Patrick Moore's Data Book of Astronomy

Packed with up-to-date astronomical data about the Solar System, our Galaxy and the wider universe, this is a one-stop reference for astronomers of all levels.

It gives the names, positions, sizes and other key facts of all the planets and their satellites; discusses the Sun in depth, from sunspots to solar eclipses; lists the dates for cometary returns, close-approach asteroids, and significant meteor showers; and includes 88 star charts, with the names, positions, magnitudes and spectra of the stars, along with key data on nebulæ and clusters.

Full of facts and figures, this is the only book you need to look up data about astronomy. It is destined to become the standard reference for everyone interested in astronomy.

PATRICK MOORE CBE, FRS, is an astronomer and author. He has received numerous awards and prizes in recognition of his work, including the CBE in 1988 and knighthood in 2001 'for services to popularisation of science and to broadcasting'. A former President of the British Astronomical Association, he is now honorary Life Vice President, and is the only amateur ever to have held an official post at the International Astronomical Union.

ROBIN REES, FRAS, is Director of Canopus Publishing and has produced a number of best-selling astronomy books, and under the Canopus Academic Publishing imprint he publishes academic physics titles.

Patrick Moore's Data Book of Astronomy

Edited by
Patrick Moore and Robin Rees

CAMBRIDGE
UNIVERSITY PRESS

CAMBRIDGE UNIVERSITY PRESS
Cambridge, New York, Melbourne, Madrid, Cape Town, Singapore,
São Paulo, Delhi, Dubai, Tokyo, Mexico City

Cambridge University Press
The Edinburgh Building, Cambridge CB2 8RU, UK

Published in the United States of America by Cambridge University Press, New York

www.cambridge.org
Information on this title: www.cambridge.org/9780521899352

First published 2011

Printed in the United Kingdom at the University Press, Cambridge

A catalogue record for this publication is available from the British Library

Library of Congress Cataloguing-in-Publication Data

Moore, Patrick.
    Celestia, the data book of astronomy / Patrick Moore, Robin Rees. – 2nd ed.
        p. cm.
    Rev. ed. of: The data book of astronomy : Bristol, UK ; Philadelphia, PA : Institute of Physics Pub., c2000.
    ISBN 978-0-521-89935-2 (Hardback)
    1. Astronomy–Handbooks, manuals, etc.   I. Rees, Robin.   II. Moore, Patrick. Data book of astronomy   III. Title.
    QB64.M623 2011
    520–dc22
                                                                    2010031372

ISBN 978-0-521-89935-2 Hardback

Contents

Foreword

Patrick Moore has inspired generations of astronomers. He has done unparalleled service, through his handbooks, lectures and articles – not to mention his BBC programme *The Sky at Night*.

Over his prolific career, Patrick has witnessed, recorded and expounded a huge enlargement of our cosmic knowledge. To see this, one need only compare the present book with one of its precursors: the *Guinness Book of Records in Astronomy* published more than 50 years ago, at the dawn of the space age.

We owe this progress to sophisticated telescopes on the ground, and to a flotilla of instruments launched into space. The planets and moons of our Solar System are now better mapped that some parts of our Earth were before the twentieth century. An unsuspected population of *trans-Neptunian objects* has been revealed – telling us that the Solar System is more complex and extensive than thought hitherto. Even more important, planets have been detected around hundreds of other stars. The study of *'extra-solar'* planets is proceeding apace: within a decade we will have discovered thousands of planetary systems, and will for the first time have evidence on just how unusual our Solar System is.

Novel technology has not only led to more powerful optical telescopes, but also to space telescopes that observe the cosmos in other wavebands out to distances exceeding 10 billion light years. We inhabit a much vaster Universe than was envisaged 50 years ago; we understand a surprising amount about how it evolved and what it contains.

This latest *Data Book of Astronomy* conveys the fascination and vibrancy of our subject – and the wonder of the skies. All astronomers should be grateful to Patrick Moore, to his co-author Robin Rees and to their team of consultants, for the immense labour that went into this book: it is surely unique in gathering such a wide and eclectic range of information into a single volume.

It will be an invaluable reference work for serious observers – but it is equally suitable for armchair browsers, and indeed for anyone who is curious about what lies beyond the Earth.

Martin Rees
Professor of Cosmology and Astrophysics,
University of Cambridge

Preface

The ancestor of this book was published more than half a century ago as the *Guiness Book of Records in Astronomy*. It went through five editions, and was then transformed into the *Astronomy Data Book* published by the Institute of Physics. By this time it had ceased to be merely lists of facts and had become much more general, and many observatories began using it as a book of quick reference. Now, 10 years later, there has been another transformation. The essential basic plan has been retained, but the text has been largely rewritten with all new data, and the tables have been enlarged and brought up to date. I pay tribute here to Robin Rees without whom I am quite certain that this book would never have seen the light of day. Invaluable help has also been given by Iain Nicolson who read the entire manuscript very carefully – though I hasten to add that any remaining errors are entirely my own.

So far as bringing the text up to date is concerned, the cut-off date is 1 December, 2010. I hope that will be acceptable.

Patrick Moore, Selsey, 1 December, 2010

Acknowledgements

Quite apart from Robin Rees and Iain Nicolson, I have had help from many friends in the preparation of this book, and none more so than from Peter Cattermole. Especial thanks to Chris Dascalo-poulos for stepping in at short notice to help with the final proofs. Very valuable administrative help has been provided by Ian Makins, and the help and encouragement of the staff at Cambridge University Press has been unfailing.

My most grateful thanks are due to the Astronomer Royal, Lord Rees of Ludlow, for writing a Foreword to the book, it is indeed a great honour for me.

The various chapters have also been read by astronomical friends who specialise in particular subjects. I am truly grateful to all to those listed below.

Paul Abel
Allan Chapman
Chris Davis
Gilbert Fielder
Alan Fitzsimmons
John Fletcher
Gerry Gilmore
Peter Gill
Monica Grady
The late Richard Gregory
Noah Hardwicke
Garry Hunt
Guy Hurst
Tom Kerss
Bruce Kingsley
Pete Lawrence
Gain Lee
Chris Lintott
Michael Maunder

John Mason
Brian May
Richard Miles
Martin Mobberley
Ian Morison
Terry Moseley
Carl Murray
Chris North
Greg Parker
Roger Prout
Martin Rees
John Rogers
David Rothery
Carl Vetterlein
Derek Ward-Thompson
Iwan Williams
John Zarnecki

Notes about units

The Celsius temperature scale is due to the Swedish astronomer Anders Celsius in 1741. The Fahrenheit scale was due to the German physicist Daniel Fahrenheit in 1724. There are other scales, now virtually obsolete. One of these is the Réaumur, due to Rene Réaumur in 1730; another is the Rankine, due to the British engineer and physicist William Rankine in 1859. The Kelvin scale is named in honour of the great British physicist Lord Kelvin.

	Réaumur (°Ré)	Kelvin (K)	Celsius (°C)	Fahrenheit (°F)	Rankine (°R)
Absolute zero		0	−273.5	−459.67	0
Water freezes	0	273.15	0	32	491.67
Water boils	80	373.14	99.98	211.97	671.64

$1\,°C = 2.25\,°F = 1\,K$

Convenient equivalents:

By definition the triple point of water is $273.6\,K = 0.01\,°C = 32.018\,°F$.

Temperature conversions are as follows:

to find °C from K: $°C = K − 273.15$;
to find K from °C: $K = °C + 273.15$;
to find °F from K: $°F = (K × 1.8) − 459.67$;
to find K from °F: $K = (°F + 459.67)/1.8$;
to find °F from °C: $°F = (9/5)°C + 32$;
to find °C from °F: $°C = (5/9)(°F − 32)$.

The old Centigrade scale is equal to the Celsius to within a degree. The Celsius and Kelvin scales are always used in science.

In everyday life the Fahrenheit scale is used in the United States. Efforts by the European Union to bully Britain into changing from Fahrenheit to Celsius for civil use have so far been mainly unsuccessful.

THE METRIC CONVERSION TABLE

The current practice of giving lengths in metric units rather then Imperial ones has been followed. To help in avoiding confusion, the following table may be found useful.

Centimetres	To	Inches	Kilometres	To	Miles
2.54	1	0.39	1.61	1	0.62
5.08	2	0.79	3.22	2	1.24
7.62	3	1.18	4.83	3	1.86
10.16	4	1.58	6.44	4	2.49
12.70	5	1.97	8.05	5	3.11
15.24	6	2.36	9.66	6	3.73
17.78	7	2.76	11.27	7	4.35
20.32	8	3.15	12.88	8	4.97
22.86	9	3.54	14.48	9	5.59
25.40	10	3.94	16.09	10	6.21
50.80	20	7.87	32.19	20	12.43
76.20	30	11.81	48.28	30	18.64
101.6	40	15.75	64.37	40	24.86
127.0	50	19.69	80.47	50	31.07
152.4	60	23.62	96.56	60	37.28
177.8	70	27.56	112.7	70	43.50
203.2	80	31.50	128.7	80	49.71
228.6	90	35.43	144.8	90	55.92
254.0	100	39.37	160.9	100	62.14

1 · The Solar System

The Solar System is made up of one star (the Sun), the eight planets with their satellites (Table 1.1) and various minor members such as asteroids, comets and meteoroids, plus a vast amount of thinly spread interplanetary matter. The Sun contains 99.86% of the total mass of the System, while Jupiter and Saturn account for 90% of what is left. Jupiter is the largest member of the planetary family, and is in fact more massive than all the other planets combined. Mainly because of Jupiter, the centre of gravity of the Solar System lies just outside the surface of the Sun.

The Solar System is divided into two parts. There are four comparatively small, rocky planets (Mercury, Venus, the Earth and Mars), beyond which comes the zone of the Main-Belt asteroids, of which only one (Ceres) is over 900 km in diameter. Next come the four giants (Jupiter, Saturn, Uranus and Neptune), plus a swarm of trans-Neptunian objects, of which the largest known are Eris and Pluto. For many years after its discovery, in 1930, Pluto was regarded as a true planet, but in August 2006 the International Astronomical Union, the controlling body of world astronomy, introduced a new scheme of classification, as follows:

A **planet** is any body in orbit round the Sun which is massive enough to assume a spherical shape, and has cleared its immediate neighbourhood of all smaller objects. All these criteria are met by the eight familiar planets, from Mercury to Neptune.

A **dwarf planet** is spherical, but has not cleared its neighbourhood. Three were listed: Eris, Pluto and Ceres.

Small solar system bodies (SSSBs) are other bodies orbiting the Sun.[1]

Natural satellites are objects in orbit round planets, dwarf planets or SSSBs rather than directly round the Sun itself.

Distances from the Sun are conventionally given in **astronomical units** (a.u.). The a.u. is defined as the mean distance between the Earth and the Sun: in round numbers 149 600 000 km (93 000 000 miles in Imperial measure). Jupiter is approximately 5.2 a.u. from the Sun; one light-year, used for interstellar distances, is equal to 63 240 a.u.

It now seems that the distinctions between the various classes of bodies in the Solar System are much less clear-cut than used to be thought. For example, it may well be that some 'near-Earth' asteroids, which swing inward away from the main swarm, are ex-comets which have lost all their volatiles, and many of the small planetary satellites are certainly captured SSSBs.

All planets, dwarf planets and SSSBs move round the Sun in the same sense, and (with one exception) so do the larger satellites orbiting their primary planets, though many of the tiny 'asteroidal' satellites move in the opposite (retrograde) sense. The orbits of the planets are not strongly eccentric, and are not greatly inclined to that of the Earth, so that to draw a plan of the main Solar System on a flat piece of paper is not grossly inaccurate. However, dwarf planets and SSSBs may have paths which are more eccentric and inclined, and comets come into a different category altogether. Those with periods of a few years or a few tens of years have direct motion, but brilliant comets come from the depths of space, and often travel in a retrograde sense. Their periods may amount to centuries, or to thousands or even millions of years.

It is also notable that six of the planets rotate in the same sense as the Earth, though the axial periods are different – over 58 Earth days for Mercury, less than 10 hours for Jupiter. The exceptions are Venus, which has retrograde rotation, and Uranus, where the rotational axis is tilted to the orbital plane by 98 degrees, more than a right angle. The cause of these anomalies is unclear.

ORIGIN AND EVOLUTION OF THE SOLAR SYSTEM

In investigating the past history of the Solar System, we do at least have one important piece of information: the age of the Earth is 4.6 thousand million years, and the Sun, in some form or other, must be older than this. We are entitled to be confident about the Earth's age, because there are several reliable methods of research, and all give the same value. There are no modern dissentients, apart of course from the Biblical Fundamentalists.

Many theories have been proposed. Of particular note is the 'Nebular Hypothesis', usually associated with the name of the eighteenth-century French astronomer Pierre Simon de Laplace, though he was not actually the first to describe it; the original idea was put forward in 1734 by Emanuel Swedenborg, of Sweden, who carried out useful scientific work but who is best remembered today for his later somewhat eccentric theories (he was on excellent terms with a number of angels, and gave graphic accounts of life on all the planets!). Swedenborg's suggestion was elaborated by Thomas Wright in England and Immanuel Kant in Germany, but the Nebular Hypothesis in its final form was due to Laplace, in 1796.

Laplace started with a vast hydrogen gas-cloud, disc-shaped and in slow rotation; it shrank steadily and threw off rings, each of which produced a planet, while the central part of the cloud – the so-called solar nebula – heated up as the atoms within it began to collide with increasing frequency. Eventually, when the temperature had risen sufficiently, the Sun had been born, and the planets were in orbits which were more or less in the same plane. All seemed well – until mathematical analysis showed that

Table 1.1 *Basic data for the planetary system*

Name	Mean distance from Sun (km)	Orbital period	Orbital eccentricity	Orbital inclination	Equatorial diameter (km)	Equatorial rotation period	Number of satellites
Mercury	57 900 000	87.97 d	0.206	7° 0′ 15″ .5	4878	58.6 d	0
Venus	108 200 000	224.7 d	0.007	178°	12 104	243.2 d	0
Earth	149 598 000	365.25	0.017	0	12 756	23h 56m 4s	1
Mars	227 940 000	687.0 d	0.093	1° 51′	6794	24h 37m 23s	2
Jupiter	778 340 000	11.86 y	0.048	1° 18′ 16″	143 884	9h 50m 30s	63
Saturn	1427 000 000	29.5 y	0.056	2° 29′ 21″	120 536	10h 14m	61
Uranus	2869 600 000	84.0 y	0.047	0° 46′ 23″	51 118	17h 14m	27
Neptune	4496 700 000	164.8 y	0.009	1° 34′ 20″	50 538	16h 6m	13

a thrown–off ring would not condense into a planet at all; it would merely disperse. There were other difficulties, too. Most of the angular momentum of the system would reside in the Sun, which would be in rapid rotation; actually, most of the angular momentum is due to the planets, and the Sun is a slow spinner (its axial rotation period amounts to several Earth weeks). In its original form, the Nebular Hypothesis had to be given up.

In 1901, T. C. Chamberlin and F. R. Moulton proposed an entirely different theory, according to which the planets were pulled off the Sun by a passing star. The visitor's gravitational pull would tear out a cigar-shaped tongue of material, and this would break up into planets, with the largest planets (Jupiter and Saturn) in the middle part of the system, where the thickest part of the 'cigar' would have been. Again there were fatal mathematical objections, and a modification of the idea by A.W. Bickerton (New Zealand), involving 'partial impact,' was no better. However, the theory in its original form remained in favour for some time, particularly as it was supported by Sir James Jeans, a leading British astronomer who was also the author of popular books on astronomy which were widely read (and in fact still are). Had it been valid, planetary systems would have been very rare in the Galaxy, because close encounters between stars seldom occur. As we now know, this is very far from being the truth. Another modification was proposed later by G. P. Kuiper, who believed that the Sun somehow acquired enough material to produce a binary companion, but that this material never formed into a true star; the planets could be regarded as stellar débris. This idea never met with much support.

In many ways our current theories are not too unlike the old Nebular Hypothesis. We do indeed begin with a gas-and-dust cloud, which began to collapse, and also to rotate, possibly because of the gravitational pull of a distant supernova. The core turned into what we call a proto-star, and the solar nebula was forced into the form of a flattened disc. As the temperature rose, the proto-star became a true star – the Sun – and for a while went through what is called the T Tauri stage, sending out a strong 'stellar wind' into the cloud and driving out the lightest gases, hydrogen and helium. (The name has been given because the phenomenon was first found with a distant variable star, catalogued as T Tauri.) The planets

built up by accretion. The inner, rocky planets lacked the gas which had been forced out by the stellar wind but, further away from the Sun, where the temperature was much lower, the giant planets were able to form and accumulate huge hydrogen-rich atmospheres. Jupiter and Saturn accreted first; Uranus and Neptune built up later, when much of the hydrogen had been dispersed. This is why they contain less hydrogen and more icy materials than their predecessors. It is fair to say that Jupiter and Saturn are true gas-giants, while Uranus and Neptune are better described as ice-giants.

In the early history of the Solar System there was a great deal of 'left-over' material. Jupiter's powerful pull prevented a planet from being formed in the zone now occupied by the Main-Belt asteroids; further out there were other asteroid-sized bodies which make up the Kuiper Belt. All the planets were subjected to heavy bombardment, and this is very evident; all the rocky planets are thickly cratered, and so are the satellites – including our Moon, where the bombardment went on for several hundreds of millions of years. (Earth was not immune, but by now most of the terrestrial impact craters have been eroded away or subducted.) It is widely believed that the gas-giants, particularly Jupiter, have acted as shields, protecting the inner planets from even more devastating bombardment. See Table 1.2 for planetary and satellite feature names.

In other ways, too, the young Solar System was very different from that of today. The Sun was much less luminous, so that, for example, Venus may well have been no more than pleasantly tropical. It is also likely that there was an extra planet in the inner part of the System, which collided with the proto-Earth and produced the Moon (though there are differing views about this). The outer planets at least may not have been in their present orbits, and interactions with each other and with general débris is thought to have caused 'planetary migration'; it has even been suggested that at one stage Uranus, not Neptune, was the outermost giant. We cannot pretend that we know all the details about the evolution of the Solar System, but at least we can be confident that we are on the right track.

How far does the Solar System extend? It is difficult to give a precise answer. The main System ends at the orbit of Neptune (unless there is a still more remote giant, which is unlikely though

Table 1.2 *Planetary and satellite feature names*

Arcus (arcus)	Arc-shaped feature
Catena (catenæ)	Chain of craters
Cavus (cavi)	Hollows; irregular steep-sided depressions
Chaos	Irregular area of broken terrain
Chasma (chasmata)	Deep, elongated, steep-sided depression
Colles	Small hills
Corona (coronæ)	Ovoid-shaped feature
Dorsum (dorsa)	Ridge
Facula (faculæ)	Bright spot
Farrum (farra)	Pancake-shaped structure
Flexus (flexûs)	Low curvilinear ridge
Fluctus (fluctûs)	Flow terrain
Flumen (flumina)	Channel that might carry liquid
Fossa (fossæ)	'Ditch'; long, narrow depression
Insula (insulæ)	Island
Labes (labes)	Landslide
Labyrinthus (labyrinthi)	Complex of intersecting ridges or valleys
Lacus	Lake
Lenticula (lenticulæ)	Small dark spot
Linea (lineæ)	Dark or bright elongated marking, either curved or straight
Macula (maculæ)	Dark spot or patch
Mare (maria)	'Sea'; large, comparatively smooth plain
Mensa (mensæ)	Flat-topped prominence with cliff-like edges
Mons (montes)	Mountain
Oceanus	'Ocean'; very large dark plain
Palus (paludes)	'Marsh'; small, often irregular plain
Patera (pateræ)	Irregular crater-like structure with scalloped edges
Planitia (planitiæ)	Low-lying plain
Planum (plana)	Plateau, or high plain
Promontorium (promontoria)	'Cape' (promontory)
Regio (regiones)	Large area, clearly different from adjacent areas
Rill (rills)	Crack-like feature (also spelled 'rille')
Rima (rimæ)	Fissure
Scopulus (scopuli)	Lobate or irregular scarp
Sinus (sinus)	'Bay'
Sulcus	Groove or trench
Tessera (tesseræ)	'Parquet' (tile-like, polygonal terrain)
Tholus (tholi)	Small, dome-like hill
Undæ	Dunes
Vallis (valles)	Valley
Vastitas	Extensive plain
Virga (virgæ)	Coloured streak

not impossible), but comets and many trans-Neptunians recede to much greater distances, and the Oort Cloud lies well over a light year away. The nearest stars beyond the Sun, those of the α Centauri group, are just over four light years away. Therefore, it seems fair to say that the effective border of the Solar System is of the order of two light years from us.

At present the Solar System is essentially stable, but this state of affairs cannot last for ever. The Sun is becoming steadily more luminous, and in no more than four thousand million years will have swelled out to become a red giant star, far more powerful than it is today. Mercury and Venus will be destroyed; Earth may survive, because the Sun's loss of mass will weaken its gravitational pull, and the planets will spiral outward to a limited degree. Yet even if our world does survive, it will be in the form of a red-hot, seething mass. Next, the Sun will collapse to become a tiny, feeble, super-dense white dwarf star, and scorching heat will be replaced by numbing cold. In the end the Sun will lose the last of its power, and will become a dead black dwarf, perhaps still attended by the ghosts of its remaining planets. It is even possible that following the merger between our Galaxy and the Andromeda Spiral, the Solar System, or rather, what is left of it – may end up in the outer part of the Milky Way, or in the depths of intergalactic space.

However, for us, all these crises lie so far ahead that we cannot predict them really accurately. We know that the Solar System has a limited lifetime, but as yet it is no more than middle-aged.

ENDNOTE

1 I was a member of that IAU Commission for many years, but felt bound to retire in 2001 as I was no longer able to travel to meetings. Had I been present at the 2006 meeting I would have put forward some alternative proposals, because the Resolution, as passed, seems to be unclear. Of the two largest Main-Belt asteroids, why should Ceres be a dwarf planet and Pallas an SSSB? Of the trans-Neptunians, a good many, such as Quaoar and Varuna, are considerably larger than Ceres. I would have retained the main planets and their satellites, and lumped the rest together as 'planetoids'. Moreover, one can hardly regard a comet as a 'small' body; the coma of Holmes' Comet of 2007 was larger than the Sun, though admittedly its mass was negligible. I suspect that the 2006 Resolution will be revised before long, but meanwhile it must be accepted.

2 · The Sun

The Sun, the controlling body of the Solar System, is the only star close enough to be studied in detail. It is 270 000 times closer than the nearest stars beyond the Solar System, those of the α Centauri group. Data are given in Table 2.1.

DISTANCE

The first known estimate of the distance of the Sun was made by the Greek philosopher Anaxagoras (500–428 BC). He assumed the Earth to be flat, and gave the Sun's distance as 6500 km (using modern units), with a diameter of over 50 km. A much better estimate was made by Aristarchus of Samos, around 270 BC. His value, derived from observations of the angle between the Sun and the exact half Moon, was approximately 4 800 000 km; his method was perfectly sound in theory, but the necessary measurements could not be made with sufficient accuracy. (Aristarchus also held the belief that the Sun, not the Earth, is the centre of the planetary system.) Ptolemy (c. AD 150) increased the distance to 8 000 000 km, but in his book published in AD 1543 Copernicus reverted to only 3 200 000 km. Kepler, in 1618 gave a value of 22 500 000 km.

The first reasonably accurate estimate of the Earth–Sun distance (the astronomical unit) was made in 1672 by Giovanni Cassini, from observations of the parallax of Mars. Some later determinations are given in Table 2.2.

One early method involved transits of Venus across the face of the Sun, as suggested by J. Gregory in 1663 and extended by Edmond Halley in 1678; Halley rightly concluded that transits of Mercury could not give accurate results because of the smallness of the planet's disc. In fact, the transit of Venus method was affected by the 'Black Drop' – the apparent effect of Venus drawing a strip of blackness after it during ingress on to the solar disc, thus making precise timings difficult. (Captain Cook's famous voyage, during which he discovered Australia, was made in order to take the astronomer C. Green to a suitable site (Tahiti) in order to observe the transit of 1769.)

Results from the transits of Venus in 1874 and 1882 were still unsatisfactory, and better estimates came from the parallax measurements of planets and (particularly) asteroids. However, Spencer Jones' value as derived from the close approach of the asteroid Eros in 1931 was too high. The modern method – radar to Venus – was introduced in the early 1960s by astronomers in the United States. The present accepted value of the astronomical unit is accurate to a tiny fraction of 1%.

THE SUN IN THE GALAXY

The Sun lies close to the inner ring of the Milky Way Galaxy's Orion Arm. It is contained within the Local Bubble, an area of rarefied high-temperature gas (caused by a supernova outburst?). The distance between our local arm and the next one out, the Perseus Arm, is about 6500 light-years. The Sun's orbit is somewhat elliptical, and passes through the galactic plane about 2.7 times per orbit. The Sun has so far completed from 20 to 25 orbits (20 to 25 'cosmic years').

ROTATION

The first comments about the Sun's rotation were made by Galileo, following his observations of sunspots from 1610. He gave a value of rather less than one month.

The discovery that the Sun shows differential rotation – i.e. that it does not rotate as a solid body would do – was made by the English amateur Richard Carrington in 1863; the rotational period at the equator is much shorter than that at the poles. Synodic rotation periods for features at various heliographic latitudes are given in Table 2.3. Spots are never seen either at the poles or exactly on the equator, but from 1871 H. C. Vogel introduced the method of measuring the solar rotation by observing the Doppler shifts at opposite limbs of the Sun.

THE SOLAR CONSTANT

The solar constant may be defined as being the amount of energy in the form of solar radiation per second which is vertically incident per unit area at the top of the Earth's atmosphere: it is roughly equal to the amount of energy reaching ground level on a clear day. The first measurements were made by Sir John Herschel in 1837–8, using an actinometer (basically a bowl of water; the estimate was made by the rate at which the bowl was heated). He gave a value which is about half the actual figure. The modern value is 1.95 cal cm^{-2} min^{-1} (1368 Wm^{-2}).

SOLAR PHOTOGRAPHY

The first photograph of the Sun – a Daguerreotype – seems to have been taken by Lerebours, in France, in 1842. However, the first good Daguerreotype was taken by Fizeau and Foucault, also in France, on 2 April 1845, at the request of F. Arago. In 1854 B. Reade used a dry collodion plate to show mottling on the disc.

Table 2.1 *The Sun: data*

Distance from Earth:
  mean   149 597 893 km (1 astronomical unit (a.u.))
  max.   152 103 000 km
  min.   147 104 000 km
Mean parallax: 8″.794
Distance from centre of the Galaxy: ~26 000 light-years
Velocity round centre of Galaxy: ~250 km s^{-1}
Period of revolution round centre of Galaxy: ~225 000 000 years
(1 'cosmic year')
Velocity toward solar apex: 19.5 km s^{-1}
Apparent diameter: mean 32′ 01″
                   max. 32′ 25″
                   min. 31′ 31″
Equatorial diameter: 1 391 980 km
Density, water = 1: mean 1.409
Volume, Earth = 1: 1 303 600
Mass, Earth = 1: 332 946
Mass: 2×10^{27} tonnes (>99% of the mass of the entire Solar System)
Surface gravity, Earth = 1: 27.90
Escape velocity: 617.7 km s^{-1}
Luminosity: 3.85×10^{23} kW
Solar constant (solar radiation per second vertically incident at unit
  area at 1 a.u. from the Sun); 1368 W m^{-2}
Mean apparent visual magnitude: −26.78 (600 000 times as bright
  as the full Moon)
Absolute magnitude: +4.82
Spectrum: G2
Temperature: surface 5500 °C
             core ~15 000 000 °C
Rotation period: sidereal, mean: 25.380 days
                 synodic, mean: 27.275 days
Time taken for light to reach the Earth, at mean distance: 499.012 s
  (8.3 min)
Age: ~4.6 thousand million years

The first systematic series of solar photographs was taken from Kew (outer London) from 1858 to 1872, using equipment designed by the English amateur Warren de la Rue. Nowadays the Sun is photographed daily from observatories all over the world, and there are many solar telescopes designed specially for this work. Many solar telescopes are of the 'tower' type, but the largest solar telescope now in operation, the McMath Telescope at Kitt Peak in Arizona, looks like a large, white inclined tunnel. At the top is the upper mirror (the heliostat), 203 cm in diameter; it can be rotated, and sends the sunlight down the tunnel in a fixed direction. At the bottom of the 183 m tunnel is a 152 cm mirror, which reflects the rays back up the tunnel on to the halfway stage, where a flat mirror sends the rays down through a hole into the solar laboratory where the analyses are carried out. This means that the heavy equipment in the solar laboratory does not have to be moved at all.

SUNSPOTS

The bright surface of the Sun is known as the photosphere, composed mainly of hydrogen and helium and it is here that we see the dark patches which are always called sunspots. Really large spot-groups may be visible with the naked eye, and a precisely dated Chinese record from as far back as 28 BC describes a patch which was 'a black vapour as large as a coin'. There is a Chinese record of an 'obscuration' in the Sun, which may well have been a spot, as early as 800 BC.

The first observer to publish telescope drawings of sunspots was J. Fabricius, from Holland, in 1611, and although his drawings are undated he probably saw the spots toward the end of 1610. C. Scheiner, at Ingoldstädt, recorded spots in March 1611, with his pupil J. Cysat. Scheiner wrote a tract which came to the notice of Galileo, who claimed to have been observing sunspots since November 1610. No doubt all these observers recorded spots telescopically at about the same time (the date was close to solar maximum when spot groups should have been frequent) but their interpretations differed. Galileo's explanation was basically correct. Scheiner regarded the spots as dark bodies moving round the Sun close to the solar surface; Cassini, later, regarded them as mountains protruding through the bright surface. Today we know that they are due to the effects of bipolar magnetic field lines below the visible surface.

Direct telescopic observation of the Sun through any telescope is highly dangerous, unless special filters or special equipment is used. The first observer to describe the projection method of studying sunspots may have been Galileo's pupil B. Castelli. Galileo himself certainly used the method, and said (correctly) that it is 'the method that any sensible person will use'. This seems to dispose of the legend that he ruined his eyesight by looking straight at the Sun through one of his primitive telescopes.

A major spot consists of a darker central portion (umbra) surrounded by a lighter portion (penumbra); with a complex spot there may be many umbræ contained in one penumbral mass. Some 'spots' at least are depressions, as can be seen from what is termed the Wilson effect, announced in 1774 by A. Wilson of Glasgow. He found that with a regular spot, the penumbra toward the limbward side is broadened, compared with the opposite side, as the spot is carried toward the solar limb by virtue of the Sun's rotation. From these observations, dating from 1769, Wilson deduced that the spots must be hollows. The Wilson effect can be striking, although not all spots and spot-groups show it.

Some spot-groups may grow to immense size. The largest group on record is that of April 1947; it covered an area of 18 130 000 000 km^2, reaching its maximum on 8 April. To be visible with the naked eye, a spot-group must cover 500 millionths of the visible hemisphere. (One millionth of the hemisphere is equal to 3 000 000 km^2.)

A large spot-group may persist for several rotations. The present record for longevity is held by a group which lasted for 200 days, between June and December 1943. On the other hand, very small spots, known as pores, may have lifetimes of less than an hour. A pore is usually regarded as a feature no more than 2500 km in diameter.

The darkest parts of spots – the umbræ – have temperatures of around 4000 °C, while the surrounding photosphere is at well over

Table 2.2 *Selected estimates of the length of the astronomical unit*

Year	Authority	Method	Parallax (arcsec)	Distance (km)
1672	G. D. Cassini	Parallax of Mars	9.5	138 370 000
1672	J. Flamsteed	Parallax of Mars	10	130 000 000
1770	L. Euler	1769 transit of Venus	8.82	151 225 000
1771	J. de Lalande	1769 transit of Venus	8.5	154 198 000
1814	J. Delambre	1769 transit of Venus	8.6	153 841 000
1823	J. F. Encke	1761 and 1769 transits of Venus	8.5776	153 375 000
1867	S. Newcomb	Parallax of Mars	8.855	145 570 000
1877	G. Airy	1874 transit of Venus	8.754	150 280 000
1877	E. T. Stone	1874 transit of Venus	8.884	148 080 000
1878	J. Galle	Parallax of asteroids Phocæa and Flora	8.87	148 290 000
1884	M. Houzeau	1882 transit of Venus	8.907	147 700 000
1896	D. Gill	Parallax of asteroid Victoria	8.801	149 480 000
1911	J. Hinks	Parallax of asteroid Eros	8.807	149 380 000
1925	H. Spencer Jones	Parallax of Mars	8.809	149 350 000
1939	H. Spencer Jones	Parallax of asteroid Eros	8.790	149 670 000
1950	E. Rabe	Motion of asteroid Eros	8.798	149 526 000
1962	G. Pettengill	Radar to Venus	8.794 0976	149 598 728
1992	Various	Radar to Venus	8.794 148	149 597 871

Table 2.3 *Synodic rotation period for features at various heliographic latitudes*

Latitude (°)	Period (days)
0	24.6
10	24.9
20	25.2
30	25.8
40	27.5
50	29.2
60	30.9
70	32.4
80	33.7
90	34.0

Table 2.4 *Zürich sunspot classification*

A Small single unipolar spot, or a very small group of spots without penumbræ.
B Bipolar sunspot group with no penumbræ.
C Elongated bipolar sunspot group. One spot must have penumbræ.
D Elongated bipolar sunspot group with penumbræ on both ends of the group.
E Elongated bipolar sunspot group with penumbræ on both ends. Longitudinal extent of penumbræ exceeds 10° but not 15°.
F Elongated bipolar sunspot group with penumbra on both ends. Longitudinal extent of penumbræ exceeds 15°.
H Unipolar sunspot group with penumbræ.

5000 °C. This means that a spot is by no means black, and if it could be seen shining on its own the surface brightness would be greater than that of an arc-lamp. The accepted Zürich classification of sunspots is given in Table 2.4.

Sunspots are essentially magnetic phenomena, and are linked with the solar cycle. Every 11 years or so the Sun is at its most active, with many spot-groups and associated phenomena; activity then dies down to a protracted minimum, after which activity builds up once more toward the next maximum. A typical group has two main spots, a leader and a follower, which are of opposite magnetic polarity.

The magnetic fields associated with sunspots were discovered by G. E. Hale, from the United States, in 1908. This resulted from the Zeeman effect (discovered in 1896 by the Dutch physicist P. Zeeman), according to which the spectral lines of a light source are split into two or three components if the source is associated with a magnetic field. It was Hale who found that the leader and the follower of a two-spot group are of opposite polarity – and that the conditions are the same over a complete hemisphere of the Sun, although reversed in the opposite hemisphere. At the end of each cycle the whole situation is reversed, so that it is fair to say that the true cycle (the 'Hale cycle') is 22 years in length rather than 11.

The magnetic fields of spots are very strong, and may exceed 4000 G. With one group, seen in 1967, the field reached 5000 G. The preceding and following spots of a two-spot group are joined by loops of magnetic field lines which rise high into the solar atmosphere above. The highly magnetised area in, around and above a bipolar sunspot group is known as an *active region*.

The modern theory of sunspots is based upon pioneer work carried out by H. Babcock in 1961. The spots are produced by bipolar magnetic regions (i.e. adjacent areas of opposite polarity) formed where a bunch of concentrated field lines (a 'flux tube') emerges through the photosphere to form a region of outward-directed or positive field; the flux tube then curves round in a loop, and re-enters to form a region of inward-directed or negative field. This, of course, explains why the leader and the follower are of opposite polarity.

Babcock's original model assumed that the solar magnetic lines of force run from one magnetic pole to the other below the bright surface. An initial polar magnetic field is located just below the photosphere in the convective zone. The Sun's differential rotation means that the field is 'stretched' more at the equator than at the poles. After many rotations, the field has become concentrated as toroids to either side of the equator, and spot-groups are produced. At the end of the cycle, the toroid fields have diffused poleward and formed a polar field with reversed polarity, and this explains the Hale 22-year cycle.

Each spot-group has its own characteristics, but in general the average two-spot group begins as two tiny specks at the limit of visibility. These develop into proper spots, growing and also separating in longitude at a rate of around $0.5 \, \text{km s}^{-1}$. Within two weeks the group has reached its maximum length, with a fairly regular leader together with a less regular follower. There are also various minor spots and clusters; the axis of the main pair has rotated until it is roughly parallel with the solar equator. After the group has reached its peak, a decline sets in; the leader is usually the last survivor. Around 75% of groups fit into this pattern, but others do not conform, and single spots are also common.

ASSOCIATED PHENOMENA

Plages are bright, active regions in the Sun's atmosphere, usually seen around sunspot groups. The brightest features of this type seen in integrated light are the faculæ.

The discovery of faculæ was made by C. Scheiner, probably about 1611. Faculæ (Latin, 'torches') are clouds of incandescent gases lying above the brilliant surface; they are composed largely of hydrogen, and are best seen near the limb, where the photosphere is less bright than at the centre of the disc (in fact, the limb has only two-thirds the brilliance of the centre, because at the centre we are looking down more directly into the hotter material). Faculæ may last for over two months, although their average lifetime is about 15 days. They often appear in areas where a spot-group is about to appear, and persist after the group has disappeared.

Polar faculæ are different from those of the more central regions, and are much less easy to observe from Earth; they are most common near the minimum of the sunspot cycle, and have latitudes higher than 65° north or south, with lifetimes ranging from a few days to no more than 12 min. They may well be associated with coronal plumes.

Even in non-spot zones, the solar surface is not calm. The photosphere is covered with granules, which are bright, irregular polygonal structures; each is around 1000 km across, and may last from 3 to 10 min (8 min is about the average). They are vast convective cells of hot gases, rising and falling at average speeds

Table 2.5 *Classification of solar flares*

Area (square degrees)	Classification
Over 24.7	4
12.5–24.7	3
5.2–12.4	2
2.0–5.1	1
Less than 2	s

F = faint, N = normal, B = bright.
Thus the most important flares are classified as 4B.

of about $0.5 \, \text{km s}^{-1}$; the gases rise at the centre of the granule and descend at the edges, so that the general situation has been likened to a boiling liquid, although the photosphere is of course entirely gaseous. They cover the whole photosphere, except at sunspots, and it has been estimated that at any one moment the whole surface contains about 4 000 000 granules. At the centre of the disc the average distance between granules is of the order of 1400 km. The granular structure is easy to observe; the first really good pictures of it were obtained from a balloon, Stratoscope II, in 1957.

Supergranulation involves large organised cells, usually polygonal, measuring around 30 000 km across; each contains several hundreds of individual granules. They last from 20 h to several days, and extend up into the chromosphere (the layer of the Sun's atmosphere immediately above the photosphere). Material wells up at the centre of the cell, spreading out to the edges before sinking again.

Spicules are needle-shaped structures rising from the photosphere, generally along the borders of the supergranules, at speeds of from 10 to $30 \, \text{km s}^{-1}$. About half of them fade out at peak altitude, while the remainder fall back into the photosphere. Their origin is not yet completely understood.

Flares are violent, short-lived outbursts, usually occurring above active spot-groups. They emit charged particles as well as radiations ranging from very short gamma-rays up to long-wavelength radio waves; they are most energetic in the X-ray and EUV (extreme ultraviolet) regions of the electromagnetic spectrum. They produce shock waves in the corona and chromosphere, and may last for around 20 min, although some have persisted for 2 h and one, on 16 August 1989, persisted for 13 h. They are most common between 1 and 2 years after the peak of a sunspot cycle. They are seldom seen in visible light. The first flare to be seen in 'white' light was observed by R. Carrington on 1 September 1859, but generally flares have to be studied with spectroscopic equipment or the equivalent. Observed in hydrogen light, they are classified according to area. The classification is given in Table 2.5.

It seems that flares are explosive releases of energy stored in complex magnetic fields above active areas. They are powered by magnetic reconnection events, when oppositely directed magnetic fields meet up and reconnect to form new magnetic structures. As the field lines snap into their new shapes, the temperature rises to tens of millions of degrees in a few minutes, and this can result in clouds of plasma being sent outward through the solar atmosphere into space; the situation has been likened to the sudden snapping of a tightly wound elastic band. These huge 'bubbles' of plasma, containing thousands of millions of tonnes of material, are known as Coronal Mass Ejections (CMEs). The particles emitted by the CME travel at a slower speed than flare's radiations and reach Earth a day or two later, striking the ionosphere and causing 'magnetic storms' – one of which, on 13 March 1989, caused power blackouts for over nine hours in Quebec, while on 20 January 2005 a flare caused widespread disruption of communications, and scrambled detectors on space-craft. And, in October 2007, material from a CME stripped the tail off Encke's periodical comet, which was then at about the distance of the orbit of Mercury.

Research carried out by A.G. Kosovichev and V. V. Zharkova has shown that flares produce seismic waves in the Sun's interior. They also cause shock waves in the solar chromosphere, known as Moreton waves, which propagate outwards at speeds of from 500 to 1000 km s^{-1}, and have been likened to solar tsunamis! They were first described by the American astronomer G. Moreton in 1960, though there had in fact been earlier observations of them by Japanese solar observers. However, nothing matches the brilliance and the effects of Carrington's flare of 1859. This was exceptional in every way.

A major CME is very likely to produce brilliant displays of auroræ. Cosmic rays and energetic particles sent out by CMEs are dangerous to astronauts moving above the protective screen of the Earth's atmosphere and, to a much lesser extent, passengers in very high-flying aircraft.

Flares are, in fact, amazingly powerful and a major outburst may release as much energy as 10 000 million one-megaton nuclear bombs. Some of the ejected particles are accelerated to almost half the velocity of light.

THE SOLAR CYCLE

The first suggestion of a solar cycle seems to have come from the Danish astronomer P. N. Horrebow in 1775–1776, but his work was not published until 1859, by which time the cycle had been definitely identified. In fact the 11-year cycle was discovered by H. Schwabe, a Dessau pharmacist, who began observing the Sun regularly in 1826 – mainly to see whether he could observe the transit of an intra-Mercurian planet. In 1851 his findings were popularised by W. Humboldt. A connection between solar activity and terrestrial phenomena was found by E. Sabine in 1852, and in 1870 E. Loomis, at Yale, established the link between the solar cycle and the frequency of auroræ.

The cycle is by no means perfectly regular. The mean value of its length since 1715 has been 11.04 years, but there are marked fluctuations; the longest interval between successive maxima has been 17.1 years (1788 to 1805) and the shortest has been 7.3 years (1829.9 to 1837). Since 1715, when reasonably accurate records began, the most energetic maximum has been that of 1957.9; the least energetic maximum was that of 1816. (See Table 2.6.) The numbered solar cycles are given in Table 2.7.

There are, moreover, spells when the cycle seems to be suspended, and there are few or no spots. Four of these spells have been identified with fair certainty: the Oort Minimum (1010–1050), the Wolf Minimum (1280–1340), the Spörer

Table 2.6 *Sunspot maxima and minima, 1718–2000*

Maxima	Minima
1718.2	1723.5
1727.5	1734.0
1738.7	1745.0
1750.5	1755.2
1761.5	1766.5
1769.7	1777.5
1778.4	1784.7
1805.2	1798.3
1816.4	1810.6
1829.9	1823.3
1837.2	1833.9
1848.1	1843.5
1860.1	1856.0
1870.6	1867.2
1883.9	1878.9
1894.1	1899.6
1907.0	1901.7
1917.6	1913.6
1928.4	1923.6
1937.4	1933.8
1947.5	1944.2
1957.8	1954.3
1968.9	1964.7
1979.9	1976.5
1990.8	1986.8
2000.1	1996.8
	2008.9

Minimum (1420–1530) and the Maunder Minimum (1645–1715). Of these the best authenticated is the last. Attention was drawn to it in 1894 by the British astronomer E. W. Maunder, based on earlier work by F. G. W. Spörer in Germany.

Maunder found, from examining old records, that between 1645 and 1715 there were virtually no spots at all. It is significant that this coincided with a very cold spell in Europe; during the 1680s, for example, the Thames froze every winter, and frost fairs were held on it. Auroræ too were lacking; Edmond Halley recorded that he saw his first aurora only in 1716, after forty years of watching.

Since then there has been a cool period, with low solar activity; it lasted between about 1790 and 1820, and is known as the Dalton Minimum.

Records of the earlier prolonged minima are fragmentary, but some evidence comes from the science of tree rings, dendrochronology, founded by an astronomer, A. E. Douglass. High-energy cosmic rays which pervade the Galaxy transmute a small amount of atmospheric nitrogen to an isotope of carbon, carbon-14, which is radioactive. When trees assimilate carbon dioxide, each growth ring contains a small percentage of carbon-14, which decays exponentially with a half-life of 5730 years. At sunspot maximum, the magnetic field ejected by the Sun deflects some of the cosmic rays away from the Earth, and reduces the level of carbon-14 in the

Table 2.7 *Numbered solar cycles*

Cycle	Began	Ended	Duration, years	No. of spotless days (throughout cycle)
1	Mar 1755	June 1766	11.3	
2	June 1766	June 1775	9.0	
3	June 1775	Sept 1284	9.3	
4	Sept 1784	May 1798	13.7	
5	May 1798	Dec 1810	12.6	
6	Dec 1810	May 1823	12.4	
7	May 1823	Nov 1833	10.5	
8	Nov 1833	July 1843	9.8	
9	July 1843	Dec 1855	12.4	
10	Dec 1855	Mar 1867	11.3	~654
11	Mar 1867	Dec 1878	11.8	~406
12	Dec 1878	Mar 1890	11.3	~736
13	Mar 1890	Feb 1902	11.9	~938
14	Feb 1902	Aug 1913	11.5	~1019
15	Aug 1913	Aug 1923	10.0	534
16	Aug 1923	Sept 1933	10.1	568
17	Sept 1933	Feb 1944	10.4	269
18	Feb 1944	Apr 1954	10.2	446
19	Apr 1954	Oct 1964	10.5	227
20	Oct 1964	June 1976	11.7	272
21	June 1976	Sept.1986	10.3	273
22	Sept 1986	May 1996	9.7	309
23	May 1996	Dec 2008	12.6	>730
24	Dec 2008			

Solar minimum of ~2009 (Cycle 24). This was the deepest for many years. In 2008 there were no spots on 266 days (73%), and 2009 was even lower. This recalls 1913 (311 spotless days).

atmosphere, so that the tree rings formed at sunspot maximum have a lower amount of the carbon-14 isotope. Careful studies were carried out by F. Vercelli, who examined a tree which lived between 275 BC and AD 1914. Then, in 1976, J. Eddy compared the carbon-14 record of solar activity with records of sunspots, auroræ and climatic data, and confirmed Maunder's suggestion of a dearth of spots between 1645 and 1715. Yet strangely, although there were virtually no records of telescopic sunspots during this period, naked-eye spots were recorded in China in 1647, 1650, 1655, 1656, 1665 and 1694; whether or not these observations are reliable must be a matter for debate. There is strong evidence for a longer cycle superimposed on the 11-year one.

The law relating to the latitudes of sunspots (Spörer's law) was discovered by the German amateur Spörer in 1861. At the start of a new cycle after minimum, the first spots appear at latitudes between 30° and 45° north or south. As the cycle progresses, spots appear closer to the equator, until at maximum the average latitude of the groups is only about 15° north or south. The spots of the old cycle then die out (before reaching the equator), but even before

they have completely disappeared the first spots of the new cycle are seen at the higher latitudes. This was demonstrated by the famous 'Butterfly Diagram', first drawn by Maunder in 1904.

The Wolf or Zürich sunspot number for any given day, indicating the state of the Sun at that time, was worked out by R. Wolf of Zürich in 1852. The formula is $R = k(10g + f)$, where R is the Zürich number, g is the number of groups seen, f is the total number of individual spots seen and k is a constant depending on the equipment and site of the observer (k is usually not far from unity). The Zürich number may range from zero for a clear disc up to over 200. A spot less than about 2500 km in diameter is officially classed as a pore.

Rather surprisingly, the Sun is actually brightest at spot maximum. The greater numbers of sunspots do not compensate for the greater numbers of brilliant plages.

SPECTRUM AND COMPOSITION OF THE SUN

The first intentional solar spectrum was obtained by Isaac Newton in 1666, but he never took these investigations much further, although he did of course demonstrate the complex nature of sunlight. The sunlight entered the prism by way of a hole in the screen, rather than a slit.

In 1802 W. H. Wollaston, in England, used a slit to obtain a spectrum and discovered the dark lines, but he merely took them to be the boundaries between different colours of the rainbow spectrum. The first really systematic studies of the dark lines were carried out in Germany by J. von Fraunhofer, from 1814. Fraunhofer realised that the lines were permanent; he recorded 5740 of them and mapped 324. They are still often referred to as the Fraunhofer lines.

The explanation was found by G. Kirchhoff, in 1859 (initially working with R. Bunsen). Kirchhoff found that the photosphere yields a rainbow or continuous spectrum; the overlying gases produce a line spectrum, but since these lines are seen against the rainbow background they are reversed, and appear dark instead of bright. Since their positions and intensities are not affected, each line may be tracked down to a particular element or group of elements. In 1861–1862 Kirchhoff produced the first detailed map of the solar spectrum. (His eyesight was affected, and the work was actually finished by his assistant, K. Hofmann.) In 1869 Anders Ångström, of Sweden, studied the solar spectrum by using a grating instead of a prism, and in 1889 H. Rowland produced a detailed photographic map of the solar spectrum. The most prominent Fraunhofer lines in the visible spectrum are given in Table 2.8.

By now many of the known chemical elements have been identified in the Sun. The list of elements which have now been identified is given in Table 2.9. The fact that the remaining elements have not been detected does not necessarily mean that they are completely absent; they may be present, although no doubt in very small amounts.

So far as relative mass is concerned, the most abundant element by far is hydrogen (71%). Next comes helium (27%). All the other elements combined make up only 2%. The numbers of atoms in the Sun relative to one million atoms of hydrogen are given in Table 2.10.

Table 2.8 *The most prominent Fraunhofer lines in the visible spectrum of the Sun*

Letter	Wavelength (Å)	Identification	Letter	Wavelength (Å)	Identification
A	7593	O_2			
a	7183	H_2O			
B	6867	O_2			
(These three are telluric lines – due to the Earth's intervening atmosphere.)					
C(Hα)	6563	H	b4	5167	Mg
D1	5896		F(Hβ)	4861	H
D2	5890	Na	f(Hγ)	4340	H
E	5270	Ca, Fe	G	4308	Fe, Ti
	5269	Fe	g	4227	Ca
b1	5183	Mg	h(Hδ)	4102	H
b2	5173	Mg	H	3968	Ca1[1]
b3	5169	Fe	K	3933	

Note: one Ångström (Å) is equal to one hundred-millionth part of a centimetre; it is named in honour of Anders Ångström. The diameter of a human hair is roughly 500 000 Å. To convert Ångströms into nanometres, divide all wavelengths by 10, so that, for instance, Hα becomes 656.3 nm.

Helium was identified in the Sun (by Norman Lockyer, in 1868) before being found on Earth. Lockyer named it after the Greek ηλιος, the Sun. It was detected on Earth in 1894 by Sir William Ramsay, as a gas occluded in cleveite.

For a time it was believed that the corona contained another element unknown on Earth, and it was even given a name – coronium – but the lines, described initially by Harkness and Young at the eclipse of 1869, proved to be due to elements already known. In 1940 B. Edlén, of Sweden, showed that the coronium lines were produced by highly ionised iron and calcium.

SOLAR ENERGY

Most of the radiation emitted by the Sun comes from the photosphere, which is no more than about 500 km deep. It is easy to see that the disc is at its brightest near the centre; there is appreciable limb darkening – because when we look at the centre of the disc we are seeing into deeper and hotter layers. It is rather curious to recall that there were once suggestions that the interior of the Sun might be cool. This was the view of Sir William Herschel, who believed that below the bright surface there was a temperature region which might well be inhabited – and he never changed his view (he died in 1822). Few of his contemporaries agreed with him, but at least his reputation ensured that the idea of a habitable Sun would be taken seriously. And as recently as 1869 William Herschel's son, Sir John, was still maintaining that a sunspot was produced when the luminous clouds rolled back, bringing the dark, solid body of the Sun itself into view[1].

Spectroscopic work eventually put paid to theories of this kind. The spectroheliograph, enabling the Sun to be photographed in the light of one element only, was invented by G. E. Hale in 1892; its visual equivalent, the spectrohelioscope, was invented in 1923, also by Hale. In 1933 B. Lyot, in France, developed the Lyot filter, which is less versatile but more convenient, and also allows the Sun to be studied in the light of one element only.

But how did the Sun produce its energy? One theory, proposed by J. Waterson and, in 1848, by J. R. Mayer, involved meteoritic infall. Mayer found that a globe of hot gas the size of the Sun would cool down in 5000 years or so if there were no other energy source, while a Sun made up of coal, and burning furiously enough to produce as much heat as the real Sun actually does, would be turned into ashes after a mere 4600 years. Mayer therefore assumed that the energy was produced by meteorites striking the Sun's surface.

Rather better was the contraction theory, proposed in 1854 by H. von Helmholtz. He calculated that if the Sun contracted by 60 m per year, the energy produced would suffice to maintain the output for 15 000 000 years. This theory was supported later by the great British physicist Lord Kelvin. However, it had to be abandoned when it was shown that the Earth itself is around 4600 million years old – and the Sun could hardly be younger than that. In 1920 Sir Arthur Eddington stated that atomic energy was necessary, adding 'Only the inertia of tradition keeps the contraction hypothesis alive – or, rather, not alive, but an unburied corpse.'

The nuclear transformation theory was worked out by H. Bethe in 1938, during a train journey from Washington to Cornell University. Hydrogen is being converted into helium, so that energy is released and mass is lost; the decrease in mass amounts to 4 000 000 tonnes per second. Bethe assumed that carbon and nitrogen were used as catalysts, but C. Critchfield, also in America, subsequently showed that in solar-type stars the proton–proton reaction is dominant.

Slight variations in output occur, and it is often claimed that it is these minor changes which have led to the ice ages which have affected the Earth now and then throughout its history, but for the moment at least the Sun is a stable, well-behaved Main Sequence star.

The core temperature is believed to be around 15 000 000 °C, and the density about 10 times as dense as solid lead. The core

Table 2.9 *The chemical elements and their occurrence in the Sun. The following is a list of elements 1 to 92. * = detected in the Sun. R = included in H. A. Rowland's list published in 1891. For elements 43, 61, 85–89 and 91 the mass number is that of the most stable isotope*

Atomic number (superscript) and symbol	Name	Atomic weight	Occurrence in the Sun
¹H	Hydrogen	1.008	*R
²He	Helium	4.003	*
³Li	Lithium	6.939	*(in sunspots)
⁴Be	Beryllium	9.013	*R
⁵B	Boron	10.812	*(in compound)
⁶C	Carbon	12.012	*R
⁷N	Nitrogen	14.007	*
⁸O	Oxygen	16.000	*
⁹F	Fluorine	18.999	*(in compound)
¹⁰Ne	Neon	20.184	*
¹¹Na	Sodium	22.991	*R
¹²Mg	Magnesium	24.313	*R
¹³Al	Aluminium	26.982	*R
¹⁴Si	Silicon	28.090	*R
¹⁵P	Phosphorus	30.975	*
¹⁶S	Sulphur	32.066	*
¹⁷Cl	Chlorine	35.434	
¹⁸A	Argon	39.949	*(in corona)
¹⁹K	Potassium	39.103	*R
²⁰Ca	Calcium	40.080	*R
²¹Sc	Scandium	44.958	*R
²²Ti	Titanium	47.900	*R
²³V	Vanadium	50.944	*R
²⁴Cr	Chromium	52.00	*R
²⁵Mn	Manganese	52.94	*R
²⁶Fe	Iron	55.85	*R
²⁷Co	Cobalt	58.94	*R
²⁸Ni	Nickel	58.71	*R
²⁹Cu	Copper	63.55	*R
³⁰Zn	Zinc	65.37	*R
³¹Ga	Gallium	69.72	*
³²Ge	Germanium	72.60	*R
³³As	Arsenic	74.92	
³⁴Se	Selenium	78.96	
³⁵Br	Bromine	79.91	
³⁶Kr	Krypton	83.80	
³⁷Rb	Rubidium	85.48	*(in spots)
³⁸Sr	Strontium	87.63	*R
³⁹Y	Yttrium	88.91	*R
⁴⁰Zr	Zirconium	91.22	*R
⁴¹Nb	Niobium	92.91	*R
⁴²Mo	Molybdenum	95.95	*R
⁴³Tc	Technetium	99	
⁴⁴Ru	Ruthenium	101.07	*
⁴⁵Rh	Rhodium	102.91	*R
⁴⁶Pd	Palladium	106.5	*R
⁴⁷Ag	Silver	107.87	*R
⁴⁸Cd	Cadmium	112.41	*R
⁴⁹In	Indium	114.82	*(in spots)
⁵⁰Sn	Tin	118.70	*R
⁵¹Sb	Antimony	121.76	*
⁵²Te	Tellurium	127.61	
⁵³I	Iodine	126.91	
⁵⁴Xe	Xenon	131.30	
⁵⁵Cs	Cæsium	132.91	
⁵⁶Ba	Barium	137.35	*R
⁵⁷La	Lanthanum	138.92	*R
⁵⁸Ce	Cerium	140.13	*R
⁵⁹Pr	Praseodymium	140.91	*
⁶⁰Nd	Neodymium	144.25	*R
⁶¹Pm	Promethium	147	
⁶²Sm	Samarium	150.36	*
⁶³Eu	Europium	151.96	*
⁶⁴Gd	Gadolinium	157.25	*
⁶⁵Tb	Terbium	158.93	*
⁶⁶Dy	Dysoprosium	162.50	*
⁶⁷Ho	Holmium	164.94	
⁶⁸Er	Erbium	167.27	*R
⁶⁹Tm	Thulium	168.94	*
⁷⁰Yb	Ytterbium	173.04	*
⁷¹Lu	Lutecium	174.98	*
⁷²Hf	Hafnium	178.50	*
⁷³Ta	Tantalum	180.96	*
⁷⁴W	Tungsten	183.86	*
⁷⁵Re	Rhenium	186.3	
⁷⁶Os	Osmium	190.2	*
⁷⁷Ir	Iridium	192.2	*
⁷⁸Pt	Platinum	195.1	*
⁷⁹Au	Gold	197.0	*
⁸⁰Hg	Mercury	200.6	
⁸¹Tl	Thallium	204.4	
⁸²Pb	Lead	207.2	*R
⁸³Bi	Bismuth	209.0	
⁸⁴Po	Polonium	210	
⁸⁵At	Astatine	211	
⁸⁶Rn	Radon	222	
⁸⁷Fr	Francium	223	
⁸⁸Ra	Radium	226	
⁸⁹Ac	Actinium	227	
⁹⁰Th	Thorium	232	*
⁹¹Pa	Protoactinium	231	
⁹²U	Uranium	238	

The remaining elements are 'transuranic' and radioactive, and have not been detected in the Sun. They are:

⁹³Np	Neptunium	237
⁹⁴Pu	Plutonium	239
⁹⁵Am	Americium	241
⁹⁶Cm	Curium	242
⁹⁷Bk	Berkelium	243
⁹⁸Cf	Californium	244

Table 2.9 (cont.)

Atomic number (superscript) and symbol	Name	Atomic weight	Occurrence in the Sun
^{99}Es	Einsteinium	253	
^{100}Fm	Fermium	254	
^{101}Md	Mendelevium	254	
^{102}No	Nobelium	254	
^{103}Lw	Lawrencium	257	
^{104}Rf	Rutherfordium	–	
^{105}Ha	Hahnium	–	
^{106}Sg	Seaborgium	–	
^{107}Ns	Neilsborium	–	
^{108}Hs	Hassium	–	
^{109}Mt	Meitnerium	–	
^{110}Ds	Darmstadtium		
^{111}Rg	Röntgenium		
^{112}Cn	Copernicium		
^{113}Uut	Ununtrium		
^{114}Uuq	Ununquadium		
^{115}Uup	Ununpentium		
^{116}Uuh	Ununhexium		
^{117}Uus	Ununseptium		
^{118}Uuo	Ununoctium (a noble gas)		

Table 2.10 *Relative frequency of numbers of atoms in the Sun*

Hydrogen	1 000 000
Helium	85 000
Oxygen	600
Carbon	420
Nitrogen	87
Silicon	45
Magnesium	40
Neon	37
Iron	32
Sulphur	16
Aluminium	3
Calcium	2
Sodium	2
Nickel	2
Argon	1

extends one-quarter of the way from the centre of the globe to the outer surface; about 37% of the original hydrogen has been converted to helium. Outside the core comes the radiative zone, extending out from 25% to 70% of the solar radius between the centre and the surface; here, energy is transported by radiative diffusion. In the outer layers it is convection which is the transporting agency.

It takes radiation about 170 000 years to work its way from the core to the bottom of the convective zone, which starts at about 70% of the solar radius and extends to just below the surface, where the temperature is over 2 000 000 °C.

One problem which baffled solar astronomers for a very long time concerned neutrinos. These are particles with virtually no rest mass and no electrical charge, so that they are extremely difficult to detect. Theoretical considerations indicate that the Sun should emit vast quantities of them, and in 1966 efforts to detect them were begun by a team from the Brookhaven National Laboratory in the USA, led by R. Davis. The 'telescope' is located in the Homestake Gold Mine in South Dakota, inside a deep mineshaft, and consists of a tank of 454 600 litres of cleaning fluid (tetrachloroethylene). Only neutrinos can penetrate so far below ground level (cosmic rays which would otherwise confuse the experiment, cannot do so). The cleaning fluid is rich in chlorine, and if a chlorine atom is struck by a neutrino it will be changed into a form of radioactive argon – which can be detected. The number of 'strikes' would therefore provide a key to the numbers of solar neutrinos.

In fact, the observed flux was much smaller than had been expected, and the detector recorded only about one-third the anticipated numbers of neutrinos. The same result was obtained by a team in Russia, using 100 tonnes of liquid scintillator and 144 photodetectors in a mine in the Donetsk Basin. Further confirmation came from Kamiokande in Japan, using light-sensitive detectors on the walls of a tank holding 3000 tonnes of water. When a neutrino hits an electron it produces a spark of light, and the direction of this, as the electron moves, tells the direction from which the neutrino has come – something which the Homestake detector cannot do. Another sort of detector, in Russia, uses gallium-71; if hit by a neutrino, this gallium will be converted to germanium-71. Another gallium experiment was set up in Gran Sasso, deep in the Apennines, and yet another detector equally deep in the Caucasus Mountains.

At Hid, in Japan, neutrino-hunters have set up their equipment 1000 metres underground in the Kamioke Mining Company's tunnels. The detector here is ultra-pure water – 50 000 tonnes of it contained in a stainless steel superstucture 41 metres high and 39 metres in diameter. When a neutrino interacts with an electron in water it produces a charged particle that moves faster than the speed of light in water. This in turn creates a cone of light known as Čerenkov radiation (the optical equivalent of a sonic boom) which can be identified by a detector at the inner wall of the container. The Japenese neutrino detector SuperKAMIOKANDE was active by the late 1980s, but the numbers of solar neutrinos detected remained much fewer than predicted.

The problem was finally solved by researchers at the Sudbury Neutrino Observatory (SNO) at the Ceighton mine in Ontario, Canada. The detector consists of 1000 tonnes of 'heavy water' (D_2O) held in a 121-metre acrylic container, 2 km below the Earth's surface. There are 10 000 light-sensitive phototubes, which produce flashes of Čerenkov radiation when neutrinos interact with the water.

The neutrinos produced in the Sun are of the type known as 'electron-neutrinos'; it transpired that during the journey from

Sun to Earth some of the electron-neutrinos changed into other types – muon-neutrinos and tau-neutrinos – which could not be recorded before SNO was ready. When the new information was taken into account, all was well, and the total number of solar neutrinos was after all in agreement with the theoretical predictions.

Predictably, the Sun emits radiation over the whole range of the electromagnetic spectrum. Infrared radiation was detected in 1800 by William Herschel during an examination of the solar spectrum; he noted that there were effects beyond the limits of red light. In 1801 J. Ritter detected ultraviolet radiation by using a prism to produce a solar spectrum and noting that paper soaked in $NaCl$ was darkened if held in a region beyond the violet end of the visible spectrum. Cosmic rays from the Sun were identified by S. Forbush in 1942, and in 1954 he established that cosmic-ray intensity decreases when solar activity increases (Forbush effect).

The discovery of radio emission from the Sun was made by J. S. Hey and his team, on 27–28 February 1942. Initially, the effect was thought to be due to German jamming of radar transmitters. The first radar contact with the Sun was made in 1959, by V. Eshleman and his colleagues at the Stanford Research Institute in the United States.

Solar X-rays are blocked by the Earth's atmosphere, so that all work in this field has to be undertaken by space research methods. The first X-ray observations of the Sun were made in 1949 by investigators at the United States Naval Research Laboratory.

SOLAR PROBES

The first attempt at carrying out solar observations from high altitude was made in 1914 by Charles Abbott, using an automated pyrheliometer launched from Omaha by hydrogen-filled rubber balloons. The altitude reached was 24.4 km, and in 1935 a balloon, Explorer II, took a two-man crew to the same height. The initial attempt at solar research using a modern-type rocket was made in 1946, when a captured and converted German V.2 was launched from White Sands, New Mexico; it reached 55 km and recorded the solar spectrum down to 2400 Å. The first X-ray solar flares were recorded in 1956, from balloon-launched rockets, although solar X-rays had been identified as early as 1948.

Many solar probes have now been launched. (In 1976 one of them, the German-built Helios 2, approached the Sun to within 45 000 000 km.) The first vehicle devoted entirely to studies of the Sun was OSO 1 (Orbiting Solar Observatory 1) of 1962: it carried 13 experiments, obtaining data at ultraviolet, X-ray and gamma-ray wavelengths.

Extensive solar observations were made by the three successive crews of the first US space-station, Skylab, in 1973–1974. The equipment was able to monitor the Sun at wavelengths from visible light through to X-rays. The last of the crews left Skylab on 8 February 1974, although the station itself did not decay in the atmosphere until 1979. Solar work was also undertaken by many of the astronauts on the Russian space-station Mir, from 1986, until the end of the mission.

One vehicle of special note was the Solar Maximum Mission, launched on 14 February 1980 into a circular, 574 km orbit. It was designed to study the Sun through the peak of a cycle. Following a breakdown, the vehicle was repaired in April 1984 by a crew from the Space Shuttle, and functioned until 2 December 1999. The robotic space probe Ulysses, sent up from Cape Canaveral on 6 October 1990, concentrated on the poles of the Sun, which can never be seen from Earth because our view is always more or less broadside-on. The only way to obtain a good view of the solar poles was to send a probe out of the ecliptic, which is not so easy as might be thought. Ulysses was first sent out to Jupiter, and on 8 February 1992 flew past the Giant Planet, which used its strong gravitational pull to put Ulysses into the required orbit. It flew over the Sun's south pole on 26 June 1994, and over the north pole on 31 July 1995. Some of its discoveries were unexpected; in particular the south magnetic pole was found to be very dynamic and without any fixed clear location. The south magnetic pole exists, of course, but it is more diffuse than its northern counterpart.

Ulysses is still in solar orbit, but it will never fly close to the Sun, and in fact it will always stay outside the orbit of the Earth. Its own orbital period is 6 years. It has now been switched off.

The Japanese space-craft Yohkoh, sent up from the Kagoshima Space Centre (South Japan) on 30 August 1991, was designed to study flares and kindred phenomena at X-ray and gamma-ray wavelengths; it worked well, and lasted until 16 December 2001 – by now the Japanese were very much to the fore in space research. Then, on 2 December 1995, came NASA's Solar and Heliospheric Observatory – SOHO – which as been an outstanding success.

The observatory SOHO was put into an unusual orbit. It remains 1 500 000 km from the Earth, exactly on a line joining the Earth to the Sun; it lies at a stable point, known as a Lagrangian point, so that from Earth it is effectively motionless, and is in sunlight all the time. There was an alarm on 25 June 1996, when contact was lost and it was feared that the mission had come to a premature end, but SOHO was reactivated on 27 July, and was still operating well over 10 years later.

This observatory has been immensely informative. For example, it has detected vast solar tornadoes whipping across the Sun's surface, with gusts up to 500 000 km h^{-1}. There are jet streams below the visible surface, and definite 'belts' in which material moves more quickly than the gases to either side. There has been a major surprise, too, with regard to the Sun's general rotation. On the surface, the rotation period is 25 days at the equator, rising to 25.7 days at latitudes of 40 degrees north or south and as much as 34 days at the poles. This differential rotation persists to the base of the convection zone, but here the whole situation changes: the equatorial rotation slows down and the higher-latitude rotation speeds up. The two rates become equal at a distance about half-way between the surface and the centre of the globe; deeper down the Sun rotates in the same way as a solid body. As yet we have to admit that we do not know the reason for this bizarre behaviour.

Using data from SOHO it has also been found that the entire outer layer of the Sun, down to about 24 000 km, is slowly but steadily flowing from the equator to the poles. The polar flow rate is no more than 80 km h^{-1}, but this is enough to transport an object from the equator to the pole in little over a year.

The Genesis mission was launched on 8 August 2001, and made for the first Lagrangian point between the Earth and the Sun. It stayed there, and between 3 December 2001 and 1 April 2004 it used exposed collector arrays to pick up particles of the solar wind. It then came home, and a capsule bringing the precious samples re-entered the atmosphere on 8 September 2004. As it parachuted down it should have been caught by a waiting helicopter but, owing to a constructional error this did not happen, and the capsule crash-landed in the Utah desert. Fortunately many of the particles were retrieved. The 'bus' is still orbiting the Sun.

The Hinode satellite – Japanese for 'Sunrise' – was sent up on 22 September 2006. It carries optical and X-ray telescopes, and equipment to carry out studies in the extreme ultraviolet range. The images sent back are probably the best obtained up to the present time. Another recent mission is STEREO (Solar TErrestrial RElations Observatory), which consists of two space-craft and was launched by NASA on 25 October 2006 to study the Sun and solar wind in three dimensions.

The Solar Dynamics Observatory (SDO) was launched on 11 February 2010. It was put into a geosynchronous orbit, and carries a variety of instruments. It is a three-axis stabilised space-craft, with two solar arrays and two high-gain antennæ. The Helioseismic and Magnetic Imager (HMI) studies solar variability and the various aspects of magnetic variability, while the Extreme Ultraviolet Variability Experiment (EVE) measures the Sun's extreme ultraviolet irradiance with improved special resolution. This is of special importance, because the Sun's output of energetic extreme ultraviolet photons is mainly responsible for heating the Earth's upper atmosphere and creating the ionosphere. The output varies with the 11-year cycle, and this affects atmospheric heating, satellite drag, and interference with communications systems. Testing and assembly were carried out at NASA's Goddard Space Flight Center in Greenbelt, Maryland.

A selected list of solar probes is given in Table 2.11.

HELIOSEISMOLOGY

The first indications of solar oscillation were detected as long ago as 1960; the period was found to be 5 min, and it was thought that the effects were due to a surface 'ripple' in the outer 10 000 km of the Sun's globe. More detailed results were obtained in 1973 by R. H. Dicke, who was attempting to make measurements of the polar and equatorial diameters of the Sun to see whether there was any appreciable polar flattening. Dicke found that the Sun was 'quivering like a jelly', so that the equator bulges as the poles are flattened, but the maximum amplitude is only 5 km, and the velocities do not exceed 10 m s^{-1}.

This was the real start of the science of helioseismology. Seismology involves studies of earthquake waves in the terrestrial globe, and it is these methods which have told us most of what we know about the Earth's interior. Helioseismology is based on the same principle. Pressure waves – in effect, sound waves – echo and resonate through the Sun's interior. Any such wave moving inside the Sun is bent or refracted up to the surface, because of the increase in the speed of sound with increased depth. When the

wave reaches the surface it will rebound back downward and this makes the photosphere move up and down. The amplitude is a mere 25 m, with a temperature change of 0.005 °C, but these tiny differences can be measured by the familiar Doppler principle involving tiny shifts in the positions of well-defined spectral lines. Waves of different frequencies descend to different depths before being refracted up toward the surface – and the solar sound waves are very low-pitched; the loudest lies about 12 ½ octaves below the lowest note audible to human beings. There are, of course, a great many frequencies involved, so that the whole situation is very complex indeed.

Various ground-based programmes are in use – such as GONG, the Global Oscillation Network Group, made up of six stations spread out round the Earth so that at least one of them can always be in sunlight. However, more spectacular results have come from solar space-craft, such as SOHO.

THE SOLAR ATMOSPHERE

With the naked eye, the outer surroundings of the Sun – the solar atmosphere – can be seen only during a total solar eclipse. With modern-type equipment, or from space, they can however be studied at any time, although the outer corona is more or less inaccessible except by using space research methods. The structure of the Sun is summarised in Table 2.12.

Above the photosphere, rising to 5000 km, is the chromosphere ('colour-sphere'), so named because its hydrogen content gives it a strong red colour as seen during a total eclipse. The temperature rises quickly with altitude (remembering that the scientific definition of temperature depends upon the speeds at which the atomic particles move around, and is by no means the same as the ordinary definition of 'heat'; the chromosphere is so rarefied that it certainly is not 'hot'). In the chromosphere we find *spicules* (long, thin 'fingers' of luminous gas which rise to the top of the layer and sink back down again over a period of about 10 minutes, and *fibrils*, horizontal wisps of gas; in extent, they are similar to spicules, but last for considerably longer (up to about 20 min). The dark Fraunhofer lines in the solar spectrum are produced in the chromosphere.

Rising from the chromosphere are the *prominences*, structures with chromospheric temperatures embedded in the corona. They were first described in detail by the Swedish observer Vassenius at the total eclipse of 1733, although he believed them to belong to the Moon rather than to the Sun. (They may have been recorded earlier, in 1706, by Stannyan at Berne.) It was only after the eclipse of 1842 that astronomers became certain that they are solar rather than lunar.

Prominences (once, misleadingly, known as Red Flames) are composed of hydrogen. Quiescent prominences may persist for weeks or even months, but eruptive prominences show violent motions, and may attain heights of several hundreds of thousands of kilometres. Following the eclipse of 19 August 1868, J. Janssen (France) and Norman Lockyer (England) developed the method of observing them spectroscopically at any time. By observing at hydrogen wavelengths, prominences may be seen against the bright

Table 2.11 *Solar missions*

Name	Launch data	Nationality	Experiments
Pioneer 4	3 Mar 1959	American	Lunar probe, but in solar orbit: solar flares.
Vanguard 3	18 Sept 1959	American	Solar X-rays.
Pioneer 5	11 Mar 1960	American	Solar orbit, 0.806×0.995 a.u. Flares, solar wind. Transmitted until 26 June 1960, at 37 000 000 km Earth.
OSO 1	7 Mar 1962	American	Orbiting Solar Observatory 1. Earth orbit, 553×595 km. Lost on 6 Aug 1963.
Cosmos 3	24 Apr 1962	Russian	Earth orbit, 228×719 km; decayed after 176 days. Solar and cosmic radiation.
Cosmos 7	28 July 1962	Russian	Earth orbit, 209×368 km. Monitoring solar flares during Vostok 3 and 4 missions. Decayed after 4 days.
Explorer 18–IMP	26 Nov 1963	American	Interplanetary Monitoring Platform 1. Earth orbit, $125\,000 \times 202\,000$ km. Provision for flare warnings for manned missions.
OGO 1	4 Sept 1964	American	Orbiting Geophysical Observatory 1. Earth–Sun relationships.
OSO 2	3 Feb 1965	American	Earth–Sun relationships. Flares.
Explorer 30–Solrad	18 Nov 1965	American	Solar radiation and X-rays; part of the IQSY programme (International Year of the Quiet Sun).
Pioneer 6	16 Dec 1965	American	Solar orbit, 0.814×0.985 a.u. First detailed space analysis of solar atmosphere.
Pioneer 7	17 Aug 1966	American	Solar orbit, 1.010×1.125 a.u. Solar atmosphere. Flares.
OSO 3	8 Mar 1967	American	Earth–Sun relationships. Flares.
Cosmos 166	16 June 1967	Russian	Earth orbit, 260×577 km. Solar radiation. Decayed after 130 days.
OSO 4	18 Oct 1967	American	Earth–Sun relationships. Flares.
Pioneer 8	13 Dec 1967	American	Solar orbit, 1.00×1.10 a.u. Solar wind; programme with Pioneers 6 and 7.
Cosmos 215	19 Apr 1968	Russian	Solar orbit, 260×577 km. Solar radiation. Decayed after 72 days.
Pioneer 9	8 Nov 1968	American	Solar orbit, 0.75×1.0 a.u. Solar wind, flares etc.
HEOS 1	5 Dec 1968	American	High-Energy Orbiting Satellite. Earth orbit, $418 \times 112\,400$ km. With HEOS 2, monitored 7 years of the 11-year solar cycle.
Cosmos 262	26 Dec 1968	Russian	Earth orbit, 262×965 km. Solar X-rays and ultraviolet.
OSO 5	22 Jan 1969	American	Earth orbit, 550 km, inclination $32° .8$. General solar studies.
OSO 6	9 Aug 1969	American	Earth orbit, 550 km, inclination $32° .8$. General solar studies, as with OSO 5.
Shinsei SS1	28 Sept 1971	Japanese	Earth orbit, 870×1870 km. Operated for 4 months.
OSO 7	29 Sept 1971	American	Earth orbit, 329×575 km. General studies: solar X-ray, ultraviolet, EUV. Operated until 9 July despite having been put into the wrong orbit.
HEOS 2	31 Jan 1972	American	Earth orbit. High-energy particles, in conjunction with HEOS 1.
Prognoz 1	14 Apr 1972	Russian	First of a series of Russian solar wind and X-ray satellites. (Prognoz = Forecast.) Earth orbit, $965 \times 200\,000$ km.
Prognoz 2	29 June 1972	Russian	Earth orbit, $550 \times 200\,000$ km. Solar wind and X-ray studies.
Prognoz 3	15 Feb 1973	Russian	Earth orbit, $590 \times 200\,000$ km. General solar studies, including X-ray and gamma-rays.
Intercosmos 9	19 Apr 1973	Russian–Polish	Earth orbit, 202×1551 km, inclination $48°$. Solar radio emissions.
Skylab	14 May 1973	American	Manned missions. Three successive crews. Decayed 11 July 1979.
Intercosmos 11	17 May 1974	Russian	Earth orbit, 484×526 km. Solar ultraviolet and X-rays.
Explorer 52–Injun	3 June 1974	American	Solar wind.
Helios 1	1 Dec 1974	German	American-launched. Close-range studies; went to 48 000 000 km from the Sun.
Aryabhāta	19 Apr 1975	Indian	Russian-launched. Solar neutrons and gamma-radiation.
OSO 8	18 Jun 1975	American	General studies, including solar X-rays.
Prognoz 4	22 Dec 1975	Russian	Earth orbit, $634 \times 199\,000$ km. Continuation of Prognoz programmes.
Helios 2	15 June 1976	German	American-launched. Close-range studies: went to 45 000 000 km from the Sun.
Prognoz 5	25 Nov 1976	Russian	Earth orbit, $510 \times 199\,000$ km. Carried Czech and French experiments.
Prognoz 6	22 Sept 1977	Russian	Earth-orbit. Effects of solar X-rays and gamma-rays on Earth's magnetic field.
Solar Maximum Mission	14 Feb 1980	American	Long-term vehicle. Repaired in orbit April 1984; decayed December 1989.
Ulysses	6 Oct 1990	European	American-launched. Solar polar probe.

Table 2.11 (cont.)

Name	Launch data	Nationality	Experiments
Yohkoh	30 Aug 1991	Japanese	X-ray studies of the Sun.
Koronos-1	3 May 1994	Russian	Long-term studies: carries coronagraph and X-ray telescope.
Wind	1 Nov 1994	American	Solar–terrestrial relationships.
SOHO	2 Dec 1995	European	Wide range of studies.
Polar	24 Feb 1996	American	Solar–terrestrial relationships. Polar orbit.
Cluster	4 June 1996	American	Failed to orbit.
TRACE	1 Apr 1998	American	Studies of solar transition region.
Cluster ii	16 July 2000 and 7 Aug 2000	European	Four space-craft to study small-scale structure of the Earth's magnetosphere in three dimensions.
GENESIS	8 Aug 2001	American	Solar wind collection and return.
Hinode	22 Sept 2006	Japanese	Optical, X-ray and EUV studies.
STEREO	25 Oct 2006	American	Two space-craft to study the Sun and solar wind in three dimensions.
Solar Dynamics Observatory	11 Feb 2010	American	Solar variability and magnetic studies. EUV observations. Altitude 35 880 km.

In addition, some satellites (such as the SPARTAN probes) have been released from the space shuttles and retrieved a few days later.

Table 2.12 *Structure of the Sun*

Core	The region where energy is being generated. The outer edge lies about 175 000 km from the Sun's centre. The core temperature is about 15 000 000 °C; the density 150 g cm^{-3} (10 times the density of lead). The temperature at the outer edge is about half the central value.
Radiative zone	Extends from the outer edge of the core to the interface layer, i.e. from 25 pc to 70 pc of the distance from the centre to the surface. Temperatures range from 7 000 000 °C at the base to 2 000 000 °C at the top: the density decreases from 20 g cm^{-3} (about the density of lead) to 0.2 g cm^{-3} (less than the density of water).
Interface layer	Separates the radiative zone from the convective zone. The solar magnetic field is generated by a magnetic dynamo in this layer.
Convective zone	Extends from 200 000 km to the visible surface. At the bottom of the zone the temperature is low enough for heavier ions to retain electrons; the material then inhibits the flow of radiation, and the trapped heat leads to 'boiling' at the surface. Convective motions are seen as granules and supergranules.
Photosphere	The visible surface: temperature 5700 °C, density 0.000 0002 g cm^{-3} (1/10 000 of that of the Earth's air at sea level). Sunspots are seen here. Faculæ lie on and a few hundred kilometres above the bright surface.
Chromosphere	The layer above the photosphere, extending to 5000 km above the bright surface. The temperature increases rapidly with altitude, until the chromosphere merges with the transition layer. The Fraunhofer lines in the solar spectrum are produced in the chromosphere, which acts as a 'reversing layer'. During a total eclipse the chromosphere appears as a red ring round the lunar disc (hence the name: colour-sphere).
Transition region	A narrow layer separating the chromosphere from the higher-temperature corona.
Corona	The outer atmosphere; temperature up to 2 000 000 °C, density on average about 10–15 g cm^{-3}. The solar wind originates here.
Heliosphere	A 'bubble' in space produced by the solar wind and inside which the Sun's influence is dominant.
Heliopause	The outer edge of the heliosphere, where the solar wind merges with the interstellar medium and loses its identity; the distance from the Sun is probably about 150 a.u.

disc of the Sun as dark filaments, sometimes termed flocculi. (Bright flocculi are due to calcium.)

Above the chromosphere, and the thin transition region, comes the *corona*, the 'pearly mist', which extends outward from the Sun in all directions. It has no definite boundary; it simply thins out until its density is no greater than that of the interplanetary medium. The density is in fact very low – less than one million millionth of that of the Earth's air at sea level, so that its 'heat' is negligible even though the temperature reaches around 2 000 000 °C. Because of its high temperature, it is brilliant at X-ray wavelengths.

Seen during a total eclipse, the corona is truly magnificent. The first mention of it may have been due to the Roman writer Plutarch, who lived from about AD 46 to 120. Plutarch's book *On the Face in the Orb of the Moon* contains a reference to 'a certain splendour' around the eclipsed Sun, which could well have been the corona. The corona was definitely recorded from Corfu during the eclipse of 22 December 968. The astronomer Clavius saw it at the eclipse of 9 April 1567, but regarded it as merely the uncovered edge of the Sun; Kepler showed that this could not be so, and attributed it to a lunar atmosphere. After observing the eclipse of 16 June 1806 from Kindehook, New York, the Spanish astronomer Don José Joaquin de Ferrer pointed out that if the corona were due to a lunar atmosphere, then the height of this atmosphere would have to be 50 times greater than that of the Earth, which was clearly unreasonable. However, it was only after careful studies of the eclipses of 1842 and 1851 that the corona and the prominences were shown unmistakably to belong to the Sun rather than to the Moon.

The shape and extent of the corona varies according to the amount of activity on the Sun; at eclipses seen during the Maunder Minimum (1645–1715) it was apparently very inconspicuous. Certainly the shape at spot-maximum is more symmetrical than at spot-minimum, when there are streamers and 'wings' – as was first recognised after studies of the eclipses of 1871 and 1872.

The high temperature of the corona was for many years a puzzle. It now seems that the cause is to be found in what is termed 'magnetic reconnection'. This occurs when magnetic fields interact to produce what may be termed short circuits; the fields 'snap' to a new, lower-energy state, rather reminiscent of the snapping of a twisted rubber band. Vast amounts of energy are released, which can produce flares and other violent phenomena as well as causing the unexpectedly high coronal temperature. A reconnection event was actually recorded, on 8 May 1998, from a space-craft, TRACE (the Transition Region and Coronal Explorer), which had been launched on 1 April 1998 specifically to study the Sun at a time when solar activity was starting to rise toward the peak of a new cycle.

In 2009, Sheffield and Belfast astronomers used the Swedish Solar Telescope, in La Palma, to detect Alfvén waves in the Sun's lower atmosphere. It is very probable that these elusive waves play a major role in raising the temperature of the corona to over a million degrees, though of course this does not correspond to the everyday meaning of "heat".

Coronal loops form the basic structure of the lower corona; they are highly structured, because of the twisted lines of magnetic force inside the Sun. A loop is a magnetic phenomenon, fixed at both ends, threading its way through the Sun's globe and protruding into the atmosphere. They appear in both active and quiet regions of the surface.

A coronal mass ejection is, as the name indicates, an outburst in the corona which results in the violent ejection of material. A CME is usually observed with a white-light coronagraph. The first CME was identified on 14 December 1971 by R. Tusey, from data supplied by the Orbiting Solar Observatory OSO-7. Ejecta reaching the Earth may disrupt the magnetosphere, and cause displays of auroræ; together with flares, they cause power surges and interfere with radio transmissions, as well as damaging satellite equipment. And as we have noted, an outburst in October 2007 ripped off the tail of Encke's Comet – though the comet soon recovered.

THE SOLAR WIND

The corona is the source of what is termed the *solar wind* – a stream of particles being sent out from the Sun all the time. The first suggestion of such a phenomenon was made in the early 1950s, when it was realised that the Sun's gravitational pull is not strong enough to retain the very high-temperature coronal gas, so that presumably the corona was expanding and was being replenished from below. L. Biermann also drew attention to the fact that the tails of comets always point away from the Sun, and he concluded that the ion or gas tails are being 'pushed outward' by particles from the Sun. In this he was correct. (The dust tails are repelled by the slight but definite pressure of solar radiation.) In 1958 E. N. Parker developed the theory of the expanding corona, and his conclusions were subsequently verified by results from space-craft. One of these was Mariner 2, sent to Venus in 1962. En route, Mariner not only detected a continuously flowing solar wind, but also observed fast and slow streams which repeated at 27 day intervals, suggesting that the source of the wind rotated with the Sun.

The solar wind consists of roughly equal numbers of protons and electrons, with a few heavier ions. It leads to a loss of mass of about 10^{12} tons per year (which may sound a great deal, but this is negligible by solar standards). As the wind flows past the Earth its density is of the order of 5 atoms cm^{-3}; the speed usually ranges between 200 and 700 $km\,s^{-1}$, with an average value of 400 $km\,s^{-1}$, although the initial speed as it leaves the Sun may be as high as 900 $km\,s^{-1}$.

The slow component of the solar wind comes from low solar latitudes. The fast component, which has an average velocity of around 800 $km\,s^{-1}$, comes from coronal holes, where the density is below average; coronal holes are often found near the poles, and here the magnetic field lines are open, making it easier for wind particles to escape. The wind is 'gusty', and when at its most violent the particles bombard the Earth's magnetosphere, producing magnetic storms and displays of auroræ.

The solar wind blows a 'bubble' in the interstellar medium, which is known as the *heliosphere*. Out to a distance of about 15 000 million kilometres from the Sun, the wind travels at several million kilometres per hour. As it starts to collide with the rarefied hydrogen and helium of the interstellar medium it slows down, becoming subsonic at the 'termination shock' and finally ceases altogether; this point – where the solar wind become inappreciable – is the *heliopause*; the solar wind turns back and flows down the tail of the heliosphere. As the heliopause moves through interstellar space it produces a *bow shock*. By 2005 the two Voyager probes, launched 1977 and still transmitting, had reached this outer part of the Solar System.

IBEX

NASA's Interstellar Boundary Explorer (IBEX), launched in October 2008, has mapped the invisible interactions occurring at the

edge of the Solar System, finding them to be surprisingly structured and intense. For example, IBEX data show the pile-up of the
solar wind in the magnetopause.

Pressure causes the wind to form a comet-like tail behind the
Sun; this is termed the *heliosheath*. It marks the boundary of
the heliosphere, and may even be said to form the boundary of
the true Solar System. Its distance from the Sun has been given as
150 astronomical units, though it is difficult to give a definite value.

ECLIPSES OF THE SUN

A solar eclipse occurs when the Moon passes in front of the Sun;
strictly speaking, the phenomenon is an occultation of the Sun by
the Moon. Eclipses may be total (when the whole of the photosphere is hidden), partial or annular (when the Moon's apparent
diameter is less than that of the Sun, so that a ring of the photosphere is left showing round the lunar disc: Latin annulus, a ring).
Recent and future solar eclipses are listed in Tables 2.13, 2.14, 2.15
and 2.16. On 24 October 2098 the obscuration amounts to no more
than 0.004%.

The solar corona can be well seen from Earth only during a
total eclipse. In 1930 B. Lyot built and tested a coronograph,
located at the Pic du Midi Observatory (altitude, 2870 m); this
instrument produces an 'artificial eclipse' inside the telescope.
With it Lyot was able to examine the inner corona and its spectrum,
but the outer corona remained inaccessible.

The greatest number of eclipses possible in one year is seven;
thus in 1935 there were five solar and two lunar eclipses, and in
1982 there were four solar and three lunar. The least number
possible in one year is two, both of which must be solar, as in 1984.

The length of the Moon's shadow varies between 381 000 km
and 365 000 km, with a mean of 372 000 km. As the mean distance
of the Moon from the Earth is 384 000 km, the shadow is on average
too short to reach the Earth's surface, so that annular eclipses are
more frequent than total eclipses in the ratio of five to four. On
average there are 238 total eclipses per century. During the twenty-
first century there will be 224 solar eclipses: 68 total, 72 annular,
seven annular/total (that is to say, annular along most of the track)
and 77 partial.[2]

The track of totality across the Earth's surface can never be
more than 272 km wide, and in most cases the width is much less
than this. A partial eclipse is seen to either side of the track of
totality, although some partial eclipses are not total or annular
anywhere on Earth.

The longest possible duration of totality is 7 min 31 s. This has
never been observed, but at the eclipse of 20 June 1955 totality over
the Philippines lasted for 7 min 8 s.

The longest totality during the twenty-first century was on
22 July 2009 (6 min 30 s). The shortest possible duration of totality
can be less than 1 s. This happened at the eclipse of 3 October 1986,
which was annular along most of the central track, but total
for about 1/10 s over a restricted area in the North Atlantic Ocean.
(So far as I know, totality was not observed.) The shortest total
eclipse of the twenty-first century will be that of 6 December 2067:
a mere 8 s.

Table 2.13 *Solar eclipses 1923–1999. T = total, P = partial,
A = annular*

Date	Type	Area
1923 Mar 16	A	South Africa
1923 Sept 10	T	California, Mexico
1924 Mar 5	P	South Africa
1924 July 31	P	Antarctic
1924 Aug 26	P	Iceland, Northern Russia, Japan
1925 Jan 24	T	Northeastern USA
1925 July 20/1	A	New Zealand, Australia
1926 Jan 14	T	East Africa, Indian Ocean, Borneo
1926 July 9/10	A	Pacific
1927 Jan 3	A	New Zealand, South America
1927 June 29	T	England, Scandinavia
1927 Dec 24	P	Polar zone
1928 May 19	T	South Atlantic
1928 June 17	P	Northern Siberia
1928 Nov 12	P	England to India
1929 May 9	T	Indian Ocean, Philippines
1929 Nov 1	A	Newfoundland, Central Africa, Indian Ocean
1930 Apr 28	T	Pacific
1930 Oct 21/2	T	South Pacific to South America
1931 Apr 17/18	P	Arctic
1931 Sept 12	P	Alaska, North Pacific
1931 Oct 11	A	South America, South Pacific, Antarctic
1932 Mar 7	A	Antarctic
1932 Aug 31	T	USA
1933 Feb 24	A	South America, Central Africa
1933 Aug 21	A	Iran, India, Northern Australia
1934 Feb 13/14	T	Pacific
1934 Aug 10	A	South Africa
1935 Jan 5	P	No land surface
1935 Feb 3	P	North America
1935 June 30	P	Britain
1935 July 30	P	No land surface
1935 Dec 25	A	New Zealand, southern South America
1936 June 19	T	Greece, Turkey, Siberia, Japan
1936 Dec 13/14	A	Australia, New Zealand
1937 June 8	T	Pacific, Chile
1937 Dec 2/3	A	Pacific
1938 May 29	T	South Atlantic
1938 Nov 21/2	P	East Asia, Pacific coast of North America
1938 Apr 19	A	Alaska, Arctic
1939 Oct 12	T	Antarctic
1940 Apr 7	A	USA, Pacific
1940 Oct 1	T	Brazil, South Atlantic, South Africa
1941 Mar 27	A	South Pacific, South America
1941 Sept 21	T	China, Pacific
1942 Mar 16/17	P	South Pacific, Antarctic
1942 Aug 12	P	Antarctica
1942 Sept 10	P	Britain
1943 Feb 4/5	T	Japan, Alaska

Table 2.13 (cont.)

Date	Type	Area
1943 Aug 1	A	Pacific
1944 Jan 25	T	Brazil, Atlantic, Sudan
1944 July 20	A	India, New Guinea
1945 Jan 14	A	Australia, New Zealand
1945 July 9	T	Canada, Greenland, Northern Europe
1946 Jan 3	P	Antarctica
1946 May 30	P	South Pacific
1946 June 29	P	Arctic
1946 Nov 23	P	North America
1947 May 20	T	Pacific, Equatorial Africa, Kenya
1947 Nov 12	A	Pacific
1948 May 8/9	A	East Asia
1948 Nov 1	T	Kenya, Pacific
1949 Apr 28	P	Britain
1949 Oct 21	P	New Zealand, Australia
1950 Mar 18	A	South Atlantic
1950 Sept 12	T	Siberia, North Pacific
1951 Mar 7	A	Pacific
1951 Sept 1	A	Eastern USA, central and southern Africa
1952 Feb 25	T	Africa, Arabia, Russia
1952 Aug 20	A	South America
1953 Feb 13/14	P	East Asia
1953 July 11	P	Arctic
1953 Aug 9	P	Pacific
1954 Jan 5	A	Antarctic
1954 June 30	T	Iceland, Norway, Sweden, Russia, India
1954 Dec 25	A	South Africa, Southern Indian Ocean
1955 June 20	T	Southern Asia, Pacific, Philippines
1955 Dec 14	A	Sudan, Indian Ocean, China
1956 June 8	T	South Pacific
1956 Dec 2	P	Europe, Asia
1957 Apr 29/30	A	Arctic
1957 Oct 23	T	Antarctica
1958 Apr 19	A	Indian Ocean, Pacific
1958 Oct 12	T	Pacific
1959 Apr 8	A	Southern Indian Ocean, Pacific
1959 Oct 2	T	North Atlantic, North Africa
1960 Mar 27	P	Australia, Antarctica
1960 Sept 20/1	P	North America, Eastern Siberia
1961 Feb 15	T	France, Italy, Greece, Yugoslavia, Russia
1962 Aug 11	A	South Atlantic, Antarctica
1963 Feb 4/5	T	Pacific
1963 July 31	A	South America, central Africa
1964 Jan 25	A	Pacific, South Africa
1964 July 20	T	Japan, northern North America, Pacific
1965 Jan 14	P	Tasmania, Antarctica
1965 July 9	P	Northern Canada, Arctic
1965 Dec 3/4	P	Northeast Asia, Alaska, Pacific
1966 May 30	T	Pacific, New Zealand, Peru coast
1966 Nov 23	A	Russia, Tibet, East Indies
1967 May 20	A	Greece, Russia
1967 Nov 12	T	South America, Atlantic
1968 May 9	P	North America, Iceland, Scandinavia
1968 Nov 2	T	South Atlantic
1969 Mar 28/9	P	Pacific, Antarctica
1969 Sept 22	T	Arctic, Mongolia, Siberia
1970 Mar 18	A	Indian Ocean, Pacific
1970 Sept 11	A	Peru, Bolivia
1971 Mar 7	T	Mexico, USA, Canada
1971 Aug 31/ Sept 1	T	East Indies, Pacific
1972 Feb 25	P	Europe, northwest Africa
1972 July 22	P	Alaska, Arctic
1972 Aug 20	P	Australasia, South Pacific
1973 Jan 16	A	Antarctica
1973 July 10	T	Alaska, Canada
1974 Jan 4	A	Pacific, South Atlantic
1974 June 30	T	Atlantic, North Africa, Kenya, Indian Ocean
1974 Dec 24	A	Brazil, Atlantic, North Africa
1974 June 20	T	Indian Ocean
1974 Dec 13	P	North and central America
1975 May 11	P	Europe, northern Asia, Arctic
1975 Nov 3	P	Antarctic
1976 Apr 29	A	Northwest Africa, Turkey, China
1976 Oct 23	T	Tanzania, Indian Ocean, Australia
1977 Apr 18	A	Atlantic, southwest Africa, Indian Ocean
1977 Oct 12	T	Pacific, Peru, Brazil
1978 Apr 7	P	Antarctic
1978 Oct 2	P	Arctic
1979 Feb 26	T	Pacific, USA, Canada, Greenland
1980 Aug 10	A	South Pacific, Brazil
1981 Feb 4	A	Pacific, southern Australia, New Zealand
1981 July 31	T	Russia, North Pacific
1982 Jan 25	P	Antarctic
1982 June 21	P	Antarctic
1982 July 20	P	Arctic
1982 Dec 15	P	Arctic
1983 June 11	T	Indian Ocean, East Indies, Pacific
1983 Dec 4	A	Atlantic, equatorial Africa
1984 May 30	A	Pacific, Mexico, USA, North Africa
1984 Nov 22/3	T	East Indies, South Pacific
1985 May 19	P	Arctic
1985 Nov 12	T	South Pacific, Antarctica
1986 Apr 9	P	Antarctic
1986 Oct 3	T	North Atlantic
1987 Mar 29	T	Argentina, central Africa, Indian Ocean
1987 Sept 23	A	Russia, China, Pacific
1988 Mar 18	T	Indian Ocean, East Indies, Pacific
1989 Mar 7	P	Arctic
1989 Aug 31	P	Antarctic

Table 2.13 (cont.)

Date	Type	Area
1990 Jan 26	A	Antarctic
1990 July 22	T	Finland, Russia, Pacific
1991 Jan 15	A	Pacific, New Zealand, southwestern Australia
1991 July 11	T	Pacific, Mexico, Hawaii
1992 Jan 4	A	Pacific
1992 June 30	T	Atlantic
1992 Dec 24	P	Arctic
1992 May 21	P	Arctic
1993 Nov 13	P	Antarctic
1994 May 10	A	Pacific, Mexico, USA, Canada
1994 Nov 3	T	Peru, Brazil, South Atlantic
1995 Apr 29	A	South Pacific, Peru, South Atlantic
1995 Oct 24	T	Iran, India, Borneo, Pacific
1996 Apr 17	P	Antarctic
1997 Mar 9	T	Siberia, Arctic
1997 Sept 2	P	Antarctic
1998 Feb 26	T	Pacific, Venezuela, Atlantic
1998 Aug 22	A	Indian Ocean, East Indies, Pacific
1999 Feb 16	A	Indian Ocean, Australia, Pacific
1999 Aug 11	T	Atlantic, England, Turkey, India

Annularity can last for longer; the maximum is as much as 12 min 24 s. The annular eclipse of 15 January 2010 lasted for 11 min 8 s – that of 16 December 2085 will last for only 19 s. The largest partial eclipse of the twenty-first century will be that of 11 April 2051, when the Sun will be 98.5% obscured.

The longest totality ever observed was during the eclipse of 30 June 1973. A Concorde aircraft, specially equipped for the purpose, flew underneath the Moon's shadow, keeping pace with it so that the scientists on board (including the British astronomer John Beckman) saw a totality lasting for 72 min. They were carrying out observations at millimetre wavelengths, and at their height of 55 000 feet were above most of the water vapour in our atmosphere which normally hampers such observations. They were also able to see definite changes in the corona and prominences during their flight. The Moon's shadow moves over the Earth at up to 3000 km h^{-1}, so that only Concorde could easily match it.

The first recorded solar eclipse seems to have been that of 2136 BC, seen in China during the reign of the Emperor Chung K'ang. A famous story is attached to it. The Chinese believed that during an eclipse the Sun was being attacked by a hungry dragon, and the only remedy was to beat drums, bang gongs, shout and wail, and in general make as much noise as possible in order to scare the dragon away. Not surprisingly this procedure always worked. It was the duty of the court astronomers to give warning of a forthcoming eclipse, and it has been said that on this occasion the astronomers, who rejoiced in the names of Hsi and Ho, forgot – with the result that they were executed for negligence. Alas, there can be no doubt that this story is apocryphal. The next eclipse

which may be dated with any certainty is that of 1375 BC, described on a clay tablet found at Ugarit in Syria.

Predictions were originally made by studies of the Saros period. This is the period after which the Sun, Moon and node (intersection of the planes of the orbits) arrive back at almost the same relative positions. It amounts to 6 585 321 solar days, or approximately 18 years 11 days. Therefore, an eclipse tends to be followed by another eclipse in the same Saros series 18 years 11 days later, although conditions are not identical, and the Saros is at best a reasonable guide. (For example, the eclipse of June 1927 was total over parts of England, but the 'return', in July 1945, was not.) One Saros series lasts for 1150 years; it includes 64 eclipses, of which 43 or 44 are total, while the rest are partial eclipses seen from the polar zones of the Earth.

The first known predictions about which we have reasonably reliable information were made by the Greeks. There does seem good evidence that the eclipse of 25 May 585 BC was predicted by Thales of Miletus, the first of the great Greek philosophers. It occurred near sunset in the Mediterranean area, and is said to have put an end to a battle between the forces of King Alyattes of the Lydians and King Cyraxes of the Medes; the combatants were so alarmed by the sudden darkness that they concluded a hasty peace.

Eclipse stories and legends are plentiful. Apparently the Emperor Louis of Bavaria was so frightened by the eclipse of 840 that he collapsed and died, after which his three sons engaged in a ruinous war over the succession. There was also the curious case of General William Harrison (later President of the United States) when he was Governor of Indiana Territory, and was having trouble with the Shawnee prophet Tenskwatawa. He decided to ridicule him by claiming that he could make the Sun stand still and the Moon 'alter its course'. Unluckily for him, the prophet knew more astronomy than the General, and he was aware that an eclipse was due on 16 July 1806. He therefore said that he would demonstrate his own power by blotting out the Sun. A crowd gathered at the camp, and the prophet timed his announcement at just the right moment, so that Harrison was nonplussed (although in 1811 he did destroy the Shawnee forces at the Battle of Tippecanoe).

The first total solar eclipse recorded in the United States was that of 24 June 1778, when the track passed from Lower California to New England. Two years later, on 21 October 1780, a party went to Penobscot, Maine, to observe an eclipse; it was led by S. Williams of Harvard and had been given 'free passage' by the British forces. Unfortunately, a mistake in the calculations meant that the astronomers went to the wrong place, and remained outside the track of totality. The first American expedition to Europe was more successful: on 28 July 1851 G. P. Bond took a party to Scandinavia, and obtained good results.

Astronomers have always been ready to run personal risk to study eclipses, and one man who demonstrated this in 1870 was Jules Janssen, a leading French expert concerning all matters relating to the Sun. The eclipse was due on 22 December; Janssen was in Paris, but the city was surrounded by German forces, and there was no obvious escape. Janssen's solution was to fly out in a hot-air balloon. He arrived safely at Oran, only to be met with overcast skies.

Table 2.14 *Solar eclipses 2000–2020. T = total, P = partial, A = annular*

Date	Mid-eclipse (GMT)	Type	Maximum length of totality/annularity min	s	Obscuration (percent)	Area
2000 Feb 5	13	P	–		56	Antarctic
2000 July 31	02	P	–		60	Arctic
2000 Dec 25	18	P	–		72	Arctic
2001 June 21	12	T	4	56	–	Atlantic, South Africa
2001 Dec 14	21	A	3	54	–	Central America, Pacific
2002 June 10	24	A	1	13	–	Pacific
2002 Dec 4	08	T	2	04	–	South Africa, Indian Ocean, Australia
2003 May 31	04	A	3	37	–	Northern Scotland, Iceland
2003 Nov 23	23	T	1	57	–	Antarctic
2004 Apr 19	14	P	–		74	Antarctic
2004 Oct 14	03	P	–		93	Arctic
2005 Apr 8	21	T	0	42	–	Pacific, northern South America
2005 Oct 3	11	A	4	32	–	Atlantic, Spain, Africa, Indian Ocean
2006 Mar 29	10	T	4	07	–	Atlantic, North Africa, Turkey, Russia
2006 Sept 22	12	A	7	09	–	Northeastern South America, Atlantic, southern Indian Ocean
2007 Mar 19	03	P	–	–	88	North America, Japan
2007 Sept 11	13	P	–	–	75	South America, Antarctic
2008 Feb 7	04	A	2	14	–	South Pacific, Antarctica
2008 Aug 1	10	T	2	27	–	Northern Canada, Greenland, Siberia, China
2009 Jan 26	08	A	7	56	–	Southern Atlantic, Indian Ocean, Sri Lanka, Borneo
2009 July 22	03	T	6	40	–	India, China, Pacific
2010 Jan 15	07	A	11	11	–	Africa, Indian Ocean
2010 July 11	20	T	5	20	–	Pacific
2011 Jan 4	08.51	P	–		86	North Africa, Middle East, Europe, western Asia
2011 June 1	21.16	P	–		60	Northern Canada, eastern Russia
2011 Nov 25	06.20	P	–		91	Antarctica
2012 May 20	23.53	A	5	46	–	China, Japan, western USA
2012 Nov 13	22.12	T	4	2	–	Northern Australia, South Pacific
2013 May 10	00.25	A	6	3	–	Northern Australia, central Pacific
2013 Nov 3	12.46	A/T	1	40	–	Atlantic, central Africa
2014 Apr 29	06.03	A	0	12	–	Antarctica
2014 Oct 23	21.44	P	–		81	USA, western Canada, Mexico
2015 Mar 20	09.46	T			–	Faroe Islands, Arctic
2015 Sept 13	06.54	P	–		79	South Africa, Antarctica
2016 Mar 9	01.57	T	4	9	–	Indian Ocean, Pacific
2016 Sept 1	09.07	A	3	6	–	Madagascar, Indian Ocean
2017 Feb 26	14.52	A	0	44	–	Southern South America, Atlantic, southern South Africa
2017 Aug 21	18.25	T	2	40	–	Right across USA
2018 Feb 15	20.51	P	–		60	Chile, Argentina, Antarctica
2018 July 13	03.01	P	–		34	Ocean south of Tasmania
2018 Aug 11	09.46	P	–		73	Greenland, Scandinavia
2019 Jan 6	01.41	P	–		72	Northeastern China, Japan, eastern Russia
2019 July 2	19.23	T	4	33	–	South Pacific, Chile, Argentina
2019 Dec 26	05.18	A	3	30	–	Southern India
2020 June 21	06.40	A	0	38		Mid-Africa, northern India, Pacific
2020 Dec 14	16.13	T	2	10		Pacific, Atlantic, Chile, Argentina

There will be total eclipses on 2021 Dec 4, 2023 Apr 20, 2024 Apr 8, 2026 Aug 12, 2027 Aug 2, 2028 July 22, 2030 Nov 25, 2031 Nov 14, 2033 Mar 30, 2034 Mar 20, 2035 Sept 2, 2037 July 13, 2038 Dec 26 and 2039 Dec 15.

Table 2.15 *British annular eclipses, 1800–2200*

Date	Location
1820 Sept 7	Shetland
1836 May 15	Northern Ireland, southern Scotland
1847 Oct 9	Southern Ireland, Cornwall
1858 Mar 15	Dorset to the Wash
1921 Apr 8	Northwest Scotland, Orkney, Shetland
2003 May 21	Scotland
2173 Apr 12	Hebrides

Table 2.16 *British total eclipses, 1–2200* [a]

Date		Location
21	June 19	Cornwall, Sussex
28	July 10	Southern Ireland, Cornwall
118	Sept 3	Cornwall, Sussex
122	June 21	Faroe Islands; between Shetland and Orkney
129	Feb 6	Wales to Humberside
143	May 2	Annular/total; total in southern Ireland, annular in Wales
158	July 13	London
183	Mar 11	Northern Ireland, northern England, southern Scotland
228	Mar 23	Almost all Ireland, England, Wales
303	Sept 27	Scotland
319	May 6	London
364	June 16	Northern Scotland, Orkney
393	Nov 20	London
413	Apr 16	Southern Ireland, north Wales, west Midlands
458	May 28	Wales to Lincolnshire
565	Feb 16	Channel Islands
594	July 23	Ireland, northern England, southern Scotland
639	Sept 3	Wales, Midlands
664	May 1	Northern Ireland, northern England, southern Scotland
758	Apr 12	Kent
849	May 25	Shetland Islands
865	Jan 1	Central Ireland, Cumberland
878	Oct 29	London
885	June 16	Northern Ireland, Scotland
968	Dec 22	Scilly, Cornwall, Jersey
1023	Jan 24	Cornwall, Wales, southern Scotland
1133	Aug 2	Scotland, northeastern England
1140	Mar 20	Wales to Norfolk
1185	May 1	Scotland
1230	May 14	Almost all England
1330	July 16	Northern Scotland
1339	July 7	Between Shetland and Orkney
1424	June 28	Orkney, Shetland
1433	June 17	Scotland
1440	Feb 3	Near miss of Outer Hebrides
1598	Feb 25	Wales, southern Scotland
1630	June 10	Cork, Scilly Isles
1652	Apr 8	Anglesey, Scotland
1654	Aug 12	Grampian, Aberdeen
1679	Apr 10	Western Ireland
1699	Sept 23	Southeastern tip of Scotland
1715	May 3	Cornwall, London, Norfolk
1724	May 22	South Wales, Hampshire, London
1925	Jan 24	Near miss of Outer Hebrides
1927	June 29	Wales, Preston, Giggleswick
1954	June 30	Northernmost Scotland (Unst)
1999	Aug 11	Cornwall, Devon, Alderney
2015	Mar 20	Faroes: misses Scotland
2081	Sept 3	Channel Islands
2090	Sept 23	Southern Ireland, Cornwall
2133	June 3	Hebrides, Scotland
2135	Oct 7	Southern Scotland, northern England, northern Wales
2142	May 25	Channel Islands
2151	June 14	Scotland, North London, Kent
2160	June 4	Cork, Land's End
2189	Nov 8	Cork, Cornwall
2200	Apr 14	Northern Ireland, Isle of Man, Lake District

[a] Calculations by Sheridan Williams, whom I thank for allowing me to quote them.

In Britain, eclipse records go back a long way. The first account comes from the *Anglo-Saxon Chronicle*; the eclipse took place on 15 February 538, four years after the death of Cerdic, the first King of the West Saxons. The Sun was two-thirds eclipsed from London.

The celebrated chronicler William of Malmesbury gave a graphic description of the eclipse of 1133: the Sun 'shrouded his glorious face, as the poets say, in hideous darkness, agitating the hearts of men by an eclipse and on the sixth day of the week there was so great an earthquake that the ground appeared to sink down; a horrid noise being first heard beneath the surface'. In fact there can be no connection between an eclipse and a ground tremor, but William was again busy at the eclipse of 1140: 'It was feared that Chaos had come again ... it was thought and said by many, not untruly, that the King [Stephen] would not continue a year in the government.' (In fact, Stephen reigned until 1154.) Several Scottish eclipses were given nicknames; Black Hour (1433), Black Saturday (1598), Mirk Monday (1652).

The eclipse of 1715 was well observed over much of England. Edmond Halley saw it, and gave a vivid description of the corona: 'A luminous ring of a pale whiteness, or rather pearl colour, a little tinged with the colours of the Iris, and concentric with the Moon'. He was also the first to see Baily's Beads – brilliant spots caused by the Sun's rays shining through valleys on the lunar limb immediately before and immediately after totality. They can sometimes be seen during an annular eclipse (as by Maclauin, from Edinburgh, on 1 March 1737) but the first really detailed description of them was given in 1836, at the annular eclipse of 15 May, by Francis Baily, after whom they are named. (They were first photographed at the eclipse of August 1869 by C. F. Hines and members of the Philadelphia Photographic Corps, observing from Ottuma in Iowa.)

The last British mainland totality before 1927 was that of 1724. Unfortunately the weather was poor and the only good report came from a Dr Stukeley, from Haraden Hill near Salisbury. The spectacle, he wrote, 'was beyond all that he had ever seen or could picture to his imagination that most solemn'. The eclipse was much better seen from France.

In 1927, the track crossed parts of Wales and northern England, but there was a great deal of cloud and the best results came from Giggleswick, where the Royal Astronomical Society party was stationed. Totality was brief – only 24 s – but the clouds cleared away at the vital moment, and useful photographs were obtained. On 30 June 1954 the track brushed the tip of Unst, northernmost of the Shetland Islands, but most observers went to Norway or Sweden. On 11 August 1999 the track crossed Devon and Cornwall, but most of the area was cloudy, though the partial phase was well seen from most of the rest of Britain. Turkey and Iran had good views; the prominences were particularly striking – not at all surprising as the Sun was rising to the peak of its 11-year cycle.

The maximum theoretical length of a British total eclipse is 5.5 min. That of 15 June 885 lasted for almost 5 min, and so will the Scottish total eclipse of 20 July 2381.

Another phenomenon seen at a total eclipse is that of shadow bands, wavy lines crossing the landscape just before and just after totality; they are, of course, produced in the Earth's atmosphere. They were first described by H. Goldschmidt at the eclipse of 1820.

The first attempt to photograph a total solar eclipse has made by the Austrian astronomer G. A. Majocci on 8 July 1842. He failed to record totality, although he did manage to photograph the partial phase. The first real success, showing the corona and prominences, was due to M. Berkowski on 28 July 1851, using the 6.25 Königsberg heliometer with an exposure time of 24 s. The flash spectrum was first photographed by the American astronomer C. Young, on 22 December 1870. (The flash spectrum is the sudden change in the Fraunhofer lines from dark to bright, when the Moon blots out the photosphere in the background and the chromosphere is left shining 'on its own'.) The flash spectrum was first observed during an annular eclipse by N. R. Pogson, in 1872.

Nowadays, of course, total eclipses are shown regularly on television. The first attempt to show totality on television from several stations spread out along the track was made by the BBC at the eclipse of 15 February 1961. All went well, and totality was shown successively from France, Italy and what was then Jugoslavia. There was, however, one bizarre incident. I was stationed atop Mount Jastrebač, in Jugoslavia, and with our party were several oxen used to haul the equipment up to the summit. It is quite true that animals tend to go to sleep as darkness falls, and, unknown to me, the Jugoslav director decided to show this as soon as totality began – so he trained the cameras on to the oxen and to make sure that the viewers were treated to a good view, he switched on floodlights! I made a gesture which could not possibly be misinterpreted even in Serbo Croat, and the floodlights were promptly switched off.

The last total eclipse will probably occur in about 700 million years from now. By then the Moon will have receded to about 29 000 km further away from the Earth, and the disc will no longer appear large enough to cover the Sun.

EVOLUTION OF THE SUN

The Sun is a normal Main Sequence star. It is in orbit round the centre of the Galaxy; the period is of the order of 225 000 000 years – sometimes known as the 'cosmic year'. One cosmic year ago, the most advanced creatures on Earth were amphibians; even the dinosaurs had yet to make their entry. (It is interesting to speculate as to conditions here one cosmic year hence!) The apex of the Sun's way – i.e. the point in the sky toward which it is moving – is RA 18h, declination $+34°$, in Hercules; the antapex is at RA 6h, declination $-34°$, in Columba.

The age of the Earth is about 4600 million years, and the Sun is certainly older than this, so that perhaps 4800 million years to around 5000 million years is a reasonable estimate. The Sun was born inside a giant gas cloud, perhaps 50 light-years in diameter, which broke up into globules, one of which produced the Sun. The first stage was that of a proto-star, surrounded by a cocoon of gas and dust which may be termed a solar nebula (an idea first proposed by Immanuel Kant as long ago as the year 1755). Contraction led to increased heat; there was a time when the fledgling star varied irregularly, and sent out an energetic 'wind' (the so-called T Tauri stage), but eventually the cocoon was dispersed, and the Sun became a true star. When the core temperature reached around 10 000 000 °C, nuclear reactions began. Initially the Sun was only 70% as luminous as it is now but eventually it settled on to the Main Sequence, and began a long period of comparatively steady existence.

The supply of available hydrogen 'fuel' is limited, and as it ages the Sun is bound to change. Over the next thousand million years there will be a slow but inexorable increase in luminosity, and the Earth will become intolerably hot from our point of view. Worse is to come. Four thousand million years from now the Sun's luminosity will have increased threefold, so that the surface temperature of the Earth will soar to 100 °C and the oceans will be evaporated. Another thousand million years, and the Sun will leave the Main Sequence to become a giant star, with different nuclear reactions in the core. There will be a period of instability, with swelling and shrinking (the 'asymptotic giant' stage) but eventually the Sun's diameter will grow to 50 times its present size; the surface temperature will drop, but the overall luminosity will increase by a factor of at least 300, with disastrous results for the inner planets. The temperature at the solar core will once again reach 100 000 000 °C, and helium will react to produce carbon and oxygen. A violent solar wind will lead to the loss of the outer layers, so that for a relatively brief period on the cosmical scale the Sun will become a planetary nebula. Finally, all that is left will be a very small, dense core made up of degenerate matter; the Sun will have become a white dwarf, with all nuclear reactions at an end. After an immensely long period – perhaps several tens of thousands of millions of years – all light and heat will depart, and the end product will be a cold, dead black dwarf, perhaps still circled by the ghosts of the remaining planets.

It does not sound an inviting prospect, but at least it need not alarm us. The Sun is no more than halfway though its career on the Main Sequence; it is no more than middle-aged.

ENDNOTES

1 It may be worth recalling that in 1952 a German lawyer, Godfried Büren, stated that the Sun had a vegetation-covered inner globe, and offered a prize of 25 000 Marks to anyone who could prove him wrong. The leading German astronomical society took up the challenge, and won a court case, although whether the prize was actually paid does not seem to be on record! So far as I know, the last serious protagonist of theories of this sort was an English clergyman, the Reverend P. H. Francis, who held a degree in mathematics from Cambridge University. His 1970 book, *The Temperate Sun*, is indeed a remarkable work.

2 The calculations were made by Fred Espenak of NASA. I thank him for allowing me to quote them.

3 · The Moon

The Moon is officially ranked as the Earth's satellite. Relative to its primary, it is however extremely large and massive, and it might well be more appropriate to regard the Earth–Moon system as a double planet. Data are given in Table 3.1.

The *synodic period* (i.e. the interval between successive new moons or successive full moons) is 29d 12h 44m, so that generally there is one full moon every month. However, it sometimes happens that there are two full moons in a calendar month and one month (February) may have none. Thus in 1999 there were two full moons in January (on the 2nd and the 31st), none in February and two again in March (on the 2nd and the 31st as with January). By tradition a second full moon in a month is known as a *blue moon*, but this has nothing whatsoever to with a change in colour. (This is not an old tradition. It comes from the misinterpretation of comments made in an American periodical, the *Maine Farmers' Almanac*, in 1937.) Yet the Moon can occasionally look blue, due to conditions in the Earth's atmosphere. For example, this happened on 26 September 1950, because of dust in the upper air raised by vast forest fires in Canada. A blue moon was seen on 27 August 1883 caused by material sent up by the volcanic outburst at Krakatoa, and green moons were seen in Sweden in 1884 – at Kalmar, on 14 February, for 8 min, and at Stockholm on 12 January, also for 3 min.

Other full moons have nicknames (Table 3.2), but of these only two are in common use. In the northern hemisphere, the full moon closest to the autumnal equinox, which falls around 22 September, is called Harvest Moon. This is because the ecliptic then makes its shallowest angle with the horizon, and then the time lapse – that is to say, the interval between moonrise on successive nights – is at its minimum; maybe no more than 15 min, although for most of the year this amounts to at least 30 min. It was held that this was useful to farmers gathering in their crops. Harvest Moon looks the same as any other full moon – and it is worth noting that the full moon is no larger when low down than when high in the sky. Certainly it does give this impression, but the 'Moon Illusion' *is* an illusion and nothing more.

In Islam, the calendar follows a purely lunar cycle, so that over a period of about 33 years the months slowly regress through the seasons. Each month begins with the first sighting of the crescent Moon, and this is important in Islamic religion. An early sighting was made on 15 March 1972 by R. Moran of California, who used 10×50 binoculars and glimpsed the Moon 14h 53m past conjunction; on 21 January 1996 P. Schwann, from Arizona, used 25×60 binoculars to glimpse the Moon only 12h 30m after conjunction.

(As an aside: in 1992 a British political party, the Newcastle Green Party, announced that it would meet at new moon to discuss ideas and at full moon to act upon them. It has not, so far, won any seats in Parliament!)

There is no conclusive evidence of any link between the lunar phases and weather on Earth, or of any effect upon living things – apart from aquatic creatures, since the Moon is the main agent in controlling the ocean tides.

During the crescent stage the 'night' part of the Moon can usually be seen shining faintly. This is known as the *Earthshine* and is due solely to light reflected on to the Moon by the Earth – as was first realised by Leonardo da Vinci [1452–1519].

MOON LEGENDS AND MOON WORSHIP

Every country has its own Moon legends – and who has not heard of the Man in the Moon? According to a German tale, the Old Man was a villager caught stealing cabbages, and was placed in the Moon as a warning to others: he was also a thief in Polynesian lore. Frogs and toads have also found their way there, and stories about the hare in the Moon are widespread. From China comes another delightful story. A herd of elephants made a habit of drinking at the Moon Lake and trampled down the local hare population. The chief hare then had an excellent idea; he told the elephant that by disturbing the waters they were angering the Moon goddess, by destroying her reflection. The elephants agreed that this was most unwise, and made a hasty departure.

There is another Chinese story – that of the 'Moon Maiden' Chang-ē. It is connected with the Sun legend of the ten suns which moved round the world and were ordered to shine one at a time. Eventually they decided to shine together, with the result that the Earth was scorched, and nine of the suns were shot down by the archer Hou Yi. One day Yi was given the elixir of immortality, and asked his wife Chang-ē to guard it for him. Alas, when Yi was out hunting, Chang-ē was threatened by the evil archer Peng Meng, and rather then give him the elixir she swallowed it. Instantly she began to float to the Moon – and landed safely, accompanied by a jade rabbit; as a Moon goddess she can still be seen there. Yi was sad at the loss of his beautiful wife, but at least he can worship her, particularly on the 15th day of the 8th lunar month each year, when the Moon is particularly brilliant. This is one of the most popular of all Chinese legends – and when the first Chinese lunar space-probes were launched, on 24 October 2007, it was named Chang-ē 1.

To the people of Van, in Turkey, the Moon was a young bachelor who was engaged to the Sun. Originally the Moon had shone in the daytime and the Sun at night, but the Sun, being feminine, was afraid of the dark – and so they changed places.

25

Table 3.1 *The Moon: data*

Distance from the Earth, centre to centre (km):
  mean 384 400
  max. 406 697 (apogee)
  min. 356 410 (perigee)
Distance from the Earth, surface to surface (km):
  mean 376 284
  max. 398 581 (apogee)
  min. 348 294 (perigee)
Revolution period: 27.321 661 days
Synodic period: 29.53 days (29d 12h 44m 2s .9)
Mean orbital velocity: 1.023 km s^{-1} (3682 km h^{-1})
Mean sidereal daily motion: 47434″ .8899 = 13° .17636
Mean transit interval: 24h 50m .47
Orbital eccentricity: 0.0549
Mean orbital inclination: 5° 9′
Axial rotation period: 27.321661 days (synchronous)
Inclination of lunar equator: to ecliptic 1° 32′ 30″, to orbit 6° 41′
Rate of recession from Earth: 3.8 cm year^{-1}
Diameter: equatorial 3476 km
          polar 3470 km
Oblateness: 0.003
Equatorial circumference, km: 10,921
Apparent diameter from Earth:
  max. 33′ 31″
  mean 31′ 5″
  min. 29′ 22″
Reciprocal mass, Earth = 1: 81.301 (= 7.350 × 1025 g)
Density, water = 1: 3.342
Escape velocity: 2.38 km s^{-1}
Volume, Earth = 1: 0.0203
Surface gravity, Earth = 1: 0.1653
Mean albedo: 0.067
Atmospheric density: 10^{-14} that of the Earth's atmosphere at sea level
Surface temperature range (°C): −184 to +101
Optical libration, selenocentric displacement: longitude ± 7° .6
                                               latitude ± 6° .7
Nutation period, retrograde: period 18.61 tropical years

Table 3.2 *Legendary names of full moons*

January	Winter Moon, Wolf Moon
February	Snow Moon, Hunger Moon
March	Lantern Moon, Crow Moon
April	Egg Moon, Planter's Moon
May	Flower Moon, Milk Moon
June	Rose Moon, Strawberry Moon
July	Thunder Moon, Hay Moon
August	Grain Moon, Green Corn Moon
September	Harvest Moon, Fruit Moon
October	Hunter's Moon, Falling Leaves Moon
November	Frosty Moon, Freezing Moon
December	Christmas Moon, Long Night Moon

In many mythologies the Sun is female and the Moon male, although this is not always the case. For example, in Greenland it is said that the Sun and Moon were brother and sister, Anninga and Malina. When Malina smeared her brother's face with soot, she fled to avoid his anger: reaching the sky, she became the Sun. Anninga followed and became the Moon, but he cannot fly equally high, and so he flies round the Sun hoping to surprise her. When he becomes tired at the time of lunar First Quarter, he leaves his house on a sled towed by four dogs, and hunts seals until he is ready to resume the chase.

From the Maori of New Zealand comes the legend of Rona, daughter of the sea-god Tangaroa. One evening she took a bucket and walked along to a brook, where she filled the bucket with fresh water to take back to her children. Then, without warning, the Moon dipped behind a cloud, and in the sudden darkness Rona stumbled and fell, dropping the bucket and spilling the water. Instinctively she made some very unkind remarks about the Moon – and the Moon, hearing them, grabbed both her and the bucket, swinging them up into the sky. Rona can still be seen in the Moon; every time she tips her bucket, rain falls.

Of course there were many lunar deities, such as Isis (Egypt), Tsuki-yomi-nokami (Japan) and Diana (Rome). In early Greek mythology the Moon goddess Selene is the sister of the Sun god Helios. When Helios completes his daily journey across the sky, Selene, freshly washed in the great Ocean, which surrounds the world, takes his place and starts her own journey, to provide light during the hours of night.

Moon worship continued until a surprisingly late stage, at least in Britain; from the Confessional of Ecgbert, Archbishop of York, we learn that in the eighth century AD homage was still being paid to the Moon as well as to the Sun.

ROTATION OF THE MOON

The Moon's rotation is synchronous (captured), i.e. the axial rotation period is the same as the orbital period. This means that the same area of the Moon is turned Earthward all the time, although the eccentricity of the lunar orbit leads to libration zones which are brought alternately in and out of view. From Earth, 59% of the Moon's surface can be studied at one time or another; only 41% is permanently out of view. There is no mystery about this behaviour; tidal forces over the ages have been responsible. Most other planetary satellites also have synchronous rotation with respect to their primaries.

The barycentre, or centre of gravity of the Earth–Moon system, lies 1707 km beneath the Earth's surface, so that the statement that 'the Moon moves round the Earth' is not really misleading.

The fact that the Moon has synchronous rotation was noted by Cassini in 1693; Galileo may also have realised it. The libration zones are so foreshortened that from Earth they are difficult to map, and good maps were not possible until the advent of spacecraft. The first images of the averted 41% were obtained in 1959 by the Russian vehicle Luna (or Lunik) 3.

Because of tidal effects, the Moon is receding from the Earth at a rate of 3.83 cm year^{-1}; also, the Earth's rotation period is lengthening, on average, by 0.000 000 2 s day^{-1}, although motions of material inside the Earth mean that there are slight irregularities superimposed on the tidal increase in period.

Ancient eclipse records are helpful here. We know when these eclipses should have occurred, assuming that the Earth's rotation is constant, but there are discrepancies, and old records are quite clear. It is true that we now have atomic clocks which are better timekeepers than the Earth itself, and we know that the rotation period sometimes changes very slightly, but on average each day is 0.000 000 02 of a second longer than its predecessor. Let us see how this 'secular acceleration of the Moon' shows up.

As each day is 0.000 000 2 of a second longer than the previous day, then a century (36 525 days) ago the length of the day was shorter by 0.00073 of a second. Taking the average between then and now, the length of the day was half of this value, or 0.00036 of a second shorter than at present. But since 36 525 days have passed by, the total error is 36 525 × 0.00036 = 13 seconds. Therefore the position of the Moon, when calculated back, will be in error; it will seem to have moved too far, i.e. too fast. It follows that eclipses of many centuries ago will not occur at the moments they would have done if there were no secular acceleration.

ORIGIN OF THE MOON

Various theories have been put forward to describe the origin of the Moon. At least we have one definite fact to help us – the Earth and the Moon are of the same age, 4.6 thousand million years, as we know from analyses of the rocks brought back by the Apollo astronauts and the Russian unmanned probes. We also know that the overall density of the Moon's globe is appreciably lower than that of the Earth.

The first theory to gain widespread acceptance was that of G. H. Darwin, proposed in 1878 and extended by O. Fisher four years later. The Earth and Moon were originally combined and rotated quickly – so quickly, in fact, that part of the globe broke away, and became the Moon. Fisher believed that the basin now filled by the Pacific Ocean was left by the departing Moon. All this sounds plausible enough, but mathematical analysis shows that there are fatal objections. A thrown-off mass would not form a lunar-sizd globe, the high angular momentum of the Earth–Moon system cannot be explained, and there are other problems too. The Moon's diameter is roughly one-third that of the Earth, but the depth of the Pacific basin is absolutely negligible compared with the diameter of the Earth itself.

H. C. Urey proposed the 'capture' theory: the Moon was formed from the solar nebula in the same way as the Earth, and was originally an independent planet, but after a while the two became gravitationally linked. However, this would require a set of very special circumstances, and the theory cannot easily explain why the Moon is so obviously less dense than the Earth. These problems remained even if the Earth and the Moon had formed close together, and at the same time. At one stage Urey commented that since all theories of the Moon's origin were so unsatisfactory, science had proved that the Moon does not exist!

Then, in 1974, came the Giant Impact theory, proposed by W. Hartmann and D. R. Davies. This involved a collision about 4000 million years ago between the proto-Earth and a body about the size of Mars – in fact, an extra member of the inner group of planets. The cores of the two bodies merged, and the mantle debris was spread round, subsequently accreting to form the Moon. At the time of the collision so much energy was liberated that the outer part of the fledgling Moon was melted to form a deep global magma ocean. Over the next 100 million years this ocean crystallised and differentiated; iron silicates such as olivine and pyroxene sank, while the less dense feldspar rose up and floated. When the process was completed, the Moon was left with a crust and a mantle. The stage was set for the Heavy Bombardment.

The Hartmann–Davies theory avoids most of the difficulties faced by older hypotheses, and most astronomers now accept it, but we cannot pretend that we know all the details, and some people still believe that the Moon was originally independent.

MINOR SATELLITES

No minor Earth satellites of natural origin seem to exist. Careful searches have been made for them, notably in 1957 by Clyde Tombaugh (discoverer of the dwarf planet Pluto), but without result. A small satellite reported in 1846 by F. Pettit, Director of the Toulouse Observatory in France, was undoubtedly an error in observation – although Jules Verne found it very useful in his great novel *From the Earth to the Moon* and its sequel *Round the Moon* (1865) – and very faint clouds of loose material at the Lagrangian points L4 and L5 were reported visually in 1956 by the Polish astronomer K. Kordylewski; in 1961 he obtained decidedly inconclusive photographs. Little is known about them; their very existence has been questioned, and it is fair to say that any true satellite more than a few metres in diameter would certainly have been discovered by now.

Several small asteroids have been referred to as minor Earth satellites, but they are nothing of the kind; they are moving round the Sun, though they are strongly influenced by the gravitational pull of the Earth. The best-known asteroid of these 'quasi-satellites' is 3753 Cruithne, which was discovered in 1986 by D. Waldron. Its mean distance from the Sun is 149 260 million kilometres (0.998 a.u.) and its orbital period is 364.02 days, but its orbit is more eccentric ($e = 0.515$) than ours and there is no danger of collision; it cannot come within 12.5 million kilometres of Earth (the next 'close' approach will be in the year 2292). Relative to the Earth, it has been said that the shape of its path resembles that of a kidney bean! Several other quasi-satellites are known, but they will not keep company with us indefinitely.

MAPPING THE MOON

The first known map of the Moon is prehistoric. It was identified in 1999 by Dr Philip Stooke, of the University of Western Ontario, carved into a rock at one of Ireland's most famous Neolithic tombs,

Knowth, in County Meath, and is thought to be about 5000 years old. Stooke commented: 'I was amazed when I saw it. Place the markings over a picture of the full moon, and you will see that they line up ... You can see the overall pattern of the lunar features, from the Mare Humorum through to the Mare Crisium. The people who carved this Moon map knew a great deal about the movements of the Moon; they were not primitive at all.' The Knowth site consists of a large mound containing two passages in an east–west line. At certain times of the year the Moon's light shines down the eastern passage, and falls upon the Neolithic map.

The first suggestion that the Moon is mountainous was made by the Greek philosopher Democritus (460–370 BC). Earlier, Xenophanes (*c.* 450 BC) had supposed that there were many suns and moons according to the regions, divisions and zones of the Earth! Certainly the main maria and some other features can be seen with the naked eye, and the first historic map which has come down to us was that of W. Gilbert, drawn in 1600, although it was not published until 1651 (Gilbert died in 1603).

Telescopes became available in the first decade of the seventeenth century. The first known telescopic map was produced in July 1609 by Thomas Harriot, one-time tutor to Sir Walter Raleigh. It shows a number of identifiable features, and was more accurate than Galileo's map of 1610. Another very early telescopic observer of the Moon was Sir William Lower, an eccentric Welsh baronet. His drawings, made in or about 1611, have not survived, but he compared the appearance of the Moon with a tart that his cook had made – 'Here some bright stuffe, there some darke, and so confused lie all over.'

Galileo did at least try to measure the heights of some of the lunar mountains, from 1611, by the lengths of their shadows. He concentrated on the lunar Apennines, and although he overestimated their altitudes his results were of the right order. Much better results were obtained by J. H. Schröter, from 1778.

The first systems of nomenclature were introduced in 1645 by van Langren (Langrenus) and in 1647 by Hevelius, but few of their names have survived; for example, the crater we now call Plato was named by Hevelius 'the Greater Black Lake'. At that time, of course, it was widely although not universally believed that the bright areas were lands, and the dark areas were watery. The modern-type system was introduced in 1651 by the Jesuit astronomer G. Riccioli, who named the features in honour of scientists – plus a few others. He was not impartial; for instance, he allotted a major formation to himself and another to his pupil Grimaldi, and he did not believe in the Copernican theory that the Earth moves round the Sun – so he 'flung Copernicus into the Ocean of Storms'. Riccioli's principle has been followed since, although clearly all the major craters were used up quickly and later distinguished scientists had to be given formations of lesser importance, at least until it became possible to map the Moon's far side by using space research methods.

Other maps followed, some of which are listed in Table 3.3. Tobias Mayer in 1775 was the first to introduce a system of lunar coordinates, although the first accurate measurements with a heliometer were not made until 1839, by the German astronomer F. W. Bessel.

Table 3.3 *Selected list of pre-Apollo lunar maps*

Date	Diameter (cm)	Author
1610	7	Galileo
1634	21	Mellan
1645	40	van Langren
1647	29	Hevelius
1651	11.3	Riccioli
1662	38	Montanari
1680	53	Cassini
1775	21	Mayer
1797	30 (globe)	Russell
1824	95	Lohrmann (unfinished)
1837	95	Beer and Mädler
1859	30	Webb
1873	30	Proctor
1876	61	Neison
1878	187	Schmidt
1895	46	Elger
1898	43	König
1910	196	Goodacre
1927	46	Lamèch
1930	508	Wilkins
1934	156	Lamèch
1935	100	International Astronomical Union
1936	86	Fauth
1946	762	Wilkins[a]

[a] Revised and re-issued to one-third scale in 1959.

Undoubtedly the first really great lunar observer was J. H. Schröter, whose astronomical career extended from 1778, when he set up his private observatory at his home in Lilienthal, near Bremen in Germany, until 1813, when his observatory was destroyed by invading French troops (the soldiers even plundered his telescopes, which were brass-tubed and were taken to be made of gold). Schröter made many drawings of lunar features and was also the first to give a detailed description of the rills,[1] although some of these had been seen earlier by the Dutch observer Christiaan Huygens.

In 1837–8 came the first really good map of the Moon, drawn by W. Beer and J. H. Mädler from Berlin. Although they used a small telescope (Beer's 3.75 inch or 9.50 cm refractor) their map was a masterpiece of careful, accurate work and it remained the standard for several decades. They also published a book, *Der Mond*, which was a detailed description of the whole of the visible surface. A larger map completed in 1878 by Julius Schmidt was based on that of Beer and Mädler, so too was the 1876 map and book written by E. Neison (real name, Nevill). Other useful atlases were those of Elger (1895) and Goodacre (1910, revised 1930); in 1930 the Welsh observer H. Percy Wilkins published a vast map, 300 inches (over 500 cm) in diameter; it was re-issued, to one-third the scale, in 1946.

The first good photographic atlas was published in 1899 by the Paris astronomers Loewy and Puiseux, but the first actual photographs date back much further; a Daguerreotype was taken on 23 March 1840 by J. W. Draper, using a 12.0 cm reflector, but the image was less than 3 cm across and required an exposure time of 20 min.

Nowadays, of course, there are photographic atlases of the entire surface, obtained by space-craft, and it is fair to say that the Moon is better charted than some regions of the Earth.

A useful lunar atlas in the 1960s was the *Photographic Lunar Atlas* edited by G. P. Kuiper in Chicago (1960). It was based on selected plates from the great observatories of the world, and showed most areas of the Moon in morning, highlight and evening illuminations.

However, special mention should be made of an Earth-based photographic atlas produced by H. R. Hatfield, using his 32 cm reflector. It was re-issued in 1999, and is ideal for use by the amateur observer, as it shows all areas of the Moon under different conditions of illumination. A more detailed map of immense value is that of the Czech observer Antonín Rükl updated and published by Cambridge University Press in 2004.

There were, of course, some oddities. No less a person than the great Sir William Herschel, who died in 1822, never wavered in his belief that the Moon must be inhabited, and in 1822 the German astronomer F. von Paula Gruithuisen described a structure with 'dark gigantic ramparts', which he was convinced was a true city built by the local populace whereas in fact the area shows nothing but low, haphazard ridges. There was also the famous Lunar Hoax of 1835. A daily paper, the New York *Sun*, published some quite fictitious reports of discoveries made by Sir John Herschel in the Cape of Good Hope. The reports were written by a reporter, R. A. Locke, and included descriptions of batmen and quartz mountains. The first article appeared in August, and was widely regarded as authentic; only in September did the *Sun* confess to a hoax. One religious group in New York City even started to make plans to send missionaries to the Moon in an attempt to convert the batmen to Christianity.

This sounds very strange, but as late as the 1930s one eminent astronomer, W. H. Pickering, was maintaining that certain dark patches on the Moon might be due to the swarms of insects or even small animals. Only since dawn of the Space Age have we been sure that the Moon is, and always has been, totally sterile.

SURFACE FEATURES

The most prominent features are of course the *maria* (seas). Although they have never contained water (as one eminent authority, H. C. Urey, believed as recently as 1966) they are undoubtedly old lava-plains, and the Moon was once the scene of violent volcanic activity. Some, such as Mare Crisium and Mare Humorum, are more or less regular in outline, and their basins were due originally to impacts; others, such as Mare Frigoris and Oceanus Procellarum, are very irregular. Details of the maria are given at the end of the chapter in Table 3.13, and selected craters on the far and near side of the Moon are given in Tables 3.14 and 3.15, respectively.

The largest of the regular 'seas' is Mare Imbrium, 1300 km across; it is also the largest to be associated with an impact basin.

It is bounded in part by the Apennines, Alps and Carpathians; its area is about the same as that of Pakistan, and surrounding the basin is a region blanketed by ejecta from the impact. Its age is thought to be about 3850 million years – and its formation had marked effects over the entire lunar surface. At its antipodal region there is an area of chaotic terrain thought to have been produced when seismic waves from the Imbrian impact were focused there after passing around the Moon's outer layers.

Oceanus Procellarum is considerably larger, 2500 km across its north–south axis and covering an area of 4 million square kilometres – greater than that of our Mediterranean. However, it is not associated with a definite impact basin; it was formed by basaltic flooding, covering the region in solidified lava. It is separated from Imbrium by the Carpathian Mountains.

Most of the main seas make up a connected system; the main exception is Mare Crisium. Though it looks elongated in a north–south direction, this is an effect of foreshortening; the east–west diameter is actually greater (590 km, against 490 km). In general the regular maria are the more depressed; the Mare Crisium lies about 4 km below the mean sphere, whereas the depth of the Oceanus Procellarum is on average no more than about 1 km.

The most striking feature on the Moon's far side is the Mare Orientale (Eastern Sea), a vast, multi-ringed impact structure probably the youngest on the Moon – younger even than the Mare Imbrium. The main central area is about 300 km across, but the outer rings extend to over 900 km. Only a very small part of it can be seen from Earth, and even then only under favourable conditions of libration; on classical maps these outer rings are named the Rook and Cordillera Mountains. The central area is covered with a layer of mare basalt, which may be no more than 1 km thick, but much of Orientale is unflooded. There are no other comparable seas on the far side; those named, such as Mare Moscoviense and Mare Ingenii, are relatively minor. Moscoviense is 277 km across, but this is much smaller than some of the far-side impact craters, such as Hertzsprung (591 km).

The largest lunar basin by far is the South Pole-Aitken Basin. It is 2400 km across, and over 8 km deep. The impact punched into layers of the lunar crust, scattering material across the Moon and into space. Near its edge is the 537-km Apollo Basin. The effect may be likened to going into a hollow and digging a deeper hole. The central part of the Apollo Basin is probably one of a very few places where we can see an exposed portion of the Moon's deep lower crust, elsewhere covered by volcanic material. It is barely visible from Earth, but the peaks on its outer rim show up as a chain of peaks along the southern limb of the Moon, and were once known as the Leibnitz Mountains. The crust below the basin is thought to be no more than about 15 km deep, because of the enormous amount of material excavated at the time of the impact.

A selected list of crater depths is given in Table 3.4. They are of many types; very often 'walled plains' would be a better term. In profile, a lunar crater is more like a shallow saucer than a mineshaft; the walls rise to only a modest height above the outer surface, while the floors are depressed. Central mountains or mountain groups are very common, but never attain the height of the outer ramparts. The depths of some craters are given in Table 3.4, but it

Table 3.4 *Crater depths. Depth values for lunar craters may carry large standard errors, and the figures given here are no more than approximate. The following are some typical mean values of the ramparts above the floor. Formations whose walls are particularly irregular in height are marked *. Depths are in metres*

6100	Newton	2570	Kepler
5220	Werner	2400	Ptolemæus*
4850	Tycho	2530	Pytheas
4730	Maurolycus	2510	Halley
4400	Theophilus	2450	Almanon
4130	Walter	2400	Proclus
3900	Alpegragius	2320	Plinius
3850	Theon Junior	2300	Posidonius
3830	Alfraganus	2200	Hell
3770	Herschel	2150	Archimedes
3770	Copernicus	2100	Le Verrier
3750	Stiborius	2080	Campanus
3730	Abenezra	2000	Brayley
3650	Aristillus	1960	Fauth
3620	Arzachel	1860	Aratus
3570	Eratosthenes	1850	C Herschel
3510	Bullialdus	1850	Ammonius
3470	Theon Senior	1850	Gassendi
3430	Autolycus	1810	Feuillée
3320	Hipparchus*	1770	Boscovich*
3270	Thebit	1760	Mercator
3200	Godin	1740	Bessel
3140	Catharina	1730	Regiomontanus*
3130	Cayley	1730	Vitello
3120	Kant	1700	Manners
3110	Timocharis	1650	Beer
3110	Abulfeda	1550	Vitruvius
3110	Lansberg	1490	D' Arrest
3050	Manilius	1240	Cassini*
3010	Menelaus	1230	Tempel
3000	Aristarchus	1180	Agatharchides*
2980	Purbach	1040	Birt
2970	Diophantus	850	Kunowsky
2890	Lambert	750	Encke
2830	Theætetus	750	Hyginus
2800	Ukert	650	Stadius
2760	Stöfler*	600	Linné
2760	Mösting	380	Kies
2760	Triesnecker	310	Spörer
2720	Thebit A	0	Wargentin

must be remembered that these are at best no more than approximate. Some craters near the poles are so deep that their floors are always in shadow and therefore remain bitterly cold; one of these craters, Newton, has a depth of over 6.1 km below the crest of the wall.

Some craters, such as Grimaldi and Plato, have dark floors, which make them identifiable under any conditions of illumination.

There are also some very bright craters; the most brilliant of these is Aristarchus, which often appears prominent even when lit only by Earthshine. It has terraced walls and a prominent central peak. Some craters are the centres of systems of bright rays which stretch for long distances over the surface; the most prominent of these ray systems are associated with Copernicus, in the Mare Nubium, and Tycho in the southern uplands. Other important ray centres include Kepler, Olbers, Anaxagoras and Thales. The rays are not visible under low illumination, but near the time of full moon they dominate the entire scene.

One remarkable crater, Wargentin, is lava-filled, so that it is a large plateau. There are other plateaux here and there, but none even remotely comparable with Wargentin.

The most conspicuous rills (Table 3.5) on the Moon are those on the Mare Vaporum area (Hyginus, Ariadæus) and the Hadley Rill in the Apennine area, visited by the Apollo 15 astronauts. The Hadley Rill is 80 km long, 1–2 km wide and 370 m deep. Other famous rills are associated with Sirsalis, Bürg, Hesiodus, Triesnecker, Ramsden and Hippalus. Extending from the crater Herodotus, near Aristarchus, is the imposing valley known as Schröter's Valley in honour of its discoverer; in a way this is misleading, since the crater named after Schröter is in a completely different area. It is worth noting that some rills are, in part, crater chains; the Hyginus Rill is an example of this, as it consists of a chain of small confluent craters. A much larger crater valley is to be found near Rheita, in the southeast quadrant of the Moon.

Domes, up to 80 km in diameter, are found in various parts of the Moon – for instance near the crater Arago in the Mare Tranquillitatis, in the Aristarchus area, and on the floor of Capuanus. Many domes have symmetrical summit pits; their slopes are gentle.

Mountain ranges are merely the ramparts of the large maria: the Apennines, bordering the Mare Imbrium, are particularly impressive. Isolated peaks and clumps of peaks are to be found all over the surface (Table 3.6).

LOBATE SCARPS

Lobate scarps are thrust faults found mainly in the highlands. They were imaged in the last three Apollo missions, but many more were shown in the LRO pictures in 2010. They seem to be globally distributed, not clustered near the equator. They are small – no more than 100 metres high and a few kilometres long – and cut through very small craters; they are young, formed less than 1000 million years ago and perhaps only 100 million years ago. As the lunar interior contracted and cooled over recent geologic time, the entire Moon shrank by about 100 metres; its brittle crust ruptured and thrust (compression) faults produced lobate scarps. Slow, slight shrinking may continue even now.

Lobate scarps are common on Mercury, and are much larger than those of the Moon.

In 1945 the American geologist and selenographer J. E. Spurr drew attention to the 'lunar grid' system, made up of families of linear features aligned in definite directions. It is also obvious that the distribution of the craters is not random; they form groups,

Table 3.5 *Rills and valleys*
(a) Valleys (valles)

Name	Lat. (°)	Long. (°)	Length (km)	
Alpine Valley	49 N	3 E	166	Very prominent; cuts through Alps; there is a delicate central rill.
Capella	7 S	35 E	49	Really a crater chain; cuts through Capella.
Reichenbach	31 S	48 E	300	Southeast of Reichenbach, narrowing to the south. Really a crater chain, much less prominent than that of Rheita.
Rheita	43 S	51 E	445	Major crater chain, starting in Mare Nectaris and abutting on Rheita.
Schröter's Valley	20 N	51 W	168	Great winding valley, extending from Herodotus. It starts at a 6 km crater to the north and widens to 10 km to form what is nicknamed the Conra-Head. The maximum depth is about 1000 m. The crater named after Schröter is nowhere near; it lies in quite another part of the Moon (Mare Nubium area).
Snellius	31 S	56 E	592	Very long valley, directed toward the centre of the Nectaris basin.

Table 3.5 *(b) rills (rimæ)*

Name	Lat. (°)	Long. (°)	Length (km)
Agatharcides	20 S	28 W	50
Archytas	53 N	3 E	90
Ariadæus	6 N	14 E	250
Birt	21 S	9 W	50
Brayley	21 N	37 W	311
Cauchy	10 N	38 E	140
Conon	19 N	2 E	30
Gay-Lussac	13 N	22 W	40
Hadley	25 N	3 E	80
Hesiodus	30 S	20 W	256
Hyginus	7 N	8 E	219
Marius	17 N	49 W	121
Sheepshanks	58 N	24 E	200

Table 3.5 *(c) rill systems (rimæ)* *(* = within crater)*

Name	Lat. (°)	Long. (°)	Length (km)
Alphonsus*	14 S	2 W	80
Archimedes	27 N	4 W	169
Arzachel*	18 S	2 W	50
Atlas*	47 N	46 E	60
Boscovich*	10 N	11 E	40
Bürg	44 N	24 E	147
Doppelmayer	26 S	45 W	162
Gassendi*	18 S	40 W	70
Gutenberg	5 S	38 E	330
Hevelius	1 N	68 W	182
Hippalus	25 S	29 W	191
Hypatia	0 S	22 E	206
Janssen*	46 S	40 E	114
Littrow	22 N	30 E	115
Menelaus	17 N	18 E	131
Mersenius	21 S	49 W	84
Petavius*	26 S	59 E	80
Pitatus	18 N	24 E	94
Posidonius*	32 N	29 E	70
Prinz	27 N	43 W	80
Ramsden	34 S	31 W	108
Repsold	51 N	82 W	166
Riccioli	2 N	74 W	400
Ritter	3 N	18 E	100
Sirsalis	16 S	62 W	426
Taruntius*	6 N	46 E	25
Triesnecker	4 N	5 E	215
Zupus	15 S	53 W	120

chains and pairs, and when one crater intrudes into another it is almost always the smaller feature which breaks into the larger.

ORIGIN OF THE LUNAR FORMATIONS

This is a problem which has caused considerable controversy. Eccentric theories have not been lacking; for example P. Fauth, who died in 1943, supported the idea that the Moon is covered with ice. J. Weisberger, who died in 1952, denied the existence of any mountains or craters, and attributed the effects to storms and cyclones in a dense lunar atmosphere. The Spanish engineer Sixto Ocampo claimed, in 1951, that the craters were the result of an atomic war between two races of Moon men (the fact that some craters have central peaks while others have not proves, of course, that the two sides used different types of bombs; the last detonations on the Moon fired the lunar seas, which fell back to Earth and caused the Biblical Flood). However, in modern times the only serious question has been as to whether the craters were produced by internal action – that is to say, vulcanism – or whether they were due to impact. By now all authorities support the impact theory, which was proposed by Franz von Paula Gruithuisen in 1824,

revived by G. K. Gilbert in 1892 and put into its present-day form by Ralph Baldwin in 1949.

According to this scenario, the sequence of events may have been more or less as follows (Table 3.7). The Moon was formed at about the same time as the Earth (4600 million years ago). The heat generated

Table 3.6 *Mountain ranges (Montes)*

Name	Mid-Lat. (°)	Long. (°)	Length (km)	
Alps	46 N	1 W	281	Borders Mare Imbrium. Contains Mont Blanc, Alpine Valley.
Apennines	19 N	4 W	401	Borders Mare Imbrium. Contains Hadley, Huygens, Ampere, Serao, Wolf.
Carpathians	14 N	24 W	361	Borders Mare Imbrium.
Caucasus	38 N	10 E	445	Borders Mare Serenitatis. Fairly high peaks.
Cordilleras	17 S	82 W	574	Forms outer wall of Orientale.
Hæmus	20 N	9 E	560	Part of the border of Mare Serenitatis, separating it from Mare Vaporum.
Harbingers	17 N	41 W	90	Clumps of peaks east of Aristarchus.
Juras	47 N	34 W	422	Borders Sinus Iridum ('Jewelled Handle' effect).
Pyrenees	16 S	41 E	164	Not a true range, but a collection of moderate hills, roughly between Gutenberg and Bohnenberger.
Riphæans	8 S	28 W	189	Low range on the Mare Nubium, close to the bright crater Euclides. The northern section is sometimes called the Ural Mountains.
Rook	21 S	82 W	791	One of the inner circular mountain chains surrounding the Orientale basin.
Rupes Cauchy	9 N	37 E	120	A fault, changing into a rill; in some ways not too unlike the Straight Wall.
Scarps (Rupes) Altai	24 S	23 E	427	Often called the Altai Mountains; really a scarp, on the edge of the Nectaris basin.
Spitzbergen	35 N	5 W	60	Bright little hills north of Archimedes, lying on the edge of a ghost ring. So named because in shape they resemble the terrestrial island group.
Straight Range	48 N	20 W	90	Remarkable line of peaks in the Mare Imbrium, west of Plato (Montes Recti).
Taurus	28 N	41 E	172	Not a true range; mountainous region near Rømer.
Teneriffes	47 N	12 W	182	Mountainous region between Plato and the Straight Range.
Straight Wall	22 S	8 W	134	In Mare Nubium, west of Thebit; very prominent, appearing dark before full moon because of the shadow and bright after full moon. The angle of slope is no more than 40°.
Mons (Mountain)				
Ampère	19 N	4 W	30	Mountain massif in the Apennines.
Bradley	22 N	1 E	30	Mountain massif in the Apennines, close to Conon.
Hadley	26.5 N	5 E	25	Mountain massif in the northern Apennines.
Huygens	20 N	3 W	40	5400 m mountain massif in the central part of the Apennines.
La Hire	28 N	25 W	25	Isolated mountain in Mare Imbrium, northwest of Lambert.
Mont Blanc	45 N	1 E	25	3600 m mountain in the Alps, southwest of Cassini.
Pico	46 N	9 W	25	Triple-peaked mountain on the Mare Imbrium, south of Plato, over 2400 m high. It is bright and prominent. The area between it and Plato is occupied by a ghost ring, once named Newton although this name has since been transferred to a deep crater in the far south of the Moon.
Piton	41 N	1 W	25	Prominent mountain in Mare Imbrium, between Cassini and Piazzi Smyth.
Capes (Promontoria)				
Agarum	14 N	66 E	70	On the eastern border of Mare Crisium.
Agassiz	42 N	2 E	20	Edge of the Alps, northwest of Cassini.
Archerusia	17 N	22 E	10	Edge of Mare Serenitatis, between Plinius and Tacquet.
Deville	43 N	1 E	20	Edge of the Alps, between Cape Agassiz and Mont Blanc.
Fresnel	29 N	5 E	20	Northern cape of the Apennines.
Heraclides	40 N	33 W	50	Western cape of Sinus Iridium.
Kelvin	27 S	33 W	50	In Mare Humorum, southwest of Hippalus.
Laplace	46 N	26 W	50	Eastern cape of Sinus Iridium.
Tænarium	19 S	8 W	70	Edge of Mare Nubium, north of the Straight Wall. Sometimes spelled Ænarium.

Table 3.7 *Lunar systems*

System	Age (thousand million years)	Events
pre-Nectarian	>3.92	Basins and craters formed before the Nectaris basin (multi-ring basins), e.g. Grimaldi.
Nectarian	3.92–3.85	Post-Nectaris, pre-Imbrian; includes some multi-ring basins, e.g. Clavius.
Imbrian	3.85–3.1	Extends from the formation of the Imbrian basin to the youngest mare lavas (Orientale, Schrödinger most basaltic maria, craters such as Archimedes and Plato).
Eratosthenian	3.1–1.0	Youngest craters and mare lavas (e.g. Eratosthenes).
Copernican	1.0–present	Begins with formation of Copernicus. Youngest craters (e.g. Tycho); ray systems.

during the formation made the outer layers melt down to a depth of several hundred kilometres; less dense materials then separated out to the surface and in the course of time produced a crust. Then, between 4400 million and about 4000 million years ago, came the Great Bombardment, when meteorites rained down to produce the oldest basins such as the Mare Tranquillitatis and Mare Fœcunditatis. The Imbrium basin dates back perhaps 3850 million years, and as the Great Bombardment ceased there was widespread vulcanism, with magma pouring out from below the crust and flooding the basins to produce structures such as the Mare Orientale and ringed formations of the Schrödinger type. Craters with dark floors, such as Plato, were also flooded at this time. The lava flows ended rather suddenly, by cosmical standards, and for the last few thousand million years the Moon has seen little activity, apart from the occasional formation of impact craters such as Copernicus and Tycho. It has been claimed that Copernicus is no more than a thousand million years old, and Tycho even younger. The ray systems are certainly late-comers, since the rays cross all other formations.

The Moon has experienced synchronous rotation since early times, and there are marked differences between the Earth-turned and the averted hemispheres. The crust is thicker on the far side, and some of the basins are unflooded, which is why they are not classed as maria *palimpsests*. The prominent feature Tsiolkovskii seems to be between a rare-type structure and a crater: it has a flooded, mare-type floor, but high walls and a central peak. It adjoins a formation of similar size, Fermi, which is unflooded.

One thing is certain; the Moon is today essentially inert. On 4 May 1783, and again on 19 and 20 April 1787, Sir William Herschel reported seeing active volcanoes, but there is no doubt that he observed nothing more significant than bright areas (such as Aristarchus) shining by Earthlight. In modern times, transient lunar phenomena (TLP) have been reported on many occasions; they take the form of localised obscurations and glows. On 3 November 1958, N. A. Kozyrev, at the Crimean Astrophysical Observatory, obtained a spectrum of an event inside the crater Alphonsus, and on 30 October 1963 a red event in the Aristarchus area was recorded from the Lowell Observatory by J. Greenacre and J. Barr. In 1967, NASA published a comprehensive catalogue of the many TLP reports, compiled by Barbara Middlehurst and myself; I issued a subsequent supplement (Barbara, sadly, has died). Over 700 TLP events were listed, and although no doubt many of these are due to observational error it seems that others are genuine,

presumably due to gaseous emissions from below the crust. They occur mainly round the peripheries of the regular maria and in areas rich in rills, and are most common near lunar perigee, when the Moon's crust is under maximum strain.

Until recently, the reality of TLP was questioned, perhaps because so many of the reports (though by no means all) came from amateur observers. However, full professional confirmation has now been obtained. Using the 83 cm telescope at the Observatory of Meudon, Audouin Dollfus has detected activity in the large crater Langrenus. He wrote: 'Illuminations have been photographed on the surface of the Moon. They appeared unexpectedly on the floor of Langrenus. Their shape and brightness was considerably modified in the following days, and they were simultaneously recorded in polarised light. They are apparently due to dust grain levitation above the lunar surface, under the effect of degassing from the interior. The Langrenus observations indicate that the Moon is not a completely 'dead body'. Degassing occasionally occurs in areas that are particularly fractured or fissured. Clouds of dust are lifted off the ground by the gas pressure.

The brilliant crater Aristarchus has long been known to be particularly subject to TLP phenomena, and images from the Clementine space-craft in the 1990s have shown that there have been recent colour changes in the area; patches of ground have darkened and reddened. Winifred Cameron of the Lowell Observatory in Arizona, who has made a long study of TLP, considers that these changes are due to gaseous outbreaks stirring up the ground material, and it is indeed difficult to think of any other explanation. By terrestrial standards the lunar outbreaks are of course very mild indeed, but there is no longer any serious doubt that they do occur.

It has also been suggested that TLP may be due to dust-storms; this was proposed in 2005 by G. Olhoeft (Colorado), who examined records from an experiment carried out on Apollo 17: LEAM (Lunar Ejecta And Meteorites). When the Sun rises over the lunar surface, dust begins to stir, and the dust storm extends along the terminator. The LEAM experiment recorded a large number of particles every lunar morning, coming mainly from east to west and moving more slowly than would be expected from lunar ejecta. The storm swirls across the surface, following the shifting of the terminator. Apollo crews also sketched 'bands' or 'twilight rays', where sunlight was apparently filtering through dust above the Moon's surface, and the earlier Surveyor craft photographed twilight glows over the lunar horizon, which persisted after the Sun had set there, leading to the suggestion that the Moon may have 'a tenuous atmosphere of moving dust particles'.

The most recent claim concerning the formation of a large impact crater related to a report dating from July 1178, by Gervase of Canterbury. 'The crescent Moon was seen to split in two ... a flaming torch sprung spewing out over a considerable distance fire, hot coals and sparks. Meanwhile, the body of the Moon which was below, writhed ... and throbbed like a wounded snake.' This indicates a terrestrial cloud phenomenon, if anything; nevertheless, it has been seriously suggested that the phenomenon was the result of an impact on the far side which led to the formation of the ray-crater Giordano Bruno. In fact this is absurd, and in any case the altitude of the Moon at the time of the observation, as seen from Canterbury, was less than 5°. Therefore there is no doubt that the claim must dismissed as merely a 'Canterbury tale'.

Major structural changes do not now occur. There have been two cases which have caused widespread discussion. But neither stands up to close examination. In the Mare Fœcunditatis there are two small craters. Messier and Messier A, which were said by W. B. Beer and J. H. von Mädler, in 1837, to be exactly alike, with a curious comet-like ray extending from them to the west; in fact A is the larger of the two and is differently shaped, but changes in solar illumination mean that they can often appear identical. In 1866 J. Schmidt, at the Athens Observatory, reported that a small, deep crater in the Mare Serenitatis, Linné, had been transformed into a white patch. Many contemporary astronomers, including Sir John Herschel, believed that a moonquake had caused the crater walls to collapse. Today, Linné is a small impact crater standing on a white nimbus, and there seems no possibility of any real change having occurred – particularly as Mädler observed it in the 1830s and again after 1866, and reported that it looked exactly the same as it had always done.

Presumably there is a certain amount of exfoliation, because the temperature range is very great. Surface temperatures were first measured with reasonable accuracy by the fourth Earl of Rosse, from Birr Castle in Ireland. His papers from 1869 indicated that near noon the temperature rose to about $100\,°C$, although S. P. Langley later erroneously concluded that the temperature never rose above $0\,°C$. It is now known that the noon equatorial temperature is about $101\,°C$, falling to $-184\,°C$ at night; the poles remain fairly constant at $-96\,°C$.

There are some meteorites which are believed to have been blasted away from the Moon; by the end of 2007, fifty of these had been listed. The evidence for lunar origin is very strong indeed. Most have been found in Antarctica, Northern Africa and the Sultanate of Oman; the first to be identified as such was ALH.89001, found in 1981 in the Allan Hills region of Antarctica. Lunar origin is established by comparing the mineralogy, chemical composition and other characteristics with samples brought back from the Moon by the Apollo astronauts and the Russian robots.

MISSIONS TO THE MOON

The idea of reaching the Moon is very old; as long ago as the second century AD, a Greek satirist, Lucian of Samosata, wrote a story about a lunar voyage (his travellers were propelled on to the Moon by the force of a powerful waterspout!). The first serious idea was due to Jules Verne, in his novel *From the Earth to the Moon* (1865); he planned to use a space-gun, but neglected the effects of friction against the atmosphere, quite apart from the shock of starting at a speed enough to break free from Earth (escape velocity: $11.2\,\mathrm{km\,s^{-1}}$). Before the end of the ninth century the Russian theoretical rocket pioneer, K. E. Tsiolkovskii, realised that the only way to achieve space travel is by using the power of the rocket.

The Space Age began on 4 October 1957, with the launch of Russia's first artificial satellite, Sputnik 1. Less than a year later the Americans made their first attempt to send a rocket vehicle to the Moon. It failed, as did others in the succeeding months, and the Russians took the lead; on 4 January 1959 their probe Luna 1 flew past the Moon at less than $6000\,\mathrm{km}$ and sent back useful information (such as the fact that the Moon has no detectable overall magnetic field). The first lander, again Russian (Luna 2), came down on the Moon on 13 September 1959, and in the following month the Soviet scientists achieved a notable triumph by sending Luna 3 round the Moon, obtaining the first pictures of the areas which are always turned away from Earth.

In the period from 1961 to 1965, the Americans launched their Ranger probes, which impacted the Moon and sent back valuable data before crash-landing. These were followed by the Surveyors (1966–8), which sent back a great deal of information as well as images. However, the first controlled landing was made by Russia's Luna 9, on 31 January 1966, which came down in the Oceanus Procellarum and finally disposed of T. Gold's curious theory that the maria, at least, would be covered by deep layers of soft, treacherous dust.

Both the USA and the then USSR were making efforts to achieve manned lunar landings. The Russian plans had to be abandoned when it became painfully obvious that their rockets were not sufficiently reliable, but the American Apollo programme went ahead, and culminated in July 1969 when Neil Armstrong and Buzz Aldrin stepped out on to the bleak rocks of the Mare Tranquillitatis. By the end of the Apollo programme, in December 1972, our knowledge of the Moon had been increased beyond all recognition. Meanwhile the Russians had used unmanned sample-and-return probes and had also dispatched two movable vehicles, the Lunokhods, which could crawl around the lunar surface under guidance from their controllers on Earth.

There followed a long hiatus in the programme of lunar exploration, but new probes were sent to the Moon during the 1990s, including one Japanese vehicle (Hagomoro, carrying the small satellite Hiten). The American Clementine (1994) and Prospector (1998) provided maps of the entire surface which were superior to any previously obtained, as well as making some surprising claims; such as the possibility of locating ice inside the deep polar craters whose floors are always in shadow.

The European Space Agency probe SMART-1, designed and constructed in Sweden, was launched from Kourou, in French Guiana, on 27 September 2003. It was one metre across, with a total launch weight of 367 kg. This is not much by conventional standards, but Smart was certainly not conventional; instead of using a rocket to send it on its way, it used a solar-powered Hall-effect thruster, using xenon as propellant. (This is a type of thruster in which the propellant is accelerated by an electric field. Though its initial thrust is modest, it has many clear advantages over conventional rocket power.)

The SMART probe was in no hurry. It spiralled its way outward, and not until November 2004 did it pass into the area dominated by the Moon's gravitational pull rather than that of the Earth. It then spiralled moonward, and on 26 January 2005 sent back its first close-range images of the lunar surface. It entered its final orbit round the Moon on 27 February 2006, and began its main programme, using mainly X-ray and infrared instruments. Its aims included a search for ice inside polar craters – which, predictably, proved to be unsuccessful – and studies of the 'Mountains of Eternal Light', summits of high polar peaks which are always in sunlight.

The mission came to its end at 05.12 hours on 3 September 2006, when SMART was deliberately de-orbited and hit the surface at a speed of 2000 m s^{-1}. It had done all that had been expected of it, and above all it had shown that ion propulsion really is effective.

The next important launch was Japanese, the Kaguya probe. Launched from the Tageshima Space Centre on 14 September 2007 by JAXA (Japan Aerospace and Exploration Agency), the Japanese equivalent of NASA. The probe's official name was Selene (Selenological and Engineering Explorer); the nickname Kaguya was chosen by members of the public. The name honours a lunar princess in Japanese folklore. The space-craft carried two small sub-satellites, which were called Okina and Ouma after the princess's foster-parents.

Kaguya was launched in the conventional way, and by 3 October had entered a polar orbit round the Moon. By 19 October it was moving in an approximately circular path 100 km above the lunar surface. Its shape was that of a rectangular box measuring 4.2 × 2.1 m, with a launch weight of 2914 kg. Okina and Ouma were 1 m octagonal prisms, each with a weight of 53 kg. They were detached on 9 and 12 October respectively, and put into elliptical orbits, Okina ranging from 100 to 2400 km and Ouna from 100 to 800 km. (As they were higher up than Kaguya, it was said that the two parents were keeping careful watch upon their daughter!) Their main rôle was to measure the Moon's gravity field, to investigate the distribution of mass in the interior of the lunar globe; each used a mere 70 watts of power, just enough to power an ordinary household electric light bulb. Kaguya itself was packed with instruments of all kinds, and was soon starting to send back very valuable information. By now Japan was well and truly a senior participant in the exploration of space, and predictably China was quick to follow; the first Chinese lunar mission, the Chang-ē 1 orbiter, was sent up from the Xichang Satellite Launch Centre in Sichuan Province, atop a Long March 3A rocket, on 24 October 2007.

A list of lunar missions is given in Table 3.8.

Chang-ē 1, named after a Chinese mythological figure, was launched and began with three orbits around the Earth. A series of burns put it on course for the Moon on 31 October 2007; further burns placed it in its final polar orbit round the Moon, at a distance of 200 km from the surface; the orbit was virtually circular. Instruments carried included a stereo camera, an imaging spectrometer, a gamma- and X-ray spectrometer, a microwave radiometer high-energy particle detector, and two solar wind detectors. Studies of all kinds were planned, notably improved mapping of the Moon's polar regions, and valuable data were collected concerning the space-environment between 40 000 and 400 000 km from the Earth, the recording of the primitive solar wind and the study of the effects of solar activity on both the Earth and Moon. The satellite weighed 2350 kg, with 130 kg of payload.

All went well. The first picture of the Moon was relayed on 26 November 2007, and by November 2008 Chang-ē 1 had produced a map of the entire lunar surface. Beijing was well satisfied.

Not to be outdone, ISRO – the Indian Space Research Organisation – launched its first satellite, Chandrayaan-1, on 22 October 2008 from the Satish Dhawan Space Centre at Sriharikota, around 80 km north of Chennai. The mass at launch was 523 kg, of which the payload accounted for 90 kg; the satellite was cuboid in shape, approximately 1.5 m. Power was provided by a solar array. Instruments included a mapping camera and spectrometers of various kinds. Some non-Indian instruments were carried, notably a Moon Mineralogy Mapper from the Brown University and the Jet Propulsion Laboratory in America, funded by NASA, and instruments from the European Space Agency and the Bulgarian Space Agency.

The final orbit round the Moon, achieved on 11 November, was polar and almost circular, 100 km above the lunar surface, with a period of 2h 9m. There were no problems, and data transmission began at once.

One important feature was that Chandrayaan-1 carried a small impact probe (MIP), which was separated on 14 November and crash-landed in or close to Shackleton Crater. Data were sent back all through its headlong plunge; after free fall for 30 min, rocket braking was used to soften the impact. Shackleton was chosen as the landing site because of NASA's continued obsession – shared by other space agencies – that water ice might exist on the floors of polar craters. Predictably, the MIP impact showed no trace of anything of the kind.

A list of man-made objects on the Moon is given in Table 3.9.

STRUCTURE OF THE MOON

The results from the Apollo missions and the various unmanned probes have led to a change in many of our ideas about the Moon. Moreover, one professional geologist has been there – Dr Harrison ('Jack') Schmitt, with Apollo 17 – and his expertise was naturally invaluable.

The upper surface is termed the regolith. This is a loose layer or débris blanket, continually churned by the impacts of micrometeoroids. (It is often referred to as 'soil', but this is misleading, because there is nothing organic about it.) It is made up chiefly of very small particles ('dust'), but with larger rocks, a few metres across, here and there; it contains many different ingredients. In the maria it is around 2–8 m deep, but it is thicker over the highlands, and may in places go down to 10 m or even more.

Table 3.8 *Missions to the Moon*

Name	Launch date	Landing date	Lat. (°)	Long. (°)	Results
(a) American					
Pre-Ranger					
Pioneer 0 (Able 1)	17 Aug 1958	–	–	–	Failed after 77 s (explosion of lower stage of launcher).
Pioneer 1	11 Oct 1958	–	–	–	Reached 113 000 km. Failed to achieve escape velocity.
Pioneer 2	9 Nov 1958	–	–	–	Failed when third stage did not ignite.
Pioneer 3	6 Dec 1958	–	–	–	Reached 106 000 km. Failed to achieve escape velocity.
Pioneer 4	3 Mar 1959	–	–	–	Passed within 60 000 km of the Moon on 5 March. Now in solar orbit.
Able 4	26 Nov 1959	–	–	–	Failure soon after take-off.
Able 5A	25 Oct 1960	–	–	–	Total failure.
Able 5B	15 Dec 1960	–	–	–	Exploded 70 s after take-off.
Rangers (intended hard landers)					
Ranger 1	23 Aug 1961	–	–	–	Launch vehicle failure.
Ranger 2	18 Nov 1961	–	–	–	Launch vehicle failure.
Ranger 3	26 Jan 1962	–	–	–	Missed Moon by 37 000 km on 28 Jan. No images returned. Now in solar orbit.
Ranger 4	23 Apr 1962	26 Apr 1962	?	?	Landed on night side; instruments and guidance failure.
Ranger 5	18 Oct 1962	–	–	–	Missed Moon by over 630 km. No data received. Now in solar orbit.
Ranger 6	30 Jan 1964	2 Feb 1964	0.2 N	21.5 E	Landed in Mare Tranquillitatis. Camera failed; no images received.
Ranger 7	28 July 1964	31 July 1964	10.7 S	20.7 W	Landed in Mare Nubium. 4306 images returned.
Ranger 8	17 Feb 1965	20 Feb 1965	2.7 N	24.8 E	Landed in Mare Tranquillitatis. 7137 images returned.
Ranger 9	21 Mar 1965	24 Mar 1965	12.9 S	2.4 W	Landed in Alphonsus. 5814 images returned.
Surveyors (controlled landers)					
Surveyor 1	30 May 1966	2 June 1966	2.5 S	43.2 W	Landed in Mare Nubium, near Flamsteed. 11 150 images returned. Transmitted until 13 July; contact regained until January 1967.
Surveyor 2	20 Sept 1966	22 Sept 1966	Southeast of Copernicus		Guidance failure; crash-landed, site uncertain; no images returned.
Surveyor 3	17 Apr 1967	19 Apr 1967	2.9 S	23.3 W	Landed in Oceanus Procellarum, 612 km east of Surveyor 1, close to site of later Apollo 12 landing. 6315 images returned. Soil physics studied.
Surveyor 4	14 July 1967	16 July 1967	0.4 N	1.3 W	Crashed in Sinus Medii. No data returned.
Surveyor 5	8 Sept 1967	10 Sept 1967	1.4 N	23.2 E	Landed in Mare Tranquillitatis, 25 km from later Apollo 11 site. 18 000 images returned; soil physics studied. Contact lost on 16 December.
Surveyor 6	7 Nov 1967	9 Nov 1967	0.5 N	1.4 W	Landed in Sinus Medii. 29 000 images returned, soil physics studied. Restarted and moved 3 m. Contact lost on 14 December.
Surveyor 7	7 Jan 1968	9 Jan 1968	40.9 S	11.5 W	Landed on northern rim of Tycho. 21 274 images returned and much miscellaneous information. Contact lost on 20 Feb 1968.
Orbiters (Mapping vehicles; no data attempted from the lunar surface)					
Orbiter 1	10 Aug 1966	29 Oct 1966	6.7 N	162 E	207 images returned. Controlled impact on far side at end of mission.
Orbiter 2	7 Nov 1966	11 Oct 1967	4.5 N	98 E	422 images returned. Controlled impact on far side at end of mission.
Orbiter 3	4 Feb 1967	9 Oct 1967	14.6 N	91.7 W	307 images returned. Controlled impact on far side at end of mission.
Orbiter 4	4 May 1967	6 Oct 1967	Far side		326 images returned; first images of polar regions. Impact on far side at end of mission; location uncertain.
Orbiter 5	1 Aug 1967	31 Jan 1968	0	70 W	212 images returned. Controlled impact at end of mission.

Table 3.8 (cont.)

Apollo (manned missions)

No	Command Module name	Lunar Module (LM) name	Launch	Land	Splash-down	Lat. (°)	Long. (°)	Area	Crew	Extra vehiculr schedule	Schedule
7	–	–	11 Oct 1968	–	22 Oct 1968	–	–	–	W. Schirra, D. Eisele, R. Cunningham	–	Test orbiter (10 days 20 h)
8	–	–	21 Dec 1968	–	27 Dec 1968	–	–	–	F. Borman, J. Lovell, W. Anders	–	Flight round Moon (6 days 3 h)
9	Gumdrop	Spider	3 Mar 1969	–	13 Mar 1969	–	–	–	J. McDivett, D. Scott, W. Schweickart	–	LM test in Earth orbit (10 days 2 h)
10	Charlie Brown	Snoopy	18 May 1969	–	26 May 1969	–	–	–	T. Stafford, J. Young, E. Cernan	–	LM test in lunar orbit (8 days 0 h)
11	Columbia	Eagle	16 July 1969	19 July 1969	24 July 1969	0° 40' N	23° 49' E	Mare Tranquillitatis	N. Armstrong, E. Aldrin, J. Collins	2.2 h	Landed: Apollo Lunar Surface Experiental Package (ALSEP)
12	Yankee Clipper	Intrepid	14 Nov 1969	19 Nov 1969	24 Nov 1969	3° 12' S	23° 24' W	Oceanus Procellarum, near Surveyor 3	C. Conrad, A. Bean, R. Gordon	7.6 h (1.4 km)	Landing: ALSEP
13	Odyssey	Aquarius	11 Apr 1970	–	17 Apr 1970	–			J. Lovell, F. Haise, J. Swigert	–	Aborted landing
14	Kitty Hawk	Antares	31 Jan 1971	5 Feb 1971	9 Feb 1971	3° 40' S	17°28' W	Fra Mauro Formation, Mare Nubium	A. Shepard, E. Mitchell, S. Roosa	9.2 h (3.4 km)	Exploration: lunar cart
15	Endeavour	Falcon	26 July 1971	30 July 1971	7 Aug 1971	26° 06' N	3° 39' E	Hadley-Apennine region, near Hadley Rill	D. Scott, J. Irwin, A Worden	18.3 h (28 km)	Exploration: LRV
16	Casper	Orion	16 Apr 1972	21 Apr 1972	27 Apr 1972	8° 36' S	15°31' E	Descartes formation. 50 km W of Kant	J. Young, B. T. Mattingly	20.1 h (26 km)	Various experiments: Lunar Roving Vehicle (LRV)
17	America	Challenger	7 Dec 1972	11 Dec 1972	19 Dec 1972	20° 12' N	30°45' E	Taurus-Littrow in Mare Serenitatis, 750 km east of Apollo 15	E. Cernan, H. Schmitt, R. Evans	22 h (29 km)	Geology: LRV

Apollo 1 (21 Feb 1967) exploded on the ground, killing the crew (G. Grissom, E. White and R. Chaffee). Apollos 2 and 3 were not used. Apollos 4 (9 Nov 1967), 5 (22 Jan 1968) and 6 (4 Apr 1968) were unmanned test Earth orbiters.

Post-Apollo missions

Clementine	Launch 25 Jan 1994	Entered lunar orbit 19 Feb 1994; mapping programme, surveying 38 000 000 km^2 of the Moon at 11 different wavelengths. Left lunar orbit on 3 May to rendezvous with asteroid Geographos, but failed to achieve this (on-board malfunction).

Table 3.8 (cont.)

Prospector	Launch 6 Jan 1998	Extensive and prolonged lunar mapping and analysis programme; crashed into polar crater 31 July 1999 – unsuccessful search for water ice.
Lunar Reconnaissance Orbiter	LRO 18 June 2009	In orbit.
Lunar Crater and Sensing Satellite	Lcross 18 June 2009	9 Oct 2009 Impacted in Cabæus

(b) Russian

Name	Launch	Landing	Lat. (°)	Long. (°)	
Luna 1	2 Jan 1959	–	–	–	Passed Moon at 5955 km on 4 Jan, proving that the Moon lacks a magnetic field. Contact lost after 62 h. Studied solar wind. Now in solar orbit.
Luna 2	12 Sept 1959	13 Sept 1969	30 N	1 W	Crash-landed in Mare Imbrium, probably near Archimedes (uncertain).
Luna 3	4 Oct 1959	–	–	–	Went round Moon, imaging the far side. Approached Moon to 6200 km.
Luna 5	9 May 1965	12 May 1965	1.5 S	25 W	Unsuccessful soft lander. Crashed in Mare Nubium.
Luna 6	8 June 1965	–	–	–	Passed Moon at 161 000 km on 11 July; failure. Now in solar orbit.
Zond 3	18 July 1965	–	–	–	Approached Moon to 9219 km. Photographic probe; 25 images returned, including some of the far side. Images returned on 27 July from 2 200 000 km. Now in solar orbit.
Luna 7	4 Oct 1965	7 Oct 1965	9.8 N	47.8 W	Unsuccessful soft-lander. Crashed in Oceanus Procellarum.
Luna 8	3 Dec 1963	6 Dec 1963	9.6 N	62 W	Unsuccessful soft-lander. Crashed in Oceanus Procellarum.
Luna 9	31 Jan 1966	3 Feb 1966	7.1 N	64 W	Successful soft-lander; 100 kg capsule landed in Oceanus Procellarum. Images returned. Contact lost on 7 Feb.
Luna 10	31 Mar 1966	–	–	–	Lunar satellite; approached Moon to 350 km; gamma-ray studies of lunar surface layer. (Entered lunar orbit on 3 Apr.) Contact lost on 30 May, after 460 orbits. Now in lunar orbit.
Luna 11	24 Aug 1966	–	–	–	Lunar satellite; entered lunar orbit on Aug 28, and approached Moon to 159 km. Radiation, meteoritic and gravitational studies. Contact lost on 1 Oct.
Luna 12	22 Oct 1966	–	–	–	Lunar satellite. Entered lunar orbit on 25 Oct, and imaged craters to a resolution of 15 m. Contact lost on 19 Jan 1967.
Luna 13	21 Dec 1966	23 Dec 1966	18.9 N	63 W	Soft landing in Oceanus Procellarum. Images returned; soil and chemical studies. Contact lost on 27 December.
Luna 14	7 April 1968	–	–	–	Lunar satellite; approached Moon to 160 km. Valuable data obtained.
Zond 5	14 Sept 1968	–	–	–	Went round the Moon, approaching to 1950 km, and returned to Earth on 21 September. Plants, seeds, insects and tortoises carried.
Zond 6	10 Nov 1968	–	–	–	Went round the Moon, approaching to 2420 km, and filmed the far side. Returned to Earth on 17 November.
Luna 15	13 July 1969	21 July 1969	17 N	60 E	Unsuccessful sample and return probe. Crashed in Mare Crisium.
Zond 7	7 August 1969	–	–	–	Went round the Moon, approaching to 2000 km, and took colour images of both Moon and Earth. Returned to Earth.
Luna 16	12 Sept 1970	15 Sept 1970	0.7 S	56.3 E	Landed in Mare Fœcunditatis, secured 100 g of material, and after $26\frac{1}{2}$ hours lifted off and returned to Earth (24 September).
Zond 8	20 Oct 1970	–	–	–	Circum-lunar flight; colour pictures of Earth and Moon. Returned to Earth on 27 October, splashing down in Indian Ocean.
Luna 17	10 Nov 1970	17 Nov 1970	30.2 N	35 W	Carried Lunokhod 1 to Mare Imbrium.
Luna 18	2 Sept 1971	15 Sept 1971	3.6 N	56.5E	Unsuccessful soft-lander. Contact lost during descent manœuvre to Mare Fœcunditatis.
Luna 19	28 Sept 1971	–	–	–	Lunar satellite. Contact kept for 4000 orbits. Studies of mascons, lunar gravitational field, solar flares, etc.
Luna 20	14 Feb 1972	17 Feb 1972	3.5 N	56.6 E	Landed near Apollonius, south of Mare Crisium (120 km north of Luna 16 site), drilled into the lunar surface, and returned with samples on 25 February.
Luna 21	8 Jan 1973	16 Jan 1973	25.9N	30.5E	Carried Lunokhod 2 to a site near Le Monnier, 180 km from Apollo 17 site.

Table 3.8 (cont.)

Name	Launch	Landing	Lat. (°)	Long. (°)	
Luna 22	29 May 1974	–	–	–	Lunar satellite. Television images, gravitation and radiation studies. Contact maintained until 6 November 1975.
Luna 23	28 Oct 1974	–	Mare Crisium	–	Unsuccessful sample and return mission; crashed.
Luna 24	9 Aug 1976	18 Aug 1976	12.8 N	62.2 E	Landed in Mare Crisium, drilled down to 2 m, collected samples, lifting off on 19 August and landing back on Earth on 22 August.

Lunokhods

Name	Carrier	Weight (kg)	Site	
Lunokhod 1	Luna 17	756	Mare Imbrium	Operated for 11 months after arrival on 17 Nov 1970. Area photographed exceeded 80 000 m^2. Over 200 panoramic pictures and 20 000 images returned. Distance travelled, 10.5 km.
Lunokhod 2	Luna 21	850	Le Monnier	Operated until mid May 1973. 86 paranoramic pictures and 80 000 television images obtained. Distance travelled, 37 km. On 3 June the Soviet authorities announced that the programme had ended.

(c) Japanese

Name	Impact dates		
Hagomoro	24 Jan 1990		Launched by Muses-A vehicle. The satellite Hiten ('Flyer') was ejected and put into lunar orbit on 15 Feb 1992; it had a mass of 180 kg and carried a micrometeoroid detector. Hiten crash-landed on the Moon on 10 Apr 1993, at lat. 34 S, long. 55 E, near Furnerius.
Nozomi	3 July 1998		Mars mission. Imaged Moon from 514 000 km on 18 July 1998.
Kaguya	14 Sept 2007	20 June 2009	Images. General data. Two sub-satellites, Okina and Ouma.

(d) Chinese

Chang- ē 1	24 Oct 2007	3 January 2009	Images. General data. Search for water.

Chang- ē 2. Launch 2010 Oct. 1. by a Long March 3C carrier rocket from the Xichang Satellite Center at Xichang, Sichuan, China. The distance from the Moon will range fom 15 km to 100 km). This was the first time that a Chinese probe has entered an Earth–Moon transfer orbit without first orbiting the Earth. Chang-ē 2 caries an improved stereo camera, a laser altimeter, gamma/X-ray spectrometers and a microwave detector. It will carry out a careful survey of Sinus Iridum, the intended landing site of Chang-ē 3.

(e) Indian

Chandrayaan-1	22 Oct 2008	–29 Aug 2009	Mission ended. In lunar orbit

(f) US/European

Galileo	Launch 18 Oct 1989. Galileo, en route to Jupiter, flew past the Earth–Moon system on 8 Dec 1990; surveyed the Moon's far side, and on 9 Dec imaged the far hemisphere from 550 000 km. The closest approach to Earth was 960 km. A second flyby occurred on 8 Dec 1992, when Galileo passed Earth at 302 km, and on 9 Dec imaged the Earth and Moon together.
NEAR–Shoemaker	(Near Earth Asteroid Rendevous Spacecraft) named in honour of E. Shoemaker. Swung past Earth on 23 Jan 1998, en route for the asteroid Eros, and obtained a view of Earth and Moon from above their south poles.

On 18 August 1999 the US Saturn probe Cassini flew past the Earth-Moon system, and imaged the Moon from a range of 377 000 km.

The highland crust averages 61 km in depth, but ranges from an average of 55 km on the near or Earth-turned side of the Moon to up to 67 km on the far side. The maria are of course volcanic: they cover 17% of the surface, mainly on the near side (they are much less common on the far side, because of the greater thickness of the crust), and they are in general no more than 1–2 km deep, except near the centres of the large basins. At a fairly shallow level there are areas of denser material, which have been located because an orbiting space-craft will speed up when affected by them: these are known as mascons (*mas*s concentrations). They lie under the large basins, such as Imbrium and Orientale.

Table 3.9 *Selected list of man-made objects on the Moon*

Name	Nationality	Launched	Mass, kg	Landing site: Lat. (°)	Long. (°)
Luna 2	USSR	1959	390.2	29.1 N	0.0
Ranger 4	USA	1962	331	12.9 S	129,1 W
Ranger 6	USA	1964	381	9.4 N	21.5 E
Ranger 7	USA	1974	365.7	10.6 S	20.6 W
Luna 5	USSR	1965	1474	1.6 S	25 W
Luna 7	USSR	1965	1504	9.8 N	47.8 W
Luna 8	USSR	1965	1550	9.6 N	62 W
Ranger 8	USA	1965	367	2.64 N	24.77 E
Ranger 9	USA	1965	367	12.79 S	2.36 W
Luna 9	USSR	1966	1589	7.13 N	64.37 W
Luna 10	USSR	1966	1600	?	?
Luna 11	USSR	1966	1640	?	?
Luna 12	USSR	1966	1670	?	?
Luna 13	USSR	1966	1700	18.87 N	63.05 W
Surveyor 1	USA	1966	270	2.45 S	43.22 W
Orbiter 1	USA	1966	386	6.35 N	160.72 E
Surveyor 2	USA	1966	292	4.0 S	11.0 W
Orbiter 2	USA	1966	385	2.9 N	119.1 E
Orbiter 3	USA	1966	386	14.6 N	97.7 W
Surveyor 3	USA	1967	281	2.99 S	23.34 W
Orbiter 4	USA	1967	386	?	?
Surveyor 4	USA	1967	283	0.45 N	1.39 W
Explorer 35	USA	1967	104.3	?	?
Orbiter 5	USA	1967	386	2.8 S	83.1 W
Surveyor 5	USA	1967	281	1.42 N	23.2 E
Surveyor 6	USA	1967	282	0.53 N	1.4 W
Surveyor 7	USA	1967	290	40.86 S	11.47 W
Luna 14	USSR	1968	1670	?	?
Apollo 10 LM*	USA	1969	2211	?	?
Luna 15	USSR	1969	2718	?	?
Apollo 11 LM*	USA	1969	2034	0.40.25.7 N	23.28.22.679 E
Apollo 12 LM*	USA	1969	2164	2.99 S	23.24 W
Luna 16*	USSR	1970	5727	0.78 S	56.3 E
Luna 17/Lunokhod 1	USSR	1970	5600	38.28 N	35.0.W
Luna 18	USSR	1971	5600	3.57 N	56.5 E
Luna 19	USSR	1971	5600	?	?
Apollo 14*	USA	1971	2144	3.38.44 S	28.16.9 W
Apollo 15*	USA	1971	2132	26.87.56 N	3.38.02 E
Luna 20*	USSR	1972	5727	3.53 N	56.55 E
Apollo 16*	USA	1972	2765	8.58.23 S	15.30.1 E
Apollo 17*	USA	1972	2798	20.11.27 N.	30.46.19
Luna 21/Lunokhod 2	USSR	1973	4850	25.85 N	30.45 E
Explorer 49	USA	1973	328	?	?
Luna 22	USSR	1974	4000	?	?
Luna 23	USSR	1974	5600	12 N	62 E
Luna 24*	USSR	1976	5800	12.75 N	62.2 E
Hagomoro	Japan	1990	12	?	?
Hiten	Japan	1993	143	34.3 S	55.6 E

Table 3.9 (cont.)

| Name | Nationality | Launched | Mass, kg | Landing site: | |
				Lat. (°)	Long. (°)
Prospector	USA	1998	126	87.7 S	42.1 E
Smart-1	USA	2006	307	34.24 S	46.12 W

*Descent stage.

Apollo 13, of course, did not land (1970); the only jettisoned part (13454 kg) landed at lat. 2.75 °S, long. 27.86 °W. To date, the total known mass of man-made material on the Moon is 180 000 kg.

(It is reported that some lunar artefacts have been offered for sale and, according to the London *Times* in December 1993, a buyer paid 678 000 US dollars for Luna 17 and Lunokhod 1. Presumably, buyer collects!)

Table 3.10 *Materials in the lunar crust*

Pyroxene, a calcium–magnesium–iron silicate, is the most common mineral in lunar lavas, making up about half of most specimens; it forms yellowish–brown crystals up to a few centimetres in size.

Plagioclase or feldspar, a sodium– or calcium–aluminium silicate, forms elongated white crystals.

Anorthosite is a rock type containing the minerals plagioclase, pyroxene and/or olivine in various proportions.

Basalt is a rock type containing the minerals plagioclase, pyroxene and ilmenite in varying proportions.

Olivine, a magnesium–iron silicate, is made up of pale green crystals a few millimetres in size; it is not uncommon in the anorthosites.

Ilmenite, present in the basalts, is an iron–titanium oxide.

On the highlands, the rock fragments are chiefly anorthosites, with minerals such as plagioclase, pyroxene and ilmenite. Details of these materials are given in Table 3.10.

Much of our knowledge about the lunar interior comes from seismic investigations – in fact, moonquakes – just as we depend upon earthquakes for information about the interior of our own world. Of course, moonquakes are very mild by terrestrial standards, and never exceed class 3 on the Richter scale, and they are of two main types. Most originate from a zone 800–1000 km below the surface, and are common enough; there is a definite correlation between moonquake frequency and lunar perigee. Shallow moonquakes, at depths of 50–100 km, also occur; although they are much less frequent, it is worth noting that the epicentres of moonquakes seem to be linked with areas particularly subject to TLP – mainly although not entirely around the peripheries of the regular maria.

There have also been man-made moonquakes, caused by the impacts of discarded lunar modules. These show that the outer few kilometres of the Moon are made up of cracked and shattered rock, so that signals can echo to and fro; it was even said that after the impact of an Apollo module the Moon 'rang like a bell'!

Below the crust comes the mantle, the structure of which seems to be relatively uniform. A one tonne meteorite which hit the Moon in July 1972 indicated, from its seismic effects, that there is a region 1000–1200 km below the surface where the rocks are hot enough to be molten (Apollo measurements disposed of an earlier theory that the Moon's globe could be cold and solid all the way through). Finally, there may be a metallic core, although its existence has not been confirmed, and it cannot be much more than 1000 km in diameter. Results from the Lunar Prospector mission of 1998/9 have led to an estimate of an iron-rich core between 440 and 900 km in diameter. Certainly the Moon's core is much smaller than that of the Earth, both relatively and absolutely. It is significant that there is no overall magnetic field now, although the remnant magnetism of some rocks indicates that between 3.6 and 3.9 thousand million years ago a definite field existed – which was not evident either before or after that period. There are, however, locally magnetised areas, notably the curious Reiner Gamma, a 'swirl' in the near side, and the crater Van de Graaff on the far side.

All the rocks brought home for analysis are igneous, or breccias produced by impact processes; the Apollo missions recovered 2196 samples, with a total weight of 381.69 kg, now divided into 35 600 samples. The youngest basalt (No 12022) was given an age of 3.08 thousand million years, while the oldest (No 10003) dated back 3.85 thousand million years. There were no water-laid sedimentary rocks. The famous 'orange soil', found by the Apollo 17 astronauts and at first thought to indicate recent vulcanism, proved to be small glassy orange beads, sprayed out some 3.7 thousand million years ago in erupting fountains of basaltic materials.

In the lavas, basalts are dominant. They contain more titanium than terrestrial lavas: over 10% in the Apollo 11 samples, for example, as against 1–3% in terrestrial basalts. Small amounts of metallic iron were found. Many lunar rocks have much less sodium and potassium than do terrestrial rocks. A new mineral – an opaque oxide of iron, titanium and magnesium, not unlike ilmenite – has been named armalcolite, in honour of *Arm*strong, *Al*drin and *Col*lins. There is also a different type of basalt, KREEP; this name comes from the fact that it is rich in potassium (chemical symbol

K), rare earth elements and phosphorus. The average age of the highland rocks is from 4 to 4.2 thousand million years; over 99% of the surface dates go back for over 3 thousand million years and 90% go back for more than 4 thousand million years. One interesting anorthosite rock, 4 thousand million years old, was collected by the Apollo 15 astronauts; it is white and was at once nicknamed the Genesis Rock.

Apollo 12 sample 12013 (collected by Conrad from the Oceanus Procellarum) is unique. It is about the size of a lemon, and contains 61% of SiO_2, whereas the associated lavas have only 35–40% of SiO_2. It also contains 40 times as much potassium, uranium and thorium as most other rocks, making it one of the most radioactive rocks found anywhere on the Moon. It is composed of a dark grey breccia, a light grey breccia and a vein of solidified lava.

Unexpected results came from one of the most recent lunar probes, Clementine, named after the character in the old mining song who was 'lost and gone forever'. Although Clementine was among the cheapest of all probes (it cost $55 000 000) it was remarkably successful insofar as the Moon is concerned. It was a joint NASA–US Air Force venture, and was launched not from Canaveral, but from the Vandenberg Air Force Base, on 23 January 1994. On 21 February it entered lunar orbit and continued mapping until 23 April; by the time it left lunar orbit, on 3 May, the whole of the surface had been mapped. Unfortunately a fault developed, making it impossible to go on to an encounter with an asteroid (Geographos), as had been hoped. The last lunar flyby was on 20 July, the 25th anniversary of the Apollo 11 landing, after which Clementine entered a solar orbit. The minimum distance from the Moon had been 425 km.

Surprisingly, some investigators claimed that the neutron spectrometer on Clementine had been used to detect ice in some of the deep polar craters, whose floors are always in shadow and where the temperature is always very low. This seemed to be inherently unlikely, since none of the materials brought home by the astronauts had shown any sign of hydrated substances, and in any case it was not easy to see how the ice could have got there. It could hardly have been deposited by an impacting comet, because the temperature at the time of the collision would have been too high; and there is no evidence of past water activity, as there is for instance upon Mars. Yet it was even suggested that there might be enough ice to provide a useful water supply for future colonists; a thousand million gallons of water was one estimate.

Then came Prospector, launched on 6 January 1998, and put into a stable orbit which took it round the Moon once in every 118 min at a distance of 96 km from the surface. Prospector was designed not only to continue with the mapping programme, but also to make a deliberate search for ice deposits – and before long the results seemed to confirm those of Clementine. Yet it was not claimed that ice, as such, had been detected. All that had been found were apparent indications of hydrogen.

Prospector, like Clementine, carried a neutron spectrometer. Neutrons are ejected when cosmic rays from space strike atoms in the Moon's crust, and these neutrons can be detected from the space-craft. Collisions between cosmic-ray particles and atoms heavier than hydrogen produce 'fast' neutrons; if hydrogen atoms are hit, the 'slow' neutrons are much less energetic and the spectrometers can distinguish between the two types. It was found that the neutron energy coming from the polar regions was reduced and from this the presence of hydrogen was inferred, which in turn could suggest the presence of water ice. On the other hand, the hydrogen could be due to the solar wind, which bombards the lunar surface all the time.

Efforts were made to confirm the presence of ice by using the large radio telescope at Arecibo in Puerto Rico. Radar studies did indeed give the same indications – but these were also found in regions which are not in permanent shadow and where frozen material could not possibly exist, so that very rough ground might well be responsible. From the outset there were many sceptics about the 'ice' idea. As for one thing, how could the ice have arrived there? All the samples brought home so far are completely lacking in hydrated materials. A major sceptic was Harrison Schmitt, the only professional geologist who has been to the Moon.

Then, on 31 July 1999, a test was made. At the end of its active career Prospector was deliberately crashed on to the Moon, landing inside a polar crater where ice, if it existed at all, would be present. It was hoped that the cloud of material thrown up would show traces of water. In fact the results were completely negative.

During the first part of the twenty-firstst century the search for lunar water has continued unabated. The first claim of success was due to the Indian satellite Chandrayaan-1 launched on 22 October 2008 from Sriharikota Space Centre. On the following 12 November it entered lunar orbit and began transmitting data. Its original altitude ranges between 100 and 200 m. It carried NASA's M3M (Moon Mineralogy Mapper) to continue the water hunt and it was claimed that tiny water or hydroxyl particles existed in the top 2 mm of lunar 'soil'; this could be due to hydrogen particles in the solar wind combining with oxygen particles in the lunar material, which is not the same thing as saying there is water 'in' or 'on' the Moon.

On 14 November 2008 the Moon Impact Probe separated from Chandrayaan and made a controlled impact near the polar crater Shackleton; it had been hoped that the resulting plume of debris would show traces of water – as usual the results were negative. Signals from the probe ceased on 29 August 2009, shortly after which the mission was officially declared over. Chandrayaan had operated for 312 days, and had completed 95% of its planned objectives. The probe will eventually crash on to the lunar surface, probably some time during 2012. It had been an undoubted success, and the M3M mapper confirmed the theory that the Moon had once been completely molten.

The Lunar Reconnaissance Orbiter (LRO) was launched from Cape Canaveral on an Atlas V 401 rocket (18 June 2009). It was given a polar orbit, and was soon successful in sending back clear images of the Apollo equipment left on the Moon, as well as carrying equipment of all kinds.

It was launched together with Lcross, the lunar crater and sensing satellite, NASA's next attempt to find water. The plan was to crash the launch vehicle's spent Centaur upper stage into the polar crater Cabæus. One minute later Lcross itself would fly through the débris plume, analyse it in the hope of finding water, and transmit back its results before it too crashed into the crater. Cabæus was certainly a suitable target as part of its floor is always shadowed.

The plan was followed. Lcross was separated from the Centaur on 9 October and flew along the same path to the Moon. It impacted in the pre-arranged position on 9 October at 11:31 GMT. The Centaur weighed 2249 kg and impacted at a speed of over 10000 km h^{-1}. Lcross itself followed 6 min later and all its equipment worked perfectly. Unfortunately the expected large plume was conspicuous only by its absence. Nothing of it was seen by observers using the 200-inch reflector on Mount Palomar or the team in Hawaii; other observers in various parts of the world were equally disappointed (at my modest observatory in Selsey the sky was overcast). Some predictions had claimed that the plume might be within the range of my 15-inch reflector, and though I was decidedly sceptical, there was every reason to make the attempt.

A small plume was later reported by the Lcross team. A further statement from a group of lunar scientists at Brown University (USA) in October 2010 gave a more elaborate analysis. Peter Schultz and Brendan Hermalyn, together with NASA, announced that the regolith was far more complex than previously believed. It harboured not only water but also other compounds such as carbon monoxide, carbon dioxide, ammonia, free sodium and silver.

Schultz, lead author of the report, considers that the volatiles originated with the long, very heavy bombardment of the Moon by comets, asteroids and meteoroids. They could have been quickly liberated by later small impacts or could have been heated by the Sun, supplying them with energy to escape and move around until they reached the poles – where they became trapped in crater-floors that are permanently shadowed.

There can be no doubt that the volatiles found by Lcross were brought to the Moon from space and cannot be classed as true lunar water. In any case, the Moon is far drier than the driest deserts on Earth. Future astronauts will never be able to paddle in the Mare Crisium!

I was always sceptical about true 'lunar water', and I maintained that any surface water droplets were caused by particles in the solar wind combining with particles in the Moon's crust. This view seems to have been confirmed by researches carried out in 2010 by Zachary Sharp and his team at the University of New Mexico, USA. They have compared the composition of Earth rocks, primitive meteorite samples and volcanic rocks with particular reference to two isotopes of chlorine, Cl-35 and Cl-37. In Earth rocks the ratio between the two isotopes is almost constant, but for Moon rocks it varies wildly – by as much as 25 times the variation found on Earth.

It is now generally believed that the Moon was formed by a collision between the proto-Earth and a body possibly as large as Mars. The Moon coalesced from the débris of the collision, and as it cooled a magma ocean covered its surface and started to crystallise. Both isotopes of chlorine were present. Cl-35 has fewer neutrons in its nucleus than Cl-37, so that was more prone to vaporise out from the magma ocean. But there is another factor to be considered. If the magma contained a good deal of hydrogen, perhaps in the form of water, it would bond with the Cl-37 and

vaporise out as hydrogen chloride, so that more Cl-37 would escape from the magma along with Cl-35. Actually, as the New Mexico team found, the wide range of the ratios of Cl-35 to Cl-37 in the lunar samples indicates that the Moon's magma ocean – unlike that of the Earth – contained virtually no hydrogen, and therefore no water. If the ocean had contained large amounts of hydrogen, the separation of the chlorine into different isotopes could never have occurred.

The most plausible picture, therefore, is of water droplets produced at or near the surface by solar wind particles and, possibly, impacting comets; the globe is bone-dry. There is no true lunar water.

ATMOSPHERE

The Moon's low escape velocity means that it cannot be expected to retain much in the way of atmosphere. Initially it was believed that the atmosphere must be dense; this was the view of Schröter (1796) and also Sir William Herschel, who always believed the habitability of the Moon to be 'an absolute certainly'. W. H. Pickering (1924) believed the atmosphere to be dense enough to support insects or even small animals, but in 1949 B. Lyot searched for lunar twilight effects and concluded that the atmosphere must have a density less than 1/10 000 of that of the Earth at sea level. In the former USSR, V. Fesenkov and Y. N. Lipski made similar investigations and came to the final conclusion that the density was indeed in the region 1/10 000 that of our air.

The first reliable results came from the Apollo missions. The orbiting sections of Apollos 15 and 16 traced small quantities of radon and polonium seeping out from below the surface, and this was no surprise, because these gases are produced by the radioactive decay of uranium, which is not lacking in the lunar rocks. The Lunar Atmospheric Composition Experiment (LACE,), taken to the Moon by Apollo 17, did detect an excessively tenuous atmosphere, mainly helium (due to the solar wind) and argon (seeping out from below the crust). Later D. Potter and T. Morgan, at the McDonald Observatory in Texas, identified two more gases, sodium and potassium. Traces of silicon, aluminium and oxygen have also been detected in the excessively tenuous upper atmosphere. The sodium seems to surround the Moon rather in the manner of a cometary corona. The lunar atmosphere seems to be in the nature of a collisionless gas: the total weight of the lunar atmosphere can be no more than about 30 tonnes. The density is of the order of 10^{-14} that of the Earth's atmosphere. If the entire lunar atmosphere were condensed to the density of the Earth's air at sea level, it could be packed inside a box with a diameter of 65 m.

So what is its source? As we have noted, outgassing from the interior is a major factor. So is sputtering, because the Moon's surface is unceasingly bombarded by meteorites and solar-wind ions; some of the gases released in this way will be re-absorbed into the regolith, while others will be lost to space. Neither must we forget the gases due to man-made lunar probes. At least it is patently obvious that the dense atmosphere of the Earth and the negligible atmosphere of the Moon are in every way different.

Table 3.11 *Lunar eclipses, 2008–2020*

Date	Time mid-eclipse	Type	Totality, UT	Area of observation
2008 Feb 21	03.26	Total	03.10–03.52	America, Europe, Africa
2008 Aug 16	21.10	Partial, 81%	–	Australasia, Europe, Africa
2009 Dec 31	19.23	Partial, 8%	–	Asia, Australasia, Europe
2010 June 26	11.38	Partial, 54%	–	Australasia, Pacific
2010 Dec 21	08.17	Total	07.40–08.53	Australasia, northwest Europe,
2011 June 15	20.13	Total	19.22–21.03	Europe, Africa, Australia
2011 Dec 10	14.32	Total	14.06–14.58	East Asia, Australia, Europe
2012 June 4	11.03	Partial, 38%	–	America, Pacific, Australasia
2013 Apr 25	20.07	Partial, 2%	–	Australasia, Europe, Africa
2014 Apr 15	07.46	Total	07.06–08.24	Americas, eastern Australia
2014 Oct 8	10.55	Total	10.24–11.24	East Asia, North America
2015 Apr 4	07.46	Total	07.06–08.24	East Asia, Australia
2015 Sept 28	02.47	Total	02.11–03.23	Americas, Europe, Africa
2017 Aug 7	18.20	Partial, 25%	–	Europe, Africa, Australia
2018 Jan 31	11.30	Total	13.51–14.08	Asia, Australia, North America
2018 July 27	20.22	Total	19.30–21.14	Europe
2019 Jan 21	05.12	Total	04 40 – 05.44	Americas, Europe, Africa
2019 July 16	21.31	Partial, 66%	–	Europe, Africa, Australia

Penumbral eclipses

Date	Mid-eclipse	Penumbral obscuration
2009 Feb 9	14.38	92
2007 July 7	09.19	18
2009 Aug 6	00.39	43
2012 Nov 28	14.33	94
2013 May 25	04.19	4
2013 Oct 18	23.50	79
2016 Mar 23	11.47	80
2016 Aug 18	09.42	2
2016 Sept 16	18.54	93
2017 Feb 11	00.44	Total
2020 Jan 10	19.10	92

Table 3.12 *The Danjon scale for lunar eclipses*

0	Very dark; Moon almost invisible.
1	Dark; grey or brownish colour; details barely identifiable.
2	Dark or rusty-red, with a dark patch in the middle of the shadow; brighter edges.
3	Brick-red: sometimes a bright or yellowish border to the shadow.
4	Coppery or orange-red: very bright, with a bluish cast and varied hues.

ECLIPSES OF THE MOON

Eclipses of the Moon are caused by the Moon's entry into the cone of shadow cast by the Earth. At the mean distance of the Moon, the diameter of the shadow cone is approximately 9000 km: on average the shadow is 1 380 000 km long. Totality may last for up to 1 h 44 min.

Lunar eclipses may be either total or partial. If the Moon misses the main cone and merely enters the zone of 'partial shadow' or penumbra to either side, there is light dimming, but a penumbral eclipse is not easy to detect with the naked eye. Of course, the Moon must pass through the *penumbra* before entering the main cone or umbra (Table 3.11).

During an eclipse the Moon becomes dim and often looks a coppery colour. The colour and brightness during an eclipse depend upon the conditions in the Earth's atmosphere; thus the eclipse of 19 March 1848 was so 'bright' that lay observers refused to believe that an eclipse was happening at all. On the other hand, it is reliably reported that during the eclipses of 18 May 1761 and 10 June 1816 the Moon became completely invisible with the naked eye. The French astronomer A. Danjon

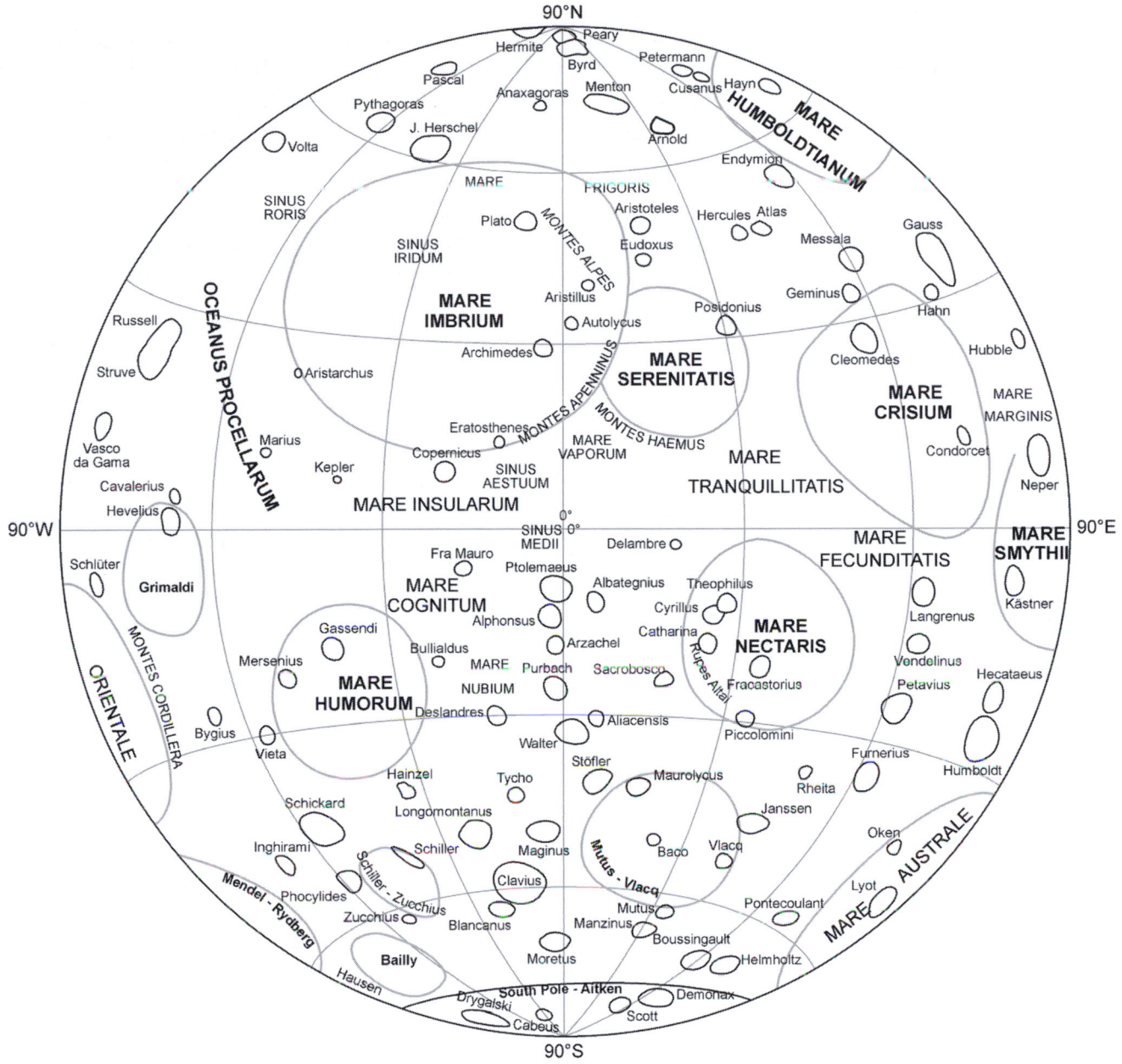

Figure 3.1 Outline map of the Moon

has given an 'eclipse scale' from 0 (dark) to 4 (bright), and has attempted to correlate this with solar activity, although the evidence is far from conclusive. The Danjon scale is given in Table 3.12.

The Greek astronomer Anaxagoras (*c.* 500–428 BC) gave a correct explanation of lunar eclipses, but in early times eclipses caused considerable alarm. For example, the Californian Indians believed that a monster was attacking the Moon, and had to be driven away by making as much noise as possible (as with the Chinese at the time of a solar eclipse), while in an old Scandinavian poem, the Edda, it is said that the monster Managarmer is trying to swallow the Moon, and staining the air and ground with blood.

During an eclipse the Orinoco Indians would take their hoes and labour energetically in their cornfields, as they felt that the Moon was showing anger at their laziness.

Ancient eclipse records are naturally uncertain. It has been claimed that an eclipse seen in the Middle East can be dated back to 3450 BC; the eclipse of 1361 BC is more definite. Ptolemy gives the dates of observed eclipses as 721 BC and 720 BC. In *The Clouds*, the Greek playwright Aristophanes alludes to an eclipse seen from Athens on 9 October 425 BC. There was certainly a lunar eclipse in August 413 BC, which had unfortunate results for Athens, since it persuaded Nicias, the commander of the Athenian expedition to Sicily, to delay the

Figure 3.2 Patrick Moore's outline Moon map. For telescopic
observes south is up.

Table 3.13 *Named mare, lacus, palus and sinus areas. Since these cover wide areas, the coordinates are given for their centres, but are of course approximate only. Diameter values are also approximate, since many of the maria have very irregular boundaries*

Name		Lat. (°)	Long. (°)	Diameter (km)	
Mare Anguis	Serpent Sea	23 N	67 E	150	Area 10 000 km^2. Narrow darkish area northeast of Mare Crisium.
Mare Australe	Southern Sea	40 S	93 E	603	Irregular, patchy area in the southeast. Area about 149 000 km^2.
Mare Cognitum	Known Sea	10 S	23 W	376	Part of Mare Nubium. East of the Riphæans. Landing site of Ranger 7 in 1964.
Mare Crisium	Sea of Crises	17 N	59 E	505	Well defined; separate from main system. Area about 200 000 km^2 (similar to Great Britain).
Mare Fœcunditatis	Sea of Fertility	8 S	31 E	909	Irregular; confluent with Mare Tranquillitatis. Area 344 000 km^2.
Mare Frigoris	Sea of Cold	56 N	1 E	1596	Elongated, irregular; in places narrow. Area about 441 000 km^2 (including Lacus Mortis). Bounded in part by the Alps.
Mare Humboldtianum	Humboldt's Sea^a	57 N	81 E	273	Limb sea, beyond Endymion; fairly regular.
Mare Humorum	Sea of Humours	24 S	39 W	389	Regular; leads off Mare Nubium. Area 118 000 km^2.
Mare Imbrium	Sea of Showers	33 N	16 W	1123	Largest regular sea; area 863 000 km^2 (equal to Britain and France combined). Bounded by the Alps, Apennines and Carpathians. Contains Palus Nebularum and Palus Putredinis, as well as major craters such as those of the Archimedes group.
Mare Insularum	Sea of Islands	7 N	31 W	513	Part of Mare Nubium; ill defined; south of Copernicus.
Mare Marginis	Marginal Sea	13 N	86 E	420	Limb sea beyond Mare Crisium; fairly well defined. Area 62 000 km^2.
Mare Nectaris	Sea of Nectar	15 S	36 E	333	Leads off Mare Tranquillitatis. Area 100 000 km^2 regular. Central part of a very ancient basin, whose border is marked by the Altai Scarp.
Mare Nubium	Sea of Clouds	21 S	17 W	715	Ill-defined northern border. Area (with Mare Cognitum) 265 000 km^2.
Mare Orientale	Eastern Sea	19 S	93 W	327	Limb sea, beyond Corderillas. Vast ringed structure. Only the eastern part visible from Earth under favourable libration.
Oceanus Procellarum	Ocean of Storms	18 S	57 W	2568	Area 2 290 000 km^2. Irregular; contains Aristarchus.
Mare Serenitatis	Sea of Serenity	28 N	17 E	707	Regular; area 314 000 km^2. Contains few conspicuous craters; Bessel is the most prominent and Linné is also on the mare. Crossed by a long ray coming from the south, and wrinkle-ridges also cross it.
Mare Smythii	Smyth's Sea^b	1 N	87 E	373	Well-defined limb sea; area 104 000 km^2.
Mare Spumans	The Foaming Sea	1 N	65 E	139	Darkish area south of Mare Crisium. Area 16 000 km^2.
Mare Tranquillitatis	Sea of Tranquillity	9 N	31 E	873	Confluent with Mare Serenitatis, but is lighter, patchier and less regular. Area 440 000 km^2; rather irregular. Nectaris and Fœcunditatis lead off it.
Mare Undarum	Sea of Waves	7 N	64 E	243	Darkish irregular area near Firmicus. Area 21 000 km^2.
Mare Vaporum	Sea of Vapours	13 N	4 E	245	Area 55 000 km^2; southeast of the Apennines. Contains some very dark patches; also the Hyginus Rill and part of the Ariadæus Rill.
Mare Moscoviense	Moscow Sea	27 N	148 E	277	Far side.
Mare Ingenii	Sea of Ingenuity	34 S	163 E	318	Far side.
Lacus Æstatis	Summer Lake	15 S	69 W	90	Two dark areas north of Crüger. Combined area about 1000 km^2.
Lacus Autumns	Autumn Lake	10 S	84 W	183	Dark patches in the Corderillas.
Lacus Bonitas	Lake of Goodness	23 N	44 E	92	Small darkish area near Macrobius.
Lacus Doloris	Lake of Grief	17 N	9 E	110	Darkish area north of Manilius.

Table 3.13 (cont.)

Name		Lat. (°)	Long. (°)	Diameter (km)	
Lacus Excellentiæ	Lake of Excellence	35 S	44 W	184	Vague darkish area near Clausius.
Lacus Felicitatis	Lake of Happiness	19 N	5 E	90	Small darkish area north of Mare Vaporum.
Lacus Gaudii	Lake of Joy	16 N	13 E	50	Darkish area between Manilius and the Hæmus Mountains.
Lacus Hiemis	Winter Lake	15 N	14 E	50	Small darkish area southwest of Menelaus.
Lacus Lenitatis	Lake of Tenderness	14 N	12 E	80	Darkish area east of Manilius.
Lacus Mortis	Lake of Death	45 N	27 E	151	Dark, adjoining Lacus Somniorum. Area 21 000 km^2. Contains the Bürg rills.
Lacus Odii	Lake of Hate	19 N	7 E	70	Darkish patch west of the Hæmus Mountains.
Lacus Perseverantinæ	Lake of Perseverance	8 N	62 E	70	Darkish patch adjoining Firmicus to the west.
Lacus Somniorum	Lake of the Dreamers	38 N	29 E	384	Irregular darkish area leading off the Mare Serenitatis. Area 70 000 km^2.
Lacus Spei	Lake of Hope	43 N	65 E	80	Dark strip between Messala and Zeno.
Lacus Temporis	Lake of Time	45 N	58 E	117	Irregular area between Mercurius and Atlas.
Lacus Timoris	Lake of Fear	39 S	27 W	117	Narrow darkish patch east of Hainzel.
Lacus Veris	Spring Lake	16 S	86 W	396	Narrow irregular area in Rook Mountains. Total area 12 000 km^2.
Lacus Luxuriæ	Lake of Luxury	19 N	176 E	50	Far side.
Lacus Oblivionis	Lake of Forgetfulness	21 S	168 W	50	Far side.
Lacus Solitudinis	Lake of Solitude	28 S	104 E	384	Far side.
Palus Epidemiarum	Marsh of Epidemics	32 S	28 W	286	Darkish area adjoining Mercator and Campanus. Area 27 000 km^2.
Palus Nebularum	Marsh of Clouds	140 N	6 W	150	Eastern part of Mare Imbrium. Name deleted from some maps.
Palus Putredinis	Marsh of Decay	16 N	0.4 E	161	Part of Mare Imbrium, near Archimedes.
Palus Somnii	Marsh of Sleep	14 N	45 E	143	Curiously coloured area bounded by bright rays from Proclus.
Sinus Æstuum	Bay of Heats	11 N	9 W	290	Fairly regular dark area leading off the Mare Nubium. east of Copernicus; area 40 000 km^2.
Sinus Amoris	Bay of Love	18 N	39 E	130	Part of Mare Tranquillitatis, east of Maraldi.
Sinus Asperitatis	Bay of Asperity	4 S	27 E	206	Part of Mare Nectaris; rough area between Theophilus and Hypatia.
Sinus Concordiæ	Bay of Harmony	11 N	42 E	142	Bay south of Palus Somnii.
Sinus Fidei	Bay of Faith	18 N	2 E	70	Outlet of Mare Vaporum, to the north.
Sinus Honoris	Bay of Honour	12 N	18 E	109	Edge of Mare Tranquillitatis, northwest of Maclear.
Sinus Iridum	Bay of Rainbows	44 N	31 W	236	Beautiful bay, extending from Mare Imbrium.
Sinus Lunicus	Luna Bay	32 N	1 W	50	In Mare Imbrium, between Aristillus and Archimedes; probable landing site of Luna 2 in 1959.
Sinus Medit	Central Bay	2 N	2 E	260	Small bay near the apparent centre of the disc; area 22 000 km^2.
Sinus Roris	Bay of Dew	54 N	57 W	400	Area joining Mare Frigoris to Oceanus Procellarum.
Sinus Successus	Bay of Success	1 N	59 E	132	Ill-defined darkish area west of Mare Spumans.

[a]Alexander von Humboldt. German natural historian (1769–1859).
[b]Admiral William Henry Smyth. British astronomer (1768–1865).

evacuation of his army; the astrologers advised him to stay where he was 'for thrice nine days'. When he eventually tried to embark his forces, he found that he had been blockaded by the Spartans. His fleet was destroyed and the expedition annihilated – a reverse which led directly to the final defeat of Athens in the Peloponnesian War. According to Polybius, an eclipse in September 218 BC so alarmed the Gaulish mercenaries in the service of Attalus I of Pergamos that they refused to continue a military advance. On the other hand, Christopher Columbus turned the eclipse of AD 1504 to his advantage. He was anchored off Jamaica and the local inhabitants refused to supply his men with food; he threatened to extinguish the Moon, and when the eclipse took place the natives were so alarmed that there was no further trouble.

Table 3.14 *Selected craters on the near side of the Moon*

Name	Lat. (°)	Long. (°)	Diameter (km)	Notes (former names in parentheses)	Named after	
Abbot	5.6 N	54.8 E	10	Uplands, east of Taruntius (Apollonius K)	Charles; American	1872–1973
Abel	34.5 S	87.3 E	122	Flooded walled plain, east of Furnerius	Niels; Norwegian mathematician	1802–1829
Abenezra	21.0 S	11.9 E	42	West of Nectaris; well-formed pair with Azophi	Abraham ben Ezra; Spanish–Jewish astronomer	1092–1167
Abetti	19.9 N	27.7 E	65	Obscure; dark floor; in Serenitatis, northwest of Argæus	Antonio; Italian astronomer	1846–1928
Abulfeda	13.8 S	13.9 E	65	Abenezra area; pair with Almanon	Abu'L fida, Ismail; Syrian geographer	1273–1331
Acosta	5.6 S	60.1 E	13	North of Langrenus (Langrenus C)	Cristobal; Portuguese natural historian	1515–1580
Adams	31.9 S	68.2 E	66	East of Vendelinus; irregular walls	John Couch; English astronomer	1819–1892
Agatharchides	19.8 S	30.9 W	48	Humorum area; remains of central peak; irregular walls	Greek geographer	?–150 BC
Agrippa	4.1 N	10.5 E	44	Vaporum area; regular walls; pair with Godin	Greek astronomer	c. 92 AD
Airy	18.1 S	5.7 E	36	Pair with Argelander; irregular walls	George Biddell; English Astronomer Royal	1810–1892
Al-Bakri	14.3 N	20.2 E	12	Northwest of Plinius (Tacquet A)	Al-Bakri; Spanish–Arab astronomer	1010–1094
Al-Biruni	17.9 N	92.5 E	77	Libration zone; Marginis area	Persian mathematician/ geographer	973–1048
Al-Marrakushi	10.4 S	55.8 E	8	West of Langrenus (Langrenus D)	Moroccan geographer/ astronomer	c. 1261
Albategnius	11.7 S	4.3 E	114	Companion to Hipparchus; irregular, terraced walls	Al-Battani; Iraqi astronomer	850–929
Aldrin	1.4 N	22.1 E	3	On Tranquillitatis, east of Sabine (Sabine B)	Buzz; American astronaut; Apollo 11	1930–
Alexander	40.3 N	13.5 E	81	North end of Caucasus; darkish floor; low walls	Alexander the Great of Macedon	356–323
Alfraganus	5.4 S	19.0 E	20	Northwest of Theophilus; very bright; minor ray-centre	Al Fargani; Persian astronomer	?–840
Alhazen	15.9 N	71.8 E	32	Near border of Crisium (not Schröter's Alhazen)	Abu Ali Ibn Al Haitham; Iraqi mathematician	987–1038
Aliacensis	30.6 S	5.2 E	79	Pair with Werner; Walter area; high walls	D'Ailly Pierre; French geographer	1350–1420
Almanon	16.8 S	15.2 E	49	Regular; pair with Abulfeda	Al Mamun; Persian astronomer	786–833
Alpetragius	16.0 S	4.5 W	39	Outside Alphonsus; high, terraced walls; huge central peak with summit pit	Nur Ed-Din Al Betrugi; Moroccan astronomer	?–c. 1100
Alphonsus	13.7 S	3.2 W	108	Ptolemæus chain; low central peak; rills on floor	Alfonso X; Spanish astronomer	1223–1284
Ameghino	3.3 N	57.0 E	9	Uplands southwest of Apollonius	Fiorino; Italian natural historian	c. 1854–1911
Ammonius	8.5 S	0.8 W	8	Prominent; in Ptolemæus (Ptolemæus A)	Greek philosopher	?–c. 517
Amontons	5.3 S	46.8 W	2	Craterlet of Fœcunditatis, south of Messier	Guillaume; French physicist	1663–1705

Table 3.14 (cont.)

Name	Lat. (°)	Long. (°)	Diameter (km)	Notes (former names in parentheses)	Named after	
Amundsen	84.3 S	85.6 S	101	Libration area; well-formed; pair with Scott	Roald; Norwegian explorer	1872–1928
Anaxagoras	73.4 N	10.1 W	50	Northern polar area; distorts Goldschmidt; ray-centre	Greek astronomer	500–428 BC
Anaximander	66.9 N	51.3 W	67	Pythagoras area; pair with Carpenter; no central peak	Greek astronomer	c. 611–547 BC
Anaximenes	72.5 N	44.5 W	80	Near Philolaus; rather low walls	Greek astronomer	585–528 BC
Andĕl	10.4 S	12.4 E	35	Highlands west of Theophilus; low, irregular walls	Karel; Czech astronomer	1884–1947
Andersson	49.7 S	95.3 W	13	Libration zone; south of Guthnick and Rydberg	Leif; American astronomer	1943–1979
Ångström	29.9 N	41.6 W	9	In Imbrium, north of Harbingers	Anders; Swedish physicist	1814–1874
Ansgarius	12.7 S	79.7 E	94	Distinct; east of Fœcunditatis; pair with La Peyrouse	St Ansgar; German theologian	801–864
Anuchin	49.0 S	101.3 E	57	Libration zone; beyond Australe	Dimitri; Russian geographer	1843–1923
Anville	1.9 N	49.5 E	10	East of Secchi. In Fœcunditatis (Taruntius G)	Jean-Baptiste; French cartographer	1697–1782
Apianus	26.9 S	7.9 E	63	Aliacensis area; high walls	Bienewitz; German astronomer	1495–1552
Apollonius	4.5 N	61.1 E	53	Uplands; south of Crisium; well formed	Greek mathematician	Third century BC
Arago	6.2 N	21.4 E	26	On Tranquillitatis; domes nearby	François; French astronomer	1786–1853
Aratus	23.6 N	4.5 E	10	In Apennines; very bright; not regular in outline	Greek astronomer	c. 315–245 BC
Archimedes	29.7 N	4.0 W	82	On Imbrium; very regular; darkish floor; no central peak	Greek mathematician/physicist	c. 287–212 BC
Archytas	58.7 N	5.0 E	31	On Frigoris; bright, distinct; central peak	Greek mathematician	c. 428–347 BC
Argelander	16.5 S	5.8 E	34	Albategnius area; pair with Airy; central peak	Friedrich; German astronomer	1799–1875
Ariadæus	4.6 N	17.3 E	11	Vaporum area; associated with great rill	King of Babylon; chronologist	?–317 BC
Aristarchus	23.7 N	47.4 W	40	Brilliant; terraced walls; central peak; inner bands	Greek astronomer	?310–230 BC
Aristillus	33.9 N	1.2 E	55	Archimedes group; fine central peak	Greek astronomer	c. 280 BC
Aristoteles	50.2 N	17.4 E	87	High walls; pair with Eudoxus	Greek astronomer	383–322 BC
Armstrong	1.4 N	25.0 E	4	Tranquillitatis; east of Sabine (Sabine E)	Neil; American astronaut (Apollo 11)	1930–
Arnold	66.8 N	35.9 E	94	Northwest of Democritus; north of Frigoris; low walls	Christoph; German astronomer	1650–1695
Arrhenius	55.6 S	91.3 W	40	Libration zone; beyond Inghirami	Svante; Swedish chemist	1859–1927
Artemis	25.0 N	25.4 W	2	Pair with Verne; between Euler and Lambert	Greek moon goddess	–
Artsimovich	27.6 N	36.6 W	8	On Imbrium (Diophantus A)	Lev; Russian physicist	1909–1973
Aryabhāta	6.2 N	35.1 E	22	Between Maskelyne and Cauchy; flooded; irregular (Maskelyne E)	Indian astronomer	476–c. 550
Arzachel	18.2 S	1.9 W	96	Ptolemæus group; high walls; central peak	Al Zarkala; Spanish–Arab astronomer	c. 1028–1087
Asada	7.3 N	49.9 E	12	Northwest of Taruntius; edge of Fœcunditatis; distinct (Taruntius A)	Goryu; Japanese astronomer	1734–1799

Table 3.14 (cont.)

Name	Lat. (°)	Long. (°)	Diameter (km)	Notes (former names in parentheses)	Named after	
Asclepi	55.1 S	25.4 E	42	Southern uplands; west of Hommel; distinct	Giuseppe; Italian astronomer	1706–1776
Aston	32.9 N	87.7 W	43	Limb; beyond Ulugh Beigh	Francis; British chemist	1877–1945
Atlas	46.7 N	44.4 E	87	Pair with Hercules; high walls; much interior detail	Mythological Greek; Titan	
Atwood	5.8 S	57.7 E	29	Northwest of Langrenus; trio with Biharz and Naonobu (Langrenus K)	George; British mathematician	1745–1807
Autolycus	30.7 N	1.5 E	39	Archimedes group; regular, distinct	Greek astronomer	?–c. 330 BC
Auwers	15.1 N	17.2 E	20	Foothills of Hæmus; not very bright	Georg Friedrich; German astronomer	1838–1915
Auzout	10.3 N	64.1 E	32	Outside Crisium; low central peak	Adrien; French astronomer	1622–1691
Avery	1.4 S	81.4 E	9	Limb; north of Mare Smythii (Gilbert U)	Oswald; Canadian doctor	1877–1955
Avicenna	39.7 N	97.2 W	74	Libration zone; north of Lorentz	Abu Ali Ibn Sina; Persian doctor	980–1037
Azophi	22.1 S	12.7 E	47	West of Nectaris; well-formed pair with Abenezra	Al-Sûfi; Persian astronomer	903–986
Baade	44.8 S	81.8 W	55	Limb; beyond Schickard and Inghirami	Walter; German astronomer	1893–1960
Babbage	59.7 N	57.1 W	143	Irregular enclosure near Pythagoras	Charles; British mathematician	1792–1871
Babcock	4.2 N	93.9 E	99	Libration zone; Smythii area, beyond Neper	Harold; American astronomer	1882–1968
Back	1.1 N	80.7 E	35	Limb; south of Schubert (Schubert B)	Ernst; German physicist	1881–1959
Bacon	51.0 S	19.1 E	69	Licetus area; high walls; low central peak	Roger; British natural philosopher	1214–1294
Baillaud	74.6 N	37.5 E	89	Uplands northeast of Meton; rather low walls	Benjamin; French astronomer	1848–1934
Bailly	66.5 S	69.1 W	287	'Field of ruins'; uplands in the far south	Jean Sylvain; French astronomer	1736–1793
Baily	49.7 N	30.4 E	26	Frigoris area, north of Bürg	Francis; British astronomer	1774–1844
Balboa	19.1 N	83.2 W	69	Southeast of Otto Struve	Vasco de; Spanish explorer	1475–1517
Ball	35.9 S	8.4 W	41	On edge of Deslandres; high walls	William; British astronomer	?–1690
Balmer	20.3 S	69.8 E	138	Ruined wall plain west of Hekatæus	Johann; Swiss mathematician	1825–1898
Banachiewicz	5.2 N	80.1 E	92	Libration zone; north of Schubert	Tadeusz; Polish astronomer	1882–1954
Bancroft	28.0 N	6.4 W	13	Deep; northwest of Archimedes (Archimedes A)	William, American chemist	1867–1953
Banting	26.6 N	16.4 E	5	In Serenitatis, east of Linné (Linné E)	Frederick Grant; Canadian doctor	1891–1941
Barkla	10.7 S	67.2 E	42	Complex between Langrenus and Kapteyn (Langrenus A)	Charles; British physicist	1877–1944
Barnard	29.5 S	85.6 E	105	Limb; beyond Ansgarius and Legendre	Edward; American astronomer	1857–1923
Barocius	44.9 S	16.8 E	82	Outside Maurolycus; high but broken walls	Francesco; Italian mathematician	c. 1570
Barrow	71.3 N	7.7 E	92	Northwest of W C Bond; low, broken walls	Isaac; British mathematician	1630–1677
Bartels	24.5 N	89.8 W	55	Limb; beyond Otto Struve	Julius; German geophysicist	1899–1964
Bayer	51.6 S	35.0 W	47	Outside Schiller; high, terraced walls	Johann; German astronomer	1572–1625
Beals	37.3 S	86.5 E	48	Limb; beyond Gauss	Carlyle F.; Canadian astronomer	1899–1979
Beaumont	18.0 S	28.8 E	53	Bay on Nectaris	Leonce; French geologist	1798–1874

Table 3.14 (cont.)

Name	Lat. (°)	Long. (°)	Diameter (km)	Notes (former names in parentheses)	Named after	
Beer	27.1 N	9.1 W	9	On Imbrium; twin with Feuillée	Wilhelm; German selenographer	1797–1850
Behaim	16.5 S	79.4 E	55	East of Vendelinus; high walls; central craters	Martin; German geographer	1436–1506
Belkovich	61.1 N	90.2 E	214	Humboldtianum area; high walls with two craters; central peaks	Igor; Russian astronomer	1904–1949
Bell	21.8 N	96.4 W	86	Libration zone; beyond Einstein	Alexander; Scottish inventor	1847–1922
Bellot	12.4 S	48.2 E	17	Edge of Fœcunditatis; bright floor	Joseph; French explorer	1826–1853
Bernouilli	35.0 N	60.7 E	47	East of Geminus; fairly regular	Jacques; Swiss mathematician	1667–1748
Berosus	33.6 N	69.9 E	74	East of Cleomedes; terraced walls; pair with Hahn	Babylonian astronomer	c. 250 BC
Berzelius	36.6 N	50.9 E	50	Taurus area; darkish area; central peak	Jons; Swedish chemist	1779–1848
Bessarion	14.9 N	37.3 W	10	Deep craterlet on Imbrium; south of Brayley	Johannes; Greek scholar	c. 1369–1472
Bessel	21.8 N	17.9 E	15	On Serenitatis; associated with a long ray	Friedrich Wilhelm; German astronomer	1784–1846
Bettinus	63.4 S	44.8 W	71	Bailly area; one of a line; high walls	Mario; Italian mathematician	1582–1657
Bianchini	48.7 N	34.3 W	38	In Jura Mountains; central peak; rather irregular	Francesco; Italian astronomer	1662–1729
Biela	54.9 S	51.3 E	76	Vlacq area; high walls; central peak	Wilhelm von; Austrian astronomer	1782–1856
Biharz	5.8 S	56.3 E	43	Trio with Naonobu and Atwood (Langrenus F)	Theodor; German doctor	1825–1862
Billy	13.8 S	50.1 W	45	Pair with Hansteen; very dark floor; southern edge of Procellarurr	Jacques de; French mathematician	1602–1679
Biot	22.6 S	51.1 E	12	In Fœcunditatis; very bright	Jean-Baptiste; French astronomer	1774–1862
Birmingham	65.1 N	10.5 W	92	North of Frigoris; low-walled, irregular	John; Irish astronomer	1829–1884
Birt	22.4 S	8.5 W	16	On Nubium, west of Straight Wall; rill to the west; profile irregular	William; British selenographer	1804–1881
Black	9.2 S	80.4 E	18	Northeast of La Peyrouse (Kästner F)	Joseph; French chemist	1728–1799
Blagg	1.3 N	1.5 E	5	Quite distinct; in Sinus Medii	Mary; British astronomer	1858–1944
Blancanus	63.8 S	21.4 W	117	Near Clavius; pair with Scheiner; high walls	Giuseppe Biancani; Italian mathematician	1566–1624
Blanchard	58.5 S	94.4 W	40	Libration zone; north of Hausen, beyond Pingré	Jean P. F., formerly Arrhenius P.; French aeronaut	1753–1809
Blanchinus	25.4 S	2.5 E	61	North of Werner; uneven walls; rough floor	Giovanni Blanchini; Italian astronomer	c. 1458
Bobillier	19.6 N	15.5 E	6	In Serenitatis; east of Sulpicius Gallus	Etienne, formerly Bessel E.; French geometer	1798–1840
Bode	6.7 N	2.4 W	18	Outside Æstuum; bright; minor ray-centre	Johann Elert; German astronomer	1747–1826
Boethius	5.6 N	72.3 E	10	Well formed (Dubiago U)	Greek physicist	c. 480–524
Boguslawsky	72.9 S	43.2 E	97	Southern uplands; high walls	Palon von; German astronomer	1789–1851
Bohnenberger	16.2 S	40.0 E	33	Edge of Nectaris; low walls	Johann von; German astronomer	1765–1831
Bohr	12.4 N	86.6 W	71	Limb; beyond Vasco da Gama	Niels; Danish physicist	1885–1962
Boltzmann	74.9 S	90.7 W	76	Libration zone; closely north of Drygalski	Ludwig; Austrian physicist	1844–1906

Table 3.14 (cont.)

Name	Lat. (°)	Long. (°)	Diameter (km)	Notes (former names in parentheses)	Named after	
Bombelli	5.3 N	56.2 E	10	East of Apollonius (Apollonius T)	Raphael; Italian mathematician	1526–1572
Bond, G P	33.3 S	35.7 W	23	East of Posidonius; fairly regular	George P.; American astronomer	1825–1865
Bond, W C	65.4 N	3.7 E	156	North of Frigoris; old and broken	William Cranch; American astronomer	1789–1859
Bonpland	8.3 S	17.4 W	60	In Nubium; Fra Mauro group; fairly regular	Aimé; French botanist	1773–1858
Boole	63.7 N	87.4 W	63	Limb; beyond Pythagoras	George; British mathematician	1815–1864
Borda	25.1 S	46.6 E	44	West of Petavius; low walls	Jean; French astronomer	1733–1799
Borel	22.3 N	26.4 E	4	In Serenitatis; southwest of Le Monnier (Le Monnier C)	Felix; French mathematician	1871–1956
Born	6.0 S	66.8 E	14	Distinct; southeast of Langrenus (Maclaurin Y)	Max; German physicist	1882–1970
Boscovich	9.8 N	11.1 E	46	Edge of Vaporum; low walls; irregular; very dark floor	Ruggiero; Italian physicist	1711–1787
Boss	45.8 N	89.2 E	47	Limb; beyond Mercurius	Lewis; American astronomer	1846–1912
Bouguer	52.3 N	35.8 W	22	In Jura uplands; very distinct	Pierre; French hydrographer	1698–1758
Boussingault	70.2 S	54.6 E	142	Southern uplands; made up of three large rings	Jean; French chemist	1802–1887
Bowen	17.6 N	9.1 E	8	North of Manilius; flattish floor (Manilius A)	Ira; American astronomer	1898–1973
Brackett	17.9 N	23.6 E	8	In Serenitatis; north of Plinius; inconspicuous	Frederick; American physicist	1896–1988
Bragg	42.5 N	102.9 W	84	Libration zone; beyond Gerard	William; Australian physicist	1862–1942
Brayley	20.9 N	36.9 W	14	On Procellarum; low central peak	Edward; British geographer	1801–1870
Breislak	48.2 S	18.3 E	49	Southeast of Maurolycus; fairly regular; central peak	Scipione; Italian chemist	1748–1826
Brenner	39.0 S	39.3 E	97	Broken; adjoins Janssen (southern uplands)	'Leo' (assumed name); Austrian astronomer	1855–1928?
Brewster	23.3 N	34.7 E	10	Between Rømer and Littrow (Rømer L)	David; Scottish optician	1781–1868
Brianchon	75.0 N	86.2 W	134	Large limb enclosure; beyond Carpenter	Charles; French mathematician	1783–1864
Briggs	26.5 N	69.1 W	37	On Procellarum; Otto Struve area; similar to Seleucus	Henry; British mathematician	1556–1630
Brisbane	49.1 S	68.5 E	44	Australe area; fairly regular	Sir Thomas; Scottish geographer	1770–1860
Brown	46.4 S	17.9 W	34	Northeast of Longomontanus; irregular	Ernest; British mathematician	1866–1938
Bruce	1.1 N	0.4 E	6	Distinct; in Sinus Medii	Catherine Wolfe; American philanthropist	1816–1900
Brunner	9.9 S	90.9 E	53	Libration zone; beyond Kästner; Hirayama area	William; Swiss astronomer	1878–1958
Buch	38.8 S	17.7 E	53	Maurolycus area; adjoins Büsching	Christian von; German geologist	1774–1853
Bullialdus	20.7 S	22.2 W	60	On Nubium; massive walls; terraced; central peak	Ismael Bouillaud; French astronomer	1605–1694
Bunsen	41.4 N	85.3 W	52	Between Gerard and La Voisier	Robert; German physicist	1811–1899
Burckhardt	31.1 N	56.5 E	56	North of Cleomedes; member of a complex group	Johann; German astronomer	1773–1825

Table 3.14 (cont.)

Name	Lat. (°)	Long. (°)	Diameter (km)	Notes (former names in parentheses)	Named after	
Burnham	13.9 S	7.3 E	24	Albategnius area; low walls	Sherburne; American astronomer	1838–1921
Bürg	45.0 N	28.2 E	39	North of Lacus Mortis; large central peak with pit; major rill system nearby	Johann; Austrian astronomer	1766–1834
Büsching	38.0 S	20.0 E	52	Adjoins Buch. But is less regular	Anton; German geographer	1724–1793
Byrd	85.3 N	9.8 E	93	Northern polar area; walled plain adjoining Gioja	Richard; American explorer	1888–1957
Byrgius	24.7 S	65.3 W	87	West of Humorum; Byrgius A, on its eastern crest, is a ray-centre	Joost Burgi; Swiss horologist	1552–1632
Cabæus	84.9 S	35.5 W	98	Southern polar uplands; fairly high walls	Niccolo Cabeo; Italian astronomer	1586–1651
Cajal	12.6 N	31.1 E	9	On Tranquillitatis, east of Jansen (Jansen F)	Santiago; Spanish doctor	1852–1934
Calippus	38.9 N	10.7 E	32	Northern end of Caucasus; irregular	Greek astronomer	c. 330 BC
Cameron	6.2 N	45.9 E	10	Intrudes into northwestern wall of Taruntius (Taruntius C)	Robert; American astronomer	1925–1972
Campanus	28.0 S	27.8 W	48	Western edge of Nubium; pair with Mercator, but with a lighter floor	Giovanni Campano; Italian astronomer	c. 1200–?
Cannizarro	55.6 N	99.6 W	56	Libration zone; in Poczobut	Stanislao; Italian chemist	1826–1910
Cannon	19.9 N	81.4 E	56	Limb north of Marginis; light floor	Annie Jump; American astronomer	1863–1943
Capella	7.5 S	35.0 E	90	Uplands north of Nectaris; large central peak; cut by crater valley	Martianus; Roman astronomer	c. 400–?
Capuanus	34.1 S	26.7 W	59	Edge of Epidemiarum; domes in floor	Francesco; Italian astronomer	c. 1400–?
Cardanus	13.2 S	72.4 W	49	On Procellarum; central peak; pair with Krafft	Girolamo Cardano; Italian mathematician	1501–1570
Carlini	33.7 N	24.1 W	10	Bright craterlet on Imbrium	Francesco; Italian astronomer	1783–1862
Carmichael	19.6 N	40.4 E	20	Well formed; west of Macrobius, on Tranquillitatis	Leonard; American psychologist	1898–1973
Carpenter	69.4 N	50.9 W	59	North of Frigoris; adjoins Anaximander	James; British astronomer	1840–1899
Carrel	10.7 N	26.7 E	15	On Tranquillitatis; central crater; southwest of Jansen (Jansen B)	Alexis; French doctor	1873–1944
Carrillo	2.2 S	80.9 E	16	North of Kästner	Flores; Mexican soil engineer	1911–1967
Carrington	44.0 N	62.1 E	30	Messala group	Richard; British astronomer	1826–1875
Cartan	4.2 N	59.3 E	15	Regular; closely west of Apollonius (Apollonius D)	Elié; French mathematician	1869–1951
Casatus	72.8 S	29.5 W	108	South of Clavius; high walls; intrudes into Klaproth	Paolo Casani; Italian mathematician	1617–1707
Cassini	40.2 N	4.6 E	56	Edge of Nebularum; low walls; contains a deep crater, A	Giovanni; Italian astronomer	1625–1712
Cassini, J J	68.0 N	16.0 W		Near Philolaus; irregular ridge-bounded area	Jacques J.; Italian–French astronomer	1677–1756
Catalán	45.7 S	87.3 W	25	Limb; beyond Schickard and Bade	Miguel; Spanish spectroscopist	1894–1957
Catharina	18.1 S	23.4 E	104	Theophilus group; rough floor; no central peak	Greek theologian (St Catherine)	?– c. 307
Cauchy	9.6 N	38.6 E	12	Bright crater in Tranquillitatis	Augustin; French mathematician	1789–1857
Cavalerius	5.1 N	66.8 W	57	In Hevel group; central ridge	Buonaventura Cavalieri; Italian mathematician	1598–1647

Table 3.14 (cont.)

Name	Lat. (°)	Long. (°)	Diameter (km)	Notes (former names in parentheses)	Named after	
Cavendish	24.5 S	53.7 W	56	West of Humorum; fairly high walls	Henry; British chemist	1731–1810
Caventou	29.8 N	29.4 W	3	On Imbrium; fairly prominent (Lahire D)	Joseph; French chemist	1795–1877
Cayley	4.0 N	15.1 E	14	Very bright; uplands west on Tranquillitatis	Arthur; British astronomer	1821–1895
Celsius	34.1 S	20.1 E	36	Rabbi Levi group; rather elliptical	Anders; Swedish astronomer	1701–1744
Censorinus	0.4 S	32.7 E	3	Brilliant; uplands southeast of Tranquillitatis	Roman astronomer	238–?
Cepheus	40.8 N	45.8 E	39	Somniorum area; forms a pair with Franklin	Mythological character	–
Chacornac	29.8 N	31.7 E	51	Edge of Serenitatis; adjoins Posidonius	Jean; French astronomer	1823–1873
Chadwick	52.7 N	101.3 W	30	Libration zone; beyond Pingré, near de Roy	James; British physicist	1891–1974
Challis	79.5 N	9.2 E	55	Northern polar area; contact pair with Main	James; British astronomer	1803–1862
Chamberlin	58.9 S	95.7 E	58	Libration zone; beyond Hanno	Thomas; American geologist	1843–1928
Chapman	50.4 N	100.7 W	71	Libration zone; beyond Galvani	Sydney; British geophysicist	1888–1970
Chappe	61.2 S	91.5 W	59	Libration zone	Jean-Baptiste d'Auteroche; French astronomer	1728–1769
Chevallier	44.9 N	51.2 E	52	Atlas area; low walls	Temple; British astronomer	1794–1873
Ching-te	20.0 N	30.0 E	4	Craterlet on Tranquillitatis, southwest of Littrow	Chinese male name	–
Chladni	4.0 N	1.1 E	13	Sinus Medii area; bright; abuts on Murchison	Ernst; German physicist	1756–1827
Cichus	33.3 S	21.1 W	40	Just south of Nubium; well formed	Francesco; Italian astronomer	1257–1327
Clairaut	47.7 S	13.9 E	75	Maurolycus area; broken walls	Alexis; French mathematician	1713–1765
Clausius	36.9 S	43.8 W	24	Schickard area; bright and distinct	Rudolf; German physicist	1822–1888
Clavius	58.8 S	14.1 W	245	Southern uplands; massive walls; no central peak; curved line of craters on floor	Christopher; German mathematician	1537–1612
Cleomedes	27.7 N	56.0 E	125	Crisium area; distorted by Tralles	Greek astronomer	?– c. 50 BC
Cleostratus	60.4 N	77.0 W	62	Pythagoras area; distinct	Greek astronomer	?– c. 500 BC
Clerke	21.7 N	29.8 E	6	On Tranquillitatis; west of Littrow (Littrow B)	Agnes; British astronomer	1842–1907
Collins	1.3 N	23.7 E	2	On Tranquillitatis; east of Sabine (Sabine D)	Michael; American astronaut (Apollo 11)	1930–
Colombo	15.1 S	45.8 E	76	In Fœcunditatis; irregular; broken by large crater, A	Christopher Columbus; Spanish explorer	1446–1506
Compton	55.3 N	103.8 E	182	Libration zone; beyond Humboldtianum; crossed by rill	Arthur Holly; American physicist	1892–1962
Condamine	53.4 N	28.2 W	48	Edge of Frigoris; fairly regular	Charles de la; French physicist	1701–1774
Condon	1.9 N	60.4 E	34	Dark floor; low walls; north of Webb (Webb R)	Edward; American physicist	1902–1974
Condorcet	12.1 N	69.6 E	74	Outside Crisium; regular; no central peak	Jean; French mathematician	1743–1794
Conon	21.6 N	2.0 E	21	In Apennine uplands; fairly distinct.	Greek astronomer	c. 260 BC
Cook	17.5 S	48.9 E	46	Edge of Fœcunditatis; darkish floor	James; British explorer	1728–1779
Copernicus	9.7 N	20.1 W	107	Great ray-centre; massive terraced walls; central peak	Mikołaj Kopernik; Polish astronomer	1473–1543

Table 3.14 (cont.)

Name	Lat. (°)	Long. (°)	Diameter (km)	Notes (former names in parentheses)	Named after	
Couder	4.8 S	92.4 W	21	Libration zone; north of Orientale	Andre; French astronomer	1897–1978
Cremona	67.5 N	90.6 W	85	Libration zone; beyond Pythagoras	Luigi; Italian mathematician	1830–1903
Crile	14.2 N	46.0 E	9	Distinct; south of Proclus (Proclus F)	George; American doctor	1864–1943
Crozier	13.5 S	50.8 E	22	Fœcunditatis area; southeast of Colombo; central peak	Francis; British explorer	1796–1848
Crüger	16.7 S	66.8 W	45	Southwest of Prodellarum; regular; very dark floor; no central peak	Peter; German mathematician	1580–1639
Curie	22.9 S	91.0 E	151	Libration zone; beyond Wilhelm Humboldt; pair with Sklodowska	Pierre; French physicist	1859–1906
Curtis	14.6 N	56.6 E	2	On Crisium; east of Picard (Picard Z)	Heber; American astronomer	1872–1942
Curtius	67.2 S	4.4 E	95	Moretus area; massive, terraced walls; regular	Albert Curtz; German astronomer	1600–1671
Cusanus	72.0 N	70.8 E	63	Limb beyond Democritus; well formed; no central peak	Nikolas Krebs; German mathematician	1401–1464
Cuvier	50.3 S	9.9 E	75	Licetus area; high walls; central peak	Georges; French palæontologist	1769–1832
Cyrillus	13.2 S	24.0 E	98	Theophilus group; low central peak; rather irregular	St Cyril; Egyptian theologian	?–444
Cysatus	66.2 S	6.1 W	48	Moretus area; high walls	Jean-Baptiste Cysat; Swiss astronomer	1588–1657
D'Arrest	2.3 N	14.7 E	30	East of Godin; low, broken walls	Heinrich; German astronomer	1822–1875
da Vinci	9.1 N	45.0 E	37	North of Taruntius; irregular; low, broken walls	Leonardo; Italian artist and inventor	1452–1519
Daguerre	11.9 S	33.6 E	46	In Nectaris; very low walls	Louis; French photographer	1789–1851
Dale	9.6 S	82.9 E	22	Northeast of La Peyrouse; trio with Black and Kreiken	Sir Henry; British physiologist	1875–1968
Dalton	17.1 N	84.3 W	60	Limb crater; east of Einstein and west of Krafft	John; British chemist	1766–1844
Daly	5.7 N	59.6 E	17	Well formed; twin with Apollonuis F (Apollonius P)	Reginald; Canadian geologist	1871–1957
Damoiseau	4.8 S	61.1 W	36	East of Grimaldi; very irregular	Marie; French astronomer	1768–1846
Daniell	35.3 N	31.1 E	29	Adjoins Posidonius; no central peak	John; British physicist/meteorologist	1790–1845
Darney	14.5 S	23.5 W	15	Bright crater in Nubium; Fra Mauro area	Maurice; French astronomer	1882–1958
Darwin	20.2 S	69.5 W	120	Grimaldi area; low walls; contains large dome	Charles; British naturalist	1809–1882
Daubrée	15.7 N	14.7 E	14	Southwest of Menelaus; darkish floor (Menelaus S)	Gabriel; French geologist	1814–1896
Davy	11.8 S	8.1 W	34	Edge of Nubium; irregular walls	Humphry; British physicist	1778–1829
Dawes	17.2 N	26.4 E	18	Distinct; between Serenitatis and Tranquillitatis	William Rutter; British astronomer	1799–1868
De Gasparis	25.9 S	50.7 W	30	West of Humorum; south of Mersenius; fairly regular	Annibale; Italian astronomer	1819–1892
De la Rue	59.1 N	52.3 E	134	Very low, broken walls; adjoins Endymion to the northwest	Warren; British astronomer	1815–1889
De Morgan	3.3 N	14.9 E	10	Bright craterlet; uplands west of Tranquillitatis	Augustus; British mathematician	1806–1871
De Roy	55.3 S	99.1 W	43	Libration zone; beyond Pingré; pair with Chadwick	Felix; Belgian astronomer	1883–1942
De Sitter	80.1 N	39.6 E	64	Libration zone; north of Euctemon	Willem; Dutch astronomer	1872–1934

Table 3.14 (cont.)

Name	Lat. (°)	Long. (°)	Diameter (km)	Notes (former names in parentheses)	Named after	
De Vico	19.7 S	60.2 W	20	Deep crater; west of Gassendi	Francesco; Italian astronomer	1805–1848
Debes	29.5 N	51.7 E	30	Outside Cleomedes; fusion of two rings	Ernst; German cartographer	1840–1923
Dechen	46.1 N	68.2 W	12	In northwest of Procellarum; not bright	Ernst von; German geologist	1800–1889
Delambre	1.9 S	17.5 E	51	Tranquillitatis area; high walls	Jean-Baptiste; French astronomer	1749–1822
Delaunay	22.2 S	2.5 E	46	Albategnius area, near Faye; irregular	Charles; French astronomer	1816–1872
De l'Isle	29.9 N	34.6 W	25	On Procellarum; pair with Diophantus; central peak	Joseph; French astronomer	1688–1768
Delmotte	27.1 N	60.2 E	32	North of Crisium; not prominent	Gabriel; French astronomer	1876–1950
Deluc	55.0 S	2.8 W	46	Clavius area; walls of moderate height	Jean; Swiss geologist	1727–1817
Dembowski	2.9 N	7.2 E	26	Low walls; east of Sinus Medii	Baron Ercole; Italian astronomer	1815–1881
Democritus	62.3 N	35.0 E	39	Highlands north of Frigoris; very deep	Greek astronomer	c. 460–360 BC
Demonax	77.9 S	60.8 E	128	Boguslawski area; fairly regular	Greek philosopher	?– c. 100 BC
Desargues	70.2 N	73.3 W	85	Limb formation; beyond Anaximander	Gerard; French mathematician	1593–1662
Descartes	11.7 S	15.7 E	48	Northeast of Abulfeda; low, broken walls	René; French mathematician	1596–1650
Deseilligny	21.1 N	20.6 E	6	Distinct craterlet on Serenitatis	Jules; French selenographer	1868–1918
Deslandres	33.1 S	4.8 W	256	Ruined enclosure west of Walter	Henri; French astrophysicist	1853–1948
Dionysius	2.8 N	17.3 E	18	Brilliant crater on edge of Tranquillitatis	Greek astronomer	9–120
Diophantus	27.6 N	34.3 W	17	On Procellarum; central peak; pair with de l'Isle	Greek mathematician	?– c. 300
Doerfel	69.1 S	107.9 W	68	Libration zone; beyond Hansen	Georg; German astronomer	1643–1688
Dollond	10.4 S	14.4 E	11	West of Theophilus; borders a large 'ghost'	John; British optician	1706–1761
Donati	20.7 S	5.2 E	36	South of Albategnius; irregular; central peak; pair with Faye	Giovanni; Italian astronomer	1826–1873
Donner	31.4 S	98.0 E	58	Libration zone; beyond western Humboldt	Anders; Finnish astronomer	1873–1949
Doppelmeyer	28.5 S	41.4 W	63	Bay in Humorum; remnant of central peak	Johann; German astronomer	1671–1750
Dove	46.7 S	31.5 E	30	Janssen area; low walls	Heinrich; German physicist	1803–1879
Draper	17.6 N	21.7 W	8	In Imbrium; south of Pytheas; one of a pair	Henry; American astronomer	1837–1882
Drebbel	40.9 S	49.0 W	30	Schickard area; well formed	Cornelius; Dutch inventor	1572–1634
Dreyer	10.0 N	96.9 E	61	Libration zone; beyond Marginis	Johann Ludwig Emil; Danish astronomer	1852–1926
Drude	38.5 S	91.8 W	24	Libration zone; beyond Orientale	Paul; German physicist	1863–1906
Drygalski	79.3 S	84.9 W	149	Cabæus area; irregular	Erich von; German geophysicist	1865–1949
Dubiago	4.4 N	70.0 E	51	Smythii area; regular	Dimitri; Russian astronomer	1850–1918
Dugan	64.2 N	103.3 E	50	Libration zone; beyond Humboldtianum; beyond Belkovich	Raymond; American astronomer	1878–1940
Dunthorne	30.1 S	31.6 W	15	Edge of Epidemiarum; broad walls	Richard; British astronomer	1711–1775
Dziewulski	21.2 N	98.9 E	63	Libration zone; beyond Marginis	Wladyslaw; Polish astronomer	1878–1962
Eckert	17.3 N	58.3 E	2	Craterlet on Crisium; northeast of Picard	Wallace; American astronomer	1902–1971

Table 3.14 (cont.)

Name	Lat. (°)	Long. (°)	Diameter (km)	Notes (former names in parentheses)	Named after	
Eddington	21.3 N	72.2 W	118	Flooded plain between Seleucus and Otto Struve	Sir Arthur; British astronomer	1882–1944
Edison	25.0 N	99.1 E	62	Libration zone; Margins area; adjoins Lomonosov	Thomas; American inventor	1847–1931
Egede	48.7 N	10.6 E	37	Near Alpine Valley; low walls; lozenge-shaped	Hans; Danish natural historian	1686–1758
Eichstädt	22.6 S	78.3 W	49	Orientale area; regular	Lorente; German mathematician	1596–1660
Eimmart	24.0 N	64.8 E	46	Near edge of Crisium; regular	Georg; German astronomer	1638–1705
Einstein	16.3 N	88.7 W	198	Otto Struve area; contains central crater	Albert; German physicist/mathematician	1879–1955
Elger	35.3 S	29.8 W	21	Edge of Epidemiarum; low walls; imperfect	Thomas Gwyn; English selenographer	1838–1897
Ellison	55.1 N	107.5 W	36	Libration zone; between Xenophanes and Poczobut	Mervyn; Irish astronomer	1909–1963
Elmer	10.1 S	84.1 E	16	Limb crater; beyond La Peyrouse	Charles; American astronomer	1872–1954
Encke	4.7 N	36.6 W	28	On Procellarum; Kepler area	Johann; German mathematician/astronomer	1791–1865
Endymion	53.9 N	57.0 E	123	Humboldtianum area; darkish floor; no central peak	Greek mythological character	–
Epigenes	67.5 N	4.6 W	55	North of Frigoris; broad walls	Greek astronomer	?– c. 200 BC
Epimenides	40.9 S	30.2 W	27	One of a pair east of Hainzel	Greek philosopher	c. 596 BC
Eppinger	9.4 S	25.7 W	6	Deep craterlet northeast of Riphæans (Euclides D)	Hans.; Czech doctor	1879–1946
Eratosthenes	14.5 N	11.3 W	58	End of Apennines; terraced; very deep; central peak	Greek astronomer/geographer	c. 276–196 BC
Erro	5.7 N	98.5 E	61	Libration zone; Smythii area; beyond Babcock	Luis; Mexican astronomer	1897–1955
Esclangon	21.5 N	42.1 E	15	On Tranquillitatis; west of Macrobius; low walls (Mercurius L)	Ernest; French astronomer	1876–1954
Euclides	7.4 S	29.5 W	11	Near Riphæans; lies on bright nimbus	Euclid; Greek mathematician	?– c. 300 BC
Euctemon	76.4 N	31.3 E	62	Walled plain beyond Meton	Greek astronomer	?– c. 432 BC
Eudoxus	44.3 N	16.3 E	67	South of Frigoris; pair with Aristoteles	Greek astronomer	c. 408–355 BC
Euler	23.3 N	29.2 W	27	On Imbrium; minor ray-centre	Leonhard; Swiss mathematician	1707–1783
Fabbroni	18.7 N	29.2 E	10	Edge of Mare; northwest of Vitruvius (Vitruvius E)	Giovanni; Italian chemist	1752–1822
Fabricius	42.9 S	42.0 E	78	Intrudes into Janssen; rough floor; central peak	David; Dutch astronomer	1564–1617
Fabry	42.9 N	100.7 E	184	Libration zone; in Harkhebi	Charles; French physicist	1867–1945
Fahrenheit	13.1 N	61.7 E	6	On Crisium; west of Agarum (Picard X)	Gabriel; Dutch physicist	1686–1736
Faraday	42.4 S	8.7 E	69	Intrudes into Stöfler; irregular	Michael; British chemist	1791–1867
Faustini	87.3 S	77.0 E	39	Libration zone	Arnaldo; Italian polar geographer	1874–1944
Fauth	6.3 N	20.1 W	12	On Procellarum; south of Copernicus; double crater	Philipp; German selenographer	1867–1941
Faye	21.4 S	3.9 E	36	South of Albategnius; irregular; central peak; pair with Donati	Hervé; French astronomer	1814–1962
Fedorov	28.2 N	37.0 W	6	On Imbrium; west of Diophantus	Alexei; Russian rocket engineer	1872–1920

Table 3.14 (cont.)

Name	Lat. (°)	Long. (°)	Diameter (km)	Notes (former names in parentheses)	Named after	
Fényi	44.9 S	105.1 W	38	Libration zone; beyond Rydberg and Guthnick	Gyula; Hungarian astronomer	1845–1927
Fermat	22.6 S	19.8 E	38	Altai area; distinct	Pierre de; French mathematician	1601–1665
Fernelius	38.1 S	4.9 E	65	North of Stöfler; rather irregular	Jean; French astronomer/doctor	1497–1558
Feuillée	27.4 N	9.4 W	9	On Imbrium; twin with Beer	Louis; French natural scientist	1660–1732
Finsch	23.6 N	21.3 E	4	In Serenitatis; northeast of Bessel; darkish floor	Otto; German zoologist	1839–1987
Firmicus	7.3 N	63.4 E	56	South of Crisium; dark floor; no central peak	Julius; Italian astronomer	?– c. 330
Flammarion	3.4 S	3.7 W	74	Northwest of Ptolemæus; irregular enclosure	Camille; French astronomer	1842–1925
Flamsteed	4.5 S	44.3 W	20	On Procellarum; associated with 100-km 'ghost'	John; British astronomer	1646–1739
Focas	33.7 S	93.8 W	22	Libration zone; well formed; beyond Rook Mountains	Ionnas; Greek astronomer	1908–1959
Fontana	16.1 S	56.6 W	31	Between Billy and Crüger; well-formed central peak	Francesco; Italian astronomer	c. 1585–1656
Fontenelle	63.4 N	18.9 W	38	Northern edge of Frigoris; deep and distinct	Bernard de; French astronomer	1657–1757
Foucault	50.4 N	39.7 W	23	Jura area; bright and deep	Leon; French physicist	1819–1858
Fourier	30.3 S	53.0 W	51	West of Humorum; terraced, with central crater	Jean-Baptiste; French mathematician	1768–1838
Fox	0.5 N	98.2 E	24	Libration zone; Smythii area; regular	Philip; American astronomer	1878–1944
Fra Mauro	6.1 S	17.0 W	101	On Nubium; low walls; trio with Bonpland and Parry	Italian geographer	?–1459
Fracastorius	21.5 S	33.2 E	112	Great bay at south of Nectaris	Girolamo Fracastoro; Italian astronomer	1483–1553
Franck	22.6 N	35.5 E	12	On Mars; south of Rømer (Rømer K)	James; German physicist	1882–1964
Franklin	38.8 N	47.7 E	56	Regular; southeast of Atlas	Benjamin; American inventor	1706–1790
Franz	16.6 N	40.2 E	25	Edge of Somnii; low walls	Julius; German astronomer	1847–1913
Fraunhofer	39.5 S	59.1 E	56	South of Furnerius; northwestern wall broken by craters	Joseph von; German optician	1787–1826
Fredholm	18.4 N	46.5 E	14	Highlands south of Macrobius (Macrobius D)	Erik; Swedish mathematician	1866–1927
Freud	25.8 N	52.3 W	2	Craterlet adjoining Schröter's Valley	Sigmund; Austrian psychoanalyst	1856–1939
Froelich	80.3 N	109.7 W	58	Libration zone; beyond Mouchez; pair with Lovelace	Jack; American rocket scientist	1921–1967
Fryxell	21.3 S	101.4 W	18	Libration zone; beyond Orientale	Roald; American geologist	1934–1974
Furnerius	36.0 S	60.6 E	135	In Petavius chain; rather broken walls	Georges Furner; French mathematician	c. 1643
Galen	21.9 N	5.0 E	10	South of Aratus; east of Conon (Aratus A)	Claudius; Greek doctor	c. 129–200
Galilaei	10.5 N	62.7 W	15	On Procellarum; obscure	Galileo; Italian scientist	1564–1642
Galle	55.9 N	22.3 E	21	On Frigoris; distinct	Johann; German astronomer	1812–1910
Galvani	49.6 N	84.6 W	80	Limb; beyond Repsold	Luigi; Italian physicist	1737–1796
Gambart	1.0 N	15.2 W	25	On Procellarum, south-southeast of Copernicus; regular; low walls	Jean; French astronomer	1800–1836

Table 3.14 (cont.)

Name	Lat. (°)	Long. (°)	Diameter (km)	Notes (former names in parentheses)	Named after	
Ganswindt	79.6 S	110.3 E	74	Libration zone; beyond Demonax; adjoins Schrödinger	Hermann; German rocket inventor	1856–1934
Gardner	17.7 N	34.6 E	18	Distinct; east of Vitruvius (Vitruvius A)	Irvine; American physicist	1889–1972
Gärtner	59.1 N	34.6 E	115	Bay on Frigoris; 'seaward' wall barely traceable	Christian; German geologist	c. 1750–1823
Gassendi	17.6 S	40.1 W	101	Edge of Humorum; 'seaward' wall low; central peak; many rills on floor	Pierre; French astronomer	1592–1655
Gaudibert	10.9 S	37.8 E	34	Edge of Nectaris, low walls	Casimir; French astronomer	1823–1965
Gauricus	33.8 S	12.6 W	79	Pitatus group; irregular outline	Luca Gaurico; Italian astronomer	1476–1556
Gauss	35.7 N	79.0 E	177	High walls; central peak; along limb from Humboldtianum	Karl; German mathematician	1777–1855
Gay-Lussac	13.9 N	20.8 W	26	North of Copernicus; irregular enclosure	Joseph; French physicist	1778–1850
Geber	19.4 S	13.9 E	44	Regular; between Almanon and Abulfeda	Jabir ben Aflah; Arab astronomer	c. 1145
Geissler	2.6 S	76.5 E	16	Small crater west of Smythii (Gilbert D)	Heinrich; German physicist	1814–1879
Geminus	34.5 N	56.7 E	85	Crisium area; broad, terraced walls; central hill	Greek astronomer	?– c. 70 BC
Gemma Frisius	34.2 S	13.3 E	87	North of Maurolycus; high but broken walls	Reinier; Dutch doctor	1508–1555
Gerard	44.5 N	80.0 W	90	West of Sinus Roris; fairly distinct	Alexander; Scottish explorer	1792–1839
Gernsback	36.5 S	99.7 E	48	Libration zone; beyond Australe	Hugo; American writer	1884–1967
Gibbs	18.4 S	84.3 E	76	Limb; northwest of Hekatæus	Josiah; American physicist	1839–19&$$$;
Gilbert	3.2 S	76.0 E	112	Walled plain east of Smythii; northwest of Kästner	Grove; American geologist	1843–1913
Gill	63.9 S	75.9 E	66	Beyond Rosenberger	Sir David; Scottish astronomer	1843–1914
Ginzel	14.3 N	97.4 E	55	Libration zone; beyond Marginis	Friedrich; Austrian astronomer	1850–1935
Gioja	83.3 N	2.0 E	41	Polar crater; fairly regular and distinct	Flavio; Italian inventor	c. 1302
Glaisher	13.2 N	49.5 E	15	Western border of Crisium; obscure	James; British meteorologist	1809–1955
Goclenius	10.0 S	45.0 E	72	Edge of Fœcunditatis; lava-flooded; low central peak	Rudolf Göckel; German mathematician	1572–1612
Goddard	14.8 N	89.0 E	89	Limb beyond Marginis; dark floor	Robert; American rocket scientist	1882–1945
Godin	1.8 N	10.2 E	34	South of Vaporum; central peak; pair with Agrippa	Louis; French astronomer	1704–1750
Goldschmidt	73.2 N	3.8 W	113	East of Anaxagoras; north of Frigoris; low, broken walls	Hermann; German astronomer	1802–1866
Golgi	27.8 N	60.0 W	5	Small distinct craterlet north of Schiaparelli	Camillo; Italian doctor	1843–1926
Goodacre	32.7 S	14.1 E	46	East of Aliacensis; low central peak	Walter; British selenographer	1856–1938
Gould	19.2 S	17.2 W	34	'Ghost' in Nubium; east of Bullialdus	Benjamin; American astronomer	1824–1896
Graff	42.4 S	88.6 W	36	Limb formation, beyond Schickard	Kasimir; Polish astronomer	1878–1950
Greaves	13.2 N	52.7 E	13	Deep craterlet on Crisium; north of Lick (Lick D)	William; British astronomer	1897–1955
Grimaldi	5.5 S	68.3 W	172	West of Procellarum; no central peak; very dark floor; irregular, low walls	Francesco; Italian physicist/astronomer	1618–1667
Grove	40.3 N	32.9 E	28	In Somniorum; bright and deep	Sir William; British physicist	1811–1896
Gruemberger	66.9 S	10.0 W	93	Moretus area; high walls	Christoph; Austrian astronomer	1561–1636

Table 3.14 (cont.)

Name	Lat. (°)	Long. (°)	Diameter (km)	Notes (former names in parentheses)	Named after	
Gruithuisen	32.9 N	39.7 W	15	Bright craterlet on Procellarum contains a very deep crater	Franz von; German astronomer	1774–1852
Guericke	11.5 S	14.1 W	63	Fra Mauro group; broken, irregular walls	Otto von; German physicist	1602–1656
Gum	40.4 S	88.6 E	54	Australe area beyond Marinus; shallow, flooded	Colin; Australian astronomer	1924–1961
Gutenberg	8.6 S	41.2 E	74	Edge of Fœcunditatis; near Goclenius; irregular	Johann; German inventor	c. 1398–1458
Guthnick	47.7 S	93.9 W	36	Libration zone; pair with Rydberg	Paul; German astronomer	1879–1947
Gyldén	5.3 S	0.3 E	47	North of Ptolemæus; partly lava-filled; crater valley to west	Hugo; Swedish astronomer	1841–1896
Hagecius	59.8 S	46.6 E	76	Vlacq group; wall broken by craters	Thaddeus Hayek; Czech astronomer	1525–1600
Hahn	31.3 N	73.6 E	84	Crisium area; regular; central peak; pair with Berosus	Friedrich von; German astronomer	1879–1968
Haidinger	39.2 S	25.0 W	22	South of Epidemiarum; inconspicuous	Wilhelm von; Austrian geologist	1795–1871
Hainzel	41.3 S	33.5 W	70	North of Schiller; compound; two coalesced rings	Paul; German astronomer	c. 1570
Haldane	1.7 S	84.1 E	37	Limb formation; Smythii area	John; British biochemist	1892–1964
Hale	74.2 S	90.8 E	83	Libration zone; beyond Boussingault; terraced	George Ellery; American astronomer	1868–1938
Hall	33.7 N	37.0 E	35	Somniorum area; east of Posidonius	Asaph; American astronomer	1829–1907
Halley	8.0 S	5.7 E	36	Hipparchus group; regular; pair with Hind	Edmond; British astronomer	1656–1742
Hamilton	42.8 S	84.7 E	57	Limb, beyond Oken; deep and regular	Sir William; Irish mathematician	1805–1865
Hanno	56.3 S	71.2 E	56	Australe area; darkish floor	Roman explorer	c. 500 BC
Hansen	14.0 N	72.5 E	39	Regular; similar to Alhazen; Crisium area near Agarum	Peter; Danish astronomer	1795–1874
Hansky	9.7 S	97.0 E	43	Libration zone; Smythii area, southeast of Hirayama	Alexei; Russian astronomer	1870–1908
Hansteen	11.5 S	52.0 W	44	Regular; southern edge of Procellarum; pair with Billy	Christopher; Norwegian astronomer	1784–1873
Harding	43.5 N	71.7 W	22	Sinus Roris; low walls	Karl; German astronomer	1765–1834
Hargreaves	2.2 S	64.0 E	16	Irregular; east of Maclaurin (Maclaurin S)	Frank; British optician	1891–1970
Harkhebi	39.6 N	98.3 E	237	Libration; eroded, incomplete; north of Marginis; contains Fasry	Egyptian astronomer	c. 300 BC
Harpalus	52.6 N	43.4 W	39	Edge of Frigoris; deep, prominent	Greek astronomer	c. 460 BC
Hartwig	6.1 S	80.5 W	79	West of Grimaldi; adjoins Schlüter to the east	Carl; German astronomer	1851–1923
Hase	29.4 S	62.5 E	83	South of Petavius; rather irregular	Johann; German mathematician	1684–1742
Hausen	65.0 S	88.1 W	167	Bailly area; central peak	Christian; German astronomer	1693–1743
Hayn	64.7 N	85.2 E	87	Limb; beyond Strabo	Friedrich; German astronomer	1863–1928
Hédervári	81.8 S	84.0 E	69	Libration zone	Peter; Hungarian astronomer	1931–1984
Hekatæus	21.8 S	79.4 E	167	Southeast of Vendelinus; irregular walls; central peak	Greek geographer	c. 476 BC
Heinrich	24.8 N	15.3 W	6	In Imbrium; northwest of Timocharis (Timocharis A)	Wladimir; Czech astronomer	1884–1965

Table 3.14 (cont.)

Name	Lat. (°)	Long. (°)	Diameter (km)	Notes (former names in parentheses)	Named after	
Heinsius	39.5 S	17.7 W	64	Tycho area; irregular; three craters on its southern wall	Gottfried; German astronomer	1709–1769
Heis	32.4 N	31.9 W	14	Bright craterlet on Imbrium	Eduard; German astronomer	1806–1877
Helicon	40.4 N	23.1 W	24	In Imbrium; pair with Le Verrier	Greek astronomer	?– c. 400 BC
Hell	32.4 S	7.8 W	33	In west of Deslandres; low central peak	Maximilian; Hungarian astronomer	1720–1792
Helmart	7.6 S	87.6 E	26	Libration zone; Smythii area; adjoins Kao	Friedrich; German astronomer	1843–1917
Helmholtz	68.1 S	64.1 E	94	Fairly regular; southern uplands; Boussingault area	Hermann von; German scientist	1821–1894
Henry, Paul	23.5 S	58.9 W	42	West of Humorum; distinct pair with Prosper Henry	Paul Henry; French astronomer	1848–1905
Henry, Prosper	23.5 S	58.9 W	42	Pair with Paul Henry	Prosper; French astronomer	1849–1903
Heraclitus	49.2 S	6.2 E	90	Very irregular; Cuvier–Licetus group, south of Stöfler	Greek philosopher	c. 540–480
Hercules	46.7 N	39.1 E	69	Bright walls; deep interior crater; pair with Atlas	Greek mythological hero	–
Herigonius	13.3 S	33.9 W	15	Northeast of Gassendi; bright; central peak	Pierre Hérigone; French astronomer	c. 1644
Hermann	0.9 S	57.0 W	15	On Procellarum; east of Lohrmann; bright	Jacob; Swiss mathematician	1678–1833
Hermite	86.0 N	89.9 W	104	Well formed; limb; beyond Anaxagoras	Charles; French mathematician	1822–1901
Herodotus	23.2 N	49.7 W	34	Fairly regular; pair with Aristarchus; great valley nearby	Greek historian	c. 484–408 BC
Herschel	5.7 S	2.1 W	40	North of Ptolemæus; terraced walls; large central peak	William; Hanoverian/British astronomer	1738–1822
Herschel, Caroline	34.5 N	31.2 W	13	On Imbrium; bright; group with Carlini and de l'Isle	Caroline; Hanoverian/British astronomer	1750–1848
Herschel, John	62.0 N	42.0 W	165	North of Frigoris; ridge-bordered enclosure	John; British astronomer	1792–1871
Hesiodus	29.4 S	16.3 W	42	Companion to Pitatus; rill runs southwest from it	Hesiod; Greek author	c. 735 BC
Hevel	2.2 N	67.6 W	115	Grimaldi chain; convex floor; with low central peak and several rills	Johann Hewelcke; Polish astronomer	1611–1687
Heyrovsky	39.6 S	95.3 W	16	Libration zone; beyond Cordilleras	Jaroslav; Czech chemist	1890–1967
Hill	20.9 N	40.8 E	16	Edge of Mare, west of Macrobius (Macrobius B)	George; American astronomer	1838–1914
Hind	7.9 S	7.4 E	29	Hipparchus group; pair with Halley; regular	John Russell; British astronomer	1823–1895
Hippalus	24.8 S	30.2 W	57	Bay on Humorum; remnant of central peak; associated with rills	Greek explorer	?– c. 120
Hipparchus	5.1 S	5.2 E	138	Low-walled, irregular; pair with Albategnius	Greek astronomer	c. 140 BC
Hirayama	6.1 S	93.5 E	132	Libration zone; Smythii area; regular	Kiyotsugu; Japanese astronomer	1874–1943
Hohmann	17.9 S	94.1 W	16	Libration zone; small crater in Orientale	Walter; German space engineer	1880–1945
Holden	19.1 S	62.5 E	47	South of Vendelinus; deep	Edward; American astronomer	1846–1914
Hommel	54.7 S	33.8 E	126	Southern uplands; two large craters in floor	Johann; German astronomer	1518–1562

Table 3.14 (cont.)

Name	Lat. (°)	Long. (°)	Diameter (km)	Notes (former names in parentheses)	Named after	
Hooke	41.2 N	54.9 E	36	West of Messala; fairly regular	Robert; British scientist	1635–1703
Hornsby	23.8 N	12.5 E	3	Craterlet in Serenitatis; between Linné and Sulpicius Galles	Thomas; British astronomer	1733–1810
Horrebow	58.7 N	40.8 W	24	Pair with Robinson; outside John Herschel; deep	Peder; Danish astronomer	1679–1764
Horrocks	4.0 S	5.9 E	30	Within Hipparchus; regular	Jeremiah; British astronomer	1619–1641
Hortensius	6.5 N	28.0 W	14	On Procellarum; bright; domes to north	Hove, Martin van den; Dutch astronomer	1605–1639
Houtermans	9.4 S	87.2 E	29	Limb; southeast of Kästner	Friedrich; German physicist	1903–1966
Hubble	22.1 N	86.9 E	80	Limb; southeast of Plutarch; partly flooded	Edwin; American astronomer	1889–1953
Huggins	41.1 S	1.4 W	65	Between Nasireddin and Orontius; irregular	Sir William; British astronomer	1824–1910
Humason	30.7 N	56.6 W	4	On Procellarum; east of Lichtenberg (Lichtenberg G)	Milton; American astronomer	1891–1972
Humboldt, Wilhelm	27.0 S	80.9 E	189	East of Petavius; rills on floor	Wilhelm von; German philologist	1767–1835
Hume	4.7 S	90.4 E	23	Libration zone; Smythii area; bordering Hirayama	David; Scottish philosopher	1711–1776
Huxley	20.2 N	4.5 W	4	Apennine region; east of Imbrium (Wallace B)	Thomas; British biologist	1825–1895
Hyginus	7.8 N	6.3 E	9	Depression in Vaporum; great crater-rill	Caius; Spanish astronomer	?–100?
Hypatia	4.3 S	22.6 E	40	South of Tranquillitatis; low walls; irregular	Egyptian mathematician	?–415
Ibn Battuta	6.9 S	50.4 E	11	Prominent; northeast of Goclenius (Goclenius A)	Moroccan geographer	1304–1377
Ibn Rushd	11.7 S	21.7 E	32	Between Cyrillus and Kant (Cyrillus B)	Spanish astronomer/ philosopher	1126–1198
Ibn Yunis	14.1 N	91.1 E	58	Libration zone; Marginis area; adjoins Goddard	Averrdes; Egyptian astronomer	950–1009
Ideler	49.2 S	22.3 E	38	Southeast of Autolycus; distinct	Christian; German astronomer	1766–1846
Idelson	81.5 S	110.9 E	60	Libration zone; Demonax area	Naum; Russian astronomer	1855–1951
Ilyin	17.8 S	97.5 W	13	Libration zone; in Orientale	Andrei; Russian rocket scientist	1901–1937
Inghirami	47.5 S	68.8 W	91	Schickard area; regular; central peak	Giovanni; Italian astronomer	1779–1851
Isidorus	8.0 S	33.5 E	42	Outside Nectaris; deep; pair with Capella	St Isidore; Roman astronomer	570–636
Ivan	26.9 N	43.3 W	4	Small craterlet northeast of Prinz (Prinz B)	Russian male name	–
Jacobi	56.7 S	11.4 E	68	Heraclitus group; fairly distinct	Karl; German mathematician	1804–1851
Jansen	13.5 N	28.7 E	23	In Tranquillitatis; low walls, darkish floor	Janszoon; Dutch optician	1580– c. 1638
Janssen	45.4 S	40.3 E	199	Southern uplands; great ruin, broken in the north by Fabricius	Pierre Jules; French astronomer	1842–1907
Jeans	55.8 S	91.4 E	79	Beyond Hanno; libration zone; between Chamberlin and Lyot	Sir James; British astronomer	1877–1946
Jehan	20.7 N	31.9 W	5	On Imbrium; southwest of Euler (Euler K)	Turkish female name	–
Jenkins	0.3 N	78.1 E	38	West of Smythii	Louise; American astronomer	1888–1970
Jenner	42.1 S	95.9 E	71	Libration zone; Australe area	Edward; British doctor	1749–1823

Table 3.14 (cont.)

Name	Lat. (°)	Long. (°)	Diameter (km)	Notes (former names in parentheses)	Named after	
Joliot	25.8 N	93.1 E	164	Libration zone; northeast of Marginis; interior detail	Fréderic Joliot-Curie; French physicist	1900–1958
Joy	25.0 N	6.6 E	5	Craterlet in Hæmus foothills	Alfred, American astronomer	1882–1973
Julius Cæsar	9.0 N	15.4 E	90	Vaporum area; low, irregular walls; very dark floor	Roman emperor	c. 102–44 BC
Kaiser	36.5 S	6.5 E	52	North of Stöfler; well marked; no central peak	Frederik; Dutch astronomer	1808–1872
Kane	63.1 N	26.1 E	54	North of Frigoris; fairly regular	Elisha; American explorer	1820–1857
Kant	10.6 S	20.1 E	33	West of Theophilus; large central peak with summit pit	Immanuel; German philosopher	1724–1804
Kao	6.7 S	87.6 E	34	Adjoins Helmert; between Kästner and Kiess	Ping-Tse; Taiwan astronomer	1888–1970
Kapteyn	10.8 S	70.6 E	49	East of Langrenus; inconspicuous	Jacobus; Dutch astronomer	1851–1922
Kästner	6.8 S	78.5 E	108	Smythii area; distinct	Abraham; German mathematician	1719–1800
Keldysh	51.2 N	43.6 E	33	In Hercules; bright, regular (Hercules A)	Mstislav; Russian mathematician	1911–1978
Kepler	8.1 N	38.0 W	31	In Procellarum; pair with Encke; major ray-centre	Johannes; German mathematician/astronomer	1571–1630
Kies	26.3 S	22.5 W	45	On Nubium; Bullialdus area; very low walls	Johann; German mathematician/astronomer	1713–1781
Kiess	6.4 S	84.0 E	63	Limb; beyond Kästner	Carl; American astrophysicist	1887–1967
Kinau	60.8 S	15.1 E	41	Jacobi group; high walls; central peak	Adolf Gottfried; German selenographer/botanist	1814–1888
Kirch	39.2 N	5.6 W	11	Bright craterlet in Imbrium	Gottfried; German astronomer	1639–1710
Kircher	67.1 S	45.3 W	72	Bettinus chain; Bailly area; very high walls	Athanasius; German humanitarian	1601–1680
Kirchhoff	30.3 N	38.8 E	24	Cleomedes area; larger of two craters west of Newcomb	Gustav; German physicist	1824–1887
Klaproth	69.8 S	26.0 W	119	Darkish floor; southern uplands; contact pair with Casatus	Martin; German mineralogist	1743–1817
Klein	12.0 S	2.6 E	44	In Albategnius; regular; central peak	Hermann; German selenographer	1844–1914
Knox-Shaw	5.3 N	80.2 E	12	Abuts on Banachiewicz (Banachiewicz F)	Harold; British astronomer	1855–1970
Kopff	17.4 S	89.6 W	41	Limb; Orientale area; west of Crüger	August; German astronomer	1882–1960
Krafft	16.6 N	72.6 W	51	On Procellarum; pair with Cardanus; darkish floor	Wolfgang; German astronomer	1743–1814
Kramarov	2.3 S	98.8 W	20	Libration zone; Rook area	G. M.; Russian space scientist	1887–1970
Krasnov	29.9 S	79.6 W	40	Irregular; limb; west of Doppelmayer	Aleksander; Russian astronomer	1866–1907
Kreiken	9.0 S	84.6 E	23	Northeast of La Peyrouse; trio with Black and Dale	E. A.; Dutch astronomer	1896–1964
Krieger	29.0 N	45.6 W	22	Aristarchus area; distinct, but walls broken by craterlets	Johann; German selenographer	1865–1902
Krishna	24.5 N	11.3 E	3	Craterlet in Serenitatis; south of Linné	Indian male name	–
Krogh	9.4 N	65.7 E	19	Well formed; southeast of Auzout (Auzout B)	Schack; Danish zoologist	1874–1949
Krusenstern	26.2 S	5.9 E	47	Werner area; not prominent	Adam; Russian explorer	1770–1846

Table 3.14 (cont.)

Name	Lat. (°)	Long. (°)	Diameter (km)	Notes (former names in parentheses)	Named after	
Kugler	53.8 S	103.7 E	65	Libration zone; beyond Brisbane and Jeans	Franz Xaver; German chronologist	1862–1929
Kuiper	9.8 S	22.7 W	6	Prominent; on Mare Cognitum; west of Bonpland (Bonpland E)	Gerard; Dutch astronomer	1905–1973
Kundt	11.5 S	11.5 W	10	Prominent; Guericke and Davy (Guericke C)	August; German physicist	1839–1894
Kunowsky	3.2 N	32.5 W	18	On Procellarum; east of Encke; low central ridge	Georg; German astronomer	1786–1846
La Peyrouse	10.7 S	76.3 E	77	East of Fœcunditatis; well formed; pair with Ansgarius	Jean, Comte de; French explorer	1741–1788
Lacaille	23.8 S	1.1 E	67	Werner area; irregular	Nicholas de; French astronomer	1713–1762
Lallemand	14.3 S	84.1 W	18	West of Rocca (Kopff A)	Andre; French astronomer	1904–1978
Lacroix	37.9 S	59.0 W	37	North of Schickard; regular; central peak	Sylvestre; French mathematician	1765–1843
Lade	1.3 S	10.1 E	55	Highlands south of Godin; low walls	Heinrich von; German astronomer	1817–1904
Lagalla	44.6 S	22.5 W	85	Abuts on Wilhelm 1; low-walled; irregular	Giulio; Italian philosopher	1571–1624
Lagrange	32.3 S	72.8 W	225	West of Humorum; low walls; rough floor	Joseph; Italian mathematician	1736–1813
Lalande	4.4 S	8.6 W	24	Ptolemæus area; low central peak	Joseph de; French astronomer	1732–1807
Lamarck	22.9 S	69.8 W	100	Ruined plain; south of Darwin and west of Byrgius	Jean; French natural historian	1744–1829
Lamb	42.9 S	100.1 E	106	Libration zone; Australe area	Sir Horace; British mathematician	1849–1934
Lambert	25.8 N	21.0 W	30	In Imbrium; bright, central crater	Johann; German astronomer	1728–1777
Lamé	14.7 S	64.5 E	84	On northeastern wall of Vendelinus; well-formed	Gabriel; French mathematician	1795–1870
Lamèch	42.7 N	13.1 E	13	Eudoxus area; very low, irregular walls	Felix; French selenographer	1894–1962
Lamont	4.4 N	23.7 E	106	On Tranquillitatis, Arago area; low walls	John; Scottish astronomer	1805–1879
Landsteiner	31.3 N	14.8 W	6	Craterlet on Imbrium; east of Carlini	Karl; Austrian pathologist	1868–1943
Langley	51.1 N	86.3 W	59	Limb beyond Repsold; north of Galvani	Samuel; American astronomer/ physicist	1834–1906
Langrenus	8.9 S	61.1 E	127	Petavius chain; massive walls; central peak	Michel van Langren; Belgian selenographer	c. 1600–1675
Lansberg	0.3 S	26.6 W	38	On Nubium; massive walls; central peak	Philippe van; Belgian astronomer	1561–1632
Lassell	15.5 S	7.9 W	23	In Nubium area; Alphonsus area; low walls	William; British astronomer	1799–1880
Laue	28.0 N	96.7 W	87	Libration zone; beyond Ulugh Beigh; intrudes into Lorentz	Max von; German physicist	1879–1960
Lauritsen	27.6 S	96.1 E	52	Beyond Humboldt; abuts on Curie	Charles; Danish physicist	1892–1968
La Voisier	38.2 N	81.2 W	70	West of Procellarum; well-formed	Antoine; French chemist	1743–1794
Lawrence	7.4 N	43.2 E	24	Flooded; northwest of Taruntius (Taruntius M)	Ernst; American physicist	1901–1958
Le Monnier	26.6 N	30.6 E	60	Bay on Serenitatis; smooth floor	Pierre; French astronomer	1715–1799
Le Verrier	40.3 N	20.6 W	20	Distinct crater on Imbrium; pair with Helicon	Urbain; French astronomer	1811–1877
Leakey	3.2 S	37.4 E	12	Obscure; near Censorinus (Censorinus F)	Louis; British archæologist	1903–1972

Table 3.14 (cont.)

Name	Lat. (°)	Long. (°)	Diameter (km)	Notes (former names in parentheses)	Named after	
Lebesque	5.1 S	89.0 E	11	Smythii area; near Warner	Henri; French mathematician	1875–1941
Lee	30.7 S	40.7 W	41	Southern edge of Humorum; damaged by lava	John; British astronomer	1783–1866
Legendre	28.9 S	70.2 E	78	Wilhelm Humboldt area; central ridge	Adrien; French mathematician	1752–1833
Legentil	74.4 S	76.5 W	113	South of Bailly; fairly distinct	Guillaume; French astronomer	1725–1792
Lehmann	40.0 S	56.0 W	53	Schickard area; irregular	Jacob; German astronomer	1800–1863
Lepaute	33.3 S	33.6 W	16	Distinct small crater at edge of Epidemiarum	Nicole Reine; French astronomer	1723–1788
Letronne	10.8 S	42.5 W	116	Bay at southern edge of Procellarum; low central peak; northern wall destroyed by lava	Jean; French archæologist	1787–1848
Lexell	35.8 S	4.2 W	62	Edge of Deslandres; northern wall reduced; remnant of central peak	Anders; Finnish astronomer	1740–1784
Licetus	47.1 S	6.7 E	74	Cuvier-Heraclitus group, near Stöfler; fairly regular	Fortunio Liceti; Italian physicist	1577–1657
Lichtenberg	31.8 N	67.7 W	20	Edge of Procellarum; minor ray-centre	Georg; German physicist	1742–1799
Lick	12.4 N	52.7 E	31	Edge of Crisium; incomplete	James; American benefactor	1796–1876
Liebig	24.3 S	48.2 W	37	West of Humorum; moderate walls	Justus, Baron von; German chemist	1803–1873
Lilius	54.5 S	6.2 E	61	Jacobi group; high walls; central peak	Luigi; Italian philosopher	?–1576
Lindbergh	5.4 S	52.9 E	12	Distinct; on Fœcunditatis; southeast of Messier (Messier G)	Charles; American aviator	1902–1974
Lindblad	70.4 N	98.8 W	66	Libration zone; beyond Pythagoras; southwest of Brianchon	Bertil; Swedish astronomer	1895–1965
Lindenau	32.3 S	24.9 E	53	Rabbi Levi group; terraced walls	Bernhard von; German astronomer	1780–1854
Lindsay	7.0 S	13.0 E	32	In highlands; north of Åndel (Dollond C)	Eric; Irish astronomer	1907–1974
Linné	27.7 N	11.8 E	2	In Serenitatis; craterlet on a light nimbus	Carl von; Swedish botanist	1707–1778
Liouville	2.6 N	73.5 E	16	Distinct crater west of Schubert (Dubiago S)	Joseph; French mathematician	1809–1882
Lippershey	25.9 S	10.3 W	6	In Nubium; Pitatus area; distinct craterlet	Hans (Jan); Dutch optician	?–1619
Littrow	21.5 N	31.4 E	30	Edge of Serenitatis; irregular	Johann von; Czech astronomer	1781–1840
Lockyer	46.2 S	36.7 E	34	Intrudes into Janssen; bright walls	Sir (Joseph) Norman; British astronomer	1836–1920
Loewy	22.7 S	32.8 W	24	Edge of Humorum; distinct	Moritz; French astronomer	1833–1907
Lohrmann	0.5 S	67.2 W	30	Between Grimaldi and Hevel; fairly regular; central peak	Wilhelm; German selenographer	1796–1840
Lohse	13.7 S	60.2 E	41	On western wall of Vendelinus; deep; central peak	Oswald; German astronomer	1845–1915
Lomonosov	27.3 N	98.0 E	92	Libration zone; north of Marginis; Joliot group	Mikhail; Russian astronomer	1711–1765
Longomontanus	49.6 S	21.8 W	157	Clavius area; complex walls; much floor detail	Christian; Danish astronomer	1562–1647
Lorentz	32.6 N	95.3 W	312	Libration zone; huge enclosure; contains Nernst	Hendrik; Dutch mathematician	1853–1928
Louise	28.5 N	34.2 W	2	Craterlet between Diophantus and de l'Isle	French female name	–

Table 3.14 (cont.)

Name	Lat. (°)	Long. (°)	Diameter (km)	Notes (former names in parentheses)	Named after	
Louville	44.0 N	46.0 W	36	In Jura Mtns; low walls, darkish floor	Jacques; French astronomer	1671–1732
Lovelace	82.3 N	106.4 W	54	Libration zone; pair with Froelich; south of Hermite	William; American space scientist	1907–1965
Lubbock	3.9 S	41.8 E	13	Western edge of Fœcunditatis; fairly bright	Sir John; British astronomer	1803–1865
Lubiniezky	17.8 S	23.8 W	43	In Nubium; Bullualdus area; low walls	Stanislaus; Polish astronomer	1623–1675
Lucian	14.3 N	36.7 E	7	On Tranquillitatis; west of Lyell (Maraldi B)	Greek writer	125–190
Ludwig	7.7 S	97.4 E	23	Libration zone; beyond Smythii and Hirayama	Carl; German physiologist	1816–1895
Luther	33.2 N	24.1 E	9	Distinct craterlet in northern part of Serenitatis	Robert; German astronomer	1822–1900
Lyell	13.6 N	40.6 E	32	Western edge of Somnii; darkish floor	Sir Charles; Scottish geologist	1797–1875
Lyot	49.8 S	84.5 E	132	Flooded; irregular wall; dark floor; north of Australe	Bernard; French astronomer	1897–1952
Maclaurin	1.9 S	68.0 E	50	West of Smythii; concave floor; uneven walls	Colin; Scottish mathematician	1698–1746
Maclear	10.5 N	20.1 E	20	On Tranquillitatis; darkish floor	Thomas; Irish astronomer	1794–1879
MacMillan	24.2 N	7.8 W	7	In Imbrium; southwest of Archimedes	William; American astronomer	1871–1948
Macrobius	21.3 N	46.0 E	64	Crisium area; high walls; compound central peak	Ambrosius; Roman writer	?– c. 140
Mädler	11.0 S	29.8 E	27	On Nectaris; irregular	Johann von; German selenographer	1794–1874
Mæstlin	4.9 N	40.6 W	7	On Procellarum; near Encke; obscure	Michael; German mathematician	1550–1631
Magelhæns	11.9 S	44.1 E	40	Edge of Fœcunditatis; pair with A; darkish floor	Fernao de (Magellan); Portuguese explorer	1480–1521
Maginus	50.5 S	6.3 W	194	Clavius area; irregular walls; obscure near full moon	Giovanni Magini; Italian astronomer	1555–1617
Main	80.8 N	10.1 E	46	Northern polar area; contact twin with Challis	Robert; British astronomer	1808–1878
Mairan	41.6 N	43.4 W	40	Jura area; bright, regular	Jean de; French geophysicist	1678–1771
Malapert	84.9 S	12.9 E	69	Irregular form; near southern pole, east of Cabæus	Charles; Belgian astronomer	1581–1639
Mallet	45.4 S	54.2 E	58	Rheita Valley area; inconspicuous	Robert; Irish seismologist	1810–1881
Manilius	14.5 N	9.1 E	38	On edge of Vaporum; brilliant walls; central peak	Marcus; Roman writer	c. 50 BC
Manners	4.6 N	20.0 E	15	On Tranquillitatis; Arago area; bright	Russell; British astronomer	1800–1870
Manzinus	67.7 S	26.8 E	98	Boguslawsky area; high terraced walls	Carlo Manzini; Italian astronomer	1599–1677
Maraldi	19.4 N	34.9 E	39	In north of Tranquillitatis; distinct	Giovanni; Italian astronomer	1709–1788
Marco Polo	15.4 N	2.0 W	28	Apennines area; irregular; darkish floor	Italian explorer	1254–1324
Marinus	39.4 S	76.5 E	58	Australe area; distinct; central peak	Greek geographer	c. 100
Markov	53.4 N	62.7 W	40	On Sinus Roris; sharp rim	Aleksandr; Russian astrophysicist	1897–1968
Marth	31.1 S	29.3 W	6	In Epidemiarum; concentric crater	Albert; German astronomer	1828–1897
Maskelyne	2.2 N	30.1 E	23	In Tranquillitatis; low central peak	Nevil; British astronomer	1732–1811
Mason	42.6 N	30.5 E	33	Bürg area; pair with Plana	Charles; British astronomer	1730–1787

Table 3.14 (cont.)

Name	Lat. (°)	Long. (°)	Diameter (km)	Notes (former names in parentheses)	Named after	
Maunder	14.6 S	93.8 W	55	Libration zone on northern edge Orientale; regular; central peak	Annie; British astronomer	1868–1947
					Edward; British astronomer	1851–1928
Maupertuis	49.6 N	27.3 W	45	In Juras; irregular mountain enclosure	Pierre de; French mathematician	1698–1759
Maurolycus	42.0 S	14.0 E	114	East of Stöfler; rough floor; central mountain group	Francesco Maurolico; Italian mathematician	1494–1575
Maury	37.1 N	39.6 E	17	Atlas area; bright and deep	Matthew; American oceanographer	1806–1873
Maxwell	30.2 N	98.9 E	107	Libration zone; beyond Gauss; intrudes into Richardson	James Clerk; Scottish physicist	1831–1879
Mayer, C	63.2 N	17.3 E	38	North of Frigoris; rhomboidal	Charles; German astronomer	1719–1783
Mayer, T	15.6 N	29.1 W	50	In Carpathians; central peak	Johann Tobias; German astronomer	1723–1762
McAdie	2.1 N	92.1 E	45	Libration zone; east of Smythii; low walls	Alexander; American meteorologist	1863–1943
McClure	15.3 S	50.3 E	23	Edge of Fœcunditatis; east of Colombo; regular	Robert; British explorer	1807–1873
McDonald	30.4 S	20.9 W	7	On Imbrium; southeast of Carlini (Carlini B)	Thomas; Scottish selenographer	?–1973
McLaughlin	47.1 N	92.9 W	79	Libration zone; beyond Galvani; rather irregular	Dean; American astronomer	1901–1965
Mee	43.7 S	35.3 W	126	Abuts on Hainzel; low, broken walls	Arthur; Scottish astronomer	1860–1926
Mees	13.6 N	96.1 W	50	Libration zone; beyond Einstein	Kenneth; English photographer	1882–1960
Mendel	48.8 S	109.4 W	138	Libration zone; beyond Orientale	Gregor; Austrian biologist	1822–1884
Menelaus	16.3 N	16.0 E	26	In Hæmus Mountains; brilliant; central peak	Greek astronomer	c 98
Menzel	3.4 N	36.9 E	3	On Tranquillitatis; east of Maskelyne	Donald; American astronomer	1901–1976
Mercator	29.3 S	26.1 W	46	Pair with Campanus; dark floor	Gerard de; Belgian cartographer	1512–1594
Mercurius	46.6 N	66.2 E	67	Humboldtianum area; low central peak	Roman messenger of the gods	
Merrill	75.2 N	116.3 W	57	Libration zone; beyond Brianchon	Paul; American astronomer	1887–1961
Mersenius	21.5 S	49.2 W	84	West of Humorum; convex floor; rills nearby	Marin Mersenne; French mathematician	1588–1648
Messala	39.2 N	60.5 E	125	Humboldtianum area; oblong; broken walls	Ma-Sa-Allah; Jewish astronomer	762–815
Messier	1.9 S	47.6 E	11	On Fœcunditatis; twin with Messier A; 'comet rays' (appearance of a comet) to the west	Charles; French astronomer	1730–1817
Metius	40.3 S	43.3 E	87	Janssen group; distinct; pair with Fabricius	Adriaan; Dutch astronomer	1571–1635
Meton	73.6 N	18.8 E	130	Northern polar area; smooth floor; compound formation	Greek astronomer	c. 432 BC
Milichius	10.0 N	30.2 W	12	On Procellarum; bright; dome to the west	Jacob Milich; German mathematician	1501–1559
Miller	39.3 S	0.8 E	61	Orontius group; fairly distinct	William; British chemist	1817–1870
Mitchell	49.7 N	20.2 E	30	Distinct; abuts on Aristoteles	Maria; American astronomer	1818–1889
Moigno	66.4 N	28.9 E	36	West of Arnold; contains central crater; no central peak	François; French mathematician	1804–1884

Table 3.14 (cont.)

Name	Lat. (°)	Long. (°)	Diameter (km)	Notes (former names in parentheses)	Named after	
Möltke	0.6 S	24.2 E	6	Distinct; northern edge of Tranquillitatis	Helmuth; German benefactor	1800–1891
Monge	19.2 S	47.6 E	36	Edge of Fœcunditatis; rather irregular	Gaspard; French mathematician	1746–1818
Montanari	45.8 S	20.6 W	76	Longomontanus area; distorted	Geminiano; Italian astronomer	1633–1687
Moretus	70.6 S	5.8 W	111	Southern uplands; very high walls; massive central peak	Theodore Moret; Belgian mathematician	1602–1667
Morley	2.8 S	64.6 E	14	Northwest of Maclaurin (Maclaurin R)	Edward; American chemist	1838–1923
Moseley	20.9 N	90.1 N	90	Libration zone; beyond Einstein	Henry; British physicist	1887–1915
Mösting	0.7 S	5.9 W	24	In Medii; A, to the north, is used as a reference point	Johan; Danish benefactor	1759–1843
Mouchez	78.3 N	26.6 W	81	Ruined plain near Philolaus	Ernest; French astronomer	1821–1892
Moulton	61.1 S	97.2 E	49	Libration zone; beyond Hanno	Forest Ray; American astronomer	1871–1952
Müller	7.6 S	2.1 E	22	Ptolemæus area; fairly regular	Karl; Czech astronomer	1866–1942
Murchison	5.1 N	0.1 W	57	Edge of Medii; low walled; irregular	Sir Roderick; Scottish geologist	1792–1871
Mutus	63.6 S	30.1 E	77	Southern uplands; two large craters on floor	Vincente; Spanish astronomer	?–1673
Nansen	80.9 N	05.3 E	104	Libration zone; beyond Einstein	Fridtjof; Norwegian explorer	1861–1930
Naonobu	4.6 S	57.8 E	34	Trio with Biharz and Atwood (Langrenus B)	Ajima; Japanese mathematician	c. 1732–1798
Nasireddin	41.0 S	0.2 E	52	Orontius group; fairly distinct	Nasir al-Din; Persian astronomer	1201–1274
Nasmyth	50.5 S	56.2 W	76	Phocylides group; fairly regular; no central peak	James; Scottish engineer	1808–1890
Natasha	20.0 N	31.3 W	12	On Imbrium; southwest of Euler (Euler P)	Russian female name	–
Naumann	35.4 N	62.0 W	9	In the north of Procellarum; bright walls	Karl; German geologist	1797–1873
Neander	31.3 S	39.9 E	50	Rheita Valley area; well-formed	Michael; German mathematician	1529–1581
Nearch	58.5 S	39.1 E	75	Vlacq area; craterlets on floor	Greek explorer	c. 325 BC
Neison	68.3 N	25.1 E	53	Meton area; regular; no central peak	Edmond Neville; British selenographer	1849–1940
Neper	8.5 N	84.6 E	137	Marginis/Smythii area; deep	John; Scottish mathematician	1550–1617
Neumayer	71.1 S	70.7 E	76	Boussingault area; distinct	Georg; German meteorologist	1826–1909
Newcomb	29.9 N	43.8 E	41	Cleomedes area; south wall broken by crater	Simon; Canadian astronomer	1835–1909
Newton	76.7 S	16.9 W	78	Moretus area; very deep; irregular	Sir Isaac; British mathematician	1643–1727
Nicholson	26.2 S	85.1 W	38	In Rook Mountains	Seth; American astronomer	1891–1963
Nicolai	42.4 S	25.9 E	42	Janssen area; regular	Friedrich; German astronomer	1793–1846
Nicollet	21.9 S	12.5 W	15	In Nubium; west of Birt; distinct	Jean; French astronomer	1788–1843
Nielsen	31.8 N	51.8 W		On Procellarum; between Lichtenberg and Gruithuisen (Wollaston C)	Axel; Danish astronomer	1902–1980
Niépce	71.7 N	119.1 W	57	Libration zone; beyond Brianchon	Joseph; French photographer	1765–1833
Nobile	85.2 S	53.5 E	73	Libration zone	Umberto; Italian explorer	1885–1978
Nobili	0.2 N	75.9 E	42	Trio with Jenkins and X (Schubert Y)	Leopoldo; Italian physicist	1784–1835
Nöggerath	48.8 S	45.7 W	30	Schiller area; low walls	Johann; German geologist	1788–1877
Nonius	34.8 S	3.8 E	69	Stöfler area; fairly regular	Pedro Nunez; Portuguese mathematician	1492?–1578

Table 3.14 (cont.)

Name	Lat. (°)	Long. (°)	Diameter (km)	Notes (former names in parentheses)	Named after	
Nunn	4.6 N	91.1 E	19	Libration zone; on northern edge of Smythii; low walls	Joseph; American engineer	1905–1968
Œnopides	57.0 N	64.1 W	67	Limb area near Sinus Roris; high walls	Greek astronomer	?500–430 BC
Œrsted	43.1 N	47.2 E	42	Edge of Somniorum; rather irregular	Hans; Danish chemist	1777–1851
Oken	43.7 S	75.9 E	71	Australe area; prominent, darkish floor	Lorenz Okenfuss; German biologist	1779–1851
Olbers	7.4 N	75.9 W	74	Grimaldi area; major ray-centre	Heinrich; German astronomer/ doctor	1758–1840
Omar Khayyám	58.0 N	102.1 W	70	Libration zone; in Poczobut	Al-Khayyami; Persian astronomer/poet	c. 1050–1123
Opelt	16.3 S	17.5 W	48	'Ghost' in Nubium; east of Bullialdus	Friedrich; German astronomer	1794–1863
Oppolzer	1.5 S	0.5 W	40	Edge of Medii; low walls	Theodor von; Czech astronomer	1841–1886
Orontius	40.6 S	4.6 W	105	Irregular; one of a group northeast of Tycho	Finnæus Oronce; French mathematician	1494–1555
Palisa	9.4 S	7.2 W	33	Alphonsus area; on edge of larger ring	Johann; Czech astronomer	1848–1925
Palitzsch	28.0 S	64.5 E	64×91×32	Outside Petavius to the east; really a crater-chain	Johann; German astronomer	1723–1788
Pallas	5.5 N	1.6 W	46	Medii area; adjoins Murchison; central peak	Peter; German geologist	1741–1811
Palmieri	28.6 S	47.7 W	40	Humorum area; darkish floor	Luigi; Italian physicist	1807–1896
Paneth	63.0 N	94.8 W	65	Libration zone; beyond Xenophanes; north of Smoluchewski	Friedrich; German chemist	1887–1958
Parkhurst	33.4 S	103.6 E	96	Libration zone; beyond Australe; irregular	John; American astronomer	1861–1925
Parrot	74.5 S	3.3 E	70	Albategnius area; very irregular, compound structure	Johann; Russian physicist	1792–1840
Parry	7.9 S	15.8 W	47	On Nubium; Fra Mauro group; fairly regular	William; British explorer	1790–1855
Pascal	74.6 N	70.3 W	115	Limb formation beyond Carpenter	Louis; French mathematician	1623–1662
Peary	88.6 N	33.0 E	73	Northern polar area; beyond Gioja	Robert; American explorer	1856–1920
Peirce	18.3 N	53.5 E	18	In Crisium; conspicuous	Benjamin; American astronomer	1809–1880
Peirescius	46.5 S	67.6 E	61	Australe area; rather irregular	Nicolas Peiresc; French astronomer	1580–1637
Pentland	64.6 S	11.5 E	56	Soutern uplands; near Curtius; high walls	Joseph; Irish geographer	1797–1873
Petermann	74.2 N	66.3 E	73	Limb beyond Arnold	August; German geographer	1822–1878
Petavius	25.1 S	60.4 E	188	Great crater; central peak; major rill on floor	Denis Petau; French chronologist	1583–1652
Peters	68.1 N	29.5 E	15	Plain; no central peak; between Neison and Arnold	Christian; German astronomer	1806–1880
Petit	2.3 N	65.3 E	5	Small crater east of Spumans (Apollonius W)	Alexis; French physicist	1771–1820
Petrov	61.4 S	88.0 E	49	Flooded; beyond Pontécoulant	Evgenii; Russian rocket scientist	1900–1942
Pettit	27.5 S	86.6 W	35	Limb formation; pair with Nicholson	Edison; American astronomer	1889–1962
Petzval	62.7 S	110.4 W	90	Libration zone; beyond Hausen	Joseph von; Austrian optician	1807–1891
Phillips	26.6 S	75.3 E	122	West of Wilhelm Humboldt; central ridge	John; British astronomer/ geologist	1800–1874

Table 3.14 (cont.)

Name	Lat. (°)	Long. (°)	Diameter (km)	Notes (former names in parentheses)	Named after	
Philolaus	72.1 S	32.4 W	70	Frigoris area; deep, regular; pair with Anaximenes	Greek astronomer	c. 400
Phocylides	52.7 S	57.0 W	121	Schickard group; much interior detail	Johannes Holwarda; Dutch astronomer	1618–1651
Piazzi	36.6 S	67.9 W	134	Schickard area; very broken walls	Giuseppe; Italian astronomer	1746–1826
Piazzi Smyth	41.9 N	3.2 W	13	On Imbrium; bright	Charles; Scottish astronomer	1819–1900
Picard	14.6 N	54.7 E	22	Largest crater on Crisium; central hill	Jean; French astronomer	1620–1682
Piccolomini	29.7 S	32.2 E	87	End of Altai scarp; high walls; central peak	Alessandro; Italian astronomer	1508–1578
Pickering	2.9 S	7.0 E	15	Hipparchus area; distinct	Edward; American astronomer	1846–1919
					William; American astronomer	1858–1938
Pictet	43.6 S	7.4 W	62	Closely east of Tycho; fairly regular	Marc Pictet-Turretin; Swiss physicist	1752–1825
Pilâtre	60.2 S	86.9 W	50	Beyond Pingré; low, irregular walls	de Rozier; French aeronaut	1753–1785
Pingré	58.7 S	73.7 W	88	Limb formation beyond Phocylides; no central peak	Alexandre; French astronomer	1711–1796
Pitatus	29.9 S	13.5 W	106	Southern edge of Nubium; passes connect it with Hesiodus	Pietri Pitati; Italian astronomer	?–c. 1500
Plana	42.2 N	28.2 E	44	Bürg area; pair with Mason; darkish floor	Baron Giovanni; Italian astronomer	1781–1864
Plaskett	82.1 N	174.3 E	109	Libration zone; large walled plain; northern polar area	John; Canadian astronomer	1865–1941
Plato	51.6 N	9.4 W	109	Edge of Imbrium; very regular; very dark floor	Greek philosopher	c. 428 – c. 347 BC
Playfair	23.5 S	8.4 E	47	Abenezra area; fairly regular	John; Scottish mathematician/ geologist	1748–1819
Plinius	15.4 N	23.7 E	43	Between Serenitatis and Tranquillitatis; central craters	Gaius; Roman natural scientist	23–79
Plutarch	24.1 N	79.0 E	68	Northeast of Crisium; distinct; central peak	Greek biographer	c. 46 – c. 120
Poczobut	57.1 N	98.8 W	195	Libration zone; large plain broken by several craters	Martin; Polish astronomer	1728–1810
Poisson	30.4 S	10.6 E	42	Aliacensis area; compound; very irregular	Simeon; French mathematician	1781–1840
Polybius	22.4 S	25.6 E	41	Theophilus/Catharina area	Greek historian	?204–?122 BC
Pomortsev	0.7 N	66.9 E	23	Distinct (Dubiago P)	Mikhail; Russian rocket scientist	1851–1916
Poncelet	75.8 N	54.1 W	69	Limb, beyond Anaximenes and Philolaus	Jean; French mathematician	1788–1867
Pons	25.3 S	21.5 E	41	Altai area; very thick walls	Jean; French astronomer	1761–1831
Pontanus	28.4 S	14.4 E	57	Altai area; regular; no central peak	Giovanni Pontano; Italian astronomer	1427–1503
Pontécoulant	58.7 S	66.0 E	91	Southern uplands; high walls.	Philippe; Comte de; French mathematician	1795–1874
Popov	17.2 N	99.7 E	65	Libration zone; beyond Marginis	Aleksandr; Russian physicist	1859–1905
					Cyril; Bulgarian astronomer	1880–1966
Porter	56.1 S	10.1 W	51	On wall of Clavius; well-formed; central peak	Russell; American telescope designer	1871–1949
Posidonius	31.8 N	29.9 E	95	Edge of Serenitatis; narrow walls; much interior detail	Greek geographer	?135–?51 BC

Table 3.14 (cont.)

Name	Lat. (°)	Long. (°)	Diameter (km)	Notes (former names in parentheses)	Named after	
Priestley	57.3 S	108.4 E	52	Libration zone; beyond Hanno and Chamberlin	Joseph; British chemist	1733–1804
Prinz	25.5 N	44.1 W	46	Northeast of Aristarchus; incomplete; domes nearby	Wilhelm; Belgian astronomer	1857–1910
Proclus	16.1 N	46.8 E	28	West of Crisium; brilliant; low central peak; minor ray-centre	Greek mathematician/ astronomer	410–485
Proctor	46.4 S	5.1 W	52	Maginus area; fairly regular	Mary; British astronomer	1862–1957
Protagoras	56.0 N	7.3 E	21	In Frigoris; bright and regular	Greek philosopher	?481–?411 BC
Ptolemæus	9.3 S	1.9 W	164	Trio with Alphonsus and Arzachel; has Ammonius; darkish floor	Greek astronomer, geographer/ mathematician	c 120–180
Puiseux	27.8 S	39.0 W	24	On Humorum; near Doppelmeyer; very low walls	Pierre; French astronomer	1855–1928
Pupin	23.8 N	11.0 W	2	Small but distinct; southeast of Timocharis (Timocharis K)	Michael; Jugoslav physicist	1858–1935
Purbach	25.5 S	2.3 W	115	Walter group; rather irregular	Georg von; Austrian mathematician	1423–1461
Purkyně	1.6 S	94.9 E	48	Libration zone; beyond Smythii	Jan; Czech doctor	1787–1869
Pythagoras	63.5 N	63.0 W	142	Northwest of Iridum; high, massive walls; high central peak	Greek philosopher and mathematician	c. 532 BC
Pytheas	20.5 N	20.6 W	20	On Imbrium; bright; central peak; minor ray-centre	Greek navigator and geographer	c. 308 BC
Rabbi Levi	34.7 S	23.6 E	81	One of a group southwest of Piccolomimi.	Ben Gershon; Jewish philosopher/astronomer	1288–1344
Raman	27.0 N	55.1 W	10	Irregular; northwest of Herodotus (Herodotus D)	Chandrasekhara; Indian physicist	1888–1970
Ramsden	0.0 N	0.0 E	24	Edge of Epidemiarum; rills nearby	Jesse; British instrument maker	1735–1800
Rankine	3.9 S	71.5 E	8	Craterlet east of Gilbert	William; Scottish physicist	1820–1872
Rayleigh	29.3 N	89.6 E	114	Limb walled plain northeast of Seneca	John (Lord); British physicist	1842–1919
Réaumur	2.4 S	0.7 E	30	Medii area; low walls	René; French physicist	1683–1757
Regiomontanus	28.3 S	1.0 W	129 × 105	Between Walter and Purbach; distorted; central peak	Johann Muller; German astronomer	1436–1476
Regnault	54.1 N	88.0 W	46	Limb; near Xenophanes	Henri; French chemist	1810–1878
Reichenbach	30.3 S	48.0 E	71	Rheita area; irregular; crater valley to southeast	Georg von; German optician	1722–1826
Reimarus	47.7 S	60.3 E	48	Rheita Valley area; irregular	Nicolai Reymers; German mathematician	1550– c. 1600
Reiner	7.0 N	54.9 W	29	On Procellarum; pair with Marius; central peak	Vincento Reinieri; Italian astronomer	?–1648
Reinhold	3.3 N	22.8 W	42	Southwest of Copernicus; pair with B to the northeast	Erasmus; German astronomer	1511–1553
Repsold	51.3 N	78.6 W	109	West of Roris; one of a group	Johann; German inventor	1770–1830
Respighi	2.8 N	71.9 E	18	Southeast of Dubiago; distinct (Dubiago C)	Lorenzo; Italian astronomer	1824–1890
Rhæticus	0.0 N	4.9 E	45	Medii area; low walls	Georg von; Hungarian astronomer	1514–1576
Rheita	37.1 S	47.2 E	70	Sharp crests; associated with great crater valley	Anton; Czech astronomer	1597–1660
Riccioli	3.3 S	74.6 W	139	Companion to Grimaldi; low walls; very dark patches on floor	Giovanni; Italian astronomer	1598–1671
Riccius	36.9 S	26.5 E	71	Rabbi Levi group; broken walls; rough floor	Matteo Ricci; Italian mathematician	1552–1610

Table 3.14 (cont.)

Name	Lat. (°)	Long. (°)	Diameter (km)	Notes (former names in parentheses)	Named after	
Richardson	31.1 N	180.5 E	141	Limb beyond Gauss. Vestine; broken by Maxwell	Sir Owen; British physicist	1879–1959
Riemann	38.9 N	86.8 E	163	Ruined walled plain; beyond Gauss	Georg; German mathematician	1826–1866
Ritchey	11.1 S	8.5 E	24	East of Albategnius; broken walls	George; American astronomer/optician	1864–1945
Rittenhouse	74.5 S	106.5 E	26	Libration zone beyond Neumayer west of Schrödinger	David; American astronomer	1732–1796
Ritter	2.0 N	19.2 E	29	On Tranquillitatis; central peak; pair with Sabine	Karl; German geographer	1779–1859
Ritz	15.1 S	92.2 E	51	Libration zone; beyond Ansgarius	Walter; Swiss physicist	1878–1909
Robinson	59.0 N	45.9 W	24	Frigoris area; distinct; similar to Horrebow	(John) Romney; Irish astronomer	1792–1882
Rocca	12.7 S	72.8 W	89	South of Grimaldi; irregular walls	Giovanni; Italian mathematician	1607–1656
Rocco	28.0 N	45.0 W	4	Craterlet closely east of Krieger (Krieger D)	Italian male name	–
Rosenberger	55.4 S	43.1 E	95	Vlacq group; darkish floor; central peak	Otto; German mathematician	1800–1890
Ross	11.7 N	21.7 E	24	On Tranquillitatis; central peak	James Clark; British explorer	1800–1862
					Frank; American astronomer	1874–1966
Rosse	17.9 S	35.0 E	11	On Nectaris; bright	3rd Earl of Rosse; Irish astronomer	1800–1867
Röst	56.4 S	33.7 W	48	Schiller area; regular; pair with Weigel	Leonhard; German astronomer	1688–1727
Rothmann	30.8 S	27.7 E	42	Altai area; fairly deep and regular	Christopher; German astronomer	?–1600
Rozhdestvensky	85.2 N	155.4 W	177	Libration zone; polar area; contains two craters	Dimitri; Russian astronomer	1876–1940
Rømer	25.4 N	36.4 E	39	Taurus area; massive central peak with summit pit	Ole; Danish astronomer	1644–1710
Röntgen	33.0 N	91.4 W	126	Libration zone; in Lorentz	Wilhelm; German physicist	1845–1923
Rümker	40.8 N	58.1 W	70	Very irregular structure; part-plateau; near Harding	Karl; German astronomer	1788–1862
Runge	2.5 S	86.7 E	38	Smythii area	Carl; German mathematician	1856–1927
Russell	26.5 N	75.4 W	103	Extension of Otto Struve	Henry Norris; American astronomer	1877–1957
Ruth	28.7 N	45.1 W	3	Craterlet adjoining Krieger to the northeast	Hebrew female name	–
Rutherfurd	60.9 S	12.1 W	48	On wall of Clavius; distinct central peak	Lewis; American astronomer	1816–1892
Rydberg	46.5 S	96.3 W	49	Libration zone; pair with Guthnick	Johannes; Swedish physicist	1854–1919
Rynin	47.0 N	103.5 W	75	Libration zone, beyond Galvani and McLaughlin	Nikolai; Russian rocket scientist	1877–1942
Sabatier	13.2 N	79.0 E	10	East of Condorcet; low walls	Paul; French chemist	1854–1941
Sabine	1.4 N	20.1 E	30	On Tranquillitatis; central peak; pair with Ritter	Sir Edward; Irish physicist/astronomer	1788–1883
Sacrobosco	23.7 S	16.7 E	98	Altai area; irregular	Johannes Sacrobuschus; British astronomer	c. 1200–1256
Sampson	29.7 N	16.5 W	1	On Imbrium; distinct craterlet northwest of Timochans	Ralph; British astronomer	1866–1939
Santbech	20.9 S	44.0 E	64	East of Fracastorius; darkish floor	Daniel; Dutch mathematician	c. 1561?
Santos–Dumont	27.7 N	4.8 E	8	Southern end of Apennines (Hadley B)	Alberto; Brazilian aeronaut	1873–1932

Table 3.14 (cont.)

Name	Lat. (°)	Long. (°)	Diameter (km)	Notes (former names in parentheses)	Named after	
Sarabhai	24.7 N	21.0 E	7	On Serenitatis; norheast of Bessel (Bessel A)	Vikram; Indian astrophysicist	1919–1971
Sasserides	39.1 S	9.3 W	90	Irregular enclosure north of Tycho	Gellius Sascerides; Danish astronomer	1562–1612
Saunder	4.2 S	8.8 E	44	East of Hipparchus; low walls	Samuel; British selenographer	1852–1912
Saussure	43.4 S	3.8 W	54	North of Maginus; interrupts larger ring	Horace de; Swiss geologist	1740–1799
Scheele	9.4 S	37.8 W	4	Distinct; on Procellarum; south of Wichmann (Letronne D)	Carl; Swedish chemist	1742–1786
Scheiner	60.5 S	27.5 W	110	Clavius area; high walls; floor craterlet; pair with Blancanus	Christopher; German astronomer	1575–1650
Schiaparelli	23.4 N	58.8 W	24	On Procellarum; distinct	Giovanni; Italian astronomer	1835–1910
Schickard	44.3 S	55.3 W	206	Great walled plain; rather low walls	Wilhelm; German mathematician/astronomer	1592–1635
Schiller	51.9 S	39.0 W	180 × 97	Schickard areal fusion of two rings	Julius; German astronomer	c. 1627
Schlüter	5.9 S	83.3 W	89	Beyond Grimaldi; prominent; terraced walls	Heinrich; German astronomer	1815–1844
Schmidt	1.0 N	18.8 E	11	In Tranquillitatis; Sabine/Ritter area; bright	Julius; German astronomer	1825–1884
					Bernhard; Estonian optician	1879–1935
					Otto; Russian astronomer	1891–1956
Schömberger	76.7 S	24.9 E	85	Regular; Boguslawsky area	Georg; Austrian astronomer	1597–1645
Schönfeld	44.8 N	98.1 W	25	Libration zone; regular; beyond Gerard	Eduard; German astronomer	1828–1891
Schorr	19.5 S	89.7 E	53	Limb formation; beyond Gibbs	Richard; German astronomer	1867–1951
Schrödinger	67.0 S	132.4 E	312	Libration zone; beyond Hanno; associated with great valley	Erwin; Austrian physicist	1887–1961
Schröter	2.6 N	7.0 W	35	Medii area; low walls	Johann; German astronomer	1745–1816
Schubert	2.8 N	81.0 E	54	In Smythii area; distinct	Theodor; Russian cartographer	1789–1865
Schumacher	42.4 N	60.7 E	60	Messala area; fairly distinct	Heinrich; German astronomer	1780–1850
Schwabe	65.1 N	45.6 E	25	Northeast of Democritus; dark floor	Heinrich; German astronomer	1789–1875
Schwarzschild	70.1 N	121.2 E	212	Libration zone; beyond Petermann; interior detail	Karl; German astronomer	1873–1916
Scoresby	77.7 N	14.1 E	55	Polar uplands; deep and prominent; central peak	William; British explorer	1789–1857
Scott	82.1 S	48.5 E	103	Beyond Schömberger; pair with Nansen	Robert Falcon; British explorer	1868–1912
Secchi	2.4 N	43.5 E	22	In Fœcunditatis; bright walls; central peak	(Pietro) Angelo; Italian astronomer	1818–1878
Seeliger	2.2 S	3.0 E	8	North of Hipparchus; irregular	Hugo von; German astronomer	1849–1924
Segner	58.9 S	48.3 W	67	Schiller area; well-formed; prominent; with Zucchius	Johann; German mathematician	1704–1777
Shackleton	89.9 S	0.0 E	19	South polar	Ernest; British explorer	1874–1922
Shaler	32.9 S	85.2 W	48	Limb beyond Lagrange; pair with Wright	Nathaniel; American geologist	1841–1906
Shapley	9.4 N	56.9 E	23	Off northern edge of Crisium; dark floor (Picard H).	Harlow; American astronomer	1885–1972
Sharp	45.7 N	40.2 W	39	In Jura Mountains; deep; small central peak	Abraham; British astronomer	1651–1742
Sheepshanks	59.2 N	16.9 E	25	North of Frigoris; fairly regular	Anne; British benefactor	1789–1876

Table 3.14 (cont.)

Name	Lat. (°)	Long. (°)	Diameter (km)	Notes (former names in parentheses)	Named after	
Shi Shen	76.0 N	104.1 E	43	Libration zone; beyond Nansen	Chinese astronomer	c. 300 BC
Short	74.6 S	7.3 W	70	Moretus group; deep; high walls	James; Scottish mathematician/ optician	1710–1768
Shuckburgh	41.6 N	52.8 E	38	Cepheus group; fairly regular	Sir George; British geographer	1751–1804
Shuleykin	27.1 S	92.5 W	15	Libration zone; just beyond Orientale; Rook area	Mikhail; Russian radio engineer	1884–1939
Sikorsky	66.1 S	103.2 E	98	Libration zone; crossed by Schrödinger Valley	Igor; Russian aeronautical engineer	1889–1972
Silberschlag	6.2 N	12.5 E	13	Near Ariadæus; bright	Johann; German astronomer	1721–1791
Simpelius	73.0 S	15.2 E	70	Moretus area; deep	Hugh Sempill; Scottish mathematician	1596–1654
Sinas	8.8 N	31.6 E	11	On Tranquillitatis; not prominent	Simon; Greek benefactor	1810–1876
Sirsalis	12.5 S	60.4 W	42	Contact pair with A; associated with rill	Gerolamo Sirsale; Italian astronomer	1584–1654
Sklodowska	18.2 S	95.5 E	127	Libration zone; well-formed, beyond Hekatæus	Marie (Curie); Polish physicist/ chemist	1867–1934
Slocum	3.0 S	89.0 E	13	Smythii area	Frederick; American astronomer	1873–1944
Smithson	2.4 N	53.6 E	5	In east of Tranquillitatis (Taruntius N)	James; British chemist	1765–1829
Smoluchowski	60.3 N	96.8 W	83	Libration zone; intrudes into Poczobut	Marian; Polish physicist	1872–1917
Snellius	29.3 S	55.7 E	82	Furnerius area; high walls; central peak; pair with Stevinus	Willibrord Snell; Dutch mathematician	1591–1626
Somerville	8.3 S	64.9 E	15	Distinct; east of Langrenus (Langrenus J)	Mary; Scottish physicist/ mathematician	1780–1872
Sömmering	0.1 N	7.5 W	28	Medii area; low walls	Samuel; German doctor	1755–1830
Sosigenes	8.7 N	17.6 E	17	Edge of Tranquillitatis; bright, low central peak	Greek astronomer/chronologist	c. 46 BC
South	58.0 N	50.8 W	104	Frigoris area; ridge-bounded enclosure	James; British astronomer	1785–1867
Spallanzani	46.3 S	24.7 E	32	West of Janssen; low walls	Lazzaro; Italian biologist	1729–1799
Spörer	4.3 S	1.8 W	27	North of Ptolemæus; partly lava-filled	Friedrich; German astronomer	1822–1895
Spurr	27.9 N	1.2 W	11	In Putredinis; southeast of Archimedes.	Josiah; American geologist	1870–1950
Stadius	10.5 N	13.7 W	60	Pitted ghost ring east of Copernicus	Jan Stade; Belgian astronomer	1527–1579
Steinheil	48.6 S	46.5 E	67	Southwest of Janssen; contact twin with Watt	Karl von; German astronomer	1801–1870
Stevinus	32.5 S	54.2 E	74	Furnerius area; high walls; central peak; pair with Snellius	Simon Stevin; Belgian mathematician	1548–1620
Stewart	2.2 N	67.0 E	13	Distinct; southwest of Dubiago (Dubiago Q)	John Quincy; American astrophysicist	1894–1972
Stiborius	34.4 S	32.0 E	43	Altai area; south of Piccolomimi; central peak	Andreas Stoberl; German astronomer	1465–1515
Stöfler	41.1 S	6.0 E	126	South of Walter; dark floor; broken by Faraday	Johann; German astronomer	1452–1531
Stokes	52.5 N	88.1 W	51	Limb; Regnault–Repsold area	Sir George; British mathematician	1819–1903
Strabo	61.9 N	54.3 E	55	Near De la Rue; minor ray-centre	Greek geographer	54 BC–AD 2-
Street	46.5 S	10.5 W	57	South of Tycho; fairly regular; no central peak	Thomas; British astronomer	1621–1689

Table 3.14 (cont.)

Name	Lat. (°)	Long. (°)	Diameter (km)	Notes (former names in parentheses)	Named after	
Struve	43.0 N	65.0 E	18	Adjoins Messala; lies on dark patch	Friedrich G. W.; Russian astronomer	1793–1864
Struve, Otto	22.4 N	77.1 W	164	Western edge of Procellarum; pair of two ancient rings	Otto; Russian astronomer	1819–1905
					Otto; American astronomer	1897–1963
Suess	4.4 N	47.6 W	8	On Procellarum; west of Encke; obscure	Eduard; Austrian geologist	1831–1914
Sulpicius Gallus	19.6 N	11.6 E	12	On Serenitatis; very bright	Gaius; Roman astronomer	c. 166 BC
Sundman	10.8 N	91.6 W	40	Libration zone; beyond Vasco da Gama and Bohr	Karl; Finnish astronomer	1873–1949
Sven Hedin	2.0 N	76.8 W	150	West of Hevel; irregular; broken walls	Swedish explorer	1865–1962
Swasey	5.5 S	89.7 E	23	Smythii area	Ambrose; American inventor	1846–1937
Swift	19.3 N	53.4 E	10	On Crisium; north of Peirce; prominent (Peirce B)	Lewis; American astronomer	1820–1913
Sylvester	82.7 N	79.6 W	58	Libration zone; beyond Philolaus	James; British mathematician	1814–1897
Tacchini	4.9 N	85.8 E	40	Limb beyond Banachiewicz (Neper K)	Pietro; Italian astronomer	1838–1905
Tacitus	16.2 S	19.0 E	39	Catharina area; polygonal; two floor craters	Cornelius; Roman historian	c 55–120
Tacquet	16.6 N	19.2 E	7	Just on Serenitatis, near Menelaus; bright	André; Belgian mathematician	1612–1660
Talbot	2.5 S	85.3 E	11	Smythii area	William Fox; British photographer	1800–1877
Tannerus	56.4 S	22.0 E	28	Southern uplands, near Mutus; central peak	Adam Tanner; Austrian mathematician	1572–1632
Taruntius	5.6 N	46.5 E	56	On Fœcunditatis; concentric crater; central peak; low walls	Lucius Firmanus; Roman philosopher	c. 86 BC
Taylor	5.3 S	16.7 E	42	Delambre area; rather elliptical	Brook; British mathematician	1685–1731
Tebbutt	9.6 N	53.6 E	31	Flooded; foothills of Crisium (Picard G)	John; Australian astronomer	1834–1916
Tempel	3.9 N	11.9 E	45	Uplands west of Tranquillitatis; bright	(Ernst) Wilhelm; German astronomer	1821–1889
Thales	61.8 N	50.3 E	31	Near Strabo; major ray-centre	Greek philosopher/astronomer	c. 636–546 BC
Theætetus	37.0 N	6.0 E	24	On Nebularum; low central peak	Greek geometrician	c. 380 BC
Thebit	22.0 S	4.0 W	56	Edge of Nubium; wall broken by Thebit A, which is itself broken by Thebit F	Thabit ibn Qurra; Arab astronomer	836–901
Theiler	13.4 N	83.3 E	7	Craterlet west of Marginis	Max; South African bacteriologist	1899–1972
Theon Junior	2.3 S	15.8 E	17	Near Delambre; bright; pair with Theon Senior	Greek astronomer	? – c. 380 BC
Theon Senior	0.8 S	15.4 E	18	Near Delambre; bright; pair with Theon Junior	Greek mathematician	?– c. 100
Theophilus	11.4 S	26.4 E	110	Very deep; massive walls, central peak complex; trio with Cyrillus and Catharina	Greek astronomer	?–412
Theophrastus	17.5 N	39.0 E	9	On Tranquillitatis; northwest of Franz (Maraldi M)	Greek botanist	c. 372–287 BC
Timæus	62.8 N	0.5 W	32	Edge of Frigoris; bright	Greek astronomer	?– c. 400 BC
Timocharis	26.7 N	13.1 W	33	On Imbrium; central crater; minor ray-centre	Greek astronomer	c. 280 BC

Table 3.14 (cont.)

Name	Lat. (°)	Long. (°)	Diameter (km)	Notes (former names in parentheses)	Named after	
Timoleon	35.0 N	75.0 E	130	Adjoins Gauss; fairly distinct	Greek general and statesman	*c.* 337 BC
Tisserand	21.4 N	48.2 E	36	Crisium area; regular; no central peak	François; French astronomer	1845–1896
Titius	26.8 S	100.7 E	73	Libration zone; beyond Wilhelm Humboldt and Lauritsen	Johann; German astronomer	1729–1796
Tolansky	9.5 S	16.0 W	13	Between Parry and Guericke; flat floor (Parry A)	Samuel; British physicist	1907–1973
Torricelli	4.6 S	28.5 E	22	On Nectaris; irregular, compound structure	Evangelista; Italian physicist	1608–1647
Toscanelli	27.4 N	47.5 W	7	Distinct; in highlands north of Aristarchus (Aristarchus C)	Paolo; Italian cartographer/doctor	1397–1482
Townley	3.4 N	63.3 E	18	South of Apollonius; distinct (Apollonius G)	Sidney; American astronomer	1867–1946
Tralles	28.4 N	52.8 E	43	On wall of Cleomedes; very deep	Johann; German physicist	1763–1822
Triesnecker	4.2 N	3.6 E	26	Vaporum area; associated with great rill system	Franz; Austrian astronomer	1745–1817
Trouvelot	49.3 N	5.8 E	9	Alpine Valley area; rather bright	Etienne; French astronomer	1827–1895
Tucker	5.6 S	88.2 E	7	Craterlet in Smythii area	Richard; American astronomer	1859–1952
Turner	1.4 S	13.2 W	11	On Nubium, near Gambart; deep	Herbert Hall; British astronomer	1861–1930
Tycho	43.4 S	11.1 W	102	Terraced walls; central peak; brightest ray-centre	Tycho Brahe; Danish astronomer	1546–1601
Ukert	7.8 N	1.4 E	23	Edge of Vaporum; rills nearby	Friedrich; German historian	1780–1851
Ulugh Beigh	32.7 N	81.9 W	54	West of Procellarum; high walls; central peak	Mongolian astronomer	1394–1449
Urey	27.9 N	87.4 E	38	Limb formation; beyond Seneca	Harold; American chemist	1893–1981
Väisälä	25.9 N	47.8 W	8	Distinct craterlet; north of Aristarchus	Yrjo; Finnish astronomer	1891–1971
Van Albada	9.4 N	64.3 E	21	Distinct; closely south of Auzout (Auzout A)	Gale; Dutch astronomer	1912–1972
Van Biesbroeck	28.7 N	45.6 W	9	Closely south of Krieger (Krieger B)	Georges; Belgian astronomer	1880–1974
Van Vleck	1.9 S	78.3 E	31	Dark floor; north of Kästner (Gilbert M)	John; American astronomer	1833–1912
Vasco da Gama	13.6 N	83.9 W	83	West of Procellarum; central ridge	Portuguese navigator	1469–1524
Vashakidze	43.6 N	93.3 E	44	Libration zone; outside Harkhebi	Mikhail; Russian astronomer	1909–1956
Vega	45.4 S	63.4 E	75	Australe area; deep	Georg von; German mathematician	1756–1802
Vendelinus	16.4 S	61.6 E	131	Petavius chain; broken and irregular	Godefroid Wendelin; Belgian astronomer	1580–1667
Vera	26.3 N	43.7 W	2	Aristarchus area; origin of long rill (Prinz A)	Latin female name	–
Verne	24.9 N	25.3 W	2	Between Euler and Lambert; pair with Artemis	Latin male name	–
Very	25.6 N	25.3 E	5	Craterlet in Serenitatis; west of Le Monnier (Le Monnier B)	Frank; American astronomer	1852–1927
Vestine	33.9 N	93.9 E	61	Libration zone; beyond Gauss	Ernest; American physicist	1906–1968
Vieta	29.2 S	56.3 W	87	West of Humorum; low; central peak	François; French mathematician	1540–1603
Virchow	9.8 N	83.7 E	16	Adjoins Neper to the north (Neper G)	Rudolph; German pathologist	1821–1902
Vitello	30.4 S	37.5 W	42	Southern edge of Humorum; concentric crater	Erazmus Witelo; Polish physicist	1210–1285

Table 3.14 (cont.)

Name	Lat. (°)	Long. (°)	Diameter (km)	Notes (former names in parentheses)	Named after	
Vitruvius	17.6 N	31.3 E	29	Between Serenitatis and Tranquillitatis; low but rather bright walls	Marcus; Roman engineer	c. 25 BC
Vlacq	53.3 S	38.8 E	89	Janssen area; deep; central peak; one of a group of six	Adriaan; Dutch mathematician	c. 1600–1667
Vögel	15.1 S	5.9 E	26	Southest of Albategnius; chain of four craters	Hermann; German astronomer	1841–1907
Volta	53.9 N	84.4 W	123	Limb formation near Repsold	Count Allessandro; Italian physicist	1745–1827
von Behring	7.8 S	71.8 E	38	Distinct; west of Kästner (Maclaurin F)	Emil; German bacteriologist	1854–1917
Voskresensky	28.0 N	88.1 W	49	Flooded; beyond Otto Struve	Leonid; Russian rocket scientist	1913–1965
Wallace	20.3 N	8.7 W	26	In Imbrium; imperfect ring; very low walls	Alfred Russel; British natural historian	1823–1913
Walter	33.1 S	1.0 E	128	Massive walls; interior peak and craters; trio with Regiomontanus and Purbach	Bernard Walther; German astronomer	1430–1504
Wargentin	49.6 S	60.2 W	84	Schickard group; the famous plateau	Per; Swedish astronomer	1717–1783
Warner	4.0 S	87.3 E	35	Regular; Smythii area	Worcester; American inventor	1846–1929
Watt	49.5 S	48.6 E	66	Southwest of Janssen; contact twin with Steinheil	James; Scottish inventor	1736–1819
Watts	8.9 N	46.3 E	15	North of Taruntius; low walls; darkish floor (Taruntius D)	Chester; American astronomer	1889–1971
Webb	0.9 S	60.0 E	21	Edge of Fœcunditatis; darkish floor; central peak; minor ray-centre	Thomas; British astronomer	1806–1885
Weierstrass	1.3 S	77.2 E	33	Fairly regular; east of Maclaurin (Gilbert N)	Karl; German mathematician	1815–1897
Weigel	58.2 S	38.8 W	35	Schillar area; fairly regular; pair with Röst	Erhard; German mathematician	1625–1699
Weinek	27.5 S	37.0 E	32	Northeast of Piccolomimo; darkish floor	Ladislaus; Czech astronomer	1848–1913
Weiss	31.8 S	19.5 W	66	Pitatus group; very irregular	Edmund; German astronomer	1837–1917
Werner	28.0 S	3.3 E	70	Very regular; high walls; central peak; pair with Aliacensis	Johann; German mathematician	1468–1528
Wexler	69.1 S	90.2 E	51	Libration zone; beyond Neumayer; regular	Harry; American meteorologist	1911–1962
Whewell	4.2 N	13.7 E	13	Bright craterlet in Tranquillitatis uplands	William; British astronomer	1794–1866
Wichmann	7.5 S	38.1 W	10	On Procellarum; associated with large 'ghost'	Moritz; German astronomer	1821–1859
Widmanstätten	6.1 S	85.5 E	46	Limb; east of Maclauri	Aloys; German physicist	1753–1849
Wildt	9.0 N	75.8 E	11	Distinct; west of Neper (Condorcet K).	Rupert; German astronomer	1905–1976
Wilhelm I	43.4 S	20.4 W	106	Lomgomintanus area; uneven walls	Landgrave of Hesse; German astronomer	1532–1592
Wilkins	29.4 S	19.6 E	57	Rabbi Levi group; irregular	(Hugh) Percy; Welsh selenographer	1896–1960
Williams	42.0 N	37.2 E	36	Somniorum area; not prominent	Arthur; British astronomer	1861–1938
Wilson	69.2 S	42.4 W	69	Bettinus chain; near Bailly; deep, regular; no central peak	Alexander; Scottish astronomer	1714–1786
					Charles; Scottish physicist	1869–1959

Table 3.14 (cont.)

Name	Lat. (°)	Long. (°)	Diameter (km)	Notes (former names in parentheses)	Named after	
Winthrop	10.7 S	44.4 W	17	Ruined crater on west wall of Letronne (Letronne P)	John; American astronomer	1714–1779
Wöhler	38.2 S	31.4 E	27	East of Riccius; fairly regular	Friedrich; German chemist	1800–1882
Wolf	22.7 S	16.6 W	25	In south of Nubium; irregular; low walls	Maximilian; German astronomer	1863–1932
Wollaston	30.6 N	46.9 W	10	Bright crater in Harbinger Mountains	William Hyde; British physicist/chemist	1766–1828
Wright	31.6 S	86.6 W	39	Beyond Lagrange; pair with Shaler	Thomas; British philosopher	1711–1786
					William; American astronomer	1871–1959
					Frederick; American astronomer	1878–1953
Wrottesley	23.9 S	56.8 E	57	Petavius group; twin-peaked central mountain	John (Baron); British astronomer	1798–1867
Wurzelbauer	33.9 S	15.9 W	88	Pitatus group; irregular walls, much floor detail	Johann von; German astronomer	1651–1725
Wyld	1.4 S	98.1 E	93	Libration zone; beyond Smythii	James; American rocket scientist	1913–1953
Xenophanes	57.5 N	82.0 W	125	Limb near Sinus Roris; high walls; central peak	Greek philosopher	?560–?478 BC
Yakovkin	54.5 S	78.8 W	37	Limb; beyond Phocylides (Pingré H)	Avenir Aleksandrovich; Russian astronomer	1887–1974
Yangel	17.0 N	4.7 E	8	In highlands northwest of Manilius (Manilius F)	Mikhail; Russian rocket scientist	1911–1971
Yerkes	14.6 N	51.7 E	36	On western edge of Crisium; low walls; irregular	Charles; American benefactor	1837–1905
Young	41.5 S	50.9 E	71	Rheita Valley area; irregular	Thomas; British doctor/physicist	1773–1829
Zach	60.9 S	5.3 E	70	East of Clavius; fairly deep and regular	Freiherr von; Hungarian astronomer	1754–1832
Zagut	32.0 S	22.1 E	84	Rabbi Levi group; irregular	Abraham; Jewish astronomer	?– c. 1450
Zähringer	5.6 N	40.2 E	11	Deep; west of Taruntius (Taruntius F)	Josef; German physicist	1929–1970
Zasyadko	3.9 N	94.2 E	11	Libration area; Smythii area; in Babcock	Alexander; Russian rocket scientist	1779–1837
Zeeman	75.2 S	133.6 W	190	Libration zone; beyond Drygalski	Pieter; Dutch physicist	1865–1943
Zeno	45.2 N	72.9 E	65	East of Mercurius; deformed	Greek philosopher	c. 335–263 BC
Zinner	26.6 N	58.8 W	4	Distinct craterlet; north of Schiaparelli (Schiaparelli B).	Ernst; German astronomer	1886–1970
Zöllner	8.0 S	18.9 E	47	Northwest of Theophilis; elliptical	Johann Karl; German astronomer	1834–1882
Zsigmondy	59.7 N	104.7 W	65	Libration zone; beyond Poczobut	Richard; Austrian chemist	1865–1929
Zucchius	61.4 S	50.3 W	64	Schiller area; distinct; pair with Segner	Niccolo Zucchi; Italian astronomer	1586–1670
Zupus	17.2 S	52.3 W	38	South of Billy; low walls; irregular; very dark floor	Giovanni Zupi; Italian astronomer	1590–1650

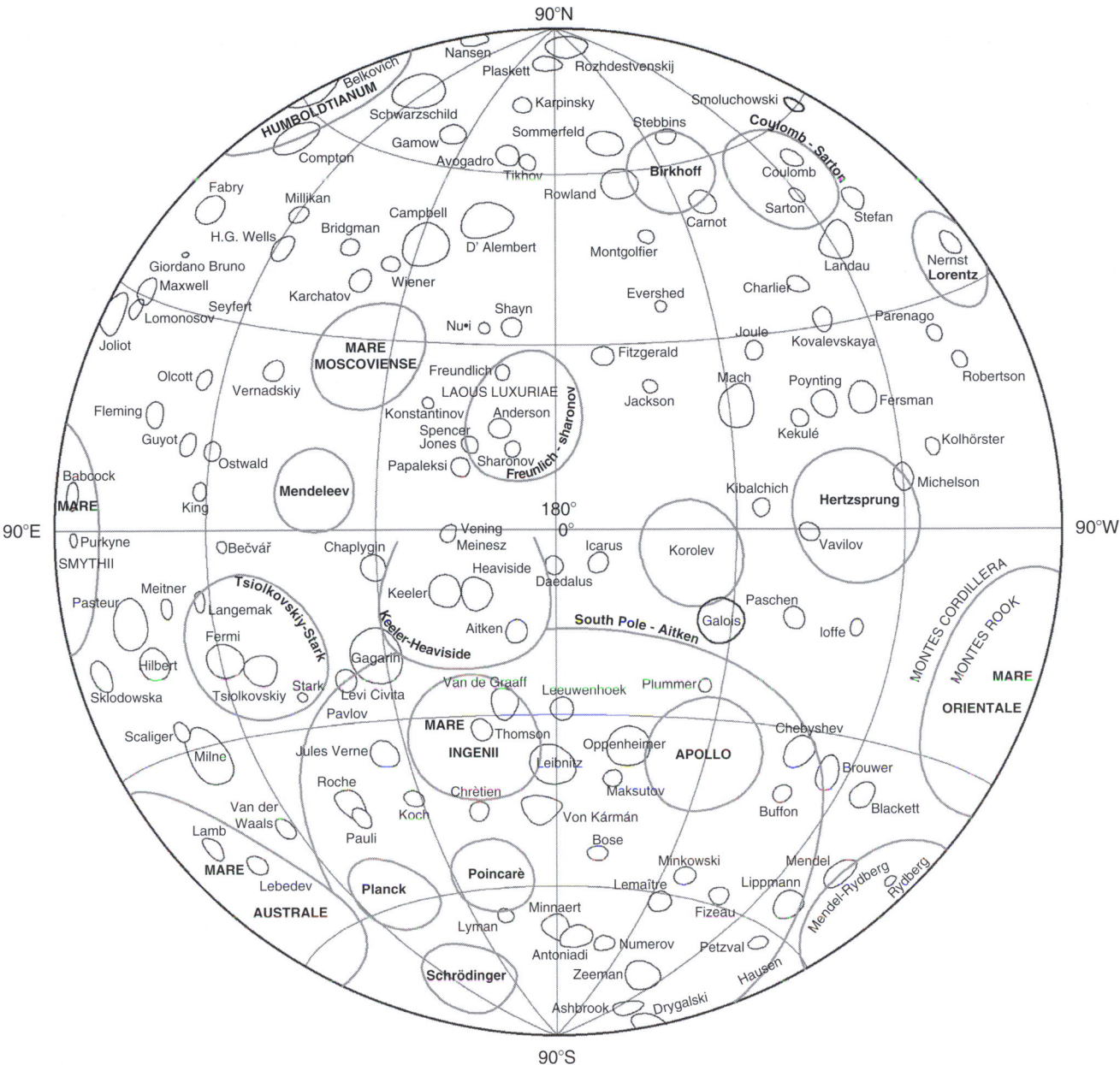

Figure 3.3 Outline map of the far side of the Moon

Obviously, a lunar eclipse can happen only at full moon. In the original edition of the famous novel *King Solomon's Mines*, H. Rider Haggard described a full moon, a solar eclipse and another full moon on successive days. When the mistake was pointed out he altered the second edition, turning the solar eclipse into a lunar one!

OCCULTATIONS

Occultations of stars by the Moon are common, and were formerly of great value for positional purposes. When a star is occulted, it shines steadily up to the instant of immersion; the lunar atmosphere is far too rarefied to have an effect. A planet will take some time to disappear. W. H. Pickering maintained that the lunar atmosphere could produce effects during an occupation of Jupiter, but this has long since been discounted.

On 23 April 1998 four Brazilian observers, led by E. Karkoschka, witnessed an exceptional event. Using a 10-cm refractor, they saw Venus and Jupiter occulted simultaneously; the site was 100 km north of Recife. The last occultation when Venus and Jupiter were occulted at the same time was AD 567.

Table 3.15 *Selected craters on the far side of the Moon*

Name	Lat. (°)	Long. (°)	Diameter (km)	Name origin	
Abbe	57.3 S	175.2 E	66	Ernst; German astronomer	1840–1905
Abul Wafa	1.0 N	116.6 E	55	Persian astronomer	940–998
Aitken	16.8 S	173.4 E	135	Robert; American astronomer	1864–1951
Alden	23.6 S	110.8 E	104	Harold; American astronomer	1890–1964
Alekhin	68.2 S	131.3 W	70	Nikolai; Russian rocket scientist	1913–1964
Alter	18.7 N	107.5 W	64	Dinsmore; American astronomer	1888–1968
Amici	9.9 S	172.1 W	54	Giovanni; Italian astronomer	1787–1863
Anders	41.3 S	142.9 W	40	William; American astronaut	1933–
Anderson	15.8 N	171.1 E	109	John; American astronomer	1876–1959
Antoniadi	69.7 S	172.0 W	143	Eugenios; Greek astronomer	1870–1944
Apollo	36.1 S	151.8 W	537	Named in honour of Apollo lunar missions	
Appleton	37.2 N	158.3 E	63	Sir Edward; British physicist	1892–1965
Artamonov	25.5 N	101.5 E	60	Nikolai; Russian rocket scientist	1906–1965
Artemev	10.8 N	144.4 W	67	Vladimir; Russian rocket scientist	1885–1962
Ashbrook	81.4 S	112.5 W	156	Joseph; American astronomer	1918–1980
Avicenna	39.7 N	97.2 W	74	Ibn Sina; Persian doctor	980–1037
Avogadro	63.1 N	164.9 E	139	Amedeo (Comte de); Italian physicist	1776–1856
Backlund	16.0 S	103.0 E	75	Oscar; Russian astronomer	1846–1916
Baldet	53.3 S	151.1 W	55	François; French astronomer	1885–1964
Barbier	23.8 S	157.9 E	66	Daniel; French astronomer	1907–1965
Barringer	28.9 S	149.7 W	68	Daniel; American engineer	1860–1929
Becquerel	40.7 N	129.7 E	65	Antoine; French physicist	1852–1908
Bečvar	1.9 S	125.2 E	67	Antonin; Czech astronomer	1901–1965
Beijerinck	13.5 S	151.8 E	70	Martinus; Dutch botanist	1851–1931
Bellingshausen	60.6 S	164.6 W	63	Faddey; Russian explorer	1778–1852
Belopolsky	17.2 S	128.1 W	59	Aristarch; Russian astronomer	1854–1934
Belyayev	23.3 N	143.5 E	54	Pavel; Russian cosmonaut	1925–1970
Bergstrand	18.8 S	176.3 E	43	Carl; Swedish astronomer	1873–1948
Berkner	25.2 N	105.2 W	86	Lloyd; American geophysicist	1905–1967
Berlage	63.2 N	162.8 W	92	Hendrik; Dutch geophysicist	1896–1968
Bhabba	55.1 S	164.5 W	64	Homi; Indian physicist	1909–1966
Birkeland	30.2 S	173.9 E	82	Olaf; Norwegian physicist	1867–1917
Birkhoff	58.7 N	146.1 W	345	George; American mathematician	1884–1944
Bjerknes	38.4 S	113.0 E	48	Vilhelm; Norwegian physicist	1862–1951
Blazhko	31.6 N	148.0 W	54	Sergei; Russian astronomer	1870–1956
Bobone	26.9 N	131.8 W	3	Jorge; Argentine astronomer	1901–1958
Boltzmann	74.9 S	90.7 W	76	Ludwig; Austrian physicist	1844–1906
Bolyai	33.6 S	125.9 E	135	Janos; Hungarian mathematician	1802–1860
Borman	38.8 S	147.7 W	50	Frank; American astronaut	1928–
Bose	53.5 S	168.6 W	91	Jagadis; Indian botanist	1858–1937
Boyle	53.1 S	178.1 E	57	Robert; British chemist	1627–1691
Brashear	73.8 S	170.7 W	55	John; American astronomer	1840–1920
Bredikin	17.3 N	158.2 W	59	Fedor; Russian astronomer	1831–1904
Brianchon	75.0 N	86.2 W	134	Charles; French mathematician	1783–1864
Bridgman	43.5 N	137.1 E	80	Percy; American physicist	1882–1961
Brouwer	36.2 S	126.0 W	158	Dirk; American astronomer	1902–1966
Brunner	9.9 S	90.9 E	53	William; Swiss astronomer	1878–1958
Buffon	40.4 S	133.4 W	106	Georges; French natural historian	1707–1788
Buisson	1.4 S	112.5 E	56	Henri; French astronomer	1873–1944
Butlerov	12.5 N	108.7 W	40	Alexander; Russian chemist	1828–1886
Buys–Ballot	20.8 N	174.5 E	55	Christoph; Dutch meteorologist	1817–1890

Table 3.15 (cont.)

Name	Lat. (°)	Long. (°)	Diameter (km)	Name origin	
Cabannes	60.9 S	169.6 W	80	Jean; French physicist	1885–1959
Cajori	47.4 S	168.8 E	70	Florian; American mathematician	1859–1930
Campbell	45.3 N	151.4 E	219	Leon; American astronomer	1862–1938
Cantor	38.2 N	118.6 E	80	Georg; German mathematician	1845–1918
Carnot	52.3 N	143.5 W	126	Nicholas; French physicist	1796–1832
Carver	43.0 S	126.9 E	59	George; American botanist	1864?–1943
Cassegrain	52.4 S	113.5 E	55	Giovanni; French astronomer	1625–1712
Ceraski	49.0 S	141.6 E	56	Witold; Polish astronomer	1849–1925
Chaffee	38.8 S	153.9 W	49	Roger; American astronaut	1935–1967
Champollion	37.4 N	175.2 E	58	Jean; French Egyptologist	1790–1832
Chandler	43.8 N	171.5 E	85	Seth; American astronomer	1846–1913
Chang Heng	19.0 N	112.2 E	43	Chinese astronomer	78–139
Chant	40.0 S	109.2 W	33	Clarence; Canadian astronomer	1865–1956
Chaplygin	6.2 S	150.3 E	137	Sergei; Russian mathematician	1869–1942
Chapman	50.4 N	100.7 W	71	Sydney; British geophysicist	1888–1970
Chappell	54.7 N	177.0 W	80	James; American astronomer	1891–1964
Charlier	36.6 N	131.5 W	99	Carl; Swedish astronomer	1862–1934
Chaucer	3.7 N	140.0 W	45	Geoffrey; British writer/astronomer	c. 1340–1400
Chauvenet	11.5 S	137.0 E	81	William; American astronome	1820–1870
Chebyshev	33.7 S	131.1 W	178	Pafnutif; Russian astronomer	1821–1894
Chrétien	45.9 S	162.9 E	88	Henri; French astronomer/mathematician	1870–1956
Clark	38.4 S	118.9 E	49	Alvan; American astronomer/optician	1804–1887
				Alvan G.; American astronomer/optician	1832–1897
Coblentz	37.9 S	126.1 E	33	William; American astronomer	1873–1962
Cockcroft	31.3 N	162.6 W	93	Sir John; British nuclear physicist	1897–1967
Comrie	23.3 N	112.7 W	59	Leslie; New Zealand astronomer	1893–1950
Comstock	21.8 N	121.5 W	72	George; American astronomer	1855–1934
Congreve	0.2 S	167.3 W	57	Sir William; British rocket pioneer	1772–1828
Cooper	52.9 N	175.6 E	36	John; American humanitarian	1887–1967
Coriolis	0.1 N	171.8 E	78	Gaspard de; French physicist	1792–1843
Coulomb	54.7 N	114.6 W	89	Charles de; French physicist	1736–1806
Crocco	47.5 S	150.2 E	75	Gætano; Italian aeronautical engineer	1877–1968
Crommelin	68.1 S	146.9 W	94	Andrew; Irish astronomer	1865–1939
Crookes	10.3 S	164.5 W	49	Sir William; British physicist	1832–1919
Cyrano	20.5 S	156.6 E	80	Cyrano de Bergerac; French writer	1615–1655
Dædalus	5.9 S	179.4 E	93	Greek mythological character	–
D'Alembert	50.8 N	163.9 E	248	Jean; French mathematician	1717–1783
Danjon	11.4 S	124.0 E	71	Andre; French astronomer	1890–1967
Dante	25.5 N	180.0 E	54	Alighieri; Italian poet	1265–1321
Das	26.6 S	136.8 W	38	Smil; Indian astronomer	1902–1961
Davisson	37.5 S	174.6 W	87	Clinton; American physicist	1881–1958
Dawson	67.4 S	134.7 W	45	Bernhard; Argentinian astronomer	1890–1960
Debye	49.6 N	176.2 W	142	Peter; Dutch physicist	1884–1966
De Forest	77.3 S	162.1 W	57	Lee; American inventor/physicist	1873–1961
Dellinger	6.8 S	140.6 E	81	John; American physicist	1886–1962
Delporte	16.0 S	121.6 E	45	Eugene; Belgian astronomer	1882–1955
Denning	16.4 S	142.6 E	44	William; British astronomer	1848–1931
Deutsch	24.1 N	110.5 E	66	Armin; American astronomer	1918–1969
de Vries	19.9 S	176.7 W	59	Hugo; Dutch botanist	1848–1935
Dewar	2.7 S	165.5 E	50	Sir James; British chemist	1842–1923
Dirichlet	11.1 N	151.4 W	47	Peter; German mathematician	1805–1859

Table 3.15 (cont.)

Name	Lat. (°)	Long. (°)	Diameter (km)	Name origin	
Doppler	12.6 S	159.6 W	110	Christian; Austrian physicist	1803–1853
Douglass	35.9 N	122.4 W	49	Andrew; American astronomer	1867–1962
Dryden	33.0 S	155.2 W	51	Hugh; American physicist	1898–1965
Dufay	5.5 N	169.5 E	39	Jean; French astronomer	1896–1967
Dugan	64.2 N	103.3 E	50	Raymond; American astronomer	1878–1940
Dunér	44.8 N	179.5 E	62	Nils; Swedish astronomer	1839–1914
Dyson	61.3 N	121.2 W	63	Sir Frank; British astronomer	1868–1939
Ehrlich	40.9 N	172.4 W	30	Paul; German doctor	1854–1915
Eijkman	63.1 S	143.0 W	54	Christian; Dutch doctor	1858–1930
Einthoven	4.9 S	109.6 E	69	Willem; Dutch physiologist	1879–1955
Ellerman	25.3 S	120.1 W	47	Ferdinand; American astronomer	1869–1940
Elvey	8.8 N	100.5 W	74	Christian; American astronomer	1899–1970
Emden	63.3 N	177.3 W	111	J. Robert; Swiss astrophysicist	1862–1940
Engelhardt	5.7 N	159.0 W	43	Vasili; Russian astronomer	1828–1915
Eötvös	35.5 S	133.8 E	99	Roland von; Hungarian physicist	1848–1919
Esnault-Pelterie	47.7 N	141.4 W	79	Robert; French rocket engineer	1881–1957
Espin	28.1 N	109.1 E	75	Thomas; British astronomer	1858–1934
Evans	9.5 S	133.5 W	67	Sir Arthur; British archæologist	1851–1941
Evdokimov	34.8 N	153.0 W	50	Nikolai; Russian astronomer	1868–1940
Evershed	35.7 N	150.5 W	66	John; British astronomer	1864–1956
Fechner	59.0 S	124.9 E	63	Gustav; German physicist	1801–1887
Feoktistov	30.9 N	140.7 E	23	Konstantin; Russian cosmonaut	1926–2009
Fermi	19.3 S	122.6 E	183	Enrico; Italian physicist	1901–1954
Fersman	18.7 N	126.0 E	151	Alexander; Russian geochemist	1883–1945
Firsov	4.5 N	112.2 E	51	Georgi; Russian rocket engineer	1917–1960
Fitzgerald	27.5 N	171.7 W	110	George; Irish physicist	1851–1901
Fizeau	58.6 S	133.9 W	111	Armand; French physicist	1819–1896
Fleming	15.0 N	109.6 E	106	Alexander; British doctor	1881–1955
				Williamina; American astronomer	1857–1911
Foster	23.7 N	141.5 W	33	John; Canadian physicist	1860–1964
Fowler	42.3 N	145.0 W	146	Alfred; British astronomer	1868–1940
Freundlich	25.0 N	171.0 E	85	Erwin Finlay; German astronomer	1885–1964
Fridman	12.6 S	126.0 W	102	Alexander; Russian physicist	1885–1925
Frost	37.7 N	118.4 W	75	Edwin; American astronomer	1866–1935
Gadomski	36.4 N	147.3 W	65	Jan; Polish astronomer	1889–1966
Gagarin	20.2 S	149.2 E	265	Yuri; Russian cosmonaut	1934–1968
Galois	14.2 S	151.9 W	222	Evariste; French mathematician	1811–1832
Gamow	65.3 N	145.3 E	129	George; Russian astronomer/physicist	1904–1968
Ganswindt	79.6 S	110.3 E	74	Hermann; German inventor	1856–1934
Garavito	47.5 S	156.7 E	74	José; Colombian astronomer	1865–1920
Geiger	14.6 S	158.5 E	34	Johannes; German physicist	1882–1945
Gerasimovič	22.9 S	122.6 W	86	Boris; Russian astronomer	1889–1937
Giordano Bruno	35.9 N	102.8 E	22	Italian astronomer/philosopher	1548–1600
Glasenapp	1.6 S	137.6 E	43	Sergei; Russian astronomer	1848–1937
Golitzyn	25.1 S	105.0 W	36	Boris; Russian physicist	1862–1916
Golovin	39.9 N	161.1 E	37	Nicholas; American rocket scientist	1912–1969
Grachev	3.7 S	108.2 W	35	Andrei; Russian rocket scientist	1900–1964
Green	4.1 N	132.9 E	65	George; British mathematician	1793–1841
Gregory	2.2 N	127.3 E	67	James; Scottish astronomer	1638–1675
Grigg	12.9 N	129.4 W	36	John; New Zealand astronomer	1838–1920
Grissom	47.0 S	147.4 W	58	Virgil; American astronaut	1926–1967

Table 3.15 (cont.)

Name	Lat. (°)	Long. (°)	Diameter (km)	Name origin	
Grotrian	66.5 S	128.3 E	37	Walter; German astronomer	1890–1954
Gullstrand	45.2 N	129.3 W	43	Allvar; Swedish opthalmologist	1862–1930
Guyot	11.4 N	117.5 E	92	Arnold; Swiss geographer	1807–1884
Hagen	48.3 S	135.1 E	55	Johann; Austrian astronomer	1847–1930
Harriot	33.1 N	114.3 E	56	Thomas; British astronomer/mathematician	1560–1621
Hartmann	3.2 N	135.3 E	61	Johannes; German astronomer	1865–1936
Harvey	19.5 N	146.5 W	60	William; British doctor	1578–1657
Heaviside	10.4 S	167.1 E	165	Oliver; British physicist/mathematician	1850–1925
Helberg	22.5 N	102.2 W	62	Robert; American aeronautical engineer	1906–1967
Henderson	4.8 N	152.1 E	47	Thomas; Scottish astronomer	1798–1844
Hendrix	46.6 S	159.2 W	18	Don; American optician	1905–1961
Henyey	13.5 N	151.6 W	63	Louis; American astronomer	1910–1970
Hertz	13.4 N	104.5 E	90	Heinrich; German physicist	1857–1894
Hertzsprung	2.6 N	129.2 W	591	Ejnar; Danish astronomer	1873–1967
Hess	54.3 S	174.6 E	88	Victor; Austrian physicist	1883–1964
Heymans	75.3 N	144.1 W	50	Corneille; Belgian physiologist	1892–1968
Hilbert	17.9 S	168.2 E	55	Johann; Austrian astronomer	1847–1930
Hippocrates	70.7 N	145.9 W	60	Greek doctor	c. 140 BC
Hoffmeister	15.2 N	136.9 E	45	Cuno; German astronomer	1892–1968
Hogg	33.6 N	121.9 E	38	Arthur; Australian astronomer	1903–1966
				Frank; Canadian astronomer	1904–1951
Holetschek	27.6 S	150.9 E	38	Johann; Austrian astronomer	1846–1923
Houzeau	17.1 S	123.5 W	71	Jean; Belgian astronomer	1820–1888
Hutton	37.3 N	168.7 E	50	James; Scottish geologist	1726–1797
Icarus	5.3 S	173.2 W	96	Greek mythical aviator	–
Idelson	81.5 S	110.9 E	60	Naum; Russian astronomer	1885–1951
Ingalls	26.4 N	153.1 W	37	Alnert; American optician	1888–1958
Innes	27.8 N	119.2 E	42	Robert; Scottish astronomer	1861–1933
Izsak	23.3 S	117.1 E	30	Imre; Hungarian astronomer	1929–1965
Jackson	22.4 N	163.1 W	71	John; Scottish astronomer	1887–1958
Joffe	14.4 S	129.2 W	86	Abram; Russian physicist	1880–1960
Joule	27.3 N	144.2 W	96	James; British physicist	1818–1889
Jules Verne	35.0 S	147.0 E	143	French writer	1828–1905
Kamerlingh Onnes	15.0 N	115.8 W	66	Heike; Dutch physicist	1853–1926
Karpinsky	73.3 N	166.3 E	92	Alexei; Russian geologist	1846–1936
Kearons	11.4 S	112.6 W	23	William; American astronomer	1878–1948
Keeler	10.2 S	161.9 E	160	James; American astronomer	1857–1900
Kekulé	16.4 N	138.1 W	94	Friedrich; German chemist	1829–1896
Khwolson	13.8 S	111.4 E	54	Orest; Russian physicist	1852–1934
Kibaltchich	3.0 N	146.5 W	92	Nikolai; Russian rocket scientist	1853–1881
Kidinnu	35.9 N	122.9 E	56	Or Cidenas; Babylonian astronomer	?–c. 343 BC
Kimura	57.1 S	118.4 E	28	Hisashi; Japanese astronomer	1870–1943
King	5.0 N	120.5 E	76	Arthur; American physicist	1876–1957
				Edward; American astronomer	1861–1931
Kirkwood	68.8 N	156.1 W	67	Daniel; American astronomer	1814–1895
Kleimenov	32.4 S	140.2 W	55	Ivan; Russian rocket scientist	1898–1938
Klute	37.2 N	141.3 W	75	Daniel; American rocket scientist	1921–1964
Koch	42.8 S	150.1 E	95	Robert; German doctor	1843–1910
Kohlschütter	14.4 N	154.0 E	53	Arnold; German astronomer	1883–1969
Kolhörster	11.2 N	114.6 W	97	Werner; German physicist	1887–1946
Komarov	24.7 N	152.2 E	78	Vladimir; Russian cosmonaut	1927–1967

Table 3.15 (cont.)

Name	Lat. (°)	Long. (°)	Diameter (km)	Name origin	
Kondratyuk	14.9 S	115.5 E	108	Yuri; Russian rocket pioneer	1897–1942
Konstantinov	19.8 N	158.4 E	66	Konstantin; Russian rocket scientist	1817–1871
Korolev	4.0 S	157.4 W	437	Sergei; Russian rocket scientist	1906–1966
Kostinsky	14.7 N	118.8 E	75	Sergei; Russian astronomer	1867–1937
Kovalevskaya	30.8 N	129.6 W	115	Sofia; Russian mathematician	1850–1891
Kovalsky	21.9 S	101.0 E	49	Marian; Russian astronomer	1821–1884
Kramers	53.6 N	127.6 W	61	Hendrik; Dutch physicist	1894–1952
Krasovsky	3.9 N	175.5 W	59	Feodosii; Russian geodetist	1878–1948
Krylov	35.6 N	165.8 W	49	Alexei; Russian mathematician	1863–1945
Kulik	42.4 N	154.5 W	58	Leonid; Russian mineralogist	1883–1942
Kuo Shou Ching	8.4 N	133.7 W	34	Chinese astronomer	1231–1316
Kurchatov	38.3 N	142.1 E	106	Igor; Russian nuclear physicist	1903–1960
Lacchini	41.7 N	107.5 W	58	Giovanni; Italian astronomer	1884–1967
Lamarck	22.9 S	69.8 W	100	Jean; French natural historian	1744–1829
Lamb	42.9 S	100.1 E	106	Sir Horace; British mathematician	1849–1934
Lampland	31.0 S	131.0 E	65	Carl; American astronomer	1873–1951
Landau	41.6 N	118.1 W	214	Lev; Russian physicist	1908–1968
Lane	9.5 S	132.0 E	55	Jonathan; American astrophysicist	1819–1880
Langemak	10.3 S	118.7 E	97	Georgi; Russian rocket scientist	1898–1938
Langevin	44.3 N	162.7 E	58	Paul; French physicist	1872–1946
Langmuir	35.7 S	128.4 W	91	Irving; American physicist	1881–1957
Larmor	32.1 N	179.7 W	97	Sir Joseph; British mathematician	1857–1942
Leavitt	44.8 S	139.3 W	66	Henrietta; American astronomer	1868–1921
Lebedev	47.3 S	107.8 E	102	Petr; Russian physicist	1866–1912
Lebedinsky	8.3 N	164.3 W	62	Alexander; Russian astrophysicist	1913–1967
Leeuwenhoek	29.3 S	178.7 W	125	Antony van; Dutch microscopist	1632–1723
Leibnitz	38.3 S	179.2 E	245	Gottfried; German mathematician	1646–1716
Lemaître	61.2 S	149.6 W	91	Georges; Belgian mathematician	1894–1966
Lenz	2.8 N	102.1 W	21	Heinrich Emil; Estonian physicist	1804–1865
Leonov	19.0 N	148.2 E	33	Alexei; Russian cosmonaut	1934–
Leucippus	29.1 N	116.0 W	56	Greek philosopher	c. 440 BC
Levi-Civita	23.7 S	143.4 E	121	Tullio; Italian mathematician	1873–1941
Lewis	18.5 S	113.8 W	42	Gilbert; American chemist	1875–1946
Ley	42.2 N	154.9 E	79	Willy; German rocket scientist	1906–1969
Lobachevsky	9.9 N	112.6 E	84	Nikolai; Russian mathematician	1793–1856
Lodygin	17.7 S	146.8 W	62	Alexander; Russian inventor	1847–1923
Lomonsov	27.3 N	98.0 E	92	Mikhail; Russian cartographer	1711–1765
Lorentz	32.6 N	95.3 W	312	Hendrik; Dutch physicist	1853–1928
Love	6.3 S	129.0 E	84	Augustus; British mathematician	1863–1940
Lovelace	82.3 N	106.4 W	54	William; American space scientist	1907–1965
Lovell	36.8 S	141.9 W	34	James; American astronaut	1928–
Lowell	12.9 S	103.1 W	66	Percival; American astronomer	1855–1916
Lucretius	8.2 S	120.8 W	63	Titus; Roman philosopher	c. 95–55 BC
Lundmark	39.7 S	152.5 E	106	Knut; Swedish astronomer	1889–1958
Lyman	64.8 S	163.6 E	84	Theodore; American physicist	1874–1954
Mach	18.5 N	149.3 W	180	Ernst; Austrian physicist	1838–1916
Maksutov	40.5 S	168.7 W	83	Dimitri; Russian optician	1896–1964
Malyi	21.9 N	105.3 E	41	Alexander; Russian rocket scientist	1907–1961
Mandelstam	5.4 N	162.4 E	197	Leonid; Russian physicist	1879–1944
Marci	22.6 N	167.0 W	25	Jan; Czech physicist	1595–1667
Marconi	9.6 S	145.1 E	73	Guglielmo; Italian radio pioneer	1874–1937

Table 3.15 (cont.)

Name	Lat. (°)	Long. (°)	Diameter (km)	Name origin	
Mariotte	28.5 S	139.1 W	65	Edme; French physicist	1620–1684
McKellar	15.7 S	170.8 W	51	Andrew; Canadian astronomer	1910–1960
McMath	17.3 N	165.6 W	86	Francis; American engineer/astronomer	1867–1938
				Robert; American astronomer	1891–1962
McNally	22.6 N	127.2 W	47	Paul; American astronomer	1890–1955
Mees	13.6 N	96.1 W	50	Kenneth; British photographer	1882–1960
Meggers	24.3 N	123.0 E	52	William; American physicist	1888–1966
Meitner	10.5 S	112.7 E	87	Lise; Austrian physicist	1878–1968
Mendeleev	5.7 N	140.9 E	313	Dimitri; Russian chemist	1834–1907
Merrill	75.2 N	116.3 W	57	Paul; American astronomer	1887–1961
Mesentsev	72.1 N	128.7 W	89	Yuri; Russian rocket scientist	1929–1965
Meshcerski	12.2 N	125.5 E	65	Ivan; Russian mathematician	1859–1935
Michelson	7.2 N	120.7 W	123	Albert; German physicist	1852–1931
Milankovič	77.2 N	168.8 E	101	Milutin; Jugoslav astronomer	1879–1958
Millikan	46.8 N	121.5 E	98	Robert; American physicist	1868–1953
Mills	8.6 N	156.0 E	32	Mark; American physicist	1917–1958
Milne	31.4 S	112.2 E	272	Arthur; British mathematician/astronomer	1896–1950
Mineur	25.0 N	161.3 W	73	Henri; French mathematician/astronomer	1899–1954
Minkowski	56.5 S	146.0 W	113	Hermann; German mathematician	1864–1909
				Rudolph; American astronomer	1895–1976
Mitra	18.0 N	154.7 W	92	Sisir Kumar; Indian physicist	1890–1963
Möbius	15.8 N	101.2 E	50	August; German mathematician	1790–1868
Mohorovičič	19.0 S	165.0 W	51	Andrija; Jugoslav geophysicist	1857–1936
Moiseev	9.5 N	103.3 E	59	Nikolai; Russian astronomer	1902–1955
Montgolfier	47.3 N	159.8 W	88	Jacques; French balloonist	1745–1799
				Joseph; French balloonist	1740–1810
Moore	37.4 N	177.5 W	54	Joseph; American astronomer	1878–1949
Morozov	5.0 N	127.4 E	42	Nikolai; Russian natural scientist	1854–1945
Morse	22.1 N	175.1 W	77	Samuel; American inventor	1791–1872
Nagaoka	19.4 N	154.0 E	46	Hantaro; Japanese physicist	1865–1940
Nassau	24.9 S	177.4 E	76	Jason; American astronomer	1892–1965
Nernst	35.3 N	94.8 W	116	Walther; German physical chemist	1864–1941
Neujmin	27.0 S	125.0 E	101	Grigori; Russian astronomer	1885–1946
Nièpce	72.7 N	119.1 W	57	Joseph N; French photographer	1765–1833
Nijland	33.0 N	134.1 E	35	Albertus; Dutch astronomer	1868–1936
Nikolayev	35.2 N	151.3 E	41	Andrian; Russian cosmonaut	1929–
Nishina	44.6 S	170.4 W	65	Yoshio; Japanese physicist	1890–1951
Nobel	15.0 N	101.3 W	48	Alfred; Swedish inventor	1833–1896
Nöther	66.6 N	113.5 W	67	Emmy; German mathematician	1882–1935
Numerov	70.7 S	160.7 W	113	Boris; Russian astronomer	1891–1941
Nušl	32.3 N	167.6 E	61	Frantisek; Czech astronomer	1867–1925
Obruchev	38.9 S	162.1 E	71	Vladimir; Russian geologist	1863–1956
O'Day	30.6 S	157.5 E	71	Marcus; American physicist	1897–1961
Ohm	18.4 N	113.5 W	64	Georg; German physicist	1787–1854
Olcott	20.6 N	117.8 E	81	William; American astronomer	1873–1936
Omar Kháyyám	58.0 N	102.1 W	70	Al Khayyami; Persian mathematician/poet	c. 1050–1123
Oppenheimer	35.2 S	166.3 W	208	J. Robert; American physicist	1904–1967
Oresme	42.5 S	169.2 E	76	Nicole; French mathematician	1323?–1382
Orlov	25.7 S	175.0 W	81	Alexander; Russian astronomer	1880–1954
				Sergei; Russian astronomer	1880–1958
Ostwald	10.4 N	121.9 E	104	Wilhelm; German chemist	1853–1932
Pannekoek	4.2 S	140.5 E	71	Antonie; Dutch astronomer	1873–1960

Table 3.15 (cont.)

Name	Lat. (°)	Long. (°)	Diameter (km)	Name origin	
Papaleski	10.2 N	164.0 E	97	Nikolai; Russian physicist	1880–1947
Paracelsus	23.0 S	163.1 E	83	Theopnrastus von Hohenheim; Swiss chemist	1493–1541
Paraskevopoulos	50.4 N	149.9 W	94	John; Greek astronomer	1889–1951
Parenago	25.9 N	108.5 W	93	Pavel; Russian astronomer	1906–1960
Parkhurst	33.4 S	103.6 E	96	John; American astronomer	1861–1925
Parsons	37.3 N	171.2 W	40	John; American astronomer	1913–1952
Paschen	13.5 S	139.8 W	124	Friedrich; German physicist	1865–1940
Pasteur	11.9 S	104.6 E	224	Louis; French chemist	1822–1895
Pauli	44.5 S	136.4 E	84	Wolfgang; Austrian physicist	1900–1958
Pavlov	28.8 S	142.5 E	148	Ivan; Russian physiologist	1849–1936
Pawsey	44.5 N	143.0 E	60	Joseph; Australian radio astronomer	1908–1962
Pease	12.5 N	106.1 W	38	Francis; American astronomer	1881–1938
Perelman	24.0 S	106.0 E	46	Yakov; Russian rocket scientist	1882–1942
Perepelkin	10.0 S	129.0 E	97	Evgeny; Russian astrophysicist	1906–1940
Perkin	47.2 N	175.9 W	62	Richard; American telescope maker	1906–1969
Perrine	45.2 N	127.8 W	86	Charles; American astronomer	1867–1951
Petrie	45.3 N	108.4 E	33	Robert; Canadian astronomer	1906–1966
Petropavlovsky	37.2 N	114.8 W	63	Boris; Russian rocket engineer	1898–1933
Petzval	62.7 S	110.4 W	90	Joseph von; Austrian optician	1807–1891
Pirquet	20.3 S	139.6 E	65	Baron Guido von; Austrian space scientist	1867–1936
Pizzetti	34.9 S	118.8 E	44	Paolo; Italian geodetist	1860–1918
Planck	57.9 S	136.8 E	324	Max; German physicist	1858–1947
Plaskett	82.1 N	174.3 E	109	John; Canadian astronomer	1865–1941
Plummer	25.0 S	155.0 W	73	Henry; British astronomer	1875–1946
Pogson	42.2 S	110.5 E	50	Norman; British astronomer	1829–1891
Poincaré	56.7 S	163.6 E	319	Jules; French mathematician	1854–1912
Poinsot	79.5 N	145.7 W	68	Louis; French mathematician	1777–1859
Polzunov	25.3 N	114.6 E	67	Ivan; Russian heat engineer	1728–1766
Poynting	18.1 N	133.4 W	128	John; British physicist	1852–1914
Prager	3.9 S	130.5 E	60	Richard; German astronomer	1884–1945
Prandtl	60.1 S	141.8 E	91	Ludwig; German physicist	1875–1953
Priestley	57.3 S	108.4 E	52	Joseph; British chemist	1733–1804
Quetelet	43.1 N	134.9 W	55	Lambert; Belgian astronomer	1796–1874
Racah	13.8 S	179.8 W	63	Giulio; Israeli physicist	1909–1965
Raimond	14.6 N	159.3 W	70	J. J.; Dutch astronomer	1903–1961
Ramsay	40.2 S	144.5 E	81	Sir William; British chemist	1852–1916
Rasumov	39.1 N	114.3 W	70	Vladimir; Russian rocket engineer	1890–1967
Rayet	44.7 N	114.5 E	27	George; French astronomer	1839–1906
Rayleigh	29.3 N	89.6 E	114	John, Lord Rayleigh; British physicist	1842–1919
Riccò	75.6 N	176.3 E	65	Annibale; Italian astronomer	1844–1911
Riedel	48.9 S	139.6 W	47	Klaus; German rocket scientist	1907–1944
				Walter; German rocket scientist	1902–1968
Riemann	38.9 N	86.8 E	163	Georg; German mathematician	1826–1866
Rittenhouse	74.5 S	106.5 E	26	David; American astronomer/inventor	1732–1796
Roberts	71.1 N	174.5 W	89	Alexander; South African astronomer	1857–1938
				Isaac; British astronomer	1829–1904
Robertson	21.8 N	105.2 W	88	Howard; American physicist	1903–1961
Roche	42.3 S	136.5 E	160	Edouard; French astronomer	1820–1883
Rowland	57.4 N	162.5 W	171	Henry; American physicist	1848–1901
Rozhdestvensky	85.2 N	155.4 W	177	Dimitri; Russian physicist	1876–1940
Rumford	28.8 S	169.8 W	61	Benjamin; Count Rumford; British physicist	1753–1814

Table 3.15 (cont.)

Name	Lat. (°)	Long. (°)	Diameter (km)	Name origin	
Šafárik	16.6 N	176.9 E	27	Vojtech; Czech astronomer	1829–1902
Saha	1.6 S	102.7 E	99	Meghrad; Indian astrophysicist	1893–1956
Sänger	4.3 N	102.4 E	75	Eugen; Austrian rocket engineer	1905–1964
St John	10.2 N	150.2 E	68	Charles; American astronomer	1857–1935
Sanford	32.6 N	138.9 W	55	Roscoe; American astronomer	1883–1958
Sarton	49.3 N	121.1 W	69	George; Belgian historian of science	1844–1956
Scaliger	27.1 S	108.9 E	84	Joseph; French chronologist	1540–1609
Schaeberle	26.7 S	117.2 E	62	John; American astronomer	1853–1924
Schjellerup	69.7 N	157.1 E	62	Hans Carl; Danish astronomer	1827–1887
Schlesinger	47.4 N	138.6 W	97	Frank; American astronomer	1871–1943
Schliemann	2.1 S	155.2 E	80	Heinrich; German archæologist	1822–1890
Schneller	41.8 N	163.6 W	54	Herbert; German astronomer	1901–1967
Schrödinger	67.0 S	132.4 E	312	Erwin; Austrian physicist	1887–1961
Schuster	4.2 N	146.5 E	108	Sir Arthur; British mathematician	1851–1934
Schwarzschild	70.1 N	121.2 E	212	Karl; German astronomer	1873–1916
Seares	73.5 N	145.8 E	110	Frederick; American astronomer	1873–1964
Sechenov	7.1 S	142.6 W	62	Ivan; Russian physiologist	1829–1905
Segers	47.1 N	127.7 E	17	Carlos; Argentine astronomer	1900–1967
Seidel	32.8 S	152.2 E	62	Ludwig von; German astronomer	1821–1896
Seyfert	29.1 N	114.6 E	110	Carl; American astronomer	1911–1960
Shajn	32.6 N	172.5 E	93	Grigeri; Russian astrophysicist	1892–1956
Sharanov	12.4 N	173.3 E	74	Vsevolod; Russian astronomer	1901–1964
Shatalov	24.3 N	141.5 E	21	Vladmir; Russian cosmonaut	1927–
Shi Shen	76.0 N	104.1 E	43	Chinese astronomer	?–c. 300 BC
Siedentopf	22.0 N	135.5 E	61	Heinrich; German astronomer	1906–1963
Siepinski	27.2 S	154.5 E	69	Waclaw; Polish mathematician	1882–1969
Sisakian	41.2 N	109.0 E	34	Noran; Russian doctor	1907–1966
Sklodowska	18.2 S	95.5 E	127	Marie Curie; Polish physicist	1867–1934
Slipher	49.5 N	160.1 E	69	Earl; American astronomer	1883–1964
				Vesto; American astronomer	1875–1969
Sniadecki	22.5 S	168.9 W	43	Jan; Polish astronomer	1756–1830
Sommerfeld	65.2 N	162.4 W	169	Arnold; German physicist	1868–1951
Spencer Jones	13.3 N	165.6 E	85	Sir Harold; British astronomer	1890–1960
Stark	25.5 S	134.6 E	49	Johannes; German physicist	1874–1957
Stebbins	64.8 N	141.8 W	131	Joe; American astronomer	1878–1966
Stefan	46.0 N	108.3 W	125	Josef; Austrian physicist	1835–1893
Stein	7.2 N	179.0 E	33	Johan; Dutch astronomer	1871–1951
Steklov	36.7 S	104.9 W	36	Vladimir; Russian mathematician	1864–1926
Steno	32.8 N	161.8 E	31	Nicolaus; Danish doctor	1638–1686
Sternfeld	19.6 S	141.2 W	100	Ari; Russian space scientist	1905–1980
Stetson	39.6 S	118.3 W	64	Harian; American astronomer	1885–1964
Stoletov	45.1 N	155.2 W	42	Alexander; Russian physicist	1839–1896
Stoney	55.3 S	156.1 W	45	(George) Johnstone; Irish physicist	1826–1911
Størmer	57.3 N	146.3 E	69	Carl; Norwegian astronomer	1874–1957
Stratton	5.8 S	164.6 E	70	Frederick; British astronomer	1881–1960
Strömgren	21.7 S	132.4 W	61	Elis; Danish astronomer	1870–1947
Subbotin	29.2 S	135.3 E	67	Milhail; Russian astronomer	1893–1966
Sumner	37.5 N	108.7 E	50	Thomas; American geographer	1807–1876
Swann	52.0 N	112.7 E	42	William; British physicist	1884–1962
Szilard	34.0 N	105.7 E	122	Leo; Hungarian physicist	1898–1964
Teisserene de Bort	32.2 N	135.9 W	62	Leon; French meteorologist	1855–1913

Table 3.15 (cont.)

Name	Lat. (°)	Long. (°)	Diameter (km)	Name origin	
ten Bruggencate	9.5 S	134.4 E	59	Paul; German astronomer	1901–1961
Tereshkova	28.4 N	144.3 E	31	Valentina; Russian cosmonaut	1937–
Tesla	38.5 N	124.7 E	43	Nikola; Jugoslav inventor	1856–1943
Thiel	40.7 N	134.5 W	32	Walter; German space scientist	1910–1943
Thiessen	75.4 N	169.0 W	66	Georg; German astronomer	1914–1961
Thomson	32.7 S	166.2 E	117	Sir Joseph; British physicist	1856–1940
Tikhomirov	25.2 N	162.0 E	65	Nikolai; Russian chemical engineer	1860–1930
Tikhov	63.3 N	171.7 E	83	Gavriil; Russian astronomer	1875–1960
Tiling	53.1 S	132.6 W	38	Reinhold; German rocket scientist	1890–1933
Timiryazev	5.5 S	147.0 W	53	Kliment; Russian botanist	1843–1920
Titov	28.6 N	150.5 E	31	German; Russian cosmonaut	1935–2000
Trümpler	29.3 N	167.1 E	77	Robert; Swiss astronomer	1866–1956
Tsander	6.2 N	149.3 W	181	Friedrich; Russian rocket scientist	1887–1933
Tsiolkovskii	21.2 S	128.9 E	185	Konstantin; Russian rocket engineer	1857–1935
Tsu Chung-chi	17.3 N	145.1 E	28	Chinese mathematician	430–501
Tyndall	34.9 S	117.0 E	18	John; British physicist	1820–1893
Valier	6.8 N	174.5 E	67	Max; German rocket engineer	1895–1930
Van de Graaff	27.4 S	172.2 E	233	Robert; American physicist	1901–1967
Van den Bergh	31.3 N	159.1 W	42	George; Dutch astronomer	1890–1966
Van der Waals	43.9 S	119.9 E	104	Johannes; Dutch physicist	1837–1923
Van Gent	15.4 N	160.4 E	43	Hendrik; Dutch astronomer	1900–1947
Van Maanen	35.7 N	128.0 E	60	Adriaan; Dutch astronomer	1884–1946
Van Rhijn	52.6 N	146.4 E	46	Pieter; Dutch astronomer	1886–1960
Van't Hoff	62.1 N	131.8 W	92	Jacobus; Dutch astronomer	1852–1911
Van Wijk	62.8 S	118.8 E	32	Uco; Dutch astronomer	1924–1966
Vavilov	0.8 S	137.9 W	98	Nicolai; Russian botanist	1887–1943
Vening Meinesz	0.3 S	162.6 E	87	Felix; Dutch geophysicist	1887–1966
Ventris	4.9 S	158.0 E	95	Michael; British archæologist	1922–1956
Vernadsky	23.2 N	130.5 E	91	Vladimir; Russian mineralogist	1863–1945
Vesalius	3.1 S	114.5 E	61	Andreas; Belgian doctor	1514–1564
Vetchinkin	10.2 N	131.3 E	98	Vladimir; Russian physicist engineer	1888–1950
Vilev	6.1 S	144.4 E	45	Mikhail; Russian chemist	1893–1919
Volterra	50.8 N	132.2 E	52	Vito; Italian mathematician	1890–1940
von der Pahlen	24.8 N	132.7 W	56	Emanuel; German astronomer	1882–1952
von Kármán	44.8 S	175.9 E	180	Theodor; Hungarian aeronautical scientist	1881–1963
von Neumann	40.4 N	153.2 E	78	John; American mathematician	1903–1957
von Zeipel	42.6 N	141.6 W	83	Eduard Hugo; Swedish astronomer	1873–1959
Walker	26.0 S	162.2 W	32	American pilot	1921–1966
Waterman	25.9 S	128.0 E	76	Alan; American physicist	1892–1967
Watson	62.6 S	124.5 W	62	James; American astronomer	1838–1880
Weber	50.4 N	123.4 W	42	Wilhelm; German astronomer	1804–1891
Wegener	45.2 N	113.3 W	88	Alfred. Austrian meteorologist	1880–1930
Wells	40.7 M	122.8 E	114	H. G. (Herbert) Wells; English writer	1866–1946
Weyl	17.5 N	126.2 W	108	Hermann; German mathematician	1885–1955
White	44.6 S	158.3 W	39	Edward; American astronaut	1930–1967
Wiechert	84.5 S	163.0 E	41	Emil; German geophysicist	1861–1928
Wiener	40.8 N	146.6 E	120	Norbert; American mathematician	1894–1964
Wilsing	21.5 S	155.2 W	73	Johannes; German astronomer	1856–1943
Winkler	42.2 N	179.0 W	22	Johannes; German rocket scientist	1897–1947
Winlock	35.6 N	105.6 W	64	Joseph; American astronomer	1826–1875
Woltjer	45.2 N	159.6 W	46	Jan; Dutch astronomer	1891–1946

Table 3.15 (cont.)

Name	Lat. (°)	Long. (°)	Diameter (km)	Name origin	
Wood	43.0 N	120.8 W	78	Robert; American physicist	1868–1955
Xenophon	22.8 S	122.1 E	25	Greek natural philosopher	c. 430–354 BC
Yablochkov	60.9 N	128.3 E	99	Pavel; Russian electrical engineer	1847–1894
Yamamoto	58.1 N	160.9 E	76	Issei; Japanese astronomer	1889–1959
Zeeman	75.2 S	133.6 W	190	Pieter; Dutch physicist	1865–1943
Zelinsky	28.9 S	166.8 E	53	Nikolai; Russian chemist	1860–1953
Zernike	18.4 N	168.2 E	48	Frits; Dutch physicist	1888–1966
Zhiritsky	24.8 S	120.3 E	35	Georgi; Russian rocket scientist	1893–1966
Zhukovsky	7.8 N	167.0 W	81	Nikolai; Russian physicist	1847–1921
Zsigmontly	59.7 N	104.7 W	65	Richard; Austrian chemist	1865–1929
Zwicky	15.4 S	168.1 E	150	Swiss astrophysicist	1898–1974

THE FUTURE OF THE MOON

We know that at the present time tidal effects mean that the Moon is receding from the Earth at the rate of 3 cm per year, and that the Earth's rotation is slowing down. If this recession were maintained, then in 15 000 million years' time the Moon would move out to a distance of 540 000 km, and its orbital period would be equal to 47 of our present days; this would also be the length of the Earth's axial rotation period, so that the Moon would be motionless in our sky. Sir James Jeans, the famous twentieth-century astronomer, described this in graphic terms: 'The inhabitants of one of the hemispheres of the Earth will never see the Moon at all, while the other side will be lighted by it every night .... After this, tidal friction will no longer operate in the sense of driving the Moon further away from the Earth. The joint effect of solar and lunar tides will be to slow the Earth's rotation still further, the Moon at the same time gradually lessening its distance from the Earth. When it has finally, after unthinkable ages, been dragged down to within about 12 000 miles (19 000 km) of the Earth, the tides raised by the Earth in the solid body of the Moon will shatter the latter into fragments. These will form a system of tiny satellites revolving round the Earth in the same way as the particles of Saturn's rings revolve around Saturn. ...'

In fact this will not happen, simply because within a few thousand million years the Sun will have evolved into a red giant star, with fatal consequences for both the Earth and Moon. Their survival is unlikely, and their survival in their present form is impossible. Fortunately, there is no cause for immediate alarm; in our time, and perhaps as long as humanity lasts, both the Earth and Moon will remain very much as they are now.

ENDNOTE

1 Often spelled rilles: I have kept to the original spelling. They can also be known as clefts.

4 · Mercury

Mercury, the innermost planet, is also the smallest. It always stays in the same part of the sky as the Sun, and can therefore never be seen against a really dark background, and is not a conspicuous naked-eye object, though at its best it is actually brighter than any star. Its quick movements led to its being named after Hermes (Mercury), the fleet-footed Messenger of the Gods.

Data for Mercury are given in Table 4.1.

VULCAN

It was once thought that a planet existed closer to the Sun than the orbit of Mercury. It was even given a name – Vulcan, after the blacksmith of the gods. Only in the twentieth century was it finally found to be non-existent and relegated to the status of a ghost.

The story of Vulcan really goes back to 1781, when William Herschel discovered a new planet, Uranus, moving far beyond the orbit of Saturn. Over the years it was found that Uranus was not moving quite as it was expected to do; something was perturbing it, and mathematicians began to suspect that there might be yet another planet still further from the Sun. From these tiny perturbations a leading French astronomer, U. J. J. Le Verrier, worked out the position of the unknown world, and in 1846 J. Galle and H. D'Arrest, at the Berlin Observatory, discovered Neptune, very close to the position given by Le Verrier.

Mercury, too, was straying slightly from its predicted path, and Le Verrier decided the perturbations must be due to an inner planet. Obviously it would be very difficult to observe, because it would be so close to the Sun, but it might be caught in transit, crossing the solar disc. In 1859 Le Verrier published his first paper about the movements of Mercury, and a French amateur astronomer, a Dr Lescarbault, claimed that he had see the planet in transit. Le Verrier made haste to visit him – to find that Lescarbault doubled as the village carpenter; he used a small telescope, recorded his observations on planks of wood, planing them off when they were no longer needed. His timekeeper was a watch minus one of its hands. Amazingly, Le Verrier decided that the observation was genuine, and that the inner planet really existed. 'Vulcan' found its way into the books. It was said to be 1 000 000 km from the Sun, with a period of 19 days 7 hours; it was thought to be comparable in size with the Moon.

However, other astronomers who had been watching the Sun at the time of Lescarbault's observation had seen nothing at all, and Vulcan was generally discounted – though Le Verrier continued to believe in it (he died in 1877). It had been suggested that there might be a chance of success during a total solar eclipse, when the sky darkens, and there was a determined search during the eclipse of 29 July 1878, carried out by two well-known American astronomers, Lewis Swift and J. Watson. They reported some faint stars that they could not identify, but it is certain that these were ordinary stars, probably θ and ζ Cancri. Future searches were equally fruitless. Then, in the first two decades of the new century, came Einstein' theory of relativity, which at once accounted for the behaviour of Mercury. There was no need for Vulcan!

There may well be asteroids in the innermost part of the Solar System – 'Vulcanoids' – and if they do exist they will probably be found before long, but the search for Le Verrier's Vulcan has finally been given up. There is no substantial body closer-in than the orbit of Mercury.

EARLY OBSERVATIONS

Mercury has been known from very early times. The oldest observation that has come down to us is dated as 15 November 265 BC, when, according to Ptolemy, Mercury lay one lunar diameter away from a line joining the stars δ and ß Scorpii. Plato (*Republic*, ch. X, 14) commented upon the yellowish colour of the planet, though most naked-eye observers will probably call it white.

There is an oft-quoted story that the great astronomer Copernicus never saw Mercury in his life because of mists rising from the River Vistula, near his home in Toruń. I am quite sure that this story is wrong. Mercury is easy to find when well placed, and the skies are much more polluted now than they were in Copernicus's time (he died in 1543). In 1974, I spent three weeks at Toruń University, and saw Mercury clearly on eight consecutive evenings.

The phases of Mercury are easy to see telescopically. They were probably suspected in the first half of the seventeenthth century by Galileo, Simon Marius, and Martin van der Hove (= Hortensius), but we cannot be sure. They were definitely seen by Giovanni Zupus in 1639, and confirmed by Hevelius in 1644. Unlike Venus, Mercury is brightest when gibbous.

Mercury is by no means hard to identify when well placed, and can actually become brighter than any star, but can never be seen against a really dark sky. The maximum elongation from the Sun is 28 degrees. Elongations for the period 2010–2020 are given in Table 4.2.

TRANSITS OF MERCURY

Mercury, like Venus, can pass in transit across the face of the Sun and does so more frequently than does Venus, although during a transit it is not visible with the naked eye.

Table 4.1 *Mercury: data*

Distance from the Sun:
  aphelion 69.82 million km (0.47 a.u.)
  mean   57.91 million km (0.39 a.u.)
  perihelion 46.00 million km (0.31 a.u.)
Orbital eccentricity: 0.206
Orbital inclination: 7.00°
Orbital period: 58.6 days
Synodic period: 115.88 days
Diameter   4879 km
Surface area: 7.475×10^7 km^2
Polar compression: <0.0006
Mass, Earth = 1: 0.055
Volume, Earth = 1: 0.056
Density, water = 1: 5.427
Escape velocity: 4.25 km s^{-1}
Surface gravity, Earth = 1: 0.38
Rotation period; 87.97 days (synchronous)
Axial inclination: 2.1′
Albedo: 0.11
Mean surface temperature, °C: day +350, night −170
Extremes of surface temperature, °C: day +427, night −183
Maximum apparent magnitude: −1.9
Apparent diameter seen from Earth: max. 13.0″, min. 4.5″
Mean diameter of the Sun, seen from Mercury: 1° 22′ 40″
Maximum elongation from the Sun: 28.3°
Pole stars: north, o Draconis (mag. 4.7); south, α Pictoris (mag. 3.3)

Table 4.2 *Elongations of Mercury, 2010–2020*

Eastern
2010 Apr 8, Aug 7, Dec 1
2011 Mar 23, July 20, Nov 14
2012 Mar 5, July 1, Oct 26
2013 Feb 16, June 12, Oct 9
2014 Jan 31, May 25, Sept 21
2015 Jan 14, May 7, Sept 4, Dec 29
2016 Apr 18, Aug 16, Dec 11
2017 Apr 1, July 30, Nov 24
2018 Mar 15, July 12, Nov 6
2019 Feb 27, June 23, Oct 20
2020 Feb 10, June 4, Oct 1

Western
2010 Jan 27, May 26, Sept 19
2011 Jan 9, May 7, Sept 3, Dec 23
2012 Apr 18, Aug 16, Dec 4
2013 Mar 31, July 30 Nov 18
2014 Mar 14, July 12, Nov 1
2015 Feb 24, June 24, Oct 16
2016 Feb 7, June 5, Sept 28
2017 Jan 19, May 17, Sept 12
2018 Jan 1, Apr 29, Aug 26, Dec 15
2019 Apr 11, Aug 9, Nov 28
2020 Mar 24, July 22, Nov 10

Transits can occur only in May and November. May transits occur with Mercury near aphelion; at November transits Mercury is near perihelion, and November transits are the more frequent in the ratio of seven to three. The longest transits (those of May) may last for almost nine hours.

The first transit predictions were made by Johannes Kepler. He said (correctly) that both Mercury and Venus would transit in 1631, Mercury on 7 November and Venus on 8 December, The Mercury transit was seen by four observers, Pierre Gassendi, Jean-Baptiste Cysatus, Johann Quietanus and a Bavarian whose name has not been preserved. (The Venus transit was missed; it occurred during night-time in Europe, and there were virtually no observers in the opposite hemisphere.)

In 1677, Edmond Halley was at St Helena, making the first telescopic survey of the southern stars, and with his '24-foot' telescope observed a transit of Mercury. It occurred to him that, in theory, transits could be used to measure the length of the astronomical unit. Mercury's disc was inconveniently small, but the method was used for later transits of Venus, though the results were disappointing.

The transit of 6 November 1993 was observed at X-ray wavelengths. The observations were made from the Japanese satellite Yohkoh; Mercury blocked X-ray emissions from the corona, so appearing as a tiny dark hole in the X-ray corona. The transit of 15 November 1999 was exceptional. It was a 'grazing transit':

Mercury followed a short chord across the Sun's northeastern limb. In fact, over some parts of the Earth there was only a partial transit, as Mercury did not pass wholly on to the solar disc. The transit was total from Papua New Guinea, northeastern Australia, Hawaii, western South America and most of North America: partial from Antarctica and most of Australia. In New Zealand, the transit was total from North Island but partial from South Island. This situation will not recur for several centuries.

From my Selsey observatory I had a splendid view of the transit of 7 May 2003. I was impressed at the blackness of Mercury compared with a sunspot which happened to be on view.

Transits of Mercury could be seen from other planets – if you could get there; from Venus (3 June 2011, 18 December 2012, 17 December 2016); from Mars (10 May 2013, 4 June 2014, 15 April 2015); from Jupiter (28 November 2011, 11 January 2018); from Saturn (30 December 2011, 28 March 2012, 25 June 2012 and 22 September 2012). The year 2012 would be a good one for a Saturnian observer, but the apparent diameter of the Sun will be only 3.5 arc-minutes, and that of Mercury a mere 0.75 of an arc-second. However, at least the transit will last for about 8 hours! Transit dates for the period 1631–2100 are shown in Table 4.3.

OCCULTATIONS AND CONJUNCTIONS

Mercury can of course be occulted by the Moon. Occasionally Mercury itself may occult another planet; this last happened on 9 December 1808, when Mercury passed in front of Saturn. The

Table 4.3 *Transits of Mercury*

(a) 1631–2000

1631 Nov 7
1644 Nov 9
1651 Nov 3
1661 May 3
1664 Nov 4
1677 Nov 7
1690 Nov 10
1697 Nov 3
1707 May 5
1710 Nov 6
1723 Nov 9
1736 Nov 11
1740 May 2
1743 Nov 5
1753 May 6
1756 Nov 7
1769 Nov 9
1776 Nov 2
1782 Nov 12
1786 May 4
1789 Nov 5
1799 May 7
1802 Nov 9
1815 Nov 12
1822 Nov 5
1832 May 5
1835 Nov 7
1845 May 8
1848 Nov 9
1861 Nov 12
1868 Nov 5
1878 May 6
1881 Nov 8
1891 May 10
1894 Nov 10
1907 Nov 14
1914 Nov 7
1924 May 8
1927 Nov 10
1937 May 11
1940 Nov 11
1953 Nov 14
1957 May 6
1960 Nov 7
1970 May 9
1973 Nov 10
1986 Nov 13
1993 Nov 6
1999 Nov 15

(b) 2000–2100

Date	Mid-transit (GMT)
2003 May 7	07.53
2006 Nov 8	21.42
2016 May 9	14.59
2019 Nov 11	15.21
2032 Nov 13	08.55
2039 Nov 7	08.48
2049 May 7	14.26
2052 Nov 9	02.32
2062 May 10	21.39
2065 Nov 11	20.08
2078 Nov 14	13.44
2085 Nov 7	13.37
2095 May 8	21.09

Table 4.4 *Planetary occultations and close conjunctions involving Mercury*

	Date	GMT	Separation (″)	Elongation (°)
Neptune	1914 Aug 10	08.11	−30	18 W
Mars	1942 Aug 19	12.36	−20	16 E
Mars	1985 Sept 4	21.00	−46	16 W
Mars	1989 Aug 5	21.54	+47	18 E
Mars	2032 Aug 23	04.26	+16	13 W
Saturn	2037 Sept 15	21.32	+18	15 W
Neptune	2039 May 5	09.42	−39	13 W
Neptune	2050 June 4	06.46	−46	17 W
Neptune	2067 July 15	12.04	+13	18 W
Mars	2079 Aug 11	01.31	Occ.	11 W
Venus	2084 Dec 24	05.11	+48	17 W

next occasion will be on 11 August 2079, when Mercury will occult Mars. Mercury will occult Jupiter on 27 October 2088 and 7 April 2094, but the elongations will be less than 6°. Close planetary conjunctions involving Mercury are given in Table 4.4 for the period 1900–2100; all these occur more than 10° from the Sun and the separations are below 60 arcsec.

MAPS OF MERCURY

Mercury is a difficult object to study from Earth (and, incidentally, it cannot be studied at all with the Hubble Space Telescope, because it is too close to the Sun in the sky). The first serious telescopic observations were made in the late eighteenth century by Sir William Herschel, who, however, could make out no surface detail. At about the same time observations were made by J. H. Schröter, who recorded some surface patches and who believed that he had detected high mountains. It is certain that these patches were illusory.

The first attempt to produce a proper map was made by G. V. Schiaparelli, from Milan, using 21.8-cm and 49-cm refractors between 1881 and 1889. His method was to study the planet in daylight, when both it and the Sun were high above the horizon. Schiaparelli recorded various dark markings, and concluded that the rotation period must be synchronous – that is to say equal to Mercury's orbital period. This would mean that part of the planet would be in permanent sunlight and another part in permanent darkness, with an intervening 'twilight zone' over which the Sun would rise and set, always keeping fairly close to the horizon. Percival Lowell, at the Flagstaff Observatory in Arizona, drew a map in 1896 showing canal-like linear features, but these were completely non-existent.

The best pre-Space Age map was drawn by E. M. Antoniadi, using the 83-cm refractor at Meudon, near Paris; like Schiaparelli, he observed in daylight. The map was published in 1934, together with a book dealing with all aspects of the planet. (Surprisingly, the book was not translated into English until 1974, when I did so – by which time its interest was mainly historical. Although Antoniadi was Greek, he spent much of his life in France and wrote his book in French.) Antoniadi agreed with Schiaparelli that the rotation period must be synchronous, and he also believed the atmosphere to be dense enough to support obscurations. Both these conclusions are now known to be wrong.

Antoniadi's map showed various dark and bright features, and to these he gave names; there was a degree of agreement with Schiaparelli's map, although there were very marked differences. Antoniadi's names are given Table 4.5. Yet they referred only to albedo features, and the map is not accurate enough for the names to be retained – which was not Antoniadi's fault; he was certainly the best planetary observer of his time.

Modern maps are based almost entirely on images sent by space-craft which have flown past Mercury; the first of which was Mariner 10, launched from Cape Canaveral on 3 November 1973.

Mercury has proved to be a world of craters, mountains, low plains (planitiæ), scarps (rupes), ridges (dorsa) and valleys (valles). The craters are named after famous artists, musicians and authors: planitiæ after the names for Mercury in different languages; the rupes after ships of discovery or scientific expeditions, and valleys after radio telescopes. Only three astronomers are commemorated on Mercury; Antoniadi and Schaparelli, and also G. P. Kuiper, who was closely connected with planetary space research. A selected list of named features on Mercury is given in Table 4.6. Figure 4.1 gives a map of the surface.

AXIAL ROTATION

For many years the synchronous rotation period favoured by Schiaparelli and Antoniadi was accepted, but in 1962, W. E. Howard and his colleagues at Michigan made observations at infrared wavelengths, and found that the dark side was much warmer than it would be if it never received any sunlight. In 1965 the non-synchronous period was confirmed by R. Dyce and G. Pettengill, using the large radio telescope at Arecibo in Puerto Rico. The true period is 58.6 days – two-thirds of the orbital period. There is no

Table 4.5 *Albedo features on Antoniadi's map*

Name	Lat. (°)	Long. (°W)
Apollonia	45.0 N	315.0
Aurora (Victoria Rupes region)	45.0 N	90.0
Australia (Bach region)	72.5 S	0.0
Borea (Borealis region)	75.0 N	0.0
Caduceata (Shakespeare region)	45.0 N	135.0
Cyllene	41.0 S	270.0
Heliocaminus	40.0 N	170.0
Hesperis	45.0 S	355.0
Liguria	45.0 N	225.0
Pentas	5.0 N	310.0
Phæthontias (Tolstoy region)	0.0 N	167.0
Pieria	0.0 N	270.0
Pleias Gallia	25.0 N	130.0
Sinus Argiphontæ	10.0 S	335.0
Solitudo Admetei	55.0 N	90.0
Solitudo Alarum	15.0 S	290.0
Solitudo Atlantis	35.0 S	210.0
Solitudo Criophori	0.0 N	230.0
Solitudo Helii	10.0 S	180.0
Solitudo Hermæ Trismegisti (Discovery Rupes region)	45.0 S	45.0
Solitudo Horarum	25.0 N	115.0
Solitudo Iovis	0.0 N	0.0
Solitudo Lycaonis (Beethoven region)	0.0 N	107.0
Solitudo Maiæ	15.0 S	155.0
Solitudo Martis	35.0 S	100.0
Solitudo Neptuni	30.0 N	150.0
Solitudo Persephones	41.0 S	225.0
Solitudo Phœnicis	25.0 N	225.0
Solitudo Promethei (Michelangelo region)	45.0 S	142.5
Tricrena (Kuiper region)	0.0 N	36.0

area of permanent daylight, no region of permanent night, and no twilight zone. The axial inclination is negligible, so that Mercury spins in an almost 'upright' position with respect to its orbital plane. The planet's calendar is frankly weird, and there is one relationship which, understandably, misled Schiaparelli and Antoniadi. It is best explained by making six points:

1. The *synodic period* of Mercury – that is to say, the time that elapses between successive appearances at the same phase – is on average, 116 Earth days. If Mercury is 'new' on a particular date it will again be new 116 days later.
2. The rotation period (58.7 Earth days) is equal to two thirds of the revolution period (88 Earth days).
3. It follows that to an observer at a fixed point on Mercury the interval between one sunrise and the next will be 176 Earth days, or 2 Mercurian years.

Table 4.6 *Geological features on Mercury*

Feature	Lat. (°)	Long. (° W)	Named after
Montes			
Caloris Montes	39.4 N	187.2	Latin for 'mountains of heat'
Dorsa (ridges)			
Antoniadi Dorsum	25.1 N	30.5	E. M. Antoniadi
Schiaparelli Dorsum	23 N	164.1	Giovanni Schiaparelli
Fossæ			
Pantheon Fossæ	30.5 N	197.0	The Pantheon, Rome
Valles (valleys)			
Arecibo Vallis	27.5 S	28.4	Arecibo Observatory
Goldstone Vallis	15.8 S	31.7	Goldstone Observatory
Haystack Vallis	4.7 N	46.2	Haystack Observatory
Simeiz Vallis	13.2 S	64.3	Simeiz Observatory
Planes (plains)			
Borealis Planitia	73.4 N	79.5	Latin for 'northern plain'
Budh Planitia	22 N	150.9	Hindu word for Mercury
Caloris Planitia	30.5 N	189.8	Latin for 'heat's plain'
Odin Planitia	23.3 N	171.6	Norse god Odin
Sobkou Planitia	39.9 N	129.9	Messenger god
Suisei Planitia	59.2 N	150.8	Japanese for Mercury
Tir Planitia	0.8 N	176.1	Norse for Mercury
Rupes (escarpments)			
Adventure Rupes	−65.1	65.5	HMS *Adventure*, ship of Captain Cook
Astrolabe Rupes	−42.6	70.7	*Astrolabe*, ship of Jules Dumont d'Urville
Beagle Rupes	−1.9	258.89	HMS *Beagle*, ship on which Charles Darwin sailed
Discovery Rupes	−56.3	38.3	HMS *Discovery*, ship of Captain Cook
Endeavour Rupes	37.5	31.3	HM Bark *Endeavour*, ship of Captain Cook
Fram Rupes	−56.9	93.3	*Fram*, ship of Fridtjof Nansen, Otto Sverdrup and Roald Amundsen
Gjøa Rupes	−66.7	159.3	*Gjøa*, ship of Roald Amundsen
Heemskerck Rupes	25.9	125.3	Ship of Abel Tasman
Hero Rupes	−58.4	171.4	Ship of Nathaniel Palmer
Mirni Rupes	−37.3	39.9	Ship of Fabian von Bellingshausen
Pourquoi-Pas Rupes	−58.1	156	*Pourquoi Pas? IV*, ship of Jean-Baptiste Charcot
Resolution Rupes	−63.8	51.7	HMS *Resolution*, ship of Captain Cook
Santa Maria Rupes	5.5	19.7	*Santa María*, ship of Christopher Columbus
Victoria Rupes	50.9	31.1	*Victoria*, ship of Ferdinand Magellan

Crater	Latitude (°)	Longitude (°)	Diameter (km)	Named after
Abedin	61.7 N	10.2 W	110.0	Zainul Abedin, Bangladeshi painter
Abu Nuwas	17.4 N	20.4 W	116.0	Abu Nuwas, Arabic poet
Africanus Horton	51.5 S	41.2 W	135.0	Africanus Horton, Sierra Leonean writer
Ahmad Baba	58.5 N	126.8 W	127.0	Ahmad Baba al Massufi, West African writer
Al-Akhtal	59.2 N	97.0 W	102.0	Akhtal, Arab poet
Al-Hamadhani	38.8 N	89.7 W	186.0	Badi' az-Zaman al-Hamadhani, Arab writer
Al-Jāhiz	1.2 N	21.5 W	91.0	Al-Jāhiz, Arab author
Alencar	63.5 S	103.5 W	120.0	José de Alencar, Brazilian novelist
Amaral	26.4 S	242.3 W	106.0	Tarsila do Amaral, Brazilian artist
Amru Al-Qays	12.3 N	175.6 W	50.0	Imru Al-Qays Ibn Hujr, Arabic poet
Andal	47.7 S	37.7 W	108.0	Aandaal, Tamil writer
Apollodorus	30.58 N	197.01 W	41.0	Apollodorus of Damascus, Ancient Greek architect
Aristoxenes	82.0 N	11.4 W	69.0	Aristoxenus, Ancient Greek writer
Aśvaghosa	10.4 N	21.0 W	90.0	Aśvaghosa, Sanskrit, poet

Table 4.6 (cont.)

Crater	Latitude (°)	Longitude (°)	Diameter (km)	Named after
Atget	25.65 N	193.93 W	100.0	Eugène Atget, French photographer
Bach	68.5 S	103.4 W	214.0	Johann Sebastian Bach, German composer
Balagtas	22.6 S	13.7 W	98.0	Francisco Balagtas, Filipino poet
Balzac	10.3 N	144.1 W	80.0	Honoré de Balzac, French writer
Bartók	41.3 S	162.8 W	128.0	Béla Bartók, Hungarian composer
Barma	29.6 S	134.6 W	112.0	Postnik 'Barma' Yakovlev, Russian architect
Bashō	32.7 S	169.7 W	80.0	Matsuo Bashō, Japanese poet
Beckett	40.1 S	248.8 W	57.0	Clarice Beckett, Australian painter
Beethoven	20.8 S	123.6 W	643.0	Ludwig van Beethoven, German composer
Belinskij	76.0 S	103.4 W	70.0	Vissarion Belinsky, Russian literary critic
Bello	18.9 S	120.0 W	129.0	Andrés Bello, South American writer
Benoit	7.6 N	256.2 W	43.0	Rigaud Benoit, Haitian artist
Berkel	13.6 S	333.5 W	21.0	Sabri Berkel, Turkish painter
Bernini	79.2 S	136.5 W	146.0	Gianlorenzo Bernini, Italian, sculptor
Bjørnson	73.1 N	109.2 W	88.0	Bjørnstjerne Bjørnson, Norwegian poet
Boccaccio	80.7 S	29.8 W	142.0	Giovanni Boccaccio, Italian writer
Boethius	0.9 S	73.3 W	129.0	Anicius Manlius Severinus Boethius, Roman philosopher
Botticelli	63.7 N	109.6 W	143.0	Sandro Botticelli, Italian artist
Brahms	58.5 N	176.2 W	96.0	Johannes Brahms, German composer
Bramante	47.5 S	61.8 W	159.0	Donato Bramante, Italian architect
Bronte	38.7 N	125.9 W	60.0	The Brontë family, English writers and artists
Bruegel	49.8 N	107.5 W	75.0	Pieter Bruegel the Elder, Flemish painter
Brunelleschi	9.1 S	22.2 W	134.0	Filippo Brunelleschi, Italian architect
Burns	54.4 N	115.7 W	45.0	Robert Burns, Scottish poet
Byron	8.5 S	32.7 W	105.0	Lord Byron, English poet
Callicrates	66.3 S	32.6 W	70.0	Kallicrates, Ancient Greek architect
Camões	70.6 S	69.6 W	70.0	Luís de Camões, Portuguese writer
Carducci	36.6 S	89.9 W	117.0	Giosuè Carducci, Italian poet
Calvino	3.9 S	56.0 N	68.0	Italo Calvino, Italian writer
Cervantes	74.6 S	122.0 W	181.0	Miguel de Cervantes, Spanish writer
Cézanne	8.5 S	123.4 W	75.0	Paul Cézanne, French painter
Chaikovskij	7.4 N	50.4 W	165.0	Pyotr Ilyich Tchaikovsky, Russian composer
Chao Meng-Fu	87.3 S	134.2 W	167.0	Zhao Mengfu, Chinese artist
Chekov	36.2 S	61.5 W	199.0	Anton Chekhov, Russian writer
Chiang K'ui	13.8 N	102.7 W	35.0	Jiang Kui, Chinese poet
Chong Ch'ol	46.4 N	116.2 W	162.0	Jeong Cheol, Korean poet
Chopin	65.1 S	123.1 W	129.0	Frédéric Chopin, Polish composer
Chu Ta	2.2 N	105.1 W	110.0	Zhu Da, Chinese painter
Coleridge	55.9 S	66.7 W	110.0	Samuel Taylor Coleridge, English poet
Copley	38.4 S	85.2 W	30.0	John Singleton Copley, American painter
Couperin	29.8 N	151.4 W	80.0	The Couperin family of French musicians
Cunningham	30.48 N	203.07 W	37.0	Imogen Cunningham, American photographer
Dali	45.3 N	240.6 W	175.0	Salvador Dali, Spanish painter
Darío	26.5 S	10.0 W	151.0	Rubén Darío, Nicaraguan writer
de Graft	22.1 N	358.0 W	65.0	Joe de Graft, Ghanaian playwright
Degas	37.4 N	126.4 W	45.0	Edgar Degas, French artist
Delacroix	44.7 S	129.0 W	146.0	Eugène Delacroix, French artist
Derain	8.7 S	340.3 W	190.0	André Derain, French artist
Derzhavin	44.9 N	35.3 W	159.0	Gavril Romanovich Derzhavin, Russian poet
Desprez	80.8 N	90.7 W	50.0	Josquin Desprez, Franco–Flemish composer
Dickens	72.9 S	153.3 W	78.0	Charles Dickens, English novelist
Donne	2.8 N	13.8 W	88.0	John Donne, English poet

Table 4.6 (cont.)

Crater	Latitude (°)	Longitude (°)	Diameter (km)	Named after
Dostœvskij	45.1 S	176.4 W	411.0	Fyodor Dostoyevsky, Russian novelist
Dowland	53.5 S	179.5 W	100.0	John Dowland, English composer
Dürer	21.9 N	119.0 W	180.0	Albrecht Dürer, German artist
Dvořák	9.6 S	11.9 W	82.0	Antonín Dvořák, Czech composer
Eastman	9.6 N	234.3 W	80.0	Charles Eastman, Sioux author
Eitoku	42.7 N	19.2 W	75.0	Kano Eitoku, Japanese artist
Equiano	22.1 S	156.9 W	100.0	Olaudah Equiano, West African writer
Eminescu	10.79 N	245.87 W	125.0	Mihail Eminescu, Romanian poet
Enwonwu	9.9 S	238.4 W	38.0	Ben Enwonwu, Nigerian painter
Equiano	40.2 S	30.7 W	99.0	Olaudah Equiano, Benin writer
Fet	4.9 S	179.9 W	24.0	Afanasy Fet, Russian poet
Flaubert	13.7 S	72.2 W	95.0	Gustave Flaubert, French author
Futabatei	16.2 S	83.0 W	66.0	Futabatei Shimei, Japanese author
Gainsborough	36.1 S	183.3 W	100.0	Thomas Gainsborough, English painter
Gauguin	66.3 N	96.3 W	72.0	Paul Gauguin, French artist
Ghiberti	48.4 S	80.2 W	123.0	Lorenzo Ghiberti, Italian sculptor
Gibran	35.5 N	110.4 W	102.0	Khalil Gibran, Lebanese American poet and artist
Giotto	12.0 N	55.8 W	150.0	Giotto di Bondone, Italian painter
Glinka	14.8 N	111.7 W	86.0	Mikhail Glinka, Russian composer
Gluck	37.3 N	18.1 W	105.0	Christoph Willibald Gluck, Austrian composer
Goethe	78.5 N	44.5 W	383.0	Johann Wolfgang von Goethe, German writer
Gogol	28.1 S	146.4 W	87.0	Nikolai Gogol, Russian playwright
Goya	7.2 S	152.0 W	135	Francisco Goya, Spanish artist
Grieg	51.1 N	14.0 W	65.0	Edvard Grieg, Norwegian composer
Guido d'Arezzo	38.7 S	18.3 W	66.0	Guido of Arezzo, Italian music theorist
Hals	54.8 S	115.0 W	100.0	Frans Hals, Dutch painter
Han Kan	71.6 S	143.8 W	50.0	Han Gan, Chinese painter
Handel	3.4 N	33.8 W	166.0	George Frederic Handel, German composer
Harunobu	15.0 N	140.7 W	110.0	Suzuki Harunobu, Japanese artist
Hauptmann	23.7 S	179.9 W	120.0	Gerhart Hauptmann, German playwright
Hawthorne	51.3 S	115.1 W	107.0	Nathaniel Hawthorne, American novelist
Haydn	27.3 S	71.6 W	270.0	Joseph Haydn, Austrian composer
Heine	32.6 N	124.1 W	75.0	Heinrich Heine, German poet
Hemingway	17.5 N	2.9 W	130.0	Ernest Hemingway, American writer
Hesiod	58.5 S	35.0 W	107.0	Hesiod, Ancient Greek poet
Hiroshige	13.4 S	26.7 W	138.0	Ando Hiroshige, Japanese artist
Hitomaro	16.2 S	15.8 W	107.0	Kakinomoto Hitomaro, Japanese poet
Hodgkins	29.2 N	341.9 W	20.0	Frances Hodgkins, New Zealand painter
Holbein	35.6 N	28.9 W	113.0	Hans Holbein the Younger, German artist
Holberg	67.0 S	61.1 W	61.0	Ludvig Holberg, Danish writer
Homer	1.2 S	36.2 W	314.0	Homer, Ancient Greek poet
Horace	68.9 S	52.0 W	58.0	Horace, Roman poet
Hovnatanian	7.6 S	187.5 W	34.0	Hakop Hovnatanian, Armenian painter
Hugo	38.9 N	47.0 W	198.0	Victor Hugo, French writer
Ibsen	24.1 S	35.6 W	159.0	Henrik Ibsen, Norwegian playwright
Ictinos	79.1 S	165.2 W	119.0	Iktinos, Ancient Greek architect
Imhotep	18.1 S	37.3 W	159.0	Imhotep, Ancient Egyptian architect
Ives	32.9 S	111.4 W	20.0	Charles Ives, American composer
Izquierdo	1.6 S	253.1 W	170.0	María Izquierdo, Mexican painter
Janáček	56.0 N	153.8 W	47.0	Leoš Janáček, Czech composer
Jókai	72.4 N	135.3 W	106.0	Mór Jókai, Hungarian writer
Judah Ha-Levi	10.9 N	107.7 W	80.0	Yehuda Halevi, Spanish–Jewish writer

Table 4.6 (cont.)

Crater	Latitude (°)	Longitude (°)	Diameter (km)	Named after
Kālidāsa	18.1 S	179.2 W	107.0	Kālidāsa, Sanskrit writer
Keats	69.9 S	154.5 W	115.0	John Keats, English poet
Kenkō	21.5 S	16.1 W	99.0	Yoshida Kenkō, Japanese writer
Kertész	27.44 N	214.06 W	33.0	Andre Kertesz, Hungarian photographer
Khansa	59.7 S	51.9 W	111.0	Al-Khansa, Arabic poet
Kōshō	60.1 N	138.2 W	65.0	Kōshō, Japanese sculptor
Kuan Han-ch'ing	29.4 N	52.4 W	151.0	Guan Hanqing, Chinese playwright
Kuiper	11.3 S	31.1 W	62.0	Gerard Kuiper, American astronomer
Kunisada	1.8 N	247.6 W	280.0	Utagawa Kunisada, Japanese woodblock printmaker
Kurosawa	53.4 S	21.8 W	159.0	Kinko Kurosawa Japanese musician
Lange	6.4 N	260.0 W	180.0	Dorothea Lange, American photographer
Leopardi	73.0 S	180.1 W	72.0	Giacomo Leopardi, Italian writer
Lermontov	15.2 N	48.1 W	152.0	Mikhail Lermontov, Russian writer
Lessing	28.7 S	89.7 W	100.0	Gotthold Ephraim Lessing, German dramatist
Li Ch'ing-Chao	77.1 S	73.1 W	61.0	Li Qingzhao, Chinese writer
Li Po	16.9 N	35.0 W	120.0	Li Bai, Chinese poet
Liang K'ai	40.3 S	182.8 W	140.0	Liang Kai, Chinese artist
Liszt	16.1 S	168.1 W	85.0	Franz Liszt, Hungarian pianist
Lu Hsun	0.0	23.4 W	98.0	Lu Xun, Chinese writer
Lysippos	0.8 N	132.5 W	140.0	Lysippos, Ancient Greek sculptor
Ma Chih-Yuan	60.4 S	78.0 W	179.0	Ma Zhiyuan, Chinese writer
Machaut	1.9 S	82.1 W	106.0	Guillaume de Machaut, French poet and composer
Mahler	20.0 S	18.7 W	103.0	Gustav Mahler, Bohemian composer
Mansart	73.2 N	118.7 W	95.0	Jules Hardouin Mansart, French architect
Mansur	47.8 N	162.6 W	100.0	Ustad Mansur, Mughal artist
March	31.1 N	175.5 W	70.0	Ausias March, Catalan poet
Mark Twain	11.2 S	137.9 W	149.0	Mark Twain, American novelist
Martí	75.6 S	164.6 W	68.0	José Martí, Cuban writer
Martial	69.1 N	177.1 W	51.0	Martial, Roman poet
Matabei	39.7 S	13.9 W	24.0	Iwasa Matabei, Japanese painter
Matisse	24.0 S	89.8 W	186.0	Henri Matisse, French painter
Melville	21.5 N	10.1 W	154.0	Herman Melville, American novelist
Mena	0.2 S	124.4 W	52.0	Juan de Mena, Spanish poet
Mendes Pinto	61.3 S	17.8 W	214.0	Fernão Mendes Pinto, Portuguese writer
Michelangelo	45.0 S	109.1 W	216.0	Michelangelo, Italian artist
Mickiewicz	23.6 N	103.1 W	100.0	Adam Mickiewicz, Polish writer
Milton	26.2 S	174.8 W	186.0	John Milton, English poet
Mistral	4.5 N	54.0 W	110.0	Gabriela Mistral, Chilean poet
Mofolo	37.7 S	28.2 W	114.0	Thomas Mofolo, Lesotho writer
Molière	15.6 N	16.9 W	132.0	Molière, French playwright
Monet	44.4 N	10.3 W	303.0	Claude Monet, French artist
Monteverdi	63.8 N	77.3 W	138.0	Claudio Monteverdi, Italian composer
Moody	13.1 S	215.4 W	80.0	Ronald Moody, Jamaican painter
Mozart	8.0 N	190.5 W	225.0	Wolfgang Amadeus Mozart, Austrian composer
Munch	40.6 N	207.3 W	54.0	Edvard Munch, Norwegian painter
Munkácsy	21.9 N	259.1 W	180.0	Mihály Munkácsy, Hungarian painter
Murasaki	12.6 S	30.2 W	130.0	Murasaki Shikibu, Japanese writer
Mussorgskij	32.8 N	96.5 W	125.0	Modest Mussorgskij, Russian composer
Myron	70.9 N	79.3 W	31.0	Myron, Ancient Greek sculptor
Nampeyo	40.6 S	50.1 W	52.0	Nampeyo, Hopi potter
Navoi	59.0 N	200.0 W	66.0	Ali-Shir Nava'i, Uzbek poet
Nawahi	36.1 N	214.9 W	34.0	Joseph Nawahi, Hawaiian painter

Table 4.6 (cont.)

Crater	Latitude (°)	Longitude (°)	Diameter (km)	Named after
Neruda	52.47 S	234.55 W	110.0	Pablo Neruda, Chilean poet
Nervo	43.0 N	179.0 W	63.0	Amado Nervo, Mexican poet
Neumann	37.3 S	34.5 W	120.0	Johann Balthasar Neumann, German architect
Nizami	71.5 N	165.0 W	76.0	Nezami, Persian poet
Ōkyo	69.1 S	75.8 W	65.0	Maruyama Ōkyo, Japanese painter
Oskison	60.6 N	215.3 W	120.0	John Milton Oskison, Cherokee author
Ovid	69.5 S	22.5 W	44.0	Ovid, Roman poet
Petrarch	30.6 S	26.2 W	171.0	Petrarch, Italian poet
Phidias	8.7 N	149.3 W	160.0	Phidias, Ancient Greek artist and architect
Philoxenus	8.7 S	111.5 W	90.0	Philoxenus of Cythera, Ancient Greek poet
Pigalle	38.5 S	9.5 W	154.0	Jean-Baptiste Pigalle, French sculptor
Po Chu-I	7.2 S	165.1 W	68.0	Bai Juyi, Chinese poet
Po Ya	46.2 S	20.2 W	103.0	Bo Ya, Chinese musician
Poe	44.0 N	201.2 W	75.0	Edgar Allan Poe, American poet
Polygnotus	0.3 S	68.4 W	133.0	Polygnotus, Ancient Greek painter
Praxiteles	27.3 N	59.2 W	182.0	Praxiteles, Ancient Greek sculptor
Proust	19.7 N	46.7 W	157.0	Marcel Proust, French novelist
Puccini	65.3 S	46.8 W	70.0	Giacomo Puccini, Italian composer
Pushkin	66.3 S	22.4 W	231	Pushkin, Russian writer
Purcell	81.3 N	146.8 W	91.0	Henry Purcell, English composer
Qi Baishi	4.2 S	196.0 W	15.0	Qi Baishi, Chinese painter
Rabelais	61.0 S	62.4 W	141.0	François Rabelais, French writer
Rachmaninoff	27.6 N	302.4 W	290.0	Sergei Rachmaninoff, Russian composer
Raden Saleh	2.2 N	201.3 W	25.0	Raden Saleh, Javanese painter
Raditladi	27.28 N	240.93 W	257.0	Leetile Disang Raditladi, Botswanan writer
Rajnis	4.5 N	95.8 W	82.0	Rainis, Latvian writer
Rameau	54.9 S	37.5 W	51.0	Jean Philippe Rameau, French composer
Raphael	19.9 S	75.9 W	343.0	Raphael, Italian artist
Ravel	12.0 S	38.0 W	75.0	Maurice Ravel, French composer
Rembrandt	33.2 S	271.8 W	720.0	Rembrandt, Dutch artist
Renoir	18.6 S	51.5 W	246.0	Pierre-Auguste Renoir, French artist
Repin	19.2 S	63.0 W	107.0	Ilya Yefimovich Repin, Russian artist
Riemenschneider	52.8 S	99.6 W	145.0	Tilman Riemenschneider, German sculptor
Rilke	45.2 S	12.3 W	86.0	Rainer Maria Rilke, German poet
Rimbaud	62.0 S	148.0 W	85.0	Arthur Rimbaud, French poet
Rodin	21.1 N	18.2 W	229.0	Auguste Rodin, French sculptor
Rubens	59.8 N	74.1 W	175.0	Peter Paul Rubens, French artist
Rublev	15.1 S	156.8 W	132.0	Andrei Rublev, Russian icon painter
Rudaki	4.0 S	51.1 W	120.0	Rudaki, Persian poet
Rude	32.8 S	79.6 W	75.0	François Rude, French sculptor
Rumi	24.1 S	104.7 W	75.0	Mawlana Rumi, Persian poet
Saadi	78.6 S	56.0 W	68.0	Saadi, Persian poet
Saikaku	72.9 N	176.3 W	88.0	Ihara Saikaku, Japanese poet
Sander	42.59 N	205.6 W	50.0	August Sander, German photographer
Sarmiento	29.8 S	187.7 W	145.0	Domingo Faustino Sarmiento, Argentinian writer
Sayat-Nova	28.4 S	122.1 W	158.0	Sayat-Nova, Armenian poet
Scarlatti	40.5 N	100.0 W	129.0	Domenico Scarlatti, Alessandro Scarlatti, Italian composers
Schoenberg	16.0 S	135.7 W	29.0	Arnold Schoenberg, Austrian composer
Schubert	43.4 S	54.3 W	185.0	Franz Schubert, Austrian composer
Scopas	81.1 S	172.9 W	105.0	Scopas, Ancient Greek sculptor and architect
Sei	64.3 S	89.1 W	113.0	Sei Shōnagon, Japanese writer
Shakespeare	49.7 N	150.9 W	370.0	William Shakespeare, English writer

Table 4.6 (cont.)

Crater	Latitude (°)	Longitude (°)	Diameter (km)	Named after
Shelley	47.8 S	127.8 W	164.0	Percy Bysshe Shelley, English poet
Sher-Gil	45.1 S	225.5 W	73.0	Amrita Sher-Gil, Indian painter
Shevchenko	53.8 S	46.5 W	137.0	Taras Shevchenko, Ukranian poet
Sholem Aleichem	50.4 N	87.7 W	200.0	Sholom Aleichem, Yiddish writer
Sibelius	49.6 S	144.7 W	90.0	Jean Sibelius, Finnish composer
Simonides	29.1 S	45.0 W	95.0	Simonides of Ceos, Greek poet
Smetana	48.5 S	70.2 W	190.0	Bedřich Smetana, Czech composer
Snorri	9.0 S	82.9 W	19.0	Snorri Sturluson, Icelandic poet
Sophocles	7.0 S	145.7 W	150.0	Sophocles, Ancient Greek dramatist
Sor Juana	49.0 N	23.9 W	93.0	Sor Juana Inez de la Cruz, Mexican writer
Sōseki	38.9 N	37.7 W	90.0	Natsume Sōseki, Japanese novelist
Sōtatsu	49.1 S	18.1 W	165.0	Tawaraya Sōtatsu, Japanese artist
Spitteler	68.6 S	61.8 W	68.0	Carl Spitteler, Swiss poet
Stravinsky	50.5 N	73.5 W	190.0	Igor Stravinsky, Russian composer
Strindberg	53.7 N	135.3 W	190.0	August Strindberg, Swedish writer
Sullivan	16.9 S	86.3 W	145.0	Louis Sullivan, American architect
Sur Das	47.1 S	93.3 W	132.0	Surdas, Hindu poet
Surikov	37.1 S	124.6 W	120.0	Vasily Surikov, Russian artist
Sveinsðottir	2.58 S	259.96 W	220.0	Júlíana Sveinsðottir, Icelandic artist
Takanobu	30.8 N	108.2 W	80.0	Fujiwara Takanobu, Japanese poet
Takayoshi	37.5 S	163.1 W	139.0	Takayoshi, Japanese painter
Tansen	3.9 N	70.9 W	34.0	Tansen, Hindustani composer
Thakur	−2.5	64	115	Rabindranath Tagore, Indian writer
Theophanes	3.0 S	63.5 W	118.0	Theophanes the Greek, icon painter
Thoreau	5.9 N	132.3 W	80.0	Henry David Thoreau, American poet
Tintoretto	48.1 S	22.9 W	92.0	Tintoretto, Italian artist
Titian	3.6 S	42.1 W	121.0	Titian, Italian artist
To Ngoc Van	52.3 N	110.7 W	63.0	To Ngoc Van, Vietnamese painter
Tolstoj	16.3 S	163.5 W	390.0	Leo Tolstoy, Russian writer
Ts'ai Wen-chi	22.8 N	22.2 W	119.0	Cai Wenji, Chinese poet and composer
Ts'ao Chan	13.4 S	142.0 W	110.0	Cao Xueqin, Chinese novelist
Tsurayuki	63.0 S	21.3 W	87.0	Tsurayuki Kino, Japanese writer
Tung Yuan	73.6 N	55.0 W	64.0	Dong Yuan, Chinese artist
Turgenev	65.7 N	135.0 W	116.0	Ivan Turgenev, Russian writer
Tyāgarāja	3.7 N	148.4 W	105.0	Tyāgarāja, Indian composer
Unkei	31.9 S	62.7 W	123.0	Unkei, Japanese sculptor
Ustad Isa	32.1 S	165.3 W	136.0	Ustad Isa, architect
Valmiki	23.5 S	141.0 W	221.0	Valmiki, Indian poet
Van Dijk	76.7 N	163.8 W	105.0	Anthony van Dijk, Flemish artist
Van Eyck	43.2 N	158.8 W	282.0	Jan van Eyck, Dutch artist
Van Gogh	76.5 S	134.9 W	104.0	Vincent van Gogh, Dutch artist
Velázquez	37.5 N	53.7 W	129.0	Diego Velázquez, Spanish artist
Verdi	64.7 N	168.6 W	163.0	Giuseppe Verdi, Italian composer
Vincente	56.8 S	142.4 W	98.0	Gil Vicente, Portuguese writer
Vivaldi	13.7 N	85.0 W	213.0	Antonio Vivaldi, Italian composer
Vlaminck	28.0 N	12.7 W	97.0	Maurice de Vlaminck, French painter
Vyāsa	48.3 N	81.1 W	290.0	Vyāsa, Indian poet
Wagner	67.4 S	114.0 W	140.0	Richard Wagner, German composer
Wang Meng	8.8 N	103.8 W	165.0	Wang Meng, Chinese artist
Wergeland	38.0 S	56.5 W	42.0	Henrik Wergeland, Norwegian writer
Whitman	41.4 N	110.4 W	70.0	Walt Whitman, American poet
Wren	24.3 N	35.2 W	221.0	Christopher Wren, English architect

Table 4.6 (cont.)

Crater	Latitude (°)	Longitude (°)	Diameter (km)	Named after
Xiao Zhao	10.64 N	236.21 W	23.0	Xiao Zhao, Chinese artist
Yeats	9.2 N	34.6 W	100.0	William Butler Yeats, Irish poet
Yun Son-Do	72.5 S	109.4 W	68.0	Yun Sondo, Korean poet
Zeami	3.1 S	147.2 W	120.0	Zeami Motokiyo, Japanese playwright
Zola	50.1 N	177.3 W	80.0	Emile Zola, French novelist

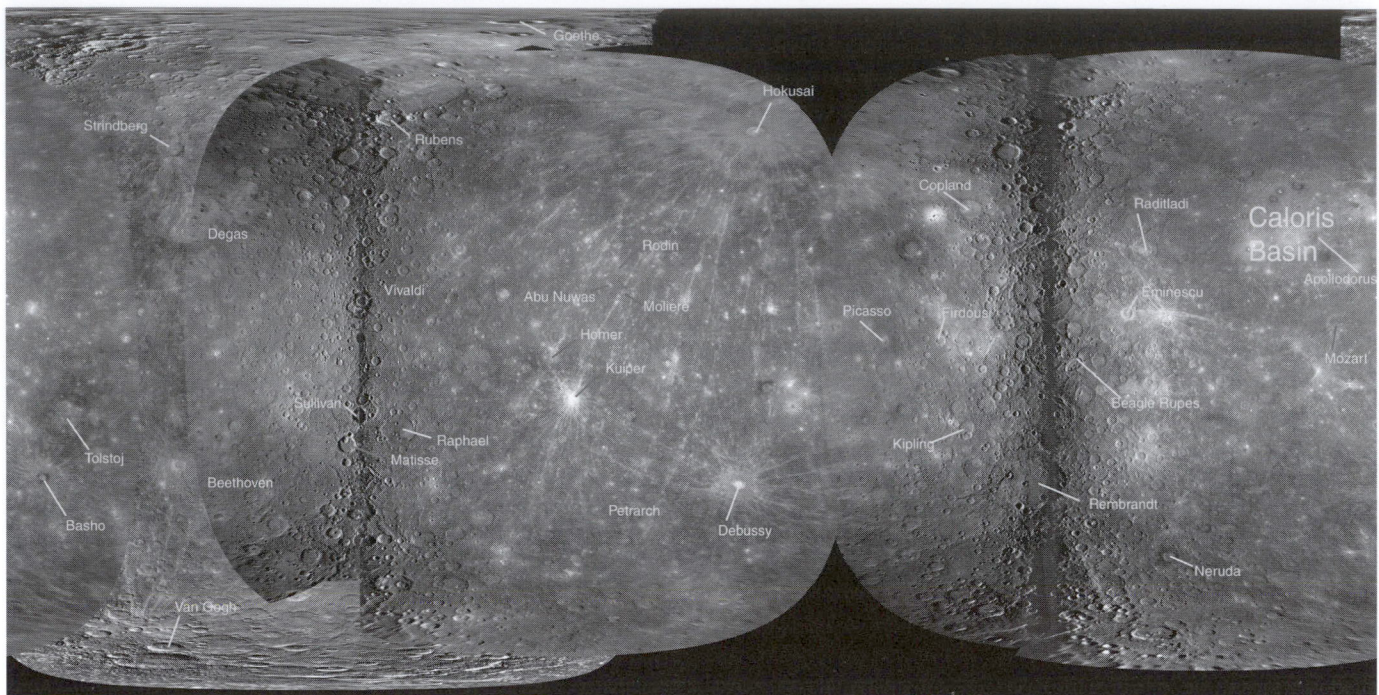

Figure 4.1 Surface of Mercury

4. This interval, 176 Earth days, is approximately equal to 1½ synodic periods.

5. From this, it will be found that after every three synodic periods, the same face of Mercury will be seen at the same phase.

6. Now for the coincidence – if coincidence it be! Three synodic periods of Mercury add up to approximately one Earth year. Consequently, the most favourable times for looking at Mercury recur every three synodic periods. Glance back at point 5. You will realise that every time Mercury is best placed for observation, we see the same hemisphere, with the same markings, at the same positions on the disc. Antoniadi made the best use of his opportunities, but even with the great Meudon refractor at the Paris Observatory the surface features are hard to make out. He could see them best every third synodic period – and each time, he saw the same markings: in fact, those drawn on his map. He could not be expected to look for so curious a relationship, and it was natural for him to think that the rotation period must be synchronous. There was nothing wrong with the observations themselves; it was the interpretation which was faulty. True, the relationship is not exact, but it was good enough to deceive all observers for many years.

We are now in the position to work out the weird calendar of Mercury, which is unlike any other planet in the Solar System and is far from being straightforward. Because Mercury's orbit is so eccentric the orbital speed varies considerably between perihelion and aphelion; however, the rate of axial rotation spin remains constant. Near perihelion, the orbital angular velocity exceeds the constant spin angular velocity, so that an observer on Mercury would see the Sun slowly retrograde, or 'move backwards' through rather less than its own apparent diameter for eight Earth days around each perihelion passage. The Sun would then almost hover over what may be called a 'hot pole'. There are two hot poles, one or the other will always receive the full blast of solar radiation when Mercury reaches perihelion; the intensity will be 2 ½ time greater than for regions of the surface 90 degrees away. Bear in mind, too, that from Earth the Sun has a mean apparent diameter of only about half a degree of arc; from Mercury the apparent diameter ranges between about 1.1 and 1.6 degrees.

Let us consider two observers, both of whom are placed on Mercury's equator, but who are 90 degrees in longitude away from each other. Observer A is at a 'hot pole' so that the Sun is at the zenith, or overhead position, at perihelion. This means that the Sun will rise when Mercury is near aphelion, and the solar disc will be at smallest. As the Sun nears the zenith, it will slow down and grow in size. It will pass the zenith, and then stop and move backwards for 8 Earth days before resuming its original direction of motion. As it drops towards the horizon it will shrink, finally setting 88 Earth days after having risen.

Observer B, 90 degrees away, will see the Sun at its largest near the time of rising, which is also Mercury's perihelion. Sunrise itself will be curiously erratic, because the Sun will come into view and then sink again until it has almost vanished. Then it will climb into the sky, shrinking as it nears the zenith. There will be no 'hovering' as it passes overhead, but sunset will be protracted; Mercury is back at perihelion, so that the Sun will disappear, rise again briefly as though bidding adieu, and then finally depart, not to rise again for another 88 Earth days.

Another odd fact is that the stars will move across the sky at roughly three times the average rate of the Sun. One hates to think what a Mercurian would make of all this!

The temperature at the hot pole reaches 427 °C at maximum, but the night temperature drops to −183 °C. The temperature range is greater than for any other planet in the Solar System. Near the poles there are some craters whose floors are always in shadow, and which remain bitterly cold: in 1991, radar measurements made with the VLA (Very Large Array) in New Mexico led to the suggestion that ice might exist in these craters. Results from Arecibo, and from Mariner 10 were later quoted in support. However, the evidence is at best very suspect and the existence of ice on a world such as Mercury would be very surprising indeed. Moreover, the same radar results have been found in areas on Mercury which do receive sunlight, and where ice could not possibly survive.

SPACE-CRAFT TO MERCURY

Mariner 10 showed the way. It was launched on 3 November by an Atlas-Centaur rocket. Its structure consists of an eight-sided magnesium and aluminium framework; it measured 1.39 m diagonally and 0.457 m in depth. There were two solar panels, each 2.69 m long and 0.97 m wide; fully deployed, the space-craft measured 8 m across the solar panels. It carried a motor-driven high-gain dish antenna, 1.37 m across, and a low-gain antenna mounted at the end of a 2.85-m boom. Instruments were carried to investigate the atmosphere, surface and physical characteristics of Venus. Mariner 10 was the first probe to use the gravity-assist technique, by-passing Venus and using the gravitational pull of that planet to send it on to rendezvous with Mercury.

It by-passed Venus on 5 February 1974, at a range of 4200 km, after which it went on to three active passes of Mercury: 29 March 1974 (at 705 km from the surface), 21 September 1974 (at 47 000 km) and 16 March 1975 (at 327 km). By that time the equipment was deteriorating, and contact was finally lost on 24 March 1975. No doubt Mariner is still in solar orbit, and still

making periodical passes of Mercury, but we have no hope of contacting it again. The same areas of the planet were in sunlight during all three active passes, so that less than half the total surface was surveyed, but the mission had been an outstanding success, and gave us the first detailed view of the surface features. It also found that Mercury, unlike Venus, has a detectable magnetic field. This was a surprise even though the field is only 1% as strong as that of the Earth.

The well-known MESSENGER (the MErcury Surface, Space ENvironment, GEochemistry and Ranging) probe was launched from Cape Canaveral on 3 August 2004, on a Delta II rocket. It was shaped like a squat box, measuring 1.27 × 1.42 × 1.85 m, with two solar panel wings and a 3.6-m magnetometer boom. The total mass was 1093 kg, of which 608 kg was propellant. Instruments of all kinds were carried. The cameras were capable of resolving surface features down to 18 m – a great improvement on Mariner 10. It was also planned to cover wide areas which had been inaccesible to Mariner 10.

The Boeing Delta II 4 lifted off at 02.15.56 EDT (Eastern Daylight Time). An hour later MESSENGER separated from the third-stage booster, and began its circuitous journey to Mercury. The plan was:

3 Aug 2004	Launch
2 Aug 2005	Pass of Earth, at 2347 km
12 Dec 2005	Deep-space manœuvre (DSM) to adjust for Venus pass
24 Oct 2006	Pass of Venus, at 2992 km
5 June 2007	Second pass of Venus, at 338 km
17 Oct 2007	DSM to adjust for rendezvous with Mercury
14 Jan 2008	Pass of Mercury, at 200 km
6 Oct 2008	Second pass of Mercury
29 Sept. 2009	Third pass of Mercury.
18 Mar 2011	Enter orbit round Mercury
2012	Scheduled end of mission

By 2012, it was hoped that as well as providing maps of the whole Mercurian surface, MESSENGER would prove the presence of ice on the permanently shadowed floors of some of the polar craters; NASA was particularly anxious about this.

In fact all three passes of Mercury were successful, and by the time that MESSENGER was put into a path that would enter Mercurian orbit, over 90% of the planet had been mapped. There was only one temporary 'glitch', when MESSENGER went unexpectedly into what is termed a safe mode, and close-range images during the outbound leg of the encounter were lost.

Quiet though it is today, Mercury has had a violently volcanic past; and lava dominates the scene. There are features of special interest such as the vast impact crater Rembrandt, over 700 km in diameter and around 3.9 thousand million years old; the 257 km Raditladi is much younger and has its floor partially filled with smooth, reddish plains material, with an age of no more than a thousand million years. Like some other large formations it has a complete inner ring.

MESSENGER has carried out its programme well, but it still has much to do; as I write these words (March 2010) it is on its way to enter orbit round Mercury. Meanwhile another mission, Bepi-Colombo, is well into the planning stage.

This BepiColombo mission, named after the Italian scientist Giuseppe (Bepi) Colombo, will be the joint work of the European Space Agency and the Japan Aerospace Exploration Agency, JAXA. There will be three components launched together in late 2014 on a Soyuz-2 rocket from Kourou (in French Guyana, South America), and due to arrive in orbit about Mercury in 2020. Observations of all kinds will be carried out in greater detail and with greater sophistication than with MESSENGER. Unfortunately a planned lander has been cancelled, being deemed too expensive.

SURFACE EVOLUTION

The main forces responsible for moulding Mercury's surface have been vulcanism, tectonics driven by thermal contraction, and meteoritic bombardment. Superficially the surface looks very like that of the Moon, but there are important differences in detail. The geological timescale for Mercury is given in Table 4.7.

Mercury was formed at the same time as the Earth, 4.6 thousand million years ago. Its geological history is usually divided into five periods: pre-Tolstojian, Tolstojian, Calorian, Mansurian and Kuiperian. ('Geological' really should be 'hermological', and why Tolstoy has become 'Tolstoj' I know not!) There was a long early period of bombardment, so that there was extensive cratering, with global vulcanism and re-surfacing. The Late Heavy Bombardment ended about 3.8 thousand million years ago.

The most prominent surface feature is the Caloris Basin. It lies at one of the hot poles – hence its name; it was formed at the end of the Late Heavy Bombardment, and was first shown by Mariner 10, when only part of it was in sunlight. MESSENGER imaged the whole structure, which is 1550 km in diameter, and is ringed by mountains up to 2 km high. Its floor is filled by lava plains. Near the centre of the floor there is a 40-km crater, from which radiates a pattern of troughs which looked so uncannily like a spider's web that the crater itself was always known as the Spider – until the IAU decided that this was far too undignified. The crater is now known as Apollidorus, after a second-century Greek architect, while the 'web' has become Pantheon Fossæ, after the domed Roman building begun in 27 BC by Marcus Agrippa and re-built by Hadrian between AD 118 and 128. The origin of the 'web' is unclear, but may be related to uplift and stretching of the basin floor.

The Caloris impact was so violent that its effects are evident over wide areas of Mercury. At the exact opposite of the globe – Caloris 'antipodes' – we find an area known as Chaotic Terrain (nickname, 'weird terrain'), hilly and grooved; it was very probably created as seismic waves from the impact converged at the antipode. Other large basins are much older, and less well defined. Tolstoj, 390 km across, is very ancient; two discontinuous rings around it are poorly developed in the north and northeast. A radially delineated ejecta blanket surrounds it for up to two-thirds of its circumference. It may well correspond to Solitudo

Table 4.7 *Geological timescale for Mercury*

Period	Age, thousands of millions of years	Lunar counterpart
Pre-Tolstojian	4.2–4.0 Intercrater plains; obliteration of oldest craters; cratering; extensive global vulcanism	Pre-Nectarian
Tolstojian	4.0–3.8 Tolstoj and other basins; cratering; large highland craters	Nectarian
Calorian	~3.85 Smooth lowland plains (Suisei Planitia, etc.); Caloris Basin impact; lobate scarps formed	Imbrian
Mansuarian	3–5–3.0 Cratering, but on a reduced scale	Eratosthenian
Kuiperian	~ 1 Reduced cratering; ray-craters	Copernican

Maiæ on Antoniadi's albedo map. Also very ancient is the Goethe Basin (diameter 383 km); part of its wall is buried by the surrounding smooth plains. On its north and east sides it is bounded by a gently sloping wall with material which may well be ejecta. The suggested Skinakas basin, thought to be large and double-ringed, does not exist.

There are numerous ray-craters on Mercury. The first to be identified on the Mariner images – in fact, the first crater of any kind to be identified – was named Kuiper, after Gerard Kuiper, one of the pioneers of planetary space-research astronomy; it is 60 km across, with high walls and central peaks. It is the brightest crater on Mercury, just as Aristarchus is the brightest on the Moon, so that it must be one of the youngest craters on the planet. It overlays the rim of a much larger formation, Murasaki. Other ray-craters include Snorri and Degas.

To illustrate the variety of craters on Mercury, it may be useful to give brief mention of a few typical examples. Beethoven (643 km in diameter) is indeed huge, but it is not multi-ringed, part of its wall is buried by its ejecta blanket and by plains material. Homer (314 km) is notable because deposits around it indicate past explosive vulcanism. Molière (132 km) is nearly circular, but on its southern and southwestern ends the rim is cut off into a flat edge, and the eastern rim is broken by a smaller crater. To its southeast is the 116 km Abu Nuwas, and to the north is Rodin (229 km), which has two well-marked craterlets on its floor. Mozart (225 km), just south of Caloris, has a stream of darkish hills on its floor, and round it are long chains of secondary craters. Petrarch (171 km) is low-walled, with a fairly level floor; it lies in the 'weird terrain' antipodal to Caloris. Vivaldi (213 km across) has a prominent, nearly complete

Table 4.8 *Craters over 200 km in diameter*

Rembrandt	720
Beethoven	643
Doestovskij	411
Tolstoj	390
Goethe	383
Shakespeare,	370
Raphael	343
Homer	314
Monet	303
Vyāsa	290
Van Eijk	282
Kunisada	280
Mozart	225
Haydn	270
Raditladi	257
Renoir	246
Pushkin	231
Rodin	229
Valmiki	221
Wren	221
Sveinsdóttir	220
Michelangelo	216
Mendes Pinto	214
Bach	214
Vivaldi	213
Sholem Aleichem	200

inner ring about half the diameter of the outer rampart, but there are no signs of exterior additional rings as there are with some of the Moon's concentric craters. Apollodorus – 'the Spider' (41 km) on the floor of the Caloris Basin, has walls which are more or less continuous, but broken in places, particularly in the south; there are low peaks in its centre, plus a probably more recent craterlet; the floor is flooded with volcanic material. The system of troughs (Pantheon Fossæ) is thought to have been formed at the same time as the impact which produced the crater. The 45-km ray crater Degas is well formed, with a central peak; it breaks into a rather larger and less definite crater. Hun Kal, chosen to mark the 20th meridian, is only 1.5 km across, but is deepish, bowl-shaped and easy to identify. Table 4.8 names the craters with a diameter of 200 km or more.

Like the Moon, Mercury has a surface upon which there are formations of all kinds. Mariner and MESSENGER have provided a great deal of information, but really complete maps must await the results from MESSENGER when the space-craft enters orbit round the planet.

THE ATMOSPHERE OF MERCURY

Because of its low escape velocity – a mere 4.25 km s^{-1} – Mercury would not be expected to have much of an atmosphere, and indeed this is so. There is a very tenuous atmosphere, but the ground density is 10^{-15} of a bar, which we would regard as a good laboratory vacuum, and it is more appropriate to call it an exosphere. Its total mass can be no more than 1000 kg. The main constituents are oxygen (42%), sodium (29%) and hydrogen (22%).

The exosphere is not stable; atoms are continuously lost and replenished from various sources. Thus hydrogen and helium probably come from the solar wind, diffusing into Mercury's magnetosphere for a while and then escaping back into space. Radioactive decay of elements in the crust can produce helium, potassium and sodium, but atoms can also be 'sputtered' from the surface by micrometeorite and cosmic-ray bombardment. MESSENGER found that there is a long 'tail' of sodium behind the planet; calcium was detected for the first time, plus a small amount of water vapour.

MERCURY

Interesting STEREO results in 2010 record escaping gas from the Mercurian atmosphere. It has even been said that the planet has a comet-like tail. This is not to suggest, however, that Mercury is really a comet!

MAGNETIC FIELD

Unlike Venus and Mars, Mercury has a significant magnetic field. It was first detected by Mariner 10, and fully confirmed by MESSENGER. The strength is about 1.1% of the Earth's field (about 350 gammas); it is dipolar, and the magnetic axis is virtually coincident with the axis of rotation. It is strong enough to deflect the solar wind, so that there is a small but very definite magnetosphere; there is a bow shock at about 1.5 Mercurian radii from the centre of the globe. In all probability it is generated by a dynamo effect, as with the Earth's field, and this means that Mercury must have an outer liquid core.

COMPOSITION OF MERCURY

The mean density of Mercury (5.4 times that of water) is greater than that of any other planet apart from the Earth. There is an iron-rich core larger than the whole globe of the Moon, above which is a solid mantle about 600 km thick; the silicate crust seems to be 100 to 200 km thick. There is an outer regolith, going down to a few metres or a few tens of metres. There have been suggestions that in the early history of the Solar System Mercury was struck by a large impactor and had its outer layers ripped off, leaving behind the heavy iron-rich core; this sounds plausible, but of course there is no proof. By weight Mercury is 70% iron and only 30% rocky material, so that it contains twice as much iron per unit volume as any other planet or satellite.

Mercury was once thought to be solid throughout its globe, but this would make the presence of a magnetic field hard to explain. In an attempt to solve the problem, a team at Cornell University, led by J. L. Margot, used radar to make very accurate measurements of Mercury's rotation. After five years' work they found that the planet does not spin exactly as it would do if it were solid throughout; slight variations in the rotation rate indicate that the mantle must be decoupled from the core (a raw egg will spin less regularly than a hard-boiled egg). The iron core almost certainly contains a few per cent of sulphur. The discovery of a magnetic field shows that the outer part of the core must be liquid.

SATELLITE?

Mercury has no satellite. This seems certain. There was, however, an alarm on 27 March 1974, two days before the Mariner 10 pass; one instrument reported bright emissions in the extreme ultraviolet. They vanished, but then reappeared, and a satellite was suspected. Disappointingly, it turned out that the object was an ordinary star, 31 Crateris. If Mercury had a satellite of appreciable size, it would certainly have been found by now.

Of course, many comets cross Mercury's orbit on their way to perihelion, and there are also some Mercury-crossing asteroids,

Table 4.9 *Selected list of Mercury-crossing asteroids*

True crossers	1566 Icarus	3200 Phæthon
	5786 Talos	40267
	66063	66253
Outer crossers	2101 Adonis	2212 Hephaistos
	2340 Hathor	3838 Epona
	5143 Heracles	37655 Illapa

True crosser; perihelion inside Mercury's orbit, aphelion outside.
Outer crosser; perihelion closer-in than Mercury' aphelion.

such as, 1566 Icarus, 3200 Phæthon and 5786 Talos (see Table 4.9). No doubt an automatic observatory on Mercury will eventually be set up to study these objects – but Mercury itself is a solitary traveller in space.

No doubt more space-craft will be sent to Mercury during the twenty-first century, but manned landings there seem to be most unlikely. Mercury is a fascinating world, but it is not a welcoming one.

5 · Venus

Venus, the second planet in order of distance from the Sun, is almost a twin of the Earth in size and mass; it is only very slightly smaller and less dense. However, in all other respects it is quite unlike the Earth. Only during the past 40 years have we been able to find out what Venus is really like; its surface is permanently hidden by its thick, cloudy atmosphere, and before the Space Age Venus was often referred to as 'the planet of mystery'. Data are given in Table 5.1.

Venus is the brightest object in the sky apart from the Sun and the Moon. At its best it can even cast shadows – as was noted by the Greek astronomer Simplicius, in his *Commentary on the Heavens of Aristotle*, and by the Roman writer Pliny around 60 AD. Venus must have been known since prehistoric times. The most ancient observations which have come down to us are Babylonian, and are recorded on the Venus Tablet found by Sir Henry Layard at Konyunjik, now to be seen in the British Museum. Homer (*Iliad*, XXII, 318) refers to Venus as 'the most beautiful star set in the sky' and the name is, of course, that of the Goddess of Love and Beauty.

It may have been Pythagoras, in the sixth century BC, who first realised that the evening and morning apparitions of Venus relate to the same body – though like almost everyone else at that period, he believed Earth to be the centre of the universe.

Of course, old references to Venus are found everywhere. The Greeks were the first to name it after the Goddess of Beauty, and the month of April was sacred to her. Our name Friday is derived from the Anglo-Saxon 'Frigedæg' (*Friga*, or Venus, and *dæ*, day). Even the Aborigines of Australia paid attention to it, notably the Yolngu people of Arnhem Land. When people die, they are taken by a mystical canoe, Larrpart, to the spirit-island Baralku in the sky, where you can see their camp-fires burning along the bank of the great river of the Milky Way. The canoe returns to Earth as a shooting-star, letting their family on Earth know that they have arrived safely. After sunset the Yolngu people gather to await the rising of Barnumbirr, the Morning Star – Venus. As she approaches, before dawn, she draws behind her a rope of light attached to Earth, along which the people are able to communicate with their loved ones.

The great civilisations of the New World, before the Spanish invasion, were the Mayas, the Aztecs and the Incas, all of whom paid great attention to Venus. Indeed, the Mayas, in Yucatán, developed an elaborate calendar; it depended upon several cycles involving the movements of Venus together with those of the Sun and the Moon. They found that every eight years Venus rises as far south in the sky as it can ever do, and this eight-year period was very important; it was referred to as the Grand Cycle. The Mayan calendar, known as the Tzolkin, had a year 260 days long. And far away, in Polynesia, it has been claimed that human sacrifices were offered to the Morning Star, though the evidence for this is decidedly vague.

As an interesting aside, Venus was once referred to by Napoleon Bonaparte. According to the French astronomer F. Arago, Napoleon was visiting Luxembourg when he saw that the crowd was paying more attention to the sky than to him; it was noon, but Venus was easily visible and Napoleon saw it. Not surprisingly, his followers referred to it as being the star 'of the Conqueror of Italy'. In more recent times, Venus has been responsible for innumerable UFO (unidentified flying object) reports – one of them, indeed, from President Carter of the United States!

MOVEMENTS

Venus moves round the Sun in a practically circular orbit. Its elongation can be as much as 47°, so that it can be above the horizon from as much as 5½ hours after sunset or before sunrise; phenomena for the period 2000–2020 are given in Table 5.2.

In 1721, Edmond Halley was the first to note that Venus, unlike Mercury, is at its brightest during the crescent stage, when about 30% of the daylight hemisphere is turned in our direction. When full, Venus is of course on the far side of the Sun; at inferior conjunction, when it is closest to the Earth, its dark side faces us, and the planet cannot be seen at all except during the rare occasions of a transit.

The observed and theoretical phases do not always agree. This is particularly evident during the time of dichotomy or half-phase. When Venus is waning, in the evening sky, dichotomy is earlier than predicted; when Venus is waxing, in the morning sky, dichotomy is late. The discrepancy may amount to several days, although it is true that timing the exact time of an observed dichotomy is not easy. This effect was first noted by J. H. Schröter in 1793 – and is now generally referred to as the Schröter effect, a term which I introduced about 50 years ago. It is due to the effects of Venus' atmosphere.

The phases were first observed telescopically by Galileo, in 1610. This was important, because according to the old Ptolemaic theory, with the Earth in the centre of the planetary system, Venus could never show a full cycle of phases. The observation strengthened Galileo's conviction in the Copernican or Sun-centred system. (The phases had not previously been mentioned specifically, although very keen-sighted people can see the crescent form with the naked eye.)

Table 5.1 *Venus: data*

Distance from the Sun:
mean 108.2 million km = 0.723 a.u.
max. 109.0 million km = 0.728 a.u.
min. 107.4 million km = 0.718 a.u.
Sidereal (orbital) period : 224.701 days
Synodic period: 583.9 days
Rotation period: 243.018 days
Mean orbital velocity: 35.0 km s^{-1}
Axial inclination: 177° .33
Orbital inclination: 3° 23′ 39″ .8
Orbital eccentricity: 0.0068
Diameter: 12.104 km
Oblateness: negligible
Apparent diameter from Earth:
mean 37″ .3
max. 65″ .2
min. 9″ .5
Mass: 4.868 × 10^{24} kg
Reciprocal mass, Sun = 1: 408.520
Mass, Earth = 1: 0.815
Density, water = 1: 5.2
Volume, Earth = 1: 0.86
Escape velocity 10.46 km s^{-1}
Surface gravity, Earth = 1: 0.909
Mean surface temperature:
cloud-tops −33 °C
surface 467 °C
Albedo: 0.76
Maximum magnitude: −4.4
Mean diameter of Sun, as seen from Venus: 44′ 15″

TRANSITS

Transits occur in pairs, separated by eight years, after which no more occur for over a century. Dates of past and future transits are given in Table 5.3. Unlike Mercury, Venus is easy so see with the naked eye during transit.

The first prediction of a transit was made by Johannes Kepler, who in 1627 found that a transit was due on 6 December 1631. It was not actually observed, because it happened during the night over Europe. The first transit to be observed was that of 1639, 24 November O.S. (Old Style; before the calendar changed) [4 December N.S. (New Style)] and was seen by two amateurs, Jeremiah Horrocks and William Crabtree; independent calculations had been made by Horrocks (Kepler had not predicted a transit for 1639). It has been claimed that the Eastern scholar Al-Farabi saw a transit in AD 910, from Kazakhstan, and this may well be true, but there is no proof, and neither is it clear that Al-Farabi had any idea about the cause – even if he did really seen Venus against the Sun. (It might even have been a large sunspot.)

Early in the eighteenth century, Edmond Halley suggested using transits of the inferior planets to measure the length of the

Table 5.2 *Phenomena of Venus, 2000–2020*

Eastern elongation	Inferior conjunction	Western elongation	Superior conjunction
2001 Jan 17	2001 Mar 29	2001 June 8	2000 June 11
2002 Aug 22	2002 Nov 1	2003 Jan 11	2002 Jan 14
2004 Mar 29	2004 June 8	2004 Aug 17	2003 Aug 18
2005 Nov 3	2006 Jan 13	2005 Mar 25	2005 Mar 31
2007 June 9	2007 Aug 18	2007 Oct 28	2006 Oct 27
2009 Jan 14	2009 Mar 27	2009 June 5	2008 June 9
2010 Aug 20	2010 Oct 29	2011 Jan 8	2010 Jan 11
2012 Mar 27	2012 June 5	2012 Aug 15	2011 Aug 16
2013 Nov 1	2014 Jan 10	2014 Mar 22	2013 Mar 28
2015 June 6	2015 Aug 16	2015 Oct 26	2014 Oct 25
2017 Jan 12	2017 Mar 25	2007 June 3	2016 June 6
2018 Aug 17	2018 Oct 27	2019 Jan 6	2018 Jan 9
2020 Mar 24	2020 June 3	2020 Aug 13	2019 Aug 14

The maximum elongation of Venus during this period is 47° 07′, but all elongations range from 45° 23′ to 47° 07′. At the superior conjunctions of 11 June 2000 and 9 June 2008 Venus was actually occulted by the Sun.

Table 5.3 *Transits of Venus, 1631–2200*

Date	Mid-transit (GMT)
1631 Dec 7	05.21 (not observed)
1639 Dec 4	18.27
1761 June 6	05.19
1769 June 3	22.26
1874 Dec 9	04.07
1882 Dec 6	17.06
2004 June 8	08.21
2012 June 6	01.31
2117 Dec 11	02.52
2125 Dec 8	16.06

These are followed by transits on 2247 June 11, 2255 June 8, 2360 Dec 13 and 2368 Dec 10. Earlier transits occurred in 1032, 1040, 1153, 1275, 1283, 1396, 1518 and 1526.

astronomical unit or Earth–Sun distance. Transits of Mercury could not be observed with sufficient accuracy, but those of Venus seemed more promising, and Halley, following up an earlier comment by James Gregory, recommended careful studies of the transits of 1761 and 1769. Unfortunately the method proved to be disappointing and is now obsolete, so that future transits will be regarded as of academic interest only. The trouble was due to an effect termed the Black Drop. As Venus passes on to the solar disc it seems to draw a strip of blackness after it, and when this strip disappears the transit has already begun; again the atmosphere of Venus is responsible. However, the 1769 transit had one other important consequence. Captain Cook was detailed to take the

Table 5.4 *Planetary conjunctions involving Venus*

(a) 2000–2015

Planet	Date	Closest approach (GMT)	Distance (″)
Uranus	2000 Mar 4	00.37	234
Jupiter	2000 May 17	10.30	42
Uranus	2003 Mar 28	12.46	157
Mercury	2005 June 27	16.01	233
Uranus	2006 Feb 14	15.31	83
Saturn	2006 Aug 26	23.36	257
Uranus	2015	18.41	317

(b) 2010–2100 (close conjunctions: separations below 60″; elongation at least 10° from the Sun)

Planets	Date	GMT	Separation (″)	Elongation (°)
Neptune	2022 Apr 27	19.21	−26	43 W
Neptune	2023 Feb 15	15.35	−42	28 E
Uranus	2077 Jan 20	20.01	−43	11 W
Mercury	2084 Dec 24	05.11	+48	17 W

Venus occulted Regulus on 7 July 1959, at 14.28 GMT, and will do so again on 1 October 2044, at 22.02 GMT. On 17 November 1981 Venus occulted the second-magnitude star Nunki (σ Sagittarii).

astronomer Charles Green to Tahiti, to observe the transit; the observations were duly made – and Cook then continued upon his voyage of discovery to Australia.

The last transit, that of 2004, was well seen; at my observatory at Selsey, in Sussex, we had a 'transit party', with well over a hundred watchers, both professional and amateur. Conditions were ideal; excellent results were obtained, and the Black Drop was very evident. All those taking part regretted that in our time there can be no repeat; the last transit of the century, that of 2012, will not be seen from Selsey!

OCCULTATIONS

Venus can of course be occulted by the Moon, and can itself occult stars and, occasionally, planets. An occultation of Mars by Venus was observed on 3 October 1590 by M. Möstlin, from Heidelberg, and Mercury was occulted by Venus on 17 May 1737; this was seen by J. Bevis from Greenwich. The last occultation of a planet by Venus was on 3 January 1818, when Venus passed in front of Jupiter. The next occasion will be on 22 November 2065, when Venus will again occult Jupiter, but the elongation from the Sun will be only 8° W. Close planetary conjunctions involving Venus for the period 2000–2100 are given in Table 5.4.

When Venus occults a star, the light from the star dims appreciably as it passes through Venus' atmosphere before the actual occultation takes place. This was very evident when Regulus was occulted on 7 July 1959, and proved to be very useful in estimating the density of Venus' atmosphere, which was not then well known. (I was able to make good estimates with a 30-cm reflector at Selsey, in Sussex.)

TELESCOPIC OBSERVATIONS

Dark markings on the disc were reported by F. Fontana in 1645, but he was using a small-aperture, long-focus refractor, and there is no doubt that his 'markings' on Venus were illusory. In 1727 F. Bianchini, from Rome, went so far as to produce a map of the surface, and even gave names to the features he believed that he had recorded – such as 'the Royal Sea of King John', 'the Sea of Prince Constantine' and 'the Strait of Vasco da Gama'. Again these markings were illusory; Bianchini's telescope was of small aperture, and as the focal length was about 20 m it must have been very awkward to use.

J. H. Schröter, using better telescopes (including one made by William Herschel) observed Venus from 1779, at his observatory at Lilienthal, near Bremen. He recorded markings, which he correctly interpreted as being atmospheric, but also claimed to have seen high mountains protruding above the atmosphere.

In fact, no Earth-based optical telescope will show surface details; all that can be made out are vague, impermanent, cloudy features. Neither is conventional photography more helpful, but in 1923 F. E. Ross, at Mount Wilson, took good photographs at infrared and ultraviolet wavelengths. The infrared pictures showed no detail, but vague features were shown in ultraviolet, indicating high-altitude cloud phenomena.

THE ASHEN LIGHT

The Ashen Light – the faint visibility of the night side of Venus – was first reported in 1643 by the Italian astronomer Johannes Riccioli. It was very faint, but we do know that Riccioli, best remembered for his 1651 map of the Moon, was a competent observer, and since then the Light has been seen on countless occasions, though it is always very dim and can be observed only when Venus is in the crescent stage. Attempts have been made to dismiss it as a mere contrast effect, but this is certainly wrong.

I tried a new experiment when Venus was a slender crescent, and was as well placed as it can be, against a reasonably dark sky. Using my 15-inch reflector, I hid the crescent behind a specially made curved occulting bar – and the Light could still be seen, so that contrast was ruled out. In any case, the Light can sometimes be so clear that its reality is not in question. But what causes it?

There can be no analogy with the Earthshine on our Moon; Venus has no satellite. All sorts of explanations have been proposed. The nineteenth-century German astronomer Franz von Paula Gruithuisen believed it to be due to brilliant illuminations lit by the local inhabitants to celebrate the election of a new Emperor (!) while William Herschel attributed it to phosphorescent oceans. Much more plausible is the theory that it might be the effect of electrical phenomena in Venus' upper atmosphere – in other words auroræ, though we must add that the planet lacks a true magnetic field. (My attempts to link the visibility of the Light with activity on the Sun were completely unsuccessful.) The latest idea is that a certain amount of glow from the hot surface can percolate through the clouds in the atmosphere. This is by far the likeliest explanation, and even if it not absolutely conclusive, it is now generally accepted.

ROTATION PERIOD

Until fairly recent times the axial rotation period of Venus was unknown. Efforts were made to determine it by observing the drifts of surface markings across the disc, as is easy enough with Mars or Jupiter, but fails for Venus because the markings are too ill-defined. The first attempt, by G. D. Cassini in 1666, gave 23h 21m. Many other estimates followed, by visual, photographic and spectroscopic methods, but they were no better. In a monograph published in 1962, I listed all the estimates of rotation periods published between 1666 and 1960. There were over 100 of them – and every one turned out to be wrong.

In 1890, G. V. Schiaparelli proposed a synchronous rotation period. This would mean that the rotation period and the orbital period would be equal at 224.7 days, and Venus would keep the same hemisphere turned toward the Sun all the time. However, this did not seem to fit the facts, and in 1954 G. P. Kuiper proposed a period of 'a few weeks'.

An estimate of the rotation period found by spectroscopic methods (the Doppler shift) was made by R. S. Richardson, at Mount Wilson, in 1956. He concluded that the rotation was very slow and retrograde – that is to say, opposite in sense to that of the Earth. During the 1950s, French observers, using photography, claimed that the upper clouds had a retrograde rotation period of four days. Surprisingly, both these results proved to be correct.

Venus was first contacted by radar in 1961, by a team at the Lincoln Laboratory in the United States (an earlier result, in 1958, proved to be erroneous). Several other groups made radar contact with the planet at about the same time, and it became possible to obtain a reliable value for the rotation period.

The true period is 243.02 days, retrograde, so that Venus is the only planet to have a rotation period longer than its orbital period. The solar day on Venus is equal to 118 Earth days, and if it were possible to see the Sun from the surface it would rise in the west and set in the east. (In fact, the Sun could never be seen through the clouds.) However, the upper clouds do indeed rotate in only four days, retrograde, so that the atmospheric structure is very unusual.

The reason or this curious state of affairs is unclear. According to a popular theory, Venus was once struck by a large impactor, and literally tipped over, but this does not sound very plausible. Moreover, we know that Uranus also has an exceptional tilt (98 degrees), and to visualise knocking over an ice-giant 50 000 km in diameter really does strain one's credulity. Longer-term interactions between members of the early Solar System seem much more plausible, but we have to admit that we do not really know.

ATMOSPHERE

The atmosphere of Venus was first reported by the Russian astronomer, M. V. Lomonosov, during the transit of 1761; he saw that the outline of the planet was hazy rather than clear-cut, and correctly interpreted this as being due to a dense atmosphere. Subsequently, the existence of a substantial atmosphere was not seriously questioned except by Percival Lowell, at Flagstaff, who in 1897 published a map showing linear features radiating from a dark central patch which he named 'Eros'. These features were, however, illusory.

In 1923–28 E. Pettit and S. B. Nicholson, using a thermocouple attached to the 2.5-m (100-inch) Hooker reflector at Mount Wilson, made the first reliable measurements of the temperatures of the upper clouds of Venus. They gave a value of -38 °C for the day side and -33 °C for the dark side, which is in excellent agreement with modern values. Then, in 1932, W. S. Adams and T. Dunham, also at Mount Wilson, used spectroscopic methods to analyse the upper atmosphere and identified carbon dioxide – now known to make up almost the whole of the atmosphere. Carbon dioxide has a strong greenhouse effect, and it followed that the surface of Venus must be very hot indeed. A high surface temperature was also indicated by the first microwave measurements of Venus at centimetre wavelengths, made by T. Mayer and his team in the United States in 1958.

The composition of the clouds remained uncertain. In 1937 R. Wildt suggested that they were made up of formaldehyde, and this remained the favoured theory until space-craft results showed it to be incorrect. Neither was the nature of the surface known. In 1954, F. L. Whipple and D. H. Menzel proposed that Venus was mainly water-covered, and that the clouds were composed chiefly of H_2O, but this attractive 'marine' theory was disproved by the Mariner 2 results of 1962. Venus is far too hot for liquid water to exist on its surface.

Virtually all our detailed knowledge of Venus' atmosphere, beneath the cloud-tops, comes from the various space-craft launched since 1961; a list of these is given later in Table 5.7. Some of the earlier Venera probes designed to land on the surface were actually crushed during their descent, because the atmosphere was even thicker than had been expected.

Below an altitude of 80 km the atmosphere is made up of over 96% of carbon dioxide and about 3% of nitrogen (N_2), which does not leave much room for anything else: there are minute traces of carbon monoxide, helium, argon, sulphur dioxide, oxygen and water vapour and other gases such as krypton and xenon. The clouds are rich in sulphuric acid, and at some levels there must be sulphuric acid 'rain', which however evaporates well before reaching the surface. The troposphere extends from the surface to an altitude of 65 km; above, the mesosphere extends to 120 km, and then comes the upper atmosphere, which reaches out to at least 400 km. Venus has no detectable overall magnetic field (it must be at least 25 000 times weaker than that of the Earth), but the dense atmosphere and magnetic eddy currents induced in its ionosphere produce a well-marked bow shock, and prevent the solar wind particles from reaching the surface.

Despite the great differences between the atmospheres of Venus and the Earth, very similar mechanisms produce lightning on the two planets. The rates of discharge, intensity and spatial distribution are comparable.

There is evidence of a Hadley cell, which means that hot air rises at the equator and is transported polewards in the upper atmosphere. This produces a polar dipole, i.e. a vast vortex with an unusual double-eye feature, which rotates around the pole and

Table 5.5 *Pressure and temperature in Venus' atmosphere*

Altitude (km)	Temperature (°C)	Pressure, Earth = 1
0	460	92
5	425	67
10	380	47
20	300	23
30	220	10
40	140	3
50	75	1
60	−10	0.2
70	−40	0.04
100	−110	0.00002

Table 5.6 *Composition of Venus' atmosphere*

	%	ppm
Carbon dioxide	96.5	
Nitrogen	3.5	
Sulphur dioxide	0.015	150
Argon	0.007	70
Water vapour	0.020	200
Carbon monoxide	0.017	170
Helium	0.0012	12
Neon	0.0007	7

features reduced cloud cover associated with rapidly descending air. This polar vortex is bounded by a polar collar, a wide (*c.* 1000-km) shallow river of cold air which circulates around the pole at about 70 degrees. Results from Venus Express in September 2010 show that the double eye in the giant vortex at the planet's south pole has disappeared. An Italian team led by Giuseppe Piccioni commented that 'the polar region of Venus has peculiar dynamics, quite different from the rest of the planet'.

The wind structure is remarkable: the whole atmosphere may be said to be super-rotating. The winds decrease from 100 m s^{-1} at the cloud-top level to only 50 m s^{-1} at 50 km, and only a few metres per second at the surface, although even a slow wind will have tremendous force in that dense atmosphere. It is notable that there is little wind erosion on the surface, although there are obvious signs of æolian depositional activity such as dunes. The atmospheric pressure at the surface is about 90 times greater than that of the Earth's air at sea level – roughly equivalent to being under water on the sea floor of Earth at a depth of 9 km. The greenhouse effect of the carbon dioxide leads to a surface temperature of around 467 °C, and this is practically the same for the day and night hemispheres of the planet.

Details of the composition and pressure and temperature are given Tables 5.5 and 5.6.

SPACE-CRAFT TO VENUS

The first of all interplanetary probes was Russia's Venera 1, launched on 12 February 1961. It lost contact when it had receded to 7 500 000 km from Earth, and we cannot be sure what happened to it; it may have by-passed Venus in May 1961 at around 100 000 km, and is presumably still in solar orbit. In July 1962 the Americans made their first attempt, with Mariner 1, but the result was disastrous – Mariner 1 plunged into the sea, apparently because someone had forgotten to feed a minus sign into a computer (a slight mistake which cost approximately $4 280 000). But then came the triumphant Mariner 2 and the era of direct planetary exploration had well and truly begun.

During its encounter with Venus, Mariner 2 revolutionised many of our ideas about the planet. The Whipple–Menzel marine theory was killed at once; the high surface temperature and long rotation were confirmed, as was the absence of any detectable magnetic field. Venus and Earth were indeed non-identical twins.

At this stage Soviet plans were concentrated mainly upon controlled landings. The density of the atmosphere had been under-estimated; the first landers were crushed during descent, and the first real success came with Venera 7, which came down on 15 December 1970 in the area now known as Navka Planitia and sent back signals for 23 minutes after arrival, so becoming the first man-made object to transmit from another planet. Venera 8 (July 1972) did even better and after landing, also in Navka Planitia, transmitted for 50 minutes before losing contact. The high surface pressures and temperatures were fully confirmed, and it was also possible to measure the surface illumination, estimated to be about the same as that in Moscow on a cloudy winter day. It was also found that the wind speeds on the surface were very low.

The sole American effort around this time was Mariner 10, launched from Cape Canaveral on 3 November 1973. The main target was not Venus, but the innermost planet Mercury, which up to that stage had remained unvisited. En route, Mariner 10 passed Venus at a range of 3600 miles, sending back good images of the cloud-tops as well as obtaining useful data; it used Venus' gravitational pull to send it on its way to Mercury – which was encountered in March 1974.

So far as Venus was concerned, the next major advances came in 1975, with two more Soviet spacecraft: Veneras 9 and 10. Both were dispatched in the summer, and landed on 21 and 25 October respectively, about 240 km apart. Before starting their descent through Venus' atmosphere they were chilled below freezing-point; after arrival each operated for about an hour (to be precise, 55 minutes for Venera 9 and 65 minutes for Venera 10), and each sent back a picture of the surface, relayed to Earth by the orbiting sections of the probes which had been left in closed paths around Venus.

The pictures were of amazingly good quality. Those from Venera 9 showed a rock-strewn landscape; the rocks were sharp-edged and most of them were 1–3 m across. One Soviet astronomer, M. Marov, described the scene as 'a stony desert', and there was no marked evidence of erosion, which was surprising in view of the fact that even a slow wind would have tremendous force in the thick atmosphere of Venus. Measurements indicated a wind speed of about 7 knots, but this would be as devastating as the effect of a sea-wave on a terrestrial cliff during a 40-knot gale. The scene from Venera 10 was much the same, but it seemed that the large rocks were outcrops from below the surface rather than simple scattering, as on the Venera 9 site. Also, the rocks were rather smoother, and the general outlook was of a landscape which was appreciably older.

There had been earlier suggestions that the surface of Venus might be too dark to allow any picture-transmission at all. Both the Veneras took floodlights with them, but there was no need to use them. It had also been thought that the great density of the atmosphere might bend light-waves strongly enough to produce what is called 'super-refraction' – so that an observer on the surface would seem to be standing in a huge bowl, with the horizon curving upward on all sides of him. This would have been weird in the extreme, but it proved not to be the case.

Veneras 11 and 12 (December 1978) were of essentially the same type as their predecessors, carrying both orbiters and landers. The instruments had been improved, and both probes sent back signals after landing, but the television cameras failed to work.

American efforts were concentrated upon fly-by missions and orbiters. The Pioneer Venus mission in 1978 was complex; it consisted of a 'bus' carrying four smaller probes, which were released well before the rendezvous and made 'hard' landings on the surface, leaving the 'bus' to burn away in Venus' upper atmosphere. The landers were not designed to transmit after arrival, although in fact one of them (the 'Day' probe, which came down in the area now called Themis Regio on the sunlit hemisphere) did so for over an hour.

The NASA Magellan probe, which operated for four years (1990–1994) was an outstanding success, and greatly improved our knowledge of the surface. It was finally de-orbited in October 1994, and burned away in Venus' atmosphere. No more intentional Venus missions were launched for some time, though data were obtained by Galileo (1990) on its way to Jupiter, and Cassini (1998 and 1999) en route for Saturn.

Magellan's work was continued by the European probe Venus Express, launched on 9 November 2005 from Baikonur Cosmodrome, Kazakhstan, by a Soyuz-Fregat rocket. It entered Venus' orbit on 11 April 2006; the distance from the surface ranged between 250 km and 66 000 km. The orbit was polar, and the orbital period approximately 24 hours. High-quality images were received almost immediately. Instruments of various kinds were carried – notably VIRTIS, the Visible and Infrared Thermal Imaging Spectrometer, and VMC, the Venus Monitoring Camera; VIRTIS detected nightglow in the unlit side of Venus, due to nitric oxide. A nitric oxide nightglow has never been detected in the atmospheres of either Earth or Mars; on Venus the glow occurs at altitudes from 110 to 120 km.

The polar atmosphere of Venus has been found to be thinner than expected. Venus Express flew through its upper layers three times in 2009–2010, and found that the density above the poles was 60% less than predicted.

The atmosphere extends up to 250 km. In April 2010, Venus Express skimmed briefly down to 175 km. The orbit at that time took the probe from 250 km to 66 000 km from the planet; period 24 hours.

Akatsuki1 ('Dawn' in Japanese) was launched by JAXA from Kagoshima on a H-HA 202 rocket on 20 May 2010. (Akatsuki is also known as Venus Climate Orbiter VCO.) The total mass of the space-craft was 640 kg, including 320 kg of propellants and 34 kg of scientific instruments; the rocket itself was conventional. The main bus is a 1.6 m $\times$ 1.6 m $\times$ 1.25 m box with two solar arrays. The main instruments are an ultraviolet imager (UVI), a long-wave infrared camera (LIR), a 1-μm camera (IR2) and the radio science experiment (RS).

The main objective is to study Venus' atmosphere, where violent winds in the upper layers drive storms and clouds up to speeds of 360 km h^{-1}, 60 times faster than the planet itself rotates. We do not yet understand the cause of this super-rotation. Akatsuki has been described as the first meteorological satellite of a planet other than the Earth.

The Interplanetary Kite-craft Accelerated by Radiation Of the Sun (IKAROS) was launched on the same rocket as Akatsuki, and had the same initial trajectory, but for propulsion relies entirely upon the pressure of sunlight. Its membrane sail – half the thickness of a human hair – carries thin-film, electricity-generating solar cells. The name recalls the mythological flyer Icarus, who flew so close to the Sun that the heat melted the wax fastening his wings – with unfortunate results!

See Table 5.7 for a list of missions to Venus.

SURFACE FEATURES

In every way Venus is an intensely hostile planet. The first pictures from the surface, sent back by Veneras 9 and 10, were obtained under a pressure of about 90 bar and an intolerably high temperature – which had been expected, in view of the greenhouse effects of the atmospheric carbon dioxide. The Venera 9 landscape was described as 'a heap of stones', several dozen centimetres in diameter and with sharp edges; the Venera 10 landing site was smoother, as though it were an older plateau.

The first attempts at analysis of the surface materials were made in March 1982 by Veneras 13 and 14, which landed in the general region of the area now known as Phœbe Regio. Venera 13 dropped a lander which continued to transmit for a record 60 minutes after arrival; the temperature was +457 °C and the pressure 89 bar. Venera 14 came down in a plain near Navka Planitia; there were fewer of the sharp, angular rocks of the Venera 13 site. The temperature was given as 465 °C, and the pressure 94 atmospheres. In both cases it was reported that highly alkaline potassium basalts were much in evidence.

The first attempts at mapping Venus were made by using Earth-based radar, but only with the Pioneer mission in 1978 did it become possible to obtain really reliable information. This radar map is Figure 5.1. Then came the Magellan mission. Magellan could resolve features down to 120 m; the orbital period was 3.2 h, and the high inclination of the orbit meant that the polar zones could be studied as well as the rest of the planet. The main dish (3.7 m across) transmitted downwards a radar pulse at an oblique angle to the space-craft, striking the surface below much as a beam of sunlight will do on Earth. The surface rocks modified the pulse before it was reflected back to the antenna; rough areas are radar-bright, while smooth areas are radar-dark. A smaller antenna sent

Table 5.7 *Missions to Venus, 1961–2000*

Name	Launch date	Encounter date	Closest approach (km)	Capsule landing area			Results
				Lat. (°)	Long. (°)		
Venera 1	12 Feb 1961	19 May 1961	100 000	–	–		Contact lost at 7 500 000 km from Earth.
Mariner 1	22 July 1962	–	–	–	–		Total failure; fell in sea.
Mariner 2	27 Aug 1962	14 Dec 1962	34 833	–	–		Fly-by. Contact lost on 4 Jan 1963.
Zond 1	2 Apr 1964	?	100 000?	–	–		Contact lost in a few weeks.
Venera 2	12 Nov 1965	27 Feb 1966	24 000	–	–		In solar orbit.
Venera 3	16 Nov 1965	1 Mar 1966	Landed	?	?		Lander crushed during descent.
Venera 4	12 June 1967	18 Oct 1967	Landed	+19	038	Eistla Regio	Data transmitted during descent.
Mariner 5	14 June 1967	19 Oct 1967	4100	–	–		Fly-by. Data transmitted.
Venera 5	5 Jan 1969	16 May 1969	Landed	−03	018	East of Navka Planitia	Lander crushed during descent.
Venera 6	10 Jan 1969	17 May 1969	Landed	−05	023	East of Navka Planitia	Lander crushed during descent.
Venera 7	17 Aug 1970	15 Dec 1970	Landed	−05	351	East of Navka Planitia	Transmitted for 23 min after landing.
Venera 8	26 Mar 1972	22 July 1972	Landed	−10	335	East of Navka Planitia	Transmitted for 50 min after landing.
Mariner 10	3 Nov 1973	5 Feb 1974	5800	–	–		Data transmitted. En route to Mercury.
Venera 9	8 June 1975	21 Oct 1975	Landed	+31.7	290.8	Beta Regio	Transmitted for 55 min after landing. One picture.
Venera 10	14 June 1975	25 Oct 1975	Landed	+16	291	Beta Regio	Transmitted for 65 min after landing. One picture.
Pioneer Venus 1	20 May 1978	4 Dec 1978	145	–	–		Orbiter, 145 to 66 000 km, period 24 h Contact lost, 9 Oct 1992.
Pioneer Venus 2	8 Aug 1978	4 Dec 1978	Landed (9 Dec)				Multiprobe. 'Bus' and four landers.
			Large Probe	+04.4	304.0	Beta Regio	No transmission after landing.
			North Probe	+59.3	004.8	Ishtar Regio	No transmission after landing.
			Day Probe	+31.7	317.0	Themis Regio	Transmitted for 67 min after landing.
			Night Probe	−28.7	056.7	North of Aino Planitia	No transmission after landing.
			Bus	−37.9	290.9	Themis Regio	Crash-landed.
Venera 11	9 Sept 1978	25 Dec 1978	Landed	−14	299	Navka Planitia	Transmitted for 95 min after landing.
Venera 12	14 Sept 1978	22 Dec 1978	Landed	−07	303.5	Navka Planitia	Transmitted for 60 min after landing.

Table 5.7 (cont.)

Name	Launch date	Encounter date	Closest approach (km)	Capsule landing area Lat. (°)	Long. (°)		Results
Venera 13	30 Oct 1981	1 Mar 1982	Landed	−07.6	308	Navka Planitia	Transmitted for 60 min. Soil analysis.
Venera 14	4 Nov 1981	5 Mar 1982	Landed	−13.2	310.1	Navka Planitia	Transmitted for 60 min. Soil analysis.
Venera 15	2 June 1983	10 Oct 1983	1000	–	–		Polar orbiter, 1000–65 000 km. Radar mapper.
Venera 16	7 June 1983	16 Oct 1983	1000	–	–		Polar orbiter, 1000–65 000 km. Radar mapper.
Vega 1	15 Dec 1984	11 June 1985		–	–		Fly-by; en route to Halley's Comet.
			Lander	+08.5	176.9	Rusalka Planitia	Lander transmitted for 20 min after arrival. Balloon dropped into Venus' atmosphere.
Vega 2	20 Dec 1984	15 June 1985		–	–		Fly-by; en route to Halley's Comet.
			Lander	−07.5	179.8	Rusalka Planitia	Lander transmitted for 21 min after arrival. Balloon dropped into Venus' atmosphere.
Magellan	5 May 1989	10 Aug 1990	294	–	–		Orbiter, 294–8450 km. Radar mapper. Burned away in Venus' atmosphere, 11 Oct 1994.
Galileo	18 Oct 1989	10 Feb 1990	16 000	–	–		Fly-by. En route to Jupiter.
Cassini	18 Oct 1997	26 Apr 1998	284	–	–		Fly-by. En route for Saturn.
Venus Express	9 Nov 2005	11 Apr 2006	250	–	–		Polar orbiter; 250–66 000 km.
Messenger	3 Aug 2004	24 Oct 2006 5 Jun 2007	2992 338			Fly-bys	En route for Mercury
Akatsuki	20 May 2010						Launched by an H-HA 202 rocket from Tanegashima Space Center, Japan. Orbiter.
IKAROS	20 May 2010						Launched by same rocket as Akatsuki. It has a solar sail.

down a vertical signal and the time lapse between transmission and return gave the altitude of the surface below to an accuracy of 30 m. Altogether, Magellan studied about 98% of the total surface of Venus.

Venus is a world of volcanic plains, highlands and lowlands. The plains cover 65–70% of the surface, with lowlands accounting for less than 30% and highlands for only 8%. About 60% of the surface lies within 500 m of Venus' mean radius, with only 5% at more than 2 km above it; the total range of elevations is 13 km. The highest mountains are the Maxwell Mountains, which rise to 11 km above the mean radius or 8.2 km above the adjacent plateau in Ishtar Terra. The lowest point is Diana Chasma, in the Aphrodite area, 2 km below the mean radius.

The features now mapped on Venus are given in Table 5.8 and Figure 5.2 is the identification map. But this is of course a selected list, since many more features have not been given names. It was laid down that all names on the planet should be female – the only exception being the Maxwell Mountains, named after the

nineteenth-century Scottish physicist James Clerk Maxwell; this name was given before the official policy was formulated. Many of the names are familiar, such as Florence Nightingale and Marie Curie, but not everyone will know that, for example, Auralia was Julius Cæsar's mother, Heng-O was a Chinese Moon goddess, Marie Vigier Lebrun a French painter and Vellamo a Finnish mermaid!

Vulcanism dominates the surface, and there are lava flows everywhere. There are two very large highland areas, Ishtar Terra in the northern hemisphere and Aphrodite area, which lies mainly in the south but is crossed by the equator. Ishtar is about the size of Australia (2900 km in diameter) and consists of western and eastern components separated by the Maxwell Mountains, the highest elevations on Venus and which have steep slopes of up to 35° in places. Maxwell forms the eastern edge of a high plateau, Lakshmi Planum, which is bounded to the south, west and north respectively by the Freyja, Akna and Danu Mountains. Lakshmi is relatively smooth, covered with lava which has flowed from the

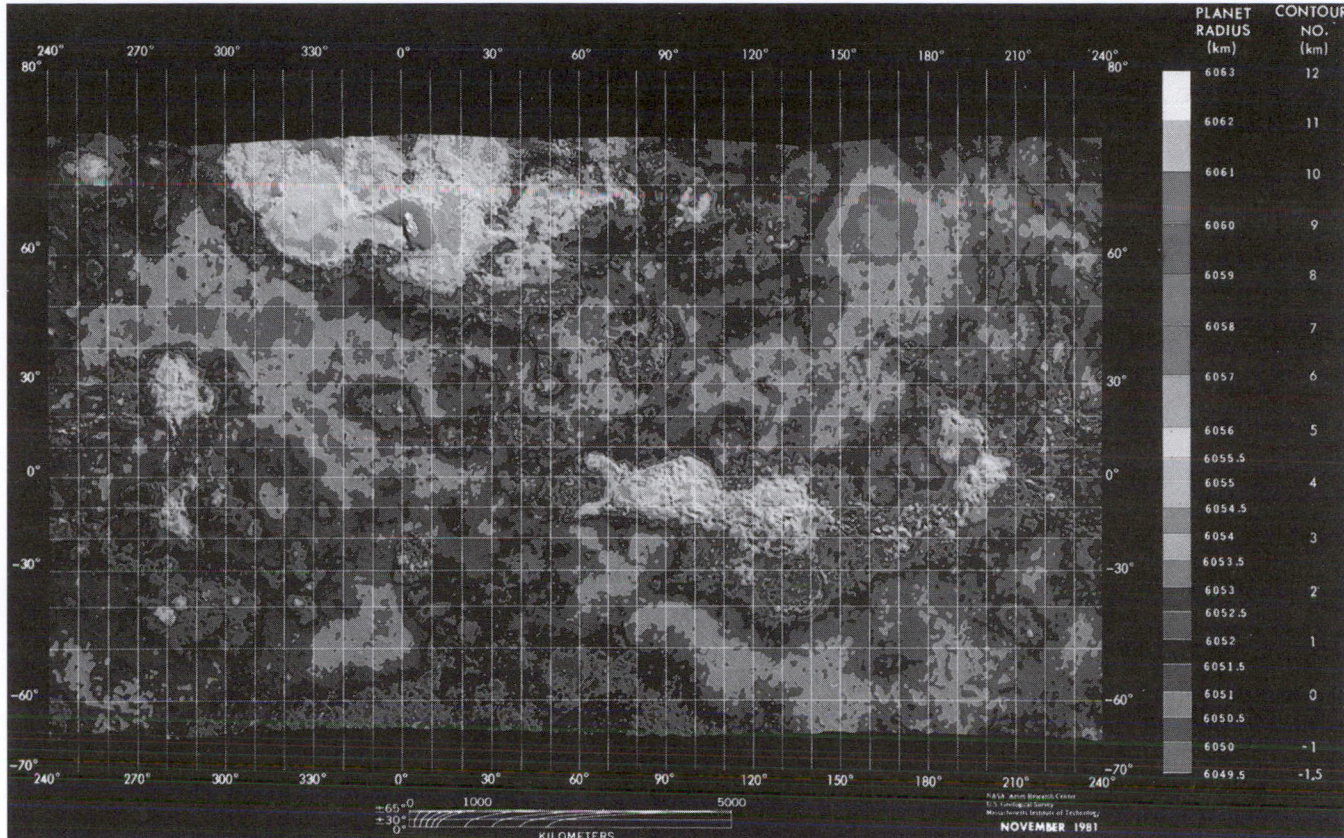

Figure 5.1 Radar map of Venus

caldera–like structure Colette; Colette itself has collapsed to 3 km below the adjacent surface.

Aphrodite is larger – 9700 × 3200 km – and consists of eastern and western elevated areas separated by a lower area. Western Aphrodite is made up of two regions, Thetis and Ovda, which lie from 3 to 4 km above the mean radius of Venus. These regions are dominated by what used to be called 'parquet' terrain; this term was abandoned as being too unscientific, and was replaced by 'tesseræ'. Tessera terrain is characterised by extreme roughness, and covers about 8% of the total surface of Venus. It seems to be unique to Venus – at least so far as we know; this also applies to the coronæ and arachnoids. Eastern Aphrodite is dominated by chasmata, which are deep, narrow canyons. One end of Aphrodite has been nicknamed the Scorpion's Tail, although known officially as Atla Regio; it is thought to be one of the main volcanic areas.

Also of note is the highland area of Beta Regio and Phœbe Regio. Here we have what is partly a large shield volcano and partly a tessera-type highland, cut by a rift valley similar to a large-scale version of the terrestrial East African Rift. There is also the southern highland of Lada Terra, first noted by the US probe Magellan.

The crust of Venus seems to be less mobile than that of the Earth, so that terrestrial-type plate tectonics do not apply; a volcano which forms over a 'hot spot' will not drift away, as Mauna Kea in Hawaii has done, but will remain active over a very long period. There has been extensive flooding from lavas sent out both from volcanic calderæ and from fissures at lower levels, probably from 300 to 500 million years ago. Current activity is probable, perhaps in the smaller highland area of Beta Regio, which has two massive peaks – Theia Mons, which is certainly a shield volcano, and its neighbour Rhea Mons.

Radar sounding from the Magellan probe revealed evidence for comparatively recent volcanic activity at Maat Mons, which is the highest volcano on Venus and towers to 11 km above the mean planetary radius; it has a large summit caldera (31 by 28 km) inside which are at least five smaller craters with diameters' up to 10 km. We do not yet know whether Maat Mons shows any activity now.

Impact craters abound, but are rather different from those on Mercury or the Moon; the dense atmosphere means that only impactors capable of producing >30 km craters can reach the surface with undiminished speed, and impact craters below 3 km across are absent, although there are some vast structures. Mead, the largest, is 280 km in diameter. The smaller craters are not lunar-type bowls, but are less regular. For the last few hundred million years the surface activity has been dominated by rift-associated vulcanism, and the fact that impact craters are less crowded than those of the Moon or Mercury indicates that the overall age of the surface features cannot be more than about 750 million years – perhaps considerably less. The largest circular lowland is Atalanta Regio, east of Ishtar; it is on average 1.4 km below the level of the mean radius and is about the area of the Gulf

Table 5.8 *Features on Venus (bold numbers indicate map references)*

Name	Lat. (°)	Long. (° E)	Diameter (km)
Craters: selected list			
Addams	56.1 S	98.0	85
Aglaonice	26.51 S	339.9	65
Alcott	59.5 S	354.5	63
Andreianova	3.0 S	68.8	70
Aurelia	20.3 N	331.8	31
Baker	62.6 N	40.5	105
Barsova	61.3 N	223.0	79
Barton	27.4 N	337.5	54
Boleyn	24.5 N	220.0	70
Bonnevie	36.1 S	127.0	91
Boulanger	26.5 S	99.3	62
Cleopatra	65.9 N	7.0	105
Cochran	51.8 N	143.2	100
de Beauvoir	2.0 N	96.1	58
Dickinson	74.3 N	177.3	69
Dix	36.9 S	329.1	68
Ermolova	60.2 N	154.2	64
Erxleben	59.9 S	39.4	30
Fedorets	59.6 N	65.1	54
Gautier	26.5 N	42.8	60
Graham	6.0 S	6.0	75
Greenaway	22.9 N	145.0	92
Henie	51.9 S	145.8	70
Hepworth	5.1 N	94.6	61
Isabella	29.8 S	204.2	165
Jhirad	16.8 S	105.6	50
Joliot-Curie	1.6 S	62.5	100
Kenny	44.3 S	271.1	50
Klenova	78.1 N	104.2	140
Langtry	17.0 S	155.0	50
Marie Celeste	23.5 N	140.2	95
Markham	4.1 S	155.6	69
Mead	12.5 N	57.2	280
Meitner	55.6 S	321.6	150
Millay	24.4 N	111.1	50
Mona Lisa	25.5 N	25.1	86
Nevelson	35.3 S	307.8	75
O'Keeffe	24.5 N	228.7	72
Ponselle	63.0 S	289.0	52
Potanina	31.6 N	53.1	90
Sanger	33.8 N	288.6	85
Sayers	67.5 S	230.0	90
Seymour	18.2 N	326.5	65
Stanton	23.3 S	199.1	104
Stowe	43.3 S	233.2	78
Tubman	23.6 N	204.5	50
Vigier Lebrun	17.3 N	141.1	59
Warren	11.8 S	176.5	50
Wheatley	16.6 N	268.1	72
Yablochkina	48.2 N	195.1	63
Zhilova	66.3 N	125.5	56

Chasmata			Length (km)
Aranyani	69.3 N	74.4	718
Artemis	41.2 S	138.5	3087
Baba-Jaga	53.2 N	49.5	580
Dali	17.6 S	167.0	2077
Daura	72.4 N	53.8	729
Devana **22**	9.6 N	284.4	1616
Diana	14.8 S	154.8	938
Ganis	16.3 N	196.4	615
Hecate	18.2 N	254.3	3145
Heng-O	6.6 N	355.5	734
Ix Chel	10.0 S	73.4	503
Juno	30.5 S	111.1	915
Kaygus	49.6 N	52.1	503
Kottravey	30.5 N	76.8	744
Kozhla-Ava	56.2 N	50.6	581
Kuanja	12.0 S	99.5	890
Lasdona	69.3 N	34.4	697
Medeina	46.2 N	89.3	606
Mežas Male	51.0 N	50.7	506
Misne	77.1 N	316.5	610
Morana	68.9 N	24.0	317
Mots	51.9 N	56.1	464
Parga	24.5 S	271.5	1870
Quilla	23.7 S	127.3	973
Varz	71.3 N	27.0	346
Vir-Ava	14.7 S	124.1	416
Vires-Akka	75.6 N	341.6	742
Colles			Diameter (km)
Akkruva	46.1 N	115.5	1059
Jurate	56.8 N	153.5	418
Mena	52.5 S	160.0	850
Coronæ			
Artemis	35.0 S	135.0	2600
Atete	16.0 S	243.5	600
Beiwe	52.6 N	306.5	600
Ceres	16.0 S	151.5	675
Copia	42.5 S	75.5	500
Heng-O	2.0 N	355.0	1060
Lilwani	29.5 S	271.5	500
Maram	7.5 S	221.5	600
Quetzalpetlatl	64.0 S	354.5	400
Tacoma	37.0 S	288.0	500
Dorsa			Length (km)
Ahsonnutli	47.9 N	194.8	1708
Aušrā	49.4 N	25.3	859
Bezlea	30.4 N	36.5	807
Breksta	35.9 N	304.0	700
Dennitsa	85.6 N	205.9	872
Frigg	51.2 N	148.9	896
Hera	36.4 N	29.5	813
Iris	52.7 N	221.3	2050
Juno	31.0 S	95.6	1652
Laūma	64.8 N	190.4	1517
Mardezh-Ava	32.4 N	68.6	906

Table 5.8 (cont.)

Name	Lat. (°)	Long. (° E)	Diameter (km)
Nambi	72.5 S	213.0	1125
Nephele	39.7 N	139.0	1937
Okipeta	66.0 N	238.5	1200
Saule	58.0 S	206.0	1375
Sel-Anya	79.4 N	81.3	975
Semuni	75.9 N	8.0	514
Tezan	81.4 N	47.1	1079
Uni	33.7 N	114.3	800
Varma-Ava	62.3 N	267.7	767
Vedma	49.8 N	170.5	3345
Zorile	39.9 N	338.4	1041
Fluctûs			
Eriu	35.0 S	358.0	1200
Kaiwan	48.0 S	1.5	1200
Mylitta	56.0 S	353.5	1250
Fossæ			Diameter (km)
Arionrod	37.0 S	239.9	715
Bellona	38.0 N	222.1	855
Enyo	61.0 S	344.0	900
Hildr	45.4 N	159.4	677
Nike	62.0 S	347.0	850
Linæ			
Antiope	40.0 S	350.0	1240
Guor	17.0 N	2.6	1050
Kalaipahoa	60.5 S	338.0	2400
Kara	44.0 S	306.0	700
Molpadia	48.0 S	359.0	1350
Morrigan	54.5 S	311.0	3200
Penardun	54.0 S	344.0	975
Mons			
Gula	21.9 N	359.1	276
Hathor	38.7 S	324.7	333
Innini	34.6 S	328.5	339
Maat	0.5 N	194.6	395
Mbokomu	15.1 S	215.2	460
Melia	62.8 N	119.3	311
Nephthys	33.0 S	317.5	350
Ozza	4.5 N	201.0	507
Rhea 20	32.4 N	282.2	217
Sapas	8.5 N	188.3	217
Sekmet	44.2 N	240.8	338
Sif	22.0 N	352.4	200
Tefnut	38.6 S	304.0	182
Theia 21	22.7 N	281.0	226
Tuulikki	10.3 N	274.7	520
Ushas	24.3 S	324.6	413
Venilia	32.7 N	238.8	320
Xochiquetzal	3.5 N	270.0	80
Montes			
Akna 17	68.9 N	318.2	830
Danu	58.5 N	334.0	808
Freyja 18	74.1 N	333.8	579
Maxwell 19	65.2 N	3.3	797
Nokomis	18.9 N	189.9	486
Pateræ			
Anning	66.5 N	57.8	135
Aspasia	56.4 N	189.1	150
Boadicea	56.0 N	96.0	220
Colette	66.5 N	322.8	149
Eliot	39.0 N	79.0	116
Hatshepsut	28.1 N	64.5	118
Hiei Chu	48.3 N	97.4	139
Hroswitha	35.8 N	34.8	163
Kottauer	36.7 N	39.6	136
Nzingha	69.0 N	206.0	143
Raskova	51.0 S	222.8	80
Razia	46.2 N	197.8	157
Sacajawea	64.3 N	335.4	233
Sand	42.0 M	15.5	181
Sappho	14.1 N	16.5	92
Schumann-Heink	74.0 N	215.0	120
Stopes	42.5 N	47.0	169
Tarbell	58.2 S	351.5	80
Tipporah	38.9 N	43.0	99
Tituba	42.5 N	214.0	163
Trotula	41.3 N	18.9	146
Yaroslavna	38.8 N	21.2	112
Planitiæ			
Aino 9	40.5 S	94.5	4983
Atalanta 10	70.0 N	165.8	2048
Audra	61.5 N	71.5	1861
Bereghinya	28.6 N	23.6	3902
Ganiki	25.9 N	189.7	5158
Guinevere 11	24.0–60.0 N	325.0	7519
Helen 12	51.7 S	263.9	4362
Kawelu	32.8 N	246.5	3910
Lavinia 13	47.3 S	347.5	2820
Leda 14	44.0 N	65.1	2890
Louhi	80.5 N	120.5	2441
Navka	8.1 S	317.6	2100
Niobe 15	21.0 N	112.3	5008
Nsomeka	55.0 S	170.0	7000
Rusalka	9.8 N	170.1	3655
Sedna 16	42.7 N	340.7	3572
Snegurochka	86.6 N	328.0	2773
Vellamo	45.4 N	149.1	2154
Vinmara	53.8 N	207.6	1634
Plana			
Lakshmi 23	68.6 N	339.3	2343
Regiones			
Alpha 1	25.5 S	1.3	1897
Asteria 2	21.6 N	267.5	1131
Atla	9.2 N	200.1	3200
Bell	32.8 N	51.4	1778
Beta 3	25.3 N	282.8	2869

Table 5.8 (cont.)

Name	Lat. (°)	Long. (° E)	Diameter (km)
Dione	31.5 S	328.0	2300
Eistla	10.5 N	21.5	8015
Hyndla	22.5 N	294.5	2300
Imdr	43.0 S	212.0	1611
Metis 4	72.0 N	256.0	729
Ovda	2.8 S	85.6	5280
Phœbe 5	6.0 S	282.8	2852
Tethus 7	66.0 N	120.0	2410
Themis 8	37.4 S	284.2	1811
Thetis	11.4 S	129.9	2801
Ulfrun	20.5 N	223.0	3954
Rupes			
Fornax	30.3 N	201.1	729
Gabie	67.5 N	109.9	350
Hestia	6.0 N	71.1	588
Uorsar	76.8 N	341.2	820
Ut	55.3 N	321.9	676
Vesta	58.3 N	323.9	788
Terræ			
Aphrodite	5.8 S	104.8	9999
Ishtar	70.4 N	27.5	5609
Lada	54.4 S	342.5	8614
Tessera			
Ananke	53.3 N	133.3	1060
Atropos	71.5 N	304.0	469
Clotho	56.4 N	334.9	289
Dekla	57.4 N	71.8	1363
Fortuna	69.9 N	45.1	2801
Itzpapalotl	75.7 N	317.6	380
Kutue	39.5 N	108.8	653
Lachesis	44.4 N	300.1	664
Laima	55.0 N	48.5	971
Manzan-Gurme	39.5 N	359.5	1354 (Tesseræ)
Meni	48.1 N	77.9	454
Meshkenet	65.8 N	103.1	1056
Moira	58.7 N	310.5	361
Nemesis	45.9 N	192.6	355
Shimti	31.9 N	97.7	1275
Tellus 6	42.6 N	76.8	2329
Virilis	56.1 N	239.7	782
Tholi			
Ale	68.2 N	247.0	87
Ashtart	48.7 N	247.0	138
Bast	57.8 N	130.3	83
Brigit	49.0 N	246.0	115
Mahuea	37.5 S	164.7	110
Nertus	61.2 N	247.9	66
Semele	64.3 N	202.0	194
Upunusa	66.2 N	242.4	223
Wurunsemu	40.6 N	209.9	83
Zorya	9.4 S	335.3	22

Undæ			
Al-Uzza	67.7 N	90.5	150
Menat	24.8 S	339.4	25
Ningal	9.0 N	60.7	225
Valles			Length (km)
Anuket	66.7 N	8.0	350
Avfruvva	2.0 N	70.0	70
Baltis	37.3 N	161.4	6000
Bayara	45.6 N	16.5	500
Belisama	50.0 N	22.5	220
Bennu	1.3 N	341.2	710
Citlalpul	57.4 S	185.0	2350
Kallistos	51.1 S	21.5	900
Lo Shen	12.8 S	80.6	224
Samundra	24.1 S	347.1	85
Sati	3.2 N	334.4	225
Saga	76.1 N	340.6	450
Sinann	49.0 S	270.0	425
Ta'urua	80.2 S	247.5	525
Vakarine	5.0 N	336.4	625
Ymoja	71.6 S	204.8	390

of Mexico. There are long lava channels, such as Hildr Chasm, some of which are longer than the Nile.

The volcanoes range from huge shield volcanoes such as Sapas Mons (217 km across, height 1.5 km), with a large summit caldera and the even loftier Maat Mons down to much smaller volcanic structures. A 'large volcano' is defined as having a base diameter over 100 km across; there are 156 known shield volcanoes from 100 to 600 km across. Most have summit pits or calderæ, and probably erupt in rift or fault zones, producing widespread lava-flows. There are very few explosive volcanoes; the high atmospheric pressure on Venus inhibits the flow of gas coming out in solution as the molten rock rises, making Hawaiian-type fire fountains unlikely.

Coronæ are complex features, usually more or less circular, up to 2 km high and 400 km across, surrounded by ridges and troughs. They seem to be due to hot, buoyant 'blobs' of material – known to geologists as diapirs – which originate below the surface of Venus; they rise upward, and as they approach the surface they lift the crust above them, fracturing the surface in a radial pattern. When a diapir nears the surface, it cools and loses its buoyancy, and the crust sags under its own weight, developing concentric faults as it does so. The usual result is an easily identifiable system of faults, ridges and fractures. Also easy to identify are the features termed novæ, closely-spaced graben radiating from a central area. Over fifty novæ are known.

We also find Pancake Dome volcanoes; the name is appropriate, because they do look remarkably like pancakes! They are abundant in the plains regions, often near coronæ. Generally they are about 700 m high, and 25 km across; they may occur in pairs, in clusters or in isolation. Their origin is unclear. One theory is that they consist of andesite, a volcanic lava which flows less easily than

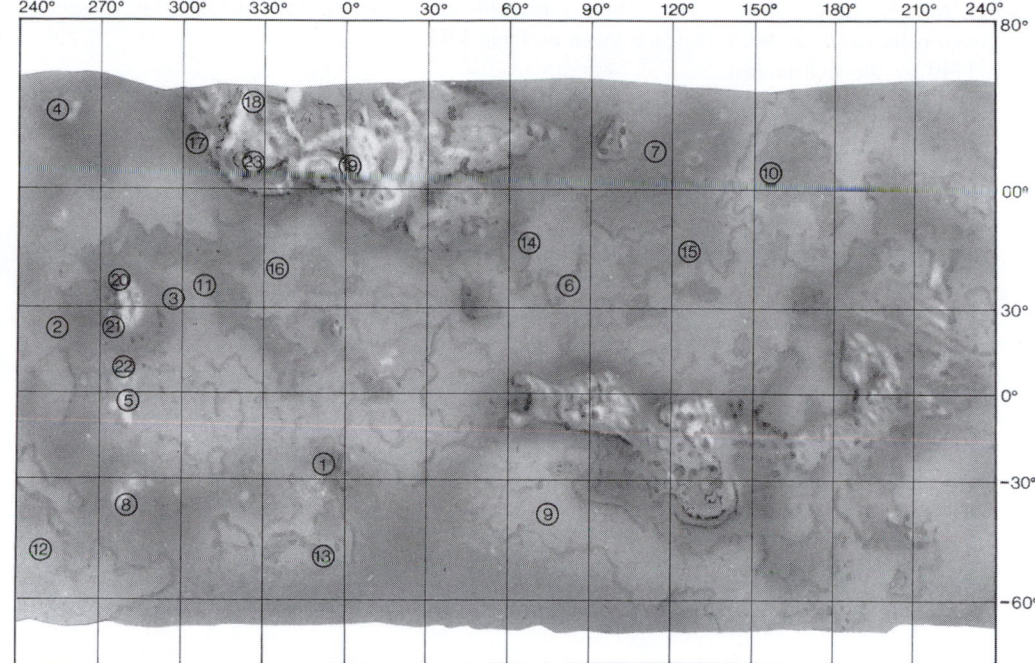

Figure 5.2 Map of Venus showing surface features

basalt, and so piles up into steep-sided domes. So far as we know, they are found only on Venus.

Also unique to Venus are arachnoids, so named because of their spider-web appearance. They are concentric ovals surrounded by complex networks of fractures, and can span 200 km. Over thirty have been identified, but their origin remains unclear.

THE INTERIOR OF VENUS

Our knowledge of the structure is obviously limited, but in some ways it may not be too unlike that of the Earth. The lithosphere is dominated by basalt, which is only to be expected. The thickness of the crust is thought to be from about 20 to 50 km, and this may be fairly uniform over much of the planet, though probably rather thicker above the highlands, possibly down to 60 km. As we have noted, plate tectonics do not operate on Venus. Below the crust comes the rocky mantle, and then the metallic core, the outer border of which is thought to lie around 3000 km from the centre of the globe. True, the lack of an intrinsic magnetic field means that we have no absolute proof of the existence of a metal-rich core, but we can hardly doubt that it does exist.

SATELLITE?

Venus has no true satellite. This seems certain; a satellite of appreciable size would certainly have been found by now. There were various early reports – the first by G. D. Cassini, discoverer of four of the satellites of Saturn on 1686 August: 'Using a telescope of 14 feet focal length … I saw at a distance of 3/5 of her diameter, a luminous appearance that seemed to have the

Table 5.9 *Selected list of Venus-crossing asteroids*

1566	Icarus
1862	Apollo
1864	Dædalus
1865	Cerberus
1981	Midas
2063	Bacchus
2100	Ra-Shalom
2101	Adonis
2201	Oljato
2212	Hephaistos
2340	Hathor
3200	Phæthon
3360	Syrinx
3362	Khufu
3554	Amun
3753	Cruithne
3838	Epona
4341	Poseidon
4450	Pan
4581	Asclepius
4769	Castalia
5143	Heracles
5381	Sekhmet
6063	Jason
6239	Minos
37655	Illapa
69230	Hermes

same shape as Venus ... the diameter of this object was exactly one quarter that of Venus.' He definitely believed it to be a satellite. A similar report was made in 1740 by the well-known telescope maker James Short. Various other observations of the same kind were made during the eighteenth century; the last came in March 1764. After that the satellite was seen no more. Some astronomers were disinclined to reject all the reports; in 1884 the French astronomer P. Houzeau suggested that they might be due to a separate object with an orbit similar to that of Venus, and he even proposed a name: Neith. However, there is no doubt that the observers had been deceived by telescopic 'ghosts', which in view of the great brilliancy of Venus is not surprising.

Venus does, however, have a temporary 'quasi-satellite', asteroid 2002 VE$_{68}$, discovered at the Lowell Observatory at Flagstaff, Arizona. It has a diameter of less than 6 km, and has an orbital period almost the same as that of Venus, though its path is much more eccentric, and actually crosses that of Mercury. It seems to travel once round Venus' sky every Venus year. It will not remain a quasi-satellite indefinitely.

Various asteroids are known to cross Venus' orbit. A selected list is given in Table 5.9.

LIFE ON VENUS?

It is very obvious that Venus is today totally unsuited to life of our kind. This may not always have been the case. In the early period of the Solar System, the Sun was about 40% less luminous than it is now, and Venus and Earth may well have started to evolve in similar fashion, with lands, seas and atmospheres of much the same type. Then, however, the Sun became hotter. Earth was moving at a safe distance; Venus was not, so that the oceans boiled away, the carbonates were driven out of the rocks, and Venus became the hostile world we know. Any life there was ruthlessly snuffed out.

It has been suggested that very primitive organisms might survive in the clouds, but even this must be regarded as improbable. Manned expeditions to Venus lie in the far future, if they can ever be attempted at all. We must be content to view our sister world from a respectful distance.

6 · Earth

The Earth is the largest and most massive of the inner group of planets. Data are given in Table 6.1. In the Solar System, only the Earth is suited for advanced life of our kind; it lies in the middle of the 'ecosphere', the region round the Sun where temperatures are neither too high nor too low. Venus lies at the extreme inner edge of the ecosphere, and Mars at the extreme outer edge.

The Earth–Moon system is often regarded as a double planet rather than as a planet and a satellite. The effect of tidal friction increases the Earth's axial rotation period by an average of 1.7 m s^{-1} per century.

STRUCTURE OF THE EARTH

The rigid outer crust and the upper mantle of the Earth's globe make up what is termed the lithosphere; below this comes the æsthenosphere, where rock is partially melted. Details of the Earth's structure are given in Table 6.2. The crust has an average depth of 10 km below the oceans, but down to around 50 km below the continents. The base of the crust is marked by the Mohorovičić discontinuity (the Moho) named after the Jugoslav scientist Andrija Mohorovičić, who discovered that the velocity of seismic waves changes abruptly at this depth, indicating a sudden change in density. Between 50 and 100 km below the surface the lithospheric rocks become hot and structurally weak.

The outer shell is divided into 8 major 'plates' and over 20 minor ones; the boundaries between the plates are associated with transform faults, spreading axes, subduction zones, earthquakes, volcanoes and mountain ranges. The continents drift around relative to each other; this was first proposed in 1915 by the Austrian meteorologist Alfred Wegener, and has led on to the science of plate tectonics.

Below the lithosphere comes the mantle, which extends down to 2890 km and contains 67% of the Earth's mass. Partial melting of mantle material produces basalts, which issue from volcanic vents on the ocean floor. The base of the mantle is marked by the Gutenberg discontinuity, where the rock composition changes from silicate to metallic and its state from solid to liquid. The outer liquid core extends down to 5150 km and contains 31% of the Earth's mass. The inner core, down to the centre of the globe, is solid and has 1.7% of the total mass; it has been said to 'float' in the surrounding liquid core. The core is iron-rich. The solid inner core, approximately 2400 km in diameter, is thought to have a central temperature of about 4530 °C with a density of 13.1 g cm^{-3}. Currents in the liquid core containing iron are responsible for the Earth's magnetic field. The outer boundary of the solid core is known as the Lehmann discontinuity, first identified by the Danish scientist Inga Lehmann in 1936.

Most of our knowledge of the Earth's interior comes from studies of earthquakes. Seismic waves are of three types: surface, primary (p-waves), also known as push waves, and secondary (s-waves), also known as shake waves. The p-waves can travel through liquid; the s-waves cannot; and it was this that gave the first definite proof that the Earth does have a liquid core.

AGE OF THE EARTH

Many estimates of the age of the Earth have been made. One of these, widely accepted for many years, was due to James Ussher (1581–1686), Archbishop of Armagh. By using Biblical chronology, and other similar calculations, which were in fact about as relevant as the flowers that bloom in the spring, he announced in 1650 that the Earth was created on 23 October 4004 BC; one source gives his exact moment as 9 pm (it is not known whether he made any allowance for Summer Time). However, after 1700 it became clear that the world is much older than this, and Ussher's date was generally abandoned, though the idea of a very young Earth is still accepted by followers of Creation Theory, now re-named Intelligent Design. (One supporter is former US President George W. Bush!)

In 1774, the Comte de Buffon adopted a new approach. He knew that the original Earth was a molten mass, and worked out how long it would take to cool down, arriving at a value of 75 000 years, though he admitted that this might be much too short. In 1788, James Hutton, usually regarded as 'the father of modern geology', wrote that 'we can find no vestige of a beginning – no prospect of an end', while at about the same time Mikhail Lomonosov, the first great Russian scientist, proposed that the Earth was created several hundred thousand years before the rest of the universe.

In 1862, Lord Kelvin, the leading physicist of the time, re-calculated the cooling period and increased the Earth's age to 98 million years; by 1897, he had amended this to 'anything between 20 and 400 million years', though in 1892 Simon Newcomb still preferred a modest 18 million years, based upon the time that the Sun would take to contract from its original size, as a molten mass, down to its present diameter.

The whole situation was changed by the discovery of radioactivity. Radioactive dating methods showed that the Earth was indeed ancient: 1.6 to 3 thousand million years (Arthur Holmes, 1927), 3.3 thousand million years (Holmes, 1946), 4.5 thousand million years (J. G. Houtermans, 1953), 4.55 thousand million years (C. C. Patterson, 1956). All the estimates made since then are in close agreement. The value now accepted is 4.54 thousand million years, with an uncertainty of no more than 0.02 thousand million years either way.

Table 6.1 *Earth: data*

Distance from Sun:
   mean 149 597 900 km (1 a.u.)
   max. 152 096 200 km (1.0167 a.u.)
   min. 147 099 600 km (0.9833 a.u.)
Perihelion (2010): 3 January
Aphelion (2010): 6 July
Equinoxes (2010): 20 March, 13h 33m 18s;
   23 September, 03h 10m 7s
Solstices (2010): 21 June, 11h 29m 31s;
   21 December, 23h 39m 33s
Obliquity of the ecliptic:
   23° .43942 (2010), 23° .43929 (2010)
Sidereal period: 365.256 days
Rotation period: 23h 56m 04s
Mean orbital velocity: 29.79 km s^{-1}
Orbital inclination: 0° (by definition)
Orbital eccentricity: 0.01671
Diameter: equatorial 12 756 km; polar 12 714 km
Oblateness: 1/298.25
Circumference: 40 075 km (equatorial)
Surface area: 510 565 500 km^2
Mass: 5.974 × 10^{24} g
Reciprocal mass, Sun = 1: 328 900.5
Density, water = 1: 5.517
Escape velocity: 11.18 km s^{-1}
Albedo: 0.37
Mean surface temperature: 22 °C

Table 6.2 *Structure of the Earth*

Depth (km)		% of Earth's mass
0–50	Continental crust	0.374
0–10	Oceanic crust	0.099
10–400	Upper mantle	10.3
400–650	Transition zone	7.5
650–2890	Lower mantle	49.2
2890–5150	Outer core	30.8
5150–6370	Inner core	1.7

HISTORY OF THE EARTH

When the Earth had formed from the solar nebula, it eventually became hot and viscous. The early atmosphere, consisting of hydrogen and helium, was soon lost. Cooling was steady; the solid crust formed after only 150 million years or so. Then, between 4 and 3.8 thousand million years ago, came the period of Heavy Bombardment; steam and other volatiles escaped from below the crust, gases were sent out by volcanos, and a second atmosphere was produced, containing ammonia, methane, carbon dioxide, nitrogen and water vapour, but virtually no free oxygen. According to the Giant Impact Theory, the Moon was formed at an early stage of Earth history.

Clouds formed; there was torrential rainfall, and by 3.8 thousand million years ago the oceans had filled the lower-lying areas; impacts of comets may have added to the water, though it is not now thought that the water in our oceans is chiefly of cometary origin. Life began perhaps 4000 million years ago; we do not know how or where. According to the panspermia theory, due originally to Svante Arrhenius in the early twentiethth century and later revived, in modified form, by Fred Hoyle and Chandra Wickramasinghe, life did not begin here, but was brought here by bodies from outer space. This cannot be disproved, but it is fair to say that no panspermia theories have ever been widely supported by the majority of scientists.

A timetable of Earth history is given in Table 6.3.

There have been various extinctions, when whole classes of living creatures have disappeared; the last of these occurred 65 million years ago, when the dinosaurs, which had ruled the world for so long, died out, together with 50% of all species. However, the Permian extinction, 248 million years ago (the 'Great Dying') was even more devastating; 96% of marine species and 76% of land species vanished. There have also been other extinctions, not so complete but still very marked. The reasons for these are not definitely known, and there may have been several processes, operating either independently or together. Extensive volcanic activity could be involved; flood-basalt events can cause widespread devastation, for instance. There is also the clathrate-gas theory, involving the release of large amounts of methane, for example. Clathrates are structures in which a lattice of one substance forms a cage round another; e.g. methane clathrates, in which water molecules make up the cage. These structures form on continental shelves, and can be broken up by a comparatively sudden event such as an earthquake. Methane is a much more effective greenhouse gas than carbon dioxide, so that a major 'clathrate gun' methane eruption would be catastrophic to many species of living creatures. An event of this kind has often been suggested as a prime cause of the global 'Great Dying' which occurred at the end of the Permian Period.

Recently there has been great support for the asteroid-impact hypothesis. There is no doubt that the Earth has been hit many times, even since the end of the Great Bombardment, and a collision with a large impactor (asteroid, meteoroid or even comet) could certainly trigger off a global climate change with associated side effects. The extinction 65 million years ago, at the end of the Cretaceous Period and ushering in the Tertiary (the K–T extinction) has been linked with the fall of an impactor which produced an 80-km crater, now submerged by shallow water off the coast of Yucatán (Mexico), near the town of Chicxulub. The crater has been well studied, and was caused by an impactor around 10 km in diameter. It must have changed world climate, and the popular view is that the dinosaurs were unable to adapt to the new conditions. However, it is fair to say that this is only a very plausible theory, without definite proof; some authorities believe that the dinosaurs died out much more gradually. The Permian extinction has been linked with impact craters in Wilkes Land (Antarctica) and Bedout off the coast of North Australia, but the evidence is rather slender – which is not surprising, in view of the fact that we are looking back over 250 million years.

Table 6.3 *Geological periods (ages in millions of years)*

	From	To		
Pre-Cambrian era				
Hadean	~4500	3950	Heavy bombardment.	No life.
Archæan	3950	2500		No life.
			First crustal formations.	
Proterozoic era				
Palæproterozoic	2500	1600	Preserved rocks.	No life.
				Stromatolites
			Shallow seas.	
Mesoproterozoic	1600	1000		Marine algæ, jellyfish
Neoproterozoic era				
Ionian	1000	950	Widespread seas.	Marine life.
Cryogenian	950	630	Ice ages. 'Snowball Earth'.	
Edicaran	630	542	End of severe ice ages.	Fossils.
Palæozoic era				
Cambrian	542	505	Climate probably fairly mild in the northern hemisphere.	Graptolites, trilobites.
Ordovician	505	438	Probably moderate to warm. Volcanic activity.	Brachiopods, trilobites. Shelled invertebrates.
Silurian	438	408	Probably warm. Continental movements.	Armoured fishes. Scorpions.
Devonian	408	360	Warm, sometimes arid. Greenland, northwestern Scotland, and North America probably joined.	Land plants, Amphibians. Insects. Spiders. Graptolites die out.
Carboniferous (Mississippian)	360	320	Africa moves against a joined Europe and North America. Climate warm. Swamps and shallow seas.	Amphibians, winged insects. Coal measures laid down. First reptiles at end of the period.
(Pennsylvanian)	320	286		
Permian	286	248	Widespread deserts. Gondwanaland (South America, Africa, India, Australia, Antarctica) near south polar regions. Formation of great continent of Pangæa.	Last trilobites. Spread of reptiles. Conifers. Cold period at end; major extinction of species.
Mesozoic era				
Triassic	248	213	Pangæa comprised Eurasia to the north and Gondwanaland to the south, separated by the Tethys Ocean. Hot climate.	Ammonites. Large marine reptiles. First dinosaurs. First small and primitive mammals.
Jurassic	213	144	Pangæa breaking up; rupture between Africa and North America began in the Gulf of Mexico.	Dinosaurs. Archæopteryx. Ammonites. Small mammals.
Cretaceous	144	65	Continental shifts; separation of South Africa and South America. Cooler than in the Jurassic; ice cap over Antarctica.	Dinosaurs, dying out at the end of the period – K–T extinction.
Cenozoic era (Tertiary)				
Palæocene	65	55	Continental drifting. Climate warm.	Rise of mammals. First modern-type plants.
Eocene	55	38	Continental drifting. Australia separates. Climate warm. Volcanic activity.	Widespread forests. Mammals. Snakes.
Oligocene	38	25	North Europe moves northward. Climate warm to temperate.	Rise of modern-type mammals.

Table 6.3 (cont.)

	From	To		
Miocene	25	5	Spread of grasslands, at the expense of forests. Climate temperate.	Grazing mammals. Whales. First primates.
Pliocene	5	2.5	Continents approaching present form. Climate cooler.	Primates. Apes.
(Quaternary)				
Pleistocene	2.5	10 000 yr	Periodical ice ages, with inter-glacials	First men.
Holocene	10 000 yr	Present	End of Ice Ages. Modern world.	Civilisation.

Table 6.4 *Structure of the atmosphere*

	Height (km)	
Troposphere	0 to 8–15 (high and middle latitudes)	Normal
	0 to 16–18 (low latitudes)	Clouds
Tropopause	Upper boundary of the troposphere	
Stratosphere	18 to 50	Ozone layer
Stratopause	Upper boundary of the stratosphere	
Mesosphere	50 to 80	Meteors
Mesopause	Upper boundary of the mesosphere	
Thermosphere	80 to 690	Aurora
Exosphere	Over 690	Collisionless gas

Table 6.5 *Composition of the lower atmosphere*

		Volume (%)
Nitrogen	N_2	78.08
Oxygen	O_2	20.95
Argon	Ar	0.93
Carbon dioxide	CO_2	0.03
Neon	Ne	18.18×10^{-4}
Helium	He	5.24×10^{-4}
Krypton	Kr	1.14×10^{-4}
Xenon	Xe	0.09×10^{-4}
Hydrogen	H_2	0.5×10^{-4}
Methane	CH_4	2.0×10^{-4}
Nitrous oxide	N_2O	0.5×10^{-4}

Very slight, variable traces of sulphur dioxide (SO2) and carbon monoxide (CO).
The amount of water vapour is variable, in the range of 1%.

A controversial theory was advanced by a Princeton Universiity team in 2009. From studies of stromatolites they propose what they term 'snowball Earth'. They believe that it relates to the Neoproterozoic era, which lasted from 1000 million to 542 million years ago (Table 6.3) and is divided into the Tonian, Cryogenian and Ediacaran periods. During the Cryogenian there were repeated ice ages, one of which – the Sturtian, at the start of the period –.was global, covering the planet with ice and disturbing the normal carbon cycle. It lasted for 1.5 million years, and since the Cryogenian there have been ice ages about once every 100 to 200 million years. The whole concept is of course highly speculative.

ATMOSPHERE

The Earth's atmosphere is divided into various layers. The structure is given in Table 6.4, and the composition in Table 6.5.

The *troposphere* is the lowest layer, containing all normal clouds and all 'weather'. It extends up to between 7 km at the poles and 17 km at the equator, though these limits vary to some extent. The temperature falls steadily with height; the average decrease (known as the *lapse rate*) is 1.6 °C per 300 m. The average temperature at the Earth's surface is 15 °C. 50% of the total mass of the atmosphere lies below an altitude of 5 km, and 80% of the total mass is included in the troposphere.

The *stratosphere* extends above the troposphere to an altitude of 50 km. From 1904, it was studied by T. de Bort, using unmanned balloons; the temperature is stable up to 25 km, but then increases – bearing in mind that scientific 'temperature' is defined as the speeds at which the atoms and molecules move around, and is not the same as conventional 'heat'. It rises to about 0 °C at the stratopause, which marks the upper boundary of the stratosphere. The ozone layer lies in the stratosphere, at an altitude between 15 and 35 km; it absorbs 97 to 99% of the Sun's ultraviolet light, which would otherwise be very dangerous to life. Ozone, O_3, is a form of oxygen. The ozone layer is somewhat variable, due presumably to activity on the Sun. (There have been recent suggestions that it is affected by human activities, but this is no more than speculation.)

The *mesosphere* (a term introduced by S. Chapman in 1950), ranges from 50 km to 80–85 km. Temperature decreases with altitude. Above comes the region often called the thermosphere (up to about 690 km), within which the temperature rises again. It is within the range of Shuttles and artificial satellites. The *ionosphere* is the part of the atmosphere ionised by solar radiation. It extends from 50 to 1000 km, overlapping both exosphere and thermosphere. It affects radio waves and produces auroræ.

CLIMATIC VARIATIONS

Earth temperatures depend on the Sun, and slight changes in the amount of radiation sent to us could have major effects. There have been periodical cold spells, and it has been proposed that one of these, between about 850 and 630 million years ago, resulted in almost global coating of ice ('Snowball Earth'). However, for a very long time now temperatures have remained tolerable insofar as life is concerned.

The last major ice age, well documented, was originally divided into four glaciations, known by the names of Gunz, Mindel, Riss and Würm, separated by warmer periods (interglacials). However, recent data indicate that glacial periods have been much more frequent than this. In fact, climate changes appear to follow a distinct 'sawtooth' curve. The last glaciation, the Wurm, ended 10 000 years ago; in all probability we are now in the midst of an interglacial. The cause may be due to several factors, including slight variations in the Earth's orbit (Milankovič cycles) as well as fluctuations in solar output. Shorter-term cool and warm periods have occurred during the last few thousand years, such as the Mediæval Maximum (c. AD 800–1300). The latest well-documented cold spells are the Spörer Minimum (c. AD 1420–1570) and the Maunder Minimum (1645–1715) named after the astronomers who drew attention to them from examining historical records. These, and other cold spells, are associated with periods of low solar activity. During the Maunder Minimum, nicknamed the Little Ice Age, the River Thames froze every winter, and Londoners enjoyed frost fairs on it. There were apparently few or no sunspots, and during total solar eclipses the corona was very inconspicuous. When the sunspots returned, and the solar cycle reverted to normal, the world warmed up. One probable explanation is that during low solar activity the Earth is less shielded from cosmic rays; more of these penetrate the atmosphere and increase cloudiness, so that a greater amount of solar radiation is reflected back into space.

MAGNETIC FIELD

The Earth has a fairly strong magnetic field; at the equator it amounts to 0.305 G. It is essentially a magnetic dipole, so that its shape is like that of a bar magnet and there are two well-defined poles. The magnetic poles are not coincident with the poles of rotation, so that a compass does not point due north; at present the angle between the two axes is just over 11°.

The magnetic poles are not fixed in position. The north magnetic pole is currently in Ellef Ringnes in the Canadian Arctic, while the south magnetic pole is off the coast of Antarctica, south of Australia. At the north magnetic pole the magnetic field is directed vertically downward relative to the Earth's surface, so that the inclination (magnetic dip) is 90°. The magnetic pole is drifting in a north-westerly direction at a rate of about 40 km per year; according to measurements made by the Geological Survey of Canada, the 2001 position was lat. 81.3°N, long. 110.8° W. The south magnetic pole is drifting more slowly, at 5 km per year; according to the Australian Antarctic Division, the 2007 position was lat. 64.5° S., long. 137.7° E. Here of course the magnetic dip is 90°.

The strength of the magnetic field varies over time, and at the moment it seems to be weakening; the magnetic patterns found in volcanic rocks means that the changes in the field can be traced far back into the past. These variations are presumably cyclic. There also seem to be geomagnetic reversals, so that the north magnetic pole becomes a south magnetic pole, and vice versa. The pioneer work here was undertaken in the 1920s by the Japanese scientist Motonori Matuyama; initially it was believed that reversals happened every million years or so, but it is now clear that the intervals are erratic. There have been many reversals throughout Earth history. There were times (as in the Cretaceous Period) when there were no reversals for tens of millions of years, while others have occurred much more rapidly. The last reversal seems to have taken place about 780 000 years ago.

There are sharp divisions of opinion about causes of the reversals. They are widely believed to be due to chaotic motions of liquid metal in the Earth's core, but we do not really know. Further researches are needed.

Before a reversal there may be a time when the dipole field temporarily collapses; if the present weakening continues this may next happen around AD 3000–4000, but there are not likely to be any adverse effects on life. After all, living creatures – including humankind – have survived many magnetic reversals in the past. Having said that, recent work indicates that several recent reversals can be correlated with cold periods.

MAGNETOSPHERE

Magnetic field lines run between the magnetic poles, and charged particles become trapped, forming the magnetosphere (a term introduced by Thomas Gold in 1959). The field is due to dynamo action in the liquid part of the Earth's iron-rich core.

The impact of the solar wind on the magnetopause (the outer boundary of the magnetosphere) compresses the magnetosphere on the day side of the Earth, while the field lines facing away from the Sun stream back to form the magnetotail. In the day side the magnetosphere extends out to about 60 000 km, while on the night side it trails out to well over 300 000 km. Not surprisingly, the magnetosphere is strongly affected by events on the Sun.

In the magnetosphere lie the Van Allen belts, discovered by equipment carried in the first US satellite, Explorer 1, launched by Wernher von Braun's team in 1958. The equipment was designed by James Van Allen of Iowa University. The outer belt extends from an altitude of 13 000 km up to 65 000 km, and consists mainly of trapped high-energy electrons; the intensity is greatest between 14 500 km and 19 000 km. The inner belt lies between 700 and 10 000 km, and consists of energetic protons plus electrons. These belts are distinctly dangerous, and astronauts would be unwise to stay in them for long.

Because the Earth's magnetic axis is offset to the axis of rotation, there is an area where the inner Van Allen belt dips down toward the Earth's surface; this happens above the South Atlantic, off the Brazilian coast, and is known as the South Atlantic Anomaly. It allows charged particles to penetrate deeper into the atmosphere, and this can affect artificial satellites.

FUTURE OF THE EARTH

At present the Earth is in a stable state; there are only relatively slight changes in the amount of heat that we receive from the Sun. However, these will not last indefinitely. The Sun is becoming more luminous as it ages. In about 1000 million years hence the Earth will have become too hot to support advanced life-forms, and when the Sun becomes a red giant it is not likely that the Earth can survive. Its distance from the Sun will have increased, because the Sun will have lost mass, but this may not be enough to save our world. If it does continue to exist, it will be as merely a lump of molten rock.

This may sound depressing, but the end of the Earth lies so far ahead that we can do no more than speculate. If humanity survives, our remote descendants may find a way to save themselves. We will never know!

7 · Mars

Mars, the fourth planet in order of distance from the Sun, must have been known since very ancient times, since when at its best it can outshine any other planet or star apart from Venus. Its strong red colour led to its being named in honour of the God of War, Ares (Mars): the study of the Martian surface is still officially known as 'areography'.

Mars was recorded by the ancient Egyptian, Chinese and Assyrian star-gazers, and the Greek philosopher Aristotle (384–22 BC) observed an occultation of Mars by the Moon, although the exact date of the phenomenon is not known. According to Ptolemy, the first precise observation of the position of Mars dates back to 27 January 272 BC, when the planet was close to the star β Scorpii.

Data for Mars are given in Table 7.1. Oppositions occur at a mean interval of 779.9 days, so that, in general, they fall in alternate years (Table 7.2). The closest oppositions occur when Mars is at or near perihelion, as in 2003 when the minimum distance was only 56 000 000 km. The greatest distance between Earth and Mars, with Mars at superior conjunction, may amount to 400 000 000 km. The least favourable oppositions occur with Mars at aphelion, as in 1995 (minimum distance 101 000 000 km).

Mars shows appreciable phases, and at times only 85% of the day side is turned toward us. At opposition, the phase is of course virtually 100%. At times Mars may be occulted by the Moon, and there are also close conjunctions with other planets (Table 7.3). Planetary occultations of by Mars are very rare: the next occasion will be on 11 August 2079 when Mars will be occulted by Mercury. Occultations of Mars by the Moon are reasonably frequent (Table 7.4).

THE MARTIAN SEASONS

The seasons on Mars are of the same general type as those of Earth, since the axial tilt is very similar and the Martian day (sol) is not a great deal longer (1 sol = 1.029 days). The lengths of the seasons are given in Table 7.5.

Southern summer occurs near perihelion. Therefore, climates in the southern hemisphere of Mars show a wider range of temperature than those in the north. The effects are much greater than for Earth, partly because there are no seas on Mars and partly because of the greater eccentricity of the Martian orbit. At perihelion, Mars receives 44% more solar radiation than at aphelion.

TELESCOPIC OBSERVATIONS

The best pre-telescopic observations of the movements of Mars were made by the Danish astronomer Tycho Brahe, from his island observatory on Hven between 1576 and 1596. It was these observations which enabled Kepler, in 1609, to publish his first laws of planetary motion, showing that the planets move round the Sun in elliptical rather than circular orbits.

The first telescopic observations of Mars were made by Galileo, in 1610. No surface details were seen. However, Galileo did detect the phase, as he recorded in a letter written to Father Castelli on 30 December of that year. The first telescopic drawing of the planet was made by F. Fontana, in Naples, in 1636 (the exact date has not been recorded): Mars was shown as spherical and 'in its centre was a dark cone in the form of a pill'. This feature was, of course, an optical effect. Fontana's second drawing (24 August 1638) was similar.

On 28 November 1659, at 7 pm, Christiaan Huygens made the first telescopic drawing to show genuine detail. His sketch shows the Syrtis Major in easily recognisable form, although exaggerated in size. This sketch has been very useful in confirming the constancy of Mars' rotation period. It was Huygens himself who gave the first reasonably good estimate of the length of the rotation period; on 1 December 1659 he recorded that the period was 'about 24 hours'. In 1666, Giovanni Cassini gave a value of 24h 40m, which is very close to the truth; in the same year he made the first record of the polar caps. It has been claimed that a cap was seen by Huygens in 1656, but his surviving drawing is very inconclusive. However, Huygens undoubtedly saw the south polar cap in 1672.

The discovery that the polar caps do not coincide with the rotational (areographical) poles was made in 1719 by G. Maraldi – a year in which Mars was at perihelic opposition and was so bright that it caused a mild panic: some people mistook it for a red comet which was about to collide with the Earth! In 1704, Maraldi had made a series of observations of the caps, and had given a value for the rotation period of 24h 39m.

William Herschel observed Mars between 1777 and 1783, and suggested that the polar caps were made of ice and snow. Herschel also measured the rotation period, and his observations were later re-worked by W. Beer and J. H. Mädler, yielding a period of 24h 37m 23s .7, which is only one second in error.

Herschel also made the first good determination of the axial inclination of Mars; he gave a value of 28° which is only 4° too great. At present the north pole star of Mars is Deneb (α Cygni), but the inclination ranges between 14.9° and 35.2° over a cycle of 51 000 Earth years, which has important long-term effects. In 25 000 years from now it will be the northern hemisphere which is turned sunward when Mars is at perihelion. These greater precessional effects are due to the fact that the globe of Mars is considerably more oblate than that of the Earth.

Table 7.1 *Mars: data*

Distance from the Sun:
  max. 249 100 000 km (1.666 a.u.)
  mean 227 940 000 km (1.524 a.u.)
  min. 206 700 000 km (1.381 a.u.)
Sidereal period: 686.980 days (= 668.60 sols)
Synodic period: 779.9 days
Rotation period: 24h 37m 22s .6 (= 1 sol)
Mean orbital velocity: 24.1 km s^{-1}
Axial inclination: 23° 59′
Orbital inclination: 1° 50′ 59″
Orbital eccentricity: 0.093
Diameter: equatorial 6794 km
  polar 6759 km
Apparent diameter from Earth:
  max. 25″ .7
  min. 3″ .3
Reciprocal mass, Sun = 1: 3098 700
Mass, Earth = 1: 0.107
Mass: 6.421 × 1026 g
Density, water = 1: 3.94
Volume, Earth = 1: 0.150
Escape velocity: 5.03 km s^{-1}
Surface gravity, Earth = 1: 0.380
Oblateness: 0.009
Albedo: 0.16
Surface temperature:
  max. +26 °C
  mean −23 °C
  min. −137 °C
Maximum magnitude: −2.8
Mean diameter of Sun, seen from Mars: 21′
Maximum diameter of Earth, seen from Mars: 46″ .8

Table 7.2 *Oppositions of Mars 1900–2010*

1900–2000
1901 Feb 22
1903 Mar 29
1905 May 8
1907 July 6
1909 Sept 24
1911 Nov 25
1914 Jan 5
1916 Feb 9
1918 Mar 15
1920 Apr 21
1922 June 10
1924 Aug 23
1926 Nov 4
1928 Dec 21
1931 Jan 27
1933 Mar 1
1935 Apr 6
1937 May 19
1939 July 23
1941 Oct 10
1943 Dec 5
1946 Jan 13
1948 Feb 17
1950 Mar 23
1952 Apr 30
1954 June 24
1956 Sept 11
1958 Nov 17
1960 Dec 30
1963 Feb 4
1965 Mar 9
1967 Apr 15
1969 May 31
1971 Aug 10
1973 Oct 25
1975 Dec 15
1978 Jan 22
1980 Feb 25
1982 Mar 31
1984 May 11
1986 July 10
1988 Sept 28
1990 Nov 27
1993 Jan 7
1995 Feb 12
1997 Mar 17
1999 Apr 24
2001 June 13
2003 Aug 28
2005 Nov 7
2007 Dec 23

Useful drawings of Mars were made by J. H. Schröter, from Lilienthal in Germany, between 1785 and 1814. But Schröter never produced a complete map, and the first reasonably good map was due to Beer and Mädler, from Berlin, in 1830–32; the telescope used was Beer's 9.5-cm refractor. Beer and Mädler were also the first to report a dark band round the periphery of a shrinking polar cap. This band was seen by almost all subsequent observers, and in the late nineteenthth century Percival Lowell attributed it – wrongly – to moistening of the ground by the melting polar ice.

Telescopic observations also revealed the presence of a Martian atmosphere. In 1783, Herschel observed the close approach of Mars to a background star, and from this concluded that the atmosphere could not be very extensive. Clouds on Mars were first reported by the French astronomer H. Flaugergues in 1811, who also suggested that the southern polar cap must have a greater range of size than that in the north, because of the more extreme temperature range – a comment verified observationally in 1811 by F. Arago. 'White' clouds in the Martian atmosphere were first seen

Table 7.2 (cont.)

2010–2020

Date	Closest to Earth	Distance, millions of km	Apparent diameter,	Magnitude	Constellation
2010 Jan 29	Jan 27	99.3	14.1	−1.3	Cancer
2012 Mar 3	Mar 9	100.8	13.9	−1.2	Leo
2014 Apr 8	Apr 14	92.4	15.2	−1.3	Virgo
2016 May 22	May 30	75.3	18.6	−2.1	Scorpius
2018 July 27	July 31	57.6	24.3	−2.8	Capricornus
2020 Oct 13	Oct 6	52.2	22.6	−2.5	Pisces

The interval between successive oppositions of Mars is not constant; it may be as much as 810 days or as little as 764 days. Perihelion of Mars, 2000–2020: 2001 Oct 22, 2003 Aug 30, 2005 July 17, 2007 June 4, 2009 Apr 21, 2011 Mar 9, 2013 Jan 24, 23014 Dec 12, 2016 Oct 29, 2018 Sept 16, 2020 Aug 3.

Table 7.3 *Close planetary conjunctions involving Mars, 1900–2100*

	Date	UT	Separation (″)	Elongation (°)
Mercury–Mars	1942 Aug 19	12.36	−20	16 E
Mars–Uranus	1947 Aug 6	01.59	+43	48 W
Mercury–Mars	1985 Sept 4	21.00	−46	16 W
Mars–Uranus	1988 Feb 22	20.48	+40	63 W
Mercury–Mars	1989 Aug 5	21.54	+47	18 E
Mercury–Mars	2032 Aug 23	04.26	+16	13 W
Mercury–Mars	2079 Aug 11	01.31	Occultation	11 W

Table 7.4 *Occultations of Mars by the Moon, 2010–2020*

Date	Time, UT	Magnitude of Mars	Elongation (°)	Visible from
2010 Dec 6	22h	+1·5	15 E	Pacific, Hawaii, North America
2011 July 27	17h	+1·6	39 W	Pacific, South America
2012 Sept 19	21h	+1·4	21 E	Pacific, South America
2013 May 9	14h	+1·5	5 W	America, Iceland, Europe
2014 July 6	1h	+0·3	97 E	Pacific, Hawaii, South America
2015 Mar 21	23h	+1·5	22 E	South Pacific
2015 Dec 6	3h	+1·7	60 W	Arabia, Indian Ocean, Australia
2017 Jan 3	7h	+1·1	58 E	Indian Ocean, Indonesia, northwestern Australia
2017 Sept 18	20h	+1·9	18 W	Pacific, Hawaii, central America
2018 Nov 16	5h	−0·1	96 E	Antarctica
2019 July 4	6h	+1·9	19 E	Arabia, Japan, Pacific
2020 Feb 18	13h	+1·4	58 W	America, Cuba, Atlantic

by Angelo Secchi (Italy) in 1858. Secchi's sketches also show surface features, notably the Syrtis Major, which he called the 'Atlantic Canal' – an inappropriate name, particularly because there was at that time no suggestion that it might be artificial.

During the 1850s, good drawings were made by the British amateur Warren de la Rue, using his 33-cm reflector, and useful maps were subsequently compiled by Sir Norman Lockyer, F. Kaiser. R. A. Proctor and others, although it was not until the work of G. V. Schiaparelli, from 1877, that really detailed maps were produced.

For some time it was tacitly assumed that the bright areas on Mars must be lands, while the dark areas were regarded as seas (although, strangely, Schröter believed that all the observed

Table 7.5 *The Martian seasons*

	Days	Sols
Southern spring (northern autumn)	146	142
Southern summer (northern winter)	160	156
Southern autumn (northern spring)	199	194
Southern winter (northern summer)	182	177
Total	687	669

SEASONS ON MARS 2010–2020

Spring Equinox	Summer Solstice	Autumn Equinox	Winter Solstice
2011 Sept 13	2012 Mar 30	2012 Sept 29	2013 Feb 23
2013 July 31	2014 Feb 15	2014 Aug 17	2015 Jan 11
2015 June 18	2016 Jan 3	2016 July 4	2016 Nov 28
2017 May 5	2017 Nov 20	2018 May 22	2018 Oct 16
2019 Mar 23	2019 Oct 8	2020 Oct 8	2020 Sept 2

Table 7.6 *Martian nomenclature*

Proctor	Schiaparelli
Beer Continent	Aeria and Arabia
Herschel II Strait	Sinus Sabæus
Arago Strait	Margaritifer Sinus
Burton Bay	Mouth of the Indus canal
Mädler Continent	Chryse
Christie Bay	Auroræ Sinus
Terby Sea	Solis Lacus
Kepler Land and Copernicus Land	Thaumasia
Jacob Land	Noachis and Argyre I
Phillips Island	Deucalionis Regio
Hall Island	Protei Regio
Schiaparelli Sea	More Sirenum, Lacus Phœnicis
Maraldi Sea	Mare Cimmerium
Hooke Sea and Flammarion Sea	Mare Tyrrhenum and Syrtis Minor
Cassini Land and Dreyer Island	Ausonia and Iapygia
Lockyer Land	Hellas
Kaiser Sea (or the Hourglass Sea)	Syrtis Major

Table 7.7 *Old and new nomenclature*

Old	New
Mare Acidaliurus	Acidalia Planitia
Amazonis	Amazonis Planitia
Aonium Sinus	Aonium Terra
Arabia	Arabia Terra
Arcadia	Arcadia Planitia
Argyre I	Argyre Planitia
Ascræus Lacus	Ascræus Mons
Auroræ Sinus	Auroræ Planum
Mare Australe	Australe Planum
Mare Boreum	Boreum Planum
Mare Chronium	Chronium Planum
Chryse	Chryse Planitia
Mare Cimmerium	Cimmeria Terra
Elysium	Elysium Planitia
Mare Hadriacum	Hadriaca Patera
Hellas	Hellas Planitia
Hesperia	Hesperia Planum
Icaria	Icaria Planum
Isidis Regio	Isidis Planitia
Lunæ Lacus	Lunæ Planum
Margaritifer Sinus	Margaritifer Terra
Meridiani Sinus	Meridiani Terra
Nix Olympica	Olympus Mons
Noachis	Noachis Terra
Nodus Gordii	Arsia Mons
Ophir	Ophir Planum
Pavonis Lacus	Pavonis Mons
Promethei Sinus	Promethei Terra
Mare Sirenum	Sirenum Terra
Solis Lacus	Solis Planum
Syria	Syria Planum
Syrtis Major	Syrtis Major Planum
Tempe	Tempe Terra
Tharsis	Tharsis Planum
Mare Tyrrhenum	Tyrrhenum Terra
Utopia	Utopia Planitia
Xanthe	Xanthe Terra

features were atmospheric in nature). Then, in 1860, E. Liais, a French astronomer living in Brazil, suggested that the dark areas were more likely to be vegetation tracts than oceans, and in 1863 Schiaparelli pointed out that the dark areas did not show the Sun's reflection, as they would be expected to do if they were made of water. Agreement was by no means universal; Secchi wrote that 'the existence of continents and seas has been conclusively proved', and in 1865 Camille Flammarion wrote that 'in places the water must be very deep', although he later modified this view and suggested that the dark areas might be composed of material in an intermediate state, neither pure liquid nor pure vapour. It was only it the late nineteenthth century that the concept of major oceans on Mars was definitely abandoned.

NOMENCLATURE

With the compilation of better maps, thought was given to naming the various markings. In 1867, the British astronomer R. A. Proctor produced a map in which he named the features after famous observers such as Cassini and Mädler. His system was followed by other British observers, but was widely criticised. Various modifications were introduced, but, in 1877, the whole system was overthrown in favour of a new one by Schiaparelli. The Proctor and Schiaparelli names are compared in Table 7.6.

Basically, Schiaparelli's system has been retained, although the space-probe results have meant that in recent years it has had to be drastically amended. The old and new systems are compared in Table 7.7. The system has also been extended to take into account features such as craters, which are not identifiable with Earth-based telescopes.

THE CANALS

Who has not heard of the canals of Mars? Not so very many decades ago they were regarded as well-established fixtures, quite possibly of artificial origin.

The first detection of an alleged canal network was due to Schiaparelli in 1877, when he recorded 40 features; he called them *canali* (channels), but this was inevitably translated as canals. Streaks had been recorded earlier by various observers, including Beer and Mädler (1830–32) and W. R. Dawes (1864), but Schiaparelli's work marked the beginning of the 'canal controversy'. Schiaparelli himself maintained an open mind, and wrote 'The suggestion has been made that the channels are of artificial origin. I am very careful not to combat this suggestion, which contains nothing impossible.' In 1879, he reported the twinning, or gemination, of some canals. The first observers to support the network were H.-J.-A. Perrotin and L. Thollon, at the Nice Observatory, in 1886, and subsequently canals became fashionable; they were widely reported even by observers using small telescopes. Percival Lowell, who built the observatory at Flagstaff in Arizona mainly to observe Mars, was convinced of their artificiality, and wrote 'That Mars is inhabited by beings of some sort or other is as certain as it is uncertain what these beings may be.' In 1892, W. H. Pickering observed canals and other features in the dark regions as well as the bright areas, and thus more or less killed the theory that the dark areas might be seas, but it was often claimed that a canal was a narrow watercourse, possibly piped, surrounded to either side by strips of irrigated land. As recently as 1956, G. de Vaucouleurs, a leading observer of Mars, still maintained that though the canals were certainly not artificial, they did have 'a basis of reality'.

In fact, this is not so. The canals were due to tricks of the eye, and do not correspond to any true features on Mars. Equally illusory is the 'wave of darkening': it had been claimed that when a polar cap shrank, releasing moisture, the plants near the cap became more distinct, and that this effect spread steadily from the polar regions to the Martian equator. Only since the Space Age has it been shown that the dark areas are not old sea-beds, and are not coated with organic material.

COMMUNICATING WITH MARS

Radio came back into the story in 1906, when Guglielmo Marconi set up a telegraph station at Cape Clear; British operators reported a strange regular signal of three dots (the Morse letter S) which they could not explain, while in 1921 Marconi himself reported receiving the Morse letter V at 1500 metres. Not to be outdone, the astronomer David Todd and balloonist L. Stevens planned to take a sensitive radio receiver up in a balloon, reducing terrestrial interference and making it easier to detect messages from Mars. At about the same time (1909) the well-known physicist R. W. Wood suggested building a large array of cylinders, blacked on one side, so that they could be set up in a desert and swung round to send signals to the Martians.

A concerted effort was made in August 1924, when Mars was at its closest. Radio transmitters in various parts of the United States were temporarily shut down so that signals from Mars could be picked up, and ready to translate them was W. F. Friedman, head of the code section of the US Army Signals Corps. At Dulwich, in Outer London, radio listeners reported possible Martian signals. In 1926, a Dr Mansfield Robinson went to the Central Telegraph Office in London and dispatched a telegram to Mars – for which he was charged the standard 18 pence per word. Tactfully, the postal authorities noted it as 'Reply not guaranteed'.

When I was making lunar observations with the Lowell refractor, I also turned the telescope toward Mars. I saw nothing remotely resembling a canal, and neither could I follow the alleged 'wave of darkening'. Under the circumstances, I am delighted that I failed!

It is interesting to look back at some of the suggestions made about signalling to the Martians. The first idea seems to have been due to the great German mathematician K. F. Gauss, about 1802; his plan was to draw vast geometrical patterns in the Siberian tundra. In 1819, J. von Littrow, of Vienna, proposed to use signal fires lit in the Sahara. Later, in 1874, Charles Cros, in France, put forward a scheme to focus the Sun's heat on to the Martian deserts by means of a huge burning-glass; the glass could be swung around to write messages in the deserts.

The first (and only) prize offered for communicating with extraterrestrial beings was the Guzman Prize, announced in Paris on 17 December 1900. The sum involved was 100 000 francs – but Mars was excluded, because it was felt that calling up the Martians would be too easy.

Quite apart from the usual eccentrics, there have been at least two very eminent scientists who have given attention to possible contact with Martians. In particular there was Nikola Tesla (1856–1943), the brilliant if unconventional Croatian electrical engineer. In 1899, while living at Colorado Springs in America, he constructed what he called a 'Magnifying Transmitter', and used it to pick up radio signals which, he wrote, could only have come from Mars: 'The feeling is growing in me that I had been the first to hear the greeting of one planet to another'. He gave no details; we know little about his Magnifying Transmitter, and after his death his notebooks could not be found.

Perhaps the last word was said in 1992 by Mr C. Cockell, who fought the General Election on behalf of the Forward to Mars Party. The consituency he selected was Huntingdon, where the sitting member was the Prime Minister, Mr John Major. It is sad to relate that Mr Cockell was not elected!

EARTH-BASED OBSERVATIONS, PRE-1964

Even a small telescope will show considerable detail on Mars when the planet is well placed. The albedo map given here – Figure 7.1 – was drawn solely with the aid of the telescopes in my own observatory in Selsey, Sussex. It shows the main features that are prominent. It is intended for the observer, so south is at the top. I used the Schiaparelli nomenclature.

Energetic observations of Mars were continued during the years before the opening of the Space Age. The behaviour of the polar caps was carefully followed; it was generally believed that the caps were very thin – probably no more than a few millimetres thick, so that they would be in the nature of hoar-frost; there was considerable support for a theory due to A. C. Ranyard and Johnstone Stoney (1898) that the material was solid carbon dioxide rather than water ice. The first useful measurements of the surface temperature of the planet were made in 1909 by S. Nicholson and P. Petit at Mount Wilson and by Coblentz and Lampland at Flagstaff; they found that Mars has a mean surface temperature of $-28\,^{\circ}C$, as against $+15\,^{\circ}C$ for Earth. Lampland also announced the detection of what became known as the 'violet layer', supposed to block out shortwave solar radiations, and prevent them from reaching the Martian surface, except on occasions when it

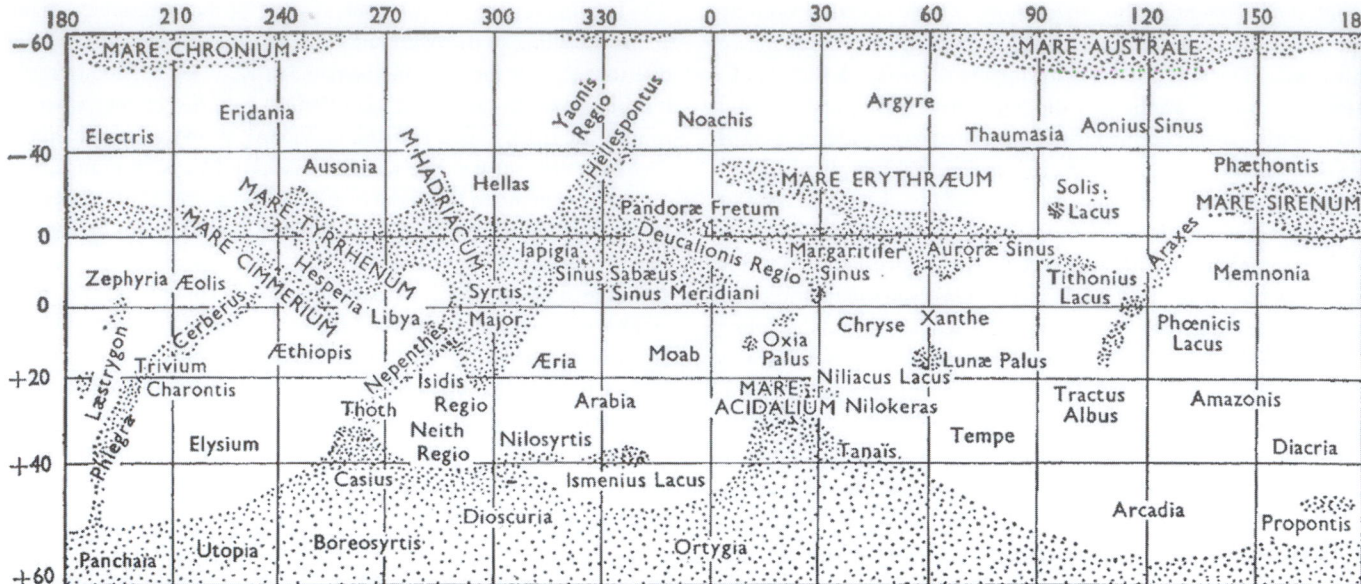

Figure 7.1 Albedo map of Mars by Patrick Moore, drawn using his 12.5-inch reflector. The old nomenclature is used.

temporarily cleared away. We now know that it does not exist – it is as unreal as the canal network and the wave of darkening.

In 1933, W. S. Adams and T. Dunham analysed the Martian atmosphere by using the Doppler method. When Mars is approaching the Earth, the spectral lines should be shifted to the shortwave end of the band; when Mars is receding, the shift should be to the red. It was hoped that in this way any lines due to gases in the Martian atmosphere could be disentangled from the lines produced by these same gases in the Earth's atmosphere. From their results, Adams and Dunham concluded that the amount of oxygen over Mars was less than 0.1% of the amount existing in the atmosphere of the Earth. In 1947, G. P. Kuiper reported spectroscopic results indicating carbon dioxide in the Martian atmosphere, but it was generally believed that most of the atmosphere was made up of nitrogen – whereas we now know that it is mainly composed of carbon dioxide. In 1934, the Russian astronomer N. Barabaschev estimated the atmospheric pressure at the surface, giving a value of 50 mbar; he later increased this to between 80 and 90 mbar – a gross overestimate, since the real pressure is nowhere as high as 10 mbars. In 1954, W. M. Sinton claimed to have detected organic matter in the spectra of the dark areas, but these results were later found to be spurious. One intriguing theory was due to E. J. Öpik, an Estonian astronomer resident at Armagh in Northern Ireland. He maintained that the dark areas had to be made up of material which could grow and push aside the red, dusty stuff blown from the 'deserts' – otherwise they would soon be covered up. Certainly there were major dust storms, usually when Mars was near perihelion.

Before 1964 it was believed that Mars was a world without major mountains or valleys; that the caps were thin, and very probably made up of solid carbon dioxide; that the atmosphere was composed chiefly of nitrogen, with a ground pressure of the order of 87 mbar; that the dark areas were old sea-beds, covered with primitive vegetation; and that the red regions were 'deserts',

not of sand but of reddish minerals. Then came the flight of Mariner 4 and in a very short time all these conclusions, apart from the last, were found to be completely wrong.

EARLY SPACE MISSIONS TO MARS

Once the Moon had been contacted, the space-planners began to look toward the planets. Only two were within 'reasonable' range, Mars and Venus. Venus was the nearer of the two, and was regarded as the friendlier; in those days; remember, it was widely believed that Venus might not be too unlike the Earth, possibly with fairly advanced life-forms. Predictably, the Russians took the lead, but as well as sending probes to Venus they also considered Mars, and it is believed that they launched several space-craft toward the Red Planet between 1960 and 1962. However, the USSR authorities were notoriously secretive, and no details were released at the time.

The first Mars mission about which we have any definite information is Mars 1, launched on 1 November 1962; contact with it was lost on 21 March 1963, at a range of about 105 000 km, and though the probe may have passed fairly close to its target contact with it was never regained. The first successful Mars mission was not Russian, but NASA's space-craft Mariner 4, which by-passed Mars in July 1965.

Since then many space-craft have been sent to the Red Planet – some successful, others not. Most have been either American or Russian, but Japan has launched one mission (Nozomi, 1998 and the European Space Agency joined in with Mars Express (2003). Up to the end of the twentieth century almost all the information had come from NASA; the Russians had had no luck at all, which is curious in view of the fact that Venus, with its dense atmosphere and its searing heat, would be expected to be a much more difficult target.

Data for all the space-craft launched before 2009 are given in Table 7.8.

Table 7.8 *Missions to Mars*

Name	Nationality	Launch date	Encounter date	Closest approach of orbiter (km)	Landing site of capsule	Results
Mars 1	USSR	1 Nov 1962	?	190 000?	–	Contact lost at 106 000 km.
Mariner 3	USA	5 Nov 1964	–	–	–	Shroud failure. In solar orbit, but contact lost soon after launch.
Mariner 4	USA	28 Nov 1964	14 July 1965	9789	–	Returned 21 images, plus miscellaneous data. Contact lost on 21 Dec 1967.
Zond 2	USSR	30 Nov 1964	Aug 1965?	?	–	Contact lost on 2 May 1965.
Mariner 6	USA	24 Feb 1969	31 July 1969	3392	–	Returned 76 images; flew over Martian equator. In solar orbit.
Mariner 7	USA	27 Mar 1969	4 Aug 1969	3504	–	Returned 126 images, mainly over the southern hemisphere. In solar orbit.
Mariner 8	USA	8 May 1971	–	–	–	Total failure: fell in the sea.
Mars 2	USSR	19 May 1971	27 Nov 1971	In orbit, 2448 × 24 400	44° S, 213° W (Eridania)	Capsule landed, with Soviet pennant, but no images received.
Mars 3	USSR	28 May 1971	2 Dec 1971	In orbit, 1552 × 212 800	45° S, 158° W (Phæthontis)	Orbiter returned data. Contact with lander lost 20 s after arrival.
Mariner 9	USA	30 May 1971	13 Nov 1971	In orbit, 1640 × 16 800	–	Returned 7329 images. Contact lost on 27 Oct 1972.
Mars 4	USSR	21 July 1973	10 Feb 1974	Over 2080	–	Missed Mars; some fly-by data returned. Failed to orbit.
Mars 5	USSR	25 July 1973	12 Feb 1974	In orbit, 1760 × 32 500	–	Failure; contact lost.
Mars 6	USSR	5 Aug 1973	12 Mar 1974	?	?24° S, 25° W (Erythræum)	Contact lost during landing sequence.
Mars 7	USSR	9 Aug 1973	9 Mar 1974	1280	–	Failed to orbit; missed Mars.
Viking 1	USA	20 Aug 1975	19 June 1976	In orbit	22.4° N, 47.5° W (Chryse)	Landed 20 July 1976.
Viking 2	USA	9 Sept 1975	7 Aug 1976	In orbit	48° N, 226° W (Utopia)	Landed 3 Sept 1976.
Phobos 1	USSR	7 July 1988	?	?	–	Contact lost, 29 Aug 1988.
Phobos 2	USSR	12 July 1988	–	In orbit, 850 × 79 750	–	Contact lost, 27 Mar 1989. Some images and data from Mars and Phobos returned.
Mars Observer	USA	25 Sept 1992	24 Aug 1993	–	–	Contact lost, 25 Aug 1993.
Mars 96	Russia	16 Nov 1996	–	–	–	Total failure. Fell in the sea.
Pathfinder	USA	4 Dec 1996	4 July 1997	–	19.33° N, 33.55° W	Landed in Ares Vallis. Carried Sojourner rover. Contact lost 6 Oct 1997.
Global Surveyor	USA	7 Nov 1996	11 Sept 1997	In orbit	–	Data returned.
Nozomi	Japan	3 July 1998	–	–	–	Intended orbiter.
Mars Climate Orbiter	USA	11 Dec 1998	–	–	–	Intended orbiter. Contact lost, 23 Sept 1999.
Mars Polar Lander	USA	3 Jan 1999	3 Dec 1999	–	76.25° S, 195.3° W (probable)	Intended orbiter/lander. Contact lost, 3 Dec 1999.

Table 7.8 (cont.)

Name	Nationality	Launch date	Encounter date	Closest approach of orbiter (km)	Landing site of capsule	Results
Mars Odyssey	USA	Launch 7 Apr 2001; entered Mars orbit 24 Oct 2001		In orbit	No lander	Mapping and surface analysis.
Mars Express	European	Launch 2 June 2003; entered Mars orbit 25 Dec 2003		In orbit	No lander	Carried Beagle 2, unsuccessful lander.
Spirit rover (MER 1)	USA	Launch 10 June 2003; landed 4 Jul 2004			14.82° S, 184.85° W	Rover (MER 1).
Opportunity rover (MER 2)	USA	Launch 7 July 2003; landed 25 Jan 2004			80.07° S, 06.08° W	Rover (MER 2).
Mars Reconnaissance Orbiter	USA	Launch 12 Apr 2005; entered Mars orbit 10 Mar 2006		In orbit		Mapping and analysis.
Phoenix	USA	Launch 4 Aug 2007; landed 25 May 2008			68.22° N, 234.25° E	Surface analysis, search for water. Final signals received 2 Nov 2010.

The USSR launched six probes between 1971 and 1974, but their story is not a happy one, and all were either partial or total failures; either they missed Mars or lost contact before they could transmit useful data. Mars 2 (1971) did at least land a capsule in the southern region of Eridania, but apparently carried nothing apart from a Soviet pennant. Mars 3 (later in 1971) achieved a safe landing in Phæthontis, but transmitted for only twenty seconds after arrival; Moscow claimed that it had been blown over by a dust-storm on the surface.

In 1988, there were also two missions to the inner satellite, Phobos, but the first of these was lost during the outward journey because it had been sent a faulty command; the second sent back a few images of Mars, but for unknown reasons 'went silent' before it had started its main programme – most disappointing, because it had been scheduled to land on the surface. There was also the loss of the elaborate Mars 96 space-craft, which carried instruments of all kinds. Great things were expected of it, but unfortunately the fourth stage of the rocket launcher failed, and Mars 96 fell back Earthward, breaking up during descent and ending its career in the sea.

Japan's Nozomi ('Hope') went on its way in 1999, scheduled to orbit Mars and send back data, plus images. However, in-flight problems, aggravated by damage due to a solar storm, meant that Mars orbit could not be achieved, and on 9 December 2003 JAXA (the Japanese Space Agency) announced that the attempt had been abandoned. Nozomi did at least enter solar orbit, with a period of two years, and sent back some useful data about solar activity, but so far as Mars was concerned it was a failure.

In America, things were very different, where NASA had planned seven Mariner probes, three to Venus (Nos. 1, 2 and 5) and the other four to Mars. Of the Mars missions, all except the first were successful, and made us revise many of our ideas about the planet.

THE MARINERS

Mariner 3 (5 November 1964) was a prompt failure; control was lost, and although the probe must have entered solar orbit there is no hope of contacting it. However, its twin, Mariner 4, was a triumphant success. It flew past Mars, making its closest approach at 01h 0m 57s on 14 July 1965 and sent back the first close-range images, showing unmistakable craters. Many years earlier, it had been claimed that craters had been seen from Earth, by E. E. Barnard, in 1892, using the Lick Observatory refractor, and by J. Mellish, in 1917, with the Yerkes refractor: but these observations were never published, and are decidedly dubious. Craters had been predicted in 1944 by D. L. Cyr (admittedly for the wrong reasons), but it was not until the flight of Mariner 4 that they were definitely found. Mariner 4 also confirmed the thinness of the atmosphere, and demonstrated that the dark areas were not depressed sea-beds; indeed some, such as Syrtis Major, are plateaux. The idea of vegetation tracts was abandoned, and the final nail driven into the coffin of the canals. Mariner 4 remained in contact until 21 December 1967; it is now in solar orbit, with a period of 587 days. Its perihelion distance from the Sun is 165 000 000 km, while at aphelion it swings out to 235 000 000 km. Of course, all track of it has since been lost.

Mariners 6 and 7, of 1969, were also successful and sent back good images – notably of Hellas, a bright feature which was once thought to be a snow-covered plateau but is now known to be a deep basin; when filled with cloud it can become so brilliant that it can easily be mistaken for an extra polar cap. Yet it was only later that the true character of Mars was revealed. By ill chance, Mariners 4, 6 and 7 passed over the least spectacular areas, and it was only in 1971 that the picture changed. Mariner 8 failed – the second stage of the rocket launcher failed to ignite – but Mariner 9 entered a closed path round Mars and for the first time provided views of the towering volcanoes and the deep valleys. When it first reached the neighbourhood of Mare, a major dust storm was in progress, but this soon cleared, and the full-scale photographic coverage of the surface could begin.

The most widespread dust-storms occur when Mars is near perihelion, as was the case when Mariner 9 reached the planet (perihelion had fallen in the previous September). If the wind speed exceeds a certain critical value – 50 to 100 m s^{-1} – grains of surface material, about 100 mm across, are whipped up and given a 'skipping' motion, known technically as saltation. On striking the surface they propel smaller grains, a few micrometres across, into the atmosphere, where they may remain suspended for weeks. Over 100 localised storms and 'dust devils' occur every Martian year, and at times a storm becomes global, so that for a while all surface details are hidden. The maximum wind speeds may reach 400 km h^{-1}, but in that tenuous atmosphere they will have little force.

THE ATMOSPHERE OF MARS

The Mariners told us a great deal about the Martian atmosphere, but new data were obtained by the Vikings, which were launched in 1975 and reached Mars in 1976. The Vikings were virtually twins; each consisted of an orbiter and a lander. The orbiters continued the surveys started by Mariner 9, and then acted as relays for the landers.

The landers, separated during Mars orbit, came down gently, braked partly by parachute and partly by retro-rockets; the touchdown speed was no more than 9.6 km h^{-1}. Fortunately, both landers came down clear of the rocks which littered the landscape, and by the end of 1976 our knowledge of the planet and its atmosphere had been vastly improved.

The main constituent of the atmosphere is carbon dioxide, which accounts for more than 95% of the total; nitrogen accounts for 2.7% and argon for 1.6%, which does not leave much room for other gases (Table 7.9). The highest atmospheric pressure so far measured is 8.9 mbar, on the floor of the deep impact basin Hellas, while the pressure at the top of the lofty Olympus Mons is below 3 mbar. When Viking 1 landed in the golden plain of Chryse (latitude 22° .4 N) the pressure was approximately 7 mbar. A decrease of 0.012 mbar per sol was subsequently measured, due to carbon dioxide condensing out of the atmosphere to be deposited on the south polar cap, but this is a seasonal phenomenon, and ceased while Vikings 1 and 2 were still operating. The maximum temperature of the aeroshell of Viking 1 during the descent to Mars was 1500 °C; that of Viking 2 was similar.

Table 7.9 *Composition of the Martian atmosphere, at the surface (ppm = parts per million)*

Carbon dioxide, CO_2	95.32%
Nitrogen, N_2	2.7%
Argon, ^{40}Ar	1.6%
Oxygen, O_2	0.03%
Carbon monoxide, CO	0.07%
Water vapour, H_2O	0.03% (variable)
Neon, Ne	2.5 ppm
Krypton, Kr	0.3 ppm
Xenon, Xe	0.08 ppm
Ozone, O_3	0.03 ppm

The atmospheric pressure is at present too low for liquid water to exist on the surface, but there is no doubt that water did once exist; Mariner 9 and the Viking orbiters showed clear evidence of old riverbeds and even islands; while confirmation was obtained from the Pathfinder mission of 1997 that the area where Pathfinder landed (Ares Vallis) was once water-covered. Mars may well go through very marked climatic changes. This may be due to the effects of the changing axial inclination (between 14.9° and 35.5° over a cycle of 51 000 years) and the changing orbital eccentricity (from 0.004 to 0.141 in a cycle of 90 000 years). When one of the poles is markedly tilted sunward when Mars is at perihelion, some of the volatiles may sublime, temporarily thickening the atmosphere and even causing rainfall. It has also been suggested that every few tens of millions of years Mars goes through spells of intense volcanic activity, when tremendous quantities of gases and vapours (including water vapour) are sent out from beneath the crust.

Clouds are common. Some are due to ice crystals, such as the lee clouds formed downwind of large obstacles such as mountains, ridges and craters; wave clouds are seen at the edges of the polar caps; there are cloud sheets, linear rows of clouds looking rather like our cumulus; and streaky cloud formations. There are fogs, usually seen in low-lying areas near dawn or dusk; ground hazes, due to dust, and also what are called plumes, caused by rising material. Night-time clouds have been detected, initially by the orbiting probe Mars Global Surveyor; the probe's laser altimeter bounced pulses of light off the surface of the planet, and this light was scattered when it encountered wispy clouds. Needless to say, rainfall on Mars is unknown now.

Quite apart from the dust-storms, ice-crystal clouds occur; they are made up of water ice, and lie from about 11 to 15 km above the surface. Localised white clouds may be seen anywhere, together with the sunrise and sunset fogs and hazes.

Because the Martian atmosphere is heated from below, temperatures in the troposphere decrease with altitude, as on Earth; the lapse-rate has been given as 1.5 degrees per kilometre up to the tropopause at about 40 km, above which comes the mesosphere, where the temperature remains fairly constant at −130 °C. The mesosphere extends to about 80 km, and then comes the excessively tenuous ionosphere, where the temperature rises again because of the effects of ultraviolet radiation from the Sun, reaching a maximum of −27 °C. The ionosphere extends out to several hundred kilometres, and is really a collisionless gas; unlike the Earth's ionosphere it is not shielded from the solar wind by a strong magnetic field.

Space-craft have given information about atmospheric circulation on a global scale. For example, it has been found that large temperature gradients at the edges of the polar caps produce strong high-altitude jet-streams, particularly during autumn and winter.

MARTIAN AIRCRAFT?

There have been proposals to design aircraft for use on Mars. Aircraft would certainly be welcomed by future colonists, and it seems that they cannot be ruled out, but the tenuity of the atmosphere means that they will have to be very different from ours. A proposal to fly a winged robotic aircraft over the Valles Marineris, in 2003, was put forward by NASA's Ames Research Centre, but did not survive onslaughts from official accountants. Solar powered or rocket aircraft may of course be used.

(My own view is that if centres are established in areas all over Mars, as will quite probably happen in the foreseeable future, the main travel will be on a global railway network. One can only hope that British Rail will not be involved in running it!)

TOPOGRAPHY OF MARS: THE POLAR CAPS

The polar caps are arguably the most striking features on the surface. Both are composed mainly of water ice, and each has a seasonal coating of carbon dioxide ice. They are layered, and at least 3 km thick; it is interesting to look back at theories supported not so many decades ago, when some astronomers believed them to be made entirely of solid carbon dioxide, while others dismissed them as wafer-thin deposits of hoarfrost! When they are warmed they do not melt; they sublime.

Because the climate of the southern hemisphere is more extreme than that in the north, the southern cap is the more extensive of the two at maximum, when it may reach to latitude 50° S. In spring it may recede by 1° of latitude every Martian day, and becomes very small in the summer. When it begins to increase again, observations are hampered by the development of clouds making up what is called the polar hood, which is made up of particle of water ice and carbon dioxide ice; it may extend to within 35° of the equator. When it dissipates, the shrinking of the cap can be followed; the border becomes very ragged. The northern cap never becomes as large as its southern counterpart, and in summer becomes hard to see at all; during shrinking its outline is much more regular, because the terrain is comparatively level.

The caps are chilly places; in the south the temperature may drop to −130 °C, while temperatures in the residual northern cap can rise to −68 °C, which is well above the frost point of carbon dioxide and not far from the frost point of water in an atmosphere as thin as that of Mars.

It has been estimated that if all the water in the caps were allowed to condense, it would produce a layer at least 10 m deep over the entire planet.

BASINS, VALLEYS AND VOLCANOES

There are high peaks and deep valleys on Mars. On Earth, altitudes are reckoned from sea level, but there are no seas on Mars and it has been agreed to use a datum line where the average atmospheric pressure is 6.2 mbar. This means that all values of altitudes and depressions must be regarded as somewhat arbitrary.

The two hemispheres are not alike. Generally speaking, the southern part of the planet is heavily cratered, and much of it lies up to 3 km above the datum line; the northern part is lower – mainly below the datum line – and is more lightly cratered, so that it is presumably younger. The very ancient craters which once existed there have been eroded away, and in general the slopes are lower than those in the south. However, the demarcation line does not follow the Martian equator; instead, it is a great circle inclined to the equator at an angle of 35°.

Rather surprisingly, the two deepest basins on Mars, Hellas and Argyre, are in the south; the floor of Hellas is almost 5 km below the datum line. Argyre is about 3 km below. Both are relatively smooth, and both can become brilliant when cloud-filled.

The main volcanic area is the Tharsis bulge. This is a crustal upwarp, centred at latitude 14° S longitude 101° W; it straddles the demarcation line between the two 'hemispheres' and rises to a general altitude of around 9 km. Along it lie the three great volcanoes of Arsia Mons, Pavonis Mons and Ascræus Mons, which are spaced out at intervals of between 650 and 720 km; only Arsia Mons is south of the equator. Olympus Mons lies 1500 km to the west of the main chain, and is truly impressive, with a 600-km base and a complex 80-km caldera. The slopes are fairly gentle (6° or less); lava flows are much in evidence, and there is an extensive 'aureole', made up of blocks and ridges, around the base. It is just over 24 km high – three times the height of our Everest – and is a shield volcano, essentially similar to those of Hawaii, but on a much grander scale. Summit calderæ are also found on Arsia Mons (diameter 110 km), Ascræus Mons (65 km) and Pavonis Mons (45 km), Images from the Mars Reconnaissance Orbiter show several dark spots on the flanks of Arsia Mons, which seem to be the entrances to caves so deep that nothing can be seen inside them. It has been suggested that there could be a system of connecting tunnels, but as yet we know little about these features. A spiral dust-cloud forms regularly over the volcano, and can rise to a height of 30 km above the summit.

On the northern flank of the Tharsis rise is the remarkable Alba Patera, which is no more than 3.2 km high but is over 1500 km across, with a central caldera.

The second major volcanic area is Elysium, centred at latitude 25° N, longitude 210° W. It is considerably smaller than Tharsis, but is still well over 1900 km across; on average it reaches to about 4 km above the datum line. Volcanoes there include Elysium Mons, Albor Tholus and Hecates Tholus. Isolated volcanos are also found elsewhere – for example the Hellas area, as well as near Syrtis Major and Tempe. The 'tholi' (domes) are smaller and steeper than the 'montes', so that the material from which they formed may have been relatively viscous; like the montes, most of them have summit calderæ. There are also the 'pateræ', scalloped, collapsed shields

Table 7.10 *Altitudes of some Martian volcanoes. (These altitudes are bound to be rather arbitrary, as there is no sea level on Mars. They are reckoned from the 'datum line' where the average atmospheric pressure is 6.2 mbar)*

Name	Altitude (m)
Olympus Mons	24 000
Ascræus Mons	18 000
Pavonis Mons	18 000
Arsia Mons	9 100
Elysium Mons	9 000
Tharsis Tholus	6 000
Hecates Tholus	6 000
Albor Tholus	5 000
Uranius Tholus	3 000
Ceraunius Tholus	2 000

Table 7.11 *Estimated volcano ages*

	Age (thousands of millions of years)
Tempe Patera	3.4
Ceraunius Tholus	2.4
Uranius Tholus	2.3
Elysium Mons	2.2
Alba Patera	1.7
Hecates Tholus	1.7
Tharsis Tholus	1.4
Arsia Mons	0.7
Pavonis Mons	0.3
Ascræus Mons	0.1
Olympus Mons	0.03

with shallow slopes and complex summit cadcræ: some are symmetrical, others less regular, with radial channels running down their flanks. They may be composed of relatively loose material, with ash flows very much in evidence.

There are some features that are decidedly mysterious. One of these is Orcus Patera, which lies between Elysium Mons and Olympus Mons. It is a depression, measuring 380 × 140 km in a north-northeast–south-southwest direction; the rim rises to 1800 metres above the surrounding plains, while the floor is 400 to 600 km below its surroundings. A volcanic origin has been suggested, and the rim is cut by graben-like structures up to 2.5 km wide. Alternatively, it may have been formed by an oblique impact, perhaps less than 5° from the horizontal. Opinions differ. The altitudes of some of the Martian volcanoes are given in Table 7.10.

Researches by Dorothy Oehler and Carlton Allen of NASA, in 2010, indicate that a region in the Acidalia Planitia may contain mud volcanoes, spewing out muddy sediments from underground; these might well contain organic materials – possibly the biosignatures of past or even present life. Using images obtained by MRO, Allen has mapped 18 000 of these circular mounds, and the total number may be as much as 40 000.

Plate tectonics do not apply to Mars, so that when a volcano forms over a 'hot spot' it remains there for a very long time – which accounts of the great size of the major structures. They are generally assumed to be extinct, although a certain amount of doubt must remain. Some, such as Uranius Tholus and Elysium Mons, seem to be over 2000 million years old, but some of the Tharsis volcanoes may have been active much more recently and Olympus Mons may have last erupted a mere 30 million years ago. Estimated volcano ages, based on crater-counting statistics, are given in Table 7.11.

The greatest canyon system is the Valles Marineris, a huge gash in the surface over 4500 km long, with a maximum width of 600 km and a greatest depth of about 7 km below the rim. It begins at the complex Noctis Labyrinthus, often nicknamed the Chandelier, where we find canyons which dwarf our own Grand Canyon in Colorado. The Valles Marineris extends eastward toward Auroræ Planum, ending in the blocky terrain not far from the well-known Margarifer Terra (once known as Margaritifer Sinus, the Gulf of Pearls). For most of its course it runs roughly parallel to the line of demarcation.

Melas Chasma, part of the Valles Marineris rift valley, is one of the deepest known parts on the Martian surface; the height difference between the floor and the plateau is over 9 km, and lie 5 km below the areoid. (The areoid is the shape and size of the planet based on space observations and may be used as the Martian equivalent of sea level.) The surrounding plateau rises to 4 km above the areoid. Around Melas Chasma there is abundant evidence of past flowing water, and the sides of the valley show evidence of multiple large landslides that have created vast fan-shaped areas of material.

Craters abound, all over Mars; many of them have central peaks similar to those of the lunar craters, and the distribution laws are much the same. They have in general been named after astronomers. There are also features which look so like dry riverbeds that they can hardly be anything else.

There is evidence of past flash-floods, so that some of the old craters have been literally sliced in half. Raging torrents must have carried rocks down into the low-lying areas, and some of the 'rivers', such as the Kasei Vallis, are hundreds of kilometres long.

There can be no doubt that Mars, like the Earth, has experienced both warm periods and 'ice ages' throughout its long history, and there must also be short-term spells of global warming and cooling. Recently (from around 1999) shrinkage of the south polar cap indicates global warming, and it is logical to conclude that these warming and cooling spells coincide with similar climatic fluctuations on Earth – the climates of both planets are affected by events in the Sun.

Tables 7.17 and 7.18 at the end of this chapter detail the surface features and craters, and Figure 7.2 is the latest topographic map of the surface from the US Geological Survey. In this map, in contrast to Figure 7.1, north is up. The official current nomenclature is used.

SURFACE EVOLUTION

Our knowledge of the evolution of the Martian surface is far from complete, but there are several fairly well-defined epochs, listed in

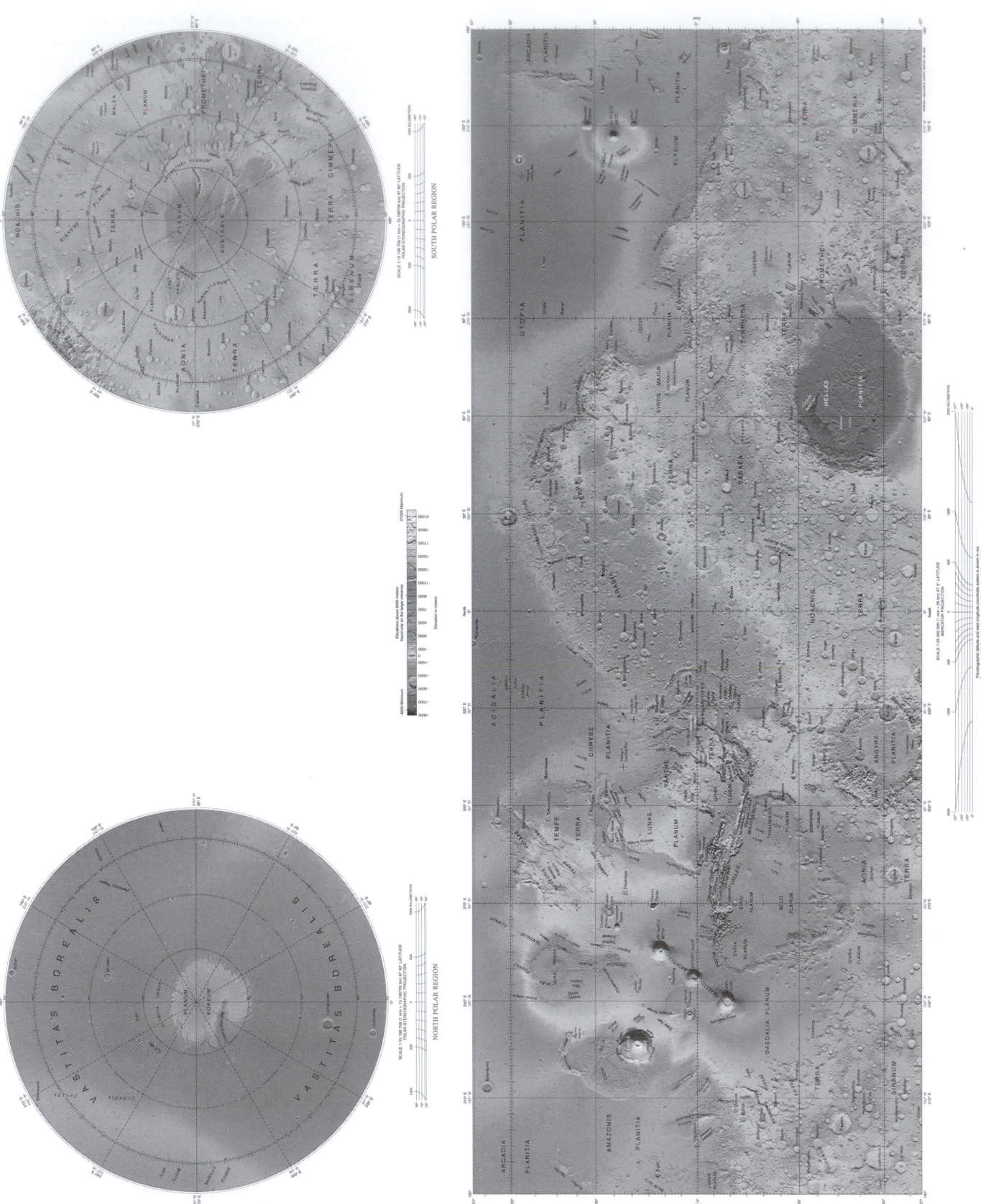

Figure 7.2 Topographic map of Mars

Table 7.12 *Martian epochs. Ages are given in thousands of millions of years, but are bound to be somewhat uncertain*

Epoch	From	To	
Early Noachian	4.5	4.4	Intense bombardment; impact basins (Hellas, Argyre)
Middle Noachian	4.4	4.3	Cratering; highland vulcanism
Late Noachian	4.3	3.8	Intercrater plains; lava flows; sinuous channels
Early Hesperian	3.8	3.7	Initial faulting of Valles Marineris; complex ridged plains; lava flows; declining cratering rate
Late Hesperian	3.7	3.6	Vulcanism in Tharsis; long sinuous channels
Early Amazonian	3.6	2.3	Smooth plains such as Acidalia; extensive vulcanism; Lava flows
Middle Amazonian	2.3	0.7	Continued vulcanism; major lava flows (Tharsis)
Late Amazonian	0.7	Present	Residual ice caps; major vulcanism in Tharsis and elsewhere, dying out at a relatively late stage; disappearance of surface water

Table 7.12. The main bombardment came during the Noachian epoch; cratering declined during the Hesperian, and virtually ceased toward the end of the Amazonian.

INTERNAL STRUCTURE

Our knowledge of the interior of Mars is very limited. It is thought that at the end of the accretion period, about 4.5 thousand million years ago, the globe differentiated into a metal-rich core and a rocky mantle; the core has a radius of between 1300 and 2000 km. The crustal thickness may be only a few kilometres in some areas, but much more (several tens of kilometres) in others, such as the southern end of the Tharsis rise.

Areas of magnetised rock have been found in the oldest terrains, possibly indicative of an ancient magnetic field, which has now become very weak. An overall field was detected in 1997 by the Mars Global Surveyor orbiter, but it is no more than 1/800 as strong as the Earth's field; whether it is produced by a weak dynamo action in the core is unclear. The orientation is the same as ours, so that in theory a compass needle would point north, but there are localised magnetic areas in the crust (for instance, near Syrtis Major and Hesperia) so that using a magnetic compass on Mars would be far from easy.

THE SEARCH FOR LIFE

Viking landers

Both Viking landers came down in the red plains of Mars, Viking 1 in Chryse and Viking 2 in the more northerly Utopia. The first picture from Viking 1, taken immediately after touchdown, showed a rockstrewn landscape and the overall impression was that of a barren, rocky desert, with extensive dunes as well as pebbles and boulders. The colour was formed by a thin veneer of red material, probably limonite (hydrated ferric oxide) covering the dark bedrock. The sky was initially said to be salmon-pink, although later pictures modified this to yellowish pink. Temperatures were low, ranging between $-96\,°C$ after dawn to a maximum of $-31\,°C$ near noon. Winds were light. They were strongest at about 10 am local time, but even then were no more than 22 km h^{-1} breezes: later in the sol they dropped to around 7 km h^{-1}: coming from the south-west rather than the east. The pattern was fairly regular from one sol to the next.

The Viking 2 site, in Utopia, was not unlike Chryse. But there were no large boulders, and the rocks looked 'cleaner'. There were no major craters in sight; the nearest large formation, Mie, was over 200 km to the west. Small and medium rocks were abundant, most of them vesicular. (Vesicles are porous holes, formed as a molten rock cools at or near the surface of a lava flow, so that internal gas bubbles escape.) There were breccias and one good example of a xenolith – a 'rock within a rock' probably formed when a relatively small rock was caught in the path of a lava flow and was coated with a molten envelope which subsequently solidified. The temperatures were very similar to those at Chryse.

One main aim of the Viking missions was to search for life. Soon after arrival, Viking 1 collected samples by using a scoop, drew them inside the space-craft, analysed them and transmitted the results to Earth. There were three main experiments:

- *Pyrolytic release.* Pyrolysis is the breaking up of organic compounds by heat. The experiment was based on the assumption that any Martian life would contain carbon, one species of which, carbon-14, is radioactive, so that when present it is easy to detect. The sample was heated sufficiently to break up any organic compounds which were present.
- *Labelled release.* This also involved carbon-14, and assumed that the addition of water to a Martian sample would trigger off biological processes if any organisms were present.
- *Gas exchange.* It was assumed that any biological activity on Mars would involve the presence of water, and the idea was to see whether providing a sample with suitable nutrients would persuade any organisms to release gases, thereby altering the composition of the artificial atmosphere inside the test chamber.

However, the results of all three experiments were inconclusive, and this was also the case when they were repeated from Viking 2. The investigators had to admit that they still could not say definitely whether there was or was not any trace of life on Mars.

PATHFINDER AND SOJOURNER

The next successful lander was Pathfinder, which came down on Mars on 4 July 1997 – America's Independence Day. This time there was no attempt to make a 'soft' landing or to put the space-craft into an initial orbit round Mars. Pathfinder was encased in airbags, and on impact it literally bounced in the manner of a beach ball. The landing speed was over 90 km h^{-1}, and the first bounce sent Pathfinder up to more than 160 m; altogether there were 15 to 17 bounces before the space-craft came to rest in an upright position, having rolled along for about a minute after the final touchdown. On Sol 2 – the second day on Mars – the tiny Sojourner rover could emerge, crawling down a ramp on to the Martian surface. Like Viking 1, Pathfinder had landed in Chryse: the distance from the Viking 1 lander was 800 km. On arrival, the main base was renamed the Sagan Memorial Station, in honour of the American astronomer Carl Sagan.

The site had been carefully chosen. It lay at the mouth of a large outflow channel, Ares Vallis, which had been carved by a violent flood in the remote past. Ares Vallis had once been a raging torrent of water, and it was thought that rocks of many different types would have been swept down into the area, which did indeed prove to be the case; there was no reasonable doubt that the whole region had once been a flood plain of standing water. Mars had not always been as arid as it is now.

Sojourner itself was a miniature vehicle 65 cm long by 18 cm high; there were six 13-cm wheels. Sojourner could be guided from Earth, and was able to analyse the surface rocks and make general observations to supplement the panoramic views from the Sagan station. Sojourner could climb over small rocks and skirt round larger ones. As it moved around it left a track in the 'soil', exposing the darker material below; the soil itself was finer than talcum powder, and was likened to the fine-grained silt found in regions such as Nebraska in the United States. The soil density was 1.2–2.0 g cm^{-3}, much the same as dry soils on Earth.

Rock analysis was carried out by an instrument on Sojourner known as the Alpha Proton X-ray Spectrometer, or APXS for short. It carried a small quantity of radioactive curium-244, which emits alpha-particles (helium nuclei). When Sojourner came up to a rock, the alpha-particles from APXS bombarded the rock; in some cases the particles interacted with the rock and bounced back, while in other cases protons or X-rays were generated. The back-scattered alpha-particles, protons and X-rays were counted, and their energies determined. The numbers of particles counted at each energy level gave a clue as to the abundance of the various elements in the rocks and also to the rock types.

Most of the rocks were, as expected, volcanic. The first to be examined, because it was nearest to the Sagan Station, was nicknamed Barnacle Bill, because of its outward appearance,[1] and proved to be similar to terrestrial volcanic rocks known as andesites; another rock, Yogi, was basaltic and less rich in silicon. However, several Martian rocks proved to have a higher silicon content than the common basalt, and it was reasoned that the original molten lava must have modestly differentiated, so that the heavy elements, such as iron, sank to the bottom and the lighter silicon compounds rose to the top of the crust, which was the source of the lavas. There was evidence in Yogi of layering or bedding, suggesting a sedimentary origin, and the pebbles and cobbles found in hollows of some of the rocks also suggested conglomerates formed in running water. Wind speeds were light during the operational life of Pathfinder, seldom exceeding 10 m s^{-1}. Rather surprisingly, dust devils – miniature tornadoes – were common, although in that tenuous atmosphere they had little force. Ice-crystal clouds were recorded, at altitudes of around 13 km.

Altogether, the lander sent back 16 000 images, and 550 were received from Sojourner before the mission ended. The last really good transmission was received on 27 September 1997. There were brief fragmentary signals on 2 and 6 October, but then Pathfinder lost contact – and of course Sojourner no longer had a relay to send data back to Earth.

Mars global surveyor

The next US probe, Mars Global Surveyor (MGS), entered Mars orbit on 11 September 1997. It was purely an orbiter. The initial orbit round Mars was highly elliptical, but at each closest approach the probe dipped into the upper Martian atmosphere and was slowed by friction, so that successive closest approaches would be lower and lower; this would continue, until eventually MGS would be moving in a circular path at an altitude of around 400 km above the surface of the planet. Achievement of the final orbit was delayed because of technical problems on the space-craft itself, but the method proved in the end to be very satisfactory.

A weak magnetic field (with a strength about 1/800th of that of the Earth's field) was confirmed on 15 September 1997, and before long excellent images were being received. (One picture, sent back on 5 April 1998, showed a rock which had earlier been imaged by the Viking orbiters, and gave a strange resemblance to a human face, although the MGS images showed, once and for all, that the rock was quite ordinary and the curious appearance had been due to nothing more significant than light and shadow patterns. Another crater, Galle, imaged by MGS, has been nicknamed the 'Happy Smiling Face', because of the arrangement of details on its floor.) Views of the canyons in the Noctis Labyrinthus area showed clear evidence of layering; flooded craters were identified, and it was also evident that many of the surface features had been shaped by winds. The polar caps were found to be thick; the northern ice cap goes down to a depth of at least 5 km, and large areas of it are surprisingly smooth. The probe's laser altimetre (MOLA) provided the best altimetric data for the planet to date.

Less successful missions

Following Japan's unsuccessful Nozomi came NASA's Mars Climate Orbiter, which would, it was hoped, operate for a full Martian year after entering its final orbit 240 km above the surface of the planet. Unfortunately, two teams were involved in the approach manœuvre; one team was working in Imperial units and the other in metric. The engine was programmed to fire using thrust data in pounds rather than in metric newtons. Five minutes after the firing of the orbital insertion engine, the space-craft entered the denser atmosphere, at an altitude 80 to 90 km lower than had been intended – with the result that it either burned up or else crashed on to the surface. It was a blunder which cost 125 million dollars.

Mars Polar Lander was launched on 3 January 1999, and was designed to come down on the following 3 December at latitude 76° S, longitude 195° W, 800 km from the Martian south pole.

It carried microphones, so that it was hoped to pick up actual sounds from Mars. All went well until ten minutes before the scheduled landing on 3 December. Two microprobes (Amundsen and Scott) were released; these were intended to send down penetrators and search for subsurface water. They would impact at over 600 km h^{-1}, and penetrate to a depth of several metres. Unfortunately, no signals were received after landing, either from the main probe or from the microprobes. The cause of this failure is unclear.

TWENTY-FIRST CENTURY MISSIONS

Though the twentieth century had ended on a rather depressing note, the first years of the twenty-first were much more encouraging. Between 2000 and 2008 three highly successful orbiters were launched, plus one fly-by (Rosetta) and four landers, including the remarkable rovers, Spirit and Opportunity, which far exceeded all expectations. Of the landers, only Beagle 2 failed. Mars Global Surveyor, which had been orbiting Mars ever since 1997 and had sent back results of the utmost value, finally lost contact on 2 November 2006. Its work had indeed been well done. A list of Mars probes is given in Table 7.8; meanwhile it will be convenient to note here those of 2000–2008.

Mars Odyssey: launched 2001 Apr; entered Mars orbit on 24 Oct 2001;

Mars Express: launched June 2 2003; entered Mars orbit on 25 Dec 2003;

Spirit rover: launched 10 June 2003, landed on 4 Jan 2004;

Opportunity rover: launched 7 July 2003, landed on 24 Jan 2003;

Mars Reconnaissance Orbiter: launched 12 Aug 2005, entered Mars orbit on 10 Mar 2006;

Phoenix: launched on 4 Aug 2007, landed 25 May 2008.

Beagle 2 was carried by Mars Express. The Rosetta comet probe flew past Mars on 25 February 2007, on its way to rendezvous with Comet Churyumov-Gerasimenko in 2011, and sent back some useful information about conditions in space in the area. At its closest point Rosetta was a mere 250 km from Mars.

Odyssey used the aerobraking technique to put it into its planned path; the aerobraking ended in January 2002, and the main mapping mission began in the following months; the orbit was almost circular, and the height above the surface was approximately 400 km, with a period of just under two days. Odyssey was designed to search for water, either past or present, and to survey promising sites for later landers. In May 2002, its spectrometer detected large amounts of hydrogen, indicating that there should be ice only a metre or two below the surface. It also acted as a relay for data obtained by the rovers Spirit and Opportunity, and discovered the strange 'cave entrances' on the flanks of the great volcano Arsia Mons.

Mars Express was the next orbiter and was put into a path which ranged between 298 and 10 107 km from the surface. It was scheduled to undertake high-resolution mapping as well as carry out a variety of other tasks, ranging from investigations of the atmospheric circulation to radar sounding of the sub-surface structure. (One picture, taken in September 2006, showed the famous 'face' in Cydonia, described on page 139, and claimed by some people to provide evidence of a past Martian civilisation; alas for these hopes! Express gave a far clearer and more detailed view than Viking 1 had done, and all semblance of a face, human or otherwise, was lost. No doubt this rocky mass will eventually become a popular tourist attraction.) Almost all its instruments worked well, notably the Sub-Surface Sounding Radar Altimeter (MARSIS) and the High Resolution Stereo Camera (HRSC). The Fourier spectrometer detected methane in the atmosphere, possibly indicative of either sub-surface micro-organisms or some kind of active vulcanism, while results from MARSIS confirmed the presence of ice not far below the surface.

All this was excellent, but was offset to some extent by the loss of Beagle 2, designed and master-minded by Colin Pillinger of Britain's Open University. It was carried by Mars Express, and was successfully released on 19 December 2003: it was expected to land on 25 December in the basin of Isidis Planitia (lat. 10.6° N, long. 270° W). It was not a rover, but among other activities it was expected to undertake a serious search for any sign of life. Unfortunately, nothing was heard from it after it left Express. It may have broken up during descent; it may have crashed and disintegrated; it may have landed intact but was unable to transmit – we simply do not know. Efforts to locate it were finally abandoned on 6 February 2004. Perhaps future explorers will find it.

We come next to the two Mars Exploration Rovers, MER-1 and MER-2, better known as Spirit and Opportunity. Each travelled to Mars on its own, rather than being attached to any other space-craft; each was protected by an effective aeroshell during the journey and, after entering the Martian atmosphere, rocket and parachute braking led to a gentle touch-down. Each lander was encased in airbags, so that after bouncing several times and rolling along the rover could be released. This always sounds a dangerous procedure, and indeed it is, but it had worked before, and with Spirit and Opportunity it worked again.

The two rovers were perfect twins, six-wheeled and solar-powered; 1.5 m high, 2.3 m wide and 1.6 m long. Each wheel had its own motor, and the rovers were highly manoeuvrable, capable of turning in one place and guiding round any obstacle likely to be encountered. The average speed was 10 mm s^{-1}, but if need be they could hurtle along at the dizzy rate of a full 50 mm s^{-1}. It was expected that each rover would have an active lifetime of only 90 sols (remember that a sol is equal to 24 hours 40 minutes, over half an hour longer than our day). This proved to be very pessimistic. Both the rovers were still in full operation at the beginning of 2010. They have had problems aplenty – for instance Spirit has become stuck in a dust drift, while Opportunity has also spent some time stuck in a drift – but they have been quite remarkably durable.

Spirit was first down, landing inside the crater Gusev (lat. 14.6° S, long. 184° E); the crater was named in honour of the nineteenth-century Russian astronomer Matvei Gusev, and is fairly regular in outline. It was chosen because it seems to be an ancient lake, with thick sediments on its floor; an extensive channel system, Ma'adim Vallis, drains into it – dry now but once, no doubt, a raging torrent. As Spirit moved around the crater, guided by NASA operators over 300 000 kilometres away, it transmitted a continuous flow of data of all kinds – for instance it used its 'rock abrasion tool' to grind away the surface of a convenient rock, and find out a great deal about the rock's composition; this was first done on 6 February 2004, when Spirit had been in the crater for 14 sols. The hole made was 2.7 mm deep, and what lay below was then inspected by Spirit's microscopic imager. Colour and monochrome images were taken regularly, and the landscape was varied; Spirit drove along the rim of the 140-m crater Bonneville, skirted the smaller crater Lahontan, and then set out to reach the summit of Husband Hill, in the Columbia Mountains. (These names are used by NASA and indeed by all investigators; no doubt they will eventually become official.)

Spirit reached the summit on 22 August (2005). It spent nearly two months there, examining the various outcops as well as Cumberland Ridge, where there are rocks containing more than the usual amount of phosphorus. Spirit then descended, and on 7 February 2006 reached Home Plate, spending the next winter there; Home Plate is thought to be an explosive volcanic deposit, surrounded by basalt. Spirit next moved on to McCool Hill, actually the highest of the Columbias, and spent the next winter on a north-facing slope, Low Ridge Haven. One problem was that on 13 March the right front wheel ceased working. By 25 October 2006 Spirit had been exploring Gusev crater for 1000 sols (Martian days).

A violent dust storm caused trouble in June 2007, and there was fear that dust covering the solar panels would cut off communication, bringing the mission to a premature end. Fortunately this did not happen, and on 6 February 2009 an obliging wind blew off the worst of the dust. On 1 May, the wheels were stuck in soft 'soil' but Spirit was still transmitting data as NASA engineers worked out the best way of extricating the rover and moving it on to its next target.

Opportunity was aimed at the opposite side of Mars: Meridiani Planum, just south of the equator (lat. 0.2° S, long. 357.5° W). It contains the 40-km crater Airy, inside which is the craterlet Airy-O, selected as the zero of longitude in fact, the Martian Greenwich (George Biddell Airy was Britain's Astronomer Royal, and it was he who obtained international agreement to use Greenwich Observatory as the zero, during the 1880s.) Crater Airy is 375 km southwest of Opportunity's landing site, Meridiani Planum was selected because it was known to contain appreciable amounts of hæmatite, an iron oxide mineral which on Earth is usually associated with water and, if this is true for Earth, why not for Mars also? In Meridiani the hæmatite seems to be part of a sedimentary rock formation several hundreds of metres thick. There are also volcanic basalts, plus the inevitable impact craters. There seems no doubt that the whole area was once covered with salty water, probably of high acidity; there are considerable amounts of magnesium sulphate and other sulphur-rich minerals such as jarosite, and there are vugs inside rocks. (Vugs are formed when crystals form inside rocks and are subsequently eroded away, leaving voids.) Rather fortunately, Opportunity landed inside a 20-m crater, at once christened Eagle, which was found to contain interesting rock outcrops and to have a floor coated with fine reddish grains and dark grey grains. The layered rocks ('Opportunity Ledge') might be either volcanic ash deposits, possibly including both ash-fall and ash-flow layers, or else sediments laid down by the action of wind and water. Rocks of all kinds were imaged and analysed, and a shallow trench was dug to examine the material immediately below the surface. The rover then began to treck across the landscape, collecting data all the time. By the end of April 2004 Endurance Crater had been reached, and in early June the decision was made to risk going down or to the floor, braving slopes of up to 30°. A hundred and eighty sols were spent inside the crater, after which Opportunity backed out (December 2004); shortly afterwards, it paused to examine its own discarded heat shield, which had protected it during the landing manœuvre, and also found a genuine meteorite, the first to be identified on another world ('Heat Shield Rock'). Various other craters were examined as Opportunity moved along, but there was trouble in April 2005, when several of the wheels were trapped in a sand-dune and the rover extricated itself only with considerable difficulty. Free from the dune, Opportunity made its way southward past a large, shallow crater, Erebus, and on 26 September 2006 reached the rim of the 730-m, well-formed crater Victoria (lat. 2.05° S, long 5.50° W) which had been well imaged by orbiters. The whole journey from the landing site in Eagle had taken 21 months, and Opportunity was still functioning well-nigh perfectly; nobody at NASA would have dared to predict that it would last for anything like as long as this. Victoria was circumnavigated, but Opportunity was in no hurry. Like Spirit, it was affected by the dust-storms; also like Spirit, it survived. On 11 September 2007, it entered the crater at a point that had been named Duck Bay, carried out some surveys and then, on 29 August 2008, reversed out again by the same route.

Opportunity spent two years examining Victoria, and then, in September 2008, set off for Endeavour, 21 km across – 25 times wider than Victoria. On 3 May 2010 it had its first glance of Endeavour's rim, 13 km away, and also the rim of a smaller crater, Iazu, 38 km away. Iazu (lat. 2.7° S, long. 5.2° W) is 6.8 km in diameter. (Its name is Romanian.)

Sadly, on 1 May 2009, 5years, 3 months, 27 Earth days after landing – almost 22 times the planned mission duration – Spirit became hopelessly stuck in soft soil. All efforts to free it failed, and on 26 January 2010 NASA had to admit that its travels were over, though it would continue to transmit data for as long as possible. We know exactly where it is, and it will wait there until astronauts rescue it and take it to a Martian museum.

Opportunity, on the other side of Mars, was more fortunate, and in December 2010 (when these words are being written) it was still functioning perfectly. On 11 July 2010, Opportunity, en route for Endeavor, had dust on its solar array cleared by an obliging solar wind. By that time the rover had travelled 21.76 km. Both the rovers have had asteroids named for them: Spirit is Asteroid 37452; Opportunity is 39382.

Mars reconnaisance orbiter

Mars Reconnaissance Orbiter (MRO) entered Mars' orbit on 10 March 2006, and was thus operational while Spirit and Opportunity were exploring the surface; one image clearly showed Opportunity at the rim of Victoria Crater. It was programmed to undertake investigations of all kinds, including a survey of the Martian Arctic (Vastitas Borealis) where the next lander was scheduled to touch

Table 7.13 *SNC ('Martian') meteorites – selected list*

Name	Location found	Date found	Mass (g)
Chassigny	Chassigny, France	3 Oct 1815	±4000
Shergotty	Shergotty, India	25 Aug 1865	±5000
Nakhla	Nakhla, Egypt	28 June 1911	±40 000
Lafayette	Lafayette, Indiana	1931	±800
Governador Valadares	Governador Valadares, Brazil	1958	158
Zagami	Zagami, Nigeria	3 Oct 1962	±18 000
ALHA 77005	Allan Hills, Antarctica	19 Dec 1977	482
Yamato 793605	Yamato Mtns, Antarctica	1979	16
EETA 79001	Elephant Moraine, Antarctica	13 Jan 1980	7900
ALH 84001	Allan Hills, Antarctica	27 Dec 1984	1940
LEW 88516	Lewis Cliff, Antarctica	22 Dec 1988	13
QUE 94201	Queen Elizabeth Range, Antarctica	16 Dec 1994	12
–	Northern Africa	May 1998	2200
Los Angeles	Mojave Desert, California	Oct 1999	452.5 and 254.4

down; it also searched – unsuccessfully – for Beagle 2 (or what remained of it) and the lost Polar Landers. It was put into its final path using the now well-tested aerobraking technique; the distance above the surface ranges between 250 and 316 km, and all the on-board instruments worked well. The images from HiRISE (the High Resolution Imaging Experiment) were probably better than any previously obtained. Thus, MRO paved the way for NASA's Phoenix.

Phoenix

Phoenix was not a rover. It was designed to land in the Vastitas Borealis, the Martian Arctic, during the summer; it would then have constant sunlight, and would continue sending data until the Sun finally set. Phoenix could not hope to survive the long winter, so that it would have no time to spare.

It was not particularly large; it measured 5.5 m long with its solar panels extended, with a 3.5-m diameter science deck and a height of 2.2 m to the top of its transmitting mast. The initial weight was 350 kg, and power for the solar array panels was generated by gallium arsenide. The expected active life was 90 sols, which is just over 92 Earth days, but in the end it was 150 sols before contact was lost. There were two main objectives: to study the geologic history of water, which would give the story of past climate changes, and to find out whether the ice layer expected to be present close to the surface was suited to life, either past or present. The position of the intended landing area was lat. 68.22° N, long. 234.25° E, in what was nicknamed the 'Green Valley' of the Vastitas Borealis. Phoenix would land on 25 May; the first sunset would be delayed until 89 sols later.

The launch on a Delta 7925 rocket, on 4 August 2007, was faultless, and so was the journey. Phoenix entered the Martian atmosphere at a velocity of 21 000 km h^{-1}, and there followed a nerve-racking seven minutes as the speed was reduced to a mere 8 km s^{-1}. All went well, and it was even possible for Mars Reconnaissance Orbiter to photograph Phoenix during its descent. Eventually the long journey was over; Phoenix touched down. It had travelled a total of 680 million km since leaving Cape Canaveral.

The landing area was flat, and the first pictures sent back showed a surface littered with small pebbles and cut by small troughs into polygons about 5 m across and 10 cm high. As expected, there were no hills or large rocks. Green Valley had been well chosen.

Phoenix wasted no time. A flow of data began almost at once, and went on unceasingly. Soil blown away during the landing uncovered bright patches which looked icy. As quickly as possible the all-important robotic arm was extended, and on 31 May started to scoop up material, providing 'soil' that could be analysed; the arm could extend 2.35 m from the base of the lander, and could dig down to half a metre below the surface. Water ice was detected on 31 July, so that one of the main objectives had been achieved. Detailed on-board analyses of the scooped-up material were carried out, and the surface soil turned out to be moderately alkaline. The SSI (the Surface Stereo Imager) took high-resolution pictures; the TEGA (Thermal and Evolved Gas Analyzer), a combination of a high-temperature furnace with a mass spectrometer, baked the samples and determined their content; the MET Meteorological Station) was hard at work, making measurements of air temperatures and pressures, dust content, wind velocities and so on . . . All the systems worked even better than had been hoped.

It was too good to last. From Phoenix the Sun first dipped behind a rock and then, on sol 89, set for 75 minutes. Autumn had come to the Green Valley; the nights grew longer, the numbing cold increased, and before long the Sun could not provide enough power to keep the solar batteries charged. The last signals were received on 2 November. Efforts to re-contact Phoenix in May 2010 were unsuccessful. New images showed definite ice damage and the mission – reluctantly – was declared over.

Though Phoenix is now silent, it still exists and has been photographed by the orbiters. Perhaps it will be finally taken to a Martian museum; certainly it has an honoured place in scientific history.

Table 7.14 *Selected list of named rocks on Mars (* = meteorite)*

VIKING PROGRAMME

Big Joe	Bonneville	Mr Badger	Mr Mole	Mr Rat
Mr Toad	Patch Rock			

PATHFINDER

Anthil	Barnacle Bill	Blackhawk	Garfield	Kitten
Piglet	Pinocchio	Jedi	Poptart	Pumpkin
Soufle	Squeeze	Spock	Stripe	T. Rex

SPIRIT

Adirondack	*Allan Hills	Cheyenne	Clovis	Concordia
Halley	Home Plate	Humphrey	King George Island	
Macquarie	Korolev	Melchior	Pot of Gold	Prat
Sejong	Sushi	Vernadsky	White Boat	*Zhong Shan

OPPORTUNITY

*Block Island	Ellesmere	Last Chance	Shark's Tooth	*Shelter Island
Vidiera	Wave	*Heat Shield Rock	*Mackinac	Wopmay
*Santa Catarina	Kalavrita	Shoemaker	Xanxer	Yuri

THE CONTINUING SEARCH FOR LIFE; METEORITES FROM MARS

One way of establishing whether or not there is any trace of life on Mars is to obtain samples of Martian material which we can analyse in our laboratories; a sample-and-return mission should be launched within the next few years. Meanwhile, it is believed by most investigators (not all) that we already have samples: meteorites of Martian origin. These are the SNC meteorites, because the first three to be identified came from Shergotty in India, Nakhla in Egypt and Chassigny in France. Over 30 have been listed; a selection of these is given in Table 7.13.

Of special note is the meteorite ALH 84001, found in the Allan Hills of Antarctica. It is shaped rather like a potato, and measures 15 cm × 10 cm × 7.6 cm; when found, it was covered with a fusion crust made of black glass. It is thought to be very ancient – of the order of 4.5 thousand million years old.

Careful examination revealed the presence of tiny features which some investigators regarded as being due to primitive forms of microscopic life. They were, moreover, accompanied by minerals which are often associated with micobacteria on Earth. Yet they were very small indeed; even the largest of them was no more than 500 nm long (1 nm is one thousand-millionth of a metre). Today it is almost universally agreed that the features are either inorganic or due to terrestrial contamination.

The official view is that ALH 84001 was blasted away from Mars about 16 000 000 years ago by a violent impact. It could not have been put straight into an Earth-crossing path, but entered an orbit round the Sun. At first the orbit was similar to that of Mars itself, but there were various perturbations; the two bodies separated, and ALH 84001 continued in its own orbit, until by chance it encountered the Earth. Around 13 000 years ago it plumped

conveniently down in Antarctica, and remained there until the meteorite collectors found it. True, this seems rather glib, but the Martian origin of it and the other SNC meteorites is supported by a mass of convincing evidence.

This raises another point. If Martian material has arrived here, could not terrestrial material have been similarly transported to Mars? It would be more difficult (breaking free from Earth is not so easy as breaking free from the less massive Mars) but if we do find living organisms there, we must carry out very careful tests to prove that the organisms really are indigenous to the Red Planet.

METEORITES ON MARS

Martian meteorites *in situ* have been found, mainly by Opportunity; the Meridiani area is clearly more meteorite-prone than the floor of Gusev, which is why Spirit has not made an equal number of discoveries.

The first to be identified was the Merediani Planum Meteorite, found by Opportunity in January 2005 (it was officially called Heat Shield Rock, because purely by chance it lay near the rover's discarded heat-shield). It is 67 cm long. Analysis with the rover's APXS (Alpha Particle X-ray Spectrometer) showed that it is 93% iron and 7% nickel, with traces of germanium and gallium; the surface shows the *regmaglypts* (pits formed by the ablation of a meteorite as it plunges through the atmosphere). Shortly afterwards Spirit managed to identify two nickel–iron meteorites in Gusev: the Allan Hills and Zhong Shan meteorites.

The largest meteorite so far found (by Opportunity in August 2009) is Block Island, in Meridiani Planum, 60 cm long and probably weighing 1½ tonnes; it is shaped rather like a melon, and shows the Widmanstätten pattern so characteristic of

Table 7.15 *The satellites of Mars: data*

	Phobos	Deimos
Mean distance from Mars (km):	9378	23 459
Mean angular distance from Mars, at mean opposition:	24″.6	1′01″.8
Mean sidereal period (days):	0.3189	1.2624
Mean synodic period:	7h 39m 26s.6	1d 5h 21m 15s.7
Orbital inclination (°):	1.068	0.8965
Orbital eccentricity:	0.01515	0.0003
Diameter (km):	27 × 22 × 19	15 × 12 × 10
Density, water = 1:	2.0	1.7
Mass (g):	1.08×10^{19}	1.8×10^{18}
Escape velocity (m s^{-1}):	3–10	6
Magnitude at mean opposition:	11.6	12.8
Maximum apparent magnitude, seen from Mars:	−3.9	−0.1
Apparent diameter seen from Mars:		
max.	12′.3	2′
min.	8′	1′.7

Table 7.16 *Features on Phobos and Deimos*

	Lat. (°)	Long. (° W)	Diameter (km)	
PHOBOS				
Craters				
Clustril	+60.0	91.0	3.4	
D'Arrest	−39.0	719.0	2.1	H., German astronomer, 1822–1875
Drunio	+36.5	92.0	4.2	
Flimnap	+60.0	350.0	1.5	
Grildrig	+81.0	195.0	2.8	
Gulliver	+62.0	163.0	5.5	
Hall	−80.0	210.0	5.4	Asaph, US astronomer, 1829–1907
Limtoo	−11.0	54.0	2.0	
Reldresal	+41.0	39.0	2.9	
Roche	+53.0	183.0	2.3	E., French astronomer, 1820–1883
Sharpless	−27.5	154.0	1.8	B., USA astronomer, 1904–1960
Skyresh	+52.5	320.0	1.5	
Stickney	+1.0	49.0	9.0	Asaph Hall's wife, 1830–1892
Todd	−9.0	153.0	2.9	D., US astronomer, 1855–1939
Wendell	−1.0	132.0	1.7	O., USA astronomer, 1845–1912
Dorsum				
Kepler	−45.0	356.0	15.0	J., German astronomer, 1571–1630
DEIMOS				
Craters				
Swift	+12.5	358.2	1.0	J., English writer, 1667–1745
Voltaire	+22.0	3.5	1.9	F., French writer, 1694–1778

All names, unless otherwise stated, are of characters in Swift's *Voyage to Laputa*.

nickel–iron meteorites. Around 700 m from it Opportunity found the smaller Shelter Island Meteorite, 47 cm long. And there is also the striking Mackinac Meteorite, found by Opportunity as the rover started its trek to Endeavour Crater.

On 16 September 2010, its 2363rd sol on Mars, Opportunity made a careful study of the Oilean Ruaidh meteorite (the name is that of an island off the northwest coast of Ireland). The meteorite is of the nickel–iron type. Opportunity then resumed its journey to Endeavour Crater.

A selected list of named rocks on Mars is given in Table 7.14.

MARTIAN METEORS

Shooting-stars will be seen in the Martian sky. The first image of a meteor was obtained on 7 March 2004, from the Spirit rover. Meteors will become visible at a height of about 120 km above the ground, where the atmospheric density is approximately equal to that of the Earth at the same altitude, but the entry velocities will be lower, and on average the meteors will be half a magnitude fainter than ours.

The parent comet of the meteor imaged by Spirit was probably 114 P/ Wiseman-Skiff, which has a period of 6.5 years. It may well have been a member of a shower, with the radiant in Cepheus; Martian Cepheids may be seen every Martian year. Other regular showers may be associated with Halley's Comet (radiant near λ Geminorum), Comet P/13 Olbers (radiant near β Canis Majoris) and even the enigmatical asteroid 5335 Damocles, which has 'dual nationality' as both a comet and an asteroid (radiant in Draco).

In 2003, Apostolis Christou, of Armagh Observatory, predicted a shower associated with Comet P/79 du Toit-Hartley. The shower was confirmed visually, and by related effects in the Martian ionosphere detected by the orbiting MGS satellite. Other regular showers have been predicted, though of course all this is, at present, highly speculative and uncertain.

At least future observers will not have to contend with moonlight; Phobos and Deimos can never be bright enough to flood the sky with radiance, and there is no interference from man-made light pollution – at least, not yet!

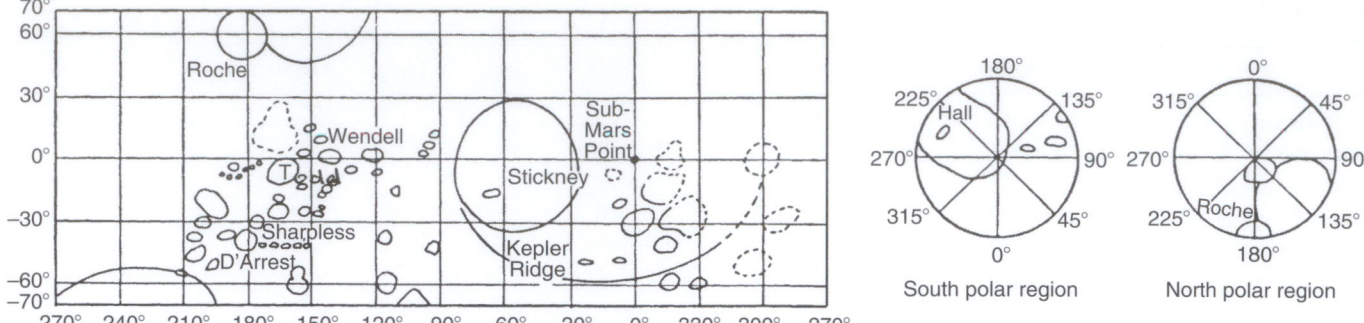

Figure 7.3 Phobos: craters are indicated by solid lines, ridges by dashed lines

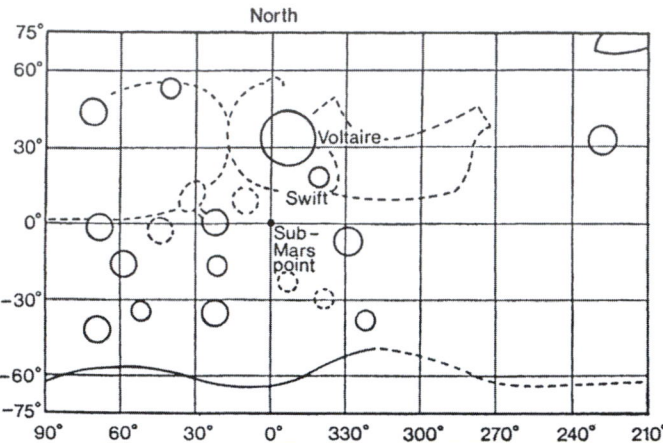

Figure 7.4 Deimos: craters are indicated by solid lines, ridges by dashed lines

THE SATELLITES OF MARS

Mars has two dwarf satellites, Phobos and Deimos, named after the attendants of the mythological war god. Their basic data are given in Table 7.15 and their features in Table 7.16. Maps of their surfaces are shown in Figures 7.3 and 7.4.

The first mention of possible Martian satellites was fictional – by Jonathan Swift, in Gulliver's *Voyage to Laputa* (1727), Swift described two satellites, one of which had a revolution period shorter than the rotation period of its primary, but at that time there was no telescope which could have shown either Phobos or Deimos. Two satellites were also described in another novel, Voltaire's *Micromegas* (1750). The reasoning was, apparently, that since Earth had one satellite and Jupiter was known to have four, Mars could not possibly manage with less than two!

A satellite reported telescopically in 1645 by A. Schyrle was certainly nothing more than a faint star. The first systematic search for satellites was made in 1783 by William Herschel. The result was negative, and H. D'Arrest, from Copenhagen, in 1862 and 1864, was similarly unsuccessful.

The first satellite to be discovered was Deimos, on 10 August 1877, by Asaph Hall; the telescope used was the 66-cm Clark refractor at the US Naval Observatory (Washington, DC). On 16 August he discovered Phobos. The names were suggested by Henry Madan, Science Master at Eton College; they commemorate Terror and Fear, two attendants of the mythological war god, Ares (Mars). In 1952, 1954 and 1956 G.P. Kuiper, using the McDonald 208-cm reflector, made a systematic search for extra satellites, but with no success, and he concluded, no doubt rightly, that no new satellite as much as a kilometre across could exist.

Both moons are elusive telescopic objects; they are not particularly faint (about magnitude 12.8 for Deimos and 11.6 for Phobos under the best possible observing conditions) but they are inconveniently close to Mars. In 1930, E. M. Antoniadi, using the Meudon 83-cm refractor, reported that Phobos was white and Deimos bluish, but these results were certainly wrong; it would be virtually impossible to see such colours on objects as small and faint as Phobos and Deimos.

The origin of the satellites is not certainly known. It is tempting to suggest they are ex-asteroids, captured by Mars long ago; certainly they seem to be similar to carbonaceous (C-type) asteroids. The fact that their orbits are virtually circular, and lie in Mars' equatorial plane, means that their capture would involve some special set of circumstances; all the same, the capture theory does seem to be the most likely answer. Neither could raise appreciable tides in Martian oceans, if they existed, and neither can stabilise the tilt of Mars' axis of rotation, as the Moon does for the Earth.

Tidal effects indicate that Phobos is gradually spiralling downward, so that the satellite may impact Mars at some time between 30 million and 100 million years hence. Deimos, much further out, is in a stable orbit, and will not suffer a similar fate.

Both are cratered. Phobos is covered with a layer of fine-grained regolith, probably about 100 m deep – it has been said that an astronaut would find himself 'hip-deep in dust'! The whole surface has been pounded by constant impacts of meteoroids, some of which have started landslides leaving dark trails on the slopes of the main craters. This is particularly evident in the largest crater, Stickney, which is 9.6 km in diameter; boulders on the rim are up to 50 m across, and rocks rolling down the slopes have left obvious tracks, showing that the gravity on Phobos, weak though it is, is not inappreciable. The gravity field has about 1/1000th of the strength

Table 7.17 *Selected list of Martian formations*

	Lat.	Long. (° W)	Diameter (km)
(a) Craters			
Adams	31.3 N	197.1	100
Agassiz	70.1 S	88.4	104
Airy	5.2 S	0.0	56
Alexei Tolstoy	47.6 S	246.4	94
Alitus	34.5 S	49.0	29
Aniak	32.1 S	69.6	58
Antoniadi	21.7 N	299.0	381
Arago	10.5 N	330.2	154
Arandas	42.6 N	15.1	22
Arkhangelsky	41.3 S	24.6	119
Arrhenius	40.2 S	237.0	132
Azul	42.5 S	42.3	20
Azusa	5.5 S	40.5	41
Babakin	36.4 S	71.4	78
Bakhuysen	23.1 S	344.3	162
Balboa	3.5 S	34.0	20
Baldet	23.0 N	294.5	195
Baltisk	42.6 S	54.5	48
Bamba	3.5 S	41.7	21
Bamberg	40.0 N	3.0	57
Barabashov	47.6 N	68.5	126
Barnard	61.3 S	298.4	128
Becquerel	22.4 N	7.9	167
Beer	14.6 S	8.2	80
Bernard	23.8 S	154.2	129
Berseba	4.4 S	37.7	36
Bianchini	64.2 S	95.1	77
Bjerknes	43.4 S	188.7	89
Boeddicker	14.8 S	197.6	107
Bond	33.3 S	35.7	104
Bouguer	18.6 S	332.8	106
Bozkir	44.4 S	32.0	89
Brashear	54.1 S	119.2	126
Briault	10.1 S	270.2	100
Bunge	34.2 S	48.4	78
Burroughs	72.5 S	243.1	104
Burton	14.5 S	156.3	137
Byrd	65.6 S	231.9	122
Camiling	0.8 S	38.1	21
Camiri	45.1 S	41.9	20
Campbell	54.0 S	195.0	123
Cartago	23.6 S	17.8	33
Cassini	23.8 N	327.9	415
Cerulli	32.6 N	337.9	120
Chamberlin	66.1 S	124.3	125
Charlier	68.6 S	168.4	100
Chekalin	24.7 S	26.6	87
Chia	1.6 N	59.5	94
Chimbote	1.5 S	39.8	65
Chincoteague	41.5 N	236.0	35
Choctaw	41.5 S	37.0	20
Clark	55.7 S	133.2	93
Coblentz	55.3 S	90.2	111
Cobres	12.1 S	153.7	89
Columbus	29.7 S	165.8	114
Comas Solà	20.1 S	158.4	132
Concord	16.6 N	34.1	20
Copernicus	50.0 S	168.6	292
Crommelin	5.3 N	10.2	111
Cruls	43.2 S	196.9	83
Curie	29.2 N	4.9	98
Da Vinci	1.5 N	39.1	98
Daly	66.4 S	23.0	99
Dana	72.6 S	33.1	95
Darwin	57.2 S	19.2	166
Dawes	9.3 S	322.3	191
Dein	38.5 N	2.4	24
Dejnev	25.7 S	164.5	156
Denning	17.5 S	326.6	165
Dia-Cau	0.3 S	42.8	28
Dison	25.4 S	16.3	20
Dokuchaev	60.8 S	127.1	73
Douglass	51.7 S	70.4	97
Du Martheray	5.8 N	266.4	94
Du Toit	71.8 S	49.7	75
Edam	26.6 S	19.9	20
Eddie	12.5 N	217.8	90
Eiriksson	19.6 S	173.7	56
Escanalte	0.3 N	244.8	83
Eudoxus	45.0 S	147.2	92
Fesenkov	21.9 N	86.4	86
Flammarion	25.7 N	311.7	160
Flaugergues	17.0 S	340.9	235
Focas	33.9 N	347.2	82
Fontana	63.2 S	71.9	78
Foros	34.0 S	28.0	23
Fournier	4.25 S	287.5	112
Gale	5.3 S	222.3	172
Gali	44.1 S	36.9	24
Galilaei	5.8 N	26.9	124
Galle	50.8 S	30.7	230
Gilbert	68.2 S	273.8	115
Gill	15.8 N	354.5	81
Glazov	20.8 S	26.4	20
Gledhill	53.5 S	272.9	72
Globe	24.0 S	27.1	45
Graff	21.4 S	206.0	157
Green	52.4 S	8.3	184
Grójec	21.6 S	30.6	37
Guaymas	26.2 N	44.8	20
Gusev	14.6 S	184.6	166
Hadley	19.3 S	203.0	113
Haldane	53.0 S	230.5	72
Hale	36.1 S	36.3	136

Table 7.17 (cont.)

	Lat.	Long. (° W)	Diameter (km)
Halley	48.6 S	59.2	81
Hartwig	38.7 S	15.6	104
Heaviside	70.8 S	94.8	103
Heinlein	64.6 S	243.8	83
Helmholtz	45.6 S	21.1	107
Henry	11.0 N	336.8	165
Herschel	14.9 S	230.1	304
Hilo	44.8 S	35.5	20
Hipparchus	45.0 S	151.1	104
Holden	26.5 S	33.9	141
Holmes	75.0 S	293.9	109
Hooke	45.0 S	44.4	145
Huancayo	3.7 S	39.8	25
Huggins	49.3 S	204.3	82
Hussey	53.8 S	126.5	100
Hutton	71.9 S	255.5	99
Huxley	62.9 S	259.2	108
Huygens	14.0 S	304.4	456
Ibragimov	25.9 S	59.5	89
Innsbrück	6.5 S	40.0	64
Janssen	2.8 N	322.4	166
Jarry-Desloges	9.6 S	276.1	97
Jeans	69.9 S	205.5	71
Joly	74.6 S	42.5	81
Kaiser	46.6 S	340.9	201
Kakori	41.9 S	29.6	25
Kansk	20.8 S	17.1	34
Kantang	24.8 S	17.5	64
Karpinsk	46.0 S	31.8	28
Karshi	23.6 S	19.2	22
Kashira	27.5 S	18.3	68
Kasimov	25.0 S	22.8	92
Keeler	60.7 S	151.2	92
Kepler	47.2 S	218.7	219
Kipini	26.0 N	31.5	75
Knobel	6.6 S	226.9	127
Korolev	72.9 N	195.8	84
Kovalsky	30.0 S	141.4	299
Krishtofovich	48.6 S	262.6	111
Kuba	25.6 S	19.5	25
Kufra	40.6 N	239.7	32
Kuiper	57.3 S	157.1	86
Kunowsky	57.0 N	9.0	60
Kushva	44.3 S	35.4	39
Labria	35.3 S	48.0	60
Lamas	27.4 S	20.5	21
Lambert	20.1 S	334.6	87
Lamont	58.3 S	113.3	72
Lampland	36.0 S	79.5	71
Lassell	21.0 S	62.4	86
Lasswitz	9.4 S	221.6	122
Lau	74.4 S	107.3	109
Le Verrier	38.2 S	342.9	139
Lebu	20.6 S	19.4	20
Li Fan	47.4 S	153.0	103
Liais	75.4 S	252.9	128
Libertad	23.3 N	29.4	31
Liu Hsin	53.7 S	171.4	129
Lockyer	28.2 N	199.4	74
Lohse	43.7 S	16.4	156
Lomonosov	64.8 N	8.8	151
Lorica	20.1 S	28.3	67
Loto	22.2 S	22.3	22
Lowell	52.3 S	81.3	201
Luga	44.6 S	47.2	42
Luki	30.0 S	37.0	20
Luzin	27.3 N	328.8	86
Lyell	70.0 S	15.6	134
Lyot	50.7 N	330.7	220
Mädler	10.8 S	357.3	100
Magadi	34.8 S	46.0	57
Magelhæns	32.9 S	174.5	102
Maggini	28.0 N	350.4	146
Main	76.8 S	310.9	102
Maraldi	62.2 S	32.1	119
Marbach	17.9 N	249.2	20
Marca	10.4 S	158.2	83
Mariner	35.2 S	164.3	151
Marth	13.1 N	3.6	104
Martz	35.2 S	215.8	91
Maunder	50.0 S	358.1	93
McLaughlin	22.1 N	22.5	90
Mellish	72.9 S	24.0	99
Mena	32.5 S	18.5	31
Mendel	59.0 S	198.5	82
Mie	48.6 N	220.4	93
Milanković	54.8 N	146.6	113
Millman	54.4 S	149.6	82
Millochau	21.5 S	274.7	102
Mitchel	67.8 S	284.0	141
Molesworth	27.8 S	210.6	175
Moreux	42.2 N	315.5	138
Müller	25.9 S	232.0	120
Murgoo	24.0 S	22.3	24
Mutch	0.6 N	55.1	200
Nansen	50.5 S	140.3	82
Nardo	27.8 S	32.7	23
Peridier	25.8 N	276.0	99
Navan	26.2 S	23.2	25
Newcomb	24.1 S	359.0	259
Newton	40.8 S	157.9	287
Nicholson	0.1 N	164.5	114
Niesten	28.2 S	302.1	114
Nitro	21.5 S	23.8	28
Noma	25.7 S	24.0	38

Table 7.17 (cont.)

	Lat.	Long. (° W)	Diameter (km)
Nordenskiold	53.0 S	158.7	87
Ochakov	42.5 S	31.6	30
Oraibi	17.4 N	32.4	31
Ostrov	26.9 S	28.0	67
Ottumwa	24.9 N	55.7	55
Oudemans	10.0 S	91.7	121
Pāros	22.2 N	98.1	40
Pasteur	19.6 N	335.5	114
Perepelkin	52.8 N	64.6	112
Perrotin	3.0 S	77.8	95
Pettit	12.2 N	173.9	104
Phillips	66.4 S	45.1	183
Pickering	34.4 S	132.8	112
Playfair	78.0 S	125.5	68
Podor	44.6 S	43.0	25
Polotsk	20.1 S	26.1	23
Poona	24.0 N	52.3	20
Porter	50.8 S	113.8	113
Poynting	8.4 N	112.8	80
Priestly	54.3 S	229.3	40
Proctor	47.9 S	330.4	168
Ptolemæus	46.4 S	157.5	184
Pulawy	36.6 S	76.7	51
Pylos	16.9 N	30.1	29
Quenisset	34.7 N	319.4	127
Rabe	44.0 S	325.2	99
Radau	17.3 N	4.7	115
Rayleigh	75.7 S	240.1	153
Redi	60.6 S	267.1	62
Renaudot	42.5 N	297.4	69
Reuyl	9.5 S	193.1	74
Revda	24.6 S	28.3	26
Reynolds	75.1 S	157.6	91
Richardson	72.6 S	180.3	82
Ritchey	28.9 S	50.9	82
Roddenberry	49.9 S	4.5	140
Ross	57.6 S	107.6	88
Rossby	47.8 S	192.2	82
Ruby	25.6 S	16.9	25
Rudaux	38.6 N	309.0	52
Russell	55.0 S	347.4	138
Rutherford	19.2 N	10.6	116
Ruza	34.3 S	52.8	20
Salaga	47.6 S	51.0	28
Sangar	27.9 S	24.1	28
Santa Fe	19.5 N	48.0	20
Sarno	44.7 S	54.0	20
Schaeberle	24.7 S	309.8	160
Schiaparelli	2.5 S	343.4	461
Schmidt	72.2 S	77.5	194
Schöner	20.4 N	309.5	185
Schröter	1.8 S	303.6	337
Secchi	57.9 S	257.7	218
Semeykin	41.8 N	351.2	71
Seminole	24.5 S	18.9	21
Shambe	20.7 S	30.5	29
Sharanov	27.3 N	58.3	95
Shatskii	32.4 S	14.7	69
Sibu	23.3 S	19.6	31
Sigli	20.5 S	30.6	31
Sklodowska	33.8 N	2.8	116
Slipher	47.7 S	84.5	129
Smith	66.1 S	102.8	71
Sögel	21.7 N	55.1	29
Sokol	42.8 S	40.5	20
Soochow	16.8 N	28.9	30
South	77.0 S	338.0	111
Spallanzani	58.4 S	273.5	72
Spencer Jones	19.1 S	19.8	85
Stege	2.6 N	58.4	72
Steno	68.0 S	115.3	105
Stokes	56.0 N	189.0	70
Stoney	69.8 S	138.4	177
Suess	67.1 S	178.4	72
Sumgin	37.0 S	48.6	83
Sytinskaya	42.8 N	52.8	90
Tabor	36.0 S	58.5	20
Tara	44.4 S	52.7	27
Tarakan	41.6 S	30.1	37
Taza	44.0 S	45.1	22
Teisserenc de Bort	0.6 N	315.0	118
Terby	28.2 S	286.0	135
Tikhonravov	13.7 N	324.1	390
Tikhov	51.2 S	254.1	107
Timbuktu	5.7 S	37.7	63
Timoshenko	42.1 N	63.9	84
Trouvelot	16.3 N	13.0	168
Trümpler	61.7 S	150.6	77
Turbi	40.9 S	51.2	25
Tuskegee	2.9 S	36.2	69
Tycho Brahe	49.5 S	213.8	108
Tyndall	40.1 N	190.4	79
Valverde	20.3 N	55.8	35
Verlaine	9.4 S	295.9	42
Very	49.8 S	176.8	127
Viana	19.5 N	255.3	29
Victoria	2.05 S	5.50	730
Vik	36.0 S	64.0	25
Vinogradov	56.3 S	216.0	191
Vishniac	76.7 S	276.1	76
Vivero	49.4 N	241.3	64
Voeykov	32.5 S	76.1	67
Vogel	37.0 S	13.2	124
von Kármán	64.3 S	58.4	100
Wabash	21.5 N	33.7	42

Table 7.17 (cont.)

	Lat.	Long. (° W)	Diameter (km)
Wallace	52.8 S	249.2	159
Waspam	20.7 N	56.6	40
Wegener	64.3 S	4.0	70
Weinbaum	65.9 S	245.5	86
Wells	60.1 S	237.4	94
Wicklow	2.0 S	40.7	21
Wien	10.5 S	220.1	105
Williams	18.8 S	164.1	125
Windfall	2.1 S	43.5	20
Wirtz	48.7 S	25.8	128
Wislencius	18.3 S	348.7	138
Wright	58.6 S	150.8	106
Zilair	32.0 S	33.0	43
Zongo	32.0 S	42.0	23
Zulanka	2.3 S	42.3	47
Zuni	19.3 N	29.6	25

(b) The features listed here are not visible with ordinary Earth-based telescopes, apart from the main dark areas and some of the smaller features such as Olympus Mons (formerly Nix Olympica, the Olympic Snow). The length or diameter in km is given. Positions refer to the centres of the features.

Catenæ (chains of craters)

Acheron Catena	38.2 N	100.7	554
Alba Catena	35.2 N	114.6	148
Ceraunius Catena	37.4 N	108.1	51
Elysium Catena	18.0 N	210.4	66
Ganges Catena	2.6 S	69.3	221
Labeatis Catenæ	18.8 N	95.1	318
Phlegethon Catena	40.5 N	101.8	875
Tithoniæ Catena	5.5 S	71.5	380
Tractus Catena	27.9 N	103.2	1234

Chasmata (large recilinear chains)

Chasma Australe	82.9 S	273.8	491
Chasma Boreale	83.2 N	21.3	318
Capri Chasma	8.7 S	42.6	1498
Candor Chasma	6.5 S	71.0	816
Eos Chasma	12.6 S	45.1	963
Gangis Chasma	8.4 S	48.1	541
Hebes Chasma	1.1 S	76.1	285
Ius Chasma	7.2 S	84.6	1003
Juventæ Chasma	1.9 S	61.8	495
Melas Chasma	10.5 S	72.9	526
Ophir Chasma	4.0 S	72.5	251
Tithonium Chasma	4.6 S	86.5	904

Fossæ (ditch)

Acheron Fossæ	38.7 N	136.6	1120
Alba Fossæ	43.4 N	103.6	2077
Amenthes Fossæ	10.2 N	259.3	1592
Ceraunius Fossæ	24.8 N	110.5	711
Claritas Fossæ	34.8 S	99.1	2033
Coloe Fossæ	37.1 N	303.9	930
Coracis Fossæ	34.8 S	78.6	747
Elysium Fossæ	27.5 N	219.5	1114
Icaria Fossæ	53.7 S	135.0	2153
Ismeniæ Fossæ	38.9 N	326.1	858
Labeatis Fossæ	20.9 M	95.0	388
Mareotis Fossæ	45.0 N	79.3	795
Memnonia Fossæ	21.9 S	154.4	1370
Nili Fossæ	24.0 N	283.0	709
Noctis Fossæ	3.3 S	99.0	692
Olympica Fossæ	24.4 N	115.3	573
Sirenum Fossæ	34.5 S	158.2	2712
Tantalus Fossæ	44.5 N	102.4	1990
Tempe Fossæ	40.2 N	74.5	1553
Thaumasia Fossæ	45.9 S	97.3	1118
Ulysses Fossæ	11.4 N	123.3	720
Uranius Fossæ	25.8 N	90.1	438
Zephyrus Fossæ	23.9 N	214.4	452

Labyrinthus (valley complex)

Adamas Labyrinthus	36.5 N	255.0	664
Noctis Labyrinthus	7.2 S	101.3	976

Mensæ

Aeolis Mensæ	3.7 S	218.5	841
Bætis Mensa	5.4 S	72.4	181
Cydonia Mensæ	37.0 N	12.8	854
Deuteronilus	45.7 N	337.9	627
Galaxias Mensæ	36.5 N	212.7	327
Nepenthes Mensæ	9.3 N	241.7	1704
Nilokeras Mensæ	32.9 N	51.1	327
Nilosyrtis Mensæ	35.4 N	293.6	693
Protonilus Mensæ	44.2 N	309.4	592
Sacra Mensæ	25.0 N	69.2	602
Zephyria Mensæ	10.1 S	188.1	230

Mons / Montes (mountain or volcano)

Arsia Mons	9.4 S	120.5	485
Ascræus Mons	11.3 N	104.5	462
Charitum Montes	58.3 S	44.2	1412
Elysium Mons	25.0 N	213.0	432
Erebus Montes	39.9 N	170.5	594
Hellespontus Montes	45.5 S	317.5	681
Libya Montes	2.7 N	271.2	1229
Nereidum Montes	41.0 S	43.5	1677
Olympus Mons	18.4 N	133.1	624
Pavonis Mons	0.3 N	112.8	375
Phlegra Montes	40.9 N	197.4	1310
Tartarus Montes	25.1 N	188.7	1011
Tharsis Montes	2.8 N	113.3	2105

Pateræ (shallow crater, volcanic structure)

Alba Patera	40.5 N	109.9	464
Amphitrites Patera	59.1 S	299.0	138
Apollinaris Patera	8.3 S	186.0	198
Biblis Patera	2.3 N	123.8	117
Diacria Patera	34.8 N	132.7	75
Hadriaca Patera	30.6 S	267.2	451

Table 7.17 (cont.)

	Lat.	Long. (° W)	Diameter (km)
Meroe Patera	7.2 N	291.5	60
Nili Patera	9.2 N	293.0	70
Orcus Patera	14.4 N	181.5	381
Peneus Patera	58.0 S	307.4	123
Tyrrhena Patera	21.9 S	253.2	597
Ulysses Patera	2.9 N	121.5	112
Uranius Patera	26.7 N	92.0	276
Planitiæ (low, smooth plain)			
Acidalia Planitia	54.6 N	19.9	2791
Amazonis Planitia	16.0 N	158.4	2816
Arcadia Planitia	46.4 N	152.1	3052
Argyre Planitia	49.4 S	42.8	868
Chryse Planitia	27.0 N	36.0	1500
Elysium Planitia	14.3 N	241.1	3899
Hellas Planitia	44.3 S	293.8	2517
Isidis Planitia	14.1 N	271.0	1238
Utopia Planitia	47.6 N	277.3	3276
Planæ (high plain or plateau)			
Auroræ Planum	11.1 S	50.2	590
Planum Australe	80.3 S	155.1	1313
Planum Boreum	85.0 N	180.0	1066
Bosporus Planum	33.4 S	64.0	330
Planum Chronium	62.0 S	212.6	1402
Dædalia Planum[a]	13.9 S	138.0	2477
Hesparia Planum	18.3 S	251.6	1869
Icaria Planum	42.7 S	107.2	840
Lunæ Planum	9.6 N	66.6	1845
Malea Planum	65.9 S	297.4	1068
Ophir Planum	9.6 S	62.2	1068
Planum Angustum	80.0 S	83.9	208
Sinai Planum	12.5 S	87.1	1064
Syria Planum	12.0 S	103.9	757
Syrtis Major Planum	9.5 N	289.6	1356
Thaumasia Planum	22.0 S	65.0	930
Rupes (scarp)			
Amenthes Rupes	1.8 N	249.4	441
Argyre Rupes	63.3 S	66.7	592
Arimanes Rupes	10.0 S	147.4	203
Avernus Rupes	8.9 S	186.4	200
Bosporus Rupes	42.8 N	57.2	500
Cerberus Rupes	8.4 N	195.4	1254
Chalcoporus Rupes	54.9 S	338.6	380
Claritas Rupes	26.0 S	105.5	–
Cydnus Rupes	59.6 N	257.5	430
Elysium Rupes	25.4 N	211.4	180
Morpheos Rupes	36.2 S	234.1	321
Ogygis Rupes	34.1 S	54.9	225
Olympus Rupes	17.2 N	133.9	1819
Phison Rupes	26.6 N	309.4	149
Pityusa Rupes	63.0 S	328.0	290
Promethei Rupes	76.8 S	286.8	1491
Rupes Tenuis	81.9 N	65.2	–
Tartarus Rupes	6.6 S	184.4	81
Thyles Rupes	73.2 S	205.6	269
Ulyxis Rupes	63.2 S	198.8	303
Scopuli (lobate or irregular scarps)			
Charybdis Scopulus	24.9 S	339.9	513
Coronæ Scopulus	33.9 S	294.1	306
Eridania Scopulus	53.4 S	217.9	510
Nilokeras Scopulus	31.5 N	57.2	1064
Œnotria Scopulus	11.2 S	283.2	1438
Scylla Scopulus	25.5 S	342.0	474
Tartarus Scopulus	6.4 S	181.7	127
Sulci (subparallel furrows or ridges)			
Amazonis Sulci	3.3 S	145.1	243
Apollinaris Sulci	11.3 S	183.4	300
Cyane Sulci	25.5 N	128.4	286
Gigas Sulci	9.9 N	127.6	467
Gordii Sulci	17.9 N	125.6	316
Lycus Sulci	29.2 N	139.8	1639
Medusæ Sulci	5.3 S	159.8	91
Memnonia Sulci	6.5 S	175.6	361
Terræ (land masses)			
Aonia Terra	58.2 S	94.8	3372
Arabia Terra	25.0 N	330.0	6000
Terra Cimmeria	34.0 S	215.0	2285
Margaritifer Terra	16.2 S	21.3	2049
Terra Meridiani	7.2 S	356.0	1622
Noachis Terra	45.0 S	350.0	3500
Promethei Terra	52.9 S	262.2	2761
Terra Sabæa	10.4 S	330.6	1367
Terra Sirenum	37.0 S	160.0	2165
Tempe Terra	41.3 N	70.5	2055
Tyrrhena Terra	14.3 S	278.5	2817
Xanthe Terra	4.2 N	46.3	3074
Tholi (domed hills)			
Albor Tholus	19.3 N	209.8	165
Australis Tholus	59.0 S	323.0	40
Ceraunius Tholus	24.2 N	97.2	108
Hecates Tholus	32.7 N	209.8	183
Iaxartes Tholus	72.0 N	15.0	53
Jovis Tholus	18.4 N	117.5	61
Kison Tholus	73.0 N	358.0	50
Ortygia Tholus	70.0 N	8.0	100
Tharsis Tholus	13.4 N	90.8	153
Uranius Tholus	26.5 N	97.8	71
Undæ (dune)			
Abalos Undæ	81.0 N	83.1	80 090
Hyperboreæ Undæ	77.5 N	46.0	80 050
Valles (valleys)			
Al-Qahira Vallis	17.5 S	196.7	546
Ares Vallis	9.7 N	23.4	1690
Auqakuh Vallis	28.6 N	299.3	195
Bahram Vallis	21.4 N	58.7	403
Brazos Valles	6.3 S	341.7	494

Table 7.17 (cont.)

	Lat.	Long. (° W)	Diameter (km)
Dao Vallis	36.8 S	269.4	667
Evros Vallis	12.6 S	345.9	335
Granicus Valles	28.7 N	227.6	445
Harmakhis Vallis	39.1 S	267.2	585
Hebrus Valles	20.0 N	233.9	299
Hrad Vallis	37.8 N	221.6	719
Huo Hsing Vallis	31.5 N	293.9	340
Indus Vallis	19.2 N	322.1	253
Kasei Valles	22.8 N	68.2	2222
Loire Valles	18.5 S	16.4	670
Louros Valles	8.7 S	81.9	423
Ma'adim Vallis	21.0 S	182.7	861
Maja Valles	15.4 N	56.7	1311
Mamers Vallis	41.7 N	344.5	945
Mangala Valles	7.4 S	150.3	880
Marti Vallis	11.0 N	182.0	1700
Mawrth Vallis	22.4 N	16.1	575
Naktong Vallis	5.2 N	326.7	807
Nanedi Valles	5.5 N	48.7	470
Nirgal Vallis	28.3 S	41.9	511
Ravi Vallis	0.4 S	40.7	169
Samara Valles	24.2 S	18.9	575
Scamander Vallis	16.1 N	331.1	272
Shalbatana Vallis	5.6 N	43.1	687
Simud Vallis	11.5 N	38.5	1074
Tinjar Valles	38.2 N	235.7	390
Tisia Valles	11.6 S	313.9	390
Tiu Vallis	8.6 N	34.8	970
Uzboi Vallis	30.0 S	35.6	340
Valles Marineris	11.6 S	70.7	4128
Vastitas (widespread lowland)			
Vastitas Borealis	67.5 N	180.0	9999

[a]Formerly Erythræum Planum.

of the Earth's field at sea level, and the escape velocity is a modest 11 m s^{-1}. The impact which produced Stickney must have come close to breaking up the whole satellite. Both satellites are irregular in shape, in which respect they are similar to the small asteroids which have been examined from close range.

Conspicuous grooves are to be seen on Phobos, typically up to 20 km long, l00 to 200 m wide and up to 30 m deep. Superficially they look as though they are radiating from Stickney, and it was at first thought that they were formed at the same time as the Stickney impact, but this is not so; they are not radial to Stickney, but are centred on the leading apex of Phobos in its orbit. Presumably they are due to material hurled up from Mars following a crater-producing impact there. There are no similar features on Deimos, because of the much greater distance from the Martian surface, and neither are there any major craters; the two largest, Swift and Voltaire, are no more than 3 km in diameter.

The close fly-bys of Phobos confirmed that it shares many surface characteristics with C-type asteroids, but its origin is still not finally settled. It is not easy to explain either the capture mechanism or the subsequent evolution of the orbit into the equatorial plane of Mars. An alternative hypothesis is that it formed around Mars and is therefore a remnant from the planetary formation period. In addition, Mars Express studied the intended landing sites for the Phobos-GRUNT mission (see below). The closest approach was 67 km.

The unmanned Russian sample-and-return mission to Phobos, Phobos-GRUNT, was due to be launched in October 2009, on a Zenit rocket with a Fregat upper stage. (The word GRUNT means 'soil' – there are no vocal associations.) It was designed to land, collect samples, and return with up to 160 g of Phobos soil. Lift off from Phobos would be gentle – required velocity a mere 35 km h^{-1}. General studies of Mars and both satellites were also scheduled to be carried out from orbit, over a period of several months, before touchdown on Phobos, and the landers *in situ* experiments would continue for a year. The whole mission would last for 3 years.

Technical problems meant that the 2009 launch window was missed, and the mission was rescheduled for 2011. As originally planned the Zenit also carries the Chinese 10 kg Mars Orbiter – Yinghuo-1 ('Firefly'), entering Martian orbit with periapsis 80 000 km, period 3 days, inclination 5°; it will be active for 1 year and will have a full programme of research.

Deimos, with its gravity field only 1/3000th of that of the Earth, has a more subdued surface; the largest crater, Voltaire, is less than 3 km in diameter, though it has been suggested that an 11 km-wide depression near the south pole might be an impact scar. There are nearly as many craters per unit area as there are on Phobos, but those on Deimos have been more eroded and filled in.

Neither satellite would be of much use in lighting up Martian nights. Moreover, Phobos would be invisible to any observer on the planet above latitude 69° north or south: for Deimos the limiting latitude would be 82° for a Martian observer. Phobos would appear less than half the diameter of the Moon as seen from Earth, and could give only about as much light as Venus does to us: from Phobos, Mars would subtend a mean angle of 42°. The maximum apparent diameter of Deimos as seen from Mars would be only about twice that of Venus seen from Earth, and with the naked eye the phases would be almost imperceptible. Deimos would remain above the Martian horizon for 2½ sols consecutively: Phobos would cross the sky in only 4½ h, moving from west to east, and the interval between successive risings would be a mere 11 h. Total solar eclipses could never occur. Phobos would transit the Sun 1300 times a year, taking 19 s to cross the disc, while Deimos would show an average of 130 transits, each taking 1 min 48 s. To an observer on Mars, eclipses of the satellites by the shadow of the planet would be very frequent.

Predictably, both Phobos and Deimos have synchronous rotation periods. This leads to great ranges in temperature, particularly in the case of Phobos. During its 7½ hour 'day' the temperature may rise to −4 °C, while at night it falls to −112 °C. There will, of course, be pronounced libration effects.

Table 7.18 *Martian crater names of astronomers*

Adams	Walter S.	American	1876–1956
Airy	George Biddell	British	1801–1892
Arago	Dominique François	French	1786–1853
Bakhuysen	Henricus van de Sande	Dutch	1838–1923
Baldet	Ferdinand	French	1885–1964
Barabashov	Nikolai	Russian	1894–1971
Barnard	Edward Emerson	American	1857–1923
Beer	Wilhelm	German	1797–1850
Bianchini	Francesco	Italian	1662–1729
Bond	George P.	American	1825–1865
Boeddicker	Otto	German	1853–1937
Briault	P.	French	?–1922
Burton	Charles E.	British	1846–1882
Campbell	William Wallace	American	1862–1938
Cassini	Giovanni	Italian	1625–1712
Cerulli	Vicenzio	Italian	1859–1927
Charlier	Carl V. L.	Swedish	1862–1934
Clark	Alvan	American	1804–1887
Coblentz	William	American	1873–1962
Comas Solá	Jose	Spanish	1868–1937
Copernicus	Nikolas (Mikołaj)	Polish	1473–1543
Crommelin	Andrew Claude de la C.	Irish	1865–1939
Darwin	George	British	1845–1912
Dawes	William Rutter	British	1799–1868
Denning	William Frederick	British	1848–1931
Douglass	Andrew E.	American	1867–1962
Du Martheray	Maurice	Swiss	1892–1955
Du Toit	Alexander	South African	1878–1948
Eddie	Lindsay	South African	1845–1913
Escalante	F.	Mexican	*c.* 1930
Eudoxus	–	Greek	408–355 BC
Fesenkov	Vasili	Russian	1889–1972
Flammarion	Camille	French	1842–1925
Flaugergues	Honoré	French	1755–1835
Focas	Ioannas	Greek	1909–1969
Fontana	Francesco	Italian	1585–1685
Fournier	Georges	French	1881–1954
Gale	Walter F.	Australian	1865–1945
Galilei	Galileo	Italian	1564–1642
Galle	Johann	German	1812–1910
Gill	David	Scottish	1843–1914
Gledhill	Joseph	British	1836–1906
Graff	Kasimir	German	1878–1950
Green	Nathan	British	1823–1899
Gusev	Matwei	Russian	1826–1866
Hale	George Ellery	American	1868–1938
Halley	Edmond	British	1656–1742
Hartwig	Ernst	German	1851–1923
Henry	Paul and Prosper	French	1848–1905 1849–1903
Herschel	William	Hanoverian/ British	1738–1822
Hipparchus	–	Greek	*c.* 160–125 BC
Holden	Edward	American	1846–1914
Hooke	Robert	British	1635–1703
Huggins	William	British	1824–1910
Ibragimov	Nadir	Russian	1932–1977
Janssen	Pierre	French	1824–1910
Jarry-Desloges	René	French	1868–1951
Jeans	James Hopwood	British	1877–1946
Kaiser	Frederik	Dutch	1808–1872
Keeler	James	American	1857–1900
Kepler	Johannes	German	1571–1630
Knobel	Edward	British	1841–1930
Kovalsky	Marian	Russian	1821–1884
Kunowsky	George	German	1786–1846
Lamont	Johann von	German	1805–1879
Lampland	Carl O.	American	1873–1951
Lau	Hans E.	Danish	1879–1918
Le Verrier	Urbain	French	1811–1877
Li Fan	–	Chinese	*c.* 85 AD
Liais	Emmanuel	French	1826–1900
Liu Hsin	–	Chinese	?–22
Lockyer	J. Norman	British	1836–1920
Lohse	Oswald	German	1845–1915
Lowell	Percival	American	1855–1916
Lomonosov	Mikhail	Russian	1711–1765
Lyot	Bernard	French	1897–1952
Mädler	Johann Heinrich von	German	1794–1874
Maggini	Mentore	Italian	1890–1941
Main	Robert	British	1808–1878
Maraldi	Giacomo	Italian	1665–1729
Marth	Albert	British	1828–1897
Maunder	Edward W.	British	1851–1928
McLaughlin	Dean B.	American	1901–1965
Mellish	John	American	1886–1970
Milankovič	Milutin	Jugoslav	1879–1958
Millman	Peter	Canadian	1906–1990
Millochau	Gaston	French	1866–?
Mitchel	Ormsby	American	1809–1862
Moreux	Theophile	French	1867–1954
Müller	Carl	German	1851–1925
Mutch	Thomas	American	1931–1980
Newcomb	Simon	American	1835–1909
Newton	Isaac	British	1643–1727
Nicholson	Seth	American	1891–1963
Niesten	Louis	Belgian	1844–1920
Oudemans	Jean	Dutch	1827–1906

Table 7.18 (cont.)

Perepelkin	Evgenii	Russian	1906–1940
Perrotin	Henri	French	1845–1904
Pettit	Edison	American	1890–1962
Phillips	Theodore	British	1868–1942
Phillips	John	British	1800–1874
Pickering	Edward C.	American	1846–1919
Pickering	William H.	American	1858–1938
Porter	Russell W.	American	1871–1949
Poynting	John Henry	British	1852–1914
Proctor	Richard A.	British	1837–1888
Ptolemæus	Claudius	Greek	c. 120–180
Quenisset	Ferdinand	French	1872–1951
Radau	Rudolphe	French	1835–1911
Rabe	Wilhelm	German	1893–1958
Renaudot	Gabrielle	French	1877–1962
Ritchey	George W.	American	1864–1945
Ross	Frank E.	American	1874–1966
Rudaux	Lucien	French	1874–1947
Russell	Henry Norris	American	1877–1957
Schaeberle	John	American	1853–1924
Schiaparelli	Giovanni	Italian	1835–1910
Schmidt	Johann	German	1825–1884
Schmidt	Otto	Russian	1891–1956
Schröter	Johann Hieronymus	German	1745–1816
Secchi	Angelo	Italian	1818–1878
Semeykin	Boris	Russian	1900–1937
Sharonov	Vesvolod	Russian	1901–1964
Slipher	Vesto	American	1875–1969
South	James	British	1785–1867
Spencer Jones	Harold	British	1890–1960
Sytinskaya	Nadezhda	Russian	1906–1974
Terby	François	Belgian	1846–1911
Tikhonravov	Gavriil	Russian	1851–1916
Trouvelot	Etienne	French	1827–1895
Tycho Brahe	–	Danish	1546–1601
Very	Frank W.	American	1852–1927
Vogel	Hermann Carl	German	1841–1907
Vishniac	Wolf	American	1922–1974
von Kármán	Theodor	Hungarian	1881–1963
Wirtz	Carl Wilhelm	German	1876–1939
Wislensius	Walter	German	1859–1905
Wright	William H.	American	1871–1959
Williams	Arthur Stanley	British	1861–1938

It has been suggested that the Kaidun Meteorite, which fell on a Soviet military base on 12 March 1980, may have come from Phobos. This is based on the presence of alkaline-rich material in the meteorite, but the idea that there is an association with Phobos is, to put it mildly, uncertain!

PLANETARY TRANSITS VISIBLE FROM MARS

From Mars, the Earth is an inferior planet, and can pass in transit across the face of the Sun; it will be easy enough to observe, and neither will the Moon be a difficult object. The last transit took place on 11 May 1984; the next will be on 10 November 2084 – possibly this event will be observed from Martian bases, which may well have been established by then! The following transits will be on 15 November 2163, 10 May 2189 and 13 May 2268. Venus will next transit on 19 August 2030, 18 June 2032 and 5 November 2059; its apparent diameter will be just over 18 seconds of arc. Mercury will transit on 10 May 2013, 4 June 2014, 15 April 2015, and 25 October 2023. These events may well be recorded by unmanned space-craft near or on Mars, but as Mercury will have an angular diameter of a mere 6 seconds of arc it will be by no means conspicuous.

MARTIAN TROJANS

There are several small Martian 'Trojan' asteroids, which move in the same orbit as the planet, though they keep either well ahead of Mars or else well behind, and are in no danger of being swallowed up. The brightest of them is 647 Eureka, discovered by David Levy in 1990; it is about 3 km in diameter, and trails the planet; its orbital period is 686.8 days, virtually the same as that of Mars.

As an aside: in 1958, the eminent Soviet astronomer Iosif Shklovskii published a paper in which he suggested that because Phobos was being 'braked' by the upper limits of the Martian atmosphere it must be of negligible mass, and either hollow or else 'an iron sphere 16 km across and less than 6 cm thick'. It would therefore be a satellite built by the Martians. The idea was not well received by the USSR Academy of Sciences but, in February 1960, no less a person than S. F. Singer, science adviser to the US President Dwight D. Eisenhower, wrote that 'My conclusion is that I back Shklovskii.' He added that the purpose of Phobos would probably be 'to sweep up radiation in Mars' atmosphere, so that the Martians could safely operate round their planet'. Another supporter was Robert H. Wilson, Chief of Applied Mathematics at NASA, who opined that Phobos might be 'a colossal base orbiting Mars'.

Naturally, all this was welcomed by Flying Saucer enthusiasts, who maintained that Phobos and Deimos had not been discovered before 1877 because they had not then existed.

Years later, when Phobos had been studied from close range, I asked Shklovskii whether he had changed his mind. He replied that his original paper had been nothing more than a practical joke. No comment!

MANNED FLIGHT TO MARS?

Serious consideration is now being given to the possibility of sending astronauts to Mars. Technically this seems feasible, but there are very obvious extra problems compared with flights to the Moon. Radiation is probably the worst of these. The crew will be in space, unprotected by any natural radiation screen, for several months, and as yet we do not know how to cope with this. No doubt unmanned vehicles will have deposited materials on Mars to await the first travellers, but there is no need to emphasise the dangers facing the pioneers.

My personal view – which may well be completely wrong! – is that several spacecraft will take part, taking off probably from the Lunar Base and going first to Deimos, making use of the tiny outer satellite as a convenient natural space-station, before making the final descent to Mars. Setting up bases on the planet itself lies further in the future, but eventually there may be two inhabited worlds in the Solar System instead of only one. The 'new Martians' will have arrived.

ENDNOTE

1 Nicknames included Hassock, Sausage, Mermaid, Squash, Wedge, Stripe, Soufflé and Desert Princess. Nobody can accuse the NASA teams of being lacking in imagination; Yogi was so named because from some angles it bore a slight resemblance to the celebrated Yogi Bear.

8 · Minor members of the Solar System

'The asteroids are of no interest to the amateur observer; in fact they are of very little interest to anybody.' So wrote G. F. Chambers, a well-known astronomical author, in 1898. A German professional astronomer, whose name does not seem to have been preserved, even described them as 'vermin of the skies', because his photographs were often found to be streaked with unwanted asteroid tracks. The modern view is very different. The asteroids are of immense interest and importance, and indeed one major observatory, the Klět Observatory in the Czech Republic, is devoted mainly to studies of them.

These minor 'planets' are now divided into two classes: dwarf planets, and Small Solar System Bodies (SSSBs). Between the orbits of Mars and Jupiter lies what is termed the Main Belt of what are still universally called asteroids, the largest of which, Ceres, has been elevated to dwarf planet status. All dwarf planets and SSSBs are given numbers, usually (not always) in order of discovery. 1 Ceres was found in 1801, and three more Main Belt objects were discovered before the end of 1807: 2 Pallas, 3 Juno and 4 Vesta. With Ceres they make up the so-called 'Big Four', though in fact the fourth largest asteroid is 10 Hygiea, with Juno only fourteenth in order of size. Of the entire Main Belt swarm, only Vesta is ever easily seen with the naked eye.

Some relatively small asteroids have orbits which swing them out of the Main Belt, and may bring them close to the Earth, while others move in orbits wholly within that of Mars. There are asteroids with eccentric orbits which take them out far beyond the Main Belt swarm, even beyond Uranus; there are the Jupiter 'Trojans', which move in the same orbit as the Giant Planet, though they keep prudently well ahead or else well behind Jupiter itself, and are in no danger of being engulfed. But there are also many substantial bodies moving beyond Neptune, and of these four, 136 199 Eris, 134 340 Pluto, 136 108 Haumed and 136 472 Makemake are classed as dwarf planets. For the moment let us deal with minor 'planets' whose mean distances from the Sun are less than that of Neptune, and deal later with the trans-Neptunians in what is known as the Kuiper Belt. (In fact the first idea of a belt of small bodies beyond the orbit of Neptune seems to have come from F. C. Leonard as early as 1930, as much more detailed proposals were made in 1943 by an English amateur, Colonel Kenneth Edgeworth. These were so similar to Kuiper's, made independently much later, that it is unfair to ignore his contribution, and some authorities still refer to the Edgeworth–Kuiper Belt.)

DISCOVERY

A systematic search for a planet moving round the Sun between the orbits of Mars and Jupiter was started in 1801 by six astronomers who met at Lilienthal, where Johann Schröter had his observatory. They based their search on Bode's law (actually first described by J. Titius of Wittemberg; it was popularised by J. E. Bode in 1772). It runs as follows:

Take the numbers 0, 3, 6, 12, 24, 48, 96, 192 and 384, each of which, apart from the first, is double its predecessor. Add 4 to each. Taking the Earth's mean distance from the Sun as 10 units, the remaining figures give the distances of the planets with reasonable accuracy out as far as Saturn, the remotest planet known in 1772. Uranus, discovered in 1781, fitted in well. Neptune does not, but it was not discovered until 1846. Bode's law is given in Table 8.1.

It is now believed that the 'law' is nothing more than coincidence, and it breaks down completely for Neptune, the third most massive planet in the Solar System. However, it seemed good enough for the self-styled 'Celestial Police' to begin a hunt for a planet corresponding to the Bode number 28. Schröter became President of the Police, with the Hungarian astronomer Baron Franz Xaver von Zach as Secretary. They very logically began their search in the region of the ecliptic.

Ceres was discovered on 1 January 1891, the first day of the new century, by G. Piazzi from Palermo. He was working on a new star catalogue, and was not a member of the Police (though he joined later). Ceres was found to have a Bode distance of 27.7. Pallas, Juno and Vesta were identified by the time that the Police disbanded in 1813, and the fifth asteroid, Astræa, was not found until 1845. The discovery was made by the German amateur K. Hencke, after a long and painstaking search which he began, unaided, in 1830.

When Ceres was discovered it was regarded as a planet, and given a planetary symbol. Pallas, Juno and Vesta were also given symbols. The practice was continued after the discovery of Astræa and down to Hygiea., but not with any real enthusiasm. Symbols for the first ten asteroids are given here, but are hardly ever used.

Since 1847, no year has passed without the discovery of at least one new asteroid. By 1868, 100 were known; by 1921 the number had risen to 1000; by 1946, to 2000; by 1985, to 3000; by 1999, to over 10 000. By 2010, the number had soared to well over 130 000. Some of these high-numbered asteroids have unexpected names; for example 108 153 is Uccle, while 108 912 is Emerald Lane.

Yet the combined mass of all the inner asteroids is much less than that of the Earth. Moreover, most of the mass of the Main Belt asteroids is concentrated in a few of the largest bodies; Ceres accounts for 32% of the total; add Vesta (9%), Pallas (7%) and Hygiea (3%) and we reach 52%. Table 8.2 list these first four. It is clear that the rest of the Main Belt members make no more than a tiny contribution.

Table 8.1 *Bode's law*

Planet	Distance by Bode's law	Actual distance
Mercury	4	3.9
Venus	7	7.2
Earth	10	10.0
Mars	16	15.2
–	28	–
Jupiter	52	52.0
Saturn	100	95.4
Uranus	196	191.8
Neptune	–	300.7
Pluto	388	394.6

Table 8.2 *The first four asteroids*

Asteroid number and name	Diameter (km)	Approximate mass (10^{15} kg)	Rotation period (h)	Orbital period (years)	Type
1 Ceres	960×932	870 000	9.075	4.60	C
2 Pallas	$571 \times 525 \times 482$	318 000	7.811	4.61	U
3 Juno	288×230	20 000	7.210	4.36	S
4 Vesta	525	300 000	5.342	3.63	V

The first asteroid to be discovered photographically was 323 Brucia, on 21 December 1891, by Max Wolf. Systematic searches are now being made with dedicated telescopes, such as the 0.91m Spacewatch Telescope and the 1-m Linear Telescope.

NOMENCLATURE

When an asteroid is found, it is given a provisional designation, such as 1999 DE. The first four characters give the year of discovery; the fifth, the time of year reckoned in two-week periods in an alphabetical sequence – A the first two weeks in January, B the second two weeks in January, C the first two weeks in February and so on (I is omitted); the sixth character indicates the order of discovery. Thus 1999 DE indicates that the asteroid was the fifth to be discovered during the second half of February. If a month has over 24 discoveries, a subscript is added: thus the 25th asteroid to be found during the first half of February 1999 is 1999 CA_1. When a reliable orbit has been worked out, the asteroid is assigned a number. Data for the first hundred known asteroids are given in Table 8.3, and Table 8.4 gives data on further interesting examples.

The discoverer of an asteroid is entitled to suggest a name, and this recommendation is almost always accepted by the International Astronomical Union. The first asteroid to be named was 1 Ceres, discovered by Piazzi from Palermo and named by him after the patron goddess of Sicily. All early names were mythological, but when the supply of deities became exhausted more 'modern' names were introduced, some of them decidedly bizarre: the first was that

of 125 Liberatrix, which seems to have been given by the discoverer, P. M. Henry of Paris, in honour of Joan of Arc. Asteroids may be named after countries (136 Austria, 1125 China, 1197 Rhodesia) and people (1123 Shapleya, after Dr Harlow Shapley; 1462 Zamenhof, after the inventor of Esperanto; 1486 Marilyn, after the daughter of Paul Herget, Director of the Cincinnati Observatory).

I feel very honoured that 2602 Moore, discovered by E. Bowell from Flagstaff in Arizona, was named by him after me!

The first asteroid to be named after an electronic calculator is 1625 The NORC – the Naval Ordnance Research Calculator at Dahlgre, Virginia. Plants are represented (978 Petunia, 973 Aralia), as are universities (694 Ekard – 'Drake' spelled backwards), musical plays (1047 Geisha), foods (518 Halawe – the discoverer, R. S. Dugan, was particularly fond of the Arabian sweet *halawe*), social clubs (274 Philagoria – a club in Vienna) and shipping lines (724 Hapag–Haniburg–Amerika line). One name has been expunged; 1267 was named Vladimirov in honour of a Soviet philanthropist, Anatoli Vladimirov, but was withdrawn when Vladimirov was exposed as a financial swindler. One name was auctioned: 250 Bettina, whose discoverer Palisa offered to sell his right of naming for £50. The offer was accepted by Baron Albert von Rothschild, who named the asteroid after his wife. Names of politicians and military leaders are in general not allowed, although we do have 1581 Abanderada – someone who carries a banner – associated with Eva Peron, wife of the former ruler of Argentina. Asteroid 1000 is, appropriately enough, named Piazzia.

(My own favourite names are 2309 Mr Spock named after a ginger cat which was itself named after the Vulcanian space-traveller in the fictional starship *Enterprise*, and 6042 Cheshirecat!)

There have been some strange misidentifications. There was for instance the object discovered on 5 December 1991 by J. V. Scotti, who was carrying out a routine patrol with the 0.9m Spacewatch telescope. It passed within 460 000 km of the Earth, and varied in brightness; the estimated diameter was 6 m. The orbital inclination was 0.4°, and the period only slightly longer than that of the Earth. It was given an asteroid designation, 1991 VG, but was not seen again, and was probably a piece of man-made satellite débris. And, on 7 November 2007, astronomers in Arizona reported discovering a near-Earth asteroid, designated 2007 VN84, which according to the Minor Planet Center would have a close brush with Earth. An observant Russian, Denis Denisenko, then pointed out that 2007 VN84 was actually the Rosetta space-craft. These mistakes will happen!

One oddity is 2001 QE 24, diameter 0.9 km. Its rotation period is only 29.2 minutes, so that any loose object on the surface would be thrown off. Landing a space probe there would be a decidedly interesting manœuvre.

719 Albert was discovered in 1911, and then lost. It was finally recovered in 2000, so that no numbered asteroids are now 'unaccounted for'. Asteroid 330, Adalberta, never existed at all; it was recorded by Max Wolf in 1892, but was photographed only twice, and it was then found that the images were of two separate stars. One named asteroid, Hermes, was 'lost' for 66 years. It was discovered on 28 October 1937 by K. Reinmuth, and flew by us at only 800 000 km – a record at that time. It reached the 8th magnitude and moved at 5° h^{-1}, so that it crossed the sky in 9 days.

Table 8.3 *The first 100 Main Belt asteroids*

		Discoverer	Date	Mean distance from Sun (×10⁶ km)	q (a.u.)	Q (a.u.)	e	P(y)	i (°)
1	Ceres	Piazzi	1 Jan 1801	414.7	2.55	2.99	0.08	4.60	10.59
2	Pallas	Olbers	28 Mar 1802	414.9	2.13	3.41	0.23	4.62	34.84
3	Juno	Harding	1 Sept 1804	399.2	1.98	3.36	0.26	4.36	12.97
4	Vesta	Olbers	29 Mar 1807	353.3	2.15	2.57	0.09	3.63	7.13
5	Astræa	Hencke	8 Dec 1845	384.9	2.08	3.07	0.19	4.13	5.37
6	Hebe	Hencke	1 July 1847	362.9	1.94	2.91	0.20	3.78	14.75
7	Iris	Hind	13 Aug 1847	357.0	1.83	2.94	0.23	3.68	56.53
8	Flora	Hind	18 Oct 1847	329.4	1.86	2.55	0.16	3.27	5.89
9	Metis	Graham	25 Apr 1848	357.1	2.10	2.68	0.12	3.69	5.58
10	Hygiea	de Gasparis	12 Apr 1849	469.3	2.76	3.51	0.12	5.56	3.84
11	Parthenope	de Gasparis	11 May 1850	367.0	2.21	2.70	0.10	3.84	4.62
12	Victoria	Hind	13 Sept 1850	349.2	1.82	2.84	0.22	3.57	8.36
13	Egeria	de Gasparis	2 Nov 1850	383.4	2.36	2.79	0.08	4.13	16.34
14	Irene	Hind	19 May 1851	386.7	2.15	3.02	0.17	4.16	9.10
15	Eunomia	de Gasparis	29 July 1851	395.4	2.15	3.14	0.19	4.30	11.74
16	Psyche	de Gasparis	17 Mar 1852	436.9	2.51	3.33	0.14	4.99	3.09
17	Thetis	Luther	17 Apr 1852	369.5	2.13	2.80	0.13	3.88	5.59
18	Melpomene	Hind	24 June 1852	343.4	1.79	2.80	0.22	3.48	10.13
19	Fortuna	Hind	22 Aug 1852	365.2	2.05	2.83	0.16	3.81	1.57
20	Massalia	de Gasparis	16 Sept 1852	360.3	2.06	2.75	0.14	3.74	0.70
21	Lutetia	Goldschmidt	15 Nov 1852	364.3	2.04	2.83	0.16	3.80	3.06
22	Kalliope	Hind	16 Nov 1852	435.2	2.61	3.21	0.10	4.96	13.71
23	Thalia	Hind	15 Dec 1852	393.1	2.02	3.24	0.23	4.26	10.14
24	Themis	de Gasparis	5 Apr 1853	468.2	2.71	3.54	0.13	5.54	0.76
25	Phocæa	Chacorna	6 Apr 1853	359.0	1.79	3.01	0.26	3.72	21.58

		D (km)	Density	R (h)	Axial tilt (°)	Albedo	Type	Magnitude
1	Ceres	970	2.08	9.07	3	0.09	C	6.7
2	Pallas	570 × 525 × 500	2.8	7.82	60	0.16	B	6.4
3	Juno	290 × 240 × 190	3.4	7.21	51	0.40	S	7.5
4	Vesta	578 × 560 × 458	3.4	5.28	29	0.42	V	5.1
5	Astræa	167 × 123 × 82	2.7	16.80	33	0.23	S	8.7
6	Hebe	205 × 185 × 170	4.1	7.20	42	0.27	S	7.5
7	Iris	225 × 190 × 190	2.9	7.20	85	0.28	S	6.7
8	Flora	145 × 145 × 120	2.7	12.87	78	0.24	S	7.9
9	Metis	235 × 195 × 140	2.7	5.08	72	0.24	S	8.1
10	Hygiea	500 × 385 × 350	2.4	27.60	60	0.07	C	9.1
11	Parthenope	153	2.7	9.43	–	0.18	S	8.6
12	Victoria	113	2.7	8.64	–	0.18	S	8.5
13	Egeria	217 × 196	2.7	6.96	–	0.08	C	6.7
14	Irene	182	2.7	15.12	–	0.16	S	8.8
15	Eunomia	330 × 245 × 205	3.8	6.08	165	0.21	S	7.9
16	Psyche	280 × 230 × 190	3.3	4.19	95	0.12	M	9.3
17	Thetis	90	2.0	12.27	–	0.17	S	10.7
18	Melpomene	150 × 125	2.0	11.57	–	0.22	S	8.9
19	Fortuna	225	2.1	7.44	–	0.04	C	8.9
20	Massalia	160 × 145 × 130.	3.2	8.10	45	0.21	S	8.3
21	Lutetia	120 × 100 × 80	2.7	8.16	89	0.21	M	9.2
22	Kalliope	215 × 180 × 150	2.0	4.15	103	0.14	M	10.6
23	Thalia	108	2.0	12.31	–	0.25	S	9.2
24	Themis	198	?	8.40	–	0.07	C	11.8
25	Phocæa	75	2.0	9.94	–	0.23	S	10.3

Table 8.3 (cont.)

		Discoverer	Date	Mean distance from Sun (×10⁶ km)	q (a.u.)	Q (a.u.)	e	P(y)	i (°)
26	Proserpina	Luther	5 May 1853	397.6	2.42	2.89	0.09	4.33	3.56
27	Euterpe	Hind	8 Nov 153	351.3	1.95	2.75	0.17	3.60	1.58
28	Bellona	Luther	1 Mar 1854	415.6	2.37	3.19	0.15	4.63	0.40
29	Amphitrite	Marth	1 Mar 1854	382.1	2.37	2.74	0.07	4.08	6.10
30	Urania	Hind	22 July 1854	354.1	2.01	2.67	0.13	3.64	2.10
31	Euphrosyne	Ferguson	1 Sept 1854	471.2	2.44	3.86	0.23	5.59	26.32
32	Pomona	Goldschnidt	26 Oct 1854	387.1	2.37	2.80	0.08	4.16	5.53
33	Polyhymnia	Chacornac	24 Oct 1854	428.6	1.81	3.83	0.34	4.85	1.87
34	Circe	Chacornac	6 Apr 1855	401.7	2.39	2.98	0.11	4.40	5.50
35	Leukothea	Luther	19 Apr 1855	447.2	2.31	3.67	0.23	5.17	7.94
36	Atalante	Goldschmidt	5 Oct 1855	410.9	1.91	3.58	0.30	4.55	18.43
37	Fides	Luther	5 Oct 1855	395.1	2.18	3.11	0.18	4.29	3.07
38	Leda	Chacornac	12 Jan 1856	410.4	2.33	3.16	0.15	4.54	6.96
39	Lætitia	Chacornac	8 Feb 1856	414.2	2.45	3.09	0.11	4.61	10.38
40	Harmonia	Goldschmidt	11 Mar 1856	339.3	3.16	2.37	0.05	3.42	4.25
41	Daphne	Goldschmidt	22 Mar 1856	413.7	2.01	3.52	0.27	4.60	15.77
42	Isis	Pogson	23 May 1856	365.3	1.90	2.99	0.22	3.82	8.53
43	Ariadne	Pogson	15 Apr 1857	329.6	1.83	2.57	0.17	3.27	3.46
44	Nysa	Goldschmidt	27 May 1857	3462.5	2.06	2.78	0.15	3.77	3.70
45	Eugenia	Goldschmdt	27 June 1857	406.9	2.50	2.94	0.08	4.49	6.61
46	Hestia	Pogson	16 Aug 1857	377.8	2.09	2.96	0.17	4.01	2.34
47	Aglaia	Luther	15 Sept 1857	430.5	2.49	3.27	0.14	4.88	4.99
48	Doris	Mayer & Goldschmidt	19 Sept 1857	465.3	2.88	3.34	0.08	5.49	6.55
49	Pales	Goldschmidt	19 Sept 1857	461.5	2.37	3.09	0.23	5.42	3.18
50	Virginia	Luther	4 Oct 1857	396.6	1.89	3.41	0.29	4.32	2.83

		D (km)	Density	R (h)	Axial tilt (°)	Albedo	T	m
26	Proserpina	95	2.1	10.60	–	0.19	S	11.3
27	Euterpe	124 × 75	2.0	10.41	–	0.16	S	8.3
28	Bellona	121	2.0	15.69	–	0.18	S	10.8
29	Amphitrite	212	2.0	5.39	–	0.18	S	8.6
30	Urania	100	2.0	13.69	–	0.17	S	9.4
31	Euphrosyne	256	1.9	5.53	–	0.05	C	10.2
32	Pomona	81	2.0	9.45	–	0.10	S	11.0
33	Polyhymnia	62	2.0	18.60	–	0.10	S	11.0
34	Circe	114	2.0	12.15	–	0.05	C	12.3
35	Leukothea	103	2.0	31.20	–	0.07	C	12.4
36	Atalante	106	2.0	9.93	–	0.07	C	11.6
37	Fides	108	2.0	7.33	–	0.18	S	10.7
38	Leda	116	2.0	12.84	–	0.06	C	12.2
39	Lætitia	149	2.0	5.14	–	0.29	S	8.9
40	Harmonia	108	2.0	8.91	–	0.24	S	9.3
41	Daphne	174	2.?	5.99	–	0.08	C	10.3
42	Isis	100	2.?	13.59	–	0.17	S	9.2
43	Ariadne	95 × 60 × 50	2.7	5.76	105	0.27	S	8.8
44	Nysa	71	2.0	6.43	–	0.55	E	8.5
45	Eugenia	305 × 220 × 145	1.1	5.70	117	0.03	F	11.5
46	Hestia	124	2.0	21.04	–	0.05	C	11.1
47	Aglaia	127	2.0	13.0	–	0.08	C	12.1
48	Doris	222	2.0	11.90	–	0.06	C	11.9
49	Pales	150	2.0	10.60	–	0.06	C	11.4
50	Virginia	100	2.0	14.31	–	0.04	C	11.6
51	Nemausa	148	2.0	7.8	–	0.09	C	10.8

Table 8.3 (cont.)

		D (km)	Density	R (h)	Axial tilt (°)	Albedo	T	m
52	Europa	360 × 313 × 240	3.6	5.62	14 or 54	0.06	C	11.3
53	Kalypso	115	2.0	26.6	–	0.04	C	11.7
54	Alexandra	160 × 135	2.0	7.0	–	0.06	C	10.0
55	Pandora	67	2.0	4.8	–	0.30	M	11.6
56	Melete	113	2.0	16.0	–	0.07	P	11.5
57	Mnemosyne	113	2.0	12.3	–	0.21	S	12.2
58	Concordia	93	2.0	?	–	0.06	C	13.1
59	Elpis	165	2.0	13.7	–	0.04	C	11.4
60	Echo	60	2.0	26.2		0.25	S	11.5
61	Danæ	82	1.1	11.45		0.22	S	11.7
62	Erato	95	2.0	?		0.06	C	13.0
63	Ausonia	103	2.0	8.8		0.16	S	10.2
64	Angelina	48 × 53	2.0	8.8		0.16	E	11.3
65	Cybele	237	2.0	6.1		0.07	C	11.8
66	Maia	73	2.0	?		0.06	C	12.5
67	Asia	58	2.0	15.89		0.26	S	11.6
68	Leta	123	2.0	14.84		0.23	S	10.8
69	Hesperia	138	2.0	5.66		0.14	M	11.3
70	Panopæa	122	2.0	14.0		0.07	C	11.5
71	Niobe	83	2.0	28.8		0.31	S	11.3
72	Feronia	86	2.0	8.1		0.06	U	12.0
73	Klytia	44	2.0	?		0.23	S	13.4
74	Galatea	119	2.0	9.0		0.04	C	12.0
75	Eurydike	56	2.0	8.9		0.15	M	11.1

		Discoverer	Date	Mean distance from Sun ($\times 10^6$ km)	Q (a.u.)	e	P (y)	i (°)	
51	Nemausa	Laurent	22 Jan 1858	363.9	3.21	2.52	0.07	3.64	9.97
52	Europa	Goldschmidt	4 Feb 1858	463.9	2.79	3.41	0.10	5.46	7.47
53	Kalypso	Luther	4 Apr 1858	391.9	2.09	3.15	0.20	4.24	5.15
54	Alexandra	Goldschmidt	10 Sept 1858	405.8	3.18	3.25	0.20.	4.47	11.80
55	Pandora	Searle	10 Sept 1858	412.6	2.16	3.16	0.15	4.58	7.19
56	Melete	Goldschmidt	9 Sept 1859	388.2	2.00	3.21	0.24	4.18	8.07
57	Mnemosyne	Luther	23 Sept 1859	471.1	2.78	3.52	0.12	5.59	15.20
58	Concordia	Luther	24 Mar 1860	404.0	2.58	2.82	0.04	4.44	5.06
59	Elpis	Chacornac	12 Sept 1860	406.2	2.40	3.03	0.12	4.47	8.63
60	Echo	Ferguson	14 Sept 1860	358.1	1.99	2.83	0.18	3.70.	3.60
61	Danæ	Goldschmidt	9 Sept 1860	446.1	2.48	3.48	0.17	5.15	18.22
62	Erato	Forster	14 Sept 1860	467.1	2.57	3.68	0.18	5.52	2.22
63	Ausonia	de Gasparis	10 Feb 1861	358.3	2.10	2.70	0.13	3.71	5.79
64	Angelina	Tempel	4 Mar 1861	401.6	2.35	3.02	0.12	4.40	1.31
65	Cybele	Tempel	8 Mar 1861	513.6	3.07	3.80	0.11	6.36	3.55
66	Maia	Tuttle	9 Apr 1861	395.7	2.19	3.10	0.17	4.30	3.05
67	Asia	Pogson	17 Apr 1861	362.2	1.97	2.87	0.19	3.77	6.03
68	Leto	Luther	29 Apr 1861	416.3	2.27	3.30	0.19	4.64	7.97
69	Hesperia	Schiaparelli	26 Apr 1869	445.8	2.49	3.47	0.17	5.14	8.59
70	Panopæa	Goldschmidt	5 May 1861	391.4	2.14	2.62	0.18	4.23	11.59
71	Niobe	Luther	13 Aug 1861	412.2	2.27	3.24	0.18	4.57	23.26
72	Feronia	Peters	29 May 1861	339.1	1.99	2.53	0.12	3.41	5.42
73	Klytia	Tuttle	7 Apr 1862	398.7	2.55	2.78	0.04	4.35	2.37
74	Galatea	Tempel	29 Aug 1862	415.7	2.11	3.45	0.24	4.63	4.07
75	Eurydike	Peters	22 Sept 1862	399.9	1.86	3.49	0.31	4.37	5.00
76	Freia	D'Arrest	21 Oct 1862	511.3	2.86	3.98	0.16	6.32	2.12
77	Frigga	Peters	12 Nov 1862	399.2	2.31	3.02	0.13	4.36	2.43

Table 8.3 (cont.)

		Discoverer	Date	Mean distance from Sun (×10⁶ km)		Q (a.u.)	e	P (y)	i (°)
78	Diana	Luther	15 Mar 1865	391.9	2.08	3.16	0.21	4.24	8.69
79	Eurynome	Watson	14 Sept 1863	365.7	1.98	2.91	0.19	3.82	4.62
80	Sappho	Pogson	2 May 1864	343.6	1.84	2.76	0.20	3.48	8.66
81	Terpsichore	Tempel	30 Sept 1864	427.0	2.25	3.46	0.21	4.82	7.81
82	Alkmene	Luther	27 Nov 1864	412.9	2.14	3.38	0.22	4.59	2.83
83	Beatrix	de Gasparis	26 Apr 1863	363.8	2.23	2.63	0.08	3.79	5.00
84	Klio	Luther	25 Aug 1865	353.4	1.80	2.92	0.24	3.63	9.33
85	Io	Peters	19 Sept 1865	396.8	2.14	3.16	0.19	4.32	11.97
86	Semele	Tietjen	4 Jan 1866	465.9	2.46	3.76	0.21	5.50	4.82
87	Sylvia	Pogson	16 May 1866	522.1	3.21	3.77	0.08	6.52	10.86
88	Thisbe	Peters	15 June 1866	414.0	2.31	3.22	0.17	4.60	5.22
89	Julia	Stephan	6 Aug 1866	381.5	2.08	3.02	0.18	4.07	16.14
90	Antiope	Luther	1 Oct 1866	472.1	2.66	3.65	0.16	5.61	2.22
91	Aegina	Stephan	4 Nov 1866	387.6	2.32	2.59	0.03	4.17	2.11
92	Undina	Peters	7 July 1867	477.2	2.87	3.51	0.10	5.70	9.92
93	Minerva	Watson	24 Aug 1867	412.0	2.36	3.14	0.14	4.57	8.56
94	Aurora	Watson	6 Sept 1867	472.6	2.88	3.43	0.09	5.61	7.96
95	Arethusa	Luther	23 Nov 1867	459.0	2.61	3.52	0.15	5.37	13.00
96	Ægle	Coggia	17 Feb 1868	457.4	2.66	3.46	0.13	5.35	15.94
97	Klotho	Tempel	17 Feb 1868	399.2	1.98	3.35	0.26	4.36	11.78
98	Lanthe	Peters	18 Apr 1868	401.6	2.18	3.19	0.19	4.40	15.61
99	Dike	Borrelly	26 May 1868	398.5	2.14	3.19	0.20	4.25	13.88
100	Hekate	Watson	11 July 1868	462.8	2.58	3.60	0.17	5.44	6.43

		D (km)	Density	R (h)	Axial tilt (°)	Albedo	T	m
76	Freia	184	2.0	9.8	–	0.04	P	12.3
77	Frigga	69	2.0	9.0	–	0.14	M	12.1
78	Diana	121	2.0	7.2	–	0.07	C	10.9
79	Eurynome	67	2.0	8.8	–	0.26	S	10.8
80	Sappho	78	2.0	14.1	–	0.19	S	12.8
81	Terpsichore	119	2.0	?	–	0.05	C	12.1
82	Alkmene	61	2.0	13.0	–	0.21	S	11.5
83	Beatrix	81	2.0	10.2	–	0.09	C	12.0
84	Klio	79	2.0	?	–	0.05	C	11.2
85	Io	180 × 160 × 160	1.4	6.88	125	0.07	C	11.2
86	Semele	121	2.0	16.6	–	0.05	C	12.6
87	Sylvia	385 × 265 × 230	1.2	5.18	29	0.04	P	12.6
88	Thisbe	232	2.0	6.0	–	0.07	B	10.7
89	Julia	152	2.0	11.4	–	0.18	S	10.3
90	Antiope	93.0 × 87.0 × 83.6	1.3	16.50	–	0.06	C	12.4
91	Ægina	110	2.0	6.0	–	0.04	C	12.2
92	Undina	126	2.0	15.9	–	0.25	M	12.0
93	Minerva	143	2.0	6.0	–	0.09	C	11.5
94	Aurora	205	2.0	7.2	–	0.04	C	12.5
95	Arethusa	136	2.0	8.7	–	0.07	C	12.1
96	Ægle	170	2.0	13.82	–	0.05	U	12.4
97	Klotho	83	2.0	35.0	–	0.23	M	10.5
98	Lanthe	105	2.0	?		0.05	C	12.6
99	Dike	72	2.1	10.35	–	?	C	12.1
100	Hekate	89	2.7	13.20		0.19	S	12.1

Abbreviations and symbols are as follows: q, perihelion distance; Q, aphelion distance; e, orbital eccentricity; P, orbital period; R, rotation period; i, orbital inclination; D, diameter; T, type; m, mean apparent magnitude at opposition.

Table 8.4 *Some further asteroid data (symbols as Table 8.3). The following asteroids are of note in some way or other – either for orbital peculiarities, physical characteristics or because they are the senior members of 'families' – or even because they have very fast or very slow rotation periods, for example*

No.	Name	Discoverer, date	q (a.u.)	Q (a.u.)	P (y)	e	i (°)	T	D (km)	R(h)	m
128	Nemesis	Watson, 25 Nov 1872	2.40	2.75	4.56	0.126	6.25	CEU	116	39.0	11.6
132	Æthra	Watson, 13 June 1873	1.61	2.61	4.22	0.389	25.07	SU	38	?	11.9
135	Hertha	Peters, 18 Feb 1874	1.93	2.42	3.78	0.204	2.29	M	80	8.4	10.5
153	Hilda	Palisa, 2 Nov 1875	3.40	3.97	7.91	0.142	7.83	P	171	?	13.3
158	Koronis	Knorre, 4 Jan 1876	2.71	2.87	4.86	0.053	1.00	S	36	14.2	14.0
221	Eos	Palisa, 18 June 1882	2.72	3.01	5.23	0.098	10.87	CEU	104	10.4	12.4
243	Ida	Palisa, 29 Sept 1884	2.74	2.86	4.84	0.042	1.14	S	58 × 23	4.6	14.6
253	Mathilde	Palisa, 12 Nov 1885	1.94	3.35	5.61	0.262	6.70	C	66 × 48 × 46	417.7	10.0
279	Thule	Palisa, 25 Oct 1888	4.22	4.27	8.23	0.011	2.33	D	130	7.4	15.4
288	Glauke	Luther, 20 Feb 1890	2.18	2.76	4.58	0.210	4.33	S	30	1500	13.2
311	Claudia	Charlote, 11 Jan 1891	2.89	2.90	4.94	0.003	3.23	S	28	11.5	14.9
324	Bamberga	Palisa, 23 Feb 1892	1.77	2.68	4.39	0.341	11.14	C	252	29.4	9.2
349	Dembowska	Charlote, 9 Dec 1892	2.66	2.92	5.00	0.092	8.26	R	164	4.7	10.6
423	Diotima	Charlote, 7 Dec 1896	2.97	3.07	5.38	0.031	11.22	C	208	4.6	12.4
434	Hungaria	Wolf, 11 Sept 1898	1.80	1.94	2.71	0.074	22.51	E	20	26.5	13.5
444	Gyptis	Coggia, 31 Mar 1899	2.29	2.77	4.61	0.173	10.26	CS	166	6.2	11.8
451	Patientia	Charlote, 4 Dec 1899	2.85	3.06	5.36	0.070	15.23	C	280	9.7	11.9
511	Davida	Dugan, 30 May 1903	2.61	3.18	5.66	0.177	15.93	C	324	5.2	10.5
532	Herculina	Wolf, 20 Apr 1904	2.28	2.77	4.62	0.176	16.35	S	220	9.4	10.7
704	Interamnia	Cerulli, 2 Oct 1910	2.61	3.06	5.36	0.148	17.30	F	338	8.7	11.0
747	Winchester	Metcalf, 7 Mar 1913	1.97	3.00	5.19	0.342	18.17	C	204	9.4	10.7
878	Mildred	Nicholson, 6 Sept 1916	1.82	2.36	3.63	0.231	2.02	?	26	?	15.0
903	Nealley	Palisa, 13 Sept 1918	3.16	3.24	5.84	0.025	11.69	?	42	?	15.1
944	Hidalgo	Baade, 31 Oct 1920	2.01	5.84	13.77	0.656	42.40	MEU	38	10.0	14.9
951	Gaspra	Neujmin, 30 July 1916	1.82	2.20	3.28	0.173	4.10	S	19 × 12 × 11	20.0	14.1
1269	Rollandia	Neujmin, 20 Sept 1930	2.76	3.90	7.69	0.097	2.76	D	110	55	15.3
1929	Kollaa	Väisälä, 20 Jan 1939	2.19	2.36	3.63	0.08	7.8	V	12	13.5	
1300	Marcelle	Reisa, 10 Feb 1934	2.76	2.78	4.64	0.009	9.54	?	21	?	15.9
1578	Kirkwood	Cameron, 10 Jan 1951	3.02	3.94	7.82	0.231	0.81	D	56	?	15.8
2602	Moore	Bowell, 24 Jan 1982	2.14	2.38	3.69	0.10	6.6	UX	14	?	17.0
9969	Braille	Helin and Laurence, 27 May 1992	1.32	3.35	3.58	0.43	–	V	2.2 × 0.6	6	–

It was given the provisional designation of 1937 UB, but was not seen again until 15 October 2003, when it was rediscovered by B. Skiff of the Lowell Near-Earth Object Search (LONEOS) project. It could then be given a number: 69 230. It has proved to be a binary asteroid; each component is 600 m in diameter, and the separation is 1200 km. The rotation period is 13.9 hours. The orbit crosses that of Mars, but not that of the Earth. A close approach to Earth in 1942 passed unobserved; Hermes then passed by us at only 1.7 times the distance of the Moon.

Asteroid-hunting has always been a popular pastime, and some nineteenth century observers were particularly successful; for instance K. Reinmuth discovered 246 asteroids; Max Wolf, 232; and J. Palisa, 121. Professional teams are now very active. Among these are:

1. LONEOS , at the Lowell Observatory in Arizona, which began work in 1993, and is now directed by E. Bowell. The telescope used is a 0.6-m, f/1.8 Schmidt.

2. LINEAR (Lincoln Near-Earth Asteroid Research), by NASA, the US Air Force, and MIT's Lincoln Laboratory. Over 68 000 asteroids have been found, plus over 100 comets. The principal investigator is G. Stokes.

3. Spacewatch (University of Arizona), led by R. S. McMillan. As well as asteroids and comets, one satellite of Jupiter (Callirrhoë) was found – it was originally mistaken for an asteroid.

4. ADAS (Asiago–DLR Asteroid Survey); the University of Padua (Italy) together with theDeutschen Zentrum für Luft- und Raumfahrt (DLR), the German Aerospace Centre. The principal investigators are C. Barbieri (Padua/Asiago) and G. Hahn (Berlin). There is collaboration also with UDAS, the Uppsala–DLR Asteroid Survey, in (Sweden).

5. NEAT (Near-Earth Asteroid Tracking), run by NASA and the Jet Propulsion Laboratory (JPL); the original principal investigator was Eleanor Helin. From April 2001, the Oschin Telescope – the 1.2-m Schmidt at Palomar – was used to produce

Table 8.5 *Asteroids also classed as comets (symbols as in Table 8.3)*

Name	Discovered	Q (a.u.)	q (a.u.)	P (y)	e	i (°)	D (km)
2060 Chiron = 95 P/Chiron	1 Oct 1977 by C. Kowal	18.891	8.51	50.6	0.379	6.93	233
4015 Wilson–Harrington = 107 P/Wilson–Harrington	19 Nov 1949 by A. Wilson and R. Harrington	4.285	0.993	4.29	0.624	2.785	~3
60558 Echeclus = 174 P/Echeclus	3 Mar 2000 by Spacewatch	2346.7	876.28	35.4	0.456	4.335	84
7968 Elst–Pizarro = P/133 Elst–Pizarro	Asteroid: 1979 by M. Hawkins et al.; comet: 1996 by W. Elst and G. Pizarro	550.5	395.1	5.62	0.164	1.386	~5
118401 LINEAR = 176 P/LINEAR	7 Sept 1999	3.81	2.58	5.71	0.19	0.238°	<6

images which led to the discoveries of, among others, the large trans-Neptunians Quaoar (in 2002) and Sedna (in 2004).

Amateurs also contribute; in England, Brian Manning now has five numbered asteroids to his credit.

ASTEROID ORBITS

Asteroid orbits are of various definite types (for the moment let us discuss them together with dwarf planets; the distinction is in any case arbitrary and unsatisfactory):

6. Main Belt: orbits between those of Mars and Jupiter.
7. Inner asteroids, including NEAs (near-Earth asteroids).
8. Vulcanoids: orbits wholly within that of Mercury. These are hypothetical only, as yet; no Vulcanoid has been found, despite careful searches. If they exist, they must be very small.
9. Aten class: mean distance from the Sun less than 1 a.u., although some may cross the Earth's orbit. All are small.
10. Apollo class: mean distance from the Sun greater than 1 a.u., but their orbits do cross that of the Earth.
11. Amor class: orbits cross that of Mars, but not that of the Earth. (One, 251 Lick, has an orbit entirely between those of Earth and Mars; inclination 39°, period 1.63 years.)
12. Trojans: a Trojan moves in the same orbit as a planet, though keeping on average either 60° ahead of the planet (the L4 position) or 60° behind (L5) and are in no danger of being engulfed. Jupiter has hundreds of Trojans, some of which are large by asteroidal standards; Neptune has several, though no Saturnian or Uranian Trojans have yet been found. Mars has several small Trojans, such as 5261 Eureka (diameter 3 km), but Earth, Venus and Mercury apparently do not.
13. 'Wanderers' (my term!): asteroids with very eccentric orbits, taking them from the inner to the outer parts of the Solar System. The first to be found was 944 Hidalgo, whose path takes it from the Main Belt out to beyond Saturn; it is larger than the inner asteroids (diameter 50 km). The tiny 5335 Damocles has an orbit crossing those of Mars, Jupiter, Saturn and Uranus; other 'Damocloids' are known.
14. Centaurs: orbits with perihelia greater than that of Jupiter but less than that of Neptune. The first to be found was 2060 Chiron, which moves mainly between the orbits of Saturn and Uranus (period 50 years); diameter well over 100 km. Another Centaur, 5145 Pholus, has an orbit crossing those of Saturn, Uranus and Neptune.

COMET/ASTEROID ORBITS

It seems that the link between asteroids and comets may be much closer than was believed until recently. There are now five objects included in both categories (see Table 8.5): 2060 Chiron (De95), 60558 Echeclus (P/74), 4015 Wilson–Harrington (P/107), 7968 Elst–Pizarro (P/133 and 118401 LINEAR P/176). In fact it does not seem logical to class Chiron and Echeclus as comets; they are far too large.

It is also rather difficult to class objects such as 1996 PW, discovered by Eleanor Helin on 9 August 1996. It seems to have a period of about 5800 years, and an aphelion distance of 645 a.u. – almost 100 000 million km; the diameter is probably about 10 km.

These curious bodies may well bridge the gap between the Kuiper Belt and the Oort Cloud of comets, and indeed the distinction between the various small members of the Solar System seems to be blurred. 1996 TL$_{66}$ and 1996 PW look like asteroids, but their orbits are cometary. On the other hand, Comet Wilson–Harrington, first seen in 1951 and which had a perceptible tail, was lost for many years and recovered in 1979, this time in the guise of an asteroid: it has been given an asteroid number, 4015, and now shows no sign of cometary activity. Comet P/133 Elst–Pizarro, discovered in August 1996, also has a tail, but moves entirely within the main asteroid belt and has been given an asteroid number 7968. The NEA asteroid 3200 Phæthon, discovered on 11 December 1983 from the IRAS satellite, has an orbit so like that of the Geminid meteor stream that it could be the parent of that stream – in which case Phæthon is an ex-comet which has lost all its volatiles.

A larger body with comet-like characterstics is 60558 Echeclus, which is classed as a Centaur; it may be over 80 km in diameter, it was discovered in 2000 by Spacewatch, and looked completely asteroidal, but in December 2005 it developed a coma, and was also listed as a comet, P/174 Echeclus. Its orbital period is 35.4 years, its perihelion distance is 5.8 a.u., and at aphelion it recedes to 15.7 a.u., well beyond the orbit of Saturn. No doubt there are many more similar cases.

CLASSIFICATION OF ASTEROIDS

Asteroids are divided into various types according to their spectral and surface characteristics. The Tholen system, devised in 1984 by David J. Tholen, has been extended with the SMASS (Small Main Belt Asteroid Spectroscopic Survey) by S. J. Bus and R. P. Binzel (2002). In any system there are three main categories: carbonaceous, silicaceous and metallic.

C (carbonaceous)

These are the commonest of all asteroids, increasing in number from 10% at a distance of 2.2 a.u. up to 80% at 3 a.u. They are dark, with albedoes of from 0.03 to 0.1. Their spectra show ultraviolet absorption at short wavelengths, but are rather featureless at longer wavelengths. 10 Hygiea is of type C. There are various subgroups; B asteroids lack marked ultraviolet absorption, and the albedoes are higher than for the 'pure C' asteroids; 2 Pallas is of this type. F asteroids are very similar to B-type, with only minor differences in spectra. G objects are relatively uncommmon; the dwarf planet 1 Ceres belongs here, but is listed as being of type C in the SMASS system, which has no type G.

S (silicaceous)

These are most numerous in the inner part of the main zone, making up 609 of the total at 2.2 a.u., but only 15% at 3 a.u. Their albedoes range from 0.1 to 0.22; they seem to consist largely of iron and magnesium silicates. The largest S-asteroids are 15 Eunomia, 3 Juno and 29 Amphitrite; there seems to be strong affinity with stony meteorites. There are various sub-classes. Type A: rare – only about twenty are known; strong indications of olivine. K: uncommon – low albedoes, spectra very like those of some stony meteorites. L: similar to K, though with difference in spectra; examples are 387 Aquitania and 980 Anacostia. P: low albedo, reddish, found in the outer part of the Main Belt. Q: uncommon inner-belt asteroids, spectra showing olivine and pyroxene, and indicative metals. The leading examples are 1862 Apollo and 2063 Bacchus; there is no known large Q-asteroid. R: moderately reflective inner-belt asteroids, showing indications of olivine, pyroxene and perhaps plagioclase. Examples are 148 Gallia, 246 Asporina and 349 Dembowska.

M (metallic)

Moderate albedoes (0.1 to 0.2); most are mainly nickel–iron, possibly from the cores of differentiated meteoroids that were broken up by collision. They could well be the source of iron meteorites. 16 Psyche is the largest M-asteroid, and may be almost pure nickel–iron; others are 21 Lucretia and the strangely low-density 22 Kallkope. 26 Kleopatra, imaged by radar from Arecibo, is shaped uncannily like a dog's bone!

In addition to the main types listed above, the following types are also recognised:

E (enstatite): Relatively rare; high albedoes, up to 0.4. They may resemble some types of chondrites, in which ensatite ($MgSO_4$) is a major constituent; they are common only in the inner part of the Main Belt. Most are small; the largest is 64 Angelina (diameter 60 km). 434 Hungaria is of type F.

D: Low albedo, reddish. Their surfaces seem to be 90% clays, with magnetite carbon-rich material. Most D-asteroids are a long way out, including many Trojans, but there are also some in the Main Belt, including the largest example, 704 Interamnia; others are 336 Lacadiera and 361 Bononia.

T: Reddish. Intermediate in type between D and S.

V (igneous rock surfaces): Very rare; rather bright. Vesta is the only large example. Other V-type asteroids may well be fragments of Vesta. One well-studied V-type asteroid is 1459 Magnya.

U: Otherwise unclassifiable – for example 2201 Oljato, discovered in 1947 by H. Giclas and recovered in 1979 by E. Helin. Its spectrum is unlike any other in the Solar System; it is no more than 3 km in diameter, and shows no sign of cometary activity.

X: There is some possible confusion here; in the original Tholen system the X group contains the types E, M and P.

ASTEROID FAMILIES

There are distinct 'families' of asteroids, whose members probably result from the breaking-up of a larger body by collision; about one-third of Main Belt asteroids belong to families, named after the member with the lowest number. They are called Hirayama families, because they were first noted by the Japanese astronomer Kiyotsugu Hirayama, in 1918. The Flora family has more than 800 known members. The main families are listed in Table 8.6.

The Hungaria family lies in the innermost part of the Main Belt, in a 9:2 resonance with Jupiter (and 3:2 with Mars). Most are of E type, and all are small; the largest, 434 Hungaria, is only about 20 km in diameter. They orbit on the inside of the 4:1 Kirkwood Gap.

The Maria group is separated from the main swarm by inclinations over 15°. Most members are of type S, including 170 Maria itself, which is 40 km in diameter.

Phocæa is a group of which the largest member, the S-type 24 Phocæa, is 72 miles in diameter. The orbits lie at distances of between 2.25 and 2.5 a.u. from the Sun, but this is a rather ill-defined family.

In Nysa, the distances are between 2.41 and 2.5 a.u., eccentricities between 0.12 and 0.2; inclinations from 1.4 to 4.3°. 44 Nysa itself is 71 km in diameter, and of type E; other major members are 135 Hertha and 750 Oskar, The members of the associated family of 142 Polana are of type F.

Koronis is one of the most populous families, with over 300 members, mainly S-type, at distances between 2.82 and 2.95 a.u. from the Sun. It is thought that they are due to the break-up of a planetesimal about 100 km across, perhaps 2000 million years ago. Members include 158 Koronis (mean diameter 35.4 km), 167 Urda (39.9), 208 Lacrimosa (41.0) and 243 Ida (31.3); Ida and its tiny

Table 8.6 *The main asteroid families*

Family	Distance from the Sun (a.u.)	
Hungaria	1.8–2.0	Just inside the Main Belt. Most are of type E, with some of type S. Inclinations exceed 16°. Members include 434 Hungaria, 1103 Sequoia and 1025 Riema. Hungaria is 20 km in diameter; most of the rest are much smaller.
Flora	2.2	Most populous family; well over 150 members. Most are reddish. 8 Flora itself is of type S, 162 km in diameter and is virtually spherical.
Maria	2.25	Separated from the Main Belt by inclinations of around 15°. Most members are of type S, including 170 Maria itself, which is 40 km in diameter.
Phocæa	2.4	Rather ill–defined. 25 Phocæa is of type S, with a diameter of 72 km.
Koronis	2.8–2.9	The only rich family with low inclination and low eccentricity; about 60 members. Almost all are of type S. 158 Koronis has a diameter of 36 km.
Budrosa	2.9	Only six members; inclinations about 6°. 338 Budrosa, diameter 80 km, is of type M; most of the others have to be classified as U.
Eos	3.0	The most compact of all the families; high inclinations, 11° for Eos itself. Almost all are of type S. The largest members are 221 Eos (diameter 104 km), 579 Sidonia (80 km) and 639 Latona (68 km).
Themis	3.1	Over 100 members; mainly of type C. Inclinations and eccentricities are low. 24 Themis is 198 km in diameter – large by asteroidal standards. Other members include 90 Antiope (128 km), 222 Lucia (55 km) and 171 Ophelia (113 km).
Hilda	4.0	In the outer region; over 30 members mainly of types P, C and D. 153 Hilda is 171 km in diameter, and of type P.

279 Thule, apparently 'on its own', lies beyond the Main Belt, at a mean distance of 4.3 a.u. The orbit is of low eccentricity (0.011); diameter 130 km, type D. The period is 8.82 years.

satellite Dactyl were imaged in 1993 by the Galileo space-craft en route for Jupiter.

Eos is another populous group, with around 500 known members; distances 2.99 to 3.03 a.u., low eccentricities (0.01 to 0.15) and moderate inclinations (8 to 12 degrees). 221 Eos is of type K; diameter 104 km.

Themis, at distances from 3.08 to 3.24 a.u., is a very large family, with over 530 known members, mainly of type C. 24 Themis is 198 km in diameter; other major members are 62 Erato, 90 Antiope, 104 Klymene and 171 Ophelia.

The Griqua group of asteroids is found at distances from 3.1 to 3.3 a.u.,with eccentricities greater than 0.35; these asteroids are at 2:1 resonance with Jupiter, and over long periods orbits in this family will be unstable. 1362 Griqua has the lowest number; the next is 8373 Stephengould.

Members of the Cybele family lie at distances between 3.27 and 3.7 a.u., with eccentricities below 0.3 and inclinations less than 25 degrees. 65 Cybele itself is a dark C-type asteroid 237 km across: its diameter is known accurately, because on 17 October 1979 it occulted a star.

Water, ice and organic molecules were found on 65 Cybele by scientists at the University of Central Florida in 2010. Humberto Campins of the team commented that water on asteroids may be much more common than expected.

The members of the Hilda group are in 3:2 resonance with Jupiter, but they may not be a true family due to the break-up of a parent body; distances 3.7 to 4.2 a.u. 153 Hilda itself has a diameter of 171 km, and an orbital period of almost 8 years. On 31 December 2002, Japanese astronomers saw it occult a star.

There are some references to a Thule family, but so far as we know at present 279 Thule moves in splendid isolation at a distance of 4.3 a.u. It is large – diameter 127 km – and of type D, probably made up of organic-rich silicates. It is in 4:3 resonance with Jupiter. Lonely Thule seems to mark the outpost of the Main Belt. If it really is part of a family, the other members must be very faint.

There are other 'families' which are very often sparse and difficult to pin down. An example of this relates to 298 Baptistina, which can be no more than about 30 km in diameter, and moves in an orbit taking it between 2.05 and 2.26 a.u. from the Sun; the eccentricity is 0.096, the inclination 6.3°, and the period 3.41 years (1224.2 days). Several smaller bodies seem to be related to it. In 2007, Czech and American studies suggested that it is the largest remnant of a 170-km body which was broken up 160 million years ago, and that fragments landed on both the Earth and the Moon. The impact on Earth resulted in the Chicxulub crater allegedly linked with the disappearance of the dinosaurs and the formation of the lunar crater Tycho, but to link all these events seems to be highly speculative.

MAIN BELT ASTEROIDS

The Main Belt asteroids move in orbits between those of Mars and Jupiter. Many tens of thousands have now been numbered, but most of them are small, and only four – 1 Ceres, 2 Pallas, 4 Vesta and 10 Hygiea – are over 400 km in diameter; these make up more than half the mass of the entire Main Belt. (3 Juno, though one of the traditional 'Big Four', is in fact considerably smaller.) The Belt was formed from the original solar nebula as a group of planetesimals, but Jupiter was already in existence, and its immense gravitational pull prevented any large planets from being formed in this region of the Solar System. Jupiter is also responsible for the Kirkwood Gaps, where an asteroid would be in orbital resonance with Jupiter and perturbing effects would be cumulative; for instance a 3:1 resonance would mean that the asteroid would make three orbits in the time that Jupiter would make to complete one. (They are named in honour of Daniel Kirkwood, who noted them and correctly interpreted them in 1857.)

The orbits of Main Belt asteroids show a wide range of inclinations; for example 52.0° for 1580 Betulia, but only 0.014° for 1383 Limburgia, while 2 Pallas has an inclination of almost 35°. Some orbits are almost circular; 311 Claudia has an eccentricity of only 0.0031, and for 903 Neally it is 0.0039. 132 Æthra has an orbital eccentricity of 0.389 and can actually come within the orbit of Mars. Mutual perturbations are measurable; thus the mass of Vesta was determined largely by the effects on the movements of 197 Arete, an S-type asteroid 42 km in diameter and therefore much smaller and less massive than Vesta. Telescopically, most asteroids appear starlike, but surface markings have been mapped on some or them, and others have been encountered by space-craft. Small asteroid diameters have been measured by 'thermal radiometry', balancing the intensity of absorbed infrared radiation with that of the radiation at visual wavelengths. Some asteroids have been observed to occult stars, and of course the duration of the occultation at once gives a clue to the minimum diameter of the asteroid.

Ceres

This is the giant of the Main Belt, and the only member classed as a dwarf planet. It is almost spherical, since the difference between the equatorial and polar diameters is only 28 km, and it is virtually in hydrostatic equilibrium. It contains one-third the mass of the whole Main Belt, and is almost certainly a proto-planet which has survived its battering. It is undoubtedly differentiated, and the best model gives a rocky core overlaid by an icy mantle 60 to 200 km deep; if this is composed of water ice, it could contain 200 million cubic kilometres of water. The surface layer seems to be a mix of water ice and hydrated minerals (clays and carbonates). The temperature may rise to almost −40 degrees °C. The rotation period is just over 9 hours, and is prograde; the axial tilt is between 3° and 5° – the north polar star is probably ι Draconis.

As yet, we know little about the surface features of Ceres. Images taken in 1995 with the Hubble Space Telescope showed what was thought to be a crater, and it was even given a name – Olbers – but subsequently the Keck telescope could not find it. However, the Keck telescope did show two dark patches which showed movement as Ceres rotated, and these probably are craters.

It is quite likely that Ceres once had an ocean below its crust, and this has led to a suggestion that primitive life might have appeared. There is even a slight chance that an ocean survives now. We may know more when the Dawn probe, launched on 27 September 2007, reaches Ceres in 2015.

It is interesting to note that Piazzi, its discoverer, named it Cerere Ferdinandea; Cerea after the goddess of the Earth, and Ferdinandea after Ferdinand III, King of Sicily. Unsurprisingly, Ferdinand was soon forgotten.

Pallas

Pallas is always regarded as the second largest member of the Main Belt swarm, but its geometric mean diameter is only marginally greater than that of Vesta, though it is rather less massive. It accounts for 7% of the total mass of the swarm, but is not regular in shape, measuring 571 × 525 × 482 km. However, like Ceres and Vesta, it may well be a planetesimal which has survived disruption by bombardment. Its rotation period is rather over 7 hours, and the axial inclination seems to be at least 60°, so that most areas of the surface must have long periods of alternate constant sunlight and constant darkness. The orbit is highly inclined and decidedly eccentric; when Pallas is closest to us its magnitude rises to 6.4, and I have two exceptionally keen-sighted friends who have then been able to see it with the naked eye.

Very little is known about the surface, but spectroscopic observations indicated material which is a silicate, and there may be similarities to carbonaceous chondrite meteorites. Unfortunately, the orbital inclination of almost 35° means that the Dawn space-craft will not be able to go anywhere near it. The asteroid is named for Pallas Athena, the goddess of wisdom.

Juno

Though third in order of discovery, Juno is not so large or so massive as several other Main Belt asteroids, though it and 15 Eunomia are the largest of the S-type bodies. It has been suggested that many of the ordinary chondrite meteorites come from Juno.

No missions to it are planned as yet, but in 1996 careful studies were carried out with the Hooker 100-inch telescope at Mount Wilson (still in the forefront of research despite its age!), and the images revealed what seems to be a crater, 96 km in diameter, which is huge with respect to Juno's small size, and distorts the whole shape of the asteroid. Quite possibly, the crater was produced by an impact no more than a thousand million years ago. One astronomer, Sallie Baliunas, commented that 'the recent large impact on Juno gives us an opportunity to see through the regolith and study excavated material from beneath the surface – a rare look into the material out of which the early Earth was formed'.

Juno was named for the queen of Olympus, wife of Zeus (Jupiter).

Vesta

Vesta is the second most massive member of the Main Belt, and easily much the brightest as seen from Earth. It has been more intensively studied than any other member of the swarm, and is the first target of the Dawn space-craft, now (2010) on its way. Its orbit is of low eccentricity and reasonably low inclination; the rotation period is short (5.34 hours) and its axial tilt is 29°, so that the north pole star is Deneb. With a zenithal Sun, the temperature may rise to a balmy −20 °C, though a polar night may drop to a frigid −190 °C. The escape velocity is 0.35 km s^{-1}; there is no sign of any atmosphere.

Vesta is certainly a surviving planetesimal with a differentiated interior. It is thought that there is a metallic nickel–iron core overlying a rocky olivine mantle, above which lies the thin crust; there is abundant evidence of past vulcanism, and there seems to have been resurfacing. Images taken with the Hubble Space Telescope show a huge crater, 460 km across, near the south pole; the width is 80% of Vesta's entire diameter, and the crater is well-formed, with walls rising to almost 12 km above the outer country, while the floor is 13 km below – a majestic structure indeed; to complete the picture there is a central peak towering to 18 km above the crater bottom. An impact of this violence has probably excavated an appreciable percentage of Vesta's original mass, and flung out débris some of which has fallen on the Earth. There are other craters of appreciable size, plus a dark feature, provisionally named Olbers, whose nature is uncertain. There are two distinct hemispheres, with different types of solidified lava; one shows what may be called quenched lava flows, while the other has characteristics of molten rock that cooled and solidified underground before being subsequently exposed by impacts.

Many meteorites (eucrites) are believed to have come from Vesta. One of these, a 6-cm eucrite achondrite, was seen to fall at Millibillillie, Australia, in 1960. It is probable that asteroids with the rare V-type spectra may also be fragments of Vesta; these include 1999 Braille and 1929 Kollaa. The asteroid was named in honour of Vesta, the goddess of hearth and home.

Lutetia

Lutetia, a large Main Belt asteroid, has an unusual spectrum. Officially it is of type M, but does not display much metal on its surface. It is decidedly irregular and elongated, with a longest diameter of 134 km. On 10 July 2010, the Rosetta space-craft (en route to Comet 67P/Churyumov-Gerasikmenko) flew past Lutetia at 3160 km, at a speed of 15 km s^{-1}. The surface of the asteroid was found to be crater-pitted. The radar albedo is low, unlike the high radar albedos of strongly metallic asteroids such as 16 Psyche.

(En passant, the encounter took place on the anniversary of the 200-km encounter between Giotto, America's first deep-space mission, and Comet Grigg–Skjellerup in 1992.)

Themis

Themis is one of the larger Main Belt asteroids, discovered by Annibale de Gasparis on 5 April 1853. It is the senior member of its family; diameter 198 km, mean distance from the Sun 468 226 km, eccentricity 0.132, period 5.54 years, inclination 0.76°, rotation period 8h.23m, albedo 0.067, escape velocity 87m s^{-1}, density 2.78 (water = 1), spectral type C, mean temperature ~ 159 K (−114 °C), absolute magnitude 7.08.

On 7 October 2009, Professors Josh Emery (University of Tennesee) and Andrew Rivkin (Johns Hopkins University), using NASA's Infrared Telescope Facility on Mauna Kea, found that Themis surface is competely covered in water ice. Organic material

was also detected, inclding the polycyclic aromatic hydrocarbons CH_2 and CH_3. The ice is continually sublimated, and is presumably replenished by a reservoir of ice under the surface. (The only other explanation seems to be trace amounts of water impinging oxide minerals present on the asteroid's surface, as with the Moon.) This unexpected discovery indicates that asteroids may be icier than has been previously believed.

Other Main Belt asteroids

There is plenty of variety in the Main Belt. A few members of interest are listed in Table 8.3. 216 Kleopatra was discovered by Palisa in 1880; astronomers using the 3.6-m telescope at La Silla detected its unusual shape, and work with the Arecibo radio telescope showed it to be shaped very like a dog's bone. It may well be a contact binary. It seems to be loosely packed, and metallic. 288 Glauke, discovered by Robert Luther in 1890, has a very long rotation period – 50 days, for reasons that are unclear. 324 Bamberga, discovered by Palisa in 1892, can at times reach the eighth magnitude; its type is intermediate between C and P. 490 Veritas was discovered by Max Wolf in 1902; it and 92 Undina are the largest members of a family of 200, and it has been suggested that we are seeing the débris from a collision which occurred a mere 8 million years ago. 511 Davida is one of the largest of the Main Belt asteroids; it was discovered by R. S. Dugan in 1903; it is dark, with a carbonate composition, and its shape indicates that it has been heavily cratered. Not much is known about 704 Interamnia, though it is the fifth largest Main Belt asteroid; it was discovered by V. Cerulli in 1910, but is a long way out, and never becomes much brighter than the tenth magnitude. It seems to account for 2% of the mass of the Main Belt; only Ceres, Pallas and Vesta contribute more.

INNER ASTEROIDS

Vulcanoids

In a survey of asteroids in the inner Solar System, we must start with Vulcanoids, in a stable region between 0.08 and 0.21 a.u. of the Sun, that is to say around 21 000 000 km, well within the orbit of Mercury. The original suggestion was due to C. D. Perrine, around 1900, and seemed to be perfectly reasonable. The trouble was that searching for small bodies so close to the Sun would be very difficult. A total solar eclipse would provide a darkish sky, and searches have been made on several eclipses, notably when it was widely believed that a comparatively large planet (Vulcan) existed there. Trying to detect an asteroid-sized body passing in transit across the face of the uneclipsed Sun was virtually hopeless, and so far no Vulcanoids have been found.

If they do exist, they cannot be more than a few tens of kilometers in diameter, at most, but Mercury has been severely bombarded in the past, and asteroidal bodies in these torrid regions seem quite credible. There have been recent NASA searches using F-18 aircraft and Black Brant sub-orbital rockets, and the quest has certainly not been abandoned. Probably space research investigations are the most promising. The Solar and Heliospheric Observatory (SOHO) is not ideal, but it has of course detected many comets.

Against this, it has been suggested by D. Vokrouhlicky and others, that no Vulcanoids survive, as all the original members have either impacted Mercury or been drawn into the Sun. The jury is still out.

Asteroid diameters and absolute magnitudes

Asteroid diameters and absolute magnitude diameters of an asteroid are linked to its absolute magnitude, H. In the table below, the diameter is given in kilometres when H is in the left-hand column and in metres when H is to the right. Thus, $H = 7.0$ gives a diameter of 110–240 km, while $H = 26.0$ gives 17–37 m. There is bound to be some uncertainty, because asteroids have a wide range of albedoes, and for small bodies the albedoes are not known precisely.

H	Diameter	H
3.0	670–1500	18.0
3.5	530–1200	18.5
4.0	420–940	19.0
4.5	330–740	19.5
5.0	260–590	20.0
5.5	210–470	20.5
6.0	170–370	21.0
6.5	130–300	21.5
7.0	110–240	22.0
7.5	85–190	22.5
8.0	65–150	23.0
8.5	50–120	23.5
9.0	40–95	24.0
9.5	35–75	24.5
10.0	25–60	25.0
10.5	20–50	25.5
11.0	17–37	26.0
11.5	13–30	26.5
12.0	11–24	27.0
12.5	8–19	27.5
13.0	7–15	28.0
13.5	5–12	28.5
14.0	4–9	29.0
14.5	3–7	29.5
15.0	3–6	30.0
15.5	2–5	
16.0	2–4	
16.5	1–3	
17.0	1–2	
17.5	1–2	

Apohele class

Apohele asteroids have orbits wholly inside that of the Earth. (Officially, Apoheles are regarded as being members of the Aten class.) By 2009, less than a dozen were known, and only one had been numbered: the 2-km. 163963 Atira, discovered in 2003 from LINEAR. It can never make close approaches to the Earth, but its orbit does cross that of Venus. In the following year B. A. Skiff found 2004 JG$_6$ (not yet numbered or named), which has an orbital period shorter than that of Venus, and an orbit crossing those of both Venus and Mercury; its mean surface temperature must be well over 600 °C. No doubt there are many more small, faint Apoheles.

Aten class

An Aten asteroid is defined as having a mean distance from the Sun which is less than one astronomical unit. Their orbits therefore cross that of the Earth. By 2009, over forty had been numbered; the largest, 2240 Hathor, has a diameter of slightly over 5 km. They have been given names from Egyptian mythology.

2100 Ra-Shalom, discovered in 1978 by Eleanor Helin, can come to within 3 000 000 km of Mars, the Earth and Venus, and within 1 200 000 million km of Mercury, but its high orbital inclination makes a collision unlikely. 3554 Amun is one of the smallest known M-type asteroids; it has been contacted by radar. 5381 Sekhmet, discovered in 1991 by Carolyn Shoemaker, is 1 km in diameter and has a 0.3-km satellite (see Table 8.15).

Two Atens are of special interest. 3753 Cruithne, discovered by D. Waldron in 1986, is 5 km across, and never becomes much brighter than the 16th magnitude, but its orbital period is only about a day shorter that of the Earth. Relative to Earth its orbit is shaped rather like a kidney bean, and it makes regular close approaches, though the orbit is much more eccentric than ours, and the separation is always over 10 000 000 km. (Note that Cruithne is in orbit round the Sun, and is not a satellite of the Earth.) Several smaller asteroids moving in much the same way have been found.

99942 Apophis, discovered in 2004 by R. Tucker, D. Tholen and F. Bernardi, caused a sensation when its orbit was worked out, because there seemed to be a definite chance of a collision with the Earth in 2029 – and as it is about 270 metres across, it would then do an immense amount of damage; on the Torino scale (see Table 8.8) it was given a rating of 4. Further calculations removed this possibility, but Apophis will brush past at a distance less than those of geosynchronous satellites, and will be easily visible with the naked eye. Moreover, we do not know quite how it will behave, and the chance of a collision in 2036 cannot be absolutely ruled out. There have been suggestions that it might be monitored during the 2029 pass, perhaps by landing a beacon upon it, so that, if there is a real danger, we might explode a nuclear bomb close to it and 'nudge' it into a safe orbit. Fittingly, it has been named after the Egyptian god of destruction!

Near-Earth asteroids

Some asteroids (all, fortunately, small) make close passes of the Earth, and we are not immune from collision; in fact we were hit by a comparatively large body 65 million years ago (call it either a large meteorite or a small asteroid) and this is widely believed to have caused a climate change fatal to the dinosaurs. A selected list of PHAs (potentially hazardous asteroids) is given in Table 8.7. Risks

Table 8.7 *Selected list of potentially hazardous asteroids*

No.	Name	Mean distance from Sun (× 10^6km)	q (a.u.)	Q (a.u.)	H
69230	Hermes	0.003	0.616	2.662	18
1566	Icarus	0.040	0.187	1.969	16.9
1620	Geographos	0.046	0.828	1.663	15.6
1862	Apollo	0.028	0.647	2.295	16.2
1981	Midas	0.000	0.622	2.930	15.5
2101	Adonis	0.012	0.441	3.308	18.7
2102	Tantalus	0.029	0.905	1.675	16.2
2135	Aristæus	0.015	0.795	2.405	17.9
2201	Oljato	0.001	0.623	3.721	15.2
2340	Hathor	0.006	0.464	1.223	19.2
3200	Phæthon	0.026	0.140	2.403	14.6
3361	Orpheus	0.013	0.819	1.599	19.0
3362	Khufu	0.018	0.526	1.453	18.3
3671	Dionysus	0.034	1.003	3.388	16.3
3757		0.026	1.017	2.653	18.9
4015	Wilson–Harrington	0.049	1.000	4.289	16.0
4034		0.023	0.023	1.530	18.1
4179	Toutatis	0.006	0.919	4.104	15.3
3183	Cuno	0.038	0.718	3.243	14.4
4450	Pan	0.027	0.596	2.287	17.2
4486	Mithra	0.045	0.743	3.658	15.6
4581	Asclepius	0.004	0.657	1.387	20.4
4660	Nereus	0.005	0.953	2.026	18.2
4769	Castalia	0.023	0.550	1.577	16.9
4953		0.040	0.555	2.687	14.1
5011	Ptah	0.026	0.818	2.453	17.1
5189		0.044	0.810	2.292	17.3
5604		0.037	0.551	1.303	16.4
5189		0.044	0.810	2.292	17.3
5604		0.037	0.551	1.303	16.4
5693		0.008	0.527	2.016	17.0
6037		0.024	0.636	1.904	18.7
6239	Minos	0.028	0.676	1.627	17.9
6489	Golovka	0.038	1.012	4.023	19.2
7335		0.042	0.913	2.628	17.0
7482		0.017	0.904	1.788	16.8
7753		0.005	0.761	2.175	18.6
7822		0.033	0.938	1.308	17.4
8014		0.018	0.951	2.543	18.7
8566		0.017	0.857	2.156	16.5
9856		0.033	0.844	3.647	17.4
99942	Apophis	0.00237	0.746	1.099	19.7

of asteroid impacts are evaluated by what is called the Torino scale (Table 8.8), from 0 (no risk) to 10 (certain global devastation, threatening the whole of civilisation). Italian astronomers from the universities of Pisa and Valladolid find that 101955 1999

RQ36, diameter 5 km, has a one in a thousand chance of hitting Earth in 2182. (Warn your grandchildren!) A list of the closest known approaches by asteroids is given in Table 8.9.

Asteroid impacts

On 6 October 2008, R. Kowalski, at the Catalina Sky Survey, using the 1.5-m Mt. Lemmon Telescope at Tucson, discovered a small near-Earth asteroid, designated 2008 TC_3, clearly on a collision course. Impact was due within 21 hours. The Minor Planet Center issued a prompt announcement, and by the time the asteroid entered the Earth's shadow, 19 hours after discovery, 570 positions had been obtained from 26 observatories. Impact occurred in northern Sudan before dawn on 7 October. The diameter was between 2 and 5 m, and the asteroid probably exploded at a height of 37 km with an energy equivalent of ~1 kilotonne of TNT. It came in at a shallow angle at 12.4 km s^{-1}.

The entry was observed from US Government satellites, infrasound signals from at least one ground station, images from Meteosat 8, and a number of US satellites. A subsequent search for meteorites produced 280 fragments, which are classified as urelites, and contain, among other minerals, nanodiamonds.

Asteroids closer in than the Main Belt

Apollo asteroids move in orbits which cross that of the Earth, but whose mean distances from the Sun are greater than 1 a.u. Well over 2000 are known; most are small, but it has been estimated that there may be at least 600 Apollos of more than one kilometre in diameter.

The first Mars-crosser to be found was 433 Eros, in 1898. (To be precise, 132 Æthra can theoretically approach the Sun to within the orbit of Mars; its perihelion distance from the Sun is 1.61 a.u., and the aphelion distance of Mars is 1.657 a.u, but Æthra is not usually classed as an Apollo.) One asteroid, 1951 Lick, discovered by C. Wirtanen in 1949, has an orbit lying entirely between those of the Earth and Mars. Its diameter is 5 km, so that it is a faint object. The largest Apollo is 1866 Sisyphus, diameter 8.5 km; most members of the class are very small (Table 8.10.)

1862 Apollo was discovered by Karl Reinmuth in 1932, the first asteroid known to cross the Earth's orbit. It was promptly lost, and was not found again until 1973; its orbit is now well known. It is 1.7 km in diameter; in 2005, radar observations from Arecibo led to the detection of an 80-m satellite, moving round Apollo at a range of only 3 km. 1685 Toro, discovered by Wirtanen in 1948, orbits in an 8:3 resonance with the Earth – that is to say, 3 'Toro years' are equal to 8 Earth years; there are regular fairly close passes, and in 1972 some reports falsely claimed it to be a second Earth satellite. It is 3 km in diameter, and of type S. 2101 Adonis can also make reasonably close passes, as in 2036 (at just over 5 000 000 km). In 1966, radar was used to study the Q-type 2063 Bacchus; it was found to be bilobate in shape, measuring 1.1 × 1.1 × 2.6 km, and to have a rotation period of 14 hours 54 minutes.

Some asteroids pass remarkably close to the Sun. The first found to move within the orbit of Mercury was 1566 Icarus, discovered by W. Baade in 1949. It has a period of 409 days, and can approach the Sun to a distance of 28 000 000 km, so that its surface

Table 8.8 *The Torino impact hazard scale: assessing asteroid and comet impact hazard possibilities in the twenty-first century*

No hazard (White Zone)	0	The likelihood of a collision is zero, or is so low as to be effectively zero. Also applies to small objects such as meteors and bodies that burn up in the atmosphere as well as infrequent meteorite falls that rarely cause damage.
Normal (Green Zone)	1	A routine discovery in which a pass near the Earth is predicted that poses no unusual level of danger. Current calculations show the chance of collision is extremely unlikely with no cause for public attention or public concern. New telescopic observations very likely will lead to re-assignment to Level 0.
Meriting attention by astronomers (Yellow Zone)	2	A discovery, which may become routine with expanded searches, of an object making a somewhat close but not highly unusual pass near the Earth. While meriting attention by astronomers, there is no cause for public attention or public concern as an actual collision is very unlikely. New telescopic observations very likely will lead to re-assignment to Level 0.
	3	A close encounter, meriting attention by astronomers. Current calculations give a 1% or greater chance of collision capable of localised destruction. Most likely, new telescopic observations will lead to re-assignment to Level 0. Attention by public and by public officials is merited if the encounter is less than a decade away.
	4	A close encounter, meriting attention by astronomers. Current calculations give a 1% or greater chance of collision capable of regional devastation. Most likely, new telescopic observations will lead to re-assignment to Level 0. Attention by public and by public officials is merited if the encounter is less than a decade away.
Threatening (Orange Zone)	5	A close encounter posing a serious, but still uncertain threat of regional devastation. Critical attention by astronomers is needed to determine conclusively whether or not a collision will occur. If the encounter is less than a decade away, governmental contingency planning may be warranted.
	6	A close encounter by a large object posing a serious but still uncertain threat of a global catastrophe. Critical attention by astronomers is needed to determine conclusively whether or not a collision will occur. If the encounter is less than three decades away, governmental contingency planning may be warranted.
	7	A very close encounter by a large object, which, if occurring this century, poses an unprecedented but still uncertain threat of a global catastrophe. For such a threat in this century, international contingency planning is warranted, especially to determine urgently and conclusively whether or not a collision will occur.
Certain collisions (Red Zone)	8	A collision is certain, capable of causing localised destruction for an impact over land or possibly a tsunami if close offshore. Such events occur on average between once per 50 years and once per several 1000 years.
	9	A collision is certain, capable of causing unprecedented regional devastation for a land impact or the threat of a major tsunami for an ocean impact. Such events occur on average between once per 10 000 years and once per 100 000 years.
	10	A collision is certain, capable of causing global climatic catastrophe that may threaten the future of civilization as we know it, whether impacting land or ocean. Such events occur on average once per 100 000 years, or less often.

temperature must then be around 500 °C. It has a very eccentric orbit, and at aphelion recedes to almost 2 a.u. – well beyond the path of Mars – so that it has a very extreme climate. The next asteroid found to invade these torrid regions was 3200 Phæthon, discovered in 1983 by Simon Green and John Davis while they were searching data supplied by IRAS, the Infra-Red Astronomical Satellite. Phæthon is about 5 km in diameter (much larger than Icarus); its distance from the Sun ranges between 21 000 000 and 390 000 000 km, with a period of 524 days and an orbital eccentricity of 0.890. Unusually for an Apollo object, it is of type B. The orbit looks cometary, and Phæthon moves in the same path as the Geminid meteors, which are always plentiful in mid December. There is every reason to believe that Phæthon is the parent of the shower, in which case it is an extinct comet, though since its discovery it has shown no trace of a coma or tail. During each orbit it crosses the paths of Mercury, Venus, the Earth and Mars.

Another asteroid which passes well within the orbit of Mercury is 5786 Talos. Note, however, that bodies moving in this way are not Vulcanoids; for most of the time they are well beyond the orbit of Mercury.

The most famous *Amor* asteroid (see Table 8.11 for a list of Amor type asteroids) is 433 Eros, discovered in 1898 by C. G. Witt, from Berlin. Eros can approach Earth at a distance of 25 000 000 km, as it did in 1931. Its position was then intensively studied in an attempt to improve our knowledge of the length of the astronomical unit, although the final value, derived by H. Spencer Jones, is now known to have been rather too great. The last close approach was in 1975. Eros can then reach magnitude 8.3, but generally it is a very faint object. On 23 December 1998, it was surveyed from a range of 4100 km by the NEAR (Near Earth Asteroid Rendezvous) spacecraft, subsequently named in honour of the American planetary geologist E. M. Shoemaker. On 14 February 2000, at a distance of

Table 8.9

Encounters <0.006 a.u., 2010–2100

Number	Name	Encounter distance (a.u.)	Date
99942	Apophis	0.0002	2029 Apr 13
137108		0.003	2027 Aug 7
	2005 UO	0.003	2096 Oct 12
	2001 GQ2	0.003	2100 Apr 27
	1994 WR12	0.005	2080 Nov 23

Encounters <0.020 of Numbered Asteroids, 2010–2100

Number	Encounter distance (a.u.)	Date
99942 Apophis	0.0002	2029 Apr 13
2340 Hathor	0.006	2086 Oct 21
35396 1997 XF11	0.006	2028 Oct 26
2340 Hathor (again)	0.007	2069 Oct 21
171576 1999 VP11	0.008	2086 Oct 22
4660 Nereus	0.008	2060 Feb 14
35396 1997 XF11 (again)	0.010	2095 Oct 27
4581 Asclepius	0.012	2051 Mar 24
33342 1998 WT24	0.013	2099 Dec 18
41429 2000 GE2	0.013	2040 Oct 1
7842 1994 PC1	0.013	2022 Jan 18
171576 1997 VP11	0.015	2017 Oct 22
4660 Nereus (again)	0.015	2071 Feb 4
66391 1999 KW4	0.016	2036 May 25
65679 1989 UQ	0.016	2093 Aug 13
41429 2000 GE2(again)	0.016	2038 Oct 1
152637 1997 NC1	0.017	2026 June 27
66391 1999 KW4(again)	0.018	2071 May 26
163899 2003 SD220	0.018	2070 Dec 20
163899 2003 SD220(again)	0.019	2018 Dec 22
4179 Toutatis	0.020	2069 Nov 5
3200 Phæthan	0.020	2093 Dec 14

256 000 km from the Earth and 1800 km from Eros, the Shoemaker probe entered an orbit round Eros at a range of 327 km and began a long-term programme of mapping.

Detailed views of Eros were obtained. Many craters between 500 and 1000 metres in diameter were found; there is one very prominent sharp-rimmed crater associated with a huge, hollowed-out gorge. However, there was a surprising scarcity of boulders large enough to make such impacts, and small craters were also fewer than expected, There were also 'ponds', flat surfaces, at the bottoms of craters, formed by regolith deposits.

Eros is a very primitive body, and is very ancient. It is an S-type asteroid: a composition of iron- and magnesium-bearing silicates (pyroxene and olivine) mixed with metallic nickel and iron. The shape of the asteroid itself is rather reminiscent of a banana. The escape velocity is 10 m s^{-1} and it has a rotation period of 5 hours. The 'day time' temperature is about 100 °C, plunging to –150 °C at 'night'.

On 12 February 2001, the NEAR–Shoemaker space-craft made a controlled landing on Eros, just south of the saddle-shaped feature Himeros. The landing speed was only about 1.5 to 1.8 m s^{-1}, and transmissions began immediately. The probe carried a gamma-ray spectrometer, and this functioned well after the landing; when cosmic-ray particles hit the surface they shatter atomic nuclei in the 'soil', and neutrons flying away from the cosmic-ray impact sites hit other atoms in turn. These secondary neutrons can excite atomic nuclei without breaking them apart, and these excited atoms emit gamma-rays, which the spectrometer can use to reveal which elements are present. Studies of the surface features, both from the orbit and from the landing spot, indicate that most of the larger rocks scattered over Eros were ejected from a single meteoritic impact perhaps a thousand million years ago. Table 8.12 lists the craters. A map of the surface is shown in Figure 8.1

The last signals from the probe were received on 26 February 2001. NEAR–Shoemaker remains on the edge of Himeros, and there it will stay until some future astronaut collects it.

1620 Geographos is the most elongated object known in the Solar System. It is 5.1 km long and only 1.8 km broad. 4769 Castalia is dumbbell-shaped, about 1.8 km across at its widest; its two distinct lobes are each 0.75 km across and there is a narrow waist about 123 m long. The two lobes were probably separate objects which came together after a gentle collision. Another compound asteroid is 4179 Toutatis, discovered on 4 January 1989 by C. Pollas on photographic plates taken on the 0.9-m Schmidt telescope at Caussols, France, by A. Maury and D. Mulholland during astrometric observations of the faint outer satellites of Jupiter. Toutatis is about 4.6 km long, dominated by two components in contact, one twice as large as the other. It is not certain whether the components are actually separated or are joined by a very narrow 'neck'. The rotation is extraordinary and chaotic. Radar images have been obtained and the craters have been revealed. There are frequent close approaches to the Earth – for instance in 2006, when it passed by at a range of only 0.0073 of an astronomical unit.

The largest Amor asteroid is 1036 Ganymed (not to be confused with Jupiter's satellite Ganymede). It is 40 km in diameter, but fortunately there is no danger of a collision with Earth.

Another asteroid that has been contacted by radar is 3908 Nyx. Its shape is described as 'spherical, but with many protruding bumps'. It is between 1 and 2 km in diameter, and has a V-type spectrum, so that it may possibly be a fragment of Vesta.

SPACE MISSIONS TO ASTEROIDS

Several asteroids have now been examined from close range, some as definite targets and others by space-craft en route to the outer planets.

The first three Main Belt asteroids to be surveyed by space-craft were 951 Gaspra (13 November 1991), 243 Ida (28 August 1993) and 253 Mathilde (27 June 1997) – the first two by the Galileo probe en route for Jupiter and the third by the NEAR (Shoemaker) probe making for Eros. They proved to be very different from each other. Gaspra, imaged from a range of 16 000 km, proved to be wedge-shaped (not unlike a distorted potato) with a darkish, rocky surface pitted with craters. There were also

Table 8.10 *Apollo-type asteroids*

	q	Q	Mean distance from the Sun ($\times 10^6$ km)	P (d)	R (y)	e	I (°)	D (km)	T
1566 Icarus	0.187	1.969	161.3	408.8	1.12	0.827	22.85	1.4	U
1685 Toro	0.771	1.963	136.7	583.9	1.6	0.436	9.38	3	S
1862 Apollo	0.647	2.294	220.0	651.5	1.78	0.560	6.36	1.7	Q
1866 Sisyphus	0.874	2.914	283.3	952.1	2.61	0.679	41.14	8.5	U
2063 Bacchus	0.701	1.455	161.3	408.8	1.12	0.349	9.43	$1.1 \times 1.1 \times 2.6$	Q
2101 Adonis	0.441	3.307	280.3	936.7	2.56	0.765	1.34	~1	?
3200 Phæthon	0.140	2.40	190.2	523.6	1.43	0.890	22.17	5.1	B
4183 Cuno	0.721	3.243	296.5	1019.0	2.79	0.636	6.75	4.5	S
4486 Mithra	0.750	3.660	329.7	1194.8	3.27	0.66	3.03	~3	S
4581 Asclepius	0.657	1.387	151.9	377.0	1.03	0.357	4.91	~0.1	?
6489 Golevka	0.986	4.010	373.6	1441.9	3.95	0.605	2.28	0.53	?
69230 Hermes	0.622	2.688	247.6	777.7	2.13	0.624	6.07	0.4.	S

Table 8.11 *Amor type asteroids (symbols as in Table 8.3)*

No	Name	q (a.u.)	Q (a.u.)	P (years)	e	i (°)	H	D (km)	m	T	R (h)	Discoverer
433	Eros	1.133	1.783	1.76	0.223	10.83	11.1	$33 \times 13 \times 13$	11.9	S	5.270	Witt, 1898; close in 1931, 1975
719	Albert	1.191	2.637	4.28	0.550	11.31	15.8	2.6	16.8	?	?	Palisa, 1911
887	Alinda	1.087	3.884	3.97	0.558	9.25	13.8	4	15.0	S	74	Wolf, 1918
1036	Ganymed	1.227	4.090	4.35	0.537	26.45	9.4	40	10.4	S	10.3	Baade, 1924
1221	Amor	1.083	2.755	2.66	0.435	11.90	17.7	1	19.1	?	?	Delporte, 1932; well placed every 8 years
1580	Betulia	1.119	3.270	3.26	0.049	52.01	14.5	1	15.0	U	6.1	Johnson, 1950
1627	Ivar	1.124	2.603	2.54	0.397	8.44	13.2	6	14.2	S	?	Hertzsprung, 1929
1915	Quetzalcoatl	1.081	3.994	4.03	0.577	20.50	19.0	0.4	20.1	SU	4.9	Wilson, 1953; recovered in 1974
1916	Boreas	1.250	3.295	3.43	0.450	12.84	14.9	3	16.1	S	?	Arend, 1953; recovered in 1974
1917	Cuyo	1.067	3.235	3.15	0.505	23.99	13.9	3	16.5	?	?	Cesco and Samuel, 1968
1943	Anteros	1.064	1.796	1.71	0.256	8.70	15.7	4	16.5	S	?	Gibson, 1973
1951	Lick	1.304	1.390	1.63	0.062	39.09	15.3	5	17.2	?	?	Wirtanen, 1949
1980	Tezcatlipoca	1.085	2.334	2.23	0.365	26.85	13.9	6.2	15.1	U	?	Wirtanen, 1980
2059	Babuquivari	1.256	4.044	4.31	0.526	10.99	15.8	3.8	16.0	?	?	Goethe, 1963
2061	Anza	1.048	3.481	3.40	0.537	3.74	16.6	2.4	18.0	C	?	Giclas, 1960
2202	Pele	1.120	1.120	3.463	0.512	1.12	17.6	1.2	18.5	?	?	Lemola, 1972
2368	Beltrovata	1.234	2.976	3.06	0.413	5.26	15.2	4.8	16.8	DU	?	Wild, 1977
2608	Seneca	1.044	3.940	3.90	0.586	15.63	17.5	1.4	18.0	?	18.5	Schuster, 1978
3199	Nefertiti	1.128	2.021	1.98	0.283	32.97	14.8	4.4	16.3	?	?	Shoemaker, 1982
3271	UI	1.271	2.933	3.05	0.394	25.00	16.7	2	18.0	?	?	Shoemaker, 1982
3288	Seleucus	1.103	2.962	2.90	0.457	5.93	15.3	4	16.5	?	75	Schuster, 1982
3352	McAuliffe	1.185	2.573	2.58	0.369	4.78	15.8	2.4	17.5	?	?	Thomas, 1981
3671	Dionysus	1.003	3.387	3.26	0.540	13.61	16.3	2	17	?	?	Shoemaker, 1984

Other named Amors include 3122 Florence, 3551 Verenia, 3552 Don Quixote, 3553 Mera, 4055 Magellan, 4401 Aditi, 4487 Pocohontas, 4503 Cleobulus, 4947 Ninkasi, 4954 Eric, 4957 Brucemurray, 5324 Lyapunov, 5370 Taranis, 5751 Zao, 5797 Bivoj, 5863 Tara, 5869 Tanith, 6489 Golevka, 7088 Ishtar and 7480 Norwan, Comet Wilson–Harrington has been assigned an asteroid number, 4015.

Table 8.12 *Features on Eros*

	Lat. (°)	Long. (° W)	Diameter (km)
Craters			
Abelard	3.53 S	12.23	1.15
Aida	7.94 N	130.50	1.66
Avtandil	22.53 S	233.10	1.16
Bovary	61.01 S	27.30	0.82
Casanova	46.58 N	236.01	0.94
Catherine	9.13 N	171.08	1.12
Cupid	8.29 N	230.18	1.84
Don Juan	29.55 N	356.67	1.15
Don Quixote	57.70 S	250.82	0.88
Dulcinea	76.13 S	272.86	1.40
Eurydice	13.49 N	170.03	2.24
Fujitsubo	03.75 S	62.70	1.70
Galatea	10.17 S	183.05	1.38
Gamba	20.58 S	4.13	1.27
Genji	19.50 S	88.62	1.52
Heathcliff	7.35 N	167.89	1.09
Himeros	21.16 N	282.32	10.0
Hios	9.41 S	130.90	1.26
Jahan	74.16 N	293.46	2.06
Kastytis	6.76 N	161.31	1.66
Leander	25.61 N	210.34	1.39
Leylie	03.01 S	23.48	1.88
Lolita	35.17 S	197.67	1.81
Mahal	79.36 N	169.98	1.16
Majnoon	03.76 N	28.80	2.06
Narcissus	18.15 N	7.12	2.91
Orpheus	25.56 N	176.74	1.07
Pao-yu	73.20 S	105.58	0.84
Psyche	31.60 N	94.61	4.85
Pygmalion	01.85 S	191.09	1.75
Radanes	5.16 S	115.09	1.63
Selene	14.18 S	12.53	3.64
Tai-yu	47.02 S	126.06	1.40
Tutanekai	56.36 N	3.33	2.10
Valentine	14.64 N	208.43	2.21
Regiones			
Charlois Regio	16 S	330	
Vitt Regio	18 N	348	
Dorsa			
Hinks Dorsum	42 N	318	
Finsan Dorsun	48 S	350	

indications of three areas, from which pieces had been broken off, so that Gaspra may well be the survivor from a series of major collisions; it is small, measuring 19 km × 12 km × 11 km. Craters on it have been named after famous spas such as Bath, Aix and Rotorua. Surprisingly, there were magnetic effects strong enough to create a 'bubble' in the solar wind. Some named features are listed in Table 8.13, and Figure 8.2 maps the surface.

Ida is larger than Gaspra (58 km × 23 km) and is heavily cratered; many of these are larger than craters on Gaspra. Ida is a member of the Koronis family, with a rotation period of 4.6 hours; like Gaspra it is of type S and is reddish, so that it is presumably composed of a mixture of pyroxene, olivine and iron minerals. It was found to have a tiny satellite, now named Dactyl, measuring 1.2 km × 1.4 km × 1.6 km, orbiting at a distance of about 100 km from Ida's centre. Strangely, it and Ida seem to be composed of decidedly different materials. The largest crater is Afon, named after a Russian cave.

Mathilde proved to be surprising. It measured 66 km × 4.8 km × 46 km, but is very irregular in outline and is cratered; the largest crater is 30 km across. The albedo is only 3%, so that Mathilde is as black as charcoal; the surface may consist of carbon-rich material which has not been altered by planet-building processes. The mean density is only 1.3 g cm^{-2}. Fittingly, the craters are to be named after famous coal mines! The rotation period is very long indeed: 417.7 hours, over 17 days. Clearly Mathilde has had a tortured history; it was even claimed that 'there are more huge craters than there is asteroid'.

5535 Annefrank was imaged on 2002 November 2 by the Stardust space-craft bound for Comet Wild 2. Stardust passed the asteroid at 3300 km; Annefrank is darkish, 1.3 km long and 1 km wide, with a prominent impact crater. 9969 Braille was imaged by the Deep Space 1 probe on 28 July 1999; it is 2.2 km long and 1 km wide, and is suspected of being a fragment of Vesta. 25143 Itokawa (540 m × 270 m) was the target of the Japanese probe Hayabusa; the probe was designed to touch the surface, collect samples and return to Earth. It carried a small 'mini-lander', Minerva. Itokawa was reached in August 2005; unfortunately Minerva missed its target but Hayabusa made contact and has now returned. On 28 June 2010 JAXA announced that they had detected small particles inside Hayabusa's capsule, but whether they came from Itokawa or were due to Earth contamination was initially unclear.

The most successful asteroid mission has been the landing on 433 Eros. On 2007 September 27, the Dawn spacecraft was launched by NASA, to go first to Vesta (2011) and then to Ceres (2015).

ASTEROID SATELLITES

Many asteroids are now known to have satellites. Space research methods have shown a large number; they are of different types and a list of some typical satellite systems is given in Table 8.14.

One particularly interesting system is that of 87 Sylvia, a Main Belt asteroid 385 km in diameter, orbiting at a distance of 522 000 000 km from the Sun in a period of 6.5 years. It is of low density (1.2 times that of water) and is a quick spinner (rotation period 5.2 hours). There are two small satellites, named Romulus and Remus after the sons of the legendary Rhea Sylvia. Romulus, the larger of the two (diameter 18 km) orbits at 1356 km from Sylvia in a period of 87.5 hours; for Remus, 7 km across, the distance is 706 km and the period 33 hours.

From Sylvia, the satellites would appear almost the same size, 0.9° for Romulus and 0.8 for Remus; as they move in the same plane, they would occult each other every 2.2 days. From the satellites, Sylvia would loom large in the sky, with an apparent diameter of 30° as seen from Remus, 16° from Romulus. Certainly any 'Sylvians' would have an entertaining sky.

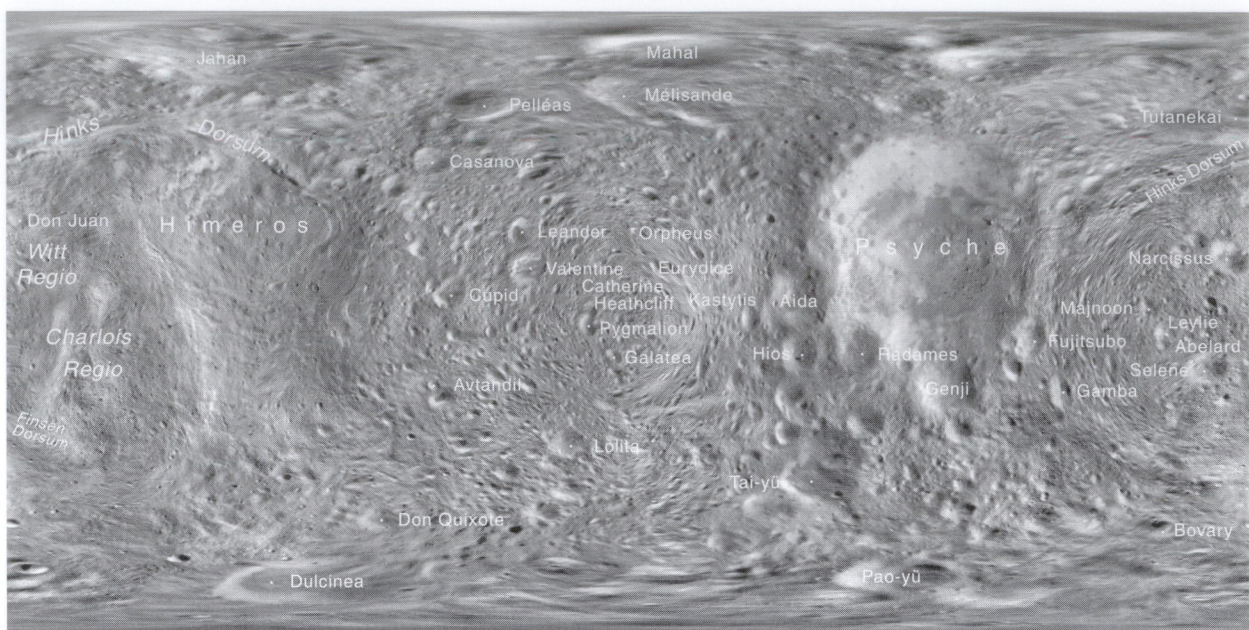

Image base courtesy of Philip Stooke, University of Western Ontario.

Figure 8.1 Surface map of Eros

Table 8.13 *Selected list of features on Gaspra*

Crater	Lat. (° N)	Long. (° W)	Diameter (km)
Aix	47.9	160.3	6
Bath	13.4	122.0	10
Charax	8.6	0.0	11
Lisdoonvara	16.5	358.1	10
Ramlösa	15.0	4.9	10
Rotorua	18.8	30.7	6

There are three named regions: Dunne (15.0° N, 15.0° W), Neujmin (2.0° N, 80.0° W) and Yeates (65.0° N, 75.0° W). These are named after persons associated with Gaspra. Grigori Neujmin discovered the asteroid, on 30 July 1916; James Dunne was the planner of the Galileo mission and Clatne Yeates was project manager of the mission.

Satellites occur in the Trojan asteroids moving in Jupiter's orbit, and also among the trans-Neptunians; Eris, the largest trans-Neptunian, has one known satellite, while Pluto has three, one large and the other two very small (see Table 8.14).

JUPITER TROJANS

In 1906 Max Wolf, at Heidelberg, discovered asteroid 588 Achilles, and found that it moved in the same orbit as Jupiter, 60° ahead at the L4 Lagrangian point. This was the first Trojan; others followed, some at the L4 point and others at the L5, trailing Jupiter. They oscillate around the main points, sometimes by as much as 25°, but they keep well clear of Jupiter, and are in no danger of collision with the Giant Planet. By asteroidal standards they are large, but their great distances mean that they appear faint.

Names commemorate the Trojan War. It was intended to make L4 the Greek camp and L5 the Trojan, but inevitably there were some errors; thus 624 Hektor is a spy in the Greek camp, while 625 Patroclus is among the Trojans.

The brightest Trojan, 617 Patroclus, is a binary system, made up of two similarly-sized objects orbiting their common centre of gravity in a period of 4.3 days. The diameters are respectively 122 and 113 km; the smaller of the two has been given a separate name, Menoetius. Both members of the pair have low density; less than that of water. 624 Hektor measures 370 km by 195 km, but appears to be a single bilobated body rather than a binary; there is however a 15-km moonlet at a separation of 1000 km. Hektor itself is of type D, darkish and reddish.

Many Jupiter Trojans have been numbered and named. No doubt many members of the swarm await discovery.

The trans-Neptunians, including KBOs (Kuiper Belt objects), have mean distances from the Sun greater than that of Neptune, though some, notably 134340 Pluto, cross Neptune's orbit. These will be discussed separately.

Table 8.15 gives a selected list of Trojan asteroids.

ASTEROIDS FURTHER OUT THAN THE MAIN BELT

Centaurs

Centaur asteroids are fairly large bodies; classical Centaurs move between the orbits of Saturn and Neptune (see Table 8.16). They take their name from the first-discovered member of the class, 2060 Chiron, found on 1 November 1977 by C. Kowal, using the Oschin Schmidt telescope at Palomar. (Chiron was the wise Centaur, half-horse and half-man, who taught Jason and other Argonauts.) Its orbit lies almost entirely between those of Saturn and

Figure 8.2 Gaspra

Table 8.14 *Selected list of asteroids with satellites (symbols as for Table 8.3; separation in km)*

Primary	Type	Mean distance from Sun (a.u.)	P (y)	D	Secondary, Name	D	Separation (km)	e	i (°)	R (h)
5381 Sekhmet	Aten	0.95	0.92	1	–	0.3	1.5	?	?	10.2
22 Kalliope	Main Belt	2.91	4.96	181	Linus	38	1065	0.00	93.4	?
87 Sylvia	Main Belt	3.50	6.52	286	Romulus	18	1356	0.001	1.7	3.65
					Remus	7	706	0.016	2.0	1.37
243 Ida	Main Belt	2.86	4.85	313	Dactyl	1.4	108	1.545	–	–
3671 Dionysus	Amor	2.20	3.26	1.5	–	0.3	3.6	0.07	?	?
69230 Hermes	Apollo	1.66	2.13	0.4	–	0.4	~1.0			
45 Eugenia	Main Belt	2.72	4.49	305	Petit-Prince	13	?	?	?	5
90 Antiope	Main Belt	3.16	5.61	93		89	170		0.687	
617 Patroclus	Jupiter Trojan	5.23	11.96	122	Menoetius	113	680	0.02	low	?

Table 8.15 *Selected list of Trojan asteroids (symbols as Table 8.3, H = absolute magnitude, m = apparent magnitude, at mean opposition). Asteroids marked * are east of Jupiter (L3); the others, west (L4)*

Number	Name	Year of discovery	q (a.u.)	Q (a.u.)	P (y)	e	i (°)	H	D (km)	R (h)	T	m
Jupiter Trojans												
588	Achilles*	1906	4.413	5.593	11.77	0.149	10.3	8.7	116	?	D	15.3
617	Patroclus	1906	4.501	5.957	11.97	0.139	22.1	8.2	164	?	P	15.2
624	Hektor*	1907	5.088	5.321	11.76	0.022	18.2	7.5	300 × 150	6.92	D	16.2
659	Nestor*	1908	4.624	5.812	12.01	0.114	4.5	9.0	110	?	C	15.8
884	Priamus	1917	4.522	5.786	11.71	0.123	8.9	8.8	94	?	D	16.0
911	Agamemnon*	1919	4.880	5.588	11.87	0.068	21.8	7.9	144	?	D	15.1
1143	Odysseus*	1930	4.771	5.743	12.01	0.092	3.1	7.9	179	?	D	15.6
1172	Æneas	1930	4.635	5.714	11.72	0.104	16.7	8.3	162	?	D	15.7
1173	Anchises	1930	4.596	6.055	12.21	0.137	6.9	8.9	162	?	C	16.0
1208	Troilus*	1931	4.744	5.698	11.85	0.091	33.6	9.0	124	?	C	16.0
1404	Ajax*	1936	4.685	5.897	12.01	0.115	18.0	9.0	92	?	?	16.0
1437	Diomedes*	1937	4.903	5.364	11.52	0.045	20.6	9.3	172	18.0	C	15.7
1583	Antilochus*	1950	4.825	5.376	11.55	0.054	28.6	8.6	158	?	D	16.3
1647	Menelaus*	1957	5.124	5.369	12.03	0.023	5.6	10.3	50	?	?	18.1
1749	Telamon*	1949	4.615	5.780	11.99	0.112	6.1	9.2	56	?	?	17.5
1867	Deiphobus	1971	4.915	5.375	11.75	0.045	26.9	8.6	140	?	D	
1868	Thersites*	1960	4.708	5.876	12.07	0.110	16.8	10.7	104	?	CFPO	
1869	Philoctetes*	1960	4.957	5.647	12.24	0.065	4.0	11.0	28	?	?	
1870	Glaukos	1971	5.083	5.426	12.00	0.033	6.6	10.5	41	?	?	
1871	Astyanax	1971	5.126	5.497	12.33	0.035	8.6	11.0	35	?	?	
1872	Helenos	1971	5.003	5.500	11.76	0.047	14.7	11.2	50	?	?	
1873	Agenor	1971	4.780	5.743	12.09	0.092	21.8	10.5	45	?	?	
2146	Stentor*	1976	4.663	5.738	11.88	0.103	39.3	10.8	50	?	?	
2148	Epeios*	1976	4.892	5.503	11.02	0.059	9.2	11.1	38	?	?	
2207	Antenor	1977	5.048	5.211	11.67	0.016	6.8	8.9	122	?	D	
2223	Sarpedon	1977	5.095	5.261	11.69	0.016	16.0	9.4	96	?	D	
2241	Alcathous	1979	4.897	5.570	12.02	0.066	16.6	8.6	132	?	D	
2260	Neptolemus*	1975	4.953	5.426	11.82	0.046	17.8	9.3	98	?	D	
2357	Phereclos	1981	4.951	5.420	11.78	0.045	2.7	8.9	96	?	D	
2363	Cebriones	1977	4.954	5.336	11.92	0.037	32.2	9.1	100	?	?	
2456	Palamedes*	1966	4.758	5.550	11.95	0.077	13.9	9.6	80	?	?	
2594	Acamas	1978	4.672	5.551	11.70	0.086	5.5	11.5	26	?	?	
2674	Pandarus	1982	4.821	5.529	11.79	0.068	1.9	9.0	80	?	?	
2759	Idomeneus*	1980	4.828	6.604	11.73	0.965	22.0	9.8	51	?	?	
2797	Teucer*	1981	4.660	5.601	11.83	0.092	22.4	8.4	101	?	?	
2893	Peiroös	1975	4.793	5.597	11.95	0.077	14.6	9.2	80	?	?	
2895	Memnon	1981	4.965	5.484	11.88	0.050	27.2	9.3	64	?	?	
2920	Automedon*	1981	4.984	5.295	11.85	0.030	21.1	8.8	80	?	?	

Other named Jupiter Trojans include:

3063	Makhaon*	3596	Meriones*	4063	Euforbo*	4708	Polydoros
3420	Laocoon	3709	Polypoites*	4068	Menestheus*	4709	Ennomos
3317	Paris	3793	Leonteus*	4086	Podalirius*	4722	Agelaos
3391	Sinon*	3794	Sthenelos*	4138	Kalchas*	4754	Panthoös
3451	Mentor	3801	Thrasymedes*	4348	Poulydamas	4791	Iphidamas
3540	Protesilaos*	4007	Euryalos*	4501	Eurypyloa*	4792	Lykaon
3548	Eurybates*	4057	Demophon*	4543	Phoinix*	4805	Asteropaios
3564	Talthybius*	4060	Deipylos*	4707	Khryses	4827	Dares

Table 8.15 (cont.)

Other named Jupiter Trojans include:

4828	Misebus	5012	Eurymedon*	5244	Amphilochos*	5637	Gyas
4829	Sergestus	5023	Agapenor*	5254	Ulysses*	5638	Deikoon
4832	Palinurus	5027	Andrigeos*	5259	Epeigeus	5652	Amphimachus*
4833	Meges*	5028	Halæsus*	5264	Telephus*	6997	Laomedon
4834	Thoas*	5041	Theotes*	5283	Pyrrhus*	6998	Tithonus
4836	Medon*	5120	Bitias	5284	Orsilocus*	7119	Hiera*
4867	Polites*	5126	Achæmenides+	5258	Krethon*	7152	Euneus*
4902	Thessandrus*	5130	Ilioneus	5436	Eumelos*	7214	Antielus*
4946	Askalaphus*	5144	Achates	5511	Cloanthus*		

Table 8.16 *Selected list of Centaur asteroids*

		q	Q	e	i	P (y)	D (km)	R (h)	Density
2060 Chiron	C. Kowal, 1977	8.45	18.89	0.38	6.94	50.54	~233	5.92	~2.0
5145 Pholus	D. Rabinowitz, 1992	8.73	32.13	0.57	24.69	92.35	~185	9.98	2.1
8405 Asbolus	J. Scotti, 1995	6.83	29.12	0.62	22.1	?	72	?	?
10199 Chariklo	J. Scotti, 1997	13.08	18.66	0.18	23.37	63.17	259	?	?
10370 Hylonome	Jewitt & Luu, 1995	18.89	31.51	0.25	4.1	126.4	?	?	?
52872 Okryhoe	Spacewatch, 1998	5.80	10.96	0.31	15.6	?	45	?	?j
60558 Echeclus	Spacewatch, 2000	5.86	15.69	0.46	4.34	35.36	70	?	2.1

Uranus. At perihelion (as on 14 February 1996) it comes to within 1 278 000 000 km of the Sun and about one-sixth of its orbit lies within that of Saturn, but at aphelion it recedes to 2 827 000 000 km, greater than the minimum distance between the Sun and Uranus. The period is 50.54 years. The orbit is unstable over a timescale of some millions of years. In 1664 BC, Chiron approached Saturn to a distance of 16 000 000 km, which is not much greater than the distance between Saturn and its outermost satellite, Phœbe, which is almost certainly a captured asteroid. At discovery the magnitude was 18, but at perihelion it rose to 15. Light-curve studies give a rotation period of just under 6 hours. Estimates of the diameter are not in perfect agreement, but by asteroid standards Chiron is large. One favoured value is 233 km, though this may be a slight overestimate.

Preliminary spectroscopic results indicated a fairly low albedo with a dusty or rocky surface, but there was a major surprise in 1988, when Chiron was found to be brightening – not spectacularly, but appreciably. Inevitably there were suggestions that it might be a huge comet rather than an asteroid, and this idea was strengthened in April 1990, when K. Meech and J. S. Belton, using electronic equipment on the 4-m reflector at Kitt Peak in Arizona, photographed Chiron and found that it appeared to be 'fuzzy'; in other words, it had developed a coma. Using the 2.24-m telescope on Mauna Kea on 29 January 1990, D. Jewitt and J. X. Luu found that the coma extended for 80 000 km and was elongated away from the Sun in comet-like fashion. The ejected material was thought to be vaporised carbon monoxide carrying away dust grains.

This is certainly comet-like behaviour, and on occasion the diameter of the coma has been known to reach almost 2 000 000 km: the brightness can vary by a factor of four over a few hours, and a gravitationally bound 'dust atmosphere' appears to be suspended in the inner 1200 km of the coma. Moreover, this dust shows evidence of structure, indicating that there may be particle plumes issuing from the nucleus. Yet Chiron is 40 times larger and 50 000 times more massive than any comet.

5145 Pholus, the second Centaur to be found, is different from Chiron. While Chiron has been given a cometary number (P/95) Pholus (in mythology, Chiron's brother) is red, and shows no sign of cometary activity; the spectrum indicates that there may be complex organic compounds on the surface. 8405 Asbolus is icy; spectral analysis of its composition by the Hubble Space Telescope revealed a fresh impact crater on the surface; Centaurs are dark, due to effects of solar wind, but fresh impact craters excavate brighter ice from below, and this is what Hubble found. 10199 Chariklo is the largest known Centaur, and has water ice on its surface; when at perihelion (as in 2004) the magnitude rises to 17.7. 60558 Echeclus can show a coma, and has been given a cometary number (P/174). On 11 April 2006, part of the body apparently broke off, and resulted in a large cloud of dust. The cause is unclear; it may be an impact, or release of volatiles from below the surface. No doubt many more Centaurs move in this part of the Solar System.

944 Hidalgo was discovered on 31 October 1920 by W. Baade. Its orbital eccentricity is 0.66; its perihelion distance is 1.95 a.u., at the inner edge of the main asteroid belt, out to 4.94 a.u., just

Table 8.17 *Numbered Damocloids*

	q (a.u.)	Q (a.u.)	e	i (°)	P (y)	H	D (km)
5335 Damocles	1.58	11.8	0.87	62.0	40.7	13.3	~10
20461 Dioretsia	2.39	23.8	0.90	160.4	116	13.8	~9
654071	2.47	55.3	0.96	119.1	412	12.3	~25

2006 WD4 has an estimated Q at 1270. a.u. and a period of 38 400 years; this is, however, very uncertain. The estimated diameter is beween 1 and 2 km.

beyond the orbit of Saturn. The period is 13.77 years; inclination 42.6°, diameter about 38 km. It is of type D, and is officially classed as a Centaur. It has been carefully monitored, but has never shown any sign of cometary activity.

Damocloids

Damocloids have eccentric, often highly inclined orbits of the 'Halley type' – that is to say, similar to the paths of long-period comets. Some have retrograde orbits. Over 30 Damocloids are known, though as yet (2010) very few have been named and numbered (Table 8.17).

The first to be found was 5335 Damocles, by R. McNaught in 1991; its perihelion distance is inside the aphelion distance of Mars, but it moves out to the distance of Uranus. Its diameter is about 10 km, the average for others of the same type. The other numbered Damocloids are 20461 Dioretsa and 654071.

COMETARY ASTEROIDS

4015 was discovered on 19 November 1949 by A. Wilson and R. Harrington at Palomar, and it showed cometary characteristics, but it was not followed long enough for a orbit to be worked out, and it was lost until 15 November 1979, when Eleanor Helin, also at Palomar, recovered it in the guise of an asteroid. Since then it has shown no sign of activity. Its distance from the Sun ranges between 0.99 and 4.29 a.u. (mean distance 2.64 a.u., 394 800 800 km); period 4.29 years. 7968 was discovered on 24 July 1979, when it had a definite tail, noted in 1996 by E. Elst and G. Pizarro. It again showed cometary activity when near perihelion in 2002. The orbit lies entirely between those of Mars and Jupiter (2.6 to 3.7 a.u., semi-major axis 472 800 000 km, period 5.62 years, eccentricity 0.16, inclination 1.34°). 118401 LINEAR was discovered on 7 September 1999; in November 2005, it developed a coma, and since it is very small (diameter around 6 km) it is very different from Chiron or Echeclus (distance between 2.6 and 3.2 a.u., semi-major axis 479 000 000 km, period 5.7 years).

P/2010 A2 (LINEAR)

This weird object was discovered on 6 January 2010 by LINEAR, using a CCD on a 1-metre reflector. Initially it was classed as a comet, and data given were:

Q = 2,57 a.u.	Eccentricity = 0.12
q = 2.00 a.u.	Inclination = 5.26°
Semi-major axis = 2.29 a.u.	Period = 3.47 years
Diameter = ~220 m	Apparent magnitude = ~18
Absolute magnitude = ~21.3	

From the outset it was recognised as being very strange. It had a tail, which, however, lacked any sign of ice. Observations with the Hubble Space Telescope and OSIRIS, the narrow-angle camera on the Rosetta space-craft, indicate that the dust tail was probably created by the impact of a small rock, a few metres across, on to the larger asteroid (possibly a member of the Flora family) in 2009; OSIRIS results give the collision date as 10 February of that year. The two bodies impacted at an estimated speed of 15 000 km h^{-1}, releasing more energy than a nuclear bomb. Hubble shows a bizarre X-shaped object at the head of a comet-like trail of material. The origin of the X-structure is still unclear.

OTHER ASTEROID BELTS?

There is no reason to doubt the existence of asteroid belts round other stars. One of these stars is the A2-type ζ Leporis; apparent magnitude 3.5, distance 70.2 light-years, luminosity 15 times that of the Sun, mass about twice that of the Sun. Infrared investigations reveal a dust-cloud, apparently coming from a belt of asteroids orbiting the star at around 12 a.u. The total mass of the belt is thought to be half that of the Moon.

9 · Jupiter

Jupiter is much the largest and most massive planet in the Solar System; its mass is greater than those of all the other planets combined. It has been suggested that it may have been responsible for preventing approaching comets invading the inner Solar System, and thereby protecting the Earth from bombardment. Data are given in Table 9.1. Figure 9.1 is a surface map.

MOVEMENTS

Jupiter is well placed for observation for several months in every year. The opposition brightness has a range of about 0.5 magnitude. Generally speaking it 'moves' about one constellation per year; thus the opposition of 2003 was in Cancer, that of 2004 in Leo, 2005 in Virgo, and so on. Opposition dates for 2008–2020 are given in Table 9.2. Some years pass without an opposition; thus that of 3 December 2012 is followed by the next on 8 January 2014, missing out 2013.

Jupiter passes perihelion on 17 March 2011 (4.95 a.u.), and aphelion (5.46 a.u.) on 17 February 2017.

Generally, Jupiter is the brightest of the planets apart from Venus; its only other rival is Mars at perihelic opposition.

OCCULTATIONS AND CONJUNCTIONS

Occultations of and by Jupiter involving other planets are rare. The last occasion when Jupiter was occulted by a planet was on 3 January 1818, when Venus occulted Jupiter; this will happen again on 22 November 2065 at 12h 46m UT, but the elongation from the Sun will be only 8° W. Jupiter last occulted a planet on 15 August 1623, when it occulted Uranus, but on 10 May 1955, at 20h 39m UT, Jupiter and Uranus were only 46″ apart; elongation from the Sun was 65° E, so that the event was easily observed (some unwary observers thought that Jupiter had acquired an extra satellite, although in fact Uranus appeared appreciably larger and dimmer than the Galilean satellites). Data for occultations by the Moon and close conjunctions with other planets are given in Tables 9.3 and 9.4.

EARLY OBSERVATIONS

Since Jupiter is generally the brightest object in the sky apart from the Sun, the Moon and Venus, it must have been known since the dawn of human history. The first telescopic observations were made in 1610 by pioneers such as Galileo and Simon Marius. The four main satellites were discovered, but at first no details were seen on Jupiter itself. Using the current small-aperture, long-focus refractors, N. Zucchi may have seen the main equatorial belts in 1630, and F. Fontana definitely recorded three belts in 1633. In 1648, F. Grimaldi showed that the belts are parallel with the Jovian equator. In 1659, C. Huygens published a good drawing showing the two equatorial belts.

More detailed drawings were made from 1665 by G. D. Cassini, first at Milan and then from Paris. He found the globe of Jupiter to be appreciably oblate, and recorded over half a dozen 'bands'; by watching the drift of the surface features, including one well-marked spot, he gave a rotation period of just over 9h 55m, which was very near the truth. Other observers of the period included G. Campani and Robert Hooke.

Careful studies of Jupiter were made in the latter part of the eighteenth century by William Herschel and J. H. Schröter, but detailed results were delayed until the nineteenth century, with observers such as J. H. Mädler, W. de la Rue, W. Lassell, W. R. Dawes and the Earl of Rosse. Rotation periods were measured; Sir George Airy, the Astronomer Royal, gave 9h 55m 21s, but it became clear that the rotation is differential, with different latitudes having different periods. Studies of Jupiter were pioneered in America by William Bond, at Harvard College. A famous drawing by Warren de la Rue, on 25 October 1865, showed that different features show differences in colour (de la Rue used a home-made 13-inch reflector). In 1890, the British Astronomical Association was founded, and ever since then its Jupiter Section has monitored the surface features. These observations are invaluable, particularly since they go back to the era before good planetary photographs became available. Nowadays, the planet is continuously monitored by amateur astronomers all around the world using high-resolution webcam imaging.

Some early theories sound bizarre today. In 1698, Huygens maintained that Jupiter must be a moist, life-bearing world and that the belts were strips of vegetation, no doubt supporting animals. Even in the early seventeenth century, W. Whewell believed Jupiter to be a globe of 'ice and water' with a cindery nucleus; he described huge gelatinous monsters 'languidly floating in icy seas'. However, it was later assumed that Jupiter must be a miniature sun, able to warm its satellite system – a theory which was still generally accepted until the 1920s.

BELTS AND ZONES

The surface is dominated by the dark belts and bright zones, all of which are variable, although in general their latitudes do not change much. There are also various striking features, notably the Great Red Spot, described below. The main belts are listed in Table 9.5.

Table 9.1 *Jupiter: data*

Distance from the Sun (km):
  max. 815 700 000 (5.455 a.u.)
  mean 778 340 000 (5.203 a.u.)
  min. 740 900 000 (4.951 a.u.)
Sidereal period: 11.86 years = 4332.59 days
Synodic period: 398.88 days
Rotation period:
  System I (equatorial) 9h 50m 30.003s
  System II (rest of planet) 9h 55m 40.623s
  System III (radio methods) 9h 55m 29.711s
Mean orbital velocity: 13.07 km s^{-1}
Axial inclination: 3° 4'
Orbital inclination: 1° 18' 15''.8
Diameter: equatorial 142 884 km
        polar 133 708 km
Oblateness: 0.065
Apparent diameter from Earth: max 50''.1
        min 30''.4
Reciprocal mass, Sun = 1: 1047.4
Density, water = 1: 1.33
Mass, Earth = 1: 317.89 (1.899 × 10^{24} kg)
Volume, Earth = 1: 1318.7 (143.128 × 10^{10} km^3)
Escape velocity: 60.22 km s^{-1}
Surface gravity, Earth = 1: 2.64
Mean surface temperature: −150 °C
Albedo: 0.43
Maximum magnitude: −2.6
Mean diameter of Sun, as seen from Jupiter: 6° 9''
Distance from Earth: max. 968 100 000 km
        min. 588 500 000 km

Jupiter has the shortest 'day' insofar as the principal planets are concerned (some of the asteroids rotate much more quickly). It is not possible to give an overall value for the rotation period, because Jupiter does not spin in the way that a solid body would do. The equatorial zone has a shorter period than the rest of the planet. Conventionally, System I refers to the region between the north edge of the South Equatorial Belt and the south edge of the North Equatorial Belt; the mean period here is 9h 50m 30s, though individual features may have periods which differ perceptibly from this. System II, comprising the rest of the surface of the planet, has a period of 9h 55m 41s, although again different features have their own periods; that of the Great Red Spot varies between 9h 55m 36s and 9h 55m 42s. In addition, there is System III, which relates not

Table 9.2 *Oppositions of Jupiter, 2008–2020*

	Distance (millions of km)	Diameter (')	Apparent magnitude	Constellation
2008 July 9	622.5	47.4	−2.3	Sagittarius
2009 Aug 14	602.6	48.9	−2.4	Capricornus
2010 Sept 21	591.5	49.9	−2.5	Virgo
2011 Oct 29	594.3	49.6	−2.5	Leo
2012 Dec 3	609.0	48.4	−2.4	Gemini
2014 Jan 8	630.0	46.8	−2.2	Gemini
2015 Feb 6	650.1	45.4	−2.1	Cancer
2016 Mar 8	663.4	44.5	−2.0	Leo
2017 Apr 2	666.3	44.3	−2.0	Virgo
2018 May 9	658.1	44.8	−2.0	Libra
2019 June 10	640.9	46.0	−2.1	Sagittarius
2020 July 14	619.4	47.6	−2.3	Capricornus

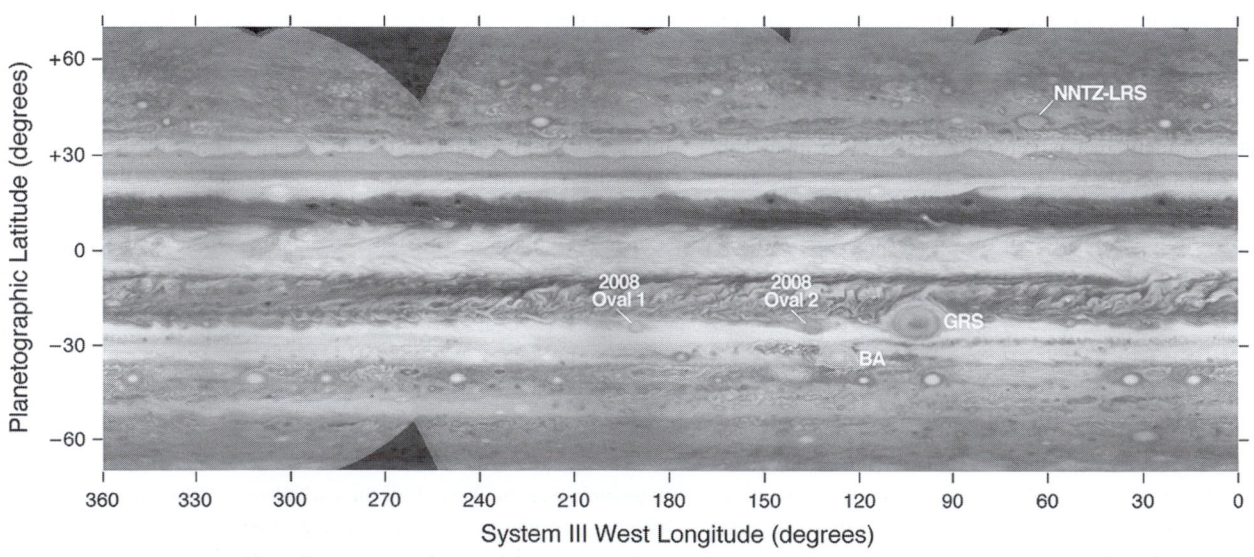

Figure 9.1 Belts and zones on Jupiter

Table 9.3 *Occultations of Jupiter by the Moon*

Date	Time (h)	Area of visibility
2012 Jan 17	08	Arctic
2012 July 15	03	Most of Europe and Asia; Japan
2012 Aug 11	21	Indonesia, northwestern Australia, Pacific, Hawaii
2012 Sept 8	11	Southeast Pacific, South America
2012 Oct 5	21	Southeast Indian Ocean, South Australia
2012 Nov 2	01	South Atlantic, South Africa
2012 Nov 29	01	Atlantic, South Africa
2012 Dec 26	06	South America, southwest Africa
2013 Jan 22	03	Pacific, South America
2013 Feb 18	12	Southern Indian Ocean, southern Australia
2016 July 9	10	Southeast Africa, Antarctica
2016 Aug 6	03	Indonesia, New Guinea, northern Australia
2016 Sept 2	22	Pacific, Hawaii, central America
2016 Sept 30	17	Northern North America, western Europe, northwest Africa
2019 Nov 28	11	Europe, Arabia, Asia
2020 Jan 23	03	Indian Ocean, Australia, New Zealand
2020 Feb 19	20	Antarctica

Table 9.4 *Planetary conjunctions, 2012–2020*

Date	Jupiter with planet	GMT (h)	Separation
2012 July 1	Venus	8	4° 48′
2016 Aug 27	Venus	21.47	24′
2016 Aug 20	Mercury	6	3° 47′
2020 May 18	Saturn	5	4° 42′

Table 9.5 *Average latitudes of the belts.*[a] *These latitudes are subject to slight variation and are given here in round numbers*

		Lat. (°)
South Polar Region	SPRn	−53
South South Temperate Belt	SSTBs	−47
	SSTBn	−42
South Temperate Belt	STBs	−33
	STB	−30
	STBn	−27
South Tropical Band	STropB	−25
South Equatorial Belt	SEBs	−21
	SEBn	−7
North Equatorial Belt	NEBs	+7
	NEBn	+18
North Tropical Band	NTropB	+23
North Temperate Belt	NTBs	+24
	NTBn	+31
North North Temperate Belt	NNTBs	+36
	NNTB	+38
	NNTBn	+39
North North North Temperate Belt	NNNTBs	+43
	NNNTB	+45
	NNNTBn	+47
North North North North Temperate Belt	NNNNTB	+49

[a]Detailed values are given by J. H. Rogers, 1995, *The Giant Planet Jupiter* (Cambridge: Cambridge University Press).

to optical features but to the decimetric radio radiation; the period is 9h 55m 29s .7. There are no true 'seasons' on Jupiter, because the axial inclination to the perpendicular of the orbital plane is only just over 3° – less than for any other planet.

The most prominent belt is generally the North Equatorial, while the South Equatorial Belt is very variable and may become obscure at times; on the other hand, at times during 1962–3 the two equatorial belts appeared to merge, and during 1988 the South Equatorial was fully equal to the North Equatorial in width and intensity. Outbreaks in the South Equatorial Belt are the most spectacular phenomena seen on Jupiter; they involve sudden outbreaks of bright and dark clouds, with intense turbulence and many spots moving on rapid currents. Changes can be surprisingly rapid. In early 2010, the SEB suddenly vanished, leaving the Great Red Spot lonely and bereft of its Hollow! This has happened before, but the Belt always returns. Other belts are also subject to marked variations in intensity, although the North Temperate and South Temperate belts are usually well marked. Around latitude 16° N may be seen brown ovals, which are known as 'barges'; they are low-lying,

and it may be that their colour is due to the fact that they are slightly warmer than the adjacent regions, so that some of the ammonia ice particles begin to melt.

We now have a sound knowledge of the way in which the Jovian winds blow. In the equatorial region, they blow west to east at about 100 m s^{-1} relative to the core, reaching maximum speed 6° to 7° north and south of the equator. In the northern hemisphere the east wind decreases with increasing latitude, until at 18° N the clouds are moving Jupiter westward at 25 m s^{-1}. North of this, the wind speed falls to zero, and then shifts in an eastward direction, reaching a maximum of about 170 m s^{-1} at 24° N. In the southern hemisphere conditions are not quite the same, due probably to the presence of the Great Red Spot, but in both hemispheres there is an alternating pattern of eastward and westward jet streams, which mark the boundaries of the visible belts. Spots such as the Great Red Spot circulate cyclonically or anticyclonically as if rolling between the jet streams.

THE GREAT RED SPOT

The Great Red Spot is undoubtedly the most famous feature on Jupiter. Together with its characteristic 'Hollow', it has certainly been in existence for nearly two centuries, and possibly much longer. It may have been recorded by Cassini as long ago as 1665, although the identification is not certain. A sketch made by

R. Hooke on 26 June 1666 may also show it, and it is also possible that he made an observation of it in 1664. It was seen several times during the nineteenth century, following the first observation of the Hollow in 1831 by H. Schwabe; it was seen to encroach into the southern part of the South Equatorial Belt. The Spot itself was seen in 1858–9 by W. Huggins as a dark ring with a light interior, and also by Lord Rosse, with his great Birr Castle reflector. Suddenly, in 1878, it became very prominent, and brick-red in colour. It remained very conspicuous until 1882, but subsequently faded. Since then it has shown variations in both intensity and colour, and at times it has been invisible, but it always returns, so that during telescopic times it has been to all intents and purposes a permanent feature – unlike any other of the spots. At its greatest extent it was said to measure 40 000 km in an east–west direction and 14 000 km north–south, giving it a greater surface area than that of the Earth, although more recently the dimensions have been 24 000 km by 12 000 km. Whether or not this shrinkage will continue remains to be seen. The latitude varies little from a value of 22.4° S, but the spot drifts around in longitude, and over the past century the total longitude drift has amounted to about 1200°.

It was once thought that the Great Red Spot might be the top of a glowing volcano, but this was soon shown to be untenable. It was then suggested that it might be a solid or semi-solid body floating in Jupiter's outer gas, in which case it would be expected to disappear if its level sank for any reason (possibly a decrease in the density of the outer gas). In 1963, R. Hide suggested that it might be the top of a 'Taylor column': a sort of standing wave above a mountain or a depression below the gaseous layer. However, the space-probe results have given us quite a different picture. The Great Red Spot is a phenomenon of Jovian meteorology – a high-level anticyclonic vortex, with wind speeds of up to 360 km h^{-1}. To the south the spot is bounded by an eastward wind, while to the north it is bounded by a strong westward wind. This means that as the winds are deflected round the spot, they set up anti-clockwise rotation. In the 1960s, the period was 12 days at the outer edge and nine days inside. By 2006, as the GRS shrank, it had decreased to 4.5 days. The vortex is a high-pressure area elevated 8 km above the adjacent cloud deck by the upward convection of warmer gases from below; smaller clouds to the northeast and northwest, beautifully shown on a 1997 image from the Galileo probe, look very like Earth's towering thunderstorms. The cause of the red colour is not definitely known. It may be due to the condensation of phosphorus or sulphur compounds at the cloud-tops. Certainly there is a great deal of interior structure.

THE OVALS

The ovals in the South Temperate Zone have proved to be of great importance. Several began to form in 1939, and eventually formed three definite features which were lettered FA, BC and DE. During the Voyager flybys, they extended about 9000 km from east to west, and 5000 km from north to south, rotating every five days.

Ovals BC and DE merged in 1998, forming Oval BE. Then, in March 2000, BE and FA merged to form Oval BA. On 24 February 2006, Filipino amateur astronomer Christopher Go found that

Oval BA had become red; he at once notified the British Astronomical Association Jupiter Section, and the change was confirmed. Predictably, it became known as the Little Red Spot. It is about half the size of the Great Red Spot, but it too is a massive hurricane-like system, and its winds are just as strong, and it may be at a similar height above the cloud-tops – around 8 km. Its latitude is south of that of the Great Red Spot; in 2006, it passed by the Great Red Spot. There were no obvious interactions between the two, and in June 2007 the second red spot (sometimes called Red Spot Junior!) passed the Great Red Spot, and was apparently unscathed.

A third smaller red spot (the Baby Red Spot) appeared in May 2008, in much the same latitude as the GRS. In July 2008, it became caught up in the circulation of the Great Red Spot, and will probably be absorbed by it. What will happen in the future is at present (2010) unclear, and neither do we know the cause of the colour; it has been suggested that phosphorus may be responsible, or perhaps deeper material dredged up by storms, and exposed to ultraviolet radiation.

THE SOUTH TROPICAL DISTURBANCE

Also in the latitude of the Great Red Spot was a feature known as the South Tropical Disturbance (STD), discovered by P. Molesworth on 28 February 1901 and last recorded by many observers during the apparition of 1939–40. It took the form of a shaded sector of the South Tropical zone between white spots. The rotation period of the STD was shorter than that of the Great Red Spot, so that periodically the STD caught up the Great Red Spot and passed it; the two were at the same latitude, and the interactions were of great interest. Nine conjunctions were observed, and possibly the beginning of the tenth in 1939–40, although by then the STD had practically vanished. Its average rotation period was 9h 55m 27s. 6. Since then it has not reappeared, but there have been several smaller, shorter-lived disturbances of same type; the most notable of these lasted from 1979 to 1981, and there were two of them in 2007. These were observed respectively by the Voyager and the New Horizons spacecraft.

INTERNAL STRUCTURE OF JUPITER

In 1923 and 1924 a classic series of papers by H. Jeffreys finally disposed of the idea that Jupiter is a miniature sun, giving off vast amounts of heat. Jeffreys proposed a model in which Jupiter would have a rocky core, a mantle composed of solid water ice and carbon dioxide, and a very deep atmosphere. Methane and ammonia – both hydrogen compounds – were identified in the atmosphere by R. Wildt in 1932, and it was proposed that Jupiter must consist largely of hydrogen. In 1934, Wildt proposed a model giving Jupiter a rocky core 60 000 km in diameter overlaid by ice shell 27 000 km thick, above which lay a hydrogen-rich atmosphere. (This was certainly more plausible than a strange theory proposed by E. Schoenberg in 1943. Schoenberg believed Jupiter to have a solid surface, with volcanic rifts along parallels of latitude; heated gases rising from these rifts would produce the belts!)

New models were proposed independently in 1951 by W. Ramsey in England and W. DeMarcus in America. According to Ramsey, the 120 000-km-diameter core was composed of hydrogen, so compressed that it assumed the characteristics of a metal. The core was overlaid by an 8000-km-deep layer of ordinary solid hydrogen, above which came the atmosphere. Today, it is believed that Jupiter is mainly liquid (a suggestion made long ago, in 1871, by G. W. Hough, who also believed the Great Red Spot to be a floating island). The latest models are based on work carried out by J. D. Anderson and W. B. Hubbard in the United States. There is no reason to think that they are very far from the truth, although it would be idle to pretend that our knowledge is at all complete.

There is probably a dense core, much more massive than the Earth, where the temperature is of the order of 25 000 °C, and the central pressure around 10 to 100 megabars (i.e. 10 to 100 million Earth atmospheres). The core diameter has been estimated at from 0.3 to 1.1 times that of the Earth. Surrounding the core there is thought to be a thick layer of hydrogen, which at such pressure will be in a metallic state. At about 46 000 km from the core the temperature is 20 000 °C (perhaps rather more) and there is a transition from liquid metallic hydrogen to liquid molecular hydrogen; in the transition region the temperature is assumed to be around 11 000 °C, with a pressure about three million times that of the Earth's air at sea level. Above the liquid molecular hydrogen comes the gaseous atmosphere, which is about 1000 km deep. The change in state is gradual; there is no hard, sharp boundary, so that we cannot say definitely where the 'atmosphere' ends and the actual body of the planet begins.

However, other models have been proposed. According to the American astronomers B. Militzer and W. B. Hubbard (2008), Jupiter has a large rocky core, 14 to 18 times the mass of the Earth, made up of layers of metals, rock and ices of methane, ammonia and water; above it is the atmosphere, composed chiefly of hydrogen and helium. To quote Militzer, of the University of California: 'Jupiter formed beyond the ice line and so accreted ice along with the rocky material. As a result, ice is part of the core and is not in the envelope.'

Jupiter radiates 1.7 times more energy than it would do if it depended only upon radiation received from the Sun. Probably this excess heat is nothing more than what remains of the heat generated when Jupiter was formed. It has been suggested that the globe is slowly contracting, with release of energy, but this explanation is not now generally favoured.

Note, incidentally, that Jupiter's core is not nearly hot enough to trigger stellar-type nuclear reactions. Jupiter is not a 'failed star' or even a brown dwarf; it is definitely a planet.

COMET COLLISION, 1994

A remarkable event occurred in July 1994, when a comet was observed to hit Jupiter.

The comet was discovered on 26 March 1993 by Eugene and Carolyn Shoemaker, working in collaboration with David Levy; because it was this team's ninth discovery the comet was known as Shoemaker–Levy 9 (S/L 9). The image was found on a plate taken three nights earlier with the Schmidt telescope at Palomar. Carolyn Shoemaker described it as a 'squashed comet' quite unlike anything previously seen. It was in orbit not round the Sun, but round Jupiter, and had probably been in this sort of path ever since 1929 – perhaps even earlier. Calculations showed that on 7 July 1992 it had passed only 21 000 km from Jupiter, and had been disrupted, literally torn apart by the Giant Planet's powerful gravitational pull. A year later, in July 1993, it reached apogee – its furthest point from Jupiter – and solar perturbations put it into a collision course. It was calculated that the chain of fragments would impact Jupiter in July 1994, and this is precisely what happened. Over 20 fragments were identified, strung out in the manner of a pearl necklace, and were lettered from A to W (I and O being omitted). The first fragment (A) was due to impact on 16 July, at 20h 11m UT and the last (W) on 22 July, at 8h 5m UT.

The largest fragments were G and Q, while J and M soon faded out altogether. Subsequently P and Q split in two; P2 then split again, while P1 disappeared. By the time they reached Jupiter, the fragments were stretched out over about 29 000 000 km, with separate 'tails' and dusty 'wings' extending ahead of and behind the main swarm.

All the fragments impacted in about the same latitude: around 47° S, well south of the Great Red Spot and in the area of the S.S.S. Temperate Belt. Unfortunately, all the impacts occurred on the side of Jupiter turned away from the Earth, but the planet's quick rotation brought the affected areas into view after only a few minutes, and the results of the impacts were very marked.[1] Great dark spots were produced; the most impressive was G, which impacted at 7h 32m on 18 July. Fragment G, probably about 0.5 km across, created a fireball at least 3000 km high and left a multi-ringed scar on the cloud deck. It was estimated that if fragment G had hit the Earth it would have made a crater 60 km in diameter. The large fragments produced vast clouds of 'smoke', which remained visible from Earth for many months. At one stage the scars seemed to link up, producing what gave the impression of an extra belt on the planet.

The impacts were observed from the Hubble Space Telescope, and also from the Galileo probe, then on its way to Jupiter. Hubble results showed the presence of ammonia and hydrogen sulphide in the G fireball as it cooled, but there was no sign of the expected water layer beneath the cloud-tops.

There were no permanent effects on Jupiter, but the whole area was violently disturbed, and dark material in the Jovian stratosphere produced by the impacts could still be traced well into 1996. There had been suggestions that the impactor might have been an asteroid rather than a comet, but it now seems definite that S/L 9 really was a comet which had spent most of its career in the Kuiper Belt, beyond the orbit of Neptune.

Could there have been any previous observations of cometary impacts? In 1690, G. D. Cassini recorded suspicious dark spots, as did J. Schröter in 1785 and 1786, using an excellent 13-cm reflector made by William Herschel; but it is quite impossible to decide whether or not these were due to an impact of a comet.

THE GALILEO ENTRY PROBE

On 18 October 1989, the Galileo space-craft was launched. It was made up of an orbiter, designed to assume a closed path round Jupiter and transmit data, and an entry probe, to plunge into the clouds and send back results until being destroyed. The experiment was highly successful. At 22h 40m UT on 7 December 1995, the entry probe met the Jovian atmosphere, at latitude 6.5° N, and transmitted for 57.6 min before losing contact. By then it had penetrated to a depth of about 600 km below the tenuous upper reaches of the Jovian atmosphere. The results were in some ways decidedly unexpected. Only one distinct cloud layer was found, apparently corresponding to the previously predicted cloud layer of ammonium hydrosulphide. However, one major surprise concerned the winds. It had been suspected that the strong Jovian winds, about 380 km h^{-1} at the entry level, were more or less confined to the upper atmosphere, but Galileo showed that this is not the case; the velocity increased to over 500 km h^{-1} below the visible level. This seems to indicate that the Jovian winds are not produced by solar heating, as on Earth, or by the condensation of water vapour; it is now more likely that the cause is heat escaping from the deep interior of the globe.

The atmosphere was found to be much dryer than had been expected. There are large, dark dry, clear areas, known as 'hot spots', near the Jovian equator, and that the entry probe plunged into one of these; they are caused by waves of up-and-down winds. A better name for these areas would be 'bright spots', since their temperature at their visible depth is only 0 °C – though this it well above the –130 °C of the surrounding cloud-tops.

LIGHTNING AND AURORÆ

Lightning is very intense on Jupiter; for example enormous bursts were recorded on the planet's night side by the Galileo orbiter in November 1996, with individual flashes hundreds of kilometres across. The flashes are of the 'cloud-to-cloud' variety, and no doubt there is thunder as well. It occurs in turbulent cyclonic regions, and may be generated by convection in water-clouds as on Earth. Auroræ are also intense; they were first detected in 1977, and were recorded by Voyager 1 during its passage across the night side of the planet in March 1979.

RADIO EMISSIONS AND MAGNETOSPHERE

Radio emissions from Jupiter were discovered in 1955 by Bernard Burke and Kenneth Franklin, of the Carnegie Institute (it has to be admitted that the discovery was fortuitous; they made it while testing their antenna array on the Crab Nebula, and at first dismissed it as terrestrial interference). The frequency of the radiation was 22 megahertz, corresponding to a wavelength of 1.38 decametres. Bursts of radiation sometimes made Jupiter the brightest source in the sky at this wavelength apart from the Sun. The bursts of noise from three distinct areas provided the first evidence for a Jovian magnetic field. It was then found that Jupiter is also a source of steady emission at decimetre wavelengths. These two types are

now always known as decametre radiation and decimetre radiation. The non-thermal component of the continuous decimetre radiation is attributed to synchrotron emission (i.e. emission from fast electrons moving within the magnetic field).

Jupiter has a very powerful magnetic field – much the strongest in the entire Solar System. The strength is 4.2 G at the Jovian equator and 10–14 G at the magnetic poles; by contrast, the strength of the Earth's magnetic field at the equator is a mere 0.3 G. With Jupiter, the magnetic axis is inclined to the rotational axis at an angle of 9.6°. The polarity is opposite to that of the Earth, so that if it were possible to use a magnetic compass on Jupiter the needle would point south. With regard to magnetic phenomena, distances from the centre of Jupiter are usually reckoned in terms of the planet's radius R_J. The volcanic satellite Io, which has such a profound effect upon these phenomena, lies at a distance of 5.9 R_J, corresponding to about 422 000 km.

The magnetic field is generated inside Jupiter, near the outer boundary of the shell of metallic hydrogen. The field is not truly symmetrical, but beyond a distance of a few R_J it more or less corresponds to a dipole.

There is a huge magnetosphere: if it could be seen with the naked eye from Earth, its apparent diameter would exceed that of the full moon. The outer boundary is formed at what is known as the magnetopause, where the incoming solar wind particles are deflected and produce a bow shock at about 10 R_J ahead of the actual magnetopause. The region between the bow shock and the true magnetopause is termed the magnetosheath. On the sunward side of Jupiter the field tends to be compressed; on the night side it is stretched out into a 'magnetotail', which may be up to 650 000 000 km long, so that at times it can even engulf Saturn.

There are zones of intense radiation (protons and electrons) at least 10 000 times more powerful than the Van Allen zones surrounding the Earth. Pioneer 10, the first space-probe to encounter these zones (in December 1972), received a total of over 250 000 rads. Since a dose of 500 rads is fatal to a human, future astronauts will be well advised to keep well clear of the danger zone.

In 1964, K. E. Bigg realised the orbital position of Io has a marked effect upon Jupiter's decametric radiation, and it is now known that the satellite is connected to Jupiter by a very strong flux tube setting up a potential difference of 400 000 V. Molecules are sputtered off Io's surface by particles in the magnetosphere, producing a 'torus' tilted to Io's orbit by 7°, so that Io passes through it twice for each rotation of Jupiter. There is also a tenuous sodium cloud round Io which extends all round the orbit of the satellite.

COMET OR ASTEROID COLLISION, 2009

Exactly 15 years after the SL9 comet crash on to Jupiter, Anthony Wesley detected a new impact – on 19 July 2009. It took the form of a dark spot with an oval core and a diffuse patchy arc on the northwestern side. The impact must have taken place on the dark side of the planet between 7.40 and 14.00 UT on 19 July. Confirmation was quickly obtained; the site was marked by a high-altitude impact cloud, similar to the S/L9 impact sites. Thus it consisted of

Table 9.6 *Rings of Jupiter*

Name	Inner radius (km)	Vertical thickness (km)	Outer radius (km)	
Halo	92 000	30 500	125 000	Merges into Main Ring
Main	122 500	30 to 300	129 100	Bounded by Adrastea
Gossamer, Amalthea	129 000–182 000	2000	53 000	Connected with Amalthea
Gossamer, Thebe	129 000–226 000	8400	97 000	Connected with Thebe

almost black 'smoke' thrown up into the stratosphere by the initial explosion and the splash-back of the ejecta. The appearance persisted over the next week, sometimes looking not unlike a satellite shadow, but from 29 July began to fade, breaking up and spreading out in longitude; by mid-August it had broken into eight small grey clouds dispersed over 50° of longitude. Traces of it lingered on into September.

Anthony Wesley's nickname is 'Bird'. Inevitably, the 2009 impact became known as the Bird Strike! The impactor was either an asteroid or a comet, less than 1 km in diameter.

Wesley had another success on 3 June 2010, at 20.31 GMT, when he observed another impact event; it was also seen by Christopher Go in the Philippines, who captured it on video. It took the form of a bright flash lasting for two seconds. It occurred on the position of the southern edge of the then-absent SEB. Observations made by amateurs with the Hubble Space Telescope indicate that there was no subsequent cloud of débris, so that presumably the impactor was a meteoroid that did not plunge deeply enough into the atmosphere to cause an explosion. Instead, it burned away during its descent.

On 20 August 2010, the Japanese amateur Masayuki Tachikawa, from Kumamoto, observed a fireball impact on Jupiter; he took a movie, showing the fireball scintillating along with other features on the planet. It was also recorded by another Japanese amateur, Aoki Kazuo, 800 km away. The site was on the northern edge of the NEB. No after-effects were visible.

Encounters of this sort may happen fairly frequently. It may also be significant that Jupiter 'captured' a comet, P/147 Kushida–Muramatsu, in 1949, and only in 1961 did the comet escape back into solar orbit; Comet P/111 Helin–Roman–Crockett became a temporary satellite from 1967 and 1985, orbiting Jupiter three times, and is expected to complete six laps of the planet between 2068 and 2086.

Jupiter is a source of cosmic radiation – and an energetic one. Jovian cosmic rays have even been detected as far out as the orbit of Mercury.

THE RINGS OF JUPITER

Jupiter's ring system was discovered on 4 March 1979, on a single image sent back by Voyager 1 as it passed through the equatorial plane of the planet. They have since been detected with the Hubble Space Telescope, and studied in detail with the Galileo space-craft. Details are given in Table 9.6. There are three main components: Halo, Main and Gossamer (there are actually two components of the Gossamer ring).

The rings are very dark, and are quite unlike the bright, icy rings of Saturn. They are caused by material coming from the small inner satellites; this dark, slightly reddish material is thrown off the surfaces of the satellites following impacts by meteoroids travelling at speeds greatly enhanced by Jupiter's strong gravitational pull – just as a cloud of chalk dust will be produced when two erasers are banged together. The small inner satellites Metis and Adrastea, which move inside the Main Ring, are the most important contributors to the Main and Halo rings, with Amalthea and Thebe responsible for the Gossamer rings. There seem to be virtually no ice particles anywhere in the ring system.

The outer (Gossamer) ring is actually composed of two faint and more or less uniform rings, one enclosing the other: they extend from the outer boundary of the Main Ring (129 100 km) and extend out to over 222 000 km, although the ring is so tenuous that it is difficult to give a precise boundary. The fainter of the two extends radially inward from the orbit of Thebe, while the denser of the two – the enclosed ring – extends radially inward from the orbit of Amalthea. In each case the centres of the rings are fainter than the edges. The Main Ring extends from the orbit of Adrastea to the edge of the Halo Ring, while the Halo Ring is toroidal, extending radially from 122 000 km to 92 000 km.

Voyager and Galileo results show that the rings are much brighter in forward-scattered light than in back-scattered or reflected light. This indicates that the ring particles are in general only 1–2 μm across. Such particles have relatively short lifetimes in stable orbits, so that the rings must be continually replenished by material produced by the small satellites.

SPACE-CRAFT TO JUPITER

Eight space-craft had encountered Jupiter by 2009: two Pioneers, two Voyagers, Galileo, the solar probe Ulysses, the Saturn probe Cassini, and New Horizons, bound for Pluto. Ulysses, Cassini and New Horizons used Jupiter's powerful pull to put them into their required orbits. Details are given in Table 9.7.

The Pioneers were virtual twins; both were successful. Pioneer 10 carried out studies of the Jovian atmosphere and magnetosphere, and returned over 300 images. It showed that the radiation zones are far stronger than had been previously believed. The first energetic particles were detected when Pioneer was still over 20 000 000 km from Jupiter, and the radiation level increased steadily as Pioneer moved inward; at the minimum distance from the upper clouds (131 400 km) the instruments were almost saturated, and if the minimum distance had been much less the mission would

Table 9.7 *Missions to Jupiter, 1972–2000*

Name	Launch date	Encounter date	Nearest approach (km)	Remarks
Pioneer 10	2 Mar 1972	3 Dec 1973	131 400	Complete success; new images and data. Now on its way out of the Solar System. Last communication in 2003; now silent.
Pioneer 11	5 Apr 1973	2 Dec 1974	46 400	Complete success. Went on to rendezvous with Saturn (1979 Sept 1). Now on its way out of the Solar System. Last communication in 1995; now silent.
Voyager 1	5 Sept 1977	5 Mar 1979	350 000	Detailed information about Jupiter and the Galilean satellites Io, Ganymede and Callisto; volcanoes on Io discovered. Went on to rendezvous with Saturn (12 Nov 1980). Now on its way out of the Solar System; still contactable.
Voyager 2	20 Aug 1977	9 July 1979	714 000	Complemented Voyager 1. Went on to fly by Saturn (1981), Uranus (1986) and Neptune (1989). Now on its way out of the Solar System.
Galileo	18 Oct 1989	7 Dec 1995	Entry	Fly-by of Venus (10 Feb 1990) and Earth (8 Dec 1990 and 8 Dec 1992); images of asteroids Gaspra (29 Oct 1991) and Ida (8 Aug 1993). Orbiter deliberately impacted Jupiter, 21 Sept 2003.
Ulysses	6 Oct 1990	8 Feb 1992		Studies of Jupiter's magnetosphere, radiation zones and general envcironment. Went on to survey the poles of the Sun. Finally deactivated on 1 July 2007.
Cassini	15 Oct 1997	30 Dec 2000	10 000 000	Fly-by imaged Jupiter from 10 000 000 km en route for Saturn.
New Horizons	19 Jan 2006	28 Feb 2007	2 300 000	Fly-by. Imaged Jupiter en route for Pluto.

have failed. The proposed orbit of Pioneer 11 was hastily altered to a different trajectory which would carry it quickly over Jupiter's equatorial zone, where the danger is at its worst. Pioneer 10 is now on its way out of the Solar System; it carries a plaque to give a clue to its planet of origin – although whether any other beings would be able to decipher the message seems rather debatable.

Pioneer 11 confirmed the earlier findings, and were then put into a path which took it out to a rendezvous with Saturn. It too is now leaving the Solar System permanently.

The third Jupiter probe, Voyager 1, was actually launched a few days later than its twin Voyager 2, but travelled in a more economical path. (To confuse matters still further, initial faults detected in the first space-craft caused a switch in numbers, so that the original Voyager 1 became Voyager 2 and *vice versa*.) The Voyagers were much more elaborate than the Pioneers, and the results obtained were of far higher quality. Voyager 1 also surveyed the satellites Io, Ganymede and Callisto. Voyager 2 followed much the same programme, and was also sent close to Europa, the only Galilean satellite not well studied by its predecessor. Voyager 1 went on to survey Saturn, while Voyager 2 was able to encounter Uranus and Neptune as well.

Galileo was made up of an orbiter and an entry probe. After a six-year journey through the Solar System, it approached Jupiter in 1995: on 13 July of that year the orbiter was separated from the entry probe, and the two reached Jupiter on different trajectories. The entry probe penetrated to a depth of about 600 km. Six hours before entry it also detected a new, very intense radiation zone round Jupiter; the orbiter acted as a relay, and, after the demise of the entry probe, began a long-continued survey of the satellite system.

When its fuel was almost exhausted, the orbiter was deliberately crashed into Jupiter, at a speed of 50 km s^{-1}, to avoid the possibility of impacting Europa and contaminating the satellite. The end came on 21 September 2003, by which time the probe had covered 4 631 778 000 km and had completed 35 orbits of Jupiter. It had been in space for 14 years, 8 of them in the Jovian system.

The Cassini space-craft, en route for Saturn, made a gravity-assist fly-by of Jupiter on 30 December 2000, at a range of 10 000 000 km, and took 26 000 images. A swirling dark oval of atmospheric haze, about the size of the Great Red Spot, was detected near the north pole, and images of very small storm-cells were seen in the dark belts. For a while, in December 2000, Jupiter was being surveyed by Cassini and Galileo simultaneously.

SATELLITES

Jupiter's satellite family is unlike any other in the Solar System. There are four main satellites, of planetary size, always known as the Galileans, although probably Marius saw them slightly before Galileo did so. Of these, three are larger than our Moon and the fourth (Europa) only slightly smaller, while Ganymede is actually larger than Mercury, although less massive.

By 2008, the total number of known satellites had risen to 63. There are four small inner satellites; then come the Galileans, while the remaining members of the system are further out. Some of these 'irregular' satellites have retrograde motion. The outer satellites are probably asteroidal; most of them are below 10 km in diameter. No doubt more outer satellites await discovery. Data for the known satellites are given in Table 9.8.

Table 9.8 *Satellites of Jupiter*

Number	Name	Diameter, D (km)	Mass ($\times 10$ kg)	Semi-major axis, SMA ($\times 1000$ km)	Orbital period, P (a minus sign indicates retrograde motion)	Orbital inclination, I (°)	Orbital eccentricity, e	Year of discovery	Discoverer	Group
1	XVI	Metis	$60 \times 40 \times 34$	-36	127 690	$+0.295$	0.06	0.00002	1979	Synott	Ir
2	XV	Adrastea	$20 \times 16 \times 24$	-0.2	128 690	$+0.298$	0.03	0.0015	1979	Jewitt (Voyager 2)	In
3	V	Amalthea	$270 \times 170 \times 150$	208	181 366	$+0.498$	$0.374°$[31]	0.0032	1892	Barnard	In
4	XIV	Thebe	$116 \times 98 \times 84$	-43	221 889	$+0.675$	$1.076°$[31]	0.0175	1979	Synnott (Voyager 1)	In
5	I	Io	$3660.0 \times 3637.4 \times 3630.6$	8 900 000	421 700	$+1.769137786$	$0.050°$[31]	0.0041	1610	Galilei	G
6	II	Europa	3121.6	4 800 000	671 034	-3.551181041	$0.471°$	0.0094	1610	Galilei	G
7	III	Ganymede	5262.4	15 000 000	1 070 412	-7.15455296	$0.204°$	0.0011	1610	Galilei	G
8	IV	Callisto	4820.6	11 000 000	1 882 709	$+16.6890184$	$0.205°$	0.0074	1610	Galilei	G
9	XVIII	Themisto	8	0.069	7 393 216	$+129.87$	$45.762°$	0.2115	1975/2000	Kowal & Roemer / Sheppard *et al.*	T
10	XIII	Leda	16	0.6	11 187 781	-241.75	$27.562°$	0.1673	1974	Kowal	H
11	VI	Himalia	170	670	11 451 971	-250.37	$30.486°$	0.1513	1904	Perrine	H
12	X	Lysithea	36	6.3	11 740 560	$+259.89$	$27.006°$	0.1322	1938	Nicholson	H
13	VII	Elara	86	87	11 778 034	$+261.14$	$29.691°$	0.1948	1905	Perrine	H
14	–	S/2000 J 11	4	0.0090	12 570 424	$+287.93$	$27.584°$	0.2058	2001	Sheppard *et al.*	H
15	XLVI	Carpo	3	0.0045	17 144 873	$+458.62$	$56.001°$	0.2735	2003	Sheppard *et al.*	C
16	–	S/2003 J 12	1	0.00015	17 739 539	-482.69	$142.680°$	0.4449	2003	Sheppard *et al.*	?
17	XXXIV	Euporie	2	0.0015	19 088 434	-538.78	$144.694°$	0.0960	2002	Sheppard *et al.*	A
18	–	S/2003 J 3	2	0.0015	19 621 780	-561.52	$146.363°$	0.2507	2003	Sheppard *et al.*	A
19	–	S/2003 J 18	2	0.0015	19 812 577	-569.73	$147.401°$	0.1569	2003	Gladman *et al.*	A
20	XLII	Thelxinoe	2	0.0015	20 453 753	-597.61	$151.292°$	0.2684	2003	Sheppard *et al.*	A
21	XXXIII	Euanthe	3	0.0045	20 464 854	-598.09	$143.409°$	0.2000	2002	Sheppard *et al.*	A
22	XLV	Helike	4	0.0090	20 540 266	-601.40	$154.586°$	0.1374	2002	Sheppard *et al.*	A
24	XXIV	Iocaste	5	0.019	20 722 566	-609.43	$147.248°$	0.2874	2001	Sheppard *et al.*	A
25	–	S/2003 J 16	2	0.0015	20 743 779	-610.36	$150.769°$	0.3184	2003	Gladman *et al.*	A
26	XXVII	Praxidike	7	0.043	20 823 948	-613.90	$144.205°$	0.1840	2001	Sheppard *et al.*	A
27	XXII	Harpalyke	4	0.012	21 063 814	-624.54	$147.223°$	0.2440	2001	Sheppard *et al.*	A
28	XL	Mneme	2	0.0015	21 129 786	-627.48	$149.732°$	0.3169	2003	Gladman *et al.*	A
29	XXX	Hermippe	4	0.0090	21 182 086	-629.81	$151.242°$	0.2290	2002	Sheppard *et al.*	A
30	XXIX	Thyone	4	0.0090	21 405 570	-639.80	$147.276°$	0.2525	2002	Sheppard *et al.*	A
31	XII	Ananke	28	3.0	21 454 952	-642.02	$151.564°$	0.3445	1951	Nicholson	A
32	–	S/2003 J 17	2	0.0015	22 134 306	-672.75	$162.490°$	0.2379	2003	Gladman *et al.*	C

Table 9.8 (cont.)

	Number	Name	Diameter, D (km)	Mass ($\times 10$ kg)	Semi-major axis, SMA ($\times 1000$ km)	Orbital period, P (a minus sign indicates retrograde motion)	Orbital inclination, I (°)	Orbital eccentricity, e	Year of discovery	Discoverer	Group
33	XXXI	Aitne	3	0.0045	22 285 161	−679.64	165.562°	0.3927	2002	Sheppard et al.	C
34	XXXVII	Kale	2	0.0015	22 409 207	−685.32	165.378°	0.2011	2002	Sheppard et al.	C
35	XX	Taygete	5	0.016	22 438 648	−686.67	164.890°	0.3678	2001	Sheppard et al.	C
36	–	S/2003 J 19	2	0.0015	22 709 061	−699.12	164.727°	0.1961	2003	Gladman et al.	C
37	XXI	Chaldene	4	0.0075	22 713 444	−699.33	167.070°	0.2916	2001	Sheppard et al.	C
38	–	S/2003 J 15	2	0.0015	22 720 999	−699.68	141.812°	0.0932	2003	Sheppard et al.	A
39	–	S/2003 J 10	2	0.0015	22 730 813	−700.13	163.813°	0.3438	2003	Sheppard et al.	C
40	–	S/2003 J 23	2	0.0015	22 739 654	−700.54	148.849°	0.3930	2004	Sheppard et al.	P
41	XXV	Erinome	3	0.0045	22 986 266	−711.96	163.737°	0.2552	2001	Sheppard et al.	C
42	XLI	Aoede	4	0.0090	23 044 175	−714.66	160.482°	0.6011	2003	Sheppard et al.	P
43	XLIV	Kallichore	2	0.0015	23 111 823	−717.81	164.605°	0.2041	2003	Sheppard et al.	C
44	XXIII	Kalyke	5	0.019	23 180 773	−721.02	165.505°	0.2139	2001	Sheppard et al.	C
45	XI	Carme	46	13	23 197 992	721.82	165.047°	0.2342	1938	Nicholson	C
46	XVII	Callirrhoe	9	0.087	23 214 986	722.62	139.849°	0.2582	2000	Gladman et al.	P
47	XXXII	Eurydome	3	0.0045	23 230 858	723.36	149.324°	0.3769	2002	Sheppard et al.	P
48	XXXVIII	Pasithee	2	0.0015	23 307 318	726.93	165.759°	0.3288	2002	Sheppard et al.	C
49	XLVIII	Cyllene	2	0.0015	23 396 269	731.10	140.148°	0.4115	2003	Sheppard et al.	P
50	XLVII	Eukelade	4	0.0090	23 483 694	−735.20	163.996°	0.2828	2003	Sheppard et al.	C
51	–	S/2003 J 4	2	0.0015	23 570 790	739.29	147.175°	0.3003	2003	Sheppard et al.	P
52	VIII	Pasiphaë	60	30	23 609 042	741.09	141.803°	0.3743	1908	Gladman et al.	P
53	XXXIX	Hegemone	3	0.0045	23 702 511	745.50	152.506°	0.4077	2003	Sheppard et al.	P
54	XLIII	Arche	3	0.0045	23 717 051	746.19	164.587°	0.1492	2002	Sheppard et al.	C
55	XXVI	Isonoe	4	0.0075	23 800 647	750.13	165.127°	0.1775	2001	Sheppard et al.	C
56	–	S/2003 J 9	1	0.00015	23 857 808	752.84	164.980°	0.2761	2003	Sheppard et al.	C
57	–	S/2003 J 5	4	0.0090	23 973 926	758.34	165.549°	0.3070	2003	Sheppard et al.	C
58	IX	Sinope	38	7.5	24 057 865	762.33	153.778°	0.2750	1914	Nicholson	P
59	XXXVI	Sponde	2	0.0015	24 252 627	771.60	154.372°	0.4431	2002	Sheppard et al.	P
60	XXVIII	Autonoe	4	0.0090	24 264 445	772.17	151.058°	0.3690	2002	Sheppard et al.	P
61	XLIX	Kore	2	0.0015	23 345 093	776.02	137.371°	0.1951	2003	Sheppard et al.	P
62	XIX	Megaclite	5	0.021	24 687 239	792.44	150.398°	0.3077	2001	Sheppard et al.	P
63	–	&$$$;	2	0.0015	30 290 846	1077.02	153.521°	0.1882	2003	Sheppard et al.	?

*In = inner, G = Galilean, H = Himalia, C = Carme, A = Ananke, P = Pasiphaë. Themisto and Carpo do not seem to be associated with any known group.

The Galileans can be seen with almost any telescope or even with good binoculars. Very keen-sighted people have even reported naked-eye sightings, and there is considerable evidence that one of them (probably Ganymede, or else two satellites close together) was seen from China by Gan De as long ago as 364 BC. The first attempted maps of the Galileans were due to A. Dollfus and his colleagues at the Pic du Midi Observatory in 1961. Some features were recorded, but, predictably, the maps were not very accurate. Today we have detailed maps obtained by space-craft, and details can also be followed with the Hubble Space Telescope.

THE SMALL INNER SATELLITES

Metis

Metis is named after the daughter of Jupiter by his first consort, Oceanus. It orbits inside the Main Ring, within a 500-km gap in the ring. It was well imaged by the Galileo space-craft, and is irregular in shape, with a longest diameter of 60 km. The surface is thickly cratered; the leading hemisphere is brighter than the trailing hemisphere (as with the other small inner satellites, the axial rotation period is synchronous). The main constituent of the globe is presumably water ice.

Adrastea

Adrastea is named after a daughter of Jupiter and Ananke (equated with Nemesis, the goddess of rewards and punishments). It lies just inside the outer edge of the Main Ring, and is probably the main source of the ring material. It is irregular in shape, with a longest diameter of 20 km, but has not been well imaged. No doubt it is composed mainly of water ice.

Amalthea

Amalthea is another mythological name; possibly the goat which suckled the infant Jupiter (Zeus) or possibly the daughter of Melisseus, King of Crete, who brought up the infant on a diet of goat's milk. It was discovered in 1892 by E. E. Barnard, using the 36-inch (91-cm) refractor at the Lick Observatory: this was the last satellite discovery to be made visually.

The satellite is irregular in form; the surface is very red, due probably to contamination from Io. It has synchronous rotation, with its longest axis pointing toward Jupiter, and is heavily cratered. The two largest craters, Gaea and Pan, are of immense size relative to the overall diameter of Amalthea. Pan is 100 km in diameter. It was well imaged in January 2000 by the Galileo probe. Gaea seems to have a depth of between 10 and 20 km; if the latter figure is correct, the slope angle of the wall will be 30°. (It would be interesting to watch a piece of material fall from the crest to the floor: the descent time would be about 10 min!) Both Pan and Gaea are deeper, relatively, than craters of similar size on the Moon – between them, from longitude 0–60° W, is a complex region of troughs and ridges, tens of kilometres long and up to at least 20 km wide. The two bright patches, Ida and Lyctos, are each about 15 km across (Table 9.9).

Amalthea is exposed to the Jovian radiation field, and also to energetic ions, protons and electrons produced in Jupiter's

Table 9.9 *Features on Amalthea*

	Lat. (°)	Long. (° W)	Diameter (km)
Craters			
Gaea	80.0 S	90.0	80
Pan	55.0 N	35.0	100
Faculæ			
Ida Facula	20.0 N	175.0	
Lyctos Facula	20.0 S	120.0	

magnetosphere; it is also bombarded by micrometeorites, and by sulphur, oxygen and sodium ions that have been blasted away from Io.

Amalthea lies near the outer edge of the Amalthea Gossamer Ring; the ring material is due to dust blasted away from the satellite. The escape velocity is only about 0.06 km s^{-1}, and the moon is presumably very ice-rich.

From Amalthea, the apparent diameter of Jupiter would be over 45°. Jupiter is stationary in the sky, and would hide the Sun for an hour and a half at each revolution; Amalthea's quick rotation gives it under six hours of sunlight at each revolution, but Jupiter would be 900 times brighter than our full moon.

Thebe

Thebe is named after the daughter of the river god Asopus; it was discovered on the Voyager images. It moves beyond the main part of the Gossamer Ring. It has synchronous rotation and low albedo; the surface is dark and reddish; the leading hemisphere is the brighter of the two. The surface is cratered; the largest crater, 40 km across, lies on the anti Jove side of the satellite, and has been named Zethus. It was discovered on images taken by the Galileo space-craft. Thebe is rather elliptical; the longest axis (116 km) always points toward Jupiter as must be the same for all these inner satellites.

THE GALILEAN SATELLITES

Io

Io is named after the daughter of Inachus, King of Argos; Jupiter was enamoured of her and Juno, Jupiter's wife, ill-naturedly changed Io into a white heifer.

The satellite is slightly larger than our Moon, and is the densest of the four Galileans. Before the space missions it was tacitly assumed to be a rocky, cratered world, but in the event nothing could have been further from the truth; it is the most volcanically active of any body in the Solar System.

In March 1979, S. Peale and his colleagues in America calculated that since Io's orbit is not perfectly circular, the interior would be 'flexed' by the gravitational pulls of Jupiter and the other Galileans, heating it sufficiently to produce active surface volcanoes. A week later, on 9 March, this prediction was dramatically verified. Linda Morabito, a member of the Voyager imaging team, was looking for a faint star, AGK-10021, as a check on Io's position when she saw what was undoubtedly a volcanic plume rising from the limb of the satellite. Subsequently nine plumes were detected, together with volcanic craters and numerous calderæ; impact craters were absent. The lack of impact craters means that the surface cannot be more than a million years old, and there must be constant 'resurfacing', with deposition of a layer 1 mm thick each year.

The surface is made up of vent regions, plains regions and mountains; the average surface temperature is −143 °C. Mountains are appreciable; the highest, Hæmus, rises to 13 km, and its steep slopes mean that it cannot be solid sulphur, even though sulphur and sulphur dioxide cover most of Io's surface. The mountains are presumably siliceous, with an outer coating of sulphur sent out by the volcanoes. The plains are crossed by yellow and brownish-yellow flows; the original Voyager images made them look redder than they really are. There are extensive deposits of sulphur dioxide (SO_2); the gas is vented from the volcanic areas and is frozen out when it reaches the bitterly cold surface.

The volcanoes seem to be of two main types. The sulphur volcanoes, such as Pele, Surt and Aten, send out material at up to 1 km s^{-1}; eruptions last for days or months (for example, Pele was erupting at the time of the Voyager 1 pass, but was quiescent when Voyager 2 flew past the planet). The sulphur dioxide volcanoes, such as Prometheus, Amirani and Volund, have lower vent velocities, but eruptions go on for months or years consecutively. (Loki, one of the most violent centres, seems to be of a hybrid type.) The temperatures of the volcanoes are very high, and the Galileo probe recorded that Pillan Patera reached over 2000 °C. This is too hot for the material to be sulphur, and it now seems that intensely heated lava, in the form of silica enriched by magnesium and sodium, may be responsible for much or all of Io's vulcanism. This hot lava vaporises sulphur near the vent, or SO_2 as the lava flows across the frozen SO_2 plains, to create the plumes. They rise to hundreds of kilometres, although the vent velocities are not sufficient to expel material from Io altogether. The black patches seen round the geysers are due to sulphur dioxide frost. Over 200 calderæ have been identified, although by no means all are active; there are probably many lava lakes. Observations from the Galileo probe in late 1999 showed over 100 active centres. Pele volcano showed a red ring of sulphur, over 1200 km in diameter, deposited by a plume of material emerging from the volcano. Loki is the most powerful volcano in the Solar System, emitting more heat than all the Earth's active volcanoes combined; the temperature of the lava attains 1027 °C. A selected list of surface features is given in Table 9.10 and Figure 9.2 gives a map of the surface.

Io seems to have a dense core, rich in iron and iron sulphide, which extends half-way from the centre of the globe to the surface, and is overlaid by a mantle of partly molten rock; above this comes the relatively thin, rocky, sulphur-coated crust. The atmosphere of sulphur dioxide is excessively tenuous, and corresponds to what we usually call a good laboratory vacuum; it may also be very variable in both density and distribution. Its highest pressure is one-hundred millionths of that of the Earth's air at sea level.

Table 9.10 *Selected list of features on Io. Heights and widths are derived from observations from space-craft and the Hubble Space Telescope. They are no doubt very variable*

	Lat. (°)	Long. (° W)	Height (km)	Width (km)
Eruptive sites				
Amirani	25.9 N	114.5	95	220 (Plume 5)
Aten	47.9 S	310.1	300	1200
Culann Patera	19.9 S	158.7	–	–
Kanehekili	18.0 S	037.0	–	–
Loki	17.9 N	302.6	200	400 (Plume 2)
Malik Patera	34.2 S	128.5	–	–
Marduk	27.1 S	207.5	70	195 (Plume 7)
Masubi	46.3 S	54.7	–	– (Plume 8)
Maui	16.5 N	124.0	90	230 (Plume 6)
Pele	18.6 S	257.8	400	1200 (Plume 1)
Pillan Patera	12.0 S	244.0	140	400
Prometheus	1.6 S	153.0	75	270 (Plume 3)
Ra Patera	8.6 S	325.3	–	400
Surt	45.5 N	337.9	300	1200
Volund	25.0 N	184.3	100	125 (Plume 4)
Zamama	18.0 N	173.0	–	–

	Lat. (°)	Long. (° W)	Diameter (km)
Pateræ			
Amaterasu	37.7 N	306.6	100
Aten	47.9 S	310.0	40
Atar	30.2 N	278.9	125
Babbar	39.5 S	272.1	95
Cataquil	24.2 S	18.7	125
Creidne	52.4 S	343.5	125
Daedalus	19.1 N	274.3	40
Discura	37.0 N	119.0	70
Emakong	3.2 S	119.1	80
Galai	10.7 S	288.3	90
Gibil	14.9 S	294.9	95
Gish Bar	17.0 N	90.0	150
Heinseb	29.7 N	244.8	60
Horus	9.6 S	338.6	125
Huo Shen	15.1 S	329.3	90
Isum	29.0 N	208.0	100
Kane	47.8 S	13.4	115
Khalla	6.0 N	303.4	80
Loki	12.6 N	308.8	250
Lu Huo	38.4 S	354.1	90
Mafuike	13.9 S	260.0	110
Malik	34.2 S	128.5	85
Masaya	22.5 S	348.1	125
Mihr	16.4 S	305.6	40
Nina	38.3 S	164.2	425

Table 9.10 (cont.)

	Lat. (°)	Long. (° W)	Width (km)
Nusku	64.7 S	4.6	90
Nyambe	0.6 N	343.9	50
Reiden	13.4 S	235.7	70
Ruwa	0.4 N	3.0	50
Shakuru	23.6 N	266.4	70
Shamash	33.7 S	152.1	110
Svarog	48.3 S	267.5	70
Taranis	70.8 S	28.6	105
Tol-Ava	1.7 N	322.0	70
Tupan	18.0 S	141.0	50
Ülgen	40.4 S	288.0	49
Vahagn	23.8 S	351.7	70
Viracocha	61.2 S	281.7	55
Zal	42.0 N	76.0	130
Catenæ			
Mazda	8.6 S	313.5	
Reshet	0.8 N	305.6	
Fluctus			
Eubœa	45.1 S	351.3	
Fjorgynn	17.5 N	358.0	300
Ionian	5.0 N	250.0	
Kanehekili	16.0 S	38.0	250
Lei-Kung	38.0 N	204.0	400
Marduk	27.0 S	209.0	150
Masubi	48.0 S	60.4	800
Tung Yo	16.4 S	357.8	
Uta	32.6 S	19.2	
Montes			
Boösaule	4.4 S	270.1	590
Eubœa	46.3 S	339.9	
Hæmus	68.9 S	46.6	
Silpium	52.6 S	272.9	
Mensæ			
Echo	79.6 S	357.4	
Epaphus	53.5 S	241.3	
Iynx	61.1 S	304.6	
Pan	49.5 S	35.4	
Plana			
Argos	47.0 S	318.2	140
Danube	20.9 S	258.7	150
Dodona 13	56.8 S	352.9	390
Ethiopia	44.9 S	27.0	105
Hybristes	54.0 S	21.1	150
Iopolis	34.5 S	333.5	125
Lyrcea	40.3 S	269.3	310
Nemea	73.3 S	275.5	500
Regiones			
Bactria 10	45.8 S	123.4	
Chalybes 11	45.5 N	83.2	
Colchis 12	5.3 N	199.8	
Illyrikon	72.0 S	160.0	700
Lerna 14	64.0 S	292.6	
Media 15	4.6 N	58.8	
Mycenæ 16	37.3 S	165.9	
Tarsus	43.7 S	61.4	
Tholi			
Apis	11.2 S	348.8	
Inachus	15.9 S	348.9	

Io plays a major rôle in the shape of Jupiter's magnetosphere which sweeps up gas and dust from the thin Ionian atmosphere at the rate of 1 tonne per second. Surrounding Io, up to a distance of 6 radii from the surface of the satellite, is a cloud made up of neutral atoms of sulphur, potassium, sodium and oxygen, and Io orbits within a belt known as the Io plasma torus, originating when the neutral atoms in the 'cloud' are ionised and carried along by the Jovian magnetosphere. Unlike the particles in the neutral cloud, these particles co-rotate with the magnetosphere revolving around Jupiter at a rate of almost 75 km s^{-1}. High-resolution images acquired when Io was in eclipse show an auroral glow, due to radiation striking the upper atmosphere.

On Io, the active volcanic areas and pateræ have been named after gods of fire, thunder, volcanoes, mythical blacksmiths and solar deities (for example, Pele is the Hawaiian goddess of fire), catenæ after Sun gods and the other features after people and places associated with myths involving Io.

Io is indeed a strange, colourful place, but since it moves well within Jupiter's radiation zones it must be just about the most lethal world in the entire Solar System.

Europa

Europa is the second and smallest of the four Galileans. In mythology she was the daughter of King Agenor of Tyre and sister of Cadmus; Jupiter (Zeus) assumed the form of a bull and carried her across the sea to Crete, where she bore him several children.

Europa is as different from Io as it could possibly be. Its surface is smooth and white, covered with water ice or snow; the average albedo is 0.7 for the white regions and 0.5–0.6 for the slightly darker areas, so that Europa is particularly reflective. It is also very cold, with a mean surface temperature of –145 °C. There are few impact craters, showing that the surface must be young – perhaps only a few millions of years old, so that there must be constant resurfacing.

The globe seems to consist mainly of silicate rock, with an upper layer of water and water ice. The outer layer of solid ice may be 10–25 km thick (although this issue is controversial), below which there is a sea which could go down to 100 km, kept warm by tidal heating. The Galileo space-craft found that magnetic interactions with Jupiter indicate the presence of a layer of highly electrically conductive material in Europa, made probably of salty water; the dark reddish streaks and patches on the surface may be

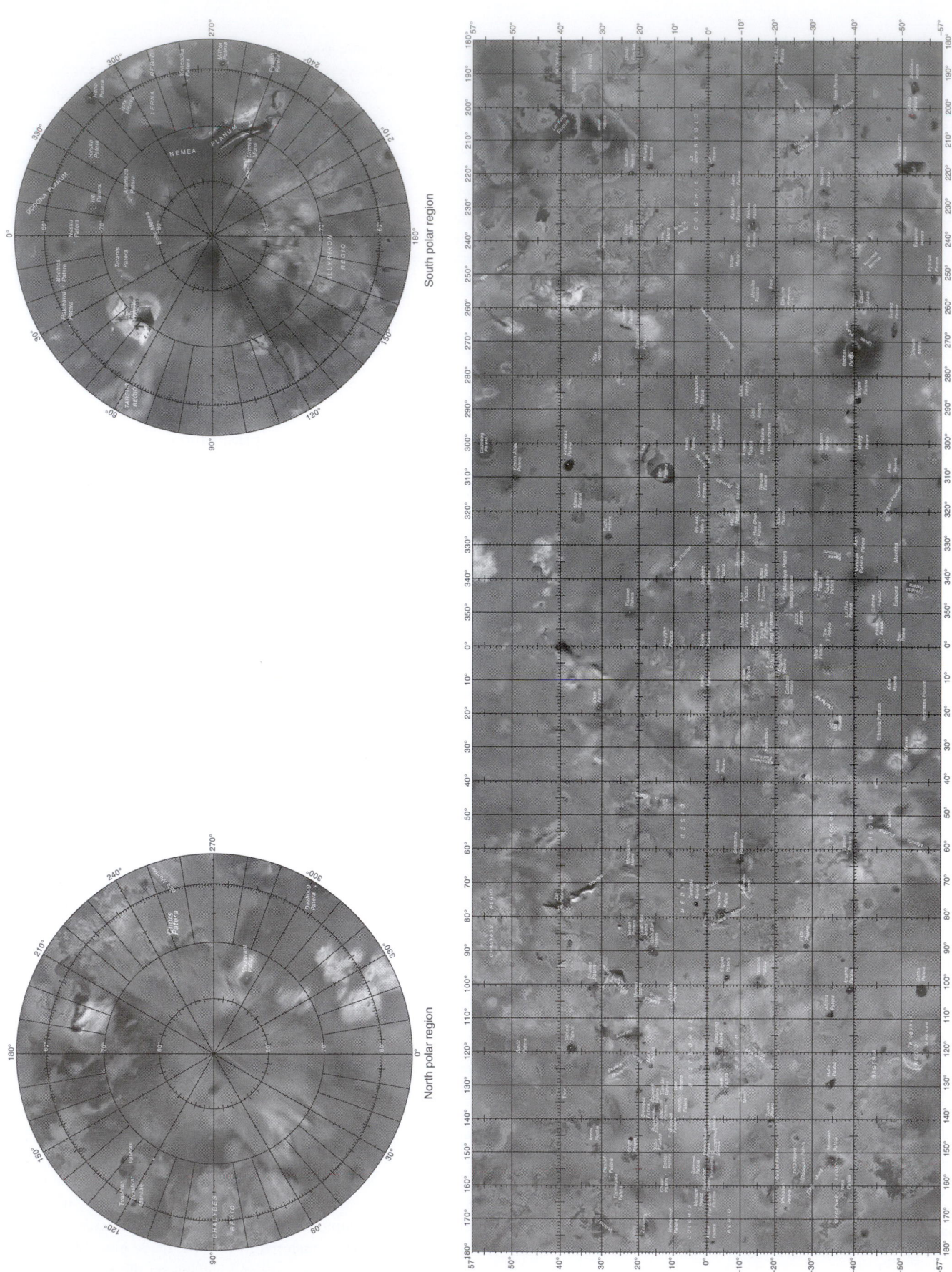

South polar region

North polar region

Figure 9.2 Io

rich in salts such as magnesium sulphate, deposited by evaporation from the ocean. If the underground sea exists – and most astronomers now believe that it does – it must be an eerie place. Life there might seem to be unlikely – but it cannot be entirely ruled but. Already there are plans for landing a space-craft and boring through the ice to reach the water below.

According to Richard Greenberg (University of Arizona) Europa's ice covered ocean may contain twice as much liquid water as all the Earth's oceans combined. He also maintains that there is plenty of oxygen in the ocean to support microfauna, assuming that their oxygen demands are similar to those of terrestrial fish!

The plains are broken into plates a few kilometres across, and it has even been proposed that they drift about above the liquid or mushy material below; along fracture lines, warm material may well up to produce the linear features (a process termed cryovulcanism). One very young crater, Pwyll, shows bright rays extending in all directions and crossing all other features; there may have been partial melting in the solid ice. Calanish and Tyre have numerous concentric fissures: possibly the impactor penetrated the crust through to the darker material below, and the crater floor is now at the same level as the outer terrain, so that it may be filled with slushy material. There are even features which look uncannily like icebergs.

Craters on Europa are named after Celtic heroes, and other features after people associated with myths involving Europa. A selected list of surface features is given in Table 9.11 and the map is shown in Figure 9.3.

There are many cycloid-shaped cracks known as *flexi*. G. Hoppa and B. Randall (University of Arizona) suggested in 1999 that they were formed as Europa's icy crust responded to tidal forces induced by Jupiter. There is certainly a tidal bulge 30 m high, and this shifts location during each revolution, since the orbit of the satellite is slightly eccentric. According to Hoppa and Randall, this causes tension cracks to open and propagate along the surface at a rate of around 3 km h^{-1}. This, of course, would indicate that the tidal bulge is sliding freely over the interior, and is further evidence in support of an underground ocean.

Observations from the Hubble Space Telescope and the Galileo space-craft have shown that Europa has an excessively tenuous oxygen atmosphere, with a density no more than one hundred thousand millionths of that of the Earth's air at sea level. If all of it were compressed to the density of our air, it would just about fill the Royal Festival Hall. The icy surface of the satellite is subject to impacts from dust and charged particles, and these processes cause the surface ice to produce water vapour as well as gaseous fragments of water molecules. These are then broken up into hydrogen and oxygen. The hydrogen escapes, while the oxygen is retained to form a thin atmosphere extending up to perhaps 200 km: obviously it must be continuously replenished from below.

Ganymede

Ganymede is the largest satellite in the Solar System. It is named after a handsome shepherd boy summoned by Jupiter to become cup-bearer to the gods. (One has to admit that Jupiter's motives were not entirely altruistic!)

Table 9.11 *Selected list of features on Europa*

Name	Lat. (°)	Long. (° W)	Diameter/length (km)
Craters			
Cilix	1.2 N	181.9	23
Govannan	37.5 S	302.6	10
Manann'an	20.0 N	240.0	30
Morvran	5.7 S	152.2	25
Pwyll	26.0 S	271.0	26
Rhiannon	81.8 S	199.7	25
Taliesin	23.2 S	137.4	48
Tegid	0.6 S	164.0	29
Flexus			
Cilicia	47.6 S	142.6	639
Delphi	69.7 S	172.3	1125
Gortyna	42.4 S	144.6	1261
Phocis	48.6 S	197.2	298
Sidon	64.5 S	170.4	1216
Lineæ			
Adonis	51.8 S	113.2	758
Agenor	43.6 S	208.2	1326
Alphesibœa	28.0 S	182.6	1642
Argiope	8.2 S	202.6	934
Asterius	17.7 N	265.6	2735
Astypalœa	76.5 S	220.3	1030
Belus	11.8 N	228.3	2580
Cadmus	27.8 N	173.1	1212
Echion	13.1 S	184.3	1217
Ino	5.0 S	163.0	1400
Katreus	39.5 S	215.5	245
Libya	56.2 S	183.3	452
Minos	45.3 N	195.7	2134
Pelagon	34.0 N	170.0	800
Pelorus	17.1 S	175.9	1770
Phœnix	14.5 N	184.7	732
Phineus	33.0 S	269.2	1984
Rhadamanthys	18.5 N	200.8	1780
Sarpedon	42.2 S	89.4	940
Tectamus	17.9 N	181.9	719
Telephassa	2.8 S	178.8	800
Thasus	68.7 S	187.4	1027
Thynia	57.9 S	148.6	398
Maculæ			
Boestia	54.0 S	166.0	22
Cycleides	64.0 S	192.0	105
Thera	47.7 S	180.9	78
Thrace	46.6 S	171.2	173
Large ring features			
Callanish	16.0 S	333.4	100
Tyre	31.7 N	147.0	148

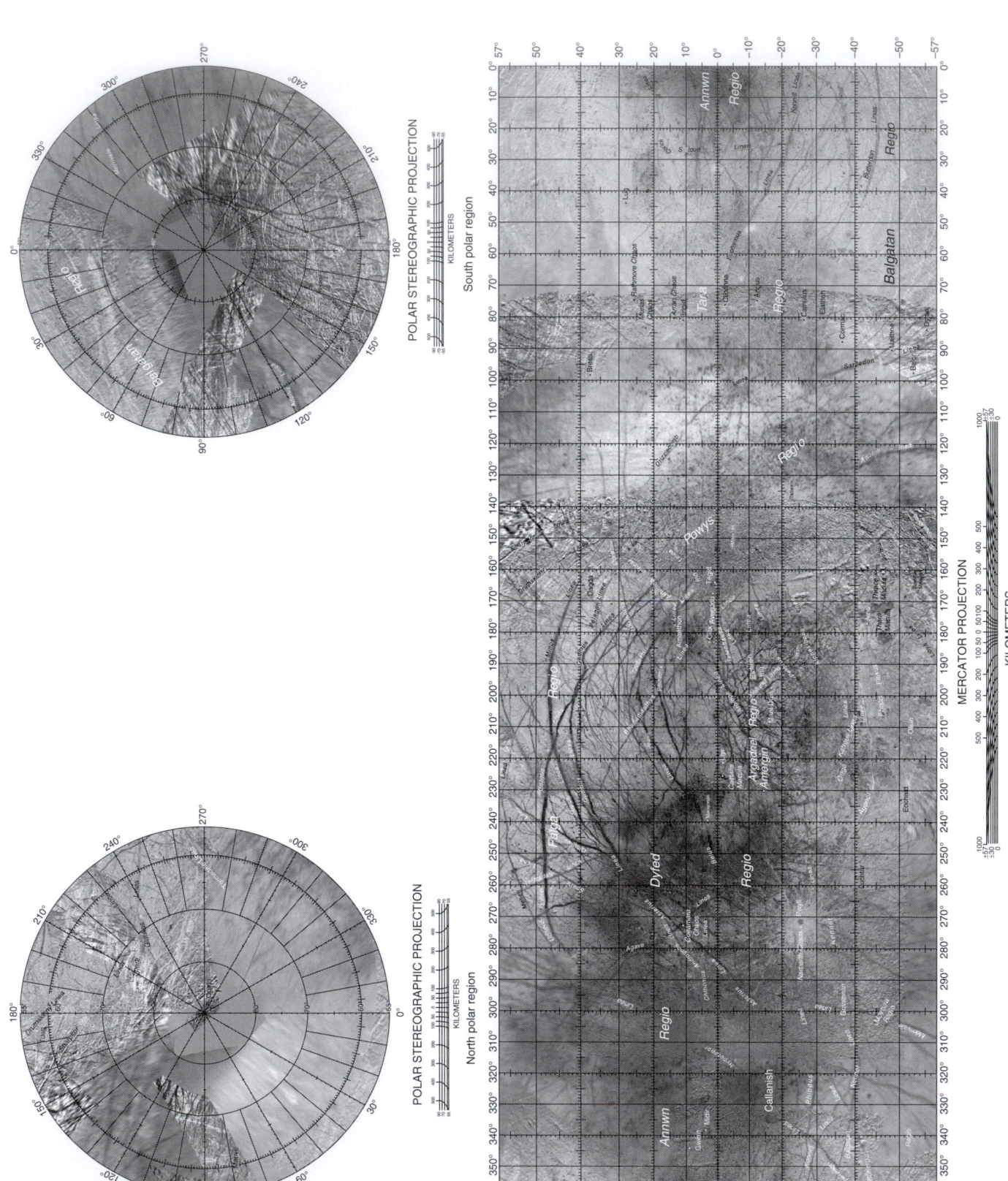

Figure 9.3 Europa

Table 9.12 *Selected list of features on Ganymede. The latitudes and longitudes are central values*

Name	Lat. (°)	Long. (° W)	Diameter/length (km)
Regiones			
Barnard	0.8 N	1.0	2547
Galileo	35.7 N	137.6	3142
Marius	12.1 N	199.3	3572
Nicholson	34.0 S	356.7	3719
Perrine	38.8 N	30.0	2145
Craters			
Achelous	60.3 N	13.5	51
Agreus	15.2 N	225.4	72
Amon	33.4 N	223.3	102
Anubis	82.7 S	118.5	97
Ashima	37.7 S	122.4	82
Bau	24.1 N	53.3	81
Enkidu	27.9 S	328.4	121
Eshmun	17.8 S	191.5	99
Gilgamesh	61.7 S	123.9	175
Halieus	35.2 N	168.0	90
Hapi	31.3 S	212.4	85
Irkilla	31.1 S	114.7	116
Ishkur	0.1 M	11.5	83
Isimu	8.1 N	2.5	90
Isis	67.9 S	197.2	68
Kadi	48.8 N	181.0	94
Khonsu	38.0 S	189.7	86
Kingu	35.7 S	227.4	91
Kulla	34.8 N	115.0	82
Melkart	10.0 S	185.8	111
Misharu	5.3 S	338.3	95
Mush	13.5 S	115.0	97
Neith	28.9 N	9.0	93
Nidaba	19.0 N	123.8	188
Ninki	6.6 S	120.9	170
Ninlil	7.6 N	118.7	91
Nunsum	13.3 S	140.1	91
Nut	60.1 S	268.0	93
Osiris	37.8 N	165.2	109
Sati	30.5 N	14.9	98
Sebek	59.5 N	178.9	70
Seker	40.8 S	351.0	117
Selket	16.7 N	107.4	140
Ta-Urt	26.5 N	306.5	85
Thoth	42.4 S	146.0	107
Tros	11.0 N	31.1	109
Zaqar	57.5 N	41.3	52
Faculæ			
Abydos	34.1 N	154.0	165
Busiris	14.9 N	216.1	348
Buto	12.6 N	204.3	236
Coptos	9.4 N	209.8	332
Dendera	0.0	257.0	114
Edfu	26.8 N	147.7	187
Memphis	15.4 N	132.5	344
Ombos	3.8 N	238.6	90
Punt	26.1 S	242.2	228
Sais	37.9 N	14.2	137
Siwah	7.5 N	143.2	220
Tettu	38.6 N	160.9	86
Thebes	4.8 N	202.4	475
Fossæ			
Lakhamu Fossa	12.5 S	228.3	392
Lakhmu Fossæ	30.3 N	142.3	2871
Zu Fossæ	53.0 N	129.4	1386
Sulci			
Aquarius Sulcus	50.0 N	11.5	1341
Arbela Sulcus	22.3 S	353.6	1896
Bubastis Sulci	79.8 S	263.1	2197
Dardanus Sulcus	39.3 S	20.2	2559
Elam Sulci	57.4 N	205.5	1866
Mashu Sulcus	31.1 N	209.2	3030
Mysia Sulci	9.6 S	28.6	4221
Nippur Sulcus	40.9 N	191.5	2158
Phrygia Sulcus	12.4 N	19.3	3205
Sippar Sulcus	15.8 S	191.0	1539
Tiamat	3.2 N	209.2	1310
Ur Sulcus	48.0 N	178.0	950
Uruk Sulcus	8.4 N	169.0	2456
Xibaltia	35.0 N	80.0	2000

Ganymede is much less dense than Io or Europa, and is of quite different type: the overall density is less than twice that of water. Ganymede is fully differentiated; the best models give an iron-rich core around 1800 km in diameter, with a silicate mantle and an upper ice layer 150 km thick. As with Europa, magnetic interactions with Jupiter suggest that there may be an ocean of salty water, but if this ocean exists it will be much deeper down than Europa's. Then comes the mantle, of which the lower part is siliceous and the upper part icy. The mantle is overlaid by the thin, icy crust. All in all, it seems that the globe is made up of a combination of rocky materials (60%) and ice (40%).

There are two types of surface: dark and bright regions. The dark regions are well defined; the most prominent has been appropriately named Galileo Regio. They are heavily cratered, and are presumably the oldest parts of the surface. Crossing them are dark furrows (fossæ), from 5 to 10 km wide; very often they indicate the outlines of distorted circles, and may well have originated from massive impacts early in Ganymede's history. There are also light-floored, rough features without walls; these are termed palimpsests (faculæ) believed to be the traces of ancient impacts before the crust was fully frozen solid, and are very ancient indeed. Their floors are relatively flat; some of them are as much as 200 km across.

The bright regions are characterised by grooves (sulci), which run in some cases for thousands of kilometres, although the vertical

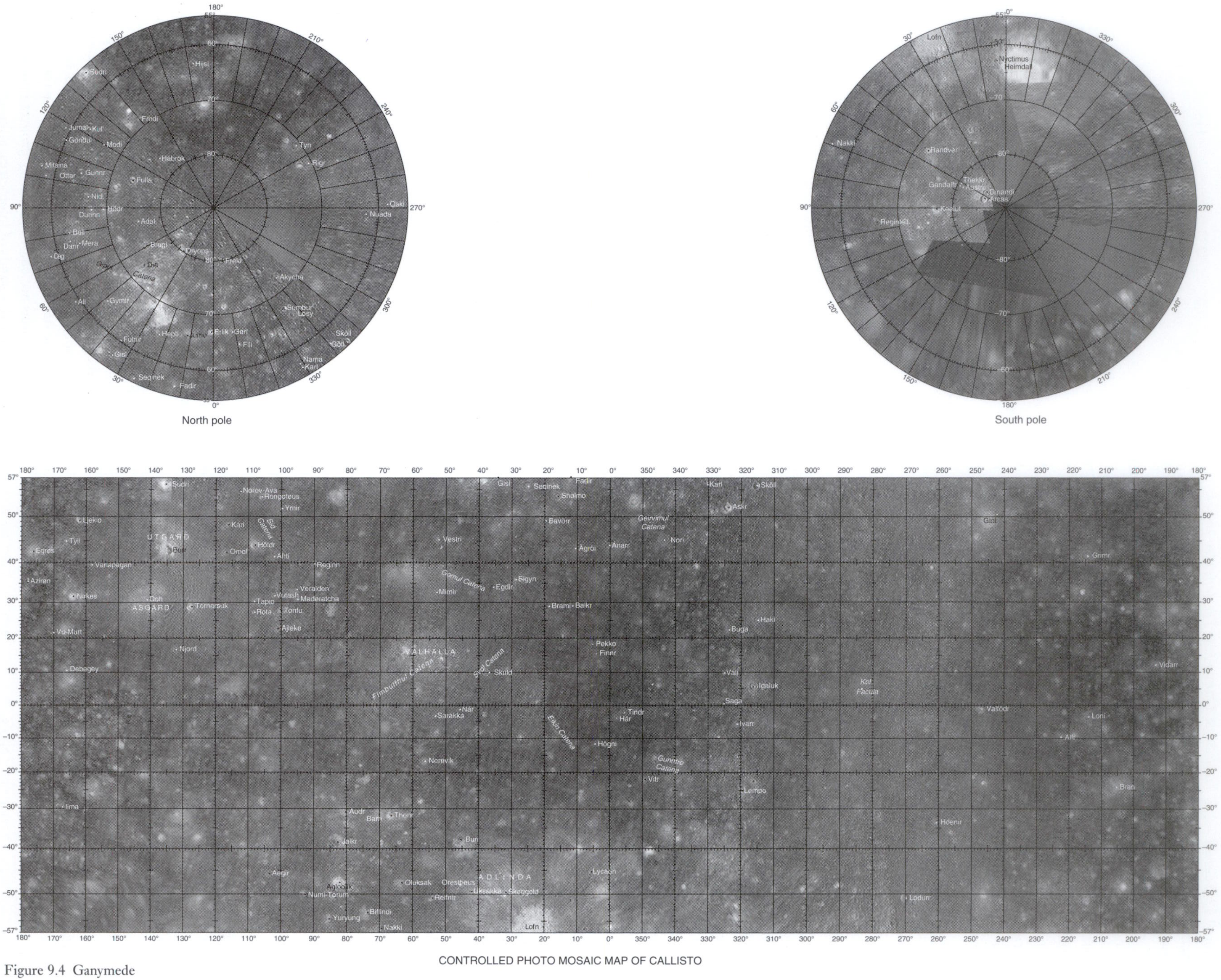

North pole

South pole

CONTROLLED PHOTO MOSAIC MAP OF CALLISTO

Figure 9.4 Ganymede

Table 9.13 *Selected list of features on Callisto*

Name	Lat. (°)	Long. (° W)	Diameter/length (km)
Basins			
Adlinda	56.6 S	23.1	900
Asgard	32.0 N	139.8	1347
Valhalla	15.9 N	56.6	2748
Catenæ			
Gipul Catena	70.2 N	48.2	588
Craters			
Adal	75.4 N	80.8	40
Ägröi	43.3 N	11.0	55
Ahti	41.8 N	103.1	52
Akycha	72.5 N	318.6	67
Ali	59.3 N	56.2	61
Anarr	44.1 N	0.6	47
Aningan	50.5 N	8.2	287
Askr	51.7 N	324.1	64
Aziren	35.4 N	178.3	64
Balkr	29.1 N	11.9	64
Bavörr	49.2 N	20.3	84
Brami	28.9 N	19.2	67
Bran	24.3 S	207.7	89
Buri	38.7 S	46.2	98
Burr	42.5 N	135.5	74
Fadir	56.4 N	12.7	81
Finnr	15.5 N	4.3	65
Gloi	49.0 N	245.7	112
Grimr	41.6 N	215.2	90
Haki	24.9 N	315.1	69
Hödr	69.0 N	91.0	76
Högni	13.5 S	4.5	65
Igaluk	5.6 N	315.9	105
Ivarr	6.1 S	321.5	68
Lodurr	51.2 S	270.8	76
Loni	3.6 S	214.9	86
Nār	1.7 S	46.4	63
Nori	45.4 N	343.5	86
Nuada	62.1 N	273.2	66
Reginn	39.7 N	90.8	51
Rigr	70.9 N	245.0	54
Sequinek	55.5 N	25.5	80
Sköll	55.6 N	315.3	55
Skuld	10.1 N	37.7	81
Sudri	55.4 N	137.1	69
Tindr	2.5 S	355.5	64
Tornarsuk	28.7 N	128.6	104
Tyll	43.3 N	165.4	65
Tyn	70.8 N	233.6	60
Valfödr	1.2 S	247.8	81
Vanapagan	38.1 N	158.0	62
Veralden	33.2 N	96.1	75
Vestri	43.3 N	52.8	75
Vidarr	11.9 N	193.6	84
Vitr	22.4 S	349.3	76
Vu-Mart	22.9 N	170.9	79
Vutash	31.9 N	102.9	55
Ymir	51.4 N	101.3	77

relief does not exceed a few hundred metres; they are often flat-topped, with gentle slopes of up to 20 degrees. These areas are essentially icy. Of the faculæ (bright spots) the most prominent is Memphis, which contains dark-floored craters which have been punched through to the darker material below – indicating that much of the bright palimpsest is a thin sheet.

Impact craters abound; the largest well-marked crater is the 175-km Gilgamesh, which is surrounded by outlying concentric escarpments with an overall diameter of 800 km. Small craters tend to have central peaks, while with larger craters (over 35 km across), central pits are more common. There are also ray-craters, such as Osiris, whose brilliant rays stretch out for over 1000 km. Unquestionably, there has been marked tectonic activity in past ages, although Ganymede today is to all intents and purposes inert.

One major surprise, due to the Galileo probe, is that Ganymede has a magnetic field. By terrestrial standards it is weak, but it is sufficient to produce a well-defined magnetosphere – so that we have a magnetosphere within a magnetosphere. The magnetic axis is inclined to the rotational axis by about 10°. Because Ganymede has a magnetic field, auroræ are not unexpected. The brighter displays coincide roughly with the boundaries of the polar caps. Ultraviolet emissions imaged by the Hubble Space Telescope are indeed striking.

The Hubble Space Telescope detected ozone on the surface, caused by the disruption of icy particles as a result of bombardment from charged particles. There is also a very tenuous oxygen atmosphere, no denser than that of Europa and presumably of the same type.

On Ganymede, craters and fossæ are named after gods and heroes of the ancient Fertile Crescent peoples, faculæ after places associated with Egyptian myths, sulci after places associated with other ancient myths, and regions after astronomers who have discovered Jovian satellites (Galileo, Simon Marius, E. E. Barnard, S. B. Nicholson and C. D. Perrine). A selected list of formations on Ganymede is given in Table 9.12 (with the associated map in Figure 9.4).

Callisto

Callisto, the fourth Galilean, is named after the daughter of King Lycaon of Arcadia, who was turned into a bear by Juno and subsequently placed in the sky as Ursa Major.

Callisto is almost as large as Mercury, but its relatively low albedo means that it is the faintest of the Galileans. It is also much further away from Jupiter, so that eclipse, transit and occultation phenomena are less frequent than with Io, Europa or Ganymede. It is also the least dense of the four, although its escape velocity is still higher than that of Europa.

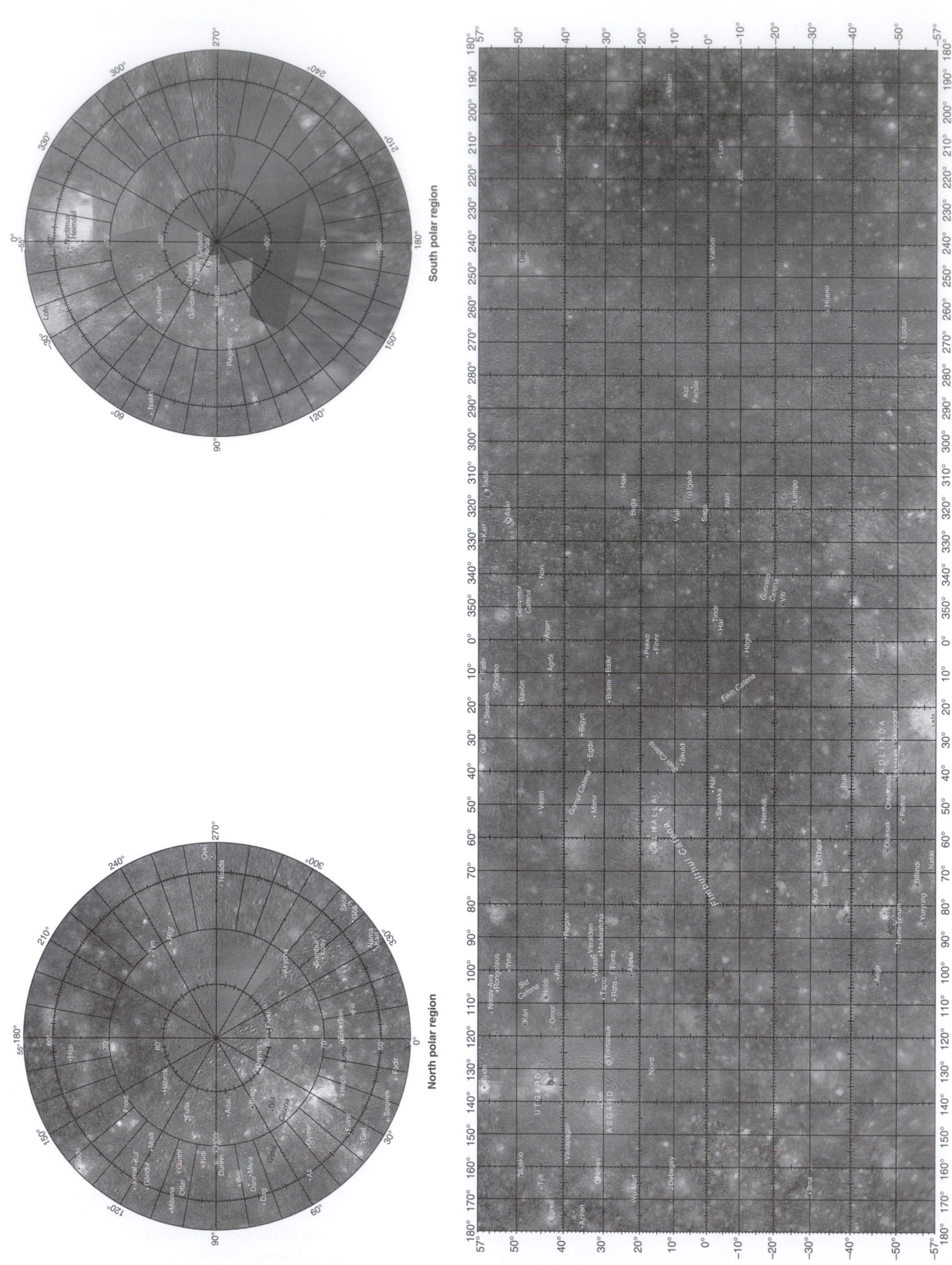

Figure 9.5 Callisto

Although Callisto is almost equal to Ganymede in size, it is different in many respects. It seems to be comparatively undifferentiated, and until recently it was thought to lack any substantial iron-rich core; we are not yet certain whether such a core exists. The surface is icy, and is saturated with craters. The dominant features are two large ringed basins, Valhalla and Asgard. Valhalla is a complex structure; around the central 600 km palimpsest there are concentric rings, and the palimpsest itself is less heavily cratered than the surrounding areas, showing that Valhalla is young by Callistan standards – although the entire surface is very ancient, and there is no evidence of past tectonic activity, as there is on Ganymede. Asgard is similar in type, although smaller; the third basin Adlinda is partly obscured by rays from a bright, fresh crater now called Lofa. Elsewhere there are craters all over the surface, some of which have dark floors with bright rims and central peaks, but very small craters seem to be less numerous than was thought before the data was sent back by the Galileo space-probe. This is because thousands of millions of years of erosion have stripped the topography on a submillimetre scale leaving icy pinnacles prostruding above dusty plains. There are various catenæ (crater-chains), due probably to the impacts of comets or asteroids which were broken up before impacting. For example, Gipul Catena is a crater-chain 588 km long; the largest crater is 40 km across. Svol, crossing the 81 km crater Skuld, is another good example of a catena. Callisto shows no Ganymede-type grooves, and there are not many craters over 100 km in diameter.

Ringed basins are named after the homes of the Norse gods and heroes; craters from heroes and heroines from Northern myths, and catenæ from mythological places in Scandinavia. A selected list of features on Callisto is given in Table 9.13, and Figure 9.5 shows a map of the surface.

Before the Galileo results it was generally assumed that Callisto was solid through to its centre, and that it had been totally inert since the very early days of the Solar System. Now, we think rather differently. There may be an upper icy lithosphere between 80 and 100 km deep, overlying a salty ocean up to 200 km deep. The lowest part of the interior consists of compressed rock and ices; the diameter of the core can hardly be much more than 1200 km. As with Ganymede, the possibility of subsurface oceans exists.

Galileo has also detected an excessively tenuous atmosphere, which seems to be made up of carbon dioxide. The density is so low that it ranks as an exosphere, i.e. a collisionless gas. It is easily lost because of the effects of ultraviolet radiation from the Sun, and so it must be constantly replenished, perhaps by venting from Calisto's interior – although this again mitigates against the assumption that the satellite is completely inert.

THE SMALL OUTER SATELLITES

Before 2000, only Himalia, Lysithea, Elara, Ananke, Carme, Pasiphaë and Sinope were known; by 2008, the total had grown to 63, and will no doubt have increased still further by the time that this book as been published. Details of the known outer satellites are given in Table 9.8. Only Himalia is as much as 100 km in diameter.

The satellites may be included in various groups, some prograde, some retrograde. Some have been imaged from great distances by the Voyager and Galileo probes. The smallest satellites at least are presumably asteroidal, and the outermost are so far from Jupiter that no two orbital cycles are alike. The main groups are Himalia (prograde), Carpo (prograde), Ananke (retrograde), Carme (retrograde) and Pasiphaë (retrograde).

Not much can yet be said about most of the satellites, but a few notes may be appropriate:

Themisto was discovered by C. Kowal and E. Roemer in 1975, lost, and recovered in 2000 by S. Sheppard *et al.* It does not belong to a group, but orbits alone between the Galileans and the inner (Himalia) prograde irregulars.

Himalia (diameter 170 km) is much the largest of the outer satellites; it appears neutral grey, similar to a C-type asteroid. It was imaged by the Cassini spacecraft, en route to Saturn, and seems to be elongated – with axes 150 and 120 km. The spectrum showed a slight indication of water.

The only other outer satellites over 10 km across are Lysithea (36 km), Elara (86), Ananke (28), Carme (46), Pasiphaë (60) and Sinope (38)

Pasiphaë is the largest retrograde satellite. The spectrum seems to be featureless.

Sinope is said to belong to the Pasiphaë group, but is red rather than grey like Pasiphaë; it seems more to resemble a D-type asteroid. Its discovery was fortuitous; it appeared on a plate exposed by S. B. Nicholson on 21 July 1914 when he was photographing Pasiphaë.

ENDNOTE

1 I was observing with the 66-cm refractor at Herstmonceux, then the site of the Royal Greenwich Observatory. The scar left by Impact A was far more prominent than I had expected. Later scars were easily visible in the 5-cm finder of my small portable refractor, and were much blacker than anything I had ever previously seen on Jupiter, apart from shadows of its satellites.

10 · Saturn

Saturn is often regarded as the most beautiful object in the entire sky. Jupiter, Uranus and Neptune also have rings, but these systems are dark and obscure; Saturn's glorious, icy rings are unrivalled.

Saturn was the outermost planet known in pre-telescopic times, and is sixth in order of distance from the Sun. It is named in honour of the second ruler of Olympus, who succeeded his father Uranus and was himself succeeded by his son Jupiter (Zeus). It moves in the sky more slowly than the other bright planets, and has been associated with the passage of time. Data are given in Table 10.1.

MOVEMENTS

Saturn reaches opposition about 13 days later every year. Opposition dates for the period 2010–2020 are given in Table 10.2.

The opposition magnitude is affected both by Saturn's varying distance and by the angle of presentation of the rings. At its best, Saturn may outshine any star apart from Sirius and Canopus, but at the least favourable oppositions from this point of view – when the rings are edgewise-on, as in 2009 – the maximum magnitude may be little brighter than Aldebaran. A list of edgewise presentations is given in Table 10.3.

The intervals between successive edgewise presentations are 13 years 9 months and 15 years 9 months. During the shorter interval, the south pole is sunward; the southern ring-face is seen, and Saturn passes through perihelion. Perihelion fell in 1944 and 1974; the next will be in 2032. The aphelion dates are 1959, 1988 and 2018.

At edgewise presentations the rings appear as a delicate line of light, particularly when the Earth passes through the plane of the rings and also when the Sun does so. Very small telescopes will not then show them at all. This was an early observational proof that the rings are very thin.

Occultations of Saturn by the Moon are not uncommon. It is interesting to note how small Saturn looks when compared with a lunar crater! There will be close conjunctions of Saturn and Venus on 9 January 2016 and 15 September 2016, and on 15 September 2037, at 21.30 UT, Saturn and Mercury will be separated by only 18 arcsec.

EARLY RECORDS

The first observations of Saturn must have been made in prehistoric times. The first recorded observations seem to have been those made in Mesopotamia in the mid-seventh century BC. About 650 BC there is a record that Saturn 'entered the Moon', which is presumably a reference to an occultation of the planet. Very careful observations were made by the Greeks and, later, the Arabs; of course it was generally assumed that Saturn, like all other celestial bodies, moved round the Earth.

Copernicus recorded an observation of Saturn on 26 April 1514, when the planet lay in line with the stars 'in the forehead of Scorpio'; he also noted the position of Saturn on 5 May 1514, 13 July 1520 and 10 October 1527. Tycho Brahe noted that, on 18 August 1563, Saturn was in conjunction with Jupiter. However, the first telescopic observation was delayed until July 1610, when Galileo examined it with his early telescope, using a magnification of ×32. He noted that 'the planet Saturn is not one alone, but is composed of three, which almost touch one another and never move nor change with respect to one another. They are arranged in a line parallel to the Zodiac, and the middle one is about three times the size of the lateral ones.' Galileo's telescope was not powerful enough to show the rings in their true guise, and subsequently he was puzzled to find that the lateral bodies had disappeared; in fact, the rings were edgewise-on in December 1612. Later, Galileo again saw the lateral bodies, but was unable to interpret them.

A drawing made by the French astronomer Pierre Gassendi on 19 June 1633 shows what seems to be a ring, but Gassendi also failed to interpret it. The correct explanation was given by Christiaan Huygens in 1659, in his book *Systema Saturnium*. He had started his telescopic observations in 1655, using a magnification of ×50; in his book he gave the answer in an anagram which he had published to ensure priority. In translation, the anagram reads: 'The planet is surrounded by a thin flat ring, nowhere touching the body of the planet, and inclined to the ecliptic.'

OBSERVATIONS OF THE GLOBE

Saturn's globe shows belts, not unlike those of Jupiter, but much less prominent, and sensibly curved. Predictably, the globe is flattened; this was first noted in 1789 by William Herschel. He gave the ratio of the equatorial to the polar diameter as 11:10, which is approximately correct. Apart from Jupiter, Saturn has the shortest rotation period of any planet in the Solar System.

Well-defined spots were seen in 1796 by J. H. Schröter and his assistant, K. Harding, from Schröter's observatory at Lilienthal, near Bremen. Prominent white spots are found occasionally. One was seen on 8 December 1876 by A. Hall; there was another in 1903, observed by A. S. Williams: and then, in 1933, a really spectacular outbreak, discovered on 3 August by W. T. Hay, using a 6-in (15-cm) refractor. (Hay may be better remembered today as

Table 10.1 *Saturn: data*

Distance from the Sun:
   max. 1 513 330 000 km (10.116 a.u)
   mean 1 433 450 000 km (9.682 a.u.)
   min. 1 353 570 000 km (9.048 a.u.)
Distance from Earth: max. 1 658 500 000km, min. 1 195 500 000 km
Orbital period: 10 832.33 days (29.657 years)
Mean synodic period: 378.09 days
Rotation period: equatorial 10h 13m 59s, internal 10h 32m
Mean orbital velocity 9.69 km s^{-1}
Axial inclination 26° 44′ (26.73°)
Orbital eccentricity: 0.055 72
Orbital inclination 2.485°
Diameter: equatorial 120 536 km, polar 108 728 km
Apparent diameter seen from Earth: max. 20′ .1, min. 14′ .5
Oblateness: 0.098
Mass, Earth = 1: 95.15
Reciprocal mass, Sun = 1: 3498.5
Volume, Earth = 1 763.6
Density, water = 1: 0.687
Escape velocity: 35.5 km s^{-1}
Surface gravity, Earth = 1: 1.19
Albedo: 0.47
Mean surface temperature: −180 °C
Opposition magnitude: max. −0.3 (rings open), min. +0.8
   (rings edge-on)
Mean diameter of the Sun, as seen from Saturn: 3′ 22′

Table 10.2 *Oppositions of Saturn, 2010–2020*

Date	Magnitude at opposition	Diameter (″)	Dec. (°)	Constellation
2010 Mar 22	+0.7	19.6	+2	Pisces
2011 Apr 3	+0.6	19.4	−3	Pisces
2012 Apr 15	+0.4	19.1	−7	Aquarius
2013 Apr 28	+0.3	18.9	−12	Aquarius
2014 May 10	+0.3	18.7	−15	Capricornus
2015 May 23	+0.2	18.6	−18	Capricornus
2016 June 3	+0.2	18.5	−21	Scorpius
2017 June 15	+0.2	18.4	−22	Scorpius
2018 June 27	+0.2	18.4	−23	Scorpius
2019 July 9	+0.3	18.4	−22	Scorpius
2020 July 20	+0.3	18.5	−21	Scorpius

Will Hay, the stage and screen comedian.) The spot remained identifiable until 13 September 1933. W. H. Wright considered it to have been of 'an eruptive nature'. Another spot was seen in 1960 and a major outbreak in 1990; on 25 September an American amateur, S. Wilber, detected a brilliant white spot in much the same latitude as the earlier ones. Within a few days the spot had been spread out by Saturn's strong equatorial winds, and by

Table 10.3 *Edgewise presentations of Saturn's rings (Sun crossing)*

1936	Dec	28
1950	Sept	21
1956	June	15
1980	Mar	3
1995	Nov	19
2009	Aug	10
2025	May	5
2039	Jan	22

10 October had been transformed into a bright zone all round the equator. Extra outbreaks were seen in it, clearly indicating an uprush of material from below.

There is an interesting periodicity here. The white spots were seen in 1876, 1903, 1933, 1960 and 1990: the intervals between outbreaks have been 27, 30, 27 and 30 years. This is very close to Saturn's orbital period of 29.6. It could be coincidental, but observers will be on watch for a new white spot around 2020.

COMPOSITION OF SATURN

Initially it was assumed that Saturn must be a sort of miniature Sun. As recently as 1882, R. A. Proctor, in his book *Saturn and its System*, wrote:

> Over a region hundreds of thousands of square miles in extent, the glowing surface of the planet must be torn by sub planetary forces. Vast masses of intensely hot vapour must be poured forth from beneath, and rising to enormous heights, must either sweep away the enwrapping mantle of cloud which had concealed the disturbed surface, or must itself form into a mass of cloud, recognisable because of its enormous extent...

The first modern-type model of Saturn was proposed in 1938 by R. Wildt. Wildt believed that there was a rocky core, overlaid by a thick layer of ice which was itself overlaid by the gaseous atmosphere.

Today it is clear that in many respects Saturn and Jupiter are similar, though Saturn's smaller size and mass means that its internal structure is different. Both are gas–giants, whereas Uranus and Neptune are better regarded as ice–giants.

ATMOSPHERE AND CLOUDS

Saturn's atmosphere contains relatively more hydrogen and less helium than is the case for Jupiter. Details are given in Table 10.4. There are three main 'cloud decks' in what is termed the troposphere. The visible cloud deck, made up of ammonia droplets, lies about 100 km below the tropopause (i.e. the top of the troposphere; the temperature here is about −250 °C.) The second deck, made up of ammonium hydrosulphide clouds, lies 170 km below the tropopause; temperature −70 °C, and the lowest deck, where the temperature has risen to 0 °C, is composed of H_2O. Wind speeds are very high, up to 1800 km h^{-1} at the equator. There is an equatorial jet

Table 10.4 *Atmospheric composition*

	%	Parts per million
Minor		
Methane, CH_4	0.4	4500
Ammonia, NH_3	0.01	125
Hydrogen deuteride, HD	0.01	110
Ethane, C_2H_6	0.00077	7
Major		
molecular hydrogen (H)	~98%	
helium (He)	~3%	

Ices: ammonia, water, ammonium hydrosulphide (NH_4 SH).

stream; the zonal winds, along the belts, are prograde (i.e. blowing east), usually at around 600 km h^{-1}.

There is a well-defined north polar vortex, and apparently associated with this is a strange hexagonal pattern at around latitude 78° N. The straight sides of the hexagon are large, with lengths of around 13 800 km.

This mysterious hexagon was first seen by Voyager space-craft, the last time spring began on Saturn. The north pole had then been in darkness for 15 years. When it emerged into the sunshine once more in 2009, it was imaged by the Cassini space-craft. The stream is believed to whip along the hexagon at around 100 m s^{-1}. The origin of this curious feature is not known. The south polar region is different, showing a jet stream but no vortex or hexagon.

Saturn is colder than Jupiter, so that ammonia crystals form at higher levels, covering the planet with 'haze' and giving it a somewhat bland appearance. Features over 1000 km in diameter are uncommon, and even the largest ovals are no more than half the size of Jupiter's Great Red Spot. Seasonal effects occur: for example, the Sun crossed into Saturn's northern hemisphere in 1980 and during the Voyager missions the northern hemisphere was still the colder of the two. The south pole was appreciably warmer than the north pole.

MAGNETIC FIELD AND AURORÆ

Saturn has a strong magnetic field, first detected in 1979 from Pioneer 11. The strength of the source of its field – that is to say, the dipole moment – is around 600 times stronger than that of the Earth, though 30 times weaker than Jupiter's. Saturn is unique inasmuch as the magnetic axis and the rotational axis are almost coincident, so that the magnetic field is reasonably straightforward and asymmetric. The polarity is opposite to that of the Earth, i.e. the same as that of Jupiter. The field seems to be slightly stronger in the north than in the south, and the centre of the field is displaced by about 2400 km northward along the planet's axis. There is no doubt that magnetic phenomena are associated with the 'spokes' seen in Ring B.

The magnetosphere is about one-fifth the size of that of Jupiter; the bow shock lies at a range of about 20 to 50 times the radius of the planet (i.e. 100 000 000 to 300 000 000 km from Saturn), and is about 2000 m thick. Depending on variations in the solar wind, the bow shock moves in and out to some extent; when the Cassini space-craft was approaching Saturn, it crossed the bow shock no less than seven times. Radiation zones exist; the magnetosphere traps radiation belt particles, extending as far as the outer edge of the Main Ring system. The numbers of electrons fall off quickly at the outer edge of Ring A; there is a well-defined magnetotail on the planet's night side.

All the inner satellites are embedded in the magnetosphere, and there are significant effects due to emissions from the active satellite Enceladus. The outer boundary is somewhat variable, and is close to the orbit of Titan, so that Titan is sometimes just inside the magnetosphere and sometimes just outside it.

Auroræ occur on Saturn. Because the magnetic axis is virtually coincident with the rotational axis, the auroræ are truly polar. They have been recorded by all fly-by missions, beginning with Pioneer 11, and from satellites such the IUE (International Ultra-violet Explorer) and the Hubble Space Telescope. They are of course due to charged solar particles that are trapped in Saturn's magnetic field and collide with atmospheric gases, mainly molecular and atomic hydrogen. Saturnian auroræ can be very lofty – for example, a huge auroral curtain was found to rise 2000 km above the cloud-tops in October 1999, and the Hubble Space Telescope recorded an equally high aurora on 17 October 2000. In November 2008, infrared detectors on the Cassini probe, orbiting Saturn, recorded a unique type of aurora, lighting up the entire polar region and covering a vast area. It was quite unlike auroral rings such as those of Earth (and Jupiter); its origin is at present unclear.

Predictably, Saturn is a radio source, though the emissions are much weaker than those of Jupiter. Saturn's kilometric radiation (SKR) has been recorded by the fly-by probes, and by the Cassini orbiter. In 2008, Cassini also detected lightning, as well as radio-wave emission from storms on the planet. In the same year, Cassini flew into the auroral zone, and made observations of the radio aurora from 247 km above the cloud-tops. The process generating radio auroræ seems to be the same for both planets.

The first 'movie' of lightning on Saturn was obtained by the Cassini space-craft in 2009, during a storm that lasted from November to mid-December. The frames in the video were obtained over 16 minutes on 30 November; each flash lasted for less than a second. The images showed a cloud 3000 km across and regions illuminated by the flashes were 300 km in diameter. Cassini also obtained a sound-track featuring the crackle of radio waves emitted by the lightning bolts. An earlier storm had lasted from January to October – a record for longevity. In 2010, a Leicester University team, led by Jonathan Nichols, found that Saturnian auroræ – the ethereal ultraviolet glows near the poles – pulse about once per Saturnian day, like the radio emissions, so that the two phenomena are associated.

RINGS

Saturn's rings were seen by very early telescopic observers, including Galileo, but their nature was not appreciated until Huygens realised that they were truly rings. However, his explanation was not universally accepted at once, and there was considerable support for Gilles de Roberval's strange idea that the planet was surrounded by 'a torrid zone giving off vapours, transparent and in small quantity, but able to reflect sunlight off the edges if of medium density, and producing an elongated appearance when thicker'. It was not until 1665 that all astronomers agreed that the rings really existed.

Subsequent telescopic improvements meant that the rings could be examined in more detail. The division between the two main rings (A and B) was discovered by G. D. Cassini in 1675, and is named after him (claims that W. Ball had seen this division ten years earlier seem to be unfounded). A narrower feature, in Ring A, was indicated by J. F. Encke from Berlin, on 28 May 1837, and has long been referred to as the Encke Division, though now known officially as the Encke Gap. The Crêpe or Dusky Ring (Ring C) was detected by W. Bond from Harvard in November 1850, using the 15-inch (38-cm) Merz refractor there; independent confirmation came from W. Lassell and W. R. Dawes in England. The transparency of Ring C was discovered in 1852, independently by Lassell and by C. Jacob from Madras. In fact the Crêpe Ring is not a difficult object when Saturn is suitably placed; indications of it may have been shown by G. Campani as long ago as 1664.

Saturn's rings can be seen with a small telescope when suitably placed, and are unrivalled, though most of our detailed knowledge of them has been derived from space-craft, particularly Cassini. Essentially, they are made up of innumerable particles of loosely packed water ice, ranging in size from tiny fragments to blocks several metres across. Outwardly they look solid, and it is not surprising that they were at first believed to be solid sheets. The first to realise that this cannot be so was J. Cassini, in 1705; in 1875, James Clerk Maxwell showed that any solid or liquid ring would quickly be disrupted by Saturn's powerful gravitational pull – even if it could have been formed at all. Spectroscopic proof was given in 1895 by J. E. Keeler, who found that the rings do not rotate in the way that a solid mass would do; the inner sections have the fastest rotation. Using the 13-inch (33-cm) refractor at the Allegheny Observatory in Pittsburgh, Keeler measured the Doppler shifts of the spectral lines at the opposite ends of the rings (ansæ) and was able to measure the rotational speeds.

The fact that the rings are made up of swarms of small particles was established well before the end of the nineteenth century, and it was realised that despite their great extent, 270 000 km from one side to the other, they are extremely thin. If a model of the system were made from material the thickness of a ten-pence coin, the overall diameter would be over 15 km. The origin of the ring system is unclear. It may have been produced from material 'left over', so to speak, when Saturn was formed, or it may be the débris of a former icy satellite which wandered too close to the planet and was literally torn apart. Certainly the rings lie wholly within Saturn's Roche limit.

It is fair to say that nobody was prepared for the revelations first from Voyager 2 and then from Cassini. Instead of being more or less homogeneous, the rings were found to be made up of thousands of small ringlets and narrow gaps. One ingenious experiment was carried out from Voyager 2, in 1981. As seen from the space-craft, the bright star δ Scorpii was occulted by the rings. It had been expected that there would be comparatively few 'winks', as the starlight would shine through the gaps but would

be blocked by ring material, but in fact thousands of 'blips' were recorded, showing that there is very little empty space anywhere in the ring system. The total mass of the ring material is very slight.

DETAILS OF THE RINGS

Table 10.5 lists the ring data. The distances are reckoned from Saturn's centre. The two main rings, A and B, are separated by the Cassini Division. These, and Ring C (the Crêpe Ring), are easy objects. The inner 'ring', D, and the outer rings F, G and E, are beyond the range of ordinary Earth-based telescopes.

Ring D

Ring D is not a true ring, and has no sharp inner edge; particles may extend downward almost to the cloud-tops. It is very faint, and was first unmistakably recorded by Voyager 1 in 1980.

Ring C

Ring C was discovered in 1850 by W. and G. Bond from Harvard, though indications of it can be made out on earlier drawings, notably by J. Galle, in 1838. It is semi-transparent, and only a few metres thick. Within, are two gaps, the Colombo Gap (width 150 km) and the Maxwell Gap (270 km), each of which contains a bright ringlet. Ring C is very easy to see with a telescope of moderate aperture (say 8 cm) when the system is suitably angled, but is out of view when the rings are almost edgewise-on to the Earth. On average the C-ring particles are probably about 2 m in diameter.

Ring B

Ring B is the brightest and most massive of the rings; its particles range in size from 10 cm to 1 m. The particles are redder than those in the C and D regions; temperatures range from –180 °C, in sunlight, down to –200 °C, in shadow. There is a cloud of neutral hydrogen extending to 60 000 km above and below the ring plane.

Curious 'spokes' are found in the ring. They were first reported in the Voyager images; they look dark in back-scattered light bright in forward-scattered light, and are puzzling because logically no such features should form; the difference in orbital period between the inner and outer edges of Ring B is over 3 hours – and yet the spokes persisted for hours after emerging from the shadow of the globe. When they were distorted and broken up, new spokes emerged from the shadow to replace them. They are not always on view, and it has been suggested that they may be seasonal phenomena, but they were shown by Cassini mission images. Presumably, they are due to particles of a certain definite size elevated away from the ring plane by magnetic or electrostatic forces, and are unique to Ring B.

It is worth noting that they were shown on earlier drawings by E. E. Barnard and E. M. Antoniadi, and I have seen them clearly with the Lowell refractor at Flagstaff and the Meudon refractor at Paris.

The Cassini Division

The Cassini Division separates Rings B and A; it is 4700 km wide and was discovered in 1675 by Cassini. Telescopically, it looks black, but in fact it is not empty, and contains ring particles which

Table 10.5 *Ring data*

Feature	Distance from centre of Saturn (km)	Period (h)	Radial width (km)
(Cloud-tops)	60 330	10.66	–
Inner edge of Ring D	66 900	4.91 ⎱	7150
Outer edge of Ring D	73 150	⎰	
Inner edge of Ring C	74 510	5.61	
Maxwell Gap	87 500	6	270
Outer edge of Ring C	92 000	7.9	17 500
Inner edge of Ring B	92 000	7.93 ⎱	25 500
Outer edge of Ring B	117 500	11.4 ⎰	
Inner edge of Cassini Division	117 500	11.4 ⎱	4700
Centre of Cassini Division	119 000	⎰	
Outer edge of Cassini Division	122 200		
Huygens Gap	117 680	11.4	258–440
Inner edge of Ring A	122 200	11.4	
(Pan)	133 583	13.7	–
Centre of Encke Gap	135 700	13.82	(Width of division, 325 km)
Centre of Keeler Gap	136 530	14	(Width of gap, 35 km)
Outer edge of Ring A	136 800	14.14	(Width of ring, 14 600 km)
(Atlas)	137 640	14.4	–
(Prometheus)	139 350	14.7	–
Centre of Ring F	140 210	14.94	30–500
(Pandora)	141 700	15.07	–
(Epimetheus)	151 422	16.65	–
(Janus)	151 472	16.7	–
Inner edge of Ring G	164 000	⎱	
Centre of Ring G	168 000	19.90 ⎰	8000
Outer edge of Ring G	172 000		
Inner edge of Ring E	180 000	21	
(Mimas)	185 520	22.6	–
Brightest part of Ring E	230 000	31.3	
(Enceladus)	238 020	32.9	–
(Tethys/Telesto/Calypso)	294 660	44.9	
(Dione/Helene)	377 400	65.9	–
Outer edge of Ring E	480 000	80	(Width of ring, 300 000 km)
(Rhea)	527 040	108.4	–
Phœbe Ring	12 870 000	545 days	

are less red than those of Ring B; they are more like the particles of Ring C. The 400-km Huygens Gap lies at the inner edge of the Cassini Division.

Ring A
Ring A is the outer bright ring, from 10 to 30 m thick, made up of particles of size ranging from about 10 m in diameter down to fine grains. There are two gaps, the 325-km Encke Gap (until recently called the Encke Division) and the 35-km Keeler Gap. The small satellite Pan orbits within the Encke Gap, and keeps it 'swept clear'; an even smaller satellite, Daphnis, performs the same function for the Keeler Gap.

The sharp outer edge of Ring A is close to the orbit of the satellite Atlas, and rings A and F are separated by the 2600-km Roche Division. (Note that this is a division, not a gap, because it separates two definite rings – the Encke Gap does not.)

Roche Division
Like the Cassini Division, the Roche Division – named only recently – is not empty, but contains very tenuous ring material. It may well contain one or more tiny satellites but on this scale it is difficult to decide at what size a clump of ring material qualifies to be classed as a satellite!

Ring F
Ring F was discovered in 1979 from Pioneer 11 images. It is thin and irregular, no more than a few hundred kilometres wide. The Cassini space-craft has shown that Ring F is made up of one core

Table 10.6 *Space missions to Saturn*

Name	Launch date	Encounter date	Nearest approach (km)	Remarks
Pioneer 11	Apr 5 1973	Sept 11 1979	20 880	Preliminary results.
Voyager 1	Sept 5 1977	Nov 12 1980	124 200	Success. Good images of Titan, Rhea, Dione and Mimas.
Voyager 2	Aug 20 1977	Aug 25 1981	101 300	Success. Good images of Iapetus, Hyperion, Tethys and Enceladus. Went on to Uranus and Neptune.
Cassini–Huygens	Oct 15 1997	July 1 2004		Cassini orbit; Huygens landed on Titan on 14 Jan 2005.

ring and a spiral strand around it. It is stabilised by two small satellites, Prometheus and Pandora, which move to either side of its centre. The satellite slightly closer to Saturn (Prometheus, at the present time) will be moving faster than the ring particles, and will speed up a particle which strays away from the main system; the outer satellite, moving more slowly, will drag any errant particles back and return them to the central region. For obvious reasons, Prometheus and Pandora are known as 'shepherd' satellites. The Cassini probe has also shown that Prometheus, despite its small size, can pull material away from the ring and produce distortions and irregularities; presumably this is also true of Pandora.

Janus-Epimetheus Ring

Beyond the Ring F and its shepherds we come to the co-orbital satellites, Janus and Epimetheus, together with a very faint dust ring formed from material blasted away from the surfaces of these two satellites by micrometeorite impacts.

Ring G

Ring G is a very thin, faint ring between the rings F and E, with its inner edge about 15 000 km inside the orbit of Mimas. It is made up of icy particles a few metres across.

Pallene Ring

Pallene Ring is a very faint dusty ring in the same orbit as the small satellite Pallene, due to material blasted away from the satellite by meteoroid impact.

Ring E

Ring E is faint and diffuse but extremely wide. It begins at the orbit of Mimas and extends well beyond the orbit of Dione, becoming undetectable somewhere near the orbit of Rhea. Its particles are very small – microscopic, rather than macroscopic – and are now believed to be sent out by cryovulcanism in the active satellite Enceladus.

Phœbe Ring

This totally unexpected outer ring was discovered in October 2009 by A. Verbiscer and M. Skrutskie (University of Virginia) and D. Hamilton (University of Maryland) using data obtained by the Spitzer Space Telescope. It extends between 6 million and 13 million kilometres from Saturn, and contains the satellite Phœbe; undoubtedly the ring particles have been sputtered off Phœbe by impacts. The particles are tiny (diameters about 10 microns) and the ring is very tenuous, probably with only 10 to 20 particles per cubic metre. At visible wavelengths it is to all intents and purposes undetectable, which is why its discovery was not made before the infrared results from Spitzer. The inclination to the main ring plane is 27°, the same as that of Phœbe, and the particles, again like Phoebe, move in the retrograde sense. The pressure of sunlight drives the smallest particles inward, and some of them encounter Iapetus – and since the satellite has direct motion, the collisions are 'head-on'. At least it explains why Iapetus has one bright hemisphere and one that is dark.

Why are the rings so complex?

The Cassini Division had been explained as being due to the gravitational effects of the larger satellite, Mimas; a particle moving in the Division will have an orbital period half that of Mimas, and cumulative perturbations should drive it away from the 'forbidden zone', leaving the Division swept clear. Yet the Voyagers have shown that this explanation is clearly inadequate, and it now seems more likely that we are dealing with a density-wave effect. A satellite such as Mimas can alter the orbit of a ring particle and make it elliptical. This causes 'bunching' in various areas; a spiral density wave will be created, and particles in it will collide, moving inward toward the planet and leaving a gap just outside the resonance orbit. This theory, due to Peter Goldreich and Scott Tremaine, does sound plausible, but it would be wrong to claim that we yet really understand the mechanics of the ring system.

SPACE MISSIONS TO SATURN

Four space-craft have so far been launched to Saturn. Data are given in Table 10.6.

The first probe was Pioneer 11. Its prime target had been Jupiter, but the success of its predecessor, Pioneer 10, and the fact that there was adequate power reserve meant that it could be swung back across the Solar System to a rendezvous with Saturn in 1979. The results were preliminary only, but involved several important discoveries. Perhaps the most significant fact was that Pioneer was not destroyed by a ring particle collision: at that time nobody knew what conditions near Saturn would be like and estimates of the probe's chances of survival ranged from 1% to 99%.

Voyager 1, launched in 1977, surveyed Jupiter in 1979 and then went on to Saturn. It was a complete success, and obtained new data about the rings and satellites as well as the planet itself. One of its prime targets was Titan, already known to have an atmosphere; it was a surprise to find that the main constituent of this atmosphere was nitrogen. Voyager 1 then began a never-ending journey out of the Solar System: as I write these words (2010) it is still in touch.

Voyager 2 came next. This time Titan was not surveyed in detail: this would have meant that the probe would have been unable to go on to Uranus and Neptune. Again the mission was a complete success, and both the outer planets were also encountered. Like its predecessor, Voyager 2 is now leaving the Solar System permanently. Its success was all the more striking because one of the main instruments failed early in the mission, and Voyager 2 operated throughout on its back-up system.

The Cassini–Huygens mission was launched on 15 October 1997 from Cape Canaveral, using a Titan- 4B/Centaur rocket. The gravity-assist technique had to be used, involving swing-bys of Venus (21 April 1998), Venus again (20 June 1999), Earth (16 August 1999) and Jupiter (20 December 2000). Carrying Huygens, the Titan lander, it was put into Saturn orbit on 1 July 2004, only 20 000 km above the cloud-tops. On 6 November 2004 it was manœuvred into an impact trajectory with Titan; on 25 December 2004 Huygens separated from Cassini, and Cassini was deflected away from a collision course, leaving Huygens to make a controlled landing on Titan (14 January 2005). Cassini acted as a relay for Huygens until it passed out of range, and then commenced a long-term survey of the planet and its satellites. It was still functioning excellently during 2010.

Cassini–Huygens was over 6.8 m high and over 4 m wide, with a mass of 2150 kg. It carried instruments of all kinds, including spectrometers, imaging systems and magnetometers. At Saturn's distance from the Sun solar arrays were not feasible as power sources, and the Cassini orbiter carries three radioisotope thermo-electric generators (RTGs), using heat from the natural decay of plutonium to generate direct current electricity.

THE SATELLITES OF SATURN

Saturn has a wealth of satellites. Over 60 are now known, and new discoveries are being made regularly. However, all these new discoveries are of what are termed 'irregular' satellites, only a few kilometres in diameter and often with retrograde motion. They are probably captured asteroids, though Phœbe, the only reasonably large irregular satellite, may well be an escapee from the Kuiper Belt.

In April 1861, H. Goldschmidt announced that he had found a satellite moving between the orbits of Hyperion and Iapetus, and named it Chiron. In 1904, W. H. Pickering claimed that he had found a satellite moving between the orbits of Titan and Hyperion, and named it Themis. However, these satellites were never confirmed, and probably the observers were misled by faint stars.

Of course Saturn's system is dominated by Titan, which is larger than Mercury and has a dense atmosphere. Before 1900, nine satellites were known; in order of distance: Mimas, Enceladus, Tethys, Dione, Rhea, Titan, Hyperion, Iapetus and Phœbe, all of which can be seen with telescopes of moderate size, though Hyperion and Phœbe are decidedly elusive. All the rest are very difficult objects.

There are three main classes: (1) the small inner satellites, involved with or close to the ring system; (2) the main satellites (Mimas out to Iapetus) with a few small prograde bodies and also the Trojans of Tethys and Dione; and (3) the outer irregular satellites, all tiny apart from the exceptional Phœbe. All the names are mythological; some of the most recently discovered satellites have not yet (2010) been given names.

It is clear that Saturn's system is quite unlike that of Jupiter, or for that matter any other satellite family in the Solar System. New small satellites are being discovered so frequently that tables soon need revision. The tables here are complete to 2009. The satellite groups are:

Inner small satellites, including ring shepherds, from Pan out to the coorbitals Janus and Epimetheus.

Main satellites, Mimas to Iapetus, all discovered before 1900. Tethys has two Trojans, Telesto and Calypso, and Dione also has two, Helene and Polydeuces. In addition the three small Alkyonides (Methone, Pallene and Anthe) move between the orbits of Mimas and Enceladus.

Outer satellites. All small, and almost all discovered on spacecraft; only Phœbe was discovered before 1900, by W. H. Pickering, in 1898. (Two satellites reported and named, Chiron (1861) and Themis (1905) do not exist.) The outer groups are:

INUIT, five prograde satellites; Ijiraq, Kiviuq, Siarnaq and Tarqeq.

NORSE, numerous retrograde satellites, including Phœbe.

GALLIC, four prograde satellites; Albiorix, Bebhionn, Erriapus and Tarvos.

The small outer satellites, often well away from Saturn and some with highly eccentric or retrograde-motion orbits, are referred to as 'irregular' satellites.

THE SMALL INNER SATELLITES

Pan

This tiny satellite was identified in 1990 by M. Showalter from Voyager 2 images, and named after the half-man, half-goat son of Hermes and Dyope. It moves inside the Encke Gap in Ring A, and keeps the Gap open; there is a ringlet coincident with its orbit. It has been described as walnut-shaped because of its equatorial ridge, shown on the best images, due to material swept up from the Gap.

Atlas

Atlas is the outer Ring A shepherd, named after the giant Atlas who carried the sky on his shoulders. R. Terrile identified it in 1980 on Voyager images. Cassini images show it to be saucer-shaped, with a large, smooth equatorial ridge.

Table 10.7 *Satellites of Saturn (Dist. = mean distance from Saturn; P = orbital period; i =.orbital inclination; e =.orbital eccentricity; m = magnitude)*

(a) INNER SMALL SATELLITES

Name	Discovered by, date	Dist. (km)	P (days)	i (°)	e
XVIII Pan	M. Showalter, 1990	133 584	0.575	0.001	~0
XXXV Daphnis		136 505	0.595	~0	~0
XV Atlas	R. Terrile, 1980	137 670	0.608	0.003	~0
XVI Prometheus	S. Collins *et al.*, 1980	139 380	0.613	0.008	0.003
XVII Pandora	S. Collins *et al.*, 1980	141 720	0.629	0.050	0.008
XI Epimetheus	R. Walker *et al.*, 1980	151 422	0.694	0.335	0.004
X Janus	A. Dollfus, 1966	151 472	0.694	0.165	0.009

	Dimensions (km)	Density (water = 1)	Escape velocity (km s^{-1})	m	Albedo	
Pan	$35 \times 35 \times 23$	?	Low	19	0.5	In Encke Gap
Daphnis	7	?	Low	20	?	In Keeler Gap
Atlas	$46 \times 38 \times 19$	?	Low	18.0	0.9	Outer Ring A shepherd
Prometheus	$119 \times 87 \times 61$	0.27	0.022	15.8	0.6	Inner Ring F shep.
Pandora	$103 \times 80 \times 64$	0.7	0.237	16.3	0.9	Outer Ring F shep.
Epimetheus	$135 \times 108 \times 105$	0.63	0.032	15.7	0.8	Janus co-orbital
Janus	$193 \times 173 \times 137$	0.67	0.052	14.5	0.8	Epimetheus co-orbital

(b) MAJOR SATELLITES (Ang. = angular distance from Saturn, at opposition)

Name	Discoverer, date	Dist. (km)	Period	Synodic period	Ang.
I Mimas.	W. Herschel, 1789	185 404	0.942 0d.	22h 37m 12s	0′.32′
XXXII Methone	Cassini team, 2004				
XLIX Anthe	Cassinii team, 2007				
XXXIII Pallene	Cassini team, 2004				
II Enceladus	W. Herschel, 1789	237 950	1.370 1d	8h 53m 22s	0′38″ .4
III Tethys	G. Cassini, 1684	264 619	1.388 1d	21h 18m 55s	0′ 47‴ .6
XIII Telesto	B. Smith *et al.*, 1980				
XIV Calypso	D. Pasco *et al.*, 1980				
IV Dione	G. Cassini, 1684	377 396	2.737 2d	17h 42m	1′ 01″.1
XII Helene	P. Laquez and P. Lacacheus, 1980				
XXXIV Polydeuces	Cassini team, 2004				
V Rhea	G. Cassini, 1672	527 108	4.518 4d	118m	1′ 28″ .1
VI Titan	C. Huygens, 1655	1 221 930	15.945 15d	23h 1m	1′17″ .3
VII Hyperion	W. Bond, 1848	1 481 010	21.277 21d	7h 39m	3′59″ .4
VIII Iapetus	G. Cassini, 1671	3 560 820	79.322 79d	22h 05m	9′ 35″

	i (°)	e	Dimensions (km)	Mass (g)	Reciprocal mass
Mimas	1.57	0.021	$415 \times 394 \times 381$ (mean 397)	3.7×10^{22}	15 000 000
Methone	0.00	0.000	3		
Anthe	0.10	0.001	2		
Pallene	0.00	0.000	4		
Enceladus	0.01	0.005	$513 \times 503 \times 497$ (mean 504)	6.5×10^{22}	7 000 000
Tethys	0.17	0.000	$1081 \times 1062 \times 1055$ (mean 1066)	6.1×10^{23}	910 000
Telesto			$29 \times 22 \times 20$ (mean 24)		
Calypso			$30 \times 23 \times 14$ (mean 21)		
Dione	0.002	0.002	$1128 \times 1122 \times 1121$ (mean 1123)	1.1×10^{23}	490 000
Helene			$36 \times 32 \times 30$ (mean 33)		
Polydeuces			3.5		
Rhea	0.327	0.001	$1535 \times 1525 \times 1526$ (mean 1529)	2.3×10^{24}	250 000
Titan	1.634	0.029	5151		4150

	i (°)	e	Dimensions (km)	Mass (g)	Reciprocal mass
Hyperion	0.568	0.018	$360 \times 280 \times 225$ (mean 292)	0.6×10^{19}	~5 000 000
Iapetus	7.570	0.028	$747 \times 749 \times 713$ (mean 1472)	1.6×10^{24}	300 000

	Density (water = 1)	Escape Velocity	m	Albedo	Apparent diameter from Saturn
Mimas	1.17	0161	14.5	0.5	10′ 54″
Methone			25		
Anthe			26		
Pallene			25		
Enceladus	1.17	0.212	11.7	0.99	10′ 36″
Tethys	1.24	0.436	10.2	0.9	17′ .36″
Telesto			18		
Calypso			18.5		
Dione	1.49	0.500	10.3	0.7	12′ 24″
Helene	0.9	0.011	18		
Polydeuces			25		
Rhea	1.33	0.659	9.7	0.7	10′ 42″
Titan	1.88	2.65	8.3	0.21	17′ 10″
Hyperion	1.4	0.11	14.2	0.3	0′ 43″
Iapetus	1.21	0.59	10.2–11.9	0.2 (mean)	1′48″

(c) OUTER SATELLITES (* = retrograde)

		Dist. (× 1000 km)	P (days)	i (°)	e	Diameter (km)	m	Group
XXIV	Kiviuq	11 295	448.1	45.7	0.334	16	22.0	Inuit
XXII	Ijiraq	11 355	451.8	46.4	0.316	12	22.6	Inuit
IX	Phœbe	12 870	*545.1	174.8	0.164	220	16.5	Norse
XX	Paaliaq	15 103	693.0	45.1	0.364	22	21.3	Inuit
XXVII	Skathi	15 672	*732.5	152.6	0.270	8	23.6	Norse
XXVI	Albiorix	15 266	774.6	34.0	0.478	32	20.5	Gallic
XXXVII	Bebhionn	17 153	838.8	35.0	0.469	6	24.1	Gallic
XXVIII	Erriapus	17 236	844.9	34.6	0.295	10	23.0	Gallic
XLVII	Skoll	17 473	*862.4	161.2	0.464	6	24.5	Norse
XXIX	Siarnaq	17 777	884.9	45.6	0.295	40	20.1	Inuit
LII	Tarqeq	17 910	894.9	44.1	0.160	7	23.9	Inuit
LI	Greip	18 065	*906.6	179.8	0.326	6	24.4	Norse
XLIV	Hyrrokkin	18 168	*914.3	151.4	0.333	8	23.5	Norse
L	Jarnsaxa	18 557	*943.8	163.3	0.216	6	24.7	Norse
XXI	Tarvos	18 563	944.2	33.8	0.531	15	22.1	Gallic
XXV	Mundilfari	18 726	*956.7	167.3	0.210	7	23.8	Norse
XXXVII	Bergelmir	19 104	*985.8	158.5	0.142	6	24.2	Norse
XXXI	Narvi	19 395	*1008.5	145.8	0.431	7	23.8	Norse
XXIII	Suttungr	19 579	*1022.8	175.8	0.114	7	23.9	Norse
XLIII	Hati	19 709	*1033.1	165.8	0.372	6	24.4	Norse
XL	Farbauti	19 984	*1054.8	156.4	0.206	5	24.7	Norse
XXX	Thrymr	20 278	*1078.1	176.0	0.470	7	23.9	Norse
XXXVI	Aegir	20 482	*1094.5	166.7	0.252	6	24.4	Norse
XXXIX	Bestia	20 570	*1101.4	145.2	0.521	7	23.8	Norse
XLI	Fenrir	21 930	*1212.5	164.9	0.136	4	25.0	Norse
XLVIII	Surtur	22 289	*1242.4	177.5	0.451	6	24.8	Norse
XLV	Kari	22 321	*1245.1	156.3	0.478	7	23.9	Norse
XIX	Ymir	22 429	*1254.2	173.1	0.335	18	21.7	Norse
XLVI	Loge	22 984	*1300.9	167.9	0.187	6	24.6	Norse
XLII	Fornjot	24 504	*1432.2	170.4	0.206	6	24.6	Norse

Table 10.8 *Features on Mimas*

	Lat. (°)	Long. (° W)	Diameter (km)
Chasmata			
Avalon	35.0 N	147.0	120.0
Camelo	43.0 S	23.4	150.0
Oeta	19.0 N	122.7	110.0
Ossa	23.6 S	303.7	95.0
Pangea	28.1 S	340.4	150.0
Pelion	25.3 S	250.1	100.0
Tintagil	51.8 S	213.3	55.0
Craters			
Accolon	70.1 S	184.4	48
Arthur	35.4 S	196.0	64
Balin	14.7 N	82.5	35
Ban	43.9 N	160.7	37
Bors	41.8 N	172.3	34
Dynas	2.3 N	80.7	35
Galahad	45.3 S	145.3	34
Gwynevere	17.6 S	323.7	42
Herschel	1.4 S	111.8	139
Igraine	42.0 S	231.2	38
Launcelot	9.5 S	328.5	30
Lucas	40.7 N	220.3	40
Marhaus	9.0 S	0.1	34
Melyodas	74.9 S	77.2	40
Merlin	38.4 S	219.0	37
Morgan	24.2 N	245.0	43
Pellinore	29.8 N	135.5	36
Uther	35.2 N	250.2	34

Prometheus and Pandora

These are the Ring F shepherds identified in 1980 by S. Collins and his colleagues from Voyager images. In mythology, Prometheus, brother of Atlas, gave humanity the gift of fire; Pandora, made of clay by Hephæstus (at the request of Zeus), opened the box which set free all the ills which plague mankind today.

Prometheus is very elongated ($119 \times 87 \times 61$ km); its surface has ridges and valleys, but is less cratered than that of Pandora. It draws material from Ring F – clearly shown on Cassini images. Its high albedo and low density indicate that it may be icy and porous. This also applies to Pandora, which is elongated to the same extent as Prometheus ($103 \times 80 \times 64$ km) but is more heavily cratered. At times Pandora and Prometheus may be no more than 1400 km apart.

Janus and Epimetheus

These are the co-orbital satellites. Janus was discovered visually by A. Dollfus in December 1966, Epimetheus by R. Walker at the same time – at first it was thought that the two observations referred to the same object. In 1966, the rings were edgewise-on, so that conditions were favourable for observing the inner satellites. (Using the 10-inch refractor at Armagh in November 1966, I actually made

two pre-discovery observations of Janus, but, as I did not realise that it was new, can claim absolutely no credit!) Janus was named after the Roman two-faced god, Epimetheus, a minor deity.

The two moons periodically approach each other at a range of about 10 000 km, and interactions mean that they exchange orbits; this happens every four years. Both satellites are of low density, and probably icy and porous, but reasonably regular in shape, rather than very elongated. Janus is extensively cratered but lacks prominent ridges; Epimetheus is also cratered, with what may be the remains of a large impact structure near its south pole. Part of the surface is bright and fractured; another region is darker and smoother. It is possible, though unproved, that Janus and Epimetheus may be the result of the break-up of a larger body. There is a faint dust ring around the region to either side of their orbits.

Craters on these two satellites have been named after characters in the legend of the Heavenly Twins. There are four named craters on Janus (Castor, Idas, Lynceus and Phœbe), and two on Epimetheus (Hilairea and Pollux).

MAIN REGION OF SATELLITES: MIMAS TO IAPETUS

This contains all the major satellites, plus the 'Alkyonides' (Methone, Anthe and Pallene) between the orbits of Mimas and Enceladus, and the Trojans of Tethys (Telesto and Calypso) and Dione (Helene and Polydeuces).

Mimas

This was discovered by William Herschel, whose son, John, named it after one of the Titans who was killed by Ares (Mars). There are craters, which have been given names from Arthurian legend, apart from the largest, named after William Herschel himself. There are also chasams (chasmata).

Mimas is the smallest body known to be fairly spherical, but its low density indicates that it is made up chiefly of water ice together with some rock. Crater Herschel is 139 km across, one-third the diameter of Mimas itself, and images of it have been likened to the Death Star in the famous film *Star Wars*! The impact that produced it must have come near to disrupting the whole satellite. Herschel is 10 km deep, with a 6-km central peak whose base measured $30 \text{ km} \times 20 \text{ km}$. Cassini probe images show that the Herschel landscape differs slightly in colour compared with the adjacent heavily cratered terrain. An impact greater than that of Herschel would have broken Mimas up. There are many other craters, such as Modred and Bors, but few are as much as 50 km across, and larger craters are lacking in the south polar region. Grooves (chasmata) are much in evidence, and some are more or less parallel, indicating that the icy crust has been subjected to considerable strain. Like all the icy satellites, Mimas is very cold; the surface temperature is given as about −200 °C. A selected list of surface features is given in Table 10.9. Figure 10.1 is the map.

The Alcyondes

These tiny bodies were discovered by the Cassini imaging team, Methone and Pallene in 2004, and Ancic in 2007. They have been

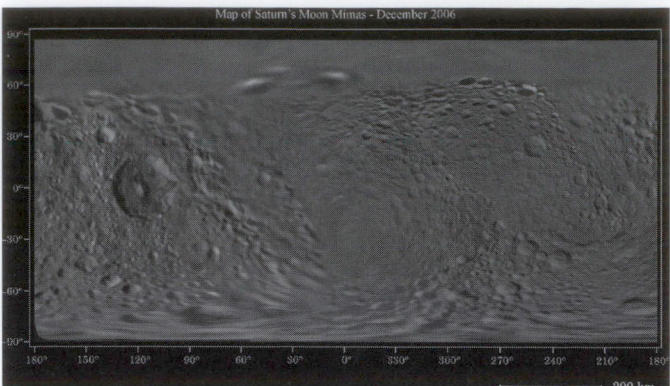

Figure 10.1 Mimas

named after three of the seven beautiful daughters of the giant Alcyoneus.

Presumably they are icy, with a small amount of rock. Their orbits are strongly influenced by the much larger Mimas and Enceladus. There is a faint dust ring in Pallene's orbit, due to particles blasted off the surface by meteoroid impacts.

Enceladus

Enceladus is the second of Saturn's 'major' satellites, but in my view is the most puzzling world in the entire Solar System. It was discovered in 1781 by William Herschel (at the same time as Mimas) and was named by John Herschel after a Titan who was crushed during a battle with the Olympian gods; earth was piled on top of him, to become the island of Sicily.

Enceladus reflects more than 99% of the sunlight falling on it, and is therefore very cold; the surface is covered with a layer of ice at a temperature of about −195 °C. Voyager 2 showed that Enceladus is very different from Mimas, and there are several distinct types of terrain. Some areas are thickly cratered, though without any really large formations; others are much smoother, giving the impression that they have been re-surfaced more recently. More detailed views were obtained from the Cassini space-craft, which made its first fly-by in July 2004. Around the south pole, and extending down to latitude 60° S, Cassini imaged a region covered with tectonic fractures and ridges; near the centre of this region were four prominent ridge-bordered fractures, which were nicknamed 'tiger stripes'. The main surprise came with the discovery of water-rich 'geysers' coming from the tiger stripes, venting icy crystals up to heights of hundreds of kilometres. The temperatures were found to be as high as −133 °C, so that the sub-crustal temperatures were presumably higher still. An excessively thin atmosphere was also detected. It has been found that as Enceladus moves along, it leaves a complex pattern of ripples and bubbles in its wake. (Remember that it lies deep within Saturn's magnetosphere.)

In 2005, Cassini actually flew through the water-plume. The icy particles were about the width of a human hair, and were vented at a speed of approximately 400 m s^{-1} – 800 miles h^{-1}; venting seemed to be continuous, producing a vast halo of ice-dust around the satellite. There is no doubt that Saturn's Ring E is due to particles vented from Enceladus; in 2008, P. Abel suggested that

Table 10.9 *Features on Enceladus*

	Lat. (°)	Long. (° W)	Diameter (km)
Craters			
Al-Kuz	18.7 S	178.2	9
Aladdin	60.7 N	26.7	37
Duban	58.4 N	282.9	19
Gharib	81.1 N	241.1	26
Musa	72.4 N	17.6	25
Shahrazad	47.3 N	199.7	20
Sindbad	67.0 N	212.1	29
Zumurrud	21.9 S	181.6	21
Dorsa			
Cufa Dorsa	3.2 N	286.2	90
Ebony Dorsum	5.7 N	280.5	70
Fossæ			
Anbar Fossæ	8.8 S	323.3	165
Daryabar Fossa	9.7 N	5.4	200
Isbanir Fossa	11.3 N	358.2	170
Khorassan Fossa	19.0 S	236.9	290
Planitia			
Diyar Planitia	13.4 S	251.9	325
Sarandib Planitia	10.2 N	311.8	165
Sulci			
Alexandria Sulcus	75.8 S	137.6	111
Baghdad Sulcus	86.9 S	230.5	176
Cairo Sulcus	81.6 S	154.5	165
Cashmere Sulci	52.1 S	296.1	260
Damascus Sulcus	80.6 S	285.9	125
Hamah Sulci	7.2 N	306.0	164
Harran Sulci	26.4 N	245.9	291
Labtayt Sulci	27.7 S	286.1	162
Lahej Sulci	10.9 S	302.0	150
Mosui Sulci	58.1 S	336.7	60
Samarkand Sulci	30.0 N	327.5	360

reported changes of brightness in the ring may be due to variations in Enceladus, and this seems plausible, though it remains speculative.

Analysis of the atmosphere showed methane and carbon dioxide as well as water, plus possibly a little ammonia and nitrogen. However, it must be remembered that the density of the atmosphere corresponds to what we would conventionally call a good laboratory vacuum.

In April 2010, Cassini flew through the water-rich plume flaring out of the south polar region, with closest approach 100 km on 27 April. A steady radio link with NASA's DSN made it possible to use the radio science instrument to measure the variations in the satellite's gravitational pull, helping to decide whether an ocean, pond or lake lies under the 'tiger stripes' that are emitting water vapour and organic particles from the south polar region. It may also help to decide whether interior bubbles of warmer ice rise to the surface in the manner of an underground lava lamp!

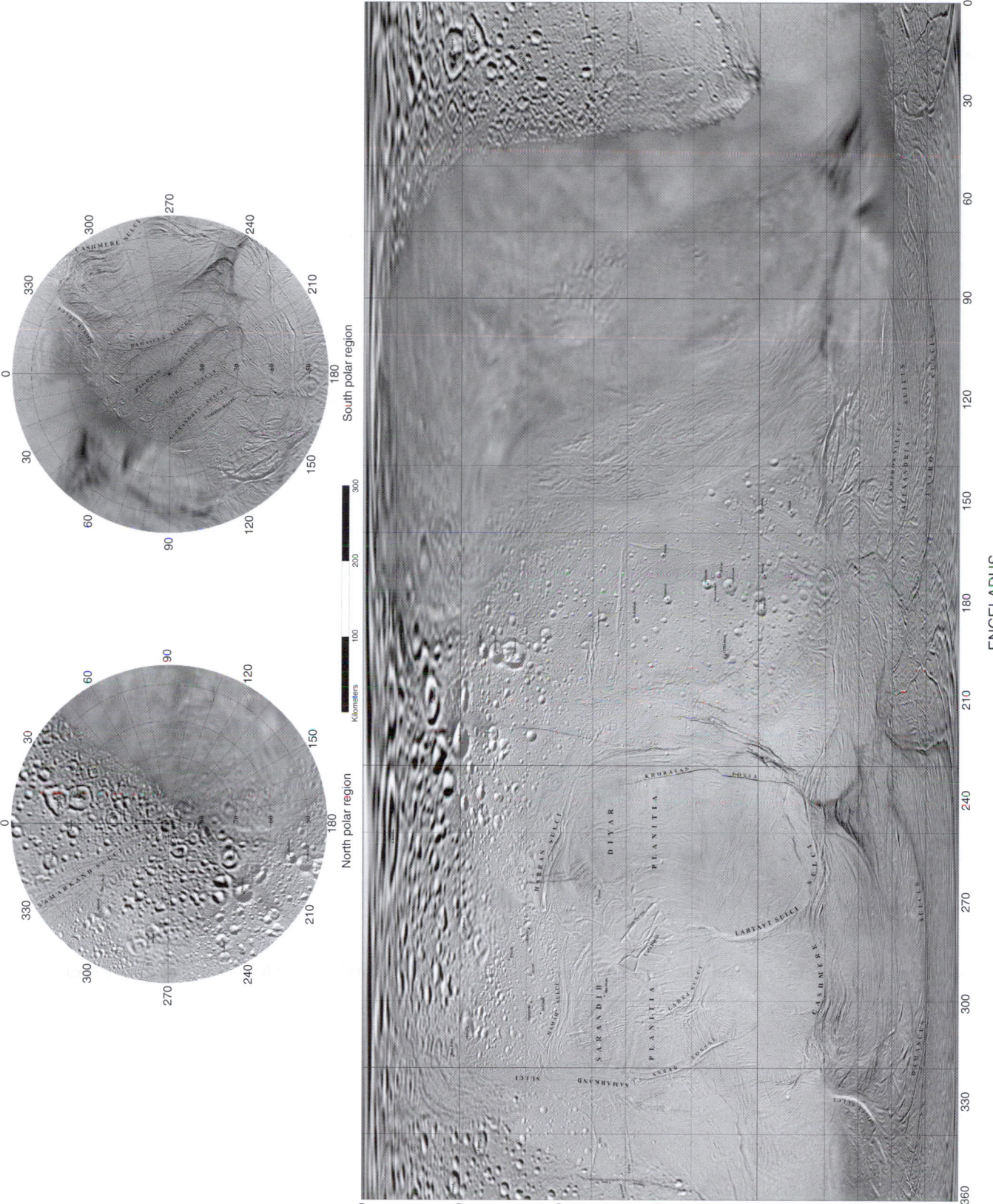

North polar region

South polar region

Kilometers

ENCELADUS

Figure 10.2 Enceladus

It seems impossible to avoid the conclusion that liquid water exists just below the surface, and cryovulcanism had certainly not been expected with a world only 500 kilometres across. The existence of an underground sea may sound surprising, but results from Cassini in 2010 seem to be best interpreted by attributing the source of the plumes to vents which channel water vapour from a warm liquid source, sending them to the surface at supersonic speeds. Apparently the water vapour forms narrow jets, and only temperatures close to the melting point of water ice could account for the high speeds of these jets. We still have much to learn about the fountains of Enceladus! A selected list of surface features is given in Table 10.9 and a map is shown in Figure 10.2

Tethys

Tethys was discovered (with Dione) by G. D. Cassini on 21 March 1684. It was named after a Titaness, wife of Oceanus and mother of the Oceanides. The names of the surface features come from Homer's 'Odyssey'. Close-range images have been obtained from two space-craft, Voyager 2 and Cassini.

Tethys seems to be made up of almost pure water ice; it is practically the same size as Dione, but considerably less massive. The surface temperature is −187 °C. Some areas are heavily cratered, but there is also a darker belt with fewer craters – presumably re-surfaced at some time in the past. The western hemisphere is dominated by a prominent crater, Odysseus, with a diameter of 400 km – larger than Mimas. It is not very deep, and the curve of its floor follows the general shape of the globe. There are other craters of considerable size, such as Penelope.

The most notable surface feature is a huge trench, Ithaca Chasma; it is 2000 km long and 3 to 5 km deep running from near the north pole across the equator and along to the south polar region. It may have been formed when the water inside Tethys froze, expanding as it did so and fracturing the crust, though there have also been suggestions that it was produced when shock-waves following the Odysseus impact travelled through the globe and fractured the surface layer on the far side.

It has also been suggested that liquid water may still exist inside Tethys, but this is decidedly speculative. No trace of atmosphere has been detected. A selected list of surface features is given in Table 10.10. Figure 10.3 is the map.

Telesto and Calypso

These are Tethys 'Trojans'. Telesto orbits 60° ahead of Tethys in the leading Lagrangian point (L1) and Calypso 60° behind (point L2). Both have been imaged by the Cassini space-craft; both have high albedoes, due to sandblasting from Saturn's Ring E. Telesto has one obvious crater, but otherwise appears to be fairly smooth; Calypso, rather larger and more elongated, has larger overlapping craters and surface material which looks loose. Like Telesto, and for the same reason, it is highly reflective.

Dione, named after the sister of Cronus and mother (by Zeus) of Aphrodite, is of special note because although little larger than Tethys it is much denser and more massive. It is indeed denser than any other Saturnian satellite apart from Titan, and there are suspicions that it may have an effect on Saturn's radio emission, since

Table 10.10 *Features on Tethys*

	Lat. (°)	Long. (° W)	Diameter/length (km)
Chasmata			
Ithaca Chasma	14.0 S	6.1	1219
Ogygia Chasma	56.0 N	95.2	120
Craters			
Achilles	0.6 N	324.4	59
Aietes	41.4 S	6.2	91
Ajax	28.4 S	282.0	88
Alcinous	30.3 N	212.6	50
Anticleia	51.3 N	32.4	101
Antinous	59.9 S	286.1	136
Circe	12.6 S	54.7	79
Demodocus	59.4 S	18.2	125
Diomedes	38.1 N	289.4	49
Dolius	30.1 S	210.3	190
Elpenor	53.4 N	263.4	60
Hermione	38.4 S	148.7	68
Icarius	5.9 S	305.6	54
Laertes	46.4 S	67.5	51
Melanthius	58.5 S	192.6	250
Mentor	0.3 N	44.2	62
Naubolos	72.2 S	305.2	54
Nausicaa	84.4 N	5.0	69
Odysseus	32.8 N	128.9	445
Penelope	10.8 S	249.2	208
Periboea	8.0 N	34.9	51
Phemius	11.3 N	286.2	76
Polyphemus	3.5 S	283.0	73
Poseidon	56.7 S	101.3	63
Salmoneus	1.8 S	335.2	93
Telemachus	54.0 N	339.4	92
Telemus	34.5 S	356.9	320
Montes			
Scheria Montes	30.0 N	131.0	185

there seems to be a radio cycle of 2.7 days – which is also the orbital period of Dione. It may also play a rôle in flexing the interior of Enceladus, whose revolution period is almost exactly half that of Dione. Surface features have been given names from Virgil's Æneaid.

The main constituent of the globe must be water ice, but there must also be silicate rock – perhaps as much as 40%. As with Tethys, there have been suggestions that there may be subcrustal water, but there is no firm evidence.

The surface is not uniform. The trailing hemisphere is relatively dark, with an albedo of 0.3, while the brightest features on the leading hemisphere have albedo 0.6. Most of the heavily cratered terrain lies on the trailing hemisphere – contrary to what had originally been expected; there are craters of considerable size, such as Æneas (diameter 161 km), Dido (122 km) and Evander (350 km). Other surface features include long 'chasms' such as

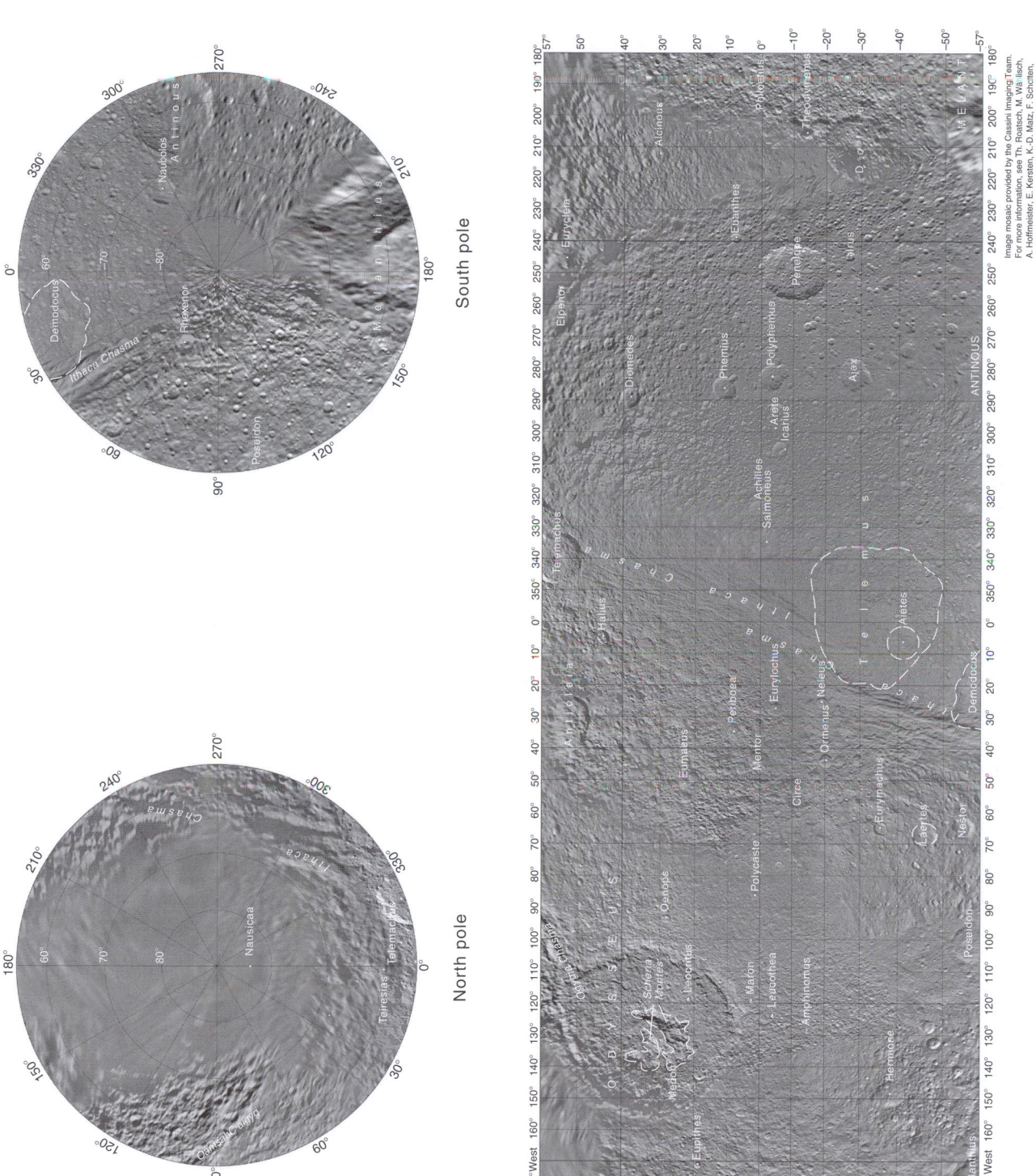

South pole

North pole

Image mosaic provided by the Cassini Imaging Team.
For more information, see Th. Roatsch, M. Wählisch,
A. Hoffmeister, E. Kersten, K.-D. Matz, F. Scholten,
R. Wagner, T. Denk, G. Neukum, P. Helfenstein, and
C. Porco, High-resolution Atlases of Mimas, Tethys,
and Iapetus derived from Cassini-ISS Images, Planetary
and Space Science, 2009.

Figure 10.3 Tethys

Table 10.11 *Features on Dione*

	Lat. (°)	Long. (° W)	Diameter/length (km)
Chasmata			
Aurunca Chasmata	11.6 N	266.7	290
Drepanum Chasma	46.0 S	265.0	360
Eurotas Chasmata	4.9 N	301.4	1000
Larissa Chasma	29.0 N	69.5	150
Latium Chasma	20.0 N	63.9	360
Padua Chasmata	17.7 N	247.2	1025
Palatine Chasmata	48.0 S	316.0	1000
Tibur Chasmata	60.0 N	69.3	156
Catenæ			
Aufidus Catena	S.78.0	296.4	275
Pactolus Catena	N.8.8	327.1	180
Pantagias Catena	S.15.3	141.7	200
Dorsa			
Janiculum Dorsa	24.8 N	144.1	900
Craters			
Acestes	50.1 N	243.4	108
Æneas	25.9 N	46.3	161
Allecto	7.7 S	224.6	106
Amastrus	10.0 S	237.2	62
Amata	5.2 S	279.8	76
Antenor	7.0 S	11.6	81
Ascanius	33.4 N	232.2	98
Aulestes	9.9 N	147.7	50
Caleta	24.7 S	79.6	50
Daucus	15.4 S	301.1	80
Dercennus	29.7 N	279.9	86
Dido	24.0 N	18.8	122
Entellus	10.9 S	210.5	63
Erulus	35.0 S	104.7	120
Evander	57.0 S	145.0	350
Herbesus	34.7 N	156.1	58
Ilia	0.5 S	346.3	52
Lagus	13.6 S	102.9	77
Latinus	52.2 N	201.0	130
Liger	24.0 N	126.6	53
Mezentius	19.2 N	183.0	51
Murranus	12.8 N	90.7	57
Pagasus	3.0 S	241.0	67
Prytanis	46.2 S	287.4	96
Remus	13.6 S	31.9	62
Romulus	8.1 S	26.8	91
Sabinus	43.7 S	186.7	88
Sagaris	4.9 N	104.2	53
Silvius	32.7 S	332.3	74
Tiburtus	29.1 N	189.7	59

Turnus	15.6 N	346.3	101
Tyrrhus	24.7 N	287.9	49
Fossæ			
Arpi Fossæ	47.5 N	130.8	330
Carthage Fossæ	11.9 N	336.7	500
Cllusium Fossæ	39.3 N	301.5	260
Fidena Fossæ	0.7 N	96.0	550
Pebelia Fossæ	8.2 N	82.4	225

Latium Chasma (length 360 km). However, before the Cassini orbiter flew within 500 km of Dione, in October 2005, the main attention was focused on the bright 'wispy streaks', together with linear troughs and ridges, which were particularly striking near Amata, whose nature was uncertain. It was generally believed that the wisps were surface deposits, making Dione look different from any other satellite. Cassini solved the mystery. Amata is a large crater, and the wisps are not mere ice deposits; they are bright ice cliffs, several hundred metres high, created by tectonic fractures. Because the nature of the original 'Amata' was uncertain, the name has (rather confusingly) been transferred to a smaller, well-defined crater (see Table 10.11).

Dione has clearly been the scene of great activity in the past, and has been riven by enormous fractures in its trailing hemisphere. As well as being denser than Tethys or Rhea, it has had a much more eventful tectonic history. Figure 10.4 is the map of the surface.

Helene and Polydeuces

These are the Dione co-orbitals. Helene, in the L1 point, is much the larger of the two (diameter over 30 km); it was named after Helen of Troy. Cassini images show that it has one fairly prominent crater and many smaller features. Polydeuces, in the trailing (Lagrangian 2) point, is a mere 3.5 km across. Polydeuces is an alterative name for Pollux, one of the Twins (Castor is the other).

Rhea

Rhea, larger than Tethys and Dione, was discovered by G. D. Cassini in 1672. It was named after the mother of Zeus (Jupiter); she was the sister of Cronus (Saturn) and also his wife – the morals of the ancient Olympians left a great deal to be desired! Surface features have been named after creation myths; see Figure 10.5.

Rhea is less dense than Dione; it is thought to consist of about 75% water ice and 25% rock. Its surface is icy; temperatures have been measured as about −174 °C in full sunlight and down to −220 °C in shade. As with Dione, the trailing hemisphere is the darker of the two, with craters of fair size (some over 40 km across); there are also wispy streaks, not nearly as prominent as Dione's. The leading hemisphere is heavily cratered, and uniformly bright. The surface looks ancient; there has certainly been no activity on Rhea for a very long time. Surface features are listed in Table 10.12.

It appears that something unknown is producing a strange symmetrical structure in the charged particle environment around Rhea, but contrary to 2008 reports there are no rings. A Cornell research team, led by Matthew Tiscareno, used the Cassini probe to search for any broad or narrow rings embedded in a circumsatellite disc of dust or cloud, but the results, announced in 2010, were negative.

North pole

Image mosaic provided by the Cassini Imaging Team.

(For more information about Dione imagery, see Th.
Roatsch, M. Wählisch, A. Hoffmeister, K.-D. Matz,
F. Scholten, E. Kersten, K. Wagner, T. Denk, G. Neukum,
and C. Porco. High-resolution Dione atlas derived from
Cassini-ISS images. Planetary and Space Sciences,
October, 2008.)

Figure 10.4 Dione

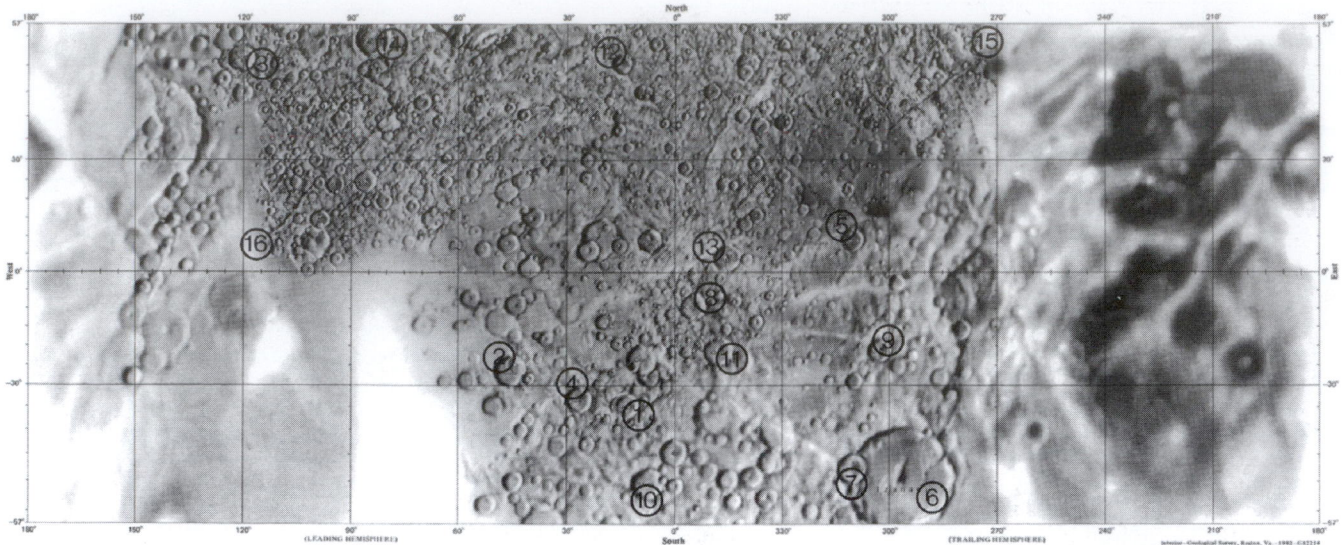

Figure 10.5 Rhea

Table 10.12 *Features on Rhea (bold numbers indicate map references)*

	Lat. (°)	Long. (° W)
Chasmata		
Kun Lun Chasma 14	46.0 N	307.5
Pu Choub Chasma 15	26.1 N	95.3
Craters		
Bulagat 1	28.2 S	15.2
Djuli 2	31.2 S	46.7
Faro 3	45.3 N	114.0
Haik 4	36.6 S	29.3
Heller 5	10.1 N	315.1
Izanagi 6	49.4 S	310.3
Izanami 7	46.3 S	313.4
Kiho 8	11.1 S	358.7
Leza 9	21.8 S	309.2
Melo 10	53.2 S	7.1
Qat 111	23.8 S	351.6
Thunapa 12	45.6 N	21.3
Xamba 13	2.1 N	349.7

A small telescope shows Rhea easily. An interesting observation was made on 8 April 1921 by six English observers independently: A. E. Levin, F. W. Hepbun, L, J. Comrie, E. A. L Atkins, P. Burnerd and C. J. Spencer. Rhea was eclipsed by the shadow of Titan, and, according to the first three observers, Rhea vanished completely for over half an hour.

No Rhea co-orbital satellites have been found.

Titan

Titan was discovered and named by Christiaan Huygens on 25 March 1655. It and its principal satellites have been named after the Titans, brothers and sisters of Cronus (Saturn). Apart from Ganymede it is the largest satellite in the Solar System, and is larger than Mercury, though less massive. It is the only satellite with a substantial atmosphere, and if it orbited the Sun rather than Saturn it would no doubt have been ranked as a planet. It is a very easy telescopic object, and when it is well placed keen-eyed observers can see it with good binoculars. It is mapped in Figure 10.6.

The first suggestion of an atmosphere was made in 1908, when the Spanish astronomer J. Comas Solá reported limb-darkening effects which could hardly be anything but atmospheric. The existence of the atmosphere was proved spectroscopically, in 1944, by G. P. Kuiper; it was tacitly assumed that the main constituent would be methane, and that the density would be low – after all, the escape velocity is only slightly greater than that of the airless Moon. As Voyager 1 approached Saturn's system, in November 1980, astronomers were still divided as to whether surface details would be seen on Titan or whether there would be too much cloud. (At Mission Control, in Pasadena, California, I well remember having an animated discussion with Professor Garry Hunt. I expected to see surface detail; he did not. He was right; surface details were conspicuous only by their absence.) Voyager 2 did not image Titan from close range, because of the need to go on to Uranus and Neptune. So far as Titan was concerned, we had to wait over twenty years – and the flight of the Cassini space-craft in 1997.

Throughout the long journey to Saturn, Huygens remained dormant except for biannual 'health checks'. It had a mass of 318 kg; its heat shield was 2.7 m across, and the probe itself was 1.3 m in diameter. On 1 July 2004, following a final check, it was separated from Cassini and began its descent through Titan's atmosphere; the descent lasted for 2 hours 30 minutes. Huygens landed at longitude 163.775° E, latitude 10.294° S, near the Xanadu region. Images and data were transmitted during the descent, and for 90 minutes after touching down – naturally, relayed via the Cassini orbiter. Surface transmissions ended when the orbiter passed below Huygens' horizon.

Instruments of all kinds were carried, because nobody knew what the surface would be like; Huygens might well have come

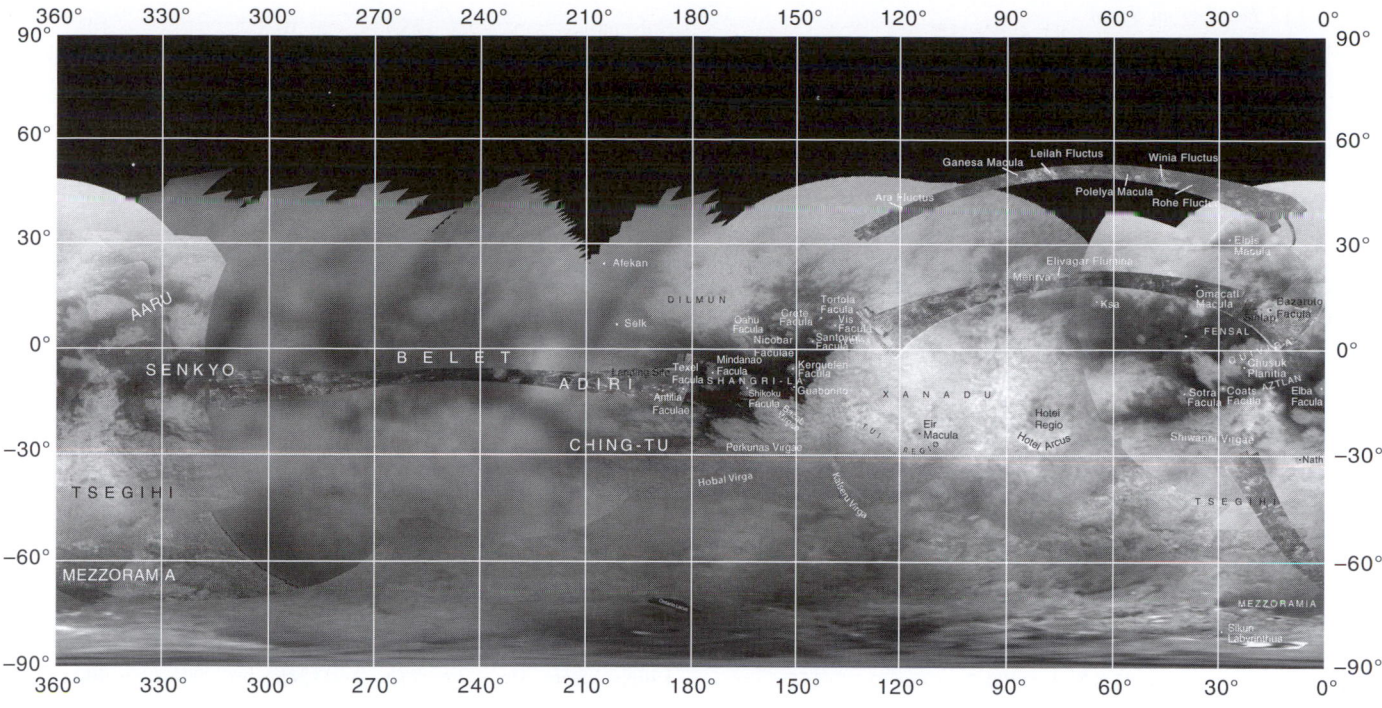

Figure 10.6 Titan

down in a chemical ocean. In fact, the images showed what looked like a flattish plain covered with 'pebbles'. There was a layer of haze around 20 km above ground level, and the 'sand-dune' surface itself seemed to have the consistency of wet clay. It was fairly clear that rain, no doubt methane rain, had fallen very recently. The dunes must be organic, and there are channels where liquid methane flows. The prevailing colour is orange, and there are indications of pieces of water ice scattered around.

Subsequent images from the Cassini orbiter showed features which cannot be anything but chemical lakes – which are, of course, smooth and therefore radar-dark. Titan has indeed a true 'lake district', and two of the features are large enough to be called 'seas', Kraken (named after a Norse sea-monster) and Ligeia (after one of the Sirens). Others are named after terrestrial lakes, such as Mývátn in Iceland and Waikare in New Zealand. A list of named lakes is given in Table 10.14.

Using one of the main instruments on Cassini, VIMS (the Visual and Infra-red Mapping Spectrometer), in 2008, ethane was identified in Ontario Lacus in the south polar area, which is 235 km long and has an area of 20 000 km^2, slightly larger than its Canadian namesake. The fact that it is liquid is shown because it reflects virtually no light at 5-micron wavelengths. Using Cassini data it has been possible to measure the changing depths of Titan's lakes, such as Ontario Lacus, and show that apart from Earth it is the only world in the Solar System to have a hydrological cycle with standing liquid on the surface. In-depth analysis of Titan's lakes were announced in 2010 by Akiva Bar-Nun of Tel Aviv University. He points out that irradiation of the methane in the atmosphere produces a variety of hydrocarbon gases, which

condense and rain down on to the surface. On reaching the surface they liquefy, flowing through the gullies and accumulating into lakes – but Bar-Nun adds 'you wouldn't want to jump into them on a summer holiday'. Further solar irradiation of the atmospheric hydrocarbons produces globules of aerosols, giving Titan its orange hue. The lakes contain enough liquid to cover Titan's entire surface to a depth of 43 m. Bar-Nun also correctly predicted that the water-ice crust has properties resembling the permafrost found in Siberia. As they are partly fluid, permafrost materials permit hills and mountains to rise no higher than 1900 m. In 2009, there was even a proposal to land a 'boat' in the far north lakes of liquid methane. This may be possible in the foreseeable future! By now we have a reasonable knowledge of the surface characteristics. There are broad and dark regions; the large reflective area of Xanadu, near the equator, was first identified in Hubble infrared images as long ago as 1994.

Xanadu, surveyed again using radar by Cassini in December 2008, is a bright, free-standing landmass surrounded by a darker region; it is about the size of Australia. At its western edge, dark sand dunes give way to land cut by river networks, hills and valleys. The rivers flow into dark areas, which are probably lakes; there is also one crater formed either by cryovulcanism or impact. Channels also cut the eastern area, but the dark outflow areas seem to lack dunes. The region is also crossed by mountain ranges, which have been compared with the Earth's Appalachians.

Xanadu is described as a land of twilight, dimmed by a haze of hydrocarbons surrounding it, and with its surface moulded by winds, rain and flowing liquids – almost certainly ethane or methane – running into a sunless sea. According to S. Wall of the

Table 10.13 *Features on Titan*

	Lat. (°)	Long. (° W)	Diameter/length (km)
Aaru	10 N	340	
Adiri	10 S	210	
Aztian	10 S	20	
Belet	56 S	255	
Ching-tu	30 S	205	
Dilmun	15 N	175	
Fensal	5 N	30	
Mezzoramia	0	0	
Quivira	0	15	
Senkyo	5 S	320	
Shangri-la	10 S	165	
Tsegihi	40 S	10	
Xanadu	15 S	100	3400
Arcûs			
Hotei Arcus	28 S	79	600
Craters			
Afekan	25.8 N	200.3	115
Ksa	14.0 N	65.4	29
Menrva	20.1 N	87.2	392
Selk	7.0 N	199.0	80
Sinlap	11.3 N	16.0	80
Faculæ			
Antilia Faculæ	11.0 S	187.0	260
Bazaruto Facula	11.6 N	16.1	215
Coats Facula	11.1 S	29.2	80
Crete Facula	9.4 N	150.1	680
Elba Facula	10.8 S	1.2	250
Kerguelen Facula	5.4 S	151.0	135
Mindanao Facula	6.6 S	174.2	210
Niobar Faculæ	2.0 N	159.0	575
Oahu Facula	5.0 N	166.7	485
Santorini Facula	2.4 N	145.8	140
Shikoku Facula	10.4 S	164.1	285
Sotra Facula	12.5 S	39.8	235
Texei Facula	11.5 S	182.6	190
Tortola Facula	8.8 N	143.1	65
Via Facula	7.0 N	138.4	215
Fluctûs			
Ara Fluctus	39.8 N	118.4	70
Leilah Fluctus	50.5 N	77.8	190
Rohe Fluctus	47.3 N	37.7	103
Winia Fluctus	49.0 N	46.0	300
Flumen			
Elivagar Flumina	19.3 N	78.6	260
Insulæ			
Mayda Insula	79.1 N	312.2	168
Lacûs (Lakes)			
Abaya	73.2 N	45.5	65
Bolsena	75.7 N	10.3	101
Faia	73.7 N	64.4	47
Kivu	87.0 N	121.0	77
Koitere	79.4 N	36.1	68
Mackay	78.3 N	97.5	180
Myvâtn	78.2 N	135.3	55
Neagh	81.1 N	32.2	98
Oneida	76.1 N	131.8	51
Ontario	72.0 S	183.0	235
Sotonera	76.7 N	17.5	63
Sparrow	84.3 N	64.7	81
Waikare	81.6 N	126.0	53
Maculæ			
Eir Macula	24.0 S	114.7	145
Elpis Macula	31.2 N	27.9	500
Ganesa Macula	50.0 N	87.3	160
Omacatl Macula	17.6 N	37.2	225
Polelya Macula	50.0 N	56.0	175
Maria (Seas)			
Kraken Mare	68 N	310	1170
Ligeia Mare	79 N	248	500
Punga Mare	85 N	340	380
Regiones			
Tui Regio	20 S	130	
Virgæ			
Bacab Virgæ	19 S	151	485
Hobal Virga	35 S	166	1075
Kaizeru Virga	36 S	137	630
Perkunas Virgæ	27 S	162	980
Shiwanni Virgæ	25 S	33	1400
Ringed feature			
Guabonito	10.9 S	150.8	55
Nath	30.5 S	7.7	95
Veles	2.0 N	137.3	45

Table 10.14 *Features on Hyperion*

	Lat. (°)	Long. (° W)
Craters		
Bahloo	36 N	195
Helios	71 N	132
Jarilo	.61 N	183
Meri	3 N	171
Dorsa		
Bond–Lassell Dorsum	48.0 N	143.5

Jet Propulsion Laboratory: 'The land is heavily tortured, convoluted and filled with hills and mountains. There are faults, valleys and deeply-cut channels. It seems to be the only area not covered with volcanic dirt; it has been washed clean. What is left underneath looks like porous water ice, maybe with caverns.'

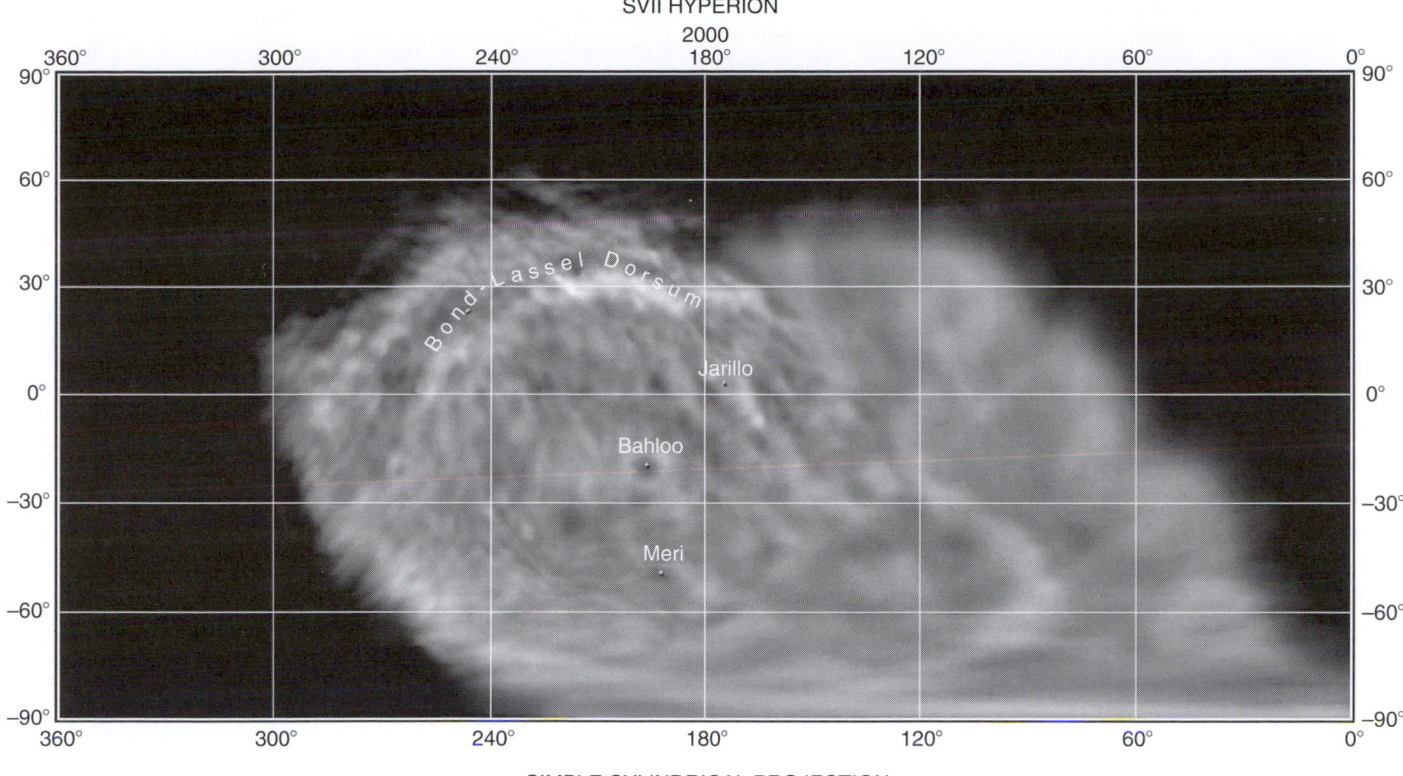

SVII HYPERION
2000

SIMPLE CYLINDRICAL PROJECTION

Figure 10.7 Hyperion

Both methane rain and dark orange hydrocarbon solids may fall from the dark skies above. Xanadu is indeed a strange, eerie place.

On Titan, all the materials needed for life are present, but the general conditions there – for example, the bitter cold – may well rule out advanced life and very possibly primitive life too. The surface temperature is around −180 °C; the atmosphere is 98.4% nitrogen, with almost all the rest made up of methane, with trace amounts of other gases such as argon, helium and cyanogen. The ground pressure is about 45% greater than that of the Earth's air at sea level. There are methane and ethane clouds, and on Titan's surface there is thought to be a persistent methane drizzle. The globe is certainly differentiated; according to one model there is a 3400-km silicate core at a fairly high temperature, surrounded by icy layers; the existence of a subcrustal sea has been suggested. The overall composition of the globe may be about half ice and half rocky material.

Further Cassini passes confirmed that Titan is unlike any other world in the Solar System. No doubt it will eventually be visited by astronauts, and since it lies at the edge of Saturn's magneto-sphere it is outside the dangerous radiation zones; it has no detect-able magnetic field of its own. Though Saturn is so much further away than Jupiter, Titan is a more attractive target than the Jovian Galileans (see Table 10.13 for details).

Hyperion

Hyperion, moving between the orbits of Titan and Iapetus, is in 3:4 orbital resonance with Titan. It is irregular in shape, measuring

$360 \times 280 \times 225$ km; it is of very low density, and seems to be composed mainly of water ice, with only a little rock. It may be porous, with only a thin dark crustal layer, and has been likened to a rubble pile – on a sponge! It may be up to 40% empty space. Surprisingly, the longer axis does not point directly to Saturn, as dynamically it would be expected to do. There are craters, named after Sun and Moon gods in various mythologies, up to 120 km in diameter and 10 km deep. There is also a long ridge or scarp, named Bond–Lassell Dorsum after the co-discoverers of Hyperion (see Table 10.14 and Figure 10.7.)

The orbital period is 21.3 days, but the rotation period is not synchronous; it is chaotic, and the orientation of its rotational axis varies unpredictably. The average rotation period is of the order of 13 days. There have been suggestions that Hyperion is half of a larger body which broke up. This is not impossible – but if so, where's the other half?

Hyperion is not a prominent telescopic object. With small telescopes, the best time to locate it is when it is in conjunction with Titan.

Iapetus

Iapetus, the outermost of Saturn's major satellites, was identified by G.D. Cassini 1671 – the first of his four discoveries (the others were Rhea, Tethys and Dione), and was named after one of the Titans; the alternative spelling, Japetus, is still sometimes used. The satellite was soon found to be very variable. When west of Saturn, it is an easy telescopic object, but when to the east it is

Table 10.15 *Features on Iapetus*

	Lat. (°)	Long. (° W)	Diameter/length (km)
Craters			
Adelroth	6.6 N	183.6	57
Astor	14.9 N	321.2	122
Baligant	16.4 N	224.9	66
Basan	33.3 N	194.7	78
Basbrun	52.0 S	111.8	80
Berenger	62.1 N	219.7	84
Bramimond	38.0 N	178.0	200
Charlemagne	55.0 N	208.8	96
Clarin	18.3 N	71.6	84
Corsablis	0.9 N	114.2	73
Engeller	40.5 S	264.7	504
Falsaron	33.8 N	82.6	424
Ganeion	44.3 S	19.8	230
Geboin	45.8 N	173.4	81
Gerin	45.6 S	233.0	445
Goderoy	71.9 N	249.1	63
Grandoyne	17.7 N	214.5	65
Hamon	10.6 N	270.0	96
Ivon	18.0 N	315.0	100
Johun	12.4 N	83.4	64
Jurfaleu	13.0 N	2.5	107
Malprimis	15.2 S	118.2	377
Malun	6.9 N	41.3	121
Margaris	27.7 N	135.8	75
Marsilion	39.2 N	176.1	136
Matthay	3.5 S	187.4	58
Milon	67.9 N	270.2	119
Naimon	9.3 N	339.3	244
Ogler	42.5 N	275.1	100
Oliver	62.5 N	200.8	113
Othon	33.3 N	347.8	86
Pinabel	39.0 S	33.0	83
Rabel	64.4 S	166.2	91
Roland	73.3 N	25.2	144
Thierry	55.0 S	8.0	110
Tibbald	57.0 N	358.0	160
Timozel	9.9 S	212.3	58
Turgia	16.9 N	28.4	580
Turpin	47.7 N	1.4	87
Valdebron	29.6 N	104.4	49
Mountains (*Montes*)			
Carcassone Montes	0	217	740
Cordova Mons	0	206	85
Gayne Mons	0	176	65
Haltil Mons	0	190	45
Seville Mons	0	346	69
Sorence Mons	0	194	46
Toledo Montes	0	136	1100
Tortelosa Montes	0	65	294
Valterne Mons	0	171	50
Terræ (*Lands*) and *Regiones* (*Regions*)			
Cassini Regio	28 S	93	
Roncevaux Terra	37 N	240	1284
Saragossa Terra	45 S	180	2300

much fainter, and at first Cassini assumed that it disappeared for a time during its 79-day orbit. However, he soon realised that it faded by around two magnitudes (10.2 to 11.8), and he gave a correct explanation; the two hemispheres have different albedo. The leading hemisphere is as black as a blackboard, while the trailing hemisphere is bright and icy. The axial rotation period is synchronous, and near western elongation it is the trailing hemisphere which is turned in our direction.

What could be the reason for this curious situation? In view of the low density of Iapetus, it had to be assumed that the globe was made up largely of ice, with a dark surface deposit. The line of demarcation between the bright and dark areas was not sharp, and the early Voyager images showed that some of the craters in the bright region were dark-floored, as though dark material had welled up from below the surface. Carl Sagan suggested that the deposit was thick and made up of organics, while other investigators believed it to be wafer-thin. Voyager showed that there are craters in both the bright and the dark areas; the names given came from the French poem 'The Song of Roland'. The dark area was named Cassini Regio, the bright area Roncevaux Terra after the Battle of Roncevaux Pass, in the Roland poem.

The mystery was finally solved in 2009, with the discovery of the Phœbe Ring. There is no doubt that dust from the ring falls upon the leading hemisphere of Iapetus, darkening it to produce the 'yin-yang' appearance. The dust particles are tiny, but they are certainly effective. On 10 September 2007, the Cassini probe flew past Iapetus at a range of only 1640 km, and made new temperature measurements; Cassini Regio is 15 °C warmer than Roncevaux.

There are other peculiarities, too. Iapetus is decidedly triaxial: $747 \times 749 \times 713$ km, with an equatorial bulge – it has been said that the shape resembles that of a walnut! A huge ridge, 1300 km long, 20 km wide and 13 km high, runs along the centre of Cassini Regio; there are peaks which tower to 20 km above the surrounding landscape. Note that it runs along Iapetus' equator, not the border between Cassini and Roncevaux Terra; it does not extend into Roncevaux, where there are 10-km isolated peaks. The origin of the ridge is unclear. There have been suggestions that it is material from a collapsed ring, either of Saturn or of Iapetus itself, but this is pure speculation, and we have to admit that Iapetus remains an enigma.

Unlike the other major satellites, its orbit is appreciably inclined to the ring-plane, so that from its surface the rings will be seen in their full glory; from the other satellites the rings will always be edgewise-on. No doubt it will become an attractive vantage-point for future space tourists. (see Table 10.15, Figure 10.8).

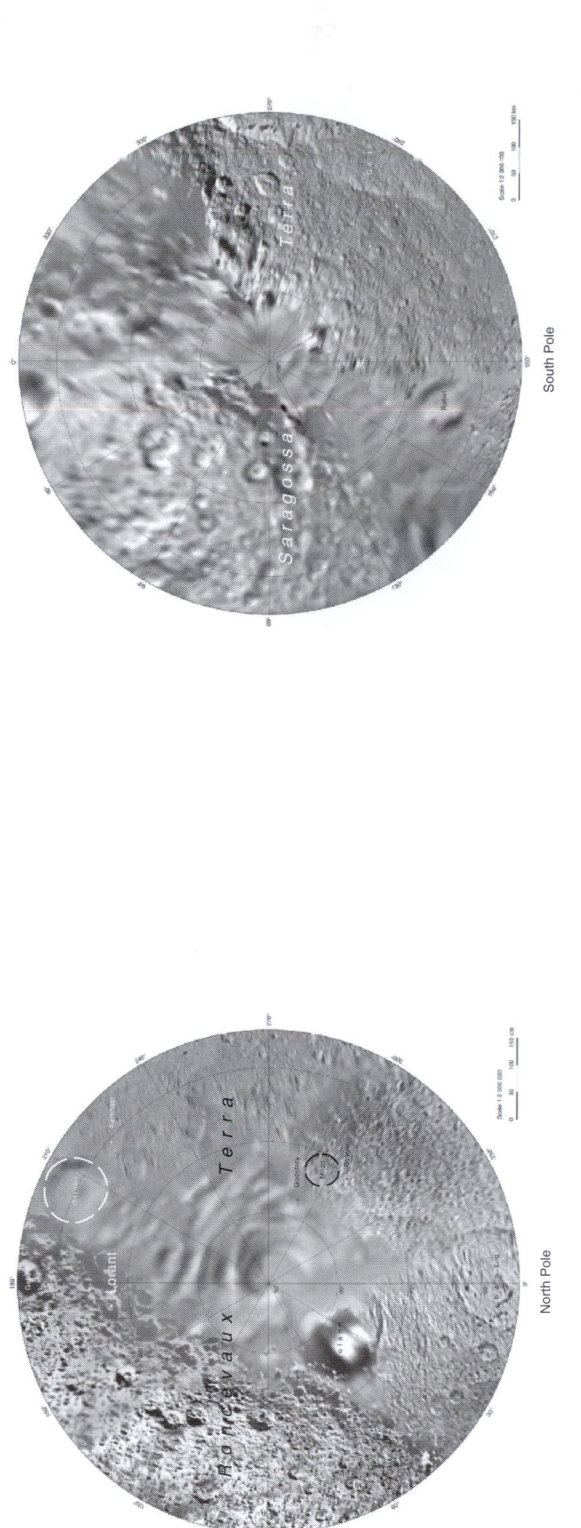

North Pole

South Pole

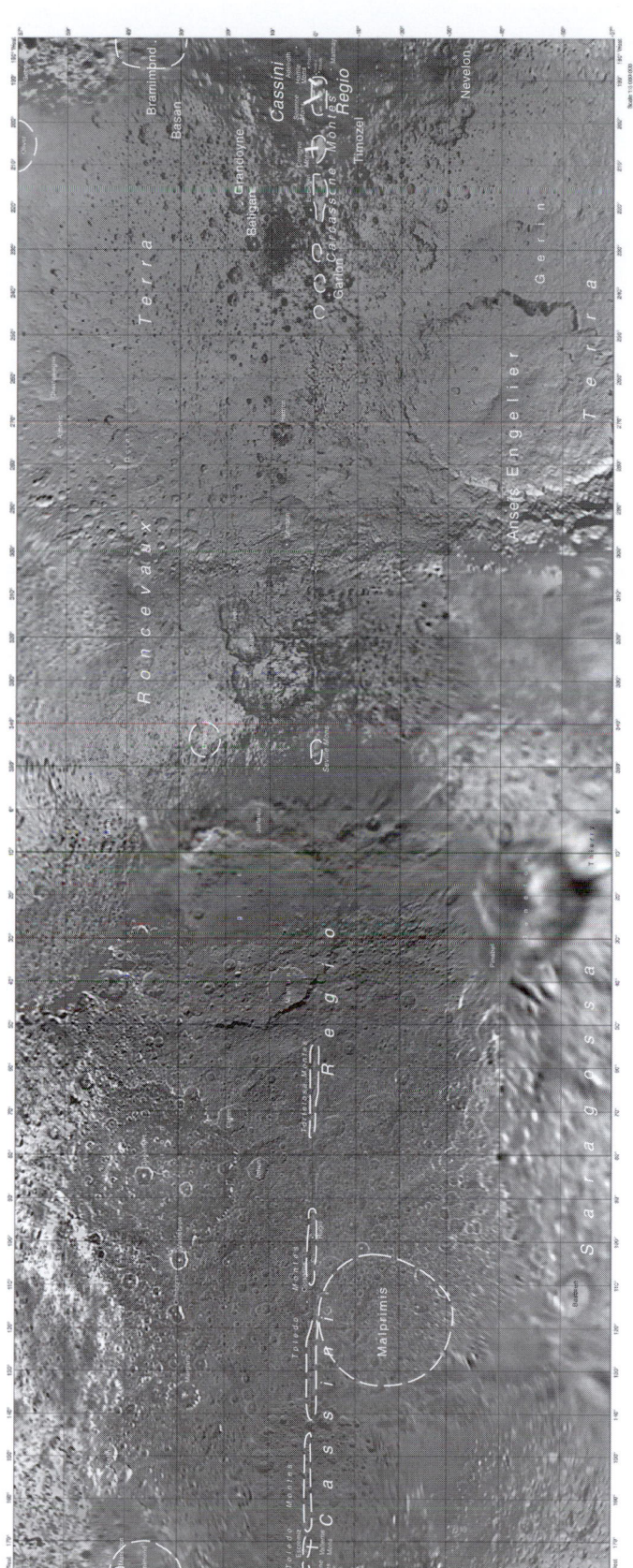

Figure 10.8 Iapetus

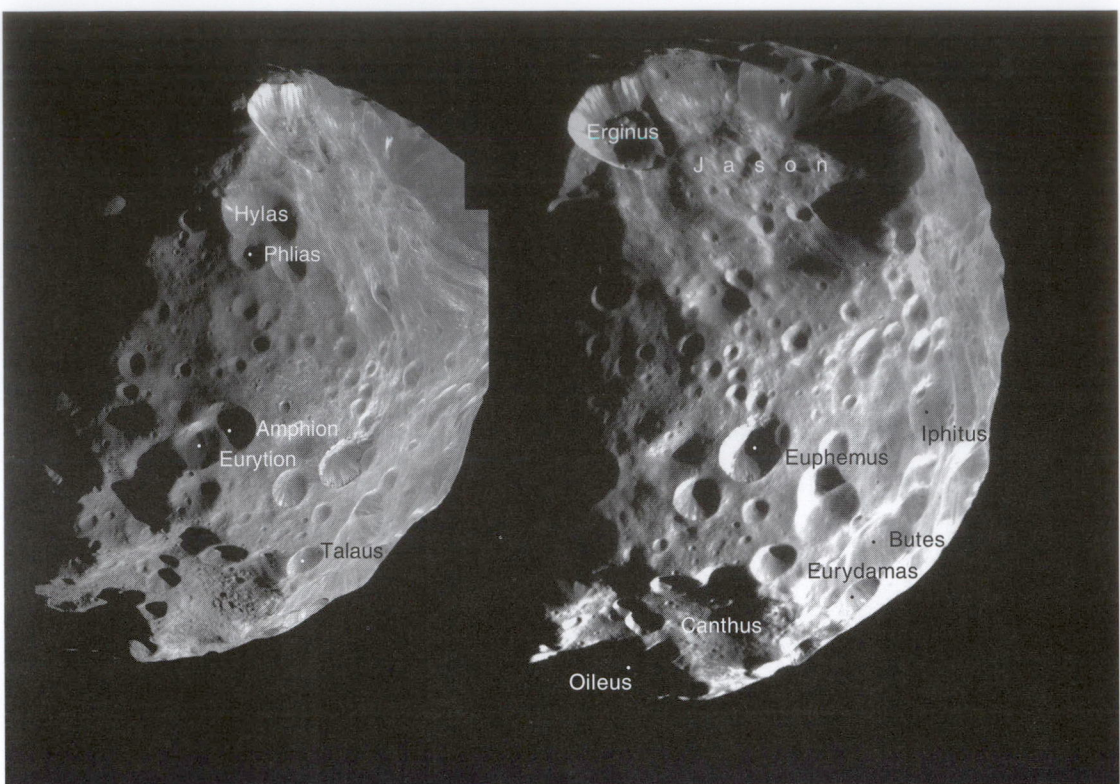

Figure 10.9 Phœbe

Table 10.16 *Features on Phœbe*

	Lat. (°)	Long. (° W)	Diameter (km)
Craters			
Acastus	9.6 N	14.6	34
Admetus	11.4 N	39.1	58
Butes	19.6 S	202.5	29
Calais	38.7 S	225.4	31
Canthus	69.6 S	342.2	44
Clytius	46.0 N	193.1	52
Eurytus	39.7 S	177.2	89
Hylas	7.9 N	354.5	30
Jason	16.2 N	317.7	101
Mopsus	6.6 N	109.1	37
Oileus	77.1 S	96.9	56
Peleus	20.2 N	102.2	44
Telamon	48.1 S	92.6	28
Zetes	20.0 S	223.0	29
Regiones (Regions)			
Leto Regio	60.0 N	20.0	95

THE OUTER SATELLITES

The outer 'irregular' satellites move beyond the orbit of Iapetus; data are given in Table 10.7. No doubt they are captured bodies; only the exceptional Phœbe is of appreciable size. Phœbe was discovered in 1898 by W. H. Pickering (the first satellite discovery to be made photographically); all the rest since 2000. There are several definite groups, appropriately named. The Inuit group comprises five prograde satellites: Ijiraq, Kiviuq, Paaliaq, Siarnaq and Tarqeq. The Norse group contains 29 retrograde satellites, from Phœbe to Fornjot. Finally, the Gallic group comprises four prograde outer satellites: Albiorix, Bebhionn, Erriapus and Tarvos.

Phœbe

This is quite different from the other outer satellites; it has no asteroidal characteristics, and is much more likely to be an escapee from the Kuiper Belt.

It moves within the outermost ring, discovered in 2009, and the ring itself must be due to particles knocked off the surface by impactors. Ring material has fallen on to the surface of Phœbe, and darkened the leading hemisphere. It was well imaged by Cassini when the space-craft drew in to Saturn in 1994, at a range of 2068 km. It is more or less spherical ($230 \times 220 \times 210$ km) and has a very dark, heavily cratered surface with craters over 50 km across, and over 10 km deep. One crater, Jason, has a diameter of 101 km. (The craters have been named after characters in the legend of the Golden Fleece.) The orbital period is 545.1 days, but the rotation period is only 9 hours 16 minutes. The density of the globe indicates a composition of about 50% rock; water ice has been detected spectroscopically, and the dark surface coating is probably thin. At magnitude 16.5, Phœbe is the only outer satellite which is within the range of telescopes of moderate aperture (see Table 10.17, Figure 10.9).

Though included with the Norse group, Phœbe must have an origin quite different from the other members of the group, and is the only one upon which surface detail has been seen.

Apart from Phœbe, the only outer satellites with estimated diameters of over 20 km are Siarnaq (40 km), Albiorix (32) and Paaliaq (22). However, the measurements are bound to be uncertain to some extent. Fenrir has an estimated diameter of only 4 km. Several as yet unnamed asteroids are almost certainly members of the Norse group: S/2004 S 7, S/2005 S 1, S/2006 S 3, S/2007 S 2 and S/2007 S 3. All are around 6 km in diameter.

The Inuit asteroids seem to have similar spectra, and to be light red; they may well be fragments of a disrupted larger body. Ijiraq is rather redder than the others (Kiviuq, Paaliaq, Siarnaq and Tarqeq). The Gallic asteroids may also have had a common parent; the largest, Albiorix, seems to have a large crater, and it has been suggested that Erriapus and Tarvos may be broken-off pieces of Albioix. There have also been suggestions that Skathi, Mundlifari, Suttungr and Thrymr may be fragments of Phœbe, but this seems less likely in view of the fact that Phœbe probably came from the Kuiper Belt rather than the asteroid zone.

All the outer satellites have been given mythological names. For example, Tarqeq is the Inuit Moon goddess, Jarnsaxa is a Norse giantess, Bestia is a frost princess, Siarnaq is an Inuit giant and Fenrir is the fearsome wolf of Norse mythology who broke his bonds at Ragnarok, the final battle between men and gods.

It has often been said that, despite its beauty, Saturn is a less rewarding telescopic object than Jupiter. There is certainly less surface activity, but one must always be prepared for the unexpected, such as the appearance of a new bright white spot and Saturn is always well worth monitoring.

11 · Uranus

Uranus, the seventh planet in order of distance from the Sun, was the first to be discovered in telescopic times, by William Herschel, in 1781. It is a giant world, but it and the outermost giant, Neptune, are very different from Jupiter and Saturn, both in size and in constitution. It is probably appropriate to refer to Jupiter and Saturn as gas giants and to Uranus and Neptune as ice giants.

Data for Uranus are given in Table 11.1.

MOVEMENTS

Since Uranus' synodic period is less than five days longer than our year, Uranus comes to opposition every year. Opposition dates for 2010–2020 are given in Table 11.2. The opposition magnitude does not vary a great deal; under good conditions the planet can just be seen with the naked eye. The most recent aphelion passage was that of 27 February 2007; Uranus was at its minimum distance from the Earth (21.09 a.u.) on 13 March of that year. The last perihelion passage was on 20 May 1966; Uranus was closest to the Earth (17.29 a.u.) on 9 March of that year. The next perihelion will be that of 13 August 2050.

In June 1989, Uranus reached its greatest southerly declination (−23.7°). Greatest northern declination had been reached in March 1950.

Close planetary conjunctions involving Uranus are listed in Table 11.3. It is interesting to note that between January and March 1610 Uranus was within 3° of Jupiter. This was the time when Galileo was making his first telescopic observations of Jupiter, but Uranus was beyond the limits of the field of his telescope.

Uranus can, of course, be occulted by the Moon. The first such record seems to be due to Captain (later Rear-Admiral) Sir John Ross, on 5 August 1924, with a power of ×500 on a reflector of focal length 25 ft (7.26 m). On 4 October 1832, Thomas Henderson, from the Cape of Good Hope, observed an occultation.

EARLY OBSERVATIONS

Uranus was seen on a number of occasions before its identification in 1781. It was recorded on 23 December 1690 by the Astronomer Royal, John Flamsteed, when it was in Taurus: Flamsteed even gave it a stellar number – 34 Tauri. Altogether 22 pre-discovery observations have been listed, as follows:

Flamsteed, 1690, 1712, four times in 1715;
J. Bradley, 1748 and 1750;
P. Le Monnier, twice in 1750, 1764, twice in 1768, six times in 1769, 1771;
T. Mayer, 1756.

It is interesting that Le Monnier failed to identify Uranus from its movement. He observed it eight times in four weeks (27 December 1768 to 23 January 1769) without realising that it was anything other than a star. He has often been ridiculed for this, but when he made his observations Uranus was near its stationary point, so it is hardly surprising that Le Monnier failed to identify it.

DISCOVERY AND NAMING

Uranus was discovered on 13 March 1781 by William Herschel, using a 6.2-inch (15.7-cm) reflector of 7 ft focal length and a magnification of × 227. Herschel realised that the object – in Gemini – was not a star, but he believed it to be a comet, and indeed his communication to the Royal Society was headed 'An Account of a Comet.'

The object was first recognised as a planet, independently but about the same time, by the French amateur astronomer Jean Baptiste de Saron – later, in 1794, guillotined during the Revolution – and by the Finnish mathematician, Anders Lexell. Lexell calculated an orbit, finding that the distance of the planet from the Sun was 19 a.u. – only slightly too small. He gave an orbital period of between 82 and 83 years, and stated that the apparent diameter was between 3 and 5 arcsec. In this case it was clear that Uranus was indeed a giant world, larger than any other planet apart from Jupiter and Saturn.

There was prolonged discussion over naming. J. E. Bode, in 1781, suggested Uranus, after the first ruler of Olympus (Uranus or Ouranos, Saturn's father). Other names were proposed – for example Hypercronius (J. Bernoulli, 1781) and 'the Georgian Planet', by Herschel himself in honour of his patron, King George III. Others called it simply 'Herschel'. Until 1850, the Nautical Almanac continued to call it the Georgian Planet, but in that year the famous mathematician and astronomer John Couch Adams suggested changing over to 'Uranus'. This was done, and the name became universally accepted.

DIAMETER AND ROTATION

The first attempt to measure the apparent diameter was made by Herschel, in 1781. His value (4″ .18) was rather too great. In 1788, he gave the diameter as 34 217 miles (55 067 km) with a mass 17.7 times that of the Earth; these values also were slightly too high. In the years 1792–4, Herschel also made an attempt to measure the polar flattening, and from his results rightly concluded that the rotation period must be short.

Table 11.1 *Uranus: data*

Distance from the Sun:
max. 3003 620 000 km (19.748 a.u.)
mean 2872 460 000 km (19.201 a.u)
min. 2741 300 000 km (18.637 a.u.)
Sidereal period: 84.01 years (30 685.4 days)
Synodic period: 369.66 days
Rotation period: 17.24 hours (17h 14.4m)
Mean orbital velocity: 6.82 km s^{-1}
Axial inclination: 97°.86
Orbital inclination: 0°.773
Orbital eccentricity: 0.04718
Diameter: equatorial 51 118 km
    polar 49 946 km
Apparent diameter, seen from Earth: max. 3″.7
    min. 3″.1
Reciprocal mass, Sun = 1: 22 869
Mass, Earth = 1: 14.6 (8.6978 × 10^{25} kg)
Volume, Earth = 1: 64
Escape velocity: 21.1 km s^{-1}
Surface gravity, Earth = 1: 1.17
Density, water = 1: 1.27
Oblateness: 0.023
Albedo: 0.51
Mean surface temperature: –214 °C
Maximum magnitude: +5.6
Mean diameter of Sun, seen from Uranus:1′ 41″
Distance from Earth: max. 3 157 300 000 km
min. 2 581 900 000 km

Table 11.2 *Oppositions of Uranus, 2010–2020*

Date	Dec. (°)	(′)	Magnitude	Apparent diameter (″)
2010 Sept 21	−1	17	6.1	3.59
2011 Sept 26	+0	19	6.1	3.59
2012 Sept 29	+1	55	6.1	3.60
2013 Oct 3	+3	30	6.1	3.60
2014 Oct 7	+5	04	6.0	3.61
2015 Oct 12	+6	38	6.0	3.61
2016 Oct 15	+8	10	6.0	3.62
2017 Oct 19	+9	40	6.0	3.62
2018 Oct 24	+11	08	6.0	3.63
2019 Oct 28	+12	34	6.0	3.64
2020 Oct 31	+13	57	6.0	3.65

During this period Uranus moves from Capricornus into Aquarius.

Table 11.3 *Planetary conjunctions involving Uranus. Close conjunctions, 1900–2100*

	Date	UT	Separation (″)	Elongation (°)
Mars–Uranus	1947 Aug 6	01.49	+43	48 W
Jupiter–Uranus	1955 May 10	20.39	−56	65 E
Mars–Uranus	1988 Feb 22	20.48	+40	63 W
Venus–Uranus	2077 Jan 20	20.01	−43	11 W

There are conjunctions with Venus on 4 Mar 2000 (234″), 28 Mar 2003 (157″), 4 Mar 2015 (317″) and 29 Mar 2018 (243″); with Mercury on 14 Feb 2006 (83″) and with Mars on 22 Mar 2013 (39″).

In 1856, J. Houzeau, in France, gave a rotation period of between 7¼ and 12½ hours. Later, before the results from spacecraft became available, the favoured period was 10 h 48 min, which is in fact decidedly too short. Uranus is unique in one respect; its axial inclination is more than a right angle, so that the rotation is technically retrograde, although not usually classed as such. From Earth, the equator of Uranus is regularly presented, as in 1923 and 1966; at other times a pole is presented – the south pole in 1901 and 1985, the north pole in 1946 and 2030. All this leads to a very peculiar Uranian calendar. Each pole has a 'night' lasting for 21 Earth years, with corresponding daylight at the opposite pole. For the rest of the orbital period conditions are less extreme.

The reason for this strange inclination is unclear. There have been suggestions that in its early career Uranus was struck by a massive impactor and literally tipped over, but this does not sound very plausible, and it may well be that interactions with other bodies in the Solar System were responsible, in which case the tilting was gradual (or even episodic) rather than sudden. There was probably some planetary migration in the young Solar System, and there have been hints that at one time Uranus was further out than Neptune, but of course all this is highly speculative.

Uranus may have had 100 000 years of disturbed movement. According to calculations by Alessandro Morbidelli (Côte d'Azur

Observatory, France, 2010), in the early history of the Solar System Uranus crossed the orbit of Saturn, and was flung inward toward Jupiter, which lobbed it back to Saturn. This may have happened three times before Uranus was finally thrown beyond Saturn. The idea of Uranus being used as a sort of cosmic pinball may sound fascinating, but most astronomers will view it with considerable scepticism!

There is, incidentally, some confusion about Uranus' poles. The International Astronomical Union has decreed that all poles above the ecliptic (i.e. the plane of the Earth's orbit) are north poles, while all poles below the ecliptic are south poles. In this case it was Uranus' south pole which was in sunlight during the pass of the Voyager 2 space probe in 1986. (However the Voyager team members reversed this and referred to the sunlit pole as the *north* pole. Take your pick!).

SURFACE MARKINGS FROM EARTH

In ordinary telescopes Uranus appears as a bland, rather greenish disc. Two bright spots were reported on 25 January 1870 by J. Buffham, using a power of ×320 on a 9-inch (23-cm) refractor; on 19 March, he described a bright streak. It seems improbable that these were genuine features, although of course one cannot be sure. From Earth, even really large telescopes show virtually no surface details on Uranus.

Spectroscopic observations were made from 1869, when Angelo Secchi, from Italy, recorded dark lines in the spectrum; the lines were photographed in 1869 by W. Huggins and, in 1902, H. Deslandres, from France, obtained spectroscopic confirmation of the retrograde rotation. Final confirmation of this was provided in 1911 by P. Lowell and V. M. Slipher, from the Lowell Observatory in Arizona, although their derived rotation period was several hours too short.

Meanwhile, very precise tables of the movements of Uranus had been compiled in 1875 by Simon Newcomb in America; earlier, in 1846, slight irregularities in the movements of Uranus had been used to identify the outer giant, Neptune, by J. Galle and H. D'Arrest, from Berlin.

Methane was identified in Uranus' atmosphere in 1933, by R. Mecke from Heidelberg; it had been suggested on theoretical grounds by R. Wildt, in 1932. Confirmation was obtained by V. M. Slipher and A. Adel from Flagstaff, in 1934. By then it had become clear that Uranus, like Jupiter and Saturn, was not a miniature sun: the outer layers at least were very cold indeed, and the visible surface was purely gaseous.

The first widely accepted model of Uranus was due to R. Wildt, who in 1934 proposed that the planet have a rocky core, overlaid with a thick layer of ice, which was in turn overlaid by a hydrogen-rich atmosphere. In 1951, W. R. Ramsey, of Manchester University, proposed an alternative model, according to which Uranus was made up largely of methane, ammonia and water. However, reliable information was delayed until the mission of Voyager 2, the only probe so far to have by-passed the planet.

VOYAGER 2

It is not too much to say that most of our detailed knowledge of Uranus comes from Voyager 2 although, in recent years, good data have also been obtained from the Hubble Space Telescope. Voyager 2 was launched on 20 August 1977 (actually before Voyager 1). It by-passed Jupiter on 9 July 1979 and Saturn on 25 August 1981. It then went on to Uranus (24 January 1986) and finally Neptune (25 August 1989). Excellent images were obtained of all four giants, together with a mass of data. Voyager 2 is now leaving the Solar System but more than 30 years after launch it was still transmitting data. By January 2008, its distance from the Sun had increased to over 12 000 000 000 km.

Because of Uranus' unusual tilt, Voyager 2 approached the planet more or less 'pole-on'. New satellites were discovered, and known satellites surveyed; studies were made of the Uranian magnetosphere, and radio waves were detected. Ultraviolet observation showed strong emissions on the day side of the planet, producing what was termed an 'electroglow'. On 'encounter day', 24 January, Voyager passed 107 100 km from the centre of the planet – that is to say just over 80 000 km from the cloud-tops. Closest approach occurred at 17h 59m UT. The Uranian equator was then twilight, and it was found that the temperatures at the poles and over the rest of the planet were much the same.

CONSTITUTION OF URANUS

One important fact is that Uranus, unlike the other giant planets, seems to have little internal heat. Jupiter radiates 1.7 times as much energy as it would do if it depended entirely upon what it receives from the Sun; Saturn radiates 1.8 times as much; and Neptune over 2. With Uranus, the upper limit is only 1.06, but with a possible uncertainty of plus or minus 1 – so that there may be no excess energy at all. Moreover, the temperatures of Uranus and Neptune as measured from Earth are almost equal, even though Neptune is so much further away from the Sun.

Uranus is made largely of 'ices', but it is to be noted that these will not be in solid form. For planetary scientists 'gas' is taken to mean hydrogen and helium, 'ices' a mixture of water (H_2O), methane (CH_4) and ammonia (NH_4) with smaller amounts of other substances. Water is the main constituent. 'Rock' is a mix of silicon dioxide (SiO_2) magnesium oxide (MgO), and either metallic iron (Fe) and nickel (Ni) or compounds of iron such as FeS and FeO. Inside Uranus and Neptune, the pressure and temperature conditions make all these materials behave as liquids, but they do not assume metallic characteristics as in the larger giants; the Jupiter/Saturn pair is very different from the Uranus/Neptune pair. Rock and ice make up 80% by mass of Uranus, as against 10% for Jupiter. In 2008, E. Schwegler (Lawrence Livermore National Laboratory) and his team proposed that at pressures above 450 000 atm ice enters a supersonic solid phrase, so that the interiors of Uranus and Neptune may contain some solid ice.

Does Uranus have a silicate core? According to one model the answer is 'yes'; a core with a radius less than 20% of the entire globe and a mass just over half that of the Earth; the central pressure would be about 8 000 000 bar. Round this would come the 'ice' mantle, made up of a fluid containing water, ammonia and other volatiles and accounting for most of the mass of the planet, and finally the atmosphere. Yet there is a good deal of speculation about this, and we have to admit that our knowledge of conditions deep inside the globe is very meagre; we have no proof that a definite core exists (bear in mind the lack or near-lack of an inner heat source), and even if it is there it may not have a well-defined boundary.

ATMOSPHERE

We know more about the atmosphere, because we can at least see its uppermost part. It is composed predominantly of molecular hydrogen (about 83% by number of molecules) and helium (13%); methane accounts for about 2%, while minor constituents include ethanol (C_2H_6) and acetylene (C_2H_2). Conventionally, the atmosphere is said to be made up of a troposphere (up to 50 km above the 1-bar pressure level), stratosphere (50 to 4000 km) and then the thermosphere and ionosphere. The immensely rarefied exosphere is very extensive, and probably spreads out as far as the inner edge of the ring system.

The troposphere is the lowest and densest part of the atmosphere; temperatures decrease with altitude, falling at the tropopause to about −225 °C. It may contain cloud layers of water and ammonium hydrosulphide; above these, in the 1 to 2 bar pressure range, lie the thin methane clouds we observe. The temperature in the stratosphere increases because of absorption of solar ultraviolet and infrared radiation by atmospheric methane and other hydrocarbons, and the temperature is still higher in the uppermost regions, the thermosphere and Uranian corona (remember that scientific 'temperature' does not imply heat!). The corona is unique to Uranus, and exerts drag on small particles orbiting the planet, causing depletion of 'dust' in the rings. The ionosphere seems to be denser than those of Neptune and Jupiter, perhaps because of the low concentration of hydrocarbons in the atmosphere. Uranus shows little detail compared with Jupiter, Saturn or Neptune, and discrete clouds are never striking; in general, there is little to be seen on the disc. I have viewed Uranus under good conditions with large telescopes, notably the Lowell and Yerkes refractors and the 60-inch reflector at Mount Wilson, with no positive results; the planet remains obstinately bland.

However, details have been made out with the Hubble Space Telescope. Good images were obtained in July to August 1997; these showed clouds in the northern hemisphere, which is now starting its 'spring' season – when Voyager 2 flew over the north pole it had been winter there, with the pole in total darkness. It has been found that features at different latitudes have rotation periods of between 14 and 17 h, so that there are winds blowing in an east–west direction: at high latitudes (around 60°) these are prograde, although there is a retrograde jet stream close to the equator. Observations made in 1998 with the Hubble Space Telescope recorded waves of massive storms on Uranus, with wind speeds in excess of 500 km h^{-1}: clearly the planet is much more dynamic than was originally thought. Obviously, the axial tilt means that wind conditions on Uranus are different from those on any other planet.

Radiation belts round Uranus exist; their intensity is similar to those round Saturn, but they differ in composition. They seem to be dominated by hydrogen ions: heavier ions are lacking.

There was unusual activity in July 2004, when some large clouds appeared, with winds of over 820 km h^{-1} – there was also what was thought to be a violent thunderstorm. Then, on 23 August, a dark spot was seen. However, none of these features lasted for very long.

MAGNETIC FIELD

Uranus has a magnetic field: the equatorial field strength at the equator is 0.25 G, as against 4.28 G for Jupiter (the value for Earth is 0.305 G). However, the magnetic axis is displaced from the rotational axis by 58.6°: neither does the magnetic axis pass through the centre of the globe – it is offset by 8000 km. The polarity is opposite to that of the Earth. The fact that the magnetic and the rotational poles are nowhere near each other means that aurorae, which were detected from Voyager 2, are a long way from the rotational pole. The magnetosphere of Uranus is relatively 'empty': it extends to 590 000 km on the day side and around 6 000 000 km on the night side.

The magnetosphere contains protons and electrons, with some H_2^+ ions. There is a bow shock at 23 Uranian radii ahead of the planet, and a magnetopause at 18 radii; the magnetotail trails for millions of kilometres, and the planet's 'sideways' rotation twists

the tail into the shape of a corkscrew! The field strength in the northern hemisphere is appreciably higher than in the southern.

With Jupiter and Saturn, the magnetic fields are generated in the cores. With Uranus and Neptune, however, the fields may well originate at much shallower levels, in the global liquid. We cannot pretend to be at all sure. The reason for the tilt of the magnetic axis is unknown. It was initially thought that Uranus might be experiencing a 'magnetic reversal', but subsequently it was found that Neptune's magnetic axis was also displaced – and to assume that two magnetic reversals were occurring simultaneously would be too much of a coincidence.

Uranian solstices and eqinoxes between 1900 and 2100 are as follows (IAU convention):

Northern hemisphere	Year	Southern hemisphere
Winter solstice	1902, 1986	Summer solstice
Vernal equinox	1923, 2007	Autumnal equinox
Summer solstice	1944, 2028	Winter solstice
Autumnal equinox	1965, 2049	Vernal equinox

THE RINGS

Uranus has a ring system, identified in 1977 and beyond the range of ordinary visual telescopes.

It is worth recalling that in 1787, a few years after his discovery of Uranus, William Herschel reported the existence of a ring: he was using his 20-ft focus reflector on 4 March, and reported the ring again on 22 February 1789. He described the ring as 'short, not like that of Saturn'. In fact he was being temporarily misled by optical effects: no telescope of that period could possibly show the true rings, and by the end of 1793 Herschel himself had realised that his 'ring' did not exist

The discovery of the ring system of Uranus was accidental. It had been predicted that, on 10 March 1977, the planet would occult the star SAO 158687, magnitude 8.9, and the occultation would provide a good opportunity to measure the diameter of Uranus. Calculations made by Gordon Taylor of the Royal Greenwich Observatory indicated that the occultation would be seen only from a restricted area in the southern hemisphere, and observations were made by J. Elliott, T. Dunham and D. Mink, flying at 12.5 km above the southern Indian Ocean in the Kuiper Airborne Observatory (KAO), which was in fact a modified C-141 aircraft carrying a 36-inch (91-cm) reflecting telescope. Close watches were also being kept from ground-based observatories, notably in South Africa. Thirty-five minutes before occultation, the star was seen by the KAO observers to 'wink' five times, so that apparently it was being temporarily obscured by material in the vicinity of Uranus. The occultation by Uranus began at 20h 52m UT, and lasted for 25 minutes. After emersion there were more winks, and these were later found to be symmetrical with the first set, indicating a system of rings. The post-emersion winks were also recorded by J. Churms from South Africa.

Subsequent observations provided full confirmation. In 1978, G. Neugebauer and his colleagues imaged the rings with the Hale

Table 11.4 *The rings of Uranus*

Ring	Distance (km)	Width (km)	Thickness, (m)	Orbital eccentricity, e	Inclination angle, i (°)	Period (h)
ζ	32 000–37 850	3500	?	?	?	Inner extension of the ζ$_c$ ring
1986 U2R	37 000–39 500	2500	?	?	?	Faint, dusty
ζ$_c$	37 850–41 350	3500	?	?	?	
6	41 837	1.6–2.2	?	1.01	0.06	6.1988
5	42 234	1.9–4.9	?	1.90	0.05	6.2875
4	42 570	2.4–4.4	?	1.10	0.03	6.3628
α	44 718	4.8–10.0	?	0.76	0.01	6.8508
β	45 661	6.1–11.4	?	0.44	0.01	7.0688
η	47 175	1.9–2.7	?	0.001	0.01	7.4239
γ	47 627	3.6–4.7	~150	0.003	0.02	7.5307
δ	48 300	4.1–6.1	?	0.001	0.00	7.6911
λ	50 023	1.0–2.0	?	0.0	0?	8.1069 Faint, dusty
ε	51 149	20–96	–150	0.079	0	6.3823
ν	66 100–69 900	3800	?	?	?	
μ	86 000–103 000	17 000	?	?	?	

The ε ring is shepherded by Cordelia and Ophelia. The γ ring lies between the orbits of Portia and Rosalind, and the κ ring at the orbit of Mab.

reflector at Palomar and in 1984, D. A. Allen, at Siding Spring, imaged them in infrared, using the Anglo–Australian Telescope. They were surveyed in detail by Voyager 2 and also studied from the Hubble Space Telescope. Details of the system are given in Table 11.4

Thirteen rings are known. Their nomenclature is chaotic, and must surely be revised sooner or later (preferably sooner!). In order of distance from the planet, they are: ζ, 6, 5, 4, α, β, η, γ, δ, λ, ε, ν and μ. Their distances range from 38 000 km for ζ out to 103 000 km for μ.

All are very thin and dark, made up mainly of water ice and radiation- processed organic particles between 0.2 and 20 m in diameter, though ζ, μ and ν rings more probably consist of small dust particles. The relative lack of dust in the main system may be due to drag in the Uranian corona.

The most prominant ring – the μ ring – is not symmetrical; and is narrowest when closest to Uranus; the satellites Cordelia and Ophelia, discovered by Voyager 2, act as 'shepherds' to it. There is little obvious dust in the main rings, but Voyager 2 took a final picture, on its outward journey from Uranus, when the planet hid the Sun, showing 200 very diffuse, nearly transparent bands of microscopic dust surrounding the system. The rings are made up of particles a few metres in diameter, with not many centimetre- and millimetre-sized particles; they are as dark as coal, and it has been suggested, although without proof, that they may be relatively young and perhaps not even permanent features of the Uranian system. Their thickness is from about 0.1 to 1 km.

The rings are not alike. Rings 6, 5 and 4 show significant internal structure. Rings α and ß lack sharp edges. Ring η does not show sensible inclination to Uranus' orbital plane, and is made up of two components – a sharp inner feature, and a much fainter one, which extends to about 55 km from the sharp feature. Both edges of the η ring are sharp while, with the δ ring, there is a faint component inside the main ring. Ring λ was discovered between the δ and ε rings; it is very thin and apparently circular. In addition to the rings, there is a dusty material closer in than Ring 6, extending from 31 000 to 41 350 km from Uranus (rings ζ, ζ$_c$, 1986 U2R.

Certainly the ring system of Uranus is interesting, but there can be no comparison with the glorious, icy rings of Saturn.

SATELLITES

Uranus has an extensive satellite system. Only five are reasonably substantial, and were known before the Voyager 2 pass in 1986: Miranda, Ariel, Umbriel, Titania and Oberon. Even Titania, the largest of them, is much smaller than our Moon. The other members of the system are very small. Table 11.5 lists the satellites of Uranus.

The first to be discovered were Oberon and Titania, by William Herschel in 1787. Both were seen on 11 January, although Herschel delayed making any announcement until he was certain of their nature. Between 1790 and 1802 Herschel claimed to have found four more satellites, but three of these are certainly non-existent; the fourth may have been Umbriel, but there is considerable uncertainty. Ariel and Umbriel were discovered on 24 October 1851 by the English amateur William Lassell (previous observations by Lassell in 1847 had been inconclusive). In 1894, 1897 and 1899, W. H. Pickering unsuccessfully searched for new satellites. The next satellite to be detected photographically from Earth was Miranda, on 16 February 1948, with the 82-inch reflector at the McDonald Observatory in Texas. Next, on 31 October 1997, P. Nicholson, J. Bums, B. Gladman and J. J. Kavelaars, using a charge-coupled device (CCD) on the 5-m Hale Telescope at

Table 11.5 *Satellites of Uranus*

(a) SMALL INNER SATELLITES

Name	Date of discovery	Distance (km)	Period, (d)	Eccentricty, e	Inclination angle, i (°)	Diameter/ dimensions (km)	Albedo	Magnitude
VI Cordelia	1986	49 751	0.335	0.0003	0.085	50 × 36 × 36	0.08	24.2
VII Ophelia	1986	53 763	0.376	0.0099	1.103	54 × 38 × 38	0.08	23.9
VIII Bianca	1986	59 166	0.434	0.0009	0.193	64 × 46 × 46	0.08	23.1
X Cressida	1986	61 767	0.464	0.0004	0.006	92 × 74 × 74	0.08	22.3
X Desdemona	1986	62 658	0.474	0.0001	0.112	90 × 54 × 54	0.08	22.5
XI Juliet	1986	64 358	0.493	0.0007	0.006	150 × 74 × 74	0.08	21.7
XII Portia	1986	66 097	0.513	0.0001	0.059	156 × 126 × 126	0.08	21.1
XIII Rosalind	1986	69 927	0.558	0.0001	0.279	72 × 72 × 72	0.08	22.5
XXVII Cupid	2003	74 392	0.618	0.0013	0.1	18	0.07	23
XIV Belinda	1986	75 255	0.624	0.0001	0.031	128 × 64 × 64	0.08	22.5
XV Puck	1985	86 004	0.762	0.0001	0.319	162	0.11	20.4
XXVI Mab	2003	97 736	0.923	0.0025	0.134	24	0.10	23

(b) PRINCIPAL SATELLITES

Name	Date of discovery	Distance (km)	Period, (d)	Eccentricty, e	Inclination angle, i (°)	Diameter/ dimensions (km)	Albedo	Magnitude
V Miranda	1948	129 390	1.413	0.001	4.232	480 × 468 × 466	0.32	17.6
I Ariel	1851	191 020	2.520	0.003	0.260	1158	0.30	14.2
II Umbriel	1851	266 000	4.144	0.004	0.205	1169	0.21	14.8
III Titania	1787	435 910	8.706	0.001	0.340	1578	0.27	13.7
IV Oberon	1787	583 520	13.463	0.001	0.058	1523	0.23	13.9

Rotation periods are synchronous, and axial inclinations very close to 0 degrees.

(c) SMALL OUTER SATELLITES

Discovery	Distance (km)	Period (d)	e	i (°)	Diameter (km)
XXII Francisco	4 276 000	266.6	0.146	147.6	22
XVI Caliban	7 231 000	579.7	0.159	139.9	72
XX Stephano	8 004 000	677.4	0.230	141.9	32
XXI Trinculo	8 504 000	749.2	0.220	166.3	18
XVII Sycorax	12 179 000	1288.3	0.522	152.5	150
XXII Margaret	14 345 000	1687.0	0.661	51.5	20
XVIII Prospero	16 256 000	1978.3	0.445	146.0	50
XIX Setebos	17 418 000	2225.2	0.591	145.9	48
XXIV Ferdinand	20 901 000	2887.2	0.368	167.3	20

Palomar, found the outermost satellites, Caliban and Sycorax. In 1999, J. J. Kavelaars and his colleagues, using the 3.58-m Canada–France–Hawaii telescope on Mauna Kea, announced the discovery of two more small outer satellites, bringing the grand total to 20. Since then, two more small inner satellites have been found, Cupid and Mab (by M. Showalter and J. Lissauer in 2003, using the Hubble Space Telescope) and four in the outer group: Francisco and Trinculo (M. Holman and J. Gladman, 2001); Ferdinand (M. Holman, J. Kavelaars, D. Milisavljevi and B. Gladman, 2001); and Margaret (S. Sheppard and D. Jewitt, 2003). The present total is 27, but no doubt more small outer satellites await discovery.

All the small satellites are presumably composed of ice with some rock. Obviously their diameters are to some extent uncertain.

All the inner satellites move more or less in the plane of Uranus' equator, and are prograde – that is to say, they move in the same sense as Uranus rotates; of the outer satellites, Margaret is prograde, while the rest are retrograde. Only the five major satellites were surveyed in any detail by Voyager. No Earth-based telescopes can show anything definite upon their surfaces.

The names of the first four satellites were suggested by Sir John Herschel: two Shakesperean, one (Ariel) from both Shakespeare and Pope's *Rape of the Lock*, and Umbriel from *Rape of the Lock*. The name 'Miranda' was proposed by Kuiper and the names of the 10 new inner and the two new outer satellites were adopted by the International Astronomical Union. Many critics feel that this departure from conventional mythology is undesirable, and should not be regarded as a precedent.

THE SMALL INNER SATELLITES

Cordelia

Cordelia is named after Lear's daughter in *King Lear*. It acts as the inner shepherd of the ε ring. Voyager images show it to be elongated (50 × 36 km), and, as with the other small elongated satellites, the major axis points to Uranus.

Ophelia

Ophelia is named after Polonius' daughter in *Hamlet*. This is the outer shepherd of the ε Ring; it is slightly brighter and larger than the inner shepherd, Cordelia, and it also is elongated (54 × 38 km).

Bianca

Bianca is named after a character in *The Taming of the Shrew*. It is longated (64 × 46 km), with a grey surface. It belongs to the Portia group, made up of Bianca, Cressida, Desdemona, Juliet, Portia, Rosalind, Cupid, Belinda and Perdita; these satellites have similar orbits and photometric characteristics.

Cressida

This satellite is named after a character in *Troilus and Cressida*. It is longated in shape (92 × 74 km) and has a grey surface.

Desdemona

Desdemona was Othello's wife in *Othello*. This satellite is grey and also elongated (90 × 54 km). It has been calculated that it may eventually collide with Cressida and Juliet.

Juliet

This satellite is named after the heroine in *Romeo and Juliet*. It is decidedly elongated (150 × 74 km).

Portia

Portia is the heroine of *The Merchant of Venice*. This is the largest member of its satellite group. It is grey and elongated (156 × 126 km). Observations of its spectrum with the Hubble Space Telescope show water-ice absorption lines.

Rosalind

This satellite is named after the character in *As you Like It*. It is also grey and apparently spherical (diameter 72 km).

Cupid

Cupid is a character in *Timon of Athens*. It is the smallest of the inner satellites, and seems to have a dark surface. It moves in an orbit only about 860 km from that of the larger satellite Belinda.

Belinda

This satellite is named after the character in Pope's poem *The Rape of the Lock*. It is elongated (128 × 64 km), with a grey surface.

Perdita

Perdita is a character in *The Winter's Tale*. This satellite seems to be spherical (diameter 30 km). It was shown on Voyager *2* images, but not recognised at the time, and its existence was not finally established until 2003.

Puck

Puck is named after the character in *A Midsummer Night's Dream*. It is the largest of the inner satellites, and was the first to be detected, on 30 December 1985, as Voyager 2 drew in toward Uranus. On 24 January 1986 – 'encounter day' – a single image of it was obtained, from a range of 500 000 km; the resolution was of the order of the 10 km. The globe is almost spherical, with a diameter of about 160 km, and the surface is thickly cratered. Observations from Earth-based telescopes and the HST have shown water-ice absorption features in the spectrum. The apparent magnitude is 20.4.

Mab

Mab is named after a fairy queen from English folklore, referred to in *Romeo and Juliet*. It moves in the same orbit as a very faint, dusty ring, and is no doubt responsible for it. The surface seems to be dark.

THE MAJOR SATELLITES

Miranda

Voyager 2 passed Miranda at only 3000 km, so that the surface features were imaged down to a resolution of 600 m. This was fortunate, because Miranda is indeed a remarkable world. Surface feature names are again Shakespearean. The measured temperature is given as –187 °C, and the surface is icy, naturally, only one hemisphere could be studied: the other was in darkness.

The landscape is amazingly varied, and there are several distinct types of terrain: old, cratered plains; brighter areas with cliffs and scarps; and 'ovoids' or coronæ all over 200 km across: Inverness, Elsinore and Arden. They are trapezoidal, and may well have been formed at the tops of diapirs (upwellings of warm ice). Inverness Corona has been nicknamed 'The Chevron', while Arden is 'The Race-track'. The surface layer is presumably water ice. The names of the surface features are of course Shakespearian. There are not many craters over 20 km across, but clearly there has been intense geological activity in the past – more so than with the other Uranian satellites. The orbital inclination with reference to the planet's equator is also much greater, indicating that Miranda could well be a captured body.

Miranda presents real problems of interpretation, it has been suggested that it has been shattered and re-formed several times, because the various types of terrain seem to have been formed at different periods, but this would involve considerable heating, which in view of Miranda's small size and icy nature does not sound probable. Features are listed in Table 11.6, and Figure 11.1 shows a view of Miranda.

Ariel

This satellite was imaged by Voyager 2 from 130 000 km, to a resolution of 2.4 km. A selected list of surface features is given in Table 11.7. The name comes from *The Tempest*: Ariel is a friendly sprite. The surface features are named after assorted spirits; thus

Table 11.6 *Features on Miranda*

	Lat. (°)	Long. (° W)	Length/diameter (km)
Coronæ			
Arden	−29.1	73.7	318
Elsinore	−24.8	257.3	323
Inverness	−66.9	325.7	234
Regiones			
Dunsinane	−31.5	11.9	244
Ephesus	−15.0	250.0	225
Mantua	−39.6	18'0.2	399
Sicilia	−30.0	317.2	174
Rupes			
Argier	−43.2	322.8	141
Vertona	−18.3	347.8	116
Sulcus			
Naples	−32.0	260.0	260
Syracusa	−15.0	293.0	40

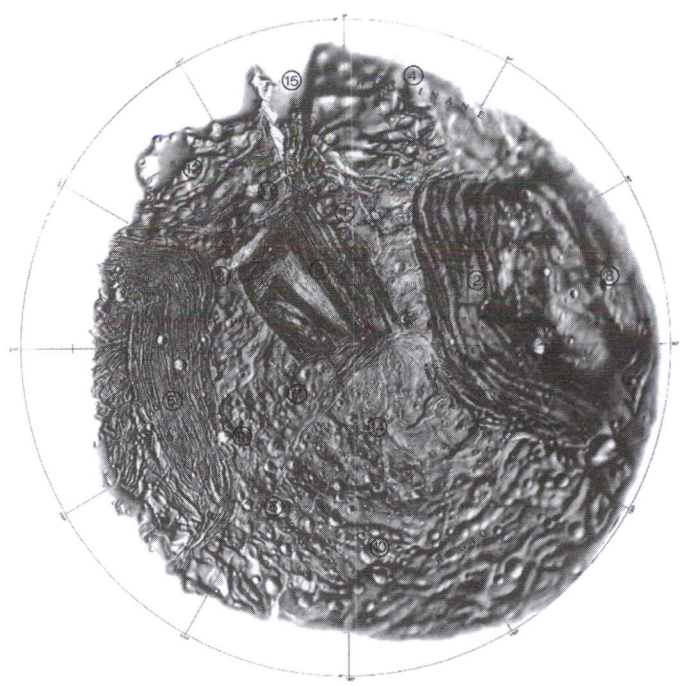

Figure 11.1 Miranda

Oonagh is the Irish queen of the fairies, Ataksak is a friendly Eskimo spirit, and British pixies live in woods.

Ariel is thought to be roughly 70% ices and 30% silicate rock, with areas of comparatively fresh frost, notably in the ejecta round young impact craters. There are plenty of craters, of which the two largest, Yangoor and Domovoy, are over 70 km in diameter (but remember that during the Voyager 2 pass Ariel's south pole was facing the Sun, so that only the southern hemisphere could be surveyed; for all we know, the north may be decidedly different). The terrain is transacted by fault scarps and graben, but the dominant features are broad, branching, steep-sided canyons ('chasmata') such as Korrigan, Sylph, Kewpie and Kachina; the longest, Kachina, runs for 622 km. They appear to have been cut by flowing liquid, possibly an ammonia–water mixture; Ariel today is inert, but must once have been the scene of great tectonic activity and icy vulanism. Certainly, its surface looks much fresher than those of Umbriel, Titania or Oberon. On 26 July 2006, a transit of Ariel across the disc of Uranus was recorded by the Hubble space Telescope. This observation was, to put it mildly, very delicate! (see Figure 11.2).

Umbriel

Umbriel was named for the dark sprite in *The Rape of the Lock*. While Ariel's surface features are named after benevolent spirits, those on Umbriel take their names from malevolent sprites. The surface is darker than those of the other main satellites; there are no ray-craters. Unfortunately the best image was taken when Voyager was still 557 000 km from Umbriel, so that the resolution is no better than 10 km (see Table 11.8, Figure 11.3).

The brightest crater, Skynd, has a central peak: other craters have darker floors, and since Umbriel is presumably icy there has been a surface coating of some kind. Wunda, at the edge of the image, is of uncertain nature: it is close to the equator (remember that Umbriel, like Uranus itself, was being imaged almost pole-on) and appears to be a ring, but it is so badly fore-shortened that its form cannot be made out, and we cannot be sure that it is a crater at all. Nothing comparable has been found anywhere else on Umbriel.

Surface features are named after trolls and gnomes (thus Vuver was an evil Finnish spirit, Wunda an Australian dark spirit). Umbriel is fractionally larger than Ariel, but has a lower albedo, and is more difficult to see.

Titania

The largest of Uranus' satellites: it is very slightly larger than Rhea and Iapetus in Saturn's system. It is named after the fairy queen in *A Midsummer Night's Dream*. Surface features are listed in Table 11.9 and shown in Figure 11.4. Voyager 2 imaged it from a range of 369 000 km.

Like Ariel, although to a lesser extent, Titania has clearly seen considerable tectonic activity in the past. There are many craters and ice cliffs, and trench-like features, notably Messina Chasmata, which extends for over 1490 km and crosses what was the boundary between the sunlit and dark sides of Titania at the time of the Voyager 2 pass. One crater, Gertrude, is over 320 km across. (Gertrude was King Claudius' wife in *Hamlet*: features on Titania are named after female Shakespearean characters.) The large crater Ursula is cut by a younger fault valley over 100 km wide. In size and mass, Titania and Oberon are virtual twins, but Oberon does not show so much evidence of past activity. On 8 September 2001, Titania was observed to occult a faint star. A careful search was made for any trace of atmosphere; the result was negative, but this was only to be expected. Though Titania's

Table 11.7 *Features on Ariel (bold numbers indicate map references)*

	Lat. (°)	Long. (° W)	Length/diameter (km)	
Chasmata				
Brownie **6**	−16.0	337.6	343	German good wood spirits
Kachina **13**	−33.7	246.0	622	Pueblo good rain spirits
Kewpie **14**	−28.3	326.9	467	British spirit babies
Korrigan **15**	−27.6	347.5	365	French wind spirits
Kra **16**	−32.1	354.2	142	Gold Coast spirits
Pixie **22**	−20.4	5.1	238	British spirits living in rocks
Sylph **25**	−48.6	353.0	349	British air spirits
Craters				
Abans **1**	−15.5	251.3	20	Spirit of the iron mines
Agape **2**	−46.9	336.5	34	Spirit in Spenser's *Fairy Queene*
Ataksak **3**	−53.1	224.3	22	Eskimo good spirit
Befanak **4**	−17.0	31.9	21	Italian good spirit
Berylune **5**	−22.5	327.9	29	Good spirit in Maeterlinck play
Deive **7**	−22.3	23.0	20	Spirit of beautiful maiden
Djadek **8**	′12.0	251.1	22	Czech good spirit
Domovoy **9**	−71.5	339.7	71	Slavic good spirit
Finvara **10**	−15.8	19.0	31	Irish spirit king
Gwyn **11**	−77.5	22.5	34	Irish war-god
Huon **12**	−37.8	33.7	40	Succeeded Oberon as spirit king
Laica **17**	−21.3	44.4	30	Inca good spirit
Mab **19**	−38.8	352.2	34	Spirit Queen
Melusine **20**	−52.9	8.9	50	French spirit heroine
Oonagh **21**	−21.9	244.4	39	Irish queen of fairies
Rima **23**	−18.3	260.8	41	Spirit in Hudson play
Yangoor **26**	−68.7	279.7	78	Spirit of day
Valles				
Leprachaun **18**	−10.4	10.2	328	Spirits or dwarfs
Sprite **24**	−14.9	340.0	305	Earth spirits

Table 11.8 *Features on Umbriel (bold numbers indicate map references)*

	Lat. (°)	Long. (° W)	Length/diameter (km)	
Craters				
Alberich **1**	−33.6	42.2	52	Dwarf guarding Niebelung gold
Fin **2**	−37.4	45.3	43	Danish troll
Gob **3**	−12.7	27.8	88	Gnome king
Kanaloa **4**	−10.8	345.7	86	Polynesian evil spirit
Malingee **5**	−22.9	13.9	164	Aboriginal night spirit
Minepa **6**	−42.7	8.2	58	Macouas evil spirit
Peri **6**	−9.2	4.3	61	Persian evil spirit
Setibos **7**	−30.8	346.3	50	Chief devil
Skynd **8**	−1.8	331.7	72	Evil troll
Vuver **9**	−4.7	311.6	98	Finnish evil spirit
Wokolo **10**	−30.0	1.8	208	Baramba devil spirit
Wunda **11**	−7.9	273.6	131	Australian dark spirit
Zlyden **12**	−23.3	326.2	44	Slavic evil spirit

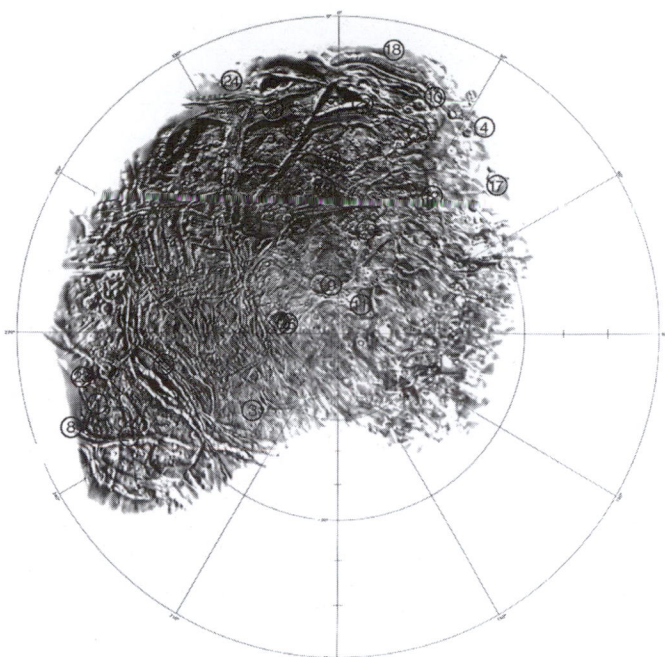

Figure 11.2 Ariel

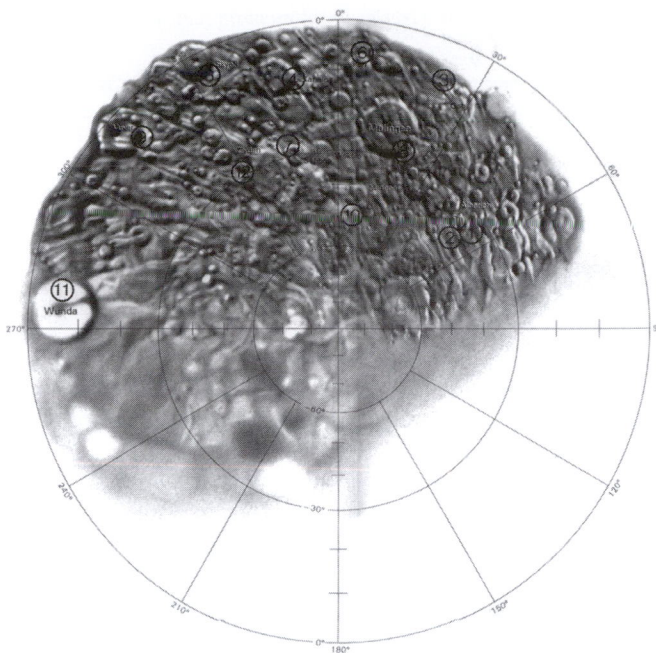

Figure 11.3 Umbriel

Table 11.9 *Features on Titania (bold numbers indicate map references)*

	Lat. (°)	Long. (° W)	Length/diameter (km)
Chasmata			
Belmont **2**	−8.5	32.6	258
Messina **13**	−33.3	335	1492
Craters			
Adriana **1**	−20.1	3.9	50
Bona **3**	−55.8	351.2	51
Calpurnia **4**	−42.4	291.4	100
Elinor **5**	−44.8	333.6	74
Gertrude **6**	−15.8	287.1	326
Imogen **7**	−23.8	321.2	28
Iras **8**	−19.2	333.8	33
Jessica **9**	−56.3	285.9	64
Katherine **10**	−51.2	331.9	75
Lucetta **11**	−14.7	277.1	58
Marina **12**	−15.5	316.0	40
Mopsa **14**	−11.9	302.2	101
Phrynia **15**	−24.3	309.2	35
Ursula **17**	−12.4	45.2	135
Valeria **18**	−34.5	4.2	59'
Rupes			
Rousillon **16**	−14.7	26.5	402

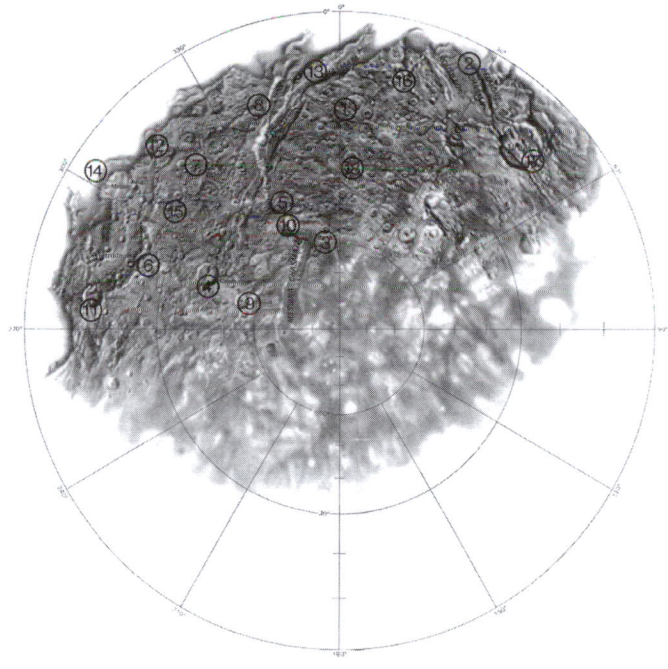

Figure 11.4 Titania

surface temperature is only about 60 °C above absolute zero, the escape velocity is a mere 0.77 km s⁻¹. Mutual satellite phenomena are excessively difficult to record, but on one occasion Oberon was observed to occult Umbriel.

Oberon

This satellite is named for the fairy king in *A Midsummer Night's Dream*. It has a brownish surface, pitted with craters; although the average albedo is low, some of the larger craters, such as Othello, are the centres of bright ray-systems. Othello and other craters, such as Falstaff and Hamlet, have dark material inside them: this may be a mixture of ice and carbonaceous material erupted from the interior.

Table 11.10 *Features on Oberon (bold numbers indicate map references)*

	Lat. (°)	Long. (°)	Length/diameter (km)
Chasmata			
Mommur **8**	−16.3	323.5	537
Craters			
Antony **1**	−27.5	65.4	47
Cæsar **2**	−26.6	61.1	76
Coriolanus **3**	−11.4	345.2	120
Falstaff **4**	−22.1	19.0	124
Hamlet **5**	−46.1	44.4	206
Lear **6**	−5.4	31.5	126
Macbeth **7**	−58.4	112.5	203
Othello **9**	−66.0	42.9	114
Romeo **10**	−28.7	89.4	159

Oberon was imaged by Voyager 2 from 660 000 km, to a resolution of 12 km. Selected features are listed in Table 11.10 and mapped in Figure 11.5.

One interesting feature is what appears to be a lofty mountain, some 6 km high, shown on the best Voyager picture, exactly at the edge of the disc, so that it protrudes from the limb (otherwise it might not be identifiable). Whether it is exceptional is something else about which we have as yet no definite information.

Oberon may be composed of about 50% water ice, 30% silicate rock, and 20% methane–related carbon–nitrogen compounds.

The names of the surface features come from Shakespeare. The only large chasm on the area imaged by Voyager 2 is Mommur – in *A Midsummer Nights Dream* this is the forest home of King Oberon.

Mutual satellite phenomena are excessively difficult to record, but on 4 May 2007 an occultation of Umbriel by Oberon was observed by M. Hidas, A. Christou and T. Brown (Las Cumbres Observatory Global Telescope). The light of Umbriel dimmed by 30%.

THE OUTER SATELLITES

All the satellites moving beyond the orbit of Oberon are very small. Margaret was named after a character in *Much Ado About Nothing*, and the others after characters in *The Tempest*. (It is reported that Nicholson wanted to name one of them Squeaker after his cat but Shakespeare prevailed!)

Sycorax, much the largest of the group, is around 150 km across, comparable with Puck, but Trinculo may have a diameter less than

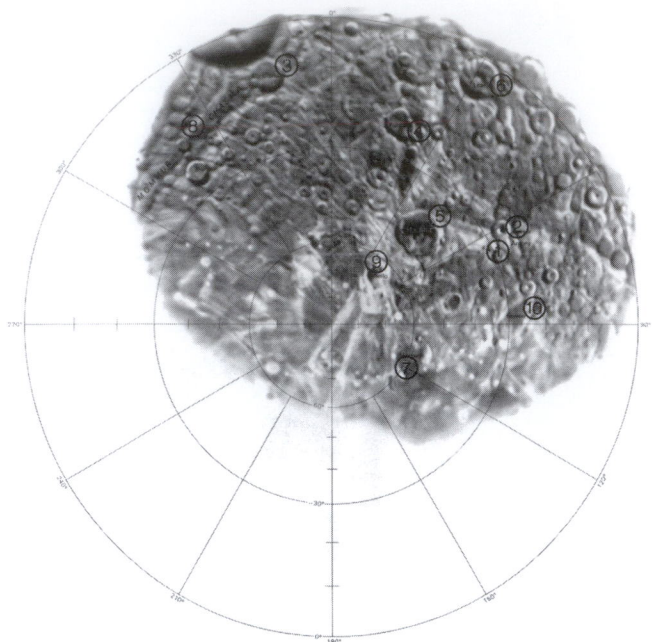

Figure 11.5 Oberon

20 km. Prosper, Sycorax and Setebos have similar orbital characteristics, but differ in colour; Sycorax is reddish, while the other two are grey. All the outer satellites apart from Margaret have retrograde motion; Margaret is prograde, with a highly eccentric orbit.

All the outer satellites are below the 20th magnitude. No doubt more very small Uranian satellites await discovery.

THE VIEW FROM URANUS

As the apparent diameter of the Sun as seen from Uranus is below 2 arcmin, all the satellites known before the Voyager 2 pass could produce total eclipses. From Uranus, Ariel would have much the largest diameter.

To an observer on Uranus – if he could get there – sunlight would be relatively strong, ranging between 1068 and 1334 times that of full moonlight on Earth. Saturn would be fairly bright when well placed (every 45½ years): Jupiter would have an apparent magnitude of 1.7, but would remain inconveniently close to the Sun in the Uranian sky. Neptune would be visible with the naked eye when near opposition.

Bland though it may appear, Uranus has proved to be of intense interest. It is very different from any other world known to us.

12 · Neptune

Neptune is the third most massive planet in the Solar System and, like Uranus, may be described as an 'ice giant'. It is too faint to be seen with the naked eye, but binoculars show it easily, and a small telescope will reveal its pale bluish disc. Data are given in Table 12.1.

MOVEMENTS

Neptune is a slow mover; it takes almost 165 years to complete one journey round the Sun so that it was discovered less than one 'Neptunian year' ago. Opposition dates are given in Table 12.2. Some close planetary conjunctions involving Neptune are listed in Table 12.3: occultations by the Moon can of course occur.

EARLY OBSERVATIONS

Neptune was observed on several occasions before being identified as a planet. The first observation seems to have been made by Galileo on 27 December 1612. While drawing Jupiter and its four satellites, he recorded a 'star' which was certainly Neptune. He again saw it twice in January 1613, and noted its movement but, not surprisingly, mistook it for a star. His telescope had a magnification of ×18 and a resolving power of 190 arcsec, with a field of view 17 arcmin in diameter. Neptune's magnitude was 7.7, and Galileo often plotted stars fainter than that. Jupiter actually occulted Neptune in 1613.

The next telescopic observation was made in May 1795 by J. J. de Lalande, but again Neptune was mistaken for a star. Further observations were made in 1845–6 by John Lamont (often referred to as Johann von Lamont: he was Director of the Munich Observatory).

DISCOVERY

Uranus was discovered in 1782 by William Herschel. Before long it became clear that it was not moving exactly as it had been expected to do; this was recognised by a mathematician, the Reverend Placidus Fixlmillner of Kremsmünster. It was reasonable to suppose that the old pre-discovery observations, used to calculate the orbit, were inaccurate, so Fixlmillner discarded them and re-worked the orbit. Within a few years new discrepancies became evident. In 1821, A. Bouvard produced new tables of Uranus' motion, but by 1832 G. Airy, then at Cambridge, found that these tables were wrong by half a minute of arc, which was unacceptable. It was around this time that there were the first definite suggestions than an unknown planet might be pulling Uranus out of position.

The idea occurred to J. E. B. Valz, Director of the Marseilles Observatory, and to F. G. B. Nicolai, Director of the Mannheim Observatory. On 17 November 1834, an English amateur, the Reverend T. J. Hussey (Rector of Hayes in Kent) wrote a letter to Airy suggesting that it might be possible to work out a position for the perturbing planet. Airy's reply was not encouraging, and Hussey took the matter no further.

In 1840, F. W. Bessel, of Königsberg, returned to the possibility of a new planet and told Sir John Herschel – son of Sir William – that he intended to search for it, in collaboration with his pupil, F. W. Flemming. This never happened. Flemming died suddenly; Bessel became ill and he too died, in 1846. Another interested astronomer was J. H. Mädler, co-author (with Beer) of the first really good map of the Moon; Mädler discussed the problem in 1841, but by then had left Germany to become Director of the new Dorpat Observatory in Estonia, and he never followed the Uranus problem through.

In 1841, John Couch Adams, then an undergraduate at Cambridge, 'formed a design of investigating as soon as possible, after taking my degree, the irregularities in the motion of Uranus', with the intention of tracking down the new planet. He began work in 1843, and by mid 1845 had calculated a position for the planet. He was in communication with James Challis, Professor of Astronomy at Cambridge, and also with Airy, now Astronomer Royal at Greenwich, but following a series of delays and misunderstandings no search was instigated. Meanwhile, U. J. J. Le Verrier, in France, had been working along similar lines, and by 1846 he had worked out a position very close to that given by Adams – about which, of course, Le Verrier was entirely ignorant.

When Airy saw Le Verrier's memoir, he realised that a search would have to be undertaken. There was no suitable telescope at Greenwich, but at Cambridge there was the 29.8-cm Northumberland refractor, and Airy instructed Challis to begin hunting. Unfortunately, Challis had no up-to-date charts of the area of the sky concerned, and he was not enthusiastic; he adopted a cumbersome method of star-checking, and was in no hurry to compare his observations. Le Verrier had failed to persuade astronomers in Paris to begin a search, but the outcome was different. Patience was never Le Verrier's strong point, and instead of waiting he contacted Johann Galle, at the Berlin Observatory. Galle was interested, and asked permission from the Observatory director, J. F. Encke, to use the 23-cm Berlin refractor for the purpose. Encke agreed: 'Let us oblige the gentleman from Paris!' Galle, together with a young astronomer, H. D'Arrest, lost no time, and on 23 September 1846, the first night of their search, they identified the planet. Galle used the telescope, while D'Arrest checked the positions of

Table 12.1 *Neptune: data*

Distance from the Sun:
max. 4553.9 million km (30.44 a.u.)
mean 4503.4 million km (30.10.a.u.)
min. 4452.9 million km (29.77 a.u.)
Sidereal period: 164.79 years (60 190 days)
Synodic period: 367.5 days
Rotation period: 16h 6m (0.671 day)
Axial inclination: 28° 48′
Orbital inclination: 1° 45′ 19″ .8
Orbital eccentricity: 0.011
Diameter (km): equatorial 49 528; polar 48 682
Apparent diameter from Earth: max. 2′ .2, min. 2″ .0
Reciprocal mass, Sun = 1: 1300
Mass, Earth = 1: 17.15
Volume, Earth = 1: 57
Escape velocity: 23.5 km s^{-1}
Surface gravity, Earth = 1: 1.2
Mean density, water = 1: 1.64
Mean surface temperature: −220 °C
Oblateness: 0.017
Albedo: 0.35
Maximum magnitude: +7.6
Mean diameter of Sun, as seen from Neptune: 1′ 04″

Table 12.2 *Oppositions of Neptune, 2010–2020*

Date	Dec (°)	Dec (′)	Distance from Earth (millions of km)
2010 Aug 20	12	52	4339
2011 Aug 22	12	09	4338
2012 Aug 24	11	26	4336
2013 Aug 27	10	42	4334
2014 Aug 29	9	56	4333
2015 Sept 1	9	10	4331
2016 Sept 2	8	24	4330
2017 Sept 5	7	36	4329
2018 Sept 7	6	48	4328
2019 Sept 10	5	59	4328
2020 Sept 11	5	10	4327

The oppsition magnitude of Neptune (in Capricornus) is +7.2; diameter 2″ .5.

Neptune was at perihelion on 28 August 1876, and will be again on 5 September 2042. Aphelion was reached on 13 July 1859.

the stars which came into view. Within minutes Galle described an eighth-magnitude star at RA 22h 53m 25s. 84. D'Arrest called out. 'That star is not on the map!'

Encke joined them in the dome, and they followed the object until it set. Next night they found that it had moved by the expected amount, and Encke wrote to Le Verrier; 'The planet whose position you have pointed out actually exists.' Le Verrier's predicted position was in error by only 55 arcmin.

It was also notable that the planet showed a definite disc; Le Verrier had predicted an apparent diameter of 3″ .3, and Encke's first measures made it 3″ .2, which is close to the correct value.

Subsequently, Challis found that he had observed the planet twice, soon after beginning his search, on 30 July and 4 August, and again on 12 August. Had Challis checked his records, he could not have failed to identify the planet, and on 12 August he even suspected a 'star' which showed a disk – yet he did not examine the suspect object with a higher magnification.[1] Following the announcement the new planet was soon seen by J. R. Hind from London, using a 17.7-cm refractor, and by the well-known amateur, William Lassell, who had set up a 61-cm reflector in Liverpool.

The first announcement of Adams' independent work (almost as accurate as Le Verrier's) was made on 3 October 1846 by Sir John Herschel, in the *Athenæum*. The announcement caused deep resentment in France, and led to acrimonious disputes, but in these Adams and Le Verrier took no part, and when they met they struck up an immediate and lasting friendship – even though Adams could not speak French and Le Verrier was equally unversed in English!

It is often said that Le Verrier and Adams should be recognised as co-discoverers of the planet, but this is not strictly correct; the true discoverers of Neptune were Johann Galle and Heinrich D'Arrest. Note also that Adams was in no way involved in the discovery, because Galle and D'Arrest knew nothing about his work.

Adams made no personal search for the planet; he was not equipped to do so. The story that he wrote to Lassell, asking for assistance and that Lassell failed to respond to because he was hors de combat with a sprained ankle, is certainly untrue. Lassell did, however, discover Neptune's main satellite, Triton, on 10 October a few weeks after Neptune itself had been found. On 14 October 1846, Lassell also reported a ring, but later observations showed that this was an optical effect, as Lassell himself later realised. His spurious ring has no connection with the ring system, which could not possibly have been detected with any telescope available at the time (in fact, even today it has never been seen with a ground-based instrument).

NAMING

A name had to be found, and the mythological tradition was followed. Challis suggested 'Occanus'; Galle, in a letter to Le Verrier, preferred 'Janus', while Verrier's own choice was 'Neptune'. Le Verrier then changed his mind, and decided that the planet should be named after himself, but this was not well received, and before long 'Neptune' was universally accepted.

PHYSICAL CHARACTERISTICS

Neptune is a twin of Uranus – but it is a non-identical twin. It is very slightly smaller, but appreciably denser and more massive. Unlike Uranus, it has an internal heat source; it sends out 2.6 times more energy than it would do if it depended entirely upon what it

Table 12.3 *Close planetary conjunctions involving Neptune, 1900–2100*

	Date	UT	Separation (″)	Elongation (°)
Mercury–Neptune	1914 Aug 10	08.11	−30	18 W
Venus–Neptune	2022 Apr 27	19.21	−26	43 W
Venus–Neptune	2023 Feb 15	12.35	−42	28 E
Mercury–Neptune	2039 May 5	09.42	−39	13 W
Mercury–Neptune	2050 June 4	06.46	−46	17 W
Mercury–Neptune	2067 July 15	12.04	+13	18 W

There was a triple conjunction of Neptune and Uranus in September 1993, but the separation was never less than 1° 08′. The previous triple conjunction of Neptune and Uranus was in 1821; the next will be in 2164.

receives from the Sun. It does not share in Uranus' unusual axial tilt; at the time of the Voyager 2 pass it was Neptune's south pole which was in sunlight. The rotation period of just over 16 h is slightly shorter than that of Uranus.

VOYAGER 2 AT NEPTUNE

Our first detailed information about Neptune was obtained from Voyager 2, in 1989; little can be seen on the planet's disc with Earth-based telescopes, and of course the Hubble Space Telescope was not launched until after the Voyager pass.

Voyager 2 had already encountered Jupiter, Saturn and Uranus. After the Saturn encounter, scientists at the Jet Propulsion Laboratory estimated that the chances of success at Uranus were about 60%, but no more than 40% at Neptune; in the event, both encounters were wellnigh faultless, and a tremendous amount of information was obtained.

At 06.50 GMT on 25 August 1989, Voyager 2 passed over the darkened north pole of Neptune, at a minimum relative velocity of 17.1 km s^{-1}. At that time the space-craft was 29 240 km from the centre of Neptune, no more than 5000 km above the cloud-tops, and 4 425 000 000 km from Earth, so that the 'light-time' was 4 h 6 min. The encounter was unlike anything previously experienced, because everything happened so quickly. The main picture-taking period was compressed into less than 12 h.

Before the closest approach over the north pole, many of the main discoveries had been made. The ring system was surveyed, and the magnetic field studied; lightning was detected from the charged particles (discharges recorded by the plasma wave equipment); auroræ were confirmed; surface features had been imaged, and six new satellites had been found. Neptune proved to be a dynamic world, very different from the bland Uranus.

After the polar pass, Voyager 2 passed through the plane of the rings, at a relative velocity of over 76 000 km h^{-1}. Impacts by ring particles were recorded 40 min before the crossing, and reached a peak of 300 per second for 10–15 min either side of the actual crossing; the Voyager team members were very relieved when the spacecraft emerged unscathed, just over 5 h later. Voyager 2 passed Triton at a range of 38 000 km, and sent back images that were as fascinating as they were unexpected. One last picture was taken,

Table 12.4 *Stars to be by-passed by Voyager 2 during the next 300 000 years*

Year (AD)	Star	Distance from Voyager to star (light-years)
8 600	Barnard's Star	4.0
20 300	Proxima Centauri	3.2
20 600	Alpha Centauri	3.5
23 200	Lalande 21185	4.7
40 100	Ross 248	1.7
44 500	DM −36°13940	5.6
46 300	AC +79°3888	2.8
129 000	Ross 154	5.8
129 700	DM +15°3364	3.5
296 000	Sirius	4.3

showing Neptune and Triton together as crescents, and then Voyager began a never-ending journey out of the Solar System. Contact was still maintained in 2010, and probably signals will be received for several years before Voyager finally passes out of range. Its fate will never be known: it carries a plaque and recordings in case any alien civilisation finds it, but the chances do not seem to be very high. Calculations indicate that it will by-pass several stars during the next 300 000 years (Table 12.4), but obviously all this is uncertain, and Voyager 2 may eventually be destroyed by a collision with some wandering body.

Since 1994, regular observations of Neptune have been made with the Hubble Space Telescope, and these have been of great value, but it is true to say that the bulk of our present knowledge of Neptune has come from that single encounter with Voyager 2.

INTERIOR OF NEPTUNE

The four major planets are often called 'gas giants', but this is strictly true only for Jupiter and Saturn: Uranus and Neptune are better called 'ice giants', since their total mass of hydrogen and helium is no more than about two Earth masses, as against 300 Earth masses for Jupiter. The interiors of Uranus and Neptune are dominated by 'ices', primarily water: the main difference between the two is that Neptune has a marked internal heat source, which Uranus lacks. It seems that Neptune has a core made up of iron, nickel and silicates, at a temperature of at least 5000 °C; the core mass is thought to be around 1.2 times that of the Earth, and the pressure is very great. The core may or may not have a well-defined boundary, but it probably extends out to around one-fifth of the planet's radius. Surrounding it is the liquid mantle, rich in water, methane, ammonia and other ices; the total mass of the mantle is between 10 and 15 times that of the Earth, and the temperature is high, so that the mantle is sometimes called a 'water–ammonia ocean'. It has also been suggested that deep down, 7000 to 8000 km below the surface, the tremendous pressure causes the methane to decompose into diamond crystals, but all this is speculative, and we must admit that our knowledge of Neptune's interior is very limited.

ATMOSPHERE

Predictably, Neptune's atmosphere consists mainly of hydrogen; it seems that the upper atmosphere is made up of 84% molecular hydrogen (H_2), 14% helium (He) and 2% methane (CH_4). Methane absorbs light of relatively long wavelength (orange and red), which is why Neptune looks blue. There are trace amounts of carbon monoxide (CO) and hydrogen cyanide (HCN), with even smaller amounts of acetylene (C_2H_2) and ethane (C_2H_6). The unusual distribution of carbon monoxide in Neptune's atmosphere, noted by German astronomers at the Max Planck Institute in 2010, led them to suggest that about 200 years ago the planet was hit by a comet. This is, however, highly speculative.

There are various cloud layers. At a level where the pressure is 3.3 bar there is a layer which seems to be made up of hydrogen sulphide, above which come layers of hydrocarbons, with a methane layer and upper methane haze. Above the hydrogen sulphide layer there are discrete clouds with diameters of the order of 100 km; these cast shadows on the cloud deck 50–75 km below. We are, in fact, dealing with clouds which may be described as methane cirrus.

Apparently, there is a definite cycle of events. First, solar ultraviolet destroys methane high in Neptune's atmosphere by converting it to other hydrocarbons such as acetylene and ethane. These hydrocarbons sink to the lower stratosphere, where they evaporate and then condense. The hydrocarbon ice particles fall into the warmer troposphere, where they evaporate and are converted back to methane. Buoyant, convective methane clouds then rise up to the base of the stratosphere or higher, thereby returning methane vapour to the stratosphere and preventing any net methane loss. In the troposphere there are variable amounts of hydrogen sulphide, methane and ammonia, all of which are involved in the creation of the cloud layers and associated photochemical processes.

Temperature measurements show that there is a cold mid-latitude region, with a warmer equator and pole (we know very little about the north pole, which was in darkness during the Voyager 2 pass). Note, *en passant*, that the general temperature is much the same as that of Uranus; at the equator it is –226 °C, as against –214 °C, even though Neptune is so much further from the Sun. The planet's internal heat source more or less compensates for the increased distance.

There are violent winds on Neptune; most of them blow in a westerly direction, or retrograde (that is to say, opposite to the planet's rotation). The winds differ from those of Jupiter and Saturn, and are distinctively zonal. At the equator they blow westward at up to 450 m s^{-1}. Further south they slacken, and beyond latitude –50° they become eastward (prograde) at up to 300 m s^{-1}, decreasing once more near the south pole. In fact, a broad equatorial retrograde jet extends from approximately latitude +45° to –50°, with a relatively narrow prograde jet at around latitude –70°. Neptune is the 'windiest' planet in the Solar System; as the heat budget is only one-twentieth of that at Jupiter, it seems that the winds are so strong because of the relative lack of turbulence.

SPOTS AND CLOUDS

At the time of the Voyager 2 encounter, the most conspicuous feature on Neptune was a huge oval, the Great Dark Spot, with a longer axis of about 10 000 km; it lay at latitude –22° and its size, relative to Neptune, was about the same as that of the Great Red Spot relative to Jupiter. It had a rotation period of 18.296 h, and drifted westward at about 325 m s^{-1} relative to the adjacent clouds; it was about 10% darker than its surroundings, while the nearby material was about 10% brighter – indicative of the altitude difference between the two regions; it was a high-pressure area, rotating counter-clockwise and showing all the characteristics of an atmospheric vortex. Hanging above it were bright cirrus-type clouds, made up of methane ice; between the cirrus and the main cloud deck there was a clear region about 50 km deep. The cirrus changed shape quite quickly, and in some cases there were shadows, cast on the cloud deck beneath – a phenomenon not observed on Jupiter or Saturn, and certainly not on Uranus.

The Great Dark Spot seemed to be migrating poleward at a rate of 15° per year, and was tacitly assumed to be a feature at least as semi-permanent as Jupiter's Red Spot. However, this proved not to be the case. By 1994, observations with the Hubble Space Telescope showed that it had disappeared, and there is no reason to expect that it will return, although we cannot be sure (after all, Jupiter's Great Red Spot vanishes sometimes, but only temporarily).

Voyager 2 also showed a smaller, very variable feature at latitude –42° with a rotation period of 15.97 h; it was nicknamed the 'Scooter', and had a bright centre. Every few revolutions it caught up with the Great Dark Spot and passed it. Still further south, at latitude –54°, there was another dark spot, D2, which also had a bright core and showed small-scale internal details which changed markedly over periods of a few hours. These too have vanished, but in June 1994 Hubble images showed a bright cloud band near latitude +30° which apparently did not exist in 1989; subsequently it faded, while two bright irregular patches appeared in the southern hemisphere. Evidently, Neptune's surface features are much more variable than had been expected.

MAGNETIC FIELD

It had been assumed that Neptune must have a magnetic field. As Voyager 2 drew inward, radio emissions were detected, but there was a delay before Voyager passed through the bow shock (that is to say, the region where the solar wind is heated and deflected by interaction with Neptune's magnetosphere). The bow shock was finally recorded at 879 000 km from the planet. The magnetic field itself was weaker than those of the other giants; the field strength at the surface is 1.2 G in the southern hemisphere but only 0.06 G in the northern. The real surprise was that the inclination of the magnetic axis relative to the axis of rotation is 47°, so that in this respect Neptune is not unlike Uranus; moreover the magnetic axis does not pass through the centre of the globe, but is displaced by 10 000 km or 0.4 Neptune radii. This indicates that the dynamo electric currents must be closer to the surface than to the centre of the globe. The magnetosphere has been described as comparatively 'empty': it gyrates dramatically as the planet rotates, and the satellites are involved. Auroræ were found, but instead of being near the rotational poles they were closer to the equator – because the rotational and the magnetic poles are so far apart. At this time, of course, the northern hemisphere was experiencing its long winter. Neptunian auroræ are considerably weaker than those of the other giants.

NEPTUNE'S RINGS

Neptune has an obscure ring system. Indications of ring material were obtained in pre-Voyager times by the occultation method, which had been used to detect the rings of Uranus: for example, on 24 May 1981, observers from Arizona observed the close approach of Neptune to a star, and found that there was a very brief drop in the star's brightness. Other similar observations led to the suggestion that there might be incomplete rings – that is to say 'ring arcs', although nobody was quite sure how or why such arcs should develop. The mystery was solved by observations from Voyager 2. There are complete rings, but the main ring is 'clumpy'. Data are given in Table 12.5; the rings have been named after astronomers involved in the discovery of Neptune (although up to now D'Arrest has been quite unfairly ignored). François Arago is included, as it was he who first drew Le Verrier's attention to the problem of the movements of Uranus.

The outer Adams Ring is the most pronounced, and contains the three 'clumps' which led to the idea of ring arcs: these 'clumps' have been named Liberté, Egalité and Fraternité, for reasons which are, at best, obscure. The 'clumps' are not evenly distributed, but are grouped together over about one-tenth of the ring circumference: their lengths range from 5000 to 10 000 km. They may be due to the effects of the small satellite Galatea, which orbits very close to the rings. It has also been suggested that the cause may be small satellites embedded in the Adams Ring, but so far these have not been seen; they may or may not exist. During the Voyager 2 pass, the Adams Ring occulted the second-magnitude star Nunki (Sigma Sagittarii), and proved to have a core only 17 km wide; the ring material is reddish, with a very low albedo. The Arago Ring is dim

Table 12.5 *The rings of Neptune*

Name	Distance from centre of Neptune (km)	Width (km)
Galle	41 900–49 000	1700
Le Verrier	53 200	50
'Plateau' (Lassell)	53 200–59 100	4000
Arago	57 600	30
Adams	62 900	50

and narrow. The Le Verrier Ring is narrow and tenuous; it is not far from the orbit of the small satellite Despina, but shows no arcs. The Plateau is a diffuse band of material, containing a high percentage of very small particles. The inner Galle ring is much broader than the Adams or Le Verrier rings: there may be 'dust' extending down almost to the cloud-tops. Unlike the Uranian rings, those of Neptune are brighter in forward-scattered light than in back-scattered light, so that they contain a larger fraction of very small particles than do the much narrower rings of Uranus. Even from Voyager, Neptune's rings were not easy to identify: they are blacker than soot! Possibly they are young, formed from the débris of small satellites.

SATELLITES

Neptune has 13 known satellites. Only one of these, Triton, is large; it is an easy telescopic object – easier than any of the satellites of Uranus – and was discovered soon after the identification of Neptune itself. Nereid, the other satellite known before the Space Age, was found by Gerard Kuiper, in 1949. The next six were all discovered on Voyager 2 images, though one of them, Larissa, had been fortuitously recorded on 24 May 1981, when it occulted a star; the observation was made by a team led by J. Reitsema, and this can surely be regarded as a discovery even though its significance was not appreciated at the time. The five small outer satellites were identified in 2002–3. Apart from Triton, all Neptune's satellites are well below the eighteenth magnitude. All are named after marine deities (see Table 12.6).

TRITON

When John Herschel heard the news about the discovery of Neptune by Galle and D'Arrest, he wrote to William Lassell and suggested that it might be worth searching for a satellite. Lassell, with his powerful telescope, wasted no time, and on 10 October 1846 he found a satellite. It was named Triton, after the son of Neptune and Amphitrite. The name was proposed (years later) by the French astronomer Camille Flammarion. It was an appropriate choice; Triton is certainly a remarkable child!

In many ways it is unique. It is the only large satellite with retrograde motion, and it has geysers unlike those found anywhere else. No Earth-based telescope can show any surface detail, so that we rely entirely upon the images sent back by Voyager 2. Up to that

Table 12.6 *Satellites of Neptune*

Name	Distance (km)	Period (d)	Inclination angle, i (°)	Eccentricity, e	Diameter (km)	Albedo	Magnitude
III Naiad	48 230	0.294	4.75	0.0004	$96 \times 60 \times 52$	0.06	25
IV Thalassa	50 075	0.311	0.21	0.0002	$108 \times 100 \times 52$	0.09	24
V Despina	52 526	0.334	0.07	0.0002	$180 \times 150 \times 130$	0.09	23
VI Galatea	61 950	0.429	0.05	0.00004	$204 \times 184 \times 144$	0.08	23
VII Larissa	73 550	0.555	0.20	0.0001	$216 \times 204 \times 164$	0.06	21
VIII Proteus	117 650	1.223	0.52	0.0005	$440 \times 416 \times 404$	0.10	20
I Triton	254 800	5.877	157.3	0.000016	2706	0.76	13.6
II Nereid	5 513 000	360.1	27.6	0.7512	340	0.16	18.7
IX Halimede	15 728 000	1880	134	0.5711	62	0.04	
XI Sao	20 420 000	2914	48.5	0.293	40	016	
XII Laomedia	23 571 000	3168	37.7	0.381	38	0.04	
X Psamathe	46 695 000	9116	137.4	0.4499	38	0.04	
XIII Neso	48 387 000	9374	132	0.4945	60	0.04	

Triton, Halimede, Psamathe and Neso have retrograde motion, the rest are prograde.

time, Triton was believed to be large – possibly larger than Mercury – perhaps with chemical oceans on its surface, but this was found to be completely wrong. Triton is much smaller than had been expected, so that in size it ranks below all four of Jupiter's Galilean satellites, Titan in Saturn's system, and our own Moon; it is also fairly reflective, with an albedo which in places rises to 0.8. It is very cold, with a temperature of –235 °C, a mere 38 K above absolute zero. The mass is 2.4×10 kg, and the density is fairly high (2.07 times that of water), so that the globe seems to be made up of a mixture of rock (two-thirds) and ice (one third). The rotation period is of course synchronous, so that the same hemisphere always faces Neptune; the rotational axis lies in the plane of Neptune's orbit, so that during each Neptunian year the poles of Triton take turns in pointing almost exactly sunward. The long 'seasons' must affect the surface features, as first one pole and then the other enjoys sunlight The orbit is almost perfectly circular – eccentricity 0.000 016 – and the escape velocity is 1.455 km s^{-1}, lower than that of the Moon, but the low temperature means that Triton can retain an extensive though very tenuous atmosphere. The ground atmospheric pressure is of the order of 14 μbar, around 1/70 000 the pressure of the Earth's air at sea level. The main constituent is nitrogen in the form of N_2, which accounts for 99%; most of the rest is methane, with a trace of carbon monoxide. There is considerable haze, seen by Voyager 2 above the surface, probably composed of microscopic ice crystals of methane or nitrogen. Winds in the atmosphere average around 5 m s^{-1} westward, though naturally they have very little force. There is a pronounced temperature inversion, since the temperature in the atmosphere rises to –173 °C at a height of 600 km. This inversion occurs at a surprisingly high altitude, for reasons which are unclear. Turbulence at Triton's surface creates what may be called a troposphere, rising to a height of about 8 km; clouds of condensed nitrogen ice particles are found at between 1 and 3 km above the surface.

In 1998, measurements with the Hubble Space Telescope found that the surface temperature had increased by a factor of about two since the Voyager pass nine years earlier, and that the atmospheric density had shown a slight but definite increase. This is certainly a seasonal effect, caused by changes in the polar cap.

The surface of Triton is very varied, and there is very marked evidence of past cryovulcanism (that is to say, icy vulcanism). There is a general coating of ice, presumably water ice overlaid by nitrogen and methane ices; water ice has not been detected spectroscopically, but it must exist, because nitrogen and methane ices are not hard enough to maintain surface relief over long periods. Not that there is much surface relief on Triton; there are no mountains or deep valleys, and the total surface relief cannot amount to more than a few hundred metres. Normal craters are scarce, and even the largest of them is no more than 27 km in diameter. Extensive flows are seen, some of them up to 80 km wide, due probably to ammonia water fluids.

The most striking feature is the southern polar cap, which is covered with pink nitrogen snow and ice. The areas surveyed by Voyager 2 have been divided into three main regions: Bubembe Region (western equatorial), Monad Regio (eastern equatorial) and Uhlanga Regio (polar). These names, and those of other features, have been allotted by the Nomenclature Committee of the International Astronomical Union. All the features have aquatic names, excluding Greek and Roman deities. Thus Bubembe is the island location of the temple of Mukasa (Uganda), Monad is a Chinese symbol of duality in nature and Uhlanga is a Zulu reed from which mankind sprang!

Uhlanga is covered with the pink cap, with some of the underlying geological units showing through in places. It is here that we find the nitrogen ice geysers, which were so totally unexpected. Apparently there is a layer of liquid nitrogen 20 to 30 m below the surface; here the pressure is high enough for nitrogen to remain liquid, but if for any reason it migrates upward it will come to a region where the pressure is about one tenth that of the Earth's air at sea level. The nitrogen will then explode in a shower of ice and vapour (about 80% ice, 20% vapour) and will travel up the nozzle of the geyser-like vent at a rate of up to 150 m s^{-1}, which is fast enough to make it rise to several kilometres before falling back. The outrush sweeps dark débris along it, producing plumes such as Mahilani and Hili. The Mahilani plume is a narrow, straight cloud 90–150 km long, while Hili is a cluster of several plumes with a length of about 100 km. The ways in which the plume clouds move indicates the force and direction of the winds in the thin Tritonian atmosphere. The wind speeds here can reach 20 m s^{-1}: the direction is northeast near the surface, east at intermediate altitude and westward at the top of the troposphere, 8 km in altitude.

The edge of the cap, separating Uhlanga from Monad and Bubembe, is sharp and convoluted. The long seasons mean that the southern pole has been in constant sunlight for over a century now, and along the cap borders there are signs of evaporation; eventually the cap may appear to shift northward, so that the entire aspect of Triton may change according to the seasons – and, as we have noted, this also affects the surface temperature and the atmospheric density. In reality, it is not the cap shifting northward but evaporation/sublimation products moving northward in the atmosphere, where at the colder northern regions they are re-deposited to form the raw materials, for the process here to be repeated during the northern summer. North of the cap there is a darker, redder region; the colour may be due to the action of solar ultraviolet upon the methane. Running across this region, more or less parallel to the edge of the cap, is a slightly bluish layer, due possibly to tiny crystals of methane ice scattering the incoming sunlight.

Monad Regio is part smooth, with knobbly or hummocky terrain; there are rimless pits (pateræ), with graben-like troughs (fossæ) and strange, mushroom-like features (guttæ) such as Zin and Akupara, whose origin is unclear. There are four walled plains or 'lakes' (Ruach, Ryugu, Sipapu, Tuonela) which have flat floors and are bounded by rougher plains units; they have been likened to calderæ. They are edged with terraces, as though the original level has been changed several times by repeated melting and re-freezing. Undulating smooth plains cover most of the higher part of Monad.

Bubembe Regio is characterised by the so-called cantaloupe terrain – a nickname given to it because of its superficial resemblance to a melon skin! Fissures cross it, meeting in elevated X or Y junctions. Liquid material, presumably a mixture of water and ammonia, seems to have forced its way up some of these fissures, so that there are central ridges; material has even flowed out on to the plain – before freezing there. The overall pattern is a network of closely spaced depressions (cavi), 25–35 km across; they do not overlap. The cantaloupe areas are probably the oldest parts of Triton's surface.

Everything indicates that Triton was not originally a satellite of Neptune, but was captured well after its formation. Initially, its orbit round Neptune would have been eccentric; over a period of perhaps 1000 million years it was forced into the present circular form, and during this time there was great internal heating, coupled with surface activity. There may also have been a dense atmosphere. A selected list of surface features is given in Table 12.7. It is mapped in Figure 12.1.

NEREID

Following the discovery of Triton, searches were made for the other satellites of Neptune. One was suspected by Lassell in 1852 and another by J. Schaeberle in 1892, using the 91-cm Lick refractor, but neither was confirmed, and a search by W. H. M. Christie, using the 152-cm reflector at Mount Wilson, was negative. Success then finally came in 1949, when G. P. Kuiper, using the 208-cm reflector at the McDonald Observatory in Texas, found a new faint satellite. It was named Nereid – and this was appropriate: the Nereids were the 50 daughters of Nereus, the sea god, and Doris.

Nereid is exceptional inasmuch as its orbit is highly eccentric, and more like that of a comet than an asteroid. Its distance from Neptune ranges between 1 372 000 km and 5 513 787 km, and it takes nearly a year (360.14 days) to complete orbit. Unfortunately, it was not well seen by Voyager, and the only reasonable image was obtained from a range of 4 500 000 km, so that little surface detail could be made out. It seems to be fairly regular in shape; it is more reflective than the small inner satellites, and is grey in colour, so that the surface may well be covered with rock or 'dirty ice'. No doubt there are craters, but for further information we must await the launch of a new space-craft to the Neptunian system – and this may well be delayed for a considerable time. Whether Nereid is a true satellite or whether it was captured is not known. It is unlikely that the rotation period is synchronous – one estimate, derived from variations in magnitude, gives 13.6 h, but it is quite probably that the period, like that of Hyperion in Saturn's system, is chaotic. To an observer on Neptune, the average apparent diameter of Nereid would be only about 19 arcsec.

MINOR SATELLITES

III Naiad
This was discovered in September 1989 by the Voyager imaging team, and named after a group of Greek water nymphs who were guardians of lakes, fountains, springs and rivers. The satellite moves only 23 500 km above the cloud-tops, and since this is below the synchronous orbit distance it may eventually enter Neptune's atmosphere or else break up to form a ring. Its low density indicates that it could well be regarded as a rubble pile.

IV Thalassa
This was iscovered in 1989 by R. Terrile and the Voyager imaging team; it is named after a daughter of Hemara and Alither in Greek mythology. Like Naiad, it seems to be an irregularly shaped rubble pile.

V Despina
Despina was discovered in 1989 by S. P. Synnott and the Voyager imaging team; it is named after a nymph who was the daughter of Poseidon and Demeter (Greek mythology). It is another irregularly shaped satellite, moving in an orbit just closer to Neptune then the Le Verrier Ring; it may have some effect on the ring. In itself it seems to be similar to Naiad and Thalassa, but is rather larger.

VI Galatea

This satellite was discovered in 1989 by S. P. Synnott and the Voyager imaging team; it named after one of the Nereids. It seems to be similar to the other small inner satellites; it acts as a shepherd for the Adams Ring, which is 1000 km further out. It is almost certainly responsible for the arcs in the ring.

Larissa

Larissa was first recorded on 24 May 1981, when it occulted a star, by a team led by J. Retisima but was not then recorded until the Voyager pass in 1989. It was named after one of the numerous lovers of Poseidon (Neptune). It is larger than the first four inner satellites, and on 24 August 1989 Voyager 2 obtained an image showing that the surface is thickly cratered.

VIII Proteus

This was discovered in 1989 by S. P. Synnott and the Voyager imaging team, and named after a sea deity in Greek mythology.

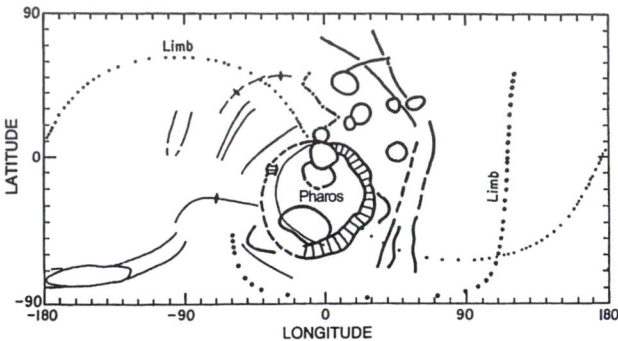

Figure 12.1 Proteus

With a mean diameter of 418 km, it is the largest of the minor satellites, but was not discovered until forty years after the smaller Nereid because it is so close to Neptune. Its density has been estimated as 1.3 times that of water, and is about as large as a body of this kind can be without assuming a spherical shape; Proteus is triaxial. The albedo is very low – about the same as that of soot. The rotation period is synchronous, 1 day 2.9 hours,

A good image was obtained by Voyager 2 on 25 August 1989. The surface shows linear streaks, which seem to be troughs from 25 to 35 km wide and several kilometres deep; they form a global network, and may be classed as graben of extensional origin. There are no cryovolcanic structures or coronæ.

The main surface feature of Proteus is a depression, Pharos (originally known as the Southern Hemisphere Depression), which dominates the southern part of the Neptune-facing hemisphere. It is basically circular, with a raised rim and a flat, rugged floor; it is 225 km in diameter, and from 10 to 15 km deep; see Figure 12.2.

IX Halimede

This was discovered on 14 August 2002 by M. Holman, J. Kavelaars, T. Grav, W. Fraser and D. Milisavljevic, and named after one of the Nereids. It has retrograde motion, and its orbit is decidedly eccentric; it is grey in colour, and may possibly be a fragment of Nereid. Its revolution period is over 5 years.

XI Sao

Sao was discovered by the Holman team, with Halimede. The name is that of a Nereid particularly associated with sailing. It has direct motion, and a period of nearly 8 years.

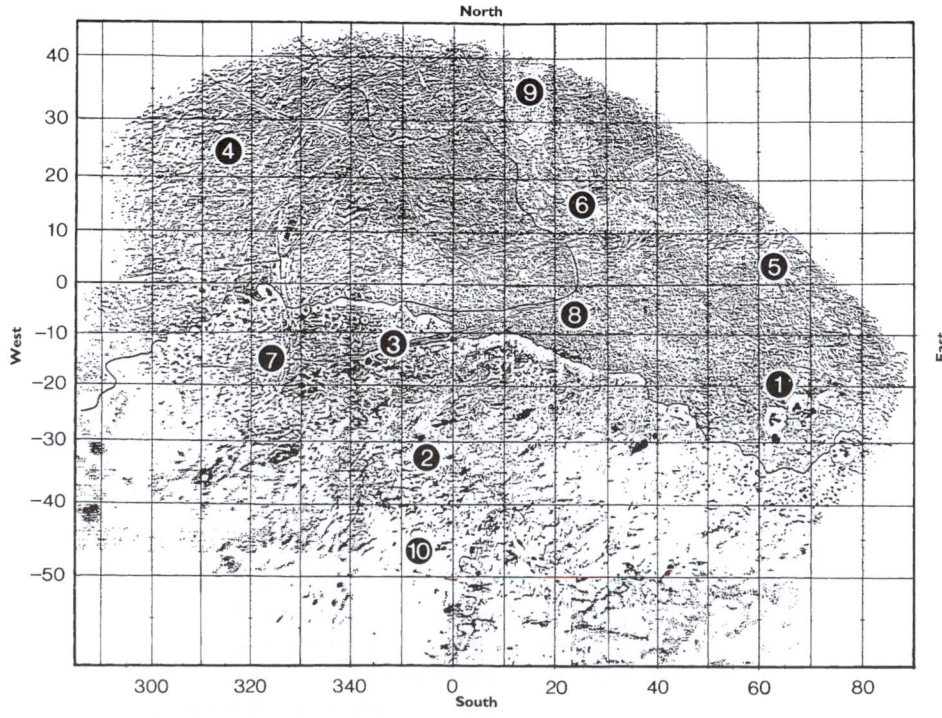

Figure 12.2 Triton

Table 12.7 *Features on Triton (bold numbers indicate map references)*

	Lat. (°)	Long. (°)	
Catenæ			
Kraken	14.0	35.5	Norse giant sea-monster
Set	22.0	33.5	Evil Egyptian water monster
Cavi			
Apep	20.0	301.5	Egyptian dragon of darkness
Bheki	16.0	308.0	Indian solar frog
Dagon	29.0	345.0	Babylonian fish god
Hekt	26.0	342.0	Egyptian frog goddess
Hirugo	14.5	345.0	Japanese jellyfish god
Kasyapa	7.5	358.0	Indian tortoise god
Kulilu	41.0	4.0	Evil Babylonian fish spirit
Mah	38.0	6.0	Persian fish that holds up the universe
Mangwe	−7.0	343.0	Ila (Zambia)
Ukupanio	35.0	23.0	Hawaiian shark god
Craters			
Amarum	26.0	24.5	Quecha (Ecuadorian water boa)
Andvari	20.5	34.0	Norse dwarf fish
Cay	−12.0	44.0	Mayan deity
Ilomba	−14.5	57.0	Lozi (evil Zambian water-snake)
Kurma	−16.5	61.0	Vishnu in tortoise form
Mazomba	−18.5	63.5	Chaga (large mythical fish)
Ravgga	−3.0	71.5	Finnish fortune-telling fish god
Tangaroa	−25.0	65.5	Maori sea god
Vodyanoy	−17.0	28.5	Slavic water spirit
Dorsa			
Awib	−7.0	80.0	Bushman word for rain
Fossæ			
Jumna	−13.5	44.0	Hindu river goddess
Raz	8.0	21.5	Breton bay of souls
Yenisey	3.0	56.2	Siberian mythical holy river
Maculæ			
Akupara	−27.5	63.0	India; tortoise upholding the world
Doro	−27.5	31.7	Nanay, marine deity (Sea of Okhotsk)
Kikimora	−31.0	78.0	Slavic swamp spirit
Namazu	−25.5	14.0	Japanese earthquake fish god
Rem	13.0	349.5	Egyptian fish god
Viviane	−31.0	36.5	Amour of Merlin (Welsh)
Zin	−24.5	68.0	Niger water spirits
Pateræ			
Dilolo	26.0	24.5	Angolan sacred lake
Gandvik	28.0	5.5	Norse tortuous sea
Kasu	39.0	14.0	Sacred lake of Zoroastrianism
Kibu	10.5	43.0	Mabuiag, island of the dead
Leviathan	17.0	28.5	Hebrew sea-monster upholding Earth
Planitiæ			
Ruach	28.0	24.0	French isle of winds
Ryugu **8**	−5.0	27.0	Japanese undersea dragon palace
Sipapu	−4.0	36.0	Pueblo lake of exit from the Underworld
Tuonela **9**	34.0	14.5	Finnish undergeound realm separation
Plana			
Abatos **1**	−21.5	58.0	Egyptian sacred isle in the Nile
Cipango	11.5	34.0	Mythical island described by Marco Polo
Medamothi **5**	3.5	69.0	'Nowhere'; French fictional island
Plumes			
Hili	−57.0	35.0	Zulu water sprite
Mahilani	−50.5	359.5	Tonga sea spirit
Regiones			
Bubembe **4**	18.0	335.0	Ugandan sacred island
Monad **6**	20.0	37.0	Chinese symbol of duality in Nature
Uhlanga **10**	−37.0	357.0	Zulu reed from which mankind sprang
Sulci			
Bia **2**	−38.0	3.0	Yoruba river
Boynne **3**	−13.0	350.0	Mythological Celtic river
Ho	2.0	305.0	Chinese sacred river
Kormet	23.0	335.6	Norse Underworld river
Leipter	7.0	9.0	Norse sacred river
Lo	3.8	321.0	Chinese sacred river
Ob **7**	−6.0	328.0	Mouth of Underworld river Ostiak
Slidr	23.5	350.0	Norse river of daggers and spears
Tano	33.5	337.0	Yoruba river
Vimur	−11.0	59.0	Elivagar river – stream of ice
Yasu	2.0	347.0	Japanese heavenly river, 'peace'

XII Laomedia

This satellite was discovered by the Holman team on 13 August 2002, and named after a Nereid. It has direct motion, with moderate orbital eccentricity and inclination.

Table 12.8. *Neptune's Trojans*

	Perihelion distance, q (a.u.)	Aphelion distance, Q (a.u.)	Inclination angle, i (°)	Eccentricity, e	Magnitude, H	Diameter (km)	Discoverer
2001 QR 322	29.36	31.25	1.3	0.031	7.0	240.	DES (Deep Ecliptic Survey) team
2004 UP 10	29.35	31.07	1.4	0.028	8.9	100	S. Sheppard, A. Trujillo
2005 TN 53	28.23	32.13	25.0	0.065	9.1	90	S. Sheppard, A. Trujillo
2005 TO 74	28.62	31.76	5.3	0.052	8.5	120	S. Sheppard, A. Trujillo
2006 RJ 103	29.25	30.91	5.2	0.028	7.4	190	SDSS (Sloan Digital Sky Survey) team
2007 VL 305	28.13	31.95	26.1	0.063	7.9	150	SDSS team
2008 LC 18	27.5	32.5	27.53	0.082	8.4	100	S. Sheppard, C. Trujillo

Diameters given here represent the maximum possible values.

X Psamathe

This is a retrograde satellite, discovered by S. Sheppard, D. Jewitt and J. Kleyna in 2003 and named after a Nereid. It and Neso may possibly be fragments of a larger body.

XIII Neso

Neso was discovered on 4 August 2002 by M. Holman and B. Gladman, and named after a Nereid (since Poseidon and Doris had 50 daughters, there were plenty of names available!). Neso has retrograde motion, and a mean distance of over 48 000 000 km from Neptune; the period is 25 years 7 months. Of all known planetary satellites, Neso has the greatest distance from its primary and also the longest period, but it is very likely that further tiny satellites of the giant planets await discovery.

Neptune Trojans

By 2008, six Neptune Trojans had been found, all at the leading Lagrangian point (L4). All are small, two have highly inclined orbits. In 2010, Scott Sheppard and Chadwick Trujillo, using the 8.2-m Japnese Subaru telescope, discovered an asteroid, 2008 LC 18, in the L5 Lagrangian point, 60° behind Neptune; all previous Neptune Trojans had been found at the L4 point, 60° ahead of the planet. The diameter of 2008 LC 18 is probably around 100 km. Like several of the L4 Trojans it has a highly inclined orbit, suggesting that it was captured at an early stage of the Solar System, when Neptune's orbit was very different from that of today. There may well be many more similar bodies, at both the L4 and L5 points. A list of known Neptune Trojans is given in Table 12.8. Observations made in 2010 with the ULT detected carbon monoxide and ethane in the atmosphere of triton. The temperature was given as $-235°$C and the atmosphere pressure 40–65 microbars. During Tritoris summer a thin surface layer of frozen nitrogen, methane and carbon dioxide sublimates, thickening and warming the atmosphere.

ENDNOTE

1 In September 1988, I decided to check Challis' observations, so from the Royal Greenwich Observatory, then at Herstmonceux in Sussex. I looked at Neptune through the telescope used by Challis and with the same magnification ($\times 117$). There was no doubt that Neptune showed a disc, and with a power of $\times 250$ the difference between the planet and a star was glaringly obvious.

13 · Beyond Neptune: the Kuiper Belt

Le Verrier's correct prediction of the position of Neptune was a personal triumph for him, there have been suggestions that there was a large element of luck about it, but it cannot be denied that the actual discovery, by Galle and D'Arrest, was due entirely to Le Verrier.

The discovery of Pluto

Despite the fact that the new planet did not move at the distance required by Bode's law, then generally believed to be significant rather than coincidental, the Solar System could well be regarded as complete. Yet there was no proof of this, and as early as 1846 Le Verrier himself suggested that there might well be a planet moving beyond the orbit of the recently discovered Neptune. The first systematic search was made in 1877 by David Peck Todd, from the US Naval Observatory. From perturbations of Uranus, he predicted a planet at a distance of 52 a.u. from the Sun, with a diameter of 80 000 km. He conducted a visual search, using the 66-cm USNO reflector with powers of ×400 and ×600, hoping to detect an object showing a definite disc. He continued the hunt for 30 clear, moonless nights between 3 November 1877 and 5 March 1878, but with negative results. A second investigation was conducted in 1879 by the French astronomer Camille Flammarion, who based his suggestion upon the fact that several comets appeared to have their aphelia at approximately the same distance, well beyond the orbit of Neptune. In 1900, G. Forbes predicted two planets, at distances of 100 and 500 a.u., respectively, with periods of 1000 and 5000 years: both were assumed to be larger than Jupiter. In 1902, T. Grigull of Minster proposed a planet the size of Uranus, moving at 50 a.u., in a period of 360 years: Grigull even gave it a name – Hades. H. E. Lau, from Denmark (1900), suggested two planets, at 46.6 and 70.6 a.u., with masses 9 and 47 times that of the Earth; G. Gaillot (1901) also believed in two planets, and T. J. J. See in America increased the number to three, naming the innermost planet 'Oceanus'. A. Garnowsky, in Russia, went further and predicted the existence of four planets. Another prediction was made by Gabriel Dalier, who believed in a single planet moving at a distance of 47 a.u. from the Sun.

Another investigator was the Indian astronomer Venkatesh Ketekar, who, in 1911, predicted two planets beyond Pluto. Using rather different mathematical methods, he gave positions for both, and his results were not very different from Lowell's (see below). He gave the period of his inner planet as 2.23 years, and its mean distance as 39 a.u. Not a great deal seems to be known about Ketekar, and there is no record of any search being made on the basis of his calculations.

THE HUNT FOR A TRANS-NEPTUNIAN PLANET

The first systematic, photographic and visual search was made by Percival Lowell, from Flagstaff, from 1905 to 1907. Lowell had worked out a position for his 'Planet X', from perturbations of Neptune and (particularly) Uranus: his planet was believed to have a mass almost seven times that of the Earth, with a period of 282 years and a rather eccentric orbit (0.202). The date of perihelion was given as 1991. A second search at Flagstaff, carried out by C. O. Lampland in 1914, was equally fruitless, and was given up on Lowell's death in 1916, a year after Lowell had published his final orbit for the planet. Also in the hunt was W. H. Pickering, who in 1898 discovered Saturn's eighth satellite, Phœbe. His planet had a mass twice that of the Earth and a period of 248 years, again with an eccentric orbit. His method – unlike Lowell's – was essentially photographical, but his conclusions were the same. On the basis of these results, Milton Humason at Mount Wilson Observatory undertook a photograpic search in 1919, but again the results were negative.

In 1929, V. M. Slipher, by then Director of Lowell Observatory, decided to make a fresh attempt. The search was entrusted to Clyde Tombaugh, using a 13-inch refractor specially acquired for the purpose. (Tombaugh was then a young, unqualified amateur; later he became one of America's most senior and respected astronomers.) Tombaugh used photographic methods, and before long success came; Pluto was detected upon plates taken on 23 and 29 January 1930, although the announcement was delayed until 13 March – 149 years after the discovery of Uranus and 78 years after Lowell's birth.

Examination of the earlier plates showed that Pluto had been recorded twice in 1915, but had been missed because it was unexpectedly faint. However, Humason's failure in 1919 was due to sheer bad luck. When his plates were examined, it was found that Pluto had been recorded twice – but once the image fell upon a flaw in the plate, and on the second occasion Pluto was masked by an inconvenient bright star.

It is interesting to compare predictions with fact:

The first orbit for Pluto issued from Flagstaff gave an eccentricity of 0.909 and a period of 3000 years, but a few months later new observations yielded a better orbit. Pluto was detected near the star δ Geminorum. Slipher's initial announcement stated that 'the object was 7 seconds of time west from δ Geminorum', agreeing with Lowell's predicted longitude.

Pluto	Actual	Planet X (Lowell)	Planet O (Pickering)
Mean distance from Sun (a.u.)	39.5	43.0	55.1
Period (years)	248	282	409
Eccentricity	0.249	0.202	0.248
Inclination (°)	17	10	15
Perihelion date	1989	1991	2129
Mass, Earth = 1	0.002	6.6	2.0

Various names for the new planet were suggested, including Odin, Persephone, Chaos, Atlas, Tempus, Lowell, Minerva, Hercules, Daisy (!), Pax, Newton, Freya, Constance and Tantalus. The final choice was Pluto, suggested by an 11-year-old Oxford schoolgirl, Venetia Burney (later Mrs Phair).

In view of later developments, one point is worth noting here. Pluto had turned up the expected position, and was thought to be comparable in size with Mars or even the Earth. No other bodies were known in those remote regions, and there was no reason to suppose that the new world was anything but a true planet. Yet before long, doubts began to creep in, quite apart from the curiously eccentric and inclined orbit, because Pluto seemed to be far too small and lightweight to exert measurable perturbations upon giants such as Uranus and Neptune. (Uranus had been used for the main investigations, because its orbit was very well known, whereas Neptune had not completed a full revolution since its identification in 1846; in fact, it has not done so even yet.) In 1936, A. C. D. Crommelin of Greenwich suggested the theory of specular reflection. If Pluto were highly reflective, the bright image of the Sun might falsify the diameter estimates, so that Pluto could be much larger – and hence more massive – than it seemed. This proved to be incorrect. In 1949, G. P. Kuiper, using the 82-inch (208-cm) reflector at the McDonald Observatory in Texas, gave the diameter as 10 200 km, with a mass eight-tenths of that of the Earth, but even this proved to be far too great. In 1950, Kuiper and Humason made new measurements with the Hale reflector at Palomar, reducing the diameter to 5800 km – smaller than Mars. In 1965, a partial occultation of a star by Pluto showed that the diameter could not be more than 5800 km, and the current value is only 2306 km. In fact, Pluto could not have been predicted by any perturbations upon Uranus or Neptune, and Lowell's accurate prediction was sheer luck. This is hard to believe, but there seems to be no alternative.

In 1936, R. A. Lyttleton, of Cambridge, suggested that Pluto might be an escaped satellite of Neptune. This idea was supported in 1956 by Kuiper, but was rejected in 1978, when Pluto was found to have a satellite, now named Charon. Data on Pluto and Charon are given in Tables 13.1 and 13.2, and Figure 13.1 (note the 2002/2003 map is rather less uncertain than the 1994 one).

DISCOVERY OF CHARON

The discovery was made on 22 June 1978 by James W. Christy, from the US Naval Observatory at Flagstaff. Pluto had been

Table 13.1 *134340 Pluto: data*

Discovery: C. W. Tombaugh, 23 and 29 January 1930 (announced 13 March)
Distance from the Sun: max. 7 375 927 931 km (49.305 a.u.)
    mean 5 906 376 272 km (39.481 a.u.)
    min. 4 436 824 613 km (29.66 a.u.)
Sidereal period: 248.09 years (90 613.3 days)
Synodic period: 366.73 days
Rotation period: 6d 9h 17m 36s (6.387 days)
Mean orbital velocity: 4.66 km s^{-1}
Axial inclination: 119.591°
Orbital inclination: 17.141°
Oribital eccentricity: 0.249
Diameter: 2390 km
Density, water = 1:   2.03
Volume, Earth = 1   0.007
Mass, Earth = 1:   0.002
Recipocal mass, Sun = 1:   135 500.00
Mean surface temperature: −233 °C
Escape velocity: 1.2 km s^{-1}
Surface gravity, Earth = 1:   0.06
Albedo: opposition   0.55 (0.49 to 0.66)
Magnitude at perihelion: 13.6
Angular diameter: 0.065 to 0.115″
Distance from Earth: max. 7 533 300 000 km
min. 4 293 700 000 km

Table 13.2 *Charon: data*

Discovery: J. W. Christy, 22 June 1978
Distance from Pluto: 19 571 km (centre to centre)
Orbital period: synchronous
Orbital eccentricity: 0.0022
Orbital inclination to Pluto's equator: 0.001°
Axial inclination: 0°
Mean orbital velocity: 0.22 km s^{-1}
Diameter: 1207 km
Mass, Pluto = 1: 0.12
Density, water = 1: 1.6
Escape velocity: 0.58 km s^{-1}
Equatorial surface gravity: 0.278 m s^{-1}
Albedo: 0.37
Surface temperature: -220 °C
Magnitude at perihelic opposition: 16.8
Apparent magnitude, seen from Pluto: -9
Elongation from Pluto, seen from Earth: 0″.6

repeatedly photographed with the 1.55-m reflector with a view to predicting occultations of stars by the planet, which could be used to give an improved value for Pluto's diameter. Images taken in April and May showed that Pluto's image seemed to be elongated.

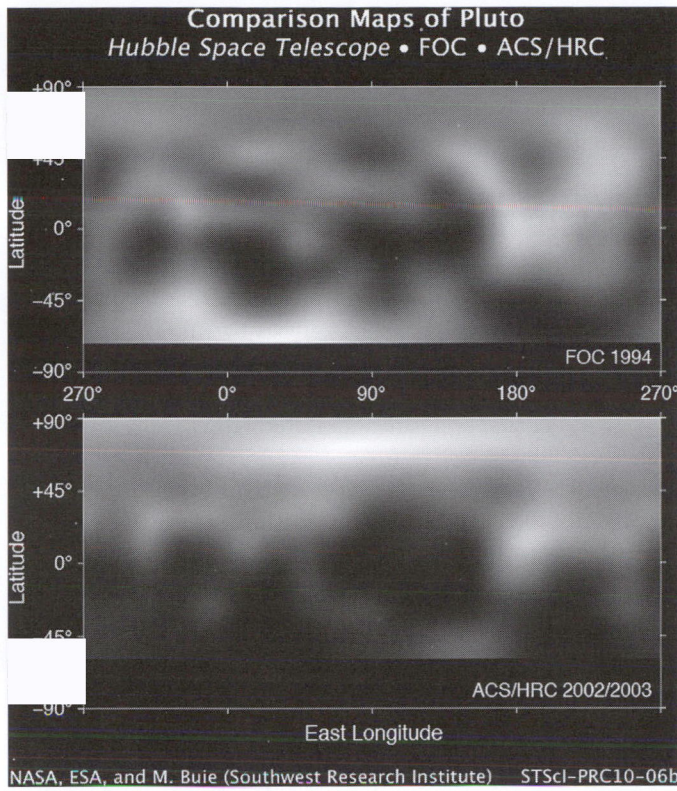

Comparison Maps of Pluto
Hubble Space Telescope • FOC • ACS/HRC

FOC 1994

ACS/HRC 2002/2003

East Longitude

NASA, ESA, and M. Buie (Southwest Research Institute)   STScI-PRC10-06b

Figure 13.1. Pluto

Table 13.3 *Minor satellites of Pluto*

Diameter (km)	Semi-major axis (km)	Period (d)	Eccentricity, e	Inclination to Pluto's orbit (°)
Nix 46	48 708	24.86	0.0030	0.195
Hydra 61	61 167	38.21	0.0051	0.212

From one hemisphere of Pluto, of course, Charon will never be seen at all.

Charon has only about one-seventh the mass of Pluto, and Pluto contributes 80% of the total light we receive from the system, partly because it is larger and also because it has a higher albedo. Charon is 'grey', whereas Pluto is reddish. Quite possibly, the two were once combined, and were separated following a major impact in the early history of the Solar System, as was suggested by R. Canup in 2005, but it is now more generally believed that Pluto and Charon may have been two separate objects which collided before going into orbit round each other.

Nix and Hydra

There are two minor satellites, Nix and Hydra, discovered in 2005 on Hubble Space Telescope images following careful examination by a special research team. They are unquestionably captured objects, no doubt from the Kuiper Belt itself (see Table 13.3).

INTERNAL STRUCTURE

We do not yet have much positive information about the interior make-up of either Pluto or Charon. Pluto's density is about twice that of water; estimates range between 1.8 and 2.1, implying that there is a greater percentage of rock and less of ice than in the satellites of the giant planets; possibly around 70% rock and 30% ice, though we cannot be sure. Below the icy crust, a few tens of kilometres thick, there may be a mantle of water ice going down to between 200 and 300 km, and then a region of partially dehydrated rock. It is likely that Pluto is differentiated, with a comparatively dense rocky core surrounded by a mantle of ice. The heat produced by radioactive substances may be significant, but we know nothing definite about the internal temperature. Charon is less dense than Pluto (about 1.6 times as dense as water), so that presumably the percentage of rock is lower, and Charon is less likely to be differentiated.

All this is very uncertain, but obviously Pluto and Charon are so small and so far away that studies with Earth-based telescopes are very difficult indeed. We may learn much more when the New Horizons space-craft, now on its way, arrives there in 2015.

SURFACE FEATURES

The apparent diameter of Pluto as seen from Earth is so small – never more than 0.11 of a second of arc – that surface features are quite beyond the reach of ordinary telescopes, and only the Hubble Space

Plates taken earlier (1965, 1970, 1971) were then checked, and the same effects were noted: further confirmation came on 6 July 1978 by J. Graham, using the 401-cm reflector at the Cerro Tololo Observatory in Chile. Either Pluto was curiously irregular in shape, or else it was attended by a satellite. The existence of the satellite as a separate body was established by D. Bonneau and P. Foy, using the 3.6-m Canada–France–Hawaii telescope on Mauna Kea: using the technique of speckle interferometry, they recorded the two bodies separately. The attendant was named Charon, after the gloomy boatman who ferried departed souls across the River Styx into the Underworld.

The Pluto–Charon system is unique, and better regarded as a binary pair (or a binary Kuiper Belt object) than as a planet and a satellite. The barycentre of the pair (that is to say, the 'balancing point') lies between the two. Charon's orbit round Pluto is almost circular, and the surface-to-surface distance is only about 18 000 km. The main point is that the orbital period of Charon is exactly the same as the rotation period of Pluto: 6.4 days, so that to a Plutonian observer Charon would remain fixed in the sky.

The rotation period of Pluto had originally been measured in 1955, by M. Walker and R. Hardie, from variations in magnitude. It was also found that the axial inclination amounts to 122.5°, even more than that of Uranus. Charon moves in the plane of Pluto's equator, so that the phase effects will be very strange – particularly as there are about 14 000 Plutonian days in every Plutonian year.

Telescope has been able to show anything at all definite. The first attempts to study the surface were made by L. Andersson and B. Fix, from 1973. They found that the amplitude of the light variations due to Pluto's rotation were increasing, but that the mean magnitude was becoming fainter. From this they inferred that there are bright poles and that one of these poles, previously presented to the Earth, was turning slowly away, so that more of the mid-latitude and equatorial zones were presented – where the ice would be 'dirtier' and older. Results from the Infra-Red Astronomical Satellite (IRAS), in 1983, indicated the presence of a band near the equator which was bright at infrared wavelengths but dark in visible wavelengths, so that it could be relatively frost-free. However, much better results came from a series of phenomena lasting from 1985 to 1990.

MUTUAL PHENOMENA

In 1978, Leif Andersson, of Sweden, pointed out that for a period occurring twice every Plutonian year – that is to say, every 124 Earth years – a situation arises when the orbits are positioned in a way which allows for mutual transits and occultations. The first observation, when the edge of Charon passed in front of Pluto, was made by R. Binzel on 17 February 1985. Total events began in 1987 and ended in 1988, while the whole series of phenomena ended in September 1990.

When Charon passed behind Pluto it was completely hidden, and Pluto's spectrum could be seen alone; when Charon passed in front of Pluto the two spectra were seen together, but that of Pluto could be subtracted. It was found that the surface of Pluto is covered with methane ice, together with large crystals of frozen nitrogen (N_2) and some of water ice and carbon monoxide. No methane ice was found on Charon, where the surface layer was apparently due to water ice.

These mutual phenomena were immensely informative. It is fortunate that they occurred when they did; the situation will not recur for well over a century.

In 1999, observations made with the 8.3-m Suburu Telescope on Mauna Kea, in Hawaii, detected solid ethane (C_2H_6), nitrogen (N_2), methane (CH_4) and carbon monoxide (CO) on Pluto; the surface temperature was found to be –233 °C. The water ice layer on Charon was confirmed.

The only attempt at drawing up a map of Pluto has been made by using images obtained with the Hubble Space Telescope. (Remember that the axial inclination is 119.4°, more than a right angle, so that in this respect Pluto is more like Uranus than the Earth.) As expected, there is a mainly dark equatorial band, and there are bright polar caps, but this tells us little about the nature of the surface features; there may well be craters, valleys, hills and ridges – again we must wait for the arrival of New Horizons. No names have yet been given to any of the features; I have suggested a system based on Underworld deities. No definite markings have been imaged on Charon, though there have been vague indications of a darkish band in one hemisphere and a brighter band in the other.

The surface layers of the two worlds are different. Spectroscopic analysis shows that Pluto's consists of over 98% nitrogen ice, with traces of methane and carbon dioxide, while most of Charon's surface is coated with water ice in crystalline form, together with

traces of ammonia ice. It has been suggested that there may be ice geysers on Charon, of the same basic nature as those of Triton, but this is highly speculative, and would involve the existence of ammonia-laced liquid water below the crust. Cryovulcanism here would indeed be surprising.

ATMOSPHERE

Pluto does have an atmosphere, albeit a very tenuous one, with a ground density of a few microbars – or a few tens of microbars at most. Preliminary spectroscopic searches by Kuiper 1943–4 were unsuccessful, but occultations of stars by Pluto have given definite proof that the atmosphere not only exists but that it is surprisingly extensive. The first occultation measures were made in 1980. On 9 June 1988, Pluto passed in front of a 12th-magnitude star; the star began to fade at a distance of 1500 km from the centre of Pluto and it seems that there is an upper transparent layer about 300 km deep, with haze below – still not opaque enough to hide the surface. The main constituent is nitrogen (N_2), together with very small amounts of methane and carbon monoxide. This means that the atmosphere is not very dissimilar to that of Triton, although the surface coating of Pluto is different.

Charon showed no trace of atmosphere, which in view of its much lower escape velocity is not surprising. It has been suggested that Charon's original atmosphere escaped and was captured by Pluto; alternatively, that an excessively tenuous atmosphere envelops both bodies, but as yet our information is very incomplete.

Charon's present lack of atmosphere was confirmed on 11 July 2005, when the satellite occulted the 15th-magnitude star UCAC2 2625713. The event had been predicted by a French team – no easy matter! – and was observed by large telescopes in South America. In addition to confirming the lack of atmosphere, the occultation also gave an improved value for Charon's diameter.

There is also the strong possibility that as Pluto moves out toward aphelion its atmosphere will freeze out, so that for part of each orbit there is no gaseous atmosphere at all.

In 2006, it was found that instead of cooling as it moved outwards from perihelion, Pluto was actually warming up, and the density of the atmosphere had slightly increased. The cause of this is unknown.

New maps of Pluto and Charon were produced in 2010 by astronomers at the Lowell Observatory, led by Marc Buie. He found dark areas on Pluto, thought to be due to dirty water ice, and brighter areas attributed to nitrogen frost. An unusual bright spot near the centre of the global map could be due to the presence of carbon monoxide, though it was decidedly puzzling.

The Hubble Space Telescope, revealed a complex-looking world with patches of white, others of dark orange or inky black. It was also confirmed that during the early twenty-first century there has been marked global warming, and the mass of the atmosphere doubled, probably because of sublimating nitrogen ice. The observations are far from easy – rather like trying to map details on a cricket ball 80 km away!

Charon's diameter was defined as 1207 km, with an uncertainty of only 1 km either way. It was confirmed that the main surface was coated with water ice, rather than nitrogen ice, as with Pluto, and no evidence of atmosphere was found. Charon is less dense than Pluto and is thought to be 55% rock and 45% ice, against 70% rock for Pluto. Apparently there are bright areas and darker markings, but obviously our knowledge is very limited.

Observations of Pluto with the Hubble Space Telescope show what has been called a 'molasses type' surface with marked changes over relatively short periods. Moreover, between 1994 and 2003 the northern hemisphere brightened, the southern hemisphere dimmed, and the atmosphere doubled in mass. Pluto is much more dynamic than expected. The New Horizons spacecraft is well on its way, and should be able to map the entire surface.

Pluto last passed perihelion in 1989 (between 1979 and 1999 it was closer-in than Neptune) and was moving outward, so that there was no time to spare. Accordingly, NASA prepared the New Horizons probe, scheduled to fly past Pluto in 2015, well before the atmosphere is expected to collapse.

The New Horizons probe, purely a NASA venture, was launched from Cape Canaveral on 16 January 2006. Its main target was Pluto (then still ranked as a planet), but en route would also survey Jupiter and an asteroid or two. After flying past Pluto and Charon it would continue in the Kuiper Belt and then leave the Solar System.

In size and shape it is comparable with a grand piano, and has been likened to a piano glued to a satellite dish. Power is provided by a radio-active thermoelectric generator (RTG); solar power cannot be used in the wastes of the Solar System. Internally, the structure is painted black, to preserve as much heat as possible, and overall the whole space-craft is blanketed. Communication is via a high-gain antenna, two low-gain antennæ and a medium-gain dish.

2006 Jan 19	Launch
2006 Apr 7	Mars orbit crossed
2006 May	Entry into the asteroid belt
2006 June 13	Fly-by of 132524 APL at a range of 101870 km; pictures taken
2006 Oct	Exit from the asteroid belt
2006 Oct 28	First image of Pluto obtained
2007 Jan 8	Start of Jupiter encounter
2007 Feb 28	Closest approach to Jupiter (2 305 000 million km, relative speed 21.219 km s^{-1})
2007 Mar 5	End of Jupiter encounter
2007 June 8	Saturn orbit crossed
So far, so good! Now for:	
2011 Mar 18	Uranus orbit crossed
2014 Aug 1	Neptune orbit crossed
2015 July 14	Fly-by of Pluto, at a range of 13 695 km (relative speed 21.2 km s^{-1})
2015 July 14.	Fly-by of Charon, at range of 24 473 km (relative speed 13.9 km s^{-1})
2016–2020	Possible encounters with other Kuiper Belt objects
2029	Exit from the Solar System
2029–2100	?

There are seven scientific instruments, covering imaging, spectro-metric research and radio science. The launch was completely successful, following a series of minor delays; this was the first probe to be put straight into a solar escape orbit. It departed at a record speed of 16.26 km s^{-1} (58 536 km h^{-1}; 10.1 miles per second). The schedule was worked out as follows:

Obviously, contact will be maintained for as long as possible, but the final fate of New Horizons will never be known.

The Jupiter encounter was highly successful, and New Horizons obtained the first close-range images of Oval BA the new Red Spot. There was no encounter with Mars or Saturn.

ILLUMINATION ON PLUTO

Pluto is certainly a remote, lonely world, but it is not shrouded in permanent darkness; indeed, sunlight there will be at least 1500 times more powerful than full moonlight on Earth. Charon will be visible from one hemisphere of Pluto, but will be very dim.

STATUS OF PLUTO

Pluto was long classed as a planet, albeit an unusual one, but the outlook was changed by the discovery of a swarm of relatively small planetary worlds moving in the outer Solar System – now known as Kuiper Belt objects (KBOs). Pluto is an ordinary member of the swarm, and is not even the largest. Eris is both larger and more massive, and very probably there are others that await discovery. Therefore, in 2006, a ruling by the International Astronomical Union officially demoted Pluto to the status of a dwarf planet.

PLANET X?

It is now definite that no sizeable planet exists closer in than the Kuiper Belt, but we must still consider the possibility of a large world at a much greater distance – often referred to a Planet X. Slight alleged discrepancies in the movements of Uranus and Neptune were quoted in support, but even if a giant planet existed it was bound to be very faint.

Comets have again been called in as evidence. In 1950, studies of eight cometary orbits led K. Schütte to assume the existence of a planet at 77 a.u., and his work was extended by H. H. Kritzinger, whose Planet X moved at 65 a.u. Later, by 'pairing' data for two of Schütte's eight comets, he amended this distance to 75.1 a.u., and the period to 650 years, with an inclination of 40° and a magnitude of 10. A photographic search was undertaken in the indicated position, but with no result.

Less convincing was a theory by M. E. Savin, who believed in a planet moving at 78 a.u. His method was to divide the known planets into two groups, inner and outer, and pair them, but for some obscure reason he included the tiny asteroid 944 Hidalgo. Pairing 'Planet X' with Mercury, he produced a planet with a period of 685.5 years, an eccentricity of 0.3 and a mass 11.6 times that of the Earth. Needless to say, no confirmation was forthcoming. The 'comet family' idea has also been discussed by J. J. Matese, D. P. Whitmire (1986), V. P. Tomanov (1986) and A. S. Guliev (1987). Guliev claimed that a new cometary family could be

Table 13.4 *Dwarf planets*

		Date	Disvoverer	Type	Distance from Sun (a.u.) Q	q	Semi-major axis (millions of km)	Eccentricity, e	Orbital period (years)	Inclination angle (°)	Diameter (km)	Rotation period	Albedo	Magnitude	Escape velocity (km s⁻¹)
90377	Sedna	2003	Brown Trujillo, Rabinowitz	SDO	975.56	525.86	7867	0.855	12 059	11.93	1600	0.42d	0.16–0.30	20.4	0.95
134340	Pluto	1930	Tombaugh	Plutino	49.31	29.00	5907	0.249	248.1	17.14	2390	6.39d	0.5–0.7	14.0	1.2
136108	Haumea	2005	Brown. Trujillo, Rabinowitz	Cubewano	51.53	39.10	6484	0.189	285.4	28.19	1500	0.16d	0.7	17.3	0.84
136199	Eris	2003	Brown, Trujillo, Rabinowitz	SDO	97.56	37.78	6767	0.441	557	44.19	2600	± 8h	0.86	18.7	±2
136472	Makemake		Brown, Trujillo, Rabinowitz	Cubewano	53.07	38.51	6850	0.159	309–9	28.96	1900	?	0.7?	16.7	± 0.84

identified, consisting of comets Halley, di Vico, Westphal, Pons–Gambart, Brorsen–Metcalf and Vaisälä 2. Projections of the aphelia of their orbits on to the celestial sphere concentrate near a large circle, indicating the existence of Planet X moving within the corresponding plane; the distance was given by Guliev as 36.2 a.u. Tomanov came to much the same conclusion as Guliev. Matese and Whitmire believed that there are 'showers' or periodical comets associated with the passage of Planet X through the Oort cloud, and they linked this with cratering and fossil records showing periods of major impacts on Earth, modulated with a period of about 30 000 000 years. Their Planet X moved between 50 and 100 a.u. In 1975, G. A. Chebotarev of what was then Leningrad used the aphelia of periodical comets to predict two outer planets, one at 53.7 a.u., and the other at 100 a.u. Much more recently (1999), J. Murray of the Armagh Observatory in Northern Ireland has used the orbits of very long-period comets to indicate the presence of a very massive planet orbiting the Sun at a distance of around one light-year. This is of course highly speculative, and the detection of a planet at so great a distance seems to be out of the question at the present time.

Halley's Comet was also involved. In 1952, R. S. Richardson made an attempt to measure the mass of Pluto by its perturbing effects on the comet. He decided that Pluto had no detectable influence, but that there might be an Earth-sized planet moving at 36.2 a.u., or 1 a.u., beyond the aphelion point of the comet; this would delay the return of the comet to perihelion by one day, while a similar planet at 35.3 a.u. would produce a delay of six days. A somewhat desultory search was put in hand, but with no result. In 1972, J. A. Brady, of the University of California, used the movements of Halley's Comet to indicate the presence of a Saturn-sized planet moving in a retrograde orbit at a distance of 59.9 a.u.; he believed the magnitude to be about 14 and indicated that the planet was situated in Cassiopeia. Searches were made, but the planet refused to show itself, and it was generally agreed that Brady's calculations were fatally flawed.

In September 1988, new predictions, based on the movements of Uranus, were made by R. Harrington, from the US Naval Observatory in Washington. His planet had a period of 600 years and a mass two to five times that of the Earth, with a present distance of around 9.6 thousand million kilometres; he gave a position in the Scorpius–Sagittarius area, but again there was no result.

A different line of approach as adopted by J. D. Anderson of the JPL (Jet Propulsion Laboratory at Pasadena, California). He claimed that there were genuine unexplained perturbations in the motions of Uranus and Neptune between 1810 and 1910, but not since. This would indicate a Planet X with a very eccentric orbit, now near its aphelion and therefore unable to produce measurable effects. Anderson gave it an inclination of about 90° and a period of between 700 and 1000 years, and a mass five times that of the Earth. Anderson suggested that the perturbations due to this planet would again become measurable about the year 2060. We must wait and see.

Uranus was again used by C. Powell, of JPL, in 1987; his planet would be located in Gemini and would move at about 39.8 a.u., with a period of 251 years – an orbit not unlike Pluto's. A brief search was made from the Lowell Observatory. Recently, it has been claimed that improved values for the masses of Uranus and (particularly) Neptune, obtained from the Voyager 2 data, show that no unexplained perturbations occur, and that Planet X does not exist. This may or may not be the case. If Planet X is real, it will no doubt be found eventually, but for the moment there is little more to be said.

However, not everyone has rejected the idea of another large planet in the depths of the the Solar System. One believer is the Japanese astronomer Patryk Lykawka (Kobe University) who bases his calculations on certain puzzling features of the Kuiper Belt. Another is Venu Malhotra, Tucson, Arizona. Their first clue is the 'Kuiper Cliff', 50 a.u. from the Sun, where the numbers of KBOs drop suddenly and dramatically for reasons that have always been unclear. Could Lykawa's planet explain this?

Assuming that Planet X comes to light, a name will have to be found for it and one favourite is 'Minerva'. In fact this was one of the names suggested for Tombaugh's planet in 1930, but the suggestion came from T. J. J. See, who was – to put it mildly – unpopular with his contemporaries. This is why Minerva is now called Pluto!

From dynamical and statistical analyses of the Oort Cloud, J. J. Matese and D. P. Whitmire suggested, in 2010, the existence of a companion to the Sun, about 1.4 times as massive as Jupiter, orbiting in the innermost region of the Oort Cloud. It may even have produced the detached KBO Sedna. Perhaps Planet 9 may exist after all!

TYPES OF TRANS-NEPTUNIANS

The discovery of the first trans-Neptunian (apart from Pluto) in 1992 caused a change in outlook, since up to that time Pluto had been regarded as a planet. By now many trans-Neptunians are known. They are of various classes:

1. Classical KBOs, known as Cubewanos. They are not controlled by an orbital resonance with Neptune. The first to be discovered was 1992 QB1, and the rest are referred to as 'QB1-o's' – Cubewanos! The largest of these is Makemake.
2. Plutinos, in 2:3 resonance with Neptune (i.e. Neptune makes three orbits in the same time that a Plutino makes two). The largest of these is Pluto itself.
3. Scattered Disc Objects (SDOs). Very eccentric and often highly inclined orbits, extending to well over 100 a.u. from the Sun. The two largest of these are Eris and Sedna. The most extreme SDOs are often called Oort objects.

The largest trans-Neptunians – Eris, Sedna, Pluto, Makemake and Haumea – are classed as dwarf planets. These are listed in Table 13.4.

THE SELECTED LIST IN TABLE 13.5

Cubewanos

This selected list includes only objects with mean distance from the Sun greater than that of Neptune. Note that the diameters (km) are the maximum likely values.

Cubewanos are also called 'classical KBOs'. These are not controlled by orbital resonance with Neptune. Semi-major axes

Table 13.5 *Selected list of Cubewanos*

		Perihelion distance, q (a.u.)	Semi-major axis (×10⁶ km)	Aphelion distance, Q (a.u.)	Period (years)	Eccentricity, e	Inclination angle, i (°)	Rotation period, R (hours)	Diameter (km)	Density (water = 1)
15760	1992 QBI	40.88	6555	46.569	289.2	0.065	2.19	?	160	?
19521	Chaos	40.93	6600	50.27	309.0	0.102	12.06	3.99	560	?
20000	Varuna	40.91	6451	61.21	283.2	0.054	17.2	3.17	874	0.99
50000	Quaoar	41.93	6494	45.29	288.0	0.038	7.99	?	1260	2.0?
53311	Deucalion	41.58	6660	47.16	295.5	0.663	0.37	?	497	?
58534	Logos	39.95	6790	50.83	305.8	0.120	2.90	?	77	?
66652	Borasisi	39.82	6540	47.41	288.0	0.087	0.56	?	166	–1
88611	Teharonhiawako	42.89	6615	45.28	201.0	0.027	2.60	?	176	73.3
136108	Haumea	35.16	6484	51.53	285.4	0.189	28.2	3.92	1500	3.3
136472	Makemake	38.51	6580	53.07	309.9	0.159	28.96	?	1900	?
148780	Altjira	41.35	6609	47.00	293.6	0.064	5.19	?	340	?
2005	UQ 513	37.00	6495	49563	285.1	0.145	25.69	?	1080?	?

range between 40 and 62 a.u.; Neptune's orbit is not crossed. The first to be discovered was 1992 QB1: hence the name! The largest Cubewanos, Makemake, and Haumea, are classed as dwarf planets.

15760 1992 QB1

This was discovered in 1992 by D. Jewitt and J. Luu; apart from Pluto, it was the first trans–Neptunian to be found. It is relatively small, and not much is known about it.

19521 CHAOS

Chaos was discovered in 1998 by the DES (Deep Ecliptic Survey) team at Kitt Peak, and named after the original state of existence in Greek mythology. It is rather bright.

20000 VARUNA

This was discovered in 2000 by R. McMillan; it is named after the Hindu god of the waters. It is reddish, with low albedo; its quick rotation may affect its shape. Small amounts of water ice have been detected on its surface.

50000 QUAOAR

Quaoar was discovered in 2002 by C. Trujillo and M. Brown; it was named after the creation deity of the North American Tongva people. It has low albedo (0.1) and is reddish, but, surprisingly, signs of crystalline ice have been detected on the surface, indicating there must have been a rise in temperature over the last million years, reaching at one time −160 °C (the present temperature is −220 °C). The reason for this is unclear; cryovulcanism has been suggested. Solid methane and ethane have also been found. The Hubble Space Telescope has been used to attempt measurements of Quaoar's diameter. A satellite, perhaps 100 km in diameter, was reported in February 2007.

53311 DEUCALION

This was discovered in 1999 by the DES team, and named after a Greek deity. Not much is known about its physical condition.

58534 LOGOS

Logos was discovered in 1997 by C. Trujillo, J. Chen, D. Jewitt and J. Luu. It has unusually high albedo (0.4). A satellite was discovered in 2001 by K. S. Noll *et al.* which orbits at a range of 8000 km in a period of 112 days (eccentricity 0.45), and is about 80 km in diameter. In Gnostic lore, Logos and Zoë are associated with the creation legend.

66652 BORASISI

This object was discovered by Trujillo, Jewitt and Luu from Mauna Kea in 1999, and named after a fictional character in a novel by K. Vonnegut. It has a 135-km satellite, Pabu, discovered in 2003 by K. S. Noll *et al.*; range 4660 km, period 46.3 days. The name also comes from Vonnegut.

88611 TEHARONHIAWAKO

This was discovered in 2001 by the DES team. It has a 135-km satellite, Sawiskera. (Try saying these names six times, quickly!) The names are of two brothers in the Iroquois creation myth.

136108 HAUMEA

Haumea was discovered in 2004 by J. L Ortiz *et al.* and M. Brown *et al.*; each team claimed priority! It has been named after a Hawaiian goddess. It has now been reclassified as a dwarf planet. It is very elliptical – around 1960 × 1500 km – and rotates in less than 4 hours. Two satellites have been found, and named after Haumea's daughters; Hi'iaka (310 km, range 50 000 km, period

49.1 days) and Namaka (l70 km, range 39 300 km, period 34.7 days). Both were discovered by Brown and his team at Caltech.

There was a rather strange episode concerning the discovery of Haumea. After Brown and his team published their results, Jose Luis Ortiz Moreno of the Astrofisica de Andalucia at Sierra Nevada Observatory, Spain, claimed to have made the discovery on plates taken as early as March 2003. Brown suspected that the Spanish had simply accessed his observation logs and had used the information to 'precover' Haumea on their own early plates! Well – I cannot possibly comment!.

Haumea is the brightest KBO apart from Pluto and Makemake; its magnitude is 17.3. Its rotational period is very short – 3.9 hours – and it seems to be denser than Pluto, its surface coated with a thin layer of crystalline water ice. It is believed to be the largest member of the only known trans-Neptunian collisional family, and is in fact the core of the differentiated progenitor. Smaller members include 24835, 19308, 120718 and 145453.

136472 MAKEMAKE

This object was discovered by M. Btown, C. Trujillo and D. Rabinowitz in 2005, and named after the creator deity of the Rapanui, the people of Easter Island (it was disovered on 31 March, close to Easter, and was a first named 'Easterbunny'). Its apparent magnitude can rise to 16, but it was missed by Tombaugh during the later searches because it is so slow moving, and then lay in a rich region of the sky (Coma Berenices). It is classed as a dwarf planet. It has high albedo (80%). Spectroscopic examination shows that its surface resembles that of Pluto, with methane grains and probably nitrogen ice. It has been suggested that there may be a transient atmosphere, similar to that of Pluto near perihelion. No satellite has been found. Makemake reaches its next aphelion in 2033.

148780 ALTJIRA

Altjira was discovered in 2001 by the DES team, and named after an Aborigine sky god. It is of moderate size; albedo about 0.1.

2005 UQ 513

The diameter of this object is uncertain; it may be from 1000 to almost 1100 km. If it does prove to be over 1000 km across, it may qualify as a dwarf planet. It next reaches perihelion in 2024.

PLUTINOS

Plutinos are KBOs in a 2:3 motion resonance with Neptune. They are very numerous. Of course Pluto itself is the largest member of the class; other large Plutinos are Ixion, Huya, Orcus and (probably) Rhadamathus. Most have orbits of low inclination and eccentricity no more than 20°, though there are a few exceptions; for example 2005 GE$_{187}$ has its perihelion well inside Neptune's path and its aphelion in the Scattered Disc region. Most, including Pluto, have periods of around 247 days – one and a half times Neptune's period.

Apart from Pluto, the first member of the class to be discovered was 1993 RO, by Jewitt and Luu. Its period is 244 years; diameter no more than 90 km. Other discoveries soon followed.

28978 Ixion

This was discovered in 2001 by the DES team, and named after a Greek deity. It is reddish, with a higher albedo (around 15%) than most red Plutinos and Cubewanos. Its apparent magnitude at opposition is 19.6; the next perihelion is due in 2070.

38628 Huya

This was discovered in 2000 by Ignacio Ferrin, and named after the Wayuu rain god. Its surface is reddish; mean opposition magnitude, 19.3.

38083 Rhadamanthus

This object was discovered in 1999 by the DES team at Kitt Peak; it is named after a son of Zeus and Europa, one of the three Judges of the Dead in Greek mythology. It is probably a Plutino, but this has been questioned. Not a great deal is known about it.

90482 Orcus

Orcus was discovered in 2004 by Brown, Trujillo and Rabinowitz; it is named after a Roman deity associated with death. It has high albedo (19%) and as its diameter is nearly 1000 km, comparable with Ceres, it could well be classed as a dwarf planet. The surface seems to show water ice. Its opposition magnitude is 19.1. A satellite, probably as much as 200 km across, was found in 2007.

134340 Pluto

Pluto, covered in depth above, was discovered by Clyde Tombaugh in 1930, and named after the God of the Underworld.

SCATTERED DISC OBJECTS

Asteroidal bodies moving in the remotest parts of the Solar System are known as Scattered Disc Objects. Orbital eccentricities may be as high as 0.8; inclinations up to 40°, with periods of many centuries. The SDOs with the greatest aphelion distances are called Distant Detached Objects (DDOs), but the boundary between these two classes is decidedly arbitrary. The two largest SDOs are Eris and Sedna. There are no SDOs with retrograde motion, and none move out as far as the Oort Cloud, 50 000 a.u. from the Sun – a quarter of the distance to α Centauri. Well over 100 SDOs are now known but this must be only a very slight percentage of the total.

All the SDOs are so far away that they are very faint, and our information about them is bound to be limited. The first to be found was 1996 TL$_{66}$, by Jewitt, Luu and Chen in 1996. 1995 TL$_8$, discovered by A. Gleason in 1995, has been found to have a large satellite, with about 10% the mass of the primary. 65489 Ceto was discovered by Trujillo and Brown, from Palomar, in 2003 and was named after a sea monster in Greek mythology; in 2006, K. Noll and his colleagues used the Hubble Space Telescope to find a satellite, christened Phorcys after a Greek sea god. It is about 130 km in diameter, large when you remember that Ceto itself is no more than 180 miles across; the orbital period is 9.6 days, and the separation 1840 km.

90377 Sedna

This was discovered by Brown, Trujillo and Rabinowitz in 2003, and named after the Inuit goddess of the sea. Its period is bound to be uncertain, but is probably around 12 000 years; it should reach perihelion in 2075 or 2076. The rotation period is about 10 hours. It is a chilly place; the temperature never rises above −240 °C. It seems to have little water ice or methane ice on its surface; the dark red colour may be due to hydrocarbons. Its apparent magnitude is at present rather below 20.

The discoverers have suggested that Sedna is too far out to be considered simply a Scattered Disc Object; it could well be the first observed member of the Oort Cloud. It and a few others, such as 2000 CR 105, have been referred to as Detached Objects.

136199 Eris

Eris is easily the largest KBO, and is over 25% more massive than Pluto. It was discovered by Brown, Trujillo and Rabinowitz in 2003, and named after the Greek goddess of strife. (Earlier it had been nicknamed Xena, after a princess in a television fiction series.) It is grey, not red like Sedna, and over the course of its 'year' the temperature must range between about −207 and −240°C. There may well be a surface layer of methane ice. The apparent magnitude can rise just above 19. As the angular diameter is only 40 milli-arcsec, studies of surface details are rather difficult! In 2005 Brown and his team, using the Keck telescopes in Hawaii, found it to be

as much 2300–2500 km in diameter, orbiting at a distance of 37 000 km in a period of 15.8 days. Recently it was discovered that Eris itself has a small moon, named Dysnomia.

2006 SQ$_{372}$

This was discovered by A. Becker and N. Kaib; it has a very eccentric path; at perihelion it actually crosses Neptune's orbit. On the other hand, 2004 XR$_{190}$, found by B. Gladman and his colleagues in 2004, has low eccentricity; the distance from the Sun ranges only between 52 300 000 000 km and 61 700 000 100 km, though the inclination is almost 47°. Depending on the albedo, the diameter may be anything between 425 and 850 km.

THE OORT CLOUD

Far beyond the Kuiper Belt we come to the Oort Cloud, named after the Dutch astronomer J. H. Oort, who in 1950 inferred its existence from studies of the orbits of long-period comets entering the main planetary system. (It is occasionally referred to as the Öpik–Oort Cloud, since a less definite proposal had been made slightly earlier by the Estonian astronomer E. Öpik.) It consists of a cloud of icy objects – essentially, cometary nuclei – moving round the Sun at distances from 20 000 to 100 000 a.u. Of course they are much too faint to be seen, and we have no final proof that the Cloud exists, but all the evidence points that way. Rather surprisingly, it is widely believed that the members of the Cloud were formed much closer to the Sun, and were driven out by the gravitational pulls of the giant planets. The distance of the inner boundary is uncertain, but the most extreme KBOs, notably Sedna and 2006 SQ 372, have been referred to as Oort objects. And if an object in the Cloud is suitably perturbed for any reason, it may swing into the inner Solar System, where we see it in its cometary guise.

Assuming that the Oort Cloud really exists, it marks the real outpost of the Solar System. Beyond comes the barren stretches of interstellar space that separate us from the nearest stars.

14 · Comets

Comets are the most erratic members of the Solar System. They may sometimes look spectacular, but they are not nearly so important as they then seem, and by planetary standards their masses are very low indeed. In most cases, though not all, their orbits round the Sun are highly eccentric. A comet has been aptly described as a dirty ice-ball.

COMET PANICS

In earlier times comets were not classed as being celestial bodies, and were put down as atmospheric phenomena, although it is true that around 500 BC the Greek philosopher Anaxagoras regarded them as being due to clusters of faint stars. They were always regarded as unlucky. Recall the lines in Shakespeare's *Julius Cæsar:*

When beggars die, there are no comets seen:

The heavens themselves blaze forth the death of princes.

In 1578, the Lutheran bishop Andreas Calichus went further, and described comets as being 'the thick smoke of human sins rising every day, every moment, full of stench and horror before the face of God'. However, his Hungarian contemporary, Andreas Dudith, sagely pointed out that in this case the sky would never be comet-free! The first proof that comets were extraterrestrial came from the Danish astronomer Tycho Brahe, who found that the comet of 1577 showed no diurnal parallax, and must therefore be at least six times as far away as the Moon (actually, of course, it was much more remote than that).

Comets were viewed with alarm partly for astrological reasons and partly because it was thought that a direct collision between the Earth and a comet might mean the end of the world. In 1696, a book by William Whiston, who succeeded Isaac Newton as Lucasian Professor of Mathematics at Cambridge, predicted that Doomsday would come on 16 October 1736, when a comet would strike the Earth. In France, in 1773, a mathematical paper by the well-known astronomer J. J. de Lalande was misinterpreted, and led to the popular belief that a comet would strike the Earth on 20 or 21 May. (Seats in Paradise were sold by members of the Clergy at inflated prices.) Another alarm occurred in 1832, when it was suggested – wrongly – that there would be a very near encounter with Biela's periodical comet. In 1843, at the time of a particularly brilliant comet, there was a widespread end-of-the-world panic in America, due to the dire prophecies of one William Miller. And in 1910, when Halley's Comet was on view, a manufacturer in the United States made a large sum of money by selling what he called anti-comet pills, and many people sealed up their windows to keep out poisonous gases. Bennett's Comet of 1970 was mistaken by some Arabs for an Israeli war weapon, and Kohoutek's Comet of 1973 was also regarded as a threat: it was expected to become brilliant but, disappointingly, failed to do so.

Then, of course, the most famous or all comets – Halley's Comet – was back in 1986, but so far as I know caused no panics. I do, however, vividly recall the demonstrators outside Wembley Stadium during a concert organised by the Halley's Comet Society in aid of handicapped children (I was involved; actually I was playing a xylophone solo!). About thirty people gathered in the street, wearing placards: 'Comet warning. The end of the world is nigh.' I have to admit that during the concert a great deal went wrong, but at least we raised several thousands of pounds for the children.

All this was harmless (in fact, decidedly funny), but the tragedy associated with the bright comet of 1997, Hale–Bopp, was quite the reverse. A Californian music teacher, Marshall Applewhite, was the leader of a cult called Heaven's Gate, whose members believed that a space-ship had been seen following the comet as it moved in toward the Earth. Thirty-nine cult members committed ritual suicide, confident that they would be taken on board and transported to a better world. (Subsequently a surviving cult sympathiser, Charles Spiegel, bought 76 acres of land near San Diego to serve as a landing site for visiting flying saucers.)

In 1994, a curious prophet named Sofia Richmond ('Sister Gabriel') achieved a degree of notoriety by predicting that a collision between Halley's Comet and the planet Jupiter (!) would result in the destruction of mankind. There were also books by an eccentric Russian-born psychoanalyst, Immanuel Velikovsky, who confused planets with comets and believed that Venus had been a comet only a few thousands of years ago; he also maintained that Biblical events, notably the Flood, were due to comets. However, Velikovsky's ignorance of astronomy was so complete that trying to argue with him was a decidedly pointless exercise.

Finally, I cannot resist mentioning one of the oddest predictions of all, due to the 'Millennium Group', which existed briefly at the end of the twentieth century. They found that in June 2000 a small, quite well-known periodical comet, West–Kohoutek–Ikemura, would approach Mars to within 70 000 kilometres. Either the comet would be pulled into Mars, hurling millions of tons of material outward to produce a catastrophic bombardment of the Earth, or else one of the two Martian satellites (Phobos or Deimos) would be wrenched away from Mars and put into a collision course with Earth. One has to admit that the members of the Millennium Group make even Sister Gabriel and Immanuel Velikovsky look fairly plausible!

No doubt there will be further comet panics in future, and it is true that a collision with a cometary nucleus would cause immense damage. We can only hope that if this does happen, the impact will be in a sparsely populated area.

THE NATURE OF COMETS

In 1948, the Cambridge astronomer R. A. Lyttleton popularised the 'flying sandbank' theory of comets. He believed that dust particles were collected by the Sun during its passage through an interstellar cloud, and that these particles collected into 'clouds', which he identified as comets. There were fatal weaknesses in this theory, and in 1950 it was abandoned in favour of the model proposed by F. L. Whipple. The nucleus of a comet – the only reasonably 'solid' part – is made up of rocky fragments held together by frozen ices such as water, methane, carbon dioxide and ammonia. When a comet is warmed as it approaches perihelion, the rise in temperature leads to evaporation, so that the comet develops a head or coma, often together with a tail or tails. Cometary tails always point more or less away from the Sun, and are of two types. There is a gas or ion tail, the molecules are repelled by the 'solar wind'. With a dust tail, the particles are driven out by the pressure of sunlight. This all means that when a comet is receding from the Sun, in travels tail-first; in general, ion tails are straight, while dust tails are curved. Many comets have tails of both types, and possibly a third, fainter tail made of sodium.

COMET NOMENCLATURE

Traditionally, comets are named after their discoverer or discoverers, thus the bright comet of 1996 was discovered by the Japanese amateur Y. Hyakutake and is named after him, while the even brighter comet of 1997 was detected independently from America by Alan Hale and Thomas Bopp and is known as Hale–Bopp. No more than three names are now allowed. Sometimes the discoverers of returns of the same comet are used; thus, in 1881, W. F. Denning discovered a comet with a period of 8 and 9 years, and it was not seen again until in 1978 by S. Fujikawa, so that it is listed as Denning–Fujikawa. Occasionally the name used is that of the first computer of the orbit (as with comets Halley and Crommelin); this applies only to periodical comets.

Up to 1994, a comet was also assigned a letter in order of discovery during the year, and then a permanent designation using Roman numerals, in order of perihelion. Thus Halley's Comet was the ninth comet to be found in 1982 and became 1982 I; it was the third comet to perihelion in 1986 and became 1986 III.

A new system was introduced in 1995, similar to that for asteroids. Each year is divided into 24 sections, each with its own letter (I and Z being omitted). Each comet is given a designation depending on the year of discovery, with a capital letter to indicate its half-month and a number to show the order of discovery in that half-month. Thus Hale–Bopp found on 23 July became 1995 O1, as it was the first comet to be discovered during the period between 16 and 30 July. A list of letter designations is given in Table 14.1.

Table 14.1 *Letter designations for comets*

A	Jan	1–15
B	Jan	16–31
C	Feb	1–15
D	Feb	16–29
E	Mar	1–15
F	Mar	16–31
G	Apr	1–15
H	Apr	16–30
J	May	1–15
K	May	16–31
L	Jun	1–15
M	Jun	16–30
N	Jul	1–15
O	Jul	16–31
P	Aug	1–15
Q	Aug	16–31
R	Sept	1–15
S	Sept	16–30
T	Oct	1–15
U	Oct	16–31
V	Nov	1–15
W	Nov	16–30
X	Dec	1–15
Y	Dec	16–31

X/indicates a comet about which so little is known that is cannot be classified.

There are also prefixes. Basically, comets are divided into two classes; periodical, with orbital periods of less than 250 years, and non-periodical, with periods so long that they cannot be predicted with any accuracy. The prefix P/ indicates a periodical comet; D/ a periodical comet that has either disintegrated or been lost; C/ a non-periodical comet. Strictly speaking, this is incorrect; all comets will return eventually unless they have been perturbed by 'planets'.

Thus Halley's Comet, with a period of 76 years, is P/ Halley; Westphal's Comet, which was seen to 'fade away' in 1913, is D/ Westphal; Hale–Bopp, with a period of over 2000 years, is C/ Hale–Bopp. A few comets which are observable all round their orbits are not given later designations; such is P/ Encke, with a period of 3.3 years.

It has also been said that the distinction between comets and asteroids has become rather blurred, and this may be true.

THE STRUCTURE OF COMETS

A comet is made up essentially of three parts: a nucleus, a head or coma, and a tail or tails. The nucleus of an average comet is surprisingly small, and even a large comet will have a nucleus which is only a few miles across.

The first comet to be well seen was Halley's Comet in 1986, from the Giotto space-craft. Earth-based observers are handicapped.

When a nucleus is unshrouded, the comet is far away; as it draws inward, the nucleus is surrounded by gas and dust, which mask it effectively. Halley's nucleus was found to be covered with a layer of blackish organic material. Under a cometary nucleus is an icy body; the ices are of various kinds, notably water, carbon monoxide and carbon dioxide. In general, these ices are shielded from sunlight, but in isolated areas jets are able to spring out to produce material for the coma and tails. Some comets are active in this respect, others are relatively inert.

A coma does not usually form until the comet is within about 3 a.u. of the Sun, when there is marked sublimation of the water ice. Gas flows outward and it is this which wrenches dust particles away from the main body of the comet. A coma may be very large (the coma of the Great Comet of 1811 was larger than the Sun and so was the coma of Holmes' Comet at its outburst in 2006) but is highly rarefied.

With comets that are poor in dust, comæ are usually round; with dusty comets the comæ are fan shaped or parabolic and there is no hard, sharp boundary between the coma and the tail. The rate of gas outflow is very high. Comet Hale–Bopp was probably losing 1000 tons of dust and 1200 tons of water per second when it was close to perihelion.

Some comets, such as P/ Tempel 2, are rich in dust grains; observations made with the Infra-Red Astronomical Satellite (IRAS) in 1983 indicated that in this comet the particles ranged from tiny grains up to 'pebbles' as much as 6 cm in diameter.

Tails are of two main types. Ion tails (otherwise known as plasma tails or gas tails) are repelled by the solar wind, and are generally straight. Our information comes mainly from spectroscopic research (the first good cometary spectrum was obtained by the Italian astronomer G. Donati as long ago as 1864), and magnetic effects are all important, particularly as the solar wind carries a magnetic current. These tails consist largely of ionised carbon monoxide (CO^+), which tends to fluoresce under the influence of sunlight; this is why plasma tails have a bluish tinge. Tails can be very extended. That of the Great Comet of 1843 was 330 000 000 km long, which is greater than the distance between the Sun and the orbit of Mars.

In April 2000, the space-probe Ulysses, which had been launched to survey the poles of the Sun, passed fortuitously through the tail of Comet C/1996 B2 (Hyakutake) and found that the length of the tail was over 500 million km. This is the longest tail ever recorded.

Plasma tails show rapid changes, due largely to variations in the solar wind; these changes are much more evident in some comets than in others. Shock waves caused by solar flares may produce 'kinks' or even spiral effects. There are also marked 'disconnection' effects, caused when the comet crosses a region where the polarity of the solar wind changes; magnetic field lines inside the tail then cross and re-connect, severing the link with the region close to the nucleus on the sunward side. The tail breaks away and a new one is formed. This happened spectacularly with Halley's Comet at the 1986 return, and also with Comet Hyakutake in 1996.

Dust tails are generally curved; the particles in them are around the size of smoke particles and the tails are yellowish, since they shine only by reflected sunlight. Occasionally, a comet may appear to have an 'anti-tail', pointing sunward – as with Comet Arend–Roland of 1957, nicknamed 'the spiked comet'. In fact, what is seen is not a tail, but material in the comet's orbit catching the sunlight at a suitable angle. Arend–Roland showed it particularly well, since it was exceptionally rich in dust. (*En passant*, this comet will never leave its hyperbolic orbit, and it has now made its permanent departure from the Solar System.)

In 1969, observations made from the OAO 2 (Orbiting Astronomical Observatory 2) led to the detection of a vast hydrogen cloud around a comet, Tago-Sago-Kosaka; the cloud was 1 600 000 km in diameter. Similar clouds were also found with other comets, notably Bennett (1970) and Kohoutek (1973), and there is no reason to doubt that they are quite common features of large comets.

The material forming the coma and tails is permanently lost to the comet. This indicates that comets are short-lived by cosmical standards; 0.1–1% of the total mass will be lost each time the comet passes through perihelion. Obviously, a comet of short period will 'waste away' much more quickly than a comet of longer period, and this explains why all the comets of really short period are faint. Very few of them ever achieve naked-eye visibility.

SHORT-PERIOD COMETS

Faint short-period comets are common, and more are discovered every year. Comets which have been observed at more than one return are given numbers, roughly in order of identification; thus, Halley, the first to have its period recognised, is P/1, while Encke, which was next identified, is P/2. The first 150 periodical comets are listed in Table 14.2. Data for selected comets are given in Table 14.3, but it must be remembered that comets are so subject to perturbations by planets that no two cycles are exactly alike, and tables of their movements become out of date very quickly. Some periodical comets seen at only one return are listed in Table 14.4; some of these may be recovered eventually. In these tables, the absolute magnitude of the comet is the magnitude it would have if seen at a distance of 1 a.u. from the Sun and 1 a.u. from the Earth.

Traditionally, it has been said that comets with a period of less than 20 years belong to the Jupiter family, while those with periods of between 20 and 200 years belong to the Halley family – though now we must presumably include Ikeya–Zhang (366 years). All the Jupiter comets have direct motion, while members of the Halley family – including Halley's Comet itself – are often retrograde. (Returns of comets with much longer periods, such as Hyakutake and Hale–Bopp, obviously cannot be predicted at all accurately.)

The following notes concern some short-period comets of special interest.

P/2 Encke

This comet was first seen in 1786 by Pierre Méchain; its first return (1822) was predicted by Johann Encke. Its orbit lies entirely within that of Jupiter, so that it can be followed all the way round. It often exhibits a tail, and has been known to reach naked-eye visibility.

Table 14.2 *Periodical comets, numbers 1–150. Of these comets, six have been lost, or at least mislaid, and have been given a D number: 3 Biela, 5 Brorsen, 20 Westphal, 25 Neujmin 2, 34 Gale and 180 Perrine-Mrkos; see Table 14.4*

Designation	Name	Discovery	Period (Years)	Designation: P	Name	Discovery	Period (Years)
1	Halley	240 BC	76.00	74	Smirnova–Chemykh	1975	8.57
2	Encke	1786	3.28	75	Kohoutek	1975	6.65
4	Faye	1843	7.34	76	West–Kohoutek–Ikemura	1975	6.41
6	D'Arrest	1851	6.30	77	Longmore	1975	6.98
7	Pons–Winnecke	1819	6.38	78	Gehrels 2	1973	7.94
8	Tuttle	1790	13.51	79	du Toit–Hartley	1945	5.21
9	Tempel 1	1867	5.50	80	Peters–Hartley	1846	8.13
10	Tempel 2	1873	5.48	81	Wild 2	1978	6.37
12	Pons–Brooks	1812	70.92	82	Gehrels 3	1975	8.11
13	Olbers	1815	69.56	83	Russell	1979	6.10
14	Wolf	1884	8.25	84	Giclas	1978	6.96
15	Finlay	1886	6.95	85	Boethin	1975	11.2
16	Brooks 2	1889	6.89	86	Wild 3	1980	6.91
17	Holmes	1892	7.09	87	Bus	1981	6.52
18	Perine–Mrkos	1896	6.72	88	Howell	1981	5.58
19	Borrelly	1904	6.88	89	Russell 2	1980	7.38
21	Giacobine–Zinner	1900	6.61	90	Gehrels 1	1972	15.1
22	Kopff	1906	6.45	91	Russell 3	1983	7.50
23	Brorsen-Metcalf	1847	70.54	92	Sangun	1977	12.50
24	Schaumasse	1911	8.22	93	Lovas 1	1980	9.09
26	Grigg–Skjellerup	1902	5.10	94	Russell 4	1984	6.57
27	Crommelin	1818	27.41	95	Chiron (Asteroid 2060)	1977	50.78
28	Neujmin 1	1913	18.21	96	Machholz	1986	5.24
29	Schwassmann–Wachmann 1	1927	14.85	97	Metcalf–Brewington	1906	7.76
30	Reinmuth 1	1928	7.31	98	Takamizawa	1984	7.22
31	Schwassmann–Wachmann 2	1929	6.39	99	Kowal 1	1977	15.02
32	Comas Solé	1926	8.83	100	Hartley 1	1985	6.02
33	Daniel	1909	7.06	101	Chemykh	1977	14.0
34	Gale	1927	11.0	102	Shoemaker	1984	7.26
35	Herschel–Rigollet	1788	155	103	Hartley 2	1986	6.26
36	Whipple	1933	8.53	104	Kowal 2	1979	6.39
37	Forbes	1929	6.13	105	Singer–Brewster	1986	6.43
38	Stephen-Oterma	1867	37.70	106	Schuster	1977	7.26
39	Oterma	1942	7.88	107	Wilson–Harrington Asteroid 4015	1949	4.29
40	Vaisala	1939	10.8				
41	Tuttle–Gracobini–Kresak	1858	5.46	108	Ciffreo	1985	7.23
42	Neujmin 3	1929	10.63	109	Swift–Tuttle	1862	135.01
43	Wolf–Harrington	1924	6.51	110	Hartley 3	1988	6.84
44	Reinmuth 2	1947	6.64	111	Helin–Roman–Crocken	1989	8.16
45	Honda–Mrkos–Pajdusakova	1948	5.30	112	Urata–Niijina	1986	6.64
46	Wirtanen	1948	5.50	113	Spitaler	1890	7.10
47	Ashbrook–Jackson	1948	7.49	114	Wiseman–Skiff	1986	6.53
48	Johnson	1949	6.97	115	Maury	1985	8.74
49	Arend–Rigaux	1951	6.82	116	Wild 4	1990	6.16
50	Arend	1951	7.99	117	Hetin–Roman–Alu 1	1989	9.50
51	Harrington	1953	8.78	118	Shoemaker–Levy 4	1991	6.51
52	Harrington–Abell	1955	7.59	119	Parker–Hartley	1989	8.89
53	Van Biesbroeck	1954	12.43	120	Mueller 1	1987	8.41
54	Di Vico–Swift	1844	6.31	121	Shoemaker–Holt 2	1989	8.05

Table 14.2 (cont.)

Designation	Name	Discovery	Period (Years)	Designation: P	Name	Discovery	Period (Years)
55	Tempel–Tuttle	1865	32.9	122	di Vico	1846	74.36
56	Slaughter–Burnham	1958	11.59	123	West–Hartley	1989	7.57
57	du Toit–Neujmin–Delporte	1941	6.39	124	Mrkos	1991	5.64
58	Jackson–Neujmin	1936	8.24	125	Spacewatch	1991	5.57
59	Kwens–Kwee	1963	8.96	126	IRAS	1983	13.29
60	Tsuchinshan 2	1965	6.82	127	Holt–Olmstead	1990	6.16
61	Shajn–Schaldach	1949	7.49	128	Shoemaker–Holt 1	1987	9.55
62	Tsuchinshan 1	1965	6.65	129	Shoemaker–Levy 3	1991	7.25
63	Wild 1	1960	13.3	130	McNaught–Hughes	1991	6.71
64	Swift–Gehrels	1889	9.21	131	Mueller 2	1990	7.05
65	Gunn	1970	6.83	132	Helin–Roman–Alu 2	1989	8.24
66	du Toit	1944	15.0	133	Elst–Pizarro (Asteroid 7968)	1996	5.61
67	Churyumov–Gerasimenko	1969	6.59	134	Kowal–Vayrova	1983	15.58
68	Klemola	1965	10.95	135	Shoemaker–Levy 8	1992	7.50
69	Taylor	1915	6.97	136	Mueller 3	1990	8.71
70	Kojima	1970	7.85	137	Shoemaker–Levy 2	1990	9.38
71	Clark	1973	5.50	138	Shoemaker– Levy 3	1991	6.73
72	Denning–Fujikawa	1881	9.01	139	Vaisala–Oterma	1979	9.55
73	Schwassmann–Wachmann 3	1930	5.34	140	Bowell–Skiff	1983	16.18
				141	Machholz 2	1994	5.22
				142	Ge–Wang	1998	11.17
				143	Kowal–Mrkos	1984	8.95
				144	Kushida	1994	7.58
				145	Shoemaker–Levy 5	1991	8.69
				146	Shoemaker–LINEAR	1984	7.88
				147	Kushida–Muramatsu	1993	7.44
				148	Anderson–LINEAR	1963	7.05
				149	Mueller 4	1992	9.01
				150	LONEOS	2000	7.66

It is associated with the Taurid meteor stream. In April 2007, Encke's Comet passed through a coronal mass ejection and its tail was completely torn off, though a new tail was soon formed and there was no permanent damage to the comet.

P/4 Faye

This was discovered in 1843, and seen at every subsequent return except those of 1903 and 1918. It has occasionally reached naked-eye visibility.

P/6 D'Arrest

D'Arrest was discovered in 1851 by H. L. D'Arrest; it is occasionally a naked-eye object.

P/7 Pons–Winnecke

This comet was discovered in 1819 by J. L. Pons, and next seen in 1858 by F. Winnecke; it is associated with the Boötid meteors of late June. The diameter of the nucleus is believed to be 2.6 km.

P/8 Tuttle

Tuttle was ddiscovered in 1858 by H. P. Tuttle, and seen at most returns since then as a conspicuous telescopic object. During the 2007 return it was green in colour. It is the parent of the December Ursid meteors.

P/9 Tempel 1

This comet was discovered in 1867 by E. W. Tempel, but following perturbations by Jupiter was lost between 1881 and 1967. It has never become brighter than magnitude 11. Measurements of is nucleus made with the Hubble Space Telescope give its dimensions as 14×4 km. In 2005 it was hit by NASA's Deep Impact Probe.

P/17 Holmes

Holmes is the strangest of the Jupiter comets. It was discovered in 1892 by the English amateur Edwin Holmes (fortuitously; he was scanning the region of the Andromeda Galaxy) but was very faint. It then brightened up suddenly to the fringe of naked-eye visibility before fading again. It proved to be a member of the Jupiter family,

Table 14.3 *Selected list of periodical comets*

Comet	Period (years)	Perihelion distance, q (a.u.)	Aphelion distance, Q (a.u.)	Eccentricity	Inclination (°)	Absolute magnitude
2 Encke	3.28	0.33	2.21	0.850	11.9	11
107 Wilson–Harrington (Asteroid 4015)	4.29	1.00	2.64	0.622	2.8	16
26 Grigg–Skjellerup	5.10	0.995	2.96	0.664	6.6	12
79 du Toit–Hartley	5.21	1.20	3.01	0.602	2.9	–
96 Machholz 1	5.24	0.12	3.02	0.959	60.1	–
10 Tempel 2	5.47	1.48	3.10	0.552	12.0	10
45 Honda–Mrkos–Pajdusakova	5.30	0.54	5.54	0.922	4.2	11
73 Schwassmann –Wachmann 3	5.35	0.93	3.06	0.695	11.4	11
41 Tuttle–Giacobini–Kresak	5.46	1.07	3.10	0.656	9.2	11
46 Wirtanen	5.46	1.07	3.10	0.657	11.7	16
9 Tempel 1	5.51	1.50	3.12	0.52	10.5	9
71 Clark	5.51	1.55	3.12	0.502	9.5	12
125 Spacewatch	5.56	1.54	3.14	0.36	10.4	
88 Howell	5.57	1.41	3.14	0.55	4.3	
133 Elst–Pizzarro (Asteroid 7968)	5.61	2.62	3.67	0.166	1.4	14
100 Hartley 1	6.02	1.82	3.31	0.450	25.7	
116 Wild	6.16	1.99	2.36	0.40	3.7	
37 Forbes	6.13	1.44	3.34	0.578	7.2	10
104 Kowal 2	6.13	1.40	3.37	0.587	15.5	
103 Hartley 2	5.28	0.95	3.40	0.720	9.3	
127 Holt–Olmstead	6.33	2.15	3.42	0.370	17.7	
81 Wild 2	6.33	1.57	3.42	0.540	3.2	6
7 Pons–Winnecke	6.37	1.26	3.44	0.634	22.3	14
57 du Ton– Neujmin–Delporte	6.39	1.72	3.44	0.501	2.9	14
31 Schwassmann–Wachmann 2	6.39	2.07	3.44	0.399	3.8	11
105 Singer–Brewster	6.44	2.03	3.46	0.413	9.2	
76 West–Kohoutek–Ikemura	6.46	1.58	3.47	0.540	30.5	10
118 Shoemaker–Levy 4	6.51	2.02	3.49	0.420	8.5	
43 Wolf–Harrington	6.51	1.61	3.49	0.539	9.3	
6 D'Arrest	6.51	1.35	3.49	0.614	19.5	6
87 Bus	6.52	2.18	3.49	0.375	2.6	
94 Russell 4	6.58	2.23	3.51	0.365	6.2	
83 Russell 1	6.10	1.61	5.06	0.517	22.7	15
67 Churyumov–Gerasimenko	6.59	1.30	3.51	0.630	7.1	10
21 Giacobini–Zinner	6.61	1.03	3.52	0.706	31.86	10
49 Arend–Rigaux	6.61	1.37	3.52	0.611	18.2	9
62 Tsuschinshan 1	6.64	1.50	3.53	0.571	10.5	14
44 Reinmuth 2	6.64	1.89	3.53	0.454	7.0	10
75 Kohoutek	6.67	1.78	3.54	0.496	5.9	
130 McNaught–Hughes	6.69	2.12	3.55	0.40	18.29	
51 Harrington	6.78	1.57	3.58	0.561	8.7	15
19 Borrelly	6.80	1.37	3.59	0.623	30.2	13
60 Tsuchinshan 2	6.82	1.78	3.60	0.504	3.6	14
65 Gunn	6.83	2.46	3.59	0.306	5.5	13
110 Hartley 3	6.88	2.48	3.62	0.314	11.7	
16 Brooks 2	6.89	1.84	3.62	0.49	5.5	13
138 Shoemaker–Levy 7	6.89	1.76	3.62	0.531	10.1	
86 Wild 3	6.91	2.30	3.63	0.366	15.5	
15 Finlay	6.95	1.09	3.64	0.699	3.7	1.3

Table 14.3 (cont.)

Comet	Period (years)	Perihelion distance, q (a.u.)	Aphelion distance, Q (a.u.)	Eccentricity	Inclination (°)	Absolute magnitude
84 Giclas	6.96	1.85	3.65	0.494	7.3	
18 Johnson	6.97	2.30	4.98	0.367	13.7	10
69 Taylor	6.97	1.95	3.65	0.466	20.6	12
77 Longmore	6.98	2.40	3.65	0.343	26.4	
131 Mueller 2	7.05	2.41	3.68	0.344	14.1	
33 Daniel	7.06	1.65	3.68	0.551	20.1	11
17 Holmers	7.09	2.16	3.68	0.412	19.2	13
113 Spitaler	7.10	1.82	5.06	0.471	12.8	16
98 Takamizawa	7.21	1.57	3.73	0.575	0.49	
102 Shoemaker 1	7.25	1.98	3.75	0.471	26.3	
108 Ciffreo	7.25	1.71	3.74	0.542	13.1	
129 Shoemaker–Levy 3	7.25	2.82	3.75	0.248	5.01	
106 Schuster	7.29	1.55	3.76	0.688	20.1	
30 Reinmuth 1	7.31	1.87	3.77	0.502	8.1	14
54 di Vico–Swift	7.32	2.15	3.77	0.431	6.1	
4 Faye	7.34	1.59	3.78	0.578	9.1	8
89 Russell 2	7.38	2.28	3.79	0.40	12.0	
61 Shajn–Schaldach	7.46	2.32	5.31	0.390	6.1	12
47 Ashbrook–Jackson	7.46	2.30	3.81	0.396	12.5	7
52 Harrington–Abell	7.53	1.75	3.84	0.542	10.2	16
123 West–Hartley	7.59	2.13	3.86	0.447	15.3	
83 Russell 1	7.64	2.18	3.88	0.437	17.7	9
78 Gehrels 2	7.94	2.37	5.62	0.409	0.9	
70 Kojima	7.85	2.41		0.39		
121 Shoemaker–Holt 2	8.05	2.66	4.02	0.337	17.7	
82 Gehrels 3	8.45	3.62	4.15	0.125	1.1	9
80 Peters–Hartley	8.12	1.62	4.02	0.598	19.9	8
111 Helin–Roman–Crockett	8.16	3.49	4.04	0.139	4.2	
50 Arend	8.24	1.91	4.08	0.530	19.2	14
58 Jackson–Neujmun	8.24	1.38	4.08	0.661	13.7	17
14 Wolf	8.21	2.41	4.07	0.407	27.5	13
24 Schaumasse	8.22	1.20	4.07	0.705	11.9	11
120 Mueller 1	8.41	2.74	4.14	0.337	8.8	
36 Whipple	8.53	3.09	4.17	0.239	9.9	
74 Smirnova–Chernykh	8.57	3.57	4.19	0.147	6.6	8
136 Muller 3	8.71	3.01	4.23	0.289	9.4	
31 Schwassmann–Wachmann 2	8.72	3.42	4.23	0.195	4.5	
32 Comas Solá	8.83	1.85	4.27	0.568	12.9	8
119 Parker–Hartley	8.89	3.05	4.29	0.290	5.2	
72 Denning–Fujikawa	9.03	0.79	4.34	0.818	9.1	11
93 Lovas 1	9.14	1.69	4.37	0.613	12.2	
64 Swift–Gehrels	9.21	1.36	4.39	0.691	9.3	15
104 Kowal–Mrkos	9.24	2.67	4.40	0.394	5.3	
137 Shoemaker–Levy 2	9.38	1.87	4.45	0.580	4.7	
128 Shoemaker–Holt 1	0.51	3.05	4.49	0.321	4.4	
139 Väisälä–Oserma	9.54	3.38	4.58	0.048	2.4	
117 Helin–Roman–Alu 1	9.57	3.71	4.51	0.176	9.7	
59 Kwens–Kwee	9.45	2.34	4.45	0.58	9.4	11
42 Neujmin 3	10.63	2.00	4.83	0.586	4.0	14
68 Klemola	10.82	1.75	4.89	0.641	11.1	
40 Väisälä 1	10.90	1.80	8.02	0.633	11.6	13

Table 14.3 (cont.)

Comet	Period (years)	Perihelion distance, q (a.u.)	Aphelion distance, Q (a.u.)	Eccentricity	Inclination (°)	Absolute magnitude
56 Slaughter–Burnham	11.6	2.54	7.71	0.504	8.2	14
85 Boethin	11.63	1.16	5.13	0.774	4.9	10
53 Van Biesbroeck	12.43	2.40	5.37	0.552	6.6	7
92 Sangum	12.50	1.81	5.39	0.663	18.7	
126 IRAS	13.30	1.70	5.61	0.697	46.0	
63 Wild 1	13.3	1.98	9.24	0.647	9.2	14
8 Tuttle	13.51	0.997	5.67	0.824	54.7	8
101 Chernykh	13.97	2.35	5.80	0.593	5.1	
99 Kowal 1	15.08	4.67	6.20	0.234	4.4	
29 Schwassmann–Wachmann 1	14.85	5.77	6.04	0.045	9.4	16
66 du Toit	15.0	1.294	10.9	0.787	18.7	16
134 Kowal–Vavrova	15.57	2.58	6.23	0.587	4.34	
140 Bowell–Skiff	16.2	1.97	6.34	0.691	3.8	
28 Neujmin 1	18.2	1.55	12.3	0.776	12.3	10
39 Orema	19.5	5.47	7.24	0.245	1.9	9

Comet	Period (years)	Perihelion distance, q (a.u.)	Aphelion distance, Q (a.u.)	Eccentricity	Inclination (°)	Next return	Absolute magnitude
27 Crommelin	27.4	0.74	17.4	0.919	19.1	2011	11
55 Tempel–Tuttle	33.2	0.98	10.3	0.905	162.5	2031	13
38 Stephan–Oterma	37.7	1.57	20.9	0.860	18.0	2018	5
13 Olbers	69.6	1.18	32.6	0.930	44.6	2024	5
23 Brorsen–Metcalf	70.6	0.48	17.8	0.972	19.3	2059	9
12 Pons–Brooks	70.9	0.77	33.5	0.955	74.2	2024	6
1 Halley	76.0	0.587	35.3	0.967	162.2	2061	4
109 Swift–Tuttle	135.0	0.96	51.7	0.964	113.4	2127	4
35 Herschel–Rigollet	155	0.75	56.9	0.974	54.2	2092	5
153 Ikeya–Zhang	366.51	0.51	101.9	0.990098	28.1	2362	~4

and returned every 6.9 years; but was always dim, and all track of it was lost between 1906 and 1964, when it was recovered by E. Roemer of the Lowell Observatory in Arizona – following calculations by Brian Marsden, who makes a habit of re-locating comets which have been mislaid. Since then it has been seen at every return, but was dim and unremarkable until 2007, when in less than fifty hours the brightness increased more than a million-fold – from a magnitude of 17 to 2.3, while the 5-km nucleus was surrounded by a coma expanding so rapidly that by the first week of November it was larger than the Sun! With the naked eye, the comet, in Perseus, looked like a fuzzy star, completely altering the aspect of that part of the sky. On 25 October, I made its magnitude 2.3, much brighter than the star ε Persei. With binoculars it could easily have been mistaken for a globular cluster. Perihelion was well past, and the comet faded as it moved outward, but showed no sign of breaking up.

Following the outburst, careful studies were made by astronomers using the Canada–France–Hawaii telescope in Hawaii. During November 2007, they found numerous small objects in the dust cloud, moving away from the nucleus at speeds of up to 25 m s^{-1}. They were too bright to be simply bare rock, but were more like minicomets, creating their own dust-clouds as ice sublimated from their surfaces.

The cause of the outburst is unknown; collision with a meteoroid can be ruled out in view of the comet's behaviour in 1892, and the best answer seems to be an outrush of gas from below the crust; but in future Holmes' Comet will be carefully monitored.

P/21 Giacobini–Zinner

This was discovered in 1900 by M. Giacobini, recovered in 1913 by E. Zinner. In 1946, it reached magnitude 5, following a series of outbursts. It is the parent of the Draconid meteors, which are sometimes called the Giacobinids. In September 1985, NASA's International Cometary Explorer (ICE) space-craft passed through its tail.

P/26 Grigg–Skjellerup

This comet was discovered in 1902 by J. Grigg, recovered in 1922 by J. F. Skjellerup (research shows that it had been observed by Pons in 1808). In 1992, the Giotto space-craft flew past at a range of

Table 14.4 *Periodical comets seen at only one return. It is unlikely that the comets in (a) will be recovered. There are, however, other short-period comets which will certainly be recovered, plus a few with longer periods. (b) All these are, of course, very uncertain.*

(a)

Comet	Year	Period (years)	Perihelion distance, q (a.u.)	Aphelion distance, Q (a.u.)	Eccentricity	Inclination (°)
Helfenzrieder	1766	4.35	0.406	4.92	0.848	7.9
Blanpain	1819	5.10	0.892	5.03	0.699	9.1
Barnard 1	1884	5.38	1.279	4.86	0.583	5.5
Brooks 1	1886	5.44	1.325	4.86	0.571	12.7
Lexell	1770	5.60	0.674	5.63	0.786	1.6
Pigott	1783	5.89	1.459	5.06	0.552	45.1
Harrington–Wilson	1951	6.36	1.664	5.10	0.515	16.4
Glacobini	1896	6.65	1.455	5.62	0.588	11.4
Schorr	1918	6.67	1.884	5.21	0.469	5.6
Swift	1895	7.20	1.298	6.16	0.652	3.0
Denning	1894	7.42	1.147	6.01	0.698	5.5
Metcalf	1906	7.78	1.631	6.22	0.584	14.6
Linear	1999	12.53	1.872	5.395	0.653	20.4
Van Houten	1961	15.6	3.957	8.54	0.367	6.7
Pons–Gambart	1827	57.5	0.807	29.0	0.946	136.5
Dubiago	1921	62.3	0.929	30.3	0.929	22.3
de Vico	1846	76.3	0.664	35.3	0.963	85.1
Väisälä	1942	85.4	1.287	37.5	0.934	38.0
Barnard 2	1889	145	1.105	54.2	0.960	31.2
Mellish	1917	145	0.198	55.1	0.993	32.7
Wilk	1937	187	0.619	64.9	0.981	26.0

(b)

Comet	Period (years)	Last perihelion	Next due
D/1889 M1 Barnard 2	145	1889	2034
D/1984 A1 Bradfield	151	1983	2134
D/1989 A3 Bradfield 2	81.9	1988	2070
P/1997 B1 Kobayashi	24.5	1997	2021
P/1997 G1 Montani	21.8	1997	2019
D/1942 EA Väisälä 2	85	1942	2027

200 km, but could not obtain images because its camera had been put out of action during the Halley encounter. It is the parent of a southern meteor shower, the π Puppids, which peak around 23 April.

P/27 Crommelin
In 1930, A. C. D. Crommelin established the identity of comets seen by Pons (1818), Coggia and Winnecke (1873) and A. Forbes (1928); the comet was named after him. Its period is 27 years; it is an easy telescopic object.

P/29 Schwassmann–Wachmann 1
This comet was discovered by these two German observers in 1925; it is usually very faint at magnitude 17, but shows outbursts to magnitude12 or 13, and has been known to brighten to 9. It is one of over 40 'Centaur' comets, whose orbits lie between those of Jupiter and Neptune.

P/35 Herschel–Rigollet
This is a Halley comet with a period of 155 years. It was discovered by Caroline Herschel in 1788, and recovered by R. Rigollet in 1939, when it was of the 8th magnitude. It is due back in 2092.

P/67 Churyumov–Gerasimenko ('Chury')
This comet was discovered by K. I. Churyumov on a photograph taken in 1969 by S. Gerasimenko at the Alma-Ata Observatory. It is the intended target of the Rosetta probe in 2014. Before 1840, its perihelion distance was over 4 a.u., making it unobservable, but perturbations by Jupiter now mean that the perihelion distance has now been reduced to 1.3 a.u.

Table 14.5 *Lost periodical comets. Comets seen at more than one return*

	Period (years)	Perihelion distance, q (a.u.)	Aphelion distance, Q (a.u.)	Eccentricity	Inclination (°)	Returns	Last seen	
D/Tempel–Swift	5.7	1,15	5.22	0.64	5.4	4	1908	Lost
D/Neujmin 2	5.4	1.34	5.43	0.57	10.6	2	1927	Probably disintegrated
D/Biela	6.6	0.86	6.19	0.76	12.6	6	1852	Broke up
D/Brorsen	5.5	0.59	5.61	0.81	29.4	5	1879	Lost: certainly disintegrated
P/4 Gale	11.3	1.21	5.02	0.76	10.7	2	1927	Lost
D/20 Westphal	61.9	1.25	30.8	0.76	12.6	2	1913	Faded out; no longer exists
Shoemaker–Levy 9	–	–	–	–	–	(1)	1994	Impacted Jupiter

P/107 Wilson–Harrington (1949) was recovered in 1979 as an asteroid and given an asteroid number, 4015.

P/133 Elst–Pizarro (1996) has been given an asteroid number, 7968. It moves wholly within the main asteroid belt.

P/73 Schwassmann–Wachmann 3

This was discovered in 1930, and is now in the process of disintegrating. In 1995, it broke into five pieces, and in 2006 dozens of fragments were recorded by the Hubble Space Telescope. These fragments passed the Earth at a range of 11.9 million km, and may come even closer in 2022 – if they still exist!

P/109 Swift–Tuttle

This comet was discovered in 1862, independently by L. Swift and H. Tuttle, but not seen again until 1992, when it was recovered by the Japanese astronomer T. Kiuchi. It has a period of 133 years, so that it belongs to the Halley family and, at the next return, in 2126, it will come very close to the Earth. A collision seems unlikely, and this is just as well, as Swift–Tuttle is about the same size as the impactor of 65 million years ago, widely believed to be linked with the disappearance of the dinosaurs. The comet is the parent of the Perseid meteor stream.

P/114 Wiseman–Skiff

This was discovered by J. Wiseman in 1987 on plates taken by B. A. Skiff at the Lowell Observatory in 1986. It is probably the parent of a meteor shower observable from Mars.

P/103 Hartley 2

Hartley 2 was discovered by M. Hartley in 1986. The estimated diameter is 1 km. It is the intended target of EPOXI, the extended mission of the Deep Impact space-craft.

P/153 Ikeya–Zhang

This comet was discovered independently in 2002 by Zhang Daqing (China) and K. Ikeya (Japan), and found to be identical with a comet seen by Chinese observers in 1661. Its period is 366.5 years, and is the longest-period comet seen at more than one return; it will be back in 2369. In 2002, it was a conspicuous naked-eye object, attaining magnitude 3.5, with a long tail.

P/177 Barnard 2

Barnard 2 was discovered in 1889 by E. E. Barnard. In 2006 (June 23), it was accidentally recovered by LINEAR (Lincoln Near-Earth Asteroid Research), and was recorded as an asteroid of magnitude 17, but L. Buzzi (Varese, Italy), using a 60-cm refractor, detected a coma, and the object was re-classified as a comet; the magnitude did not rise much above 13. The period for the 2006 apparition, computed by G. Marsden and Z. Sekanina, is 119.64 years, and the identity with Barnard's Comet of 1889 was confirmed.

These notes are very brief but may help to show that every periodical comet has a 'personality' of its own.

LOST COMETS

Some periodical comets seen at more than one return have been lost either by collision, because they have disintegrated, because they have been insufficiently observed, because they have been perturbed into new orbits that make them unobservable, or for reasons unknown. A selected list of these is given in Table 14.5.

The classic case is that of D/3 Biela, only the third comet found to be periodical (after Halley and Encke). It was discovered in 1772 by Montaigne, and seen again in 1805 (discovered by Pons), and in 1826, when it was found by the Austrian amateur W. von Biela, who computed its orbit and realised that it was identical with the comets of 1772 and 1805. It returned in 1832, was missed in 1839 because it was so badly placed, and recovered once more in 1846, when it astounded astronomers by breaking in two. The twins came back on schedule in 1852, but this was their last appearance. They were again badly placed in 1859, but should have been visible in 1866, and were eagerly awaited but despite the most careful searches they failed to show up, and have never been seen since. Undoubtedly, they have disintegrated, but on 27 November 1872 a brilliant meteor storm was seen radiating from that part of the sky where the comet ought to have been. The shower was never again as spectacular as this, but was rich in 1885 and persisted through the rest of the nineteenth century and known as the Bielids

Table 14.6 *Close-approach comets*

Comet	Name	Date	Distance (a.u.)	Magnitude
C/1491 B1	–	1491 Feb 20	0.094[a]	1?
D/1770 L1	Lexell	1770 July 1.7	0.0151	2
P/55 1366 U1	Tempel–Tuttle	1366 Oct 26.4	0.0229	3
C/1983 H1	IRAS–Araki–Alcock	1983 May 11.5	0.0312	2
P/1837 F1	Halley	837 Apr 10.5	0.0334	– 3.5
D/3 1805 V1	Biela	1805 Dec 9.9	0.0366	3
C/1743 C1	–	1743 Feb 8.9	0.0390	3
P/7	Pons Winnecke	1927 June 26.8	0.0394	3.4
C/1702 H1	–	1702 Apr 20.2	0.0437	3.5
P/73 1930 J1	Schwassmann–Wachmann 3	1930 May 31.7	0.0617	10
C/1983 J1	Sagamo–Saigasa–Fujikara	1983 June 12.8	0.0628	2
C/1760 A1	–	1760 Jan 8.2	0.0682	4
C/1853 G1	Schweizer	1853 Apr 29.1	0.0839	0
C/1797 P1	Boward–Herschel	1797 Aug 16.5	0.0879	3
L/P 374 E1	Halley	374 Apr 1.9	0.0884	0?
L/P 607 H1	Halley	607 Apr 19.2	0.0898	0?
C/1763 S1	Messier	1763 Sept 23.7	0.0934	6
C/1864 N1	Tempel	1865 Aug 8.4	0.0964	2.5
C/1862 N1	Schmidt	1862 July 4.6	0.0982	4.5
C/1996 B2	Hyakatake	1996 Mar 25.3	0.1018	0
C/1961 T1	Seki	1961 Nov 15.2	0.1019	4

[a]Very uncertain.

or Andromedids (since the radiant lies in Andromeda). A few Andromedids are still seen in late November – probably the final manifestation of Biela's Comet.

Another nineteenthth-century casualty was D/5 Brorsen, discovered in 1846 by the Danish astronomer Theodor Brorsen, and found to have a period of 5.5 years. It was seen at five returns, but vanished after 1879, and has evidently broken up. And there is (or was?) Comet D/1819 W1, discovered by W. Blanpain and found to have a period of about 5 years. It was not seen again, but David Jewitt has suggested that asteroid 2003 WY 25 may be what remains of its nucleus (around 300 m in diameter), and that the comet may be the parent of the southern Phœnicid meteor shower. True, this is highly speculative, but it is at least possible that we have not seen the last of D/ Blanpain. Much less definite is the suggested identification of Comet P2003 LINEAR with the long-lost D/1783 W1 (Pigott). (At the LINEAR facily at the US Lincoln Laboratory, the telescopes have so far discovered over 225 000 asteroids and over 230 comets.) Pigott's Comet was followed for some time in 1788, but never seen again.

We know much more about D/1771 Lexell, discovered by C. Messier. A. Lexell of St. Petersburg computed the orbit, and found a period of 5.6 years. The minimum distance from Earth was 2 200 000 km, and the comet was visible with the naked eye. A subsequent encounter with Jupiter, in 1779, changed the orbit completely; the current period is thought to be of the order of 250 years, and so far as we are concerned the comet is hopelessly lost.

Comet D/20 Westphal is lost for a different reason. It was discovered in 1852, and found to have a period of almost 62 years.

At the return of 1913 it faded out as it moved in toward perihelion, and was not seen again, despite careful searches for it in 1975; it has probably broken up. We must be wary of jumping to conclusions, however. Thus Comet Di Vico–Swift was lost for 38 years after the return of 1897, but was recovered in 1965, while Comet Barnard 3, found in 1892 and given up as lost, was tracked down again in 2008.

Comets are flimsy things, and very susceptible to perturbations by planets, particularly Jupiter. For example, in 1886, Comet P/16 Brooks 2 had a close encounter with the Giant Planet, and actually passed within the orbit of Io. The encounter was not observed, but at the return of 1889 the comet was seen to be accompanied by four minor companions, which were classified as 'splinters' and did not last for long. Comet P/82 Gehrels 3 was in orbit round Jupiter for some time in the early 1970s, but finally escaped unharmed, and returned to solar orbit.

One comet that will certainly never be seen again is Shoemaker–Levy 9, which impacted Jupiter in July 1994. We are not certain about P/85 Bothin, discovered in 1975 by the Reverend Leo Bothin. Its period was given as 11.2 years, and at the return of 2008 it was due to be contacted by a space-craft. Unfortunately, the comet cannot be found, and may have broken up, so that the space-craft has been diverted to a different target. One can almost picture NASA's plaintive request: 'Please, can we have our comet back?'

Comets can make close approaches to Earth (Table 14.6). Excluding the comet of 1491, whose orbit is highly uncertain, the record is still held by Lexell's Comet of 1770, but has been suggested that the Tunguska impact of 1908 was due to a small comet, or perhaps a fragment of one. Neither can we rule out the ideas that the disaster of 65 million years ago, which allegedly wiped out the dinosaurs, was due to a comet rather than to a meteoroid or asteroid.

All these short-period comets have direct motion, but with longer periods we begin to come to retrograde comets, One of these is P/109 Swift–Tuttle, which will next return in 2126. For a time there were fears that it could be on a collision course; it now seems that a direct hit is unlikely, but it cannot be denied that the comet will come unpleasantly close.

It is unlikely that any comet exists with a period shorter than that of P/2 Encke. The longest period comet seen at more than one return is P/153 Ikeya–Zhang (366.5 years), seen by Chinese and Japanese astronomers in 1661 and recovered in 2002 by the Chinese observer Zhang Daqing. It rose to magnitude 3.5, and developed a very respectable tail. Look out for it again in June 2369!

HALLEY'S COMET

Much the most famous of all comets is P/l Halley. It may have been recorded by the Chinese as early as 1059 BC; since 240 BC it has been seen at every return. The mean period is 76 years.

There are many historical references to Halley's Comet. In 684, Ma-tuan-lin, the Chinese historian, referred to a comet seen in the western sky during September and October; this was certainly Halley's, and the first known drawing of it relates to this return. The drawing was published in the *Nuremberg Chronicle*; this was printed in 1493, and shows woodcuts by the German artist M. Wolgemurh. In 837, the comet was at its very best; on 11 April, it was a mere 0.03 a.u. from the Earth (4 500 000 km) and the tail extended over 93°, while the brightness of the coma rivalled Venus. The return of 1066 was shown in the Bayeux Tapestry; King Harold is tottering on his throne, while his courtiers gaze up in horror.

The return of 1301 was favourable; one man who saw it was the Florentine painter Giotto di Bondone, who later used it as a model for the Star of Bethlehem in his *Adoration of the Magi*. (In fact there is no chance that the comet can be identified with the Star of Bethlehem; it returned years too early.) At the return of 1456, the comet was again bright, and as usual, was regarded as an evil omen. At that time the Turkish forces were laying siege to Belgrade, and on the night of 8 June it was said that 'a fearsome apparition appeared in the sky, with a long tail like a dragon'. The current Pope, Calixtus III went so far as to preach against the comet as an agent of the Devil, although it is unlikely that he excommunicated it, as has sometimes been claimed.

In 1682, the comet was seen by Edmond Halley (the actual discovery was made on 15 August of that year by G. Dorffel) and subsequently Halley decided that it must be identical with comets previously seen in 1607 and in 1531. He predicted a return of 1758. On Christmas night of that year the comet was duly found, by the German amateur Palitzsch, and passed through perihelion in March 1759. This was the first predicted cometary return. Since then the comet has been back in 1835, 1910 and 1986.

In 1835, the comet was recovered on 6 August by Dumouchel and di Vico, from Rome, close to the predicted position near the star ζ Tauri. It remained prominent for weeks later in the year, and was followed until 20 May 1836; the last observation of it was made by Sir John Herschel from the Cape. For 1910, very accurate predictions were made by P. Cowell and A. C. D. Crommelin, from Greenwich; the discovery was made on 12 September 1909 by Max Wolf, from Germany, and the comet was followed until 15 June 1911, by which time its distance from the Sun was over 800 000 000 km. It was brilliant enough to cause general interest – although it was not so bright as the non-periodical 'Daylight Comet', which had been seen earlier in 1910, several weeks before Halley's Comet reached its brightest magnitude. On 18–19 May 1910 the comet passed in transit across the face of the Sun. The American astronomer F. Ellerman went to Hawaii to observe under the best possible conditions, but could seen no trace of the comet.

The 1910 return was the first occasion when the comet could be studied with photographic and spectroscopic equipment, and it was fortunate that the comet was well placed. The Earth was closest to the comet on 20 May, at a range of around 21 000 000 km; the closest encounter between the Earth and the comet's tail was about 400 000 km, and there was some public unease because it had become known that comet tails contain unpleasant substances such as cyanogen. This is true enough, but the density of a tail is so low that there can be no possible ill-effects on this score. At its best, the tail was at least 140° long. The comet was indeed a magnificent sight, even if it could not equal the Daylight Comet of the preceding January.

The last perihelion occurred on 9 February 1986. The comet was recovered on 16 October 1982 by a team of astronomers at Palomar (Jewitt, Danielson and Dressler) who used the Hale reflector to detect the comet as a tiny blur of magnitude 24.3; it was a mere 8 arcsec away from its predicted position. The discovery was confirmed shortly afterwards from Kitt Peak. At the time of its recovery, the comet was still moving between the orbits of Saturn and Uranus.

Unfortunately, this was the most unfavourable return for many centuries, and although the comet became an easy naked-eye object it was never brilliant. It rose to the sixth magnitude by early December 1985, and was at its best in mid March 1986: the nucleus was then brighter than magnitude 2, and there was a very respectable tail, showing a great deal of structure. The comet was well south of the celestial equator when at its brightest; at one time it was close to the globular cluster ω Centauri, and with the naked eye the comet and the cluster looked very similar. On 24 April 1986, there was a total eclipse of the Moon and for many people (including myself) this was the last chance to see the comet without optical aid; the magnitude had by then fallen to magnitude 4.5, slightly brighter than the adjacent star α Crateris. The fan-shaped tail was still very much in evidence.

SPACE MISSIONS TO COMETS

By 1986, space-probes had been developed, and five missions were dispatched: two Russian, two Japanese and one European (Table 14.7). (The Americans withdrew on the grounds of expense.) All the

Table 14.7 *Cometary probes, 1978–2000*

Spacecraft	Launch date	Comet	Nearest to comet (km)	Closest approach to comet (km)
ISEE/ICE	12 Aug 1978	P/21 Giacobini–Zinner	11 Sept 1986	7800
Vega 1	15 Dec 1984	P/1 Halley	6 Mar 1986	8890
Vega 2	21 Dec 1984	P/1 Halley	9 Mar 1986	8030
Sakigake	8 Jan 1985	P/1 Halley	11 Mar 1986	7 000 000
Giotto	2 July 1986	P/1 Halley,	14 Mar 1986	596
		P/26 Grigg–Skjellerup	10 July 1992	200
Suisei	18 Aug 1985	P/1 Halley	8 Mar 1986	150 000
Deep Space	24 Oct 1998	P/19 Borrelly	21 Sept 2001	~2000
Stardust	7 Feb 1999	P/81 Wild 2	Jan 2004	~145
Deep Impact	12 Jan 2005	P/10 Tempel 2	4 July 2005	Impact

Halley probes were successful. The European mission, named Giotto in honour of the painter, was programmed to pass into the comet's inner coma and image the nucleus, but prior information sent back by the Japanese and Russian missions was invaluable. Giotto passed within 596 km of the comet's nucleus on the night of 13–14 March 1986. It carried a camera, the HMC (Halley Multicolour Camera) and this functioned until 14 seconds before closest approach to the nucleus, when it was made to gyrate by the impact of a dust particle probably about the size of a grain of rice and communications were temporarily interrupted; in fact the camera never worked again, and the closest image was obtained at 1675 km from the nucleus. The nucleus itself measured 15 km × 8 km × 8 km, and was shaped rather like a peanut; it had a total volume of over 500 km and a mass of from 50 000 million to 100 000 million tonnes. The mean density was 0.1–0.2 g cm^{-3}; it would take 60 000 million comets of this mass to equal the mass of the Earth.

The nucleus was dark-coated, with an albedo of 2–4%. Water ice appeared to be the main constituent of the nucleus (84%) followed by formaldehyde and carbon dioxide (each around 3%) and smaller amounts of other volatiles, including nitrogen and carbon monoxide. The shape of the terminator showed that the central region was smoother than the ends; a bright patch 1.5 km in diameter was assumed to be a hill, and there were features which appeared to be craters, around 1 km across. Dust-jets were active, although from only a small area of the nucleus on the sunward side. The sunward side was found to have a temperature of 47 °C, far higher than expected, and from this it was inferred that the icy nucleus was coated with a layer of warmer, dark dust. The icy nucleus was eroded at around 1 cm per day near perihelion, and at each return the comet must lose around 300 000 000 tonnes of material. The rotation period was found to be 53 hours with respect to the long axis of the nucleus, with a 7.3-day rotational period around the axis; the nucleus was in fact 'precessing' rather in the manner of a toppling gyroscope.

As the comet drew away from the Sun, activity naturally died down. Observations made with large telescopes – notably by R. West with the 1.54-m Danish telescope at La Silla – showed that in April to May 1988 and January 1989 the images were still diffuse, indicating some residual activity or possibly a cloud of dust. However, by February 1990, when the distance from the Sun was 12.5 a.u., and the magnitude had fallen to 24.3, the image appeared stellar. Then, on 12 February 1991, C. Hainaut and A. Smette, with the Danish telescope, recorded a major outburst; the magnitude rose to 18.9, even though the distance from the Sun had increased to 14.5 a.u.

On 22 February, Smette used the New Technology Telescope at La Silla to obtain a spectrum. The coma showed a solar-type spectrum, with no emission features, which indicated a dust composition. It was subsequently found that structures within the coma varied with time, while the central region faded by about 1 magnitude per month. It seems that a fan-like structure in the approximate direction of the Sun reached a radius of 61 000 km on 13 February, expanding to 142 000 km by 12 April. If the expansion of the coma material were about 14.5 m s^{-1}, the actual outburst would have occurred on 17 December 1990, lasting for three months or so. A short explosive event is ruled out – it would have involved higher velocities for the dust than were observed.

The cause of the outburst is uncertain. A collision with a wandering body is possible, but seems unlikely; possibly a pocket of volatile carbon monoxide ice was exposed to sunlight and the vaporizing gases carried the dust particles away from the nucleus, but this also seems improbable in view of the comet's distance from the Sun. We may have to await another return before solving the problem.

The last image of the comet was obtained on 11 January 1994. By June 1994, the comet had reached the halfway point between perihelion and aphelion. It will next reach aphelion in 2024. Unfortunately the return of 2061 will be as poor as that of 1986: for another really good view we must wait for the return of 2137.

OTHER COMETARY MISSIONS

Quite apart from the 'Halley fleet', probes have been sent to other comets; during the first decade of the new century, Comets Borrelly, Wild 2 and Tempel 1 were targeted, with considerable success. There was also one major failure: CONTOUR, the Comet Nucleus Tour, launched by NASA on 4 July 2002. It was scheduled to encounter P/2 Encke in November 2003, P/73 Schwassmann–Wachmann 3 in June 2006 and P/6 D'Arrest in August 2008, but it lost contact in August 2002, presumably because of a rocket failure, and nothing more was heard from it.

Although the Americans did not contribute to the Halley's Comet programme, they did at least send a probe to the periodical comet P/ Giacobini–Zinner. They used an older probe, ISEE (the International Sun–Earth Explorer), which had been launched in 1978 for a completely different purpose, and had been orbiting the Earth monitoring the effects of the solar wind on the Earth's outer atmosphere. It carried a full complement of instruments, and had a large fuel reserve. On 10 June 1982, it was re-named ICE (the International Cometary Explorer) and began a series of manœuvres and orbital changes, involving a sequence of 'swing-by' passages around the Moon; at the pass of 22 December 1983, ICE was a mere 196 km from the lunar surface. The closest approach to the comet's nucleus occurred on 11 September 1986 (before the Halley armada reached its target); the range was 7800 km and the relative velocity was 20.5 km s^{-1}. The probe took 20 min to cross the ion tail, and collisions with dust grains were recorded as well as magnetic effects. The distance from Earth was then 70 000 000 km.

The Giotto probe was put into 'hibernation' in April 1986 and was re-activated on 19 February 1990. On 2 July, it flew past Earth at 22 730 km, and used the gravity-assist technique to put it into a path to rendezvous with comet P/ Grigg–Skjellerup. After a further hibernation period, Giotto was again reactivated on 4 May 1992, when it was 219 000 000 km from Earth and on 10 July 1982 it encountered Grigg–Skjellerup, passing only 200 km from the nucleus. Most of the instruments on the space-craft were still working, apart from the camera, and valuable data were secured. Grigg–Skjellerup is a much older comet than Halley, and seldom produces a tail; however, the density of the gas near the nucleus was greater than expected, and there was a good deal of fine dust. It was found that the gas coma extended to at least 50 000 km beyond the visible boundary. Giotto suffered no damage, although it was hit by a particle about 3 mm across. Giotto is still in solar orbit, although it does not retain sufficient gas to send it on to another comet, as had originally been hoped.

In October 1998, NASA sent up a space-craft, Deep Space 1, to obtain close-range images of Borrelly's Comet (P/19), which had been discovered in 1904 by A. L. N. Borrelly and whose orbit was well known; the period is 6.8 years. Deep Space 1 was a great success. On its journey it obtained good pictures of asteroid 9969 Braille (29 July 1999) and flew past Borrelly on 22 September 2001 at a range of less than 2200 km.

The nucleus of the Borrelly's Comet is 8 km long and 4 km wide. The overall albedo is low (around 4%); the surface is very rough, with deep fractures and some plains. The encounter took place at a distance of 221 000 000 km from Earth and 204 000 000 km from the Sun, a week after the comet passed through perihelion. Images from the probe showed a narrow, sunward-pointing jet of vaporised ice and dust streaming into the coma. It seems that the dark material is a veneer of carbon- and organic-rich slag left behind as the comet's ices sublimate and escape into space.

Deep Space 1 had done all that its makers had expected. On 18 December 2001, it was 'retired' – that is to say switched off and left to orbit the Sun. We wish it well. The Stardust probe, NASA's next foray, was launched on 7 February 1999. The target this time was Comet P/81 Wild 2, discovered in 1978 by Paul Wild – who is a distinguished Swiss astronomer, so that his surname should be pronounced 'vilt'.

This comet has had a chequered history. It was formed in the early period of the Solar System and had a very distant, almost circular orbit. It was then perturbed, for some unknown reason, and drew inward; in 1974, it passed within a million kilometres of Jupiter, and was forced into a short-period orbit. Instead of taking more than 40 years to complete one journey round the Sun, it now takes a mere 6.4 years.

Stardust was programmed to fly past the comet, collect dust samples from the coma, and then return to the neighbourhood of the Earth and deposit the container with its precious samples. (Obtaining these samples was by no means easy; they had to collide with the probe and be embedded in aerogel, which is the least dense of all known solids, and has been nicknamed 'frozen smoke'.) During its outward journey the space-craft by-passed asteroid 5535 Annefrank at a range of 3079 km, and sent back images showing an irregular, cratered body, measuring 66 × 50 × 34 km. Its albedo was about 24%.

On 2 January 2004, Stardust flew past the comet 145 km from its surface, and thousands of particles were collected in the aerogel. The particles came from the coma, and had been released from the coma only a few hours earlier; they were of special interest, because Wild 2 had been moving in the inner Solar System for only a short time, cosmically speaking. The capsule landed back on Earth on 15 January 2006, ending its journey in Utah's Great Salt Lake desert – the entry speed was a record 12.9 km s^{-1}; a bright fireball was seen over Utah and eastern Nevada, and there was a sonic boom. The capsule was recovered intact, while Stardust itself was put into an orbit which kept it away from re-entering the atmosphere. The plan is to send it on to survey Comet Tempel 1 in February 2011, under the designation NExT (New Exploration of Tempel 1). The main object is to measure the size of the crater made in the nucleus by Deep Impact.

The samples, the largest of which is a millimetre in diameter, show a wide range of organic compounds as well as crystalline silicates such as olivine and pyroxene. These silicates indicate that the samples come partly from the Solar System and partly from interstellar space. The nucleus of Wild 2 is in the form of an oblate spheroid, 4.5 km across, with impact craters, pillars, and cliffs up to 100 metres high. Small jets were observed.

Rosetta is a European mission, launched on an Ariane 5 rocket on 2 March 2004. Its target is Comet P/67 Churyumov–Gerasimenko. There are two parts to the mission; the Rosetta space-probe and the Philæ lander (named after the Nile island where an obelisk was found, immensely helpful to scholars deciphering the Rosetta Stone). The current plan is to map the comet in 2014 and escort it round the Sun from November 2014 to December 2015; Philæ will land in November 2014. On contact with the surface two harpoons will be fired to prevent Philæ from bouncing off (remember that the comet's escape velocity is very low).

On the way, Rosetta was scheduled to fly past asteroids 2867 Šteins (5 September 2008) and 21 Lutetia (10 July 2010). The Šteins imaging was successful; Rosetta passed at a range of 800 km, showing a body with a maximum spread of 5 km, with a prominent crater. Šteins is irregular in shape, with its wide section tapering off to a point. Its rotation period was found to be 6 hours; it takes 3.6 years to complete one orbit round the Sun, at a mean distance of 353 million km.

NASA's Deep Impact probe was launched on 12 January 2005. The plan was to encounter a well-known periodical comet, P/9 Tempel 2, and deliver an impactor, which would produce a large crater and permit astronomers to learn something definite about the comet's composition. Nobody was at all sure what would happen; it was even thought possible that the impactor would simply sink down into a sponge-like surface. There seemed no chance of destroying the comet, but inevitably there were strident protesters. ('Hands off the comet!' 'Leave Tempel alone!' – and so on.) Disregarding these, NASA went ahead, and on 12 January 2005 Deep Impact began a journey of 429 000 000 km, lasting 174 days at a cruising speed of 28.6 km s^{-1}. On nearing the comet, on 3 July, the impactor was separated from the fly-by section, and 24 hours later scored a direct hit.

It was certainly dramatic. The 350-kg copper impactor slammed into Tempel 2 at a relative speed of 10.3 km s^{-1}, so that it was as effective as almost 5 tonnes of TNT would have been. A huge cloud of material was thrown up; minutes later the fly-by passed the nucleus at a range of 500 km, taking pictures of the ejecta plume and the whole of the nucleus, though the crater was masked by the débris. The event was also photographed by Earth-based telescopes and by orbital observatories, notably by Chandra, Spitzer, XMM-Newton, the Hubble Space Telescope and Rosetta, which was then about 80 000 000 km away. Rosetta's spectroscope was even able to analyse the composition of the gas-and-dust ejecta cloud.

When the débris cleared, it revealed a crater 100 m wide and 30 m deep. Outgassing from the impact continued for almost two weeks, and the comet lost at least 250 000 000 kg of water and 20 000 000 kg of dust, which had no permanent effect. The material ejected was of a consistency more like talcum powder than sand; clays and silicates were identified. It was estimated that the comet was about 75% empty space, and the outer layers were likened to a snowbank. Astronomers tended to think that Tempel 2 had been born in the outer part of the Solar System, well beyond the orbits of Uranus and Neptune.

The fly-by was still fully functional, and was due to go on to a rendezvous with Comet Boethin. However, Boethin inconveniently disappeared, and so Deep Impact's next trip (known as EPOXI: Extrasolar Planet Observation and Deep Impact Extended Investigation) was re-routed to another comet, P/ Hartley 2, which was brighter than Boethin had been, and presumably less likely to break up see p. x. Before leaving Deep Impact, I cannot resist mentioning Marina Bay, a Russian astrologer, who in 2005 sued NASA for $300 million on the basis that the impact 'ruined the natural balance of forces in the universe', upsetting her horoscopes. Alas, the Moscow Court ruled against her, and NASA was completely unfased!

BRILLIANT COMETS

Brilliant comets have been seen now and then all through the historical period, although early reports, most of them Chinese, are bound to be rather vague. A selected list of bright comets between the years 1500 and 1900 is given in Table 14.8. (In fact Sarabat's Comet of 1729 may have been the largest ever observed, but it was never less than 4.05 a.u. from the Sun and so did not become bright in our skies.)

The Great Comet of 1744 attained magnitude −7, and was visible in broad daylight when only 12° from the Sun. At perihelion it was only 33 000 000 km from the Sun, well inside the orbit of Mercury, and it had at least six bright, broad tails. The Great Comet of 1811, discovered by Honoré Flaugergues on 25 March, was also a daylight object; it had a coma about 2 000 000 km in diameter, and a tail which extended for 160 000 000 km. *En passant*, the wine crop in Portugal was particularly good, and for years afterwards 'Comet Wine' appeared in the price lists of wine merchants. A bottle was sold at Sotheby's, in London, in 1984. (It would be interesting to know what it tasted like.)

The brightest comet of modern times was probably that of 1843. According to the famous astronomer Sir Thomas Maclear it surpassed the comet of 1811, and Maclear saw both. Donati's Comet of 1858 was said to be the most beautiful of all: it was discovered by G. Donati from Florence, on 2 June 1858 and was finally lost on 4 March 1859. It had a wonderfully curved main tail and two smaller ones; the tail length was around 80 000 000 km. The period is unknown, but may be of the order of 2000 years.

Tebbutt's Comet of 1861 was brilliant, and it seems that the Earth passed through its tail on 30 June. Despite some unconfirmed reports of an unusual daytime darkness and a yellowish sky, no unusual phenomena were seen.

The first photograph of a comet (Donati's) was taken on 27 September 1858 by an English portrait artist, Usherwood, with an f/2.4 focal ratio portrait lens; but the first really good picture was taken in 1882 of Cruls' Comet, at the instigation of Sir David Gill. Many stars were also shown, and it was this picture which made David Gill, Director of the Cape Observatory, appreciate the endless potentialities of stellar photography. Earlier in 1882, on 17 May, a comet was found on an image of the total eclipse of the Sun, seen from Egypt. The comet had never been seen before and it was never seen again, so that this is the only record of it; it is generally referred to as Tewfik's Comet, in honour of the Khedive, ruler of Egypt at the time, who had made the astronomers very welcome.

THE TWENTIETH CENTURY

The last century has been relatively poor in brilliant comets; only those of 1910 and 1965 have come anywhere near to matching the splendour of the shadow-casting comets of the Victorian era. A list of selected bright twentieth-century comets is given in Table 14.9.

The Daylight Comet of 1910 was first seen on 13 January by some diamond miners in South Africa. It passed perihelion on 17 January and earlier had been seen with the naked eye when only 4.5° from the Sun. It was much brighter than Halley's – and people who claim to have seen Halley's Comet in 1910 usually saw the Daylight Comet instead. Its orbit is elliptical, but the period seems to be of the order of 4 000 000 years. The bright comet of 1948 was, like Tewfik's, discovered fortuitously during a total solar eclipse, but was subsequently followed and remained under observation until April 1949, when it had faded to the 17th magnitude. It will

Table 14.8 *Selected list of brilliant comets, 1500–1900*

Comet	Name	Discovery	Perihelion	Maximum magnitude	Naked-eye visibility	
1577	Tycho Brahe	1577 Nov 1	1577 Oct 27	−4	1577 Nov–1578 Jan	Possibly brighter than −4
1585	–	1585 Oct 13	1585 Oct 8	−4	1585 Oct–Nov	Discovered by Chinese
1665	–	1665 Mar 27	1665 Apr 24	−4?	1665 Mar–Apr	Observed by Hevelius
1677	Hevelius	1677 Apr 27	1677 May 6	−4?	1677 Apr–May	Long, thin tail
1695	Jacob	1695 Oct 28	1695 Oct 23	−3?	1695 Oct–Nov	40° tail: probably a Sun-grazer
1702	–	1702 Feb 20	1702 Feb 15?	?	1702 Feb–Mar	42° tail; discovered at Cape
1744	de Chéseaux	1743 Nov 29	1744 Mar 1	−7	1743 Dec–1744 Mar	Multi-tailed comet; Discovered by Klinkenerg; independently by de Chéseaux
C/1811 F1 (1811 I)	Flaugergues	1811 Mar 25	1811 Sept 12	0	1811 Mar–1812 Jan	Great Comet: 20′ coma, 24° ion tail
C/1819 N1 (1811 II)	Tralles	1819 July 2	1819 June 28	1	1819 July	Transited Sun (unobserved), 26 June
C/1843 D1 (1843 I)	Great Comet	1843 Feb 8	1843 Feb 27	−6	1843 Feb–Apr	Brighter than Comet of 1811; Sun-grazer.
C/1858 L1 (1858 VI)	Donati	1858 June 2	1858 Sept 30	−1	1858 June–Nov	Most beautiful of all comets: ion and dust tails, up to 60°
C/1861 N1 (1861 II)	Tebbutt	1861 May 13	1861 June 12	−2	1861 May–Aug	Earth passed through the 100° tail on June 30
C/1874 H1 (1874 III)	Coggia	1874 Apr 17	1874 July 9	−1	1874 Apr–Aug	63° tail
C/1880 C1 (1880 I)	Great Comet	1880 Feb 1	1880 Jan 28	3	1880 Feb	Southern hemisphere comet; Sun-grazer
C/1881 K1 (1881 III)	Tebbutt	1881 May 22	1881 June 16	1	1881 May–July	20° tail
C/1882 F1 (1882 I)	Wells	1882 Mar 18	1882 June 11	0	1882 May–June	Yellow colour pronounced
C/1882 R1 (1882 II)	Great Comet (Cruls)	1882 Sept 1	1882 Sept 14	−4	1882 Sept–1883 Feb	Transited Sun: perhaps as bright as magnitude −10 (transit unobserved); Sun-grazer
C/1887 B1 (1887 I)	Great Comet	1887 Jan 18	1887 Jan 11	2	1887 Jan	'Headless' comet; long, narrow tail

return in around 95 000 years. Of course, all periods of this order are very uncertain; some estimated values are given in Table 14.10.

The brightest twentieth-century comet was that of 1965 Ikeya–Seki, discovered on 18 September by two Japanese observers. It was a daylight object, and could be seen when only 2° from the Sun, but it faded quickly, and was never really well seen from Britain. The period has been given as 880 years. Other fairly conspicuous comets were Arend–Roland (1957), Bennett (1970) and West (1975). Kohoutek's Comet of 1973 was a disappointment. It was found on 7 March by L. Kohoutek from Hamburg, when it was still 700 000 000 km from the Sun. Few comets are detectable as far away as this, and the comet was expected to become a magnificent object in the winter of 1973–4, but it failed to come up to expectations even though it was visible with the naked eye. It was, however, scientifically important, and was carefully studied by the astronauts then aboard the US space-station Skylab (Carr, Gibson and Pogue). Perhaps it will do better when it next returns to the Sun, in approximately 75 000 years' time.

Two splendid comets were seen near the close of the millennium. The first, C/1996 B2, was discovered on 30 January by the Japanese amateur Yuji Hyakutake. It passed perihelion on 1 May, and was then striking in the far north of the sky; it had a long tail – the length was subsequently found to be over 500 000 000 km as found by the Ulysses space-probe, which passed through it in 2000. Its beauty was enhanced by its greenish colour. It was in fact a very small comet, and owed its brilliance to its closeness to the Earth. Its original period seems to have been about 8000 years, but its orbit was altered during its journey through the inner Solar System, and the next return is likely to be postponed for 14 000 years.

On 23 July 1995 two American astronomers, Alan Hale and Thomas Bopp independently discovered the comet which was destined to become the most celebrated of recent years. Had it come as close to us as Hyakutake had done, it would have cast obvious shadows. It was an exceptionally large comet – the nucleus was at least 40 km in diameter – and there were both ion and dust tails, plus a much fainter tail made up of sodium. It was a pity that

Table 14.9 *Bright naked-eye comets, 1900–2010*

Designation				
New	Old	Name	Naked-eye visibility	Maximum magnitude
C/1901 G1	1901 I	Viscara	1901 Apr–May	−1.5
C/1910 A1	1910 I	Daylight Comet	1910 Jan–Feb	1
1P	1910 I	Halley	1910 Feb–July	0
C/1911 S3	1911 IV	Beljawsky	1911 Sept–Oct	1
C/1911 O1	1911 V	Brooks	1911 Aug–Nov	2
C/1927 X1	1927 IX	Skjellerup–Maristany	1927 Dec–1928 Jan	−6
C/1941 B2	1941 IV	De Kock–Paraskevopoulos	1941 Jan–Feb	2
C/1947 X1	1947 XII	Southern Comet	1947 Dec	−1
C/1948 VI	1948 XI	Eclipse Comet	1948 Nov–Dec	−2
C/1956 R1	1957 III	Arend–Roland	1957 Mar–May	1
C/1957 P1	1957 V	Mrkós	1957 July–Sept	1
C/1961 O1	1961V	Wilson–Husband	1961 July–Aug	3
C/1962 C1	1962 III	Seki–Lines	1962 Feb–Apr	−2.5
C/1965 S1	1965 VIII	Ikeya–Seki	1965 Oct–Nov	−10
C/1969 Y1	1970 II	Bennett	1970 Feb–May	0.5
C/1970 K1	1970 VI	White–Ortiz–Bolelli	1970 May–June	0.5
C/1973 E1	1973 XII	Kohoutek	1973 Nov–1974 Jan	0
C/1975 VI	1976 VI	West	1976 Feb–Apr	−2
P/1	–	Halley	1986 Jan–Dec	1
C/1996 B2	–	Hyakutake	1996 Mar–May	−0.2
C/1995 O1	–	Hale–Bopp	1996 July–1997 Oct	−1
P/153		Ikeya–Zhang	2002 Mar	3.5
C/2006 P1		McNaught	2006–2007	−6.0

it never came close; however, it remained a prominent naked-eye object for over a year, and was truly beautiful – it must be the most photographed comet in history. There were marked changes in the tails, and a spiral structure in the coma. Apparently it was last at perihelion 4200 years ago, and will be back in 2360 years' time. Its orbital inclination is over 89°, so that its path lies almost at right angles to that of the Earth. Perihelion was passed on 1 April 1997. The axial rotation perihelion was given as 11.4 h.

THE TWENTY-FIRST CENTURY

During the first decade of the new century there were several fairly conspicuous comets (see Table 14.9) and one truly Great Comet, C/2006 P1. It was discovered on 7 August by R. McNaught, as a faint object in Ophiuchus – around magnitude 17. As it drew inward it brightened steadily, and by January 2007 had risen to magnitude −6, brighter than Venus, so that it was much the most brilliant since Ikeya–Seki of 1965. It reached perihelion on 12 January, when it was 0.17 a.u. from the Sun (25 500 000 km). It was closest to the Earth on 15 January, when its distance from us was 123 000 000 km. Unfortunately for northern-hemisphere observers it was best seen from south of the equator. On 3 February 2007 the Ulysses space-craft passed through the comet's long, curved tail, at around 250 000 000 km from the nucleus –even at that distance, the tail had slowed down the rate of the solar wind by about 50%. The orbit is sensibly parabolic, so that Comet McNaught will probably never return to the Sun.

SUN-GRAZING COMETS

Some Great Comets have been seen when moving very close to the Sun; such were those of 1106, 1688, 1689, 1695, 1702, 1843 (C/1843 D1) 1880 (C/1880 C1), 1882 (Tewfik, X/1882 K1) and 1965 (C/1965 S1, Ikeya–Seki). These survived, but others did not, such as C/1979 Q1 (Howard–Kooman–Michels), which fell into the Sun on 31 August of that year; its demise was clearly imaged. Kreutz suggested that these 'Sun-grazers' might be the remnants of a single giant comet which broke up near perihelion, and they are often called Kreutz comets. Recently, however, the SOHO (Solar and Heliospheric Observatory) space-craft has shown that these cometary suicides are extremely common. We might even refer to them as Kamikaze comets!

THE ORIGIN OF COMETS

Comets are very ancient objects – as old as the Solar System itself. Since they lose material at every return to perihelion, it follows that the comets we now see cannot have remained in their present orbits for thousands of millions of years. They must have come from afar. They are almost certainly bona-fide members of the Solar System. If they came from interstellar space, they would move at greater velocities than are actually found.

There is little doubt that comets formed from material that never condensed into a planet or a satellite; they may be regarded as cosmical débris. (I once defined a comet as being 'the nearest

Table 14.10 *Comets of very long period. Obviously, the periods are very uncertain!*

Comet	Year	Period (years)	Perihelion distance, q (a.u.)	Eccentricity	Inclination (°)
Great Comet	1861 II	409	0.822	0.985	85
Great Comet	1843 I	517	0.0055	0.999 91	14
Great Comet	1882 II	759	0.0077	0.999 91	14
Ikeya–Seki	1965 VIII	880	0.008	0.9999	14
Pereyra	1963 V	903	0.0051	0.999 95	14
Bennett	1970 II	1678	0.538	0.996	90
Donati	1858 VI	1951	0.578	0.996	11
Flaugergues	1811 I	3096	1.035	0.995	10
Hale–Bopp	1997	2360	0.913	0.9951	89
1680 Comet	1680	8917	0.006	0.9999	6
Hyakutake	1996	14000	0.230	0.999	12.5

Comets now in hyperbolic orbits include Morehouse (1908), Arend–Roland (1957) and Kohoutek (1973).

Table 14.11 *Selected list of Kreutz Sun-grazing comets*

Comet	Name	Perihelion date	Perihelion distance (a.u.)	Magnitude
1106[a]	–	1106 Feb 2[a]	?	−5[a]
1668[a]	–	1668 Mar 1[a]	?	0[a]
1689[a]	–	1689 Sept 2[a]	?	3[a]
1695[a]	–	1695 Oct 23[a]	?	?
1702[a]	–	1702 Feb 15[a]	?	?
C/1843 D1	Great Comet	1843 Feb 27.9	0.0055	−6[a]
C/1880 C1	–	1880 Jan 28.1	0.0055	3
X/1882 K1	Tewfik	1882 May 17.5	?	−1?
C/1882 R1	–	1882 Sept 17.7	0.0077	−4
C/1887 B1	–	1887 Jan 11.9	0.0048	2
C/1945 X1	du Toit	1945 Dec 28.0	0.0075	7
C/1963 R1	Pereyra	1963 Aug 24.0	0.0051	2
C/1965 S1	Ikeya–Seki	1965 Oct 21.2	0.0078	−10
C/1970 K1	White–Ortiz–Bolelli	1970 May 14.5	0.0089	0.5
C/1979 Q1	Howard–Kooman–Michels (SOLWIND 1)	1979 Aug 30.9	Impacted	−4

Between 1979 and 1999 SOLWIND discovered six Sun-grazers, SMM (Solar Maximum Mission satellite) discovered 10, and SOHO (the Solar and Heliospheric Observatory satellite) discovered 46. By March 2000 SOHO had discovered over 100 comets, many of them very close to the sun.

[a]Very uncertain.

approach to nothing that can still be anything'.) It is generally believed that short-period comets come from the Kuiper Belt, while long-period comets come fom the Oort Cloud, but we cannot claim to know a great deal about the early history of the Solar System.

There have often been suggestions that some comets, such as Hale–Bopp, may have been born around other stars, a theory supported recently by a Queen's University, Belfast, team led by Martin Duncan. The Sun was presumably born in a cluster of stars, each of which was associated with a disc of material; comets were flung out of these discs and became 'free-floaters' before being captured. This is, of course, speculative.

The 'panspermia' theory was due to the Swedish scientist Svante Arrhenius, whose work was good enough to win him the Nobel Prize for Chemistry in 1903. Arrhenius believed that life was brought to the Earth by way of a meteorite, but the theory never became popular, because it seemed to raise more problems than it solved. The same sort of theme has been followed up much more recently by the late Sir Fred Hoyle and his colleague Chandra Wickramasinghe, who believe that comets can deposit harmful bacteria in the Earth's upper atmosphere, thereby causing epidemics. Again there has been little support. When will the next Great Comet grace our skies? We cannot tell; it may be tomorrow, or it may not be for centuries. Time will tell.

15 · Meteors

Meteors are known commonly as shooting-stars. They are produced by small and friable particles (meteoroids), usually of no more than centimetre size, which do not reach the Earth's surface intact; they may be regarded as cosmical débris, very often of cometary origin. Strictly speaking a meteor is the visible event that occurs when a meteoroid dashes into the upper atmosphere, and is vapourised.

There are many annual well-defined showers, associated with comets, which can often be identified; other meteors are sporadic, not associated with any known comet, and so may appear from any direction at any moment. Meteors can of course occur in daylight, as was pointed out by the Roman philosopher Seneca, about AD 20, and may be tracked by radio and radar.

Meteors are not associated with meteorites, most of which come from the asteroid belt. The link with comets was first proposed in 1861 by D. Kirkwood; he believed that meteors were the remnants of comets which have disintegrated – and in some cases this is true enough. In 1862, G. V. Schiaparelli demonstrated the link between the Perseid meteor shower and the periodical comet Swift–Tuttle, and other associations were soon established.

Some well-known periodical comets are the parents of meteor showers. Halley's Comet produces two, the Aquarids of April and the Orionids of October; Comet P/Giacobini–Zinner can occasionally yield rich displays, as in 1933. Biela's Comet, which broke up and was last seen in 1852 produced 'meteor storms' in 1872 and in 1885: in recent years this shower (the Andromedids) has been almost undetectable, but it has been calculated that it may return around 2120, when the orbit of the stream will be suitably placed. The Lyrids, first recorded in 687, are linked with Thatcher's Comet of 1862, which has a period of over 400 years. The rich Geminid shower of December has an orbit very like that of asteroid 3100 Phæthon, and it is widely believed that Phæthon may be the parent of the stream, adding credibility to the suggestion that some near-Earth asteroids may be extinct comets.

EARLY THEORIES

Meteors were once regarded as atmospheric phenomena. Aristotle believed them to be due to vapours from Earth created by the warmth of the Sun; when they rose to great altitudes they caught fire, either by friction or because the column of air around them cooled, so squeezing out the hot vapours rather as toothpaste can be squeezed out of a tube. Even Newton believed that meteors were volatile gases which, when mixed with others, ignited to cause 'Lightening and Thunder and fiery Meteors'. Edmond Halley correctly maintained that they came from space and burned away in the upper air (although, curiously, he seems to have changed his

mind later and reverted to the Aristotelian picture). Myths abounded. The Mesopotamians regarded meteors as evil portents, and to the Moslems they represented artillery in a war between devils and angels. In Sparta, around 1200 BC, the priests surveyed the sky on one special night once in eight years; if a meteor were seen, it indicated the king had sinned and ought to be deposed. In mediæval Brunswick a meteor was a fiery dragon which could cause damage; however, if the observer sheltered and cried out 'Fiery Dragon, come to me', the dragon might relent, and even drop down a ham or a side of bacon!

NATURE OF METEORS

The status of meteors was solved in 1798 by two German students, H. W. Brandes and J. F. Benzenberg of the University of Göttingen. Between 11 September and 4 November they observed meteors from sites 15.2 km apart, giving them a useful 'baseline', and made 402 measurements; in 22 cases they found that the same meteor had been seen from each site, and its track plotted. This made it possible to determine the height of the meteor by the method of triangulation. The heights at which the meteors disappeared ranged between 15 km and 226 km, the mean burnout altitude was found to be 89 km – now known to be very near the truth.

The total number of meteors entering the atmosphere daily has been given as 75 000 000 for meteors of magnitude 5 or brighter. An observer under ideal conditions would expect to see between about 5 and 15 naked-eye meteors per hour (except during a shower, when the number would be higher). Meteors of magnitude −5 or brighter – that is to say, appreciably more brilliant than Venus – are conventionally termed fireballs. Very occasional fireballs, such as those of 20 November 1758 and 18 August 1783, may far outshine the Moon. The 1758 fireball was seen from England, and a contemporary eye-witness report is worth quoting:

> This night a surprising large meteor was seen at Newcastle, about 9 o'clock, which passed a little westward of the town, directly north, and illuminated the atmosphere to that degree, for a minute, that, though it was dark before, a pin might have been picked up in the streets. Its velocity was inconceivably great, and it seemed near the size of a man's head. It had a tail of between two and three yards long, and as it passed, some said that they saw sparks of fire fall from it.

A fireball which ends its career by exploding, sometimes with obvious fragmentation, is often called a bolide, though the term seems to be only semi-official. Strange green fireballs were quite widely reported by American observers in the late 1940s. Experienced meteor authorities, notably Lincoln LaPaz (University of New Mexico), believed them to be artificial. The US Department of Defense set up 'Project Twinkle' in 1950, to investigate the reports, but discontinued it after two years, concluding that the phenomena were meteors of unusual composition. However, the facts that most of the reports came from regions near US military bases, and that very few green fireballs have been seen recently indicate that LaPaz was certainly right.

A meteor may enter the atmosphere at a velocity anywhere between 11 km s^{-1} and 72 km s^{-1}; it will be violently heated as it enters the upper atmosphere at an altitude of about 150 km above sea level. It is not possible to give a precise value, because the Earth's atmosphere has no definite boundary, but simply thins out with increasing altitude until the density is no greater than that of the interplanetary medium. (The Hungarian engineer/physicist Theodore von Kármán proposed that a height of 100 km should be regarded as the edge of space, because at higher altitudes the atmosphere is too tenuous to be of use for aeronautical purposes; this Kármán Limit has never become official, but is often used.) It is often said that the incoming meteoroid is burned away by friction against the air particles, but the sequence of events is slightly more complicated then this.

What is termed 'ram pressure' must be taken into account; it is defined as the pressure exerted on a body which is moving through a resisting medium. The incoming meteor produces a shock wave generated by the very rapid compression of the air in front of the particle. It is mainly this ram pressure, rather than friction, which heats the air, and this in turn heats the particle as it flows around it. The end result is that the whole particle is vapourised; atoms from its outer surface are ablated and collide with molecules in the atmosphere, exciting and ionising them, producing a trail which may extend for many kilometres. There is little deceleration before the meteor is destroyed. What we see, therefore, is not the particle itself (the meteoroid), but the effect that it produces in the atmosphere during the final moments of its existence. Particles below about 0.1 mm in diameter are termed micrometeorites, and do not produce luminous effects. Some are cometary, while others must be classed as Zodiacal 'dust'.

Meteors are easy to photograph – the earliest really good picture, of an Andromedid, was taken by L. Weinek, from Prague, as long ago as 27 November 1885 – but meteor spectra are much more difficult, because one never knows just when or where a meteor will appear. Many spectra have been obtained (largely by amateurs) and it seems that meteors are made up of material of the type only to be expected in view of their cometary origin. Radar studies of meteor trails are now of great importance; the first systematic work was carried out in 1945 by J. S. Hey and his team, with the δ Aquarids. However, amateur observations are still very useful indeed.

METEOR RADIANTS

Because the meteors in any particular shower are moving through space in parallel paths (or virtually so), they seem to come from one set point in the sky, known as the radiant. (The effect may be likened to the view from a bridge overlooking a motorway; the parallel lanes of the motorway will seem to converge at a point near the horizon.) The shower is named after the constellation in which the radiant lies. One exception refers to the January meteors, the Quadrantids; they are named after Quadrans Muralis, a constellation added to the sky in Bode's maps of 1775 but later rejected – its stars are now included in Boötes, but the old name has been retained.

A list of the principal annual showers is given in Table 15.1. A selected list of minor showers is given in Table 15.2, although the low hourly rate of these showers means that the data are decidedly uncertain. The ZHR, or Zenithal Hourly Rate, is given by the number of naked-eye meteors which would be expected to be seen by an observer under ideal conditions, with the radiant at the zenith. In practice, these conditions are never met, so that the observed hourly rate is bound to be rather lower than the theoretical ZHR.

METEOR SHOWERS

On the night of 12–13 November 1833 there was a brilliant meteor shower; the meteors came from the constellation of Leo. It was observed from Yale, in the United States, by Denison Olmsted. H. A. Newton postulated the existence of definite showers; finding that the Leonids had appeared periodically, at intervals of 33 years, he predicted that there should be another major meteor storm in 1866. It duly appeared, although unfortunately Olmsted did not see it (he died in 1859). Subsequently, other showers were identified, initially the Perseids in 1834 (by J. Locke and A. Quetelét), the Lyrids in 1835 (by F. Arago), the Quadrantids in 1839 (by Quetelét and E. Herrick), the Orionids in 1839 (by Quetelét, Herrick and J. Benzenberg) and the Andromedids in 1838 (also by Quetelét, Herrick and Benzenberg).

Some of the recognised showers are consistent, notably the Perseids, while others, such as the Leonids, are very variable in richness. Some radiants are ill-defined, and while some showers are of brief duration – such as the Quadrantids – others spread over weeks. Material may leave a comet either in front of or behind the nucleus. Dust particles ejected from the nucleus may return to perihelion earlier than the comet itself, or may return later; gradually the material is distributed all around the comet's orbit, forming a loop. With older showers, such as the Perseids, this has had sufficient time to happen: with younger showers it has not, so that good displays are seen only when the Earth passes through the thickest part of the swarm. We must also consider what is termed the Poynting–Robertson effect. In re-radiating energy received from the Sun, a particle will lose orbital velocity and will spiral inward towards the Sun; therefore old streams are depleted in small particles, and even smaller particles are ejected altogether by radiation pressure.

The Quadrantid shower is the year's earliest – and one of the richest, but it has not been consistently well observed, partly because January is so often cloudy and damp, partly because the far-north position of the radiant means that few Quadrantids are seen in the southern hemisphere, and partly because the shower is so brief. There is no definitely known parent, but one of the world's

Table 15.1 *Principal meteor showers*

Name	Begins	Max.	Ends	ZHR	RA	Dec. (°)	Comet	
Quadrantids	1 Jan	3 Jan	6 Jan	100	15h 28m	+50	—	Sharp maximum. Can be spectacular.
Virginids	7 Apr	10 Apr	18 Apr	5	13h 36m	−11	—	Slow, long paths. Several radiants in Virgo, Mar–Apr.
Lyrids	19 Apr	22 Apr	25 Apr	10	18h 08m	+32	Thatcher	Occasionally very rich, as in 1803, 1922, 1982.
ηAquarids	24 Apr	4 May	20 May	40	22h 20m	−01	P/ Halley	Multiple radiant, broad maximum.
α Scorpiids	20 Apr	27 Apr	19 May	5	16h 32m	−24	—	Several weak radiants. One max on 12 May.
Ophiuchids	19 May	9 June	July	5	17h 56m	−23	D/Lexell?	Weak activity from several radiants.
α Cygnids	July	{ 21 July, 12 Aug }	Aug	5	21h 0m	+48	—	Weak but prolonged activity. Less rich than formerly.
Capricomids	July	{ 8 July, 15, 26 July }	Aug	5	20h 44m	−15	P/ Honda–Mrkos	Bright meteors. Three maxima. Multiple radiant.
δAquarids	15 July	{ 29 July, 6 July }	20 Aug	{ 20, 10 }	22h 36m	−17	—	Double radiant. Rich but faint meteors.
Piscis Australids	13 July	31 July	20 Aug	5	22h 40m	−30	—	Probably double maximum.
α Capricomids	15 July	2 Aug	20 Aug	5	20h 36m	−10	—	Slow, yellow fireballs. Triple maximum.
Aquarids	July	6 Aug	Aug	8	22h 10m	−15	—	Rich in faint meteors. Double radiant.
Perseids	23 July	13 Aug	20 Aug	80	03h 04m	+58	P/ Swift–Tuttle	Most reliable annual shower. Consistent.
Draconids	Sept	{ 8, 21 Sept, 13 Oct }	Sept	10	00h 36m	+07	—	Weak: multiple radiant.
Orionids	16 Oct	21 Oct	27 Oct	25	06h 24m	+15	P/ Halley	Swift with fine trains. Flat maximum.
Draconids	10 Oct	10 Oct	10 Oct	var	18h 00m	+54	P/ Giacobine–Zinner	Usually weak, but occasional storms. Also known as the Giacobinids.
Taurids	20 Oct	3 Nov	30 Nov	10	03h 44m	+14	P/ Encke	Fine display in 1988. Slow meteors.
Puppids–Velids	27 Nov	9.26 Dec	Jan	15	09h 00m	−48	—	Two of several radiants in Puppis. Vela and Carina.
Leonids	15 Nov	18 Nov	20 Nov	Var	10h 08m	+22	P/ Tempel–Tuttle	Occasional storms (1799, 1833, 1866, 1966).
Andromedids	15 Nov	20 Nov	6 Dec	very low	00h 50m	+55	D/ Biela	Now virtually extinct.
Geminids	7 Dec	14 Dec	17 Dec	75	07h 28m	+32	Phaethon? (asteroid)	Many bright meteors. Consistent. Can be even richer than the Perseids.
Ursids	17 Dec	21 Dec	25 Dec	10	14h 28m	+78	P/ Tuttle	Usually weak, but good displays 1945, 1982, 1986.

Permanent daytime showers include the Arietids (29 Mar–17 June), the ξ Perseids (1–15 June) and the β Taurids (23 June. 7 July). The β Taurids seem to be associated with Encke's Comet.

Table 15.2 *Selected list of minor annual meteor showers*

Name	Begins	Max.	Ends	
ζ Aurigids	11 Dec	31 Dec	21 Jan	Slow meteors
Boötids	9 Jan	15 Jan	18 Jan	
δ Cancrids	14 Dec	15 Jan	14 Feb	Weak shower
η Carinids	14 Jan	21 Jan	27 Jan	
η Craterids	11 Jan	16 Jan	22 Jan	Rapid meteors
ρ Geminids	1 Jan	10 Jan	15 Jan	Ill-defined
α Hydrids	15 Jan	20 Jan	30 Jan	
Aurigids	31 Jan	7 Feb	23 Feb	Some bright fireballs
α Centaurids	2 Feb	8 Feb	25 Feb	
δ Leonids	5 Feb	22 Feb	19 Mar	ZHR ~3
η Draconids	22 Mar	30 Mar	8 Apr	
β Leonids	14 Feb	20 Mar	25 Apr	ZHR 3–4
δ Mensids	14 Mar	18 Mar	21 Mar	Weak shower
γ Normids	11 Mar	16 Mar	21 Mar	
η Virginids	24 Feb	18 Mar	27 Mar	Diffuse
π Virginids	13 Feb	6 Mar	8 Apr	ZHR 2–5
θ Virginids	10 Mar	20 Mar	21 Apr	ZHR ~2
τ Draconids	13 Mar	31 Mar	17 Apr	
π Puppids	8 Apr	23 Apr	25 Apr	Weak: Comet P/ Grigg–Skjellerup
April Ursids	18 Mar	19 Apr	9 May	
α Virginids	10 Mar	13 Apr	6 May	Diffuse: complex radiants
γ Virginids	5 Apr	14 Apr	21 Apr	
May Librids	1 Apr	6 May	9 May	
Pons–Winneckeids (June Boötids)	27 June	28 June	5 July	Comet P/ Pons–Winnecke; faint: good in 1921, 1927
τ Herculids	19 May	9 June	19 June	Comet P/ Schwassmann–Wachmann 3?
June Lyrids	10 June	15 June	21 June	(magnitude 3)
θ Ophiuchids	21 May	10 June	16 June	Flat maximum: ~5 days
φ Sagittariids	1 June	18 June	15 July	Weak shower
χ and ω Scorpiids	6 May	4 June	11 July	Weak: diffuse radiants
Scutids	2 June	27 June	29 July	ZHR ~3
α Lyrids	9 July	14 July	20 July	Fast: mainly telescopic.
Phœnicids	9 July	14 July	17 July	Diffuse radiant: ZHR 2
κ Cygnids	26 July	19 Aug	1 Sept	Complex max: ZHR 6?.
υ Pegasids	25 July	8 Aug	19 Aug	ZHR 2–5; swift, yellow
α Ursæ Majorids	9 Aug	13 Aug	30 Aug	Weak shower
γ Aurigids	1 Sept	7 Sept	14 Sept	ZHR can reach 4
α Aurigids	25 Aug	1 Sept	6 Sept	ZHR may be 9; good in 1935 and 1986
October Arietids	7 Sept	8 Oct	27 Oct	One of several radiants
δ Aurigids	22 Sept	10 Oct	23 Oct	C/ Bradfield, 1972 III?
ε Geminids	10 Oct	18 Oct	27 Oct	ZHR 1–2
November Monocerotids	13 Nov	21 Nov	2 Dec	
Coma Berenicids	8 Dec	25 Dec?	23 Jan	Weak: diffuse
December Monocerotids	9 Nov	11 Dec	18 Dec	ZHR no more than 2
χ Orionids	16 Nov	10 Dec	16 Dec	Bright meteors. ZHR 3
December Phœnicids	29 Nov	3 Dec	9 Dec	Comet D/ Blanpain? Rich in 1956

All data are rather uncertain. This table is derived from several sources, but mainly from the work of Gary W. Kronk.

Table 15.3 *Leonid meteor storms*

902	Oct 12–13	Southern Europe, northern Africa
934	Oct 13–14	Europe, northern Africa, China
1002	Oct 14–15	China, Japan
1202	Oct 18–19	Japan
1238	Oct 18–19	Japan
1366	Oct 21–22	Europe, China
1533	Oct 25–27	Europe, China, Japan
1566	Oct 26–27	China, Korea
1601	Nov 5–6	China
1666	Nov 6–7	China
1698	Nov 8–9	Europe, Japan
1766	Nov 11–12	South America
1799	Nov 11–12	America
1833	Nov 12–13	North America
1866	Nov 13–14	Europe
1966	Nov 17	North America
1999	Nov 18	Europe, Middle East

Julian calendar dates before Oct 1582, Gregorian dates thereafter. It is of course possible that some storms were not recorded.

leading authorities, Peter Jenniskens, believe that the meteors are due to Comet C/1400 Y$_1$, observed by the Chinese; according to Jenniskens, asteroid 2003 EH$_1$, discovered by the Lowell Near-Earth Object Search (LONEOS) on 6 March, is the intemittently active nucleus of the comet. The stream may have been produced no more than about 500 years ago, when the débris was ejected from 2003 EH$_1$. All this is very speculative, but seems plausible particularly since the Quadrantids seem to be cometary in nature and do not penetrate the atmosphere as deeply as, for example, the higher-density Geminids.

The Perseid shower of early August is consistent, and any observer who looks up into a dark, clear sky at any time during the first part of the month will be very unlucky not to see a few Perseids. The October Draconids, associated with Comet P/Giacobini–Zinner, are usually sparse, but produced a major storm in 1933, when for a while the ZHR reached an estimated 6000; a weaker but still rich storm occurred in 1946 (this was the first occasion on which meteors were systematically tracked by radar). Nothing comparable from the Draconids has been seen since. It must be remembered that meteor streams are easily perturbed by planets, and no two orbits are exactly alike.

The Leonids can produce the most spectacular storms of all; a selected list is given in Table 15.3 (drawn from the researches carried out by John Mason). In 1833 and 1866 it was said that meteors 'rained down like snowflakes'. No major displays were seen in 1899 and 1933 because the main swarm did not intersect the Earth's orbit at the critical time, but there was another storm in 1966 – unfortunately not seen from Europe, because it occurred during European daylight, but spectacular from parts of North America, such as Arizona.

Comet Tempel–Tuttle returned to perihelion in 1998, and was expected to produce another meteor storm. The predicted date was 17 November, 258 days after the comet had passed through perihelion, but in fact the richest display was seen on 16 November – not a 'storm' but certainly striking. It was calculated that the dust stream left behind by the comet does not have uniform cylindrical structure, but consists of a number of discrete, separate arcs of dust, each released at a different return of the comet. If the Earth passes through a thin filament, the meteor shower is brief but intense. If it passes through a broader filament, the shower is less intense, but lasts longer. If the Earth passes through a gap between filaments, the display is weak. If it passes through a broad filament first, and then through the edge of a narrower filament, there will be two peaks of activity.

The great storm of 1833 was caused by a dust trail generated in 1800, 33 years earlier; the 1966 storm was due to dust released from the comet in 1899. The displays of 1998 and 1999 were due to an arc-shaped cloud of dust shed by the comet in 1366. In 1999, there was indeed a meteor storm, peaking at 02 hours GMT on 18 November; if not as splendid as the storms of 1833 and 1866, it was very spectacular, with a peak ZHR of well over 2000. It was of brief duration, but was well seen from cloud-free areas of Europe. From Oban (Scotland), Iain Nicolson found the peak activity to be from 02.00 to 02.15 GMT, and had declined markedly by 02.40. Many of the meteors were very bright, with long, sometimes persistent trains.

The Leonids were still very much in evidence in 2001 and 2002, and work by D. Asher (Armagh), R. McNaught (Australia) and E. Lyytinen (Finland) made it possible to predict bursts of activity very precisely. For instance, the double 'spikes' of activity seen in 2001 and 2002 were due to particles ejected by the comet in 1767 and in 1866. Unfortunately, perturbations by Jupiter will probably make Leonid storms unlikely in the near future.

EXTRA-TERRESTRIAL METEORS

Lunar meteors have often been reported, mainly by American observers, but the Moon's atmosphere is so tenuous that it could not possibly cause an incoming meteoroid to become luminous. The reports are almost certainly due to errors of observation or interpretation. There have also been reports of flashes on the Moon due to impacting Leonids, but whether a particle as small as a meteor could produce a visible flash seems to be questionable. A meteorite certainly could, but meteorites are not associated with comets or with meteor showers.

The Martian atmosphere, on the other hand, is quite adequate, and the first Martian meteor was imaged on 7 March 2004 by the Spirit rover on the surface of the planet. A long streak was recorded. The parent comet was undoubtedly P/114 Wiseman–Skiff, whose orbital period is 6.53 years; a Martian shower from this comet had already been predicted by F. Selsis *et al.* The radiant lies in Cepheus, and Martian Cepheids are probably seen from the planet's surface at the same time every Martian year. No doubt there are also many other annual showers.

METEOR SOUNDS?

During the Leonid shower of November 1998, the Croatian astronomer Dejan Vinkovic reported that he had recorded sounds from meteors. These have been reported before, and the deep, thunderlike or hissing noise seems to coincide with the visual appearance of the meteor. This would indicate that the sound waves could travel at the speed of light, which seems impossible. However, the phenomenon – termed electrophonic sound – can be explained by radio waves, and a plausible explanation has been given by the Australian physicist Colin Keay. Glowing meteor trails give off not only visible light, but also very low-frequency radio signals, which do travel at the speed of light and interact with objects at ground level to produce audible sound. During the Leonid meteor storm of 18 November 2001 there were many reliable reports of 'meteor noise'. Further research into this interesting problem is needed.

DANGER FROM METEORS?

From ground level, meteors – unlike meteorites – are quite harmless. It was suggested that the expected 1998–9 Leonid shower might affect space-craft, such as the Hubble Telescope, but no damage was reported, and all in all it seems that the danger from meteors is not very great.

16 · Meteorites

Meteors are cometary débris, too small and too friable to reach the surface of the Earth intact. Larger bodies, however, can survive the dash through the atmosphere, and land without being destroyed, though they may be fragmented. It may be helpful to give some definitions.

A *meteoroid* is defined by the IAU as 'a solid object moving in interplanetary space, smaller than an asteroid and considerably larger than an atom'. This is all very well, but where exactly is the boundary between a meteoroid and an asteroid? The Royal Astronomical Society gives it as 10 m, but consider then the asteroid 2008 TN$_3$, which impacted Earth on 7 October 2008. Its diameter was just about 10 metres, so that it could be classed either as a large meteoroid or a small asteroid; it was given an asteroid designation because it was followed telescopically well before it entered the atmosphere, exploded and broke into fragments. However, all the definitions could well be tightened up.

A *meteorite* is a body which has reached the Earth, or other planet, in recognisable form. If sufficiently large and dense, it may produce an impact crater. Note that the famous structure in Arizona is generally known as Meteor Crater; it really should be Meteorite Crater.

Meteorites and shooting-star meteors are very different. Most meteorites come from the asteroid belt, though some are believed to come from the Moon (see p. x) and others (the SNC meteorites) from Mars (see p. x). It has been suggested that the meteorites known as achondrites may come from Vesta, though firm evidence is lacking.

Normal meteorites are ancient; their ages are given as 4.6 thousand million years – the same as that of the Earth itself. Ages are measured chiefly by the method of radioactive decay. Meteorites contain radioactive isotopes which decay at a known rate; for instance, the half-life of uranium, U-235, is 704 million years. The oldest meteorite known was found in the Moroccan desert in 2004; it weighs 1.49 kg. Analysis of its lead isotopes give an age of 4.45682 thousand million years. (Half-life indicates the time taken for half of the original material to decay.) The isotope U-235 ends up as lead, Pb-206. Half-life periods for materials found in meteorites are given in Table 16.1.

Well over 10 000 meteorites have been identified (it is estimated that in each year the Earth sweeps up about 78 000 tonnes of extraterrestrial material). Relatively few have been seen to fall. Among famous falls which have resulted in meteorite discovery are those of the Pribrăm fireball (Czech, which was recorded as being of magnitude –19) on 7 April 1959, the Lost City meteorite (Oklahoma) in 1970, the Barwell meteorite (Leicestershire) on 24 December 1965 and the Sikhote–Alin fall in Siberia on 12 February 1947. Rather surprisingly, there are no authenticated records of any human death due to a meteorite. Reports that a monk was killed at Cremona in 1511, and another monk in Milan in 1650 are unsubstantiated. There have, however, been narrow escapes. In 1954, a woman in Alabama, USA, was disturbed by a meteorite which fell through the roof of her house, and she suffered a minor arm injury. On 21 June 1994, José Martin was driving his car from Madrid to Marbella, in Spain, when a 1.4-kg meteorite crashed through his windscreen, ricocheted off the dashboard and injured the driver's finger, fortunately not seriously. More than 50 fragments were later found within 200 m of the impact. Two boys, Brodie Spaulding and Brian Kinzie, were outdoors on 31 August 1991 in Noblesville, Indiana, when a meteorite landed 3.5 m away from them, making a crater 9 cm wide and 4 cm deep; the boys found a small black stone which was still warm – it proved to be an unusual sort of chondrite. On 15 August 1992, a piece of the Mbale meteorite (Uganda) struck a banana tree and then hit the head of a boy, again without real damage, and on 9 October of the same year a 12 kg meteorite landed on the bonnet of a car belonging to Michelle Kemp. And on the following 10 December, a house in Japan, belonging to Masaru and Maiko Matsumoto, was struck by a 6.5-kg meteorite.

There is well-known claim that one victim of a meteorite fall was an Egyptian dog, which was in the wrong place at the wrong time when the Nakhla Meteorite fell on 28 June 1911. However, careful examination seems to have disposed of the story.

One of the narrowest escapes was that of Mark Gallini, in January 2010. He was in his doctor's waiting room in North Virginia and had just got up from his chair when a meteorite the size of a tennis ball smashed through the roof and landed in the chair that he had just vacated, and where he would still have been sitting but for the fact that another patient had cancelled their appointment. The meteorite was found to be a chondrite, and if sold would probably have been worth between $25 and 50 thousand. There followed a legal battle between the doctor, the landlord and the Smithsonian Institution.

Meteorites recovered after being observed during descent are classed as *falls*; others are *finds*. Between 1000 and 2000 fall meteorites are included in private or official collections; there are well over 30 000 well-documented finds. I am 'guarding' a piece of the Barwell meteorite. This I have done. But beware! I once devoted a *Sky at Night* television programme to falls, and said that amateurs could help. Viewers then bombarded me with specimens. There were no meteorites, but there were minerals of all kinds. What really baffled me for some time was a fragment which eventually proved to be a very ancient Bath bun.

Table 16.1 *Major isotopes used to date meteorites*

Parent isotope	Daughter	Half-life (years)
Carbon, C-14	Nitrogen, N-14	5730
Aluminium, Al-26	Magnesium, Mg-26	740 000
Iodine, I-129	Xenon, Xe-129	17 000 000
Uranium, U-235	Lead, Pb-207	704 000 000
Potassium, K-40	Argon, A-40	1 300 000 000
Uranium, U-238	Lead, Pb-206	4 500 000 000
Thorium, Th-232	Lead, Pb-208	14 000 000 000
Rubidium, Rb-87	Strontium, Sr-87	49 000 000 000

The first three parents in this table are extinct; all the material present when the Earth was formed has decayed. Thorium and uranium isotopes produce helium as well as lead.

NATURE OF METEORITES: EARLY IDEAS

The earliest reports of meteoritic phenomena are recorded on Egyptian papyrus, around 2000 BC. Early meteorites falls are, naturally, poorly documented but it seems that a meteorite fell in Crete in 1478 BC, stones near Orchomenos in Boetia in 1200 BC and an iron meteorite on Mount Ida in Crete in 1168 BC. According to Livy, 'stones' fell on Alban Hill in 634 BC, and there is evidence that in 416 BC a meteorite fell at Ægospotamos in Greece. A meteorite which fell at Nogara, in Japan, in 861 AD was placed in a Shinto shrine, and the Sacred Stone at Mecca is almost certainly a meteorite. The oldest meteorite which can be positively dated fell at Ensisheim, in Switzerland, on 16 November 1492 and is now on show at Ensisheim Church.

In India, it is said that the Emperor Jahangir ordered two sword blades, a dagger and a knife to be made from the Jalandhar meteorite of 10 April 1621. A sword was made from the meteorite which fell in Mongolia in 1670, and in the nineteenth century part of a South African meteorite was used to make a sword for the Emperor Alexander of Russia. Nowadays there is a flourishing trade in meteorites; for example, in 1999, a chip of the Dar al Gani meteorite, allegedly lunar, was sold at Sothebys in London for £9200. It was 1.75 cm in diameter.

Meteorites were recognised as extraterrestrial only a few centuries ago. The original suggestion was made by E. F. Chladni in 1794, but met with considerable scepticism, and as recently as 1807 Thomas Jefferson, President of the United States, was quoted as saying 'I could more easily believe that two Yankee professors would lie than that stones would fall from heaven'. By then, however, proof had been obtained by the French astronomer J. B. Biot, following his investigation of the meteorite shower at L'Aigle on 26 April 1803.

On average, meteorites enter the Earth's atmosphere at a speed of $15 \, km \, s^{-1}$, although the extreme range is probably between $11 \, km \, s^{-1}$ and $70 \, km \, s^{-1}$. During entry the leading edge melts, and the ablation of molten material produces a smooth face. Melt droplets streaming along the sides of the meteorite collect at the opposite face and solidify, producing an oriented meteorite.

Quenching of the molten coating leads to a dark, glassy fusion crust. For stone meteorites, this crust is seldom more than 0.1 cm thick and the contrast in colour with the underlying material, which is whitish grey with specks of iron, makes it easier to establish that the object really is meteoritic.

CLASSIFICATION OF METEORITES

Meteorites are of three main types: irons (siderites), stony irons (siderolites) and stones (aerolites). Early systems of classification were due to G. Rose (1863), G. Tshcermak (1883) and A. Brezina (1904); these were extended by G. Prior (1920) and by G. J. H. McCall (1973). Stones are more commonly found than irons in the ratio of 96% to 4%, but this is misleading, as irons are much more durable and are more likely to survive. Antarctica is a particularly good area for meteorite collection, and many have been found there, initially by Japanese researchers in 1969. Siderites (irons) are made up largely of metallic iron minerals. Kamacite is essentially metallic iron with up to 7.5% nickel in solid solution; taenite is iron with more than 25% nickel in solid solution. There is also plessite, which is a mixture of tænite and finite-grained kamacite.

Siderites are divided into three main groups: hexahedrites, octahedrites and ataxites. Hexahedrites are mainly of kamacite, with between 4 and 6% of nickel. Octahedrites contain between 7 and 12% of nickel. When etched with acid and polished, these types show what are termed Widmanstätten patterns, composed of parallel bands or plates of kamacite bordered by tænite, and intersecting one another in two, three or four directions. Widmanstätten patterns are unique to these meteorites. They do not appear in ataxites, which contain more than 16% of nickel.

Aerolites (stones) are made up chiefly of silicate minerals, and are again divided into two groups: chondrites (containing chondrules) and achondrites (without chondrules). Chondrules are small spherical particles; they are fragments of minerals, and show radiating structure; their average diameter is about 1 mm. They are formed from previously melted minerals which have combined with other mineral matter to form solid rock. Chondrites account for 86% of known specimens, and are believed to be among the oldest rocks in the Solar System, with ages of around 4.5 to 4.6 thousand million years. They contain pyroxenes, which are darkish minerals also common on Earth. A stone which was seen to fall at Monahans, Texas, on 22 March 1998, was found to contain salt crystals inside which were tiny droplets of water.

Ordinary chondrites, much the commonest form, are divided into three groups. Those with 12–21% of metallic iron are known as bronzites (bronzite is usually green or brown; its chemical formula is $(MgFe)SiO_3$). Chondrites with 5–10% metallic iron are termed hyperstheres; darker than bronzite (chemical formula $(MgFe)SiO_3$). With about 2% metallic iron, the principal minerals are bronzite and olivine $(MgFe)_2SiO_4$; olivine is abundant in the mantle of the Earth. Much less common are the enstatites containing 13–25% of low nickel–iron content metal; the formula for enstatite is $MgSiO_3$–$FeSiO_3$; colour brown or yellowish.

Of special interest are the carbonaceous chondrites; on average the percentage of material by weight is 2.0 carbon, 1.8 metals, 0.2 nitrogen, 83.0 silicates and 11.0 water. There is almost no nickel–iron. Carbonaceous chondrites make up 50% of the asteroids at the inner edge of the Main Belt, and 95% at the outer edge.

Achondrites, accounting for about 7% of known specimens, contain no chondrules. Of special interest are eight meteorites known as the SNC meteorites after the regions in which they were found (Shergotty in India, Nakhla in Egypt and Chassigny in France). They seem to have crystallised only 1.3 thousand million years ago, and their composition and texture indicates that they formed on or in a planet which had a strong gravitational field. They have a concentration of volatile elements, and glassy incursions which were permanently formed in the extreme heat of whichever process ejected them from a parent body. These glassy incursions have trapped gases such as Ar, Kr, Xe and N. It has been suggested that they are of Martian origin, though this is of course highly speculative.

Siderolites (stony irons) are made up of a mixture of nickel–iron alloy and non-metallic mineral matter. Pallasites consist of a network of nickel–iron enclosing crystals of olivine; mesosiderites are heterogeneous aggregates of silicate minerals and nickel–iron alloy. There are two other groups, lodranites (iron, pyroxene, olivine) and siderophtres (iron, orthopyroxene), but these are excessively rare. Siderolites account for no more than 1.5% of known falls.

The meteorite that landed at Murchison (Australia) in 1969 is of special interest. Its age is 4.65 thousand million years; fragments were scattered over 13 km^2, some weighing up to 7 kg. In all, over 100 kg were collected, and it has been found that the meteorite contained a complex mixture of differing-sized organic chemicals, probably collected during the meteorite's journey round the early Solar System. German researchers have found signals representing over 14 000 elementary compositions, including dozens of amino acids, the building blocks of life.

CHICXULUB IMPACT

One major problem in Earth history concerns the disappearance of the dinosaurs, around 65 000 000 years ago, at the end of the Cretaceous period. Not only the dinosaurs vanished – so did many other species of living things, and there was unquestionably a great 'extinction', although there have also been others (notably toward the end of the Permian Period). In 1980, Luiz and Walter Alvarez proposed that the K–T extinction, separating the Cretaceous (K) and Tertiary (T) eras, was due to the impact of a huge asteroid, meteoroid or comet, which threw up so much material that the world climate changed abruptly. Near the mediæval town of Gubbio, in Italy, they found limestone deposits laid down at this particular time which showed an abrupt change in fossil specimens, and moreover the centimetre-thick layer was unusually rich in iridium, which is characteristic of certain meteorites. Subsequently, it was claimed that the point of impact had been found, near the

village of Chicxulub on the Yucatàn peninsula in Mexico. The crater is buried under a thick layer of sedimentary rock, and studies of the gravitational and magnetic fields indicate that the hidden crater is round 180 km in diameter; there are three major ring structures round its rim, and the whole multi-ring structure may have a diameter of at least 300 km. The Alvarez theory is now widely accepted, and is certainly plausible, but final proof is lacking, and there remain some sceptics. However, there can be no serious doubt that a massive impactor did strike the Chicxulub area about the time that the dinosaurs died out.

LARGE METEORITES

A selected list of large meteorites is given in Table 16.2. Pride of place must go to the Hoba West meteorite, near Grootfontein in Namibia (southwest Africa), which is still lying where it fell in prehistoric times; the total weight is over 60 tonnes. It is now protected, since at one stage it was being vandalised by United Nations troops who were meant to be guarding it. All known meteorites weighing more than 10 tonnes are irons; the largest known aerolite fell in Kirin Province, Manchuria, on 8 March 1976. It weighs 1766 kg.

The largest meteorite on display in a museum is the Ahnighito (Tent), found by Robert Peary in Greenland in 1897; it is now in the Hayden Planetarium, New York, along with two other meteorites found at the same time and on the same site – known, appropriately, as The Woman and The Dog. Apparently the local Eskimos were rather reluctant to let them go. The Willamette meteorite is also in the Hayden Planetarium. (This meteorite was the subject of a lawsuit. It was found in 1902 on property belonging to the Oregon Iron and Steel Company. The discoverer moved it to his own property and exhibited it; the company sued him for possession, but the Court ruled in favour of the discoverer.) Table 16.3 lists the largest meteorites found in different regions of the Earth.

Table 16.2 *Selected list of large meteorites*

	Weight (tonnes)
Hoba West, Grootfontein, southwest Africa	60
Ahnighito (The Tent), Cape York, western Greenland	30.4
Bacuberito, Mexico	27
Mbosi, Zimbabwe	26
Agpalik, Cape York, western Greenland	20.1
Armany, Outer Mongolia	20 (est.)
Willamette, Oregon, USA	14
Chupaderos, Mexico	14
Campo del Cielo, Argentina	13
Mundrabilla, Western Australia	12
Morito, Mexico	11

Table 16.3 *The largest meteorites found in different regions*

		Weight (tonnes)
Africa	Hoba West, Grootfontein	60
USA	Willamette, Oregon	14
Asia	Armanty, Outer Mongolia	20
South America	Campo del Oielo, Argentina	13
Australia	Mundrabilla	12
Europe	Magura, Czech Republic	1.5
Ireland	Limerick	48 kg
England	Barwell, Leicestershire	46 kg (total)
Scotland	Strathmore, Tayside. Perthshire	10.1 kg
Wales	Beddgelert, Gwynedd	72.3 g

Table 16.4 *British Isles meteorites*

1623 Jan 10	Stretchleigh, Devon	12 kg
1628 Apr 9	Hatford, Berkshire	Three stones about 33 kg
1719	Pettiswood, West Meath	?
1795 Dec 13	Wold Cottage, Yorkshire	25.4 kg
1804 Apr 5	High Possil, Strathclyde, Lanarkshire	4.5 kg
1810 Aug ?	Mooresfort, Tipperary	3.2 kg
1813 Sept 10	Limerick	48 kg (shower)
1830 Feb 15	Launton, Oxfordshire	0.9 kg
1830 May 17	Perth	~ 11 kg
1835 Aug 4	Aldsworth, Gloucestershire	Small shower, over 0.5 kg
1844 Apr 29	Killeter, Tyrone	Small shower
1865 Aug 12	Dundrum, Tipperary	1.8 kg
1876 Apr 20	Rowton, Shropshire	3.2 kg (iron)
1881 Mar 14	Middlesbrough	1.4 kg
1902 Sept 13	Crumlin, County Antrim	4.1 kg
1914 Oct 13	Appley Bridge, Lancashire	33 kg
1917 Dec 3	Strathmore, Tayside, Perthshire	4 stones; 13 kg
1923 Mar 9	Ashdon, Essex	0.9 kg
1931 Apr 14	Pontlyfni, Gwynedd	723 g
1949 Sept 21	Beddgelert, Gwynedd	723 g
1965 Dec 24	Barwell, Leicestershire	46 kg (total)
1969 Apr 25	Bovedy, Northern Ireland	Main mass presumably fell in the sea
1991 May 5	Glatton, Cambridgeshire	767 g
1999 Nov 28	Leighlinbridge, County Carlow	220 g

The Sikhote-Alin Meteorite fell on 12 February 1947. Observers in the Sikhote-Alin Mountains, Primorye (Russia) saw a bolide brighter than the Sun, coming from the north and descending at an angle of 40°. The impact site was 440 km northeast of Vladivostok; a 30-km smoke train in the sky persisted for several hours. The meteorite entered the atmosphere at a speed of 14 km s^{-1}, and at an altitude of 5.6 km the largest mass exploded. The object is an octahedite – truly a giant iron, with an initial mass of perhaps 28 tonnes. Many craters were produced by the fragments; the largest was 26 m across and 6 m deep.

Carbonaceous chondrites are of special interest. For example, the Orgueil Meteorite, which fell on 14 May 1864 in Southern France, has a mass of 14 kg and is of primitive composition, and contains a high concentration of the gas xenon-HL, which is carried by very ancient, fine-grained diamond dust; it was once claimed (by B. Nagy, 1862) that 'organised elements' in it were due to extraterrestrial biological structures, but Earth contamination was found to be responsible, (A seed embedded in it proved to be a hoax.) The 100-kg Murchison Meteorite, which fell in Victoria (Australia) on 28 September 1969, scattered fragments over a wide area, and contains amino acids such as glycine. Its composition includes 22.1% iron and 12% water.

BRITISH METEORITE FALLS

No really large meteorites have fallen in the British Isles in historic times, but there have been a number of small meteorites, listed in Table 16.4. The Barwell fall was well observed. Many fragments of the meteorite were found; one was detected some time later nestling coyly in a vase of artificial flowers on the windowsill of a house in Barwell village. Although it broke up during descent, the stone is the largest known to have fallen over Britain.

The Bovedy meteorite was also well observed during its descent, but the main mass was not recovered, and almost certainly fell in the sea. *En passant*, I just missed seeing it. I had been in my observatory, and went indoors to change a chart. That, of course,

was the precise moment that the meteorite flew over. (A classic case of Spode's law: if things *can* be annoying, they *are*.) No British meteorite casualties have ever been reported. The Beddgelert meteorite – a small iron – scored a direct hit on the Prince Llewellyn Hotel, but caused no damage.

The 1991 meteorite fell at Glatton, in Cambridgeshire. It was found by A. Pettifor, who heard a loud whining noise and the crash of the stone into a conifer hedge some 20 m away from him. He found the meteorite, which had made a shallow depression 2 cm deep. The meteorite was warm, not hot, when he picked it up. It has a granular structure, indicating that soon after its formation as part of an asteroid it had been hot, but did not melt, so that the mineral grains grew and interlocked. It is an ordinary chondrite of the low-iron lodranite group, with 23% by weight of iron, about

5% of which is nickel–iron metal, with 18% of stony materials – mainly pyroxene and olivine, which are common components of terrestrial basaltic lavas.

BRILLIANT FIREBALLS

The brightest fireball ever seen may have been that of 1 February 1994. It passed over the Western Pacific at 22.38 GMT; the magnitude was about –25. Presumably this was a rocky object; if moving at 15 km s^{-1}, it would have been about 7 m across, weighing 400 tonnes.

A remarkable phenomenon was seen on 9 February 1913, from the North American continent, from Toronto (Canada) through to Bermuda. C. A. Chant, an astronomer at Toronto University, recorded: 'At about 9.05 in the evening there suddenly appeared in the N.W. sky a fiery red body . . . it moved forward on a perfectly horizontal path with a peculiar, majestic, dignified deliberation . . . Before the astonishment caused by this first meteor had subsided other bodies were seen coming from the N.W., emerging from precisely the same point as the first one. Onward they moved at the same deliberate pace, in twos, threes or fours, with tails streaming behind . . . They all traversed the same path and were headed for the same point in the S.E. sky.' Because this was St Cyril's Day, the objects are remembered as the Cyrillids. The whole display lasted for perhaps three minutes.

Were the Cyrillids meteoroids, which entered the Earth's upper air and then returned to space? Unfortunately, we do not have a definite explanation. Nothing similar had ever been seen before, and nothing similar has been seen since.

On 10 August 1972, a meteoroid was seen to enter the Earth's atmosphere and then leave it again. It seems to have approached the Earth from 'behind' at a relative velocity of 10 km s^{-1}, which increased to 15 km s^{-1} as it was accelerated by the Earth's gravity. The object entered the atmosphere at a slight angle, becoming detectable at a height of 76 km above Utah, and reaching its closest point to the ground at 58 km above Montana. It then began to move outward and became undetectable at just over 100 km above Alberta after a period of visibility of 1 min 41 s; the magnitude was estimated by eye-witnesses to be at least –15 and the diameter of the object may have been as much as 80 m. After emerging from the Earth's atmosphere it re-entered solar orbit, admittedly somewhat modified by its encounter, and presumably it is still orbiting the Sun.

THE TUNGUSKA FALL

The most famous fall of recent times was that of 30 June 1908, in the Tunguska region of Siberia. As seen from Kansk, 600 km away, the decending object was said to outshine the Sun, and detonations were heard 1000 km away: reindeer were killed, and pine trees blown flat over a wide area. The first expedition to the sight was led by L. Kulik, but did not arrive before 1927. No fragments were found, and it has been suggested that the impactor was the nucleus of a small comet or even a fragment of Encke's Comet – which, if icy in nature, would presumably evaporate

during the descent and landing. (Inevitably, flying-saucer enthusiasts have claimed that it must have been an alien space-ship in trouble!)

We do have some eye-witness reports, collected by Kulik. S. Semenov was sitting outside his house at the Vanavara Trading Post, 65 km away from the site of the explosion when, looking north, he said 'I suddenly saw that over Onkoul's Tunguska Road, the sky split in two and fire appeared high and wide over the forest. The split in the sky grew larger, and the entire northern side was covered with fire. At that moment I became so hot that I couldn't bear it, as if my shirt were on fire . . . Then the sky shut closed, a strong thump sounded and I was thrown a few yards. I lost my senses for a moment, but then my wife ran out and led me to the house.' Further noises followed, plus a hot wind. One can well imagine what would have happened if the impactor had hit a city.

A new theory of the event was put forward in 2009 by Eduard Drobyshevsky of the Russian Academy of Science. He believed that an icy comet nucleus, saturated with dissolved hydrogen and oxygen, entered the Earth's atmosphere tangentially at a speed of about 20 k s^{-1}; its size was 200×500 m, and its mass from 5 to 50 million tonnes. Part of the nucleus, mass about 4 million tonnes, detonated in the upper air, causing a blast that flattened 80 million trees and devastated an area of 2150 km^2. The main part of the nucleus then went through the atmosphere and returned to space. At least this theory explains the lack of a crater and meteorite fragments.

THE CARANCAS FALL

On 15 September 2007, a large chondritic meteorite fell near the village of Carancas in Peru, near Lake Titicaca and the Bolivian border. It made a crater 13 m wide and 4.5 m deep; the ground was scorched all around. It was said that 'boiling water started to gush from the crater', with 'noxious gases'; soon afterwards 600 villagers were said to have been taken ill. Presumably this was due to contaminated water thrown up by the impact; according to one explanation, the illness was caused by the vapourisation of troilite, a sulphur-bearing compound present in the meteorite, and which would have melted on impact. A certain amount of hysteria may also have been a factor – some of the locals assumed that Peru was being attacked by Chile!

THE ONTARIO FALL

A particularly brilliant object was seen on 25 September 2008 over southern Ontario. It appeared to be about 100 times brighter than the full moon, and lit up the whole sky. It was first detected by the camera system of the University of Western Ontario, when it was moving over the Canadian town of Guelph at abut 21 km s^{-1}; its altitude was given as 100 km, and it was said to be 'about the size of a boy's tricycle'. Five small chunks of black rock, found later by Tony Garchinski in Grimsby, 25 km southeast of Hamilton, were certainly fragments of it.

Table 16.5 *Terrestrial meteorite craters*

Name	Lat. (°)	Long. (°)	Diameter (km)	Age (years)	Discovered	
Acraman, Australia	32 01 S	135 27 E	160	570 000 000	1986	Extensive ejecta: central seasonal lake
Amguid, Algeria	26 05 N	4 23 E	0.45	100 000	1980	Circular: rim rises 30 m above floor
Aouelloul, Mauritania	20 15 N	12 41 W	0.4	3 100 000	1973	Impact glass found: associated with Tenoumer?
Boxhole, Australia	22 37 S	135 12 E	0.17	30 000	1937	Rim raised 3–5 m; many fragments found
Brent, Canada	46 05 N	78 29 W	3.8	450 000 000	1951	Partly filled with Lakes Gilmour, Tecumseh
Campo del Cielo, Argentina	27 38 S	61 42 W	0.05	<40 000	1933	At least 11 craters; 1000 km northwest of Buenos Aires
Chicxulub, Mexico	21 24 N	89 31 W	54	65 000 000	1990	Buried under almost 1 km sediment; northwestern corner of Yucatan; contains Merida town
Clearwater Lake East, Quebec	56 05 N	74 07 W	20	290 000 000	1965 ⎫	Two lakes; emptying into Gulf of Richmond; probable double impact
Clearwater Lake West, Quebec	56 13 N	74 30 W	32	290 000 00	1965 ⎭	
Dalgaranga, Australia	27 43 S	117 05 E	0.021	30 000	1928	Fresh fragments of mesodiertite collected
Gosse's Bluff, Australia	23 50 S	132 19 E	22	142 000 000	1972	Very eroded; many shatter-cones; west of Alice Springs
Henbury, Australia	24 35 S	133 09 E	0.157	10 000	1931	13 craters; largest measured 180 km × 140 m
Holleford, Ontario, Canada	44 28 N	76 38 W	2.35	550 000 000	1956	Barely recognizable at surface
Kaalijarvi, Finland	58 24 N	22 40 E	0.10	40 000	1827	7 craters; main crater is lake-filled
Lappajarvi, Finland	63 12 N	23 42 E	17	77 000 000	1967	Eroded; partly exposed; contains lake; 100 km east of Vaasa
Lawn Hill, Queensland, Australia	18 40 S	138 39 E	20	540 000 000	1987	Eroded; near border with Northern Territories
Lonar, India	19 59 N	76 31 E	1.83	52 000	1970	150 m deep; rim rises 20 m above outer land; possibly volcanic
Manicouagan, Quebec, Canada	51 23 N	68 42 W	100	212 000 000	1964	Huge circle indicated by two narrow semi-circular lakes
Manson, Iowa	42 35 N	94 31 W	35	66 000 000	1940	Buried; covered by recent sediments
Meteor Crater, Arizona	35 02 N	111 01 W	1.186	49 000	1891	(Barringer Crater); tourist attraction
Morasko, Poland	52 29 N	16 54 E	0.1	10 000	1957	8 craters, 6 contain lakes; 9 km from Pozan; contains village of Morasko
New Quebec (Quebec)	61 17 N	73 40 W	3.44	1 400 000	1943	Chubb Crater; bowl-shaped filled by lake 250 m deep at centre; rim rises to 110 m; somewhat eroded
Odessa, Texas, USA	31 45 N	102 29 W	0.17	50 000	1921	Largest crater has raised rim; many meteorites found
Popigai, Russia (Siberia)	71 30 N	111 00 E	100	35 000 000	1976	300 m deep: well preserved
Ries, Germany	48 53 N	10 37 E	24	15 000 000	1978	Large; flat; circular valley surrounded by low hills; Contains town of Nordlingen

Table 16.5 (cont.)

Name	Lat. (°)	Long. (°)	Diameter (km)	Age (years)	Discovered	
Sikhote-Alin, Russia	46 07 N	134 40 E	27 000 000	53	1947	Formed 12 Feb 1947: fall observed; at least 122 craters of 0.5 m diameter or greater
Sobolev, Russia	46 18 N	138 52 E	0.053	200	1981	Asymmetrical; eastern rim lower than western rim
Steinheim, Germany	48 02 N	10 04 E	3.4	15 000 000	1905	Basin; there is a central peak, 50 m high; basin contains villages of Steinheim, Sontheim
Sudbury, Ontario. Canada	46 36 N	81 11 W	200	1 850 000 000	1968	Contains extensive nickel–copper sulphide deposits; possibly volcanic?
Tenoumer, Maurtania	22 55 N	10 24 W	1.9	2 500 000	1970	Rim 100 m high; possibly volcanic?
Tswaing, Pretoria, South Africa	23 24 S	28 05 E	1.1	200 000	1991	'Saltpan'; depth 120 m; Partially lake-filled
Vredefort, Pretoria, South Africa	27 00 S	27 30 E	140	1 970 000 000	1961	Contains Vredefort, Parys; very possibly volcanic
Waqar, Arabia	21 30 N	50 28 E	0.097	10 000	1932	Two craters; 500 km southeast of Riyadh; meteorites found
Wolf Creek, Australia	19 18 S	127 46 E	0.875	300 000	1957	Alternatively Wolfe Creek; regular; well preserved
Zhamanshin, Khazakstan	48 24 N	60 58 E	13.5	900 000	1981	Basin partly filled with lake deposits; tekites found nearby; 100 km north of the Aral Sea

IMPACT CRATERS

Impact craters are found on all the solid bodies of the Solar System – including the Earth, though many have been eroded away. Most lists give the Vredefort Ring in South Africa as the largest impact structure, 300 km in diameter and 2000 million years old, though most of the geologists who have made very exhaustive studies of it believe it to be of internal origin.

One of America's important tourist attractions is the 1200-m diameter structure in Arizona, known generally as Meteor Crater and alternatively as the Barringer Crater (after the mining engineer who first realised that it was meteoritic, in 1894) or Canyon Diablo Crater (after the nearest settlement). It is well formed, and about 170 m deep. It is easily reached from the main highway, but from afar looks unimpressive, as the wall rises to no more than 250 m above the surrounding plain. Reach the top of the wall, and the view is truly dramatic. Visitors were once allowed to scramble down to the floor (as I did), but following one serious accident this is no longer permitted. I once made a television programme from the crater floor, at dusk; I was alone, and it was quite eerie!

The impactor is thought to have been a nickel–iron meteorite, about 50 m across, which landed at a speed of between 12 and 20 km s^{-1}; it came in at an angle of 80° from the north or north-east. Much of it was vapourised; what survived is valuable, but efforts at mining were unsuccessful, and were abandoned after

1930. The crater was formed during the Pleisocene Period, about 50 000 years ago. This was before humans arrived there, and the only casualties would have been mammoths and other creatures. Nickel–iron fragments have been found littering the area; most have now been collected.

In 1891, G. K. Gilbert, head of the US Geological Survey, went to the crater and came to the conclusion that it was a 'steam explosion of volcanic origin'! The Swedish geologist and Nobel Laureate, Svante Arrhenius, described it in 1909 as being 'the most interesting place on Earth'. I tend to agree, though I could well dispense with the company of the local rattlesnakes.

Another structure of the same kind, though much smaller, is Wolfe Creek in Western Australia; 875 m across. Its rim rises to 60 m above the floor; the local Aboriginal people call it Kandimalal. The impactor struck less than 300 000 years ago, and probably had a mass of around 50 000 tonnes. It is not hard to drive to the crater from the town of Halls Creek; go 100 km along the rough Tanami Road. The best way to appreciate it is to fly over it.

Also in Australia we find the Henbury Craters, about 130 km south of Alice Springs; there are over a dozen of them, up to 73 m across, and they are 'young' – that is to say no more than 15 000 years old; the Aborigines call them 'chindu chinna waru chingi yabu' which means 'sun walk fire devil rock'. Boxhole Crater, 175 m across, is circular with a rim rising to 15 m; Gosse's Bluff is much

Table 16.6 *Main groups of tektites*

Region	Name	Geological age
Australasia	Australites	Middle/Late Pleistocene
Ivory Coast	Ivory Coast tektites	Lower Pleistocene
Czech Republic	Moldavites	Miocene
USA	Bediasites (Texas)	
	Georgiates (Georgia)	Oligocene

older – around 147 million years – and has been eroded almost beyond recognition, but must once have been a magnificent structure.

One interesting impact crater is the Haughton Crater in the Qikiqtaaluk Region on Devon Island (75° N, 88° W). This is the second largest of the Queen Elizabeth Islands (Baffin Bay, Canadian Arctic Archipelago) and is the largest uninhabited island in the world. The crater is 23 km in diameter and was formed about 39 million years ago by the impact of a meteorite 2 km in diameter. The area was then forested and for several million years the crater was a lake. Temperatures rarely exceed 19 °C and in winter can plunge to –50 °C. Haughton Crater is regarded as one of Earth's best analogies to Mars, and research stations are set up there, where geological, hydrological, botanical and microbological studies are undertaken.

There may well be a large impact crater under the ice at Wilkes Land (Antarctica). There is certainly a mascon centred at lat. 70° S long. 120° E, as first reported in 2006 by Ralph von Frese and Laramie Potts (Ohio State University). It is 300 km wide, and may be surrounded by a large crater. The estimated age is less than 500 million years, and it has been suggested that there may be a link with the Permian extinction, but obviously all this is very uncertain.

Another impact crater of comparable age, at Bedout High, 250 km off the north west coast of Australia, was reported in 1996 by the Australian geologist John Gorter. However, this idea has not met with a great deal of support. The two Clearwater Lakes, on the Canadian Shield in Quebec, are of Permian age (*c.* 290 million years) and may possibly have been formed by the impact of a binary asteroid. Tswaing Crater near Pretoria, South Africa – the 'Saltpan' – is around 220 000 years old, and contains a lake; the impactor was probably a stony meteorite about 40 m across. Britain has no obvious impact crater, but the 20-km ringed 'Silverpit' structure, deep below the surface of the North Sea, may well have been formed by a plunging meteorite.

Kamil Crater in the Egyptian desert near Sudan (latitude 22° 1′ 6.03″ N, longitude 26° 5′ 15.76″) is exceptionally 'lunar', because of its well-preserved ray structure. It is 44.8 m in diameter, and 15.8 m deep; it was discovered in 2008 by Vincenzo de Michele (Milan) during a survey of satellite images in Google Earth. It seems to be less than 5000 years old, and was produced by an iron meteorite which fragmented upon impact with the sandstone bedrock. The bulk of the recovered meteorite is now in the Egyptian

Geological Museum (Cairo). A selected list of terrestrial impact craters is given in Table 16.5.

METEORITE-HUNTING

Have you ever thought about searching for meteorites? It is a fascinating pastime, but it can be infuriating. However, I had one piece of amazing luck on 27 December 1965, and I will always remember it with glee.

Acually, it wasn't pure luck. On the previous Christmas Eve, at 4.20 p.m., a brilliant fireball shot across England, and was seen by many people in areas where the sky was clear (at Selsey, where I live, there was total cloud). Over the village of Barwell, in Leicestershire, there was a loud bang, followed by a shower of stones. Quite obviously this indicated a meteorite fall, and during Christmas Day and Boxing Day numbers of fragments were collected, showing that the object was an aerolite – that is to say, a stony meteorite rather than an iron.

I did some thinking, worked out where it had travelled on its final plunge, and decided to have a look round, so I fuelled up my elderly motor-bike (known to all and sundry as 'Vesuvius') and drove to Barwell. On arrival I knocked on the door of the largest farmhouse: 'Please could I have a look round the fields?' The farmer was extremely friendly; nothing had been planted there, so I could tramp where I liked, and if I found anything interesting I was welcome to keep it. I thanked him suitably, and began. That was where the luck came in. After only a few minutes I saw a piece of rock protruding above the ground; I went straight to it, pulled up and realised at once that it was meteoritic – a fragment of the Barwell impactor. My journey had been well worth while.

One problem about meteorite-hunting stems from the fact that it is not always easy to tell a meteorite from a non-meteorite. True meteorites are of two main types, aerolites (stones) and irons, though the division is often not clear-cut. I could classify it at once, because when a meteorite enters the atmosphere it is moving at a tremendous speed; friction against the air-particles slows it down, but its outer surface is scorched. The 7-inch long fragment that I found showed the scorching unmistakably.

I spent the rest of the day making a further search, and though I found nothing more I was well satisfied. I showed the fragment to the friendly farmer, who was totally uninterested and told me (in the nicest possible way) what I could do with it. I then went to the nearest museum, and found they already had a number of pieces, one of which had entered a house via an open window and had been found nestling coyly in a vase of artificial flowers. The curator suggested that should keep my fragment for display and leave it to the Science Museum in my will. This I have done.

One point is worth making here: there is no link between a meteorite and a shooting-star meteor. Meteors are cometary débris, and are of dust-grain size, so that they burn away at a height of over forty miles above the ground. Go out to watch a spectacular display of Leonids or Perseids, and you have no reason to feel apprehensive about being hit on the head by something falling out of the sky.

A meteorite which has been observed on its downward journey is called a 'fall'; otherwise it is a 'find'. I have seen only one fall (Barwell) but I narrowly missed another. On 25 April 1969 at 21.25 GMT, a brilliant fireball shot over Britain, all the way from Sussex to Belfast. It passed over my Selsey observatory, about five minutes after I had gone indoors in quest of a cup of coffee... Fragments were found near Belfast, but it seems that the main mass fell in the sea. And on 5 May 1991, a single small stone fell in the garden of a house at Glatton, in Cambridgeshire, a few feet away from a surprised gardener.

All these have been small, but large specimens are known; the record-holder, the Hoba West Meteorite is still lying where it fell, in prehistoric times, at Namibia in West Africa. Nobody is likely to run away with it because it weighs over 60 tons. (It was formerly in the care of United Nations troops, but, predictably, the UN troops had to be withdrawn when it was found that they were vandalising the meteorite instead of protecting it.) An impact of this kind must cause immense damage; if a missile of Hoba West size came down in Selsey, there would not be much left of Bognor Regis! There are impact craters, too, of which the most is Meteor Crater, not far short of a mile in diameter; nobody was killed, for the excellent reason that nobody lived there at the time of the impact, 50 000 years ago. There is a widely-supported theory that a fall 65 000 000 years ago caused a climatic change that wiped out the dinosaurs. The greatest impact of the last two centuries was that of 1908, when either a meteorite or a small comet hit Siberia, flattening tens of thousands of pine-trees. Luckily the area was uninhabited. There is in fact no reliable report of any human casualty from a meteorite fall, though several people have had narrow escapes.

A meteorite may be found anywhere, but beware. It's easy to be deceived, and if you find a 'meteorite' in a chalky area be suspicious at once; the object will probably turn out to be an ordinary mineral. Reports come in regularly; a few weeks ago a small object landed in a cricket field during a game and caused some excitement, but proved to be terrestrial. (Luckily, I gather that it did not damage the wicket.) So if you find an object that you think may have come from the sky, consult an expert geologist.

I have one notable memory. I presented a *Sky at Night* programme about meteorite finds, and viewers began sending me specimens – dozens of them. They came in all shapes and sizes, and I called in a friend who has an honours degree in geology. Only one 'meteorite' gave us real trouble. It was different from all the other specimens, and it baffled us. We pored over it for hours, and finally, with a flash of inspiration, found the answer. It was not a meteorite or a mineral; it was a partly fossilised Bath bun!

MICROMETEORITES

Very small particles entering from the Earth's atmosphere from space are termed micrometeorites: if they are less than about 1 mm in diameter, they cannot cause luminous effects, but end their journey to the ground in identifiable form. Many thousands of tonnes of micrometeoritic material reach the Earth's surface each year.

TEKTITES

These are glassy objects, found only in a few definite areas (Table 16.6). They are small; the largest, found in 1932 at Muong Nong in Laos, weighs 3.2 kg. They seem to have been heated twice, and are aerodynamically shaped. They were once believed to be meteoritic; in 1897 R. O. M. Verbeek even suggested that they came from the Moon, but it now seems definite that they are of terrestrial origin, being blasted away from the surface by impact and, subsequently, re-entering the atmosphere.

17 · Glows and atmospheric effects

Glows of various kinds are seen in the sky. Some, such as rainbows and haloes, are purely meteorological, while others, notably the Zodiacal Light, are purely astronomical. Of these glows, much the most famous – and the most spectacular – are the auroræ, northern and southern.

AURORÆ

Auroræ or polar lights (Aurora Borealis in the northern hemisphere, Aurora Australis in the southern) must have been observed from early times, since they are often strikingly brilliant. The Northern and Southern Lights are of the same type, but obviously the northern displays arc better known, because the Southern Lights are not well seen from inhabited countries. They are sometimes visible from the southern parts of South America and New Zealand and occasionally from South Africa, but most of the displays over the ages must have been enjoyed only by penguins.

Legends and folklore

Auroræ have a major rôle in folklore, particularly of northern peoples. Here is a description given by the Inuit of the Hudson's Bay:

> The sky is a huge dome of hard material arched over the flat Earth. On the outside there is light. In the dome there are many small holes, and through these holes you can see the light from the outside when it is dark. And though these holes the spirits of the dead can pass into the heavenly regions. The way to heaven leads over a narrow bridge which spans an enormous abyss. The spirits that, were already in heaven light torches to guide the feet of the new arrivals. These torches are called the Northern Lights.

To the North American Indians, the Lights marked a gathering of medicine men and warriors, where they held great feasts and barbecued their fallen foes in cooking-pots! The Tlingit Indians of Alaska thought that auroræ were due to battles going on in the sky, while in Norway and Sweden it was widely believed that auroræ were due to the reflection of huge shoals of herring floating near the surface of the sea. The Danes had a rather different idea; swans which flew too far north became frozen, so that as they flapped their wings they would produce disturbances in the air which resulted in auroræ. When the Lights flickered, the Greenlanders believed that the spirits of the dead were trying to contact their loved ones who were still on Earth. In Norse mythology, the Lights were due to reflections from the shields of Valkyries. Even in the far south there are legends; the New Zealand Maori believed

that the Lights were caused by people who lived in very cold climates and were doing their best to keep themselves warm.

To some of the Eskimos, an aurora represented a game of football played by spirits who were using a walrus head as a ball – but in Siberian lore it was the walruses who were playing, using a human skull as a ball! According to Scottish belief, the auroral displays were the 'Merry Dancers' of the heavens.

In some cases auroræ were regarded as harmful. The Eskimos of Alaska would take their children indoors at the start of a display, and in the Faroe Islands children would not venture out when auroræ were in the sky – because they feared that the Northern Lights would strike down and singe their hair.

Of course the ancients had no idea of the real nature of auroræ. The Greek philosopher Aneximenes, who lived from 570 to 526 BC, attributed them to a volatile vapour accumulated in the clouds; as this vapour mixed with the air, it caused luminescence. Aristotle (384–322 BC) thought that they were due to 'dry vapours' which rose upwards and caught fire, while the Roman writer Seneca (5 BC– AD 65) explained them as boiling air in the uppermost part of the sky; because these currents of air move so quickly, the stars can generate enough heat to set them alight. It is understandable that fire should be involved in many of the early ideas, and it is certainly true that in AD 37 the Roman Emperor dispatched his fire engines to the port of Ostia, because a striking red aurora led him to believe that the entire town was burning.

In Denmark and Sweden, auroræ came from an active volcano in the polar region, put there by the Creator to provide light and heat for people living in the far north. The writer of the Norse chronicle Kongespeliet (the King's Mirror), issued in around 1230, stated that, 'Otherwise it is the same with the Northern Lights as with anything else we know nothing about, that wise men put forward ideas and simple guesswork, and believe that most common and probable'. He goes on: 'Some people say that when the Sun is under the horizon, at night, some rays of light reach up to the skies over Greenland, a land so close to the edge of the Earth that the Earth's curvature, which hides the Sun, must be less there.' At that time there was still an inbred distrust of auroræ; during a display, it is said one should 'tread carefully, and in no way intimidate the Northern Lights by waving, whistling, staring or any other form of defiance'.

Auroræ and the Sun

The term 'aurora' was introduced by the French astronomer R. Gassendi, who saw a brilliant display on 12 November 1621, but properly scientific accounts had been published earlier than this. A good description of a display of Aurora Borealis was given by K.Gesner of Zürich for the aurora of 27 December 1560, and in 1731 J. J. de Mairan wrote an excellent book, *Traité physique et historique de l' auroræ boreale*. The first surviving description of the Southern Lights was written by Captain Cook for the display of 20 February 1773.

Auroræ and sunspots

The connection between auroræ and electrical discharges goes back to a suggestion made by Edmond Halley in 1716; he linked auroral displays with discharges associated with the Earth's magnetic field. (It must be remembered that Halley saw his first aurora between 1645 and 1715, following the end of the solar Maunder Minimum; between 1715 and 1745 there are no reports of auroræ from Britain.) De Mairan, in his book, saw a relationship between the frequency of sunspots and displays of auroræ while Anders Celsius (1701–44) and Olof Hiorter (1696–1750) recognised the link between aurora and disturbances of the magnetic needle, and in 1751 the Danish bishop Erik Pontoppidan came to the conclusion that the aurora was itself an electrical phenomenon.

The sunspot–aurora connection was refined in 1870 by E. Loomis of Yale, and in 1872 by the Italian astronomer G. Donati following a major display on 4–5 February of that year. The foundations of modern theory were really laid down in 1895 by the Norwegian scientist Kristian Birkeland, who found that auroræ were due to charged particles moving under the influence of the Earth's magnetic field; he also reproduced 'miniature auroræ by using a helical electromagnet and a beam of cathode rays'. In 1929, S. Chapman and V. Ferraro suggested that the auroræ were due to solar plasma and it is indeed true that auroræ are produced by electrified particles from space, originating in the Sun, colliding with atoms and molecules in the Earth's upper atmosphere.

The streams of solar plasma sent out, particularly from coronal holes, take from two to five days to reach the Earth. When they meet the boundary of the magnetosphere they are deflected, and many of the particles are forced round the Earth's globe following the magnetic field lines: the Earth's magnetic field is compressed on the day side, while on the night side a magnetic tail is produced, extending out to as much as 6 000 000 km. It may be said that the Earth's magnetic field creates a 'tunnel' in the plasma stream: the northern and southern sections are separated by a plasma sheet, and the particles in the opposite sides of the tunnel rotate in opposite directions, producing a dynamo effect. When the magnetic tail becomes unstable, due to an increase in the solar wind, the charged particles move inward to the centre of the tunnel, where they meet and cause a magnetic short circuit. This circuit closes when the particles reach the top of the atmosphere, where the tenuous gases are made up of ionised particles and are therefore electrically conducting. The energy of the dynamo it converted into light, and auroræ appear.

AURORAL OVALS

The particles from the magnetic tail stream down toward the magnetic poles, but it is wrong to assume that the poles are the best sites from which to observe auroræ: they are not. Auroral activity is more or less permanent at high latitudes round the so-called auroral ovals, which are 'rings' asymmetrically displaced around the geometrical poles; the positions of the magnetic poles shift and these changes naturally affect the positions of the auroral ovals. Basically, however, the ovals remain more or less fixed in direction, and are wider on the night side of the Earth than on the day side. They are centred on the magnetic poles, while the Earth rotates around the geographical poles. When there are major disturbances on the Sun, producing energetic solar wind particles, the ovals broaden and expand, bringing auroral displays further north and further south of the main regions. The maximum activity for auroræ is seen at latitude 68° north or south. Auroræ are far commoner there than they are at the pole itself.

On average, auroræ of one kind or another are seen on 240 nights per year in north Alaska, north Canada, Iceland, north Norway and Novaya Zemlya: 25 nights per year along the Canada/United States border and in central Scotland, only one night per year in central France. Obviously, however, this varies according to the state of the solar cycle, and in general auroræ are most common about two years after a sunspot maximum. From Mediterranean countries there are only one or two good displays per century, and close to the equator they are very rare indeed, although on one occasion, in 1909, an aurora was seen from Singapore, latitude 1.25° N.

From southern England there were brilliant displays on 25–6 January 1938, 13 March 1989 and 8–9 November 1991. The 1938 display coloured the entire sky red, while those of 1989 and 1991 were bright enough to cast shadows. On these occasions the Sun was very active – but even so, auroræ cannot be reliably predicted. A fairly bright aurora occurred on 6 April 2000, seen from much of England: from Selsey in Sussex I saw it as very conspicuous and decidedly red.

ALTITUDES

The heights of auroræ vary. In general, the sharp lower boundary lies at about 98 km above ground level, and the maximum region of auroral activity at about 110 km: the normal upper boundary is at 300 km, although in extreme cases this may reach up to 700 km or even 1000 km Extremely low auroræ have been reported now and then, but not with any certainty. Auroræ always lie far above normal clouds, so that when the sky is overcast there is no chance of seeing a display.

FORM AND BRILLIANCE

The form and brilliance of displays also vary and a definite scale of brightness has been drawn up, although it is not easy to be really precise. The accepted scale is given in Table 17.1. Auroræ may be seen in many forms. There are glows; rays or streamers; coronæ, or radiating systems of rays converging at the zenith; 'surfaces', which are diffuse patches, sometimes pulsating; flaming auroræ, made up of quickly moving sheets of light; and ghost arcs, which may persist long after the main display has ended. There are also 'flickering

Table 17.1 *Brilliance classes of several displays*

1 Rather faint, about equal in brightness to the Milky Way as seen on a clear night.
2 Brighter display, about ten times brighter than Class 1, about equal to thin cirrus clouds by moonlight.
3 Much brighter – about 100 times as bright as Class 1: comparable with cumulus clouds lit by moonlight.
4 Brilliant auroræ 100 to 2000 times as bright as those of Class 1, sometimes casting shadows *and* even giving as much light as the full moon. From England, the last Class 4 auroræ were those of 1989 and 1991.

auroræ', whose behaviour has been compared with that of a candle flame in a light breeze.

The colours of auroræ were first studied in detail by the Norwegian scientist Lars Vegard (1880–1963). The differences in hue are due to differences in composition of the atoms and molecules struck by the energetic particles which rain down from above, and each atmospheric gas produces its own characteristic colour. Oxygen atoms at the lower border of the auroral zone produce brilliant greens; the higher-altitude oxygen atoms yield the rare but spectacular all-red auroræ and ionised nitrogen atoms are responsible for blue colours, while complete nitrogen atoms give red or purple edges to the lower borders and rippled edges of the display. Sometimes all these colours can be seen over a few minutes: at other times only one colour is seen. Over Norway, for example, by far the commonest auroræ are green.

AURORAL NOISE

One interesting and frankly puzzling problem associated with auroræ is that of 'auroral noise'. There have been many reports of sounds: faint whistles, rustling, swishing, crackling and soft hissing. Many reported were listed by S. Tromholt from Norway, in 1885 (reported in *Nature*, volume 32, page 499). A typical report is that of Dr H. D. Curtis, who was in charge of the Labrador Station of the Lick Observatory in 1905. He wrote:

Auroral displays were frequent and bright during July and August. On several nights I heard faint swishing and crackling sounds, which I could only attribute to the aurora. There were times when large feint luminous patches or 'curtains' passed rapidly over our camp ... The fatal hissing and crackling sounds were much more in evidence as such luminous patches swept past us. I tried in vain to assign the sounds to some reasonable cause other than the aurora, but was forced to exclude them as possible sources. In short, I feel certain that the sounds I heard were caused by the aurora and nothing else. There was, moreover, a certain synchronism between the maxima of these sounds and the sweeping of aurora curtains over the sky.

It is extremely difficult to see how auroræ could produce sounds. Moreover, the speed of sound in the high atmosphere is such that any noise due to auroræ should take minutes to reach ground level, making it hard to reconcile the noise with visual activity; in any case auroræ are so high up that they are restricted to conditions of near vacuum, so that there is no real chance of sounds being propagated downward. Electrical phenomena have been proposed; it has also been claimed that the reports are purely psychological, but we do not really know the answer, and so far I believe that no serious attempt has been made to record the noise electronically.

Slight odours have also been reported, but with no certainty, and it must be said that smelly auroræ seem only slightly less improbable than noisy auroræ. Daytime auroræ occur, but are naturally difficult to detect because of the brightness of the sky. Moreover, the auroral belt is much narrower by day than by night. To see daytime auroræ from the Earth's surface, the observer must be within 10° to 15° of the magnetic pole, and the Sun must be at least 10° below the horizon. One favourable site for making these observations is Spitzbergen, but this is not a place which is particularly easy to reach.

A vivid description of an aurora was given in 1904 by the great Norwegian polar explorer Fridtjof Nansen, as he stood on the deck of his ship, the *Fram*, drifting over the Arctic pack-ice. He wrote: 'There is the supernatural for you – the Northern Lights flashing in matchless power and beauty over the sky in all the colours of the rainbow. Seldom or never have I seen the colours so brilliant. The prevailing one at first was yellow, but that gradually flickered over to green, and then a sparkling ruby-red began to show at the bottom of the rays on the underside of the arc. And now from the far away western horizon a fiery serpent writhed up over the sky, growing brighter and brighter as it came ...'

There are interesting phenomena known as *black auroræ* – unusually dark spaces between bright auroral forms. They were initially thought to be purely optical effects but are now known to be real, and due to negatively charged particles being sucked out of the ionosphere along ajoining magnetic-field lines. They can climb to over 20 000 km and last for several minutes. They may appear as dark rings, curls or black blobs floating upon a sea of faint, glowing auroral light. They are also referred to as 'anti-auroræ'.

Diffuse auroræ are quite different from bright auroræ. They are caused when electrons trapped in the Earth's magnetic field are funnelled towards the polar atmosphere; light is emitted when the electrons collide with neutral atoms in the upper atmosphere. The diffuse aurora is not generally visible with the naked eye, but is well shown in satellite images.

AURORÆ ON OTHER PLANETS

Obviously there can be no auroræ on the airless Moon, and this presumably applies also to Mercury, which has only an extremely tenuous atmosphere (though there must be a certain amount of doubt, because the planet does have a magnetic field). Venus lacks a detectable magnetic field, but auroræ there show up as bright, diffuse patches distributed across the full disc; they vary in shape and intensity, and are due to electrons from the solar wind entering the night-side atmosphere. The global magnetic field of Mars is negligible, but in the crust there are locally magnetised areas, and weak aurora-type effects are seen, caused by charged particles from the Sun colliding with particles in the Martian atmosphere. Strong auroræ occur on the giant planets, and have already been discussed (remember that on Uranus and Neptune the auroræ are not polar, because the magnetic poles are nowhere near the poles of rotation).

Other atmospheric phenomena

These belong strictly to meteorology, but are worth brief mention here. *Lunar haloes* – 'rings round the Moon' – are caused by the refraction of moonlight by ice crystals in high-altitude cirrostratus clouds in the Earth's atmosphere. The crystals 'bend' light at an angle of 22 °, so that the resulting halo is 22° in diameter (the same mechanism can produce a solar halo). A second, 44° ring can sometimes be seen. A *lunar corona* is similar, but is much smaller as individual crystals are involved. *Lunar and solar pillars* are produced when snow or ice crystals reflect light forward from a strong source such as the Moon or the Sun. A *Sundog* (parhelion) is a bright circular spot on a solar halo; it can be coloured from red to blue, and is caused by refraction by ice crystals which are hexagonal in shape. Two are often seen at the same time, one on each side of the Sun, usually (not always) when the Sun is low over the horizon. A *Moondog* (paraselene) is similar, but of course much fainter.

The *Belt of Venus* is seen shortly before sunrise or sunset, when the observer is surrounded by a pinkish glow or arch extending from 10° to 20 ° above the horizon. It may be separated from the horizon by a dark layer, due to the shadow of the Earth. The pink hue is caused by the back-scattering of reddened light from the rising or setting Sun.

GLOWS IN THE SKY

Other glows, much too faint to be seen with the naked eye, occur further from the poles and extend to mid latitudes. They are known as Stable Auroral Red (SAR) arcs, and are due to a steady influx of electrons coming from the Van Allen radiation zones and colliding with atoms of oxygen. Their altitudes range from 200 to 400 km, and they are very subdued near solar minimum.

SUNDOGS

A *sundog* is produced by sunlight shining through a thin layer of ice crystals in high altitude cirrus clouds. When the tiny six-sided ice crystals line up with their flat faces pointing downwards, they act in the manner of glass prisms, splitting up the solar rays into the colours of the rainbow and beaming them into a patch of light beside the Sun. The coloured light may twist upwards like a dog's tail, hence the name! If the cirrus clouds are extensive, a pair of sundogs may form at equal distance to either side of the Sun, producing what are termed *mock suns*.

Much rarer is a *Bishop's Ring*, usually seen only after major volcanic eruptions, and was first described by Bishop in 1883 after the Krakatoa eruption: 'A very peculiar corona or halo extending from 20 to 30 degrees from the Sun, which has been seen every day, of whitish haze with pinkish tint, shading off into lilac or purple against the blue'. (I have not been able to find any Bishop's Ring reports following the April 2010 of the Icelandic volcano Eynafjallajökull.)

THE GREEN FLASH (OR GREEN RAY)

This is purely an atmospheric effect. As the Sun sinks below the horizon, the last segment of it may flash brilliant green for an instant. There are vague references to this in Egyptian and Celtic folklore, but the first scientific account of it was due to W. Swan, who saw it on 13 September 1865. However, Swan's account was not published until 1883. The first published reference to the phenomenon was due to J. P. Joule, in 1869. Much more recently, superb photographs of it have been taken by D. J. O'Connell at the Vatican Observatory; these were published in book form in 1958.

There is no mystery abut the Green Flash. Light moves more slowly in the lower, denser layer of atmosphere than in the higher, thinner air above, and so the Sun's rays follow paths that curve slightly to follow the direction of the curvature of the Earth. Short wavelengths (green, blue) are hidden by the curvature of the Earth. Blue is drowned by the blueness of the sky, so that the flash is green. The phenomenon is probably best seen over a sea horizon.

The Green Flash has even been seen with Venus; the first reference to it seems to be due to Admiral Murray, who saw it from HMS *Cornwall*, off Colombo, on 28 November 1939. Venus was setting over a sea horizon; Admiral Murray was using binoculars, and described the flash as 'emerald'. A Green Flash with Jupiter was seen by G. Verschuur on 8 June 1978.

Airglow

The airglow, due to the very weak emission of light by the Earth's atmosphere, means that the night sky is never completely dark. It was first noted in 1868 by the Swedish scientist Anders Ångström. There are various reasons for it, quite apart from the diffusion of starlight. In particular, there are various processes in the upper atmosphere, such as the re-combination of ions which have been photo-ionised by the Sun during daylight, luminescence due to incoming cosmic rays, and so on. There also seems to be a sort of 'geocorona' in the exosphere (1000 to 10 000 km altitude), due to emissions from hydrogen and helium in this highly rarefied environment.

Diffused starlight is quite appreciable, coming as it does from the thousand million stars above magnitude 20, of which approximately one million are above magnitude 21.2, 4850 above magnitude 6, and 1620 above magnitude 5. Over a full visible hemisphere of the sky (20 626 square degrees) the total starlight is approximately equal to 52 stars the brilliance of Sirius or four planets the brilliance of Venus at maximum. It is, however, only a tiny fraction of the light sent to us by the Moon.

At night the airglow can be bright enough to be noticeable, and to be slightly bluish in colour; diffused starlight is quite appreciable. Dayglow is present, but in the ordinary way will obviously not be noticed.

THE ZODIACAL LIGHT

The Zodiacal Light may be seen as a faint cone of light rising from the horizon either after sunset or before sunrise. It extends away from the Sun, and is generally observable for only a fairly short period after the Sun has set or before it rises. On a clear, moonless night, under ideal conditions, it contributes about one-third of the total sky light, and may be brighter than the average Milky Way region. It is due to particles scattered in the Solar System along and near the main plane of the system; cometary débris is a major contributory factor. The diameters of the particles are of the order of 0.1 to 0.2 µm (1 µm is equal to one-millionth of a metre). Since the Zodiacal Light extends along the ecliptic, it is best seen when the ecliptic is nearly vertical to the horizon – i.e. February to March, and again in September to October. The first scientific account of the Zodiacal Light was given by G. D. Cassini in 1683; he correctly suggested that it was due to sunlight reflected from tiny particles orbiting the Sun near the plane of the ecliptic. However, it had been observed much earlier by the Persians and the Arabs, and had been clearly described by Joshua Childrey in the *Britannia Baconica* published in London in 1661.

THE GEGENSCHEIN

The Gegenschein or Counterglow is seen as a faint patch of radiance in the position in the sky exactly opposite to the Sun. It is extremely elusive, and is generally visible only under near-perfect conditions. (From England I have seen it only once, in March 1942, when the whole country was blacked out as a precaution against German air-raids.) The best opportunities occur when the anti-Sun position is well away from the Milky Way (i.e. in February to April and September to November) and when the anti-Sun position is high, at local midnight. Generally, the Gegenstein is oval in shape, measuring 10° by 22°, so that its maximum diameter is roughly 40 times that of the full Moon.

The discovery of the Gegenschein seems to have been due to Esprit Penezas, who reported it to the Paris Academy in 1731. It was named by W. Humboldt, who saw it on 16 March 1803. It was described in more detail by the Danish astronomer Theodor Brorsen, who saw it in 1854 and wrote about it in 1863. Contrary to statements made in many books. Brorsen did not claim the discovery for himself, as he was familiar with Humboldt's description.

THE ZODIACAL BAND

The Zodiacal Band is a very faint, parallel-sided band of radiance, which may extend to either side of the Gegenschein or be prolonged from the apex of the Zodiacal Light cone to join the Zodiacal Light with the Gegenschein. It may be from 5° to 10° wide and is extremely faint. Like the Zodiacal Light and the Gegenschein, it is due to sunlight being reflected from interplanetary particles near the main plane of the Solar System.

COSMIC DUST

There is a great deal of cosmic dust in the Solar System, as indeed there is all over the universe. It is responsible for the Zodiacal Light, the Zodiacal Band and the Gegenschein, but we can see visually only the dust which is illuminated by the Sun – an insignificant amount. Local sources include cometary débris, dust from asteroid collisions and from the Kuiper Belt, but there is also interstellar dust through which the Solar System happens to be passing. Remember, in the entire universe there is no such thing as truly 'empty' space!

18 · The Stars

Look at the sky on a dark, clear night and it may seem that millions of stars are visible. This is not so. Only about 5780 stars are visible with the naked eye, and this means that it is seldom possible to see more than 2500 naked-eye stars at any one time, but much depends upon the visual acuity of the observer. People with average sight can see stars down to magnitude 6, but very keen-eyed observers can reach at least 6.5. On the magnitude scale, a star of magnitude 1 is exactly 100 times as bright as a star of magnitude 6.

The proper names of stars are usually Arabic, although a few (such as Sirius) are Greek. In general, proper names are used only for the stars conventionally classed as being of the first magnitude (down to Regulus in Leo, magnitude 1.36), plus a few special stars, such as Mizar in Ursa Major, Mira in Cetus, and Polaris in Ursa Minor. The system of using Greek letters was introduced by J. Bayer in 1603; also in wide use are the numbers given in Flamsteed's catalogue.

DISTANCES OF THE STARS

It had long been known that the stars are suns, and are very remote, but early efforts to measure their distances ended in failure. William Herschel tried the method of parallax; he reasoned – quite correctly – that if a relatively nearby star is observed at an interval of six months, it will seem to shift slightly against the background of more remote stars, because in the interim the Earth will have moved from one side of its orbit to the other. At that time it was believed that double stars resulted from line of sight effects, so that the nearer component would show a parallax shift against its companion. Herschel's measurements were not sufficiently precise to show parallaxes, but in 1801 he did show that many doubles, such as Castor in Gemini, are binary systems, with the components moving around their common centre of gravity.

In 1838, F. W. Bessel, from Königsberg, successfully measured the parallax of the star 61 Cygni, finding it to be about 11 light-years away. From the Cape, T. Henderson measured the parallax of the brilliant star α Centauri, although his results were not announced until after those of Bessel; and about the same time F. G. W. Struve, from Dorpat, made a rather less accurate measurement of the parallax of Vega.

The nearest stars within 14 light-years of the Sun are listed in Table 18.1. Most are very faint dwarfs; only α Centauri, Sirius, ε Eridani, ε Indi, τ Ceti and Procyon are visible with the naked eye, and only α Centauri, Sirius and Procyon are more luminous than the Sun.

The nearest star beyond the Sun is Proxima, a member of the α Centauri system. (In 1976, O. J. Eggen reported the discovery of a red star, magnitude 10.8, in the constellation of Sculptor near

τ Sculptoris, which was regarded as being very close, but later work showed that it was well beyond the 14 light-year limit of the table.) It is most unlikely that any stars exist closer to the Sun than Proxima.

All parallax shifts are very small. A star showing a parallax of 1 arcsec would lie at a distance of 3.26 light-years, but even Proxima is well beyond this limit.

Absolute magnitude is the apparent magnitude that a star would have if seen from a standard distance of 10 parsec (32.6 light-years). These nearby stars are interesting; some are described elsewhere. *Barnard's Star* in Ophiuchus has a diameter 5% that of the Sun and some 15% of the mass, but only 0.0035 of the Sun's luminosity; at present it is approaching the Sun at a rate of about 107 km s^{-1}. It was once (wrongly) suspected of having a planetary companion. The very feeble red dwarf *Wolf 359*, with mass 0.09 of that of the Sun has only 0.0009 of the solar luminosity; its outer atmosphere is cool enough to show molecular lines in its spectrum, notably of water. *Lalande 21185* in Ursa Major has luminosity 0.025 that of the Sun, and is much brighter at infrared wavelengths than it is visually. *Kapteyn's Star* is a small orange variable of the BY Draconis type, and has been given a variable star designation; VZ Pictoris. It is approaching us at 245 km s^{-1}; in terms of the Sun its radius is 0.29, luminosity 0.004. *EZ Aquarii* (*Luyten* 789–6) is a triple-star system; and three components are red dwarfs. Luyten's Star (GL 273) is a red dwarf in Canis Minor, now receding from the Sun; it was closest to us 13 000 years ago. At present it is 1.2 light-years from Procyon, and to an observer in the Luyten's Star system Procyon would shine as a star of magnitude –7. *Teegarden's Star* (SO J 025300+16528) has luminosity only 0.000009 that of the Sun; it has a large proper motion, 5 arcsec per year. *VZ Ceti* is a flare star with mass 85% that of the Sun and 1/5000 of the Sun's luminosity; it is only 1.6 light-years away from τ Ceti. *DEN 1048–3956* is of type M9V, and is classed as a brown dwarf. *Ross 614* is another red dwarf flare star; this is also true of *DX Cancri*, which has a diameter 0.4 that of the Sun and mass 0.09 that of the Sun. *Van Maanen's Star* is the closest known solitary white dwarf; it has 70% the mass of the Sun but only 1% of the diameter; it has been calculated that the core temperature is about 6 000 000 °C. A planetary companion has been suspected, but not confirmed. α Centauri, Sirius, Procyon and τ Ceti would be naked-eye objects. It is interesting to note that if our Sun could be observed from Sirius, it would be a second-magnitude star at right ascension (RA) 18h 44m, declination (Dec.) –16° 4'.

Most of the stars in Table 18.1 are red dwarfs; the companions of Sirius and Procyon are white dwarfs. There are five flare stars; UV Ceti B, Luyten L 726–8B, Proxima, Krüger 60B and Ross 154.

Table 18.1 *The nearest stars*

Name	RA			Dec			Spectrum	Apparent magnitude, m	Absolute magnitude, M
	(h)	(m)	(s)	(°)	(')	('')			
Proxima	14	29	43.0	−62	40	46	M5.5Ve	11.1	15.5
α Centauri A	14	39	36.5	−60	50	02	G2V	0.01	4.38
B							K1V	1.44	5.71
Barnard's Star	17	57	48.5	+04	41	36	M45.0V	9.53	13.22
Wolf 359	10	56	29.2	+07	00	53	M6.0V	13.44	16.55
Lalande 21185	11	03	20.2	+35	58	12	M2.0V	7.47	10.44
Sirius A	06	45	08.9	−16	42	58	A1V	−1.43	1.47
B							DA2	8.44	11.34
BL Ceti	(01	39	01.3	−17	57	01	M5.5Ve	12.54	15.40
UV Ceti	(01	39	01.3	−17	57	01	M6.0Ve	12.99	15.85
Ross 154	18	49	49.4	−23	50	10	M3.5Ve	10.43	13.07
Ross 248	23	41	54.7	+44	10	30	M5.6Ve	12.2	14.79
ε Eridani	03	32	55.8	−09	27	30	K2V	3.73	6.19
Lacaille 9532	23	05	52.0	−35	51	11	M1.5Ve	7.34	9.75
Ross 128	11	47	44.4	−00	48	16	M4.0Vn	11.13	13.51
EZ Aquarii	22	38	31.4	−15	18	07	M.0Ve	13.33	15.64
Procyon A	07	39	18.1	+05	13	30	F5V	0.38	2.66
B							DA	10.70	12.98
61 Cygni A	21	06	53.9	+38	44	58	K5.0V	5.1	7.49
B	22	06	55.3	+38	44	31	K7.0V	6.03	8.31
Struve 2398	18	42	46.7	+59	37	49	M3.0V	8.90	11.16
Groombridge 34	00	18	22.9	+44	01	23	M.1.5V	8.08	10.32
ε Indi	22	03	21.7	−56	47	10	K5Ve	4.69	6.89
DX Cancri	08	29	49.5	+26	46	37	M6.5Ve	14.78	16.98
τ Ceti	01	44	04.1	−15	56	15	G8Vp	3.49	5.68
GJ 1061	03	35	59.7	−44	30	45	M5.5V	13.09	15.26
YZ Ceti	01	12	30.6	−16	59	56	M4.5V	12.02	14.17
Luyten's Star	07	27	24.5	+05	13	33	M3.5Vn	9.86	11.97
Teegarden's Star	02	53	00.9	−16	52	53	M6.5V	15.14	17.22
SCR 1845–6357	18	45	05.3	−63	57	48	M8.5V	17.39	19.41
Kapteyn's Star	05	11	40.6	−45	01	06	M1.5V	8.84	10.87
AX Microscopii	21	17	15.3	−38	52	03	M0.0V	6.67	8.69
Krüger 60 A	22	27	59.5	+57	41	45	M3.0V	9.79	11.76
Krüger 60 B	22	27	59.5	+57	41	15	M45.0V	11.41	13.38
DEN 1048–3956	10	48	14.7	−39	56	06	M8.5V	17.39	19.37
Ross 614	06	29	23.4	−02	48	50	M4.5V	11.15	13.09
Wolf 1061	16	30	18.1	−12	39	45	M34.0V	10.07	11.93
Van Maanen's Star	00	49	09.9	+05	23	19	DZ7	12.38	14.21

Star	Parallax ('')	Distance (light-yrs)	Radial velocity (km s^{-1})	Alternative name	Constellation
α Centauri A	0.747	4.365	−25		Cen
α Centauri B	0.747				
Barnard's Star	0.547	5.963	−111		Oph
Wolf 359	0.412	7.783	+13	CN Leonis	Leo
Lalande 21185	0.393	8.290	−84		UMa
Sirius A	0.380	8.582	−9		CMa
Sirius B	0.380				
Luyten 726–8 A	0.373	8.728	+29		Cet
Luyten 726–8 B	0.373	8.728	+32		Cet

Table 18.1 (cont.)

Name	Parallax	Distance (light-yrs)	Radial velocity (km s^{-1})	Alternative name	Constellation
Ross 154	0.337	9.681	−12	V1216 Sagittarii	Sgr
Ross 248	0.316	10.3 2	−78	HH Andromedæ	And
ε Eridani	0.310	10.52 2	+17		Eri
Lacaille 9352	0.304	10.742	+9.5		PsA
Ross 128	0.299	19.912	−31	FI Virginis	Vir
EZ Aquarii	0.290	11.266	−59.9	(triple)	Aqr
Procyon A	0.286	11.402	−4		CMi
Procyon B	0.286				
61 Cygni A	0.286	11.403	−65		Cyg
Struve 2398	0.283	11.535	−1	BD +59 1915	Dra
Groombridge	0.281	11.624	+12	GX+GQ Andromedæ	And
ε Indi	0.276	11.824	−40		Ind.
DX Cancri	0.276	11.826	+4		Cnc
τ Ceti	0.275	11.887	−17		Cet
Wolf 1061	0.272	11.991	−21.0		Hor
YZ Ceti	0.269	12.12	+28.2		Cet
Luyten's Star	0.262	12.263	−20	BD+05 1668	Cet
Teegarden's Star	0.261	12.261			Ari
1845–6357	0.259	12.571		(double)	
Kapteyn's Star	0.255	12.777	+245.5		Pic
Krüger 60	0.248	13.149	−33		
B	0.248	13.149	−32	DO Cephei	Cep
AX Microscopii	0.253	12.870	+28	Lacaille 8760	Mic
DEN 1048–3956	0.248	13.167	−10.1		Ant
Ross 614	0.244	13.349	+18.2	V577 Monocerotis	Mon
Wolf 106					
Van Maanen's Star	0.232	14.066	+6		Psc

Table 18.2 *Brightest stars in the past and future*

Years past/future	Star	Magnitude	Closest distance (light-years)	Present magnitude	Present distance (light-years)
− 4 700 000	ε Canis Majoris	−4.0	34	1.50	430
4 400 000	β Canis Majoris	−3.7	37	1.98	500
1 200 000	ζ Sagittarii	−2.7	8	2.60	89
1 000 000	ζ Leporis	−2.1	5.3	3.55	70
300 000	Aldebaran	−1.5	21.5	0.87	65
240 000	Capella	−0.8	28	0.08	42
60 000	Sirius	−1.6	7.8	−1.44	86
+ 300 000	Vega	−0.8	17	0.03	25
1 190 000	β Aurigæ	−0.4	28	1.90	82
1 250 000	δ Scuti	−1.8	9.2	4.70	187
1 500 000	γ Draconis	−1.4	28	2.24	148
2 290 000	υ Libræ	−0.5	30	3.60	195

Data are according mainly to J. Tomkin. Distances are according to Hipparcos.
Canopus was brilliant all through this period; estimates of its luminosity still do not agree well, but probably Canopus was, or will be, in prime position.

Table 18.3 *Stars that will change constellations*

Star	Magnitude	Enters	Year
ρ Aquilæ	5.0	Delphinus	1992
γ Cæli	4.6	Columba	2400
ε Indi	4.7	Tucana	2640
ε Sculptoris	5.3	Fornax	2920
λ Hydri	5.1	Tucana	3200
μ Cygni	4.8	Pegasus	4500
χ Pegasi	4.8	Pisces	5200
μ Cassiopeiæ	5.1	Perseus	5200
η Sagittarii	3.1	Corona Australis	6300
ζ Doradûs	4.7	Pictor	6400

The apparent diameters of stars are very small. For many years the largest measured diameter was that of Betelgeux (0.044 arcsec), but in 1997 measurements of the semi-regular variable R Doradûs showed that it is even larger (0.057 arcsec). The measurements were made with the New Technology Telescope at La Silla. R Doradûs is 200 light-years away; the real diameter is over 250 000 000 km. The spectrum is of type M.

STELLAR MOTIONS

All proper motions are very slight, so that to all intents and purposes the constellations we see today are the same as those visible in ancient times – although in 1718 Edmond Halley was able to show that Sirius, Arcturus, Aldebaran and Betelgeux had shifted perceptibly since Ptolemy had drawn them. The star with

Table 18.4 *The brightest stars*

Star	Name	Spectrum	Apparent magnitude	Absolute magnitude	Luminosity, Sun = 1	Distance (light-years)
α Canis Majoris	Sirius	A1	−1.44	1.15	26	8.6
α Carinæ	Canopus	FO	−0.62	−5.53	15 000	310
α Centauri	–	G2 + K1	−0.27	4.08	1.70 4s	4.4
α Boötis	Arcturus	K2	−0.05	−0.31	115	37
α Lyræ	Vega	AO	0.03	0.53	52	25
α Aurigæ	Capella	G5 + G6	0.08	−0.48	90 + 70	42
β Orionis	Rigel	B8	0.18	−0.69	40,000	770
α Canis Minoris	Procyon	F5	0.40	2.68	8	11
α Eridani	Achernar	B3	0.45	−2.77	1000	144
α Orionis	Betelgeux	M2	0.41	−5.14	11 000	430
β Centauri	Agena	S1	0.61	−5.42	13 000	530
α Aquilæ	Altair	A7	0.76	2.20	10	16.8
α Crucis	Acrux	S0 + B1	0.77	−4.19	3200 + 2000	320
α Tauri	Aldebaran	K5	0.87	−0.63	140	65
α Virginis	Spica	B1 + B2	0.98	−3.56	2200	260
α Scorpii	Antares	M1	1.06	−5.28	12 000	600
β Geminorum	Pollux	K0	1.16	1.09	33	34
α Piscis Australis	Fomalhaut	A3	1.17	1.74	20	25
α Cygni	Deneb	A2	1.25	−8.73	260 000	3000
β Crucis	Mimosa	B0	1.25	−3.92	3000	350
α Leonis	Regulus	B7	1.36	−0.52	125	78
ε Canis Majoris	Adhara	B2	1.50	−4.10	1000	430
α Geminoram	Castor	A1 + A2	1.53	0.59	53	52
γ Crucis		M3	1.59	−0.56	140	88
λ Scorpii	Shaula	B2	1.62	−6.05	9000	700
γ Orionis	Bellatrix	B2	1.64	−2.72	900	240
β Tauri	Alnath	B7	1.65	−1.37	300	130
β Carinæ	Miaplacidus	A2	1.67	−0.99	200	111
ε Orionis	Alnilam	B0	1.69	−6.38	24,000	1300
α Gruis	Alnair	B7	1.73	−0.73	150	101
ζ Orionis	Alnitak	0.95	1.74	−5.26	12,000	820
γ Velorum	Regor	W08	1.75	−5.31	12,500	840
ε Ursæ Majoris	Alioth	A0	1.76	−0.21	100	81
α Persei	Mirphak	F5	1.79	−4.50	5500	590
δ Scorpii	Dzuba	B0	1.81	−3.16	1300	400

Table 18.4 (cont.)

Star	Name	Spectrum	Apparent magnitude	Absolute magnitude	Luminosity, Sun = 1	Distance (light-years)
ε Sagittarii	Kaus Australis	B9	1.79	−3.44	300	145
α Ursæ Majoris	Dubhe	K0	1.81	−1.08	220	124
δ Canis Majoris	Wezea	F8	1.83	−6.87	50.000	1800
δ Ursæ Majoris	Alioth	B3	1.86	−0.60	150	101
ε Carinæ	Avior	F3	1.86	−4.58	6000	630
δ Scorpii	Sargas	F1	1.86	−2.75	930	270
β Aurigæ	Menkarlina	A2	1.90	0.10	95	82
α Trianguli Australis	Atria	K2	1.91	−3.62	2500	420
γ Geminorom	Alhena	AU	1.93	−0.60	145	105
δ Velorum	Koo She	A0	1.93	−0.01	85	80
α Pavonis		50	1.94	−1.81	450	180
α Ursæ Minoris	Polaris	F7	1.97	−3.64	2500	430
β Canis Majoris	Mirzam	K1	1.98	−3.95	3000	500
α Hydræ	Alphard	K3	1.99	−1.69	430	180
α Arietis	Hamal	K2	2.01	0.48	55	66
γ Leonis	Algieba	K0	2.01	−0.92	100	126
σ Sagittarii	Nunki	2.02	525	B3	209	

Excluding Mira Ceti and η Carinæ, the other stars above magnitude 2.5 are as follows:

Star	Name	Mag	Star	Name	Mag
β Ceti	Diphda	2.04	γ Draconis	Eltamin	2.25
β Andromedæ	Mirach	2.06	ι Carinæ	Tureis	2.25
α Andromedæ	Alpheratz	2.06	ε Scorpii	Wei	2.29
θ Centauri	Haratan	2.06	α Lupi	Men	2.30
κ Orionis	Saiph	2.06	ε Centauri		2.30
β Ursæ Minoris	Kocab	2.08	η Centauri		2.31
α Ophiuchi	Rasalhague	2.08	β Ursæ Majoris	Merak	2.37
ζ Ursæ Majoris	Mizar	2.09	ε Boötis	Izar	2.38
β Gruis	Al Dhanab	2.11	ε Pegasi	Enif	2.38
β Persei	Algol	2.12 max	α Phœnicis	Ankaa	2.30
γ Andromedæ	Almaak	2.14	β Pegasi	Schaet	2.4 var
β Leonis	Denebola	2.14	κ Scorpii	Girtab	2.41
γ Cassopeiæ		2.2 var	γ Ursæ Majoris	Phad	2.44
γ Cygni	Sadr	2.20	η Canis Majoris	Aludra	2.44
λ Velorum	Al Suhail al Wazn	2.21	α Cephei	Alderamin	2.44
α Coronæ Borealis	Alphekka	2.23	ε Cygni	Gienah	2.46
α Cassiopeiæ	Shedir	2.23 var?	α Pegasi	Markab	2.49
ζ Puppis	Suhai; Hadar	2.25	κ Velorum	Markeb	2.50
δ Orionis	Mintaka	2.23 var			

the greatest proper motion is Barnard's Star, discovered in June 1916 by E. E. Barnard at the Yerkes Observatory. The annual proper motion is over 10 arcsec, so that in 170 years it crosses the sky by a distance equal to the diameter of the full moon.

Radial velocities (the towards or away movements of the stars, relative to the Sun) are given as negative for a velocity of approach, positive for a velocity of recession. Barnard's Star is approaching us at 111 km s^{-1}, so that its distance is decreasing at the rate of 0.036 light-years per century. The proper motion is increasing by about 0.0013 arcsec per year, and will reach 25 arcsec by AD 11 800, when the star will be at its closest to us – 3.85 light-years, which is nearer than Proxima; the parallax will then be and the apparent magnitude will have risen to 8.5. Subsequently, the star will begin to recede.

Over sufficient periods of time, of course, the skies will alter. For example, Arcturus must first have become visible with the naked eye half a million years ago, and has brightened steadily; it is now about at its nearest and will subsequently recede, dropping below naked-eye visibility half a million years from now. Table 18.2 shows the brightest stars in the past and future, and Table 18.3 the stars which will change constellation.

STARS OF THE FIRST MAGNITUDE

The 21 brightest stars in the sky are usually said to be of the first magnitude; this takes us down to Regulus in Leo, 1.36. The Cambridge catalogue (*Sky Catalogue 2000*) was generally regarded as the best available. Today the official values for magnitudes, luminosities and distances are usually taken from the values given by the Hipparcos satellite, though it has to be admitted that in some cases Hipparcos is decidedly suspect; in any case the values for stars more than a few hundred light-years away are bound to be subject to some uncertainty. Table 18.4 shows the 50 brightest stars in the sky.

In general the Hipparcos and the Cambridge results are not very different, though there are a few exceptions – for example the Cambridge value for the luminosity of Canopus is 200 000 times that of the Sun, reduced to only 15 000 in the Hipparcos list, in this book I have followed Hipparcos, except where otherwise stated.

Some variable stars can reach the first magnitude, though in general they are fainter; thus at one time during the nineteenth century the unique η Carinæ shone as the brightest star in the sky apart from Sirius. Mira Ceti rose to 1.2 at the maximum of 1772, according to the Swedish astronomer Per Wargentin, though most maxima are below 2, and the official maximum magnitude is given as 1.7. Of course, novæ can become very brilliant for a few days, weeks or even months. The last novæ to reach magnitude zero were GK Persei, 1901 (0.0) and V603 Aquilæ, 1918 (–1.1); two others reached the first magnitude: RR Pictoris, 1925 (1.1) and DQ Herculis, 1934 (1.2).

Some stars have been suspected of permanent or secular variation. Thus Ptolemy ranked both β Leonis and θ Eridani as being of the first magnitude, although in the latter case there may well have been an error in identification. Ptolemy ranked Castor and Pollux as equal, although Pollux is now much the brighter of the two. Stars which are now decidedly brighter than as listed by Ptolemy include β Canis Majoris, β Canis Minoris, γ Cygni, δ Draconis, β Eridani, γ Geminorum, ε Sagittarii and Polaris; those which are decidedly fainter include ζ Eridani, α Sagittarii and α Microscopii. However, it is very unwise to place much reliance on those old records, and there is no proven case of a star of naked-eye brightness which has shown permanent change since Ptolemy's time.

19 · Stellar spectra and evolution

The stars show a tremendous range in luminosity, though much less in mass. Some known stars are millions of times more luminous than the Sun, while others are remarkably feeble. At its peak, in the 1840s, the erratic variable η Carinæ was estimated to be 6 000 000 times as powerful as the Sun; S Doradûs, in the Large Magellanic Cloud, has an absolute magnitude of –8.9, so that it is at least a million times as luminous as the Sun – yet because of its great distance (170 000 light-years) it cannot be seen with the naked eye. At the other end of the scale is MH18, discovered in 1990 by M. Hawkins at what was then the Royal Observatory, Edinburgh, from plates taken with the UK Schmidt telescope in Australia. It has 1/20 000 the luminosity of the Sun, and is presumably a brown dwarf (see below). Its mass is 5% that of the Sun, and its distance is 68 light-years.

The first attempt to classify the stars according to their spectra was made by the Italian Jesuit astronomer, Angelo Secchi, in 1863–7. He divided the stars into four main types:

(1) White or bluish stars: with broad, dark lines of hydrogen but obscure metallic lines. Example: Sirius.
(2) Yellow stars: hydrogen lines less prominent, metallic lines more so. Examples: Capella, the Sun.
(3) Orange stars: complicated, banded spectra. Examples: Betelgeux, Mira. The class included many long-period variables.
(4) Red stars: with prominent carbon lines; all below magnitude 5. Example: R Cygni. This class also included many variables.

Secchi's work was followed up enthusiastically by the English amateur W. Huggins, but the modern system of classification was developed in America, at the Harvard College Observatory. It was introduced by E. C. Pickering in 1890, and was extended by two famous women astronomers, Annie Jump Cannon and Wilhelmina Fleming, who produced the Draper catalogue. Conventionally, but confusingly, all elements except hydrogen and helium are classed as 'metals' from the spectroscopist's point of view. The Draper catalogue was so named because money for its development was provided by the widow of Henry Draper, who in 1872 had been the first to photograph the spectrum of a star (Vega). Secchi's catalogue had contained over 500 stars; the Draper Catalogue contained spectra of 225 000 stars down to the ninth magnitude.

The spectral types were allotted letters in order of decreasing temperature, A, B, C..., but before long it became clear that some of the original types were unnecessary and others out of order. The final sequence was alphabetically chaotic: O, B, A, F, G, K, M, R, N, S. A famous mnemonic runs 'O Be a Fine Girl Kiss Me Right Now Sweetie' (or, if you prefer it, Smack).

The original scheme depended mainly on surface temperature. Conventionally, the first in the sequence were called 'early' types and the K, M and other red stars as 'late', but this is now known to have nothing to do with evolutionary sequences.

Type W stars come at the start of the sequence (known as Wolf–Rayet stars, after the astronomers who drew attention to them). They are very hot and luminous, with prominent emission lines in their spectra, with atmospheres dominated by helium; the hydrogen layer may have been blown away by intense stellar wind, so exposing the helium shell. According to the dominance of carbon, nitrogen or oxygen emission lines, they are classed as WC, WN or WO.

In the Hertzsprung–Russell or HR diagram, drawn up in the early part of the twentieth century by E. J. Hertzsprung in Denmark and H. N. Russell in America, stars are plotted according to their luminosities and their spectral types. Most of the stars lie on a line extending from the upper left of the diagram to the lower right; this is termed the Main Sequence. Main Sequence stars (such as the Sun) are officially regarded as dwarfs. Characteristics of the main spectral types are given in Table 19.1. Stars of very 'early' type (W and O) and very 'late' type (R, N and S) are relatively rare. Types R and N are now often combined as type C.

In 1999, two new types, L and T, were added to accommodate very cool red dwarfs, plus brown dwarfs. L stars are dark red, with atmospheres cool enough for metal hydrides and alkali metals to show prominently in their spectra. T stars are cool brown dwarfs with methane prominent in their spectra. A new type, Y, has been proposed for dwarfs even cooler than T-dwarfs. No Y-star has yet been positively identified, but one candidate is CFBDS J005910.90–011.401.7, which shows ammonia absorption in the near-infrared.

Class D is now used for white dwarfs, where nuclear fusion has ceased and the bodies are of planetary size. There are various types: for example DA (strong hydrogen lines), DB and DO (helium strong), DQ (carbon lines strong) and DZ (metal-rich atmospheres). The letter is followed by a number, 1 to 9, indicating decreasing surface temperature. Examples are: Sirius B (DA2), Procyon B (DA4), Van Maanen's Star (DZ7).

The original Harvard scheme has been modified into what is known as the Yerkes or MKK classification, based on the work of W. W. Morgan, P. V. Keenan and E. Kellman. It takes into account the fact that the gravitational attraction on the surface of a giant star is much lower than for a dwarf, so that the gas pressures and densities are much lower in giants than in dwarfs. There are six luminosity classes (Table 19.2), so that the Sun would be specified as a G2V star. There are other characteristics of importance,

Table 19.1 *Stellar spectra*

Type	Surface temperature (°C)	Spectrum	Examples	Notes
W	Up to 80 000	Many bright lines; few absorption lines. Broad emission lines of hydrogen, ionised helium, carbon, nitrogen and oxygen.	γ Vel, WC7	Wolf–Rayet stars. Expanding shells, moving outward at up to 3000 km s⁻¹. All very remote. They may exist in binary systems where the companion star has stripped away the Wolf–Rayet's outer layers.
O	40 000–35 000	Both bright and dark lines; singly ionised helium lines in either emission or absorption. Strong ultraviolet continuum. Very massive and luminous.	ζ Pup, O5.8 ξ Per, O7 τ CMa, O9 ζ Ori, O9.5 10 Lac, O9	Represent a transition between W and B stars, although this does not imply any evolutionary sequence.
B	Over 26 000 for B0 to 12 000 for B9	No emission lines, but dominant absorption lines of neutral helium; hydrogen lines also prominent.	β Cru, B0 ε CMa, B2 α Eri, B5 α Leo, B7 β Ori, B8	Bluish (B0) to white (B9). Some such as Rigel (β Ori) are exceptionally luminous.
A	11 000–7500	Spectra dominated by hydrogen lines.	α Lyr, A0 α CMa, A1 α Cyg, A2 α Aql, A7	White stars, although some such as α Lyræ (Vega) are bluish.
F	7500–6000	Hydrogen lines less prominent. Metallic lines become noticeable; calcium very conspicuous.	α Car, F0 θ Sco, F0 α CMi, F5 δ CMa, F8 α UMi, F8	Yellowish, although the hue is so subdued that most F-type stars look white, Canopus being a good example.
G	Giants: 4200 Dwarfs: 5500	Solar-type spectra. Conspicuous metallic lines, hydrogen lines weaker.	η Boö, G0 δ Boö, G8 τ Cet, G8	Beginning of division between giants and Main Sequence stars (dwarfs). Yellow stars.
K	Giants: 4000–3000 Dwarfs: 5000-4000	Metallic lines dominant; hydrogen weak; weak blue continuum.	β Gem, K0 α Boö, K2 α Tau, K5 ε Eri, K2 ε Ind, K5	Orange stars. K-stars are more numerous than any other type. Division between giants and dwarfs now very marked.
M	Giants: 3000 Dwarfs: 3400	Very complicated spectra, with many bands due to molecules. Molecular bands of titanium oxide noticeable.	α Sco, M1	Orange-red stars, many of them variable. Many M-type stars are highly luminous giants, such as Antares, or supergiants, such as Betelgeux; many red dwarfs are very feeble, such as Proxima Centauri.
R	4800–2500	Carbon lines very prominent	V Ari, R	Red stars. most of them variable.
N	2800–2300	Similar to R stars, but rather cooler.	R Lep, N TX Psc,N	Types R and N are now often combined as type C.
S	2600	Prominent bands of heavier elements such as zirconium, yttrium and barium.	χ Cyg X Aqr R Cyg	Red stars: many are long-period variables.
L	1700–1000	Metal hydrides, alkali metals in atmosphere	VW Hyi	
T	1000–450	Methane dwarfs	ε Ind B	
Y	<450	Ultra-cool brown dwarfs		CFBDS J0059?

Table 19.2 *Stellar luminosity classes*

Ia	Very luminous supergiants
Ib	Less luminous supergiants
II	Luminous giants
III	Normal giants
IV	Subgiants
V	Dwarfs (Main Sequence stars)

Table 19.3 *Additional spectral classification*

comp	Composite spectrum: two stars are involved, indicating a very close binary system.
e	Emission lines are present (generally hydrogen).
m	Abnormally strong metallic lines; usually applied to A stars.
n	Broad (nebulous) lines, indicating fast rotation.
nn	Very broad lines, indicating very fast rotation.
neb	The spectrum of the star is mixed with that of a nebula.
p	Unspecified peculiarity, except with type A, where it denotes abnormally strong metallic lines.
s	Very narrow (sharp) lines.
sh	Shell star (a B to F Main Sequence star with emission lines from a shell of gas).
var	Varying spectral type.
wl	Weak lines, indicating an ancient, metal-poor star.

denoted by lower-case letters (Table 19.3); thus Alioth (Epsilon Ursæ Majoris) in the Great Bear is specified as type A0plV(CrEu), indicating that the spectrum shows strong lines of the elements chromium (Cr) and europium (Eu). The famous red variable Mira Ceti is classified as M7IIIe; this indicates that emission lines are present.

STELLAR EVOLUTION

It was originally believed that a star began its career as a large, red body, condensing out of nebular material. Its spectral type would then be M. It would contract and heat up, to become a Main Sequence star (type B or A), and then cool down while continuing to shrink, ending up as a red dwarf (type M once more). In this case the HR diagram would represent a definite evolutionary sequence, and certainly the Main Sequence is very marked, as are the giant and supergiant areas to the upper right. White dwarfs, to the lower left, were unrecognized in 1913, when the HR diagram was drawn up in its present form.

At that time the source of stellar energy was generally assumed to be gravitational, so that the star would begin at the top right of the HR diagram, cross to the Main Sequence at the top left, and then pass down the Main Sequence as it cooled. This would mean that red giants and supergiants would be very young. In fact this is not true; they are very advanced in their evolution.

Neither can gravitational contraction account for the radiation of a normal star. A star such as the Sun is several thousands of millions of years old, and simple gravitational contraction could not sustain it for anything like this period. Russell himself proposed that the energy source could be due to the annihilation of matter, so that certain types of particles were wiping each other out and releasing energy in the process. However, this would lead to a life-cycle of millions of millions of years, which was as obviously too long as previous estimates had been too short.

The key to the problem was found in 1939 by H. Bethe and, at about the same time, by G. Gamow. (Bethe actually worked it out during a train journey from Washington to Cornell University!) Normal stars shine by means of nuclear reactions. Thus, deep inside the Sun, hydrogen is being converted into helium. It takes four hydrogen nuclei to make one helium nucleus, and each time this happens a little mass is lost and a little energy is released. Each second, the Sun converts 600 million tonnes of hydrogen into helium – and loses 4 million tonnes in mass. This may sound a great deal, but the Sun is very massive, and has a great deal of hydrogen 'fuel', which lasts for a long time. The age of the Earth is

4.6 thousand million years; the Sun is certainly older than that, but even so it is no more than half-way through its main career.

What are termed 'stellar populations' were first described by W. Baade in the early 1950s. Population I stars are metal-rich, with about 2% of their mass being made up of elements heavier than helium. The most brilliant Population I stars are of types W, O and B. The disc and arms of the Galaxy (and other spirals) are mainly Population I.

Population II stars are metal-poor, and are clearly older so that their most brilliant members are red giants and supergiants which have already left the Main Sequence. They are dominant in the halo and nucleus of the Galaxy and other galaxies and in globular clusters. However, there is no hard and fast boundary between Population I and Population II regions.

There are numerous cases in which many stars of similar type, and presumably of similar age, are concentrated in a limited area. These are known as stellar associations.

STAR BIRTH

A star begins its career by condensing out of the gas and dust in the interstellar medium – that is to say, the gas and dust lying between existing stars. Vast amounts of material are concentrated in what are termed giant molecular clouds (GMCs); for example, a GMC covers almost the whole of the constellation of Orion, and the famous nebula M 42 is only a small feature of it. Nebulæ are, in fact, stellar birthplaces.

Star formation in a given area may well be triggered off by a nearby supernova explosion – the death of a very massive star, which literally blows itself to pieces, or the destruction of the white dwarf component of a binary pair. After the outburst, a shell of gas passes through the interstellar medium; if it encounters a GMC, then the GMC is compressed, and this stimulates star birth.

In nebulæ we can observe small dark 'globules', known as Bok globules in honour of the Dutch astronomer Bart Bok, who first drew attention to them in the 1940s. A typical Bok globule measures 1 to 2 light-years across, through some are larger; they are intensely cold – less than 15° above absolute zero. The denser regions inside them contract gravitationally and heat up, so that eventually they form masses of material known as proto-stars: these draw in further material by the process of accretion.

The subsequent career of the proto-star depends entirely upon its initial mass, so let us consider the different categories.

(1) Stars with mass below 0.08 that of the Sun, or 80 times that of Jupiter

The core temperature never becomes high enough to trigger off nuclear reactions such as the conversion of hydrogen into helium: for this, the temperature must rise to around 10 000 000 °C. Very low-mass stars which cannot achieve this are known as brown dwarfs, a term coined in 1975 by Jill Tarter; in fact, it is rather misleading, since visually a brown dwarf would look dull red. A young star of this kind shrinks until the pressure exerted by closely packed electrons (degenerate pressure) prevents it contracting further so that the size remains more or less constant for several thousands of millions of years: initially the dwarf shines feebly, although because no nuclear reactions are going on it gradually cools down and eventually ceases to shine at all. It becomes a black dwarf (though whether the universe in its current form is old enough for this to have happened is still problematical).

It has been said that a brown dwarf is a cross between a star and a planet, but this analogy must not be taken too far; a planet is formed by accretion from the material surrounding a young star, and its internal structure is different. Moreover, a planet has no light of its own, and depends upon reflecting the light of a nearby star, whereas a young brown dwarf may have an absolute magnitude of around +17 and a surface temperature of the order of 2000 °C. The luminosity may then be around 1/10000 that of the Sun.

Because they shine so feebly (if at all), brown dwarfs are not easy to identify, and for many years searches for them proved to be fruitless. Then, in 1995, a brown dwarf was detected in the Pleiades star-cluster: it was catalogued as PPl 15. The Pleiades cluster is only about 125 000 000 years old, so that PPl 15 was young; the surface temperature was about 2000 °C. Spectroscopic examination revealed the presence of the element lithium. Newly born stars contain small amounts of lithium (about 1 atom per thousand million), but an increased temperature will destroy lithium in about 100 million years – so that a cool object containing lithium must be a brown dwarf. Other brown dwarfs in the Pleiades, such as Teide I, were also reported.

However, even stronger evidence comes from the star Gliese 329A, a red dwarf in the constellation of Lepus; it is 19 light-years away. It has a companion, Gliese 329B, found on 27 October 1993 with the 60-inch (152-cm) and 200-inch (5-cm) telescopes at Palomar, and quickly confirmed by the Hubble Space Telescope. The distance between Gliese A and B is around 40 a.u. (much the same as the distance of Pluto from our Sun), and the mass seems to be between 20 and 50 times that of Jupiter, putting it firmly in the brown dwarf category. The surface temperature is around 700 °C; since it is between one and five thousand million

years old, it has had plenty of time to cool down. In 1997, a spectrum taken with UKIRT, the United Kingdom Infra-red Telescope on Mauna Kea in Hawaii, revealed the presence of methane and water vapour. Methane is destroyed at a temperature of 2500 °C, and so Gliese 229B cannot be as hot as this; it can only be a brown dwarf.

An isolated brown dwarf of special interest was detected in 1997 by Maria Teresa Ruiz, using the 3.6-m telescope at the La Silla Observatory in Chile; it was tracked down because of its unusually large proper motion. It seems to be 33 light-years away, and is of magnitude 22. The surface temperature is of the order of 1700 °C, and the mass is 75 times that of Jupiter, which is equivalent to 6% of the mass of the Sun; its spectrum shows unmistakable traces of lithium. It is three million times fainter than the dimmest object which can be seen with the naked eye; Ruiz named it Kelu-I, since that word means 'red' in the language of the Mapuche people, the ancient inhabitants of that part of Chile.

The coolest known methane brown dwarf, NNTDF J1205–0744, was detected in 1999 in Virgo, following a deep-field exposure with the Hubble Space Telescope. Its temperature is around 700 °C, suggesting an age of 500 to 1000 million years, and a mass 20 to 50 times that of Jupiter. It is 300 light-years away. Lacking a stable energy source at its core, it is becoming fainter and cooler, and will continue to cool. A very cool brown dwarf, Gliese 5670D, was identified in 2000 by A. Burgasser and D. Kirkpatrick with the Two Micron All Sky Survey at infrared wavelengths. A spectrum taken with the Cerro Tololo reflector (aperture 4 metres) shows lithium, which cannot survive in hotter stars. The temperature of Gliese 570D is about 750 K (480 °C). The dwarf orbits the triple star system of Gliese 570 at a distance of 1500 astronomical units, in a period of around 40 000 years. It is about the same size as Jupiter, but 50 times more massive. The distance from Earth is 19 light-years; the system lies in the constellation of Libra.

An even cooler brown dwarf – so far (2010) the record-holder – was discovered in 2008 by astronomers using infrared and near-infrared instruments on the CFH (Canada–France–Hawaii) telescope in Hawaii and the Gemini North telescope. This is CFBD 5600599; the surface temperature is only 623 K (350 °C). This means that its surface is cooler than that of Venus, and not even as hot as a standard electric stove element, and it radiates only in the near-infrared. It is 40 light-years away, and the mass is between 15 and 30 times that of Jupiter, which is not far from the lower mass limit of a brown dwarf. The composition is not yet known, but we may well have found the first definite example of a Y-type dwarf.

It has been said that a brown dwarf is a star which has failed its Common Entrance examination. This certainly seems appropriate!

Mention must also be made of quark stars, as yet hypothetical only, though at least one promising candidate has been found. Quarks are elementary particles found in nucleons (protons and neutrons), first proposed in 1964 by Murray Gell-Mann and George Zweig. It seems that quark stars may be formed inside neutron stars with over 1.5 times the mass of the Sun, where the pressures are so great that even neutrons are broken up into their constituent quarks – if so, the density must be intermediate between that of a neutron star and that of a black hole.

The best candidate so far (2010) is the object catalogued as RX J1856−35−375, detected by the orbiting Chandra X-ray Observatory, 450 light-years away. It was originally thought to be a neutron star, but the Chandra measurements indicate a diameter of only 11 km, which means that it is too small to consist of nucleons – they would be broken down into their quarks. The object cannot be proved to be a quark star, but this does seem likely.

(2) Low-mass stars

These are too massive to remain as brown dwarfs, but are below 1.4 solar masses. The processes of increasing mass in the central region of a dense cloud (accretion) continue, and eventually a proto-star develops; nuclear processes have not begun, but the protostar glows because of the heat into several stars; if the core is not rotating rapidly, it will form a single star (such as the Sun).

At first the protostar will be large and red, although by no means the same as red giants such as Arcturus or Aldebaran. As the contraction goes on, the surface temperature remains the same, so that the luminosity decreases. On the HR diagram, the star passes along what are known as the Hayashi and Henyey tracks (named after the astronomers who first described them) and finally joins the Main Sequence. T Tauri stars are still contracting toward the Main Sequence, and are irregularly variable; they are also sources of strong 'stellar winds', which blow away the original surrounding cocoon of dust. The removal of the cocoon makes the star brighten, and there is strong emission of infrared radiation.

When the cocoon of gas and dust is finally blown away, and mass accretion stops, the proto-star joins the Main Sequence at a point known as the ZAMS (Zero Age Main Sequence). Just where it joins depends upon its mass, and this also regulates the subsequent course of events. Once nuclear reactions begin – the conversion of hydrogen into helium – the star settles down to a period of stability; a star with a mass similar to that of the Sun will spend about 10 000 million years on the Main Sequence, so that the Sun is now about half-way through this stage in its evolution. At first it was only about 70% as luminous as it is now, but as the hydrogen-into-helium process continued the core temperature rose to its present value of about 15 000 000 °C. The conversion process – known, rather misleadingly, as hydrogen burning – is accomplished by a rather roundabout process known as the proton–proton reaction. With hotter stars, what is termed the carbon–nitrogen cycle is dominant, so that these two elements are used as catalysts; however, the end result is much the same – hydrogen is converted into helium, with release of energy and loss of mass. Gas radiation pressure (tending to make the star expand) and gravitation (tending to make it contract) balance each other out. In fact, the star adjusts its size to make this happen, so that the star remains stable.

This stage of evolution lasts for a long time, but not for ever. Helium is built up in the core, and eventually the supply of available hydrogen runs low. This means that less energy is generated, and finally the core can no longer support the weight of the star's outer layers pressing down on it from all directions. The inner temperature rises, with hydrogen burning continuing round the now helium-rich core. As the central temperature continues to rise, the outer layers expand and cool, so that the star becomes a red giant – as Arcturus and Aldebaran are now. It is then said to be on the asymptotic giant branch (AGB) of the HR diagram. The luminosity has increased a hundredfold, and the bloated globe is now so large that in the case of the Sun it will engulf the inner planets. Certainly the Earth cannot hope to survive as a habitable world.

When the core temperature reaches 100 000 000 °C, helium suddenly reacts to form carbon and oxygen. This is known as the 'helium flash', and is very sudden indeed; subsequently, the energy output declines, and so the outer layers again contract. The star is left smaller, hotter and dimmer, and cannot return to the Main Sequence. Very often the star becomes variable: stars of about the same mass as the Sun become what are termed W Virginis variables, while higher-mass stars become Cepheid variables. The variations are due to the fact that the stars are pulsating, changing both in luminosity and in radiation output. Following this period of instability, the outer layers of the star are puffed off; subsequently the expelled material becomes visible, excited by radiation from the star, to produce what is called a planetary nebula – a misleading term, because a planetary nebula has nothing whatsoever to do with a planet. This may dispose of 20% of the star's mass. The expelled material will not fall back, but will expand and dissipate in space; planetary nebulæ, as such, are therefore relatively short-lived, and are unlikely to last for as long as 100 000 years. Some are symmetrical, while others are irregular in form. Undeniably they are beautiful objects.

By this time nuclear reactions in the core have ceased. In most cases what remains of the old star (that is to say, the burnt-out core) will settle down as a white dwarf, such as the companion of Sirius, which is easily seen by an amateur astronomer who is equipped with an adequate telescope; it was the first white dwarf to be recognised as such. Its escape velocity and its surface gravity are very high. (Even this is not enough to produce a quark star, and in any case we cannot definitely prove that quark stars actually exist.)

As we have noted, the atoms in a white dwarf are broken up and their component parts packed tightly together with little waste of space. This explains the high density; a teaspoonful of their material would weigh several tonnes if it could be brought to the surface of the Earth. Sirius B is by no means the extreme example; we know of white dwarfs which are smaller than the Moon, but as massive as the Sun.

White dwarfs are of two main types, DA (with hydrogen-rich atmospheres) and DB (with more complex spectra). They have been referred to as 'bankrupt stars' which have used up all their energy, even though in some cases the surface temperatures may be as high as over 50 000 °C.

White dwarf material is officially termed 'degenerate'. The star cannot contract further, and it will simply cool down until it turns into a cold, dead globe – a black dwarf, sending out no energy at all. It is significant that no white dwarfs are known with surface temperatures much below 3000 °C, so that evidently the Galaxy is not yet old enough for even the most ancient white dwarfs to have cooled down below this temperature.

One very hot white dwarf, RE 1738 + 665, was discovered in 1994 by M. Barstow of Leicester University, from data supplied by

Table 19.4 *Conversion of absolute magnitude (A) to luminosity (L) in terms of the Sun. This table is no more than approximate, but serves as a general guide*

A	L	A	L
−16	200 000 000	+0.5	52
−15	80 000 000	+1	33
−14	30 000 000	+1.5	21
−13	13 000 000	+2	13
−12	5 000 000	+2.5	8.3
−11	2 000 000	+3	5.2
−10	800 000	+3.5	3.3
−9	330 000	+4	2 1
−8.5	200 000	+4.5	1.3
−8	132 000	+4.83	1 (Sun)
−7.5	83 000	+5	0.8
−7	52 500	+5.5	0.5
−6.5	33 000	+6	0.3
−6	21 000	+6.5	0.2
−5.5	13 200	+7	0.1
−5	8 300	+7.5	0.08
−4.5	5 200	+8	0.05
−4	3 300	+8.5	0.03
−3.5	2 000	+9	0.02
−3	1 300	+10	0.008
−2.5	800	+11	0.003
−2	520	+12	0.001
−1.5	330	+13	0.005
−1	200	+14	0.0022
−0.5	130	+15	0.000 08
0	83	+16	0.000 03
		+17	0.000 01
		+18	0.000 005

Table 19.5 *Evolution stage for a star 25 times as massive as the Sun*

Hydrogen burning	7 000 000 years
Helium burning	500 000 years
Carbon burning	600 years
Oxygen burning	6 months
Silicon burning	1 day
Core collapse	c. 0.1 s
Core bounce	A few milliseconds
Explosive burning	c. 10 s
Surface blows away	c. 1 h

the Röntgen Satellite (Rosat) artificial satellite. The surface temperature is of the order of 90 000 °C, and the atmosphere is pure hydrogen. The star seems to form a link between the cooler, hydrogen-rich white dwarfs and the very hot hydrogen-rich stars at the centres of planetary nebulæ.

(3) Stars more than 1.4 times as massive as the Sun

The Indian astronomer S. Chandrasekhar has shown that if a star is more than 1.4 times as massive as the Sun, it cannot become a white dwarf unless it sheds some of its mass during its evolutionary career – which happens, of course, during the unstable and the planetary nebula stages. If the mass remains above the Chandrasekhar limit, the star will meet with a very different fate; it will not subside quietly as a white dwarf, but will implode to become a neutron star via a supernova explosion.

Here, everything happens at an accelerated rate: for example, if the proto-star is 15 times as massive as the Sun it may reach the Main Sequence in a few thousand years, and will remain on the Main Sequence for a mere 10 million years before it has exhausted all its available hydrogen 'fuel'. The core heats up to 100 000 000 °C, and helium is converted to carbon, although the helium 'burns' steadily after the helium flash. By the time that helium burning has come to an end, the outer layers of the star have extended further still, and the star has become much brighter than a low-mass red giant such as Arcturus; it has turned into a supergiant, as Betelgeux in Orion is now. The absolute magnitude of a supergiant may attain −10, of the order of a million times the luminosity of the Sun (in the case of Betelgeux, the absolute magnitude is between −5 and −6). Massive though they are, supergiants are very rarefied, with average densities of less than one-millionth that of the Sun. Many are variable; these include the Cepheid variables, to be discussed below.

The most massive and most luminous known star is R 136a-1 in the Large Cloud of Magellan, 169 000 light-years away. Its mass is 265 times that of the Sun, and its luminosity is 10 000 000 times that of the Sun; I have nicknamed it 'The Monster'! It is important because it far surpasses the Eddington Limit, according to which there can be no star over 150 times as massive as the Sun. Its surface temperature is around 40 000 °C. When formed, probably about 3 000 000 years ago, it may have been well over 200 times as massive as the Sun, but it is shedding material at a furious rate, and in no more than 3 000 000 years it is bound to explode as a supernova. There are other very massive stars in this area (the Tarantula Nebula), but the Monster does seem to be exceptional.

As the core temperature continues to rise, carbon reacts in its turn; next, oxygen and silicon are produced, and then iron. This is where the whole situation changes. Iron will not react, and so energy production stops. Disaster follows. The process is out of control: in a matter of seconds the core collapses, and the electrons and protons are fused into neutrons. There is an 'implosion', followed by an explosion; shock waves race through the star, and literally blow it to pieces in what is termed a type II supernova outburst. At its peak, the luminosity may be 1000 million times that of the Sun. Heavy elements, produced in the supernova, are hurled away into space, later to form new stars: the remnant of the supernova remains as a neutron star, although it will eventually end up as a cold, dead globe.

A rough timetable for a star 25 times as massive as the Sun is given in Table 19.5.

Neutron stars

The concept of neutron stars was first proposed in 1932 by the Russian physicist L. Landau, and again in 1934 by F. Zwicky and W. Baade at Caltech (USA), but the first neutron star was not detected until 1967 – not by its visible light, but by its radio emissions. A neutron star is indeed an amazing object. The diameter is of the order of 20 km or even less, and the star cannot contract further, because of neutron degeneracy. The density is perhaps a thousand million million tonnes per cubic metre, or 1000 million million times that of water. A teaspoonful of neutron star material would weigh a thousand million tonnes, and a pin's head of the material would balance the weight of an ocean liner.

According to current theory, the mass of a neutron star is between about 1.4 to 2.1 times that of the Sun. The atmosphere is no more than a metre deep. The surface layer is crystalline, and around 100 m deep; it is iron-rich, and composed of what we may call 'normal' matter. There are 'mountains', but these cannot be more than a few centimetres high. The surface is to all intents and purposes smooth. The gravity is amazingly strong. Stand on a neutron star (admittedly rather a difficult thing to do!) and you will weigh 10 000 million times as much as you do on Earth. Drop an object from a height of 1 m, and it will land at a speed of 2000 km s^{-1}.

Neutron stars are quick spinners; the fastest known is rotating over 700 times per second. Neutron stars are created inside supernovæ, or from the collapse of other massive objects, and they slow down by a tiny fraction of a second per rotation. But now and then there are 'glitches' when the slowing-down is halted or even reversed temporarily. It is generally believed that cracks exist on the surface; these cause 'starquakes', affecting the rotation period. This may not be the full cause, but it probably contributes.

Below the crust comes the neutron-rich liquid mantle; below again we come to a superfluid core made up mainly of neutrons, and at the centre the material is made up of 'hyperons', about which we can only speculate. The inner temperature has been estimated as being 1000 million to a million degrees, but we do not really know. Over 2000 neutron stars have so far been identified; about 5% of them are members of binary systems. They can pair up with Main Sequence stars or other neutron stars; they may also be associated with black holes.

As the original star collapses, its magnetic field is concentrated and becomes very strong, reaching perhaps a hundred million tesla, as against roughly 30 millionths of a tesla for Earth. A neutron star with an even stronger field is called a *magnetar*; only a few dozens are known. A typical magnetar has a diameter of around 20 km, but the mass is greater than that of the Sun. A thimbleful of their material would weigh over 100 million tonnes. They are quick spinners, rotating several times per second. Their magnetic power does not last for long on the cosmical scale, and decays in about 10 000 years; the universe must be crowded with ex-magnetars.

At its peak the field strength may be 1 000 000 000 000 000 gauss (the Earth's is no more than 1 gauss). Luckily, all magnetars are a long way away. If one swooped by us at a range of 160 000 km, it would erase all our credit cards, and at 1000 km it would endanger life. The cause of these immensely powerful fields is unclear, but crustal starquakes may release energy in the form of gamma-rays, X-rays, and sub-atomic particles travelling at nearly the speed of light. This was confirmed in 1998, when astronomers using the VLA (Very Large Array) in New Mexico identified a short-lived 'afterglow' of particles emitted by a magnetar. The object, SGR 1900+14, is 15 000 light-years away; in August 1998 it had emitted a powerful burst of X-rays and gamma-rays, which is why attention had been drawn to it.

Magnetars and black holes

It had been assumed that a star with an initial mass 10 to 25 times that of the Sun will form a neutron star; above 25 solar masses, a black hole. However this has been challenged by studies of a star in the Westerlund cluster in Ara, 18 000 light-years away. European astronomers using the VLT (Very Large Telescope) found that the eclipsing binary system W13 has produced a magnetar, even though at 40 times the mass of the Sun it 'should' have formed a black hole. To quote Norbert Langer of the research team, 'This therefore raises the thorny question of just how massive a star must be to collapse to form a black hole if a star 40 times as heavy as the Sun cannot manage it.'

Pulsars

The first pulsar was discovered in 1967 by Jocelyn Bell, from Cambridge. She was using a special 'radio telescope', designed by A. Hewish, which was not a 'dish', but looked remarkably like a collection of barber's poles; the aim was to study the scintillation of distant radio sources. Stars twinkle, or scintillate, because of effects in the Earth's atmosphere; radio sources do so because they are affected by the clouds of electrons in the solar wind.

During her surveys, Jocelyn Bell came across a discrete radio source which 'pulsed' quickly, almost as though ticking. It was catalogued as CP 1919. Bell found the period to be 1.337 3011 s, which means that there are around 60 000 pulses per day; the period was absolutely regular, and the object was so extraordinary that Bell's colleagues were initially sceptical. However, following the initial announcement, made on 24 February 1968, other pulsars were soon found, and by now many hundreds are known.

It was some time before astronomers found out just what they were. There was even the short-lived LGM or Little Green Men theory – that the signals were artificial. Rotating white dwarfs were next suggested, but it became clear that a pulsar had to be even smaller than a white dwarf; it could be nothing other than a rotating neutron star.

A pulsar is spinning rapidly, and beams of radio radiation are sent out from its magnetic poles, which do not coincide with the poles of rotation. This leads to the 'lighthouse' effect. Each time a beam sweeps over the Earth, we receive a pulse of radiation. The 'normal' pulsar with the longest-known period (8.5 s) is PSR J2144−3933. The present holder of the short-millisecond period record (0.001396 s) is the Pulsar PSR J1748−2466 ad, which was identified in 2004 in the globular cluster Terzan 5. Next comes NP 0532, in the Crab Nebula, with a period of 0.339 s, so that it pulses 30 times in each second.

The Crab Nebula, in Taurus, is known to be the remnant of a supernova seen in the year 1054 (although since it is 6000 light-years away, the actual outburst took place in prehistoric times). It contains the first pulsar to be seen visually. This was achieved in January 1969 by a team at the Steward Observatory in Arizona, using a 36-inch reflector. They identified a faint, flashing object whose mean magnitude was about 17, and whose period was the same as that of the pulsar. Soon afterwards it was photographed from the Kitt Peak Observatory, also in Arizona. (Interestingly, R. Minkowski and W. Baade had observed it as far back as 1942, and had even suspected that it was the centre of activity in the Crab, but, not surprisingly, had failed to interpret it.) The second visual identification was that of the pulsar 0833–45 in the southern constellation of Vela; the pulsar lies in the Gum Nebula (named after its discoverer, C. S. Gum) which is unquestionably a supernova remnant. The pulsar itself was found in 1968 from the Molonglo Radio Astronomy Observatory in Australia, and in 1977 a team working with the 3.9-m Anglo–Australian Telescope at Siding Spring in New South Wales identified the pulsar as a faint, flashing object with a mean magnitude of 24.2. The period is 0.089 s, the third shortest known for a 'normal' pulsar.

As a pulsar spins and emits energy, its rate of rotation slows down by measurable amounts. Thus the period of CP 1919, Bell's original pulsar, is lengthening by a thousand millionth of a second each month, so that in 3000 years time the period will be 1.3374 s instead of the current 1.3373 s: this applies to all pulsars, including CP 1919. The period of the Crab pulsar is increasing by 3×10^{-8} s per day. Some pulsars show sudden, irregular changes in period, known as glitches; thus on 1 March 1969 the pulsar PSR 0833–45 speeded up by a full quarter of a millionth of a second. Glitches are due to disturbances in the pulsars – in other words, starquakes.

Obviously there must be many pulsars whose beams do not sweep over the Earth. For example, there is the relatively close neutron star known as Geminga; it is a radio and gamma-ray source, lying 300 light-years away. Its optical counterpart is a 25th magnitude star which has moved by 1.8 arcsec in eight years. Geminga emits light rays, X-rays and gamma-rays in all directions, but there are no pulses, so that presumably its beams never sweep across our line of sight. The closest known pulsar is JO 108–1431, in the constellation of Cetus; it is a thousand times weaker than any other known pulsar, and its age has been estimated as 160 000 000 years. The distance is 280 light-years. In the Guitar Nebula in Cepheus we find the fast-moving pulsar PSR 2224+65, 6000 light-years away; it spins 1.47 times per second, and is moving at around 800 km s^{-1}.

Binary pulsars are known; the first, PSR 1913+16, was discovered in 1974 by J. H. Taylor and R. Hulse, from Arecibo. A binary pulsar is a pulsar with a companion – usually another pulsar, a white dwarf or a neutron star. Binaries give us the best available information about pulsar masses, and the ways in which they move give an extra check on relativity theory. Their pulses have been likened to the ticks of a clock; changes indicate changes in time due to relativistic effects, and the agreement with Einstein's equations is perfect. When the two orbiting components are closest together, the gravitational field is stronger, the passage of time is lengthened, and the interval between successive ticks is also lengthened; as the 'pulsar clock' moves through the weakest part of the field (when the components are furthest apart) the time interval between successive ticks is shortened. Binary pulsars are excellent clocks!

Planets of pulsars have been reported but many people – including me – find them very hard to credit.

BLACK HOLES

(4) Stars too massive initially to form neutron stars

Here we come to the real cosmic heavyweights, with masses of the order of 40 times that of the Sun or even greater. Their fate will be different again; they will produce black holes, a term coined by Archibald Wheeler 30 years ago.

The basic principle was suggested as long ago as 1783 by the English natural philosopher John Michell, and again by the great French mathematician Laplace in 1796. Like Newton, Laplace believed light to consist of a stream of particles, and wrote that if a body were sufficiently small and dense it would be invisible, since the light particles would not be able to travel fast enough to escape from it. In fact this is reasonable enough; if the escape velocity exceeds the speed of light, then light cannot break free – and if light cannot do so, then certainly nothing else can.

Relativity theory explains the situation rather differently. A massive body distorts the curvature of space (or, more precisely, spacetime), and the paths of rays of light or particles of matter are regulated by the curvature of the space in which they are travelling. At a critical radius from the star, the curvature of space will be so great that it prevents light from escaping; this is known as the Schwarzschild radius, after the German astronomer Karl Schwarzschild, who drew up the theory in 1916. The value of the Schwarzschild radius, in kilometres, is $3.0M$, where M is the mass of the body in terms of the Sun. The Schwarzschild radius of the Sun is therefore 3 km: for a star 60 times as massive as the Sun, it will be 180 km: for the Earth, only about 9 mm. The boundary of the Schwarzschild radius determines the size of the black hole, and is termed the event horizon, because we have no positive information about events taking place inside it; the black hole is to all intents and purposes cut off from the outer universe.

A black hole is the result of the collapse of a very massive star. If the mass of the collapsing core exceeds the maximum possible for a neutron star, the collapse continues to a central point of infinite density, termed a singularity – a concept which is impossible to describe in everyday language. But before this, the collapsing star will have passed through its event horizon, and will have vanished as effectively as the hunter of the Snark.

Close to a black hole there are some very strange effects involving time dilation. A clock will run slow if it is within a strong gravity field (this has been experimentally proven). Picture an astronaut who carries a very accurate clock, and is falling towards a black hole, watched by an observer from a safe distance. The observer will conclude that the astronaut's clock is slowing down as it nears the event horizon; the interval between the ticks will increase. At the event horizon itself, the interval between the ticks will become infinitely long, so that the observer will assume that the traveller is left poised on the event horizon: but time will pass naturally insofar as the traveller is concerned, and he will simply fall through the event horizon and crash on to the singularity. By that time he will be in poor shape, because the tidal pull on his feet will be far stronger than that on his head (assuming that he is moving feet first), and he will be stretched out – a sort of cosmical Procrustean bed!

What is the true situation within a black hole, and does the original star's core crush itself out of existence altogether? We have to admit that we do not know, and since there can be no communication with the region beyond the event horizon it will be very difficult to find out.

If the black hole is rotating, the situation is rather different (a black hole of this kind is known as a Kerr black hole, after John Kerr, who first described it). A Kerr black hole is assumed to be surrounded by an ellipsoidal area or ergosphere, in which nothing can avoid being dragged round in the direction of the rotation; the singularity is a ring rather than a point of zero dimensions. There have been suggestions that an astronaut could enter a Kerr black hole, avoid the singularity, and emerge elsewhere, either in our universe or in a totally different universe, via a 'wormhole' linking one location in space time with another. It is an interesting theory, but one feels that there would be a definite shortage of volunteers willing to test it.

Stephen Hawking, one of the leaders in this field of research, believes that black holes may eventually evaporate. We know that pairs of particles and antiparticles form spontaneously in space, and almost immediately annihilate each other. If this happens close to the event horizon, one particle may fall into the black hole and the other be allowed to escape; the net result is that the mass of the black hole is slightly reduced, and eventually this process might accelerate, in which case the black hole would evaporate completely. Obviously we are theorising on the basis of very meagre data, and speculation is almost endless. In any case, the universe is not nearly old enough for any solar-mass black hole to have evaporated in this way, and though 'mini black holes' have been postulated there is no evidence that they actually exist.

Obviously we cannot see black holes, but we can detect them by their gravitational effects on visible matter. One famous example is Cygnus X-1, so called because it is an X-ray source. The system consists of a B0 type supergiant, HDE 226868, with about 30 times the mass of the Sun and a diameter 23 times that of the Sun (18 000 000 km), together with an invisible secondary with 14 times the mass of the Sun. The orbital period is 5.6 days, the distance from us is 6500 light-years, and the magnitude of the primary star is 9 (it lies at RA 19h 56m 295.3, declination $+35°$ 3' 55", near the star η Cygni). The secondary would certainly be visible if it were a normal star, and it is almost certainly a black hole which is pulling material away from the supergiant; before this material disappears over the event horizon, it is heated sufficiently to give off the

X-rays which we receive. Another famous example is V404 Cygni, which (having presumably been created by a much larger supernova explosion in the remote past) flared up as a nova in 1938, and rose from magnitude 19 to 12 before fading back to its normal brightness. In 1989, observations from the Japanese satellite Ginga showed that X-rays were being emitted, and studies by P. Charles, using the William Herschel telescope at La Palma, indicated that the system consists of a visible star 70% as massive as the Sun together with a black hole of 12 solar masses. The revolution period is 6.5 days.

There is certainly a black hole in the centre of our Galaxy, 26 000 light-years away. The centre is identified with a compact radio source, Sagittarius A* (pronounced A-star) and it is this which is apparently associated with a supermassive black hole, around 4 000 000 times as massive as the Sun; stars near it are orbiting it so quickly that they must be moving round a body so massive that it cannot be anything but a black hole. Other galaxies too are believed to have central black holes, and they may indeed be the rule rather than the exception. There also seems to be a black hole at the centre of the giant globular cluster ω Centauri.

In January 2000 it was announced that isolated black holes had been identified for the first time. Using the gravitational lensing technique, a team using Australian and Chilean telescopes found evidence of two black holes, each about six times as massive as the Sun, lensing the light of stars in the background. If the lensing objects were ordinary stars they would have outshone the background star; they seem too massive to be white dwarfs or neutron stars, but more observations are needed before any certain conclusions can be drawn.

Not all black holes are as massive as this. For instance, in January 2000 it was found that the X-ray source V4641 Sagittarii, at a distance of 1600 light-years from Earth, is almost certainly attended by a black hole companion. R. Hjellming examined the system with the VLA, and discovered twin jets shooting out from the system at over 90% of the speed of light. Only three other known systems eject material at such a velocity, and are termed 'microquasars'. Only the intense gravity of a black hole can generate so much power. This means that the companion to V4641 Sgr is the closest known black hole.

Not long ago there were many astronomers who questioned even the very existence of black holes. Now, of course, these doubts have been set at rest. Black holes are certainly widespread throughout the universe, and they are the most bizarre objects known to us.

OBSERVED EVOLUTION?

There are a few cases where we may be able to witness a stage in the evolution of a star; we may have caught a star changing from a red giant into a white dwarf. This is V4334 in Sagittarius, known as Sakurai's Object because attention was first drawn to it in February 1996, when the Japanese amateur astronomer Yukio Sakurai observed that it was rapidly brightening. This was due to a flash of helium fusion taking place in a shell surrounding the giant's carbon/oxygen core: in six months the star swelled from being a hot dwarf, with a surface temperature of 50 000 °C, into a yellow supergiant with a surface temperature of 6000 °C. The inner core contracted, and generated enough heat to start helium burning. The outburst also revealed a previously unknown planetary nebula associated with the star, and since then there have been interesting spectral changes. The nebula is expanding at the rate of 31 km s^{-1}, and now has an apparent diameter of 44 arcsec; its age has been given as between 3800 and 27 000 years. It has also been found that the rapidly evolving star showed a marked drop in temperature between 1996 and 1997. Its subsequent career will be followed with special interest.

20 · Extra-solar planets

The Sun is one of at least a hundred thousand million stars in our Galaxy, and many of these are of solar type. Therefore, most astronomers have always believed that it is unlikely to be unique in having a system of orbiting planets. If our Solar System had been formed by the near-collision between the Sun and a passing star, it would certainly have been a rarity, because close encounters seldom occur, but when this theory was abandoned there was no reason to believe that the Sun was a special case. Table 20.1 is a selected list of stars with known planets.

Proof was difficult to obtain. A planet will be very close to its parent star; it shines only by reflected light; it is much smaller than a normal star, and it will be drowned in its parent's glare, so that it will strain the power of even our largest telescopes. Only a few of these extra-solar planets ('exoplanets') have so far been actually seen, but fortunately there are other methods of detection. Of these, the most prolific are:

Astrometric. Proper motions of reasonably close stars are easy to detect with present-day equipment, and are measurable out to hundreds of light-years. If a star is attended by a planet of sufficient mass it will not move regularly, but will 'weave' its way along.

Radial velocity method. Slight variations in a star's radial velocity (i.e. the towards or away speed with respect to the Earth) can be found by shifting of the lines in the spectrum due to the Doppler effect as the planet orbits its parent star. Of course, the shifts will be detectable only when we are dealing with a massive planet moving round a comparatively lightweight star.

Transit method. If an orbiting planet passes in transit across the parent star it will block out part of the star's light, and the observed magnitude will drop. The diminution will depend on the size of the planet. An observer on, say, α Centuri would be able to see the slight fading of the Sun during a transit of Jupiter if he used equipment as sensitive as ours, but he would need much more powerful instruments to detect the transit of a planet as small as the Earth.

The main programme for the transit method is SuperWASP, the Wide-Angle Search for Planets. It consists of two robotic observatories, one at the Roque de los Muchachos (La Palma) and the other at the South African Astronomical Observatory (Sutherland). Each is equipped with an array of eight high-quality Canon 200-mm f/1.8 lenses plus CCDs, and is operated at all possible times; very slight fades due to transiting planets can be detected. The first success came in 2006, with WASP-1 in Andromeda; by the end of 2009 the number of discoveries was approaching twenty.

Space-craft are also involved. The Convection Rotation and planetary Transits (COROT) space-craft, launched in December 2006, was designed specifically to hunt for exoplanets. On 15 December 2009 NASA launched the Wide-field Infrared Survey Explorer (WISE), which carries a 40-cm diameter infrared sensitive telescope and orbits at an altitude of 525 km, period 95 minutes. It operates entirely at infrared wavelengths, and should be able to detect terrestrial-type planets; it will image at least 1.5 million stars. Its orbit is Sun-synchronous, circular and polar. It seems fitting to give it a polar orbit because, as someone at NASA pointed out, it is just about the same size as a polar bear!

The latest planet-hunting space-craft, Kepler (2010), has already proved to be extremely successful.

One question raised early in the search for extra-solar planets related to brown dwarf stars, not massive enough to trigger off nuclear reactions. They could so easily be confused with planets, even though they are basically so different. A brown dwarf does radiate, albeit very feebly, whereas no planet has any light of its own. Like normal stars, brown dwarfs form by the gravitational shrinking of a rotating mass of gas, whereas planets form by accretion in a disc of material round the parent star – and they have definite cores, whereas brown dwarfs are fully convective. Then there is the all-important question of mass. If a low-mass companion of a star has a mass greater than around 13 times that of Jupiter, the most massive planet in our Solar System, it cannot be a planet, and is almost certainly a brown dwarf; for bodies of mass between 8 and $13M_J$, the terminology becomes rather vague.

The first systematic search for a planet of another star, apart from those for the α Centauri group, was made in the mid-twentieth century by P. van de Kamp at the Sproule Observatory in the United States, using the astrometric method. In 1937, he began to monitor Barnard's Star, a faint red dwarf only 6 light-years away (in fact the nearest star apart from those of the α Centauri group). The luminosity of the star is 0.00045 that of the Sun, and it has the greatest proper motion known, so that in 170 years it crawls across the sky for a distance equal to the apparent diameter of the full moon. In 1963, he announced the detection of a planet with a mass similar to that of Jupiter, moving round the star at a distance of 4.4 a.u. He later claimed that there were at least two planets in the system. Alas, it was later found that the results were spurious, and were due to faults in van de Kamp's telescope. After these false alarms, attention was drawn to the naked-eye binary star γ Cephei (Alrai), where in 1988 slight effects were attributed to the presence of a planet. These were then found to be due to changes in the star itself, though many years later a planet was in fact orbiting the brighter member of the binary pair.

The first real clues came in 1983, from IRAS, the Infra-Red Astronomical Satellite, one of the most successful missions of the late twentieth century. While calibrating the on-board instruments, H. Aumann and F. Gillett, at the Rutherford–Appleton Laboratory

Table 20.1 *Selected list of stars with known planets*

Star	Constellation	RA (h)	(m)	(s)	Dec (°)	(')	('')	Magnitude	Distance (light-years)	Type	Planets
WASP-1	And	00	20	40	+31	59	24	11.8	1030	F7	b
ν Andromedæ	And	01	36	48	+41	24	20	4.1	44	F8	b c d
WASP-18	Phe	01	37	25	−45	40	40	9.3	330	F9	b
HD 17156	Cas	02	49	44	+71	45	12	8.2	255	G0	b
ε Eridani	Eri	03	32	55	−09	27	29	3.7	10.5	K2	b
XO-3	Cam	04	21	53	+57	49	01	9.8	850	F5	b
COROT-7b	Mon	06	43	49	−01	03	46	11.7	490	K0	b c
XO-4	Lyn	07	21	33	+58	16	05	10.7	950	F5	b
55 Cancri	Can	08	52	37	+28	20	02	5.9	41	G8	b c d e f
HD 80606	UMa	09	22	37	+50	36	13	8.9	190	G5	b
2M1207	Cen	12	07	33	−39	32	54	20.1	170	M8	b
Gliese 581	Lib	15	19	26	−07	43	20	10.6	20	M3	e b c d
HD 149026	Her	16	30	29	+38	20	50	8.1	257	G0	b
WASP-3	Lyr	18	34	32	+35	39	42	10.6	730	F7	b
HAT-P-7	Cyg	19	28	59	+47	58	10	10.5	1040	F8	b
HD 189733	Vul	20	00	43	+22	42	39	7.7	63	K2	b
Gliese 876	Aqr	22	53	13	−14	15	13	10.2	15	M4	d c b e
Fomalhaut	PsA	22	57	39	−29	37	20	1.2	25	A3	b

Planet	Mass, Jupiter = 1	Diameter, Jupiter = 1	Period (d)	Semi-major axis (a.u.)	Eccentricity, e	Inclination angle (°)	Discovery
WASP-1 b	0.86	1.42	2.520	0.382	0	83.9	2006
ν Andromedæ b	∼0.69	?	4.617	0.059	0.013	>30	1996
WASP-18	10.4	1.17	0.941	0.205	0.009	86	2009
HD 17156 b	3.21	1.02	21.22	0.162	0.675	88	2007
ε Eridani b	1.55	∼0.24	2505	3.39	0.7	30	2000
XO-3 Cam	11.79	1.22	3.191	0.045	.0.26	84	2007
COROT-7 c	0.026	?	3.698	0.046	0	?	2009
XO-4 b	1.72	1.34	4.125	0.055	0	88.7	2008
55 Cancri d	3.835	?	5218	5.77	0.025	?	2002
HD 80606 b	3.94	0.92	111.4	0.449	0.933	89.2	2001
2M1207 b	3–10	∼1.5	900 000	41	?	?	2004
Gliese 581 e	0.006	?	3.149	0.04	0	>3	2009
HD149026 b	0.359	0.6	2.876	0.03	0	85.3	2005
WASP-3 Lyr b	2.06	1.45	1.847	0.032	0	85.1	2007
HAT-P-7 b	1.8	1.421	2.205	0.038	0	84.1	2008
HD 189733 b	1.15	1.151	2.219	0.031	0	85.8	2005
Gliese 876 b	2.28	?	66.8	0.207	0.036	48.9	2000
Fomalhaut	0.05	?	320,000	∼115	∼0.11	∼66	2008

in Oxfordshire, found that Vega, one of the most brilliant stars in the sky, had 'a huge infrared excess', due presumably to solid particles moving in an extended region round the star and stretching out to about 80 a.u. The temperature of the material was around −185 °C. There were immediate suggestions that IRAS might have found a planetary system in its early stage of formation.

Other stars were subsequently found to be associated with cool, dusty material; two of these were ε Eridani, a mere 10.5 light-years away, and the first-magnitude Fomalhaut, in Piscis Australis (the Southern Fish). However, the most significant results were associated with β Pictoris, in the little constellation of the Painter, too far south in the sky to be seen in Europe. It is of magnitude 3.9, and easy to identify, fairly near Canopus. According to the Hipparcos satellite it is 63 light-years away, with a mass 1.8 times that of the Sun and with about 8 times the Sun's luminosity; the spectral type is A5, and its surface temperature is 7780 °C. It is a young star, with an age of between 8 and 20 million years, so that it has just joined the Main Sequence.

As with Vega and Fomalhaut, the IRAS satellite found a marked infrared excess, indicating a débris disc, and in 1984, R. Terrile and B. Smith, from the Las Campanas Observatory in Chile, succeeded in photographing it. It extended to nearly 150 00 000 000 km from the star, and was almost edgewise-on to us; its age can hardly be more than a few hundred million years. It is slowly rotating, and several elliptical rings have been found in its outer regions – due possibly to the system being disturbed by a passing star, perhaps the red giant β Columbæ, which passed at a range of 2 light-years about 110 000 years ago, or F-type yellow dwarf ζ Doradûs, which passed at 3 light-years 350 000 years ago.

Interesting though they were, the presence of débris discs did not necessarily show the existence of a planet. We now know that a planet almost certainly does exist, but it was not detected until November 2008, and before that there had been major developments.

In 1991, there was a startling announcement from Jodrell Bank. Using the 250-foot Lovell radio telescope, A. Lyne and his team reported the discovery of a planet orbiting a pulsar, PSR 1829–10. The method used was that of timing the pulsations, which would be affected by the pull of an orbiting body. The period was given as exactly six months. When I heard about this (a General Assembly of the International Astronomical Union was going on at the time, and I had been appointed Editor), I was incredulous, because a pulsar is a supernova remnant, and the very last place where a planet could be expected. In fact the result was spurious, because the researchers had not allowed for the fact that the Earth's path round the Sun is an ellipse, not a circle. As soon as they realised what had happened, they promptly and generously admitted their mistake – a classic example of true scientific integrity.

The real breakthrough came in September 1995, when two Swiss astronomers, Michel Mayor and Didier Queloz, made some new observations with the 30-m reflector at the Haute Province Observatory in France, and announced the detection of a planet orbiting the star 51 Pegasi, a G4-type star 54 light-years away: it is fractionally less luminous than the Sun and, at an apparent magnitude of 5.5, is easily visible with the naked eye. Mayor and Queloz used the radial velocity method. As a planet orbits a star the star itself swings round the common centre of gravity of the system. This affects its radial velocity as seen from Earth, and this can be measured by making use of the Doppler effect: the equipment used by Mayor and Queloz was sensitive enough to detect changes down to 12 m s^{-1}. The result seemed reliable, but the planet was very peculiar. The mass was about half that of Jupiter, but the distance from the parent star was a mere 7 000 000 km – about one-eighth the distance between the Sun and Mercury, the innermost planet of our Solar System. In this case, the surface should be baked to a temperature of around 1300 °C; the orbital period was 4.3 days. All in all, it seemed to be a most improbable object, but other stars were soon found to have close-in gas giants of the same kind. As they are much more like Jupiter than like the Earth, they have been nicknamed 'hot Jupiters'. As more and more discoveries were made, it became clear that planetary systems are very common in the Galaxy.

Obviously, really massive planets are the easiest to detect. Some of the gas-giants were Neptune-sized, but small exoplanets were much more elusive, though there was no reason to doubt their existence – quite possibly they are more numerous than 'hot Jupiters' or 'hot Neptunes'. The smallest exoplanet found by 2010 has less than twice the diameter of the Earth, and orbits its Sun-like parent star in a mere 20 hours. The discovery was made by equipment carried in a French-built satellite, COROT, launched on 27 December 2006 by a Russian rocket. The planet, known as COROT-7b, is not exactly welcoming; its surface temperature is probably between 1000 and 1500 °C.

Quite a number of naked-eye stars are known to have planets. They include Pollux, γ Leonis, ε Tauri, γ Cephei, υ Andromedæ, ι Draconis, τ Boötis, 70 Virginis and 51 Pegasi.

PULSAR PLANETS?

Though the first report of a planet of a pulsar, in 1991, was soon retracted, other reports have followed. In 1992, A. Wolczczan and D. Frail, at the Arecibo radio telescope, used the timing method and announced the discovery of a multi-planet system orbiting the millisecond pulsar PSR 1257+12, which is around 1000 light-years away. It was said that there were four planets (one unconfirmed), with masses ranging between 4.3 and 0.0004 times that of the Earth, periods from 1.250 to 98.211 days and distances between 0.19 and 2.6 a.u. In 2003, a planet was reported with the pulsar PSR B1620–26b, with a mass 2.5 times that of Jupiter, a distance of 23 a.u., and a period of 100 years. The pulsar was already known to have a white dwarf companion.

There is also the case of Geminga, an object which puzzled astronomers for many years (and in some ways still does). It was discovered as a gamma-ray source in 1975 by NASA's SAS 2 satellite (Small Astronomical Satellite 2) and remains the second-brightest gamma-ray source in the entire sky, but did not seem to radiate at any other wavelength. It became known as Geminga, partly because it lies in Gemini and partly because of a word in Milanese dialect meaning 'It's not there'. Then, in 1991, an X-ray pulse with a definite periodicity was detected by the Röntgen Satellite (Rosat), indicating that Geminga is a rapidly rotating neutron star – in fact a pulsar – whose beams of radio radiation do not sweep over the Earth. It seems to be the result of a supernova outburst 300 000 years ago; the distance is over 800 light-years, and may be over 1200 light-years.

In 1997, J. Maddox and his team announced the discovery of a planet orbiting Geminga; they used the method of gamma-ray timing. The planet was said to be 1.7 times as massive as the Earth, with a period of 5.1 years and a semi-major axis of 3.3 a.u. However, this remains unconfirmed, and it my well be that the results of the timing were due only to signal noise.

It is extremely difficult to see how a planet of a pulsar could have been formed. No existing planet could survive a supernova explosion, and the subsequent capture of a wandering planet seems most unlikely. In 2006, D. Chakrabarty and his team from MIT, using the Spitzer Space Telescope, reported a circumstellar disc round the magnetar 4U 0142+61, 13 000 light-years from the Earth, and suggested that it might have been due to metal-rich débris left from the supernova outburst about 100 000 years ago, so that the formation of planet-sized bodies could not be ruled out, but the whole question of pulsar planets remains controversial.

VISUAL SIGHTING

It is fair to say that astronomers searching for exoplanets would not be satisfied until one had actually been seen. There were several 'false alarms', as in 1999, when a British team using the 4.2-m William Herschel telescope in La Palma had obtained visual confirmation of a planet orbiting the star η Boötis. They believed that they had detected the reflection spectrum of the planet, using a new computer claimed to be sensitive enough to disentangle the faint light of the planet from the glare of the parent star. The result was later found to be spurious, but in November 2008 a team using the Hubble Space Telescope (HST) was successful, and photographed a planet orbiting the first-magnitude star Fomalhaut (α Piscis Australis).

Fomalhaut, an A-type star 18 times more luminous than the Sun and over twice as massive, is a mere 25 light-years away, and believed to be no more than 300 million years old. It was one of the first stars found to be surrounded by a débris disc and to emit Vega-type excess infrared radiation. The débris disc has a sharp inner edge at a distance of about 133 a.u. (2 000 000 000 km) from the star, and just inward of this, at 115 a.u., the HST photographs revealed a planet. Its mass is between 0.05 and 3 times that of Jupiter; the orbit is of low eccentricity, and the period is 872 years.

In November 2008, the Canadian astronomer Christian Maoris announced that using the Keck and Gemini North telescopes in Hawaii he had imaged three planets of the sixth-magnitude star HR 8799, in Pegasus. It is 129 light-years away, and about five times as luminous as the Sun, with a rather unusual spectrum of what is called the λ Boötis type, indicating that the outer layers at least are unusually poor in sulphur and iron. It is also slightly variable. The outermost planet, probably about seven times as massive as Jupiter, moves round the star at a distance of abut 68 a.u., just closer in than a substantial débris-ring, in a period of 460 years; the other two are more massive, though they do not seem to qualify as brown dwarfs.

The first débris disc to be imaged visually was that of β Pictoris. The system seems to be rather complex, and in 2008 the presence of a planet eight times as massive as Jupiter was indicated by 'warps' in the disc. The planet of 51 Pegasi, the first to be found, is a 'hot Jupiter' much closer to its parent star than Mercury is to the Sun; it whirls round in four days, at the giddy rate of over 130 km s^{-1}; the temperature must be about 1000 °C. Mayor and Queloz, who found it by the radial velocity method, wanted to give it a mythological name – Bellerophon – but the IAU has decided against naming exoplanets yet, and so we must accept the designation of 51 Pegasi b.

NOTABLE SYSTEMS

There are several multi-planet systems of special note. One of these, υ Andromedæ, was discovered in 1997 by Marcy and Butler, using the radial velocity method. υ Andromedæ is an F8-type Main Sequence star, of the fourth magnitude, 3.4 times as luminous as the Sun and 44 light-years away. There are four planets, with periods ranging from 4.6 days to 3.5 years, probably gas-giants of the Jupiter type, but no débris disc has been found. υ Andromedæ has an M4-type red dwarf companion, moving well beyond the primary star's system of planets.

ρ^1(55) Cancri is a binary, 41 light-years away, dimly visible with the naked eye. The primary component, a G-type dwarf, is considerably less luminous than the Sun, and is orbited by no fewer than five planets, all smaller than Jupiter, moving in periods ranging from 2.8 days out to over 14 years. The innermost planet may be comparable in mass to Uranus.

The southern yellow sub-giant star μ Aræ, 50 light-years from us, has four planets, three of which are not too unlike Jupiter, while the innermost was the first-known 'hot Neptune'. μ Aræ itself is about 1.8 times as luminous as the Sun. The first giant star found to have an orbiting planet was the orange K-type ι Draconis (Edasich), just over 100 light-years away and about 40 times as luminous as the Sun. Its planet, 8 times as massive as Jupiter, is about 1.5 a.u., from the star, with a period of 550 days.

We have certainly learned a great deal about exoplanets since Mayor and Queloz made the first discovery. There are planets of all kinds, and as we have noted, one star, ζ Leporis, seems to have an asteroid belt more populous than ours. ζ Leporis is an A-type star about twice as massive as the Sun and 18 times more luminous, easily seen with the naked eye even though it is 78 light-years away. Dust around it had been found by the IRAS satellite as long ago as 1983. In 2001, C. Chen and N. Jura, at the Keck Observatory, detected a ring of dust and débris from about 2.5 to 6.1 a.u. from the star. Because of the effects of the star's radiation, dust grains should spiral downwards in about 20 000 years, but ζ Leporis has an estimated age of 100 million years, so that the grains must be replenished from a reservoir. Collisions between asteroid-sized bodies account for this. The infrared measurements made by Chen and Jura indicated that the mass of the ζ Leporis belt is probably over 150 times that of the asteroid belt in our Solar System.

We can actually follow the destruction of one exoplanet, WASP-12b. It was discovered in 2008 by a team using SuperWASP, led by Shu-lin Li and D. N. C. Lin. The star, WASP-12, lies in Auriga; RA 06h 30m 33s, Dec +29° 40′ 20″, magnitude 11.7, type G0, distance 87 light-years. The planet has 4% more mass than Jupiter, and its radius is 79% larger; it is very close in, distance 0.02 a.u., with low eccentricity (0.049) but high inclination (83°). The gravity of the parent star is inflating the planet's size and destroying it. Tidal effects not only distort it, but create friction in its interior; this friction produces heat, and the planet expands. The planet is losing mass to its parent star at about 6000 million tonnes per second, and in about 10 million years there will be none left. The stripped-off material does not fall directly into the star, but forms a disc round the star and spirals slowly inwards. The disc may possibly contain a second, lower-mass planet.

Astronomers using the Hubble Space Telescope have recently (2010) confirmed the existence of a 'cometary planet' orbiting HD2090458, 153 light-years away; its orbital period is 3.5 days. Its heated atmosphere is escaping into space in a flow swept up by the stellar wind, producing a tail!

Table 20.2 *The Gliese 581 star system*

Companion (in order from star)	Mass ($M_\oplus$ = the mass of the Earth)	Semi-major axis (a.u.)	Orbital period (days)	Eccentricity
e	$3.1 \geq m \geq 1.94\ M_\oplus$	0.03	3.14942 ± 0.00045	0
b	$30.4 \geq m \geq 15.65\ M_\oplus$	0.04	5.36874 ± 0.00019	0
c	$10.4 \geq m \geq 5.36\ M_\oplus$	0.07	12.9292 ± 0.0047	0.17 ± 0.07
d	$13.8 \geq m \geq 7.09\ M_\oplus$	0.22	66.80 ± 0.14	0.38 ± 0.09

Table 20.3 *The υ Andromedæ system*

Companion (in order from star)	Mass (M_J = mass of Jupiter)	Semimajor axis (a.u.)	Orbital period (days)	Eccentricity
b	$\geq 0.672 \pm 0.056\ M_J$	0.0595 ± 0.0034	4.617136 ± 0.000047	0.013 ± 0.016
c	$\geq 1.92 \pm 0.16\ M_J$	0.832 ± 0.048	241.33 ± 0.20	0.224 ± 0.021
d	$\geq 4.13 \pm 0.35\ M_J$	2.53 ± 0.15	1278.1 ± 2.9	0.267 ± 0.021

THE SEARCH FOR EXTRA TERRESTRIAL INTELLIGENCE

In some cases the effects of exoplanet atmospheres during transits have made it possible to analyse the atmospheres themselves; water vapour has been found, for instance. What we have not managed to do, so far, is to find a planet so like the Earth where conditions are suitable for our kind of life. It is not for want of trying; the SETI organisation is doing its best.

SETI stands for 'the Search for Extra Terrestrial Intelligence'. If there are civilised beings within a few light-years or a few tens of light-years, contact might well be possible by radio at least. The essential first step is to locate a suitable planet. Unless we are prepared to consider aliens which we probably would not even recognise, the planet must be reasonably like Earth in size and mass, and it must be neither too hot nor too cold. Other conditions of the environment being suitable, water (an essential ingredient) must neither freeze nor boil.

Every star must have its zone of habitability, often called its 'Goldilocks' zone. The Sun's zone extends from just outside the orbit of Venus to just inside the orbit of Mars, so that the Earth sits comfortably in the middle (which is why we are here). With a less luminous star the Goldiocks zone will be closer in, while with a more powerful star it will be further out.

The best candidate found yet is Gliese 581c, the second planet in the system of this dim red dwarf star (Gliese 581, incidentally, is the number given in the well-known star catalogue drawn up by the German astronomer Wilhelm Gliese). The star he numbered 581 is 20 light-years away, with a mass no more than one-third that of the Sun; it is of type M3, and very feeble. Represent the Sun by a pocket torch, and Gliese 581 will be a glowworm.

In September 2010, astronomers at the Lick–Carnegie Exoplanet Survey, led by Steven Vogt, excitedly announced the discovery of an Earthlike, habitable planet orbiting the red dwarf Gliese 581 in Libra, already known to have several planets. It is 20 light-years away. The team used 122 radial velocity measurements from the HiRISE instrument on the Keck 1 telescope in Hawaii, and 119 from the HARPS instrument on the La Silla telescope in Chile. The planet, Gliese 581g, was said to be 1.9 times as massive as the Earth. However, in October 2010, Francesco Pepe of the Geneva Observatory, using the HARPS data, could find no trace either of it or another reported planet, Gliese 581f, and their existence must be regarded as questionable.

υ Andromedæ is also of special interest; here we have at least three planets, listed in Table 20.3. All three are likely to be gas-giants and therefore not likely to support life. The star itself is rather younger than the Sun, and does not appear to have a circumstellar belt of the same type as our Kuiper Belt. The outermost planet (d) moves within the star's Goldilocks zone, but we cannot image it, so that we know little about it; the mass may be about one-third that of Jupiter. Lower-mass planets presumably exist in the system, and in others, so that there must be vast numbers of planets suited to life. We have as yet no proof of life anywhere except on Earth, but most people (not all) are confident that it exists.

Assuming that we are not alone, how can we best try to contact 'aliens'? There is no intelligent life in the Solar System except (possibly!) on Earth, so that our nearest neighbours are light-years away. The two nearest solar-type stars, τ Ceti and ε Eridani, appeared to be particularly promising as planetary centres, and were within a dozen light-years of us; each was considerably less luminous than the Sun, though both are easily visible with the naked eye. τ Ceti turned out to be a disappointment, because it is associated with a tremendous amount of dust and rubble; there is no sign of a planet, but if one exists it will be subject to constant and merciless bombardment, and will be a most uncomfortable place. The SETI researchers were much more enthusiastic about ε Eridani.

There are certainly two asteroid-type rocky belts, one at around 3 a.u. from the star and the other at 20 a.u., and there is probably a gas-giant planet moving at a distance of 3.4 a.u. (500 000 000 km) in a period of approximately 7 years, and smaller planets of terrestrial type may well exist. So – can there be life?

In 1960, a pioneer experiment was conducted by a SETI team headed by Frank Drake, of Cornell University in the USA. Using the 26-m radio 'dish' at the Green Bank Observatory, West Virginia, they 'listened out' to both ε Eridani and τ Ceti, tuning the receiver to a wavelength of about 21 cm. This is the wavelength of radiation emitted by interstellar hydrogen in the Galaxy, so that any alien operators might well pay special attention to it. Alas, the Eridanians and the Cetians did not reply, and the pioneer experiment was given up. It was always the longest of long shots, but it was worth trying. (It was known as Project Ozma, after the main character in Baum's famous novel, though it was more commonly known as Project Little Green Men!)

It was also Drake who made the first really serious speculations about the numbers of civilisations possibly able to communicate with us. In 1961 he drew up the celebrated Drake equation, which runs as follows:

$$N = N^* f_p n_e f_l f_i f_c f_L$$

The equation can really be looked at as a number of questions:

(N^*) represents the number of stars in the Milky Way Galaxy

Question: How many stars are in the Milky Way Galaxy?

Answer: Current estimates give a hundred thousand million.

(f_p) is the fraction of stars that have planets around them

Question: What percentage of stars have planetary systems?

Answer: Current estimates range from 20% to 50%.

(n_e) is the number of planets per star that are capable of sustaining life

Question: For each star that does have a planetary system, how many planets are capable of sustaining life?

Answer: Current estimates range from 1 to 5.

(f_l) is the fraction of planets in n_e where life evolves

Question: On what percentage of the planets that are capable of sustaining life does life actually evolve?

Answer: Current estimates range from 100% (where life can evolve it will) down to close to 0%.

(f_i) is the fraction of f_l where intelligent life evolves

Question: On the planets where life does evolve, what percentage evolves intelligent life?

Answer: Estimates range from 100% (intelligence is such a survival advantage that it will certainly evolve) down to near 0%.

(f_c) is the fraction of f_i that communicate

Question: What percentage of intelligent races have the means and the desire to communicate?

Answer: 10% to 20%

(f_L) is fraction of the planet's life during which the communicating civilisations live

Question: For each civilisation that does communicate, for what fraction of the planet's life does the civilisation survive?

Strictly speaking, this should be L/T, where L is the length of time for which a communicating civilisation exists and T is the age of the Galaxy.

Answer: This depends upon the evolution of the parent star and the character of the civilisation. For example, the Earth should be able to support intelligent life for around 1000 million years before the Sun's increase in luminosity becomes intolerable – but we may well destroy ourselves in the near future unless we choose world leaders much wiser than those of today.

Drake came to the conclusion that there ought to be about 10 000 communicative civilisations in the Galaxy. Whether or not he was correct, and whether or not we will contact any of them, remains to be seen.

21 · Double stars

Double stars are of two types: optical pairs (that is to say line-of-sight effects) and binaries (physically associated pairs). Binaries are much the more frequent. They range from contact pairs, where the components are almost or quite touching, to very distant pairs separated by at least a light-year. In a binary system the components move round their common centre of gravity. For visual binaries the shortest period is that of Wolf 630 Ophiuchi (1.725 years), but shorter periods are known: the record-holder is X-1820–303, an X-ray star in the globular cluster NGC 6623, distance 30 000 light-years. Its period is 685 s or 11 min. It was discovered in 1987 by the aptly named L. Stella and collaborators with the Exosat satellite. It is impossible to say which is the binary with the longest period, and all we can say is that very widely separated components share a common motion through space. Table 21.1 lists prominent double stars.

EARLY OBSERVATIONS

The term 'double star' was first used by Ptolemy, who wrote that η Sagittarii was 'διπλυοζ'. There are of course several doubles which can be separated with the naked eye, so that presumably they have been known since antiquity; of these the most celebrated is Mizar (ζ Ursæ Majoris), which makes a naked-eye pair with Alcor (80 Ursæ Majoris). The Arabs described it – although they regarded Alcor as a rather difficult object. This is not true today, but it is most unlikely that there has been any real change.

The first double star to be discovered telescopically was Mizar itself, which is made up of two rather unequal components separated by 14″.5. The discovery was made by Castelli in 1617, and confirmed by Riccioli in 1651. Alcor is 700″ from the main pair, which is rather too wide to be classed as a recognised 'double' in the official catalogues. The duplicity of γ Arietis was discovered in 1665 by Robert Hooke, while he was searching telescopically for a comet.

The first southern double star to be discovered was Crucis by Father Guy Tachard, in 1685. Tachard was on his way to Siam, by sea, and stopped off at the Cape of Good Hope, where he was warmly welcomed by the Dutch settlers and set up a temporary observatory, mainly for navigational purposes. He recorded that 'the foot of the Crozier marked in Bayer's map is a Double Star, that is to say consisting of two bright stars distant from one another about their own diameter, only much like the most northern of the Twins, not to speak of a third much less, which is also to be seen but further from these two'.

The Crozier is the Southern Cross; 'Bayer' refers to J. Bayer's famous catalogue of 1603, and the 'northern' of the Twins is Castor,

which was already known to be double. (Like most of his contemporaries, Tachard believed the stars to show definite apparent diameters rather than being virtual point sources.)

Other doubles discovered at an early stage were α Centauri (1689), γ Virginis and the Trapezium's θ Orionis, in the Orion Nebula, which is a multiple system. In 2009, a team led by S. Kraus and G. Weigelt (Bonn, Germany) used the VLTI (Very Large Telescope Interferometer) to obtain a very sharp image of θ Orionis C. The separation of the two stars is 2 milli-arcsec, corresponding to the apparent size of a car on the Moon. The masses of the two components are relatively 38 and 9 times that of the Sun. The distance from the Sun is 1350 light-years; this is the region where massive stars are born.

DOUBLE-STAR CATALOGUES

The first true list of doubles was published in 1771 by C. Mayer of Mannheim. This list included γ Andromedæ, ζ Cancri, α Herculis and β Cygni. Most of his observations were made with an 80-foot mural quadrant, with magnifications of 60 and 80. Many catalogues have appeared since. Among them are those by F. G. W. Struve (Dorpat 1822, with later additions); E. Dembowski (Naples 1852, over 20 000 measures); S. W. Burnham (1870 and again in 1906, listing 13 665 pairs – he was personally responsible for the discovery of 1340 of them); R. Aitken (1932, 17 180 pairs) and the Lick Index Catalogue or IDS (1963, 65 000 pairs, of which 40 000 are binaries). Work of the greatest importance was carried out at the Republic Observatory, Johannesburg (formerly the Union Observatory) between 1917 and 1965, under the successive directorships of R. T. A. Innes (1917–27), H E. Wood (1927–42), W. H van den Bos (1941–56) and W. S. Finsen (1957–65): the telescope used was the 27-inch (69-cm) Innes refractor.

William Herschel, of course, discovered large numbers of double stars, and in our own time superbly accurate measurements have been made from the Hipparcos astrometric satellite. This was launched on 8 August 1989, and its main catalogue appeared seven years later. The catalogue includes details of over 12 000 double stars, of which 3000 were new discoveries. The accuracy was truly amazing. For example, images were obtained of the double star HTP 46706 in Hydra, which is 34 light-years away; the components are separated by only about 1/5000 of a degree. This is a binary system (period 18.3 years); the masses of the components are 0.42 and 0.41 that of the Sun.

The position angle of a visual double star (either an optical or a binary pair) is measured according to the angular direction of the secondary (B) from the primary (A), reckoned from 000 at north

Table 21.1 *Selected list of prominent double stars*

Star	RA (h)	(m)	Dec. (°)	(′)	Magnitudes	Separation (″)	Position angle (°)	
β Tuc	00	31.5	−62	58	4.4, 4.8	27.1	170	Both components again double.
η Cas	00	49.1	+57	49	3.4, 7.5	12.7	315	Creamy, bluish. Binary, 480 y.
ζ Psc	01	13.7	+07	35	5.6, 6.5	23.1	063	
γ Ari	01	53.6	+19	18	4.8, 4.8	7.5	001	
α Psc	02	02.0	+02	46	4.2, 5.1	1.8	274	Binary, 933 y.
γ And	02	03.9	+42	20	2.3, 5.0	9.6	062	B is double; magnitudes 5.5, 6.3; separation 0″.5, position angle 61°.
τ Cas	02	29.2	+67	25	4.9, 6.9	3.0	227	Binary, 840 y. Third star at 7″.2, position angle 118°, magnitude 8.4.
ω For	02	33.8	−28	14	5.0, 7.7	10.8	245	Fixed. Common proper motion.
γ Cet	02	43.3	+03	14	3.5, 7.3	2.6	298	
θ Eri	02	58.3	−40	18	3.4, 4.5	8.2	088	Fine pair. Both white. Acamar.
ε Ari	02	59.2	+21	20	5.2, 5.5	1.5	208	
α For	03	12.1	−28	59	4.0, 7.0	4.8	301	Binary, 314 y.
τ Pic	04	50.9	−53	28	5.6, 6.4	12.3	058	Fixed.
κ Lep	05	13.2	−12	56	4.5, 7.4	2.2	357	
β Ori	05	14.5	−08	12	0.1, 6.8	9.5	202	Rigel. Fixed. Common proper motion.
η Ori	05	24.5	−02	24	3.8, 4.8	1.8	078	Third star, magnitude 9.4, 115″.1, 051°.
λ Ori	05	35.1	+09	56	3.6, 5.5	4.3	043	Fixed.
θ Ori AB	05	35.3	−05	23	6.7, 7.9	8.8	032 ⎫	Trapezium. In M 42.
CD	'		'		5.1, 6.7	13.4	061 ⎭	
σ Ori AC	05	38.7	−02	36	4.0, 10.3	11.4	238	
ζ Ori	05	40.8	−01	57	1.9, 4.0	2.4	162	Alnitak. Binary, 1509 y
η Gem	06	14.9	+22	30	var, 6.5	1.6	257	Propus. A is orange. Binary, 474 y.
γ Vol	07	08.8	−70	30	4.0, 5.9	14.1	298	Combined magnitude 3.6.
δ Gem	07	20.1	+21	59	3.5, 8.2	5.8	225	Binary, 1200 y. Yellow, pale blue.
α Gem	07	34.6	+31	53	1.9, 2.9	3.7	067	Castor. Binary, 420 y. Widening.
κ Pup	07	38.8	−26	48	4.5, 4.7	9.8	318	Combined magnitude 3.8.
ζ Cnc AB+C	08	12.2	+17	39	5.3, 6.0	5.9	074	Binary, 1150 y.
δ Vel	08	44.7	−54	43	2.1, 5.1	2.6	153	
ε Hya	08	46.8	+06	25	3.3, 6.8	3.3	298	A is a close binary, 890 y.
υ Car	09	47.1	−65	04	3.1, 6.1	5.0	127	Fixed.
γ Leo	10	20.0	+19	51	2.4, 3.5	4.6	125	Binary, 169 y. Two distant companions.
μ Vel	10	46.8	−49	25	2.7, 6.4	2.6	057	Binary, 116 y. Closing.
ξ UMa	11	18.2	+31	32	4.3, 4.8	1.6	080	Binary, 59.8 y. Opening.
ι Leo	11	23.9	+10	32	4.0, 6.7	1.7	118	Binary, 192 y.
N Hya	11	32.3	−29	16	5.8, 5.9	9.5	210	Fixed.
D Cen	12	14.0	−45	43	5.6, 6.8	2.8	243	Orange, white, Closing.
α Cru	12	26.6	−63	06	1.4, 1.9	4.0	113	Acrux. Combined magnitude 0.8. C at 90″.1, 202°, magnitude 4.9.
γ Cru	12	31.2	−57	07	1.6, 6.7	110.6	031	C at 155″.2, 082°, magnitude 9.5.
γ Cen	12	41.5	−48	58	2.9, 2.9	1.0	347	Binary, 84.5 y. Closing
γ Vir	12	41.7	−01	27	3.5, 3.5	1.6	264	Binary, 171.4 y. Opening.
β Mus	12	46.3	−68	06	3.7, 4.0	1.2	039	
μ Cru	12	54.6	−57	11	4.0, 5.2	34.9	017	
α CVn	12	56.0	+38	19	2.9, 5.5	19.1	299	Cor Caroli. Yellow, bluish.
ζ UMa	13	23.9	+54	56	2.3, 4.0	14.4	152	Mizar. Alcor at 708″.7, magnitude 4.0, position angle 0.71°.
α Cen	14	39.6	−60	50	0.0, 1.2	14.8	221	Binary, 79.9 y. Closing.

Table 21.1 (cont.)

Star	RA (h)	RA (m)	Dec. (°)	Dec. (′)	Magnitudes	Separation (″)	Position angle (°)	
ζ Boö	14	41.1	+13	44	4.5, 4.6	0.8	300	Closing. Binary, 123.3 y.
ε Boö	14	45.0	+27	04	2.5, 4.9	2.9	341	Yellow, blue.
ξ Boö	14	51.4	+19	06	4.7, 6.8	6.7	319	Binary, 150 y.
π Lup	15	18.5	−47	53	4.6, 4.7	1.7	067	Widening. 7.2 magnitude star at 23″ .7, 130°.
γ Cir	15	23.4	−59	19	5.1, 5.5	0.8	011	Closing. Binary, 180 y.
δ Ser	15	34.8	+10	32	4.1, 5.2	4.0	175	Binary, 3168 y.
ζ CrB	15	39.4	+36	38	5.1, 6.0	6.3	305	
γ Lup	15	56.9	−33	58	5.3, 5.8	10.2	049	Fixed.
σ CrB	16	14.7	+33	52	5.6, 6.6	7.0	236	Binary, 1000 y.
α Sco	16	29.4	−26	26	1.2, 5.4	2.7	274	Antares. Red, green. Binary, 878 y.
ζ Her	16	41.3	+31	36	2.9, 5.5	0.9	029	Binary, 34.5 y. Position angle and separation change quickly.
μ Dra	17	05.3	+54	28	5.7, 5.7	2.0	017	Binary, 482 y. Closing.
α Her	17	14.6	+14	23	var, 5.4	4.6	105	Red. Green. Binary, 3600 y.
70 Oph	18	05.5	+02	30	4.2, 6.0	3.4	152	Binary, 88.1 y. Widening.
ε¹ Lyr	18	44.3	+39	40	5.0, 6.1	2.6	357 ⎫	Separation 207″ .7. Quadruple.
ε² Lyr	18	44.3	+39	40	5.2, 5.5	2.3	094 ⎭	
θ Ser	18	56.2	+04	12	4.5, 4.5	22.4	104	Fixed.
γ CrA	19	06.4	−37	04	4.8, 5.1	1.3	061	Binary, 120.4 y.
β Cyg	19	30.7	+27	58	3.1, 5.1	34.4	054	Albireo. Yellow, blue.
δ Cyg	19	45.0	+45	07	2.9, 6.3	2.5	226	Widening.
ε Dra	19	48.2	+70	16	3.8, 7.4	3.1	019	Slow binary.
γ Del	20	46.7	+16	07	4.5, 5.5	9.2	206	Both yellowish.
61 Cyg	21	06.9	+38	45	5.2, 6.0	30.5	150	Binary, 722 y.
θ Ind	21	19.9	−53	27	4.5, 7.0	6.8	271	Yellow and red. Slow binary.
μ Cyg	21	44.1	+28	45	4.8, 6.1	2.0	206	Binary, 716 y.
ξ Cep	22	03.8	+64	38	4.4, 6.5	8.0	276	White and blue.
ζ Aqr	22	28.8	−00	01	4.3, 4.5	1.9	187	Binary, 856 y. Widening.
δ Aps	16	20.3	−78	42	4.7, 5.1	102.9	012.	
α Aql	19	50.8	+08	52	0.8, 9.5	165.2	301	Altair. Optical pair.
θ Aur	05	59.7	+37	13	2.6, 7.1	3.6	313	10.6 magnitude, star at 50″, 297°.
ε CMa	06	58.6	−28	58	1.5, 7.4	7.5	161.	
α Cap	20	18.1	−12	33	3.6, 4.2	377.7	291.	
β Cap	20	21.0	−14	47	3.1, 6.0	205	267	B is a close double.
β Cep	21	28.7	+70	34	3.2, 7.9	13.3	239.	
δ Cep	22	29.2	+58	25	var, 7.5	41.0	191.	
o Cet	02	19.3	−02	59	var, 9.5	0.6	085	Mira. Binary, 400 y. Mira B is VZ Ceti.
κ CrA	18	33.4	−38	44	5.9, 5.9	21.6	359.	
δ Her	17	15.0	+24	50	3.7, 8.2	8.9	236	Optical pair.
β Hya	11	52.9	−33	54	4.7, 5.5	0.9	008	
γ Lep	05	44.5	−22	27	3.7, 6.3	96.3	350	
α Lib	14	50.9	−16	02	2.8, 5.2	231.0	314	
ζ Lyr	18	44.8	+37	36	4.3, 5.9	43.7	150	
ε Mon	06	23.8	+04	36	4.5, 6.5	13.4	027	
θ Mus	13	08.1	−65	18	5.7, 7.3	5.3	187	
ε Nor	16	27.7	−47	33	4.8, 7.5	22.8	335	
λ Oct	21	50.9	−82	43	5.4, 7.7	3.1	070	
ρ Oph	16	25.6	−23	27	5.3, 6.0	3.1	344	
β Phe	01	06.1	−46	43	4.0, 4.2	1.4	346	

Table 21.1 (cont.)

Star	RA (h)	(m)	Dec. (°)	(′)	Magnitudes	Separation (″)	Position angle (°)	
β PsA	22	31.5	−32	21		10.3	172	Optical pair.
ζ Ret	03	18.2	−62	30		310.0	218	Common proper motion.
β Sco	16	05.4	−19	48		13.6	021	A is a close double.
θ Tau	04	28.7	+15	32		337.4	346	White, orange. Optical pair.
κ + 67 Tau	04	25.4218	+22	18		339	173	Optical pair.
α UMi	02	31.8	+89	16		18.4	218	Polaris.
α + 8 Vul	19	28.7	+24	40		413.7	028	Optical pair.

Table 21.2 *Limiting magnitudes and separations for various apertures*

Aperture of object-glass (inch)	Aperture of mirror (cm)	Faintest magnitude	Smallest separation (″)
2	5.1	10.5	2.5
3	5.2	11.4	1.8
4	10.2	12.0	1.3
5	12.7	12.5	1.0
5	15.2	12.9	0.8
7	17.8	13.2	0.7
8	20.3	13.5	0.6
10	20.5	14.0	0.5
12	30.5	14.4	0.4
15	38.1	14.9	0.3

It is extremely difficult to give a definite value for limiting magnitudes and separations, since so much must depend upon individual observers. This table must be regarded as approximate only. The third column refers to stars of equal brilliancy and of about the sixth magnitude. Where the components are unequal, the double will naturally be a more difficult object, particularly if one star is much brighter than the other.

round by east (090), south (180) and west (270), back to north. With rapid binaries the position angles and separations alter quickly: a good example is the fine binary ζ Herculis, where the magnitudes are 3 and 5.6, and the period only 34 years. Some of the published catalogues are already out of date and need revision. There is scope here for the skilful and well-equipped amateur as well as for the professional.

Limiting magnitudes and separations for different aperture telescopes are given in Table 21.2.

BINARY SYSTEMS

The first suggestion that some double stars might be physically associated pairs was made by the Reverend John Michell in 1766, who wrote: 'it is highly probable, in particular, and next to a certainty in general, that such double stars as appear to consist of two or more stars placed very near together, do really consist of stars under the influence of some general law'. Michell repeated this view in 1784. However, in 1782, William Herschel commented that it was 'much too soon to form any theories of small stars revolving around large ones'. The actual proof was given by Herschel himself in 1802. From 1779 be had been attempting to measure the parallaxes of stars, since if one member of the pair were more remote than the other it followed that the closer member should show an annual parallax relative to the more distant component. He failed, because his equipment was not sufficiently sensitive, but he made the fortuitous discovery that some of the pairs under study (such as Castor) showed orbital motion, and by 1802 he was confident enough to 'publish' his findings. His classic paper actually appeared in the *Philosophical Transactions* on 9 June 1803.

The first reliable orbit for a binary pair (ξ Ursæ Majoris) was worked out by the French astronomer Felix Savary in 1830. (The period is 60 years.) Such calculations are of great importance, since they lead to a determination of the combined masses of the components – something which is much more difficult to calculate for a single star. In fact, our knowledge of stellar masses depends very largely upon the orbital movements of binaries. It has been calculated that more than 50% of all stars are members of binary systems, with a mean separation of 10–20 a.u., although this may be too high a ratio. It is important to note that both components of a binary move round their common centre of gravity, and the orbits are not in general so dissimilar as might be thought, since in mass there is a far lower spread among the stars than there is in luminosity and in size.

Incidentally, it was the measurements of a binary – 61 Cygni – which led F. W. Bessel, in 1838, to make the first successful determination of the distance of a star. 61 Cygni was selected because it had a large proper motion and was a wide binary, indicating that by stellar standards it must be comparatively close (the distance is 11.1 light-years).

The first *spectroscopic binary* – (Mizar A) – was discovered in 1889 by E. C. Pickering at Harvard: another identification (β Aurigæ) soon followed. Spectroscopic binaries have too small a separation for the components to be seen individually, but the binary nature of the system betrays itself because of the Doppler shifts in the spectra. If both spectra are visible, the absorption lines will be periodically doubled; if one spectrum is too faint to be seen, the lines due to the primary will oscillate around a mean position. There are some 'borderline' cases: thus Capella was long known to be a spectroscopic binary, but only the world's largest telescopes can just indicate that it is not a single star. The orbital period is 100 days.

The first *astrometric binaries* to be studied were Sirius and Procyon, by F. W. Bessel in 1844. In an astrometric binary, the presence of an invisible companion is inferred from slight displacements of the primary. (This, of course, is also one important method for detecting planetary or brown dwarf companions of visible stars.)

There is an interesting corollary with regard to the companion of Sirius. It was first seen in 1862 by Clark, using the Washington refractor, almost exactly where Bessel had predicted. The orbital period is 50 years, and the maximum separation is 11″.5. For many years after its discovery the companion was assumed to be faint because it was cool and red, but in 1915 W. S. Adams, at Mount Wilson, studied its spectrum and found that it was white; the surface temperature was at least 8000 °C. Since the luminosity was only 1/10 000 of that of Sirius itself, the companion had to be small – no larger than a planet such as the Earth. The companion was in fact the first known white dwarf, with a density 125 000 times that of water. The absolute magnitude is +11.4, and if we could bring a cubic inch of its material back to Earth the weight would be about two and a half tonnes. Since Sirius is commonly known as the Dog Star, the companion has predictably been nicknamed the Pup. The companion of Procyon, first seen in 1896 by J. M. Schaeberle with the aid of the Yerkes refractor, is also a white dwarf; the separation was greatest in 1990, and is now decreasing again to a minimum of no more than 2 arcsec. The real separation is 2 250 000 000 km, rather less than the distance between our Sun and the planet Uranus.

Binary systems are of many different kinds; sometimes the components are dissimilar (as with Sirius), sometimes they are identical twins – as with γ Virginis, which has a period of 171 years. Several decades ago γ Virginis was wide and easy to split with a very small telescope, but it is now much closer, though it is starting to open out again. This does not indicate any actual change in separation: everything depends upon the angle from which we see the pair. Another binary, formerly easy to split but now much less so, is Castor. Here, both components of the bright pair are spectroscopic binaries, and also associated with the system is Castor C or YY Geminorum, made up of two red dwarfs; it is an eclipsing binary. Castor therefore consists of six stars, four luminous and two very dim.

Multiple stars are not uncommon. Of special note is ε Lyræ, near Vega in the sky: it has two main components, making up a naked-eye pair, and each component is again double, making up a quadruple system. θ Orionis, in the Orion Nebula (M 42) has been nicknamed the Trapezium, for obvious reasons: it lies on the outskirts of the nebula and is responsible for making the nebulosity luminous.

Some double stars show beautiful contrasting colours. Thus β Cygni has a golden-yellow primary with a vivid blue companion. Antares and α Herculis have fainter green secondaries; with δ Geminorum the primary is yellow and the companion pale blue.

ORIGIN OF BINARY SYSTEMS

The old theory – that a binary was formed as a result of the fission of a single star – has been abandoned, and neither is it likely that binaries are due to the mutual capture of the components; it seems that the components were formed from the same cloud of interstellar material in the same region of space. When there is a marked difference in brightness between the two components, the spectra also differ. If both stars belong to the Main Sequence, the primary is usually of earlier type than the secondary, while if the primary is a giant the secondary is either a giant of earlier type or else a dwarf of similar spectral type. Novæ are binary systems (see below).

If the two components of a binary system are close together, the evolution of one component may profoundly affect the other. In such a system there is an hourglass-shaped region bounded by the points where the two stars will equally affect a small particle: each of the two segments of the 'hourglass' encloses a region termed a Roche lobe. The two Roche lobes may touch at what is termed the Inner Lagrangian Point. The giant component of a close binary will evolve more quickly than its lower-mass companion; as it expands it may fill its Roche lobe completely, and material will flow across the Inner Lagrangian Point on to the second star, so that there is actual transfer of mass from one component to the other, and the original secondary may in time become the more massive of the two. There are also cases in which the two components share a common envelope as the expansion continues, and if one component accumulates enough mass it may explode as a supernova. The evolutionary careers of the members of a close binary system are decidedly complicated.

White dwarf binaries are known: the first of these was found by W. Luyten and P. Higgins in 1973 at RA 9h 42m, Dec. +23′ 41″. The separation is 3″, the position angle 052° and the period 12 000 years; at present the components are 600 a.u. apart. X-ray binaries were spotted following the launch of the first X-ray astronomical satellite Uhuru, in 1970. A system of this kind consists of a white dwarf, neutron star or black hole orbiting another star, which may, in some cases, be a high-mass O- or B-type star, and in others a B- or K-type star of mass similar to that of the Sun.

ECLIPSING BINARIES

As the two components of a binary move round their common centre of gravity, it may happen that one component will pass in front of the other; as seen from Earth this will cause a change in brightness. Stars of such a kind are usually called eclipsing variables, although 'eclipsing binary' is much more accurate. The first to be discovered was Algol (β Persei), by G. Montanari in 1669. It is not now thought that the variability was known in ancient times – even though Algol was always called 'the Demon Star'. With Algol, one component is considerably brighter than the other, so that there is a deep minimum when the primary star is eclipsed (even though the eclipse is not total). The drop in magnitude when the fainter member of the two is hidden is too slight to be noticed with the naked eye. With stars of the β Lyræ type, the two components are much less unequal, and are almost or quite touching, so that changes in brightness are always going on. Since eclipsing binaries do show changes in brightness, they are dealt with in the next section. Incidentally, even the term 'eclipsing binary' is technically wrong. It really ought to be 'occulting binary'.

Famous doubles

It may be of interest to describe a few well-known doubles, mainly to show their great variety.

γ Arietis (Mesartim), 4.7 and 4.8, separation 7.7″. A very easy pair; the components are almost equal. They make up a binary, with a period of >5000 years; both are of type A. They were first noted by Hooke in 1664, when he was observing a comet. Orbiting the binary is a K-type star, magnitude 9.6, separation 221″.

ζ Aquarii, 4.4 and 4.6, separation 1.67″. Discovered by C. Mayer in 1777, they are a binary, period 760 years. Distance from Earth, ~250 light-years. Both are of type F, but it seems that the brighter component $ζ^2$ is a Main Sequence star and the fainter ($ζ^1$) a subgiant. Each star is about as massive as the Sun, but six times more luminous. This rather close pair is a good test for small telescopes.

δ Apodis. A fine naked-eye pair. $δ^1$ is an M-type red giant, irregularly variable between magnitude 4.66 to 4.87; $δ^2$ is a K-type orange giant, magnitude 5.27. The separation is 102.9″. Hipparcos gives their distances as 770 light-years for $δ^1$ and 663 light-years for $δ^2$; it is by no means certain that they are gravitationally bound, but they do share common motion through space.

α Canis Majoris (Sirius). The white dwarf companion would be easy to see with binoculars were it not so drowned in the glare.

α Capricorni (Al Giedi), 4.2 and 3.6. The two components are separated by 376' so that this is a naked-eye pair. It is an optical double, not a binary; the fainter component ($α^1$) is 686 light-years away, the brighter ($α^2$) only 109. Both are of Type G, and therefore yellowish. The brighter component has two faint companions.

β Capricorni (Dabih), 3.1 and 6.1, separation 205″. This is another optical double, separable with binoculars. Both components are themselves complex; the brightest ($β^1$) is an orange K-type giant, the secondary ($β^2$) is of type A, and is a close binary.

α Canum Venaticorum (Cor Caroli), 2.8/3.0 and 5.5, separation 19″.6. This is an optical pair; distances from Earth 110 light-years for the brighter component ($α^2$) and 82 light-years for the fainter ($α^1$); it's a very easy pair. The brighter component is a magnetic variable (type A); the secondary is of type F0.

ι Cancri (Descapoda), 4.0 and 6, separation 30″.5. The primary is a yellow 6-type giant; the secondary is a bluish A-type Main Sequence star, so that we have here a particularly good case of colour contrast. The distance from Earth is just under 300 light-years; the orbital period is probably about 65 000 years.

η Cassiopeiæ (Achird), 3.5 and 7.5, separation 12″. This is a binary, with a period of 480 years – at present widening. One of our nearer neighbours, only 19 light-years away. The primary is of type G and the secondary of type K, so that we have here a rather unusual yellow and red colour contrast. At periapsis (closest approach to each other) the two stars may be only 30 a.u. apart – the same as the distance between the Sun and Neptune.

α Crucis, 1.4 and 2.1, separation 4″, combined magnitude 0.77. A superb double; both components are of type B. This is a wide binary, with a revolution period of at least 1500 years; the brighter component is a spectroscopic binary. A B-class subgiant lies only 90″ away, and has a similar motion through space, but whether it is a member of the α Crucis system is unclear.

β Cygni (Albireo), 3.1 and 5.1, separation with separable with good binoculars. The K-type giant primary is golden yellow; to me, the B-type secondary appears azure-blue. It is unclear whether the two make up a binary pair; if they do, the orbital period must be at least 100 000 years. The primary, well over 1000 times as luminous as the Sun and 390 light-years away, is an excessively close binary; even the hot secondary is more than 200 times as luminous as the Sun. Albireo is certainly the loveliest coloured double in the entire sky.

γ Delphini, 4.5 and 5.5, separation 9″.2. An easy pair, made up of a yellowish F-type Main Sequence star with a K-type orange subgiant. This is probably a very wide binary.

ν Draconis (Kuma), 4.9 and 4.9, separation 62″, and easy in binoculars. Both components are of type A, and are nine times as luminous as the Sun. They make up a binary, but are around 2000 a.u. apart, with a revolution period of at least 44 000 years.

α Herculis (Rasalgethi), 3–4 and 5.4, separation 4″.7. This is a very famous binary; the primary, variable from magnitudes 3 to 4 in a very rough period of around 90 days, is a huge red supergiant, while the companion is usually described as green; no doubt effects of contrast are partly responsible for this. It is possible that expanding shells from the supergiant may actually envelop the companion, which is itself a close binary.

ζ Herculis (Rutilicus), 2.9 and 5.5, average separation ~1″.5. At a distance of about 35 light-years, Rutilicus is one of our nearer neighbours. It is a rapid binary, with a period of only 34 years, so that the separation and position angle alter quickly; their real separation ranges between 8 and 21 a.u. The primary is a G-type subgiant. This is a pair well worth monitoring with a telescope of moderate aperture. It was first noted by William Herschel in 1782.

γ Leonis (Algieba), 2.3 and 3.5, separation 4″.6 This is a fine pair of giants; a binary, types K and G, with a period of 619 years. Descriptions of the colours vary; I see them as orange and yellowish. The distance from Earth is 126 light-years.

ε Lyrae, the famous 'double-double'. The two main components may be separated with the naked eye, and a small telescope will show that each component is again double.

β Orionis (Rigel), 0.1 and 6.8, separation 9″.5. The companion of Rigel is not difficult to see with a fairly small telescope; under good conditions I can glimpse it with my 3-inch refractor. The pair are a long way apart, but are believed to be genuinely associated. The faint component is a spectroscopic binary.

ζ Reticuli, 4.8 and 5.1, separation 310″. This pair comprise two G-type dwarfs, 39 to 40 light-years away and certainly associated, though they must be at least 3700 a.u. apart. Under good conditions, this is a naked-eye pair; their luminosities are comparable with that of the Sun. They have achieved fame (or notoriety!) because of flying saucer enthusiasts.

ζ Sagittarii (Nergat), 5.1 (ζ^2) and 3.5 (ζ^1), separation 0.16°. Too wide to be classed as a conventional double, particularly as the two are not associated. ζ^2, a B-type supegiant, is >2300 light-years away, ζ^1, a G-type giant, only 372 light-years. The name is that of a Babylonian deity.

α Scorpii (Antares). The companion of Antares was probably discovered by Burg, from Vienna, during emersion from an occultation. It is of magnitude 5.4; separation 2″.7. This is a binary system with a period of 878 years. The green colour of the companion is no doubt enhanced by contrast but, on one occasion, using my 15-inch reflector, I put Antares out of view behind an occulting bar, so that I could see the companion shining on its own – it still looked green!

θ Serpentis (Alya), 4.4 and 5.0, separation 22″. Very wide and easy; the two components are A-type Main Sequence stars, about 900 a.u., apart, with a revolution period of >14 000 years. The distance from Earth is 132 light-years; the luminosities are given as 18 and 13 times that of the Sun. Telescopically, they give me the impression of being identical twins.

ζ Ursæ Majoris (Mizar), 2.3 and 4.0, separation 14″.4. The most famous of all binaries; naked-eye pair with Alcor (80 Ursæ Majoris). The Arabs described it as a test for sharp-eyed observers, whereas it is in fact very easy indeed. It is most unlikely that there has been any change in either Mizar or Alcor; between them there is an unrelated star visible in a small telescope, but this is well out of naked-eye range. The answer to this problem is unclear.

Mizar itself is a fine binary, and was the first to be discovered telescopically (by Benedetto Castelli, in 1617). The revolution period must be several thousands of years; both components are spectroscopic binaries, and all are of type A. Alcor, 709″ from the main pair, is also a spectroscopic binary. According to Hipparcos, Mizar is 78 light-years away and Alcor 81 light-years; the two are certainly associated, but the orbital period must be many tens of thousands of years.

Alcor was carefully studied in 2009 by a team using a coronagraph and adaptive optics with the Hale 200-inch reflector at Palomar. In March, a very faint star was found nearby. One of the team, Neil Zimmerman, commented that, 'Right away I spotted a faint point of light next to the star. No one had reported this object before, and it was very close to Alcor, so we realised it was probably an unknown companion star'. So it proved; the two stars form a binary system with a period of around 90 years. The faint companion is much smaller and cooler than Alcor A, and is likely to be an M-type red dwarf.

α Ursæ Minoris (Polaris), 2.0 and 9.0, separation 18″.4. This is in fact quite a complex system, but the 9th-magnitude is an easy telescopic object – though I admit that with my 3-inch refractor I find it rather elusive.

γ Virginis (Arich), both components 3.5, separation now (2010) <1″. This is a binary, with a period of 171 years. The components are of type F, and virtual twins; distance from Earth 38 light-years. When I began observing, in the 1930s, Arich was one of the best doubles for users of small telescopes; we now see it from a much less favourable angle, but it has already started to open out again, and from around 2020 will again be very easy.

22 · Variable stars

Variable star research is an important branch of modern astronomy – amateur observers make very valuable contributions. Variable stars are of many types; elaborate systems of classifying them have been proposed, and the data given here are not intended to be more than a general guide. Seven major categories are now recognised.

(1) Eclipsing stars (more properly eclipsing binaries, because they are not intrinsically variable).
(2) Pulsating variables: either radial or non-radial pulsations.
(3) Eruptive variables, where the changes are caused by flares or the ejection of shells of material.
(4) Cataclysmic variables, where the changes are due to explosions in the star or in an accretion disc round it. Novæ dwarf novæ and supernovæ come into this category.
(5) Rotating variables, where the changes are caused by star-spots, non-spherical shape or magnetic effects.
(6) X-ray variables, usually inherent in the neutron star or black hole companion of a binary.
(7) Unclassifiable stars, which do not fit into any accepted category.

We have already noted what are termed secular variables: stars which have permanently brightened or faded in historic times. Thus Ptolemy ranked β Leonis and θ Eridani as of the first magnitude, whereas today they are below magnitude 2 and 3 respectively: α Ophiuchi was ranked of magnitude 3, but is now 2.1. However, these changes must be regarded as highly suspect. It is unwise to trust the old observations too far.

EARLY IDENTIFICATIONS

The first variable to be positively identified as such was Mira (o Ceti) in 1638. It had been recorded by Fabricius (1596) and Bayer (1603), and Bayer had even allotted it a Greek letter, so that it is surprising that its fluctuations were not recognised until 1638 (by Phocylides Holwarda). In the latter part of the seventeenth century, two more variables were identified, Algol and χ Cygni. Table 22.1 lists the variables identified between 1638 and 1850.

The recognised abbreviations for the different classes of variables are given in Table 22.2, and a more detailed scheme is given in Table 22.3.

ECLIPSING BINARIES

In eclipsing binaries the light changes are due entirely to mutual eclipses (or, to be accurate, occultations) of the two components.

Algol (EA) type

The components are more or less spherical, and are unequal, so that an Algol variable remains at maximum for most of the time. The secondary minimum, when the faint component is hidden, is often very slight.

With Algol (β Persei), the prototype EA star, the primary eclipse it not total. The main component (Algol A) is of type B, 105 times as luminous as the Sun, with a diameter of about 4 000 000 km. Algol B, the companion, is not genuinely dark: it is of type G, and about three times as luminous as the Sun. It has a diameter of 5 500 000 km, so that it is larger than the primary and qualifies as a subgiant. The secondary minimum is less than 0.1 in magnitude. The distance from the Earth is 95 light-years.

EA stars are common enough, but few are naked-eye objects apart from Algol; only λ Tauri, δ Libræ and ζ Phænicis rise above the fifth magnitude. A list of bright eclipsing stars is included in the bright-variable catalogue (Table 22.4).

β Lyræ (EB) type

Here the two components are much less unequal, so that there are alternate deep and shallow minima. The prototype star β Lyræ itself has a period of 13 days. The maximum magnitude is 3.3. The star fades to magnitude 3.8, recovers and then goes through its primary minimum, which takes it down to below 4. It then returns to maximum, and the cycle is repeated. The components are almost touching each other, and are tidally distorted into egg-like shapes: the more massive component has filled its Roche lobe, and material is streaming through the Inner Lagrangian Point to form an accretion disc round the less massive star There is evidence that the components are connected by huge streamers of gas moving at over 300 km s^{-1}. The primary minimum as seen from Earth is caused by a total eclipse, the secondary minimum by a partial eclipse.

W Ursæ Majoris (EW) type

In dwarf binaries of types F or G the components are almost or quite in contact, and the two minima are more or less equal. The periods are less than one day. There are no naked-eye examples; the maximum magnitude of the prototype, W Ursæ Majoris itself, is only 7.9.

Very long-period eclipsing binaries

A few eclipsing stars have periods of years. The most celebrated of these is ε Aurigæ, one of a triangle of stars close to Capella in the Charioteer (they are often known as the Hædi, or Kids). ε Aurigæ is

Table 22.1 *The first known variable stars*

Star	Discoverer of variability	Date
Mira (o Ceti)	Holwarda	1638
Algol (β Persei)	Montanari	1669
χ Cygni	Kirch	1686
R Hydræ	Marald	1704
Rasalgethi (α Herculis)	W Herschel	1759
μ Cephei	W Herschel	1782
R Leonis	Koch	1782
δ Cephei	Goodricke	1784
β Lyræ	Goodricke	1784
η Aquilæ	Pigott	1784
R Scuti	Pigott	1795
R Coronæ Borealis	Pigott	1795
R Virginis	Harding	1809
R Aquarii	Harding	1811
ε Aurigæ	Fritsch	1821
R Serpentis	Harding	1826
η Carinæ	Burchell	1827
S Serpentis	Harding	1828
U Virginis	Harding	1831
δ Orionis	J. Herschel	1834
S Vulpeculæ	Rogerson	1837
Betelgeux (α Orionis)	J. Herschel	1840
β Pegasi	Schmidt	1847
λ Tauri	Baxendell	1848
R Orionis	Hind	1848
R Pegasi	Hind	1848
R Capricorni	Hind	1848
S Hydræ	Hind	1848
S Cancri	Hind	1848
S Geminorum	Hind	1848
R Geminorum	Hind	1848
T Geminorum	Hind	1848
R Tauri	Hind	1849
T Virginis	Boguslawsky	1849
T Cancri	Hind	1850
R Piscium	Hind	1850

Table 22.2 *Variable star types*

E	Eclipsing binary
EA	Algol type
EB	β Lyræ type
EW	W Ursæ Majoris type
M	Mira type (long period)
SR	Semi-regular
SRa	Semi-regular: well-defined periodicity
SRb	Semi-regular: poorly defined periodicity
SRc	Semi-regular disc component stars
SRd	Semi-regular: types F, G or K
RR	RR Lyræ variable
RRa	RR Lyræ: sharp asymmetrical light curve
RRa, b	RR Lyræ: asymmetrical light curve
RRc	RR Lyræ: symmetrical sinusoidal light curve
RV	RV Tauri type
RVa	RV Tauri: constant mean brightness
RVb	RV Tauri: varying mean brightness
I	Irregular
IB	Slow irregular variations
IC	Irregular supergiants
Cep	Cepheid
δ Cep	Classical Cepheid
CW	W Virginis star (Type II Cepheid)
DSCT	δ Scuti type
SXPHE	SX Phœnicis type
ACYG	Deneb (α Cygni) type
β Cep	β Cephei type
ZZ	ZZ Ceti type
FU	FUors (FU Orionis) type
GCAS	γ Cassiopeiæ type
IN	Orion variables
IT	T Tauri variables
RCB	R Coronæ Borealis variables
RS	RS Canum Venaticorum variables
SDOR	S Doradûs variables
UV	Flare stars (UV Ceti)
W	Unstable Wolf–Rayet stars
AM	Polars (AM Herculis type)
UG	Dwarf novæ (U Geminorum or SS Cygni)
UGZ	Dwarf novæ Z Camelopardalis type
ZABD	Symbiotic stars (Z Andromedæ type)
N	Novæ
RN	Recurrent novæ
NL	Nova-like variables
SN	Supernovæ
ACV	Magnetic variables (α² Canum Venaticorum)
BY	BY Draconis type
ELL	Ellipsoidal variables
FKCOM	FK Comæ variables
SXARI	SX Arietis type
Pec	Peculiar: not fitting into any class

generally just above the third magnitude, but every 27 years it fades, taking over five months to drop down by almost a magnitude. The minimum lasts for just over a year, followed by a gradual recovery. Midway through the eclipse there is a slight brightening that lasts for several weeks.

ε Aurigæ (its proper name, Almaaz, is hardly ever used) is indeed a remarkable system. Its variability seems to have been discovered by J. Fritsch in 1821, and subsequently confirmed by E. Heis and F.W. Argelander, but it was Hans Ludendorff who realised that it is an eclipsing binary. The primary is an F0-type supergiant, ~2000 light-years away and about 47 000 times as luminous as the Sun, with diameter about 100 times that of the Sun. The eclipsing component is now thought to be a massive,

Table 22.3 *The classification of variable stars*

Eclipsing variables (or eclipsing binaries)

EA	Algol	Period 0.2 day to over 27 years. Almost spherical components. Maximum for most of the time.
EB	β Lyræ	Periods over 1 day: spectra B to A. Ellipsoidal components: magnitude continuously changing.
EW	W Ursæ Majoris	Dwarfs, periods usually less than 1 day. Almost or quite in contact. Primary and secondary minima almost equal.

Pulsating variables

M	Mira	Long-period late-type giants: spectra M–C–S. Periods 80 to 1000 days. Amplitude may exceed 10 magnitudes. Periods and amplitudes vary from cycle to cycle.
SR	Semi–regular	Late-type giants (spectra M–C–S). Periods from 20 days to several years. SRa persistent periodicity (Z Aquarii). SRb: rough periodicity (RR Coronæ Borealis). SRc: red supergiants (μ Cephei). SRd: F- to K-type giants and supergiants (SX Herculis).
PVTEL	PV Telescopii	Helium supergiants: periods from a few hours to a year: Spectra Bp: small amplitudes (0.1 magnitude).
RR	RR Lyræ	Spectra A to F: periods 0.2 to 1.2 days; formerly called cluster-Cepheids. RRab steep rise to maximum. RRb: almost symmetrical light curves. RRc: sinusoidal light curves.
RV	RV Tauri	Supergiants. Usually type F to K. Periods 30 to 150 days. RVa: stars with constant mean magnitude (AC Herculls). RVb: stars with variable mean magnitude.
I	Irregular	Types K, M, C, S. IB: slow giant variables (CO Cygni). IC: slow supergiant variables (TZ Cassiopeiæ).
CEP	Cepheids	Radial pulsating stars. Periods 1 to 135 days: spectra F to K.
CW	W Virginis	Population II Cepheids. CWA: longer period (W Virginis). CWB: shorter period (BL Herculis).
DSCT	δ Scuti	Types A to F: periods less than 1 day: small amplitude.
SXPHE	SX Phoenicis	Population II subdwarfs resembling δ Scuti stars. Types A to F, periods less than 1 day, amplitude up to 0.7 magnitude.
ACYG	α Cygni	Types B to A: supergiants. With low amplitudes. Most have short periods.
BCEP	β Cephei (or β Canis Majoris)	B-type subgiants. Low amplitude and short period.
ZZ	ZZ Ceti	Non-radially pulsating white dwarfs: amplitude up to 0.2 magnitude: periods 30 seconds to 1500 seconds. ZZA: hydrogen absorption lines only (spectrum DA). ZZB: helium absorption lines only (spectrum DV). ZZO: very hot stars (spectrum DO).

Eruptive variables

FU	FUors	Prototype, FU Orionis. Types A to G: slow rise to maximum (many years) and slower decline.
GCAS	γ Cassiopeiæ	Shell stars: type B: rapid rotaters. Amplitudes usually below 2 magnitudes.
I	Irregular	Types O to M. IA: early spectral type. IB intermediate and late spectral type.
IN	Orion	INA: early type (T Orionis). INB: later type (spectra F to M).
IT	T Tauri	Similar to Orion variables: types F to M (RW Aurigæ). INT: types F to M: contained in diffuse nebulæ (T Tauri itself). IS: rapidly-varying stars, amplitudes up to 1 magnitude: ISA (early type, spectra B to A). ISB (later type. Spectra F to M).
RCB	R Coronæ Borealis	Types B to R. Occasional deep minima: large amplitude (over 9 magnitudes in some cases).
RS	RS Canum Venaticorurn	Small amplitude. Close binaries with active chromospheres.
SDOR	S Doradus	Types Bp to Fp. Very luminous supergiants with expanding shells. Often in diffuse nebulæ
UV	UV Ceti	Dwarfs of types K to M. Flare stars: amplitudes may be as much as 6 magnitudes (UV Ceti itself).
W	Wolf–Rayet	Very low amplitudes: spectra of type W. Non-stable mass outflow.

Rotating Variables

ACV	α Canum Venaticorum	Types B to A: strong magnetic fields: spectra rich in silicon, strontium, chromium and rare earth lines. Periods from 12 hours up to 160 days.
BY	BY Draconis	Types G to M: periods up to 20 days. Rotating dwarfs with starspots and active chromospheres.

Table 22.3 (cont.)

ELL	Ellipsoidal	Close binaries with no eclipses. But changing visible area. Low amplitudes (up to 0.1 magnitude).
FKCOM	FK Com	Types G to K. Periods up to several days. Amplitudes up to half a magnitude. Rapidly rotating giants with non–uniform surface brightness.
SXARI	SX Arietis	Helium stars (type B): amplitude around 0.1 Magnitude and periods of around 1 day. High-temperature versions of α^2 Canum Venatucorum stars.
Cataclysmic variables		
AM	AM Herculis	Polars: close binaries with one compact component. Amplitude up to 5 magnitudes.
UG	U Geminorum	Dwarf novæ. Periods from 10 to 1000 days. Amplitudes from 2 to 9 magnitudes. SS Cygni stars have outbursts lasting for several days.
UGSU	SU Ursæ Majoris	Dwarf novæ, with occasional supermaxima brighter and longer than normal maxima.
UGZ	Z Camelopardalis	Dwarf novæ which have occasional standstills when the normal cycle of variation is suspended.
ZAND	Z Andromedæ	Symbiotic stars: close binaries.
N	Novæ	Thermonuclear outburst on the white dwarf component of a binary system. NA: fast, fading by three magnitudes in 100 days or less (GK Persei 1901). NAB: fading at intermediate speed. NB: slow fading, no more than three magnitudes in 150 days (RR Pictoris, 1925). NC: very slow novæ with maxima which may last for years as with RR Telescopii.
NL	Nova-like	NL: poorly studied stars with nova-like outbursts (V Sagittæ). NR: recurrent novæ (such as T Coronæ Borealis) which flared in 1866 and again in 1946).
SN	Supernovæ	SNI: explosion and destruction of the white dwarf component of a binary system. SNII: collapse of a very massive star. Often leaving a neutron star or pulsar.

A few variables such as VY Canis Majoris do not seem to fit into any class and there are also very exceptional stars such as η Carinæ. Pulsars and X-ray binaries are often included in variable star classification lists.

opaque disc of dust, almost edgewise-on to us. The mid-eclipse brightening may be due to an opening in the centre of the disc. The timetable for the latest eclipse is as follows:

2009 Aug 6:	Start of partial phase
2009 Dec 21:	Start of totality
2010 May–Sept:	Mid-eclipse brightening
2011 Mar 12:	End of totality
2011 May 15:	End of partial phase.

The eclipsing component of ε Aurigæ was finally imaged in 2010 by B. Kloppenborg *et al.* using the 330-m CHARA array atop Mount Wilson (a collection of six telescopes). The images show the intrusion of a wedge-shaped structure across the face of the huge supergiant.

The primary itself is very slightly variable, but the following eclipse will not begin until 2036. Also in the triangle of the 'Kids' is the other celebrated long-period eclipsing binary, ζ Aurigæ. This is sheer coincidence because ζ is much the closer of the two. The period is 972 days, and both spectra are visible; the primary is a K4-type supergiant, while the secondary is of type B8; the distance from Earth is about 800 light-years.

As the supergiant begins to hide the secondary, at the start of the eclipse the light of the secondary comes to us through super-giant's outer layers, and there are complicated spectral changes, which are highly informative. Totality lasts for 38 days, and the magnitude falls from 3.7 to 4.1.

INTRINSIC VARIABLES

The variations are intrinsic. The star expands and contracts, changing its surface temperature and its output as it does so.

Mira variables

These are often called long-period variables: Mira (o Ceti) is the brightest member of the class. They are late-type red giants with emission lines in their spectra; their periods range from 80 days to over 1000 days, and the amplitudes are large – in some cases well over 10 magnitudes in cases (as with amplitudes χ Cygni, where the extreme range is magnitude 3.3 to 14.2). Neither the periods nor the amplitudes are constant, and no two cycles are exactly alike. At some maxima Mira may rise to the second magnitude, and it is reported that in 1779 it matched Aldebaran; but other maxima are not brighter than magnitude 4, and on average Mira is an easy naked-eye object for only a few weeks in every year. Minima are always between magnitudes 8.6 and 10.1 – usually about 10.

Estimates of the distance of Mira do not agree particularly well. Hipparcos gives 418 light-years (error margin 14%), much greater than previous estimates of ~220 light-years. Mira, an asymptotic red giant of type M7 is certainly large. Measurements with the Hubble Space Telescope gave the angular diameter as 60 milliarcsec, corresponding to a true diameter 700 times that of the Sun,

The true luminosity varies considerably. The absolute magnitude range has been given as from −2.5 to +4.7, in which case even at its peak Mira is not much more than 1000 times as powerful as

Table 22.4 *A catalogue of bright variable stars. The following list includes variable stars with a maximum of magnitude 6 or brighter, and a range of at least 0.4 magnitude*

	Max.	Min.	Period (d)	Spectrum
Eclipsing binaries (Algol type)				
R Ara	6.0	6.9	4.4	B
WW Aur	5.8	6.5	2.5	A + A
R CMa	5.7	6.3	1.1	F
RS Cha	6.0	6.7	0.1	A + F
δ Lib	4.9	5.9	2.3	B
U Oph	5.9	6.6	1.7	B + B
β Per	2.2	3.4	2.9	B + G Algol
ζ Phe	3.9	4.4	1.7	B + B
RS Sgr	6.0	6.9	2.4	B + B
λ Tau	3.3	3.8	3.9	B + A
HU Tau	5.9	6.7	2.1	A
(β Lyræ type)				
UW CMa	4.0	5.3	4.3	O7
u Her	4.6	5.3	2.0	B + B
GG Lup	5.4	6.0	2.1	B + A
β Lyr	3.3	4.3	12.9	B + A Sheliak
V Pup	4.7	5.2	1.4	B + B
(Long period)				
ε Aur	2.9	3.8	9892	F
ζ Aur	3.7	4.1	972	K + B
VV Cep	4.8	5.4	7430	M + B
Mira variables				
R And	5.8	14.9	409	S
lR Aql	5.5	12.0	284	M
R Car	3.9	10.5	309	M
S Car	4.5	9.9	149	K–M
R Cas	4.7	13.5	430	M
R Cen	5.3	11.8	546	M
o Cet	1.7	10.1	332	M Mira
S CrB	5.8	14.1	360	M
χ Cyg	3.3	14.2	407	S
U Cyg	5.9	12.1	462	N
R Gem	6.0	14.0	370	S
S Gru	6.0	15.0	401	M
R Hor	4.7	14.3	404	M
R Hya	4.0	10.0	390	M
R Leo	4.4	11.3	312	M
R Lep	5.5	11.7	432	N
V Mon	6.0	13.7	334	M
Mira variables				
X Oph	5.9	9.2	334	M + K
U Ori	4.8	12.6	372	M
RU Sgr	6.0	13.8	240	M
RT Sgr	6.0	14.1	305	M
RR Sco	5.0	12.4	279	M
S Scl	0.5	13.6	365	M
R Ser	5.1	14.4	356	M
R Tri	5.4	12.6	266	M
SS Vir	6.0	9.6	355	N
R Vir	6.0	12.1	146	M
Semi-regular variables				
UU Aur	5.1	6.8	234	N
W Boö	4.7	5.4	450	M
VZ Cam	4.7	5.2	24	M
X Cnc	5.6	7.5	195	N
TU CVn	5.6	6.6	50	M
S Cen	6.0	7.0	65	N
T Cen	5.5	9.0	90	K–M
T Cet	5.0	6.9	159	M
FS Com	5.3	6.1	58	M
W Cyg	5.0	7.6	126	M
U Del	5.7	7.6	110	M
EU Del	5.8	6.9	59	M
R Dor	4.8	6.6	338	M
μ Cep	3.4	5.1	750	M
UX Dra	5.9	7.1	168	N
RY Dra	5.6	8.0	173	N
η Gem	3.2	3.9	233	M Propus
π' Gru	5.4	6.7	150	S
g Her	5.7	7.2	70	M
α Her	3	4	±100	M Rasalgethi
R Lyr	3.9	5.0	46	M
ε Oct	4.9	5.4	55	M
α Ori	0.1	0.9	2110	M Betelgeux
W Ori	5.9	7.7	212	N
CK Ori	5.9	7.1	120	K
Y Pav	5.7	8.5	233	N
SX Pav	5.4	6.0	50	M
β Peg	2.3	2.8	38	M Scheat
ρ Per	3	4	33–55	M
TV Psc	4.6	5.4	70	M
L² Pup	2.6	6.2	140	M
R Scl	5.8	7.7	370	N
RR UMi	6.0	6.5	40?	M
Eruptive variables				
U Ant	5.7	6.8		N
η Car	−0.8	7.9		Pec
ρ Cas	4.1	6.2		F–K
				(Occasional fades)
α Cas	2.1?	2.5?		K (Suspected variable)
δ Sco	1.6	2.3		
γ Cas	1.6	3.3		B
μ Cen	2.9	3.5		B
θ Cir	5.0	5.4		B
P Cyg	3	4		Bp
T Cyg	5.0	5.5		K
BU Gem	5.7	7.5		M
RX Lep	5.0	7.0		M
S Mon	4	5		O7
BO Mus	6.0	7.7		M

Table 22.4 (cont.)

	Max.	Min.	Period (d)	Spectrum
χ Oph	4.2	5.0		B
λ Pav	3.4	4.3		B
X Per	6.0	7.0		09.5 X-ray star
d Ser	4.9	5.9		G + A
VY UMa	5.9	6.5		N
BU Tau	4.8	5.5		Bp Pleione
Cepheids				
η Aql	3.5	4.4	7.2	F–G
RT Aur	5.0	5.8	3.7	F–G
ZZ Car	3.3	4.2	36.5	F–K
U Car	5.7	7.0	38.8	F–G
SU Cas	5.7	6.2	1.9	F
δ Cep	3.5	4.4	5.4	F–G
AX Cir	5.6	6.1	5.3	F–G
X Cyg	5.9	6.9	16.4	F–G
β Dor	3.7	4.1	9.8	F–G
ζ Gem	3.7	4.1	10.1	F–G
T Mon	6.0	6.6	27.0	F–K
S Mus	5.9	6.4	9.7	F
R Mus	5.9	6.7	7.5	F
Y Oph	5.9	6.4	17.1	F–G
κ Pav	3.9	4.7	9.1	F5v
				(W Virginis type)
S Sge	5.3	6.0	8.4	F–G
X Sgr	4.2	4.8	7.0	F
W Sgr	4.3	5.1	7.6	F–G
Y Sgr	5.4	6.1	5.8	F
AH Vel	5.5	5.9	4.2	F
T Vul	5.4	6.1	4.4	F–G
RV Tauri variable				
R Sct	4.4	8.2	140	G–K
Symbiotic variables				
R Aqr	5.8	12.4	387	M + P
AG Peg	6.0	9.4	830	WN + M

	Max.	Min.	Spectrum	Outbursts
Recurrent novæ				
T CrB	2.0	10.8	M + Q	1866, 1946 Blaze Star
RS Oph	5.3	12.3	O + M	1901, 1933, 1958, 1957
R Coronæ Borealis variables				
R CrB	5.7	15	Fp	
RY Sgr	6.0	15	Gp	

football, and shows definite variations. Ultraviolet studies with NASA's GALEX (Galaxy Evolution Explorer) probe show that as it moves through the interstellar medium it sheds material from its outer envelope, creating a tail 13 light-years long.

Mira has a binary companion, which is itself variable and has been given a variable star designation (VZ Ceti). The range is from magnitude 9.5 to 12. The orbital period is 400 years; separation 70 a.u. The companion was originally thought to be a white dwarf, but in 2007 a proto-planetary disc was detected round it, presumably accreted by the stellar wind from the primary – so that the companion may really be a K1-type Main Sequence star, with mass 0.7 that of the Sun. The Hubble Space Telescope has also detected a hook-like appendage extending from the football-shaped primary in the direction of the companion – no doubt the presence of the companion is responsible for this.

Several Mira variables reach naked-eye visibility – for instance χ Cygni (which is a particularly strong infrared source) and the southern R Carinæ. All are of spectral types M or later.

Semi-regular variables (SR)

These also are of late spectral type. In some cases the periods are so rough that they are almost unrecognisable, and again no two cycles are alike. There are various subdivisions:

15. SRa. Late-type giants (types M, C or S), with small amplitude (less than 2.5 magnitudes) and relatively stable cycles. Z Aquarii is a good example.
16. SRb. Red giants (types M, C or S), with poorly-defined cycles in the range of 20 to 2300 days; small amplitudes. Occasionally all variation may temporarily stop. RR CrB is of this class.
17. SRc. Late-type supergiants such as μ Cephei, with amplitudes of one magnitude or rather more and rough periods from 30 days to several years.
18. SRd. Giants and supergiants of types F, G or K; amplitudes from 0.1 to 4 magnitudes, periods 30 to 1100 days. Good examples are SX Herculis and SV Ursæ Majoris.

The brightest semi-regular variable is Betelgeux in Orion. The official magnitude range is from 0.2 to 0.9, but there are indications that it may occasionally become as bright as Rigel. The period is given as 2110 days, but this is decidedly rough. Estimates of its distance do not agree well, and neither do estimates of its luminosity, but Betelgeux is at least 15 000 times as powerful as the Sun – perhaps much more – and the angular diameter is great enough for surface details to be made out with our largest telescopes. There are apparently large convection cells rising to the surface, and there is a vast, extended atmosphere. The mass of Betelgeux may be 20 times that of the Sun; it is well advanced in its evolutionary sequence, and eventually it will no doubt explode as a supernova. When this happens, the apparent magnitude as seen from Earth will be –9 or –10.

μ Cephei – nicknamed the 'Garnet Star' by William Herschel because of its colour – is much larger even than Betelgeux, but is further away, well over 1000 light-years. The range is from magnitude 3.4 to just below 5, and the period is officially given as 730 days but, as with Betelgeux, this is very rough indeed. Other bright semi-regulars are α Herculis, η Geminorum and β Pegasi. With β Pegasi in the 'Square', the period is reasonably well defined.

the Sun, but other estimates increase this considerably. The mass is less than twice that of the Sun. The Hubble observations show that it is not perfectly spherical, and that its shape rather resembles a

There are also many semi-regular variables whose fluctuations are too slight – <0.1 magnitude – to be detected without sensitive equipment. Antares and Aldebaran are of this type.

Cepheids

Cepheids (CEPs) take their name from the best-known member of the class, δ Cephei, whose variability was discovered by John Goodricke in 1784 – though actually the variability of η Aquilæ, which is of the same type, had been discovered a few months earlier by Edward Pigott (they might have become known as Aquilids!). They are radially pulsating yellow giants, whose spectra change between F and G over a cycle; the periods range from a few days to (in rare cases) over 100 days; the amplitudes are from 0.1 to 2 magnitudes. They are of great importance because their periods are linked with their real luminosities, and this means that once the period is known the distance can be found: Cepheids act as 'standard candles' in space, and because they are highly luminous they can be seen over vast distances. The Cepheid period–luminosity law states that the longer the period, the more powerful the star. The period of pulsation is the time taken for a vibration to travel from the surface of the star towards the centre and back again, so that these periods are longer for larger and brighter stars; at maximum a period of 3 days corresponds to a luminosity of 800 Suns, a 30-day period corresponds to 10 000 Suns. Thus η Aquilæ (7.2 days) is more powerful than Cephei (5.4 days). Other naked-eye Cepheids are ζ Geminorum, β Doradûs and ZZ Carinæ. η Aquilæ is the most conspicuous of them; when its variability was discovered it was included in the now-discarded constellation Antinous, and was referred to as η Antinoi. Even at minimum it is over 2500 times more luminous than the Sun. Its changes are particularly easy to follow because δ and θ Aquilæ, to either side of it, are ideal comparison stars.

Polaris is a Cepheid of very small amplitude. In 1899, it was found to have a range of magnitude from 1.92 to 2.07 and a period of 3.969 778 days. Subsequently the amplitude decreased, and by 1992 was down to 0.010 of a magnitude. It was thought that the pulsations might cease altogether, but since 1995 the amplitude has stabilised at 0.03 magnitude. It remains in the 'instability strip' of the Hertzsprung– Russell (HR) diagram, where every star ought to pulsate: this strip occupies a region between the Main Sequence and the red giant branch.

Unlike Mira stars, Cepheids are perfectly regular, and the cycles repeat each other, so that the magnitude at any particular moment can be predicted. Either the rise to a maximum is sharper than the subsequent decline, or else the light curve is virtually symmetrical.

δ Scuti variables

δ Scuti (DSCT) variables, also known as AI Velorum variables or as dwarf Cepheids, belong to the galactic disc (Population I) and are young A- to F-type stars. Their variations are due to both radial and non-radial surface pulsations. Amplitudes range from 0.003 to 0.9 magnitudes in periods of a few hours; many are spectroscopic binaries. β Cassiopeiæ (Chaph) is a bright δ Scuti star; it is an F-type giant, with a magnitude of 2.25 to 2.31 and a period of 2.5 hours. Sub-dwarfs showing the same characteristics are known as SX *Phoenicis stars.*

In 1999, a new class was proposed. The prototype star is γ Doradûs, a 4th-magnitude star in the southern hemisphere. These stars have spectra around type F0: they are Main Sequence stars slightly hotter and more massive than the Sun, but cooler than δ Scuti stars. Stars such as δ Scuti are expected to pulsate, since they are in the instability strip of the H–R Diagram, but γ Doradûs stars are outside the instability strip – it is suggested that they undergo non-radial pulsations: portions of their outer layers expand outward while other portions contract, as against stars such as Cepheids which expand and contract as a unit, thereby preserving spherical symmetry. The amplitudes of γ Doradûs stars are very small, several hundredths of a magnitude.

β Cephei variables

β Cephei (BCEP) variables are sometimes called β Canis Majoris variables, B0 to B3 giants or subgiants, with periods of from 0.1 to 0.7 day, and amplitudes from 0.1 to 0.3 magnitude. They are relatively massive stars which have almost exhausted their core hydrogen. Ultra-short periods of a few hundredths of a day are suffixed s; a typical example is γ Centauri.

Cygni (ACYG) variables

These are pulsating supergiants of type B or A; they have very small amplitudes. Deneb is the best-known example. Its magnitude range is 1.21 to 1.29.

ZZ Ceti (ZZ) variables

These are pulsating white dwarfs, with periods which may be as short as 30 s and never as long as half an hour. The amplitudes are below 0.2 magnitude. Suffixes A, B or O indicate spectral features such as lines of hydrogen, helium or carbon, the brightness at a particular moment can be predicted. As with the Cepheids, either the rise is steeper than the decline or else the light curve is virtually symmetrical.

W Virginis (CW) stars

These stars are associated with Cepheids, and may be termed type-2 Cepheids or Population II Cepheids. They have lower masses than the classical Cepheids and are about two magnitudes fainter: they are also metal-poor rather than metal-rich, and their periods are less precise The only naked-eye example is κ Pavonis in the southern sky.

RR Lyræ (RR) variables

These were once called Cluster Cepheids, because they are common in globular clusters: however, many of them (including the prototype RR Lyræ itself) are not cluster members. The spectra are of type A or F. They are old, with masses lower than that of the Sun, but radii four to five times greater and all are of about the same luminosity, 95 times that of the Sun; so – like the Cepheids – they can be used as 'standard candles'. The amplitude is around 1 magnitude, and the periods are from 0.2 to 1.2 days. There are various subdivisions. RRAB stars (formerly divided into two sub-classes RRa and RRb) have amplitudes of 0.3 to 1 magnitude, and periods of from 0.5 to 0.7 days. RRc stars have almost sinusoidal light-curves, with amplitudes around 0.5 magnitude and periods around 0.3 days. Some RR Lyræ stars, such as AR Herculis, have several pulsation periods, and this results in a continuous deformation of the light curve (Blazhko effect). Dwarf Cepheids were once classed with the RR Lyræ stars, as RRs: the prototype is AI Velorum. They are of types A to F with absolute magnitudes of from −1 to −5 and periods from 0.05 to 0.25 day.

RV Tauri (RV) variables
These are radially pulsating yellow to red supergiants, usually of types F to K, but occasionally M. There are alternate deep and shallow minima: the interval between successive primary minima may be from 30 to 150 days, with amplitudes up to 4 magnitudes. The subdivisions are:

RVa: the period is not constant, but is at least reasonably consistent for most of the time: the brightest member of the class. R Scuti, is of this type.

RVb: several superimposed cycles, with rough periods of from 30 to 100 days.

PV Telescopii
These variables are Bp supergiants with amplitude about 0.1 magnitude, periods from 0.1 to 1 day. Their spectra show lines of hydrogen with very strong lines of helium and carbon.

ERUPTIVE VARIABLES

FUors (FU)
FUors are named after the prototype star FU Orionis. In a star of this kind there is a very slow rise to a maximum which may last for years, followed by a slower decline. Emission lines develop in the spectrum. The types are usually A to G.

Flare stars: UV Ceti (UV) type
These are red dwarfs, of type K or (usually) M. Flare activity produces sudden outbursts of 1 to 6 magnitudes, lasting for several minutes. All flare stars show emission lines in their spectra. UV Ceti itself is normally of magnitude 13.4, but on one occasion brightened abruptly to 6.8. Many red dwarfs show flare activity of a less spectacular kind; among these is Proxima Centauri, the nearest star to the Sun. Flash variables (UVs) are of earlier spectral type, and are more luminous; they are associated with nebulosity.

T Tauri variables
These are very young stars, still contracting towards the Main Sequence and varying irregularly. They are strong infrared emitters, and send out pronounced stellar winds. T Tauri itself lies in a dark dust cloud; many T Tauri stars are found inside nebulæ. The Sun certainly passed through a T Tauri stage early in its evolution.

RS Canum Venaticorum (RS) variables
These are close binaries with chromospheric activity, causing slight light variations. Most have amplitude ~0.2 magnitude; there are

spectral types from G to M, though in some cases one component may be a white dwarf; some are X-ray and radio emitters.

R Coronæ Borealis (RCB) variables
These are hydrogen-poor, but rich in carbon. They are highly luminous, and remain at a maximum for most of the time; they then undergo sudden, unpredictable falls to a minimum, taking from several weeks to many months to recover. The amplitudes are large – at least 10 magnitudes in the case of R Coronæ Borealis (CrB) itself, which is usually on the fringe of naked-eye visibility magnitude 6 but at some minima falls below magnitude 15. The fadings are due to clouds of soot accumulating in the star's atmosphere. R CrB stars are rare.

Even more uncommon are stars of the DY Persei subclass, which are carbon-rich asymptotic giants showing both pulsational and irregular variability.

Wolf–Rayet stars (W)
These are highly luminous, and are unstable, with expanding envelopes: this causes random amplitude changes in brightness. Their spectra show emission lines of nitrogen and carbon, and there is evidence of mass outflow as a stellar wind.

S Doradûs (SDOR) variables
These are massive, very luminous blue supergiants, usually surrounded by expanding envelopes. They are often found in diffuse nebulæ: generally the amplitude is small, but there may be occasional outbursts up to 7 magnitudes, lasting for many weeks. These are due to the ejection of shells of material. S Doradûs itself lies in the Large Cloud of Magellan, at a distance of 169 000 light-years, and has a mass about 60 times that of the Sun: although it is almost a million times as luminous as the Sun, it is too faint to be seen with the naked eye (magnitude 11).

γ Cassiopeæ (GCAS) variables
These are rapidly rotating blue giants, usually of type B, which occasionally throw off shells of material. gamma Cassiopeiæ itself is the prototype. Its usual magnitude is just below 2, but in the late 1930s just rose to 1.6 before declining to 3.2; for decades now the magnitude has hovered around 2.2.

Another notable star of this type is δ Scorpii (Dzuba), which was listed as magnitude 2.3 until June 2000, when it suddenly brightened up to 1.6; it is of type B0 distance 402 light-years, luminosity about 14 000 Suns. Since then it has remained between magnitude 1.6 and 2. Spectra showed that luminous gases are sent out from its equatorial region, but the activity may be influenced by the presence of a close-in B-type binary companion orbiting in a period of 20 days, and by another companion which orbits in 10 years and was at its closest in 2000.

Slow irregular variables
These are late-type giants or supergiants with little or no detectable periodicity, and very small amplitude. Examples are Aldebaran, Antares, ε Pegasi and β Gruis.

ROTATING VARIABLES

α² *Canum Venaticorum (ACV) stars*

These are white stars, usually of type B or A, which have intense magnetic fields. It seems that there are huge 'starspots', produced by the magnetic fields: there are variable spectral lines due to silicon, strontium, chromium and rare-earth elements. Periods range between 12 h and 160 days. The spectra are variable, but the light fluctuations are very small.

BY Draconis (BY) stars

These are red dwarfs with large starspots and active chromospheres. The periods may reach 20 days.

Ellipsoidal (ELL) variables

These are tidally distorted close binaries. The changes in light are due to the varying areas presented, but there are no eclipses, as with β Lyræ stars. As fluctuations are not intrinsic, it would be more accurate to refer to these stars as ellipsoidal binaries.

FK Comæ (FKCom) variables

FK Comæ are rapidly rotating yellow giants. They are apparently ellipsoidal and the changes in brightness are due to the star's rotation, so they cannot be regarded as intrinsic variables. They are of types G to K, with periods up to several days, amplitudes of no more than half a magnitude. The surfaces are of non-uniform brightness, and it is this which causes the fluctuations, so that again it cannot really be said these stars are intrinsically variable.

SX Arietis (SXARI) variables

These are similar to the α² Canum Venaticorum stars, except that they are richer in helium and have higher temperatures. The amplitudes never exceed 0.1 magnitude.

Symbiotic variables

A symbiotic variable is a binary in which one component has an extensive outer envelope and is losing mass to its companion, which may be a white dwarf. The best-known example, R Aquarii, is made up of a Mira primary, with a period of 387 days and a magnitude range of 5.8 to 12.4, together with a white dwarf; the system incorporates a strange developed nebula, Cederblad 211.

Reflection (R) binaries

Finally there are reflection binaries, where a large, cool component is illuminated by its hotter companion leading to variations of up to a magnitude as the system rotates.

SOME REMARKABLE VARIABLES

η *Carinæ*

Unquestionably the most erratic of all variables is η *Carinæ* (before the dismemberment of the old constellation of Argo Navis it was known as η Argûs). It has a strange history. It was recorded by Halley, in 1677, as being of the fourth magnitude. It remained between magnitudes 4 and 2 until 1827, when it blazed up to the first magnitude. In 1837, John Herschel, from the Cape, made it as bright as α Centauri and following a slight fade it reached its greatest brilliance in April 1843, when it outranked Canopus and almost matched Sirius. Then a decline set in. By 1870, η was only of the 6th magnitude, and it has remained at around this level ever since. At its peak, the luminosity was of the order of 6 million times that of the Sun, and it has not really declined a great deal, since most of its emission is in the infrared. The distance is of the order of 8000 light-years and the mass is at least 100 times that of the Sun. It is associated with nebulosity, and superimposed on this is a dark, dusty region known as the Keyhole Nebula from its shape.

What apparently happened in the years following 1834 is that there was a massive explosion, which threw off a shell of material from the surface of the star. As the shell expanded, the star seemed to brighten: after 1843 the shell itself cooled dimmed and finally became opaque, hiding the light of the star beneath. After a century and a half of expansion, at a rate of 700 km s^{-1}, we now see the shell as a tiny nebula, nicknamed the Homunculus Nebula from its shape; it is orange in colour and, telescopically, η Carinæ looks quite unlike an ordinary star. Images taken with the Hubble Space Telescope show the billowing clouds of expanding material.

Between 1900 and 1940 the magnitude had fallen to about 8, but it subsequently brightened to between 6 and 7. There was a further increase in 1998, and in 2009 the magnitude was above 5, so that the star can easily be seen with the naked eye. What will happen in the near future remains to be seen, but η Carinæ is unstable, and there seems no doubt that it will eventually become a supernova – or, perhaps more probably, a hypernova.

It has been suggested that η has a binary companion, a hot but less massive star with an orbital period of 5½ years. This is possible, but unproved.

By 2010, η Carinæ had brightened to magnitude 4.7.

ρ *Cassiopeiæ*

This is a yellow hypergiant, close to β and usually a naked-eye object just above the fifth magnitude – it lies between two convenient comparison stars, σ and τ. The distance is almost 12 000 light-years. It is vast, with diameter 450 times that of the Sun. The mass is equal to 40 Suns, and the luminosity is 550 000 times that of the Sum.

At present the variations are slow and unspectacular, but in 1946 it fell temporarily to the sixth magnitude. This happened again in 2000, when the star ejected material of mass equal to 10 000 Earths, and in few months the surface temperature cooled from 7000 to only 4000 °C. This may happen every few tens of years. It seems that it is one of the most promising supernova candidates.

P *Cygni*

P Cygni is another very remote, very luminous star, but instead of being a yellow hypergiant, like ρ Cassiopeiæ, it is classed as an LBV (Luminous Blue Variable) of the S Doradûs type. It was first noted in 1600, when it was of the third magnitude, and was believed to be a nova. By 1626, it had dropped below naked-eye visibility, but brightened again in 1655. After further fluctuations it settled down in 1715 at magnitude ~5, and since then there has been little change. It is 30 times the mass of the Sun, and over 600 000 times more luminous; its distance is approximately 6000 light-years.

V838 Monocerotis

Here we have a truly weird object. It lies 20 000 light-years away, near the edge of the Galaxy, but was not noticed before 10 January 2002, when it suddenly brightened up to above magnitude 7; photographic records showed that the pre-outburst magnitude had been 15.6; peak brilliancy (magnitude 6.75) was reached on 6 February 2002, when the luminosity was at least equal to a million Suns; the brightening was due to the rapid expansion of the star's outer layers. Expansion means cooling, so that V838 became very cool and deep red. There were complicated spectral changes, and the Hubble Space Telescope took dramatic colour pictures as the surrounding dust shells were successively hit by the flash.

The true nature of V838 is unclear. It is not an ordinary variable star, and neither is it an ordinary nova, as was first assumed. Various theories have been proposed – even the destruction of a system of planets – but the star remains an enigma. It will be carefully monitored.

CATACLYSMIC VARIABLES

These are amongst the most unpredictable of all variable stars. The most impressive, by far, are the various types of novæ.

Novæ

Novæ can be spectacular in the extreme. The name is misleading because a nova is not a new star. What happens is that a formerly faint star suffers a tremendous outburst, and flares up to many times its normal brightness, remaining light for a few days, weeks or months before fading back to obscurity.

Some novæ remain bright only briefly. The rise may last only a few hours as with the bright nova V1500 Cygni 1975, which reached an absolute magnitude of –10 and an apparent magnitude of 1.8; within a week it had fallen below naked-eye visibility, whereas slow novæ, such as R Delphini, of 1967, are much more gradual in their decline. A suffix 'a' indicates a fall of three magnitudes in less than 100 days; 'b' three magnitudes in 100 to 150 days; 'c' three magnitudes in over 150 days.

To some extent novæ can be used as 'standard candles', though they are much less reliable than Cepheids. Many have a peak absolute magnitude of –7.5 and others of –8.8; novæ of both kinds have absolute magnitude –5.5, 15 days after their peak.

A list of naked-eye novæ seen since 1600 is given in Table 22.5. The brightest nova, V603 Aquilæ of 1918, briefly outshone every star in the sky apart from Sirius.

A nova is a binary system made up of a low-density red star with a white dwarf companion. The white dwarf pulls material away from the giant star, and this material builds up into an accretion disc round the white dwarf. Over long periods of time more and more material collects; it is hydrogen-rich and at a high temperature and the very strong surface gravity of the white dwarf generates tremendous pressures. Eventually, the situation becomes unstable and a runaway nuclear reaction begins, hurling material into space at

speeds up to 1500 km s^{-1}. The brilliancy may increase by a factor of at least 1000, but the outburst is brief: after a few days, weeks or months the nova returns to its old state. The mass of the ejected material is no more than 1/1000 of a solar mass. The spectra of novæ show absorption lines which are blue-shifted, indicating that gas thrown off during the outburst is moving towards us.

At the peak of the outburst a nova may be highly luminous. Nova Puppis 1942 sent out as much radiation as 1 600 000 Suns. Compared with this, the fluctuations of variable stars such as Mira seem very minor.

In some cases an old nova may be seen to be surrounded by a gas cloud. GK Persei 1901 is a case in point. It is associated with the nebulosity nicknamed the Firework Nebula. Some months after maximum it was found that nebulosity was appearing to one side of the star and to be expanding at the speed of light. This was clearly unacceptable: in fact, the nova lay in a dark nebula, and the radiation spreading out from the outburst illuminated more and more of the nebula each year. The even brighter nova V603 Aquilæ developed a tiny surrounding disc, which grew steadily in size and became fainter; by 1941, it had become too dim to be followed further. Many old novæ are now seen as eclipsing binaries: such is DQ Herculis 1934, which has a period of 4h 39m. The least luminous 'old nova' is CK Vulpeculæ of 1670, which is 2000 light-years away and has only 100X the luminosity of the Sun. Some stars have been known to undergo more than one outburst: these are the *recurrent novæ* (see Table 22.6). The best known of these is the 'Blaze Star', T Coronæ Borealis. Usually it is of around magnitude 10, but in 1866 it flared up briefly to magnitude 2: it was then regarded as a normal nova, but in 1946 it flared up again this time to magnitude 3. Several recurrent novæ are now under observation.

Faint novæ are by no means uncommon, and many of these are discovered by amateur observers. Sixteen naked-eye novæ appeared between 1970 and 2009, and one of these, V382 Velorum of 1999, reached magnitude 2.5. It was discovered on 22 May by Alan Gilmore in New Zealand and independently by Peter Williams in Australia; it was too far south in the sky to be seen from Europe. At its peak it shone 75 000 times more brightly than the Sun; the distance is believed to be 2600 light-years and before the outburst its magnitude was below 16.

Dwarf novæ

These are usually termed *U Geminorum* stars although much the brightest member of the class is SS Cygni. They show minor outbursts at roughly regular intervals. Of the two components one is a K- or M-type dwarf and the other is a white dwarf. The amplitude is of the order of 2 to 9 magnitudes (usually less), and the interval between successive outbursts may be from 10 days to several years – 103 days for U Geminorum (magnitude 14.9 to 8.2), 50 days for SS Cygni (12.4 to 8.2). The basic cause of the outbursts is the same as for true novæ but on a much reduced scale. *SU Ursæ Majoris* stars have both normal maxima and occasional 'supermaxima' of greater amplitude, while *Z Camelopardalis* stars show outbursts of from 2 to 5 magnitudes every 10 to 40 days, but with unpredictable 'standstills' when variations are temporarily suspended. *Z Andromedæ* stars or symbiotic variables are close binaries where the hot companion actually

Table 22.5 *Selected list of novæ*

Name	Year	RA (h)	RA (m)	Dec. (°)	Dec. (′)	Max. magnitude	Q	Discoverers
CK Vul	1670	19	47.6	+27	19	2.6	20.7	Anthelm
WY Sge	1783	19	32.7	+17	45	5	19.5	D'Agelet
V841 Oph	1848	16	59.5	−12	53	2	13.5	Hind
U Sco	1863	16	22.5	−17	53	8.8	19.2	Pogson
Q Cyg	1876	21	41.8	+42	51	2	15.6	Schmidt
V Per	1887	02	01.9	+56	44	4.5	11.5	Fleming
T Aur	1891	05	32.0	+30	27	4.2	15.2	Anderson
IL Nor	1893	15	29.4	−50	35	7.0	18	Fleming
RS Car	1895	11	08.1	−61	56	5.0	22	Fleming
V1059 Sgr	1898	19	01.8	−13	10	2.0	18.1	Fleming
V606 Aql	1899	19	20.4	−00	08	5.5	17.3	Fleming
GK Per	1901	03	31.2	+43	54	0.2	14.0	Anderson
DM Gem	1903	06	44.2	+29	57	4.8	16.7	Turner
OY Ara	1910	16	40.8	52	26	5.1	17.5	Fleming
DL Lac	1910	22	35.8	+52	43	4.6	14.9	Espin
DN Gem	1912	06	54.9	+32	08	3.5	15.8	Enebo
V840 Oph	1917	16	64.8	−29	38	5.5	20	Woods
V603 Aql	1918	18	49.0	+00	35	−1.1	12	Bower
HR Lyr	1919	18	53.5	+29	14	6.5	15.8	Mackie
V476 Cyg	1920	19	58.5	53	37	2.0	17.2	Denning
RR Pic	1925	05	25	−62	39	1.0	11.9	Watson
XX Tau	1937	05	19.4	16	43	5.9	18.5	Wachmann, Schwassmann
DQ Her	1934	18	07.5	45	51	1.3	14.5	Prentice
CP Lac	1939	22	15.6	55	37	13.1	16.6	Gomi, Neilson, Loreta
V630 Scr	1936	18	08	−34	20	4.5	19	Okabayasi
BT Mon	1939	06	43.8	−02	01	+5	15.5	Whipple and Wachmann
CP Pup	1942	08	11.8	−35	21	+0.5	15	Finsler and Dawson
V500 Aql	1943	19	52.4	+08	17	5.0	15.5.	Bertaud
V446 Her	1960	18	57.4	+13	14	3.0	18.0	Hasse
V533 Her	1963	18	14.4	+41	51	3.0	15.0	Dahlgren
QZ Aur	1964	05	28.7	+33	19	5.0	18.0	Sanduleak
HR Del	1967	20	42.4	+19	10	3.2	12.1	Alcock
LV Vul	1968	19	48.1	+27	11	5.2	16.9	Alcock
FH Ser	1970	18	30.8	+02	37	4.5	16.2	Honda
V3888 Sgr	1974	17	48.6	−18	46	6.5	16	Kuwano
V3964 Sgr	1975	17	49.7	−17	24	6	17	Stenholm, Lundstrom
V1500 Cyg	1975	21	11.6	+48	09	2.2	21.5	Honda, Osada
V2104 Oph	1976	18	03.4	+11	48	5.3	20.5	Kuwano
NQ Vul	1976	19	29.3	+20	28	6.0	18.5	Alcock
V1370 Aql	1982	19	23.3	+02	30	6.0	19.5	Honda
QU Vul	1984	20	26.8	+27	51	5.6	19	Collins
V842 Cen	1986	14	35.9	−57	38	4.6	20.2	McNaught
V351 Pup	1991	08	112.6	−35	07	6.4		Camilleri
V1974 Cyg	1992	20	30.5	+52	38	4.3		Collins
V705 Cas	1993	23	41.8	+57	31	6.5		Kanatsu
V382 Vel	1999	10	44.8	−52	26	3.1		Gilmore, Williams
V4743 Sgr	2002	19	09.4	−22	01	5.0		Haseda
V597 Pup	2007	08	16.2	−34	16	7.0		Pereira
V496 Sct	2009	18	43.5	−07	36	8.8		Nashimura

This list contains naked-eye novæ since 1600, plus a few others of special interest. Quiescent magnitudes (Q) are bound to be uncertain.

Table 22.6 *Selected list of recurrent novæ*

Star	Outbursts	Maximum magnitude
T Coronæ Borealis	1866, 1946	2.0
RS Ophiuchi	1901, 1933, 1958, 1967, 1985, 2006	5.1
T Pyxidis	1890, 1902, 1920, 1925, 1945, 1966	7.0
WZ Sagittæ	1913, 1946, 1976, 2001	7.0

WZ Sagittæ is an exceptional system, consisting of a white dwarf and a less massive 'normal' star. Over the ages, the white dwarf has drained material from its companion, so that the companion is now small and exceptionally cool – surface temperature about 1450 °C. It may eventually end up as a unique type of stellar end-product.

orbits within the envelope of its cool red giant companion; the variations are caused by pulsations in the red star together with interactions between the two, so that the light-curves are decidedly complicated. *RR Telescopii* stars show slow increases which may be of very long duration. *Polars* or AM Herculis stars show sudden outbursts of up to three magnitudes caused by the accretion of material on to the magnetic poles of a compact star: the light is strongly polarised – hence the name.

X-RAY NOVÆ

An X-ray nova is a binary system in which there are sudden outbursts at X-ray wavelengths. Around 200 X-ray binaries are known in our Galaxy and of these about one-third have been seen to suffer outbursts. They are of two main classes:

(a) High mass: a hot blue star (type O or B0) together with a neutron star. These are more or less regular periodic outbursts.
(b) Low mass: a cool red star (M or K type) with a compact object which may be either a neutron star or a black hole. The outbursts are unpredictable.

The strong gravitational pull of the compact member of the pair collects material from its companion to form an accretion disc. Material is then sucked down on to the surface of the compact member, and an outburst occurs when there is a sudden increase in the amount of material striking the compact object. X-rays are emitted, but optical and radio outbursts can also be detected. The term 'X-ray nova' is rather misleading; 'X-ray variable' would have been better.

GAMMA-RAY NOVÆ

In 2010, astronomers using NASA's Fermi Gamma-Ray Space Telescope detected gamma-rays from a novalike star – V407 Cygni, a symbiotic binary 9000 light-years away. The system consists of a white dwarf plus a red giant with a diameter 500 times that of the Sun. This discovery disproves the idea that novæ lack the power to emit such high-energy radiation.

SUPERNOVÆ

Supernovæ are outbursts so violent that they can be seen across the universe. Several have been observed in our Galaxy; the last was that of 1604, always known as Kepler's Star, but there have been two more recent supernovæ in our Local Group of galaxies, S Andromedæ (1885) in the Andromeda Spiral, and SN1987a in the Large Magellanic Cloud. There are several distinct types of supernovæ.

First, there is Type Ia; here we have a binary system, in which one component (A) is initially more massive than its companion (B) and therefore evolves more quickly into the red giant stage. Material is then pulled across from A to B, so that B grows in mass while A declines; there is in fact a sort of cosmic tug of war, in which B has the advantage! Eventually A is reduced to a small, dense core made up chiefly of carbon.

The tables are then turned: B evolves in the usual way, swells out and starts to lose material back to the shrunken A, which is now a carbon white dwarf. The dwarf goes on accreting matter until it reaches the Chandrasekhar limit, 1.38 solar masses. This means disaster; the star can no longer resist collapse; the carbon detonates in a vast supernova explosion, and in a matter of seconds the white dwarf is completely destroyed. There is an expanding shock wave with matter reaching velocities of 5000 to 20 000 km s^{-1}, around 3% of the speed of light. The absolute magnitude may reach -19.3, which is 5000 million times more luminous than the Sun. The most important point here is that all Type Ia supernovæ have the same peak luminosity, so they are invaluable in measuring the distances so great that we have no hope of making out objects such as supergiant stars or ordinary novæ. When the carbon dwarf explodes it creates other elements, including neon, oxygen and silicon, ending up as cobalt. Nickel decays to cobalt and then to iron, and these elements show up spectroscopically; there are no lines due to hydrogen.

A Type Ib or Ic supernova is different and is probably produced when a massive single star runs out of 'fuel' and collapses; with a Ib supernova the collapsing star may be a Wolf–Rayet.

A Type II supernova is also due to the collapse of a single star, at least eight times as massive as the Sun, which has used up its nuclear fuel. Inside a star of this kind, a succession of different fusion reactions takes place, each producing progressively heavier atomic nuclei in the star's core. Each time one nuclear 'fuel' is consumed, fusion in the core ceases and the core contracts further until the pressure and temperature within the core become sufficiently great for the next fusion reaction to commence. Successive reactions convert hydrogen to helium, helium to carbon, carbon to elements such as neon and magnesium, and so on, until eventually iron begins to form in the core. By this stage, the interior of the star is rather like an onion. Outside the core there is a layer of silicon and sulphur; next comes a layer of neon and magnesium; then a layer of carbon, neon and oxygen; then a layer of helium; and finally an outer layer of hydrogen. Each successive process produces less energy, and iron marks the end of the line.

Once the core has been converted into iron, it can no longer produce energy by means of fusion reactions. By this stage, the core

Table 22.7 *Selected supernova remnants*

Name	Observed	Maximum magnitude	Distance (light-years)	Type	Remnant
Galactic					
Vela supernova	~10,000 BC	?	800	II	Gum Nebula
Loop Nebula	~5000–8000 BC	?	1400–2600	?	Witch's Broom or Loop Nebula
SN 185	7 Dec 185	~−8	3000	Ia	(probably) RCW 86
SN 1006	1 May 1006	~−8	7200	Ia	SNR 1006, Lupus remnant
SN 1054	1054	−6	6300	II	M 1, Crab Nebula
SN 1181	1181	−1	?	?	(probably) 3C 58
SN 1572	11 Nov 1572	−4	7500	Ia	Tycho's SNR
SN 1604	8 Oct 1604	−2.5	~20 000	Ia	Kepler's SNR
Cassiopeia A	?1680	+6?	10 000	Ib	Cassiopeia SNR
Extragalactic					
S Andromedæ	20 Aug 1885	+6	In M 31	?	SNR 1885a
SN 1987a	24 Feb 1987	+2.3	In LMC	II	SNR 1987a

is only prevented from collapsing by the pressure that is exerted by fast-moving electrons (electron degeneracy pressure). As more of the star's material is converted into iron the mass of the iron core continues to grow. When the mass of the iron core exceeds the Chandrasekhar limit, degeneracy pressure can no longer resist the inward pull of gravity, and catastrophic collapse ensues. As the core collapses, electrons and protons merge to produce electrically neutral neutrons together with vast numbers of neutrinos, most of which pass through the star and into space. The collapse of the core, which takes place in milliseconds, is eventually halted by the enormous pressure exerted by closely packed neutrons (neutron degeneracy pressure), by which time the density of the compressed core has become comparable to that of an atomic nucleus (at least 3×10^{17} kg m^{-3}); put another way, about 2500 million tonnes of this material could be packed inside a matchbox. When infalling layers of the star's material encounter the unyielding core, they 'bounce' and a shock wave moves outwards, colliding with other material which is still falling inwards. The result is catastrophic and most of the star is blown away into space, leaving only the core – which may be a neutron star or, in extreme cases, a black hole. The peak luminosity is not as great as with a Ia, but the absolute magnitude may reach –17, which is 500 million times as luminous as the Sun.

Type II supernovæ show hydrogen Balmer lines in their spectra. On average they fade by 0.008 magnitude per day, much more slowly than Type Ia supernovæ. They cannot be used as reliable 'standard candles'.

Type I supernovæ may occur anywhere, but Type II seem to be confined to the spiral arms of galaxies. The light-curves differ markedly. A Type I shows a steep rise, early fading and then a long slow decline. With a Type II there is also a steep rise and an early decline, after which the magnitude drops more quickly until settling down to a gradual decline.

In 2010, a new type of supernova was discovered. In January 2005, a supernova (SN2005E) flared up in the halo of the galaxy NGC 1032, in Cetus (RA 02h 39m 23.6s, Dec. +01 05′ 38″, magnitude 12.6, distance 117 million light-years). It simply did not fit into any known type. NGC 1032 is an S0 type spiral, but there was no sign of star formation anywhere near the supernova, and the mass ejected was surprisingly low (0.3 times the mass of the Sun). This did not agree with the core collapse of a massive star. On the other hand, the theory of a self-destructing white dwarf was equally untenable, because the spectrum showed that the chemical composition was different.

The expelled material was rich in calcium, plus titanium. These elements are produced by nuclear reactions involving helium rather than the carbon and oxygen found in the centres of white dwarfs. It seems likely that SN 2005E occurred in an interacting system of two close white dwarfs when the helium shell of one component was drawn on to the other component, causing the helium to react explosively. The reactions could also produce titanium.

About half of the mass thrown out by the explosion just over 100 million years ago was calcium, so that a couple of such supernovæ every century would be enough to produce the rich amount of carbon found in galaxies such as our own. These 'helium supernovæ' could even be quite common, but because they appear faint they must be comparatively hard to detect.

Supernovæ are key sources of all elements except those produced in the original Big Bang (mainly hydrogen and helium). When they explode, they enrich the surrounding medium with 'metals' (i.e. heavier elements). It is certainly true to say that the atoms and molecules making up the Sun, the Earth, and you and me were once inside supernovæ.

Supernova remnants

A supernova remnant (SNR) is the structure left by a supernova explosion. With all types except Ia there may be a neutron star or a black hole, and there will be traceable effects as the expanding

Table 22.8 *Selected list of galactic supernovæ*

Date	Constellation	Maximum magnitude	Time visible with naked-eye (months)	Remnant name	Remnant diameter (light-years)	Estimated distance (light-years)	Maximum absolute magnitude
185	Centaurus	−8	20	RCW 86?	115	10 000	−19 ± 2
386	Sagittarius	+1.5	3	–	20?	15 000	
393	Scorpius	0	8	CTB 37 A–B?	80	35 000	
1006	Lupus	−9.5	24	PKS 1459–41	30	3000	−19.8 ± 1
1054	Taurus	−5	22	Crab, 3C 144	9	7000	
1181	Cassiopeia	0	6	3C 58	17	9000	
1572	Cassiopeia	−4	16	3C 10	18	8000	−17 ± 0.5
1604	Ophiuchus	−3	12	3C 358	12	15 000	−19.6 ± 0.5

shock-wave collides with interstellar material. A selected list of SNRs is given in Table 22.7. The most famous is undoubtedly the Crab Nebula (M 1), the remnant of the supernova of 1054. These remants are probably the main sources of galactic cosmic rays, as was first suggested in 1934 by Fritz Zwicky and Walter Baade. It seems that the frequency of supernovæ in a galaxy about the size of the Milky Way is around 1 in 50 years.

Amateurs are skilled at discovering supernovæ in external galaxies. Tom Boles in England leads the way with 136 supernovae at the end of October 2010.

Pair–Instability supernoæ

A most unusual supernova, SN 2007bi, was discovered by a robotic telescope, the Nearby Supernova Factory at Palomar. Instead of fading quickly, this one brightened, reaching its peak brightness after an unprecedented 77 days. Only in October 2008, 555 days after being first sighted, did it become really faint. Meanwhile the afterglow had been monitored using Palomar's Samuel Oschin Telescope.

An explanation was proposed by the Israeli astrophysicist Avishay Gal-Yam, at the Weizmann Institute of Science in Rehovot. A Type II supernova is produced by the collapse of a giant star, when gravity overwhelms the photon pressure. It had been suggested that with a particularly massive star (more than 200 solar masses) the temperature would become so high during nuclear fusion that photons would start to break up spontaneously into their constituent electrons and positrons. Some of the photon pressure needed to support the star's outer layers would be removed, and the resulting collapse would vapourise the star. The final burst of fusion would create vast quantitites of heavy radioactive elements, producing what became known as a *pair-instability supernova*, very much more powerful than an ordinary supernova.

Gal-Yam examined all the data, and concluded that SN 2007bi fulfilled all these qualifications. It was the first of its type to be detected, and blazed out in a fairly near dwarf galaxy; for all we know, pair-instability supernovæ may be reasonably common. Searches for them will continue energetically.

CONFIRMED GALACTIC SUPERNOVÆ

Observable SNRs show that there have been a good many outbursts in our Galaxy over the past few tens of thousands of years, and there is an SNR near the galactic centre (see Table 22.8.)

The Veil Nebula

One lovely SNR is the *Veil Nebula* or Cygnus Loop, also known as the Witch's Broom Nebula. It was discovered in 1784 by William Herschel (position RA 20h 45m 38s, Dec. +30 42′ 30″) and is 3° across; the integrated magnitude is given as 7, but telescopically it can be elusive. I find it easier to locate with good binoculars. The outburst occurred from 5000 to 8000 years ago, and the distance is somewhere between 1400 and 2600 light-years.

The Vela SNR

The *Vela SNR* is known as the Gum Nebula (not because it is sticky! – the name honours the Australian astronomer Colin Gum). The position is RA 08h 35m 21s, Dec., −45 10′ 35″; it is about 8° across, and overlaps the older and more remote Puppis SNR. It is about 11 000 years old, and contains a flashing pulsar – one of the first to be optically identified. The Vela SNR is a particularly strong X-ray source.

Several SNRs have been positively identified since reasonably reliable observations began to be made, initially by the Chinese:

SN 185

This supernova was discovered by the Chinese on 7 December between Centaurus and Circinus, not far from α Centauri. It remained visible for 8 months: approximate position RA 14h 43m, Dec. −62° 30′. The gaseous shell RCW 86 is probably its remnant, though the identification is not 100% certain.

SN 1006, the Lupus supernova

This was certainly the brightest galactic supernova ever seen; position near β Lupi. According to the Egyptian astronomer Ali ibn Ridwan, it was three times as large as 'the disc of Venus', and a quarter as bright as the Moon. The remnant, PKS 1459–41, was identified by D. Milne and F. Gardner in 1965, using the Parkes radio telescope, showing that the supernova was of Type Ia.

The Crab Nebula, M 1

This is possibly the most-studied object in the sky, as it radiates over the whole of the electromagnetic spectrum. It was discovered by John Bevis in 1731; Messier re-discovered it in 1758 and made it No.1 in his catalogue. Its nickname is due to Lord Rosse, who drew it with the great Birr reflector. I can just see it with powerful binoculars, near ζ Tauri. Its apparent size is $420'' \times 290''$, corresponding to a real diameter of around 12 to 13 light-years. The distance from Earth is about 6500 light-years. The remnant is a *plerion*, i.e. an object filling the whole area it occupies rather than producing a shell; it is expanding rapidly. The central pulsar, NP 0532, has a period of 0.032 s, and can reach magnitude 15.

In visible light the Crab Nebula appears as a roughly oval-shaped mass of filaments; $6'$ long by $4'$ wide. In 1953 Iosif Shklovsky correctly proposed that the emissions from this area are due mainly to synchrotron radiation, produced by electrons spiralling along at speeds up to half the velocity of light.

For records of the outburst we rely upon China, Japan and Korea. Apparently it was first seen in April or May 1054, peaking in July and remaining visible for two years. It has been claimed that it is shown on a cave painting made by the Anasazi Indians of Chaco Canyon, North America, and this may well be true.

SN 1181

According to Chinese and Japanese sources, a bright star appeared between β and ι Cassiopeiæ; it was almost certainly a supernova, and the SNR 3C 58 is probably its remnant.

Tycho's Star, SN 1572 (B Cassiopeiæ)

This is always known as Tycho's Star, because he left a detailed account of it, though in fact the great Dane did not discover it; it was seen by W. Schuler of Wittenberg on 6 November, though it may have been seen as early as 3 November. Tycho saw it on the 11th, and monitored it carefully as long as it remained visible.

Tycho wrote; 'In the evening, after sunset . . . I noticed that a new and unusual star, surpassing all the other stars in brilliancy, was shining almost directly above my head it was quite evident to me that there had never before been any star in that place in the sky. A miracle indeed, either the greatest of all that have occurred in the whole range of nature since the beginning of the world, or one certainly that is to be classed with those attested by the Holy Oracles.' The main point here is that, in the eyes of the Church, the sky was perfect and changeless; the appearance of a brilliant new star proved that this was not true.

The star became brighter than Venus, and then slowly faded, changing in colour. Tycho, a firm believer in astrology, put his own interpretation on this. 'The star was at first like Venus and Jupiter, giving pleasing effects; but as it then became like Mars, there will next come a period of wars, seditions, captivity and death of princes, and destruction of cities, together with dryness and fiery meteors in the air, pestilence, and venomous snakes. Lastly, the star

became like Saturn, and there will finally come a time of want, death, imprisonment and all sorts of sad things'.

The star faded from view in 1574. Radio waves from the remnant were detected in 1952 by Hanbury Brown and Hazard, and the SNR was seen visually, as a faint nebula, by astronomers at Palomar, in 1960. Clearly this was a Type Ia; it is 7500 light-years away and a shell of gas is still expanding outward from its centre at about 9000 km s^{-1}. Its radio catalogue number is 3C 10 or more correctly G120.1+1.4. A solar-type companion star was discovered in 2004; parts of its outer layers may have been stripped away by the outburst.

Kepler's Star, SNR 1604

This was seen by Johannes Kepler on 8 October, in Ophiuchus (RA 17h 30m 42s, Dec. $-2.1°$ $29'$). A few days later it rose to about magnitude -2.5, equal to Jupiter but not as bright as Venus. It faded slowly but did not drop below naked-eye visibility until March 1606. Its remnant was detected in 1941, by astronomers using the 100-inch telescope at Mount Wilson, and catalogued as 3C 358 (G4.5+6.8). It is a dim fan-shaped nebulosity, magnitude 19, so that again we have a Type Ia. The distance can be no more than 20 000 light-years.

Cassiopeia A

There must have been a galactic supernova around 1680, because its remnant is the strongest radio source in the sky beyond the Solar System It seems to have been no more than about 11 000 light-years away. Unfortunately it was hidden behind clouds of dust in the main plane of the Galaxy, and nobody actually saw it, though, just possibly, it is identical with the sixth-magnitude star 3 Cassiopeiæ in John Flamsteed's catalogue. This was a Type Ib supernova, so that the remnant is of the shell type. The progenitor star was probably a red supergiant. The optical SNR (3C 461; G111.7–2.1) was discovered in 1950, and in 1999 the Chandra X-ray Observatory found a 'hot, point-like source near the centre', which may be a neutron star or even a black hole.

In 2009, Loretta Dunne and her team, using the SCUBA polarimeter on the JCMT telescope in Hawaii, found new evidence of huge dust production in the remnant. The polarisation signal is the strongest ever found in the Milky Way; it has been suggested that much of the material may be in the form of iron needles.

When the next galactic supernova will flare up we know not, but we hope that it will not be too close. In particular they produce floods of gamma-rays, which would do us no good at all (it has even been suggested that the Ordovician extinction was due to this cause). The most famous potential supernova candidates – Betelgeux, ρ Cassiopeiæ and η Carinæ – are reassuringly remote. It seems that the only nearby SN candidate is IK Pegasi (RA 21h 26m 27s, Dec. $+19°$ $22'$ $32''$), a binary with an A8-type primary and a white dwarf secondary; separation 31 000 000 km, orbital period 21.7 days, distance from Earth a mere 150 light-years. The primary is eight times as massive as the Sun. At magnitude 6 it is just visible with the naked eye. It may well produce a Type Ia supernova eventually – but not yet, and even when it does, the Earth is just about out of harm's way.

Supernovæ in the Local Group

Two supernovæ have been seen in the Local Group. In 1885, what we now know to be a supernova flared up in M 31, the Andromeda Spiral, and despite being well over two million light years away rose to the fringe of naked-eye visibility. At the time its true nature was not recognised. It is now catalogued as S Andromedæ, and on 4 November 1988 R A. Feisen, using the 4-m Mayall telescope at Kitt Peak, Arizona, identified its remnant, which, being iron-rich, showed up as a dark patch against the background.

S Andromedæ was discovered on 20 August 1885 by Hartwig at the Dorpat Observatory in Estonia, but there is an interesting aside here. On 22 August, the Hungarian Baroness de Podmaniczky was holding a house party and for the entertainment of her guests had set up a small telescope on the lawn. Using this she saw a 'small star' in M 31. Her observation was quite independent of Hartwig's though she certainly did not appreciate its importance!

The most recent naked-eye supernova was 1987A, in the Large Cloud of Magellan at a distance of 169 000 light-years. It was discovered on 24 February by Ian Shelton, at the Las Campanas Observatory in Chile, on a routine photograph he had taken, and almost at the same time by O. Duhalde, also at Las Campanas, with the naked eye. (It had been photographed earlier by R. McNaught from Australia, but McNaught had not checked his observations.) When discovered it was of magnitude 4.5, but brightened to magnitude 2.3 before starting to fade. One surprise was that the progenitor star, Sanduleak −69 202, was identifiable and was a blue giant rather than a red star: it was thought to be about 20 000 000 years old, with a mass from five to seven times that of the Sun. The peak luminosity was low by supernova standards: only about 250 000 000 times that of the Sun – because the progenitor star was smaller, though hotter, than a red supergiant. It seems that the progenitor was originally very hot and massive; as it used up its available hydrogen it went through the red supergiant stage before shedding its outer layers and contracting. It was then, with the star, a blue supergiant, that the outburst happened.

Maximum brightness was reached in mid May 1981, and then came the expected decline. Various phenomena were seen, which were illuminated by the outburst. The supernova was observed at many wavelengths, including X-rays (by the Japanese satellite Ginga) and ultraviolet (by the International Ultra-violet Explorer satellite). Neutrinos were also recorded. So far no pulsar has been detected. In fact a pulsar was reported on 18 January 1989 by astronomers at Cerro Tololo in Chile, led by J. Middleditch: it was claimed that the pulsar was rotating at a rate of 1968.63 turns per second. Various explanations were offered, but it was then found that the observed effects were due to the mechanism of the telescope. Whether a pulsar will eventually form remains to be seen. It has been suggested that the progenitor star, the blue giant Sanduleak −69 202a, was a binary system in which the components merged about 20 000 years before the explosion; this may

or may not be true. Three neutrino bursts were recorded. The computer at Mont Blanc recorded a burst of five pulses about 8 hours before the first optical observation and, three hours before the sight reached Earth, a burst was recorded at three separate neutrino observatories. This was because the emission of neutrinos happens at the same time as the core collapse, whereas visible light emission comes later, when the shock wave has reached the star's surface. The total neutrino count was only 24, but this was the first time that neutrinos from a supernova had been observed directly.

Paul Murdin has given a very interesting timescale, linking the evolution of the progenitor star with geological periods on Earth; it is as follows:

Formation of the star. 20 000 000 years ago, during the Miocene period. Hydrogen burning continued for 15 000 000 years.
Helium burning. End of the Pleiocene period.
Carbon burning. Pleistocene period; first men on Earth.
Neon and oxygen burning. 1980: Mrs Thatcher as British Prime Minister.
Silicon burning. 1987 February 20.
Explosion of the star. 1987 February 23, 17h 35m.

(*En passant* I have to admit that I took a plane from Heathrow to Johannesburg when the supernova was at its best, simply to look at it. I reasoned that the chances of another naked-eye supernova appearing in my lifetime were, to put it mildly, slim!)

Using SINFONI (Spectrograph for Integral Field Observations in the Near Infrared) with the VLT, a team led by Karina Kjær has been able to obtain what may be called a three-dimensional picture of SN 1987A. Evidently the explosion was faster and stronger in some directions than in others, so that the supernova must have been very turbulent; this led to an irregular shape. The first material to be ejected travelled at about 100 million kilometres per hour, but even so it took 10 years to reach a previously ejected ring of dust and gas. Another wave of material is travelling much more slowly, and is being heated by radioactive elements created in the explosion.

Hypernovæ

These are outbursts far more violent even than supernovæ – perhaps by a factor of 100. What seems to happen is that the core of a hypergiant star collapses to form a black hole. The star is 100 to 150 times as massive as the Sun, and when the collapse starts a staggering amount of energy is released. In 1997, Bohdan Paczynski, of Princeton University in the United States, proposed that hypernovæ might be responsible for GRBs (gamma-ray bursters), 'the biggest bangs since the Big Bang'.

Because hypergiant stars are so rare, hypernovæ must also be rare. In a galaxy the size of the Milky Way, one hypernova might be expected every 200 million years or so.

23 · Stellar clusters

Clusters and nebulæ are among the most striking of stellar objects. Several are easily visible with the naked eye. Few people can fail to recognize the lovely star cluster of the Pleiades or Seven Sisters, which has been known since prehistoric times and about which there are many old legends. The nebula in the Sword of Orion, the Sword-Handle in Perseus, Præsepe in Cancer and the Jewel Box cluster in Crux are other objects easily visible without optical aid. Keen-sighted people have little difficulty in locating the great Andromeda Spiral and the globular cluster in Hercules, while in the far south there are the two Clouds of Magellan which cannot possibly be overlooked, as well as the bright globular clusters ω Centauri and 47 Tucanæ.

The most famous of all catalogues of nebulous objects was compiled by the French astronomer Charles Messier, and published in 1781. Ironically, Messier was not interested in the objects he listed: he was a comet-hunter, and merely wanted a quick means of identifying misty patches which were non-cometary in nature. In 1888, J. L. E. Dreyer, Danish by birth (although he spent much of his life in Ireland, and finally in England) published his New General Catalogue (NGC), augmented in 1898 and again in 1908 by his Index Catalogue (I or IC).

Messier's original catalogue included 103 objects; later it was extended to 110, although not by Messier himself. However, Messier excluded many bright clusters and nebulæ, either because they could not be confused with comets, and were therefore of no interest to him, or because they were too far south in the sky to be seen from France. In 1995, I compiled the Caldwell Catalogue of 109 objects arranged in order of declination and omitted by Messier. Much to my surprise, the C-numbers are now widely used.[1]

A few nebulous objects must have been known in prehistoric times. This certainly applies to the Pleiades cluster, which cannot be overlooked. In his great work the *Almagest*, Ptolemy records not only the Pleiades but also the Sword-Handle in Perseus, M 44 (Præsepe) and M 7 in Scorpius, now called Ptolemy's Cluster. We may also assume that early inhabitants of the southern hemisphere knew about the Magellanic Clouds, though they were not mentioned by Europeans until about 1520.

The list in Table 23.1 includes all the objects known before 1745: 21 objects in all. Some of these (and also some other objects) have popular nicknames; a selected list of these is given in Table 23.2. Many other objects had been previously listed, but subsequently found to be mere groups of stars rather than true clusters or nebulæ. One remarkable fact relates to the Great Spiral in Andromeda, which was recorded by the Persian astronomer Al-Sûfi. It was not noted again until 1612, when Simon Marius

described it. Amazingly, it was completely overlooked by the greatest observer of pre-telescope times, Tycho Brahe.

After 1745, more and more objects were found, mainly by de Chéseaux, Legentil, Lacaille, and of course the two great Frenchmen, Messier and Méchain. By 1781, when Messier published his Catalogue, 138 nebulous objects were known.

All catalogues include objects of different kinds: open or loose clusters, globular clusters, supernova remnants, diffuse or galactic nebulæ, planetary nebulæ and galaxies. The Messier Catalogue is given in Table 23.3 and the Caldwell Catalogue in Table 23.4.

OPEN CLUSTERS

Open or loose clusters are aggregations of stars arranged in no particular shape; the Pleiades, Hyades and Præsepe are good examples. Some clusters are sparse, containing no more than 10 stars, while others may include at least 10 000 stars. The diameter of an average cluster is of the order of 65 light-years, and in its richest part the distances between individual stars will be no more than a light-year. They are of various ages. The Pleiades are about 76 000 000 years old, but the lovely Jewel Box in the Southern Cross has an age of little more than 7 000 000 years, and others are even younger, at around 1 000 000 years. On the other hand, M 67, an open cluster in Cancer, has an age of 4000 million years.

Interstellar space is never completely empty, but the density is very low indeed. Here and there, denser clouds of dust and gas form, and are very cold. When a cloud reaches a mean density a thousand times greater than that of normal space, atoms combine into molecules, and we have a molecular cloud, which may be from 1 to 300 light-years across, containing enough material to produce at least 10 000 stars. If the mass of the cloud exceeds 100 000 times that of the Sun, it is classed as a giant molecular cloud (GMC). Clouds of this type are bitterly cold, at around −263 °C (only 10 above absolute zero); as they are made up chiefly of hydrogen molecules (H_2) they are hard to detect optically, and radio telescopes have to be used. However, in recent years the Hubble Space Telescope has been used to observe colour changes of background stars seen through the clouds, which leads on to information about the distribution of matter in the clouds themselves.

Still denser clumps of material accumulate in the GMCs, and star formation follows: many stars may be produced from the same cloud, resulting in a stellar cluster.

Once star formation has started, very hot, massive stars (OB stars) within any particular cluster are of the same age, and of the same initial chemical composition, so that present differences between them are due to initial differences in mass. Drawing up a

Table 23.1 *First-known clusters and nebulæ*

Number		Discoverer
M 45	Pleiades	(Prehistoric)
NGC 869/884, C 14	Sword-Handle in Perseus	(Listed by Prolemy)
M 44	Præsepe	(Listed by Prolemy)
M 7	Open cluster in Scorpius	Prolemy, *c.* 140
M 31	Andromeda Spiral	Al-Sûfi, *c.* 964
IC 2391, C 85	o Velorum cluster	Al- Sûfi, *c.* 964
Large Cloud of Magellan		1519
Small Cloud of Magellan		1519
M 42	Great Nebula in Orion	Peiresc, 1610
M 22	Globular in Sagittarius	Ihle, 1665
NGC 5139, C 80	ω Centauri	Halley, 1677
M 8	Lagoon Nebula	Flamsteed, 1680
M 11	Wild Duck Cluster	Kirch, 1681
NGC 2244	12 Monocerotis	Flamsteed, 1690
M 41	Open cluster in Canis Major	Flamsteed, 1702
M 5	Globular in Serpens	Kirch, 1702
M 50	Open cluster in Monoceros	Cassini, *c.* 1711
M 13	Hercules Globular	Halley, 1714
M 43	Part of the Orion Nebula	de Mairan, *c.* 173
M 1	Crab Nebula	Bevis, 1731

Hertzsprung–Russel (HR) diagram of a cluster shows where there is a turn-off from the Main Sequence to the giant branch, and this gives the age of the cluster. For example, the brightest stars in the Pleiades are blue and very hot, while the much older Hyades contain stars which have already evolved into red giants. In young open clusters there is still visible material which shows that star formation may sometimes be disrupted by passing 'field' stars, or by occasional encounters with molecular clouds, and will eventually disperse. The oldest open clusters, such as M 67, have maintained their separate identity, because they are well away from the main plane of the Galaxy and are not likely to encounter many non-cluster stars, while most of the open clusters are close to the galactic plane, so that encounters are frequent.

A selected list of bright open clusters is given in Table 23.5. Many of these are naked-eye objects, and all in the list are easily seen with binoculars. Fainter clusters are plentiful, and there may be some 20 000 of them in our Galaxy. Of course, open clusters are also seen in external galaxies.

Celebrated open clusters

The Pleiades (M 45, in Taurus)

This is the most famous of all clusters, and there are many old legends about it. For example, in Hindu mythology the stars are the six mothers of the war god Skanda, who developed six faces, one for each of them! They are mentioned by Homer and Hesiod, and even in the Bible. The popular name for them is the Seven Sisters, and certainly people with average eyesight can see seven stars with the naked eye. Much the brightest of them is Alcyone (η Tauri), magnitude 2.9; then come Atlas, Electra, Merope, Maia, Taygete, Pleione, Celæno and Asterope. This makes nine, but Pleione is variable as well as being very close to Atlas. A list of the main stars is given in Table 23.6. All are hot and blue, so they must be young by cosmical standards.

Table 23.2 *Some astronomical nicknames. There are many unofficial names for stellar objects in common use. The following list is far from complete, but does include some familiar nicknames*

Ant Nebula	Bipolar nebula PK 331–1.1, at RA 16h 17m .2 , Dec. −51° 59′
Antennæ	Colliding galaxies NGC 4038 and 4039; in Corvus (C 60 and 61)
Barnard's Galaxy	NGC 6822; in the Local Group, RA 19h 45m, Dec. −14° 48′,C 57
Barnard's Loop	Extensive ring of nebulosity in Orion
Barnard's Star	Munich 15040 the nearest star apart from the α Centauri group
Bear Paw Galaxy	NGC 2537; galaxy in Lynx
Becklin–Neugebauer Object	Infrared source in M 42 (Orion Nebula)
Beehive Cluster	M 44 Præsepe
Black-eye Galaxy	M 64; in Coma (spiral galaxy)
Blaze Star	The recurrent nova T Coronæ Borealis
Blinking Nebula	NGC 6826; planetary nebula in Cygnus; C 15
Blue Planetary	NGC 3918; nebula in Centaurus
Bode's Nebula	M 81 spiral galaxy in Ursa Major
Boomerang Nebula	Bipolar nebula at RA 12h 44m 8s, Dec. −54° 31'
Box Nebula	NGC 4169
Bubble Nebula	NGC 7635: nebula in Cassiopeia

Table 23.2 (cont.)

Bug Nebula	NGC 6302; plantary nebula in Scorpius. C 69
Burnham's Nebula	T Tauri nebula
Butterfly Cluster	M 6; open cluster in Scorpius
California Nebula	NGC 1499; nebula in Perseus
Cartwheel Galaxy	Ring galaxy at RA 1h 37m .4, Dec. $-33°$ 45′
Centaurus A	Radio galaxy NGC 5128; C 77
Christmas Tree Cluster	NGC 2264
Coal Sack	Dark nebula in Crux; C 99
Coat-Hanger	Cluster in Vulpecula (also known as Brocchi's Cluster)
Cocoon Nebula	IC 5146; nebula in Cygnus; C 19
Cone Nebula	IC 2264 (S Monocerotis); nebula in Monoceros
Crab Nebula	M1 supernova remnant in Taurus; NGC 1952
Crescent Nebula	NGC 6888; nebula in Cygnus; C 27
Crimson Star	The Mira variable R Leporis
Demon Star	Algol (β Persei)
Dog Star	Sirius (α Canis Majoris)
Dumbbell Nebula	M 27; planetary nebula in Vulpecula
Eagle Nebula	M 16; nebula in Serpens Cauda
Egg Nebula	CRL 2688 at RA 21h 02m. 3, Dec. 36° 42′
Eskimo Nebula	NGC 2392; planetary nebula in Gemini (also known as the Clown Face Nebula); C 38
Flaming Star Nebula	IC 405; nebula associated with AE Auria
Flying Star	61 Cygni (because of its large proper motion)
Garnet Star	μ Cephei
Geminga	Gamma-ray source in Gemini
Ghost of Jupiter	NGC 3242: plateary nebula in Hydra; C 59
Gum Nebula	Nebula in Vela (supermova remnant)
Helix Nebula	NGC 7293; planetary nebula in Aquarius
Hind's Variable Nebula	NGC 1555; nebula near T Tauri
Homunculus Nebula	Core of η Carinæ Nebula
Hourglass Nebula	The brightest part of M 8
Horse's Head Nebula	Barnard 33: dark nebula in Orion
Hubble's Variable Nebula	NGC 2261; nebula round R Monocerotis; C 46
Hyades	Mel 25, open cluster in Taurus; C 41
Innes Star	Proxima Centauri
Intergalactic Tramp	NGC 2419 globular cluster in Lynx; C 25
Jewel Box	NGC 4755: opea cluster (κ Crucis), C 94
Kepler's Star	The supernova of 1604 (in Ophiuchus)
Keyhole Nebula	Dark nebula; in the η Carinæ nebulosity
La Superba	Y Canum Venaticorum
Lacework Nebula	NGC 6960
Lagoon Nebula	M 8; nebula in Sagittarius
Little Dumbbell	M 76; planetary nebula in Perseus
Markarian's Chain	M 85 to M 88 in the Virgo Cluster
Network Nebula	NGC 6992–5 in the Veil Nebula
North America Nebula	NGC 7000 nebula in Cygnus C. 10
Omega Nebula	M 17 nebula in Sagittarius (also known as the Horseshoe Nebula)
Owl Nebula	M 97; planetary nebula in Ursa Major
Pelican Nebula	IC 5067/7; nebula in Cygnus
Pinwheel Galaxy	Triangulum Galaxy; M 33
Pinwheel Nebula	M 99 (NGC 4254)

Table 23.3 *The Messier Catalogue. Messier's original catalogue ends with M 103. The later numbers are objects which apart from the last were discovered by Méchain. (M 110 may have been discovered by Messier himself, but it is still always known as NGC 205, one of the companions to the Great Spiral in Andromeda.) M 104 was added to the Messier catalogue by Camille Flammarion in 1921, on the basis of finding a handwritten note about it in Messier's own copy of his 1781 catalogue. M 105 to 107 were listed by H. S. Hogg in the 1947 edition of the catalogue; M 108 and 109 by Owen Gingerich in 1960. The naming of NGC 205 as M 110 was proposed by K. G. Jones in 1968, but has never been accepted*

M	NGC	Constellation	Type	RA (h)	(m)	Dec. (°)	(')	Magnitude	Dimensions (')	Distance (thousands of light-years)*	
1	1952	Taurus	Supernova remnant	0.5	34.5	+22	01	8.2	6.4	6.3	Crab Nebula
2	7089	Aquarius	Globular	21	33.5	−00	49	6.5	12.9	36.2	
3	5272	Canes Venatici	Globular	13	42.2	+28	23	6.2	16.2	30.6	
4	6121	Scorpius	Globular	16	23.6	−26	32	5.6	26.3	6.8	
5	5904	Serpens	Globular	15	18.6	+02	05	5.6	17.4	22.8	
6	6405	Scorpius	Open cluster	17	40.1	−32	13	5.3	15.0	2.0	Butterfly Cluster
7	6475	Scorpius	Open cluster	17	53.9	−34	49	4.1	80.0	1.0	
8	6523	Sagittarius	Nebula	18	13.8	−24	23	6.0	60 × 35	6.5	Lagoon Nebula
9	6333	Ophiuchus	Globular	17	19.2	−18	31	7.7	9.3	26.4	
10	6154	Ophiuchus	Globular	16	57.1	−04	06	6.6	15.1	13.4	
11	6705	Scutum	Open cluster	18	51.1	−06	16	6.3	14.0	6.0	Wild Duck Cluster
12	6218	Ophiuchus	Globular	16	17.2	−01	57	6.7	14.5	17.6	
13	6205	Hercules	Globular	16	41.7	+36	28	5.8	16.6	22.2	Hercules Cluster
14	6402	Ophiuchus	Globular	17	37.6	−03	15	7.6	11.7	27.4	
15	7078	Pegasus	Globular	21	30.0	+12	10	6.2	12.3	32.6	
16	6611	Serpens	Nebula and embedded cluster	18	18.8	−13	47	6.4	7.0	7	Eagle Nebula
17	6618	Sagittarius	Nebula	18	20.8	−16	11	7.0	11.0	5.0	Omega Nebula
18	6613	Sagittarius	Open cluster	18	19.9	−17	08	7.5	9.0	6.0	
19	6273	Ophiuchus	Globular	17	02.6	−26	16	6.8	13.5	27.1	
20	6514	Sagittarius	Nebular	18	02.6	−23	02	9.0	28.0	2.2	Trifgid Nebula
21	6531	Sagittarius	Open cluster	18	04.6	−22	30	6.5	13.0	4.3	
22	6656	Sagittarius	Globular	18	36.4	−23	54	5.1	24.0	10.1	
23	6494	Sagittarius	Open cluster	17	55.8	−19	01	6.9	27.0	4.5	
24	6603	Sagittarius	Star-cloud	18	14.9	−18	29	4.6	90.0	10.0	Not a cluster
25	IC 4725	Sagittarius	Open cluster	18	31.6	−19	15	6.5	40.0	2.0	
26	6694	Scutum	Open cluster	18	45.2	−09	24	9.3	15.0	5.0	
27	6853	Vulpecula	Planetary nebula	19	54.6	+22	43	7.4	8.0 × 5.7	1.3	Dumbbell Nebula
28	6626	Sagittarius	Globular	18	24.5	−24	52	6.8	11.2	17.9	
29	6913	Cygnus	Open cluster	20	23.9	+38	32	7.1	7.0	7.2	
30	7099	Capricornus	Globular	21	40.4	−23	11	7.2	11.0	24.8	
31	224	Andromeda	Spiral galaxy	00	42.7	+41	16	4.8	178	2200	Great Spiral
32	221	Andromeda	Elliptical galaxy	00	42.7	+40	52	8.7	8 × 6	2200	Companion to M 31
33	598	Triangulum	Spiral galaxy	01	33.9	+30	39	6.7	73 × 45	2300	Triangulum spiral
34	1039	Perseus	Open cluster	02	420	+42	47	5.5	35.0	1.4	
35	2168	Gemini	Open cluster	06	03.9	+24	20	5.3	28.0	2.8	
36	1960	Auriga	Open cluster	05	35.1	+34	08	6.3	12.0	4.6	
37	2099	Auriga	Open cluster	05	52.4	+32	33	6.2	24.0	4.6	

Table 23.3 (cont.)

M	NGC	Constellation	Type	RA (h)	(m)	Dec. (°)	(′)	Magnitude	Dimensions (′)	Distance (thousands of light-years)*	
38	1912	Auriga	Open cluster	05	23.6	+35	50	7.4	21.0	4.2	
39	7092	Cygnus	Open cluster	21	32.2	+48	26	5.2	32.0	0.8	
40	–	Missing	Possibly a comet?								
41	2287	Canis Major	Open cluster	06	47.0	–20	44	4.6	38.0	2.4	
42	1976	Orion	Nebula	05	35.4	–05	27	4.0	85 × 60	1.6	Great Nebula in Orion
43	1982	Orion	Nebula	05	35.5	–05	16	9.0	20 × 15	1.6	Part of Orion Nebula
44	2632	Cancer	Open cluster	08	40.1	+19	59	3.7	95.0	0.5	Præsepe
45	–	Taurus	Open cluster	03	47.0	+24	07	1.6	110	0.4	Pleiades
46	2437	Puppis	Open cluster	07	41.8	–14	49	6.0	27.0	5.4	
47	2422	Puppis	Open cluster	07	13.8	–05	48	5.5	54.0	1.8	
48	2548	Hydra	Open cluster	08	13.8	–05	48	5.5	54.0	1.8	
49	4472	Virgo	Elliptical galaxy	12	29.8	+08	00	8.5	9 × 7.5	60 000	
50	2323	Monoceros	Open cluster	07	03.2	–08	20	6.3	16.0	3.0	
51	5194	Canes Venatici	Spiral galaxy	13	29.9	+47	12	8.1	11 × 7	37 000	Whirlpool Galaxy
52	7654	Cassiopeia	Open cluster	23	24.2	+61	35	7.3	13.0	7.0	
53	5024	Coma	Globular	13	12.9	+18	10	7.6	12.6	56.4	
54	6715	Sagittarius	Globular	18	55.1	–30	29	7.6	9.1	82.8	
55	6809	Sagittarius	Globular	19	40.0	–30	58	6.3	19.0	16.6	
56	6779	Lyra	Globular	19	16.6	+30	11	8.3	7.1	31.6	
57	6720	Lyra	Planetary nebula	18	53.6	+33	02	8.8	1.4 × 1.0	4.1	Ring Nebula
58	4579	Virgo	Spiral galaxy	12	37.7	+11	49	9.2	5.5 × 4.5	60 000	
59	4621	Virgo	Elliptical galaxy	12	42.0	+11	39	9.6	5 × 3.5	60 000	
60	4649	Virgo	Elliptical galaxy	12	43.7	–11	33	6.9	7 × 6	60 000	
61	4303	Virgo	Spiral galaxy	12	21.9	+04	28	9.6	6 × 5.5	60 000	
62	6266	Ophiuchus	Globular	17	01.2	–30	07	6.5	14.1	21.5	
63	5055	Canes Venatici	Spiral galaxy	13	15.8	+42	02	9.5	10 × 6	37 000	Sunflower Galaxy
64	4826	Coma	Spiral galaxy	12	56.7	+21	41	8.8	9.3 × 5.4	12 000	
65	2612	Leo	Spiral galaxy	11	18.9	+13	05	9.3	8 × 1.5	35 000	
66	3627	Leo	Spiral galaxy	11	20.2	+12	59	8.2	8 × 2.5	35 000	
67	2682	Cancer	Open cluster	08	50.4	+11	49	6.1	30.0	2.3	Famous old cluster
68	4590	Hydra	Globular	12	39.5	–26	45	7.8	12.0	32.3	
69	6637	Sagittarius	Globular	18	31.4	–32	21	7.6	7.1	25.4	
70	6681	Sagittarius	Globular	18	43.2	–32	18	7.9	7.8	28.0	
71	6838	Sagitta	Globular	19	53.8	+18	47	8.2	7.2	11.7	
72	6981	Aquarius	Globular	20	53.5	–12	32	9.3	5.9	52.8	
73	6994	Aquarius	Four faint stars	20	58.9	–12	38	9	3	—	Not a cluster
74	628	Pisces	Spiral galaxy	01	36.7	+15	47	9.2	10.2 × 9.5	35 000	
75	6864	Sagittarius	Globular	20	06.1	–21	55	8.5	6.0	57.7	
76	650	Perseus	Planetary nebula	01	42.4	+51	34	10.1	2.7 × 1.8	3.4	Little Dumbbell
77	1068	Cetus	Spiral galaxy	02	42.7	–00	01	8.9	7 × 6	60 000	
78	2068	Orion	Nebula	05	46.7	+00	03	8.3	8 × 6	1.6	
79	1904	Lepus	Globular	05	24.5	–24	33	7.7	8.7	39.8	
80	6093	Scorpius	Globular	16	17.0	–22	59	7.3	8.9	27.4	

Table 23.3 (cont.)

M	NGC	Constellation	Type	RA (h)	(m)	Dec. (°)	(′)	Magnitude	Dimensions (′)	Distance (thousands of light-years)*	
81	3031	Ursa Major	Spiral galaxy	09	55.6	+69	04	6.8	12 × 10	11 000	Bode's Nebula
82	3034	Ursa Major	Irregular galaxy	09	55.8	+69	41	8.4	9 × 4	11 000	
83	5236	Hydra	Spiral galaxy	13	37.0	−29	52	7.6	11 × 10	10 000	
84	4374	Virgo	Spiral galaxy	12	35.1	+12	53	9.3	5.0	60 000	
85	4382	Coma	Spiral galaxy	12	25.4	+18	11	9.3	7.1 × 5.2	60 000	
86	4406	Virgo	Elliptical galaxy	12	26.2	+12	57	9.7	7.5 × 5.5	60 000	
87	4486	Virgo	Elliptical galaxy	12	30.8	+12	24	8.6	7.0	60 000	Giant Seyfert Galaxy
88	4501	Coma	Spiral galaxy	12	32.0	+14	25	10.2	7.0 × 4	60 000	
89	4552	Virgo	Elliptical galaxy	12	35.7	+12	33	9.5	4.0	60 000	
90	4569	Virgo	Spiral galaxy	12	36.8	+13	10	10.0	9.5 × 4.5	60 000	
91	–	Not identified	Possibly a comet?								
92	6341	Hercules	Globular	17	17.1	+43	08	6.4	11.2	26.1	
93	2447	Puppis	Open cluster	07	44.6	−23	52	6.9	22.0	4.5	
94	4736	Canes Venatici	Spiral galaxy	12	30.9	+41	07	7.9	7 × 3	14 500	
95	3351	Leo	Barred spiral galaxy	10	44.0	+11	42	10.4	4.4 × 3.3	38 000	
96	3368	Leo	Spiral galaxy	10	36.8	+11	49	9.1	6 × 4	38 000	
97	3587	Ursa Major	Planetary nebula	11	14.8	+55	01	9.9	3.4 × 3.3	2.6	Owl Nebula
98	4192	Coma	Spiral galaxy	12	13.8	+14	54	1.7	9.5 × 3.2	60 000	
99	4524	Coma	Spiral galaxy	12	18.3	+14	25	10.1	5.4 × 4.8	60 000	
100	4321	Coma	Spiral galaxy	12	22.9	+15	49	10.6	7 × 6	60 000	
101	5457	Ursa Major	Spiral galaxy	14	03.2	+54	21	9.6	22.0	24 000	
102	–	Missing	Possibly a faint spiral in Draco, or else identical with M 101								
103	581	Cassiopeia	Open cluster	01	33.2	+60	42	7.4	6.0	8.0	
104	4594	Virgo	Spiral galaxy	12	40.0	−11	37	8.7	9 × 4	50 000	Sombrero Galaxy
105	3379	Leo	Elliptical galaxy	10	47.8	+12	35	9.2	2.0	38 000	
106	4528	Ursa Major	Spiral galaxy	12	19.0	+47	18	8.6	19 × 8	25 000	
107	6171	Ophiuchus	Globular	16	32.5	−13	03	9.2	10.0	19.6	
108	3556	Ursa Major	Spiral galaxy	11	11.5	+55	40	10.7	8 × 1	45 000	
109	3992	Ursa Major	Spiral galaxy	11	57.6	+53	23	10.8	7 × 4	55 000	

* Distances of galaxies in the Virgo Cluster have been rounded off to 60 000 000 light-years.

Some people can see many more Pleiads with the naked eye; the German astronomer E. Heis is said to have managed 19. The total membership is of the order of 1000; the core diameter is 16 light-years, and the whole diameter just over 80 light-years.

Quite a number of brown dwarfs have been identified in the cluster. The nineteenth-century J. H. Mädler, renowned for his lunar work, believed that Alcyone was the central star of the Galaxy, with all bodies rotating round it. How he reached this curious conclusion is unclear!

The distance of the Pleiades is thought to be about 440 light-years. The Hipparcos astrometric satellite gave a much smaller value, which is definitely wrong, because the stars in the cluster are close enough to show reliable trigonometrical parallax. Evidently Hipparcos is far from infallible.

Nebulosity is seen in the Pleiades, mainly near Maia. This a reflection nebula, not too easy to see visually but easy to photograph. It is now believed to be an unrelated cloud in the interstellar medium through which the Pleiades happen to be passing. The cluster will probably not be dispersed for at least 25 million years yet.

Table 23.4 *The Caldwell Catalogue*

C	NGC/IC	Constellation	Type	RA (2000.0) (h)	(m)	Dec. (°)	(′)	Magnitude	Size (′)	
1	188	Cepheus	Open cluster	00	44.4	+85	20	8.1	14	Very old cluster
2	40	Cepheus	Planetary nebula	00	13.0	+72	32	10.7	1.0 × 0.7	
3	4236	Draco	Sb galaxy	12	16.7	+69	27	9.6	23 × 8	
4	7023	Cepheus	Reflection nebula	21	01.8	+68	10	–	18 × 18	
5	IC 342	Camelopardalis	SBc galaxy	03	46.8	+68	06	9.2	18 × 17	
6	6543	Draco	Planetary nebula	17	58.6	+66	38	8.1	22 × 16	Cat's Eye Nebula
7	2403	Camelopardalis	Sc galaxy	07	36.9	+65	36	8.4	18 × 10	
8	559	Cassiopeia	Open cluster	01	29.5	+63	18	9.5	4	
9	Sh2–155	Cepheus	Bright nebula	22	56.8	+62	37	–	50 × 10	Cave Nebula
10	663	Cassiopeia	Open cluster	01	46.0	+61	15	7.1	16	
11	7635	Cassiopeia	Bright nebula	23	20.7	+61	12	8	15 × 8	Bubble Nebula
12	6946	Cepheus	SAB galaxy	20	34.8	+60	09	8.9	11 × 10	
13	457	Cassiopeia	Open cluster	01	19.1	+58	20	6.4	13	φ Cas Cluster
14	869/884	Perseus	Double cluster	02	20.0	+57	08	4.3	30 and 30	Sword-Handle
15	6826	Cygnus	Planetary nebula	19	44.8	+50	31	8.8	27 × 24	Blinking Nebula
16	7243	Lacerta	Open cluster	22	15.3	+49	53	6.4	21	
17	147	Cassiopeia	dE4 galaxy	00	33.2	+48	30	9.3	13 × 8	
18	185	Cassiopeia	dE0 galaxy	00	39.0	+48	20	9.2	12 × 10	
19	IC 5146	Cygnus	Bright nebula	21	53.5	+47	14	–	12 × 12	Cocoon Nebula
20	7000	Cygnus	Bright nebula	20	58.8	+44	33	–	175 × 110	North America Nebula
21	4449	Canes Venatici	Irregular galaxy	12	28.2	+44	06	9.4	6 × 5	
22	7662	Andromeda	Planetary nebula	23	25.9	+42	33	8.3	17 × 14	Blue Snowball Nebula
23	891	Andromeda	Sb galaxy	02	22.6	+42	21	9.9	14 × 3	
24	1275	Perseus	Seyfert galaxy	03	19.8	+41	31	11.6	3.5 × 2.5	Perseus A
25	2419	Lynx	Globular cluster	07	38.1	+38	53	10.4	4.1	
26	4244	Canes Venatici	Scd galaxy	12	17.5	+37	49	10.2	18 × 2	
27	6888	Cygnus	Bright nebula	20	12.0	+38	20	–	20 × 10	Crescent Nebula
28	752	Andromeda	Open cluster	01	57.8	+37	41	5.7	50	
29	5005	Canes Venatici	Sb galaxy	13	10.9	+37	03	9.5	6 × 3	
30	7331	Pegasus	Sb galaxy	22	37.1	+34	25	9.5	11 × 4	
31	IC 405	Auriga	Bright nebula	05	16.2	+34	16	–	37 × 19	
32	4631	Canes Venatici	Sc galaxy	12	42.1	+32	32	9.3	17 × 3	Whale Galaxy
33	6992/5	Cygnus	SN remnant	20	56.4	+31	42	–	60 × 8	Eastern Veil Nebula
34	6960	Cygnus	SN remnant	20	45.7	+30	43	–	70 × 6	Western Veil Nebula
35	4889	Coma Berenices	E4 galaxy	13	00.1	+27	59	11.4	3 × 2	Brightest in Coma Cluster
36	4559	Coma Berenices	Sc galaxy	12	36.0	+27	58	9.8	13 × 5	
37	6885	Vulpecula	Open cluster	20	12.0	+26	29	5.9	7	
38	4565	Coma Berenices	Sb galaxy	12	36.3	+25	59	9.6	16 × 2	Needle Galaxy
39	2392	Gemini	Planetary nebula	07	29.2	+20	55	8.6	47 × 43	Eskimo Nebula
40	3626	Leo	Sb galaxy	11	20.1	+18	21	10.9	3 × 2	
41	Melotte 25	Taurus	Open cluster	04	27	+16		0.5	330	Hyades
42	7006	Delphinus	Globular cluster	21	01.5	+16		0.5	330	
43	7814	Pegasus	Sb galaxy	00	03.3	+16	09	10.3	6 × 3	
44	7479	Pegasus	SBb galaxy	23	04.9	+12	19	10.9	4.4 × 3.4	
45	5248	Boötes	Sc galaxy	13	37.5	+08	53	10.2	7 × 5	
46	2261	Monoceros	Bright nebula	06	39.2	+08	44	–	2 × 1	Hubble's Variable Nebula
47	6934	Delphinus	Globular cluster	20	34.2	+07	24	2	5.9	

Table 23.4 (cont.)

C	NGC/IC	Constellation	Type	RA (2000.0) (h)	(m)	Dec. (°)	(′)	Magnitude	Size (′)	
48	2775	Cancer	Sa galaxy	09	10.3	+07	02	10.1	5 × 4	
49	2237–9	Monoceros	Bright nebula	06	32.3	+04	59	–	80 × 70	Rosette Nebula
50	2244	Monoceros	Open cluster	06	32.4	+04	52	4.8	24	
51	IC 1613	Cetus	Irregular galaxy	01	04.8	+02	07	9.2	12 × 11	
52	4697	Virgo	E4 galaxy	12	48.6	−05	48	9.3	6 × 4	
53	3115	Sextans	SO galaxy	10	05.2	−07	43	8.9	8 × 3	
54	2506	Monoceros	Open cluster	08	00.2	−10	47	7.6	7	
55	246	Cetus	Planetary nebula	21	04.2	−11	22	8.0	28 × 23	Saturn Nebula
56	246	Cetus	Planetary nebula	00	47.0	−11	53	8.6	4 × 3	
57	6822	Sagittarius	Irregular galaxy	19	44.9	−14	48	8.8	20 × 10	Barnard's Galaxy
58	2360	Canis Major	Open cluster	07	17.8	−15	37	7.2	13	
59	3242	Hydra	Planetary nebula	10	24.8	−18	38	8.6	40 × 35	Ghost of Jupiter
60	4038	Corvus	Sc galaxy	12	01.9	−18	52	10.7	2.6 × 2 ⎱	Antennæ
61	4039	Corvus	Sp galaxy	12	01.9	−18	53	10.7	2.6 × 2 ⎰	
62	247	Cetus	SAB galaxy	00	47.1	−20	46	9.1	20 × 7	
63	7293	Aquarius	Planetary nebula	22	29.6	−20	48	6.3	16 × 12	Helix Nebula
64	2362	Canis Major	Open cluster	07	18.8	−24	57	4.1	8	τ CMa Cluster
65	253	Sculptor	Scl galaxy	00	47.6	−25	17	7.1	25 × 7	Sculptor Galaxy
66	5694	Hydra	Globular cluster	14	39.6	−26	32	10.2	3.6	
67	1097	Fornax	SBb galaxy	02	46.3	−30	16	9.2	9 × 7	
68	6729	Corona Australis	Bright nebula	19	01.9	−36	58	–	1.0	R CrA Nebula
69	6302	Scorpius	Planetary nebula	17	13.7	−37	06	9.6	2 × 1	Bug Nebula
70	300	Sculptor	Scl galaxy	00	54.9	−37	41	8.1	20 × 15	
71	2477	Puppis	Open cluster	07	52.3	−38	33	5.8	27	
72	55	Sculptor	SB galaxy	00	15.1	−39	13	7.9	25 × 4	Brightest in Sculptor Group
73	1851	Columba	Globular cluster	05	14.1	−40	03	7.3	11	
74	3132	Vela	Planetary nebula	10	07.7	−40	26	8.2	1.4 × 0.9	Southern Ring Nebula
75	6124	Scorpius	Open cluster	16	25.6	−40	40	5.8	29	
76	6231	Scorpius	Open cluster	16	54.0	−41	48	2.6	15	
77	5128	Centaurus	Radio galaxy	13	25.5	−43	01	6.8	18 × 14	Centaurus A
78	6541	Corona Australis	Globular cluster	18	08.0	−43	42	6.6	13	
79	3201	Vela	Globular cluster	10	17.6	−46	25	6.7	18	
80	5139	Centaurus	Globular cluster	13	26.8	−47	29	3.6	36	ω Centauri
81	6352	Ara	Globular cluster	17	25.5	−48	25	8.1	7	
82	6193	Ara	Open cluster	16	41.3	−48	46	5.2	15	
83	4945	Centaurus	SBc galaxy	13	05.4	−49	28	8.7	20 × 4	
84	5286	Centaurus	Globular Cluster	13	46.4	−51	22	7.6	9	
85	IC 2391	Vela	Open cluster	08	40.2	−53	04	2.5	50	O Velorum Cluster
86	6397	Ara	Globular cluster	17	40.7	−53	40	5.7	26	
87	1261	Horologium	Globular cluster	03	12.3	−55	13	8.4	7	
88	5823	Circinus	Open cluster	15	05.7	−55	36	7.9	10	
89	6087	Norma	Open cluster	16	18.9	−57	54	5.4	12	S Normæ Cluster
90	2867	Carina	Planetary nebula	09	21.4	−58	19	9.7	12	
91	3532	Carina	Open cluster	11	06.4	−58	40	3.0	55	
92	3372	Carina	Bright nebula	10	43.8	−59	52	5.4	20	
93	6752	Pavo	Globular cluster	19	10.9	−59	59	5.4	20	
94	4755	Crux	Open cluster	12	53.6	−60	20	4.2	10	Jewel Box (κ Crucis)

Table 23.4 (cont.)

C	NGC/IC	Constellation	Type	RA (2000.0) (h)	(m)	Dec. (°)	(′)	Magnitude	Size (′)	
95	6025	Triangulum Australis	Open cluster	16	03.7	−60	30	5.1	12	
96	2516	Carina	Open cluster	07	58.3	−60	30	5.1	30	
97	3766	Centaurus	Open cluster	11	36.1	−61	37	5.3	12	
98	4609	Crux	Open cluster	12	42.3	−62	58	6.9	5	
99	–	Crux	Dark nebula	12	53	−63		–	420 × 300	Coal Sack
100	IC 2944	Centaurus	Cluster and nebulosity	11	36.6	−63	02	4.5	60 × 40	Collinder 249
101	6744	Pavo	SBb galaxy	19	09.8	−63	51	8.3	16 × 10	
102	IC 2602	Carina	Open cluster	10	43.2	−64	24	1.9	50	θ Carinæ Cluster 'Southern Pleiades'
103	2070	Dorado	Bright nebula	05	38.7	−69	06	–	30 × 20	Tarantula Nebula
104	362	Tucana	Globular cluster	01	03.2	−70	51	6.6	13	
105	4833	Musca	Globular cluster	12	59.6	−70	53	7.3	14	
106	104	Tucana	Globular cluster	00	24.1	−72	05	4.0	31	47 Tucanæ
107	6101	Apus	Globular cluster	16	25.8	−72	12	9.3	11	
108	4372	Musca	Globular cluster	12	25.8	−72	40	7.8	19	
109	3195	Chamæleon	Planetary nebula	10	09.5	−80	52	8.4	40 × 30	

The Hyades

This cluster is also in Taurus but quite unrelated to the Pleiades; it is not listed by Messier or Dreyer, but is numbered 25 in the Melotte catalogue and 41 in the Caldwell catalogue. It is rather drowned by the bright orange-red light of Aldebaran, but in fact Aldebaran is not a cluster member; at a distance of 65 light-years it lies roughly half-way between the Hyades and ourselves.

The main stars (Table 23.7) of the Hyades are arranged in a V shape. ε (3.5) and γ Tauri (3.6) are orange stars of type K, and there are some interesting pairs; δ (3.8) makes a pair with 64 Tauri (4.8), and θ is a naked-eye double, made up of a white star of magnitude 3.4 and a K-type orange companion of magnitude 3.8. The two are not particularly close, as the distance between them is 15 light-years, although of course both are condensed but of the same cloud which produced the rest of the Hyades. The proper motions of the Hyades stars show that they are converging to a point east of Betelgeux, at RA 6h, declination 9° N, so that in the far future the cluster will look more condensed than it does now.

The Hyades cluster is the closest of all the open clusters; its distance is given as 151 light-years. The four brightest stars lie within a few light-years of each other. The main core of the cluster is almost 18 light-years across, and the whole diameter is 65 light-years. Altogether the cluster contains between 300 and 400 stars.

Præsepe (Latin for 'Manger'; M 44 and NGC 2632)

This is another naked-eye cluster, lying in Cancer, more or less between γ Cancri (4.7) and δ (3.9). Since Præsepe is 'the Manger' γ

and δ are the Asses – Asellus Borealis and Asellus Australis respectively. Another name for Præsepe is 'the Beehive'. Flamsteed also gave it a stellar letter, ε Cancri, but this is never used.

Ptolemy called Præsepe 'a nebulous mass', but in fact it contains no nebulosity at all. It is a very loose cluster, and there are over a thousand stars of which about 68% are M-type dwarfs. Only a few members are hot blue stars, and there are some white dwarfs. The age of Præsepe is about the same as that of the Hyades, and the two clusters have similar motion in space. The core of Præsepe is about 22 light-years across, and the overall diameter is about 80 light-years.

M 67

This cluster is also in Cancer, near α. It is just visible with the naked eye. It contains more than 100 solar-type stars and many red giants. It is around 2700 light-years away, and is the oldest known open cluster, with an age of perhaps 4000 million years. It has been able to preserve its identity because it is a long way above the galactic plane, and there are not many passing stars to disrupt it. It was discovered by Kohler between 1772 and 1779.

The Wild Duck Cluster (M 11, NGC 6705)

This is a particularly rich and compact cluster, discovered by Gottfried Kirch in 1681. It lies in Scutum, near the 'tail' of the Eagle (Aquila), and has about 3000 members; age 220 million years. Three of its brightest stars form a triangle, and this led the ninth century observer Admiral W. H. Smyth to give it the familiar nickname: 'This object, which somewhat resembles a flight of wild

Table 23.5 *Some bright star clusters*

M	NGC	C	Constellation	Magnitude	
–	752	28	Andromeda	5.7	Open cluster
2	7089	–	Aquarius	6.5	Globular cluster
–	6193	82	Ara	5.2	Open cluster
36	1960	–	Auriga	6.0	Open cluster
37	2099	–	Auriga	5.6	Open cluster
38	1912	–	Auriga	6.4	Open cluster: Præsepe
44	2632	–	Cancer	3.1	Open cluster: Præsepe
67	2682	–	Cancer	6.9	Open cluster
41	2287	–	Canis Major	4.5	Open cluster
–	2362	64	Canis Major	4.1	Open cluster: τ Canis Majoris
–	2516	96	Carina	3.8	Open cluster
–	IC 2602	102	Carina	1.9	Open cluster: θ Carinæ
–	457	13	Cassiopeia	6.4	Open cluster: φ Cassiopeiæ
–	5139	80	Centaurus	3.6	Great globular: ω Centauri
–	4755	94	Crux	4.2	Open cluster: Jewel Box (κ Crucis)
–	6394	47	Delphinus	5.9	Globular cluster
35	2168	–	Gemini	5.0	Open cluster
13	6205	–	Hercules	5.9	Globular cluster
92	6341	–	Hercules	6.5	Global cluster
–	7243	16	Lacerta	6.4	Open cluster
–	2244	50	Monoceros	4.8	Open cluster in Rosette Nebula
50	2323	–	Monoceros	5.9	Open cluster
–	6752	93	Pavo	5.4	Globular cluster
–	869/884	14	Perseus	4.3, 4.4	Sword-Handle: two open clusters
47	2422	–	Puppis	4.4	Open cluster
25	IC 4725	–	Sagittarius	4.6	Open cluster: υ Sagittarii
23	6494	–	Sagittariuis	5.5	Open cluster
6	6405	–	Scorpius	4.2	Open cluster: Butterfly Cluster
7	6475	–	Scorpius	3.3	Open cluster: Ptolemy's Cluster
11	6705	–	Scutum	5.8	Open cluster: Wild Duck Cluster
45	1432/5	–	Taurus	1.2	Open cluster: Pleiades
–	–	41	Taurus	1	Open cluster: Hyades (Melotte 25)
–	6025	95	Triangulum Australis	5.1	Open cluster
–	104	106	Tucana	4.0	Globular cluster: 47 Tucanæ
–	IC 2391	85	Vela	2.5	Open cluster: o Velorum

Table 23.6 *Brightest stars in the Pleiades*

Name	Flamsteed	Magnitude	Spectrum	
Alcyone	25	2.86	B7IIIe	η Tauri
Atlas	27	3.62	B8III	
Electra	17	3.70	B6IIIe	
Maia	20	3.86	B7III	
Merope	23	4.17	B61ev	
Taygete	19	4.29	B6V	
Pleione	28	5.1v	B81Vep	BU Tauri
Celæno	19	5.44	B71V	
Asterope	21 and 22	5.64 and 6.41	B8Ve and B9V	
	18	5.65	B8V	

ducks in shape, is a gathering of minute stars.' Its integrated magnitude is only just below 6. But care must be taken in identifying it, because it lies in a rich part of the Milky Way.

The Jewel Box (NGC 4755, C 94)

This cluster lies in the Southern Cross. European observers never cease to regret that this superb cluster lies so far south in the sky. It surrounds the bright orange supergiant κ Crucis. It is very young for a galactic cluster, no more than about 7 million years old, and is 640 years away. It contains around 100 stars, most of which are hot and blue, contrasting beautifully with the orange glow of κ. Sir John Herschel described the cluster as 'a casket of variously coloured precious stones'. It is easily seen with the naked eye, close to β Crucis, and binoculars give a splendid view of it.

Table 23.7 *Brightest stars in the Hyades*

Bayer	Flamsteed	RA (h)	(m)	(s)	Dec. (°)	(')	('')	Magnitude	Spectrum	Name
θ^2	78	04	28	40	+15	52	15	3.4	A7	
ϵ	74	04	28	37	+19	10	50	3.5	K0	Ain
γ	54	04	19	48	+15	37	40	3.6	G8	Hyadum Primus
δ	61	04	22	56	+17	32	33	3.8	G8	
θ	77	04	28	34	+15	57	44	3.8	G7	Phaeo
κ	65	04	25	22	+22	17	38	4.2	A7	
C^2	90	04	38	09	+12	30	39	4.3	A6	
υ	69	04	26	18	+22	18	49	4.3	A8	
δ^3	68	04	25	29	+17	55	41	4.3	A2	Kleeia
	71	04	26	21	+15	37	06	4.5	F0	Polyxo
ι	102	05	03	06	+21	35	24	4.6	A7	
ρ	86	04	35	51	+41	50	40	4.6v	A8	
σ^2	92	04	39	19	+15	55	05	4.7	A5	
δ^2	64	02	24	06	+17	26	39	4.8	A7	
ψ	79	04	28	50	+13	02	52	5.0	A7	
σ^1	91	04	39	09	+15	48	00	5.1	A4	
	97	04	51	22	+18	50	24	5.1	A7	
	58	04	50	36	+15	05	44	5.3	F0	
κ^2	67	04	25	25	+22	12	00	5.3	A7	

Stellar associations

A stellar association (also termed a moving cluster) is so loose that its members are so widely scattered that the cluster is not readily identifiable as such. They are produced by the disruption of an open cluster; many of the stars originally in it will continue to move through space at similar speeds and trajectories, as was first realised in 1947 by the Armenian astronomer Viktor Ambartsumian. The nearest example, the Ursa Major moving cluster, includes five of the stars in the Plough, as well as Sirius and various other members. Spectral-type OB associations contain up to 100 very massive stars, and are short-lived; the nearest example is the Scorpius–Centaurus association, about 400 light-years away. T associations can contain young T Tauri stars while R associations illuminate reflection nebulæ.

GLOBULAR CLUSTERS

Globular clusters are quite different from open clusters. They are vast symmetrical systems, containing from 10 000 to at least a million stars, and so strongly condensed that near their centres the individual stars may be only light-months apart. An average globular cluster has a diameter of 65 light-years, but there is a wide range in both size and mass. About 160 globular clusters are known in our Galaxy, but other galaxies have their own systems of clusters.

Our Galaxy is a flattened system, and is rotating around its centre. The Sun takes about 225 000 000 years to complete one orbit and lies not far from the main plane. The globular clusters, on the other hand, are members of what is termed the galactic halo, which is roughly spherical and extends out to several hundreds of thousands of light-years. The globulars move in highly eccentric orbits which take them far outside the main Galaxy, and they do not share in the disc rotation; this means that their velocities relative to the Sun may be very high (up to 100 km s^{-1}). One globular, NGC 2419 (C 25), in Lynx, seems to be escaping from the Galaxy altogether, and has been called an Intergalactic Tramp!

Globular clusters are very ancient, and are indeed as old as the Galaxy itself. They are very poor in heavy elements, since they were formed from primitive material, whereas the open clusters were formed from 'recycled' material. The globulars are much more stable than the open clusters, and though they are subjected to disruptive forces (such as tidal effects from the parent Galaxy) they are very long-lived.

Huge though they are, most globulars are so remote that they appear faint, and only three are clearly visible with the naked-eye. They contain short-period variables (RR Lyræ stars), which enables their distances to be measured. This was first done in 1917 by Harlow Shapley, who also realised that most of the globulars lie in the southern sky; there is a heavy concentration in the area of Scorpius, Ophiuchus and Sagittarius. From this Shapley deduced, quite correctly, that we are having a lop-sided view, because the Sun is away from the centre of the Galaxy.

ω Centauri (NGC 5139, C 80) is much the brightest and largest of the globulars, and even from a range of 16 000 light-years is a prominent naked-eye object. Ptolemy recorded it as a star, and Bayer gave it a Greek letter, but it was Halley who first recognised its true nature. A small telescope will resolve its outer parts into stars, though near its centre its star density is very high. Its mass, estimated at least five million times that of the Sun, matches that of a small galaxy, and there have been suugestions that it is the remnant of a small galaxy that was torn apart and absorbed into the Milky Way system. This is unproved, but is supported by some recent observations made with the Hubble Space Telescope and the Gemini South Telescope in Chile. Movements of stars within ω Centauri seem to indicate the presence of black hole.

47 Tucanæ (NGC 104, C 106) is smaller than ω Centauri, but in a telescope is probably more striking, because all of it can be contained in the same field of view, whereas ω Centauri is too large. It is almost silhouetted against the Small Magellanic Cloud, but of course the two are in no way associated; 47 Tucanæ is 16 700 light-years away, while the Cloud is a satellite galaxy at a distance not much less than 200 000 light-years.

47 Tucanæ has a diameter of about 120 light-years, and contains around two million stars, plus over twenty known millisecond pulsars. The estimated age is ~10 000 million years. It was actually given a Greek letter ξ Tucanæ but this is never used.

In the northern hemisphere of the sky, the only naked-eye globular is M 13 (NGC 6205), between ζ and η Herculis. It was discovered by Halley in 1714, and seems to be further away than ω Centauri and 47 Tucanæ. One estimate gives its distance as 25 000 light-years. The diameter is about 140 light-years. The outer regions are fairly easy to resolve, but near the centre the star density is very high.

In 1974 a message was sent to it by using the Arecibo radio telescope in Puerto Rico. If any radio replies, we may expect an answer in the year AD 50 197 (remember to listen out for it).

Because globular clusters are so old, their leading stars are highly evolved red giants. There are, however, some stars which are hot and blue. A 'blue straggler' of this kind seems to be a rejuvenated star – either the result of a stellar collision, or the merger of the two components of a binary pair. 47 Tucanæ is particularly rich in blue stragglers.

The view from a planet orbiting a star near the centre of a globular cluster would indeed be fascinating. There would be many stars bright enough to cast shadows, and many of these would be red; there would be no darkness at night. Against this, an observer on such a world would have a very restricted view of the outer universe.

There has been a recent suggestion (2010) by a team of Australian astronomers led by Duncan Forbes (Swinburne University) that some of the Milky Way's globular clusters may be 'invaders' – born elsewhere in our Galaxy. They also suggest that the Milky Way may have accreted more dwarf galaxies then has been previously believed, so that when the dwarfs were broken up and their stars assimilated into the Milky Way their globular clusters remained intact. This is certainly an interesting theory, but much more research is required to decide whether or not it is valid.

ENDNOTE

1 Why C? Well, I have to admit that I compiled the catalogue very light-heartedly after I had finished my night's observation of Jupiter, and had amused myself by looking at some clusters and nebulæ. I wanted to allot numbers, and for obvious reasons I could not use M, so I chose C – because my surname is actually double-barrelled: Caldwell-Moore. This proved to be a mistake, as I realised when a somewhat sour writer in the *BAA Journal*, presumably irritated because the idea was not his, accused me of trying to publicise my own name! No comment.

24 · Nebulæ

Nebulæ are gas-clouds. They are of very different types. Bright nebulæ, such as M 42 in the Sword of Orion, are stellar nurseries, while planetary nebulæ are dying stars, and there are also supernova remnants, such as the Crab Nebula in Taurus. The galaxies were once called 'spiral nebulæ', but this name for them is now obsolete.

PLANETARY NEBULÆ

Planetary nebulæ are very inappropriately named. They were so called by William Herschel because their pale, often greenish discs make them look superficially not unlike Uranus or Neptune when seen through a small telescope, but they are not true nebulæ, and have absolutely nothing to do with planets.

A proto-planetary nebula marks the brief period in a star's history when it has left the asymptotic giant branch of the Hertzsprung–Russell (HR) diagram but has not reached the planetary nebula stage – it has been 'caught in the act', so to speak. These nebulæ are faint and rare; a selected list is given in Table 24.1.

(Do not confuse a proto-planetary nebula with a proto-planetary disc. 'Preplanetary nebula' has been proposed as an alternative name, and this would be a distinct improvement.)

The Red Rectangle is one of the most notable of the proto-planetaries. It takes the form of a symmetric, bipolar nebula with X-shaped spikes; the central star is a close binary, apparently surrounded by a thick dust torus which forces the otherwise spherical outflow into tip-touching cone shapes. We see the torus edgewise-on, so that the boundary edges of the cone shape seem to form an X. NGC 2359, in Canis Major, is known as Thor's Helmet, and its shape is certainly unusual; it has a bubble-like structure, about 30 light-years across, blown by extremely energetic winds from the hot Wolf–Rayet star at the centre (in some catalogues it is classed as an ordinary emission nebula rather than a planetary). The Calabash Nebula in Puppis has also been called the Rotten Egg Nebula because it contains a great deal of sulphur; the densest parts are made up of material ejected from the central star in opposite directions about 800 years ago at speeds up to 150 000 km h^{-1}. In a thousand years time it will probably become a fully developed planetary nebula. Do not confuse it with CRL 2688, the Egg Nebula in Cygnus, where there are bright arcs and circles surrounding the central star; the diameter of the nebula is almost half a light-year.

The Boomerang Nebula in Centaurus was formed from the outflow of gas from the central star; the gas is moving outwards at around 165 km s^{-1} and is expanding rapidly, and this lowers the temperature to only about one degree above absolute zero. It has been claimed that the Boomerang is the coldest known place in the entire universe. Another decidedly chilly place is the Frosty Leo Nebula, where there are two lobes separated by 2 seconds of arc; between the lobes lies an almost edgewise-on dust ring. The molecular envelope is expanding at a rate of ~25 km s^{-1} – hence the very low temperature. The long-wavelength emission spectrum is dominated by crystalline ice.

The proto-planetary stage does not last for long on the cosmic timescale, and neither do fully-developed planetary nebulæ. The dying star sheds outer layers, and what remains of the star is very hot, with a surface temperature from 20 000 to as much as 100 000 °C. It is in the process of turning into a white dwarf, and its ultraviolet radiation ionises ejected material and makes it glow. The common greenish tint is due to emissions from doubly ionised oxygen.

Planetary nebulæ cannot persist, as such, for more than about 50 000 years before the material dissipates and becomes too dim to be seen. Roughly a thousand are known in our Galaxy, but no doubt there are many more which are too faint to be detected. A selected list of planetaries given in Table 24.2.

The best-known planetary is M 57, the Ring Nebula in Lyra, which is particularly easy to locate because it lies directly between β and γ Lyræ. It can be glimpsed with good binoculars, and a telescope as modest as my 7.5-cm refractor will show the ring, while a 20-cm telescope will also reveal the central star. The ejected material forms a sphere, but we see it from long range (about 2300 light-years) and the eye observes more material at the periphery than at the centre – hence the appearance of a luminous ring; at the centre we can see 'straight through'. The material is indeed tenuous. If you take the amount of gas in an average village hall and spread it out over an area equivalent to the entire volume of the Earth, the resulting density will be about the same as that of the gas in a planetary nebula.

The first planetary to be discovered – by Messier, in 1764 – was M 27, the Dumbbell Nebula in Vulpecula, an easy binocular object which really does merit its name; the diameter is 3 light-years. In Aquarius we find two notable planetaries, the Saturn Nebula (NGC 7009) and the Helix Nebula (NGC 7293); NGC 7009 was given its name by Lord Rosse, who observed it when he had completed his 72-inch reflector in 1845 and commented that it resembled the appearance of Saturn with the rings edge-on. The Helix, sometimes called the Eye of God, is no more than 700 light-years away – the closest of the planetaries – and is thought to be about 11 000 years old in its present guise. The Owl Nebula, M 97 in Ursa Major, is so called because of 'owl-like' eyes that are seen through a powerful telescope; the Hourglass Nebula in Musca is formed by the fast stellar wind inside a slowly expanding cloud of material; the Ghost

Table 24.1 *Selected list of proto-planetary nebulæ*

Name	Constellation	RA (h)	(m)	(s)	Dec. (°)	(′)	(″)	Magnitude	Distance
Red Rectangle	Mon	06	19	58.2	−10	38	14.7	9.0	2.3 HD 44179
Thor's Helmet	CMa	07	17	36 −13 12 0			12	15	NGC 2359
Calabash	Pup	07	42	16.8 −14 42 52.1			9.5	4.2	OH231.84+4.22
Frosty Leo	Leo	09	39	54.0 +11 58 52.4			11	3	IRAS 09371+1212
Boomerang	Cen	12	44	45.5 −54 31 11.4				5	PGC 3074547
Egg	Cyg	21	02	18.8 + 36 41 37.8			14	3	RAGL 2688

Others include the Cotton Candy Nebula (IRAS 17150–3224) and the Water Lily Nebula in Ara (IRAS 16594–4656). Distances, in thousands of light-years, are approximate only in most cases.

Table 24.2 *Selected list of planetary nebulæ*

M2–9	17 05 38	−10 08 35	14.7	115 × 18″	2100	Oph	Twin Jet Nebula
M1–92	19 36 18	+29 33 00	11.7	18 × 16″	~15 000?	Cyg	Foot-print
NGC 2346	07 09 41	−00 48 56	11.6	0.9″	2000	Mon	Butterfly Nebula
NGC 3132	10 07 02	−40 26 12	9.9	1.4 × 0.9′	2000	Vel	Southern Ring
NGC 2438	07 42	−14 44	10.8 1.1′		2900	Pup	
IC 4406	14 22 26	−44 09 04	–	30″	2000	Lup	Retina Nebula
NGC 2440	07 41 55	−18 13 00	9.4	74 × 42″	4000	Pup	
NGC 5189	13 33 33	−65 58 27	–	3 × 2	3000	Mus	Spiral Planetary
Menzel 1	15 34 17	−59 09 09	12.0	76 × 23″	3400	Nor	
Menzel 3	16 17 13	−51 59 10	13.8	50 × 12″	8000	Nor	Ant Nebula
Shapley 1	15 51 43	−51 31 31	12.6		1000	Nor	
MyCn 18	13 39 35	−67 22 51	13.0		8000	Mus	Hourglass Nebula
NGC 3918	11 50 18	−57 10 57	8.5	9″	4900	Cen	Blue Planetary
Hen 3-1357	17 16 21	−5 29 24	10.7				Stingray Nebula
NGC 3242, C 39	10 24 46	−18 38 33	8.6		1400	Hya	Ghost of Jupiter
NGC 7662	23 25 54	+42 32 06	9		5000	And	Blue Snowball

of Jupiter, NGC 3242 in Hydra, shows an oval disc with a 12th-magnitude central star; the Cat's-Eye Nebula in Draco has a very complicated structure with jets, knots, and sinewy arc-like features; the central star of the Red Spider Nebula in Sagittarius is one of the hottest white dwarfs known, sending out a powerful wind which produces the spider-like waves in the nebular material; the Blue Planetary in Centaurus, discovered by John Herschel in 1834, is of 'a beautiful rich blue colour' markedly different from the pale green of most nebulæ of this class. The Eskimo Nebula in Gemini, discovered in 1787 by William Herschel, is also nicknamed the Clown Face Nebula; looking at photographs taken with adequate telescopes, it is easy to understand why! Other nebulæ with suitable nicknames are the Bug (N0 6302) and the Ant (Menzel 3). The Blinking Planetary in Cygnus, NGC 682 does not blink. When seen through a fairly small telescope the brightness of the central star masks the surrounding nebula, but the nebulosity can be seen by the use of averted vision, making the star 'blink' in and out of view as the observer's eye wanders. Details of these nebulæ are in the Messier or Caldwell catalogues in Chapter 23.

DIFFUSE NEBULÆ

Diffuse nebulæ, otherwise known as galactic or gaseous nebulæ, are among the most beautiful objects in the sky. Photographs bring out their vivid colours and bizarre forms, and even when seen through binoculars or small telescopes they are fascinating. Table 24.3 lists diffuse nebulæ.

A new nebula appeared in January 2004. It was discovered by a Kentucky amateur, J. McNeil, with a 3-inch refractor; it appeared in M 78, a reflection nebula in Orion, and surrounds a newborn star. It does not appear on images before September 2003, and is now known as McNeil's Nebula.

Diffuse nebulæ are of three main types:

1. *Emission nebulæ*: These are produced by energetic short-wave radiation from very hot stars (types O and B). Absorption of this radiation causes the gases making up the nebular material to emit visible light at certain specific wavelengths, and the result is an emission spectrum. The ultraviolet radiation ionises the hydrogen in the nebulosity and makes it glow. Because hydrogen

Table 24.3 *Selected list of diffuse nebulæ*

	Name	Constellation	RA (h)	(m)	(s)	Dec. (°)	(′)	(″)	Magnitude	Diameter (light-years)	Distance (light-years)
M 42, NGC 1976	Orion Nebula	Ori	05	37	17	−05	23	28	4	24	1270
M 8, NGC 6523	Lagoon Nebula	Sgr	18	03	37	−24	23	12	6	110 × 40	4100
M 17, NGC 6618	Omega Nebula	Sgr	18	20	23	−16	10	36	6		~5000
M 16, NGC 6611	Eagle Nebula	Ser	18	18	48	−13	49		6	140 × 110	7000
M 20, NGC 6514	Trifid Nebula	Sgr	18	02	23	−23	01	48	6.3	–	~4000
30 Dor, NGC 2070.	Tarantula	Dor	05	38	58	−69	05	42	8	40 × 25	170 000
η Car, NGC 3372	η Carinæ	Car	10	45	09	−59	52	04	1	~300	~7500
C 20, NGC 7000	North America Nebula	Cyg	20	59	17	+44	31	44	~6	–	~1800

is the major constituent of these glowing clouds, and the symbol H II is used to denote ionised hydrogen, emission nebulæ are also termed H II regions.

2. *Reflection nebulæ*: these shine only by light reflected from stars either in or very close to the nebulosity. The light is reflected from particles of dust.

3. *Dark or absorption nebulæ*: there are no suitable stars nearby, and the nebulosity remains dark, detectable only because it blots out the light of stars beyond. The most famous dark nebula is the Coal Sack in the Southern Cross (C 90). There is no basic difference between bright and dark nebulæ; for all we know there may be stars illuminating the far side of the Coal Sack, so that seen from a different vantage point in the universe it would appear bright.

The most famous nebula is M 42 in the Sword of Orion, easily visible with the naked eye. It is actually the most prominent part of a Great Molecular Cloud which covers most of Orion. It is approximately 1300 light-years away, and 24 light-years across. The material in it is almost incredibly rarefied. If you could take a 3-cm core sample right through it, the total mass of the material collected would just about balance a pound coin.

The nebula is illuminated by the Trapezium Cluster, of which the brightest members, known collectively as θ Orionis, make up the famous Trapezium pattern, which is easily seen with a 7-cm telescope. Two of the Trapezium stars are eclipsing binaries, and two of them have variable star designations; A is V1016 Orionis, and B is BM Orionis. All four are very young – only about a million years old – and very hot. The whole Orion Nebula cluster contains about 1000 stars. Inside the nebula are proto-planetary discs, brown dwarfs and Bok globules. There are also Herbig–Haro objects, named after their co-discoverers, G. Herbig, of the USA and G. Haro of Mexico; they are formed when jets of material sent out from very young stars strike existing material and compress it. In 2009, a team led by S. Kraus and G. Weigelt (Bonn, Germany) used the VLTI (Very Large Telescope Interferometer) to obtain a very sharp image of θ Orionis C, in the Trapezium. The separation of the two stars is 2 milliarcseconds, corresponding to the apparent

size of a car on the surface of the Moon. The masses of the two components are respectively 38 and 9 times that of the Sun. The distance from the Sun is 350 light-years; this is the nearest region where massive stars are born.

When a proto-star blows away its opaque cocoon cover and becomes visible at optical wavelengths, it goes through the T Tauri stage (named after the first star of this type to be discovered). Classical T Tauri stars are from 1 to 10 million years old, and vary irregularly, because they are unstable and have yet to join the Main Sequence. They send out violent stellar winds, and are usually accompanied by dark discs of material known as proplyds (proto-planetary discs), many of which have been found inside the Orion Nebula. Contraction to the Main Sequence depends upon the mass of the star – perhaps 100 million years for a star of low mass, 40 million years for a star of solar mass and as little as one million years for a star of much greater mass.

T Tauri stars are characterised by strong emission lines in their spectra, produced by interactions between the star and its disc. Once the disc dissipates, these lines vanish, but 'weak-lined' T Tauri stars produce X-rays in the hot plasma trapped in the magnetic field above the star's surface; these X-rays are very evident in observations made from space vehicles such as Chandra and Newton. Probably the dark material collects into planetesimals, which in turn collide and produce true planets.

Two objects deep inside the nebula, detectable only in infrared, are known as BN and KL (named after its discoverers, Kleinmann and Low) and may be an early cluster of proto-stars. The BN object was discovered in 1967 by Eric Becklin and Gerry Neugebauer, and seems to be a star in the early stage of its evolution; at wavelengths of less than 10 cm it is the brightest object in the sky. It is about the same size as the orbit of Neptune round the Sun, and embedded in its core is thought to be a very powerful star, at least 15 times as massive as the Sun, with a surface temperature of over 25 000 °C. We will never see BN at optical wavelengths; it is forever hidden.

Proplyds have also been detected inside the Lagoon Nebula in Sagittarius, which is much larger than M 42 but is much further away; there are also Bok globules, Herbig–Haro objects, and a

Table 24.4 *Selected list of dark nebulæ*

	Constellation	RA (h)	(m)	(s)	Dec. (°)	(′)	(″)	Dimensions (′)	Distance (thousands of light-years)
Barnard 33, Horse's Head	Ori	05	40	59	−02	27	30	8 × 6	1500
Barnard 142/3, E Nebula	Aql	19	40	42	+10	57		30	2000
Barnard 72, Snake Nebula	Oph	17	23	30	−23	38		4	650
Barnard 59, 65–67, Pipe Nebula	Oph	17	27	−26	56			stem 300 × 60	~650
								bowl 200 × 140	
NGC 3324, Keyhole Nebula	Car	10	45	09	−59	52	04	9000	
C 99, CoalSack	Cru	12	50	−62	30			7 × 5°	600

complex structure shaped like an hourglass (do not confuse it with the Hourglass Nebula in Musca). The Eagle Nebula in Serpens, M 16, is truly magnificent; a photograph taken with the Hubble Space Telescope shows three columns of gas and dust, each 1 to 3 light-years long, inside which stars are forming. Near the tops of the pillars, the radiation from the fledgling stars is evaporating the gas, so that it will eventually be dissipated: these structures are known as evaporating gaseous globules (EGGs). When the star-forming material in a nebula is exhausted, all that is left is an open star cluster.

Another interesting nebula is the North America Nebula NGC 7000 (C 20). Its shape really does recall the outline of the North American continent, and there is darker material in the position of the 'Gulf of Mexico'! Many dark rifts are seen in the Milky Way, notably in the region of Cygnus, and also in other nebulæ, notably the Trifid Nebula in Sagittarius (M 20).

Note also M17, the Omega or Swan Nebula in Sagittarius, an H_2 region discovered by P. L de Chéseaux as long ago as 1745. (An H_2 region is a cloud of glowing gas, many light-years across, in which star formation is taking place.) The cloud containing M17 is about 40 light-years in diameter. In the far south there is the complex nebula round the unique variable η Carinæ. All these are dwarfed by the immense Tarantula Nebula in the Large Magellanic Cloud. If it were as close to us as the Orion Nebula, it would cast shadows. Supernova 1987a, the only naked-eye supernova of the twentieth century, blazed up in the Tarantula's outer region.

DARK NEBULÆ

Dark nebulæ are exactly the same as diffuse nebulæ except that, to us, they are not lit up or excited by suitable stars. Obscuring matter is not uncommon; look for example at the 'rifts' in the Milky Way in Cygnus. William Herschel was wrong in calling them 'holes in the heavens'

A selected list of dark nebulæ is given in Table 24.4.

The most famous example is the Coal Sack in the Southern Cross (Caldwell 99). It extends into Centaurus and Musca, and is easily visible with the naked eye; in Aboriginal legend it was the head of 'the Emu in the Sky'. It is not totally black, but glows with about 10% of the brightness of the average Milky Way. The Snake Nebula in Ophiuchus wriggles in front of the Milky Way star-clouds extending from the north-northwestern edge of the bowl of the Pipe Nebula; it is between 2′ and 3′ wide. The E Nebula in Aquila has an obvious shape, about the size of the full moon. Finally, the Horse's Head Nebula in Orion, not far from Alnitak (ζ Orionis) in the Hunter's Belt, really does resemble the head a knight in chess. It is a photographic favourite, but strangely elusive visually.

25 · The Milky Way Galaxy

The Sun is a member of a system of stars known as the Galaxy. The system is popularly called the Milky Way, but this name is more properly restricted to the luminous band stretching across the sky. The name is the translation of the Latin *Via Lactea*; the Greek name was 'Galaxies'.

The Milky Way has been known since very ancient times – on a dark, clear night it cannot be overlooked – and there are many legends about it. In Chinese tales there are two lovers, Chuc Nu and Nguu Lang, at the court of the Jade Emperor, who neglected their official duties and were banished to the sky as the stars Vega and Altair, on opposite sides of the Milky Way river. They are allowed to meet once a year, on the seventh day of the seventh month, when flocks of birds form a bridge across the river to enable the lovers to cross. In Finland, birds are again involved; it was thought that migratory birds used the Milky Way as a guideline to their home, Lintukoto – and there may be some truth in this! To the people of the Kalahari Desert the Milky Way was due to embers from a fire lit by a girl who wanted to find her way around on a starless night. The Aborigines of South Australia had a different idea; the Milky Way was Woodlipari, a celestial river where there are caves in which live dangerous creatures called *yura* (the caves themselves are *yurakauwe*, 'homes of monsters'). Different again is a legend from North America; a dog stole some cornmeal and was chased away, scattering some corn to create the Milky Way – *Gili Ulisvsdanvyl*, 'the Dog that Ran Away'.

It seems that the Greek philosophers Anaxagoras (~500–428 BC) and Democritus (450–370 BC) were the first to propose that the Milky Way was made up of stars, but definite proof was not obtained until 1610, when Galileo made the first telescopic observations. In 1750, Thomas Wright explained the Milky Way as 'an optical effect due to our immersion in what locally approximates to a flat layer of stars', and he was certainly thinking along the right lines. In 1785, William Herschel gave the first reasonably accurate picture of the shape of the Galaxy; he described it as flattened, so that when we look along the main plane we see many stars in the same direction. This explains the Milky Way.

PATH OF THE MILKY WAY

Going westward from the centre in Sagittarius, the Milky Way runs through: Sco, Ara, TrA, Circ, Cen, Mus, Cru, Car, Vel, Pup, CMa, Mon, Ori & Gem, Tau, Aur, Per, And, Cas, Cep & Lac, Cyg, Vul, Sge, Aql, Oph, Set and back to Sgr – a total of 29 constellations.

The north galactic pole lies near β Comæ Beenice's, and the south pole near α Sculptoris.

AGE AND FORM

The Galaxy must have been formed soon after the universe itself was created in the Big Bang, 13.7 thousand million years ago. The oldest star known is HE 1523–0901: age 13.4 thousand million years. It is a metal-poor star containing radioactive uranium and thorium. These steadily decay, so that the star's age can be found by measuring the amounts remaining.

There is considerable uncertainty about the number of stars in the Galaxy. One estimate gives a total of 100 000 million, but others increase this by a factor of two or even four. We can be more positive about the shape; the Galaxy is a barred spiral of type SBc. The main disc is 100 000 light-years in diameter, with an average thickness of 1000 light-years, and there is a spherical central bulge with a diameter of 15 000 light-years. The central bar is 27 000 light-years long, inclined about 44° to the line joining the Sun to the galactic centre; the Sun's distance from the centre is given as 26 000 light-years, though with slight uncertainty. Reckoning outward from the centre, the arms are the Norma, Scutum-Crux (or Centaurus), Sagittarius, Orion, Perseus and Cygnus. The Sun lies near the edge of the Orion Arm (which is not really a major arm, but a rich region between the Sagittarius and Peruses arms). Surrounding the main system is the galactic halo, which may extend as far as the Magellanic Clouds, the largest of the Galaxy's satellite systems. The halo is made up of isolated stars as well as the globular clusters.

The disc is rotating round the centre of the Galaxy, in the bulge; the Sun takes 225 million years to complete one circuit – this is often called the 'cosmic year'. Halo objects move in very inclined orbits, and do not share in the disc rotation, so that relative to the Sun they seem to be travelling at high velocity. Mention must be made of Gould's Belt, an incomplete ring of bright, luminous stars, many of them of types O and B. It is about 3000 light-years across, tilted to the galactic plane by 15° to 20°; its origin is unknown. Attention was first drawn to it by Benjamin Gould in 1879. There is also Smith's Cloud, a high-velocity cloud between 32 000 and 49 000 light-years away, and 8000 light-years from the disc of our Galaxy.

As well as stars, the Galaxy contains a vast amount of thinly spread interstellar matter. The first proof of this was obtained by W. Hartmann in 1904, when he was studying the spectrum of the star δ Orionis. δ Orionis is a spectroscopic binary, so that its lines show Doppler shifts corresponding to the orbital motions of the components. However, Hartmann found that some of the lines remained stationary, so that clearly they were associated not with the star itself but with material lying between δ Orionis and ourselves. Further proof was obtained by J. Trümpler in 1930, when some of the Milky Way clusters were reddened and appeared fainter than logically they ought to have done. So they were being dimmed by intervening material. The gas between the stars is made up of hydrogen and helium, with smaller amounts of carbon, nitrogen, oxygen and neon. In the 1930s, astronomers at Mt Wilson found the first indications of interstellar molecules, but firm proof was postponed until 1963, when the hydroxyl radical, OH, was identified.

Many other molecules, organic as well as inorganic, have since been identified – for example ethyl alcohol (C_2H_5OH). One 'cloud' contains enough of it to fill the entire globe of the Earth with alcohol, or make 10^{28} bottles of whisky. More molecules are being detected yearly. There are also dust grains, about 0.1 μm (10^{-7} m) min radius, which are chiefly responsible for the reddening and dimming of distant objects. Quite apart from all this we have to reckon with dark matter, to be discussed below. Its total mass far exceeds the total mass of the stars, but about its nature we know absolutely nothing. The average density of the interstellar gas in the Galaxy is very low – about 500 000 hydrogen atoms per cubic metre, which corresponds to a very good laboratory vacuum.

Originally it was believed that the Sun lay in the centre of the Galaxy. This was assumed by Herschel, and also by the Dutch astronomer J. C. Kapteyn, who carried out a detailed investigation in the early part of the twentieth century. However, in 1917, H. Shapley proved that this is not so. He realised that the globular clusters are not spread symmetrically around the sky, but are concentrated in the southern hemisphere; using short-period variable stars he measured the distances of the clusters, and established that the Sun lies well away from the centre, although very close to the main plane. This, of course, explains the Milky Way band: we are looking along the main plane of the Galaxy, and seeing many stars in much the same line of sight. The centre of the galaxy lies at RA 17h 45m, Dec. −28′ 66″, in Sagittarius; the galactic pole is at RA 12h 51m .4, Dec. −27° 07′, in the star-poor region of Sculptor. Shapley gave the diameter of the Galaxy as 30 000 light-years, although this proved to be very much of an underestimate.

It is now believed that individual spiral arms are not permanent features. Density waves sweep around the Galaxy, and produce what may be called cosmic traffic jams compressing the material and triggering off star formation. Very luminous O and B stars develop, but do not shine as such for long enough to leave the spiral arms, so that these OB associations characterise the arms. In the future our Galaxy will still have spiral arms – but they will not be the same as those of the present time.

THE GALACTIC CENTRE

The centre of the Galaxy, on the far side of the lovely star-clouds in Sagittarius, cannot be studied at visible wavelengths, or in ultraviolet; there is too much dust in the way. We have to turn to gamma-ray, hard X-ray, infrared, submillimetre and radio wavelengths.

The region of the centre is decidedly crowded. In 1997, two huge star clusters, no more than 100 light-years from the centre, were identified by the NICMOS (Near Infra-red Camera and Multiprobe Spectrometer) camera on the Hubble Space Telescope. The Arches Cluster, between 2 and 4 million years old, is the densest of all known clusters in the Galaxy; it is packed with very young and hot stars far more luminous than the Sun. The Quintuplet is slightly more dispersed and seems to be older than the Arches. It contains the Pistol Star, one of the most powerful stars known, emitting as much energy in six seconds as the Sun does in a year. Its nickname comes from the shape of the nebula that it illuminates. Within a million years or so the Pistol Star is likely to explode as a supernova. It must, however, be remembered that we can never have a direct optical view of it, and it is always possible that we are dealing with a closely packed group of stars rather than a single giant sun.

Round the galactic centre comes what is termed the circumnuclear disc (CND), a torus of warm atomic molecular gas and dust. It is rotating, and extends from about 1.5 to 8 parsec from the centre. The centre is located in Sagittarius A, made up of three components. Sagittarius A East is thought to be a rather unusual supernova remnant, 25 light-years across. Sagittarius A West is composed of gas and dust clouds, which give the impression of a spiral, hence its misleading nickname, 'the Minispiral'. It is orbiting the true centre, Sagittarius A* (pronounced Sagittarius A-star), a supermassive black hole, about 44 000 000 km in diameter and as massive as almost 4 000 000 Suns. Obviously we cannot see the hole itself, so that the observed radiation comes from gas and dust heated to millions of degrees while falling into it. In 2002, a team of German astronomers found that a star, catalogued as S2, was orbiting the hole, and was moving so fast that its central body was so massive that it could not be anything other than a black hole.

SATELLITES OF THE GALAXY

The Magellanic Clouds are of course the most important of the satellites of the Galaxy. All the rest are either irregular, dwarf ellipticals, or dwarf spheroids. A list of the satellites is given in Table 25.1.

With the naked eye, the two Clouds look like broken-off parts of the Milky Way. The first surviving mention of them is due to the Arab astronomer Al-Sûfi, who in 964 referred to it as Al Bakr, the White Ox. They were described by Antonio Pigafetta during Ferdinand Magellan's voyage from 1519 to 1522 – hence the name; in 1603 Bayer called them the Nubeculæ. In the sky they are 21° apart, making the real separation distance 75 000 light-years. Streams of neutral hydrogen connect them with the Galaxy (the Magellanic Stream) and with each other (the Magellanic Bridge). A few stars can be seen in these streams.

The Large Cloud (LMC) is 169 000 light-years from the Sun. It crosses the border between Dorado and Mensa, and therefore never rises over any part of Europe. It is generally regarded as irregular, though there are definite signs of a central bar. It contains objects of all kinds, including the magnificent Tarantula Nebula, and in 1987 a supernova blazed up in it.

The Small Cloud (SMC), 197 000 light-years away, contains several hundred million stars, and it too has indications of a bar. It

Table 25.1 *Satellites of the galaxy*

Name	RA (h)	(m)	(s)	Dec. (°)	(′)	(″)	Type	Distance (light-years)	Size (°)	Apparent magnitude, *m*	Absolute magnitude, *M*
CMa Dwarf	07	12	35.0	−27	40	00	Irregular	25 000	12 × 12		
SagDEG	18	55	19.5	−30	32	44	dSph	65 000	450′ × 216′	4.5	−14
Ursa Major I Dwarf	08	51	30.3	+63	07	48	dSph?	98 000		14.3	−3.8
LMC	05	23	34.5	−69	45	22	Irregular	169 000	10.7 × 9.2	0.9	−18.1
SMC	00	52	44.8	−72	49	43	Sb	197 000	20′ × 3.5′	2.7	−16.2
UMi Dwarf	15	09	08.5	+67	13	21	dE	200 000	30′ × 19′	11.9	−8.9
Draco Dwarf	17	20	12.4	+57	54	55	Epec	260 000	35′ × 24′	10.9	−8.6
Sculptor Dwarf	01	00	09.3	−33	42	33	dE	290 000	40′ × 31	10.1	−10.7
Sextans Dwarf	10	13	02.9	−01	36	53	dSph	290 000	30′ × 12	12	
Carina Dwarf	06	41	36.7	−50	57	58	E3	330 000	23 × 55′	113	−9.2
Fornax Dwarf	02	39	59.3	−34	26	57	deO	460 000	17′ × 13′	9.3	−13.0
Ursa Major II Dwarf	10	34	52.8	+51	55	12	dSph	330 000			
Leo II Dwarf	11	13	29.2	+22	09	17	EOpec	690 000	12′ × 11′	12.6	10.2
Leo I Dwarf	10	08	27.4	+12	18	27	dSph	820 000	9.8′ × 7′	4	11.2−12.0

See the next chapter for types of galaxies. Here dSph is dwarf spherical, Epec is elliptical peculiar, EOpec is elliptical peculiar type O, deO is dwarf elliptical type O.

lies in Tucana, and it gives the misleading impression of being associated with the globular cluster 47 Tucanæ! It is not so rich as the Large Cloud, but it contains short-period variables, and it was by studying these, in 1908, that Henrietta Swan Leavitt established the vitally important Cepheid period–luminosity law. The brighter stars had the longer periods and, since they were all equally distant, it followed that they really were the more powerful.

Some of the dwarf satellites have points of special interest. The Canis Major Dwarf, discovered in 2003, is hard to detect because it lies behind the plane of the Milky Way; it contains about a thousand million stars, including an unusual number of red giants. At a mere 25 000 light-years, it is the closest galaxy known, and stars have been torn from it by the gravitational pull of our Galaxy, producing a ring of stars round the Milky Way (the Monoceros Ring). The Fornax Dwarf, at 460 000 light-years, contains six globular clusters, one of which, NGC 1049, was discovered before the dwarf itself. Some of the dwarfs are very obscure – for example

Ursa Major II, discovered by Beth Willman in 2005, is less luminous than single stars such as Deneb and Canopus.

THE FUTURE OF THE GALAXY

The Milky Way Galaxy belongs to the Local Group, in which there are three spirals (our system, M 31 in Andromeda and M 33 in Triangulum) together with over fifty dwarfs. It used to be thought that the Andromeda Spiral was much the largest and most populous member of the Group, but recent results indicate that it and our Galaxy are not very unequal.

At present M 31 is approaching us at a rate of 100 to 140 km s^{-1}. As it closes in there will be complicated movements, and finally a collision, so that the two spirals will merge producing a giant elliptical. This will indeed be a time of chaos, but it will not begin for at least 3000 million years, perhaps much longer. It will not concern us, because by that time both the Earth and the Sun will be long gone.

26 · Galaxies

Messier's catalogue contained many 'nebulæ', but spectroscopic work carried out during the 1860s by William Huggins, a brilliant English amateur, showed that there were two different types. Some, such as M 42 in Orion, were gaseous, while others, such as M 31, were apparently made up of stars. As long ago as 1755, Immanuel Kant had suggested that the starry nebulæ might be 'island universes' at vast distances, but there was no obvious way of deciding, and it was more generally believed that all nebulæ were members of the Milky Way.

There was a major development in 1845, when the third Earl of Rosse, using his remarkable home-made 72-inch reflector at Birr Castle in Ireland, found that M 51, in the constellation of Canes Venatici, was spiral in form; we now call it the Whirlpool. Other spirals were soon found, although for some years only the Birr telescope was capable of showing them as such. Other starry nebulæ were spherical, looking very like globular clusters, while others were elliptical or irregular. All were too far away to show measurable parallax, but in 1898 Agnes Clerke, a leading astronomical historian, referred to the island-universe theory as 'a half-forgotten speculation'.

In 1920 there was a famous debate between two leading American astronomers, Harlow Shapley and Heber D. Curtis. By studying short-period variables in globular clusters, Shapley had given the first reasonably accurate value for the size of the Milky Way system, but he regarded the spirals as minor features of our Galaxy, while Curtis believed them to be external. The problem was solved in 1923, when Edwin Hubble identified Cepheid variables M 31, the Andromeda Spiral. He could use the period–luminosity relationship to measure the distances of the Cepheids, and hence the distance of the spiral itself. The result was conclusive: M 31 was much too remote to be a member of our Galaxy. Hubble gave its distance as 900 000 light-years, though he later preferred 750 000 light-years. Hubble's work, carried out with the 100-inch Hooker reflector at Mount Wilson (then the only telescope of sufficient power) showed that Curtis had been right, and Shapley wrong; it must, however, be added that Shapley's value for the size of the Milky Way system was more accurate than Curtis's, so perhaps the debate can be regarded as an honourable draw!

By 1952, the 200-inch Hale reflector at Palomar had been completed, and Walter Baade used it to show that there had been a major error in the Cepheid scale. There are two different types of short-period variables, one much more luminous than the other, and those in the spirals had been wrongly identified; they were more powerful than had been believed, and hence more remote. The distance to the Andromeda Spiral is now thought to be 2 500 000 light-years, and the distances of its two principal satellite galaxies, M 32 and NGC 205, are much the same.

This was one of Walter Baade's major contributions to astronomy; another was concerned with what are known as stellar populations. Rather confusingly, cosmologists refer to all elements heavier than hydrogen and helium as 'metals' (even though in everyday terms there is nothing metal-like about, say oxygen or nitrogen). The first stars, born soon after the original Big Bang, were very massive and contained almost no metals; these were Population III stars. They had brief lives; heavier elements were formed inside them, and they then exploded as supernovæ, so that none now survive. Population II stars were formed from their debris, and contained appreciable amounts of metals. Further explosions produced relatively metal-rich Population I stars, such as the Sun. Star-forming regions – the arms of spirals – are dominated by high-metallicity Population I stars, while the brightest stars in ancient systems such as globular clusters tend to be Population II.

In 1925, Hubble drew up a simple system of classifying galaxies by their shapes, and his 'tuning fork' diagram (see Figure 26.1) is still used, though more elaborate classification systems have been proposed. Hubble's classes are:

1. *Spirals*: resembling Catherine-wheels. Sa: conspicuous, often tightly wound arms issuing from a central bulge, Sb: arms looser, central bulge less condensed. Sc: inconspicuous bulge, loose arms.
2. *Barred spirals*: the arms issue from the ends of a 'bar' through the bulge. They are divided into types SBa, SBb and SBc in the same way as for normal spirals.
3. *Ellipticals*: no signs of spirality. They range from E7. (highly flattened) to E0 (virtually spherical). Many are dwarf systems (dE) but there are also giant ellipticals, such as the immense M 87. Lenticular galaxies (SO) are apparently intermediate between the ellipticals and the spirals.
4. *Irregular* systems, with no definite shape.

There also seem to be galaxies with no stars at all, so that they consist entirely of a flat disc of dark matter, rotating in the same

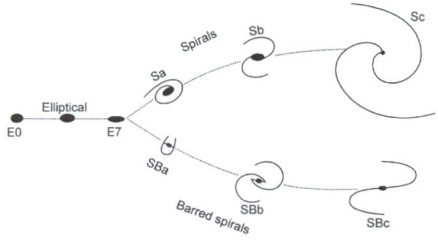

Figure 26.1 Hubble's 'tuning fork' diagram

way as a normal galaxy. Obviously a starless galaxy cannot be seen, but is detectable by its gravitational and radio effects. The first, VIRGOH121, was found in 2005 by a Cardiff University team using radio telescopes in England together with the Arecibo dish in Puerto Rico; the distance is given as 50 000 000 light-years. A second starless galaxy was reported a few months later by the Cardiff team using equipment at Mount Stromlo in Australia.

Galaxy Zoo

Spirals and irregular systems seem to make up at least 70% of all galaxies, but the total numbers are so great that even an adequate estimate of percentage numbers is very difficult indeed. The Galaxy Zoo project was initiated in 2007, mainly due to Chris Lintott of Oxford University. Images of galaxies obtained with robotic telescopes are shown on the internet, and viewers – members of the public as well as amateur astronomers – are asked to classify them. The programme is ongoing, and is a tremendous success; by August 2007, 80 000 volunteers had already classified more than a million galaxies, and the results have been found to be amazingly reliable.

Quite apart from this, unusual objects are found. One of these is 'Hanny's Voorwerp' discovered in 2007 by a Dutch schoolteacher, Hanny van Arkel, during a routine Zoo survey. It appears as a bright blob close to the spiral galaxy IC 2497 in Leo Minor, but nobody yet has any real clue about its nature ('voorwerp' is Dutch for 'unknown'). It contains hot, highly ionised gas, green in colour, and could possibly be a small galaxy reflecting an outburst in the centre or IC 2497; its distance is around 700 million light-years, the same as the neighbouring galaxy, and it has a huge central hole 16 000 light-years across.

COLLIDING GALAXIES

It was at first assumed that Hubble's tuning fork represented an evolutionary sequence, so that a spiral could evolve into an elliptical or vice versa. In fact, things are much less straightforward than this. Galaxies can however collide, such as the Mice, NGC 4676 A and B, in Coma, 290 million light-years away (RA 12h 46m, Dec. +30° 43′). The two galaxies are now colliding and merging; the name refers to the long tails produced by tidal action. Both are above magnitude 15, and can be imaged with good amateur equipment. So can the Antennæ in Corvus, NGC 4038 and 4039 (C 60 and 61), which are of magnitude 11; RA 12h 02m, Dec. –18° 52′ . 3. They are known as the Antennæ because the two long tails of stars, thrown out during the collision, look like the antennæ of an insect. About 1.2 thousand million years ago they were separate, one a spiral and the other a barred spiral, but they then passed through each other; within the next 400 million years the nuclei will collide, and the end product will be a single elliptical system.

Our own Galaxy will eventually collide with the Andromeda Spiral, M 31, to produce a giant elliptical. Also, small galaxies can be swallowed up by larger systems, a process known as 'cannibalisation'. For example, in our Local Group the dwarf SagDEG galaxy is certain to be cannibalised by the Milky Way.

In 1999, a superb picture of two interacting galaxies in Canis Major NGC 2207 and IC 2163, was captured by the Hubble Space Telescope's Wide Field Planetary Camera 2. Strong tidal forces from the larger system, NGC 2207, have distorted the shape of IC 2163, flinging out stars and gas into long streamers stretching out at least 100 000 light-years. It seems that IC 2163 is swinging past NGC 220 in a counter-clockwise direction, having made its closest approach 40 million years ago. However, IC 2163 does not have sufficient energy to escape from the gravitational pull of NGC 2207, and in the future it will be pulled back, again swinging past the larger system. Trapped in their mutual orbits round each other, several thousands of millions of years from now they will merge into a single, more massive galaxy. Table 26.1 lists some notable galaxies.

ACTIVE GALAXIES

These are galaxies with exceptional characteristics. A typical system may contain a compact, brilliant and often variable central core, termed an Active Galactic Nucleus (AGN). Whereas most of the energy radiated by a normal galaxy is simply starlight (the combined output of its constituent stars), an active galaxy radiates strongly over a much wider range of wavelengths. There are various types:

Starburst galaxies

These have exceptionally high rates of star formation, often because of a collision or interaction with another system. The archetypal starburst galaxy is M 82, due to interactions with the large neighbouring spiral M 81. Other good examples are the two Antennæ.

Seyfert galaxies

These galaxies were first identified in 1943 by the American astronomer Carl Seyfert. A system of this type has a bright, variable nucleus, probably with a supermassive central black hole, and weak spiral arms; it radiates strongly in the radio, infrared, ultraviolet and X-ray regions of the electromagnetic spectrum. Good examples are M 77 in Cetus and NGC 4151 in Canes Venatici.

Blazars

A blazar is very compact and very variable, presumably associated with a supermassive black hole at the centre of its host galaxy. They include highly variable quasars and BL Lacertæ objects. They were first described by the American astronomer E. Spiegel in 1978.

Radio galaxies

As the name indicates, these are galaxies emitting very strongly at radio wavelengths, due to the synchrotron process. Most of them are giant elliptical systems. One famous radio galaxy is Cygnus A (3C-405), discovered in 1939 by Grote Reber (then the world's only radio astronomer). It is 600 million light-years away and is a double source, containing an AGN. Images taken at radio

Table 26.1 *Selected list of notable galaxies*

M	NGC	Name	Constellation	RA (h)	(m)	(s)	Dec. (°)	(')	('')	Type	Magnitude	Dimensions (')	Distance (millions of light-years)	
31	224	Andromeda	And	00	42	43.3	+41	16	9	SAb	4.4	190 × 60	2.54	Great Spiral
32	221	–	And	00	42	41.8	+40	51	55	dE2	9.0	8.7 × 6.5	2.49	Satellite of M 31
–	205	–	And	00	40	22.1	+41	41	07	E5pec	8.9	22 × 11	2.69	Satellite of M 31
51	5194	Whirlpool	CVn	13	29	52.7	+47	11	43	SAbc	9.0	12 × 7	23	Pair of
81	3031	Bode's	UMa	09	55	33.2	+69	03	55	Saab	7.9	27 × 14	11.8	galaxies
82	3034	Cigar	UMa	09	55	52.2	+69	40	47	IO	9.3	11 × 4	11.5	
104	4594	Sombrero	Vir	12	39	59.4	–11	37	23	SA	9.0	8.7 × 3.5	29.3	
63	5055	Sunflower	CVn	13	15	49.3	+52	91	45	SAbc	9.3	13 × 31	37	
87	4486	Virgo A	Vir	12	50	49.4	+12	23	28	B	9.6	8.3 × 6.6	55	Radio galaxy
33	598	Pinwheel	Tri	01	33	50.9	+30	39	36	SA	6.3	71 × 42	2.8	Triangulum Spiral
–	253	Sculptor	Scl	00	47	33.0	+25	17	18	SAB	8.0	28 × 7	11.4	
64	4826	Black-Eye	Com	12	56	43.7	+21	40	58	Saab	9.4	11 × 4	24	(Sleeping Beauty)
–	4622	Backwards Galaxy	Cen	12	42	37.7	+40	44	35	Saab	12.6	1.7 × 1.6	200	
–	4631	Whale Galaxy	CVn	12	42	08.0	+32	32	28	SBd	9.9	16 × 3	25	
–	1275	Perseus A	Per	03	19	48.1	+41	30	42	cD	12.6	1.2 × 1.7	235	Radio Galaxy
–	–	Cartwheel	Scl	00	37	41.1	–33	42	59	Spec	15.2	1.1 × 0.9	500	Ring Galaxy
–	3115	Spindle	Sxt	10	05	14.0	–07	43	07	SO	9.9	7.2 × 2.5	31.6	
–	4565	Needle	Com	12	36.3	+25	59			SO	9.6	16 × 2	30	

wavelengths show two jets protruding in opposite directions from the galaxy's centre; at the ends of the jets there are two lobes of intense radiation, formed when material from the jets collides with the surrounding intergalactic medium. The position is: RA 19h 59m 28s .4, Dec. $+40°$ 44′ 02″. The apparent magnitude is 16.2.

The southern Centaurus A (RA 13h 25m 28s, Dec. $-43°$ 01′ 09″) is much brighter, at magnitude 7.8, and is only 13.7 million light-years away. It is a peculiar SO system, and is apparently absorbing a smaller galaxy, with energetic star formation and emissions at X-ray and radio wavelengths. There is probably a central supermassive black hole. One supernova has been seen: SN 1986G, Type Ia. It was discovered by the Reverend Robert Evans, the Australian amateur astronomer who has now found over 100 extragalactic supernovæ.

Quasars

A quasar (Quasi-Stellar Radio source) is a very powerful and remote galaxy with an AGN and a central black hole. It was formerly believed that all members of the class were strong radio emitters, but in fact this is not the case, and the term QSO (quasi-stellar object) is now widely used.

The first quasar to be discovered was 3C-273, in Virgo. (The prefix 3C indicates the third Cambridge catalogue of radio sources, published in 1962.) 3C-273 was known to be a strong radio emitter, but identifying it with a visual object proved to be difficult. However, on 5 August 1962, radio astronomers in Australia, working with the Parkes telescope followed an occultation of the radio source by the Moon, and were able to pinpoint its position very accurately. From this the source was identified with what seemed to be a bluish star of magnitude of 12.8. In 1963, M Schmidt at Palomar obtained an optical spectrum, and therefore found that the object was not a star at all: its spectrum was quite different, and showed lines which could not at first be identified. They proved to be due to hydrogen, but were tremendously red-shifted, indicating a high recessional velocity and therefore immense distance and luminosity. In fact, it is now known that the recessional velocity is 47 400 km s^{-1}; the distance is 2.2 thousand million light-years – yet even so 3C-273 is one of the closest of the quasars, and the only one which is within the range of small telescopes. It is not hard to locate, but visually it looks exactly like an ordinary dim star.

Other quasars were soon found. Their luminosities range from 10 to 10 000 times that of a normal galaxy such as our own, and they vary in luminosity over a short period so that most of their radiation must come from a small region, only about one light-week (or 1200 a.u.) in diameter. Again, the best explanation is that the power comes from a central, very massive black hole. All quasars are very remote, so that we see them as they used to be when the universe was comparatively young; it may well be that a quasar is simply one stage in the evolution of a massive galaxy.

The distances of quasars are, of course, measured by the red shifts of their spectral lines. For example, consider the quasar PC 1247 + 3406, where $z = 4.897$ (z is a measure of the red shift;

$$z = (L - L_0)/L,$$

where L is the observed wavelength and L_0 is the laboratory wavelength).

BL Lacertæ objects (BL Lacs)

A BL Lac is a powerful, remote galaxy with an AGN and a central black hole. The name comes from BL Lacertæ itself, discovered by C. Hoffmeister in 1929; it was reported as an irregular variable star; its magnitude ranges betwen 13 and 16 (the position is RA 22h 02m 43s .3, dec + 42° 16′ 40″). In 1968, John Schmitt at the David Dunlap Observatory identified it with a bright, variable radio source, with faint indications of a host galaxy. Other objects of the same type, such as AP Libræ and W Comæ, had also been catalogued as variable stars, but their almost featureless spectra meant that distance measurements were difficult to obtain. In 1973, Palomar astronomers found fuzzy 'surrounds' of the objects that did show indications of emission lines, and in 1974 Oke and Gunn were able to measure the red shift in the spectrum of BL Lac itself; the value was $z = 0.07$, corresponding to a recessional velocity of 21 000 km s^{-1} with respect to the Milky Way. In 1976, J. Wampler at the Lick Observatory found definite lines in the spectrum of the BL Lac object 0548–322, and was able to measure the red shift.

BL Lacs are classed as blazars. It may well be that they are of the same basic nature as quasars. A quasar is observed at a substantial angle to its jet, whereas with a BL Lac we are looking 'straight down' the jet so that the jet appears as a bright spot overpowering the dim gaseous surround.

Luminous infrared galaxies (LIRGs)

A LIRG emits more than $\times 10^{11}$ the luminosity of the Sun in the far-infrared part of the electromagnetic spectrum. Ultra-luminous infrared galaxies (ULIRGS) are more powerful ($\times 10^{12}$ the luminosity of the Sun), while hyper-luminous infrared galaxies (HLIRGs) are more powerful still. Most of these systems emit at least 90% of their radiation in the infrared, and show signs of continuing interactions; in 1999, observations made with the Hubble Space Telescope showed that there are 'nests' of these systems engaged in multiple collisions involving three, four or even five galaxies. ULIRGs are probably embedded in haloes of dark matter.

The closest known ULIRG is Arp 220, in Serpens (RA 15h 34m 57s .1, Dec. +23° 30′11.1″, magnitude 13.9). It is 250 million light-years away, and seems to be the product of two galaxies which are merging. The central region includes over 200 huge star clusters. Another ULIRG of note is NGC 6240, in Ophiuchus (RA 16h 42m 58.9s, Dec. +02° 24′ 03″, magnitude12.8). This is the result of a collision between two smaller galaxies, so that the original two nuclei can still be seen. The distance is 400 million light-years.

Gravitational lensing

This was proposed by Fritz Zwicky as long ago as 1937, but it was not until 1979 that there was observational confirmation. It has now become very important in astrophysical and cosmological research.

If a galaxy lies on or near the line of sight of a more remote quasar, the result will be that the quasar will show multiple images, simply because the intervening galaxy acts as a lens. The first instance of this, with the galaxy Q0956–561, was found in 1979, but many examples are now known – such as the Clover Leaf H 1413+1143, discovered from La Silla in 1988, where there are four images of the distant quasar. There is also the Einstein ring, found in 1985 by astronomers using the Canada–France–Hawaii telescope on Mauna Kea. They detected a giant luminous arc around the cluster of galaxies, Abell 370; the arc is 500 000 light-years long and 25 000 light-years wide. The effect is due to the light from a background galaxy being bent in the gravitational field of the Abell 370 cluster. This sort of effect had been predicted by Albert Einstein – hence the nickname.

However, one must be wary of jumping to conclusions. In 1987, a binary quasar was detected by S. Bjorgevskii, G. Meylan, R. Perley and P. McCarthy. The binary quasar QQ 1145–0711 is at least 10 000 million light-years away. The spectra of the two images are not identical so that we are dealing with two separate objects rather than a gravitational lens effect. There are also cases when a very remote background object has been detected only because of the 'magnifying' effect of a massive system lying in almost the same line of sight. The bending of light gathers most of the light of the distant object, causing it to appear brighter than it would do if the gravitational lens were not present.

Microlensing makes use of the same principle; when an image is gravitationally lensed by a star rather than a galaxy, it is termed microlensing. Studies were carried out from 1998 by astronomers using the Melbourne telescope at Mount Stromlo in Australia. In 1986, B. Paczynski had proposed that if what is termed a MACHO, a Massive Astronomical Compact Halo Object, passes in front of a star it will act in the manner of a lens and cause a temporary increase in the star's apparent brightness. Effects of this type have been reported, although the precise nature of MACHOs is very uncertain; brown dwarfs, black holes, neutron stars and normal stars have all been suggested as possible candidates.

Galaxies are gregarious, and collect in groups or clusters. Our Galaxy belongs to the Local Group, already described. The nearest group is the IC342 or Maffei 1 Group, whose members are around 10 000 000 light-years away; they are difficult to examine because they lie close to the ecliptic plane, and are heavily obscured by dust in the Milky Way. The nearest major cluster is the Virgo Cluster, which contains about 2000 members and is the physical centre of what is termed the Local Supercluster; the average distance from us to the galaxies in the Virgo cluster is 60 000 000 light-years. The largest member is the giant elliptical M 87, which is a powerful radio source, and which has sent out a curious jet of material. The Virgo Cluster has a pronounced effect upon our Local Group, and we are also affected by the 'Great Attractor', which may be a concentration of massive systems heavily obscured by intergalactic dust.

The galaxy IC 3418 is plunged into the massive Virgo cluster, which contains about 1500 clusters and is permeated by hot gas. IC 3418 is being drawn in at 1000 km s^{-1}, so that its gas is being forced into a 'choppy' tail. Results from the GEE (Galaxy Evolution Explorer) satellite show that the tail is speckled with massive young stars, glowing with ultraviolet light. Dense clouds of molecular hydrogen formed in the wake of the galaxy's plunge. Its distance is 54 000 000 light-years.

In October 2010, an international team of astronomers involving a team from Bristol University, UK, identified the most remote galaxy ever to be seen. The observations were made with the ESO's VLT (Very Large Telescope). The team measured the distance by analysing its feint glow of light. It is so far away that its light has taken 13.1 billion years to reach the Earth, emitted just 600 million years after the Big Bang.

EXPANSION OF THE UNIVERSE

From 1912, V. M. Slipher, at the Lowell Observatory in Arizona, examined the spectra of galaxies, using the Lowell refractor. The galaxies are composed of millions of stars, and the spectra are bound to be something of a jumble, but the absorption lines can be measured easily enough. Slipher found that apart from a few of the very nearest galaxies – those now known to belong to the Local Group – all the Doppler shifts were to the red, indicating velocities of recession. At the time the significance of this was not fully appreciated, and it was still widely believed that the 'starry nebulæ' were contained in our Galaxy.

Edwin Hubble found the answer. First he proved, by observations of the Cepheids, that the galaxies are external systems. Two years later, in 1925, he established that there is a definite link between distance and recessional velocity; the further away a galaxy lies, the faster it is receding. In fact, the entire universe is expanding, and every group of galaxies is moving away from every other group. There is no 'centre' and by now our telescopes can reach out to systems which are 13 200 million light-years away. This is indeed remarkable, bearing in mind that the universe is thought to be only 13 700 million years old.

We need to determine the value of the Hubble Constant (H_o), which gives the rate at which the universe is expanding. The latest value, derived in 2009 by Adam Riess and his team at the Space Telescope Science Institute from the Hubble Space Telescope observations, is 74.2 km s^{-1}.

SOME NOTABLE GALAXIES

The following selection includes some galaxies which are within the range of moderate-sized telescopes, plus a few others of special interest.

Excluding the Magellanic Clouds, the *Andromeda Galaxy*, M 31, is easily the brightest of the external systems. It was long thought to be the largest and most massive galaxy in the Local Group but recent estimates suggest that although it has more stars than the 100 thousand million of our Galaxy there may be less dark matter, in which case the two systems may be equal in mass.

It was first recorded in 964 by Al-Sûfi, who called it 'a small cloud'. It was first photographed in 1887 by Isaac Roberts, from his observatory at Crowborough in Sussex. Many novæ have been seen in it, and one supernova (S Andromedæ, 1885). The radial velocity was first measured in 1912 by V. M. Slipher at the Lowell

Observatory, who gave an approach value of 300 km s^{-1}, which was remarkably correct. Relative to the Earth, M 31 is inclined at an angle of 77°.

It includes objects of all kinds, and seems to be a normal spiral. The nucleus is double, with two concentrations separated by 4.8 light-years; there is a vast extended stellar halo, making the overall diameter of the system over 220 million light-years.

Like our Galaxy, M 31 has a number of satellite galaxies, of which two (M 32 and NGC 205) are visible with very small telescopes. M 32, sometimes called the *Cigar Galaxy*, is a dwarf elliptical; it was discovered in 1740 by Legentil, and consists mainly of Population 2 stars, with little interstellar dust or gas. NGC 205 is sometimes added unofficially to Messier's catalogue, as M 110; Messier himself discovered it, in 1773. Unusually for a dwarf elliptical, it does contain appreciable amounts of dust and gas.

The spiral M 33, in Triangulum, is on the fringe of naked-eye visibility; it was discovered by G. B. Hodierna around 1654. A black hole, almost 16 times the mass of the Sun, was detected in it by the Chandra X-ray Observatory, in 2007; it orbits a companion star, eclipsing it every 3.5 days.

The *Whirlpool Galaxy*, M 51, was discovered by Messier in 1774, and the companion, NGC 5194, in 1781, by Méchain. It was the first galaxy to be recognised as a spiral, by Lord Rosse in 1845, as soon as he completed his great 72-inch reflector at Birr Castle. It is the brightest member of a small group of galaxies, including M 63 (the Sunflower) and NGC 5023 and 5229.

Bode's Galaxy, M 81, in Ursa Major, pairs with its smaller neighbour M 82 (which, confusingly, has also been referred to as the 'Cigar'). M 81 was discovered by Bode in 1774, and is a normal spiral with a central black hole and an AGN. It has marked effects on the starburst galaxy M 82, and there has been at least one tidal encounter between the two. The *Sunflower Galaxy*, discovered by Méchain in 1779, has a number of short spiral arm segments; it was one of the first to be recognised as a spiral (by Lord Rosse).

M 64, the *Black Eye Galaxy* in Coma, was discovered by E. Pigott in 1789; it is a spiral characterised by the dark band of absorbing dust passing in front of the nucleus. Remarkably, the gas in the outer regions rotates in the opposite direction to the stars and gas in the inner region, which is 6000 light-years across. It may be that M 64 collided with and absorbed a satellite galaxy, so that there were violent disturbances. Another system with a prominent dark belt is the *Sombrero Galaxy*, M 104, where the central bulge is large and contains a supermassive central black hole; the galaxy has a large number of globular clusters. The *Backwards Galaxy*, NGC 4622, is a real mystery, because it is a well-developed spiral rotating with its leading arm leading 'clockwise' instead of trailing. The *Whale Galaxy*, NGC 4631, is a splendid example of a spiral edgewise-on

to us. Energetic star formation is proceeding inside it, and it has a dwarf elliptical companion NGC 4627. Another galaxy almost edgewise-on is the irregular NGC 55, in Sculptor, only just over 7 000 000 light-years away, and an easy telescopic object. Another galaxy in Sculptor is NGC 253, discovered in 1783 by Caroline Herschel; it is a starburst galaxy, but the spiral arms are not too easy to trace. The *Spindle Galaxy* (NGC 3115, C 53) is a spiral considerably larger than our Galaxy, discovered by William Herschel in 1787.

There are several prominent radio galaxies, notably *Centaurus A*, probably the product of a merger between two smaller galaxies. It is less than 14 000 000 light-years away, and has a complex structure. Observations from the Spitzer Space Telescope suggest that it is at present absorbing a small spiral. The giant elliptical M 87, the largest and brightest member of the Virgo Cluster, is 120 000 light-years in diameter; it is catalogued by radio astronomers as Virgo A, and shows a jet extending at least 5000 light-years from the nucleus. M 87 is also a source of X-rays and gamma-rays. Another radio source is the Seyfert Galaxy, NGC 1275, catalogued as Perseus A; it lies near the core of the large Perseus Cluster of galaxies.

Far away in space, 500 000 000 light-years from us, we find ESO 350–40, the *Cartwheel Galaxy* in Sculptor. It must once have been a normal spiral, 150 000 light-years across, but a smaller galaxy plunged head-on through it, producing a strong shock-wave, and a starburst region formed round the periphery of the attacked galaxy; hence the bluish ring round the bright central region. It is not in earlier catalogues because it is too faint; the apparent magnitude is below 15. Even fainter is the *Baby Boom* Galaxy (ZW296), discovered from the Spitzer Space Telescope in 2008 (it had actually been glimpsed earlier from the Hubble Space Telescope and the Japanese Subaru Telescope in Hawaii). It is 12 200 000 000 light-years away, and apparently generating over 4000 stars per year. Perhaps the most remote galaxy so far discovered (by Masanori Iye, with the Subaru Telescope) is 12 900 000 000 light-years away. This means that we see it as it was only 750 million years after the Big Bang.

There are also some curious objects not given in the older catalogues. Arp 148, Mayall's Object in Ursa Major (RA 11h 03m 54s, Dec. +40° 51′ 00″), discovered in 1940 by N. U. Mayall, seems to be the result of a collision between two systems, ending up as a ring-shaped galaxy with a tail emerging from it, while Hoag's Object in Serpens Caput, PGC 54559 (RA 15h 17m 14s .4, Dec. +21 °35′ 08″) is made up of a ring of hot, young blue stars encircling the older yellow nucleus. It was discovered by A. A. Hoag in 1950. It is at least 600 million light-years away.

The variety among the galaxies is astonishing, and they alone can guide us toward a true understanding of the formation and evolution of the universe.

27 • Evolution of the universe

When setting out to discuss the origin and evolution of the universe, we are at once confronted with three clear-cut alternatives. They are:

1. The universe began at a definite moment. This was also the beginning of time, so that there was no 'before'. It will also cease to exist at a definite moment, so that there will be no 'after'.
2. The universe has always existed, in which case we must accept a period of time which extends back for ever. It will always exist, so that there will be no 'end'.
3. The universe began at a definite instant, which was also the beginning of time, so that there was no 'before'. It will continue to exist for ever, so that there will be no 'end'.

To explain any of these precepts in plain English is not easy, even for someone with the brain of a Newton or an Einstein!

TIMESCALE

The timescale of the universe was not appreciated until comparatively modern times. The age of the Earth itself was wildly underestimated by almost all scientists of the nineteenth century, and only gradually did the evidence provided by fossils and radiometric dating show that our world must be thousands of millions of years old.

James Ussher, Archbishop of Armagh (1581–1656) held very definite views. By basing his chronology upon key events recorded in the Bible, he found that the moment of the Creation was nine o'clock on Sunday, 23 October 4004 BC. Even today we find people who genuinely believe that everything in the Bible is literally true – and Creationism, re-named Intelligent Design, is so widespread in parts of the United States that some schools teach it as a serious alternative to Darwinian evolution!

Gradually it became clear that the universe is very old indeed. By the end of the nineteenth century, Lord Kelvin, the leading physicist of the time, was estimating the age of the Earth as between 20 and 40 million years, so that the age of the universe was obviously much greater than that. However, it was not until well into the twentieth century that scientists began to think in terms of thousands of millions of years. Table 27.1 shows the timescale of the universe.

MODERN COSMOLOGY

Modern astronomers (not all) believe that the universe came into existence at one definite moment in what is termed the Big Bang (a term scornfully introduced by Fred Hoyle, who did not believe in anything of the sort). It is also assumed that the force of gravity is responsible for the overall behaviour of the universe at large scales, and therefore that Einstein's theory of general relativity can be used as a basis for cosmological models.

Accordingly, it follows that matter did not simply erupt into pre-existing space; instead space, time and matter came into existence simultaneously. It is impossible to discuss what happened before that, because there was no 'before'. Neither can we know where the Big Bang happened, because if it included the whole universe it happened everywhere. Expansion began at once and has been continuing ever since. To be accurate, it is space which is expanding, carrying all matter – and, of course, the galaxies – with it. The concept was originally described in 1927 by the Belgian Abbé Georges Lemaître.

Theory can take us back to 10^{-4} second after the Big Bang. The temperature at that time was of the order of 10^{32} °C, and the universe was dominated by radiation. Energetic particles were moving around, and some of this radiation turned into particles of matter and anti-matter – including what are termed quarks, the 'building blocks' of protons and neutrons. If a particle and an anti-particle meet, both vanish, and if the numbers had been equal there would have been nothing left of the fledgling universe. However, there were rather more particles than anti-particles, so that most of the anti-particles were rapidly annihilated. About one-millionth of a second after the beginning of time, quarks clumped together to form protons and neutrons.

There were more protons than neutrons. After about 100 seconds, nuclear reactions began, and protons and neutrons combined to form the first elements, hydrogen and helium. The universe was opaque, because photons of light could not travel far before being blocked by collisions with electrons.

About 300 000 years later, when the universe had cooled to around 3000 °C, electrons were captured by nuclei to make complete atoms. Light could now travel for vast distances without being blocked, and the universe became transparent to radiation; this is known as the decoupling stage. It followed that the radiation content of the universe was free to spread out all over the expanding volume of space. Thus expansion diluted the radiation, and the wavelength was increased, shifting into the millimetre range of the electromagnetic spectrum. We detect it as a faint glow pervading all space – we know it as the CMB (Cosmic Background Radiation), the last manifestation of the Big Bang. It indicates an overall temperature of 3 K, that is to say 3 degrees above absolute zero, the coldest temperature there can possibly be (−273 °C).

The main credit for predicting the CMB should rightfully go to the Russian cosmologist Ralph Alpher in 1948, collaborating with Robert Herman. (He was already known for his important work in collaboration with George Gamow, who had first given the

Table 27.1 *Some events in the history of the universe*

(a) TIME AFTER THE BIG BANG

0	Big Bang.
10^{-35} to 10^{-33} s	Inflationary period.
10^{-3} to 10^{-1} s	Quarks combine to form protons and neutrons. Universe opaque.
3 minutes	Nuclear fusion combines protons and neutrons to form helium and tiny quantities of other light nuclei. Still three times as much hydrogen (by mass) than helium.
370 000 years	Surface of the last scattering. Emission of CMB. Universe becomes transparent.
200 million years	First stars ignite; soon explode as supernovæ.
1 thousand million years	Quasars. Developed galaxies. Oldest stars in the Milky Way.
8 thousand million years	Start of acceleration of rate of expapnsion of the universe.
9.1 thousand million years	Formation of the Solar System.
9.9 thousand million years	First known fossils.
13 thousand million years	Present day.

(b) TIME AFTER THE PRESENT DAY

1000 million years	Earth becomes unnhabitable.
5000 million years	Sun becomes a red giant. Planetary nebula stage.
6000 million years	Sun becomes a white dwarf.
100 thousand million years	Star formation ceases. Last stars die.
10^{40} years	Protons decayed. Black holes dominate.
10^{100} years	Black holes disintegrate.
?	End of the universe.

real indication of the Big Bang. Gamow, who had a curious sense of humour, persuaded Hans Bethe to add his name to the pioneer paper: Alpher, Bethe, Gamow, $\alpha\beta\gamma$; in fact Bethe had little to do with it, but papers still refer to it as the $\alpha\beta\gamma$ theory). Alpher calculated that the overall temperature should be 5 K.

The actual detection came in 1964. The American radio astronomers Arno Penzias and Robert Wilson were using a special 'horn antenna', built for a completely different investigation, when they recorded a persistent hiss that they could not identify – at first they attributed it to pigeon droppings in the horn of the antenna! Their results came to the attention of Robert Dicke, who had been working on the problem and, apparently without knowing much about Alpher's work, had also predicted that the background temperature would be 3 K. This was exactly what Penzias and Wilson had found. Everything fell neatly into place.

Yet there was one awkward problem. Slight differences in temperature over different areas of the CMB would indicate slight density differences, but it seemed that the entire sky was uniform, and it was hard to see how a 'lumpy' universe could have formed from completely smooth expansion; how could galaxies begin to condense? To the immense relief of cosmologists, measurements obtained in 1992 from an artificial satellite, COBE (Cosmic Background Explorer), showed that there were tiny irregularities. They were confirmed in 1993 by S. S. Meyer and his team using a balloon-borne radiometer, together with results obtained by a Jodrell Bank team using a radiometer on Mount Teide in Tenerife.

In 1999 came the project known as Balloon Observations of Millimetric Extragalactic Radiation and Geophysics (BOOMERANG). The main telescope had a 1.2-m primary mirror, and weighed 2 tonnes. The equipment was carried by a giant hydrogen-filled balloon, which flew around Antarctica from 29 December 1998 to 9 January 1999 covering over 8000 km at a maximum altitude of 37 km; the launch took place from MacMurdo Base, and the landing was within 5 km of this. Over 1800 square degrees of the sky were covered. Antarctica was chosen because of the stable prevailing winds at high altitude and, of course, the constant sunshine. The scientists came from Britain, Canada, Italy and the United States.

The resolution of the BOOMERANG images was 35 times better than with COBE, and it was the first to bring the CMB into sharp focus. The images revealed hundreds of complex regions visible as tiny variations of the order of 0.0001 °C in the temperature of the CMB, thereby giving improved data for the geometry of space-time.

The Wilkinson Microwave Anisotropy probe (WMAP), named in honour of the American physicist David Wilkinson (1935–2002) was launched from Cape Canaveral on 30 June 2001. Like COBE, its aim was to measure temperature differences in the CMB, but it was 45 times more sensitive. After launch, on the Delta II rocket, it made its way to the Second Lagrangian Point, 1 500 000 km from Earth. The telescope's primary reflecting mirrors are a pair of Gregorian 14-m × 1.6-m dishes (facing opposite directions) that focus the signals on to a pair of 0.9-m × 1.0-m secondary mirrors. Following earlier data releases in 2003 and 2005, accumulated data from WMAP's first five years of operation were released on 28 February 2008, giving very precise results:

Age of the universe, 13.73 ± 0.12 thousand million years.

Hubble Constant, 70.5 km s^{-1} per megaparsec.

Content of the universe, 4.56% $\pm$ 0.15% ordinary baryonic matter, 22.8% $\pm$ 1.3% cold dark matter, 72.6% $\pm$ 1.5% dark energy.

EXPANSION, INFLATION AND ACCELERATION

Looking back at the creation and the earliest stages of the universe, we encounter concepts that seem frankly weird. For example, there is 'vacuum energy', an underlying background energy that exists even in a space that is totally devoid of matter. This involves *virtual* particles, which are created out of the vacuum in particle–antiparticle pairs, and promptly annihilate each other. There is *dark matter*, which is invisible but which makes itself evident because of its gravitational pull. And there is *dark energy*, which is thought to be a very major constituent of the universe, but about whose nature we know absolutely nothing. It all seems decidedly surreal.

Theory can take us back no further than 10^{-43} seconds after the Big Bang; this is what is termed the Planck era. Between 10^{-35} and 10^{-33} seconds after the Big Bang came the Inflationary period, when the size of the universe increased enormously and the temperature dropped. By 3 minutes after the Big Bang, the rate of expansion had slowed down, but protons and electrons were milling around, and the universe was opaque; photons of light could not travel far before being blocked. This *dark era* lasted until around 370 000 years after the Big Bang, by which time the temperature had fallen; matter had been decoupled from radiation. The dark age was over, and the universe of this age can be studied by our telescopes – we have come to 'the surface of the last scattering', and the cosmic background radiation.

The first stars formed, and ignited around 200 million to 400 million years after the Big Bang – that is to say 13.5 to 13.3 thousand million years ago. They consisted mainly of hydrogen, with virtually no metals (remember, that to an astrophysicist metals are all elements heavier than hydrogen and helium). The immensely luminous first-generation stars produced heavier elements and then exploded as supernovæ, spewing out material from which second-generation stars could form. We know a good deal about the universe as it used to be after the end of the dark age, because we can see it; quasars are particularly informative, because they are so powerful.

Expansion continued, though it seems natural to assume that the present expansion rate would slow down because of gravitational effects. We have to decide whether or not the universe will expand indefinitely. There are several alternatives. (1) The present expansion will never stop; matter will decay, leaving only extremely low-energy radiation (*an open universe*). (2) The pre-set rate of expansion will increase, until the universe literally flies apart in a Big Rip. (3) The present period of expansion will be followed by a period of contraction again in a 'Big Crunch', (a *closed universe*). (4) The Big Crunch will be in effect a new Big Bang, and we have a cyclic universe with Big Bangs happening very 80 000 million years or so. (Rather irreverently, I cannot help thinking of this as the Concertina Universe!)

We must ask whether there is enough matter in the universe to stop the expansion. The critical value seems to be about 3 atoms of hydrogen per cubic metre; if the overall density is less than this, we have an open universe. If it is greater, the universe will be closed.

The ratio of the actual mean density to the critical density is usually denoted by the Greek letter omega (Ω). If Ω is greater than 1, the universe is closed. If Ω is less than 1, the universe is open. If it is exactly 1, we have a situation in which the galaxies will have just enough energy to continue moving apart for ever; their velocities will fall closer and closer to zero, but will not actually become zero until an infinite time in the future. This is usually termed a flat universe, because space would have no curvature.

Consider first the *closed* universe, with Ω greater than 1. The clusters of galaxies will eventually start to draw together again; red shifts will be succeeded by blue shifts, and the temperature of the background radiation will rise. Around 10 000 million years before the Big Crunch the overall temperature will have climbed back to its present value (3 K). A hundred million years before the Big

Crunch, galaxies will merge and lose their separate identity. A million years before the end, the whole of space will be warmer than the present-day temperature of the surface of the Earth. About 100 000 years before the Crunch, the temperature everywhere will be around 10 000 K, hotter than the surface of the Sun; stars will explode, and the whole universe will become opaque, consisting of a mass of plasma together with radiation. A hundred seconds before the Crunch, atomic nuclei will disintegrate into protons and neutrons. When the Crunch comes it may be the end of everything, because time itself will cease.

With a *cyclic* universe the Crunch will be followed by another Big Bang, and the cycle will begin again; this could happen on an infinite number of occasions, though it has been claimed that successive cycles will become less and less energetic and will eventually die out. Frankly, we are reduced to pure speculation when we try to tackle concepts of this kind. We have to admit that we do not know enough about the forces of Nature.

But researches carried out during recent years seem to go some way toward ruling out closed or cyclic universes. As we have noted, gravitational effects ought to slow down the rate of expansion of the universe, but it now seems that the rate is actually increasing. We live in an *accelerating universe*.

The first evidence came from Type Ia supernovæ, all of which have the same peak luminosities. In 1998, two separate research projects, the Supernova Cosmology Project and the High-Redshift Supernova Search, discovered that Type Ia supernovæ in distant, high-red-shifted galaxies were systematically fainter, and hence more remote, than would have been expected if the expansion rate of the universe had been slowing down. Instead, the rate is increasing. Apparently, the expansion rate did slow down after the Big Bang, but when the universe reached approximately half its present size and age the restricting pull of gravity was overcome by a mysterious repulsive force – 'dark energy' – resulting in an accelerating rate of expansion.

In 1997, a Type Ia supernova was discovered in a dim elliptical galaxy, the distance of which was subsequently given as 11.5 thousand million light-years. Analysis of the apparent brightness and the red shift of this supernova implies that at the time when it exploded – when the universe was less than 40% of its present age – the expansion was still slowing down. This seems to imply that the rate of expansion continued to slacken until the universe was about half its present age, and only then began to accelerate. The evidence for acceleration may not be conclusive, but it is certainly very strong indeed.

Einstein had once introduced a repulsive force into his equations, calling it the cosmological constant, but later abandoned it, and even called it his 'biggest blunder'. Since dark energy seems to have the same effect as the cosmological constant it may well be that, as usual, Einstein was right.

ALTERNATIVE THEORIES

From time to time the Big Bang concept has been challenged. In 1947, Hermann Bondi and Thomas Gold, at Cambridge, proposed the Steady-State hypothesis, according to which the universe has always existed and will exist for ever. As old galaxies die they are replaced by new ones, created spontaneously out of nothingness in the form of hydrogen nuclei; the rate of creation would be so slow that it would be undetectable (go to Bognor Regis, and you would find it difficult to track one new grain of sand on the entire beach!). It followed that the universe would always look the same as it does today; a time-traveller coming back in a million million years would see the same numbers of stars and galaxies as we do, even though they would not be the same stars and galaxies. However, in looking back at objects thousands of millions of light-years away we are looking back into the past, and we find that the aspects and distribution of galaxies is not the same as it is closer to us; in fact the universe is not in a steady state. The theory was finally killed off by the discovery of the cosmic background radiation.

It is not so easy to combat the ideas of Halton Arp, a leading American cosmologist, who has produced images of quasars and galaxies which are clearly joined by luminous 'bridges' and are therefore presumably associated, but which have completely different red shifts. Arp maintains that the shifts are not pure Doppler effects, but there is an important non-velocity component, so that all our distance-measures beyond the Local Group are faulty; he has even suggested that quasars might be minor features shot out of galaxies. This is a most unpopular view with most cosmologists, perhaps because if it were correct many PhD theses would be consigned to the scrapheap, and Arp was refused time with the large American telescopes he had been using because he was obtaining embarrassing results (he is now working at the Max Planck Institute in Germany).

There is also MOND (Modified Newtonian Dynamics), initially proposed in 1981 by the Israeli cosmologist Mordehai Milgrom. This stresses that if we explain the movements of stars and galaxies by introducing dark matter and dark energy, then 95% of the universe is controlled by components about which we are completely ignorant and one cannot avoid suspecting what is often termed a fudge! MOND involves a modification to Newton's Second Law of Dynamics, and this can account for many of the observed effects without any need for ghostly forces driving them. To question Newtonian dynamics sounds like the most extreme form of heresy, but MOND has attracted considerable support, and the jury is still out.

EPILOGUE

Much of what I have written here may prove to be wrong – perhaps even by the time that the book is published. For example, at the time of going to press, Tom Shanks and his graduate student Utane Sawangwit at Durham University used new observations from WMAP to indicate that the tiny variations in the CMB are smaller than previously believed; they used astronomical objects that appear as unresolved points in radio telescopes to test the way that WMAP smooths out its maps. They find that the smoothing in much greater than as given by WMAP, so the ripples are smaller, and dark matter and dark energy are not needed.

Many cosmologiots, eyeing their Ph.D. theses, would be alarmed at any idea of this sort.

On 3 July, Planck, the European satellite dedicated to CMB measurements, sent back its first full-sky image showing the CMB, 'the oldest light in the universe', originating a mere 380 000 years after the Big Bang. Its 1.5 m telescope focused the radiation on to two arrays cooled to a temperature of $-275.05°C$, a tenth of a degree above absolute zero.

Astronomy is a fast-moving science, and each week seems to bring its own quota of new discoveries and new ideas. We live in exciting times.

28 · The constellations

From BC 4000, constellation patterns were drawn up. All these were different; the Chinese and Egyptian constellations, for example, are quite unlike ours (for example, our Draco seems to correspond with the Egyptian hippopotamus). Our system is derived from that of Ptolemy (Table 28.1); it may originally have been Cretan, though opinions differ. In all, 88 separate constellations are now in use. A list of these is given in Table 28.2.

Ptolemy gave a list of 48 constellations: 21 northern, 12 zodiacal and 15 southern (Table 28.1). All these are to be found on modern maps, although in many cases their boundaries have been altered – and the huge Argo Navis, the Ship Argo, has been chopped up into a Keel (Carina), sails (Vela) and a poop (Puppis).

Surviving post-Ptolemaic constellations are given in Table 28.3. Many of the original names have been shortened: thus Pisces Volant, the Flying Fish, has become simply Volant, while Mons Mensæ, the Table Mountain, has become Mensa. Pisces Australis may also be called Pisces Austrians, while Scorpius is often incorrectly called Scorpio. There was some confusion over two of Bayer's constellations, Apis (the Bee) and Avis Indica (the Bird of Paradise); modern maps give it as Apus. There were also two Musca, one formed by Lacaille to replace Bayer's Apis, and the other (rejected) formed by Bode out of stars near Aries. One discarded constellation, Quadrans, has at least given its name to the Quadrantid meteor shower of early January; the stars of Quadrans are now included in Boötes (near β Boötis).

The list of rejected constellations is very long (Table 28.4). Few will be regretted, though there are some people who are sad to see the demise of Noctua and Felis (the Owl and the Pussycat!). There have been occasional attempts to revive the whole constellation system, but none has met with much support.

However, the constellations are very unequal in size and importance. Apart from the dismembered Argo, the largest constellation, Hydra, covers 1303 square degrees, while the smallest, Crux, covers a mere 68 square degrees. Orion has five stars above the second magnitude and 42 above magnitude 5, while Cælum, Horologium and Sextans each have only two stars above magnitude 5.

'Star density' is also very uneven. For stars above magnitude 5, the greatest 'density' is for Crux (19.12 stars per 100 square degrees), at the other end of the scale comes Mensa (0.0), and then Sextans (0.63).

In 1603, the German astronomer Johann Bayer published his star map, allotting Greek letters to the brightest stars, in order – α β γ down to ω. The order was not always strictly followed; for example in Sagittarius the brightest stars are ϵ and σ with α and β almost two magnitudes fainter, while in Octans the brightest star

is ν. A few constellations, notably Norma and Leo Minor, have no σ at all, for reasons that are unclear. When Argo was broken up its leaders were distributed between its various parts, so that for instance the brightest star in Puppis is ζ. A few star clusters were given Greek letters; one is the great globular ω Centauri, and the famous open cluster Præsepe was catalogued as ϵ Cancri. And the globular cluster 47 Tucanæ was listed as ξ Tucanæ.

In Flamsteed's catalogue the stars are numbered in order of increasing RA, so that for example ϵ Sagittarii is Flamsteed's 20 Sagittarii and σ Sagittarii is Flamsteed's 34.

Almost all proper names of stars are Arabic, though a few, such as Sirius, are Greek. Nowadays the names are in general used for only the brightest stars, those above magnitude 2, plus a few special cases, such as Mizar (ζ Ursæ Majoris) and Polaris (α Ursæ Minoris). Some stars have been given several alternative names; thus σ Piscium may be Al Rischa, Kaitain or Okda, while the first-magnitude β Centauri may be either Agena or Hadar.

Some names are tongue-twisting; thus μ Serpentis is Leiolepidotus and δ Equulei is Pherasauval. My own favourite is

Table 28.1 *Ptolemy's original 48 constellations*

Northern	Zodiacal	Southern
Ursa Minor	Aries	Cetus
Ursa Major	Taurus	Orinon
Draco	Gemini	Eridanus
Cepheus	Cancer	Lepus
Boötes	Leo	Canis Major
Corona Borealis	Virgo	Canis Minor
Hercules	Libra	Argo Navis
Lyra	Scorpio (Scorpius)	Hydra
Cygnus	Sagittarius	Crater
Cassiopein	Capricornus	Corvus
Perseus	Aquarius	Centaurus
Auriga	Pisces	Lupus
Ophiuchus		Ara
Serpens		Corona Australis
Sagitta		Piscis Australis
Aquila		
Delphinus		
Equuleus		
Pegasus		
Andromeda		
Triangulum		

Table 28.2 *The constellations*

		1st mag	Stars to magnitude			Area (square degrees)	Number of stars above magnitude 5 per 100 square degrees (star density)
			2.00	4.00	5.00		
Andromeda	Andromeda	–	0	7	25	722	9.19
Antlia	The Airpump	–	0	0	4	239	1.67
Apus	The Bird of Paradise	–	0	2	6	206	2.91
Aquarius	The Water-bearer	–	0	7	31	980	3.16
Aquila	The Eagle	Altair	1	8	16	652	2.45
Ara	The Altar	–	0	7	10	237	4.21
Aries	The Ram	–	1	4	11	441	2.49
Auriga	The Charioteer	Capella	2	7	21	657	3.20
Boötes	The Herdsman	Arcturus	1	8	24	907	2.65
Cælum	The Graving Tool	–	0	0	2	125	1.60
Camelopardus	The Giraffe	–	0	0	11	757	1.45
Cancer	The Crab	–	0	1	6	506	1.19
Canes Venatici	The Hunting Dogs	–	0	1	7	465	1.51
Canis Major	The Great Dog	Sirius	4	10	26	380	6.84
Canis Minor	The Little Dog	Procyon	1	2	4	183	2.19
Capricornus	The Sea Goat	–	0	5	16	414	3.86
Carina	The Keel	Canopus	3	14	40	494	8.10
Cassiopeia	Cassiopeia	–	0	7	23	598	3.85
Centaurus	The Centaur	α Cen., Agena	2	14	49	1060	4.62
Cepheus	Cepheus	–	0	8	20	588	3.40
Cetus	The Whale	–	0	8	24	1232	1.95
Chamæleon	The Chameleon	–	0	0	6	132	4.55
Circinus	The Compasses	–	0	1	4	93	4.30
Columba	The Dove	–	0	4	9	270	3.33
Coma Berenices	Berenice's Hair	–	0	0	8	386	2.07
Corona Australis	The Southern Crown	–	0	0	7	128	5.47
Corona Borealis	The Northern Crown	–	0	3	10	179	5.59
Corvus	The Crow	–	0	5	6	184	3.26
Crater	The Cup	–	0	1	6	282	2.13
Crux Australis	The Southern Cross	Acrux, β Crucis	3	55	1133	68	19.112
Cygnus	The Swan	Deneb	1	11	43	804	5.35
Delphinus	The Dolphin	–	0	4	6	189	3.17
Dorado	The Swordfish	–	0	1	8	179	4.47
Draco	The Dragon	–	0	1	26	1083	2.40
Equuleus	The Foal	–	0	0	3	72	4.17
Eridanus	The River	Achernar	1	12	43	1138	3.78
Fomax	The Furnace	–	0	1	5	398	1.26
Gemini	The Twins	Pollux	3	13	23	514	4.47
Grus	The Crane	–	1	4	13	366	3.55
Hercules	Hercules	–	0	15	37	1225	3.02
Horologium	The Clock	–	0	1	2	249	0.80
Hydra	The Watersnake	–	1	9	32	1303	2.46
Hydrus	The Little Snake	–	0	3	9	243	3.70
Indus	The Indian	–	0	2	7	294	2.38
Lacerta	The Lizard	–	0	1	11	201	5.47
Leo	The Lion	Regulus	2	10	26	947	2.75
Leo Minor	The Little Lion	–	0	1	6	232	2.59

Table 28.2 (cont.)

		1st mag	Stars to magnitude			Area (square degrees)	Number of stars above magnitude 5 per 100 square degrees (star density)
			2.00	4.00	5.00		
Lepus	The Hare	–	0	7	14	290	4.83
Libra	The Balance	–	0	5	13	538	2.42
Lupus	The Wolf	–	0	8	32	334	9.58
Lynx	The Lynx	–	0	2	12	545	2.20
Lyra	The Lyre	Vega	1	4	11	280	3.85
Mensa	The Table	–	0	0	0	153	0.00
Microscopium	The Microscope	–	0	0	4	210	1.90
Monoceros	The Unicorn	–	0	2	13	482	2.70
Musca Australis	The Southern Fly	–	0	4	11	138	7.97
Norma	The Rule	–	0	0	6	165	3.64
Octans	The Octant	–	0	1	4	291	1.37
Ophiuchus	The Serpent	–	0	12	36	948	3.80
Orion	Orion	Rigel, Betelgeux	5	15	42	594	7.07
Pavo	The Peacock	–	1	4	14	378	3.70
Pegasus	The Flying Horse	–	0	9	29	1121	2.59
Perseus	Perseus	–	1	10	34	615	5.52
Phœnix	The Phœnix	–	0	7	17	469	3.62
Pictor	The Painter	–	0	2	5	247	2.02
Pisces	The Fishes	–	0	3	24	889	2.70
Piscis Australis	The Southern Fish	Fomalhaut	1	1	7	245	2.86
Puppis	The Poop	–	0	11	42	673	6.24
Pyxis	The Compass	–	0	1	7	221	3.17
Reticulum	The Net	–	0	2	7	114	6.14
Sagitta	The Arrow	–	0	2	5	80	6.25
Sagittarius	The Archer	–	1	14	33	867	3.80
Scorpius	The Scorpion	Antares	3	17	38	497	7.64
Sculptor	The Sculptor	–	0	0	6	475	1.26
Scutum	The Shield	–	0	0	6	109	5.50
Serpens	The Serpent	–	0	9	17	637	2.67
Sextans	The Sextant	–	0	0	2	314	0.63
Taurus	The Bull	Aldebaran	2	14	44	797	5.52
Telescopium	The Telescope	–	0	1	4	252	1.59
Triangulum	The Triangle	–	0	2	3	132	2.27
Triangulum Astrale	The Southern Triangle	–	1	3	6	110	5.45
Tucana	The Toucan	–	0	2	7	295	2.37
Ursa Major	The Great Bear	–	3	19	35	1280	2.73
Ursa Minor	The Little Bear	–	1	3	9	256	3.51
Vela	The Sails	–	2	10	30	500	6.00
Virgo	The Virgin	Spica	1	8	26	1294	2.01
Volans	The Flying Fish	–	0	3	7	141	4.96
Valpecula	The Fox	–	0	0	10	268	3.73
	Totals	21	50	455	1417		(Average 3.8)

These counts do not include variable stars, which can rise above the fifth magnitude, but whose average magnitude is below this limit. For instance, Mira Ceti is excluded even though its brightest maxima exceed magnitude 2.

Table 28.3 *Surviving post-Ptolemaic constellations*

Added: *By Tycho Brahe, c. 1590:*
Coma Berenices

By Bayer, 1603
Pavo
Tucana
Grus
Phœnix
Dorado
Volans (originally Piscis Volans)
Hydrus
Chamæleon
Apus (originally Avis Indica)
Triangulum Australe
Indus

By Royer, 1679:
Columba (originally Columba Noachi, Noah's Dove)
Crus Australis

By Hevelius, 1690:
Camelopardus
Canes Venatici
Vulpecula (originally Vulpecula et Anser, the Fox and Goose)
Lacerta
Leo Minor
Lynx
Scutum Sobieskii
Monoceros
Sextans (originally Sextans Uraniæ, Urania's Sextant)

By La Caille, 1752:
Sculptor (originally Apparatus Sculptoris, the Sculptor's Apparatus)
Fornax (originally Fornax Chemica, the Chemical Furnace)
Horologium
Reticulum (originally Reticulus Rhomboidalis, the Rhomboidal Net)
Cælum (originally Cæla Sculptoris, the Sculptor's Tools)
Vulpecula (originally Vulpecula et Anser, the Fox and Goose)
Pictor (originally Equuleus Pictoris, the Painter's Easel)
Pyxis (originally Pyxis Nautica, the Mariner's Compass)
Antila (originally Antila Pneumatica, the Airpump)
Octans
Circinus
Norma (or Quadra Euclidis, Euclid's Square)
Telescopium
Microscopium
Mensa (originally Mons Mensæ, the Table Mountain)

ο Piscium: Torcularis Septentrionalist. I have collected all the names I can find, and given them in the Catalogue, but I cannot claim to have tracked down the origins of most of them.

There have been a few cases of stars which have been arbitrarily transferred from one constellation to another; thus, δ Pegasi has become α Andromedæ, γ Aurigæ has become β Tauri and γ Scorpii has become σ Libræ.

The current star-map is so well established that it will certainly not be altered, but one cannot but agree with the nineteenth-century Sir John Herschel, who commented that the constellation patterns seemed to have been designed so as to cause as much confusion and inconvenience as possible.

One or two of the rejects have points of interest. *Antinoüs* was a favourite youth in the court of the Roman emperor Hadrian, but also identified with Ganymede, cup-bearer of the gods; it lay at the 'tail' of Aquila. It was shown by Tycho, and may have been on one of Ptolemy's lists. Apis is probably just an old name for Musca. *Cerberus*, captured by Hercules as one of the Twelve Labours, lies in Hercules near the border with Cancer. *Tigris*, created by Jakob Bartsch, began at Ophiuchus and meandered through Aquila, Hercules, Cygnus, Sagitta and Equuleus, ending at Pegasus, and would have added to the general confusion if astronomers had accepted it! The longest name was Augustin Royer's *Sceptrum et Manus Iustitiæ*; it was made up of a few faint stars in Lacerta, and was quickly and blissfully forgotten.

Taurus Poniatovski was created in 1777 by the Abbe Poczobut of Wilna to honour Stanislaus Poniatovski, King of Poland; it was marked by a well-defined group of stars including β Ophiuchi, but was soon rejected.

Felis was created by the French astronomer Lalande, who was particularly fond of cats; it contained the planetary nebula GC 3242, the 'Ghost of Jupiter'. *Quadrans*, near β Boötis, is of course remembered because it contains the radiant point of the January Quadrantids. The star 53 Eridani, magnitude 3.9, is named 'Sceptrum' because it was contained in *Sceptrum Brandenburgicum*, created in 1688 by the Prussian astronomer Gottfried Kirsch. The constellation itself was soon forgotten.

ASTERISMS

An asterism is a pattern of stars not regarded as a constellation. The best example is the False Cross, made up of ι and ε Carinæ with δ and κ Velorum. It does look very like the Southern Cross. True, it is larger, and not so brilliant, but it and Crux are often confused.

In Vulpecula there is Brocchi's Cluster, nicknamed the Coathanger; it was described in 964 by Al-Sūfi, and shown independently in the seventeenth century. It was mapped in the 1920s by the American amateur D. F. Brocchi, of the American Association of Variable Star Observers; in 1931, Per Collinder listed it as a genuine cluster, and it is still known as Collinder 399. The nearest reasonably bright star is α Sagittæ. It is made up of ten stars from magnitudes 5 to 7, and is visible with the naked eye as a dim patch of light (page x).

Kemble's Cascade, named after the Franciscan amateur Father Lucian Kemble, who in 1980 drew attention to it, lies in Camelopardus. It consists of a chain of 15 to 25 stars, magnitudes 5 to 9, spread over an area of 2.5 degrees. The stars are not associated; the 6th-magnitude open cluster NGC 1502 lies at the southeast end of the chain. The Cascade is a beautiful sight when seen through binoculars.

Table 28.4 *Rejected constellations. The list of rejected constellations is very long. One of these, Quadrans, has at least given its name to the Quadrantid meteor shower of early January*

Name		Introduced by	Position	Now in
Anser	The Goose	Hevelius 1690	Cygnus/Vulpecula	Vulpecula
Antinoüs	Antinoüs	Tycho 1559	Near λ Aquilæ	Aquila
Cancer Minor	The Little Crab	Lubinietzki 1650	Cancer/Gemini	Cancer
Cor Caroli	Charles' Heart	Flamsteed 1700	α Canum Venaticorum	Canes Venatrci
Cerberus	Cerberus	Hevelius 1687	Hercules	Hercules
Custos Messium	Messier's Equipment	Lalande 1776	Camelopardalis	Cepheus/Cassiopeia
Felis	The Cat	Bode 1775	Hydra/Antlia	Hydra
Gallus	The Cock	Plancius 1613	Argo Navis	Argo
Globus Ærostaticus	The Air Balloon	Lalande 1798	Capricornus	Piscis Australis/ Microscopium
Honores Frederici	The Honours of Frederick	Bode 1787	Andromeda/Pegasus	Andromeda
Jordanus Fluvius	The River Jordan	Plancius 1613	Position uncertain	
Lochium Funis	The Log Line	Bode 1787	Argo Navis	Argo
Lilium	The Lily	Royer 1679	Aries	Aries
Machina Marmor Sculptile Electrica	The Bust of Columbus Electrical Machine	Bode 1787 William Crosswell, 1810	Cetus Reticulum	Fornax/Sculptor Reticulum
Mons Mænalus	Mount Mænalus	Flamsteed 1700	Boötes	Boötes
Musca Borealis	The Northern Fly	Hevelius 1687	Aries	Aries
Noctua	The Night Owl	Burritt 1833	Hydra	Hydra/Libra/Virgo
Norma Nilotica	The Nilometer	Burritt 1833	Capricornus	Aquarius
Nubes Major	The Great Cloud	Royer 1679	Nubecula Major	Dorado
Nubes Minor	The Small Cloud	Royer 1679	Nubecula Minor	Tucana
Officina Typographica	The Printing Press	Bode 1787	Monoceros	Puppis
Polophylax	Guardian of the Pole	Plancius, 1952	Tucana/Grus	Tucana/Grus
Pomum Inperiale	Orb of Leopold I	Gottfired Kirch, 1688	Aquila	Aquila
Psalterium Georgii	George's Lute	Hell 1780	Taurus	Eridanus
Quadrans Muralis	The Mural Quadrant	Bode 1775	Boötes	Boötes
Quadratum	The Square	Allard 1700	Near Nubecula Major	Dorado
Ramus Pomifer	The Apple Tree		Hercules	Hercules
Robur Carolinum	Charles' Oak	Halley, 1680	Argo Navis	Argo
Sagitta Australis	The Southern Arrow	Hevelius 1645	Scorpius/Sagittarius	Scorpius/Sagittarius
Sceptrum Brandenburgicum	The Sceptre of Brandenburg	Kirch 1688	Eridanus/Lepus	Lepus
Sceptrum et Manus lustitiæ	The Sceptre and Hand of Justice	Augustin Royer, 1679	Lacerta	Lacerta
Scirius Volans	The Flying Squirrel		Camelopardus	Camelopardus
Solarium	The Sundial	Burritt 1833	Hydrus/Dorado	Hydrus/Dorado
Solitarius	The Solitaire	Le Monnier 1776	Hydra	Hydra
Taurus Poniatowski	The Bull of Poniatowski	Poczobut 1777	Ophiuchus	Ophiuchus
Tarandus	The Reindeer	Le Monnier 1776	Camelopardalis/ Cassiopcia	Cepheus/Cassiopeia
Telescopium Herschelii Major	Herschel's Large Telescope	Hell 1781	Lynx/Gemini	Auriga
Telescopium Herschelii Minor	Herschel's Small Telescope	Hell 1781	Taurus	Taurus
Testudo	The Turtle	Admiral W. H. Smyth, *c.* 1840	Cetus/Pisces	Pisces
Tigris	The Tigris	Bartschius 1624	Ophiuchus/Pegasus	Ophiuchus
Triangulum Minor	The Little Triangle	Hevelius 1677	Triangulum/Aries	Aries
Turdus Solitarius	The Blue Rock Thrush	Pierre-Charles le Monnier, 1776	Hydra, near Libra	Hydra, near Libra
Vespa	The Wasp	Plancius, 1612	Aries	Aries

Argo Navis was divided into Carina (the Keel), Vela (the Sails), Puppis (the Poop) and Malus (the Mast); Malus then became the constellation of Pyxis. Ophiuchus was originally called Serpentarius, and Grus was Phoenicopterus (the Flamingo).

Finally, a widely-accepted asterism for which I was unintentionally responsible. In a BBC *Sky at Night* television programme about fifty years ago, I discussed the huge triangle made up of Vega, Deneb and Altair, which to northern-hemisphere observers dominates the night sky for several months from spring to autumn. I casually referred to it as 'the Summer Triangle' – and the nickname 'stuck'; everyone now seems to use it, though the stars are in completely different constellations and in the southern hemisphere are at their best during winter.

29 · The star catalogue

Data for stars down to the fourth magnitude:

27. Flamsteed number.
28. Greek letter, or other cataogue letter.
29. RA: right ascension, hours, minutes, seconds.
30. Dec.: declination, degrees, arcmin, arcsec.
31. *m*: apparent magnitude. A v indicates slight variability, range <0.2 magnitude.
32. *M*: absolute magnitude. * indicates that the star is a spectroscopic binary.
33. *d*: distance, in light years.
34. Spectrum.
35. Name, if any.

Stars between magnitudes 4 and 4.5 are listed, with their Greek letters or F numbers.

Data for stars beyond about 300 light-years cannot be precise, and I have 'rounded them off'.

All data come mainly from the Hipparcos astrometric satellite, superseding the Cambridge catalogue data given in the last edition.

The list of double stars, variable stars, and nebular objects does not pretend to be complete, and could be extended almost *ad infinitum*, but I have included most variable stars whose maximum magnitudes are 8.0 or brighter and where the range is at least half a magnitude. For double stars, I have given pairs which are within the range of modest telescopes together with some which may be regarded as test objects. For binary stars of reasonably short period, the values of position angle and distance are for approximate date 2000.

These symbols identify the various features shown on the maps

Symbol	Feature	Symbol	Magnitude	Star Magnitudes
☼ ◉ ○	VARIABLE STARS	✦	Magnitude 1	**Star Magnitudes**
⊖ ⊕	PLANETARY NEBULÆ	✦	Magnitude 2	These are given by different
⁑	GLOBULAR CLUSTERS			symbols, though of course there
◯ ○	GALAXIES	●	Magnitude 2.5	are many
◌	OPEN CLUSTERS	●	Magnitude 3	gradations; for example, the first
🌟	GASEOUS NEBULA	•	Magnitude 4	symbol indicates stars between magnitudes −1 (Sirius) down to Regulus (+1.4).

Data given are:
VARIABLE STARS. Range, type, period in days, and spectral type.
DOUBLE STARS. Position angle (PA) in degrees, separation in seconds of arc, and magnitudes of the components.
OPEN CLUSTERS. Diameter in minutes of arc, magnitude, and approximate number of stars (though in many cases this is subject to great uncertainty).
GLOBULAR CLUSTERS. Diameter in minutes of arc, and approximate total magnitude.
PLANETARY NEBULÆ. Dimensions in seconds of arc, total magnitude, and magnitude of central star.
NEBULÆ. Dimensions in minutes of arc and, where appropriate, the magnitude of the illuminating star.
GALAXIES. Magnitude, dimensions in minutes of arc, and type.
For nebular objects I have given NGC numbers, together with Messier or Caldwell numbers where appropriate. Right ascensions and declinations for all objects are for epoch 2000.

ANDROMEDA

(Abbreviation: And)
A large and important northern constellation; one of Ptolemy's 'originals'. It contains the Great Spiral, M 31. One of the leading stars in the constellation – Alpheratz or α Andromedæ – is actually a member of the Square of Pegasus, and was formerly known, more logically, as δ Pegasi.

In mythology, Andromeda was the beautiful daughter of King Cepheus and Queen Cassiopeia. Cassiopeia offended the sea god Neptune by her boasting about Andromeda's beauty, which, she claimed, was greater than that of any sea nymph. Neptune thereupon sent a sea monster to ravage the kingdom, and the Oracle stated that the only solution was to chain Andromeda to a rock by the shore where she would be devoured by the monster.

This was duly done, but the situation was redeemed by the hero Perseus, who was on his way home after killing the Gorgon, Medusa. Mounted upon his winged sandals, Perseus arrived in the nick of time, turned the monster to stone by the simple expedient of showing it Medusa's head, and then, in the best story-book tradition, married Andromeda. Cepheus, Cassiopeia and of course Perseus are to be found in the sky; the Gorgon's head is marked by the 'Demon Star' Algol, and the sea monster has sometimes been identified with the constellation of Cetus.

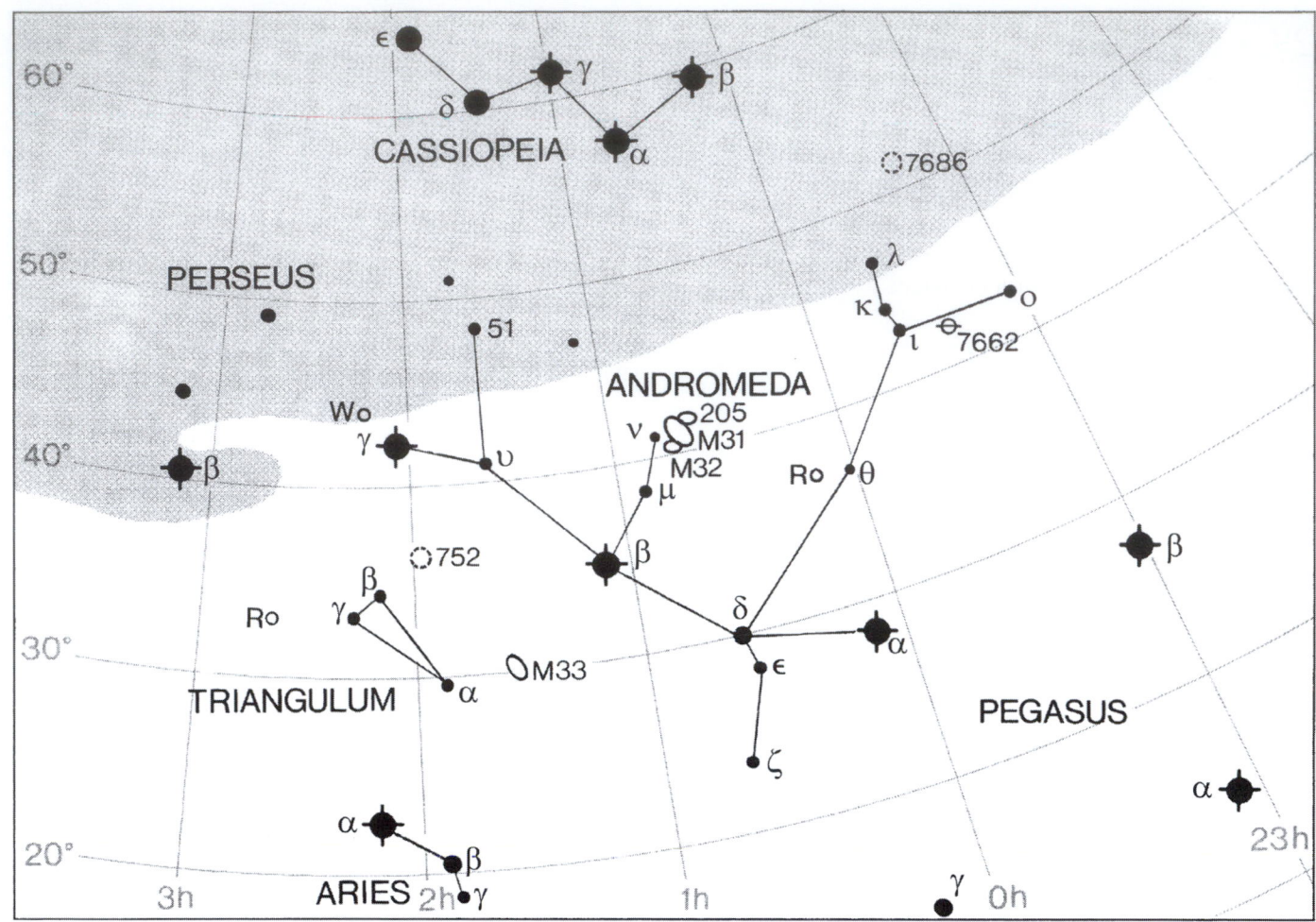

There are eight stars above the fourth magnitude:

	RA			Dec.						
	(h)	(m)	(s)	(°)	(')	('')	m	M	d (light years)	Spectrum
21 α	00	08	23.2	+29	05	27	2.07	−0.30	97	B0	Alpheratz
43 β	01	09	43.8	+35	37	15	2.07	−1.86	199	M1	Mirach
57 γ¹	02	03	43.9	+42	19	47	2.10	−3.08	355	B8	Almaak
γ²	02	03	54.7	+42	19	51	4.84			
8 δ	02	39	19.6	+30	51	40	3.27	0.81*	101	K3	
51	01	37	59.5	+48	37	42	3.59	−0.04	124	K3	Nembus
1 o	23	01	55.2	+41	19	34	3.6v	−5.01	690	B6	
16 λ	23	37	33.7	+46	27	33	3.8v.	1.75	64	G8	
37 μ	00	56	45.1	+38	29	57	3.86	0.75	136	A5	

Also above 4.5: ζ (4.0v), υ (4.10), φ (4.26), ι(4.29), π (4.34[*]), ε (4.34), and η (4.40[*]).

VARIABLE STARS

	RA (h)	RA (m)	Dec. (°)	Dec. (')	Range	Type	Period (d)	Spectral type
Z	23	>33.7	+48	49	8.0–12.4	Z And	—	M
ST	23	38.8	+35	46	7.7–11.8	Semi-regular	328	R
SV	00	04.3	+40	07	7.7–14.3	Mira	325.2	M
KU	00	06.9	+43	05	6.5–10.5	Mira	750	M
VX	00	19.9	+44	43	7.8–9.3	Semi-regular	369	N
T	00	22.4	+27	00	7.7–14.5	Mira	280.8	M
R	00	24.0	+38	35	5.8–14.9	Mira	409.3	S
TU	00	32.4	+26	02	7.8–13.1	Mira	316.8	M
RW	00	47.3	+32	41	7.9–15.7	Mira	430.3	M
W	02	17.6	+44	18	6.7–14.6	Mira	395.9	S

DOUBLE STARS

	RA (h)	RA (m)	Dec. (°)	Dec. (')	PA (°)	Separation (")	Magnitude	
π	00	36.9	+33	43	173	35.9	4.4, 8.6	
γ	02	03.9	+42	20	063	9.8	2.3, 4.8	
γ²					106	0.5	5.5, 6.3	Binary, period 61 years

OPEN CLUSTERS

M	C	NGC	RA (h)	RA (m)	Dec. (°)	Dec. (')	Diameter (')	Magnitude	No. of stars
		7686	23	30.2	+49	08	15	5.6	20
	28	752	01	57.8	+37	41	50	5.7	60

PLANETARY NEBULA

M	C	NGC	RA (h)	RA (m)	Dec. (°)	Dec. (')	Diameter	Magnitude	Magnitude of central star
	22	7662	23	25.9	+42	33	20 × 130	9.2	13.2

GALAXIES

M	C	NGC	RA (h)	RA (m)	Dec. (°)	Dec. (')	Magnitude	Dimensions (')	Type	
		205	00	40.4	+41	41	8.0	17.4 × 9.8	E6	Companion to M 31
31		224	00	42.7	+41	16	3.5	17.8 × 6.3	Sb	
32		221	00	42.7	+40	52	8.2	7.6 × 5.8	E2	Companion to M 31
	23	891	02	22.6	+42	21	9.9	13.5 × 2.8	Sb	

α (Alpheratz) was formerly δ Pegasi. 51 (Nembus) was formerly υ Persei. 54 Andromedæ was formerly φ Persei. Curiously, the great spiral galaxy has been referred to as n Andromedæ and 33 Andromedæ. ι, κ, λ and ψ made up the constellation Honores Frederici formed by Bode in 1787, but later rejected.

Alpheratz (alternative name, Sirrah) is a very close binary; the brighter component is a mercury–manganese star. γ (Almaak) is a lovely, easy double; the primary is orange, the secondary bluish. The secondary is a very close triple. δ is a long-period spectroscopic binary; orbital period 41 years. λ is also a spectroscopic binary, with a period of 20.5.days. υ is a binary; the primary, a yellow F-type star, is widely separated from the M4 red dwarf companion. The primary is attended by at least three planets. Any inhabitants of the υ Andromedæ system must have a fascinating view of the sky!

ANTLIA

(Abbreviation: Ant)

A small southern costellation, covering 239 square degrees, added to the sky by Lacaille in 1752; its original name was Antlia Pneumatica. It conains no star as bright as the 4th magnitude, and no legends are attached to it.

Andromeda is a rich constellation, but pride of place must go to the great spiral, M 31, the nearest of the large external galaxies. It was in M 31 that the supernova S Andomedæ flared up in 1885, though its nature was not recognised at the time.

See chart for Carina.

The brightest star is α, RA 17h 27m 09s .1, Dec. −41° 04′ 04″, magnitude 4.25. The only other stars above magnitude 4.7 are ε (4.51) and ι (4.60).

VARIABLE STARS

	RA (h)	RA (m)	Dec. (°)	Dec. (′)	Range	Type	Period (d)	Spectrum
S	09	32.3	−28	38	6.4–6.9	W UMa	0.65	A
U	10	35.2	−39	34	5.7–6.8	Irregular	—	N

DOUBLE STARS

	RA (h)	RA (m)	Dec. (°)	Dec. (′)	PA (°)	Separation (′)	Magnitudes
θ	09	44.2	−27	46	005	0.1	5.4, 5.6
δ	10	29.6	−30	36	226	11.0	5.6, 9.6

GALAXIES

M	C	NGC	RA (h)	RA (m)	Dec. (°)	Dec. (′)	Magnitude	Dimensions min	Type
		2997	09	45.6	−31	11	10.6	8.1 × 6.5	Sc
		3223	10	21.6	−34	16	11.8	4.1 × 2.6	Sb
		3347	10	42.8	−36	22	12.5	4.4 × 2.6	SBb

The only notable object for telescopic observers is NGC 2997, a spiral galaxy 24.8 million light-years away; its plane is inclined at an angle of 45°, so that it looks elliptical. The Antlia dwarf galaxy, magnitude 16.2, is one of the most distant members of the Local Group.

APUS

(Abbreviation: Aps)

Apus was probably introduced by Petrus Plancius around 1597, and was called 'Paradysvogel Apis Indica'; it later became Avis Indica, the Bird of Paradise, and finally Apus.

See chart or Musca.

There are two stars above the fourth magnitude:

RA (h) (m) (s)	Dec. (°) (′) (″)	m	M	d (light-years)	Spectrum
α 14 47 51.7	−79 02 41	3.83	−1.67	410	K5
γ 16 33 27.5	−78 53 49	3.86	0.41	160	K0

The only other star above magnitude 4.5 is β (4.23).

VARIABLE STARS

	RA (h)	(m)	Dec. (°)	(′)	Range	Type	Period (d)	Spectrum
θ	14	05.3	−76	48	6.4–8.6	Semi-regular	119	M
S	15	04.3	−71	53	9.5–1.5	R CrB	—	R
DW	17	23.5	−67	56	7.9–9.1	Algol	2.31	B

(R Apodis, magnitude 5.3, has been suspected of variability.)

DOUBLE STAR

	RA		Dec.		PA	Separation	Magnitudes
	(h)	(m)	(°)	(′)	(°)	(″)	
δ	16	20.3	−78	42	012	102.9	4.7, 5.1

GLOBULAR CLUSTERS

M	NGC	RA		Dec.		Diameter	Magnitude
		(h)	m	(°)	(′)	(′)	
	IC 4499	15	00.3	−82	13	7.6	10.6
	6101	16	25.8	−72	12	10.7	9.3

GALAXY

M	NGC	RA		Dec.		Magnitude	Dimensions	Type
		(h)	(m)	(°)	(′)		(′)	
	5967	15	48.1	−75	40	12.5	2.9 × 1.8	SBc

δ^1 and δ^2 make up a wide pair, separated by 100 light-years; the primary is a red M4 giant, irregularly variable between magnitudes 4.66 and 4.87, while the secondary, magnitude 5.27, is an orange K3 giant. Probably the most attractive telescopic objects in Apus are the two globular clusters.

AQUARIUS

(Abbreviation: Aqr)

A Zodiacal constellation, and of course one of Ptolemy's 'originals'. Oddly enough there are no well-defined legends attached to it though it has been associated with Ganymede, cup-bearer of the Olympian gods. There are nine stars brighter than magnitude 4. Aquarius is distinguished by the presence of two splendid planetary nebulæ, C 55 (the Saturn Nebula) and C 63 (the Helix Nebula).

There are 10 stars above magnitude 4:

	RA			Dec.			m	M	d (light-years)	Spectrum	
	(h)	(m)	(s)	(°)	(′)	(″)					
22 β	21	31	33.52	−05	34	16.2	2.90	−3.47	612	G0	Sadalsuud
34 α	22	05	47.03	−00	19	11.4	2.95	−3.88	760	G2	Sadalmelik
76 δ	22	54	39.04	−15	49	14.7	3.27	−0.18	159	A3	Skat
55 ζ	22	28	49.80	−00	01	12.2	3.65	1.14	103	F3	Altager
							4.42			F3	
88 c²	23	09	26.76	−21	10	20.9	3.68	−0.60	103	K1	
73 λ	22	52	36.86	−07	34	46.8	3.7v	−1.67	391	M2	Hydor
2 ε	20	47	40.53	−09	29	44.5	3.78	−0.46	229	A1	Albali
48 γ	22	21	39.30	−01	23	14.5	3.86	0.44	15	A0	Sadachbia
98 b′	23	22	58.30	−20	06	01.2	3.96	0.48	162	K0	

Also above magnitude 4.5: η (Hyria), 4.04; τ^2, 4.05; θ (Ancha), 4.17; φ, 4.22; ψ^1, 4.24; ι, 4.29; 99, 4.38; ψ^2, 4.4v; k, 4.4v; c^1, 4.48; ω^2, 4.49; ν (Albulaan), 4.50; o, magnitude 4.74, has a proper name, Kae Uh; π, magnitude 4.80, also has a name, Seat.

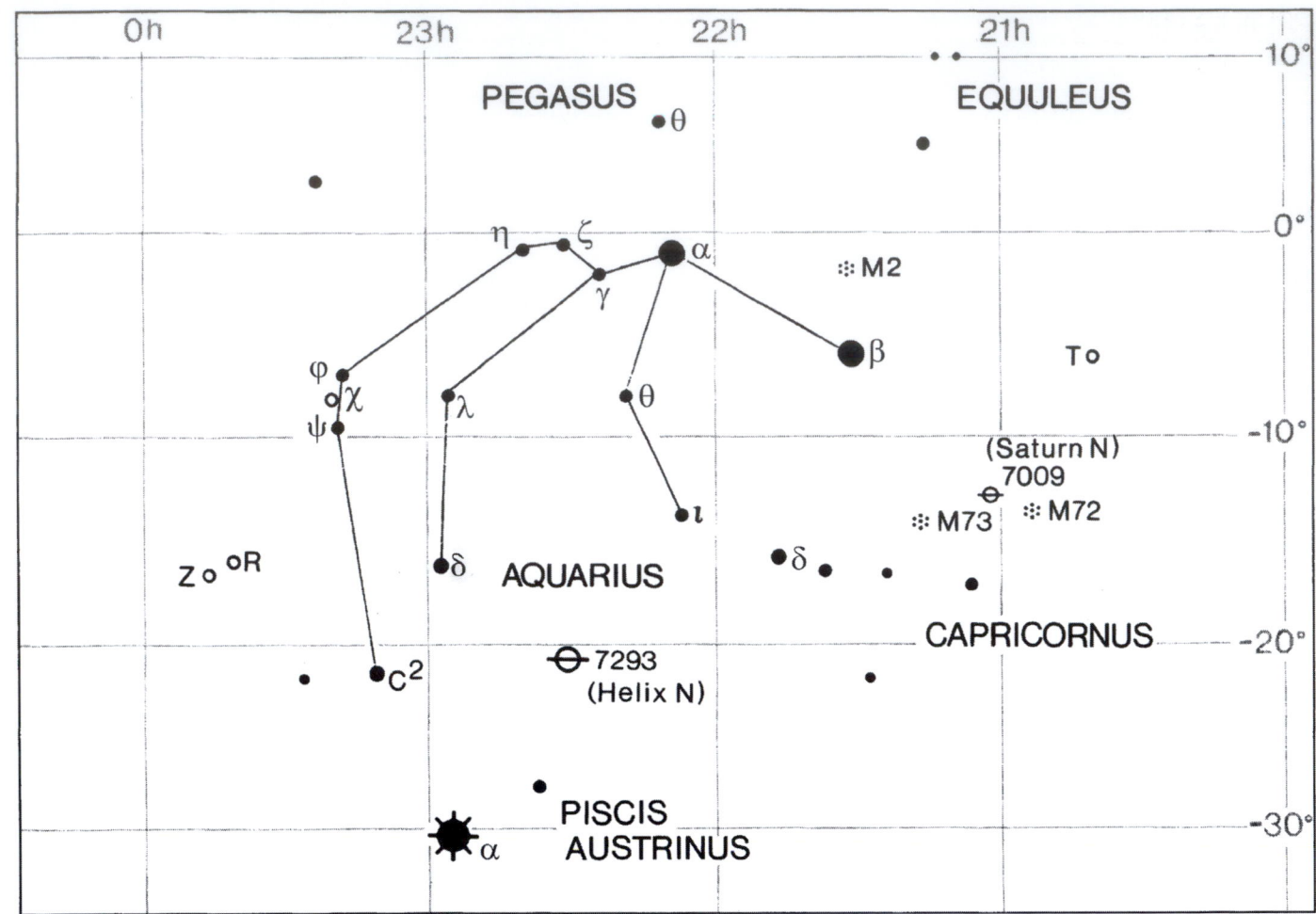

ASTERISM

M	C	NGC	RA		Dec.		
			(h)	m	(°)	(′)	
73		6994	20	58.9	−12	38	Four stars; not a true cluster

VARIABLE STARS

	RA		Dec.		Range	Type	Period (d)	Spectrum
	(h)	(m)	(°)	(′)				
W	20	46.4	−04	05	8.4–14.9	Mira	381.1	M
V	20	46.8	+02	26	7.6–9.4	Semi-regular	244	M
T	20	49.9	−05	09	7.2–14.2	Mira	202.1	M
X	22	18.7	−20	54	7.5–14.8	Mira	311.6	S
S	22	57.1	−20	21	7.6–15.0	Mira	297.3	M
R	23	43.8	−15	17	5.8–12.4	Symbiotic	387	M+Pec

DOUBLE STARS

	RA (h)	(m)	Dec. (°)	(')	PA (°)	Separation ('')	Magnitudes	
β	21	31.6	−05	34	AB 321	35.4	2.9, 10.8	
					AC 186	57.2	11.4	
41	22	14.3	−21	04	AB 114	5.0	5.6, 7.1	
					AC 043	212.1	9.0	
51	22	24.1	−04	50	AB 324	0.5	6.5, 6.5	
					AB+D 191	116.0	10.1	
					AC 342	54.4	10.2	
					AE 133	132.4	8.6	
ζ	22	28.8	−00	01	187	2.9	4.3, 4.5	Binary, period 856 y
89	23	09.9	−22	27	007	0.4	5.1, 5.9	
107	23	46.0	−18	41	136	6.6	5.7, 6.7	

ASTERISM

M	C	NGC	RA (h)	(m)	Dec. (°)	(')	
73		6994	20	58.9	−12	38	Four stars; not a true cluster

GLOBULAR CLUSTERS

M	C	NGC	RA (h)	(m)	Dec. (°)	(')	Diameter (')	Magnitude
72	6981	20	53.5	−12	32	5.9	9.3	
2		7089	21	33.5	−00	49	12.9	6.5

PLANETARY NEBULÆ

M	C	NGC	RA (h)	(m)	Dec. (°)	(')	Dimensions ('')	Magnitude	Magnitude of central star	
	55	7009	21	04.2	−11	22	2.5 × 100	8.3	11.5	Saturn Nebula
	63	7293	22	29.6	−20	48	770	6.5	13.5	Helix Nebula

GALAXIES

M	C	NGC	RA (h)	(m)	Dec. (°)	(')	Magnitude	Dimensions (')	Type
		7184	22	02.7	−20	49	12.0	5.8 × 1.8	Sb
		7606	23	19.1	−08	29	10.8	5.8 × 2.6	Sb
		7723	23	38.9	−12	58	11.1	3.6 × 2.6	Sb
		7727	23	39.9	−12	18	10.7	4.2 × 3.4	Sba

α is a yellow giant; its name, Sadalmelik, means 'luck of the king'. ζ (Altager) is an easy double; it is a slow binary, with period given as 856 years. The orbits are decidedly elliptical; the present separation is about 2.5'', but is increasing; both components are yellow dwarfs; the primary is very slightly more massive than the Sun (1.1 times) the secondary slightly less (0.9). λ is an M-type red giant, varying irregularly between magnitude 3.70 to 3.80. It has an alternative proper name, Ekkhysis – in Greek, 'outpouring water'.

Aquarius is not a very distinctive constellation, but it does have three bright clusters, the globular M 72 and two interesting planetaries, the Saturn and the Helix. Curiously, neither is in Messier's list, but M 73 is merely a grouping of four faint stars.

AQUILA

(Abbreviation: Aql)

One of the most distinctive of all the northern constellations, and, of course, an 'original'. Mythologically it represents an eagle that was sent by Jupiter to collect a Phrygian shepherd-boy, Ganymede, who was destined to become cup-bearer of the Gods – following an unfortunate episode in which the former holder of the office, Hebe, tripped and fell during a particularly solemn ceremony.

The leading star is Altair. Altogether there are eight stars above magnitude 4.

	RA			Dec.			m	M	d (light-years)	Spectrum	
	(h)	(m)	(s)	(°)	(')	('')					
53 α	19	50	46.7	+08	52	03	0.76	2.30	17	A7	Altair
50 γ	19	46	15.6	+10	36	48	2.72	−3.03	460	K3	Tarazed
17 ζ	19	05	24.6	+13	51	49	2.99	0.96	83	A0	Deneb el Okab
65 θ	20	11	18.3	+00	49	17	3.24*	−1.48	290	B9	
30 δ	19	25	29.7	+03	06	53	3.36	2.43	50	F0	Song
16 λ	19	06	15.0	+04	52	56	3.43	0.51	125	B9	Al Thalimain
60 β	19	55	18.,8	+06	24	29	3.71	3.03	45	G8	Alshain
55 η	19	52	28.4	+01	00	20	var	−3.91v	1170	F-G	Bezek

Also above magnitude 4.5; ε, 4.02; 12 (Bezed), 4.2; 71, 4.31; ι, 4.36; μ, 4.45. ν, magnitude 4.64, is a yellow supergiant, around 140 000 times as luminous as the Sun.

VARIABLE STARS

	RA		Dec.		Range	Type	Period (d)	Spectrum
	(h)	(m)	(°)	(')				
V	19	04.4	−05	41	6.6–8.4	Semi-regular	353	N
R	19	06.4	+08	14	5.5–12.0	Mira	284.2	M
TT	19	08.2	+01	18	6.4–7.7	Cepheid	13.75	F–G
W	19	15.4	−07	03	7.3–14.3	Mira	490.4	S
U	19	29.4	−07	03	6.1–6.9	Cepheid	7.02	F–G
X	19	51.5	+04	28	8.3–15.5	Mira	347.0	M
η	19	52.5	+01	00	3.5–4.4	Cepheid	7.18	F–G
RR	19	57.6	−01	53	7.8–14.5	Mira	394.8	M

DOUBLE STARS

	RA		Dec.		PA	Separation ('')	Magnitudes	
	(h)	(m)	(°)	(')	(°)			
ε	18	59.6	+15	04	187	131.1	4.0, 9.9	
23	19	18.5	+01	05	005	3.1	5.3, 9.3	
31	19	25.0	+11	57	343	105.6	5.2, 8.7	
δ	19	25.5	+03	07	271	108.9	3.4, 10.9	
ν	19	26.5	+00	20	288	201.0	4.7, 8.9	
U	19	29.4	−07	03	228	1.5	variable, 11.7	
χ	19	42.6	+11	50	077	0.5	5.6, 6.8	
γ	19	46.3	+10	37	258	132.6	2.7, 10.7	
π	19	48.7	+11	49	110	1.4	6.1, 6.9	
α	19	50.8	+08	52	301	165.2	0.8, 9.5	optical
57	19	54.6	−08	14	170	35.7	5.8, 6.5	

OPEN CLUSTERS

M	C	NGC	RA (h)	(m)	Dec. (°)	(′)	Diameter (′)	Magnitude	No. of stars
		6709	18	51.5	+10	21	13	6.7	40
		6755	19	07.8	+04	14	15	7.5	100

PLANETARY NEBULÆ

M	C	NGC	RA (h)	(m)	Dec. (°)	(′)	Diameter (″)	Magnitude	Magnitude of central star
		6741	19	02.6	−00	27	6	10.8	14.7
		6751	19	05.9	−06	00	20	12.5	13.9
		6790	19	23.2	+01	31	7	10.2	13.5
		6803	19	31.3	+10	03	6	11.3	15.2

Altair, 'the flying eagle', is rotating rapidly with a velocity at the equator of 286 km s^{-1}, so that it is flattened at the poles. It has three faint optical companions: θ has a Chinese name, Tsen Foo, and is a spectroscopic binary with a period of 17.1 days. δ is a triple; it is an astrometric binary with a period of 3.4 years, and the primary is a spectroscopic binary. η – its name, Bezek ('lightning') is Hebrew – is a Cepheid, easy to follow by using θ and δ as comparison stars.

ρ Aquilæ is now in Delphinus, because of its proper motion. Its Flamsteed number is 67 Aquilæ. The annual proper motion is +0.004″ in RA, +0.06″ in Dec. 2000 position, RA 20h 14m 16s .4, Dec. +15° 11′ 51″. Its apparent magnitude is 4.95; spectrum B2; absolute magnitude +1.4.

On 1 December 1999, A. Pereida discovered a nova at RA 19h 23m 5s. 3, Dec. +4° 57′ 20″. 1. It reached magnitude 3.5, and was the brightest northern-hemisphere nova since 1975.

Aquila is crossed by the Milky Way, and is a very rich constellation. Altair is one of the three stars of the 'Summer Triangle' (the other two are Deneb and Vega).

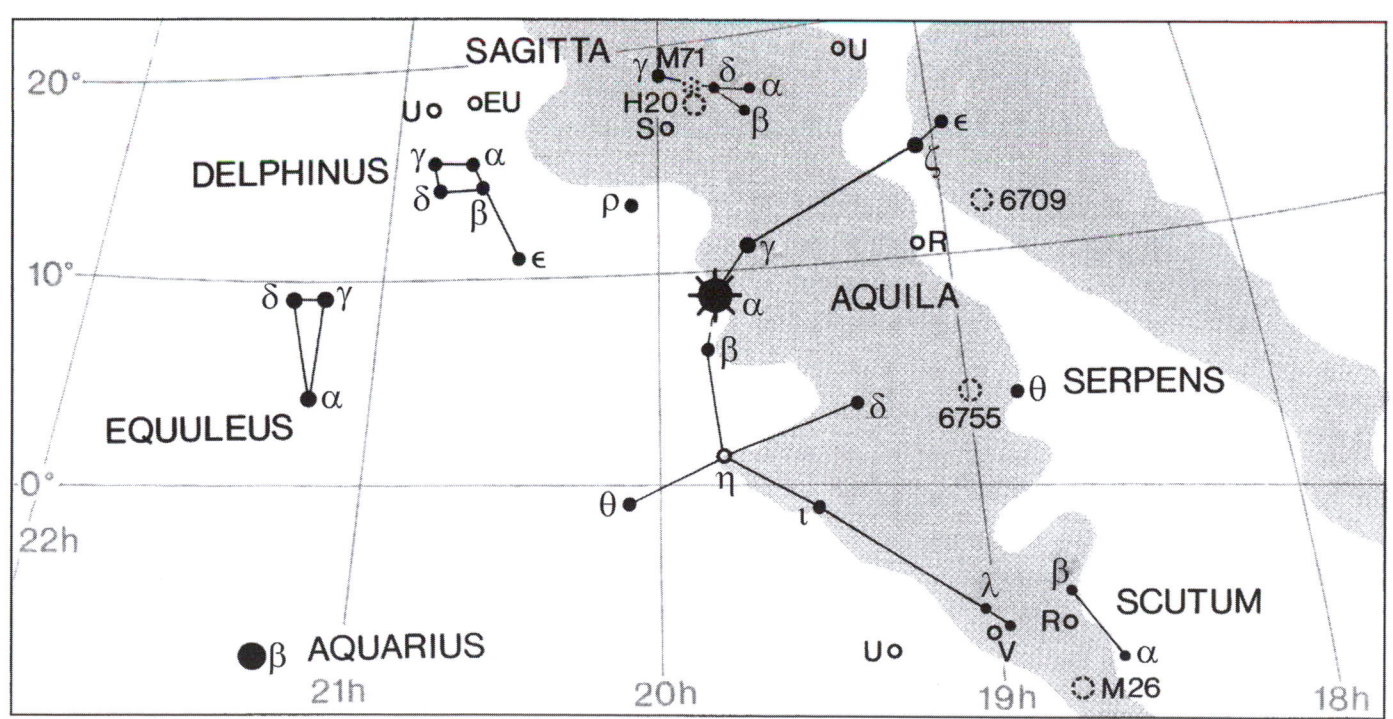

ARA

(Abbreviation: Ara)

An original constellation, though no mythological legends are attached to it.

There are seven stars above the fourth magnitude:

	RA			Dec.			m	M	d (light-years)	Spectrum	
	(h)	(m)	(s)	(°)	(′)	(″)					
β	17	25	18.0	−55	31	47	2.84	−3.40	600	K3	Karnot
α	17	31	50.5	−49	52	34	2.84	−1.51	240	B2	Choo
ζ	16	58	37.2	−55	59	24	3.12	−3.11	570	K5	Korban
γ	17	25	23.7	−56	22	40	3.31	−4.40	1130	B1	Zadok
δ	17	31	07.0	−60	41	01	3.60	−0.19	187	B8	Tseen Yin
θ	18	06	37.9	−50	05	29	3.65	−3.81	1010	B2	Tau Shou
η	16	49	47.1	−59	02	29	3.77	−1.14	310	K5	

Also above magnitude 4.5: ε (Tso Kang), 4.06.

is a rapid rotator, and is probably surrounded by a dense equatorial disc of material. μ (magnitude 5.1, RA 17h 44m 09s, Dec. −51° 50′ 03″) is a planetary centre; four planets are known.

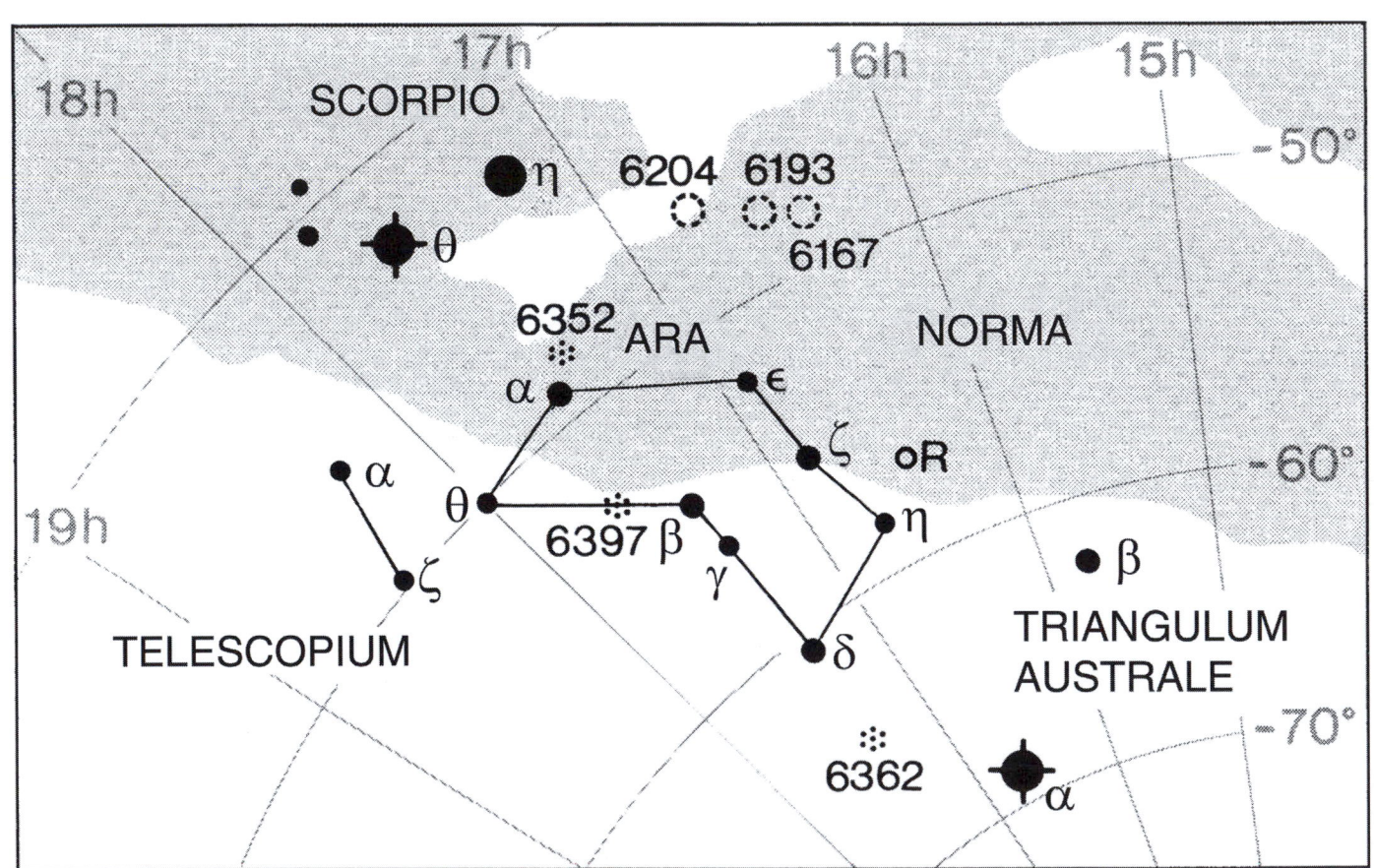

DOUBLE STAR

γ RA 17h 25m 24s Dec.e−56° 22′ 40″, separation 17″ .9, magnitudes 3.5 and 10.5, PA 328°.

VARIABLE STARS

	RA		Dec.		Range	Type	Period (d)	Spectrum
	(h)	m	(°)	(′)				
X	16	36.4	−55	24	8.0–13.5	Mira	175.8	M
R	16	39.7	−57	00	6.0–6.9	Algol	4.42	B
U	17	53.6	−51	41	7.7–14.1	Mira	225.2	M

OPEN CLUSTERS

M	C	NGC	RA		Dec.		Diameter	Magnitude	No. of stars
			(h)	(m)	(°)	(′)	(′)		
	82	6193	16	41.3	48	46	15	5.2	—
		6204	16	46.5	47	01	5	8.2	45
		6208	16	49.5	−53	49	16	7.2	60
		6250	16	58.0	−45	48	8	5.9	60
		6253	16	59.1	52	43	5	10.2	30
		H.13	17	05.4	−48	11	15	—	15
		IC 4651	17	24.7	−49	57	12	6.9	80

GLOBULAR CLUSTERS

M	C	NGC	RA		Dec.		Diameter	Magnitude
			(h)	(m)	(°)	(′)	(′)	
	81	6352	17	25.5	−48	25	7.1	8.1
		6362	17	31.9	−67	03	10.7	8.3
	86	6397	17	40.7	53	40	25.7	5.6

GALAXIES

M	C	NGC	RA		Dec.		Magnitude	Dimensions	Type
			(h)	(m)	(°)	(′)		(′)	
		6215	16	51.1	−58	59	11.8	2.0 × 1.6	Sc
		6221	16	52.8	−59	13	11.5	3.2 × 2.3	SBc

Ara has no bright stars, but its pattern makes it easy to recognise.

ARIES

(Abbreviation: Ari)

The first constellation of the Zodiac – though since the vernal equinox has shifted into Pisces. Aries should logically be classed as second! In mythology it represents a ram with a golden fleece, sent by the god Mercury (Hermes) to rescue the children of the king of Thebes from an assassination plan by their stepmother. The ram, which had the remarkable ability to fly, carried out its mission; unfortunately, the girl (Helle) lost her balance and fell to her death in that part of the sea now called the Hellespont, but the boy (Phryxus) arrived safely. After the ram's death its golden fleece was hung in a sacred grove from which it was later removed by the Argonauts commanded by Jason.

Aries is moderately conspicuous. See chart for Andromeda.

There are four stars above the fourth magnitude:

	RA (h)	(m)	(s)	Dec. (°)	(′)	(″)	m	M	d (light-years)	Spectrum	
13 α	02	07	10.3	+23	27	46	2.01	0.48	66	K2	Hamal
6 β	01	54	38.4	+20	48	30	2.64*	1.33	60	A5	Sheratan
41 c	02	49	59.0	+27	15	39	3.61	0.16	159	B8	Bharani
5 γ	01	53	31.8	+19	17	39	3.88	−0.10	204	A1p..	Mesartim

Also above magnitude 4.5: δ (Botein), 4.35.

VARIABLE STARS

	RA (h)	(m)	Dec. (°)	(′)	Range	Type	Period (d)	Spectrum
R	02	16.1	+25	03	7.4–13.7	Mira	186.8	M
T	02	48.3	+17	31	7.5–11.3	Semi-regular	317 (variable)	M
U	03	11.0	+14	48	7.2–15.2	Mira	371.1	M

DOUBLE STARS

	RA (h)	(m)	Dec. (°)	(′)	PA (°)	Separation (″)	Magnitudes
γ	01	53.5	+19	18	007	7.8	4.8, 4.8
π	02	49.3	+17	28	AB 120	3.2	5.2, 8.7
					AC 110	25.2	10.8
ε	02	59.2	+21	20	208	1.5	5.2, 5.5
υ	02	59.2	+21	20	208	1.5	5.0, 5.2

GALAXIES

M	C	NGC	RA (h)	(m)	Dec. (°)	(′)	Magnitude	Dimensions (′)	Type	
		772	01	59.3	+19	01	10.3	7.1 × 4.5	Sb	Arp 78
		976	02	34.0	+20	59	12.4	1.7 × 1.5	Sb	

α (Hamal) is a K-type giant, twice as massive as the Sun and 15 times larger in diameter. β (Sheratan) is a spectroscopic binary; the companion is probably of type G, orbiting in 107 days (orbital eccentricity 0.88). 41 c does not have a Greek letter, because it was once included in the now-rejected constellation of Musca Borealis (the Northern Fly). γ (Mesartim) is a wide, easy binary with almost equal components; magnitudes 4.75 and 4.83. The orbital period is over 5000 years. Orbiting the binary is a 9.6-magnitude K-type dwarf; separation 221″.

53 (UW) Arietis, a small-range β Cephei type variable (RA 03h 07m 25s .7, Dec. +17° 52′ 48.0″) is one of three 'runaway stars' thought to have been ejected from the Orion Nebula about two million years ago (the others are AE Aurigæ and μ Columbæ).

AURIGA (THE CHARIOTEER)

(Abbreviation: Aur)

One of the most brilliant of the northern constellations, with Capella outstanding. Mythologically it represents Erechthonius, son of Vulcan, the blacksmith of the gods; he became King of Athens, and invented the four-horse chariot. In the Cambridge catalogue there are nine stars brighter then the fourth magnitude.

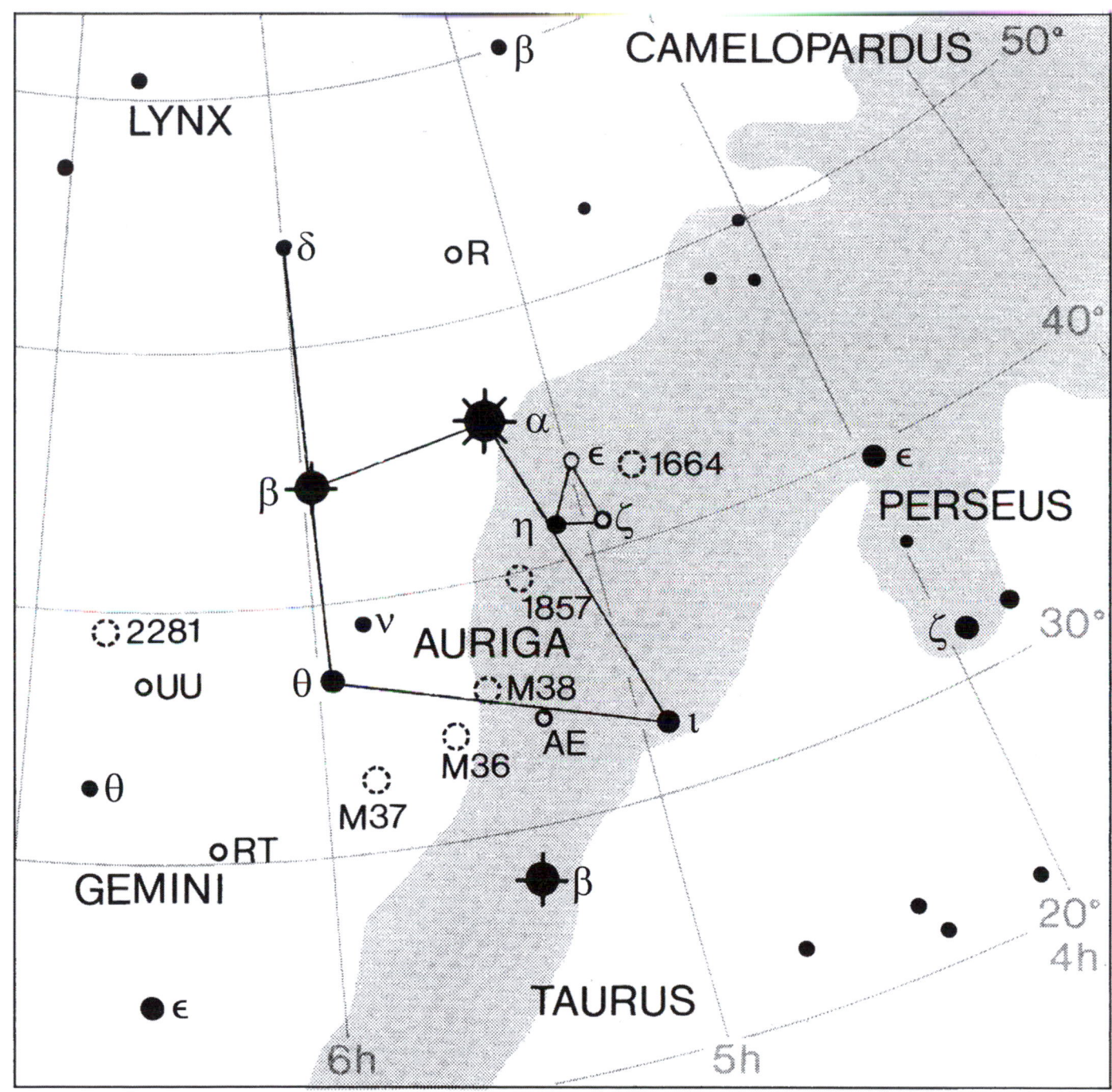

	RA			Dec.			m	M	d (light-years)	Spectrum	
	(h)	(m)	(s)	(°)	(′)	(″)					
8 α	05	16	41.3	+45	59	57	0.08*	−0.48	42	G8	Capella
34 β	05	59	32	+44	56	51	1.90	−0.10	82	A2	Menkarlina
37 θ	05	59	434.2	+37	12	46	2.65	−0.98	173	A0p	Manus
3 ι	04	56	59.6	+33	09	58	2.69	−3.29	510	K5	Hassaleh
7 ε	05	01	58.1	+43	49	24	3.0v	−5.9	2030	F0	Almaaz
10 η	05	06	30.9	+41	14	05	3.18	−0.96	219	B3	Maha-Sim
8 ζ	05	02	28.7	+41	04	34	3.7v	−3.22	790	K4	Sadatoni
33 δ	05	59	31.6	+54	17	06	3.72	0.55	140	K0	Prijipati
32 ν	05	51	29.4	+39	08	55	3.97	−0.13	215	K0	

Also above magnitude 4.5: π, 4.30; κ, 4.32.

In addition, Al Nath used to be called γ Aurigæ, but has now been given a free transfer, and is included in Taurus as β Tauri.

VARIABLE STARS

	RA		Dec.		Range	Type	Period (d)	Spectrum
	(h)	(m)	(°)	(′)				
RX	05	01.4	+39	58	7.3–8.0	Cepheid	11.62	F–G
ε	05	02.0	+43	49	2.9–3.8	Eclipsing	9892	F
ζ	05	02.5	+41	05	3.7–4.1	Eclipsing	972.1	K + B
R	05	17.3	+53	35	6.7–13.9	Mira	457.5	M
UV	05	21.8	+32	31	7.4–10.6	Mira	394.4	M
U	05	42.1	+32	02	7.5–15.5	Mira	408.1	M
X	06	12.2	+50	14	8.0–13.6	Mira	163.8	M
RT	06	28.6	+30	30	5.0–5.8	Cepheid	3.73	F–G
WW	06	32.5	+38	27	5.8–6.5	Algol	2.53	A + A
UU	06	36.5	+38	27	5.1–6.8	Semi-regular	234	N

DOUBLE STARS

	RA		Dec.		PA	Separation (″)	Magnitudes
	(h)	(m)	(°)	(′)	(°)		
ω	04	59.3	+37	53	359	5.4	5.0, 8.0
R	05	17.3	+53	35	339	47.5	variable, 8.6
ν	05	51.5	+39	09	206	54.6	4.9, 9.3
δ	05	59.5	+54	17	AB 271	115.4	3.7, 9.5
					AC 067	197.1	9.5
θ	05	59.7	+37	13	AB 313	3.6	2.6, 7.1
					AC 297	50.0	10.6

OPEN CLUSTERS

M	C	NGC	RA		Dec.		Diameter	Magnitude	No. of stars
			(h)	(m)	(°)	(′)	(′)		
		1664	04	51.1	−43	42	18	7.6	—
		1778	05	08.1	−37	7	7.7	25	

OPEN CLUSTERS

M	C	NGC	RA (h)	(m)	Dec. (°)	(')	Diameter (')	Magnitude	No. of stars
		1857	05	20.2	−39	21	6	7.0	40
		1893	05	22.7	−33	24	11	7.5	60
38	1912	05	28.7	−35	50	21	6.4	100	
36	1960	05	36.1	−34	08	12	6.0	60	
37	2099	05	52.4	−32	33	24	5.6	150	
		2126	06	03.0	−49	54	6	10.2	40
		2281	06	49.3	−41	04	15	5.4	30

NEBULA

M	C	NGC	RA (h)	(m)	Dec. (°)	(')	Dimensions (')	Magnitude of illuminating star
	31	IC 405	05	16.2	−34	16	30·19	6v

Capella is a close binary; both components are G-type giants, respectively 79 and 78 times as luminous as the Sun. The system also includes a binary pair made up of two faint red dwarfs, around 10 000 a.u. from the main pair.

Menkarlina (also called Menkarlinan) is a triple system; the two main components are A-type subgiants, and make up an eclipsing binary, so that the apparent magnitude ranges from 1.85 to 1.93 in a period of 47.5 hours. The third component is a red dwarf 130 a.u. from the main pair.

AE Aurigæ (RA 05h 16m 18s .2, Dec. +34° 18′ 44.3″), type O, distance ~1500 light-years, luminosity 30 000 times that of the Sun, lights up the Flaming Star Nebula, IC 405. It is one of three 'runaway stars' from the Trapezium in the Orion Nebula (the other two are 53 Arietis and μ Columbæ).

The two remarkable ecipsing binaries, ε and ζ, have already been dicussed. δ has three faint companions, B (9.7) C (also 9.7) and an 11th-magnitude companion to C. The name Prijipati is Sanskrit: 'Lord of Creation'.

BOÖTES

(Abbreviation: Boo)
An original constellation, dominated by Arcturus. There are various myths attached to it, but none is very defnite. According to one version, Boötes was a herdsman who invented the plough drawn by two oxen, for which service he was transferred to the sky. There are eight stars above the fourth magnitude:

	RA (h)	(m)	(s)	Dec. (°)	(')	(″)	m	M	d (light-years)	Spectrum	
16 α	14	15	40.35	+19	11	14	−0.05	−0.31	37	K2	Arcturus
36 ε	14	44	59.2	+27	04	27	2.35	−1.69	210	A0	Izar
							2.70				
8 η	13	54	41.1	+18	23	25	2.68	−2.41	37	G0	Muphrid
9 γ	14	32	04.8	+38	18	28	3.04	0.96	85	A7	Seginus
49 δ	15	15	30.1	+33	18	54	3.46	0.69	117	G8	
42 β	15	01	56.8	+40	23	26	3.49	−0.64	219	G8	Nekkar
25 ρ	14	31	49.9	+30	22	16	3.57	0.27	149	K3	Hemelein Prima
30 ζ	14	41	8.9	+13	43	42	3.78	0.06	180	A3	

Also above magnitude 4.5: θ, 4.04; υ, 4.08: λ, 4.18; μ (Alkalurops), 4.31; σ (Hemelein Secunda), 4.47, π¹, 4.49; τ, 4.50.

The discarded constellation Quadrans (the Quadrant) was made up of dim stars near Nekkar. However, the January meteors radiating from there are still called the Quadrantids.

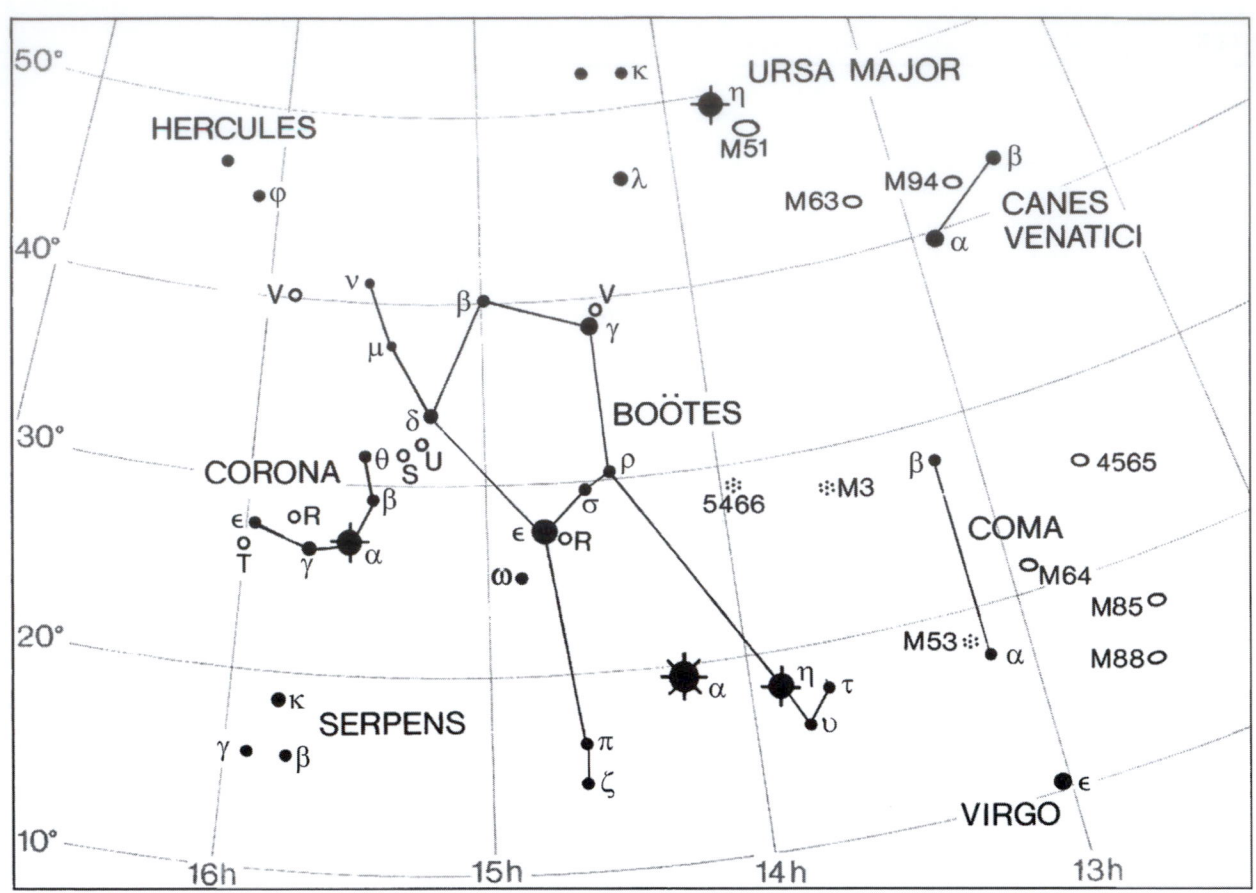

VARIABLE STARS

	RA		Dec.		Range	Type	Period (d)	Spectrum
	(h)	(m)	(°)	(′)				
ZZ	13	56.2	+25	55	5.8–6.4	W UMa	5.0	G2 + G2
S	14	22.9	+53	49	7.8–13.8	Mira	270.7	M
V	14	29.8	+38	52	7.0–12.0	Semi-reg.	258	M
R	14	37.2	+26	44	6.2–13.1	Mira	223.4	M
W	14	43.4	+26	32	4.7–5.4	Semi-regular	450	M
i (44)	15	03.8	+47	39	6.5–7.1	W UMa	0.27	G + G

DOUBLE STARS

	RA		Dec.		PA	Separation (″)	Magnitudes	
	(h)	(m)	(°)	(′)	(°)			
κ	14	13.5	+51	47	236	13.4	4.6, 6.6	
ι	14	16.2	+51	22	033	38.5	4.9, 7.5	
π	14	40.7	+16	25	108	5.6	4.9, 5.8	
ζ	14	41.1	+13	44	AB 303	1.0	4.5, 4.6	Binary, 123.3 y
					AC 259	99.3	10.9	
i (44)	15	03.8	+47	39	040	1.0	5.3v, 6.2	Binary, 225 y
μ	15	24.5	+37	23	171	108.3	4.3, 7.0	
ε	14	45.0	+27	04	341	2.9	2.5, 4.9	
ξ	14	51.4	+19	06	319	6.7	4.7, 6.8	Binary, 150 y

GLOBULAR CLUSTER

M	C	NGC	RA (h)	(m)	Dec. (°)	(')	Diameter (")	Magnitude
		5466	14	05.5	+28	32	11	9.1

GALAXIES

M	C	NGC	RA (h)	(m)	Dec. (°)	(')	Magnitude	Dimensions (')	Type
	45	5248	13	37.5	+08	53	10.2	6.5 × 4.9	Sc
		5676	14	32.8	+49	28	10.9	3.9 × 2.0	Sc

Arcturus is a K-type giant; its lovely orange colour is very different from the steely blue of Vega or the light yellow of Capella. It is three and a half times as massive as the Sun, with a diameter of about 4 000 000 km and a surface temperature of 4000 °C. It is oscillating, and the magnitude varies by 0m. 04 in a period of 8.3 days. It has relatively high proper motion, and is approaching us at a rate of 5 km s^{-1}; it is now at the point of its closest approach to the Sun.

β (Nekkar), a G-type giant, mass and diameter 1.1 times that of the Sun, which lies close to the radiant point of the Quadrantid meteors. It has an alternative proper name: Meres.

ε (Izar) also has an alternative proper name: Pulcherrima. It is a binary, made up of a K-type orange giant with a smaller Main Sequence companion. The separation is almost 3″, so that this is an easy telescopic pair.

ξ is a much closer binary; separation always < 1″. Both components are of type A; individual apparent magnitudes 4.4 and 4.8.

η (Muphrid) is a spectroscopic binary; it lies only 3.24 light-years from Arcturus, so that in the sky of any orbiting planet Arcturus would shine as a star of magnitude −5, far brighter than Venus appears to us.

τ is a binary, with an F-type primary together with a dwarf M-type secondary; the real separation is about 240 a.u., and the orbital period must be several thousands of years. The yellow primary is known to be attended by a planet, discovered in 1996 by Marcy and Butler and quickly confirmed by Mayor and Queloz. The planet is over four times as massive as Jupiter, and is presumably a gas-giant with a surface temperature of over 1000 °C; mean distance from the star 0.05 a.u., period 3.31 days, orbital eccentricity 0.02. The distance from the Sun is 50.8 light-years.

γ (Seginus) is a very small-range δ Scuti-type variable (m = 3.02 to 3.07, period 78 h). ε has a faint G-type binary companion; separation about 4000 a.u., revolution period 120 000 y. The entirely unremarkable F7-type star 38 Boötis has a proper name: Merga.

CÆLUM

(Abbreviation: Cæ)

This entirely unremarkable constellation was introduced in 1752 by Lacaille, under the name of Cæla Sculptoris. It has no star brighter than the fourth magnitude, and only two which are brighter than the fifth.

See chart for Columba.

γ Caeli has an orange giant primary (γ^1) and an F-type companion (γ-2). The F-type star is a δ Scuti-type variable, magnitude 6.28 to 6.39, period 0.13 d. Despite its very small range it has a variable star designation: X Caeli.

RA (h)	(m)	(s)	Dec. (°)	(')	(")	m	M	d	Spectrum
α 04	40	33.82	−41	51	48.9	4.44	2.92	66	F2 binary
γ 05	04	24.31	−35	28	58.3	4.55	0.78	185	K2

DOUBLE STARS

RA (h)	(s)	Dec. (°)	(')	PA (°)	Separation (")	Magnitudes
α 04	40.6	−41	52	121	6.6	4.4, 12.5
γ 05	04.4	−35	29	308	2.9	4.5, 6.3v

CAMELOPARDUS (OR CAMELOPARDALIS)

(Abbreviation: Cam)

A very barren northern constellation. It was introduced to the sky by Hevelius in 1690, and some historians have maintained that it represents the camel which carried Rebecca to Isaac! It is interesting to note that several of the apparently faint stars are in fact highly luminous and remote; for instance α Cam, which is below the fourth magnitude, is well over 20 000 times more luminous than the Sun.

There are no stars in Camelopardus above the fourth magnitude. Above 4.5 there are 2 (H), 4.21; α, 4.29; and 7, 4.47. See chart for Cassiopeia.

VARIABLE STARS

	RA (h)	RA (m)	Dec. (°)	Dec. (')	Range	Type	Period (d)	Spectrum
U	03	37.5	+62	29	7.7–8.9	Semi-regular	400	N
RV	04	30.7	+57	25	7.1–8.2	Semi-regular	101	M
T	04	40.1	+66	09	7.3–14.4	Mira	373.2	M
X	04	45.7	+75	06	7.4–14.2	Mira	143.6	K–M
S	05	41.0	+68	48	7.7–11.6	Semi-regular	327	R
V	06	02.5	+74	30	7.7–16.0	Mira	522.4	M
VZ	07	31.1	+82	25	4.8–5.2	Semi-regular	24	M
R	14	17.8	+83	50	7.0–14.4	Mira	270.2	S

OPEN CLUSTER

M	C	NGC	RA (h)	RA (m)	Dec. (°)	Dec. (')	Diameter (')	Magnitude	No. of stars
		1502	04	07.7	+62	20	8	5.7	45

PLANETARY NEBULA

M	C	NGC	RA (h)	RA (m)	Dec. (°)	Dec. (')	Diameter ('')	Magnitude	Magnitude of central star
		IC 3568	12	32.9	+82	33	6	11.6	12.3

GALAXIES

M	C	NGC	RA (h)	RA (m)	Dec. (°)	Dec. (')	Magnitude	Dimensions (')	Type
		IC 342	03	46.8	+68	06	9.2	17.8 × 17.4	SBc
		1961	05	42.1	+69	23	11.1	4.3 × 3.0	Sb
		2146	06	18.7	+78	21	10.5	6.0 × 3.8	SBb
		2366	07	28.9	+69	13	10.9	7.6 × 3.5	Irregular
	7	2403	07	36.9	+65	36	8.4	17.8 × 11.0	Sc
		2460	07	56.9	+60	21	11.7	2.9 × 2.2	Sb
		2655	08	55.6	+78	13	10.1	5.1 × 4.4	SBa
		2715	09	08.1	+78	05	11.4	5.0 × 1.9	Sc

Kemble's Cascade (Kemble 1), located in the constellation Camelopardus, is an asterism – a pattern created by unrelated stars. It is an apparent straight line of more than 20 colourful fifth to tenth magnitude stars over a distance of approximately five moon diameters, and the open cluster NGC 1502 can be found at one end.

It was named by Walter Scott Houston in honour of Father Lucian J. Kemble (1922–99), a Franciscan Friar and amateur astronomer who wrote a letter to Walter about the asterism, describing it as 'a beautiful cascade of faint stars tumbling from the northwest down to the open cluster NGC 1502' that he had discovered while sweeping the sky with a pair of 7 × 35 binoculars.

CANCER

(Abbreviation: Cnc)

Cancer is an obscure constellation, redeemed only by the presence of two famous star-clusters, Præsepe and M 67. However, it lies in the Zodiac, and it has a legend attached to it. It represents a sea-crab which Juno, queen of Olympus, sent to the rescue of the multi-headed hydra which was doing battle with Hercules. Not surprisingly, Hercules trod on the crab, but as a reward for its efforts Juno placed it in the sky!

There are only two stars above the fourth magnitude.

	RA			Dec.			m	M	d	Spectrum	
	(h)	(m)	(s)	(°)	(′)	(″)					
17 β	08	16	31.0	+09	11	08	3.53	−1.22	290	K4	Altarf
47 δ	08	44	41.1	+18	09	18	3.94	0.84	316	K0	Asellus Australis

Also above magnitude 5: ι (Decapoda), 4.03; α (Acubens), 4.26; γ (Asellus Borealis), 4.66; ζ (magnitude 4.67) is Tegmine. Præsepe (M 44) was actually given a Greek letter, ε.

ι is an easy double; a G-type primary and an A-type companion. ζ is a triple system. There are two binary pairs, 5.06″ apart, orbital period 1100 y. The name Tegmine means 'the shell of the crab'.

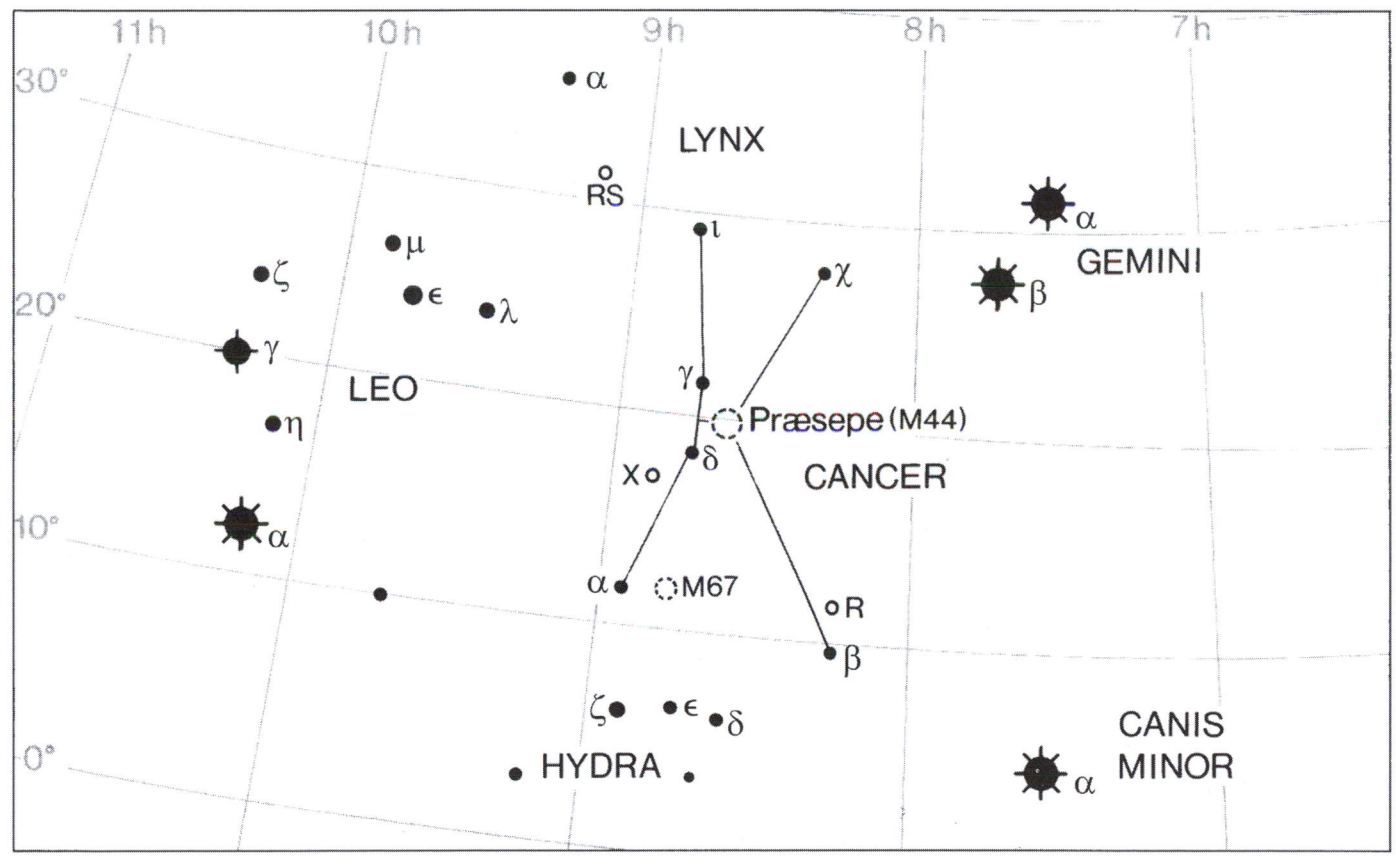

VARIABLE STARS

	RA		Dec.		Range	Type	Period (d)	Spectrum
	(h)	(m)	(°)	(′)				
R	08	16.6	+11	44	6.1–11.8	Mira	361.6	M
V	08	21.7	+17	17	7.5–13.9	Mira	272.1	S
X	08	55.4	+17	14	5.6–7.5	Semi-regular	195	N
T	08	56.7	+19	51	7.6–10.5	Semi-regular	482	R–N
W	09	09.9	+25	15	7.4–14.4	Mira	393.2	M
RS	09	10.6	+30	58	6.2–7.7	Semi-regular	120	M

The red variable X is a fine object, always within binocular range.

On the whole Cancer is a rather barren constellation, but it does contain two notable open clusters; M 44 (Præsepe) is prominent, and some very keen-sighted observes have also glimpsed M 67 with the naked eye.

OPEN CLUSTERS

	M	C	NGC	RA (h)	(m)	(°)	Dec. (')	(')	Diameter	Magnitude	No. of stars	
(ε)	44		2632	08	40.1	+19	59		95	3.1	50	Præsepe
	67		2682	08	50.4	+11	49		30	6.9	200	

GALAXY

M	C	NGC	RA (h)	(m)	(°)	Dec. (')	Magnitude	Dimensions (')	Type
		2775	09	10.3	+07	02	10.3	4.5 × 3.5	Sa

CANES VENATICI

(Abbreviation: CVn)

One of Hevelius' constellations, dating only from his maps of 1690, and evidently representing two hunting dogs (Asterion and Chara), which are being held by the herdsman Boötes – possibly to stop them from chasing the two Bears round and round the celestial pole. The name of Cor Caroli was given to α^2 CVn by Edmond Halley in honour of King Charles I. Cor Caroli (magnitude 2.90) is the only star above the fourth magnitude. It is also the prototype 'magnetic variable'. α^1 and α^2 share common proper motion in space, and must be associated, but they are a long way apart.

See chart for Ursa Major.

	RA (h)	(m)	(s)	Dec. (°)	(')	('')	m	M	d (light-years)	Spectrum	
12 α	12	56	01.9	+38	19	06	2.89	0.25	110	A0	Cor Caroli

Also above 4.5: β (Chara), 4.24.

Cor Caroli is slightly variable ($m = 2.84$ to 2.98) and has a very strong magnetic field. Chara is a solar-type star; it is 1.15 times as luminous as the Sun, its radius is 1.18 times greater and its mass 108 times greater. It is regarded as a possible planetary centre, though no planet has yet been found. Y Cvn, La Superba, is three times as massive as the Sun, and is one of the reddest stars in the sky. Its spectrum was of the old type N, C5 on the new system.

The constellation contains the bright, globular cluster M 3 and notable galaxies, including M 51 (the Whirlpool) and NGC 4631 (C 32), the Whale. There are two stars above magnitude 4.5:

Star	RA (h)	(m)	(s)	Dec. (°)	(')	('')	m	M	Spectrum	Dist (pc)	
8 β	12	33	44.4	+41	21	26	4.26	4.5	G0	9.2	Chara
12 α²	12	56	01.6	+38	19	06	2.90	0.1	A0p	20	Cor Caroli

VARIABLE STARS

	RA (h)	(m)	Dec. (°)	(')	Range	Type	Period (d)	Spectrum	
T	12	30.2	+31	30	7.6–12.6	Mira	290.1	M	
Y	12	45.1	+45	26	7.4–10.0	Semi-reg.	157	N	La Superba

VARIABLE STARS

	RA (h)	RA (m)	Dec. (°)	Dec. (')	Range	Type	Period (d)	Spectrum
U	12	47.3	+38	23	7.2–11.0	Mira	345.6	M
TU	12	54.9	+47	12	5.6–6.6	Semi-regular	50	M
V	13	19.5	+45	32	6.5–8.6	Semi-regular	192	M
R	13	49.0	+39	33	6.5–12.9	Mira	328.5	M

DOUBLE STAR

	RA (h)	RA (m)	Dec. (°)	Dec. (')	PA (°)	Separation (″)	Magnitudes
α^2	12	56.0	+38	19	229	19.1	2.9, 5.5 (α^1)

GLOBULAR CLUSTER

M	C	NGC	RA (h)	RA (m)	Dec. (°)	Dec. (')	Diameter (')	Magnitude
3		5272	13	42.2	+28	23	16.2	6.4

GALAXIES

M	C	NGC	RA (h)	RA (m)	Dec. (°)	Dec. (')	Magnitude	Dimensions (')	Type	
		4111	12	07.1	+43	04	10.8	4.8 × 1.1	S0	
		4138	12	09.5	+43	41	12.3	2.9 × 1.9	E4	
		4145	12	10.0	+39	53	11.0	5.8 × 4.4	Sc	
		4151	12	10.5	+39	24	10.4	5.9 × 4.4	Pec.	
		4618	12	41.5	+41	09	10.8	4.4 × 3.8	Sc	
		4214	12	15.6	+36	20	9.8	7.9 × 6.3	Irregular	
		4217	12	15.8	+47	06	11.9	5.5 × 1.8	Sb	
		4242	12	17.5	+45	37	11.0	4.8 × 3.8	S	
	26	4244	12	17.5	+37	49	10.7	13.0 × 10	Sb	
106		4258	12	19.0	+47	18	8.3	18.2 × 7.9	Sb	
		4395	12	25.8	+33	33	10.1	12.9 × 11.0	S	
	21	4449	12	28.2	+44	06	9.4	5.1 × 3.7	Irregular	
		4490	12	30.6	+41	38	9.8	5.9 × 3.1	Sc	
	32	4631	12	42.1	+32	32	9.3	15.1 × 3.3	Sc	
		4656–7	12	44.0	+32	10	10.4	13.8 × 3.3	Sc	
94		4736	12	50.9	+41	07	8.2	11.0 × 9.1	Sb	
	29	5005	13	10.9	+37	03	9.8	5.4 × 2.7	Sb	
		5033	13	13.4	+36	36	10.1	10.5 × 5.6	Sb	
63		5055	13	15.8	+42	02	8.6	12.3 × 7.6	Sb	
		5112	13	21.9	+38	44	11.9	3.9 × 2.9	Sc	
51		5194	13	29.9	+47	12	8.4	11.0 × 7.8	Sc	Whirlpool
		5195	13	30.0	+47	16	9.6	5.4 × 4.3	Pec	Companion to M 51
		5371	13	55.7	+40	28	10.7	4.4 × 3.6	Sb	

CANIS MAJOR

(Abbreviation: CMa)

Orion's largest dog; a brilliant constellation, containing Sirius, the Dog Star. There are 11 stars above the fourth magnitude.

Also above magnitude 4.5: ω, 4.01v; θ, 4.08; γ, Muliphein, 4.11; ξ, 4.34; 27 (EW), 4.42*; λ, 4.47.

	RA			Dec.			m	M	d (light-years)	Spectrum	
	(h)	(m)	(s)	(°)	(')	('')					
9 α	06	45	09.2	−16	42	47	−1.44	145	8.6	A0	Sirius
21 ε	06	58	37.6	−28	58	20	1.50	−4.10	430	B2	Adhara
25 δ	07	08	23.5	−26	23	36	1.83	−6.87	1790	F8	Wezea
2 β	06	22	42.0	−17	57	21	1.98	−3.95	500	B0	Mirzam
31 η	07	24	05.7	−29	18	11	2.45	−7.51	3200	B5	Aludra
1 ζ	06	20	18.8	−30	03	48	3.02	−2.05	340	B2	Phurad
24 o²	07	03	01.5	−23	49	56	3.02	−6.46	2600	B3	Menkelb Post
22 σ	07	01	43.2	−27	56	05	3.49v	−3.47	1220	K4	
13 κ	06	49	50.5	−12	30	31	3.50	−3.42v	790	B1	
16 o¹	06	54	08.0	−24	11	03	3.89	−5.02	1980	K3	Menkelb Prior
7 ν	06	36	41.0	−19	15	21	3.95	2.46	65	K1	

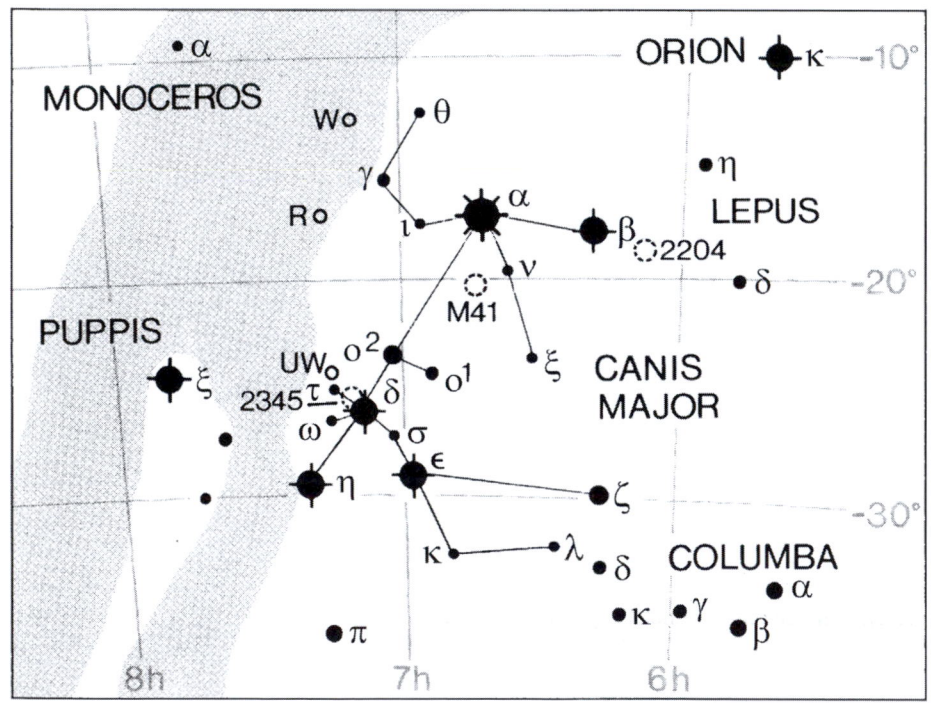

VARIABLE STARS

	RA		Dec.		Range	Type	Period (d)	Spectrum
	(h)	(m)	(°)	(')				
W	07	08.1	−11	55	6.4–7.9	Irregular	—	N
RY	07	16.6	−11	29	7.7–8.5	Cepheid	4.68	F–G
UW	07	18.4	−24	34	4.0–5.3	β Lyræ	4.39	07
R	07	19.5	−16	24	5.7–6.3	Algol	1.14	F

DOUBLE STARS

	RA		Dec.		PA	Separa-tion (")	Mag-nitudes	
	(h)	(m)	(°)	(')	(°)			
ν¹	06	36.4	−18	40	262	17 5	5 8, 8 5	
α	06	45.1	−16	43	005	4.5	−1.5, 8.3	Binary, 50 y
17	06	55.0	−20	24	AB 147	44.4	5.8, 9.3	
					AC 184	50.5	9.0	
					AD 186	129.9	9.5	
π	06	55.6	−20	08	018	11.6	4.7, 9.7	
μ	06	56.1	−14	03	AB 340	3.0	5.3, 8.6	Isis
					AC 288	88.4	10.5	
					AD 061	101.3	10.7	
ε	06	58.6	−28	58	161	7.5	1.5, 7.4	

OPEN CLUSTERS

M	C	NGC	RA		Dec.		Dia-meter (')	Mag-nitude	No. of stars	
			(h)	(m)	(°)	(')				
		2204	06	15.7	−18	39	13	8.6	80	
41		2287	06	47.0	−20	44	38	4.5	80	
		2345	07	08.3	−13	10	12	7.7	20	
	58	2360	07	17.8	−15	37	13	7.2	80	
	64	2362	07	17.8	−24	57	8	4.1	60	τ CMa

GALAXIES

M	C	NGC	RA		Dec.		Magnitude	Dimen-sions (')	Type
			(h)	(m)	(°)	(')			
		2207	06	16.4	−21	22	10.7	4.3 × 2.9	Sc
		2217	06	21.7	−27	14	10.4	4.8 × 4.4	SBa
		2223	06	24.6	−22	50	11.4	3.3 × 3.0	SBb
		2280	06	44.8	−27	38	11.8	5.6 × 3.2	Sb

Sirius, much the brightest star in the sky, is a binary system. The companion, Sirius B, is a white dwarf almost equal to the Sun in mass, but about the size of the Earth; it evolved from the red giant stage about 120 million years ago. The present surface temperature is about 25 000 °C. It is composed mainly of a carbon–oxygen mixture, with an outer atmosphere of hydrogen.

At EUV (extreme ultraviolet) wavelengths ε (Adhara) is the brightest object in the entire sky. It has a binary companion, magnitude 7.5, at a distance of 7.5″. Five million years ago it was only 34 light-years away and its apparent magnitude was −4. δ (Wezea), o² (Menkelb Post) and η (Aludra) are particularly luminous supergiants. β (Mirzam) and σ are very slightly variable.

30 (τ), an O-type blue supergiant, is an eclipsing spectroscopic binary of the β Lyræ type; $M = 4.32 - 4.37$, period 1.28 d. It is the brightest star in the open cluster NGC 2362 (Caldwell 654).

CANIS MINOR

(Abbreviation: CMi)

The second of Orion's dogs. Procyon is one of the brightest stars in the sky; the Little Dog has only one other star above the fourth magnitude.

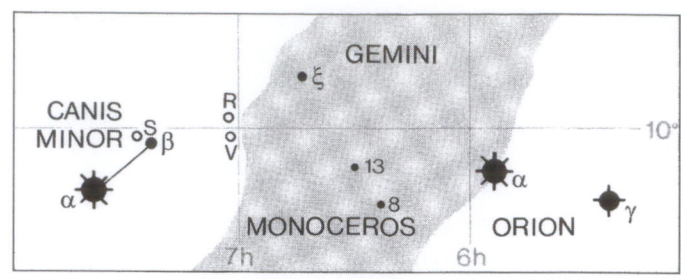

	RA			Dec.			m	M	d (light-years)	Spec-trum	
	(h)	(m)	(s)	(°)	(')	('')					
10 α	07	39	18.5	+05	13	39	0.40	2.68	11	F5	Procyon
3 β	07	27	09.1	+08	17	22	2.89	−0.70	179	B8	Gomeisa

Also above magnitude 4.5: γ, 4.33; and G, 4.39.

Procyon, like Sirius, has a white dwarf companion, which is, however, far from being an easy object.

The three Mira variables R, S and V are easy to find, near Gomeisa.

VARIABLE STARS

	RA		Dec.		Range	Type	Period (d)	Spectrum
	(h)	(m)	(°)	(')				
V	07	07.0	+08	53	7.4–15.1	Mira	366.1	M
R	07	08.7	+10	01	7.3–11.6	Mira	337.8	S
S	07	32.7	+08	19	6.6–13.2	Mira	332.9	M

DOUBLE STAR

	RA		Dec.		PA	Separation ('')	Magnitudes	
	(h)	(m)	(°)	(')	(°)			
α	07	39.3	+05	14	021	5.2	0.4, 12.9	Binary, 40.7 y

CAPRICORNUS

(Abbreviation: Cap)

A Zodiacal constellation, though by no means brilliant. It has been identified with the demigod Pan, but there seems to be no well-defined mythological associations. There are five stars above magnitude 4:

α¹ and α² make up a naked-eye pair, separated by 376' (PA 291°), but the two components are not genuinely associated, and each is itself double; the fainter component of α² is also double!

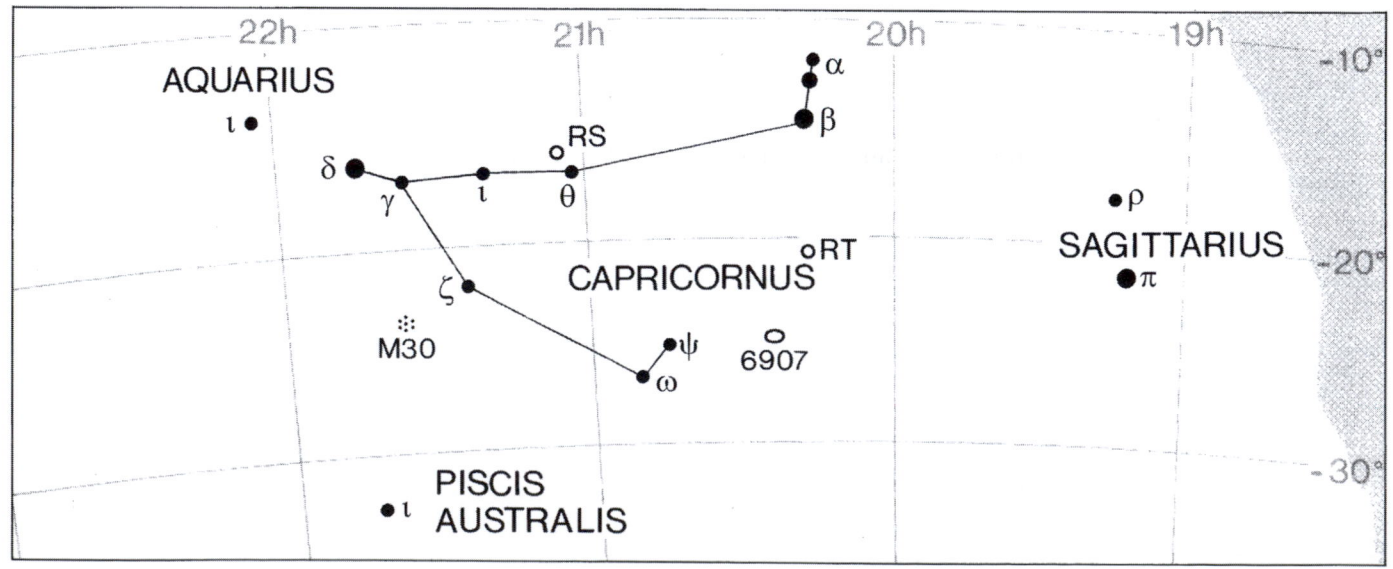

	RA			Dec.			m	M	d (light–years)	Spectrum	
	(h)	(m)	(s)	(°)	(′)	(″)					
49 δ	21	47	02.3	−16	07	36	2.85	2.49	39	A5	Deneb Algiedi
9 β	20	21	00.7	−14	46	53	3.05	−2.07	344	A5	Dabih
6 α	20	18	03.2	−12	32	41	3.58	0.97	109	G6	Al Giedi
40 γ	21	40	05.3	−16	39	44	3.69	0.54	139	A7	Nashira
34 ζ	21	26	40.0	−22	24	41	3.77	−1.66	398	G4	Marakk

Also above magnitude 4.5; θ (Dorsum), 4.08; ψ (Pazan), 4.1; ι, 4.28v; α¹, 4.30; 24, 4.49; 36, 4.50.

VARIABLE STARS

	RA		Dec.		Range	Type	Period (d)	Spectrum
	(h)	(m)	(°)	(′)				
RT	20	17.1	−21	19	6.5–8.1	Semi-regular	393	N
RR	21	02.3	−27	05	7.8–15.5	Mira	277.5	M
RS	21	07.2	−16	25	7.0–9.0	Semi-regular	340	M

DOUBLE STARS

	RA		Dec.		PA	Separation (″)	Magnitudes	
	(h)	(m)	(°)	(′)	(°)			
α	20	18.1	−12	33	291	377.7	3.6, 4.2	(α¹–α²)
α¹	20	17.6	−12	30	AB 182	44.3	4.2, 13.7	
					AC 221	45.4	9.2	
α²	20	18.1	−12	33	AB 172	6.6	3.6, 11.0	
					AD 156	154.6	9.3	
					BC 240	1.2	11.3	
σ	20	19.6	−19	07	179	55.9	5.5, 9.0	
β	20	21.0	−14	47	267	205	3.1, 6.0	B is double
π	20	27.3	−18	13	148	3.2	5.3, 8.9	
ρ	20	28.9	−17	49	158	0.5	5.0, 10.0	
ε	21	37.1	−19	28	047	68.1	4.7, 9.5	
τ	20	39.3	−14	57	118	0.3	5.8, 6.3	Binary, 200 y

GLOBULAR CLUSTER

M	C	NGC	RA		Dec.		Diameter	Magnitude
			(h)	(m)	(°)	(′)	(′)	
30		7099	21	40.4	−23	11	11.0	7.5

GALAXY

M	C	NGC	RA		Dec.		Magnitude	Dimensions	Type
			(h)	(m)	(°)	(′)		(′)	
		6907	20	25.1	−24	49	11.3	3.4 × 3.0	SBb

ζ (Marakk) is a G-type supergiant with an unusual spectrum, showing an over-abundance of the element praseodymium. Its proper name means 'the loins of the goat'. It has a white dwarf companion.

All in all, Capricornus is a rather barren constellation redeemed by the presence of the bright globular galaxy M 30.

CARINA

(Abbreviation: Car)

The keel of the ship Argo, in which Jason and his companions sailed upon their successful, if somewhat unprincipled, expedition to remove the golden fleece of the sacred ram (Aries) from its grove. Carina is the brightest part of Argo, and contains Canopus, which is second only to Sirius in brilliance. Canopus is a highly luminous star, but estimates of its power and distance vary considerably. A recent estimate makes it over 200 000 times as luminous as the Sun: other estimates are much lower. Also in this constellation is η Carinæ, which during part of the last century rivalled Sirius, but which is now well below naked-eye visibility. It is unique, and at its peak may have been the most luminous star in the Galaxy. Excluding η, there are 15 stars in Carina above the fourth magnitude.

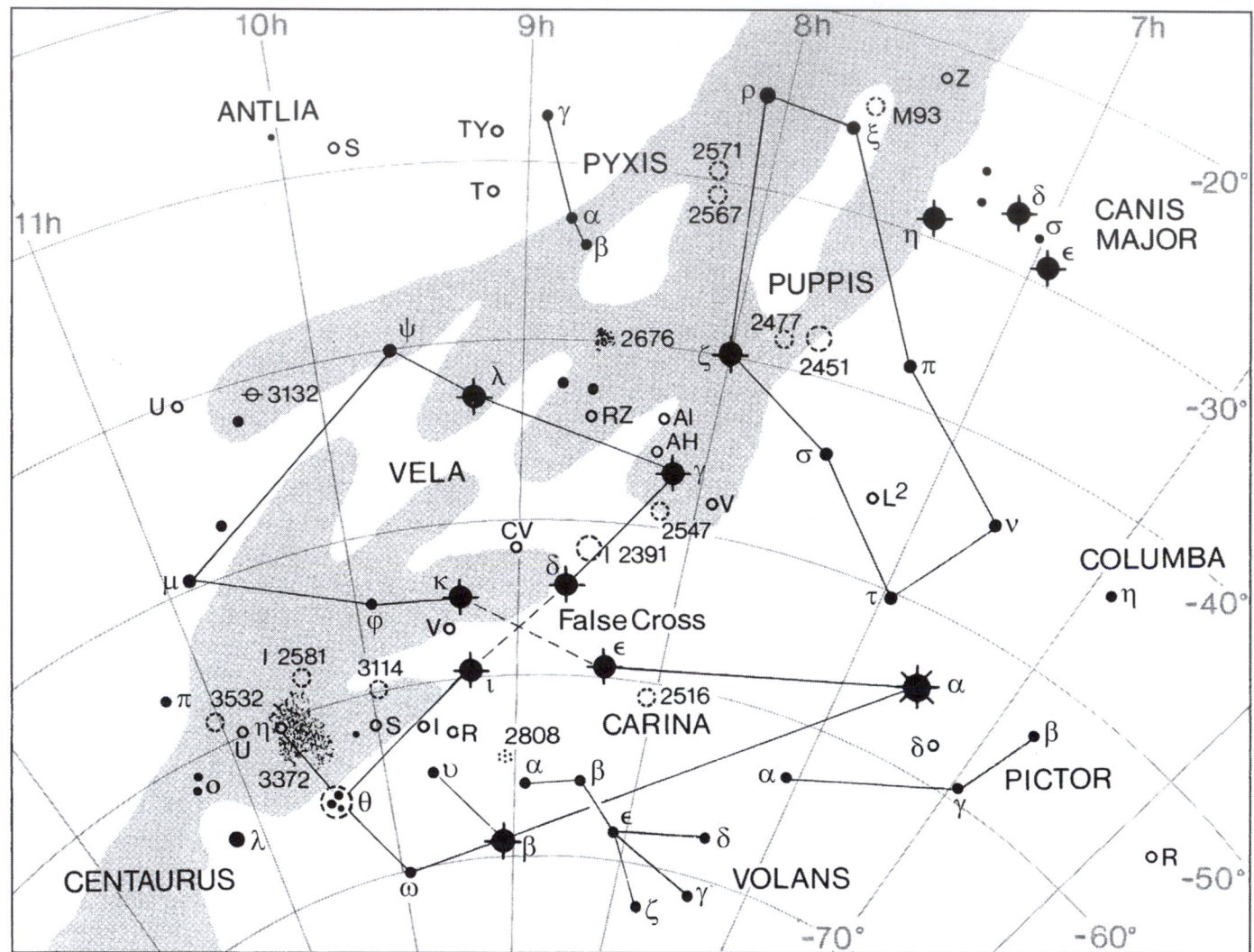

	RA			Dec.			m	M	d (light-years)	Spectrum	
	(h)	(m)	(s)	(°)	(')	(")					
α	06	23	57.1	−52	41	45	−0.62	−5.43	313	F0	Canopus
β	09	13	12.2	−69	43	01	1.67	−0.99	111	A2	Miaplacidus
ε	08	22	30.9	−59	30	34	1.86	−4.58	630	K3	Avior
ι	00	17	05.4	−59	16	31	2.21	−4.42	690	A8	Turais
θ	10	42	57.4	−64	23	40.1	2.74	−2.91	440	B0	Vathorz Post
υ	09	47	06.1	−65	04	19	2.92	−5.56	1620	A9	Vathorz Prior
ω	10	13	44.3	−70	02	17	3.29	−1.99	370	B2	Semiram
p	10	32	01.5	−61	41	07	3.30v	−2.62	500	B4	PP Car

	RA			Dec.			m	M	d (light-years)	Spectrum	
	(h)	(m)	(s)	(°)	(′)	(″)					
q	10	17	05.0	−61	19	56	3.39v	−3.38	76	K3	V357 Car
a	09	10	58.1	−58	58	01	3.43*	−2.11	420	B2	
χ	07	56	46.7	−52	58	57	3.46v	−1.91	390	B3	Drys
I (ZZ)	09	45	14.8	−62	30	29	3.69	−4.64	1500	G5	
u	10	53	29.6	−58	51	12	3.78	1.42	97	K0	
s	10	27	52.8	−58	44	22	3.81v	−3.71	1640	F2	
c	08	55	02.9	−60	38	41	3.84v	−1.06	310	B8	
x	11	08	35.4	−58	58	30.2	3.93	−7.37	·5900	G0	
i	09	11	16.8	−02	19	01	3.96	−1.97	500	B3	
l	10	24	23.7	−74	01	34	3.99	2.94	53	F2	

Also above magnitude 4.5: h, 4.08; d (V343), 4.3v; g, 4.34; N, 4.35; A*, 4.41; r , 4.45; G, 4.47.

VARIABLE STARS

	RA		Dec.		Range	Type	Period (d)	Spectrum	
	(h)	(m)	(°)	(′)					
V	08	28.7	−60	07	7.1–7.8	Cepheid	6.70	F–H	
X	08	31.3	−59	14	7.9–8.6	Beta Lyræ	1.08	A+A	
R	09	32.2	−62	47	3.9–10.5	Mira	308.7	M	
ZZ	09	45.2	−62	30	3.3–4.2	Cepheid	35.53	F–K	(e Carinæ)
S	10	09.4	−61	33	4.5–9.9	Mira	149.5	K–M	
η	10	45.1	−59	41	−0.8–7.9	Irregular	–	Pec.	
BO	10	45.8	−59	29	7.2–8.5	Irregular	–	M	
U	10	57.8	−59	44	5.7–7.0	Cepheid	38.77	F–G	

DOUBLE STARS

	RA		Dec.		PA	Separation (″)	Magnitudes
	(h)	(m)	(°)	(′)	(°)	s	
η	10	45.1	−59	41	195	0.2	var., 8.6
u	09	47.1	−65	04	127	5.0	3.1, 6.1

OPEN CLUSTERS

M	C	NGC	RA		Dec.		Diameter	Magnitude	No. of stars	
			(h)	(m)	(°)	(′)	(′)			
	96	2516	07	58.3	−60	52	30	3.8	80	
		3114	10	02.7	−60	07	35	4.2	–	
		IC 2581	10	27.4	−57	38	8	4.3	25	
	102	IC 2602	10	43.2	−64	24	50	1.9	60	θ Carinæ cluster
	91	3532	11	06.4	−58	40	55	3.0	150	
		Mel 101	10	42.1	−65	06	14	8.0	50	
		3572	11	10.4	−60	14	7	6.6	35	
		3590	11	12.9	−60	47	4	8.2	25	
		Mel 105	11	19.5	−63	30	4	8.5	70	
		IC 2714	11	17.9	−62	42	12	8.2	100	
		3680	11	25.7	−43	15	12	7.6	30	

GLOBULAR CLUSTER

M	C	NGC	RA (h)	RA (m)	Dec. (°)	Dec. (')	Diameter (')	Magnitude
		2808	09	12.0	−64	52	13.8	6.3

NEBULA

M	C	NGC	RA (h)	RA (m)	Dec. (°)	Dec. (')	Magnitude	Magnitude of illuminating star	
	92	3372	10	43.8	−59	52	6.2	variable	η Carinæ nebula

PLANETARY NEBULÆ

M	C	NGC	RA (h)	RA (m)	Dec. (°)	Dec. (')	Diametre (")	Magnitude	Magnitude of central star
		IC 2448	09	07.1	−69	57	8	11.5	12.9
	90	2867	09	21.4	−58	19	11	9.7	13.6
		IC 2501	09	38.8	−60	05	25	11.3	—
		3211	10	17.8	−62	40	12	11.8	—
		IC 2621	11	00.3	−65	15	5	—	13.6

ι and ε Carinæ are two of the stars of the False Cross (the other two stars are κ and δ Velorum). The False Cross is often mistaken for the Southern Cross, but is larger and less brilliant.

By AD 5000 ω Carinæ will be the South Pole Star.

CASSIOPEIA

(Abbreviation: Cas)
An original constellation – one of the most distinctive in the sky. Mythologically, Cassiopeia was Andromeda's mother and wife of Cepheus: it was her boasting which led to the unfortunate contretemps with Neptune's sea monster.

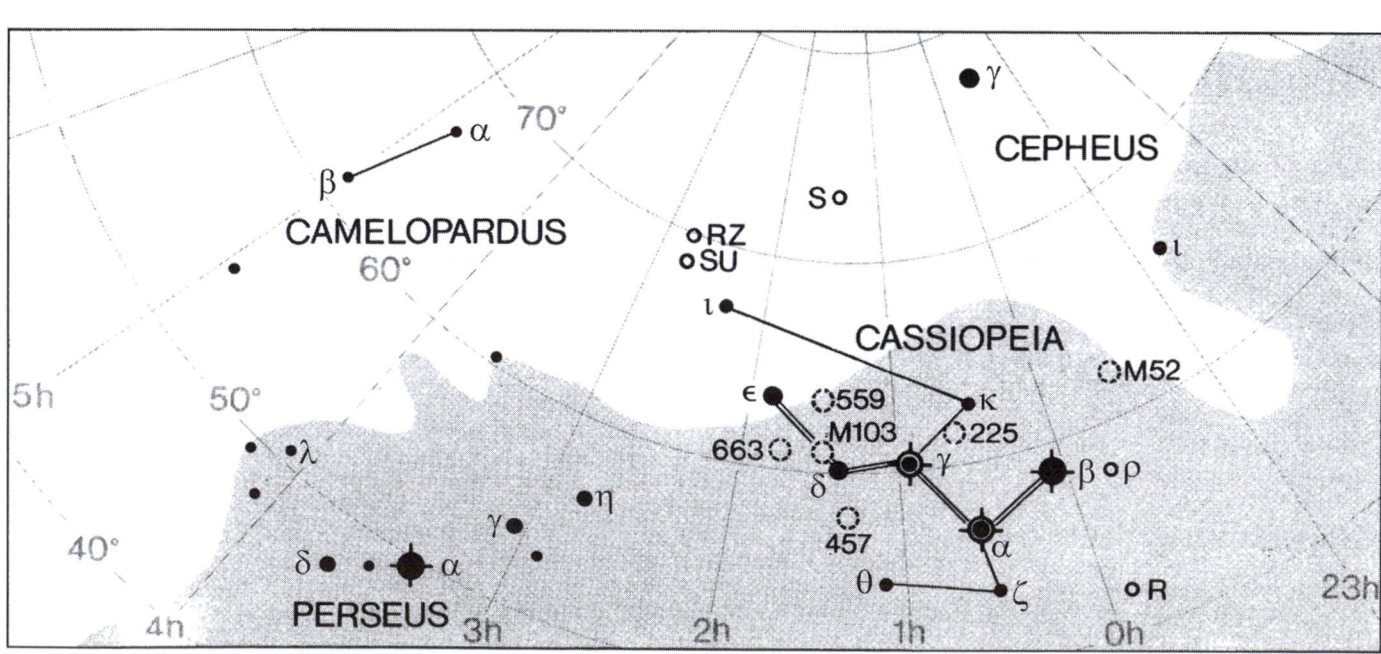

There are eight stars above the fourth magnitude:

	RA			Dec.			m	M	d (light-years)	Spectrum	
	(h)	(m)	(s)	(°)	(′)	(″)					
27 γ	00	56	42.5	+40	43	00	2.2v	−4.22	610	B0	
18 α	00	40	30.4	+56	32	15	2.24	−1.99	228	K0	Shedir
11 β	00	09	10.1	+59	09	01	2.28	1.17	54	F2	Caph
37 δ	01	25	48.6	+60	14	08	2.66[*]	0.24	99	A5	Ruchbah
45 ε	01	54	23.7	+63	40	13	3.35	−2.31	440	B2	Segin
24 η	00	49	05.1	+57	49	00	3.46	4.59	19	G0	Achird
17 ζ	00	36	58.3	+53	53	49	3.69	−2.62	600	B2	

Also above 4.5: κ (4.17v), θ (4.34), ι (4.46v), o (4.48v), 48 (4.49), ρ (5.0v).

VARIABLE STARS

	RA		Dec.		Range	Type	Period (d)	Spectrum
	(h)	(m)	(°)	(′)				
V	23	11.7	+59	42	6.9–13.4	Mira	228.8	M
ρ	23	54.4	+58	30	4.1–6.2	?	—	F–K
R	23	58.4	+51	24	4.7–13.5	Mira	430.5	M
T	00	23.2	+55	48	6.9–13.0	Mira	444.8	M
TU	00	26.3	+51	17	6.9–8.1	Cepheid	2.14	F
α	00	40.5	+56	22	2.1–2.5	Suspected	—	K
U	00	46.4	+48	15	8.0–15.7	Mira	277.2	S
RV	00	52.7	+47	25	7.3–16.1	Mira	331.7	M
W	00	54.9	+58	34	7.8–12.5	Mira	405.6	N
γ	00	56.7	+60	43	1.6–3.3	Irregular	—	B
S	01	19.7	+72	37	7.9–16.1	Mira	612.4	S
SU	02	52.0	+68	53	5.7–6.2	Cepheid	1.95	F
RZ	02	48.9	+69	38	6.2–7.7	Algol	1.19	A

DOUBLE STARS

	RA		Dec.		PA	Separation (″)	Magnitudes	
	(h)	(m)	(°)	(′)	(°)			
λ	00	31.8	+54	31	176	0.5	5.3, 5.6	
η	00	49.1	+57	49	293	12.2	3.4, 7.5	Binary, 480 y
ψ	01	25.9	+68	08	113	25.0	4.7, 9.6	
ι	02	29.1	+67	24	232	2.4	4.6, 6.9	Binary, 840 y
σ	23	59.0	+55	45	326	3.0	5.0, 7.1	

OPEN CLUSTERS

M	C	NGC	RA		Dec.		Diameter	Magnitude	No. of stars
			(h)	(m)	(°)	(′)	(′)		
52		7654	23	24.2	+61	35	13	6.9	100
		7788	23	56.7	+61	24	9	9.4	20
		7789	23	57.0	+56	44	16	6.7	300

OPEN CLUSTERS

M	C	NGC	RA (h)	(m)	Dec. (°)	(')	Diameter (')	Magnitude	No. of stars	
		H.21	23	54.1	+61	46	4	9.0	6	
		129	00	29.9	+60	14	21	6.5	35	Contains DL Cas
		133	00	31.2	+63	22	7	9.4	5	
		146	00	33.1	+63	18	7	9.1	20	
		225	00	43.4	+61	47	12	7.0	15	
		381	01	08.3	+61	35	6	9.3	50	
		436	01	15.6	+58	49	6	8.8	30	
	13	457	01	19.1	+58	20	13	6.4	80	φ Cas cluster
	8	559	01	29.5	+63	18	4.4	9.5	60	
		IC 1805	02	32.7	+61	27	22	6.5	40	
103		581	01	33.2	+60	42	6	7.4	25	
		637	01	42.9	+64	00	3.5	8.2	20	
		1027	02	42.7	+61	33	20	6.7	40	
		654	01	44.1	+61	53	5	6.5	60	
		659	01	44.2	+60	42	5	7.9	40	
	10	663	01	46.0	+61	15	16	7.1	80	

NEBULÆ

M	C	NGC	RA (h)	(m)	Dec. (°)	(')	Diameter (')	Magnitude of illuminating star	
	11	7635	23	20.7	+61	12	15 × 8	7	Bubble Nebula
		281	00	52.8	+56	36	35 × 30	8	
		IC 1805	02	33.4	+61	26	60 × 60	—	
		IC 1848	02	51.3	+60	25	60 × 30	—	

GALAXIES

M	C	NGC	RA (h)	(m)	Dec. (°)	(')	Magnitude	Dimensions (')	Type
	17	147	00	33.2	+48	30	9.3	12.9 × 8.1	dE4
	18	185	00	39.0	+48	20	9.2	11.5 × 9.8	dE0

Cassiopeia is very rich; the Milky Way passes through it. From Britain it is circumpolar.

CENTAURUS

(Abbreviation: Cen)

A brilliant southern constellation (one of Ptolemy's originals), containing many important objects, including the nearest of all the bright stars (α) and the superb globular cluster ω. There are 19 stars brighter than the fourth magnitude.

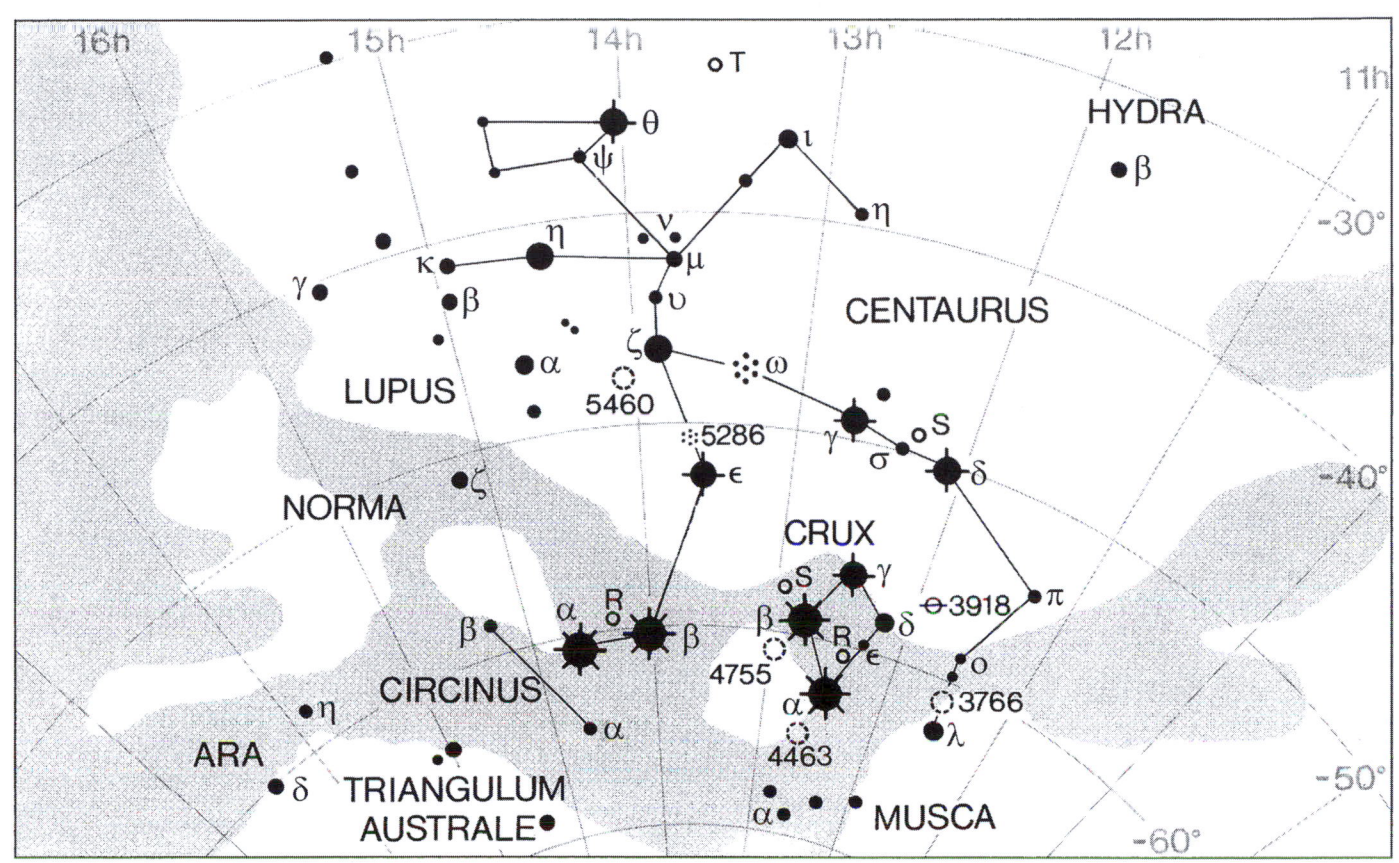

	RA			Dec.			*m*	*M*	*d* (light-years)	Spectrum	Name
	(h)	(m)	(s)	(°)	(′)	(″)					
α	14	39	40	−60	50	06.5	−0.01	4.34	4.36	G2	Toliman
	14	39	39.4	−60	50	22	1.35	5.70	4.36	K1	
β	14	03	49.4	−60	22	23	0.61	−5.42	525	B1	Agena
5 θ	14	06	41.3	−36	22	07	2.06	0.70	61	K0	Menkent
γ	12	41	31.2	−48	57	36	2.20	−0.81	130	A1	Miliphein
ε	13	39	53.27	−53	27	59	2.29	−3.02	380	B1	Birdun
η	14	35	30.45	−42	09	28	2.33	−2.55	310	B1	Marfikent
ζ	13	55	32.4	−47	17	18	2.55	−2.81	184	B2	
δ	12	08	21.5	−50	41	21	2.58	−2.84	395	B2	Wei
ι	13	20	36.1	−36	42	44	2.75	1.48	59	A2	Alkahim
λ	11	35	46.9	−63	01	11	3.11	−2.39	410	B9	Mati
κ	14	59	09.7	−42	06	15	3.13	−2.96	539	B2	Ke Kwan
ν	13	49	30.3	−41	41	16	3.41	−2.41	470	B2	Kabkent Sec.
μ	13	49	37.0	−42	28	25	3.47	−2.57	530	B2	Kabkent Prior
φ	13	58	16.3	−42	06	03	3.83	−1.94	460	B2	Kabkent Terta
τ	12	37	42.3	−48	32	29	3.85	0.82	132	A2	
υ	13	58	40.8	−44	48	13	3.87	−1.67	420	B2	

	RA			Dec.			m	M	d (light-years)	Spectrum	Name
	(h)	(m)	(s)	(°)	(′)	(″)					
G	11	21	00.5	−54	29	28	3.90	−1.97	320	B2	
D	13	31	02.7	−39	24	26	3.90	−4.03	1250	G2	
σ	12	28	02.4	−50	13	50	3.91	−1.76	440	B3	
ρ	12	11	39.1	−52	22	06	3.97	−1.1	340	B3	

Also above magnitude 4.5: b (4.01); Psi (4.05); c′ (4.06), g (4.19), i (4.23), n (4.25), ξ^2 (4.27), j (4.30v), v (4.30), 3 k (4.32), e (4.33), υ^2 (4.34), χ (4.4v) a (4.41v), V863 (4.5v), B (4.47).

VARIABLE STARS

	RA		Dec.		Range	Type	Period (d)	Spectrum
	(h)	(m)	(°)	(′)				
RS	11	20.5	−61	52	7.7–14.1	Mira	164.4	M
X	11	49.2	−41	45	7.0–13.8	Mira	315.1	M
W	11	55.0	−59	15	7.6–13.7	Mira	201.6	M
S	12	24.6	−49	26	6.0–7.0	Semi-regular	65	N
U	12	33.5	−54	40	7.0–14.0	Mira	220.3	M
RV	13	37.5	−56	29	7.0–10.8	Mira	446.0	N
XX	13	40.3	−57	37	7.3–8.3	Cepheid	10.95	F–G
T	13	41.8	−33	36	5.5–9.0	Semi-regular	60	K–M
μ	13	49.6	−42	28	2.9–3.5	Irregular	—	B
R	14	16.6	−59	55	5.3–11.8	Mira	546.2	M
V	14	32.5	−56	53	6.4–7.2	Cepheid	5.49	F–G

DOUBLE STARS

	RA		Dec.		PA	Separation (″)	Magnitudes	
	(h)	(m)	(°)	(′)	(°)			
D	12	140	−45	43	243	2.8	5.6, 6.8	
γ	12	41.5	−48	58	347	1.0	2.9, 2.9	Binary, 84.5 y
ε	13	39.9	−53	28	158	36.0	2.3, 12.7	
3	13	51.8	−33	00	108	7.9	4.5, 6.0	
4	13	53.2	−31	56	185	14.9	4.8, 8.4	
β	14	03.8	−60	22	251	1.3	0.7, 31.9	
η	14	35.5	−42	09	270	5.0	2.6, 13.5	
α	14	39.6	−60	50	221	14.8	0.0, 1.2	Binary, 79.9 y

OPEN CLUSTERS

M	C	NGC	RA		Dec.		Diameter	Magnitude	No. of stars	
			(h)	(m)	(°)	(′)	(′)			
	97	3766	11	36.1	−61	37	12	5.3	100	
	100	IC 2944	11	36.6	−63	02	15	4.5	30	λ Cen cluster, Collinder 249
		3960	11	50.9	−55	42	7	8.3	45	
		5138	13	27.3	−59	01	8	7.6	40	
		5281	13	46.6	−62	54	5	5.9	40	
		5316	13	53.9	−61	52	14	6.0	80	
		5460	14	07.6	−48	19	25	5.6	40	

OPEN CLUSTERS

M	C	NGC	RA (h)	(m)	Dec. (°)	(′)	Diameter (′)	Magnitude	No. of stars
		5617	14	29.8	−60	43	10	6.3	80
		5662	14	35.2	−56	33	12	5.5	70

GLOBULAR CLUSTERS

M	C	NGC	RA (h)	(m)	Dec. (°)	(′)	Diameter (′)	Magnitude	
	80	5139	13	26.8	−47	29	36.3	3.6	ω Centauri
	84	5286	13	46.4	−51	22	9.1	7.6	

PLANETARY NEBULA

M	C	NGC	RA (h)	(m)	Dec. (°)	(′)	Diameter (″)	Magnitude	Magnitude of central star	
		3918	11	50.3	−57	11	12	8.4	10.9	Blue Planetary

NEBULA

M	NGC	RA (h)	(m)	Dec. (°)	(′)	Diameter (′)
	5367	13	57.7	−39	59	4.3 Includes IC 4347. Double nucleus.

GALAXIES

M	C	NGC	RA (h)	(m)	Dec. (°)	(′)	Magnitude	Dimensions (′)	Type	
		4603	2	40.9	−40	59	12.0	3.8 × 2.5	Sc	
		4696	2	48.8	−41	19	10.7	3.5 × 3.2	Elp	
	83	4945	13	05.4	−49	28	9.5	20.0 × 4.4	SBc	
		4976	13	08.6	−49	30	10.2	4.3 × 2.6	E4p	
	77	5128	13	25.5	−43	01	7.0	18.2 × 14.3	S0p	Centaurus A
		5253	13	39.9	−31	39	0.6	4.0 × 1.7	E5	
		5483	14	10.4	−43	19	12.0	3.1 × 2.8	Sc	
		3557	11	10.0	−37	32	10.4	4.0 × 2.7	E3	

There are various legends about Centaurus; probably it commemorates the wise centaur Chiron, tutor to Jason and the Argonauts. α and β are the 'Pointers' to the Southern Cross. α can be called Toliman, Bungula, or (by navigators) Rigil Kent; the dim red dwarf Proxima is a flare star with a mass one eighth that of the Sun. It was discovered in 1915 by R. Innes, from the Union Observatory in Johannesburg. It is probably part of the α Centauri system. β is either Agena or Hadar; it is a variable of the β Cephei type but very small range. μ is a B-type γ Cassiopeiæ variable, with a range of magnitude 2.9 to 3.5.

ω, already described is much the largest of the Milky Way's globular clusters, and may indeed be the core of a small galaxy which was captured.

CEPHEUS

(Abbreviation: Cep)

The relatively nearby double Krüger 60 lies near δ; RA 22 h 26 m .3, Dec. +57 °27′. The mean separation is between 2 and 3 seconds of arc, but the period is only 44.5 years, so that the separation and position angle alter quite quickly. The magnitudes are 9.8 and 11.4. The real separation is about the same as that between the Sun and Saturn. Krüger 60B, the fainter component, is a flare star, and in variable star catalogues is listed as DO Cephei.

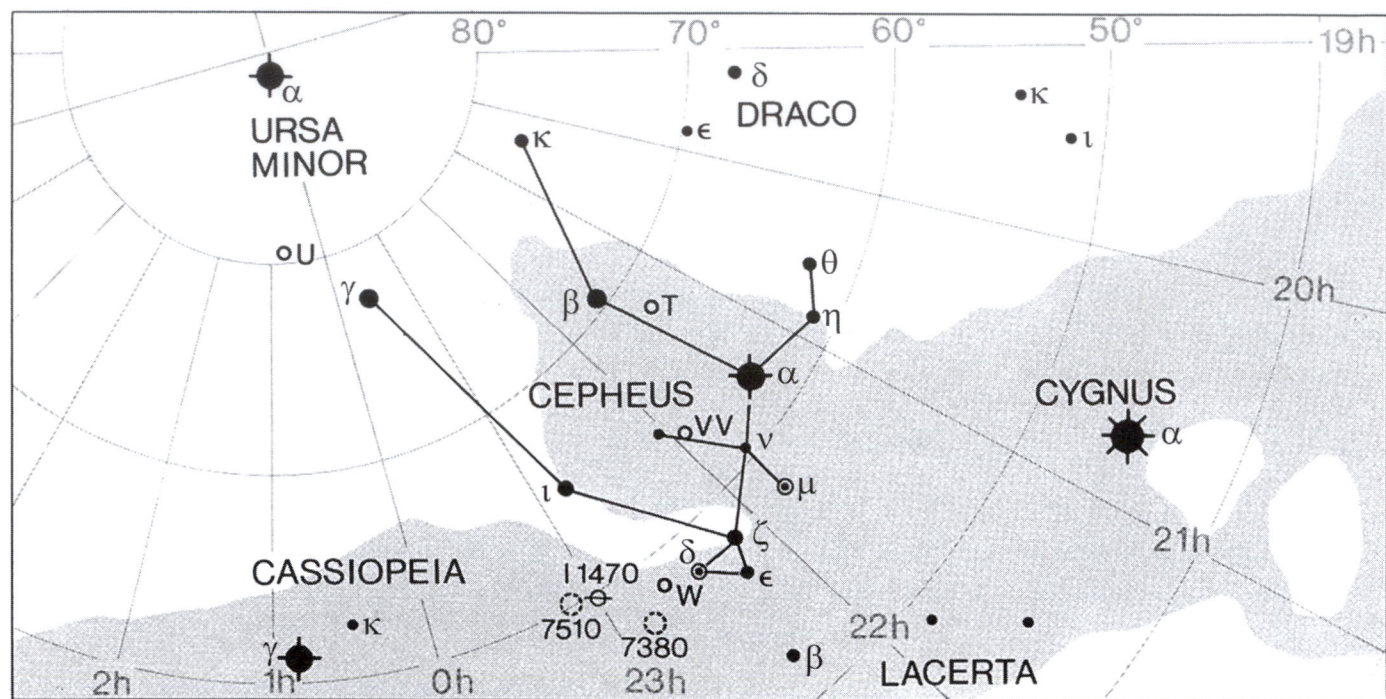

There are eight stars above the fourth magnitude, including the variables δ and μ at maximum.

	RA			Dec.			m	M	d (light-years)	Spectrum	
	(h)	(m)	(s)	(°)	(′)	(″)					
5 α	21	18	34.5	+62	18	08	2.45	1.58	49	A7	Alderamin
35 γ	23	39	21.0	+77	37	55	3.21*	2.51	45	K1	Alrai
8 β	21	28	39.6	+70	33	38	3.23v	−3.08	595	B2	Alphirk
21 ζ	22	10	51.3	58	12	05	3.30v	−3.35	730	K1	Tsao Fu
3 ε	20	45	17.3	+61	50	13	3.41	2.63	47	A0	Al Agemim
32 ι	22	49	40	+66	12	03	3.50	0.76	115	K0	Alvahet
27 δ	22	29	10.3	+58	24	55	3.7v	−3.32	982	G2	Al Radif
μ	21	43	30.5	+58	46	48	3.9v	−6.81	5260	M2	Erakis

Also above 4.5: ε (Phicares), 4.18v; θ (Al Kidr), 4.21*; 2, 4.24; ν (Cor Regis), 4.25v; χ (Alkurah), 4.26*; κ, 4.38; π, 4.41.

Also above magnitude 5:				
	Magnitude	Absolute magnitude	Spectrum	Distance (light-years)
24	4.79	0.3	G8	360
34 o	4.90	1.2	G7	

VARIABLE STARS

	RA (h)	RA (m)	Dec. (°)	Dec. (′)	Range	Type	Period (d)	Spectrum
T	21	09.5	+68	29	5.2–11.3	Mira	388.1	M
VV	21	56.7	+63	38	4.8–5.4	Eclipsing	7430	M + B
S	21	35.2	+78	37	7.4–12.9	Mira	486.8	N
μ	21	43.5	+58	47	3.4–5.1	Semi-reg.	730	M
δ	22	29.2	+58	25	3.5–4.4	Cepheid	5.37	F–G
W	22	36.5	+58	26	7.0–9.2	Semi-regular	Long	K–M
U	01	02.3	+81	53	6.7–9.2	Algol	2.49	B + G

DOUBLE STARS

	RA (h)	RA (m)	Dec. (°)	Dec. (′)	PA (°)	Separation (″)	Magnitudes	
κ	20	08.9	+72	43	122	7.4	4.4, 8.4	
β	21	28.7	+70	34	249	13.3	3.2, 7.9	
ξ	22	03.8	+64	38	277	7.7	4.4, 6.5	Binary, 3800 y
δ	22	29.2	+58	25	191	41.0	variable, 7.5	
π	23	07.9	+75	23	346	1.2	4.6, 6.6	Slow binary
o	23	18.6	+68	07	220	2.9	4.9, 7.1	Binary, 796 y

OPEN CLUSTERS

M	C	NGC	RA (h)	RA (m)	Dec. (°)	Dec. (′)	Diameter (′)	Magnitude	No. of stars
		IC 1396	21	39.1	+57	30	50	3.5	50
		7160	21	53.7	+62	36	7	6.1	12
		7235	22	12.6	+57	17	4	7.7	30
		7261	22	20.4	+58	05	6	8.4	30
		7380	22	47.0	+58	6	12	7.2	40
		7510	23	11.5	+60	34	4	7.9	60
	1	188	00	44.4	+85	20	14	8.1	120

PLANETARY NEBULA

M	C	NGC	RA (h)	RA (m)	Dec. (°)	Dec. (′)	Dimensions (″)	Magnitude	Magnitude of central star
	2	40	00	13.0	+72	32	37	10.7	11.6

NEBULÆ

M	C	NGC	RA (h)	RA (m)	Dec. (°)	Dec. (′)	Dimensions (′)	Magnitude	
	4	7023	21	01.8	+68	12	18 × 18	6.8	
	9	Sh2–155	22	56.8	+62	37	50 × 30	7.7	Cave Nebula

GALAXIES

M	C	NGC	RA (h)	RA (m)	Dec. (°)	Dec. (′)	Magnitude	Dimensions (′)	Type
	12	6946	20	34.8	+60	09	9.7	11.0 × 9.8	Sc
		6951	20	37.2	+66	06	12.2	3.8 × 3.3	Sbp

Alphirk (β) is the prototype variable of its class; the magnitude range is very small (3.15–3.21), period 0.19 day. It is a spectroscopic binary with an 8h-magnitude optical companion. Cepheus is not a prominent constellation, but it does contain two famous variable stars, δ and μ (their proper names are seldom or never used), which have already been described. There is also the eclipsing binary VV Cephei (RA 21h 56m 39s .14, Dec + 63° 37′ 32″, distance ~2400 light-years; m = 4.9, M = −9, luminosity perhaps 300 000 times that of the Sun); it is a vast M2-type red hypergiant, with a B-type companion at a separation of 17 to 34 a.u. The hypergiant, with a diameter of around 5 000 000 000 km, is one of the largest stars known. ν is a very luminous blue supergiant;

δ, ε and ζ form a distinctive triangle; this makes naked-eye or binocular estimates of δ easy to follow.

CETUS

(Abbreviation: Cet)

One of the largest of all constellations: sometimes associated with the sea monster of the Perseus legend, at others relegated to the status of a harmless whale. It contains the prototype long-period variable Mira, which can occasionally rise to magnitude 1.7, but which spends most of its period below naked-eye visibility. On average, Mira is visible with the naked eye for only a few weeks in every year. Mira is a huge star; its diameter varies, but is of the order of 650 000 000 km. The absolute magnitude varies between −2.5 and +4.7. Its companion is the flare star VZ Ceti. Cetus abounds in faint galaxies.

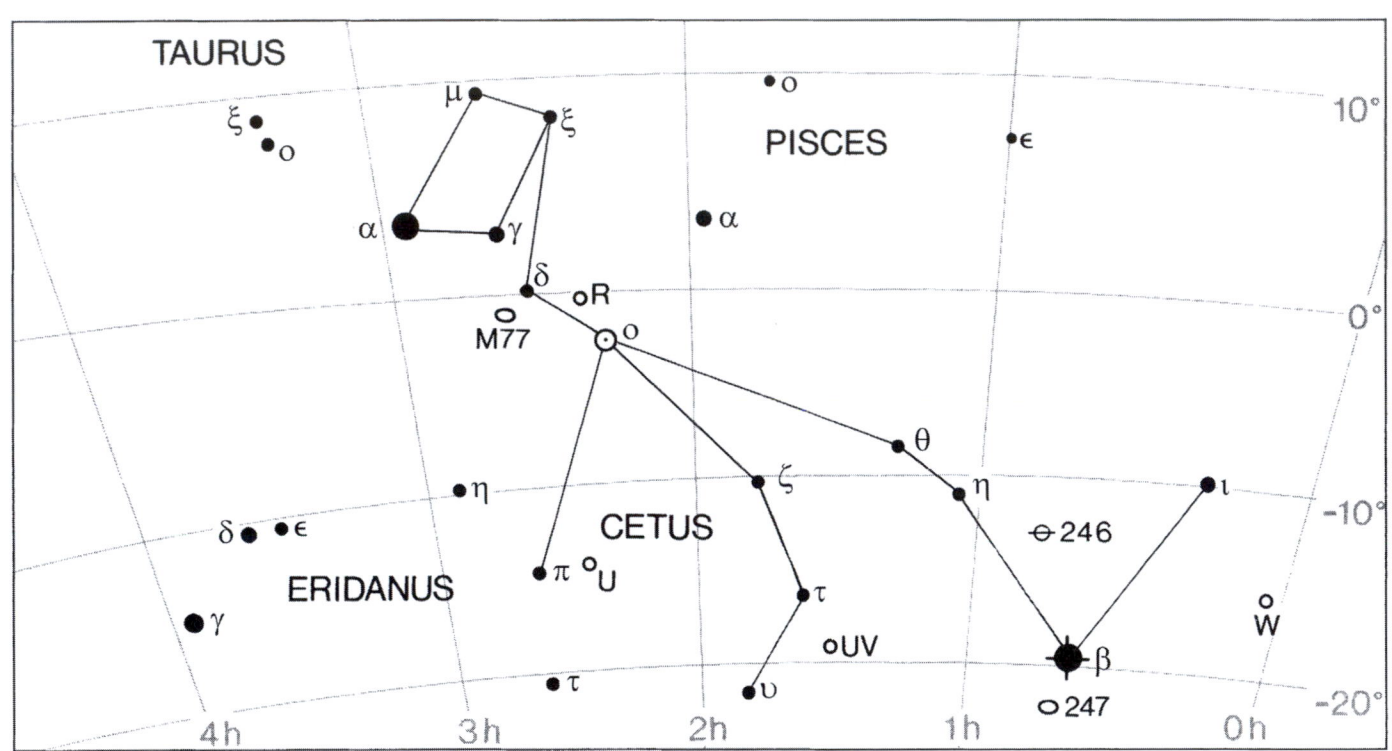

Excluding Mira, there are nine stars above the fourth magnitude.

	RA			Dec.			m	M	d (light-years)	Spectrum	Name
	(h)	(m)	(s)	(°)	(′)	(″)					
16 β	00	43	35.2	−17	59	12	2.04	−0.30	96	K0	Diphda
92 α	03	02	16.8	+04	05	23.7	2.54	−1.61	220	M2	Menkar
68 o	02	19	20.8	−02	58	37.4	var	−2.60	418	M5	Mira
6 η	01	08	35.3	−10	10	55	3.46	0.67	118	K2	Dheneb
86 γ	02	43	18.1	+03	14	10	3.47	1.47	82	A3	Kaffaejidhina
52 τ	01	44	05.1	−15	56	22	3.49	5.68	12	G8	Durre Menthor
8 ι	00	19	25.7	−08	49	26	3.56	−1.18	290	K2	Schemali
45 θ	01	24	01.5	−08	10	58	3.60	0.87	114	K0	Altawk
55 ζ	01	51	27.6	−10	20	06	3.74	−0.76	259	K2	Baten Kaitos
59 υ	02	00	00.2	−21	04	40	3.99	−0.83	300	K5	Aqueus

Also above magnitude 4.5: δ (Phycea), 4.08; π (Al Sadr al Ketus), 4.24; μ, 4.27v; ξ², 4.30; ξ¹, 4.36; 7, 4,45v

VARIABLE STARS

	RA		Dec.		Range	Type	Period (d)	Spectrum
	(h)	(m)	(°)	(′)				
W	00	02.1	−14	41	7.1–14.8	Mira	351.3	S
T	00	21.8	−20	03	5.0–6.9	Semi-reg.	159	M
S	00	24.1	−09	20	7.6–14.7	Mira	320.5	M
U	02	33.7	−13	09	6.8–13.4	Mira	234.8	M
UV	01	38.8	−17	58	6.8–13.0	Flare	–	dM
o	02	19.3	−02	59	1.7–10.1	Mira	332.0	M
R	02	26.0	−00	11	7.2–14.0	Mira	166.2	M

DOUBLE STARS

	RA		Dec.		PA	Separation (″)	Magnitudes	
	(h)	(m)	(°)	(′)	(°)			
37	01	14.4	−07	55	331	49.7	5.2, 8.7	
χ	01	49.6	−10	41	250	183.8	4.9, 6.9	
66	02	12.8	−02	24	AB 234	16.5	5.7, 7.5	
					AC 061	172.7	11.4	
o	02	19.3	−02	59	085	0.6	var., 9.5	Binary (400 y) (B is VZ Ceti)
ν	02	35.9	+05	36	081	8.1	4.9, 9.5	
ε	02	39.6	−11	52	039	0.1	5.8, 5.8	Binary, 2.7 y
γ	02	43.3	+03	14	294	2.8	3.5, 7.3	

PLANETARY NEBULA

M	C	NGC	RA		Dec.		Diameter (″)	Magnitude	Magnitude of central star
			(h)	(m)	(°)	(′)			
		246	00	47.0	−11	53	225	8.0	11.9

GALAXIES

M	C	NGC	RA		Dec.		Magnitude	Dimensions	Type
			(h)	(m)	(°)	(′)		(′)	
		45	00	14.1	−23	11	10.4	8.1 × 5.8	S
	62	247	00	47.1	−20	46	8.9	20.0 × 7.4	S
		428	01	12.9	+00	59	11.3	4.1 × 3.2	Scp
		578	01	30.5	−22	40	10.9	4.8 × 3.2	Sc
		584	01	31.3	−06	52	10.3	3.8 × 2.4	E4
		720	01	53.0	−13	44	10.2	4.4 × 2.8	E3
		864	02	15.5	+06	00	11.0	4.6 × 3.5	Sc
		895	02	21.6	−05	31	11.8	3.6 × 2.8	Sb
		908	02	23.1	−21	14	10.2	5.5 × 2.8	Sc
		936	02	27.6	−01	09	10.1	5.2 × 4.4	SBa
	51	IC1613	01	04.8	+02	07	9.9	12.0 × 11.2	Irregular
		1042	02	40.4	−08	26	10.9	4.7 × 3.9	Sc
		1055	02	41.8	+00	26	10.6	7.6 × 3.0	Sb
		1068	02	42.7	−00	01	8.8	6.9 × 5.9	SBp
		1073	02	43.7	+01	23	11.0	4.9 × 4.6	SBc
		1087	02	46.4	−00	30	11.0	3.5 × 2.3	Sc

Mira has already been described. Its overal shape has been imaged from Earth.Observations made in 2007 show a proto–planetary disc round the companion, which may be a K-type star instead of a white dwarf, as previously believed.

τ Ceti has proved to be something of a disappontment. Unlike ε Eridani, the other first Ozma-candidate, it seems to have a débris disc rather than a system of planets.

Cetus is not a rich constellation, but as well as Mira it contains some notable objects, such as the bright spiral galaxy M 77, near δ.

CHAMÆLEON

(Abbreviation: Cha)
A small southern modern constellation; no legends are attached to it. The brighest star is α RA 08h 1m 1s .3, Dec. −76° 55′ 12″; $m = 4.05$, = 2.60, $d = 63$, type F5. Also above magnitude 4.5: γ, 4.11; β, 4.24; θ, 4.34*; δ², 4.45.

See chart for Musca.

δ² makes up a pair with δ¹, magnitude 5.46, type K0 – a pleasant telescopic sight.

In 1999, a nearby open cluster was found centred on η Chamæleontis RA 08h 41m 19s, Dec. −79° 57′ 48″, magnitude −5.46. It is now known as Mamajek 1, and is only about 320 light-years away. It seems to be less than 8 million years old, and contains a high percentage of pre-Main Sequence stars. X-rays have also been detected from the cluster. Another young cluster is centred on ε (RA 12h 59m 37s, Dec. −78° 13′ 18″). Chamæleon also includes several dark molecular clouds, in which T Tauri stars are being formed.

CIRCINUS

(Abbreviation: Cir)
A very small southern constellation, in the area of α and β Centauri. No legends are attached to it. There is only one star above the fourth magnitude.

See chart for Musca.

	RA (h)	(m)	(s)	Dec. (°)	(′)	(″)	m	M	d (light-years)	Spectrum
α	14	42	30.7	−64	58	29	3.18v	2.11	53	F1

Also above magnitude 4.5: β (2.67), γ (4.48).

VARIABLE STARS

	RA (h)	(m)	Dec. (°)	(′)	Range	Type	Period (d)	Spectrum
AX	14	52.6	−63	49	5.6–6.1	Cepheid	5.27	F–G
θ	14	56.7	−62	47	5.0–5.4	Irregular, γ C	—	B

DOUBLE STARS

	RA (h)	(m)	Dec. (°)	(′)	PA (°)	Separation (″)	Magnitudes	
α	14	42.5	−64	59	232	5.7	3.2, 8.6	
δ	15	16.9	−60	57	270	50.0	5.1, 13.4	
γ	15	23.4	−59	19	011	0.8	5.1, 5.5	Binary, 180 y

OPEN CLUSTERS

M	C	NGC	RA (h)	(m)	Dec. (°)	(′)	Diameter (′)	Magnitude	No. of stars
	88	5823	15	05.7	−55	36	10	7.9	100

δ (RA 15h 16m 57s, Dec. −63° 36′ 38″) is a very powerful O8.5-type star over 6300 light-years away; its magnitude is 5.04, but it is a very small-range rotating ellipsoidal variable.

COLUMBA

(Abbreviation: Col)

A southern contellation created in 1592 by Petrus Plancius, as Columba Noachi (Noah's Dove). There are five stars above the fourth magnitude.

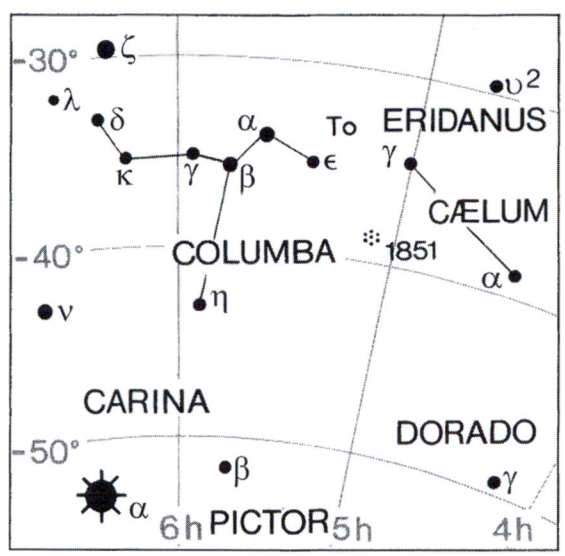

	RA			Dec.			m	M	d (light-years)	Spectrum
	(h)	(m)	(s)	(°)	(′)	(″)				
α	05	39	38.9	−34	04	27	2.65	−1.93	268	B7 Phakt
β	05	50	57.6	−35	46	10	3.12	1.02	86	K1 Wazn
δ	06	22	06.9	−33	26	11	3.85	−0.46	237	G7 Ghusn al Zaitun
ε	05	31	12.8	−35	28	14	3.86	−0.79	277	K1
η	05	59	54	−42	48	54	3.96	−2.10	530	K0

Also above 4.5: γ, 4.36; κ (Al Kurud), 4.37. δ Columbæ was formerly catalogued as 3 Canis Majoris.

μ (magnitude 56.18; RA 05h 45m 59.9s, Dec. −32 18′. 23″) is one of the three 'runaway stars' from the Orion complex (its annual proper motion is 0.025″). The other 'runaways' are 53 Arietis and AE Aurigæ.

VARIABLE STARS

	RA		Dec.		Range	Type	Period (d)	Spectrum
	(h)	(m)	(°)	(′)				
T	05	19.3	−33	42	6.6–12.7	Mira	225.9	M
R	05	50.5	−29	12	7.8–15.0	Mira	327.6	M

DOUBLE STARS

	RA		Dec.		PA	Separation (″)	Magnitudes
	(h)	(m)	(°)	(′)	(°)		
α	05	39.6	−34	04	359	13.5	2.6, 12.3
γ	05	57.5	−35	17	110	33.8	4.4, 12.7
π²	06	07.9	−42	09	150	0.1	6.2, 6.3

GLOBULAR CLUSTER

M	C	NGC	RA		Dec.		Diameter (′)	Magnitude	
			(h)	(m)	(°)	(′)			
	73	1851	05	14.1	−40	03	11.0	7.3	X-ray source

GALAXIES

M	NGC	RA		Dec.		Magnitude	Dimensions (′)	Type
		(h)	(m)	(°)	(′)			
	1792	05	05.2	−37	59	10.2	4.0 × 2.1	Sb
	1808	05	07.7	−37	32	9.9	7.2 × 4.1	SBa
	2090	05	47.0	−34	14	11.7	4.5 × 2.3	Sc

COMA BERENICES

(Abbreviation: Com)

At first glance this constellation gives the impression of being a vast, dim cluster. Coma has no star above magnitude 4.3, but it abounds in faint ones, and there are many telescopic galaxies.

Though the constellation is not 'original', there is a legend attached to it. When Ptolemy Euergetes, King of Egypt, set out in an expedition against the Assyrians, his wife Berenice vowed that if he returned safely she would cut off her lovely hair and place it in the temple of Venus. The King returned; the Queen kept her vow, and Jupiter placed the shining tresses in the sky.

See chart for Boötes.

The brightest star is β (Al Dafirah), Magnitude 4.23; RA 13h 11m 52s, Dec. +18° 31′ 45″. α (Diadem) is magnitude 4.32. 21, magnitude 5.47, has a name, Kissin; it is a close binary with almost equal components, period 26 years, separation never more than 0″.3.

Also above magnitude 5:

	Magnitude	Absolute magnitude	Spectrum	Distance (light-years)
12	4.79	0.7	F8	27
7	4.94	0.2	K0	89
11	4.74	0.3	G8	69
14	4.93		F0	24
16	4.99		A4	21
23	4.81	−0.6	A0	110
31	4.94	0.6	G0	74
36	4.78	−0.5	M1	110
37	4.90		K1	27
41	4.80	−0.3	K5	110

VARIABLE STARS

	RA (h)	RA (m)	Dec. (°)	Dec. (′)	Range	Type	Period (d)	Spectrum
R	12	04.0	+18	49	7.1–14.6	Mira	362.8	M
FS	13	06.4	+22	37	5.3–6.1	Semi-regular	58	M

GALAXIES

M	C	NGC	RA (h)	RA (m)	Dec. (°)	Dec. (′)	Magnitude	Dimensions (′)	Type
		4136	12	09.3	+29	56	11.7	4.1 × 3.9	Sc
98		4192	12	13.8	+14	54	10.1	9.5 × 3.2	Sb
		4251	12	18.1	+28	10	11.6	4.2 × 1.9	E7
99		4254	12	18.8	+14	25	9.8	5.4 × 4.8	Sc
		4278	12	20.1	+29	17	10.2	3.6 × 3.5	E1
		4314	12	22.6	+29	53	10.5	4.8 × 4.3	SBa
100		4321	12	22.9	+15	49	9.4	6.9 × 6.2	Sc
		4448	12	28.2	+28	37	11.1	4.0 × 1.6	Sb
		4450	12	28.5	+17	05	10.1	4.8 × 3.5	Sb
		4459	12	29.0	+13	59	10.4	3.8 × 2.8	E2
		4473	12	29.8	+13	26	10.2	4.5 × 2.6	E4
		4477	12	30.0	+13	38	10.4	4.0 × 3.5	SBa
88		4501	12	32.0	+14	25	9.5	6.9 × 3.9	SBb
		4548	12	35.4	+14	30	10.2	5.4 × 4.4	SBb
	36	4559	12	36.0	+27	58	9.8	10.5 × 4.9	Sc
	38	4565	12	36.3	+25	59	9.6	16.2 × 2.8	Sb
		4651	12	43.7	+16	24	10.7	3.8 × 2.7	Sop
		4689	12	47.8	+13	46	10.9	4.0 × 3.5	Sb
		4725	12	50.4	+25	30	9.2	11.0 × 7.9	SBb

GALAXIES

M	C	NGC	RA (h)	RA (m)	Dec. (°)	Dec. (′)	Magnitude	Dimensions (′)	Type	
64		4826	12	56.7	+21	41	6.6	9.3 × 5.4	Sb	Black-Eye Galaxy
	35	4889	13	00.1	+27	58	11.4	3 × 2	E4	

OPEN CLUSTER

M	C	NGC	RA (h)	RA (m)	Dec. (°)	Dec. (′)	Diam. (′)	Magnitude	No. of stars	
		Mel 111	12	25	+26		275	4	80	Coma Berenices

GLOBULAR CLUSTER

M	C	NGC	RA (h)	RA (m)	Dec. (°)	Dec. (′)	Diameter (′)	Magnitude
53		5024	13	12.9	+18	10	12.6	7.7

The Coma cluster (Melotte 111) is large and diffuse. It is only about 270 light-years away. The constellation contains the northern portion of the Virgo Cluster, including M 64, the Black-Eye Galaxy.

CORONA AUSTRALIS

(Abbreviation: CrA)
A small southern constellation, easy to identify because of its distinctive shape The brightest stars are β (RA 10h 10m 02s, magnitude 4.10), and α, magnitude 4.11.
　　See chart for Sagittarius.

Also above magnitude 5:

	Magnitude	Absolute magnitude	Spectrum	Distance (light-years)
ε	4.8v	2.6	F0	28

DOUBLE STARS

	RA (h)	RA (m)	Dec. (°)	Dec. (′)	PA (°)	Separation (″)	Magnitudes	
κ	18	33.4	−38	44	359	21.6	5.9, 5.9	
λ	18	43.8	−38	19	214	29.2	5.1, 9.7	
γ	19	06.4	−37	04	061	1.3	4.8, 5.1	Binary, 120.4 y

GLOBULAR CLUSTER

M	C	NGC	RA (h)	RA (m)	Dec. (°)	Dec. (′)	Diameter (′)	Magnitude
	78	6541	18	08.0	−43	42	13.1	6.6

NEBULA

M	C	NGC	RA (h)	RA (m)	Dec. (°)	Dec. (′)	Diameter (′)	Magnitude	Magnitude of illuminating star	
	68	6729	19	01.9	−36	57	1 (variable)	variable	9.7v	(R Coronæ Australis)

PLANETARY NEBULA

M	C	NGC	RA (h)	RA (m)	Dec. (°)	Dec. (′)	Diameter (″)	Magnitude	Magnitude of central star	
		IC 1297	19	17.4	−39	37	7	—	12.9v	RU Coronæ Australis

CORONA BOREALIS

(Abbreviation: CrB)
A small but very distinctive constellation. It represents the crown given by Bacchus to Ariadne, daughter of King Minos of Crete.

There are three stars above the fourth magnitude. Also above magnitude 4.5: θ (4.14), ε (4.14) and δ (4.59).
See chart for Boötes.

	RA			Dec.			m	M	d (light-years)	Spectrum	
	(h)	(m)	(s)	(°)	(′)	(″)					
5 α	18	34	41.2	+26	42	54	2.22	0.42	75	AO	Alphekka
3 β	15	27	49.9	+29	06	20	3.66	0.94	114	FO	Nusaka
8 γ	12	42	44.6	+26	17	44	3.81	0.57	145	A1	

Alphekka (alternative name, Gemma) is a small-range eclipsing binary; $m = 2$ to 2.32, period 17.36 days.

VARIABLE STARS

	RA		Dec.		Range	Type	Period (d)	Spec.	
	(h)	(m)	(°)	(′)					
U	15	18.2	+31	39	7.7–8.8	Algol	3.45	B + F	
S	15	21.4	+31	22	5.8–14.1	Mira	360.3	M	
R	15	48.6	+28	09	5.7–15	R CrB	–	F8p	
V	15	49.5	+39	34	6.9–12.6	Mira	357.6	N	
T	15	59.5	+25	55	2.0–10.8	Recurrent nova		M+Q	(1866, 1946)
W	16	15.4	+37	48	7.8–14.3	Mira	238.4	M	

DOUBLE STARS

	RA		Dec.		PA	Separation (″)	Magnitudes	
	(h)	(m)	(°)	(′)	(°)			
o	15	20.1	+29	37	337	147.3	5.5, 9.4	
η	15	23.2	+30	17	AB 030	1.0	5.6, 5.9	Binary, 41.6 y
					AC 012	57.7	12.5	
					AB + D 047	215.0	10.0	
ζ	15	39.4	+36	38	305	6.3	5.1, 6.0	
γ	15	42.7	+26	18	118	0.6	4.1, 5.5	Binary, 91 y
ε	15	57.6	+26	53	003	1.8	4.2, 12.6	
ρ	16	01.0	+33	18	071	89.8	5.5, 8.7	
σ	16	14.7	+33	52	236	7.0	5.6, 6.6	Binary, 1000 y

The notable variables in Corona have already been described: R and the Blaze Star, T κ (magnitude 4.8) and ρ (4.2) are each known to have a planet.

CORVUS

(Abbreviation: Crv)

Corvus is an original group. When the god Apollo became enamoured of Coronis, mother of the great doctor Æsculapius, he sent a crow to watch her and report on her behaviour. To be candid, the crow's report was decidedly adverse; but Apollo rewarded the bird with a place in the sky!

Corvus is distinctive, since its leading stars form a quadrilateral. There are four stars above magnitude 4.

See chart for Hydra.

	RA			Dec.			*m*	*M*	*d* (light-years)	Spectrum	
	(h)	(m)	(s)	(°)	(′)	(″)					
4 γ	12	15	48.8	−17	32	31	2.58	−0.94	165	B8	Gienah
9 β	12	34	23.2	−23	23	48	2.65	−0.51	140	G5	Kraz
7 δ	12	29	52.0	−16	30	54	2.94	0.79	88	B9	Algorab
2 ε	12	10	0.5	−22	37	11	3.02	−1.82	300	K2	Minkar

Also above magnitude 4.5: α (Alkhiba), 4.02; η, 4.30. All these six stars have been suspected of variability.

VARIABLE STARS

	RA		Dec.		Range	Type	Period (d)	Spectrum
	(h)	(m)	(°)	(′)				
R	12	19.6	19	15	6.7–14.4	Mira	317.0	M
SV	12	49.8	15	05	6.8–7.6	Semi-reg.	70	M

PLANETARY NEBULA

M	C	NGC	RA		Dec.		Diameter (″)	Magnitude	Magnitude of central star
			(h)	(m)	(°)	(′)			
		4361	12	24.5	−18	48	45 × 110	10.3	13.2

GALAXIES

M	C	NGC	RA		Dec.		Magnitude	Dimensions (′)	Type	
			(h)	(m)	(°)	(′)				
	60	4038	12	01.9	−18	52	11.3	2.6 × 1.8	Sc	Antennæ
	61	4039	12	01.9	−18	53	13	3.2 × 2.2	Smp	Antennæ

CRATER

(Abbreviation: Crt)
Like Corvus, this is a small constellation adjoining Hydra; it has been identified with the wine goblet of Bacchus. The only star above the fourth magnitude is δ.

	RA (h)	RA (m)		Dec. (°)	Dec. (′)		m	M	d (light-years)	Spectrum	
12 δ	11	19	20.5	−14	46	45	3.56	−0.32	195	K0	Labrum

Also above magnitude 4.5: α (Alkes), 4.06, β (Al Sharasif), 4.46.

DOUBLE STAR

	RA (h)	RA (m)	Dec. (°)	Dec. (′)	PA (°)	Separation (″)	Magnitudes
γ	11	24.9	17	41	096	5.2	4.1, 9.6

GALAXIES

M	C	NGC	RA (h)	RA (m)	Dec. (°)	Dec. (′)	Magnitude	Dimensions (′)	Type
		3511	11	03.4	−23	05	11.6	5.4 × 2.2	Sc
		3513	11	03.8	−23	15	12.0	2.8 × 2.3	SBc
		3571	11	11.5	−18	17	12.8	3.3 × 1.3	Sa
		3672	11	25.0	−09	48	11.5	4.1 × 2.1	Sb
		3887	11	47.1	−16	51	11.0	3.3 × 2.7	Sc
		3981	11	56.1	−19	54	12.4	3.9 × 1.5	Sb

The very red semi-regular R Crateris (m 8.0 to 9.5, period ~160 days) lies close to Alkes, and as Alkes is itself an orange giant of type K the two make up a notable pair, though they are not genuinely associated.

CRUX AUSTRALIS

(Abbreviation: Cru)
Though Crux is the smallest constellation in the entire sky, it is also one of the most famous. Before Royer introduced it, in 1679, it had been included in Centaurus. Strictly speaking, it is more like a kite than a cross. As well as its brilliant stars it contains the glorious 'Jewel Box' cluster, and also the dark nebula known as the Coal Sack. There are five stars above the fourth magnitude.

See chart for Centaurus.

	RA (h)	RA (m)	RA (s)	Dec. (°)	Dec. (′)	Dec. (″)	m	M	d (light-years)	Spectrum	
α	12	26	35.9	−63	05	57			321	B0	Acrux
β	12	47	43.3	−59	41	19	1.25	−3.92	352	B0	Mimosa
γ	12	31	09.9	−57	06	45	1.59	−0.56	B8	M4	
δ	12	15	08.8	−58	44	56	2.79	−2.45	364	B2	
ε	12	21	21.8	−60	24	05	3.59	−0.63	228	K3	Juxta

Also above magnitude 4.5: μ^1, 4.03; ζ, 4.06; η, 4.14; θ^1, 4.32.

VARIABLE STARS

	RA (h)	RA (m)	Dec. (°)	Dec. (′)	Range	Type	Period (d)	Spectrum
BH	12	16.3	−56	17	7.2–10.0	Mira	421	S
T	12	21.4	−62	17	6.3–6.8	Cepheid	6.73	F
R	12	23.6	−61	38	6.4–7.2	Cepheid	5.83	F–G
S	12	54.4	−58	26	6.2–6.9	Cepheid	4.69	F–G

DOUBLE STARS

	RA (h)	RA (m)	Dec. (°)	Dec. (′)	PA (°)	Separation (″)	Magnitudes
θ^1	12	03.0	−63	19	325	4.5	4.3, 13.6
η	12	06.9	−64	37	299	44.0	4.2, 11.7
α	12	26.6	−63	06	AB 113	4.0	1.4, 1.9
					AC 202	90.1	1.0, 4.9
γ	12	31.2	−57	07	AB 031	110.6	1.6, 6.7
					AC 082	155.2	9.5
ι	12	45.6	−60	59	022	26.9	4.7, 9.5
μ^1	12	54.6	−57	11	017	34.9	4.0, 5.2

OPEN CLUSTERS

M	C	NGC	RA (h)	RA (m)	Dec. (°)	Dec. (′)	Diam. (′)	Magnitude	No. of stars	
		4052	12	01.9	−63	12	8	8.8	80	
		4103	12	06.7	−61	15	7	7.4	45	
		4337	12	23.9	−58	08	3.5	8.9	–	
		4349	12	24.5	−61	54	16	7.4	30	
		H.5	12	29.0	−60	46	5	9.0	40	
		4439	12	28.4	−60	06	4	8.4	–	
	98	4609	12	42.3	−62	58	5	6.9	40	
	94	4755	12	53.6	−60	20	10	4.2	50+	Jewel Box (κ Crucis)

M	C	NGC	RA (h)	Dec. (°)	Dimensions (°)	Area, squared	
	99		12	53	−63	400 × 300	26.2	Coal Sack

Acrux is a fine binary; the magnitudes of the separate components are 1.4 and 2.1, their absolute magnitudes −3.6 and −2.9. δ is a small-range β Cephei variable; range 2.78 to 2.84, period 0.15 days.

The lovely Jewel Box cluster (κ Crucis) is probably no more than 7 100 000 years old. The three brightest stars are of type B, while the fourth is a red supergiant, magnitude 7.56. Close by is the Coal Sack (C 99), 60 to 70 light-years in diameter, and 500 to 600 light-years away.

CYGNUS

(Abbreviation: Cyg)

Cygnus is one of the richest constellations in the sky; it is often nicknamed the Northern Cross – certainly it is much more nearly cruciform than is Crux. Various legends are associated with it.

According to one, the group was placed in the sky to honour a swan into which Jupiter once transformed himself when on a visit to the wife of the King of Sparta!

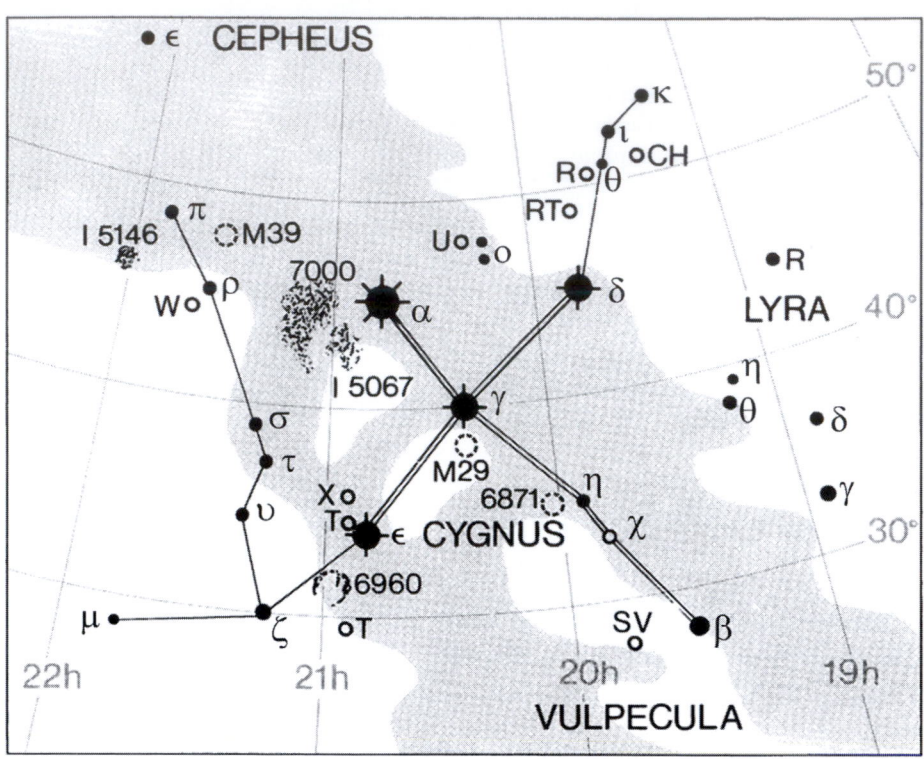

There are 14 stars above the fourth magnitude. To these must be added the red variable χ, which can rise above magnitude 4 at times.

RA			Dec.			m	M	d (light-years)	Spectrum	
(h)	(m)	(s)	(°)	(′)	(″)					
50 α	20	41	25.9	+45	16	49	1.25	−8.73	3228	A2	Deneb
37 γ	20	22	13.7	+40	15	24	2.23	−6.12	1523	F8	Sadr
53 ε	20	46	12.4	+38	58	10	2.48	0.76	72	K0	Gienah
18 δ	19	44	58.4	+45	07	51	2.86	−0.74	171	B9	Rukh
6 β	19	30	43.3	+27	57	35	3.05	−2.31	385	K3	Albireo
64 ζ	21	12	56.2	+30	13	38	3.20	−0.12	151	G8	
62 ξ	21	04	55.9	+43	55	40	3.72	−4.07	1180	K5	
65 τ	21	14	47.4	+38	02	40	3.74	2.14	68	F1	
10 ι	19	29	42.3	+51	43	46	3.76	0.89	122	A5	
1 κ	19	17	06.1	+53	22	05	3.80	0.91	123	K0	
31 o	20	13	37.9	+46	44	29	3.80	−4.29	1350	K2	
21 η	19	56	18.4	+35	05	01	3.89	0.74	139	K0	
58 υ	20	57	10.4	+41	10	02	3.94	−1.25	356	A1	
32 o	20	15	28.3	+47	42	51	3.96	−3.70	1100	K3	
73 ρ	21	33	58.9	+45	35	31	3.98	1.07	124	G8	

Also above magnitude 4.5: 41, 4.01; 52, 4.22; σ (Pennæ), 4.22; π² (Azrak), 4.23; 33, 4.2v; υ, 4.41; 39, 4.43; θ, 4.4; μ¹, 4.49.

Two fainter stars have names. π¹ (magnitude 4.69) is Azelfafage; 22 (magnitude 4.95) is Mokhtarzada.

The immense power of Deneb makes it act upon the nebulosity in its area, exciting it to luminosity. This may well apply to the famous North America Nebula (NGC 7000, C 20), which is about 70 light-years from Deneb.

P Cygni is exceptionally luminous, with an estimated absolute magnitude of −9. Its distance has been given as 5900 light-years (1800 parsec). When it rose to magnitude 3, in 1600, it was regarded as a nova, but it is now classed as a variable of the S Doradûs type. For many years its magnitude has hovered between 4.8 and 5.2.

DOUBLE STARS

	RA (h)	(m)	Dec. (°)	(′)	PA (°)	Separation (″)	Magnitudes	
β	19	30.7	+27	58	054	34.4	3.1, 5.1	Yellow, blue
δ	19	45.0	+45	07	226	2.5	2.9, 6.3	Binary, 828 y
ψ	19	55.6	+52	26	178	3.2	4.9, 7.4	
γ	20	22.2	+40	15	196	41.2	2.2, 9.9	B is a close double
61	21	06.9	+38	45	150	30.3	5.2, 6.0	Binary, 653 y
τ	21	14.8	+38	03	015	0.5	3.8, 6.4	Binary, 50 y
μ	21	44.1	+28	45	206	2.0	4.8, 6.1	Binary, 713 y

VARIABLE STARS

	RA (h)	(m)	Dec. (°)	(′)	Range	Type	Period (d)	Spec.
CH	19	24.5	+50	14	6.4–8.7	Z Andromedæ	±97	M + B
R	19	36.8	+50	12	6.1–14.2	Mira	426.4	M
RT	19	43.6	+48	47	6.4–12.7	Mira	190.2	M
SU	19	44.8	+29	16	6.5–7.2	Cepheid	3.84	F
χ	19	50.6	+32	55	3.3–14.2	Mira	406.9	S
Z	20	01.4	+50	03	7.4–14.7	Mira	263.7	M
RS	20	13.4	+38	44	6.5–9.3	Semi-regular	417	N
P	20	17.8	+38	02	3–6	S Dor	—	B2p
CN	20	17.9	+59	48	7.3–14.0	Mira	198.5	M
U	20	19.6	+47	54	5.9–12.1	Mira	462.4	N
V	20	41.3	+48	09	7.7–13.9	Mira	421.4	N
X	20	43.4	+35	35	5.9–6.9	Cepheid	16.39	F–G
T	20	47.2	+34	22	5.0–5.5	Lb?	—	K
W	21	36.0	+45	22	5.0–7.6	Semi-regular	126	M
SS	21	42.7	+43	35	8.4–12.4	SS Cygni	±50	A–G
WY	21	48.7	+44	15	7.5–14–0	Mira	304.5	M

OPEN CLUSTERS

M	C	NGC	RA (h)	(m)	Dec. (°)	(′)	Diameter (′)	Magnitude	No. of stars	
		6811	19	38.2	+46	34	13	6.8	70	
		6819	19	41.3	+40	11	5	7.3	—	
		6834	19	52.2	+29	25	5	7.8	50	
		6866	20	03.7	+44	00	7	7.6	80	
		6871	20	05.9	+35	47	20	5.2	15	27 Cygni
		6910	20	23.1	+40	47	8	7.4	50	
29		6913	20	23.9	+38	32	7	6.6	50	
		6939	20	31.4	+60	38	8	7.8	80	
		7067	21	24.2	+48	01	3	9.7	20	
39		7092	21	32.2	+48	26	32	4.6	30	

PLANETARY NEBULÆ

M	C	NGC	RA (h)	(m)	Dec. (°)	(')	Dimensions ('')	Magnitude	Magnitude of central star	
		6826	19	44.8	+50	31	30 × 140	9.8	10.4	Blinking Nebula
		7048	21	14.2	+46	16	61		11.3	18

NEBULÆ

M	C	NGC	RA (h)	(m)	Dec. (°)	(')	Dimensions (')	Magnitude of illuminating star	
	27	6888	20	12.0	+38	21	20 × 10	7.4	Crescent Nebula
	34	6960	20	45.7	+30	43	70 × 6	—	Filamentary Nebula, 52 Cygni
		IC 5067/70	20	50.8	+44	21	80 × 70	—	Pelican Nebula
	33	6992/5	20	56.4	+31	43	60 × 8	–	Veil Nebula: SNR
	20	7000	20	58.8	+44	20	120 × 100	6	North America Nebula
	19	IC 5146	21	53.5	+47	16	12 × 12	10	Cocoon Nebula, with sparse cluster

61 Cygni (RA 21h 06m 53s .9, Dec. +38° 44′ 57.9″, magnitude K5) was the first star to have its distance measured. It has been nicknamed 'the Flying Star' because of its large proper motion. It is a wide binary; orbital period 659 years, separation 44 a.u. at periapsis and 124 a.u. at apoaxis.

KY Cygni, RA 20h 25m 57″ .2, Dec. +38° 21′ 11″, is one of the largest stars known; diameter 1420 times that of the Sun, luminosity 300 000, type M3, mass 25 Suns.

The immense power of Deneb makes it act upon the nebulosity in its area, exciting it to luminosity. This may well apply to the famous North America Nebula (NGC 7000, C 20), which is about 70 light-years from Deneb.

P Cygni is exceptionally luminous, with an estimated absolute magnitude of −9. Its distance has been given as 5900 light-years (1800 parsec). When it rose to magnitude 3, in 1600, it was regarded as a nova, but it is now classed as a variable of the S Doradus type. For many years its magnitude has hovered between 4.8 and 5.2.

Cygnus is a favourite area for nova hunters; a number have flared up there, notably Q Cygni of 1875.

DELPHINUS

(Abbreviation: Del)
This is a small but compact constellation: one of Ptolemy's originals. It honours the dolphin which carried the great singer Arion to safety, after he had been thrown overboard by the crew of the ship carrying him home after winning all the prizes in a competition.

The curious names of α and β were allotted by one Nicolaus Venator, for reasons which are obvious!

There are three stars above the fourth magnitude.

See chart for Aquila.

	RA (h)	(m)	(s)	Dec. (°)	(')	('')	m	M	d (light-years)	Spectrum	
6 β	20	37	32.8	+14	35	43	3.64	1.26	07	F5	Rotanev
9 α	20	39	38.3	+15	54	44	3.77	−0.57	240	B0	Svalocin
12 γ	20	46	39.5	+16	07	29	3.9	1.8+2.7	101	K1+A2	Dulfim

Also above magnitude 4.5; ε (Deneb Dulfim), 4.02; δ, 4.4v. δ is a very small-range δ Scuti variable.

The proper motion of ρ (67) Aquilæ has carried it into Delphinus.

γ is a beautiful binary. The primary is a yellow dwarf, magnitude 5.1, 2.5 times as luminous as the Sun; the secondary, an orange subgiant, appears the brighter component at $m = 4.17$. It is 16 ½ times more luminous than the Sun.

VARIABLE STARS

	RA (h)	RA (m)	Dec. (°)	Dec. (')	Range	Type	Period (d)	Spectrum
R	20	14.9	+09	05	7.6–13.8	Mira	284.9	M
EU	20	37.9	+18	16	5.8–6.9	Semi-regular	59	M
HR	20	42.3	+19	10	3.7–12.7	Nova	—	Q
U	20	45.5	+18	05	5.7–7.6	Semi-regular	110	M
S	20	43.1	+17	05	8.3–12.4	Mira	277.2	M
V	20	47.8	+19	20	8.1–16.0	Mira	533.5	M

DOUBLE STARS

	RA (h)	RA (m)	Dec. (°)	Dec. (')	PA (°)	Separation (″)	Magnitudes	
1	20	30.3	+10	54	AB 346	0.9	6.1, 8.1	
					AC 349	16.8	14.1	
β	20	37.5	+14	36	167	0.3	4.0, 4.9	Binary, 26.7 y
α	20	39.6	+15	55	AB 224	29.5	3.8, 13.3	
					AC 272	43.4	11.8	
K	20	39.1	+10	05	286	28.8	5.1, 11.7	
γ	20	46.7	+16	07	206	9.2	4.5, 5.5	
13	20	47.8	+06	00	194	1.6	5.6, 9.2	

GLOBULAR CLUSTERS

M	C	NGC	RA (h)	RA (m)	Dec. (°)	Dec. (')	Diameter (')	Magnitude
	47	6394	20	34.2	+07	24	5.9	8.9
	42	7006	21	01.5	+16	11	2.8	10.6

PLANETARY NEBULA

M	C	NGC	RA (h)	RA (m)	Dec. (°)	Dec. (')	Dimensions (″)	Magnitude	Magnitude of central star
		6891	20	15.2	+12	42	12 × 74	11.7	12.4

DORADO

(Abbreviation: Dor)

A southern constellation, an old name for it was Xiphias. It contains most of the Large Magellanic Cloud (the rest is in Mensa) and also the South Ecliptic Pole.

See chart for Reticulum.

There are two stars above the fourth magnitude:

	RA			Dec.			m	M	d (light-years)	Spectrum
	(h)	(m)	(s)	(°)	(')	('')				
α	04	33	59.7	−55	02	42	3.3v	−0.36	176	A0
β	05	33	37.5	−62	29	24	3.5v	−3.76	1038	F4–G8

Also above magnitude 4.5: γ (4.2v), δ (4.34).

α is a small-range α² CVn variable (3.26–3.30, 2.5 days). γ is the prototype of a class of very small-range pulsating variables; m = 4.23–4.27, due to non-radial gravity wave oscillations. β is one of the brightest Cepheids in the sky, range 3.46–4.08. The diameter is over 70 times that of the Sun, and the spectum varies between F4 and G4.

R Doradûs, a Mira variable, has an angular diameter of 0'' 057, the largest of any star beyond the Solar System; its diameter is over 350 times that of the Sun, luminosity about 6500 Suns. The hypergiant S Doradûs, in the Large Magellanc Cloud, is an LBV (Luminous Blue Variable) over a million times more luminous than the Sun – but at a range of 169 000 light-years its apparent magnitude is only 8.6! Also above magnitude 5:

	Magnitude	Absolute magnitude	Spectrum	Distance
θ	4.83	−0.1	K2	80

VARIABLE STARS

	RA		Dec.		Range	Type	Period (d)	Spectrum
	(h)	(m)	(°)	(')				
R	04	36.8	−62	05	4.8–6.6	Semi-regular	338	M
β	05	33.6	−62	29	3.7–4.1	Cepheid	9.84	F–G

DOUBLE STAR

	RA		Dec.		PA	Separation ('')	Magnitudes
	(h)	(m)	(°)	(')	(°)		
α	04	34.0	−55	03	AB 182	0.2	3.8, 4.3
					AB + C 101	77.7	9.8

GALAXIES

M	C	NGC	RA		Dec.		Magnitude	Dimensions	Type
			(h)	(m)	(°)	(')		(')	
		1549	04	15.7	−55	36	9.9	3.7 × 3.2	E0
		1553	04	16.2	−55	47	9.5	4.1 × 2.8	S0
		1596	04	27.6	−55	02	11.0	3.9 × 1.2	S0
		1617	04	31.7	−54	36	10.4	4.7 × 2.4	SBa
		1672	04	45.7	−59	15	11.0	4.8 × 3.9	SBb
		LMC	05	23.6	−69	45	0.1	650 × 550	Large Cloud of Magellan. Contains 30 Doradûs and three planetary nelbulæ, NGC 1714, 1722 and 1743.
		1947	05	26.8	−63	46	10.8	3.0 × 1.6	S0p

NEBULA

M	C	NGC	RA		Dec.		Dimensions	
			(h)	(m)	(°)	(')	(')	
	103	2070	05	38.7	−69	06	40 × 25	30 Dor. In the LMC; Tarantula Nebula

DRACO

(Abbreviation: Dra)

A long, sprawling northern group. In mythology it has been identified either with the dragon killed by Cadmus before he found the city of Bœotia, or with the dragon which guarded the golden apples in the Garden of the Hesperides. Thuban (α) was the pole star in ancient times.

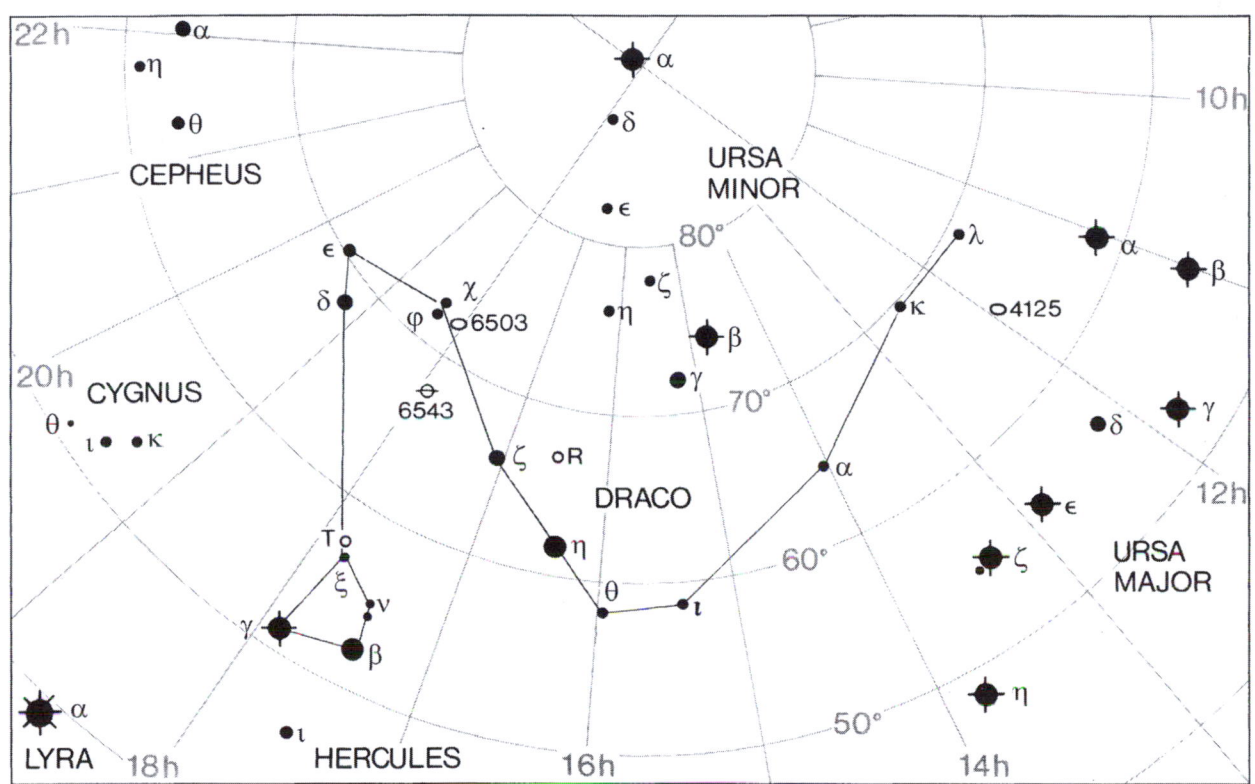

There are 12 stars above the fourth magnitude.

	RA			Dec.			m	M	d (light-years)	Spectrum	
	(h)	(m)	(s)	(°)	(')	('')					
33 γ	17	56	36.4	+51	29	20	2.24	−1.04	148	K5	Eltanin
14 η	16	23	59.5	+61	30	51	2.73	0.58	88	G8	Aldhibain
23 β	17	30	26.0	+52	18	05	2.79	2.43	360	G2	Alwaid
57 δ	19	12	33.2	+67	39	41	3.07	0.63	100	G9	Altais
22 ζ	17	08	47.2	+65	42	53	3.17	−1.92	340	B6	Kaou
12 ι	15	24	55.8	+58	57	58	3.29	0.81	102	K2	Edasich
44 χ	18	21	02.3	+72	44	01	3.55	4.02	26	F7	Bain al Thuban
11 α	14	04	23.4	+64	22	13	3.67	−1.21	309	A0	Thuban
32 ξ	17	53	31.2	+56	52	21	3.73	1.06	111	K2	Grumium
1 λ	11	31	24.3	+69	19	52	3.82	−1.23	334	M0	Giansar
63 ε	19	48	10.2	+70	16	04	3.84	0.59	146	G8	Tyl
5 κ	12	33	29.0	+69	47	18	3.85	−2.07	500	B6	Ketu

Also above 4.5: θ (4.01), φ (4.22), τ (4.45).
Also above magnitude 5:

	Magnitude	Absolute magnitude	Spectrum	Distance (light–years)	
42	4.82	−0.1	K2	96	
39	4.98	1.2	A1	53	
28 ω	4.80	3.4	F5	22	Al Dhih
24 ν¹	4.88		A8	62	Kuma
25 ν²	4.87		A4	62	
21 μ	4.92		F5		
4	4.95	−0.5	M4	120	
6	4.94	−0.1	K2	84	
18 g	4.83	−4.5	K1		
19	4.89	3.7	F6	17	
45	4.77	−4.6	F7	450	
52 υ	4.82	0.2	K0	64	
54	4.99	−0.1	K2	100	

VARIABLE STARS

	RA (h)	(m)	Dec. (°)	(′)	Range	Type	Period (d)	Spectrum
RY	12	56.4	+66	00	5.6–8.0	Semi-regular	173	N
R	16	32.7	+66	45	6.7–13.0	Mira	245.5	M
T	17	56.4	+58	13	7.2–13.5	Mira	421.2	N
UW	17	57.5	+54	40	7.0–8.0	Irregular	—	K
UX	19	21.6	+76	34	5.9–7.1	Semi-regular	168	N

DOUBLE STARS

	RA (h)	(m)	Dec. (°)	(′)	PA (°)	Separation (″) sec	Magnitudes	
η	16	24.0	+61	31	142	5.2	2.7, 8.7	
μ	17	05.3	+54	28	017	2.0	5.7, 5.7	Binary, 482 y
ν	17	32.2	+55	11	312	61.9	4.9, 4.9	
ψ	17	41.9	+72	09	015	30.3	4.9, 6.1	
ε	19	48.2	+70	16	019	3.1	3.8, 7.4	Slow binary

PLANETARY NEBULA

M	C	NGC	RA (h)	(m)	Dec. (°)	(′)	Dimensions (″)	Magnitude	Magnitude of central star
	6	6543	17	58.6	+66	38	18 × 350	8.8	9.5

GALAXIES

M	C	NGC	RA (h)	(m)	Dec. (°)	(′)	Magnitude	Dimensions (″)	Type
		3147	10	16.9	+73	24	10.6	4.0 × 3.5	Sb
		4125	12	08.1	+65	11	9.8	5.1 × 3.2	E5p
	3	4236	12	16.7	+69	28	9.7	18.6 × 6.9	Sb

GALAXIES

M	C	NGC	RA (h)	(m)	Dec. (°)	(′)	Magnitude	Dimensions (″)	Type
		5866	15	06.5	+55	46	10.0	5.2 × 2.3	E6p
		5879	15	09.8	+57	00	11.5	4.4 × 1.7	Sb
		5907	15	15.9	+56	19	10.1	12.3 × 1.8	Sb
		5985	15	39.6	+59	20	11.0	5.5 × 3.2	Sb
		6015	15	51.4	+62	19	11.2	5.4 × 2.3	Sc
		5907	15	15.9	+56	19	10.4	12.3 × 1.8	Sb
		5985	15	39.6	+59	20	11.0	5.5 × 3.2	Sb
		6503	17	49.4	+70	09	10.2	6.2 × 2.3	Sb

Thuban was the pole star in ancient times – and will be so again in AD 21 000.

γ (Eltanin) has an alternative proper name, Rastaban. In 1728, Bradley attempted to measure its parallax; he failed but his measurements led him to the discovery of the aberration of light. In 1.5 million years it will pass the Sun at only 28 light-years, and will then be the brightest star in the sky.

ε (Tyl) is an easy double. The K-type primary has been suspected of variability.

ν (Kuma) is double; the components, both of type A, are nearly equal, and the separation, 62″, means that this is a pair separable with binoculars. The distance from Earth is 100 light-years. The real separation is at least 1900 a.u., and the orbital period is probably about 44 000 years.

Probably the most spectacular object in Draco is the Cat's-Eye planetary nebula, NGC 6543.

EQUULEUS

(Abbreviation: Eql)

One of Ptolemy's original 88, but the smallest constellation in the sky apart from Crux. Mythologically it may represent the foal Celeris, offspring of Pegasus, given to Castor by Mercury. There is only one star above the fourth magnitude:

8 α, RA 21h 15m 49s ., Dec +03° 14′ 33″, m = 3.92, M = 0.14*, d = 186 light-years, spectrum G0 Kitalpha.

The only other star above 4.5 is δ (Pherasauval), 4.47*.

VARIABLE STAR

	RA (h)	(m)	Dec. (°)	(′)	Range	Type	Period (d)	Spectrum
S	20	57.2	+05	05	8.0–10.1	Algol	3.44	B + F

DOUBLE STARS

	RA (h)	(m)	Dec. (°)	(′)	PA (°)	Separation (″)	Magnitudes	
ε	20	59.1	+04	18	AB 285	1.0	6.0, 6.3	Binary, 101.4 y
					AB + C 070	10.7	7.1	
					AD 280	74.8	2.4	
γ	21	10.4	+10	08	AB 268	1.9	4.7, 11.5	
					AC 005	47.7	12.5	
δ	21	14.5	+10	00	029	0.3	5.2, 5.3	Binary, 5.7 y
β	21	22.9	+06	49	257	34.4	5.2, 13.7	

δ was for some time the visual binary with the closest known separation. The distance is 60 light-years; the components are of types G0 and F3, both over twice as luminous as the Sun. Its Arabic proper name seems to mean 'the First Horse'.

See chart for Aquila.

ERIDANUS

(Abbreviation: Eri)
An immensely long constellation, extending from Achernar in the far south as far as Kursa, near Orion. Achernar is the only brilliant star. Mythologically, Eridanus is the river Po – and this was the river into which the youth Phæthon was plunged when he had obtained permission to drive the Sun-chariot for a day, and had lost control of it, so that Jupiter was forced to strike him down with a thunderbolt.

Achernar is the closest really brilliant star to the south celestial pole. The pole lies in a very barren area, around midway between Achernar and the Southern Cross.

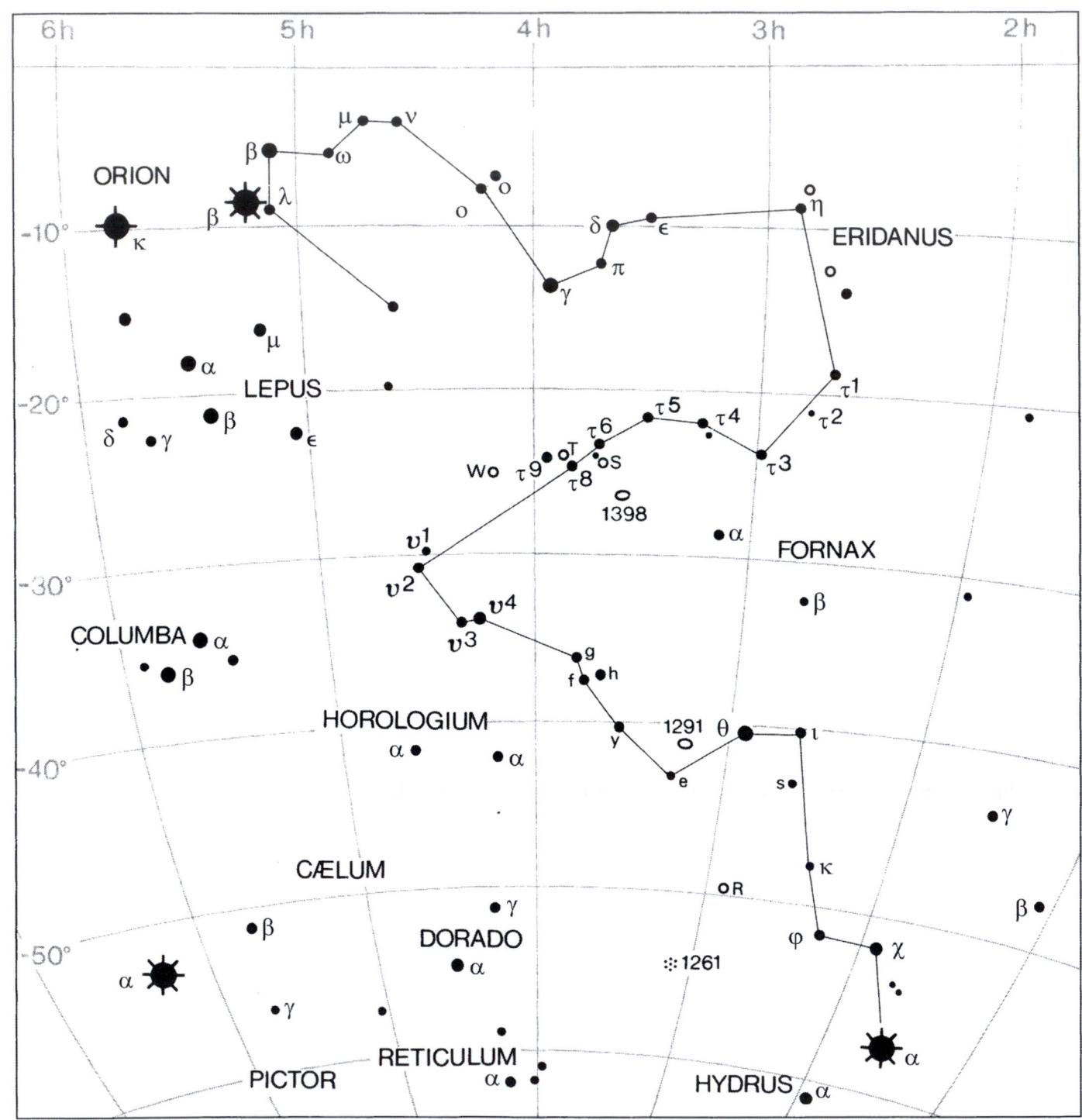

RA			Dec.				m	M	d (light-years)	Spectrum	
(h)	(m)	(s)	(°)	(′)	(″)						
α	01	37	42.8	−57	14	12	0.45	−2.77	144	B3	Achernar
67 β	05	07	41.0	−05	05	11	2.78	0.60	89	A3	Kursa
θ	02	58	15.7	−40	18	17	2.88	−0.59	161	A4	Acamar
34 γ	03	58	01.7	−13	30	30	2.97	−1.19	221	M1	Zaurak
23 δ	03	43	15.0	−09	45	55	3.52	3.74	29	KO	Rana
41 υ⁴	04	17	53.6	−33	47	54	3.55	−0.94	178	B9	
φ	02	16	30.5	−51	30	44	3.56	0.18	155	B8	
χ	01	55	56.8	−51	36	35	3.69	2.48	57	G5	
16 τ 4	03	19	31.0	−21	45	29	3.70	−0.79	258	M3	Liberflux
18 ε	03	32	56.4	−09	27	30	3.72	6.18	10	K2	Sadira
52 υ 2	04	35	33.1	−30	33	44	3.81	−0.22	209	G8	Theemin
53 I	04	38	10.9	−14	18	13	3.86	−1.23	109	K1	Sceptrum
3 η	02	56	25.6	−08	53	51	3.89	0.83	133	K1	Azha
48 ν	04	36	19.1	−03	21	09	3.93	−2.34	580	B2	
43 d	04	24	02.2	−34	01	01	3.97	−0.64	273	K4	

Also above 4.5: μ (4.01), o¹ (Beid 4.04), τ³ (4.08), ι (4.11), g (4.17), τ⁶ (4.22), κ (4.24), λ (4.25), e (4.26), τ⁵ (4.26), f (4.30), 54 (4.32), ω (4.36), π (4.43), o² (Keid 4.43), w (4.46), τ¹ (4.47) υ¹ (4.49).

VARIABLE STARS

	RA		Dec.		Range	Type	Period (d)	Spectrum
	(h)	(m)	(°)	(′)				
Z	02	47.9	−12	28	7.0–8.6	Semi-regular	80	M
RR	02	52.2	−08	16	7.4–8.6	Semi-regular	97	M
T	03	55.2	−24	02	7.4–13.2	Mira	252.2	M
W	04	11.5	−25	08	7.5–14.5	Mira	376.7	M
RZ	04	43.8	−10	41	7.8–8.7	Algol	39.28	A + G

DOUBLE STARS

	RA		Dec.		PA	Separation (″)	Magnitudes	
	(h)	(m)	(°)	(′)	(°)			
χ	01	56.0	−51	37	202	5.0	3.7, 10.7	
P	01	39.8	−56	12	194	11.2	5.5, 5.8	Binary, 484 y
θ	02	58.3	−40	18	088	8.2	3.4, 4.5	
τ⁴	03	19.5	−21	45	AB 288	5.7	3.7. 9.2	
					AC 112	39.2	10.7	
υ⁴	04	17.9	−33	48	013	49.2	3.6, 11.8	A is a close double
ρ²	03	02.7	−07	41	075	1.8	5.3, 9.5	
o²	04	15.2	−07	39	AB 104	83.4	4.4, 9.5	Binary, 248 y
					BC 279	7.6	9.5, 11.2	

PLANETARY NEBULA

M	C	NGC	RA		Dec.		Dimensions (″)	Magnitude	Magnitude of central star
			(h)	(m)	(°)	(′)			
		1535	04	14.2	−12	44	18 × 44	9.6	12.2

GALAXIES

M	C	NGC	RA (h)	(m)	Dec. (°)	(′)	Magnitude	Dimensions (′)	Type
		1084	02	46.0	−07	35	10.6	2.9 × 1.5	Sc
		1179	03	02.6	−18	54	11.8	4.6 × 3.9	Sp
		1187	03	02.6	−22	52	10.9	5.0 × 4.1	SBc
		1291	03	17.3	−41	08	8.5	10.5 × 9.1	SBa
		1300	03	19.7	−19	25	10.4	6.5 × 4.3	SBb
		1332	03	26.3	−21	20	10.3	4.6 × 1.7	E7
		1337	03	28.1	−08	23	11.7	6.8 × 2.0	S
		1395	03	38.5	−23	02	11.3	3.2 × 2.5	E3
		1407	03	40.2	−18	35	9.8	2.5 × 2.5	E0
		1532	04	12.1	−32	52	11.1	5.6 × 1.8	Sb
		1637	04	41.5	−02	51	10.9	3.3 × 2.9	Sc

Achernar, about eight times as massive as the Sun, spins so rapidly that its equatorial diameter is over 50% greater than its polar diameter – the 'flattened star'!

Acamar is a wide, splendid binary; magnitudes 2.88 and 4.35, both of type A. Separation 8″.2. Curiously, both Ptolemy and Ulugh Beigh ranked it as a first-magnitude star. Its name means 'Last of the River', and there may possibly have been some confusion between Acamar and Achernar.

The two Omicrons are not genuinely associated. o^2 or 40 Eridani (Keid) is only 16 light-years away; it is a triple, and notable because the faintest member of the trio is a white dwarf, type DA4.

FORNAX

(Abbreviation: For)

A small southern constellation, introduced by Lacaille in 1756 as Fornax Chemica, the Chemcal Furnace. There is only one star above the fourth magnitude:

α, RA 03h 12m 04s .3, Dec. −28° 59′ 21″, m = 3.80, M = 3.05, d = 46 light-years, spectrum F8; Dalim.

Also above 4.5: β (4.48). Then come ν (4.68) and ω (4.96). The sequence of Greek letters has certainly become chaotic here!

Dalim has also been included in Eridanus, as 12 Eridani. See chart for Eridanus.

VARIABLE STARS

	RA (h)	(m)	Dec. (°)	(′)	Range	Type	Period (d)	Spectrum
R	02	29.3	−26	06	7.5–13.0	Mira	387.9	N
ST	02	44.4	−29	12	7.7–9.0	Semi-regular	277	M
S	03	46.2	−24	24	?5.6–8.5?	Suspected	?	M

DOUBLE STARS

	RA (h)	(m)	Dec. (°)	(′)	PA (°)	Separation (″)	Magnitudes	
ω	02	33.8	−28	14	244	10.8	5.0, 7.7	
γ¹	02	49.8	−24	34	AB 145	12.0	6.1, 12.5	
					AC 143	40.9	10.5	
η²	02	50.2	−35	51	014	5.0	5.9, 10.1	
α	03	12.1	−28	59			4.0, 7.0	Binary, 314 y
χ³	03	28.2	−35	51	248	6.3	6.5, 10.5	

PLANETARY NEBULA

M	C	NGC	RA (h)	(m)	Dec. (°)	(′)	Diameter (″)	Magnitude	Magnitude of central star
		1360	03	33.3	−25	51	390	–	11.3

GALAXIES

M	C	NGC	RA (h)	(m)	Dec. (°)	(′)	Magnitude	Dimensions (′)	Type
		986	02	33.6	−39	02	11.0	3.7 × 2.8	SBb
	67	1097	02	46.3	−30	17	9.2	9.3 × 6.6	SBb
		1201	03	04.1	−26	04	10.6	4.4 × 2.8	Sa
		1255	03	13.5	−25	44	11.1	4.1 × 2.8	Sa
		1302	03	19.9	−26	04	11.5	4.4 × 4.2	SBa
		1316	03	22.7	−37	12	8.8	7.1 × 5.5	SB0p
		1326	03	23.9	−36	28	10.5	4.0 × 3.0	SB0
		1344	03	28.3	−31	04	10.3	3.9 × 2.3	E3
		1350	03	31.1	−33	38	10.5	4.3 × 2.4	SBb
		1365	03	33.6	−36	08	9.5	9.8 × 5.5	SBb
		1371	03	35.0	−24	56	11.5	5.4 × 4.0	SBa
		1380	03	36.5	−34	59	11.1	4.9 × 1.9	S0
		1385	03	37.5	−24	30	11.2	3.0 × 2.0	Sc
		1399	03	38.5	−35	27	9.9	3.2 × 3.1	Elp
		1398	03	38.9	−26	20	9.7	6.6 × 5.2	SBb
		1404	03	38.9	−35	35	10.2	2.5 × 2.3	E1
		1425	03	42.2	−29	54	11.7	5.4 × 2.7	Sb

The Hubble Deep Field is located within Fornax, and the constellation includes most of the Fornax Cluster of galaxes. In it lies NGC 1316, a bright elliiptical galaxy and a particularly strong radio source.

GEMINI

(Abbreviation: Gem)

A brilliant Zodiacal constellation. In mythology, Castor and Pollux were the twin sons of the King and Queen of Sparta. Pollux was immortal, but Castor was not. When Castor was killed, Pollux pleaded with the gods to be allowed to share his immortality: so Castor was brought back to life, and both youths placed in the sky.

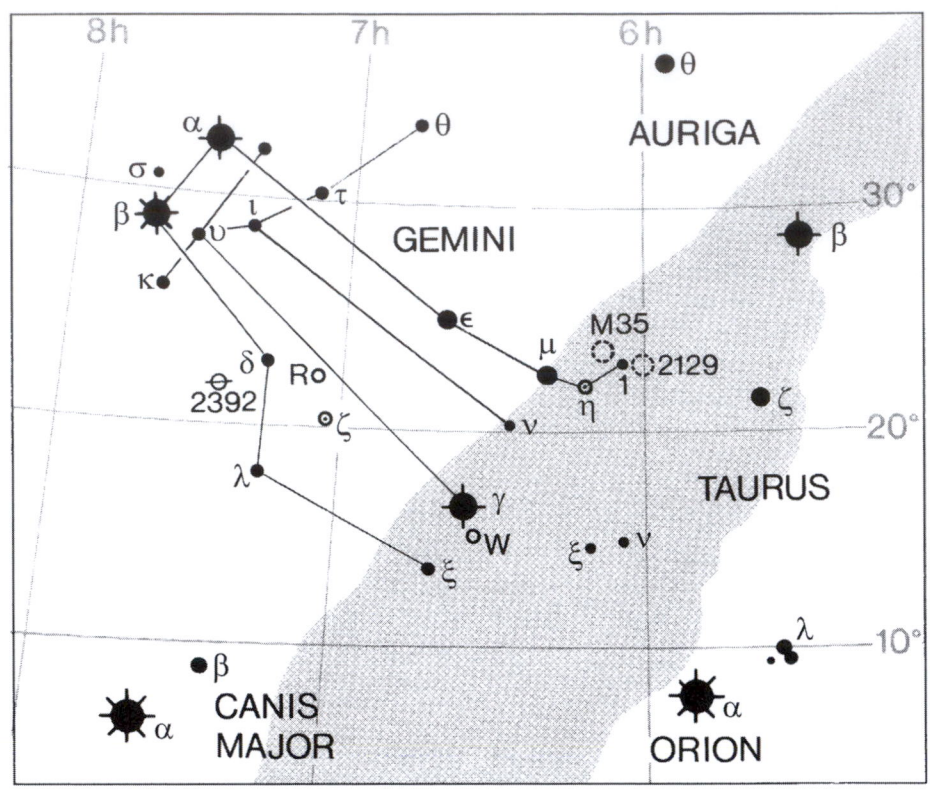

Today Pollux is the brighter of the two stars, but in ancient times Castor was recorded as being the more brilliant. If any change has occurred (and this is by no means certain) it is more likely to have been in the late-type Pollux than in Castor, which is a multiple system. There are 13 stars in Gemini above the fourth magnitude.

	RA			Dec.			m	M	d (light-years)	Spectrum	
	(h)	(m)	(s)	(°)	(')	('')					
78 β	07	45	19.4	+28	01	35	1.16	1.09	34	K0	Pollux
66 α	07	34	36.0	+31	53	19	1,58	0.59	52	A2	Castor
24 γ	06	37	42,3	+16	23	58	1.93	−0.60	105	A0	Alhena
13 μ	06	22	57.6	+22	30	50	2.87	−1.39	232	M3	Tejat Post
27 ε	06	43	55.9	+25	07	52	3.06	−4.15	900	A3	Mebsuta
7 η	06	14	52.7	+22	30	25	3.2v	−1.84	349	M3	Propus
31 ξ	06	45	17.4	+12	53	46	3.35	2.13	57	F5	Alzirr
35 δ	07	20	07.4	+21	58	56	3.50	2.22	59	F0	Wasat
77 κ	07	44	26.9	+24	23	53	3.57	0.35	143	G8	Al Kirkab
54 λ	07	18	05.6	+16	32	26	3.58	1.27	94	A3	Kebash
34 θ	06	52	47.3	+33	57	41	3.60	−0.30	197	A3	Nageba
43 ζ	07	04	13.1	+20	34	13	3.7v	−3.76	1168	G3	Mekbuda
60 ι	07	25	43.7	+27	47	54	3.78	0.85	126	G9	Yin-Yang

Also above 4.5: υ (4.06), ν (4.13), i (4.16), ρ (416), σ (4.23), τ (4.41), 30 (4.49).

VARIABLE STARS

	RA (h)	RA (m)	Dec. (°)	Dec. (')	Range	Type	Period (d)	Spectrum
BU	06	12.3	+22	54	5.7–7.5	Irregular	–	M
η	06	14.9	+22	30	3.2–3.9	Semi-regular	233	M
W	06	35.0	+15	20	6.5–7.4	Cepheid	7.91	F–G
X	06	47.1	+30	17	7.5–13.6	Mira	263.7	M
ζ	07	04.1	+20	34	3.7–4.1	Cepheid	10.15	F–G
R	07	07.4	+22	42	6.0–14.0	Mira	369.8	S
V	07	23.2	+13	06	7.8–14.9	Mira	275.1	M
T	07	49.3	+23	44	8.0–15.0	Mira	287.8	S
U	07	55.1	+22	00	8.2–14.9	SS Cygni	±103	M + WD

DOUBLE STARS

	RA (h)	RA (m)	Dec. (°)	Dec. (')	PA (°)	Separation (")	Magnitudes	
η	06	14.9	+22	30	267		variable	Binary, 474 y
μ	06	22.9	+22	31	077	72.7	3.0, 9.8	
ν	06	29.0	+20	13	329	112.5	4.2, 8.7	
ε	06	43.9	+25	08	094	110.3	3.0, 9.0	
38	06	54.6	+13	11	147	7.0	4.7, 7.7	Binary, 3190 y
ζ	07	04.1	+20	34	AB 084	87.0	variable, 10.5	
					AC 350	96.5	8.0	
λ	07	18.1	+16	32	033	9.6	3.6, 10.7	
δ	07	20.1	+21	59	255	5.8	3.5, 8.2	Binary, 1200 y
α	07	34.6	+31	53	AB 067	3.7	1.9, 2.9	Binary, 420 y
					AC 164	72.5	1.6, 8.8	
κ	07	44.4	+24	24	240	7.1	3.6, 8.1	

OPEN CLUSTERS

M	C	NGC	RA (h)	RA (m)	Dec. (°)	Dec. (')	Diameter (')	Magnitude	No. of stars	
		2129	06	01.0	+23	18	7	6.7	40	
		IC 2157	06	05.0	+24	00	7	8.4	20	
		2169	06	08.4	+13	57	7	5.9	30	
35		2168	06	08.9	+24	20	28	5.0	200	
		2266	06	43.2	+26	58	7	9.5	30	
		2355	07	16.9	+13	47	9	9.7	40	
		2395	07	27.1	+13	35	12	8.0	30	Asterism?

PLANETARY NEBULA

M	C	NGC	RA (h)	RA (m)	Dec. (°)	Dec. (')	Dimensions (")	Magnitude	Magnitude of central star	
	39	2392	07	29.2	+20	55	13 × 44	10	10.5	Eskimo Nebula

Castor has already been described it is a sextupal system. Pollux, an orange giant, is 32 times as luminous as the Sun and diameter 8 times that of the Sun and mass 1.7 times that of the Sun. It has a planet, Pollux b (nicknamed Polydeuces), discovered in 2006; orbital period 590 days at a mean distance of 1.64 a.u. from the star. It is at least 2.3 times as massive as Jupiter, and is presumably a gas-giant

GRUS

(Abbreviation: Gru)

The only conspicuous member of the Southern Birds. Its shape really can conjure up the impression of a crane in flight,

η, a spectroscopic binary. Is a semi-regular variable with a range of magnitude from 3.2.to 3.9. η Geminorum B, a G-type dwarf, orbits the main pair in a period of over 700 years. ζ is one of the brightest Cepheids in the sky.

Gemini is a very rich constellation. The open cluster M 35 and the Eskimo Nebula, NGC 2392 (C 39) – sometimes called the Clown's Face Nebula – are of special note.

and the colour contrast between Alnair and Al Dhanab is very marked.

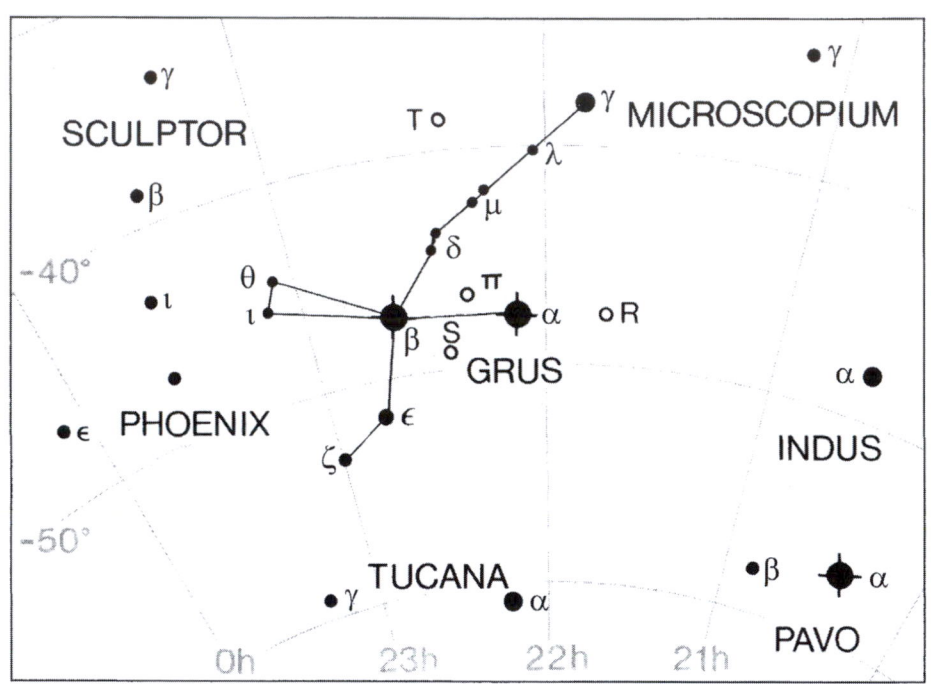

	RA			Dec.			m	M	d (light-years)	Spectrum	
	(h)	(m)	(s)	(°)	(')	('')					
α	22	08	13.9	−46	57	38	1.73	−0.73	101	B7	Alnair
β	22	42	40.0	−46	53	04	2.07	−1.52	170	M5	Al Dhanab
γ	21	53	55.7	−37	21	53	3.00	−0.97	203	B9	
ε	22	48	33.2	−51	19	00	3.49	0.49	130	A3	
ι	23	10	21.4	−45	14	47	3.88	0.11	185	K0	
δ-1	22	29	16.1	−43	29	44	3.97	−0.82	296	G6	

Also above 4.5: ζ (4.11), δ 2 (4.12), θ (4.28), λ (4.47).

δ^1 (magnitude 3.97, type G6) and δ^2 (4.12, M4) look like a very wide double, but are over 20 light-years apart. μ^1 (4.79, G8) and

μ^2 (5.11, G8) also give the impression of a wide double, but here too the real separation is ove 20 light-years.

VARIABLE STARS

	RA		Dec.		Range	Type	Period (d)	Spectrum
	(h)	(m)	(°)	(')				
RS	21	43.1	−48	11	7.9–8.5	Delta Scuti	0.15	A–F
R	21	48.5	−46	55	7.4–14.9	Mira	331.9	M
π¹	22	22.7	−45	57	5.4–6.7	Semi-regular	150	S
T	22	25.7	−37	34	7.8–12.3	Mira	136.5	M
S	22	26.1	−48	26	6.0–15.0	Mira	401.4	M

DOUBLE STARS

	RA		Dec.		PA	Separation (″)	Magnitudes
	(h)	(m)	(°)	(')	(°)		
θ	23	06.9	−43	31	075	1.1	4.5, 7.0
υ	23	06.9	−38	54	211	1.1	5.7, 8.0

GALAXIES

M	NGC	RA		Dec.		Magnitude	Dimensions	Type
		(h)	(m)	(°)	(')		(')	
	7144	22	52.7	−48	15	10.7	3.5 × 3.5	E0
	7213	22	09.3	−47	10	10.4	1.9 × 1.8	Sa
	7410	22	55.0	−39	40	10.4	5.5 × 2.0	SBa
	7412	22	55.8	−42	39	11.4	4.0 × 3.1	SBb
	7418	22	56.6	−37	02	11.4	3.3 × 2.8	SBc
	IC 1459	22	57.2	−36	28	10.0	—	E3
	IC 5267	22	57.2	−43	24	10.5	5.0 × 4.1	S0
	7424	22	57.3	−41	04	11.0	7.6 × 6.8	SBc
	IC 5273	22	59.5	−37	42	11.4	2.9 × 2.1	SBc
	7456	23	02.1	−39	35	11.9	5.9 × 1.8	Sc
	7496	23	09.8	−43	26	11.1	3.5 × 2.8	SBb
	7531	23	14.8	−43	36	11.3	3.5 × 1.5	Sb
	7552	23	16.2	−42	35	10.7	3.5 × 2.5	SBb
	7582	23	18.4	−42	22	10.6	4.6 × 2.2	SBG
	7599	23	19.3	−42	15	11.4	4.4 × 1.5	Sc

Since Grus is not one of Ptolemy's original constellations (it is too far south in the sky) no legends are attached to it. An old, obsolete name for it was 'Phoenicopterus', Latin for flamingo.

HERCULES

(Abbreviation: Her)

A large but by no means brilliant constellation, commemorating the great hero of mythology. The most interesting objects are the red supergiant α and the globular clusters M 13 and M 92;

M13 is the brightest globular in the northern hemisphere of the sky, and is surpassed only by the southern ω Centauri and 47 Tucanæ.

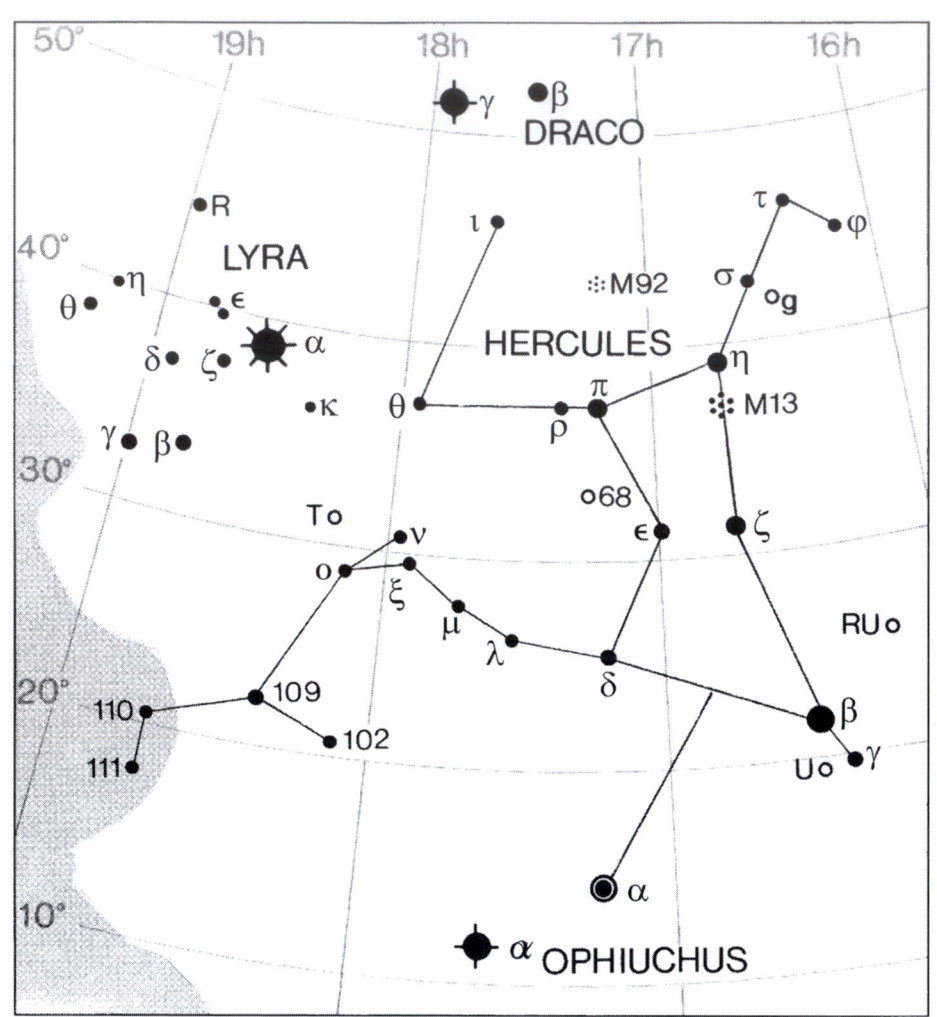

Hercules contains fifteen stars above the fourth magnitude.

	RA			Dec.			m	M	d (light-years)	Spectrum
	(h)	(m)	(s)	(°)	(′)	(″)				
27 β	16	30	13.3	+21	29	23	2.78	−0.50	143	G8
40 ζ	16	41	17.5	+31	36	07	2.81	2.64	35	F9
65 δ	17	15	01.9	+24	50	23	3.12	1.21	78	A3
67 π	17	15	02.9	+36	48	33	3.16	−2.10	367	K3
64 α	17	14	38.9	+14	23	25	3v	−2.04	182	M5
86 μ	17	46	27.7	+27	43	21	3.42	3.80	27	G5
44 η	16	42	53.7	+38	55	21	3.48	0.80	112	G3
92 χ	17	57	45.8	+29	14	53	3.70	0.61	115	K0
20 γ	16	21	55.2	+19	09	11	3.74	−0.15	1.95	A9
85 ι	17	39	28	+46	00	23	3.82	−2.09	500	B3

	RA			Dec.			m	M	d (light-years)	Spectrum
	(h)	(m)	(s)	(°)	(′)	(″)				
103 o	18	07	32.6	+28	45	45	3.84	−1.30	347	B9
109	18	23	41.8	+21	46	13	3.85	0.87	128	K2
91 θ	17	56	15.2	+37	15	02	3.86	−2.70	670	K1
22 τ	16	19	44.5	+46	18	48	3.91	−1.01	315	B5
58 g	17	00	17.4	+30	55	35	3.92	0.43	163	A0

Also above 4.5: ρ (4.15), 110 (4.19), σ (4.20), φ (4.23), 95 (4.26), 111 (4.34), 102 (4.37), λ (Masym) (4.41), μ (4.41).
The asterism sometimes called the Keystone is formed by π, η, ζ and ε.

VARIABLE STARS

	RA		Dec.		Range	Type	Period (d)	Spectrum
	(h)	(m)	(°)	(′)				
X	16	02.7	+47	14	7.5–8.6	Semi-regular	95	M
R	16	06.2	+18	22	7.13–15.0	Mira	318.4	M
RU	16	10.2	+25	04	6.8–14.3	Mira	485.5	M
U	16	25.8	+18	54	6.5–13.4	Mira	406.0	M
g (30)	16	28.6	+41	53	5.7–7.2	Semi-regular	70	M
W	16	35.2	+37	21	7.6–14.4	Mira	280.4	M
S	16	51.9	+14	56	6.4–13.8	Mira	307.4	M
α	17	14.6	+14	23	3–4	Semi-regular	±100?	M
u (68)	17	17.3	+35	06	4.6–5.3	Beta Lyræ	2.05	B + B
RS	17	21.7	+22	55	7.0–13.0	Mira	219.6	M
Z	17	58.1	+15	08	7.3–8.1	Algol	3.99	F + K
T	18	09.1	+31	01	6.8–13.9	Mira	165.0	M
AC	18	30.3	+21	52	7.4–9.7	RV Tauri	75.5	F + K
RX	18	30.7	+12	37	7.2–7.8	Algol	1.78	A + A

DOUBLE STARS

	RA		Dec.		PA	Separation (″)	Magnitudes	
	(h)	(m)	(°)	(′)	(°)			
κ	16	08.1	+17	03	012	28.4	5.3, 6.5	
γ	16	21.9	+19	09	233	41.6	3.8, 9.8	
ω	16	25.4	+14	02	AB 223	1.0	4.6, 11.6	
					AC 096	28.4	11.1	
37	16	40.6	+04	13	230	69.8	5.8, 7.0	
ζ	16	41.3	+31	36	029	1.6	2.9, 5.5	Binary, 34.5 y
54	16	55.4	+18	26	183	2.5	5.4, 12.7	
α	17	14.6	+14	23	107	4.7	variable, 5.4	Binary, 3600 y
δ	17	15.0	+24	50	236	8.9	3.7, 8.2	Optical pair
u (68)	17	17.3	+35	06	060	4.4	var., 10.2	
ρ	17	23.7	+37	09	316	4.1	4.6, 5.6	
μ	17	46.5	+27	43	247	33.8	3.4, 10.1	Binary. Very long period. B is itself a close binary.

GLOBULAR CLUSTERS

M	C	NGC	RA		Dec.		Diameter	Magnitude
			(h)	(m)	(°)	(′)	(′)	
92		6341	17	17.1	+43	08	11.2	6.5
13		6205	16	41.7	+36	28	16.6	5.9

PLANETARY NEBULÆ

M	C	NGC	RA		Dec.		Dimensions (″)	Magnitude	Magnitude of central star
			(h)	(m)	(°)	(′)			
		6058	16	04.4	+40	41	23	13.3	13.8
		IC 4593	16	12.2	+12	04	12 × 120	10.9	11.3
		6210	16	44.5	+23	49	14	9.3	12.9

α (Rasalgethi) is a huge red supergiant, about 600 000 000 km in diameter; the angular diameter, measured with an interferometer is 0.034″. It is a semi-regular variable; the official range is from magnitude 3 to 4, but I have never seen it brighter than 3.2 or fainter than 3.7. The mass is 14 times times that of the Sun and there is a vast, tenuous gaseous envelope. The companion, over 500 a.u. away, looks green when viewed through a telescope, but con-trast with the red primary may be responsible for this. The companion is actually a very close binary.

The best-known telescopic objects in Hercules are the naked-eye globular cluster M 13, and M 92, which is on the fringe of naked-eye visibility.

The apex of the Sun's way lies in Hercules, near the border with Lyra, not far from Vega.

HOROLOGIUM

(Abbreviation: Hor)
A small, obscure southern constellation. It has only one star above magnitude 4.

	RA			Dec.			m	M	d (light-years)	Spectrum
	(h)	(m)	(s)	(°)	(′)	(″)				
α	04	14	01	−42	17	38	3.85	1.07	117	K1

There is no other star above magnitude 4. Next come δ (4.3) and β (4.8).

See chart for Eridanus.

The most interesting object is ι, RA 02h 42m 33s .2, Dec. −50° 48′ 03″, magnitude 4.22, distance 56 light-years. This G3 solar-type star, with diameter 1.5 times that of the Sun and 1.8 times the luminosity of the Sun, is attended by a planet, discovered in 1998. The planet has a mass slightly more than twice that of Jupiter, moving in a low-eccentricity orbit (e =.2.2) at a range of 0.9 a.u.; period 311days. ι may have been formed in the Hyades cluster ~625 million years ago, and has drifted away more than 130 light-years from its birthplace, and is now 7 light-years from the yellow subgiant χ Eridani.

The Mira variable R Horologii has a pariculary large range (~9 magnitudes).

HYDRA

(Abbreviation: Hya)

This is the largest constellation in the sky, representing the hundred-headed monster which lived in the Lernæan marshes until it was killed by Hercules. It extends for over six hours of RA. α (Alphard) is often known as 'the Solitary One', because there are no other bright stars anywhere in the neighbourhood.

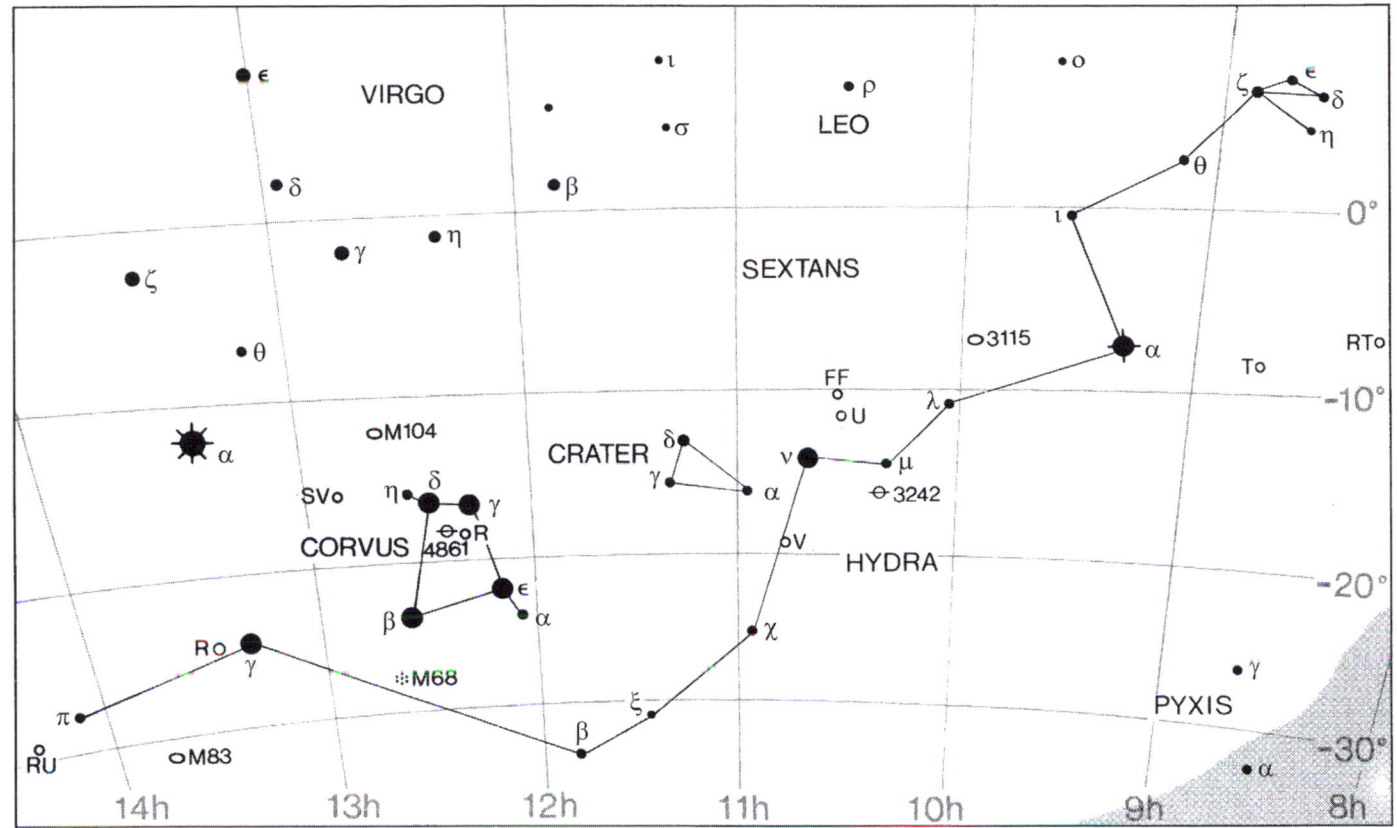

There are 12 stars above the fourth magnitude.

	RA			Dec.			m	M	d (light-years)	Spectrum	
	(h)	(m)	(s)	(°)	(')	('')					
30 α	09	27	35.3	−08	39	31	1.99	−1.69	177	K3	Alphard
46 γ	13	18	55.3	−23	10	17	2.99	−0.05	132	G8	
16 ζ	08	55	23.7	+05	56	43	3.11	−0.21	151	G8	Hydrobius
ν	10	49	37.4	−16	11	39	3.11	−0.03	138	K0	Pleura
49 η	14	06	22.8	−26	40	55	3.25	0.79	101	K2	Markeb
11 ε	08	46	46.7	+06	25	08	3.38	0.29	135	G0	Ashlesha
ξ	11	33	00.3	−31	51	27	3.54	0.55	129	G8	
41 λ	10	10	35.4	−12	21	14	3.61	0.88	115	K0	
42 μ	10	26	05.5	−16	50	10	3.83	−0.58	248	K4	
22 θ	09	14	21.8	+02	18	54	3.89	0.91	129	B9	
35 ι	09	39	51.3	−01	08	34	3.90	−0.74	276	K3	
C	08	25	39.7	+03	54	24	3.91	0.99	125	A0	

Also above 4.5: υ 1 (4.11), δ Mautina (4.14), β (4.29), η (4.32), ρ (4.35), E (4.42), σ (4.45).

VARIABLE STARS

	RA (h)	RA (m)	Dec. (°)	Dec. (')	Range	Type	Period (d)	Spectrum
RT	08	29.7	−06	19	7.0–11.0	Semi-regular	253	M
S	08	53.6	+03	04	7.4–13.3	Mira	256.4	M
T	08	55.7	−09	08	6.7–13.2	Mira	289.2	M
X	09	35.5	−14	42	8.0–13.6	Mira	301.4	M
U	10	37.6	−13	23	4.8–5.8	Semi-regular	450	N
V	10	49.2	−20	59	6.0–12.5	Mira	533	N
R	13	29.7	−23	17	4.0–10.0	Mira	389.6	M
TT	11	13.2	−26	28	7.5–9.5	Algol	6.95	A + G
HZ	11	26.3	−25	45	7.6–8.2	Semi-regular	95	M
W	13	49.0	−28	22	7.7–11.6	Semi-regular	397	M
RU	14	11.6	−28	53	7.2–14.3	Mira	333.2	M

DOUBLE STARS

	RA (h)	RA (m)	Dec. (°)	Dec. (')	PA (°)	Separation (")	Magnitudes	
ε	08	46.8	+06	25	AB 295	0.2	3.8, 4.7	Binary, 890 y
					AB + C 298	3.3	6.8	
θ	09	14.4	−02	19	197	29.4	3.9, 9.9	
α	09	17.6	−08	40	153	283.1	2.0, 9.5	
N	11	32.3	−28	16	210	9.5	5.8, 5.9	
β	11	52.9	−33	54	008	0.9	4.7, 5.5	
R	13	29.7	−23	17	324	21.2	var., 12.0	
52	14	28.2	−29	30	AB 130	0.1	5.8, 5.8	
					AB + C 279	4.2	10.0	
					AB + D 282	140.8	12.0	
59	14	58.7	−27	39	335	0.8	6.3, 6.6	

OPEN CLUSTER

M	C	NGC	RA (h)	RA (m)	Dec. (°)	Dec. (')	Diameter (')	Magnitude	No. of stars
48		2548	08	13.8	−05	48	54	5.8	80

GLOBULAR CLUSTER

M	C	NGC	RA (h)	RA (m)	Dec. (°)	Dec. (')	Diameter (')	Magnitude
68	66	5694	14	39.6	−26	32	3.6	10.2

GALAXIES

M	C	NGC	RA (h)	RA (m)	Dec. (°)	Dec. (')	Magnitude	Dimensions (')	Type
		2784	09	12.3	−24	10	10.1	5.1 × 2.3	S0
		2835	09	17.9	−22	21	11.1	6.3 × 4.4	Spectrum
		3109	10	03.1	−26	09	10.4	14.5 × 3.5	Irregular
		3585	11	13.3	−26	45	10.0	2.9 × 1.6	E5

GALAXIES

M	C	NGC	RA (h)	RA (m)	Dec. (°)	Dec. (')	Magnitude	Dimensions (')	Type
		3621	11	18.3	−32	49	9.9	10.0 × 6.5	Sc
		3923	11	51.0	−28	48	10.1	2.9 × 1.9	E3
		5078	13	19.8	−27	24	12.0	3.2 × 1.7	Sa
		5085	13	20.3	−24	26	11.9	3.4 × 3.0	Sb
		5061	13	18.1	−26	50	11.7	2.6 × 2.3	E2
		5101	13	21.8	−27	26	11.7	5.5 × 4.9	Sa
83		5236	13	37.0	−29	52	8.2	11.2 × 10.2	Sc

PLANETARY NEBULA

M	C	NGC	RA (h)	RA (m)	Dec. (°)	Dec. (')	Dimensions (")	Magnitude	
	59	3242	10	24.8	−18	38	16 × 1250	12.0	Ghost of Jupiter

Precise radial velocity measuremets show that Alphard is pulsating very slightly, and is therefore of special interest to 'astroseismologists'.

The Mira variable R Hydræ can reach naked-eye visibility. Its period has shortened from about 500 days when first discovered, by Maraldi in 1704, to about 415 days by 1925; since 1937 the period has remained more or less steady at 389.6 days. There is a distant companion, magnitude 12, separation 21"; this companion has the same movement through space as Alphard, indicating a possible physical relationship.

The Ghost of Jupiter, NGC 3242 (C 59) is a particularly atractive planetary nebula.

HYDRUS

(Abbreviation: Hyi)
This is a constellation in the far south – remarkably lacking in interesting objects. There are three stars above the fourth magnitude.

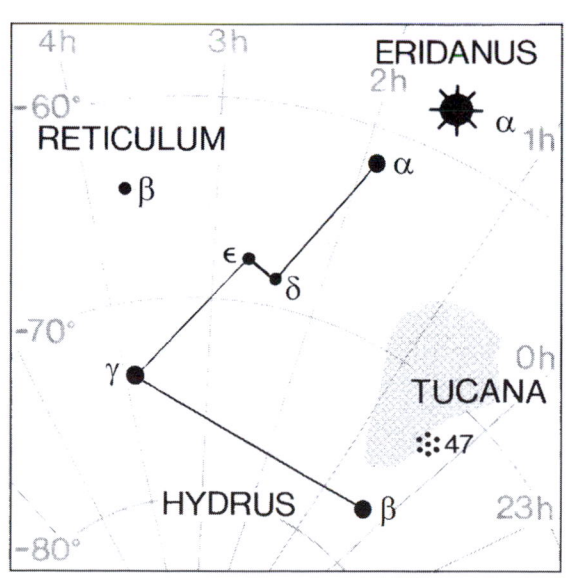

RA (h)	(m)	(s)	Dec. (°)	(')	(")	m	M	d (light-years)	Spectrum
β 00	25	39.2	−77	15	18	2.82	3.45	24	G2
α 01	58	45.9	−61	34	12	2.86	1.16	71	F0
γ 03	47	14.2	−74	14	23	3.26	−0.83	214	M2	Foo Pih

β is the closest star to the south celestial pole which is above the third magnitude.

Also above magnitude 4.5: δ (4.08), ε (4.12).

VARIABLE STAR

	RA (h)	(m)	Dec. (°)	(')	Range	Type	Period (d)	Spectrum
VW	04	09.1	−71	18	8.4–14.4	SS Cygni	100	M

INDUS

(Abbreviation: Ind)

This is a small southern constellation. There are two stars a above the fourth magnitude:

	RA			Dec.			M	m	d (light-years)	Spectrum	
	(h)	(m)	(s)	(°)	(′)	(″)					
α	22	37	34.0	−47	17	30	3.11	0.65	101	K0	Persian
β	20	54	48.6	−58	27	15	3.67	−2.66	600	K0	

Also above 4.5: θ (4.39), δ (4.40).

See chart for Grus.

VARIABLE STARS

	RA		Dec.		Range	Type	Period (d)	Spectrum
	(h)	(m)	(°)	(′)				
S	20	56.4	−54	19	7.4–14.5	Mira	399.9	M
T	21	20.2	−45	01	7.7–9.4	Semi-regular	320	N

DOUBLE STARS

	RA		Dec.		PA	Separation (″)	Magnitudes	
	(h)	(m)	(°)	(′)	(°)			
θ	21	19.9	−53	27	271	6.8	4.5, 7.0	
δ	21	57.9	−55	00	323	0.1	5.3, 5.3	Binary, 12 y

GALAXIES

M	NGC	RA		Dec.		Magnitude	Dimensions	Type
		(h)	(m)	(°)	(′)		(′)	
	7049	21	19.0	−48	34	10.7	2.8 × 2.2	S0
	7083	21	35.7	−63	54	11.8	4.5 × 2.9	Sb
	7090	21	36.5	−54	33	11.1	7.1 × 1.4	SBc
	7168	22	02.1	−51	45	12.6	2.0 × 1.6	E3
	7205	22	08.5	−57	25	11.4	4.3 × 2.2	Sb

ε Indi is a nearby K4-type dwarf. In 2007 it was found to be orbited by a pair of brown dwarfs of type T, separation from the primary ~1500 a.u., masses ~47 and ~28 that of Jupiter.

LACERTA

(Abbreviation: Lac)

This is a small northern constellation. It is not one of Ptolemy's 'originals', and no legends are associated with it. There is one star above the fourth magnitude:

RA			Dec.			m	M	d (light-years)	Spectrum
(h)	(m)	(s)	(°)	(′)	(″)				
7	22	31	1.4	+50 16 57		3.76	1.28	102	A1
α									

Also above 4.5: 1 (4.14), 5 (4.34), β (4.42).

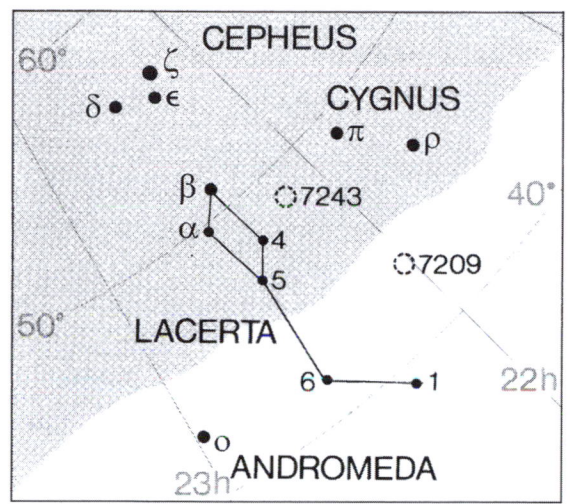

VARIABLE STARS

	RA		Dec.		Range	Type	Period (d)	Spectrum
	(h)	(m)	(°)	(′)				
S	22	29.0	+40	19	7.6–13.9	Mira	241.8	M
Z	22	40.9	+56	50	7.9–8.8	Cepheid	10.89	F–G
R	22	43.3	+42	22	8.5–14.8	Mira	299.9	M

OPEN CLUSTERS

M	C	NGC	RA (h)	(m)	Dec. (°)	(′)	Diameter (′)	Magnitude	No. of stars
	16	7243	22	15.3	+49	53	21	6.4	40
		7296	22	28.2	+52	17	4	9.7	20

BL Lacertæ was discovered by C. Hoffmeister in 1929 and assumed to be an ordinary variable star. It is, however, a highly variable extragalactic nucleus, and the host galaxy has actually been found. Position: RA 22 h 0m 43s .3, Dec. +42° 16′ 40″, magnitude range 14 to 17.

LEO

(Abbreviation: Leo)

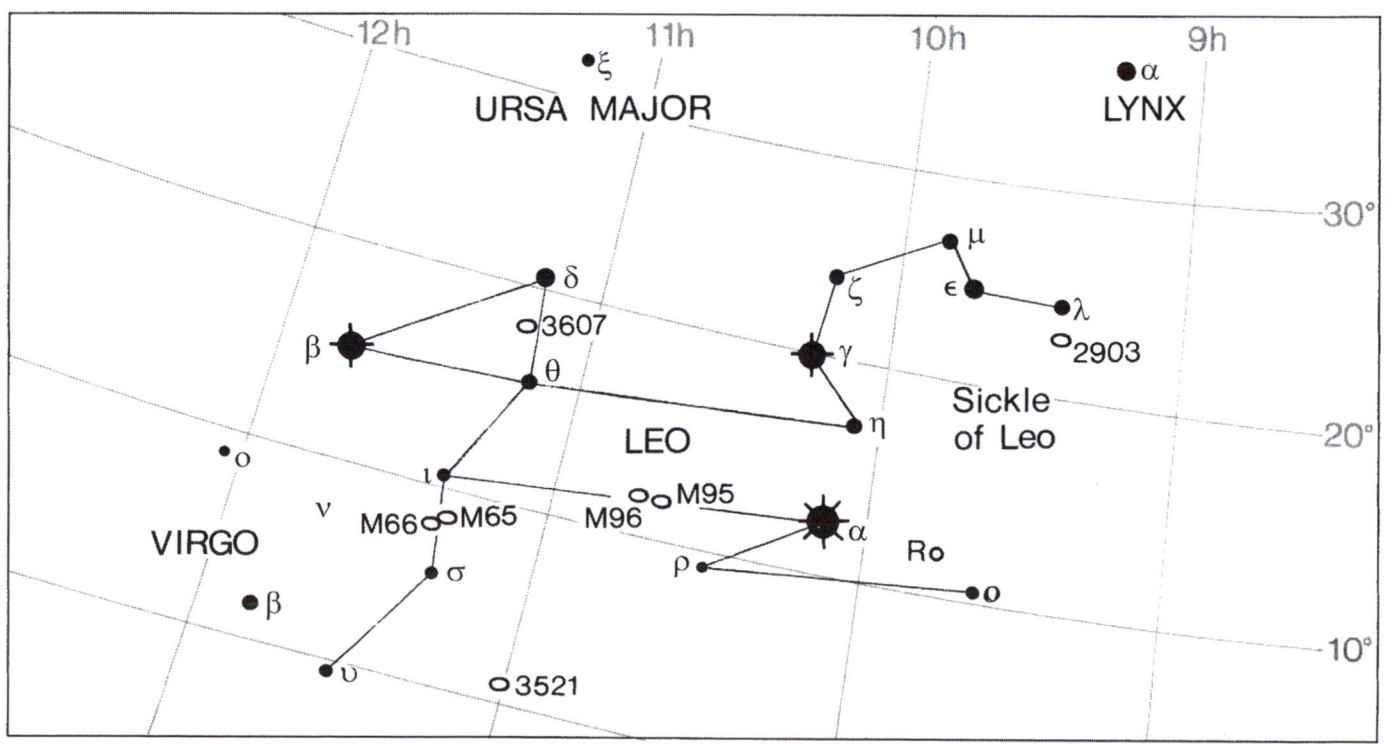

VARIABLE STAR

	RA		Dec.		Range	Type	Period (d)	Spectrum
	(h)	(m)	(°)	(')				
R	09	47.6	+11	25	4.4–11.3	Mira	312.4	M

DOUBLE STARS

	RA		Dec.		PA (°)	Separation ('')	Magnitudes	
	(h)	(m)	(°)	(')				
ω	09	28.5	+09	03	053	0.5	5.9, 6.5	Binary, 118 y
α	10	08.4	+11	58	307	176.9	1.4, 7.7	
γ	10	20.0	+19	51	AB 125	4.6	2.2, 3.5	Binary, 619 y
					AC 291	259.9	9.2	
					AD 302	333.0	9.6	
TX	10	35.0	+08	39	157	2.4	5.8, 8.5	
ι	11	23.9	+10	32	131	1.5	4.0, 6.7	Binary, 192 y
τ	11	27.9	+02	51	176	91.1	4.9, 8.0	

GALAXIES

M	C	NGC	RA		Dec.		Magnitude	Dimensions (')	Type	
			(h)	(m)	(°)	(')				
		3190	10	18.1	+21	50	11.0	4.6 × 1.8	SG	
95		3351	10	44.0	+11	42	9.7	7.4 × 5.1	SBb	
96		3368	10	46.8	+11	49	9.2	7.1 × 5.1	Sb	
		3377	10	47.7	+13	59	10.2	4.4 × 2.7	E5	
105		3379	10	47.8	+12	35	9.3	4.5 × 4.0	E1	
		3384	10	48.3	+12	38	10.0	5.9 × 2.6	E7	
		3412	10	50.9	+13	25	10.6	3.6 × 2.0	E5	
		3489	11	00.3	+13	54	10.3	3.7 × 2.1	E6	
		3521	11	05.8	−00	02	8.9	9.5 × 5.0	Sb	
		3593	11	14.6	+12	49	11.0	5.8 × 2.5	Sb	
		3596	11	15.1	+14	47	11.6	4.2 × 4.1	Sc	
		3607	11	16.9	+18	03	10.0	3.7 × 3.2	E1	
65		3623	11	18.9	+13	05	9.3	10.0 × 3.3	Sb	
	40	3626	11	20.1	+18	21	10.9	3.1 × 2.2	Sb	
66		3627	11	20.2	+12	59	9.0	8.7 × 4.4	Sb	
		3628	11	20.3	+13	36	9.5	14.8 × 3.6	Sb	Arp 317
		3630	11	20.3	+02	58	12.8	2.3 × 0.9	E7	
		3640	11	21.1	+03	14	10.3	4.1 × 3.4	E1	
		3646	11	21.7	+20	10	11.2	3.9 × 2.6	Sc	
		3686	11	27.7	+17	13	11.4	3.3 × 2.6	Sc	
		3810	11	41.0	+11	28	10.8	4.3 × 3.1	Sc	

Regulus is quadruple. The main component, Regulus A, is a bluish Main Sequence star, and is a spectroscopic binary; the secondary may be a white dwarf. Further away are B and C, faint Main Sequence stars. Regulus is so close to the ecliptic that it can be occulted by the Moon and planets.

Denebola is a very small-range δ Scuti variable; Algiba is a lovely double.

There are some bright galaxies in Leo, and the whole constellation is very rich.

LEO MINOR

(Abbreviation: LMi)

A small, entirely unremarkable northern constellation. For some unknown reason the only star to have a Greek letter is Flamsteed 20 (β) and the only star with a proper name is 46 (Præcipua). This is also the only star above the fourth magnitude.

See chart for Ursa Major.

	RA			Dec.			m	M	d (light years)	Spectrum	
	(h)	(m)	(s)	(°)	(′)	(″)					
46	10	53	18.6	+34	12	56	3.79	1.41	98	K0	Præcipua

Also above 4.5: β (4.20), 21 (4.49).

A trick question in astronomical quizzes is: 'What magnitude m is the star α Leonis Minoris?' Answer: 'Nothing'. There is no α Leonis Minoris!

VARIABLE STARS

	RA		Dec.		Range	Type	Period (d)	Spectrum
	(h)	(m)	(°)	(′)				
R	09	45.6	+34	31	6.3–13.2	Mira	371.9	M
S	09	53.7	+34	55	7.9–14.3	Mira	233.8	M
RW	10	16.1	+30	34	6.9–10.1	Mira	?	N

DOUBLE STAR

	RA		Dec.		PA	Separation (″)	Magnitudes	
	(h)	(m)	(°)	(′)	(°)			
β	10	27.9	+36	42	250	0.2	4.4, 6.1	Binary, 37.2 y

GALAXIES

M	C	NGC	RA		Dec.		Magnitude	Dimensions (′)	Type
			(h)	(m)	(°)	(′)			
		3003	09	48.6	+33	25	11.7	5.9 × 1.7	SBc
		3245	10	27.3	+28	30	10.8	3.2 × 1.9	E5
		3254	10	29.3	+29	30	11.5	5.1 × 1.9	Sb
		3294	10	36.3	+37	20	11.7	3.3 × 1.8	Sc
		3344	10	43.5	+24	55	9.9	6.9 × 6.5	Sc
		3414	10	51.3	+27	59	10.7	3.6 × 2.7	SBa
		3430	10	52.2	+32	57	11.5	3.9 × 2.3	Sc
		3432	10	52.5	+36	37	11.2	6.2 × 1.5	SB
		3486	11	00.4	+28	58	10.3	6.9 × 5.4	Sc

LEPUS

(Abbreviation: Lep)

An original constellation, probably representing a hare being chased by Orion. It is quite distinctive.

There are eight stars above the fourth magnitude.

Also above 4.5: λ (4.29), κ (4.36), ι (4.45).

ζ probably has an asteroid belt. γ is a wide, easy double. R Leporis, known as the Crimson Star, is one of the reddest stars in the sky.

M 79 is an interesting globular cluster, 41 000 light-years away. It may possibly have been formed inside the Canis Major dwarf galaxy, which is at present going through a close encounter with our Galaxy and will probably lose its separate identity, though opinions differ.

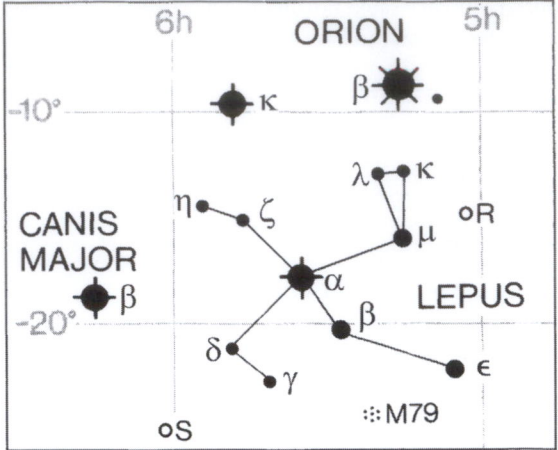

	RA			Dec.			m	M	d (light-years)	Spectrum	
	(h)	(m)	(s)	(°)	(′)	(″)					
11 α	05	32	43.8	−17	49	20	2.58	−5.40	1280	F0	Arneb
9 β	05	28	14.7	−20	45	33	2.81	−0.63	150	G5	Nihal
2 ε	05	05	27.7	−22	22	15	3.19	−1.02	227	K4	
5 μ	05	12	55.9	−16	12	20	329	0.47	184	B9	Neshmet
14 ζ	05	46	57.3	−14	49	19	3.55	1.89	70	A2	
13 γ	05	44	28.0	−22	26	51	3.59	3.83	29	F7	
16 η	05	56	24.3	−14	10	05	3.71	2.82	49	F1	
15 δ	05	51	19.2	−20	52	39	3.76	1.08	112	G8	

VARIABLE STARS

	RA		Dec.		Range	Type	Period (d)	Spectrum
	(h)	(m)	(°)	(′)				
R	04	59.6	−14	48	5.5–11.7	Mira	432.1	N
T	05	04.8	−21	54	7.4–13.5	Mira	368.1	M
RX	05	11.4	−11	51	5.0–7.0	Irregular	–	M
S	06	05.8	−24	12	7.1–8.9	Semi-regular	90	M

DOUBLE STARS

	RA		Dec.		PA (°)	Separation (′)	Magnitudes
	(h)	(m)	(°)	(′)			
ι	05	12.3	−11	52	337	12.7	4.5, 10.8
κ	05	13.2	−12	56	357	2.2	4.5, 7.4
β	05	28.2	−20	46	AB 330	2.5	2.8, 7.3
					AC 145	64.3	11.8
					AD 075	206.4	10.3
					AE 058	241.5	10.3
γ	05	44.5	−22	27	350	96.3	3.7, 6.3

GLOBULAR CLUSTER

M	NGC	RA		Dec.		Diameter (')	Magnitude
		(h)	(m)	(°)	(')		
79	1904	05	24.5	−24	33	8.7	8.0

PLANETARY NEBULA

M	NGC	RA		Dec.		Diameter (")	Magnitude	Magnitude of central star
		(h)	(m)	(°)	(')			
	IC 418	05	27.5	−12	42	12	10.7	10.7

GALAXIES

M	NGC	RA		Dec.		Magnitude	Dimensions (')	Type
		(h)	(m)	(°)	(')			
	1744	05	00.0	−26	01	11.2	6.8 × 4.1	SBc
	1964	05	33.4	−21	57	10.8	6.2 × 2.5	Sb

LIBRA

(Abbreviation: Lib)

A Zodiacal constellation, but by no means prominent. It was originally known as Chelæ Scorpionis, the Scorpion's Claws. Some Greek legends associate it, rather vaguely, with Mochis, the inventor of weights and measures.

There are six stars above the fourth magnitude

Also above 4.5: θ (5.13) and 16 (4.47).

σ, τ and υ were originally included in Scorpius; σ was γ Scorpii.

The four brightest stars were part of the Scorpius pattern: α (Zubenelgenubi) the Southern Claw, β (Zubeneschemali) the Northern Claw, γ (Zubenelakrab) the Scorpion's Claw, and σ (Brachium).

α is a very wide, easy double; the brighter member (α²) is white, the fainter (α¹) is an F4-type star, magnitude 5.2. The two make up a binary system, but the orbital period must be very long indeed.

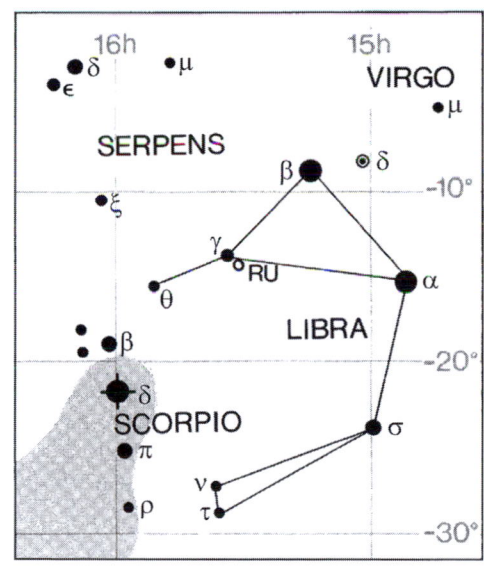

	RA			Dec.			m	M	d (light-years)	Spectrum
	(h)	(m)	(s)	(°)	(')	(")				
37 β	15	17	00.5	−09	22	58	2.61	−0.84	160	B8
9 α²	14	50	52.8	−16	02	30	2.75	0.88	77	A3
20 σ	15	04	04.3	−25	16	55	3.25	1.51	292	M5
39 υ	15	37	01.5	−28	08	06	3.60	−0.28	195	K3
40 τ	15	38	39.4	−29	46	40	3.66	−2.01	450	B2
38 γ	15	35	39.5	−14	47	22	3.91	0.56	152	K0

VARIABLE STARS

	RA (h)	RA (m)	Dec. (°)	Dec. (′)	Range	Type	Period (d)	Spectrum
δ	15	01.1	−08	31	4.9–5.9	Algol	2.33	B
Y	15	11.7	−06	10	7.6–14.7	Mira	275.0	M
S	15	21.4	−20	23	7.5–13.0	Mira	192.4	M
RS	15	24.3	−22	55	7.0–13.0	Mira	217.7	M
RU	15	33.3	−15	20	7.2–14.2	Mira	316.6	M
RR	15	56.4	−18	18	7.8–15.0	Mira	277.0	M

DOUBLE STARS

	RA (h)	RA (m)	Dec. (°)	Dec. (′)	PA (°)	Separation (″)	Magnitudes
μ	14	49.3	−14	09	355	1.8	5.8, 6.7
α	14	50.9	−16	02	314	231.0	2.8, 5.2
ι	15	12.2	−19	47	111	57.8	5.1, 9.4
κ	15	41.9	−19	41	279	172.0	4.7, 9.7

GLOBULAR CLUSTER

M	C	NGC	RA (h)	RA (m)	Dec. (°)	Dec. (°)	Diameter (′)	Magnitude
		5897	15	17.4	−21	01	12.6	8.6

β, Type B8, has often been called the only green single star, but I admit that to me it looks white. Some Greek astronomers, including Ptolemy, said that it was as bright as Antares, but any real change seems dubious,

δ (Zubenelakribi) is one of the very few Algol eclipsing binaries visible with the naked eye.

Gliese 481, centre of a particularly interesting planetary system, lies in Libra.

LUPUS

(Abbreviation: Lup)

Lupus is an original constellation, though no legends seem to be
attached to it. There are 12 stars above the fourth magnitude. α is
called either Men or Kakkab.

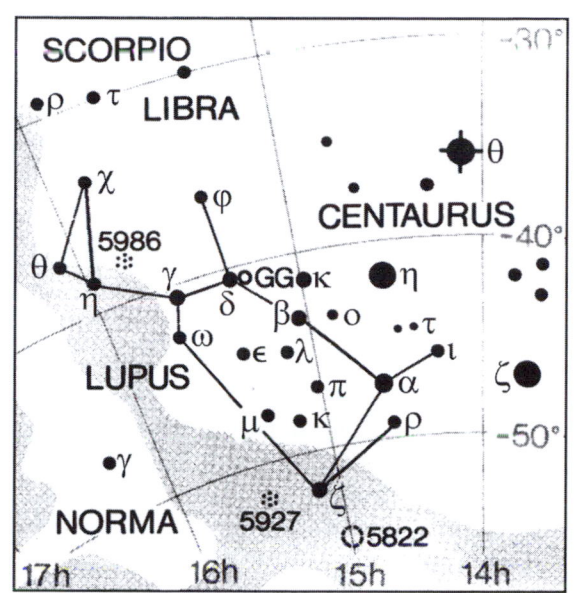

	RA			Dec.			m	M	d	Spectrum	
	(h)	(m)	(s)	(°)	(′)	(″)					
α	14	41	55.8	−47	23	17	2.30	−3.83	550	B1	Kakkab
β	14	58	32.0	−43	08	02	2.68	−3.35	522	B2	Kekouan
γ	15	35	08.5	−41	10	00	2.80	−3.40	570	B2	Thusia
δ	15	21	22.3	−40	38	51	3.22	−2.75	510	B1	Hilasmus
ε	15	22	40.9	−44	41	23	3.37	−2.58	500	B2	
ζ	15	12	17.2	−52	05	57	3.41	0.65	116	G5	
η	16	00	07.3	−38	23	48	3.42	−2.48	490	B2	
ι	14	19	24.2	−46	03	29	3.55	−1.61	152	B2	
φ	15	21	48.4	−36	15	40	3.57	−1.43	326	K5	
κ	15	11	56.2	−48	44	16	3.88	0.14	182	B9	
π	15	05	07.1	−47	03	04	3.91	−2.01	500	B5	
5 χ	15	50	57.5	−33	37	38	3.97	−0.03	206	B9	

Also above 4.5: ρ (4.05), λ (4.07), θ (4.22), μ (4.27), o (4.32),
τ 2 (4 33), ω (4.34), f (4.35) and σ (4.44).

Also above magnitude 5:

	Magnitude	Absolute magnitude	Spectrum	Distance (light-years)
l	4.91	−6.6	F0	1800
ν¹	5.00	3.4	F3	20
ξ	4.6 (5.1 + 5.6)		A + A	

VARIABLE STARS

	RA		Dec.		Range	Type	Period (d)	Spectrum
	(h)	(m)	(°)	(′)				
S	14	53.4	−46	37	7.8–13.5	Mira	342.7	S
GG	15	18.9	−40	47	5.4–6.0	βLyræ	2.16	B + A

DOUBLE STARS

	RA		Dec.		PA (°)	Separation (″)	Magnitudes
	(h)	(m)	(°)	(′)			
τ¹	14	26.1	−45	13	204	148.2	4.6, 9.3
π	15	05.1	−47	03	067	1.7	4.6, 4.7
κ	15	11.9	−48	44	144	26.8	3.9, 5.8
μ	15	18.5	−47	53	AB 142	1.2	4.6, 4.7
					AC 130	23.7	7.2
ε	15	22.7	−44	41	247	0.6	3.7, 7.2
υ	15	24.7	−39	43	038	1.4	5.4, 10.9
ξ	15	56.9	−33	58	049	10.4	5.3, 5.8
η	16	00.1	−38	24	020	15.0	3.6, 7.8

OPEN CLUSTERS

M	C	NGC	RA (h)	RA (m)	Dec. (°)	Dec. (')	Diameter (')	Magnitude	No. of stars
		5749	14	48.9	−54	31	8	8.8	30
		5822	15	05.2	−54	21	40	6.5	150

GLOBULAR CLUSTERS

M	C	NGC	RA (h)	RA (m)	Dec. (°)	Dec. (')	Diameter (')	Magnitude
		5824	15	04.0	−33	04	6.2	9.0
		5927	15	28.0	−50	40	12.0	8.3
		5986	15	46.1	−37	47	9.8	7.1

PLANETARY NEBULÆ

M	C	NGC	RA (h)	RA (m)	Dec. (°)	Dec. (')	Diameter ('')	Magnitude	Magnitude of central star
		IC 4406	14	22.4	−44	09	28	10.6	14.7
		5882	15	16.8	−45	39	7	10.5	12.0

GALAXIES

M	NGC	RA (h)	RA (m)	Dec. (°)	Dec. (')	Magnitude	Dimensions (')	Type
	5643	14	32.7	−44	10	10.7	4.6 × 4.1	SB0

LYNX

(Abbreviation: Lyn)

This is a dim northern constellation – said to have been named because to identify it an observer must be lynx-eyed! There are only two stars above the fourth magnitude.

	RA (h)	RA (m)	RA (s)	Dec. (°)	Dec. (')	Dec. ('')	m	M	d (light-years)	Spectrum	
40 α	09	21	03.5	+34	23	33	3.14	−1.02	222	K7	Alvashak
38	09	18	50.7	+36	48	10	3.82	0.96	122	A1	Maculata

Also above 4.5: 31, κ (Alsciaukat) (4.25), 15 (4.35), 2 (4.44).
See chart for Ursa Major.

VARIABLE STARS

	RA (h)	RA (m)	Dec. (°)	Dec. (')	Range	Type	Period (d)	Spectrum
RR	06	26.4	+56	17	5.6–6.0	Algol	9.95	A
R	07	01.3	+55	20	7.2–14.5	Mira	378.7	S
Y	07	28.2	+45	59	7.8–10.3	Semi-regular	110	M

DOUBLE STARS

	RA		Dec.		PA (°)	Separation ($''$)	Magnitudes
	(h)	(m)	(°)	(')			
4	06	22.1	+59	22	124	0.8	6.2, 7.7
12	06	45.2	+59	27	AB 070	1.7	5.4, 6.0
					AC 308	8.7	7.3
					AD 256	170.0	10.6
19	07	22.9	+55	17	AB 315	14.8	5.6, 6.5
					AD 003	214.9	8.9
					BC 287	74.2	10.9
38	09	18.8	+36	48	AB 229	2.7	3.9, 6.6
					BC 212	87.7	10.8
					BD 256	177.9	10.7

GLOBULAR CLUSTER

M	C	NGC	RA		Dec.		Diameter (')	Magnitude	
			(h)	(m)	(°)	(')			
	25	2419	7	38.1	+38	53	4.1	10.4	'Intergalactic Tramp'

GALAXIES

M	C	NGC	RA		Dec.		Magnitude	Dimensions	Type	
			(h)	(m)	(°)	(')		(')		
		2537	08	13.2	+46	00	12.3	1.7 × 1.5	S	Bear Paw Galaxy
		2541	08	14.7	+49	04	11.7	6.6 × 3.5	S	
		2683	08	52.7	+33	25	9.7	9.3 × 2.5	Sb	
		2776	09	12.2	+44	57	11.6	2.9 × 2.7	Sc	

α is noticeably orange. The most interesting objects in Lynx are the Bear Paw Galaxy, NGC 2537, and the globular cluster NGC 2419 (C 25), the Intergalactic Tramp.

LYRA

(Abbrevation: Lyr)

A small constellation, but a very interesting one; it is graced by the presence of the brilliant blue Vega, as well as the prototype eclipsing binary β Lyræ, the quadruple ε Lyræ, and the 'Ring Nebula' M 57. Mythologically it represents the harp which Apollo gave to the great musician Orpheus.

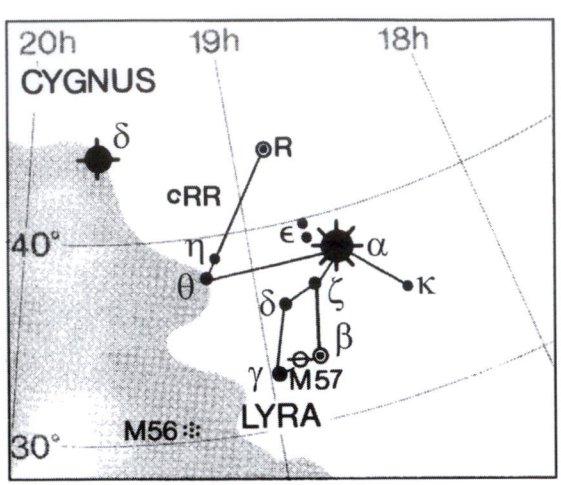

There are several stars above the fourth magnitude. Of these, one (β) is the famous variable; the combined magnitude of the quadruple ε is about 3.9, though keen-sighted people can see the two main components as separated. The brightest stars are:

	RA			Dec.			m	M	d (light-years)	Spectrum	
	(h)	(m)	(s)	(°)	(')	(")					
3 α	18	36	56.2	+38	46	59	0.03	0.58	25	A0	Vega
14 γ	18	58	56.6	+32	41	22	3.25	−3.20	634	B9	Sulaphat
10 β	18	50	04.8	+33	21	46	3.5v	−3.64	881	A8	Sheliak
5,	18	44	22.8	+39	36	45	3.7	−1.1&	161	A8 & F1	
4 ε								−1.2			

Also above 4.5: R (4.0v), δ² (4.22), κ (4.33), ζ¹(4.34), θ (4.35), η (Aladfar) (4.43). μ (5.11) has a proper name Alathfar.

Vega, 'steely blue', is the brightest star in the northern hemisphere of the sky apart from Arcturus. It rotates rapidly; it is 37 times as luminous as the Sun, just over twice as massive, and 2.3 times greater in diameter. It is surrounded by a dust-disc, and may have a planet. From Earth, we see Vega almost pole-on.

VARIABLE STARS

	RA		Dec.		Range	Type	Period (d)	Spectrum
	(h)	m	(°)	(')				
W	18	14.9	+36	40	7.3–13.0	Mira	196.5	M
T	18	32.3	+37	00	7.8–9.6	Irregular	—	R
β	18	50.1	+33	22	3.3–4.3	β Lyræ	12.94	B + A
R	18	55.3	+43	57	3.9–5.0	Semi-regular	46	M
RR	19	25.5	+42	47	7.1–8.1	RR Lyræ	0.57	A–F

DOUBLE STARS

	RA		Dec.		PA (°)	Separation (")	Magnitudes	
	(h)	(m)	(°)	(')				
ε	18	44.3	+39	40	AB + CD 173	207.7	4.7, 5.1	
					ε¹ = AB 357	2.6	5.0, 5.1	Binary, 1165 y
					ε² = CD 094	2.3	5.2, 5.5	Binary, 585 y
ζ	18	44.8	+37	36	150	43.7	4.3, 5.9	
β	18	50.1	+33	22	149	45.7	variable, 8.6	
δ¹	18	53.7	+36	58	020	174.6	5.6, 9.3	
δ²	18	54.5	+36	54	349	86.2	4.5, 11.2	
η	19	13.8	+39	09	082	28.1	4.4, 9.1	

OPEN CLUSTER

M	C	NGC	RA		Dec.		Diameter (')	Magnitude	No. of stars
			(h)	(m)	(°)	(')			
		6791	19	20.7	+37	51	16	9.5	300

GLOBULAR CLUSTER

M	C	NGC	RA		Dec.		Diameter (')	Magnitude
			(h)	(m)	(°)	(')		
56	6779	19	16.6	+30	11	7.1	8.2	

PLANETARY NEBULA

M	C	NGC	RA		Dec.		Dimensions (")	Magnitude	Magnitude of central star
			(h)	(m)	(°)	(')			
57	6720	18	53.6	+33	02		70 × 150	9.7	14.8	Ring Nebula

The eclipsing binary Sheliak has already been described. Two other stars may be associated with it, a B7-type spectroscoic bnary at a separation of 45.7″, magnitude 7.2, and an F-type star, magnitude 9.9, separation 86″.

R Lyræ, a red semi-regular variable, reaches magnitude 4 at maximum. RR Lyræ (RA 19h 25m 28s, Dec. + 42° 47′ 04″) is the prototype of its class; magnitude range 7 to 8, period 13 hours. It is 860 light-years away, and almost 50 times as luminous as the Sun.

Lyra also contains the Ring Nebula, M 57, and a bright globular cluster, M 56.

MENSA

(Abbreviation: Men)

A very dim constellation, introduced by Lacaille in 1752 under the name of Mons Mensæ (the Table Mountain). A small part of the Large Magellanic Cloud extends into it. There are no stars brighter than the fifth magnitude, and no objects to be listed. For the record, the brightest star is α (5.09). Next comes γ: RA 5h 31m 53s .1, Dec. −76° 20′ 28″, magnitude 5.19, absolute magnitude −0.3, spectrum K4, distance 101 light-years. It has an optical companion of magnitude 11, at PA 107°, separation 38″ .2.

See chart for Musca.

VARIABLE STARS

	RA		Dec.		Range	Type	Period (d)	Spectrum
	(h)	(m)	(°)	(')				
U	04	09.6	−81	51	8.0–10.9	Mira	407	M
TY	05	26.9	−81	35	7.7–8.2	W UMa	0.46	A
TZ	05	30.2	−84	47	6.2–6.9	Algol	8.57	B

MICROSCOPIUM

(Abbreviation: Mic)

A small and entirely unemarkable southern constellation. There is no star above magnitude 4.5. The brightest are:

	RA			Dec.			m	M	d (light-years)	Spectrum
	(h)	(m)	(s)	(°)	(′)	(″)				
γ	21	01c	17.5	−32	15	28	4.67	0.49	223	G8
ε	21	17	56.2	−32	10	21	4.71	1.19	165	A0

α is only of.magnitude 4.89, β even fainter at 6.06.
γ was formerly known as 1 PsA, ε as 4 PsA.
See chart for Grus.

VARIABLE STARS

	RA		Dec.		Range	Type	Period (d)	Spectrum
	(h)	(m)	(°)	(′)				
T	20	27.9	−28	16	7.7–9.6	Semi-regular	344	M
U	20	29.2	−40	25	7.0–14.4	Mira	334.2	M
S	21	26.7	−29	51	7.8–14.3	Mira	208.9	M

DOUBLE STARS

	RA		Dec.		PA (°)	Separation (″)	Magnitudes
	(h)	(m)	(°)	(′)			
α	20	50.0	−33	47	166	20.5	5.0, 10.0
θ²	21	24.4	−41	00	AB 267	0.5	6.4, 7.0
					AC 066	78.4	10.5

GALAXIES

M	C	NGC	RA		Dec.		Magnitude	Dimensions (′)	Type
			(h)	(m)	(°)	(′)			
		6923	20	31.7	−30	50	12.1	2.5 × 1.4	Sb
		6925	20	34.3	−31	59	11.3	4.1 × 1.6	Sb

AU Micoscopii (RA 20h 43m 10s, Dec. −31° 20′ 27″) is a red dwarf only 32 light-years away. Its magnitude is 8.6; it flares at all wavelengths, from radio waves to X-rays. It has a large débris disc, edge-on to us, and possibly a planet. It has only 0.029 of the luminosity and 0.6 of the diameter of the Sun. The red dwarf AX Microscopii (Lacaille 8760) is only 12.9 light-years away.

MONOCEROS

(Abbreviation: Mon)
Not an ancient constellation, and though it represents the fabled unicorn there are no definite legends attached to it. It is crossed by the Milky Way, and the general area is decidedly rich.

There are three stars above the fourth magnitude; of these, β is a double, and the magnitude is combined. The stars are (see chart for Orion):

	RA			Dec.			m	M	d (light-years)	Spectrum	
	(h)	(m)	(s)	(°)	(')	('')					
11 β	06	28	49.1	−07	01	59	4.7	−2.03	690	B3	(Triple)
26 α	07	41	14.9	−00	33	03	3.94	0.21	144	K0	
5 γ	06	14	51.3	−06	16	29	3.99	−2.49	650	K3	

Also above 4.5: δ (4.15), ζ (4.36), ε (4.39), 13 (4.47), 18 (4.48). Though lacking in bright stars, Monoceros is a rich region. It includes the Rosette Nebula and the Cone Nebula.

MUSCA AUSTRALIS

(Abbreviation: Mus)
Orginally called Apis, the Bee. It was then renamed Musca Australis, the Southern Fly, to avoid confusion with Musca Boreralis, the Northern Fly, but the Northern Fly has disappeared (no doubt someone has swatted it), and the surviving insect is generally known simply as Musca.

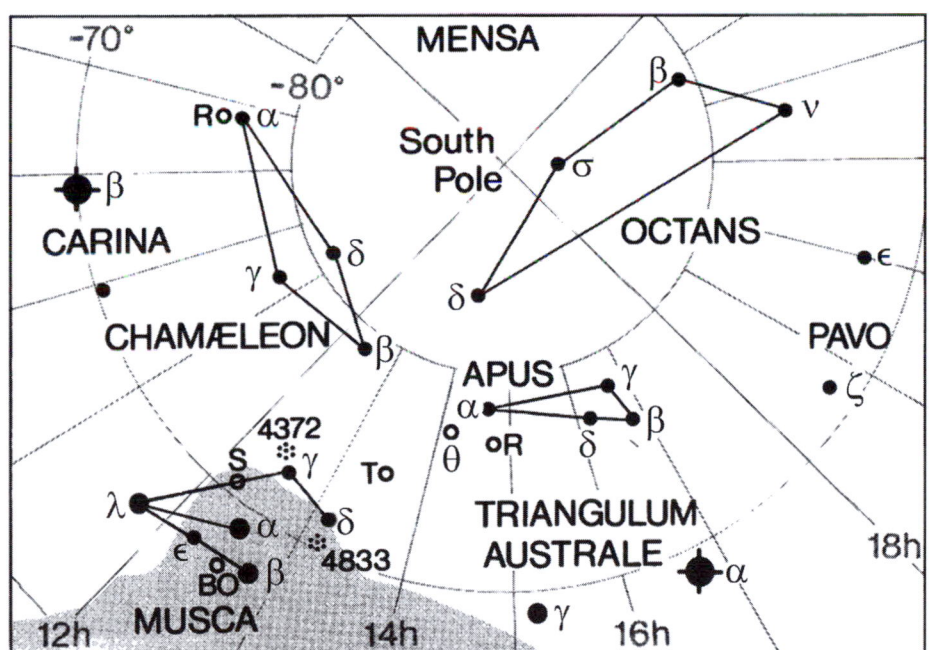

	RA			Dec.			m	M	d (light-years)	Spectrum	
	(h)	(m)	(s)	(°)	(')	('')					
α	12	37	11.1	−69	08	08	2.69	−2.17	306	B2	
δ	13	02	15.8	−71	32	56	3.61	1.39	90.8	K2	(Binary)
λ	11	45	36.6	−66	43	44	3.68	0.66	128	A2	
γ	12	32	28.1	−72	07	59	3.84	−1.14	324	B5	

Also above 4.5: ε (4.06).

VARIABLE STARS

	RA		Dec.		Range	Type	Period (d)	Spectrum
	(h)	(m)	(°)	(')				
S	12	12.8	−70	09	5.9–6.4	Cepheid	9.66	F
B0	12	34.9	−67	45	6.0–6.7	Irregular	—	M
R	12	42.1	−69	24	5.9–6.7	Cepheid	7.48	F
T	13	21.2	−74	27	7.1–9.0	Semi-regular	93	N

DOUBLE STARS

	RA		Dec.		PA	Separation (″)	Magnitudes
	(h)	(m)	(°)	(')	(°)		
ζ²	12	22.1	−67	31	130	32.4	5.2, 10.6
α	12	37.2	−69	08	316	29.6	2.7, 12.8
β	12	46.3	−68	06	039	1.2	3.7, 4.0
θ	13	08.1	−65	18	187	5.3	5.7, 7.3

OPEN CLUSTER

M	C	NGC	RA		Dec.		Diameter (')	Magnitude	No. of stars
			(h)	(m)	(°)	(')			
		4463	12	30.0	−64	48	5	7.2	30

GLOBULAR CLUSTERS

M	C	NGC	RA		Dec.		Diameter (')	Magnitude
			(h)	(m)	(°)	(')		
		4372	12	25.8	−72	40	18.6	7.8
		4833	12	59.6	−70	53	13.5	7.3

PLANETARY NEBULÆ

M	C	NGC	RA		Dec.		Diameter (″)	Magnitude	Magnitude of central star	
			(h)	(m)	(°)	(')				
		IC 4191	13	08.8	−67	39	5	12.0	—	
		5189	13	33.5	−65	59	153	10	14	Gum 47

An interesting nova appeared in Musca in 1991; it was a soft X-ray emitter, and was probably a binary consisting of a star and a black hole. Musca also contains an unusual planetary nebula, NGC 5189, which is very complex and has been likened to a miniature Crab Nebula; distance ~3000 light-years. Here too are the globular clusters NGC 4833 and NGC 4372, both partially obscured by dust clouds in the Milky Way; also the Hourglass Nebula MyCn 18, ~8000 light-years away.

NORMA

(Abbreviation: Nor)

A small southern constellation, formerly known as Quadra Euclidis (Euclid's Quadrant). Curiously, it has no α or β and the brightest star is γ²; RA 16h 19m 50s .6, Dec. −50° 09′ 20″, $m = 4.01$, $M =$ 1.05, distance =127 light-years, type G8 (it has no connection with its neighbour γ¹, $m = 4.97$, which is a yellow supergiant over 1400 light-years away). The only other star above magnitude 4.5 is ε (4.46).

See charts for Ara and Lupus.

DOUBLE STARS

C	RA (h)	(m)	Dec. (°)	(′)	PA (°)	Separation (″)	Magnitudes	
ι¹	16	03.5	−57	47	100	0.2	5.3, 5.5	Binary, 26.9 y
ε	16	27.2	−47	33	335	22.8	4.8, 7.5	

OPEN CLUSTERS

M	C	NGC	RA (h)	(m)	Dec. (°)	(′)	Diameter (′)	Magnitude	No. of stars	
		5925	15	27.7	−54	31	15	8.4	120	
		5999	15	52.2	−56	28	5	9.0	40	
		6031	16	07.6	−54	04	2	8.5	20	
	89	6067	16	13.2	−54	13	13	5.6	100	
		6087	16	18.9	−57	54	12	5.4	40	S Normæ Cluster
		H.10	16	19.9	−54	59	30	–	30	
		6134	16	27.7	−49	09	7	7.2	–	
		6152	16	32.7	−52	37	30	8.1	70	
		6167	16	34.4	−49	36	8	6.7	–	

PLANETARY NEBULA

Name	Designation	RA (h)	(m)	Dec. (°)	(′)	Diameter (″)	Magnitude	Magnitude of central star
Fine Ring Nebula	Spectrum-1	15	51.7	−51	31	76	13.6	13.8

OCTANS

(Abbreviation: Oct)

The south polar constellation. It is very obscure, and there is only one star above the fourth magnitude:

	RA			Dec.			m	M	d (light-years)	Spectrum
	(h)	(m)	(s)	(°)	(')	('')				
ν	21	41	28.5	−77	23	22	3.73	2.10	69	K0

Next come β (4.13), δ (4.31), θ (4.78) and ε (4.9 at maximum). The Greek sequence has certainly not been followed here; α is only of magnitude 5.13.

See chart for Musca.

σ (Polaris Australis) is of magnitude 5.45, so that it is of very little use to navigators. Its position is: RA 21h 08m 46s, Dec. −88° 57′ 23.4″. The absolute magnitude is 0.86, distance 270 light-years, spectrum F0. The polar distance was 45″ in 1900, but has now increased to over 1°.

VARIABLE STARS

	RA		Dec.		Range	Type	Period (d)	Spectrum
	(h)	(m)	(°)	(')				
R	05	26.1	−86	23	6.4–13.2	Mira	405.6	M
U	13	24.5	−84	13	7.1–14.1	Mira	302.6	M
S	18	08.7	−86	48	7.3–14.0	Mira	258.9	M
ε	22	20.0	−80	26	4.9–5.4	Semi-regular	55	M

DOUBLE STARS

	RA		Dec.		PA	Separation ('')	Magnitudes
	(h)	(m)	(°)	(')	(°)		
ι	12	55.0	−85	07	230	0.6	6.0, 6.5
μ²	20	41.7	−75	21	017	17.4	7.1, 7.6
λ	21	50.9	−82	43	070	3.1	5.4, 7.7

OPHIUCHUS

(Abbreviations: Oph)

This constellation is also sometimes known as Serpentarius. It commemorates Æsculapius, son of Apollo and Coronis, who became so skilled in medicine that he was even able to restore the dead to life. To avoid depopulation of the Underworld, Jupiter reluctantly disposed of Æsculapius with a thunderbolt, but relented sufficiently to place him in the sky.

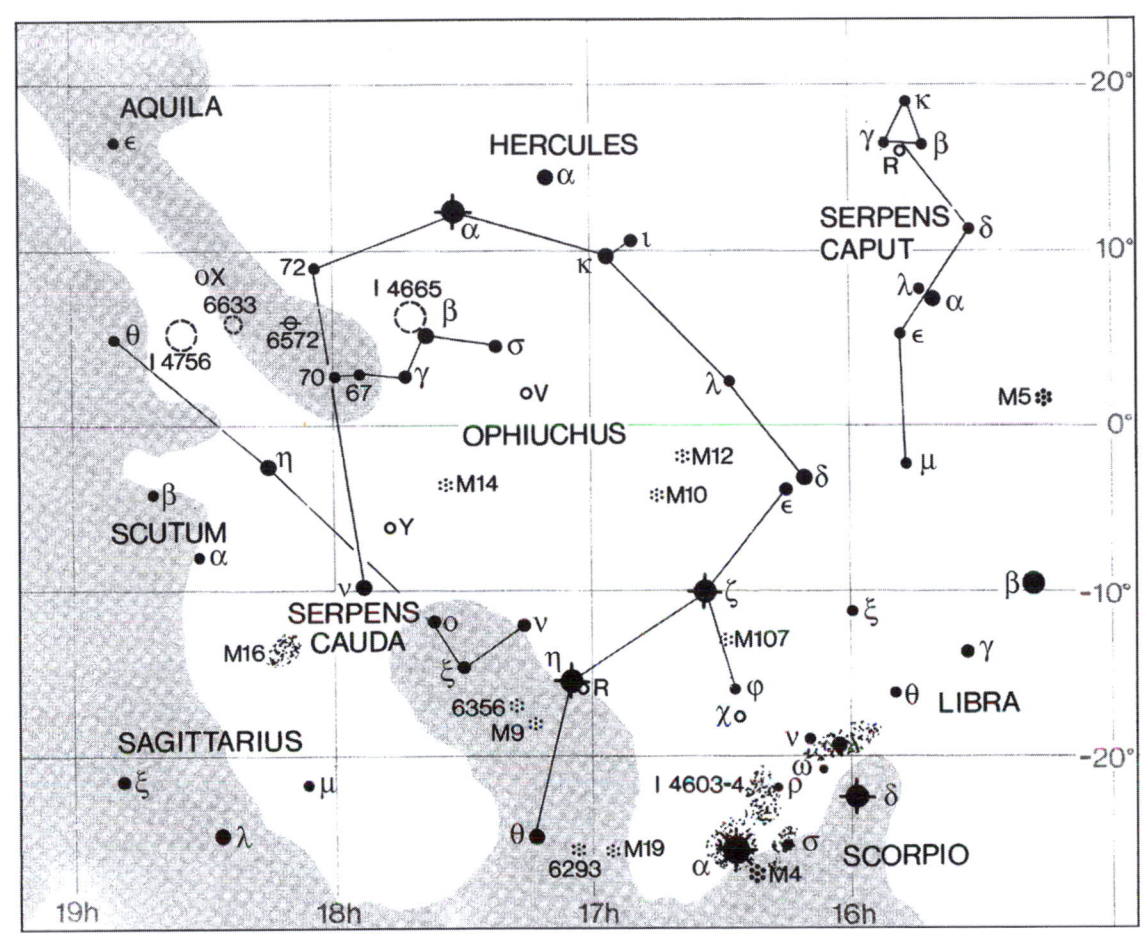

Ophiuchus is a very large constellation, with 13 stars above the fourth magnitude.

	RA			Dec.			m	M	d (light-years)	Spectrum	
	(h)	(m)	(s)	(°)	(′)	(″)					
57 α	17	34	56.0	+12	33	38	2.08	1.30	47	A5	Rasalhague
35 η	17	10	22.7	−15	43	31	2.43	0.37	84	A3	Sabik
13 ζ	16	37	09.5	−10	34	02	2.54	−3.20	460	O9	Han
1 δ	16	14	20.8	−03	41	38	2.73	0.86	170	M4	Yed Prior
60 β	17	45	28.4	+04	34	01	2.76	0.76	82	K2	Cheleb
27 κ	16	57	40.2	+09	22	20	3.19	1.09	86	K2	Helkath
2 ε	16	18	19.2	−04	41	33	3.23	0.64	107	G8	Yed Post
42 θ	17	22	00.6	−24	59	58	3.27	−2.92	560	B2	Imad
64 ν	17	59	01.6	−09	46	24	3.32	−0.03	153	K0	Sinistra
72	18	02	21.0	+09	33	49	3.71	1.69	83	A4	Phorbaceous
62 γ	17	47	53.6	+02	42	27	3.75	1.43	95	A0	Aldurajah
10 λ	16	30	54.8	+01	59	03	3.82	0.28	166	A2	Marfik
67	18	00	38.7	+02	55	54	3.91	−4.26	1400	B5	Fellah

Also above 4.5: ρ (4.03), b (4.16), χ (4.22), d (4.28), φ (4.29), A (5.44), σ (4.34), ι (4.39), ξ (4.39), 68 (4.42), ω (4.45), Psi (4.48).

VARIABLE STARS

	RA (h)	RA (m)	Dec. (°)	Dec. (′)	Range	Type	Period (d)	Spectrum
V	16	26.7	−12	26	7.3–11.6	Mira	298.0	N
χ	16	27.0	−18	27	4.2–5.0	Irreguar (γ c)	—	B
SS	16	57.9	−02	46	7.8–4.5	Mira	80.0	M
R	17	07.8	−16	06	7.0–13.8	Mira	302.6	
U	17	16.5	+01	13	5.9–6.6	Algol	1.68	B + B
Z	17	19.5	+01	31	7.6–14.0	Mira	348.7	K–M
RS	17	50.2	−06	43	5.3–12.3	Recurrent nova	—	O + M (Outbursts 1933, 1958, 1967)
Y	17	52.6	−06	09	5.9–6.4	Cepheid	17.12	F–G
RY	18	16.6	+03	42	7.5–13.8	Mira	150.5	M
X	18	38.3	+08	50	5.9–9.2	Mira	334.4	M + K

DOUBLE STARS

	RA (h)	RA (m)	Dec. (°)	Dec. (′)	PA (°)	Separation (″)	Magnitudes	
ρ	16	25.6	−23	27	344	3.1	5.3, 6.0	
υ	16	27.8	−08	22	095	1.0	4.6, 7.8	
λ	16	30.9	+01	59	AB 022	1.5	4.2, 5.2	Binary, 129.9 y
					AB + C 170	119.2	11.1	
					AD 246	313.8	9.9	
φ	16	31.1	−16	37	037	34.4	4.3, 12.8	
19	16	47.2	+02	04	089	23.4	6.1, 9.4	
η	17	10.4	−15	43	247	0.5	3.0, 3.5	Binary, 84.3 y
36	17	15.3	−26	36	150	4.7	5.1, 5.1	Binary, 549 y
41	17	16.6	−00	27	346	1.0	4.8, 7.8	
53	17	34.6	+09	35	191	41.2	5.8, 8.5	
τ	18	03.1	−08	11	AB 280	1.8	5.2, 5.9	Binary, 280 y
					AC 127	100.3	9.3	
70	18	05.5	+02	30	152	3.4	4.2, 6.0	Binary, 88.1 y
73	18	09.6	+04	00	300	0.4	6.1, 7.0	Binary, 270 y
X	18	38.3	+08	50	150	0.4	variable, 8.6	Binary, 485 y

OPEN CLUSTERS

M	C	NGC	RA (h)	RA (m)	Dec. (°)	Dec. (′)	Diameter (′)	Magnitude	No. of stars
		IC 4665	17	46.3	+05	43	41	4.2	30
		6633	18	27.7	+06	34	27	4.6	30

GLOBULAR CLUSTERS

M	C	NGC	RA (h)	RA m	Dec. (°)	Dec. (′)	Diameter (″)	Magnitude
107		6171	16	32.5	−13	03	10.0	8.1
12		6218	16	47.2	−01	57	14.5	6.6
10		6254	16	57.1	−04	06	15.1	6.6
62		6266	17	01.2	−30	07	14.1	6.6

GLOBULAR CLUSTERS

M	C	NGC	RA (h)	m	Dec. (°)	(')	Diameter (")	Magnitude
19		6273	17	02.6	−26	16	13.5	7.1
		6304	17	14.5	−29	28	6.8	8.4
		6316	17	16.6	−28	08	4.9	9.0
9		6333	17	19.2	−18	31	9.3	7.9
		6356	17	23.6	−17	49	7.2	8.4
		6355	17	24.0	−26	21	5.0	9.6
14		6402	17	37.6	−03	15	11.7	7.6
		6401	17	38.6	−23	55	5.6	9.5

PLANETARY NEBULÆ

M	C	NGC	RA (h)	(m)	Dec. (°)	(')	Dimensions (")	Magnitude	Magnitude of central star
		6309	17	14.1	−12	55	14 × 66	10.8	14.4
		6572	18	12.1	+06	51	8 × 8	9.0	13.6

GALAXY

M	C	NGC	RA (h)	(m)	Dec. (°)	(')	Magnitude	Dimensions (')	Type
		6384	17	32.4	+07	04	10.6	6.0 × 4.3	Sb

The δ and ε make up a lovely pair, the redness of δ contrasting with the yellowish hue of ε.

The celebrated binary 70 (p) Ophiuchi was once included in the rejected constellation Taurus Poniatovskii. The combined magnitude is 4.0; the primary is an orange Main Sequence star of type K0 – actually a very small-range BY Draconis variable; the secondary is a K4-type dwarf. The position is RA 18h 05m 27s, Dec. +02° 30′ 00″. It is only 16.6 light-years away; the orbital period is 88.3 years.

Barnard's Star (Munich, 15040) lies in Ophiuchus; RA 17h 57m 48s .5, Dec. +04° 41′ 36″, magnitude 9.5. Its distance is 5.98 light-years, so that is the closest star apart from the members of the α Centauri system. The absolute magnitude is 13.2, and the mass ~0.16 that of the Sun. A planet was reported by P. van de Kamp, but this turned out to be spurious. The star is appoaching the Solar System, and will pass by in AD 11 700 at a mere 3.8 light-years – but even then it will still be below the 8th magnitude, too faint to be seen with the naked eye.

Ophiuchus is crossed by the ecliptic, and should really be classed as a Zodiacal constellation – a fact that astrologers conveniently overlook!

ORION

(Abbreviation: Ori)

Orion is one of the most magnificent constellations in the sky; it represents the mythological hunter who boasted that he could kill any creature on Earth, but who was fatally stung by a scorpion. The two leading stars are Rigel, which is actually variable over a very small range (0.08 to 0.20) and the red variable Betelgeux – a name which may also be spelled Betelgeuse or Betelgeuze. The gaseous nebula M 42, in the Sword, is the most famous example of its type, and is easily visible with the naked eye.

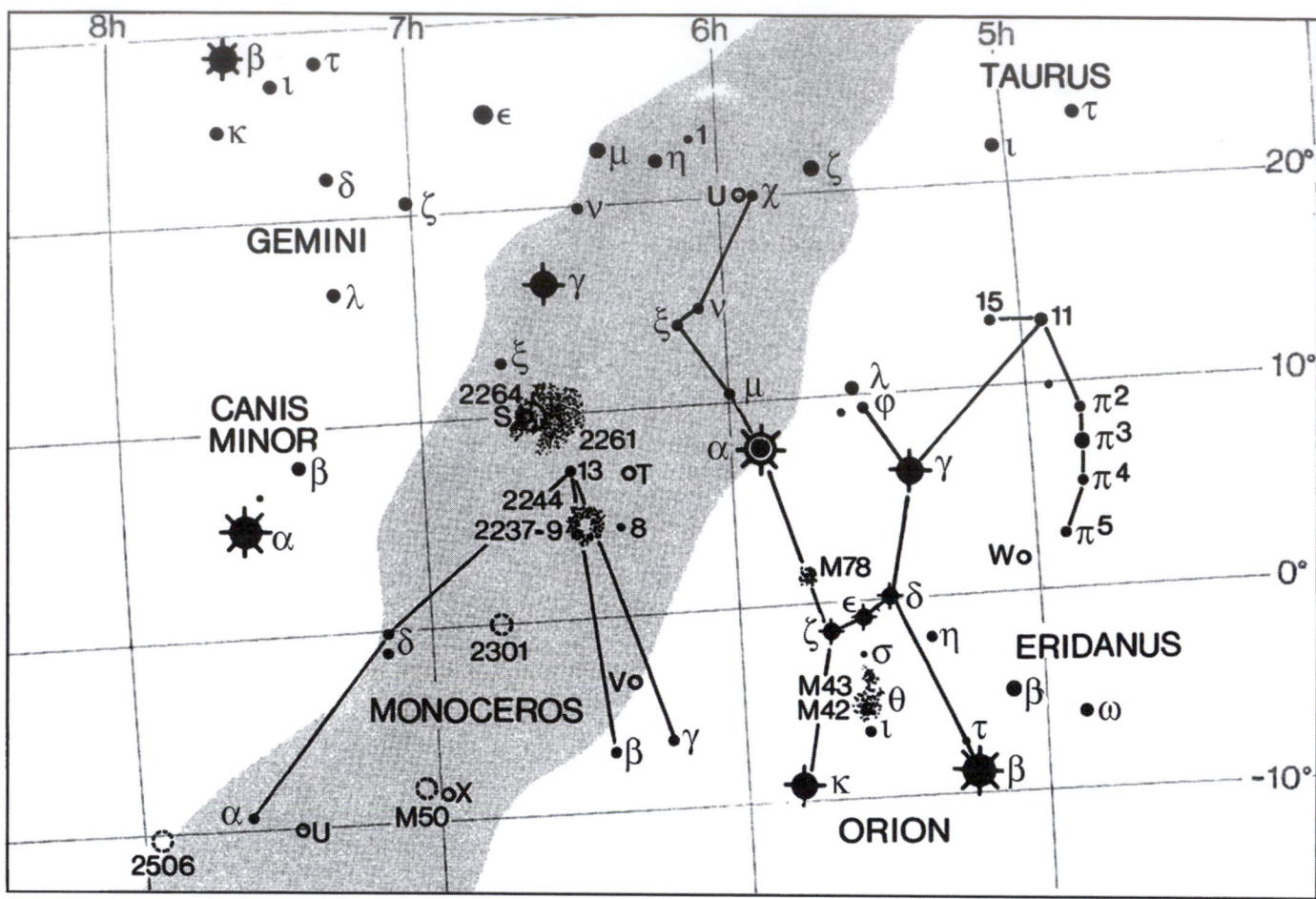

Altogether there are 15 stars above the fourth magnitude:

	RA			Dec.			m	M	d (light-years)	Spectrum	
	(h)	(m)	(s)	(°)	(′)	(″)					
19 β	05	14	32.3	−08	12	06	0.18	−6.69	773	B8	Rigel
58 α	05	55	10.3	+07	24	25	0.4v	−5.14	427	M2	Betelgeux
24 γ	05	25	07.9	+06	20	59	1.64	−2.72	243	B2	Bellatrix
46 ε	05	36	12.8	−01	12	07	1.69	−6.38	1342	B9	Alnilam
50 ζ	05	40	45.5	−01	56	33	1.74	−5.25	817	09	Alnitak
53 κ	05	47	45.4	−09	40	11	2.07	−4.65	721	BO	Saiph
34 δ	05	32	00.4	−00	17	57	2.2v	−4.99	916	O9	Mintaka
44 ι	05	15	26.0	−05	54	36	2.75	−5.30	1325	O9	Hatysa
1 π³	04	49	50.1	+06	57	41	3.19	3.67	26	F6	Tabit
28 η	05	24	28.6	−02	23	50	3.35	3.86	900	B1	Algieba
39 λ-A	05	35	08.3	+09	56	03	3.39	−4.16	1050	O8	Heka
30 τ	05	17	36.4	−06	50	40	3.59	−2.56	550	B5	

	RA			Dec.			m	M	d (light-years)	Spectrum
	(h)	(m)	(s)	(°)	(′)	(″)				
3 π^4	04	51	12.4	+05	36	18	3.68	−4.25	1260	B2
8 π^5	04	54	15.1	+02	26	26	3.71	−4.36	1340	B2
48 σ-A	05	38	24.8	−02	36	00	3.77	−3.96	1150	O9

Also above 4.5: o^2 (4.06), φ^2 (Khad Post) (4.09), μ (4.12), e (4.13), A (4.20), π^2 (4.35), φ^1 (Khad Prior) (4.39), χ^1 (4.39), ν (4.42), ξ (4.45), ρ (4.46), π^6 (4.47), ω (4.50).

DOUBLE STARS

	RA		Dec.		PA (°)	Separation (″)	Magnitudes	
	(h)	(m)	(°)	(′)				
π^3	04	49.8	+06	58	138	94.6	3.2, 8.7	
β	05	14.5	−08	12	202	9.5	0.1, 6.8	
ρ	05	15.3	+02	54	064	7.0	4.5, 8.3	
η	05	24.5	−02	24	AB 078	1.6	3.8, 4.8	
					AC 051	115.1	9.4	
δ	05	32.0	−00	18	359	52.6	2.2v, 6.3	
λ	05	35.1	+09	56	043	4.4	3.6, 5.5	
σ	05	38.7	−02	36	AB 137	0.2	4.0, 6.0	Binary, 170 y
					AB + C 238	11.4	10.3	
					AB + D 084	12.9	7.5	
					AB + E 061	42.6	6.5	
θ	05	35.3	−05	23	AB 031	8.8	6.7, 7.9	
					AC 132	12.8	5.1	
					AD 096	21.5	6.7	
ι	05	35.4	−05	55	141	11.3	2.8, 6.9	
ζ	05	40.8	−01	57	AB 162	2.4	1.9, 4.0	Binary, 1509 y
					AC 010	57.6	9.9	
μ	06	02.4	+09	39	023	0.4	4.4, 6.0	

OPEN CLUSTERS

M	C	NGC	RA		Dec.		Diameter (′)	Magnitude	No. of stars
			(h)	(m)	(°)	(′)			
		1981	05	35.2	−04	26	25	4.6	20
		2112	05	53.9	+00	24	11	9.1	5
		2175	06	09.8	+20	19	18	6.8	60
		2186	06	12.2	+05	27	4	8.7	30

NEBULÆ

M	C	NGC	RA		Dec.		Dimensions (′)	Magnitude of illuminating star	
			(h)	(m)	(°)	(′)			
42		1976	05	35.4	−05	27	66 × 60	5	Great Nebula
43		1982	05	35.6	−05	16	20 × 15	7	Extension of M 42
78		2068	05	46.7	+00	03	8 × 6	10	Nebula is magnitude 8
		IC 434	05	41.0	−02	24	60 × 10	2	(ζ). Behind Horse's Head dark nebula. Barnard 33

Orion is crossed by the celestial equator, and can therefore be seen from any inhabited country.

PAVO

(Abbreviation: Pav)

This is one of the Southern Birds. The brightest star, α (called 'Peacock' by air navigators) is rather isolated from the rest of the constellation. κ is the brightest W Virginis variable in the sky; λ is an eruptive variable of the γ Cassiopeiæ type.

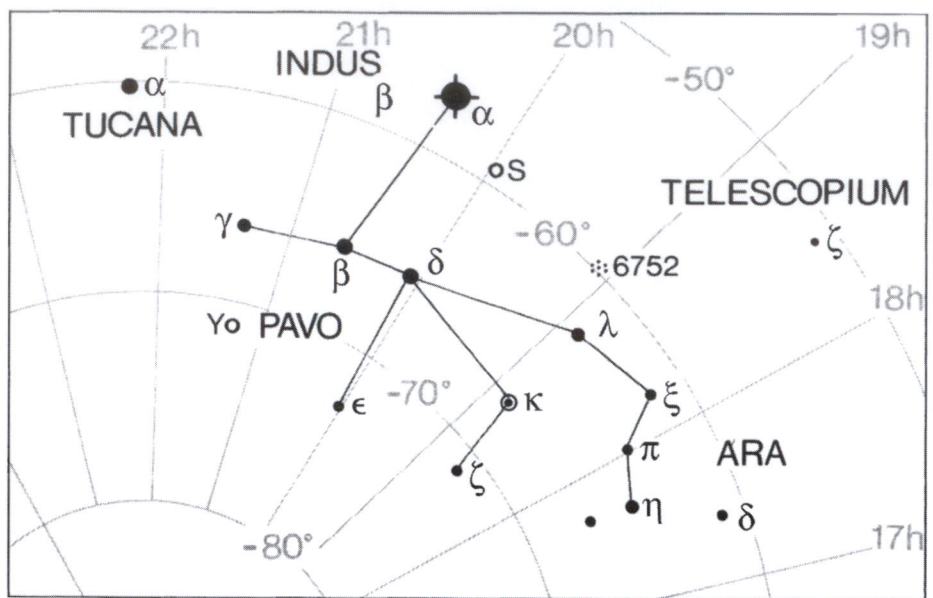

There are six stars above the fourth magnitude.

	RA			Dec.			m	M	d (light-years)	Spectrum
	(h)	(m)	(s)	(°)	(′)	(″)				
α	20	25	38.9	−56	44	06	1.94	−1.81	183	B2
β	20	44	57.6	−66	12	11	3.42	0.29	137	A5
λ	18	52	304	−18	12	14	3.4v	−4.5v	1800	B2
δ	20	08	41.9	−66	10	46	3.55	4.62	20	G5
η	17	45	44.0	−64	43	25	3.61	−1.67	370	K4
ε	20	00	35.4	−72	54	36.7	3.97	1.41	106	AO
ζ	18	43	02.1	−71	25	40	4.01	−0.03	210	K2

Also above 4.5: γ (4.21), π (4.33), ξ (4.35), κ (4.4v).

α, the brightest star in the area, makes Pavo easy to find. The most notable objects are κ and the bright globular NGC 6752 (C 93). The eruptive variable λ is usually about magnitude 4; ζ is a suitable comparison star.

VARIABLE STARS

	RA		Dec.		Range	Type	Period (d)	Spec.
	(h)	(m)	(°)	(′)				
R	18	12.9	−63	37	7.5–13.8	Mira	229.8	M
λ	18	52.2	−62	11	3.4–4.3	Irregular (γ C)	—	B
κ	18	56.9	−67	14	3.9–4.7	W Virginis	9.09	F
T	19	50.7	−71	46	7.0–14.0	Mira	244.0	M
S	19	55.2	−59	12	6.6–10.4	Semi-regular	386	M

VARIABLE STARS

	RA		Dec.		Range	Type	Period (d)	Spec.
	(h)	(m)	(°)	(′)				
Y	21	24.3	−69	44	5.7–8.5	Semi-regular	233	N
SX	21	28.7	−69	30	5.4–6.0	Semi-regular	50	M

DOUBLE STAR

	RA		Dec.		PA	Separation (″)	Magnitudes
	(h)	(m)	(°)	(′)	(°)		
ξ	18	23.2	−61	30	154	3.3	4.4, 8.6

GLOBULAR CLUSTER

M	C	NGC	RA		Dec.		Diameter	Magnitude
			(h)	(m)	(°)	(′)	(′)	
	93	6752	19	10.9	−59	59	20.4	5.4

GALAXIES

M	C	NGC	RA		Dec.		Magnitude	Dimensions	Type
			(h)	(m)	(°)	(′)		(′)	
		IC 4662	17	47.1	−64	38	11.4	2.2 × 1.4	Irregular
		6684	18	49.0	−65	11	10.4	3.7 × 2.7	SB0
	101	6744	19	09.8	−63	51	9.0	15.5 × 10.2	SBb
		6753	19	11.4	−57	03	11.9	2.5 × 2.2	Sb

PEGASUS

(Abbreviation: Peg)

One of the most distinctive of the northern constellations. It commemorates the flying horse which the hero Bellerophon rode during an expedition to slay the fire-breathing Chimæra. The main stars of Pegasus make up a square: three of these are α, β and γ. The fourth is Alpheratz, which used to be included in Pegasus as δ Pegasi, but has been officially – and frankly, illogically – transferred to Andromeda, as α Andromedæ.

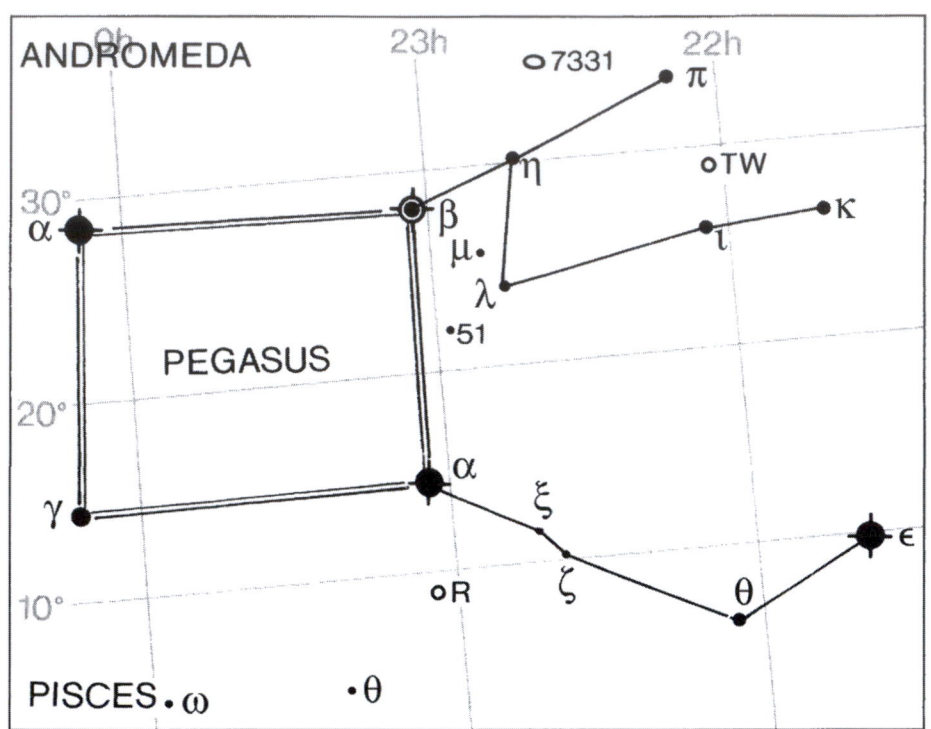

Excluding Alpheratz, there are ten stars above the fourth magnitude:

	RA			Dec.			m	M	d (light-years)	spec	
	(h)	(m)	(s)	(°)	(′)	(″)					
8 ε	21	44	11.1	+09	52	30	2.38	−4.19	670	K2	Enif
53 β	23	03	46.3	+28	04	57	variable	−1.49	199	M2	Sheat
54 α	23	04	45.6	+15	12	19	2.49	−0.67	140	B9	Markab
88 γ	00	13	14.1	+15	11	01	2.83	−2.22	330	B2	Algenib
44 η	22	43	00.1	+30	13	17	2.93	−1.16	215	G2	Matar
42 ζ	22	41	27.7	+10	49	53	3.41	−0.62	208	B8	Homan
48 μ	22	50	00.1	+24	36	06	3.51	0.74	117	M2	Sadalbari
26 θ	22	10	11.8	+06	11	52	3.52	1.16	97	A2	Biham
24 ι	22	07	00.5	+25	20	42	3.77	3.42	38	F5	
47 λ	22	46	31.8	+23	33	56	3.97	−1.45	390	G8	Sadalpheris

Above magnitude 4.5: I (4.08), κ (4.14), ξ (4.20), π (4.28), 9 (4.34), υ (4.42).

DOUBLE STARS

	RA		Dec.		PA (°)	Separation (″)	Magnitudes	
	(h)	(m)	(°)	(′)				
ε	21	44.2	+09	52	AB 325	81.8	2.4, 11.2	
					AC 320	142.5	9.4	
κ	21	44.6	+25	39	095	0.3	4.7, 5.0	Binary, 11.6 y
35	22	27.9	+04	42	AB 210	98.3	4.8, 9.8	
					AC 241	181.5	9.7	
37	22	30.0	+04	26	118	0.9	5.8, 7.1	Binary, 140 y
η	22	43.0	+30	13	339	90.4	2.9, 9.9	B is a close double
β	23	03.8	+28	05	AB 211	108.5	2v, 11.6	
					AC 098	253.1	9.4	

GLOBULAR CLUSTER

M	C	NGC	RA		Dec.		Diameter (′)	Magnitude
			(h)	(m)	(°)	(′)		
15		7078	21	30.0	+12	10	12.3	6.3

GALAXIES

M	C	NGC	RA		Dec.		Magnitude	Dimensions (′)	Type
			(h)	(m)	(°)	(′)			
	30	7331	22	37.1	+34	25	9.5	10.7 × 4.0	Sb
		7332	22	37.4	+23	48	11.8	4.2 × 1.3	E7
	44	7479	23	04.9	+12	19	11.0	4.1 × 3.2	SBb
	43	7814	00	03.3	+16	09	10.5	6.3 × 2.6	Sb

ε (Enif) is well away from the Square; it is an orange super-giant, with diameter 150 times that of the Sun. Brief brightenings have been reported, and if genuine they could possibly be due to flare activity. α (Markab) is named after the Arabic for 'the horse's saddle'.

β (Sheat, or Scheat) has a magnitude range between 2.3 and 2.8; there is a very rough period of about 38 days. Its diameter is over 9 times that of the Sun, and some catalogues give its luminosity as over 9000 times that of the Sun. Its slow variations are easy to follow with the naked eye; Markab and Algenib are convenient comparison stars.

γ (Algenib) is a variable of the β Cephei type; range 2.78 to 2.89, period 3.6 hours. Used as a comparison for β, it is good enough to take the magnitude as 2.8.

Two fainter stars in Pegasus are of special note. The yellow dwarf 51 Pegasi, the first star known to have an orbiting planet, has already been discussed. The other notable star is IK Pegasi: RA 21h 26m 27s, Dec. +19° 22′ 30″, magnitude 6.1, type A8. It has a companion, IK Pegasi B, which is a massive white dwarf; orbital period 21.7 days, average separation 31 000 000 km. When the primary evolves into a red giant stage, it is likely that the white dwarf will explode as a Type 1a supernova. This is the closest probable supernova progenitor, but it is already 150 light-years from us, and is receding, so that when the outburst occurs in the far future it could not damage the Earth.

The globular cluster M 15 is one of the finest in the sky. Several galaxies in Pegasus are easy objects. More elusive, but of special interest, are the members of Stephan's Quintet, discovered in 1877 by Eduard Stephan.

They form a compact group:

Name	RA			Dec.			Type	m
	(h)	(m)	(s)	(°)	(′)	(″)		
NGC 7317	22	35	52	+33	56	42	E4	14.6
NGC 7318a	22	35	57	+33	57	56	E2p	14.3
NGC 7818b	22	35	58	+33	57	57	Sbp	13.9
NGC 7319	22	36	04	+33	58	33	Sbp	14.1
NGC 7320c	22	.36	20	+33	59	06	SAbc	16.7

The interest here is that while four members of the quintet are associated, around 210–340 million light-years away, the fifth, NGC 7320, is in the foreground, about 40 million light-years from us. The four are on collision courses with each other. The Spitzer Space Telescope has detected a vast intergalactic shock wave caused by NGC 7818b falling into the group at several millions of miles per hour.

Pegasus is a splendid constellation, unmistakable even though maps tend to make it look smaller and brighter than it really is.

PERSEUS

(Abreviation: Per)

A prominent constellation, containing the prototype eclipsing star Algol as well as the superb Sword-Handle cluster (C 14). Mythologically, Perseus was the hero of one of the most famous of all legends; he killed the Gorgon, Medusa, and married Andromeda, daughter of Cepheus and Cassiopeia. The Gorgon's Head is marked by the winking 'Demon Star', Algol.

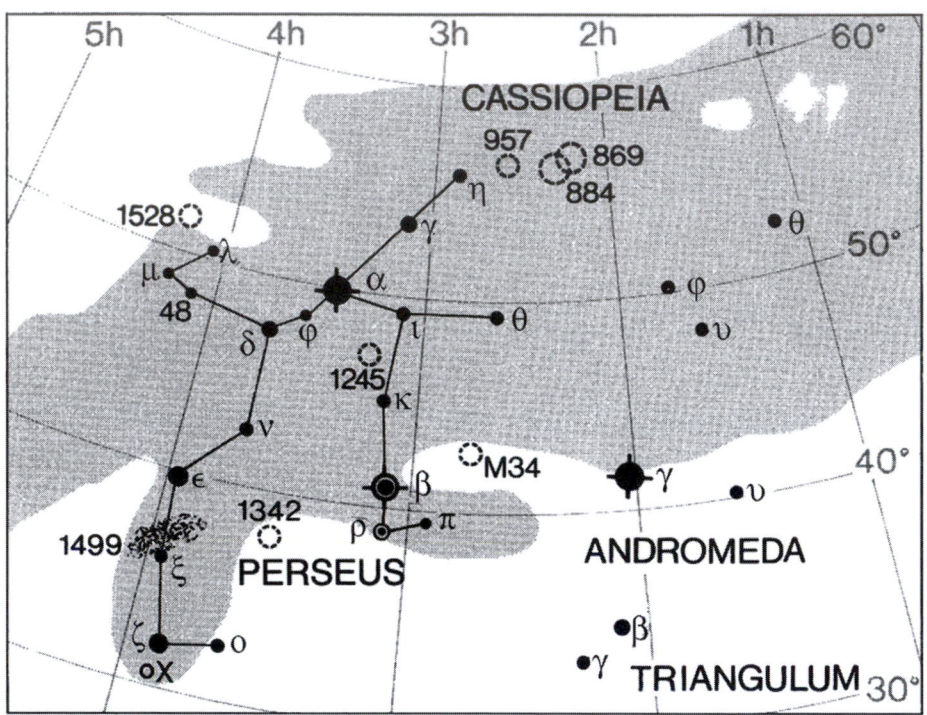

There are 14 stars above the fourth magnitude.

	RA			Dec.			m	M	d (light-years)	Spectrum	
	(h)	(m)	(s)	(°)	(')	('')					
33 α	03	24	19.4	+49	51	41	1.79	−4.50	590	F5	Mirphak
26 β	03	08	10.1	+40	57	20	2.09v	−0.18	93	B8	Algol
44 ζ	03	54	07.9	+31	53	01	2.84	−4.55	980	B1	Atik
45 ε	03	57	51.2	+40	00	37	2.90	−3.19	540	B0	Adid Aust
23 γ	03	04	47.8	+53	30	23	2.91	−1.57	256	G8	Seid
39 δ	03	42	55.5	+47	47	16	3.01	−3.04	530	B5	Basel
25 ρ	03	05	10.5	+38	50	26	3.3v	−1.67	275	M3	
15 η	02	50	41.8	+55	52	44	3.77	−4.28	1330	K3	Miram
41 ν	03	45	11.6	+42	34	43	3.77	−2.39	550	F5	Adid Media
27 κ	03	09	29.6	+44	51	28	3.79	1.11	112	K0	Misan
38 o	03	44	19.1	+32	17	18	3.84	−4.44	1470	B1	Atika
18 τ	02	54	15.5	+52	45	45	3.93	−0.48	248	G4	
48 c	04	08	39.7	+47	42	45	3.96	−2.19	550	B3	
46 ξ	03	58	57.9	+35	47	28	3.98	−4.70	1770	O7	Menkib

Also above 4.5: ψ (Alseph) (4.01), ι (4.05), θ (4.10), μ (4.12), 16 (4.22), λ (4.25), e (4.25), ψ (4.32), σ (4.36).

VARIABLE STARS

	RA (h)	RA (m)	Dec. (°)	Dec. (')	Range	Type	Period (d)	Spectrum	
U	01	59.6	+54	49	7.4–12.3	Mira	321.0	M	
S	02	22.9	+58	35	7.9–11.5	Semi-regular	Long	M	
ρ	03	05.2	+38	50	3–4	Semi-regular	33 to 55	M	
β	03	08.2	+40	57	2.1–3.4	Algol	2.87	B + G	
R	03	30.1	+35	40	8.1–14.8	Mira	210.0	M	
X	03	55.4	+31	03	6.0–7.0	Irregular	—	09.5	X-ray source
AW	04	47.8	+36	43	7.1–7.8	Cepheid	6.46	F–G	

DOUBLE STARS

	RA (h)	RA (m)	Dec. (°)	Dec. (')	PA (°)	Separation ('')	Magnitudes	
η	02	50.7	+55	54	300	28.3	3.3, 8.5	
θ	02	44.2	+49	14	215	19.8	4.1, 9.9	Binary, 2720 y
γ	03	04.8	+53	30	326	57.0	2.9, 10.6	
ζ	03	54.1	+31	53	AB 208	12.9	2.9, 9.5	
					AC 286	32.8	11.3	
					AD 195	94.2	9.5	
					AE 185	120.3	10.2	
ε	03	57.9	+40	01	010	8.8	2.9, 8.1	

OPEN CLUSTERS

M	C	NGC	RA (h)	RA (m)	Dec. (°)	Dec. (')	Diameter (')	Magnitude	No. of stars	
		744	01	58.4	+55	29	11	7.9	20	
	14	869	02	19.0	+57	09	30	4.3	200	Sword
		884	02	22.4	+57	07	30	4.4	150	Handle
		957	02	33.6	+57	32	11	7.6	30	
34		1039	02	42.0	+42	47	35	5.2	60	
		1245	03	14.7	+47	15	10	8.4	200	
		1444	03	49.4	+52	40	4	6.6	—	
		1513	04	10.0	+49	31	9	8.4	50	
		1528	04	15.4	+51	14	24	6.4	40	
		1545	04	20.9	+50	15	8	6.2	20	

PLANETARY NEBULA

M	C	NGC	RA (h)	RA (m)	Dec. (°)	Dec. (')	Diameter ('')	Magnitude	Magnitude of central star	
76		650–1	01	42.4	+51	34	65 × 290	12.2	17	Little Dumbbell

β (Algol) is the prototype eclipsing binary. The main component (A) is of type B, around 4 000 000 km in diameter, and just over 100 times as luminous as the Sun. The secondary (B) is of type G, over 5 000 000 km in diameter and 3 times as luminous as the Sun, though its mass is less than that of A. Eclipses are not total. The secondary minimum has an amplitude of no more than 0.1 magnitude.

NEBULÆ

M	C	NGC	RA		Dec.		Diameter (')	Magnitude of illuminating star	
			(h)	(m)	(°)	(')			
		1333	03	29.3	+31	25	9 × 7	9.5	(Near dark nebula B 205)
		1499	04	00.7	+36	37	145 × 40	4	California Nebula

GALAXIES

M	C	NGC	RA		Dec.		Magnitude	Dimensions (')	Type	
			(h)	(m)	(°)	(')				
		1003	02	39.3	+40	52	11.5	5.4 × 2.1	Sc	
		1023	02	40.4	+39	04	9.5	8.7 × 3.3	E7p	
	24	1275	03	19.8	+41	31	11.6	2.6 × 1.9	Pec;	Perseus A

Perseus is a rich constellation. It contains the lovely double of the Sword Handle – not to be confused with the Sword of Orion. It is not in Messier's catalogue, presumably because there was chance of mistaking it for a comet.

PHŒNIX

(Abbreviation: Phe)
This is one of the Southern Birds. Ankaa is the only bright star, but there are seven stars above the fourth magnitude.

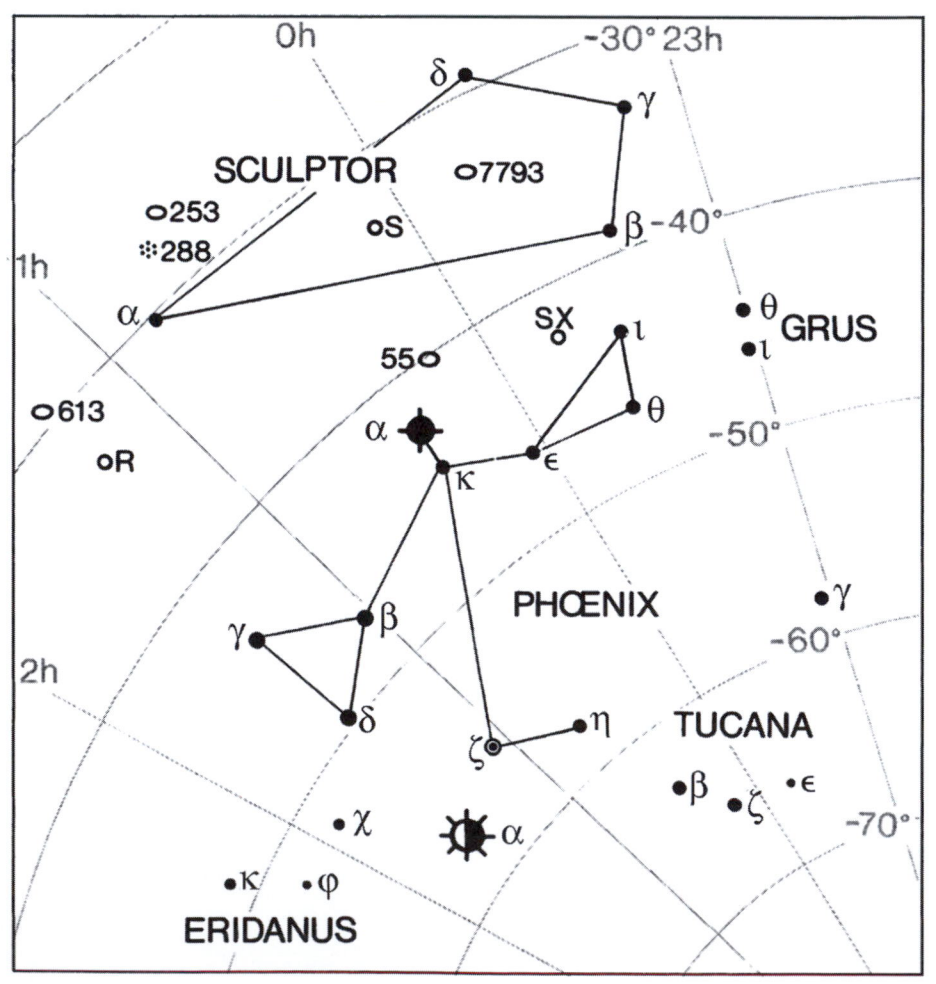

	RA			Dec.			m	M	d (light-years)	Spectrum	
	(h)	(m)	(s)	(°)	(′)	(″)					
α	00	26	16.9	−42	18	18	2.40	0.52	77	KO	Ankaa
β	01	06	05.1	−46	43	07	3.32	−0.60	198	G8	
γ	01	28	21.9	−43	19	04	3.41	−0.87	234	K5	
ε	00	09	24.5	45	11	19	3.88	0.71	140	KO	
κ	00	26	12.1	−43	40	48	3.93	2.07	77	A7	
δ	01	31	15.0	−49	04	23	3.93	0.66	147	AO	
ζ	01	08	23.1	−55	14	45	3.94	−0.73	280	B6	

Also above 4.5: η (4.36), ψ (4.39).

VARIABLE STARS

	RA		Dec.		Range	Type	Period (d)	Spectrum
	(h)	(m)	(°)	(′)				
SX	23	46.5	41	35	6.8–7.5	Delta Scuti	0.055	A
R	23	56.5	49	47	7.5–14.4	Mira	267.9	M
S	23	53.1	56	35	7.4–8.2	Semi-regular	141	M
ζ	01	08.4	55	15	3.6–4.1	Algol	1.67	B + B

DOUBLE STARS

	RA		Dec.		PA (°)	Separation (″)	Magnitudes
	(h)	(m)	(°)	(′)			
ξ	00	41.8	−56	30	253	13.2	5.8, 10.2
η	00	43.4	−57	28	217	19.8	4.4, 11.4
β	01	06.1	−46	43	346	1.4	4.0, 4.2

ζ is the brightest Algol variable in the sky, apart from Algol itself. Both components are of type B, each over 150 times as luminous as the Sun; distance 280 light-years. The system also includes a 7.2-magnitude star at a separation of 0.8″, and an 8.2-magnitude star at 0.4″.

PICTOR

(Abbreviation: Pic)

Originally Equuleus Pictoris, the Painter's Easel, this is a dim constellation near Canopus. β is of special interest. It is a pity that it has no proper name – perhaps we should gve it one!

See chart for Carina.

There are two stars above the fourth magnitude.

Also above magnitude 5: γ (4.50) and δ (4.72).

Pictor includes the nearest halo star, VZ Pictoris, better known as Kapteyn's Star (it was discovered by Jacobus Kapteyn in 1898). It is an M1-type subdwarf; RA 03h 11m 40s .6 , Dec. −46° 01′ 06″, distance 12.8 light-years, diameter 0.24 that of the Sun, mass

	RA			Dec.			m	M	d (light-years)	Spectrum
	(h)	(m)	(s)	(°)	(′)	(″)				
α	06	48	11.5	−61	56	31	3.24	0.83	99	A7
β	05	47	17.1	−51	04	00	3.85	2.43	63	A5

0.38 that of the Sun, luminosity 0.004 that of the Sun. It orbits the Galaxy in a retrograde direction. The apparent magnitude is 8.8.

PISCES

(Abbrevaiation: Psc)

A Zodiacal constellation. Its mythological associations are rather vague, but according to one Greek legend it represents the fishes into which Aphrodite and Eros changed themselves when being chased by Typhon.

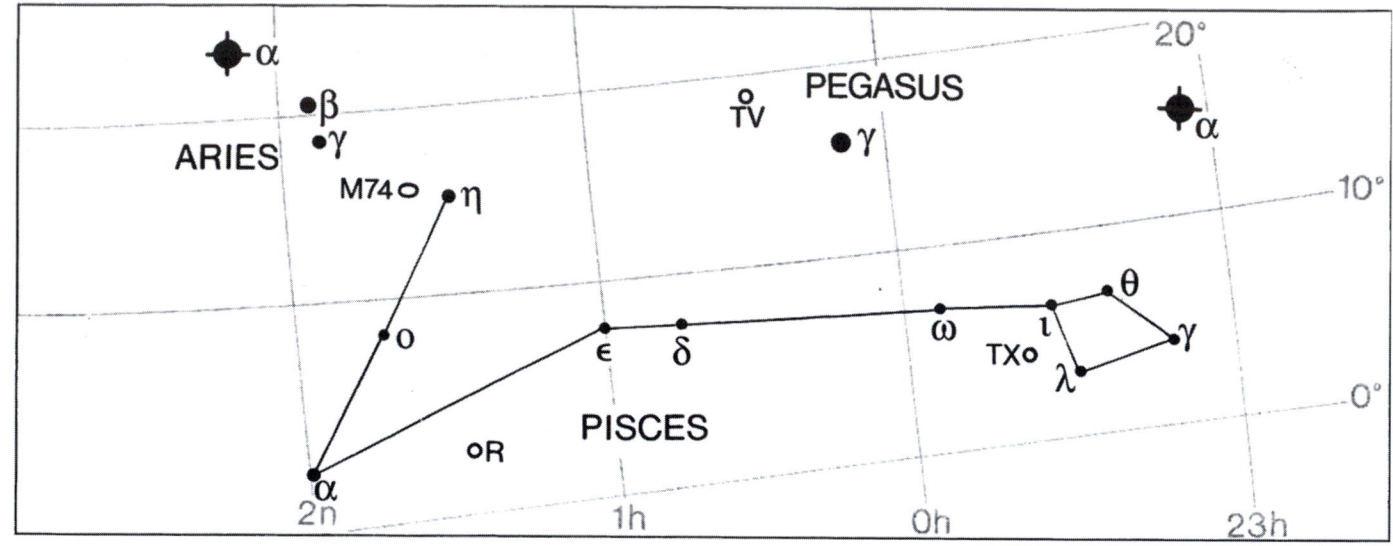

	RA			Dec.			m	M	d (light-years)	Spectrum	
	(h)	(m)	(s)	(°)	(′)	(″)					
99 η	02	31	29.0	+15	20	45	3.62	−1.16	294	G8	Alpherg
6 γ	23	17.	09.5	+03	16	56	3.70	0.68	151	G7	Simmah
113 α	02	02	02.8	+02	45	50	3.82	0.67	139	A2	Al Rischa
28 ω	23	59	18.6	+06	51	49	4.03	1.47	106	F4	Vernalis

Also down to 4.5: ι (4.13), o (Torcularis Septentrionalis) (4.26), ε (Kaht) (4.27), θ (4.27), 30 (4.37), δ (Linteum) (4.44), ν (4.45), β (Samaka) (4.48), λ (4.49), τ (Anunitum) (4.51).

Some of these proper names seem to be decidedly curious. o has the longest, but α has several alternatives: Al Rischa, Alrescha, Alrisha, El Rescha, Kaitain, Okda. Take your pick!

The vernal equinox has shifted from Aries into Pisces, and is now just south of ω; hence the star's name, Vernalis. The equinox is shifting slowly below the western fish towards Aquarius.

VARIABLE STARS

	RA (h)	(m)	Dec. (°)	(')	Range	Type	Period (d)	Spectrum
TX	23	46.4	+03	29	6.9–7.7	Irregular	–	N
TV	00	28.0	+17	24	4.6–5.4	Semi-regular	70	M
Z	01	16.1	+25	46	7.0–7.9	Semi-regular	144	N
R	01	30.6	+02	53	7.1–14.8	Mira	344.0	M

DOUBLE STARS

	RA (h)	(m)	Dec. (°)	(')	PA (°)	Separation (")	Magnitudes	
ζ	01	13.7	+07	35	063	23.0	5.6, 6.5	
ψ¹	01	05.6	+21	28	AB 159	30.0	5.6, 5.8	
					AC 123	92.6	11.2	
α	02	02.0	+02	46	279	1.9	4.2, 5.1	Binary, 933 y

GALAXIES

M	NGC	RA (h)	(m)	Dec. (°)	(')	Magnitude	Dimensions (')	Type
	470	01	19.7	+03	25	11.9	3.0 × 2.0	Sc
	474	01	20.1	+03	25	11.1	7.9 × 7.2	S0
	488	01	21.8	+05	15	10.3	5.2 × 4.1	Sb
	524	01	24.8	+09	32	10.6	3.2 × 3.2	E1
74	628	01	36.7	+15	47	9.2	10.2 × 9.5	Sc

The carbon star TX (19) Piscium is one of the reddest stars in the sky, and is always within binocular range; well worth finding. So is the fine spiral galaxy M 74, between η Piscium and γ Arietis.

Van Maanen's Star (Gliese 35), the closest known solitary white dwarf, lies in Pisces: RA 00h 49m 09s, Dec. +05° 23' 19.0", magnitude 12.4. Its distance is 14.1 light-years; absolute magnitude 14.2.

Taking the Sun as unity: mass 0.7, diameter 0.01, luminosity < 0.0002. It was discovered by Adriaan van Maanen in 1917.

The obsolete constellation of Testudo (the Turtle) lay between Pisces and Cetus; the turtle's head was marked by the star 20 Piscium.

PISCIS AUSTRALIS (OR PISCIS AUSTRINUS)

(Abbreviation: PsA)
A small southern constellation, with one brilliant star – Fomalhaut. It is one of Ptolemy's originals; no definite legends seem to be attached to it, though it has been associated with Dagon, the Assyrian fish-like water god, and with Oannes, 'Lord of the Waves'.

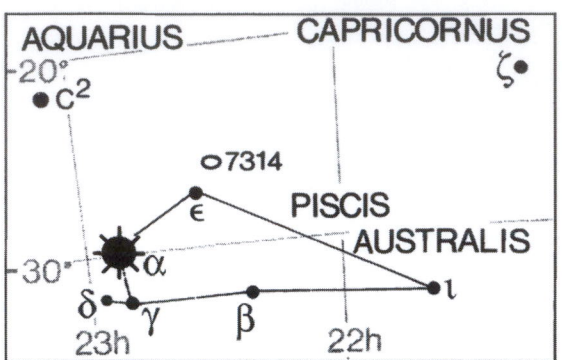

Also above 4.5: ε (4.18), δ (4.20), β (Fum al Samakah) (4.29), ι (4.35), γ (4.46), μ (4.50).

β also has a Chinese name, Tien Kang ('heavenly rope'). It is a wide, easy binary.

Fomalhaut's planet, Fomalhaut b, was the first extrasolar planet to be imaged directly. The small-range BY Draconis variable TW Piscis Australis (RA 22h 56m 3s .8, Dec. −31° 33′ 55″), Magnitude 6.5, specrum K4, lies within a light-year of Fomalhaut and may well be associated with it.

	RA			Dec.			m	M	d (light-years)	Spectrum
	(h)	(m)	(s)	(°)	(′)	(″)				
24 α	22	57	.38.8	−29.	17.	19	1.17	1.74	25	A3 Fomal-haut

VARIABLE STARS

	RA		Dec.		Range	Type	Period (d)	Spectrum
	(h)	(m)	(°)	(′)				
S	22	03.8	−28	03	8.0–14.5	Mira	271.7	M
V	22	55.3	−29	37	8.0–9.0	Semi-regular	148	M

DOUBLE STARS

	RA		Dec.		PA	Separation (″)	Magnitudes	
	(h)	(m)	(°)	(′)	(°)			
η	22	00.8	−28	27	115	1.7	5.8, 6.8	
β	22	31.5	−32	21	172	30.3	4.4, 7.9	(Optical)
γ	22	52.5	−32	53	262	4.2	4.5, 8.0	
δ	22	55.9	−32	32	244	5.0	4.2, 9.2	

GALAXIES

M	C	NGC	RA		Dec.		Magnitude	Dimensions (′)	Type	
			(h)	(m)	(°)	(′)				
		7172	22	02.0	−31	52	11.9	2.2 × 1.3	S	
		7174	22	02.1	−31	59	12.6	1.3 × 0.7	S	
		7314	22	35.8	−26	03	10.9	4.6 × 2.3	Sc	Arp 14

PUPPIS

(Abbreviation: Pup)

The poop of the good ship *Argo Navis*. The brightest star is ζ (Naos), an O-type supergiant.

There are 10 stars above the fourth magnitude. See chart for Carina.

	RA			Dec.			*m*	*M*	*d*	Spectrum	
	(h)	(m)	(s)	(°)	(′)	(″)					
ζ	08	03	35.1	−40	00	11	2.21	−5.95	1400	O5.8	Naos
π	07	17	08.6	−37	05	51	2.71	−4.92	1090	K3	Ahadi
15 ρ	08	07	32.7	−24	18	16	2.83	1.41	63	F2	Tureis
τ	06	49	56.1	−50	36	52	2.94	−0.80	183	K0	Anazitisi
ν	06	37	45.7	−43	11	45	3.17	−2.39	423	B8	Kaimana
σ	07	29	13.9	−43	18	07	3.25	−0.51	184	K5	Hadir
7 ξ	07	49	17.7	−24	51	35	3.34	−4.72	1350	G6	Asmidiske
C	07	45	15.3	−37	58	07	3.62	−4.52	1390	K4	
A	07	52	13.1	−40	34	12	3.71	−1.41	345	G5	
3 1	07	43	48.5	−28	57	17	3.94			A2	

Also above 4.5: P (4.10), II,j (4.20), o (4.40), 16.(4.40), h² (4.42), h¹ (4.44), a (4.44), HD60532 (4.46), V (4.47), I (4.49).

VARIABLE STARS

	RA		Dec.		Range	Type	Period (d)	Spectrum
	(h)	(m)	(°)	(′)				
L²	07	13.5	44	39	2.6–6.2	Semi–reg.	140	M
Z	07	32.6	−20	40	7.2–14.6	Mira	499.7	M
VX	07	32.6	−21	56	7.7–8.5	Cepheid	3.01	F
X	07	32.8	−20	55	7.8–9.2	Cepheid	25.96	F–G
W	07	46.0	−41	12	7.3–13.6	Mira	120.1	M
AP	07	57.8	−40	07	7.1–7.8	Cepheid	5.08	F
V	07	58.2	−49	15	4.7–5.2	β Lyræ	1.45	B + B
AT	08	12.4	−36	57	7.5–8.4	Cepheid	6.66	F–G
RS	09	13.1	−34	35	6.5–7.6	Cepheid	41.39	F–G

DOUBLE STAR

	RA		Dec.		PA	Separation (″)	Magnitudes
	(h)	(m)	(°)	(′)	(°)		
σ	07	29.2	−43	18	074	22.3	3.3, 9.4
κ	07	38.8	−24	48	318	9.8	4.5, 4.7

OPEN CLUSTERS

M	C	NGC	RA		Dec.		Diameter	Magnitude	No. of stars
			(h)	(m)	(°)	(′)	(′)		
		2383	07	24.8	−20	56	6	8.4	40
		2421	07	36.3	−20	37	10	8.3	70
47		2422	07	36.6	−14	30	30	4.4	30
		Mel 71	07	37.5	−12	04	9	7.1	80

OPEN CLUSTERS

M	C	NGC	RA (h)	(m)	Dec. (°)	(′)	Diameter (′)	Magnitude	No. of stars	
		Mel 72	07	38.4	−10	41	9	10.1	40	
		2432	07	40.9	−19	05	8	10.2	50	
		2439	07	40.8	−31	39	10	6.9	80	R Puppis. Asterism
46		2437	07	41.8	−14	49	27	6.1	100	
93		2447	07	44.6	−23	52	22	6.2	80	
		2451	07	45.4	−37	58	45	2.8	40	
	71	2477	07	52.3	−38	33	27	5.8	160	
		2479	07	55.1	−17	43	7	9.6	45	
		2489	07	56.2	−30	04	8	7.9	45	
		2509	08	00.7	−19	04	8	9.3	70	
		2527	08	05.3	−28	10	22	6.5	40	
		2533	08	07.0	−29	54	3.5	7.6	60	
		2539	08	10.7	−12	50	22	6.5	50	
		2546	08	12.4	−37	38	41	6.3	40	
		2567	08	18.6	−30	38	10	7.4	40	
		2571	08	18.9	−29	44	13	7.0	30	
		2580	08	21.6	−30	19	8	9.7	50	
		2587	08	23.5	−29	30	9	9.2	40	

PLANETARY NEBULÆ

M	C	NGC	RA (h)	(m)	Dec. (°)	(′)	Dimensions (″)	Magnitude	Magnitude of central star	
		2438	07	41.8	−14	44	66	10.1	17.7	In cluster NGC 2437
		2440	07	41.9	−18	13	14 × 32	10.8	14.3	Proto-planetary?

NEBULA

M	C	NGC	RA (h)	(m)	Dec. (°)	(′)	Dimensions (′)	Magnitude of illuminating star	
		2467	07	52.5	−26	24	8 × 7	9.2	Gum 9

Naos is an extremely hot O-type supergant. Its diameter is 16 times that of the Sun, mass 64 times that of the Sun. Visually it is abut 21 000 times as luminous as the Sun, but it is so hot (surface temperature over 40 000 °C) that most of its radiation is in the ultraviolet. If it were as close to us as Sirius, it would cast shadows.

PYXIS

(Abbreviation: Pyx)
Originally Pyxis Nautica, the Mariner's Compass. John Herschel's proposal to rename it Maus (the Mast) was never accepted.

See chart for Carina.
There are four stars down to the fourth magnitude.

	RA (h)	(m)	(s)	Dec. (°)	(′)	(″)	m	M	d (light-years)	Spectrum	
α	04	43	35.6	−33	11	11	3.68	−3.39	845	B1	Al Sumut
β	08	40.	06.1	−35	18	30	3.97	−1.41	388	G5	
γ	08	5	32.0	−27	42	36	4.02	−0.01	209	K4	

Next in order comes κ (4.63).

VARIABLE STARS

	RA (h)	(m)	Dec. (°)	(′)	Range	Type	Period (d)	Spectrum	
TY	08	59.7	−27	49	6.9–7.5	Eclipsing	3.20	G + G	
T	09	04.7	−32	23	6.3–14.0	Recurrent nova	—	Q	Outbursts 1890, 1902, 1920, 1944, 1966
S	09	05.1	−23	05	8.0–14.2	Mira	206.4	M	

DOUBLE STARS

	RA (h)	(m)	Dec. (°)	(′)	PA (°)	Separation (″)	Magnitudes
ζ	08	39.7	−29	34	061	52.4	4.9, 9.1
δ	08	55.5	−27	41	AB 268	23.8	4.9, 14.0
					CD 017	2.5	11.0, 11.0
ε	09	09.9	−30	22	A + BC 147	17.8	5.6, 10.5
					BC 088	0.3	10.5, 10.8
					AD 340	35.4	5.6, 13.5
κ	09	08.0	−25	52	263	2.1	4.6, 9.8

OPEN CLUSTERS

M	C	NGC	RA (h)	(m)	Dec. (°)	(′)	Diameter (′)	Magnitude	No. of stars
		2627	08	37.3	−29	57	11	8.4	60
		2658	08	43.4	−32	39	12	9.2	80

PLANETARY NEBULA

M	C	NGC	RA (h)	(m)	Dec. (°)	(′)	Diameter (″)	Magnitude	Magnitude of central star
		2818	09	16.0	−36	28	38	13.0	13.0

GALAXY

M	C	NGC	RA (h)	(m)	Dec. (°)	(′)	Magnitude	Dimensions (′)	Type
		2613	08	33.4	−22	58	10.4	7.2 × 2.1	Sb

Pyxis is not one of Ptolemy's originals, but is presumably associated with Argo.

RETICULUM

(Abbreviation: Ret)
Originally Reticulum Rhomboidalis (the Rhomboidal Net). A small but quite distinctive constellation of the far south.

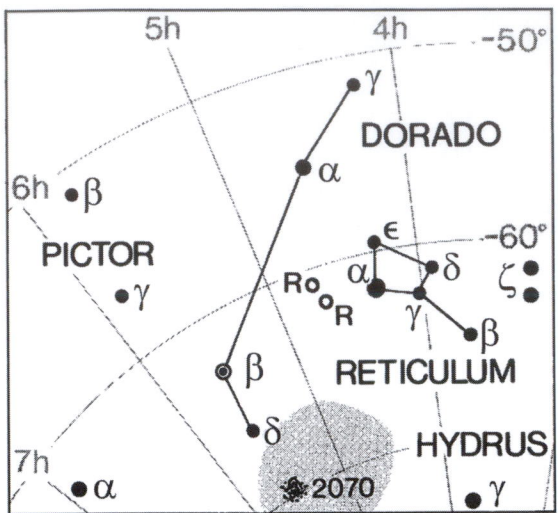

There are two stars above the fourth magnitude.

	RA			Dec.			*m*	*M*	*d* (light-years)	Spectrum
	(h)	(m)	(s)	(°)	(′)	(″)				
α	04	14.	25.4	−62	28	26	3.53	−0.17	163	G7
β	04	16	29.1	−64	48	26	3.84	1.41	100	K0

Also above magnitude 4.5. ε (4.44) and γ (4.48). Then come δ (4.56), κ (4.71) and ι (4.97).

VARIABLE STAR

	RA		Dec.		Range	Type	Period (d)	Spectrum
	(h)	(m)	(°)	(′)				
R	04	33.5	−63	02	6.5–14.0	Mira	278.3	M

GALAXIES

M	NGC	RA		Dec.		Magnitude	Dimensions (′)	Type
		(h)	(m)	(°)	(′)			
	1313	03	18.3	−66	30	9.4	8.5 × 6.6	SBd
	1559	04	17.6	−62	47	10.4	3.3 × 2.1	SBc

ε (RA, 04h 16m 29s. 02, Dec. −59° 18′ 08″) is a double. The primary is an orange subgiant, type K2, apparent magnitude 4.44, about 20% more massive than the Sun, distance 59.5 light-years; in 2000 a planet was discovered, about the size of Jupiter and rather more massive; separation 11.16 a.u., with a period of 418 days. The orbital eccentricity is low. The secondary component is a white dwarf, about 240 a.u. From the primary its magnitude is 12.5.

ζ has achieved notoriety by its association with flying saucers! There are in fact two Zetas about 9000 a.u. apart; orbital period over 1000 years.
Both are yellow dwarfs,
Data are as follows:

	ζ¹			ζ²		
RA	03h	17m	46s	03h	18m	23s
Dec.	−62	34′	31″	−62	30′	23″
Apparent magnitude	5.54			5.24		
Mass, Sun=1	0.93			0.99		
Diameter, Sun=1	0.91			0.99		
Luminosty, Sun=1	0.79			1.02		
Spectral type	G2			G1		

According to flying saucer enthusiasts ('ufologists'), ζ Reticuli is the home of intelligent aliens called Greys. In 1961, so it is said, the Greys abducted two Americans, Mr and Mrs Hill, though they did them no harm and returned them safely. Subsequently, under hypnosis, Mrs Hill drew a star map indicating the position of the Greys' home planet relative to our Sun, and another ufologist, a Miss Marjorie Fish, at once identified the star around which the Greys' planet orbted: ζ² Reticuli.

I do not feel that I can make any constructive comments. I would love to meet a Grey and issue an invitation to dinner, but somehow I have an uneasy feeling that Miss Fish's analysis of the whole episode is, well, slightly fishy!

SAGITTA

(Abbreviation: Sge)

An original constellation; small though it is, it is quite distinctive. It has been identified with Cupid's bow, and also with the arrow used by Apollo against the one-eyed Cyclops. There are two stars above the fourth magnitude.

	RA			Dec.			m	M	d (light-years)	Spectrum
	(h)	(m)	(s)	(°)	(′)	(″)				
12 γ	10	58	45.4	+19	29	32	3.51	−1.11	274	K57
7 δ	19	47	23.3	+18	32	03	3.82	−2.01	448	M2+B6

Also above magnitude 4.5: α (Alsahm) (4.39), β (4.39).

δ is a very close binary: magnitudes 3.68 and 3.80.

See chart for Aquila.

VARIABLE STARS

	RA		Dec.		Range	Type	Period (d)	Spectrum	
	(h)	(m)	(°)	(′)					
U	19	18.8	+19	37	6.6–9.2	Algol	3.38	B–K	
S	19	56.0	+16	38	5.3–6.0	Cepheid	8.38	F–G	
X	20	05.1	+20	39	7.9–8.4	Semi-regular	196	N	
WZ	20	07.6	+17	42	7.0–15.5	Recurrent nova	—	Q	Outbursts 1913, 1946, 1978, 2001

DOUBLE STAR

	RA		Dec.		PA		Separation (″)	Magnitudes
(h)	(m)	(°)	(′)	(°)				
ζ	19	49.0	+19	09	AB + C 311		8.6	5.5, 8.7
					AB 163		0.3	5.5, 6.2	Binary, 22.8 y

OPEN CLUSTER

M	C	NGC	RA		Dec.		Diameter	Magnitude	No. of stars
			(h)	(m)	(°)	(′)	(′)		
		H 20	19	53.1	+18	20	7	7.7	15

GLOBULAR CLUSTER

M	C	NGC	RA		Dec.		Diameter	Magnitude
			(h)	(m)	(°)	(′)	(′)	
71		6838	19	53.8	+18	47	7.2	8.3

PLANETARY NEBULÆ

M	C	NGC	RA		Dec.		Diameter	Magnitude	Magnitude of central star
			(h)	(m)	(°)	(′)	(″)		
		6879	20	10.5	+16	55	5	13.0	15
		IC 4997	20	20.2	+16	45	2	11.6	13 (v?)

SAGITTARIUS

(Abbreviation: Sgr)

The southernmost of the Zodiacal constellations, and not wholly visible from England. Mythologically, it has been associated with Chiron, the wise centaur who was tutor to Jason and many others; but it would certainly be more logical to associate Chiron with Centaurus, and another version states that Chiron merely invented the constellation Sagittarius to help in guiding the Argonauts in their quest of the Golden Fleece. The centre of the Milky Way lies behind the star-clouds here, and the whole area is exceptionally rich; it abounds in Messier objects. It is worth commenting that the stars lettered α and β are relatively faint. There are 16 stars above the fourth magnitude. μ is an Algol binary with a very small range (3.8 to 3.9).

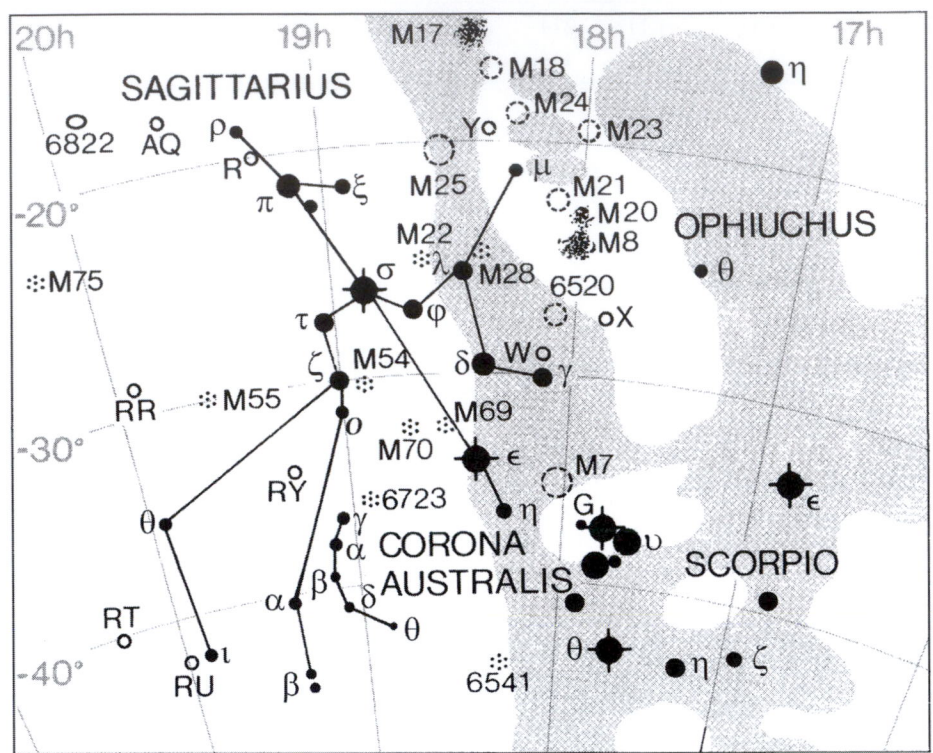

The Milky Way is very rich in Sagittarius, and the 'star-clouds' mask our view of the centre of the Galaxy.

	RA			Dec.			m	M	d (light-years)	Spectrum	
	(h)	(m)	(s)	(°)	(')	('')					
20 ε	18	24	10.4	−34	23	04	1.79	−1.44	145	B9	Kaus Australis
34 σ	18	55	15.9	−26	17	48	2.05	−2.14	224	B2	Nunki
38 ζ	19	02	36.7	−29	52	48	2.60	0.42	89	A3	Ascella
19 δ	18	20	59.6	−29	49	41	2.72	−2.14	306	K3	Kaus Media
22 λ	18	27	58.3	−25	25	17	2.82	0.95	77	K1	Kaus Borealis
41 π	19	09	45.8	−23	01	25	2.88	−2.77	440	F2	Albaldah
10 γ	18	05	48.5	−30	25	25	2.98	0.61	96	K0	Nasi
η	18	27	37.7	−36	45	41	3.10	−0.20	149	M2	Saghdar
27 φ	18	45	39.4	−26	59	27	3.17	−1.08	230	B8	Nanto
40 τ	19	06	56.4	−27	40	11	3.32	0.48	120	K1	Hekatebolos
37 ξ²	18	57	43.8	−21	06	24	3.52	−1.77	372	G8	Nergal
39 o	19	04	40.9	−21	44	29	3.76	0.61	139	K0	Manubrag
13 μ	18	13	45.8	−21	03	32	3.84	>3000		B2	Polis
44 ρ¹	19	21	40.4	−17	50	50	3.92	1.06	122	F0	Cappa
β¹	19	22	38.3	−44	27	32	3.96	−1.36	380	B9	Arkab
α	19	23	53.2	−40	36	56	3.96	0.38	1.70	B8	Rukbat

Also above 4.5: ι (4.12), μ (4.27), θ (4.37) and c (4.43).

VARIABLE STARS

	RA (h)	RA (m)	Dec. (°)	Dec. (')	Range	Type	Period (d)	Spectrum
X	17	47.6	−27	50	4.2–4.8	Cepheid	7.01	F
W	18	05.0	−29	35	4.3–5.1	Cepheid	7.59	F–G
VX	18	08.1	−22	13	6.5–12.5	Semi-regular	732	M
RS	18	17.6	−34	06	6.0–6.9	Algol	2.41	B + A
Y	18	21.4	−18	52	5.4–6.1	Cepheid	5.77	F
RV	18	27.9	−33	19	7.2–14.8	Mira	317.5	M
U	18	31.9	−19	07	6.3–7.1	Cepheid	6.74	F–G
YZ	18	49.5	−16	43	7.0–7.7	Cepheid	9.55	F–G
UX	18	54.9	−16	31	7.6–8.4	Semi-regular	100	M
ST	19	01.5	−12	46	7.6–16.0	Mira	395.1	S
T	19	16.3	−16	59	7.6–12.9	Mira	392.3	S
RY	19	16.5	−33	31	6.0–15	R Coronæ	—	Gp
R	19	16.7	−19	18	6.7–12.8	Mira	268.8	M
AQ	19	34.3	−16	22	6.6–7.7	Semi-reg.	200	N
RR	19	55.9	−29	11	5.6–14.0	Mira	334.6	M
RU	19	58.7	−41	51	6.0–13.8	Mira	240.3	M
RT	20	17.7	−39	07	6.0–14.1	Mira	305.3	M

DOUBLE STARS

	RA (h)	RA (m)	Dec. (°)	Dec. (')	PA (°)	Separation ('')	Magnitudes	
21	18	25.3	20	32	289	1.8	4.9, 7.4	
ζ	19	02.6	−29	53	320	0.3	3.3, 3.4	Binary, 21.2 y
η	18	17.6	−36	46	105	3.6	3.2, 7.8	
π	19	09.8	−21	01	AB 150	0.1	3.7, 3.7	
					AB + C 122	0.4	5.9	
β¹	19	22.6	−44	28	077	28.3	3.9, 8.0	Wide naked-eye pair with β²
κ²	20	23.9	−42	25	234	0.8	6.9, 6.9	

OPEN CLUSTERS

M	C	NGC	RA (h)	RA (m)	Dec. (°)	Dec. (')	Diameter (')	Magnitude	No. of stars	
		6469	17	52.9	−22	21	12	8.2	50	
23	6494		17	56.8	−19	01	27	5.5	150	
		6520	18	03.4	−27	54	6	7.6	60	In M 20
21		6531	18	04.6	−22	30	13	5.9	70	
		6530	18	04.8	−24	20	15	4.6	—	In M 20
		6546	18	07.2	−23	20	13	5.9	70	
		6568	18	12.8	−21	36	13	8.6	50	
	24	—	18	16.9	−18	29	90	4.5	—	Star-cloud; not a true cluster
18		6613	18	19.9	−17	08	9	6.9	20	
25	IC 4725		18	31.6	−19	15	32	4.6	30	υ Sagittarii cluster
		6645	18	32.6	−16	54	10	8.5	40	
		6716	18	54.6	−19	53	7	6.9	20	

GLOBULAR CLUSTERS

M	C	NGC	RA (h)	RA (m)	Dec. (°)	Dec. (′)	Diameter (′)	Magnitude
		6522	18	03.6	−30	02	5.6	8.6
		6544	18	07.3	−25	00	8.9	8.2
		6553	18	09.3	−25	54	8.1	8.2
		6558	18	10.3	−31	46	3.7	—
		6569	18	13.6	−31	50	5.8	8.7
		6624	18	23.7	−30	22	5.9	8.3
28		6626	18	24.5	−24	52	11.2	6.9
69		6637	18	31.4	−32	21	7.1	7.7
		6638	18	30.9	−25	30	5.0	9.2
		6652	18	35.8	−32	59	3.5	8.9
22		6656	18	36.4	−23	54	24.0	5.1
54		6715	18	55.1	−30	29	9.1	7.7
70		6681	18	43.2	−32	18	7.8	8.1
55		6809	19	40.0	−30	58	19.0	6.9
75		6864	20	06.1	−21	55	6.0	8.6

PLANETARY NEBULÆ

M	C	NGC	RA (h)	RA (m)	Dec. (°)	Dec. (′)	Diameter (″)	Magnitude	Magnitude of central star
		6567	18	13.7	−19	05	8	11.1	15.0
		6629	18	25.7	−23	12	15	11.6	12.8
		6644	18	32.6	−25	08	3	12.2	15.9
		6818	19	44.0	−14	09	17	9.9	13.0

NEBULÆ

M	C	NGC	RA (h)	RA (m)	Dec. (°)	Dec. (′)	Dimensions (′)	Magnitude of illuminating star	
20		6514	18	02.6	−23	02	29 × 27	7.6	Trifid Nebula
8	6523		18	03.8	−24	23	90 × 40	6.0	Lagoon Nebula
17	6618		18	20.8	−16	11	46 × 37	7.0	Omega Nebula

GALAXY

M	C	NGC	RA (h)	RA (m)	Dec. (°)	Dec. (′)	Magnitude	Dimensions (′)	Type	
	57	6822	19	44.9	14	48	9.3	10.2 × 9.5	Irregular	Barnard's Galaxy

τ, ζ, σ, φ, λ, ε, δ, Q and γ make up the asterism known as the Teapot.

Sagittarius is a magnificent constellation, though this cannot be properly appeciated from the latitude of Britain.

SCORPIUS

(Abbreviation: Sco)

Alternatively, and less correctly, known as Scorpio. Mythologically, it is usually associated with the scorpion which Juno caused to attack and kill the great hunter Orion. Note that Orion and Scorpius are now on opposite sides of the sky – placed there, it is said, so that the creature could do Orion no further damage!

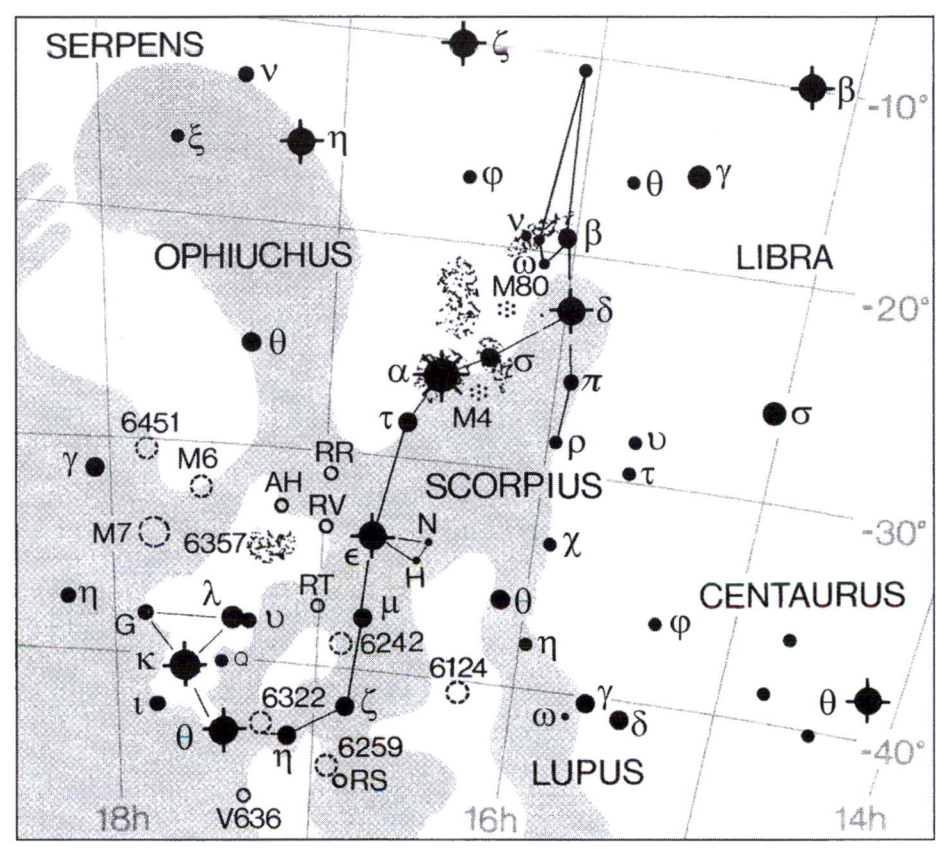

Scorpius is one of the most magnificent of all constellations, and one of the few which gives at least a vague impression of the creature it is meant to represent. It is dominated by Antares, but the whole area is exceptionally rich. The 'sting', which includes Shaula – only just below the first magnitude – is to all intents and purposes invisible from England.

	RA			Dec.			m	M	d (light-years)	Spectrum	
	(h)	(m)	(s)	(°)	(')	('')					
21 α	16	29	24.5	−26	25	55	1.06	−5.28	604	M1	Antares
15 λ	17	33	36.5	−37	06	14	1.62	−5.05	703	B1	Shaula
θ	17	37	19.1	−42	59	22	1.86	−2.75	272	F7	Sargas
7 δ	16	00	20.1	−22	17	18	2v	−3.16	400	B0	Dzuba
26 ε	16	50	10.2	−34	17	11	2.29	0.78	65	K2	Wei
κ	17	42	29.3	−39	01	48	2.39	−3.38	460	B1	Girtab
8 β	16	05	26.2	−19	48	19	2.56	−3.50	530	B9	Graffias
34 υ	17	30	45.8	−37	37	45	2.70	−3.31	520	B2	Lesath
23 τ	16	35	53.0	−28	12	58	2.82	−2.78	430	B0	Alniyat
6 P	15	58	51.1	−26	06	51	2.89	−2.85	459	B1	Vrischika
20 σ	16	21	11.3	−25	35	34	2.90	−3.86	730	B1	
ι¹	17	47	35.1	−40	07	37	2.99	−5.71	1800	F3	Apollyon
μ	16	51	52.2	−38	02	50	3.00v	−4.01	820	B1	Denchakrab

	RA			Dec.			m	M	d (light-years)	Spectrum	
	(h)	(m)	(s)	(°)	(′)	(″)					
G	17	49	51.5	−37	02	36	3.19	0.24	127	K0	Basanimus
η	17	12	09.2	−43	14	19	3.32	1.61	72	F3	
μ	16	32	20.2	−38	01	03	3.56	−2.44	510	B2	
ζ²	16	54	35.1	−42	21	39	3.62	0.30	150	K4	
5 ρ	15	56	53.1	−29	12	50	3.87	−3.62	400	B2	
9 ω¹	16	06	48.4	−20	40	09	3.91	−1.64	423	B5	
14 ν	16	11	56.7	−19	27	18	4.00	−1.63	426	B2	

Also above 4.5: ξ (4.16), H (4.18), N (4.24), Q (4.26) and ω 2 (4.31).

VARIABLE STARS

	RA		Dec.		Range	Type	Period (d)	Spectrum
	(h)	(m)	(°)	(′)				
RT	17	03.5	−36	55	7.0–16.0	Mira	449.0	M
FV	17	13.7	−32	51	7.9–8.6	Algol	5.72	B
RY	17	50.9	−33	42	7.5–8.4	Cepheid	20.31	F–G
RR	16	55.6	−30	35	5.0–12.4	Mira	279.4	M
RS	16	56.6	−45	06	6.2–13.0	Mira	320.0	M
RV	16	58.3	−33	37	6.6–7.5	Cepheid	6.06	F–G
BM	17	41.0	−32	13	6.8–8.7	Semi-regular	850	K
RU	17	42.4	−43	45	7.8–13.7	Mira	369.2	M

DOUBLE STARS

	RA		Dec.		PA	Separation (″)	Magnitudes	
	(h)	(m)	(°)	(′)	(°)			
2	15	53.6	25	20	274	2.5	4.7, 7.4	
π	15	58.9	−26	07	132	50.4	2.9, 12.1	
ξ	16	04.4	−11	22	AB 040	0.8	4.8, 5.1	
					AC 051	17.6	7.3	
β	16	05.4	−19	48	AC 021	3.6	2.6, 4.9	A is a close double
11	16	07.6	−12	45	257	3.3	5.6, 9.9	
ν	16	12.0	−19	28	AB 003	0.9	4.3, 6.8	Binary, 45.7 y
					AC 337	41.1	6.4	
12	16	12.3	−28	25	073	4.0	5.9, 7.9	
σ	16	21.2	−25	36	273	20.0	2.9, 8.5	
α	16	29.4	−26	26	274	2.7	1.2, 5.4	Binary, 878 y

OPEN CLUSTERS

M	C	NGC	RA		Dec.		Diameter (′)	Magnitude	No. of stars
			(h)	(m)	(°)	(′)			
	75	6124	16	25.6	−40	40	29	5.8	100
		6178	16	35.7	−45	38	4	7.2	12
		6192	16	40.3	−43	22	8	8.5	60
	76	6231	16	54.0	−41	48	15	2.6	−

OPEN CLUSTERS

M	C	NGC	RA (h)	(m)	Dec. (°)	(′)	Diameter (′)	Magnitude	No. of stars	
		6242	16	55.6	−39	30	9	6.4	–	
		6259	17	00.7	−44	40	10	8.0	120	
		6268	17	02.4	−39	44	6	9.5	–	
		6281	17	04.8	−37	54	8	5.4	–	
		6383	17	34.8	−32	34	5	5.5	40	(Nebulosity)
		6400	17	40.8	−36	57	8	8.8	60	
6		6405	17	40.1	−32	13	15	4.2	50	Butterfly Cluster
		6416	17	44.4	−32	21	18	5.7	40	
		6451	17	50.7	−30	13	8	8.2	80	
7		6475	17	53.9	−34	49	80	3.3	80	Ptolemy's Cluster
		6322	17	18.5	−42	57	10	6.0	30	

GLOBULAR CLUSTERS

M	C	NGC	RA (h)	(m)	Dec. (°)	(′)	Diameter (′)	Magnitude
80		6093	16	17.0	22	59	8.9	7.2
4		6121	16	23.6	26	32	26.3	5.9
		6388	17	36.3	44	44	8.7	6.8

PLANETARY NEBULÆ

M	C	NGC	RA (h)	(m)	Dec. (°)	(′)	Diameter (″)	Magnitude	Magnitude of central star	
		6153	16	31.5	−40	15	25	11.5	–	
	69	6302	17	13.7	−37	06	50	12.8	–	Bug Nebula
		6337	17	22.3	−38	29	48	–	14.7	

Antares ('the Rival of Mars') is a vast red supergiant, 16 times as massive as the Sun and as large as the orbit of Jupiter. It is very slightly variable, m −0.86 to +1.06. Its companion, magnitude 5.4, is of type B2 and though actually bluish many observers see it as green (I do!) because of the contrast with the fiery hue of the primary. Antares is one of the four first-magnitude stars close enough to the ecliptic to be occulted by the Moon and planets (the others are Aldebaran, Spica and Regulus).

λ (Shaula) is actually a triple system, made up of two B-type stars and one star still contracting toward the Main Sequence; there are three faint stars, much further out, which may or may not be genuinely associated. Shaula and υ (Lesath) in the 'sting' form a conspicuous pair, but Shaula is much the more distant and more luminous.

β (Graffias, or Akrab) is a wide, easy binary, separable with a small telescope. δ (Dzuba, or Dschubba) is a γ Cassiopeiæ variable, which in June 2000 flared up to magnitude 1.6; previously it had not been known to be variable. Spectrscopic observations showed that during the 2000 outburst Dzuba was throwing off luminous gases from its equatorial region.

18 Scorpii (RA 16h 15 17s, Dec. −08° 22′ 06″, magnitude 5.50, type G2) is a solar-type star mass 1.01 times that of the Sun, luminosity 1.08, diameter 1.02 compared with the Sun. Searches for planets have been unsuccessful.

ν (Jabbah) is a complex system. There are two main components, separated by 41″; both are binary. The brighter pair (A and B) consists of two B-type subgiants; magnitudes 4.4 and 6, separation 1.3″. The fainter (C and D) consists of two Main Sequence stars; types B8 and B9, separation 2.4″, magnitudes 6.5 and 7.9. A is itself a spectroscopic binary; the second component is of type B; separation ~0.0003″. ν Scorpii illuminates the reflection nebula IC 4592.

The whole of Scorpus is very rich. In particular, look for the globular clusters M 4 and M 80, and the two lovely open clusters, M 6 (the Butterfly) and M 7 (Ptolemy's Cluster).

SCULPTOR

(Abbreviation: Scl)

Originally Apparatus Sculptoris. There is no star above the fourth magnitude.

See chart for Phœnix.

The south galactic pole lies in Sculptor, not far from α.

The brightest star is α, RA 00h 58m 36s .4, Dec. −21° 29′ 27″, magnitude 4.3, distance 860 light-years, type B7 (it is actually an SX Arietis-type variable, but the range is only 0m. 01). Also above magnitude 4.5: β (4.38) and γ (4.41).

VARIABLE STARS

	RA (h)	RA (m)	Dec. (°)	Dec. (′)	Range	Type	Period (d)	Spectrum
Y	23	09.1	−30	08	7.5–9.0	Semi-regular	300	M
S	00	15.4	−32	03	5.5–13.6	Mira	365.3	M
R	01	27.0	−32	33	5.8–7.7	Semi-regular	370	N

DOUBLE STARS

	RA (h)	RA (m)	Dec. (°)	Dec. (′)	PA (°)	Separation (″)	Magnitudes	
δ	23	48.9	−28	08	AB 243	3.9	4.5, 11.5	
					AC 297	74.3	9.3	
ζ	00	02.3	−29	43	320	3.0	5.0, 13.0	
κ¹	00	09.3	−27	59	265	1.4	6.1, 6.2	
λ¹	00	42.7	−38	28	003	0.7	6.7, 7.0	
ε	01	45.6	−25	03	028	4.7	5.4, 8.6	Binary, 1192 y

GLOBULAR CLUSTER

M	C	NGC	RA (h)	RA (m)	Dec. (°)	Dec. (′)	Diameter (′)	Magnitude
		288	00	52.8	−26	35	13.8	8.1

GALAXIES

M	C	NGC	RA (h)	RA (m)	Dec. (°)	Dec. (′)	Magnitude	Dimensions (′)	Type
		IC 5332	23	34.5	−36	06	10.6	6.6 × 5.1	Sd
		7713	23	36.5	−37	56	11.6	4.3 × 2.0	SBd
		7755	23	47.9	−30	31	11.8	3.7 × 3.0	SBd
		7793	23	57.8	−32	35	9.1	9.1 × 6.6	Sd
		24	00	09.9	−24	58	11.5	5.5 × 1.6	Sb
	72	55	00	14.9	−39	11	8.2	32.4 × 6.5	SB
		134	00	30.4	−33	15	10.1	8.1 × 2.6	SBb
	65	253	00	47.6	−25	17	7.1	25.1 × 7.4	Scp
	70	300	00	54.9	−37	41	8.7	20.0 × 14.8	Sd
		613	01	34.3	−29	25	10.0	5.8 × 4.6	SBb

See chart for Phœnix. The Sculptor Dwarf, a member of the Local Group, lies here. There is also the Sculptor Group, the nearest group to our Local Group. Its brightest member is the barred spiral NGC 253, always called the Sculptor Galaxy. Another bright member is the irreglar galaxy NGC 55.

SCUTUM

(Abbreviation: Sct)

Originally Clypleus Sobieskii (Sobieski's Shield). There is only one star brighter than the fourth magnitude.

	RA			Dec.			m	M	d (light-years)	Spectrum	
	(h)	(m)	(s)	(°)	(')	('')					
α	18	35	12.4	−08	14	36	3.85	0.21	174	K2	Ioannina

Also above magnitude 4.5: β (4.22).

See chart for Aquila.

VARIABLE STARS

	RA		Dec.		Range	Type	Period (d)	Spectrum
	(h)	(m)	(°)	(')				
RZ	18	26.6	−09	12	7.3–8.8	Algol	15.19	B
R	18	47.5	−05	42	4.4–8.2	RV Tauri	140	G–K
S	18	50.3	−07	54	7.0–8.0	Semi-regular	148	N

OPEN CLUSTERS

M	C	NGC	RA		Dec.		Diameter (')	Magnitude	No. of stars	
			(h)	(m)	(°)	(')				
		6664	18	36.7	−08	13	16	7.8	50	EV Scuti Cluster
26		6694	18	45.2	−09	24	15	8.0	30	
		6704	18	50.9	−05	12	6	9.2	30	
11		6705	18	51.1	−06	16	14	5.8	500	Wild Duck Cluster

GLOBULAR CLUSTER

M	C	NGC	RA		Dec.		Diameter (')	Magnitude
			(h)	(m)	(°)	(')		
		6288	00	52.8	−26	35	13.8	8.1
		6712	18	53.1	−08	42	7.2	8.2

NEBULA

M	C	NGC	RA		Dec.		Dimensions (')	Magnitude of illuminating star
			(h)	(m)	(°)	(')		
		IC 1287	18	31.3	−10	50	4.4 × 3.4	5.5

δ (RA 18h 42m 16s .4, Dec. −09° 03′ 09″) is a triple star, 187 light-years away. The primary (A) is an F-type giant, pulsating regularly in a period of 4.65 hours, magnitude range 4.60 to 4.79; it is the prototype star of δ Scuti variables. It has two companions: B (magnitude 12.2, separation 15″ 2) and C (magnitude 10, separation 53″).

R Scuti is the brightest of the RV Tauri-type variables; absolute magnitude ~ −4.95, distance 3000 light-years. It is always within binocular range, though at every fourth minimum it drops to just below the eighth magnitude.

Scutum is a rich little constellation; it contains M 11, the lovely Wild Duck Cluster.

SERPENS

(Abbreviation: Ser)

A curious constellation in as much as it is divided into two parts, Caput (the Head) and Cauda (the Body). Presumably it represents the serpent with which Ophiuchus has been struggling.

Caput has stars above the fourth magnitude and Cauda three.

	RA			Dec.			m	M	d (light-years)	Spectrum	
	(h)	(m)	(s)	(°)	(′)	(″)					
Caput											
34 α	15	44	16.0	+06	25	32	2.63	0.87	73	K2	Unukalhai
32 μ	15	49	37.2	−03	25	49	3.54	0.14	156	A0	Leolepis
28 β	15	46	11.2	+15	46	11	3.65	0.29	153	A3	Chow
37 ε	15	50	48.9	+04	28	39	3.71	0.24	70	A2	Nulla Pambu
13 δ	15	34	48.1	+10	32	20	3.80	−0.24	216	F0	Qin
41 γ	15	56	27.0	+15	39	53	3.85	3.62	36	F6	Ainalhai
35 κ	15	45	44.4	+18	08	30	4.09	−1.05	350	M1	
Cauda											
58 η	18	21	08.9	−02	53	50	3.23	1.84	62	K0	Tang
61/63 θ	18	56	13.1	+04	12	13	3.4(c)	1.57	+1.77	A5+A5	Alya
55 ξ	17	37	25.2	−15	23	54	3.54	0.99	105	F0	Nehustan

Also above magnitude 4.5: λ (4.42) (Caput), o (4.2v) (Cauda), ν (4.32) (Cauda).

θ is a very wide, easy double with almost equal components. o is a very small-range δ Scuti variable.

VARIABLE STARS

	RA		Dec.		Range	Type	Period (d)	Spectrum
	(h)	(m)	(°)	(′)				
S	15	21.7	+14	19	7.0–14.1	Mira	368.6	M
τ^4	15	36.5	+15	06	7.5–8.9	Irregular (Lb)	–	M
R	15	50.7	+15	08	5.1–14.4	Mira	356.4	M
U	16	07.3	+09	56	7.8–14.7	Mira	237.9	M
d	18	27.2	+00	12	4.9–5.9	?	?	G + A

DOUBLE STARS

	RA		Dec.		PA (°)	Separation (″)	Magnitudes	
	(h)	(m)	(°)	(′)				
δ	15	34.8	+10	32	175	4.0	4.1, 5.2	Binary, 3168 y
β	15	46.2	+15	25	265	30.6	3.7, 9.9	
ν	17	20.8	−12	51	028	46.3	4.3, 8.3	
d	18	27.2	+00	12	318	3.8	5.3v, 7.6	
θ	18	56.2	+04	12	104	22.4	4.5, 4.5	Common proper motion

OPEN CLUSTERS

M	C	NGC	RA (h)	(m)	Dec. (°)	(')	Diameter (')	Magnitude	No. of stars
		6611	18	18.8	−13	47	7	6.0	In M 16
		6604	18	18.1	−12	14	2	7.0	30

GLOBULAR CLUSTER

M	C	NGC	RA (h)	(m)	Dec. (°)	(')	Diameter (')	Magnitude
5		5904	15	18.6	+02	05	17.4	5.8

NEBULA

M	C	NGC	RA (h)	(m)	Dec. (°)	(')	Dimensions (')	Magnitude star
16		6611	18	18.8	−13	47	35 × 28 Eagle Nebula	6.4

GALAXIES

M	C	NGC	RA (h)	(m)	Dec. (°)	(')	Magnitude	Dimensions (')	Type
		6118	16	21.8	−02	17	12.3	4.7 × 2.3	Sb

R Serpentis is a bright Mira variable, easy to find because it lies directly between β and γ. As the period is so close to a year, there are several consecutive years when maxima are difficult to follow.

Note the splendid globular cluster M 5.

SEXTANS

(Abbreviation: Sxt)

A very obscure southern constellation. There is no star as bright as the fourth magnitude.

Brightest star: α, RA 10h 07m 56s .3, Dec. −00° 22′ 18″, m = 4.45, M= −0.25, d = 287 light-years, type A0.
Next come β (5.1v) and δ (5.10).
See chart for Hydra.

BRIGHTEST STAR

Star	RA			Dec.			Magnitude	Absolute magnitude	Spectrum	Distance (light-years)
	(h)	(m)	(s)	(°)	(′)	(″)				
15 α	10	07	56.2	−00	22	18	4.49	−1.1	B5	326

VARIABLE STAR

Star	RA		Dec.		Range	Type	Period (d)	Spectrum
	(h)	(m)	(°)	(′)				
S	10	34.9	−00	20	8.2–13.5	Mira	261.0	M

DOUBLE STAR

Star	RA		Dec.		PA (°)	Separation (″)	Magnitudes	
	(h)	(m)	(°)	(′)				
γ	09	52.5	−08	06	AB 067	0.6	5.5, 6.1	Binary, 75.6 y
					AC 325	35.8	12.0	

GALAXIES

M	C	NGC	RA		Dec.		Magnitude	Dimensions (′)	Type	
			(h)	(m)	(°)	(′)				
		2967	09	42.1	+00	20	11.6	3.0 × 2.9	Sc	
	53	3115	10	05.2	−07	43	9.1	8.3 × 3.2	E6	Spindle Galaxy
		3166	10	13.8	+03	26	10.6	5.2 × 2.7	SBa	
		3169	10	14.2	+03	28	10.4	4.8 × 3.2	Sb	

The most notable object in Sextans is the Spindle Galaxy, NGC 3115 (C 53), discovered by William Herschel in 1787. It is 32 million light-years away. It is thought to contain a supermassive black hole.

TAURUS

(Abbreviation: Tau)

Taurus is one of the brightest of the Zodiacal constellations. Mythologically, it has been said to represent the bull into which Jupiter transformed himself when he wished to carry off Europa, daughter of the King of Crete. Taurus includes the reddish first-magnitude star Aldebaran, and also the two most famous open clusters in the sky: the Pleiades and the Hyades. Altogether there are 16 stars above the fourth magnitude.

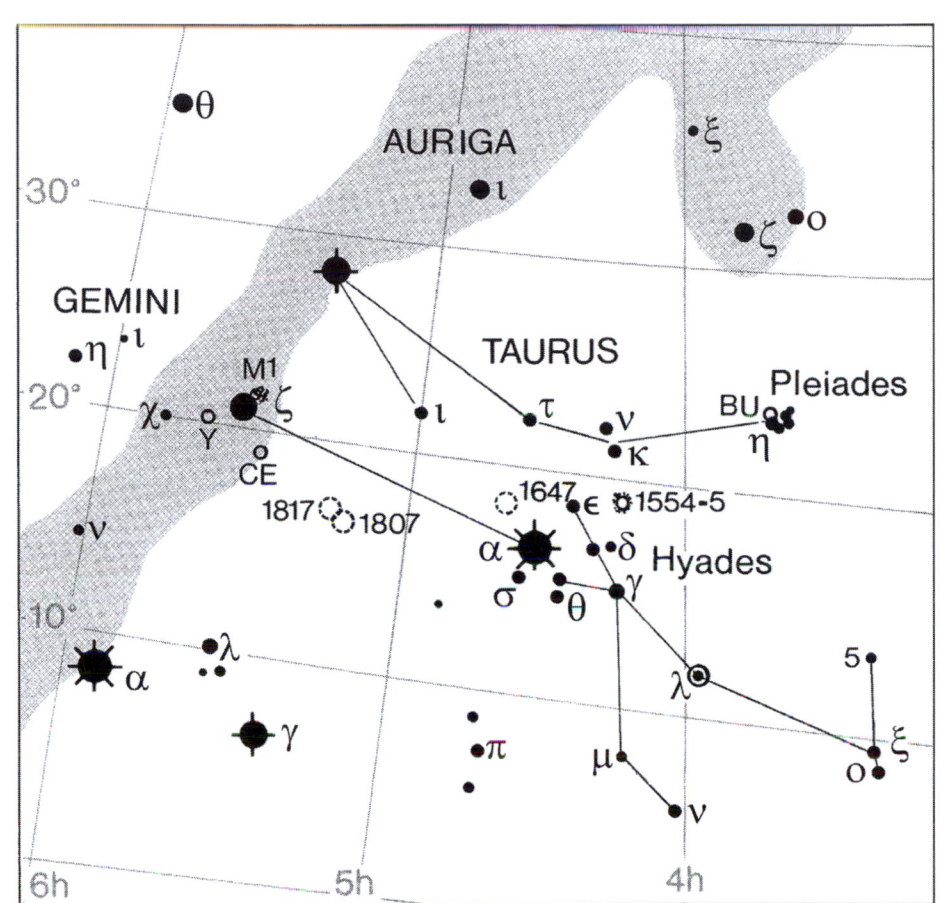

	RA			Dec.			m	M	d (light-years)	Spectrum	
	(h)	(m)	(s)	(°)	(')	('')					
87 α	04	35	55.2	+16	30	35	0.87	−0.63	65	K5	Aldebaran
112 β	05	26	17.5	+28	36	28	1.65	−1.37	131	B7	Alnath
25 η	03	47	29.1	+24	06	19	2.85	−2.41	368	B7	Alcyone
123 ζ	05	37	38.7	+21	08	33.3	2.97	−2.56	417	B4	Alheka
78 θ 2	04	28	39.7	+15	52	15	3.40	0.10	149	A7	Phæsyla
35 λ	04	00	40.8	+12	29	25	3.4v	−1.87	370	B3	Althor
74 ε	04	28	36.9	+19	10	50	3.53	0.15	155	K0	Ain
1 o	03	24	48.8	+09	01	45	3.61	−0.45	211	G8	Atirsague
27	03	49	09.7	+24	03	13	3.62	−1.72	380	B8	Atlas
54 γ	04	19	47.5	+15	37	40	3.65	0.28	154	G8	Hyadum
17	03	44	52.5	+24	08	48	3.72	−1.56	370	B6	Electra
2 ξ	03	27	10.1	+09	43	58	3.73	−0.44	222	B9	Ushakaron
61 δ¹	04	22	56.0	+17	32	33	3.77	0.41	153	G5	
77 θ¹	04	28	34.4	+15	57	44	3.84	0.42	158	G7	Phæo
20	03	45	49.6	+24	22	04	3.87	−1.34	360	B8	Maia
38 ν	04	03	09.4	+05	59	22	3.91	0.92	129	A1	Furibundus

Also above magnitude 4.5: f (4.14), 23 (Merope) (4.14), 65 κ 1 (4.21), 88 d (4.25), 49 Kattupothu (4.27), c¹ (4.27), τ (4.27), υ (4.28), 10 (4.29), q (Taygete) (4.30), δ² (4.30), 119 (4.32), 37 (4.36), 71 (Polyxo) (4.48).

VARIABLE STARS

	RA (h)	(m)	Dec. (°)	(′)	Range	Type	Period (d)	Spectrum	
BU	03	49.2	+24	08	4.8–5.5	Irregular	—	Bp	Pleione
λ	04	00.7	+12	29	3.3–3.8	Algol	3.95	B + A	
T	04	22.0	+19	32	8.4–13.5	T Tauri	Irreg.	G–K	
R	04	28.3	+10	10	7.6–14.7	Mira	323.7	M	
HU	04	38.3	+20	41	5.9–6.7	Algol	2.06	A	
ST	05	45.1	+13	35	7.8–8.6	W Virginis	4.03	F–G	
TU	05	45.2	+24	25	5.9–8.6	Semi-regular	190	N	
SU	05	49.1	+19	04	9.1–16.0	R Coronæ	—	G0p	

DOUBLE STARS

	RA (h)	(m)	Dec. (°)	(′)	PA (°)	Separation (″) sec	Magnitudes	
φ	04	20.4	+27	21	250	52.1	5.0, 8.4	
χ	04	22.6	+25	38	024	19.4	5.5, 7.6	
66	04	23.9	+09	28	265	0.1	5.8, 5.9	Binary, 51.6 y
κ + 67	04	25.4	+22	18	173	339	4.2, 5.3	
θ	04	28.7	+15	32	346	337.4	3.4, 3.8	
σ	04	39.3	+15	55	193	431.2	4.7, 5.1	
126	05	41.3	+16	32	238	0.3	5.3, 5.9	

OPEN CLUSTERS

M	C	NGC	RA (h)	(m)	Dec. (°)	(′)	Diameter (′)	Magnitude	No. of stars
45		1432/5	03	47.0	+24	07	110	1.2	300 + Pleiades
	41	—	04	27	+16		330	1	200 + Hyades
		1647	04	46.0	+19	04	45	6.4	200
		1746	05	03.6	+23	49	42	6.1	20
		1807	05	10.7	+16	32	17	7.0	20 Asterism?
		1817	05	12.1	+16	42	16	7.7	60

PLANETARY NEBULA

M	C	NGC	RA (h)	(m)	Dec. (°)	(′)	Diameter (″)	Magnitude	Magnitude of central star
		1514	04	09.2	+30	47	114	10	9.4

NEBULÆ

M	C	NGC	RA (h)	(m)	Dec. (°)	(′)	Dimensions (′)	Magnitude of illuminating star	
		1554–5	04	21.8	+19	32	variable	9v	Hind's Variable Nebula (T Tauri)
1		1952	05	34.5	+22	01	6 × 4	16	Crab Nebula: SNR

β (Alnath) was formerly included in Auriga, as γ Aurigæ.

λ is a bright Algol eclipsing binary, 370 light-years away. The primary is a B-type Main Sequence star 400 times as luminous as the Sun and diameter almost 7 times that of the Sun; the companion is of type A, 0.95 times as luminous as the Sun and diameter 5.5 times that of the Sun. Their separation is ~0.1.a.u.

TELESCOPIUM

(Abbreviation. Tel)

A dim southern constellation, made even dimmer by the fact that the International Astronomical Union has transferred several of its stars. β — η Sagittarii, γ is now G Scorpii, 0 is d Ophiuchi and σ is HD168905 Coronæ Australis.

There are only two stars above magnitude 4.5:

	RA			Dec.			m	M	d (light-years)	Spectrum
	(h)	(m)	(s)	(°)	(′)	(″)				
α	18	26	58.4	−45	58	06	3.49	−0.93	249.	B3
ζ	18	28	49.7	−49	04	12	4.10	1.14	127	G8

See chart for Ara.

VARIABLE STARS

	RA		Dec.		Range	Type	Period (d)	Spectrum
	(h)	(m)	(°)	(′)				
BL	19	06.6	−51	25	7.7–9.8	Eclipsing	778.1	F + M
RR	20	04.2	−55	43	6.5–16.5	Z Andromedæ	—	F5p
R	20	14.7	−46	58	7.6–14.8	Mira	461.9	M

PLANETARY NEBULA

M	C	NGC	RA		Dec.		Diameter (″)	Magnitude
			(h)	(m)	(°)	(′)		
		IC 4699	18	18.5	−45	59	10	11.9

TRIANGULUM

(Abbreviation: Tri)

Triangulum is a small northern constellation. Its three main stars really do make up a triangle. See chart for Andromeda.

	RA (h)	(m)	(s)	Dec. (°)	(′)	(″)	m	M	d (light–years)	Spectrum	
4 β	02	09	32.5	+34	59	15	3.00	0.09	124	A5	Deltotron
2 α	01	53	04.9	+29	34	46	3.42	1.95	64	F6	Rasalmothallah
9 γ	02	17	18.8	+33	50	50	4.03	1.24	118	A1	

Next comes δ (4.84).

VARIABLE STAR

	RA (h)	(m)	Dec. (°)	(′)	Range	Type	Period (d)	Spectrum
R	02	37.0	+34	16	5.4–12.6	Mira	266.5	M

DOUBLE STARS

	RA (h)	(m)	Dec. (°)	(′)	PA (°)	Separation (″)	Magnitudes
6	02	12.4	+30	18	071	3.9	5.3, 6.9
ι	02	15.9	+33	21	240	2.3	5.4, 7.0

GALAXIES

M	C	NGC	RA (h)	(m)	Dec. (°)	(′)	Magnitude	Dimensions (′)	Type	
33		598	01	33.9	+30	39	5.7	62 × 39	Sc	Pinwheel Galaxy
		925	02	27.3	+33	35	10.0	9.8 × 6.0	SBc	

6 Trianguli, magnitude 4.9, was the brightest star in the rejected constellaton Triagulum Minor (the other members were 10 and 12 Trianguli). It is an easy double; the primary is a G-type giant, the secondary an F5-type dwarf; each component is itelf a close binary. The G-type giant is a small-range rotating variable, and has a variable star designation: TZ Trianguli.

M 33, the face-on Pinwheel Galaxy, is on the fringe of naked-eye visibility.

TRIANGULUM AUSTRALE

(Abbreviation TrA)

This constellation also is appropriately named. There are four stars above the fourth magntude.

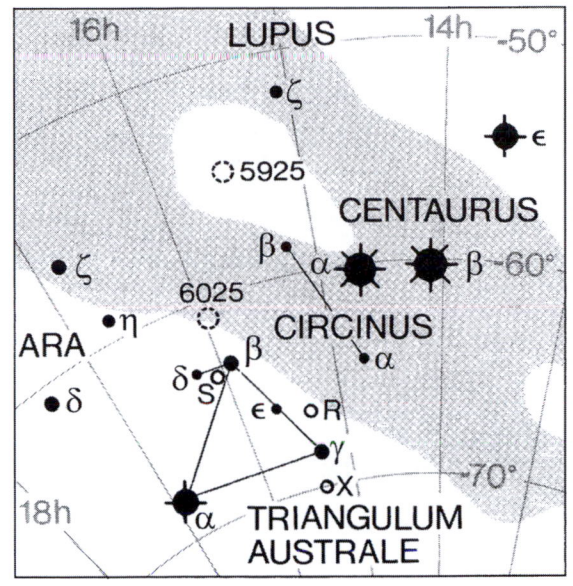

	RA			Dec.			m	M	d (light-years)	Spectrum	
	(h)	(m)	(s)	(°)	(')	('')					
α	16	48	39.9	−69	01	40	1.91	3.62	415	K2	Atria
β	15	55	08.8	−63	25	47	2.83	2.38	40	F2	
γ	15	18	54.7	−68	40	46	2.87	−0.87	183	A1	
δ	16	15	26.3	−63	41	08	3.86	−2.54	620	G5	

Also above magnitude 4.5: ε (4.11).

VARIABLE STARS

	RA		Dec.		Range	Type	Period (d)	Spectrum
	(h)	(m)	(°)	(')				
X	15	14.3	−70	05	8.1–9.1	Irregular	–	N
R	15	19.8	−66	30	6.4–6.9	Cepheid	3.39	F–G
S	16	01.2	−63	47	6.1–6.8	Cepheid	6.32	F
U	16	07.3	−62	55	7.5–8.3	Cepheid	2.57	F

DOUBLE STARS

	RA		Dec.		PA (°)	Separation ('')	Magnitudes
	(h)	(m)	(°)	(')			
ε	15	36.7	−66	19	218	83.2	4.1, 9.5
ι	16	28.0	−64	03	016	29.6	5.3, 10.3

OPEN CLUSTER

M	C	NGC	RA		Dec.		Diameter (')	Magnitude	No. of stars
			(h)	(m)	(°)	(')			
	95	6025	16	03.7	−60	30	12	5.1	60

α is obviously orange-red. The intensely red carbon star X, near γ,
is a fine sight in binoculars.

TUCANA

(Abbreviation: Tuc)

The dimmest of the Southern Birds, but graced by the presence of two bright clusters: 47 Tucanæ, inferior only to ω Centauri, and NGC 362. There are only two single stars above the fourth magnitude, but the combined magnitude of β^1 and β^2 is 3.7.

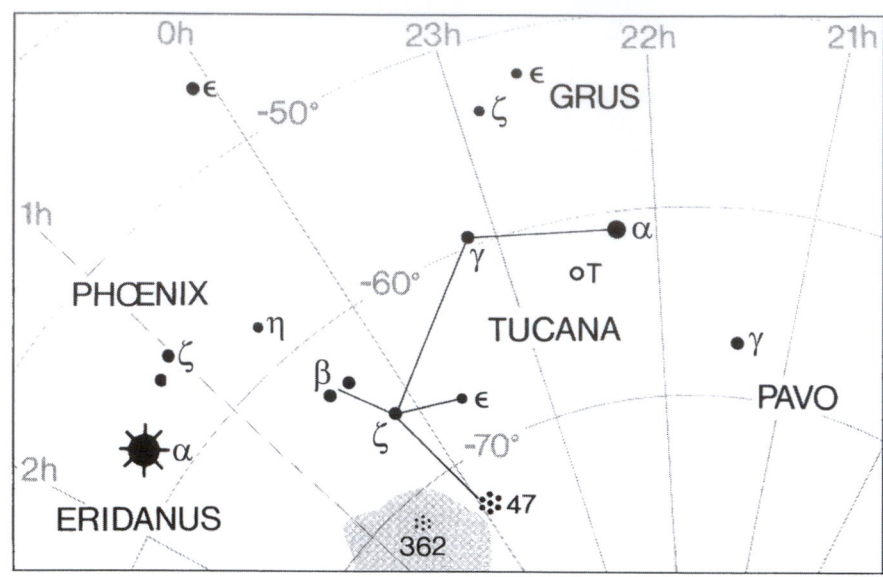

	RA			Dec.			m	M	d (light-years)	Spectrum
	(h)	(m)	(s)	(°)	(′)	(″)				
α	22	18	30.2	−60	15	34	2.87	−1.05*	199	K3
γ	23	17	25.8	−58	14	09	3.99	2.28	72	F1

Also above magnitude 4.5: ζ (4.23), κ (4.25), β^1 (4.36) and ε (4.49). Then follow δ (4.51) and β^2 (4.53).

VARIABLE STARS

	RA		Dec.		Range	Type	Period (d)	Spectrum
	(h)	(m)	(°)	(′)				
T	22	40.6	−61	33	7.7–13.8	Mira	250.8	M
S	00	23.1	−61	40	8.2–15.0	Mira	240.7	M
U	00	57.2	−75	00	8.0–14.8	Mira	259.5	M

DOUBLE STARS

	RA		Dec.		PA (°)	Separation (″)	Magnitudes	
	(h)	(m)	(°)	(′)				
δ	22	27.3	−64	58	282	6.9	4.5, 9.8	
β^1	00	32.7	−62	58	169	27.1	4.4, 4.8	
β^2	00	33.6	−62	58	295	0.6	4.8, 6.0	Binary, 44.4 y
κ	01	15.8	−68	53	336	5.4	5.1, 7.3	

GLOBULAR CLUSTERS

M	C	NGC	RA		Dec.		Diameter (′)	Magnitude	
			(h)	(m)	(°)	(′)			
	106	104	00	24.1	−72	05	30.9	4.0	47 Tucanæ
	104	362	01	03.2	−70	51	12.9	6.6	

GALAXY

	RA		Dec.		Magnitude	Dimensions (')
	(h)	(m)	(°)	(')		
Small Cloud of Magellan	00	53	−72	50	2.3	280 × 160

β seems to be a system of six stars. β^1 is a bluish B-type star, while β^2 is of type A; their real separation is about 1100 a.u. β^3 is a much fainter binary, made up of two A-type Main Sequence stars of magnitude 5.8 and 6.0 at least 23,000 a.u. from the main pair – almost a quarter of a light-year.

ζ is a relatively close star, not unlike the Sun; type F9.5 and, relative to the Sun, luminosity 1.3, mass 0.99, diameter 0.9. It seems to have a débris disc at just over 2 a.u. from the star.

URSA MAJOR

(Abbreviation: UMa)
The most famous of all northern constellations; circumpolar in England and the northern United States. Mythologically, it represents Callisto, daughter of King Lycaon of Arcadia. Her beauty surpassed that of Juno, which so infuriated the goddess that she ill-naturedly changed Callisto into a bear. Years later, Arcas, Callisto's

son, found the bear while out hunting, and was about to shoot it when Jupiter intervened, swinging both Callisto and Arcas – also transformed into a bear – up to the sky: Callisto as Ursa Major, Arcas as Ursa Minor. The sudden jolt explains why both bears have tails stretched out to decidedly un-ursine length!

The two globular clusters haave already been described. 47 Tucanæ seems to be almost silhouetted against the southernmost part of the Small Magellanic Cloud, which does extend into Tucana. 47 Tucanæ has a never-used designation, ξ Tucanæ. Of course, they are not in Messier's catalogue, but are Caldwell 106 and 104.

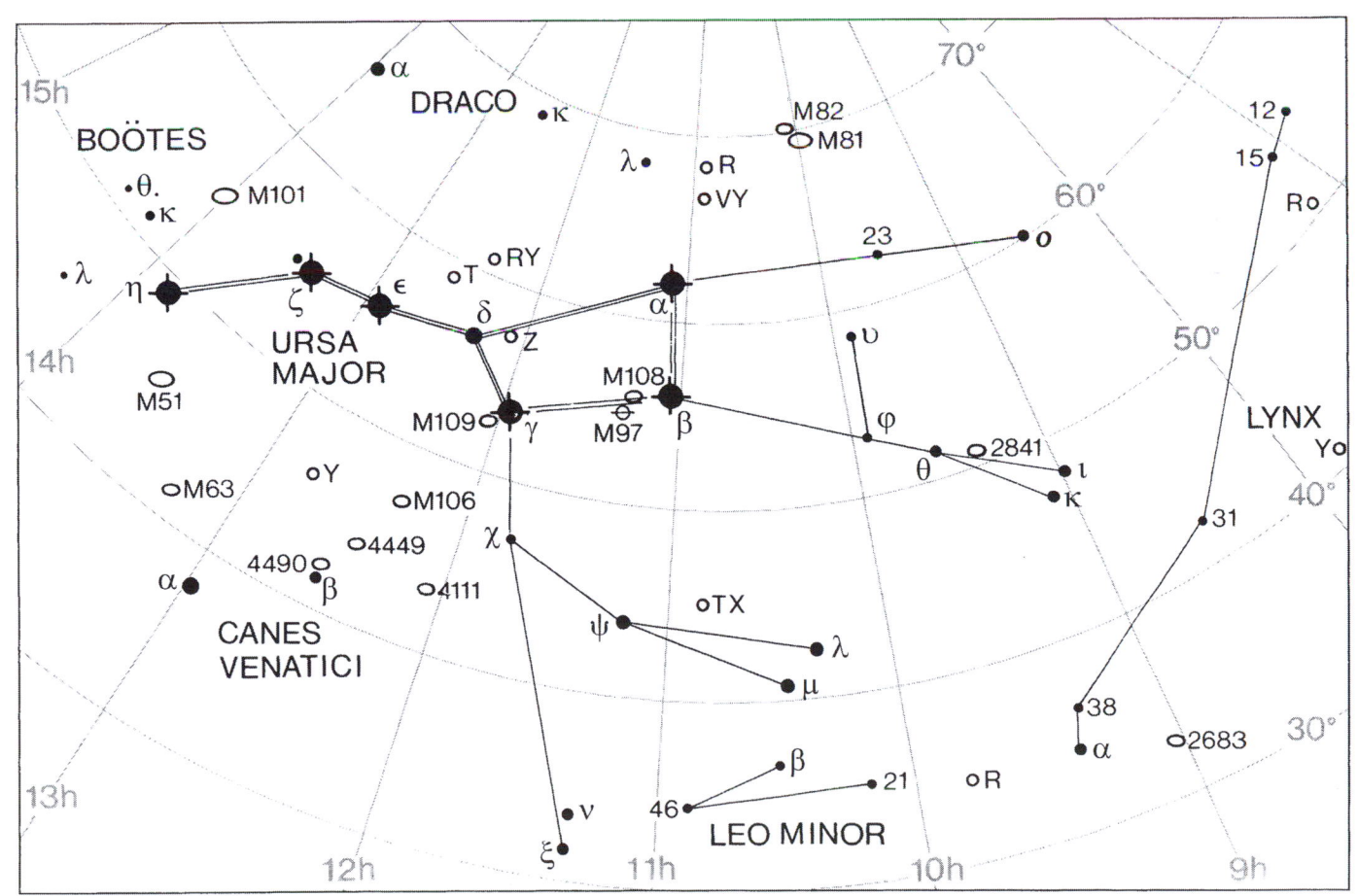

The seven main stars of Ursa Major are often called the Plough; sometimes King Charles' Wain and, in America, the Big Dipper. One of the Plough stars is Mizar, the most celebrated naked-eye double in the sky since it makes a pair with Alcor; Mizar is itself a compound system, and the two main components are easily separable with a small telescope.

Ursa Major contains 19 stars above the fourth magnitude. ξ Ursæ Majoris was the first binary to have its orbit accurately computed (by F. Savary, in 1827).

Five of the stars in the 'Plough' pattern make up a moving cluster; the two exceptions are α and η. β and α are known as the Pointers, because they show the way to the Pole Star.

	RA			Dec.			m	M	d (light-years)	Spectrum	
	(h)	(m)	(s)	(°)	(')	('')					
77 ε	12	54	01.6	+55	57	35	1.76	−0.21	81	AO	Alioth
50 α	11	03	43.8	+61	45	04	1.81	−1.08	124	F7	Dubhe
85 η	13	47	32.4	+49	18	48	1.85	−0.60	101	B3	Alkaid
79 ζ	13	23	55.4	+54	55	31	2.23*	0.33	78	A2	Mizar
48 β	11	01	50.3	+56	22	56	2.34	0.41	79	A1	Merak
64 γ	11	53	49.7	+53	41	41	2.41	0.36	84	A0	Phad
52 ξ	11	09	39.9	+44	29	55	3.00	−0.27	147	K1	Ta Tsun
34 μ	10	22	19.8	+41	29	58	3.06	−1.35	249	M0	Tania Aust
9 ι	08	59	12.8	+48	02	33	3.12	2.29	48	A7	Talita
25 θ	09	32	52.3	+51	40	43	3.17	2.52	44	F6	Sarir
69 δ	12	15	25.5	+57	01	57	3.32	1.33	81	A3	Megrez
1 o	08	30	16.0	+60	43	06	3.35	−0.40	184	G4	Muscika
33 λ	10	17	05.9	+42	54	52	3.45	0.38	134	A2	Tania Borealis
54 ν	11	18	28.8	+33	05	39	3.49	−2.07	420	K3	Alula Bor
12 κ	09	03	37.6	+47	09	24	3.57	−1.99	420	A1	Talitha
23 h	09	31	31.6	+63	03	43	3.65	1.83	75	F0	
63 χ	11	46	03.1	+47	46	46	3.69	−0.20	196	K0	Alkafzah
29 υ	09	50	59.7	+59	02	21	3.78	1.04	115	F0	
53 ξ	11	18	11.2	+31	31	51	3.79		27	G0	Alula
80 g	13	25	13.4	+54	59	17	3.99	2.01	81	A5	Alcor

Also above magnitude 4.5: 15 (4.46), 26 (4.47).

Gamma (Phad) has also been called Phekda and Phecda; η (Alkaid) also has an alternative name, Benetnasch.

There are 20 stars above the fourth magnitude.

DOUBLE STARS

	RA		Dec.		PA (°)	Separation ('')	Magnitudes	
	(h)	(m)	(°)	(')				
ι	08	59.2	+48	02	100	1.8	3.1, 10.2	Binary, 818 y
κ	09	03.6	+47	09	258	0.1	4.2, 4.4	Binary, 70 y
σ²	09	10.4	+67	08	000	3.4	4.8, 8.2	Binary, 1067 y
φ	09	52.1	+54	04	188	0.2	5.3, 5.4	Binary, 105.5 y
α	11	03.7	+61	45	283	0.7	1.9, 4.8	Binary, 44.7 y
ξ	11	18.2	+31	32	060	1.6	4.3, 4.8	Binary, 59.8 y
ν	11	18.5	+33	06	147	7.2	3.5, 9.9	
78	13	00.7	+56	22	057	1.5	5.0, 7.4	Binary, 116 y
ζ	13	23.9	+54	56	AB 152	14.4	2.3, 4.0	
					AC 071	708.7	2.1, 4.0	

VARIABLE STARS

	RA		Dec.		Range	Type	Period (d)	Spectrum
	(h)	(m)	(°)	(')				
X	08	40.8	+50	08	8.0–14.8	Mira	248.8	M
W	09	43.8	+55	57	7.9 8.6	W UMa	0.33	F + F
R	10	44.6	+68	47	6.7–13.4	Mira	301.7	M
TX	10	45.3	+45	34	7.1–8.8	Algol	3.06	B + F
VY	10	45.7	+67	25	5.9–6.5	Irregular	–	N
VW	10	59.0	+69	59	6.8–7.7	Semi-regular	125	M
ST	11	27.8	+45	11	7.7–9.5	Semi-regular	81	M
CF	11	53.0	+37	43	8.5–12	Flare	–	B
Z	11	56.5	+57	52	6.8–9.1	Semi-regular	196	M
RY	12	20.5	+61	19	6.7–8.5	Semi-regular	311	M
T	12	36.4	+59	29	6.6–13.4	Mira	256.5	M
S	12	43.9	+61	06	7.0–12.4	Mira	225.0	S

PLANETARY NEBULA

M	NGC	RA		Dec.		Diameter (″)	Magnitude	Magnitude of central star	
		(h)	(m)	(°)	(')				
97	3587	11	14.8	+55	01	194	12.0	15.9	Owl Nebula

GALAXIES

M	C	NGC	RA		Dec.		Magnitude	Dimensions (')	Type	
			(h)	(m)	(°)	(')				
		2681	08	53.5	+51	19	10.3	3.8 × 3.5	Sa	
		2685	08	55.6	+58	44	11.0	5.2 × 3.0	Sbp	
		2768	09	11.6	+60	02	10.0	6.3 × 2.8	E5	
		2787	09	19.3	+69	12	10.8	3.4 × 2.3	Sap	
		2841	09	22.0	+50	58	9.3	8.1 × 3.8	Sb	
		2976	09	47.3	+67	55	10.1	4.9 × 2.5	Scp	
		2985	09	50.4	+72	17	10.5	4.5 × 3.4	Sb	
81		3031	09	55.6	+69	04	6.9	25.7 × 14.1	Sb	Bode's Nebula
82		3034	09	55.8	+69	41	8.4	11.2 × 4.6	Pec.	
		3077	10	03.3	+68	44	9.8	4.6 × 3.6	E2p	
		3079	10	02.0	+55	41	10.6	7.6 × 1.7	Sb	
		3184	10	18.3	+41	25	9.7	6.9 × 6.8	Sc	
		3198	10	19.9	+45	33	10.4	8.3 × 3.7	Sc	
		3310	10	38.7	+53	30	10.9	3.6 × 3.0	SBc	
108		3556	11	08.7	+55	57	10.0	8 × 1	Sc	
		3359	10	46.6	+63	13	10.4	6.8 × 4.3	SBc	
		3610	11	18.4	+58	47	10.7	3.2 × 2.5	E2p	
		3631	11	21.0	+53	10	10.4	4.6 × 4.1	Sc	
		3675	11	26.1	+43	35	10.9	5.9 × 3.2	Sb	
		3687	11	28.0	+29	31	12.6	2.0 × 2.0	Sb	
		3718	11	32.6	+53	04	10.5	8.7 × 4.5	SBap	
		3726	11	33.3	+47	02	10.4	6.0 × 4.5	Sc	
		3877	11	46.1	+47	30	11.6	5.4 × 1.5	Sb	
		3898	11	49.2	+56	05	10.8	4.4 × 2.6	Sb	
		3945	11	53.2	+60	41	10.6	5.5 × 3.6	SBa	
		3949	11	53.7	+47	52	11.0	3.0 × 1.8	Sb	

GALAXIES

M	C	NGC	RA (h)	(m)	Dec. (°)	(′)	Magnitude	Dimensions (′)	Type	
		3953	11	53.8	+52	20	10.1	6.6 × 3.6	Sb	
109		3992	11	55.0	+53	39	95	7 × 0.4	Sb	
		3998	11	57.9	+55	27	10.6	3.1 × 2.5	E2p	
		4026	11	59.4	+50	58	11.7	5.1 × 1.4	S0	
		4036	12	01.4	+61	54	10.6	4.5 × 2.0	E6	
		4041	12	02.2	+62	08	11.1	2.8 × 2.7	Sc	
		4051	12	03.2	+44	32	10.3	5.0 × 4.0	Sc	
		4062	12	04.1	+31	54	11.2	4.3 × 2.0	Sb	
		4088	12	05.6	+50	33	10.5	5.8 × 2.5	Sc	
		4096	12	06.0	+47	29	10.6	6.5 × 2.0	Sc	
		4100	12	06.2	+49	35	11.5	5.2 × 1.9	Sb	
		4605	12	40.0	+61	37	11.0	5.5 × 2.3	SBcp	
		5308	13	47.0	+60	58	11.3	3.5 × 0.8	S0	
		5322	13	49.3	+60	12	10.0	5.5 × 3.9	E2	
101		5457	14	03.2	+54	21	7.7	26.9 × 26.3	Sc	Pinwheel
		5475	14	05.2	+55	45	12.4	2.2 × 0.6	Sa	

There is a good colour contrast between the red μ (Tania Australis) and its orange neigbour ξ; well seen in binoculars. The galaxies M 81 and M 82 have already been described; also the Owl Planetary Nebula (M 97).

From a site near the equator – outside Singapore, for example – it is fascinating to look upward on a dark night and see the Southern Cross in one side of the sky and the Great Bear in the other!

URSA MINOR

(Abbreviation: UMi)

The north polar constellation. Mythologically, it represents Arcas, son of Callisto (see Ursa Major).

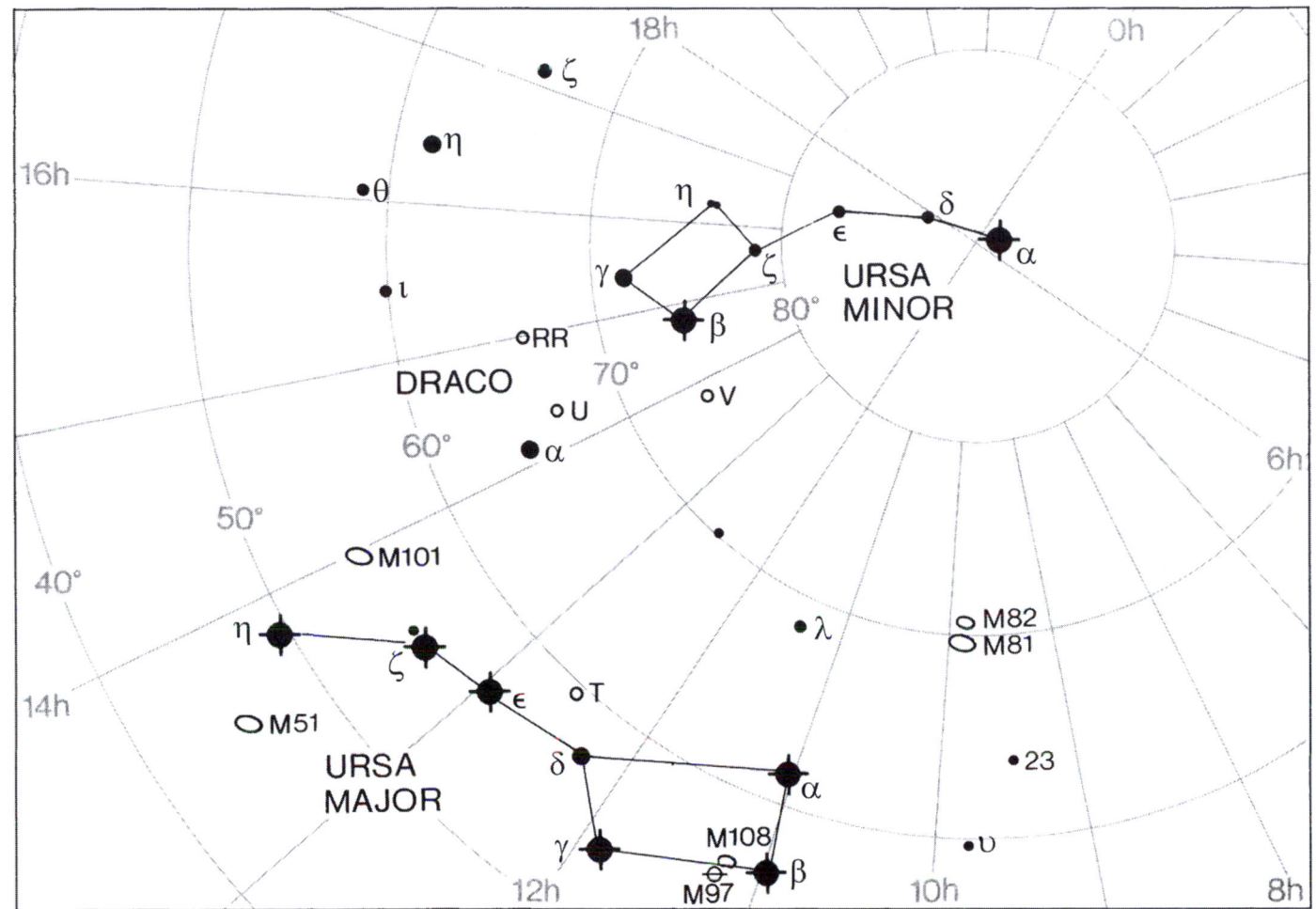

There are three stars above the fourth magnitude:

	RA			Dec.			m	M	d (light-years)	Spectrum	
	(h)	(m)	(s)	(°)	(′)	(″)					
1 α	02	31	47.1	+89	15	51	2.0v	−3.64	431	F7	Polaris
7 β	14	50	42.4	+74	09	20	2.07	−0.87	126	K4	Kocab
13 γ	15	20	43.8	+71	50	02	3.00	−2.84	480	A3	Pherkad Major

Also in the main pattern:

22 ε	16	45	58.2	+82	02	14	4.21v	−0.92	346	G5	Urodelus
16 ζ	15	44	03.5	+77	47	40	4.20	−1.02	376	A3	Alifa
23 δ	17	32	12.9	+86	35	11	4.35	0.61	183	A1	Yildun
21 η	16	17	30.5	+75	45	17	4.95	2.58	97	F5	Alasco

VARIABLE STARS

	RA (h)	RA (m)	Dec. (°)	Dec. (′)	Range	Type	Period (d)	Spectrum
T	13	34.7	+73	26	8.1–15.0	Mira	313.9	M
V	13	38.7	+74	19	7.4–8.8	Semi-regular	72	M
U	14	17.3	+66	48	7.4–12.7	Mira	326.5	M
RR	14	57.6	+65	56	6.0–6.5	Semi-regular?	40?	M
S	15	29.6	+78	38	7.7–12.9	Mira	326.2	N
R	16	30.0	+72	17	8.8–11.0	Semi-regular	324	M

DOUBLE STAR

	RA (h)	RA m	Dec. (°)	Dec. (′)	PA (°)	Separation (″)	Magnitudes
α	02	31.8	+89	16	218	18.4	2.0, 9.0

The Ursa Minor dwarf galaxy is a member of the Local Group; RA 15h 09m 09s, Dec. +67° 13′ 21″. It is of apparent magnitude 11.9; dimensions 30′×19′. It seems to be made up chiefly of old stars; with no evidence of star formation going on now. It was discovered in 1954 by A. G. Wilson at the Lowell Observatory.

Polaris will be at its closest to the pole on 24 March 2100; its declination then will be +89° 32′ 51″. It is a Cepheid variable; the range is from magnitude1.92 to 2.07, period 3.9698 days. The range decreased during the twentieth century, but now seems to be altering very little, if at all. Its companion, a ninth-magnitude Main Sequence star, is an easy object.

Kocab and Pherkad Major are often called 'the Guardians of the Pole'.

VELA

(Abbreviation: Vel)

The Sails of Argo Navis, the Ship Argo. There are 13 stars above the fourth magnitude.

	RA (h)	RA (m)	RA (s)	Dec. (°)	Dec. (′)	Dec. (″)	m	M	d (light-years)	Spectrum	
γ²	08	09	36.0	−47	20	12	1.75	−5.31	840	WC8+09	Regor
δ	08	44	42.2	−54	42	31	1.93	−0.01	80	A1	Koo She
λ	09	07	59.8	−43	25	57	2.23	−3.99	570	K4	Suhail
κ	09	22	06.8	−55	00	39	2.47	−3.62	540	B2	Markab
μ	10	46	46.1	−49	25	13	2.69	−0.06	116	O5	Al Haram
N	09	31	13.4	−57	02	04	3.16	−1.15	238	K5	Marut
φ	09	56	51.8	−54	34	04	3.52	−5.34	1929	B5	Tseen Ke
o	08	40	17.6	−52	55	19	3.60	−2.31	495	B3	Xestus
ψ	09	30	42.1	−40	28	01	3.60	2.26	60	F2	
c	09	04	09.3	−47	05	52	3.75	−1.13	309	K2	
b	08	40	37.6	−46	38	56	3.77	−6.12	3100	F3	
p	10	37	18.3	−48	13	32	3.84	1.72	86	A3	
a	08	46	01.7	−46	02	30	3.87	−4.52	1550	A1	

Also above magnitude 4.5: d (4.05), e (4.11) × (4.29), M (4.34), i (4.37), w (4.45), J (4.50).

δ and κ make up the 'False Cross' with ε and ι Carinæ. See chart for Carina.

VARIABLE STARS

	RA (h)	RA (m)	Dec. (°)	Dec. (')	Range	Type	Period (d)	Spectrum
AH	08	12.0	−46	39	5.5–5.9	Cepheid	4.23	F
Al	08	14.1	−44	34	6.4–7.1	δ Scuti	0.11	A−F
RZ	08	37.0	−44	07	6.4–7.6	Cepheid	20.40	G
T	08	37.7	−47	22	7.7–8.3	Cepheid	4.64	F
SW	08	43.6	−47	24	7.4–9.0	Cepheid	23.47	K
SX	08	44.9	−46	21	8.0–8.6	Cepheid	9.55	G
CV	09	00.6	−51	33	6.5–7.3	Algol	6.89	B + B
SY	09	12.4	−43	47	7.6–8.1	Semi-regular	63	M
RW	09	20.3	−49	31	7.8–12.0	Mira	451.7	M
V	09	22.3	−55	58	7.2–7.9	Cepheid	4.37	F
S	09	33.2	−45	13	7.7–9.5	Algol	5.93	A + K
U	09	33.2	−45	31	7.9–8.2	Semi-regular	37	M
Z	09	52.9	−54	11	7.8–14.8	Mira	421.6	M
SV	10	44.9	−56	17	7.9–9.1	Cepheid	14.10	F−G

DOUBLE STARS

	RA (h)	RA (m)	Dec. (°)	Dec. (')	PA (°)	Separation (")	Magnitudes	
γ	08	09.5	−47	20	AB 220	41.2	1.9, 4.2	
					AC 151	62.3	8.2	
					AD 141	93.5	9.1	
					DE 146	1.8	12.5	
b	08	40.6	−46	39	058	37.5	3.8, 10.2	
δ	08	44.7	−54	43	AB 153	2.6	2.1, 5.1	
					AC 061	69.2	11.0	
					CD 102	6.2	13.5	
μ	10	46.8	−49	25	057	2.6	2.7, 6.4	Binary, 116 y

OPEN CLUSTERS

M	C	NGC	RA (h)	RA (m)	Dec. (°)	Dec. (')	Diameter (')	Magnitude	No. of stars	
		2547	08	10.7	−49	16	20	4.7	80	
	85	IC 2391	08	40.2	−53	04	50	2.5	30	o Velorum Cluster
		IC 2395	08	41.1	−48	12	8	4.6	40	
		2669	08	44.9	−52	58	12	6.1	40	
		2670	08	45.5	−48	47	9	7.8	30	
		IC 2488	09	27.6	−56	59	15	7.4	70	
		2910	09	30.4	−52	54	5	7.2	30	
		2925	09	33.7	−53	26	12	8.3	40	
		2972	09	40.3	−50	20	4	9.9	25	
		3033	09	48.8	−56	25	5	8.8	50	
		3228	10	21.8	−51	43	18	6.0	15	

GLOBULAR CLUSTER

M	C	NGC	RA (h)	RA (m)	Dec. (°)	Dec. (′)	Diameter (′)	Magnitude
		3201	10	17.6	−46	25	18.2	6.7

PLANETARY NEBULA

M	C	NGC	RA (h)	RA (m)	Dec. (°)	Dec. (′)	Diameter (″)	Magnitude	Magnitude of central star
	74	3132	10	07.7	−40	26	47	8.2	10.1

γ Velorum has a traditional name, Suhail; 'Regor' is modern, but is widely used (it was invented by astronaut Gus Grissom for his fellow astronaut Roger Chaffee; 'Roger' spelled backwards). It is at least a six-star system; the brightest component (A) is a spectroscopic binary, consisting of an O9-type supergiant with a massive Wolf–Rayet companion, separation 1 a.u., period 78.5 days. The Wolf–Rayet star is at least 10 times as massive as the Sun; there are several fainter companions.

The o Velorum Cluster (IC 2391, Caldwell 85) is a conspicuous object, containing at least 30 stars.

The Gum Nebula (Gum 12) is the remnant of a supernova that exploded perhaps a million years ago. It contains a pulsar. There is nothing sticky about it! It is named after the Australian astronomer, Colin Gum, who disovered it in 1955.

VIRGO

(Abbreviation: Vir)

A very large Zodiacal constellation, representing Astræa, the goddess of justice, daughter of Jupiter and Themis. The autumnal equinox. ('First Point of Libra') has now precessed to a position close to Zaniah (η Virginis). The constellaiton includes one first-magnitude star, Spica. The 'bowl' of Virgo, bounded on its opposite side by Denebola in Leo, abounds in galaxies.

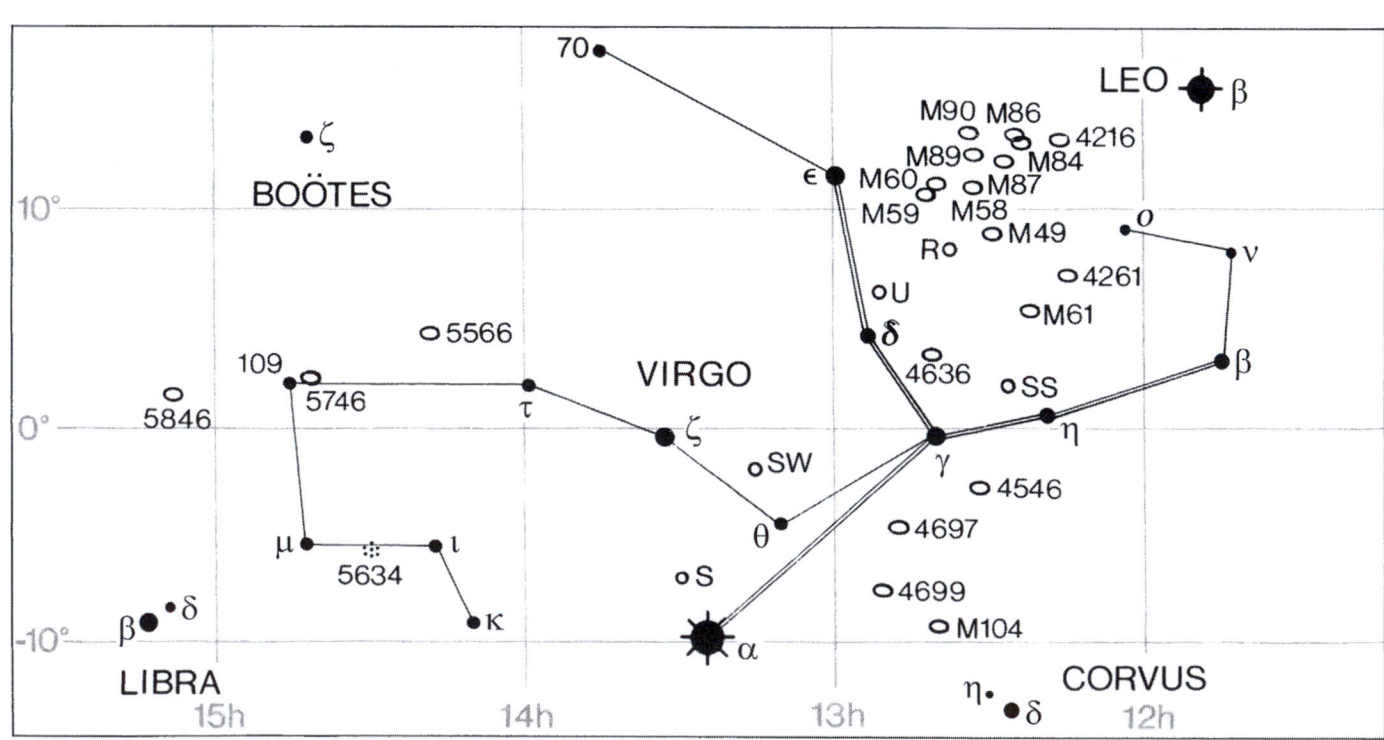

There are nine stars above the fourth magnitude.

	RA (h)	(m)	(s)	Dec. (°)	(′)	(″)	m	M	d (light-years)	Spectrum	
67 α	13	25	11.6	−11	09	41	0.98	−3.55	262	B1	Spica
29 γ	12	41	40.0	−01	36	58	2.74$_{(com)}$d	2.38	39	FO	Arich
47 ε	13	02	10.8	+10	57	33	2.85	0.37	102	G8	Vindemiatrix
79 ζ	13	34	41.8	−00	35	45	3.38	1.62	73	A3	Heze
43 δ	12	55	36.5	+03	23	51	3.39	−0.57	200	M3	Minelauva
5 β	11	50	41.3	+01	45	55	3.59	3.40	36	F8	Zavijava
109	14	46	15.0	+01	53	35	3.73	0.75	129	AO	
107 μ	14	43	03.6	−05	39	27	3.87	2.51	61	F2	Rigl Al Awwa
15 η	12	19	54.4	−00	40	00	3.89	−0.53	290	A2	Zaniah

Also above magnitude 4.5: ν (4.04), ι (Syrma) (4.07), o (4.12), κ (4.18), τ (4.23), θ (4.38), 110 (Khambalia) (4.39).

VARIABLE STARS

	RA (h)	(m)	Dec. (°)	(′)	Range	Type	Period (d)	Spectrum
X	12	01.9	+09	04	7.3–11.2	?	—	F
SS	12	25.3	+00	48	6.0–9.6	Mira	354.7	N
R	12	38.5	+06	59	6.0–12.1	Mira	145.6	M
U	12	51.1	+05	33	7.5–13.5	Mira	206.8	M
S	13	33.0	−07	12	6.3–13.2	Mira	377.4	M
RS	14	27.3	+04	41	7.0–14.4	Mira	352.8	M

DOUBLE STARS

	RA (h)	(m)	Dec. (°)	(′)	PA (°)	Separation (″)	Magnitudes	
17	12	22.5	+05	18	337	20.0	6.6, 9.4	
γ	12	41.7	−01	27	287	3.0	3.5, 3.5	Binary, 171.4 y
θ	13	09.9	−05	32	343	7.1	4.4, 9.4	
73	13	32.0	−18	44	183	0.1	6.7, 6.9	
84	13	43.1	+03	32	229	2.9	5.5, 7.9	
τ	14	01.6	+01	33	290	80.0	4.3, 9.6	
φ	14	28.2	−02	14	110	4.8	4.8, 9.3	

GALAXIES

M	NGC	RA (h)	(m)	Dec. (°)	(′)	Magnitude	Dimensions (′)	Type
	4216	12	15.9	+13	09	10.0	8.3 × 2.2	Sb
	4261	12	19.4	+05	49	10.3	3.9 × 3.2	E2
61	4303	12	21.9	+04	28	9.7	6.0 × 5.5	Sc
85	4382	12	22.9	+08	28	9.3	3.0 × 3.0	S0
84	4374	12	25.1	+12	53	9.3	5.0 × 4.4	E1
	4429	12	27.4	+11	07	10.2	5.5 × 2.6	S0

GALAXIES

M	NGC	RA (h)	(m)	Dec. (°)	(′)	Magnitude	Dimensions (′)	Type	
	4438	12	27.8	+13	01	10.1	9.3 × 3.9	Sap	
	4442	12	28.1	+09	48	10.5	4.6 × 1.9	E5p	
86	4406	12	26.2	+12	57	9.2	7.4 × 5.5	E3	
49	4472	12	29.8	+08	00	8.4	8.9 × 7.4	E4	
87	4486	12	30.8	+12	24	8.6	7.2 × 6.8	E1	Virgo A
	4699	12	49.0	−08	40	9.6	3.5 × 2.7	Sa	
	4535	12	34.3	+08	12	9.8	6.8 × 5.0	SBc	
	4546	12	35.5	−03	48	10.3	3.5 × 1.7	E6	
	4527	12	34.1	+02	39	11.3	6.3 × 2.3	Sb	
	4536	12	34.5	+02	11	11.0	7.4 × 3.5	Sc	
	4546	12	35.5	−03	48	11.3	3.5 × 1.7	E6	
89	4552	12	35.7	+12	33	9.8	4.2 × 4.2	E0	
90	4569	12	36.8	+13	10	9.5	9.5 × 4.7	Sb	
58	4579	12	37.7	+11	49	9.8	5.4 × 4.4	Sb	
104	4594	12	40.0	−11	37	8.3	8.9 × 4.1	Sb	Sombrero Hat
	4596	12	39.9	+10	11	10.5	3.9 × 2.8	SBa	
59	4621	12	42.0	+11	39	9.8	5.1 × 3.4	E3	
60	4649	12	43.7	+11	33	8.8	7.2 × 6.2	E1	
	4654	12	44.0	+13	08	10.5	4.7 × 3.0	Sc	
	4636	12	42.8	+02	41	9.6	6.2 × 5.0	E1	
	4660	12	44.5	+11	11	11.9	2.8 × 1.9	E5	
	4697	12	48.6	−05	48	9.3	6.0 × 3.8	E4	
	4699	12	49.0	−08	40	9.6	3.5 × 2.7	Sa	
	4753	12	52.4	−01	12	9.9	5.4 × 2.9	Pec	
	4762	12	52.9	+11	14	10.2	8.7 × 1.6	SB0	
	4856	12	59.3	−15	02	10.4	4.6 × 1.6	SBa	
	5247	13	38.1	−17	53	10.5	5.4 × 4.7	Sb	
	5363	13	56.1	+05	15	10.2	4.2 × 2.7	Ep	
	5364	13	56.2	+05	01	10.4	7.1 × 5.0	SB + p	
	5068	13	18.9	−21	02	10.8	6.9 × 6.3	SBc	
	5850	15	07.1	+01	33	11.7	4.3 × 3.9	SBb	

Spica is a very small-range rotating ellipsoidal variable (magnitude +0.92 to 1.04, period 4.010 days). In 1889, H. Vogel found it to be a spectroscopic binary, but visually the pair are so close that resolution is very difficult indeed. The real centre-to-centre separation is about 18 000 000 km. Spica is also a variable of the β-Cephei type.

Spica is one of only four first-magnitude stars sufficiently close to the ecliptic to be occulted (the others are Aldebaran, Antares and Regulus).

Arich (γ; alternative names Postvarta and Porrimer) is a binary; the two F0-type components are virtually equal (magnitudes 3.48 and 3.49), period 168.9 years. When I first

remember it (around 1930!) it was very wide and easy, but by about 1990 we were seeing it from a much less favouable angle, and was beyond the reach of most telescopes; the orbital eccentricity is 0.88. The separation is now increasing again; by 2010, it was almost 1″, and before long it will again be an easy pair.

Minelauva (δ) is an M-type orange-red giant, well over 600 times as luminous as the Sun and with a diameter 65 times as great; a fine sight in binoculars. It is very slightly variable (magnitude 3.32 to 3.40, semi-regular). Zaniah (η) looks single visually, but is actually a very close triple, with two stars only 0.5 a.u. apart and a third slightly further out.

VOLANS

(Abbreviation: Vol)

Originally Piscis Volans. A small southern constellation, intruding into Carina. There are five stars down to the fourth magnitude.

	RA (h)	(m)	(s)	Dec. (°)	(′)	(″)	m	M	d (light-years)	Spectrum
γ	07	08	44.8	−70	29	57	3.6	0.59+5.08	142 + 147	G8 + F2
β	08	25	44.2	−66	08	12	3.77	1.17	108	K2
ζ	07	41	49.2	−72	36	22	3.93	0.86	134	K0
δ	07	16	49.8	−67	57	26	3.97	−2.56	660	F6
α	09	02	26.8	−60	23	45	4.00	1.09	124	A

Also above magnitude 4.5: ε (4.35).

See chart for Carina.

VARIABLE STAR

	RA (h)	(m)	Dec. (°)	(′)	Range	Type	Period (d)	Spectrum
S	07	29.8	−73	23	7.7–13.9	Mira	395.8	M

DOUBLE STARS

	RA (h)	(m)	Dec. (°)	(′)	PA (°)	Separation (″)	Magnitudes
γ²	07	08.8	−70	30	300	13.6	4.0, 5.9
ζ	07	41.8	−72	36	116	16.7	4.0, 9.8
ε	08	07.9	−68	37	024	6.1	4.4, 8.0
κ	08	19.8	−71	31	AB 057	65.0	5.4, 5.7
					BC 030	37.7	8.5
θ	08	39.1	−70	23	108	45.0	5.3, 10.3

GALAXY

M	NGC	RA (h)	(m)	Dec. (°)	(′)	Magnitude	Dimensions (′)	Type
	2442	07	36.4	−69	32	11.2	6.0 × 5.5	SBb

γ² (magnitude 3.78) and γ¹ (5.68) make up a wide binary; they are at least 700 a.u. apart and the orbital period cannot be less than 7500 years.

VULPECULA

(Abbreviation: Vul)

Originally Vulpecula et Anser, the Fox and Goose. The Goose has disappeared, but the brightest star, α, has been named Anser.

There are no stars above the fourth magnitude. The brightest is the red giant α (Anser); RA 10h 28m 42s .4, Dec. + 24° 39′ 55″,

$m = 4.44$, $M = -0.35$, = 296 light-years, type M0. Next come 23 (4.50) and 31 (4.56). α has a 5.8-magnitude companion, visible with binoculars.

See chart for Cygnus.

VARIABLE STARS

	RA (h)	RA (m)	Dec. (°)	Dec. (′)	Range	Type	Period (d)	Spectrum
RS	19	17.7	+12	26	6.9–7.6	Algol	4.46	B + A
Z	19	21.7	+25	34	7.4–9.2	Algol	2.45	B + A
U	19	36.6	+20	20	6.8–7.5	Cepheid	7.99	F–G
T	20	51.5	+28	15	5.4–6.1	Cepheid	4.44	F–G
SV	19	51.5	+27	28	6.7–7.7	Cepheid	45.03	F–K
R	21	04.4	+23	49	7.0–14.3	Mira	136.4	M

DOUBLE STARS

	RA (h)	RA (m)	Dec. (°)	Dec. (′)	PA (°)	Separation (″)	Magnitudes
2 (ES)	19	17.7	+23	02	127	1.8	5.4, 9.2
α–8	19	28.7	+24	40	028	413.7	4.4, 5.8

OPEN CLUSTERS

M	C	NGC	RA (h)	RA (m)	Dec. (°)	Dec. (′)	Diameter (′)	Magnitude	No. of stars	
		Cr 399	19	25.4	+20	11	60	3.6	40	Coat-hanger (Brocchi's Cluster)
		6823	19	43.1	+23	18	12	7.1	30	
		6830	19	51.0	+23	04	12	7.9	20	
	37	6885	20	12.0	+26	29	7	5.7	30	
		6940	20	34.6	+28	18	31	6.3	60	

PLANETARY NEBULA

M	NGC	RA (h)	RA (m)	Dec. (°)	Dec. (′)	Dimensions (″) sec	Magnitude	Magnitude of central star	
27	6853	19	59.6	+22	43	350 × 910	7.6	13.9	Dumbbell Nebula

Several novæ have been seen in Vulpecula, notably Anthelm's Star of 1571 it is catalogued as CK Vulpeculæ.

The most celebrated object in the constellation is M 27, the Dumbbell planetary nebula.

Collinder 399, Brocchi's Cluster – known popularly as the Coathanger – lies in Vulpecula, RA 19h 25m 24s, Dec. +10° 11′ 00″. It consists of 10 stars from magnitude 5 to 7, of which six are in a straight line with four more making up the 'hook'. It is easily vsible with the naked eye, and was recorded by Al-Sûfi in 964; it was again mentioned by Hodierna in the seventeenth century, and mapped in 1920 by D. F. Brocchi, an amateur member of the American Association of Variable Star Observers. In 1931, the Swedish astronomer Per Collinder included it in his catalogue of open clusters, but data obtained by the Hipparcos astrometric satellite show that it is merely an asterism, not a genuine cluster.

30 · Telescopes and observatories

OBSERVATORIES

Strictly speaking, an observatory is any place from which astronomical studies are carried out. It is even possible to claim that Stonehenge was an observatory, because there is little doubt that it is astronomically aligned. The oldest observatory building now standing seems to be that at Chomsong-dae in Kyingju, South Korea; it dates from AD 632. The name means 'Star-gazing Tower'. Apparently, it was constructed under the reign of Queen Seondeok (632–647). It is 5.7 m wide at the base and 9.4 m high. Later, elaborate measuring instruments were built by the Arabs and the Indians; some of these still exist such as the great observatory at Delhi. In 1576, Tycho Brahe erected his elaborate observatory at Hven, in the Baltic, and used the equipment to draw up an amazingly accurate star catalogue. In the modern sense, observatories are of course associated with telescopes of some kind or another. A list of some great modern observatories is given in Table 30.4.

National observatories date back for centuries; the oldest seems to be that of Leiden in Holland (1632). The oldest truly national observatory was that at Copenhagen in Denmark, although unfortunately the original buildings were destroyed by fire.

The national British observatory, at Greenwich, was founded in 1675 by order of King Charles II, mainly so that a new star catalogue could be drawn up for the use of British seamen. The original buildings were designed by Wren and are now known as Flamsteed House (after the first Astronomer Royal). Greenwich became the 'timekeeping centre' of the world and the zero for longitude passes through it. Until 1971 the Director of the Royal Greenwich Observatory also held the post of Astronomer Royal: lists of Astronomers Royal are given in Tables 30.1, 30.2 and 30.3. When Greenwich Park became unsuitable as an observing site in the mid-twentieth century, the main equipment was transferred to Herstmonceux in Sussex and the largest telescope was later taken to La Palma in the Canary Islands, where observing conditions are far better than in England. In 1992, Herstmonceux was closed, and the RGO transferred to an office block in Cambridge. Finally, in October 1998, the Labour Government closed the RGO completely, ostensibly as an economy measure – probably the twentieth century's worst act of scientific vandalism.

Most modern observatories are sited at high altitude, where seeing conditions are good: the summit of Mauna Kea, for example, lies at over 4000 m. The summit of Los Muchachos, at La Palma in the Canary Islands, is also excellent as are sites in the Atacama Desert of Northern Chile. The United States national observatory is at Kitt Peak in Arizona.

Aircraft have been used; the Kuiper Airborne Observatory, brought into use in 1975, was a Lockheed C141 jet transport aircraft, in which was mounted a 0.915-m Cassegrain telescope. It was an outstanding success, and functioned until 1995, when it was abandoned for the usual reasons of economy. It could fly at 12 km, above 85% of the Earth's atmosphere and 99% of the water vapour. Its successor, SOFIA (the Stratospheric Observatory for Infared Astronomy) is a combined project of NASA and USRA, the German Universities Space Research Association, and was launched in 2007; it is based on a Boeing 747SP wide-body aircraft. The telescope, a 2.5-m Cassegrain reflector, looks out of a large door in the side of the fuselage near the aircraft's tail. The operating altitude is 12.5 km, where 85% of the full infrared range will be available.

Telescopes have of course been flown in space. The most famous of these, the Hubble Space Telescope, was launched in 1990, and was still in full operation in 2010.

In contrast, there are 'observatories' below ground level! The Homestake Mine Observatory, in South Dakota, is located in a gold-mine at a depth of 1.5 km. The 'telescope' is a 100 000-gallon tank of cleaning fluid designed to trap solar neutrinos; at a lesser depth the results would be ruined by cosmic rays. The project was started in 1966. The Super-Kamiokande at Hida, in Japan, is sited 1000 m underground, in Mozami Mine; the neutrino detector is a cylindrical stainless steel tank, 41 m tall and 39.3 m in diameter, holding 50 000 tonnes of ultra-pure water. A neutrino interacting with the electrons or nuclei of the water can produce a charged particle that moves faster than the speed of light in water, and creates a beam of light known as Čerenkov radiation (the optical equivalent of a sonic boom). This can be detected, and the results have been excellent. Operations began in 1996.

An underground laboratory has even been set up in England – in Boulby salt and potash mine near Whitby in Yorkshire. The equipment here attempts to detect WIMPS, weakly interacting massive particles. Boulby is the deepest mine in Britain, with over 1400 km of tunnels. The detector, ZEPLIN, uses a tank filled with 4 kg of liquid xenon. If a WIMP collides with a xenon nucleus in the tank, the energy from the recoil of the nucleus should produce a flash of light detectable by photomultiplier tubes, which then amplify it. So far there have been no definite results, but the experiment is certainly of great value.

TELESCOPES

We cannot be sure when the first telescopes were made. It has been suggested that Leonardo da Vinci may have had something of the kind, and the British scientific historian C. A. Ronan believed that a

Table 30.1 *Astronomers Royal*

John Flamsteed	1675–1719
Edmond Halley	1720–1742
James Bradley	1742–1762
Nathaniel Bliss	1762–1764
Nevil Maskelyne	1765–1811
John Pond	1811–1835
Sir George Airy	1835–1881
Sir William Christie	1881–1910
Sir Frank Dyson	1910–1933
Sir Harold Spencer Jones	1933–1955
Sir Richard Woolley	1956–1971
Sir Martin Ryle	1972–1982
Sir Francis Graham-Smith	1982–1990
Sir Arnold Wolfendale	1991–1995
Lord Rees of Ludlow	1995–

Table 30.3 *Royal Astronomers of Ireland*

Reverend Henry Ussher	1783–1792
John Brinkley	1792–1827
Sir William Rowan Hamilton	1827–1865
Friedrich Ernst Brunnow	1865–1874
Sir Robert Stawell Ball	1874–1892
Arthur Alcock Rambaut	1892–1897
Charles Jasper Joly	1897–1906
Sir Edmund Taylor Whittaker	1906–1912
Henry Plummer	1912–1921

(The title of Royal Astronomer for Ireland originally went with the Andrews Professorship of Astronomy in Dublin. The title lapsed in 1921, but the Andrews Chair was re-established in 1984 on an honorary basis: Patrick Arthur Wayman 1984–1998; Luke Drury 1999–.)

Table 30.2 *Astronomers Royal for Scotland*

Thomas Henderson	1834–1844
Professor Charles Piazzi Smyth	1846–1888
Ralph Copeland	1889–1905
Frank Watson Dyson	1905–1910
Ralph Allen Sampson	1910–1937
Professor William Michael Herbert Greaves	1938–1955
Professor Hermann Alexander Brück	1957–1975
Professor Vincent Cartledge Reddish	1975–1980
Professor Malcolm Sim Longair	1980–1990
Vacant	1991–1995
Professor John Campbell Brown	1995–

(The title of Astronomer Royal for Scotland was originally the title of the Director of the Royal Observatory Edinburgh, but since 1995 it has been an honorary title.)

telescope involving both lenses and mirrors was made by Leonard Digges, in England, some time between 1545 and 1559, but the evidence is very slender. (In 2000 Ronan and I made a telescope on the 'Digges pattern' and demonstrated it on BBC television. It did work, but it was amazingly clumsy!)

The world's largest neutrino 'telescope' is the aptly named 1 cubic kilometre Ice Cube in Antarctica. Construction began in 2005, and involved drilling over seventy 2.5-km deep holes in the Antarctic ice, using a novel hot-water drill and then lowering long strings of detectors into the holes, where they could be frozen in place. It will far surpass AMANDA, the smaller Antarctic Muon and Neutrino Detector Array.

The south pole area is in fact very favourable for astronomical work, despite the obvious problems. In particular, there is a site known as Ridge A, several hundred kilometres from the pole at an altitude of 4050 m, where the air is so calm that stars hardly appear to twinkle at all. There is hardly any 'weather' on the plateau, the air is 100 times drier than the Sahara, there is hardly a wisp of wind and no falling snow. Against this, it is somewhat chilly, with a winter temperature of –70 °C (–94 °F), and each year there are six months of darkness (apart from the Southern Lights), during which telescope maintenance will be difficult. However, there is little doubt that an observatory will be established there in the foreseeable future, and the Chinese have already set up a telescope at Dome Argus, one of the coldest, driest and loftiest places even in Antarctica.

The first refracting telescope whose existence can be definitely proved was made by H. Lippershey, in Holland, in 1608. Before long, telescopes were used for astronomical work; for example, a telescopic map of the Moon was drawn by Thomas Harriot in 1609, but the first great telescopic observer was of course Galileo who made a telescope for himself and began his work in January 1610. With it, he made a series of spectacular discoveries: the craters of the Moon, the phases of Venus, the strange appearance of Saturn, the four principal satellites of Jupiter and the 'myriad stars' of the Milky Way.

The first reflecting telescope was made by Isaac Newton and was presented to the Royal Society in 1671; it was probably made in 1668 or 1669. Telescopes of both kinds were steadily improved, and really large instruments became possible. In 1789, William Herschel produced a reflector with a 49-inch (124.5-cm) mirror. This was surpassed in 1845 by the 182.9-cm (72-inch) reflector made by the third Earl of Rosse, and set up at Birr Castle in Ireland, it was a strange instrument, mounted between two massive stone walls and capable of swinging for only estimated distances to either side of the meridian, but with it Rosse discovered the spiral forms of the objects we know to be galaxies. The 'Leviathan', as it was nicknamed, was in fact the largest telescope in the world until 1917, it is now (2010) in full operation again.

Early reflectors, including those of Herschel and Rosse, had metal mirrors, but glass mirrors coated with a reflective substance

Table 30.4 *Some of the world's great observatories*

Location	Name	Latitude °	′	″	Longitude °	′	″	Altitude (m)
Aarhus, Denmark	Ole Rømer Observatory	56 N	07	40.0	10 E	11	48	50
Alma Ata, Kazakhstan	Mountain Observatory of Academy of Sciences	43 N	11	16.9	76 E	57	24	1450
Ann Arbor, Michigan, USA	University of Michigan	42 N	16	48.7	83 E	43	48	282
Arcetri (Florence), Italy	Astrophysical Observatory	43 N	45	44.7	11 E	15	18	184
Armagh, N Ireland	Armagh Observatory	54 N	21	11.1	06 W	38	54	64
Athens, Greece	National Obsservatory	37 N	58	19.7	23 E	43	0	110
Auckland, New Zealand	Auckland Public Observatory	36 S	54	28.0	174 E	46	36	80
Barcelona, Spain	Fabra Observatory	41 N	24	59.3	02 E	07	36	415
Beijing, China	Beijing University Observatory	39 N	57	24	116 E	21	36	76
Beirut, Lebanon	American University Observatory	33 N	54	22.0	35 E	28	12	38
Belgrade, former Yugoslavia	Observatory of Academy of Science	44 N	48	13.2	20 E	30	48	253
Belo Horizonte, Brazil	Piedade Observatory	19 S	49	18	43 W	30	42	1746
Berlin, Germany	Wilhelm Förster Observatory	52 N	28	30.0	13 E	25	30	40
Berlin-Treptow, Germany	Archenhold Observatory	52 N	29	07.0	13 E	28	36	38
Berne, Switzerland	Astronomy Institute of University	46 N	57	12.7	07 E	25	42	563
Big Bear Lake, California, USA	Big Bear Solar Observatory	34 N	15	12	116 W	54	54	2067
Bloemfontein, South Africa	Boyden Observatory	29 S	02	18.0	26 E	24	18	1387
Bochum Germany	Astronomical Station	51 N	27	54.8	07 E	13	24	132
Bogotá, Colombia	National Astronmical Observatory	04 N	35	54	74 W	04	54	2640
Bologna, Italy	San Vittore Observatory	44 N	28	06	11 E	20	30	280
Bombay (now Mumbai), India	Government Observatory	18 N	53	36.2	72 E	48	54	14
Bonn, Germany	University Observatory	50 N	43	45.0	07 E	05	48	62
Bordeaux, France	Observatory of University of Bordeaux	44 N	50	07	00 E	31	36	73
Bosque Alegro, Argentina	Cordoba Observatory Station	31 S	35	54	64 W	32	48	1250
Boulder, Colorado, USA	Sommers-Bausch Observatory	40 N	00	13.0	105 W	15	42	1648
Brno, Czech Republic	Nicolaus Copernicus Observatory	49 N	12	18	16 E	35	18	310
Bro, Sweden	Kvistaberg Observatory	59 N	30	06	17 E	36	24	0
Brussels, Belgium	Astronomy and Astrophysics Institute	50 N	48	48	04 E	23	00	147
Bucharest, Romania	National Observatory	44 N	24	49.4	26 E	05	48	83
Budapest, Hungary	Konkoly Observatory	47 N	29	58.6	18 E	57	54	474
Buenos Aires, Argentina	Naval Observatory	34 S	37	18.3	58 W	21	18	6
Calar Alto, Spain	German-Spanish Ast. Centre	37 N	13	48	02 E	32	12	2168
Cambridge, England	University Observatory	52 N	12	51.6	00 E	05	42	28
Cambridge, Massachusetts, USA	Harvard College Observatory	42 N	22	47.6	71 W	07	48	24
Cape, South Africa	South African Astronomical Observatory	33 S	56	02.5	18 E	28	36	10
Caracas, Venezuela	Cagigal Observatory	10 N	30	24.3	66 W	55	42	1042
Castleknock, Ireland	Dunsink Observatory	53 N	23	18	06 W	20	12	85
Catania, Sicily	Astrophysical Observatory	37 N	31	42.0	15 E	04	42	193
Cerro Las Campanas, Chile	Las Campanas Observatory	29 S	00	30	70 W	42	00	2282
Cerro Le Silia, Chile	European Southern Observatory	29 S	15	24	70 W	43	48	2347
Cerro Parañal, Chile	VLT	24 S	38	00	70 W	24	00	2635
Cerro, Tololo, Chile	Cerro Tololo Inter-American Observatory	30 S	09	54	70 W	48	54	2215
Chung-li, Taiwan	National Central University Observatory	24 N	58	12	121 E	11	12	152
Cincinnati, Ohio, USA	Cincinnati Observatory	39 N	08	19.8	84 W	45	42	17
Cocoa, Florida, USA	Brevard Community Astronomical Observatory	28 N	33	06	80 W	45	42	17

Table 30.4 (cont.)

Location	Name	Latitude °	′	″	Longitude °	′	″	Altitude (m)
Copenhagen, Denmark	Urania Observatory	55 N	41	19.2	12 E	32	18	10
Corboda, Argentina	National Observatory	31 S	25	16.4	64 W	11	48	434
Crimea, Ukraine	Crimean Astronomical Observatory	44 N	43	42.0	34 E	01	00	550
Edinburgh, Scotland	Royal Observatory	55 N	55	30.0	03 W	11	00	146
Flagstaff, Arizona, USA	US Naval Observatory	35 N	11	00.0	111 W	44	24	2310
Flagstaff, Arizona, USA	Lowell Observatory	35 N	12	06.0	111 W	39	48	2210
Göttingen, Germany	University Observatory	51 N	31	48.2	09 E	56	36	161
Greenbelt, Maryland, USA	Goddard Research Observatory	39 N	01	11.5	76 W	49	36	49
Groningen, Holland	Kapteyn Astronomical Laboratory	53 N	13	13.8	06 E	33	48	4
Haleakala, Hawaii, USA	University of Hawaii	20 N	42	22.0	156 W	15	24	3054
Hamburg, Germany	Bergedorf Observatory	53 N	28	46.9	10 E	14	24	41
Hartebeespoort, South Africa	Republic Observatory Annexe	25 S	46	22.4	27 E	52	36	1220
Helsinki, Finland	University Observatory	60 N	09	42.3	24 E	57	18	33
Helwan, Egypt	Helwan Observatory	29 N	51	31.1	31 E	20	30	115
Hyderabad, India	Nizamiah Observatory	17 N	25	53.0	78 E	27	12	293
Jena, Germany	Karl Schwarzschild Observatory	50 N	58	51.0	11 E	42	48	331
Johannesburg, South Africa	Republic Observatory	26 S	10	55.3	28 E	04	30	1806
Juvisy, France	Flammarion Observatory	48 N	41	37.0	02 E	22	18	92
Kharkov, Ukraine	Kharkov Univ. Observatory	50 N	00	02	36 E	13	54	138
Kiev, Ukraine	Kiev University Observatory	50 N	27	12	30 E	29	54	184
Kiso, Japan	Kiso Observatory	35 N	47	36	137 E	37	42	1130
Kitt Peak, Arizona, USA	Kitt Peak National Observatory	31 N	57	48	111 W	36	00	2120
Kitt Peak, Arizona, USA	Steward Observatory Station	31 N	57	48	111 W	36	00	2071
Kodaikanal, India	Astrophysical Observatory	10 N	13	50.0	77 E	28	06	2343
Kracow, Poland	University Observatory	50 N	03	52.0	19 E	57	36	221
Kyoto, Japan	Kwasan Observatory	34 N	59	40.8	135 E	47	36	234
La Palma, Canary Islands	Roque de los Muchachos Observatory	28 N	45	36	17 W	52	54	2326
La Plata, Argentina	La Plata Observatory	34 S	54	30	57 W	55	54	17
Las Cruces, New Mexico, USA	Corralitos Observatory	32 N	22	48	107 W	02	36	1453
Leiden, Holland	University Observatory	52 N	09	19.8	04 E	29	00	6
Lisbon, Portugal	Lisbon Astronomical Observatory	38 N	42	42	09 N	11	12	111
Los Angeles, California, USA	Griffith Observatory	34 N	06	46.8	118 W	18	06	357
Lund, Sweden	Royal University Observatory	55 N	41	51.6	13 E	11	12	34
Madison, Wisconsin, USA	Washburn Observatory	43 N	04	36	89 W	24	30	292
Madrid, Spain	Astronomical Observatory	40 N	24	30.0	03 W	41	18	655
Mauna Kea, Hawaii, USA	Mauna Kea Observatory	19 N	46	36	155 W	28	18	4220
Meudon, France	Observatory of Physical Astronomy	48 N	48	18.0	02 E	13	54	162
Milan, Italy	Brera Observatory	45 N	27	59.2	09 E	11	30	120
Mill Hill, London, England	University of London Observatory	51 N	36	46.3	00 W	14	24	82
Minneapolis, Minnesota, USA	University of Minnesota Observatory	44 N	48	40.0	93 W	14	18	260
Mitaka, Japan	National Astronomical Observatory	35 N	40	18	139 E	32	30	58
Montevideo, Uruguay	National Observatory	34 S	54	33.0	56 W	12	42	24
Montreal, Quebec, Canada	McGill University Observatory	45 N	30	20.0	73 W	34	42	57
Moscow, Russia	Sternberg Institute Observatory	55 N	45	19.8	37 E	34	12	166
Mt Aragatz, Armenia	Byurakan Astrophysical Observatory	40 N	20	06	44 E	17	30	1500
Mt Bigelow, Arizona, USA	Steward Observatory, Catalina Station	32 N	25	00	110 W	43	54	2510
Mt Fowlkes, Texas, USA	Hobby-Eberly Observatory	30 N	40	00	104 W	01	00	2072
Mt Hamilton, California, USA	Lick Observatory	37 N	20	25.3	121 W	38	42	1283
Mt Hopkins, Arizona, USA	Fred L. Whipple Observatory	31 N	41	18	110 W	53	06	2608
Mt John, New Zealand	Mt John University Observatory	43 S	59	12	170 E	27	54	1027

Table 30.4 (cont.)

Location	Name	Latitude			Longitude			Altitude (m)
		°	′	″	°	′	″	
Mt Lemmon, Arizona, USA	Steward Observatory, Catalina Stn	32 N	26	30	110 W	47	30	2776
Mt Locke, Texas, USA	McDonald Observatory	30 N	40	18	104 W	01	06	2075
Mt Semirodriki, Caucasus, Russia	Zelenchukskaya	43 N	49	32.0	41 E	35	24	973
Mt Stromlo, Canberra, Australia	Mount Stromlo Observatory	35 S	19	12	149 E	00	30	767
Mt Wilson, California, USA	Mt Wilson Observatory	34 N	12	59.5	118 W	03	36	1742
Naples, Italy	Capodimonte Observatory	40 N	51	45.7	14 E	15	24	164
Nice, France	Nice Observatory	43 N	43	17.0	07 E	18	00	376
Ondřejov, Czech Republic	Astrophysical Observatory	49 N	54	38.1	14 W	47	00	533
Ottawa, Ontario, Canada	Dominio Observatory	45 N	23	38.1	75 W	43	00	87
Palermo, Sicily	University Astronomical Observatory	38 N	06	43.6	13 E	21	30	72
Palomar, California, USA	Palomar Observatory	33 N	21	22.4	116 W	51	48	1706
Pic du Midi, France	Observatory of University of Toulouse	42 N	56	12.0	00 E	08	30	2862
Piikkiö, Finland	Turku University Observatory	60 N	25	00	22 E	26	48	40
Pittsburgh, Pennsylvania, USA	Allegheny Observatory	40 N	20	58.1	80 W	01	18	370
Potsdam, Germany	Astrophysical Observatory	52 N	22	56.0	13 E	04	00	107
Pulkovo, Russia	Astronomical Observatory of Academy of Sciences	59 N	46	18.5	30 E	19	36	75
Purple Mountain, China	Purple Mountain Observatory	32 N	04	00	118 E	39	18	367
Quezon City, Philippines	Manila Observatory	14 N	38	12	121 E	04	36	58
Quito, Ecuador	National Observatory	00 S	14	00.0	78 W	29	36	2908
Richmond Hill, Ontario, Canada	David Dunlap Observatory	43 N	51	46.0	79 W	25	18	244
Riga, Latvia	Latvian State University Astronomical Observatory	56 N	57	06	24 E	07	00	36
Rio de Janeiro, Brazil	National Observatory	22 S	53	42.2	43 W	13	24	33
Rome, Italy	Vatican Observatory	41 N	44	48	12 E	39	06	450
St Andrews, Scotland	University Observatory	56 N	20	12.0	02 W	48	54	30
St Michel, France	Observatory of Haute-Provence	43 N	55	54	05 E	42	48	665
St Petersburg, Russia	Pulkovo Observatory	59 N	46	05.5	30 E	19	24	70
São Paulo, Brazil	Astronomical and Geophysical Institute	23 S	39	06.9	46 W	37	24	800
Siding Spring, New South Wales, Australia	Siding Spring Observatory	31 S	16	37.3	149 E	04	00	1165
Stockholm, Sweden	Saltsjöbaden Observatory	59 N	16	18.0	18 E	18	30	55
Sunspot, New Mexico, USA	Sacramento Peak National Solar Observatory	32 N	47	12	105 W	49	12	2811
Sutherland, South Africa	South African Astronomical Observatory	32 S	22	46.5	20 E	48	36	1830
Sydney, Australia	Government Observatory	33 S	51	41.1	151 E	12	18	44
Tartu, Estonia	Wilhelm Struve Astronomical Observatory	58 N	16	00	26 E	28	00	0
Tautenberg, Germany	Karl Schwarzschild Observatory	50 N	58	54	11 E	42	48	331
Tokyo, Japan	Tokyo Observatory at Mitaka	35 N	40	21.4	139 E	32	30	59
Tonantzintla, Mexico	National Astronomical Observatory	19 N	01	57.9	98 W	18	48	2150
Toruń, Poland	Copernicus University Observatory	53 N	05	47.7	18 E	33	18	90
Trieste, Italy	Trieste Astronomical Observatory	45 W	38	30	13 E	52	30	400
Tucson, Arizona, USA	Steward Observatory	32 N	14	00	110 W	56	54	757
Uccle, Belgium	Royal Observatory	50 N	47	55.0	04 E	21	30	105
Uppsala, Sweden	University Astronomical Observatory	59 N	51	29.4	17 E	37	30	21
Utrecht, Holland	Sonnenborgh Observatory	52 N	05	09.6	05 E	07	48	14
Victoria, BC, Canada	Dominion Astronomical Observatory	48 N	31	15.7	123 W	25	00	229

Table 30.4 (cont.)

Location	Name	Latitude			Longitude			Altitude (m)
		°	′	″	°	′	″	
Vienna, Austria	University Observatory	48 N	13	55.1	16 E	20	18	240
Vilnius, Lithuania	Vilnius Astronomical Observatory	54 N	41	00	25 E	17	12	122
Washington, District of Columbia, USA	US Naval Observatory	38 N	55	14.0	77 W	03	54	86
Wellington, New Zealand	Carter Observatory	41 S	17	03.9	174 E	45	54	129
Williams Bay, Wisconsin, USA	Yerkes Observatory	42 N	34	13.4	88 W	33	24	334
Zürich, Switzerland	Swiss Federal Observatory	47 N	22	36	08 E	33	06	469

such as silver or aluminium were better. Large refractors were built during the latter part of the nineteenth century, the largest of all, the Yerkes Observatory 40-inch (101-cm), was brought into use in 1895. However, a lens has to be supported around its edges, and if too large it will distort under its own weight: probably the Yerkes refractor will never be surpassed. A 124.5-cm object glass was made and shown at the Paris Exposition of 1901, but was never used for serious research. A 41-inch object-glass destined for the Pulkovo Observatory in Russia was never even mounted. Clearly, the future belonged to the reflector. Master-minded by George Ellery of the United States, a 60-inch reflector was set up in 1908 at Mount Wilson, in California; in 1917 came the Hooker 100-inch (254-cm) reflector, also at Mount Wilson. This remained in a class of its own until 1948, with the completion of the Hale 200-inch (508-cm) on Mount Palomar, also in California.

HISTORY

The Anglo–Australian Observatory (AAP) was set up in 1974 at Siding Spring, near Coonabarabran in the Warrumbungle Mountains. It was a joint venture by the UK and the Australian Governments; 'first light' with the AAT (Anglo–Australian Telescope) was achieved on 27 April 1974. Another major instrument was the UKST (United Kingdom Schmidt Telescope) which had been opened on 17 August 1973. Very valuable researches were carried out, including advances in fibre optics.

In 2002, the UK Labour Government, following its policy of abandoning valuable scientific programmes, announced its intention to withdraw from the project. On 30 June 2010 the AAO became the Australian Astronomical Observatory, while the AAT became the AAO.

During its history the original AAO had five Directors:

Joe Wampler	1974–1976
Don Morton	1976–1986
Russell Cannon	1986–1996
Bruce Boyle	1996–2003
Matthew Colless	2003–2010.

Today there are reflecting telescopes of many kinds, some have segmented mirrors: for example the first Keck Telescope, on Mauna Kea in Hawaii, has a mirror 9.8 m in diameter, made of 36 segments fitted together to form the perfect optical curve. Keck II, by its side, is identical; when used together as an interferometer, they could in theory distinguish the headlights of a car separately at a range of 25 000 km. The first multiple-mirror telescope (MMT) was set up at the Whipple Observatory at Mount Hopkins, Arizona; it used six 183-cm mirrors in conjunction, so that total light-grasp was equal to a single 442-cm mirror. It performed well, but the six mirrors have now been replaced by a single 650-cm mirror constructed by Roger Angel, using his new spin-casting technique.

A selected list of large optical telescopes are shown in Table 30.5.

The largest single telescope operating in 2010 was the Spanish 10.4-m Gran Telescopio Canarias, in the Canary Islands; it has of course a segmented mirror, and was based on the Kecks. Kecks I and II on Mauna Kea in Hawaii both have a 10-metre mirror composed of 36 segments; they can be operated either separately or else together as an interferometer. The Large Binocular Telescope at Mount Graham, Arizona, has a pair of 8.4-m mirrors on one mount, equalling the light-collecting power of an 11.8-m instrument. The Very Large Telescope at Cerro Paranâl, in the Atacama Desert of Northern Chile, has four 8.2-m mirrors – Antu, Kueyen, Melipal and Yepun – which also can be used either separately or together, giving them resolution of a 16-m instrument. As early as 2001, Antu and Melipal were used together to measure the diameters of five stars (β Doradûs, ζ Geminorum, Kapteyn's Star, Achernar, HD 217987 and HD 36395); the apparent diameter of Achernar was given as 0.00192″, corresponding to 13 000 000 km.

Until the 1970s, all really large telescopes were mounted equatorially (with the obvious exception of the Rosse). The first large modern altazimuth telescope was the 6-m Russian reflector, made in 1975. Optically it has never been a real success, but its mounting was very satisfactory, and almost all modern large reflectors are altazimuths; today's computers can deal with guidance with no trouble at all.

There are some very novel designs. Some telescopes have mirrors of liquid mercury, spinning so as to produce a perfect parabola; one of these is the LZT (Liquid Zenthal Telescope) in British Columbia, which can point only to the zenith.

The Hobby–Eberly Telescope has a spherical main mirror whose optical axis is tilted to the zenith at an angle of 35°; the mirror and telescope are mounted on a frame which turns 360° in the azimuthal direction, so that during an observation the telescope is fixed in the

Table 30.5 *Large optical telescopes*

Aperture				Latitude		Longitude		Elevation (m)	Completion
(m)	(inch)	Name	Observatory	°	′	°	′		
16.0	630	Very large Telescope[a]	Cerro Parañal, Chile	24	38 S	70	24 W	2635	1998 (Antu), 1999 (Kueyen, Melipal)
10.4	410	Gran telescopic Canarias	Oservatorio del Roque de los muchachos	17	52	52	34 W	2267	2009
10.0	387	Keck I	Manuna Kea, Hawaii	19	50 N	155	28 W	4123	1991; mirror composed of 36 segments
10.0	387	Keck II	Mauna Kea, Hawaii	19	50 N	155	28 W	4123	1996; twin of Keck I
9.2	362	Hobby-Eberly	Mount Fowlkes, Texas	30	40 N	104	01 W	2072	1986; segmented mirror; fixed elevation
8.3	327	Subaru	Manuna Kea, Hawaii	19	50 N	155	28 W	4100	1999; Japanese
8.0	315	Gemini North	Manuna Kea, Hawaii	19	50 N	155	28 W	4100	1999; twin of Gemini South
6.5	256	Mono–Mirror Telescope	Mount Hopkins, Arizona	31	04 N	110	53 W	2608	1999; succeeds 4.5-m Multiple Mirror Telescope
6.0	236	Bolshoi Teleskop Azimutalnyi	Nizhny Arkhyz, Russia	49	39 N	41	26 E	2070	1975; Large Altazimuth Telescope
5.08	200	Hale Telescope	Palomar, California	33	21 N	116	52 W	1706	1948; 200-inch reflector
4.2	165	William Herschel Telescope	La Palma, Roque de los Muchachos	28	46 N	17	53 W	2332	1987; British
4.0	158	Victor Blanco Telescope	Cerro Tololo, Chile	30	10 S	70	49 W	2215	1976; Inter-American Observatory
3.89	153	Anglo-Australian Telescope	Siding Spring, New South Wales	31	17 S	149	04 E	1149	1975; Coonabarabran, Australia
3.81	150	Nicholas U Mayall Reflector	Kitt Peak, Arizona	31	57 N	111	37 W	2120	1973
3.80	150	UKIRT	Manuna Kea, Hawaii	19	50 N	155	28 W	4194	1978; United Kingdom Infra-Red Telescope
3.58	141	Canada–France–Hawaii Telescope	Mauna Kea, Hawaii	19	49 N	155	28 W	4200	1979; CFH Telescope
3.57	141	3.6-m telescope	Calar Alto, Spain	37	13 N	02	32 W	2168	1984; Spanish
3.50	141	Astrophysical Research Consortium	Apache Point, New Mexico	32	47 N	105	49 W	2788	1993; mainly remote controlled
3.50	141	WIYN Telescope	Kitt Peak, Arizona	31	57 N	111	63 W	2089	1994; Wisconsin–Indiana–Yale–NOAO Telescope
3.50	141	NTT	La Silla, Chile	29	16 S	70	44 W	2353	1989; New Technology Telescope
3.50	141	Starfire	Kirtland AFB, New Mexico	—		—		1900	— Military
3.50	141	Telescopio Nazionle Galileo	La Palma, Canary Islands	28	45 N	17	53 W	2370	1998 Italian
3.05	120	C. Donald Shane Telescope	Lick Observatory, Mt Hamilton, California	37	21 N	121	38 W	1290	1959
3.00	118	NASA Infra-Red Telescope Facility	Mauna Kea, Hawaii	19	50 N	155	28 W	4208	1979; IRTF
3.00	118	NODO	Laval University, Quebec	32	59 N	105	44 W	2758	1999; liquid mirror telescope

Table 30.5 (cont.)

Aperture (m)	(inch)	Name	Observatory	Latitude °	′	Longitude °	′	Elevation (m)	Completion
2.72	107	Harlan Smith Telescope	McDonald Observatory, Mount Locke, Texas	30	40 N	104	01 W	2075	1969; 107-inch telescope
2.70	106	UBC–Laval Telescope	Malcolm Knapp Research Forest, British Columbia	49	07 N	122	35 W	50	1992; liquid mirror; not steerable
2.64	104	Shajn Reflector	Crimean Astrophysical Observatory, Ukraine	44	44 N	34	00 E	550	1960
2.64	104	Byurakan Reflector	Byurakan, Armenia	40	20 N	44	18 E	1500	1976; Byurakan Astrophysical Observatory
2.56	101	Nordic Optical Telescope	La Palma, Canary Islands	28	45 N	17	53 W	2382	1989; Los Muchachos
2.56	101	Issac Newton Telescope	La Palma, Canary Islands	28	46 N	17	53 W	2336	1984; Los Muchachos
2.54	100	Irénée du Pont Telescope	Las Campanas, Chile	29	00 S	70S	42 W	2282	1976
2.50	100	Sloan Digital Sky Survey	Apache Point, New Mexico	32	47 N	105	49 W	2788	1999; wide-field detector
2.50	100	Hooker Telescope	Mount Wilson, California	34	13 N	118	03 W	1742	1917; 100-inch reflector
2.40	94	Hubble Space Telescope	Orbital	—		—		—	1990; average 600 km altitude
2.34	92	Hiltner Telescope	Michigan–Dartmouth–MIT Observatory, Kitt Peak, Arizona	31	57 N	111	37 W	1938	1986
2.30	91	Vainu Bappu Telescope	Kavalur, Tamil Nadu, India	12	35 N	78	50 E	725	1986
2.30	91	2.3-m Telescope	Mt Stromlo, Australia	31	16 S	149	03 E	1149	1984; Mount Stromlo and Siding Spring Observatory
2.30	91	Bok Telescope	Kitt Peak, Arizona	31	57 N	111	37 W	2100	1991; Steward Observatory
2.30	91	Wyoming Infrared Telescope	Jelm Mountain, Wyoming	41	06 N	105	59 W	2943	1977
2.20	87	University of Hawaii Telescope	Mauna Kea, Hawaii	19	50 N	155	28 W	4200	—
2.20	87	ESO Telescope	La Silla, Chile	29	15 S	70	44 W	2200	1977; European Southern Observatory
Refractors									
1.01	40	Yerkes 40-inch Telescope	Yerkes Observatory, Williams Bay, Wisconsin, USA	42	34 N	88	33 W	334	1897
0.89	36	36-inch Refractor	Lick Observatory, Mount Hamilton, California, USA	37	20 N	121	39 W	1290	1888
0.83	33	33-inch Meudon Refractor	Paris Observatory, Meudon, France	48	48 N	02	14 E	162	1889
0.80	31	Potsdam Refractor	Potsdam Observatory, Germany	52	23 N	13	04 E	107	1899

Table 30.5 (cont.)

Aperture (m)	(inch)	Name	Observatory	Latitude °	′	Longitude °	′	Elevation (m)	Completion
0.76	30	Thaw Refractor	Allegheny Observatory, Pittsburgh, Pennsylvania, USA	40	29 N	80	01 W	380	1985
0.74	29	Lunette Bischoffschei	Nice Observatory, France	43	43 N	07	18 E	372	1886
0.68	27	Grosser Refractor	Archenold Observatory, Treptow, Germany	52	29 N	13	29 E	41	1896
0.67	26	Grosser Refractor	Vienna Observatory, Austria	48	14 N	16	20 E	241	1880
0.67	26	McCormick Refractor	Leander McCormick Observatory, Charlottesville, Virginia, USA	38	02 N	78	31 W	264	1883
0.66	26	26-inch Equatorial	US Naval Observatory, Washington, DC, USA	38	55 N	77	04 W	92	1873
0.66	26	Thompson Refractor	Herstmonceux, England	51	29 N	00	00	50	1897
0.66	26	Innes Telescope	Republic Observatory, Johannesburg, South Africa	26	10 S	55	33 E	1806	1926

Schmidt telescopes

Aperture (m)	(inch)	Name	Observatory	Latitude °	′	Longitude °	′	Elevation (m)	Completion
1.34	53	2-m Telescope	Karl Schwarzschild Observatory, Tautenberg, Germany	50	59 N	11	43 E	331	1960
1.24	49	Oschin Telescope	Palomar Observatory, California, USA	33	21 N	116	51 W	1706	1948
1.24	49	United Kingdom Schmidt Telescope (UKS)	Siding Spring, Australia	31	16 S	149	04 E	1145	1973
1.05	41	Kiso Schmidt Telescope	Kiso Observatory, Kiso, Japan	35	48 N	137	38 E	1130	1975
1.00	39	3TA-10 Schmidt Telescope	Byurakan Astrophysics Observatory, Mt Aragatz, Armenia	40	20 N	44	30 E	1450	1961
1.00	39	Kvistaberg Schmidt Telescope	Uppsala University Observatory, Kvistaberg, Sweden	59	30 N	17	36 E	33	1963
1.00	39	ESO 1-m Schmidt Telescope	European Southern Observatory, La Silla, Chile	29	15 S	70	44 W	2318	1972
1.00	39	Venezuela 1-m Schmidt Telescope	Centro F J Duarte, Merida, Venezuela	08	47 N	70	52 W	3610	1978
0.90	35	Télescope de Schmidt	Observatoire de Calern, Calern, France	43	45 N	06	56 W	1270	1981
0.84	33	Télescope Combiné de Schmidt	Royal Observatory, Brussels, Belgium	50	48 N	04	21 E	105	1958
0.80	31	Schmidt Telescope	Radiophysical Observatory, Riga, Latvia	56	47 N	24	24 E	75	1968
0.80	31	Calar-Alto-Schmidtspiegel	Calar Alto Observatory, Spain	37	13 N	02	32 W	2168	1980

[a]Four 8.2-m mirrors, working together: Antu (the Sun), Kueyen (the Moon), Melipal (the South Cross) and Yepun (Sirius); the names come from the Mapuche language of the area; installed in 2001.

azimuth and objects are tracked by moving a spherical aberration corrector to follow the reflected light. The main mirror is composed of 91 segments, each hexagonal in shape and 1 m across the flats.

South Africa now has a 10-m telescope, SALT (South African Large Telescope), completed in 2009. It follows the Hobby–Eberly pattern.

Telescopes now make use of active optics (rapid adjustments of the shape of the mirror to allow for changes in elevation) and adaptive optics (to counteract rapid variations in the atmosphere). Electronic devices, such as CCDs (charge-coupled devices) have superseded photography; indeed, the last photographic plate taken with the Hale reflector was exposed as long ago as September 1989. Fibre optics are also widely used. There are, of course, specialist telescopes, such as solar telescopes: the largest of these is at Kitt Peak in Arizona. It uses a 203-cm heliostat, and has an inclined tunnel 146 m long. It was completed in 1962, and although as signed solely for solar research it proved capable of being used for other types of observation as well – an unexpected bonus.

SPACE TELESCOPES

Observing from above the atmosphere is clearly ideal in many ways, and space telescopes have been used successfully. The first of these was the Hubble Space Telescope, launched from the Space Shuttle on 25 April 1990. It has a 340-cm (94-inch) mirror, and orbits at an altitude of 600 km in a period of 95 minutes; the orbital inclination is 28.5°. The total weight is 11 360 kg, and the length is 13.3 m; the diameter is 12 m with the solar arrays extended. Initially, the telescope had a faulty mirror, but a servicing mission by astronauts introduced corrective optics, and since then the telescope has been performing even better than originally expected. Regular servicing missions have been carried out.

The COROT telescope (COnvection ROtation and planetary Transits) was launched on 27 December 2006 from Baikonur Cosmodrome, on a Soyuz 2 carrier rocket. It was designed to search for extrasolar planets with short orbital period, and moved in a polar orbit at a height of 827 km. It detected its first planet, Corot 1b, in May 2007, and six more before the end of 2009. It carries a 27-cm diameter off-axis afocal telescope, with six CCDs cooled to –40 °C. The planned active lifetime was 2 ½ years, but in fact this was extended by several months.

The Hipparcos astrometric satellite (HIgh Precision PARallax COllecting Satellite) was launched from Kourou on an Ariane 4 on 8 August 1989 and operated until March 1993. It was designed to measure the parallaxes, distances, positions and proper motions of over 100 000 stars. The Hipparcos Catalogue of 100 000 stars was published in 1997, together with the lower-precision Tycho Catalogue of over a million stars. Hipparcos carried an all-reflective Schmidt telescope of aperture 29 cm. The orbital distance from Earth ranged between 507 and 35 888 km; orbital period 10h 40m.

The Kepler mission, launched by NASA on a Delta II rocket from Cape Canaveral on 6 March 2009 with an expected active lifetime of 3½ years, has reached an Earth-trailing heliocentric orbit at a distance of 1 a.u. Its aim is to detect extrasolar planets by the transit method, monitoring the brightness of 100 000 stars in a fixed field of view.

PLANETARIA

A planetarium is purely an educational device; an artificial sky is projected onto the inside of a large dome, by means of a very complex projector. The planetarium is really a development of the orrery, a device to show the movements of the planets. The first orrery was made in the eighteenth century by George Graham, for Prince Eugene of Savoy; the first American orrery was made by David Ritlenhouse in 1772.

The true ancestor of the modern planetarium was the Gottorp Globe, made by H. Busch in Denmark about 1654–6. It was 4 m in diameter. The audience sat inside it; the stars were painted on the inside of the globe. However, the first modern-type planetarium, due to Dr W. Bauersfeld of the Zeiss Optical Works, was opened at the Deutsches Museum in Bonn in 1923. Today most major cities have planetaria; the largest is that of Tokyo, at the Miyazumi Science Center, which has a 27 m dome, and was opened in 1987. English planetaria include the South Downs Planetarium in Chichester and the Peter Harrison Planetarium in Greenwich; there are various other planetaria in the British Isles.

31 · Non-optical astronomy

Our atmosphere acts as an effective screen against most space radiations outside the optical range, so that in most cases space research methods have to be used. All major nations have space agencies. Those capable of launch ability in 2010 are listed in Table 31.1.

Non-optical astronomy really began in 1931, with the detection of radio waves from the Milky Way, though infrared emissions from the Sun had been demonstrated by William Herschel as long ago as 1801. It is now possible to examine almost all regions of the electromagnetic spectrum, from long radio waves down to the very short gamma rays.

Wavelengths are measured in nanometres or Ångstroms – one nanometre being one thousand-millionth of a metre, and one Ångström being one ten thousand millionth of a metre we, so that 1 nm = 10 Å. The accepted regions are as follows:

Below 0.01 nm	Gamma-rays
0.01–10 nm	X-rays
	hard: 0.01–0.1
	soft: 0.1–10
10–400 nm	Ultraviolet
	EUV (Extreme ultraviolet)
400–700 nm	Visible light
(=4000–7000 Å)	
700 nm–5000 nm	Near infrared
500 nm–0.3 mm	Mid to far infrared
0.1 mm–1 mm	Submillimetre
1 mm–0.3 in	Microwaves
over 0.3 m	Radio waves

COSMIC-RAY ASTRONOMY

Cosmic rays are the only particles we can detect which have crossed the Galaxy, moving at almost the velocity of light. Unfortunately, the particles are affected by the magnetic fields in the Galaxy and this means that ordinarily it is not possible to determine the directions from which they come.

The origins of cosmic rays seem to be varied. Those of the highest energy may have come from quasars or active galactic nuclei (AGN); those of lower energy originate within the Galaxy in supernova remnants, supernova outbursts and pulsars. Those of the lowest energy come from solar flares. However, the Sun is not a major source of cosmic rays. The flux we receive decreases at the time of solar maximum, because of the increase in the strength of the interplanetary magnetic field, which acts as a screen (the Forbush effect).

This variation in the cosmic-ray flux affects conditions in our atmosphere, and is mainly responsible for periods of global warming and cooling.

Cosmic rays were discovered in 1912 by the Austrian physicist Victor Hess, who was anxious to find out why electrometers at ground level always recorded a certain amount of background. He therefore flew electrometers in a balloon, rising to a height of 4800 m, and discovered evidence that cosmic rays are coming constantly from all directions in space.

They are not rays at all, but particles, broadly divided into two categories, primary and secondary. Those produced in astrophysical sources are primaries, which interact with interstellar matter to produce secondaries. Almost 90% of cosmic ray particles entering the Earth's atmosphere are protons, 10% are helium nuclei (alpha particles) and slightly less than 1% are electrons and some heavier atomic nuclei. The primaries cannot pass directly through the Earth's atmosphere but break up both themselves and the air particles, so that all the cosmic rays reaching ground level are secondaries. The collisions in the upper air produce what are termed muons. These muons decay so quickly that they should not last long enough to reach ground level – but they do, because they are travelling so rapidly that their timescale, relative to ours, is slowed down.

GAMMA-RAY ASTRONOMY

Gamma-rays represent the most energetic form of electromagnetic radiation, at wavelengths shorter than those of X-rays – that is to say, below 0.01 nm. They are absorbed in the Earth's upper atmosphere, and only the most energetic can reach ground level, so that virtually all gamma-ray astronomy depends upon equipment carried in satellites or space-craft (although balloons have also been used). Moreover, care must always be taken to distinguish gamma-rays from the much more plentiful cosmic rays.

The first detection of high-energy gamma-rays from space was achieved by the Explorer X1 satellite, in 1961; it picked up fewer than 100 gamma-ray photons. They seemed to come from all directions, indicating more or less uniform gamma-ray background; this has been fully confirmed since. Discrete gamma-ray sources were also identified: the first of these, in the constellation of Sagittarius, was found in 1969. Detectors on OSO 3 (Orbiting Solar Observatory 3) recorded gamma-rays coming from the direction of the centre of the Galaxy.

The first really important gamma-ray satellites were SAS 2 (Small Astronomical Satellite 2), launched in 1972, and Cos-B, launched in 1975; SAS 2 failed after six months, but Cos-B

Table 31.1 *List of space agencies with launch capability*

Country	Acronym	Founded
China (China National Space Administration)	CNSA	1993
Europe (European Space Agency)	ESA	1975
Iran (Iranian Space Agency)	ISA	2004
Israel (Israeli Space Agency)	ISA	1983
India (Indian Space Research Organisation)	ISRO	1969
Japan (Japan Aerospace Exploration Agency)	JAXA	2003
USA (National Aeronautics and Space Administration)	NASA	1958
France (Centre National d'Etudes Spatiales)	CNES	1961
Ukraine (National Space Agency of Ukraine)	NSAU	1992
Russian Federal Space Agency (Roscosmos)	RFSA	1992

(Members of the ESA are Austria, Belgium, Czech Republic, Denmark, Finland, France, Germany, Greece, Ireland, Italy, Luxembourg, the Netherlands, Norway, Portugal, Spain, Sweden, Switzerland and the United Kingdom. The Soviet Space Programme (CCCP) operated in the USSR between ~1955 and ~1991.)

operated until 1982. It was confirmed that the most intense gamma radiation came from the galactic plane, due to interactions between cosmic rays and interstellar gas, and new discrete sources were identified, including the Vela pulsar, the Crab pulsar and a strange object in Gemini (already noted), which was not visible optically and which did not seem to radiate except at gamma-ray wavelengths. It became known as Geminga, partly as an abbreviation for 'Gemini gamma-ray source' and partly because in the Milanese dialect. Geminga means 'it is not there'; only in 1999 was its true nature established. The Röntgen Satellite (Rosat) detected a period of 0.237 seconds in soft X-ray emission, so that Geminga is a neutron star, the remnant of a supernova outburst about 300 000 years ago. It is only 552 light-years away, and it has been suggested that the outburst was responsible for the low density of the interstellar medium in our neighbourhood (the Local Bubble). It is a powerful gamma-ray source, although the two strongest gamma-ray sources in the sky are the Vela pulsar and the Crab pulsar. The Vela pulsar lies in the southern constellation of Vela, the Sails of the dismantled constellation of Argo Navis. The name has nothing to do with the Vela series of artificial satellites.

One of the most successful vehicles involved with gamma-ray research is the satellite originally called GRO (the Gamma-Ray Observatory), but now named after the great pioneer of this research, Arthur Holly Compton. It was launched on the Shuttle on 5 April 1991; it weighed nearly 17 tonnes, and filled two-thirds of the Shuttle bay. The instruments weighed 6 tonnes; all were capable of detecting gamma-ray photons, measuring their

energies, and determining the directions from which they came. It was found that the Milky Way glows at gamma-ray wavelengths; in 1993, the Compton satellite completed the first all-sky map of gamma-ray sources, including AGN and quasars. It was deliberately crashed into the Pacific on 4 June 2000, as the equipment had started to fail.

Very-high-energy gamma-rays may be detected indirectly from observations made at ground level. In passing through the atmosphere they generate electrons, which travel faster than the speed of light in air, resulting in blue Čerenkov radiation, which may be detected with special equipment. This does not mean that Einstein's theory of relativity is wrong, because it is still the case that nothing can travel faster than the speed of light in a vacuum. The largest gamma-ray 'telescope' of this kind is at the Whipple Observatory on Mount Hopkins, Arizona; the 'dish' is 10 m in diameter. Intensive research is being carried out by scientists at the University of Durham, where the main equipment has been set up at Narrabri in Australia.

Of special interest are the gamma-ray bursters; the first of these was detected by the Vela satellite in 1967, but many hundreds have been recorded since. They are of immense power, and last only briefly – from a few seconds to a few minutes; they may appear in any direction at any moment. For a long time they remained a complete mystery, but a great advance was made in February 1997, when an Italian satellite, BeppoSAX, pinpointed a burst in the constellation of Orion. Telescopes were promptly trained on the area, and detected a rapidly fading star, the aftermath of a titanic explosion. This was the first proof that gamma-ray bursters are both very remote and very violent. On the 31 January 2001 a burster was detected coming from a small area in Carina, about 50″ in diameter. It was designated as GRB 000131. The afterglow was observed both by the Very Large Telescope (VLT) and by the 1.54-m Danish telescope at La Silla. It was found that the distance was of the order of 11 000 million light-years and that the brightness was about a million times that of the Sun. This was the most remote gamma-ray burster observed up to that time. Since then we have learned a great deal from the Swift gamma-ray mission, launched by NASA from Cape Canaveral on a Delta II carrier on 20 November 2004, into an almost circular orbit (586 × 601 km altitude). It carries its Burst Alert Telescope (BAT), X-ray telescope (XRT) and Ultraviolet/Optical Telescope (UVOT). The BAT detects bursts, computes the positions to an accuracy of about 4′ in no more than 15 seconds, and relays the data to ground-based observatories. The XRT obtains images and performs spectral analysis of the afterglows, and the UVOT attempts to detect optical afterglows. Hundreds of GRBs have been detected. The most remote seems to be about 12.6 thousand million light-years away. (The estimated age of the Únwere is 13.7 thousand milion years.).

Gamma-ray bursts are of two main types, 'long' (duration over 2 seconds) and 'short' (less than 2 seconds). A long burst is probably due to the collapse of a very massive star into a neutron star or a black hole, and a short burst to a collision between two neutron stars. Fortunately, GRBs are rare – in our Galaxy perhaps no more than one per million years – but a GRB could have disastrous effects on life on Earth even if many thousands of light-years away.

NASA's Fermi Gamma-ray Space Telescope (originally known as the Gamma-ray Large Area Space Telescope, GLAST) was launched fom Cape Canaveral on a Delta II rocket on 11 June 2008, and put into a low-Earth circular orbit, altitude 550 km, inclination 28.5°. It was designed to study blazers, gamma-ray bursts and background radiation, neutron stars, cosmic rays and supernova remnants, solar gamma-rays, and join the search for weakly interacting massive particles (WIMPS).

Research carried out at Washington University, St. Louis, notably by M. Beilicke and H. Krawczynski, has resulted in the discovery of an outburst of very high-energy gamma-rays from the radio galaxy M 87. Particles are accelerated to extremely high energies in the immediate vicinity of a supermassive black hole, and then emit gamma-radiation.

Pulsars have been detected because of the gamma-rays they emit, thanks largely to NASA's Fermi Space Telescope and the Lovell Telescope at Jodrell Bank. It has also been found that gamma-rays pulsate in a similar fashion to millisecond pulsars, which rotate from 100 to 1000 times per second. On 10 May 2009, Fermi detected a short gamma-ray burst, GRB 090510, in a galaxy 7.3 thousand million light-years away, probably due to a collision between two neutron stars.

The gamma-ray burst GRB 09043 detected in April 2009 by NASA's Swift satellite, was about 13 000 million light-years away, making it the most remote object ever seen. The afterglow was recorded by several ground-based telescopes. It is also the oldest object ever seen – near the end of the 'Dark Age', succeeded by the period of reionisation.

A selected list of gamma-ray telescopes is given in Table 31.2.

X-RAY ASTRONOMY

X-rays were discovered in 1895 by the German scientist Wilhelm Röntgen (it must be admitted that the discovery was accidental); in the electromagnetic spectrum they lie between gamma-rays and ultraviolet. Cosmic X-ray sources range from the Sun to the hot outer atmospheres of normal stars, white dwarfs, active galaxies (including quasars) and accretion discs around neutron stars and black holes.

For obvious reasons no optical telescope can detect X-rays because the X-ray photons would simply penetrate the telescope mirror. However, just as a bullet will ricochet off a wall when striking at a grazing angle, so X-rays will ricochet off a mirror when striking it at a narrow angle. An X-ray telescope is therefore quite unlike an optical telescope; the mirrors have to be aligned almost parallel to the incoming X-rays, so that the telescope is shaped like a barrel. Moreover, X-rays are absorbed by the Earth's upper atmosphere, so that observations from ground level are virtually impossible.

The first observations of X-rays from the sky were made on 5 August 1948, when R. Burnright of the US Naval Research Laboratory detected solar X-rays from the darkening of a photographic emulsion carried to altitude on a V2 rocket. On 29 September 1948, H. Friedman and E. O. Hulbert, also using a V2, detected intense X-radiation from the Sun. In 1956, Friedman's team recorded results which could have been due to celestial X-rays, but they could not be sure, as they always had the Sun in their field of view. Then, in 1959, R. Giacconi and his colleagues published a paper in which they predicted X-rays from very hot stars and supernova remnants. They referred particularly to the Crab Nebula.

On 24 October 1961, a rocket was launched from White Sands to search for X-rays from the Moon. It was thought that these could be due to incident X-rays striking the lunar surface and causing X-ray fluorescence, together with X-rays due to the surface being struck by energetic electrons from the solar wind. The equipment failed, as the protective covers refused to open. A second rocket – an Aerobee – was launched on 18 June 1962. No lunar X-rays were found, but a discrete source was detected, later found to be Scorpius X-1. On 12 October 1962, a new launch recorded X-rays from the Crab Nebula, with an intensity of 15% of those from Scorpius X-1. By 1966, 30 sources had been identified, including the first galaxy, M 87. The first imaging X-ray telescope was designed by Giacconi and his team at Cambridge, Massachusetts, and was flown on a small rocket. In 1965, it managed to pick up images of hot spots in the upper atmosphere of the Sun.

Many X-ray satellites have been launched: a selected list is given in Table 31.3. The first satellite designed purely for X-ray work was Uhuru, launched from Kenya an 12 December 1970 (the

Table 31.2 *Selected list of gamma-ray telescopes (all are Earth orbiters)*

Name	Agency	Launch	Termination	Distance (km)
HEAO3	NASA	29 Sept 1979	29 May 1981	48 635 065
Cos-B	ESA	9 Aug 1975	25 Apr 1982	340 399 876
Gamma	RSA	1 July 1990	1992	375 393 153 000
Compton (CGRO)	NASA	5 Apr 1991	4 June 2000	3 623 457
HETE-2	NASA	9 Oct 2000	–	5 903 650
INTEGRAL	ESA	17 Oct 2002	–	6 303 153 000
Fermi	NASA	11 June 2008	–	555
Swift	NASA	20 Nov 2004	–	5 853 604

(HEAO = High Energy Astronomy Observatory. HETE = High Energy Transient Explorer. INTEGRAL = International Gamma-Ray Astrophysics Laboratory. Swift is not an acronym, but indicates that the space-craft has the speed and agility of a swift!)

Table 31.3 *A selected list of X-ray satellites*

Name	Re-named	Launch	End of mission	Nationality	Notes
Vela 58		23 May 1969	19 June 1969	US	X-ray burster detected.
SAS-1	Uhuru	12 Dec 1970	March 1973	US	Small Astronomy Satellite 1. First X-ray satellite.
OSO-7		19 Sept 1971	9 July 1974	US	Orbiting Solar Observatory. All-sky survey.
OAO 3	Copernicus	21 Aug 1972	Late 1980	US-UK	Orbiting Astronomical Observatory 3
	Ariel-5	15 Oct 1974	14 Mar 1980	UK	X-ray transients. Spectra.
SAS-3		May 1975	1979	US	Burster. Locations of sources. X-rays from Algol.
HEAO-1		12 Aug 1977	9 Jan 1979	US	High Energy Astronomy Observatory 1. All-sky survey.
HEAO-2	Einstein	12 Nov 1978	Apr 1981	US	First fully imaging X-ray telescope.
Corsa-B	Hakucho	21 Feb 1979	15 Apr 1985	Japan	Transient phenomena. Cygnus X-1 study.
Astro-B	Tenma	20 Feb 1983	Oct 1985	Japan	'Pegasus'. Four high-energy experiments.
	Exosat	26 May 1983	9 Apr 1986	ESA	AGN; X-ray binaries; SNR.
Astro-C	Ginga	5 Feb 1987	1 Nov 1991	Japan	X-ray transient phenomena.
	Granat	1 Dec 1989	27 Nov 1998	Russia	Deep imaging of galacitic centre area.
Röntgen satellite	Rosat	1 June 1990	12 Feb 1999	US	Observations of all types.
	Yohkoh	31 Aug 1991	–	US	Solar X-ray and gamma-rays.
Astro-D	ASCA	20 Feb 1993	–	Japan	Advances Satellite for Cosmology and Astrophysics. First use of charge-coupled devices in X-ray astronomy.
RXTE	Rossi X-ray Timing Explorer	30 Dec 1995	–	US	Variability in X-ray sources.
	BeppoSAX	30 Apr 1996	–	Italy	Afterglow of gamma-ray bursts.
AXAF	Chandra	23 July 1999	–	US	Observations of all kinds.
XMM	Newton	10 Dec 1999	–	ESA	Detailed X-ray spectroscopy.

name means Freedom in Swahili). Uhuru carried out a comprehensive all-sky survey and located 339 sources, including binary systems, supernova remnants, Seyfert galaxies and clusters of galaxies: it also discovered diffuse X-radiation from the material contained in clusters of galaxies.

In 1971, Uhuru showed that one source, Centaurus X-3, is a binary system; it was found that the intensity of the X-radiation varied rapidly in a period of 4.8 s, suggesting a neutron star, and a secondary period of 2.087 days indicating that the neutron star was being eclipsed by a binary companion. The larger component was then optically identified. And a second binary, Hercules X-l, was identified in 1971.

X-ray novæ are temporary phenomena. The first, Centaurus X-4, was discovered in May 1969 by an X-ray detector carried in a Vela satellite. An X-ray nova seen in December 1974 near the radio galaxy Centaurus A (although certainly unconnected with it) reached its maximum on Christmas Day and was inevitably nick-named CenXmas; another in May 1975 close to the Crab Nebula (again certainly unconnected with it) was equally inevitably nick-named Fresh Crab! X-ray novæ seem to be close binaries with low-mass optical companions.

One very successful X-ray satellite was HEAO-2, the Einstein Observatory, which was launched on 13 November 1978 and operated until 1981. It carried the first really large X-ray telescope, consisting of four nested paraboloids and four nested hyperboloids,

with an outer diameter of 58 cm. The ambitious Japanese satellite Yohkoh ('sunlight'), sent up in 1991, was designed mainly for solar studies. The Röntgen satellite, Rosat, was launched in 1990 and operated until finally switched off in February 1999: it achieved more than 9000 observations of objects including comets, quasars, black holes, clusters of galaxies, proto-stars and supernovæ, as well as carrying out the first high-resolution sky surveys at X-ray and extreme ultraviolet (EUV) wavelengths. Rosat was truly international; it used a German X-ray telescope and a British wide-field camera designed for EUV work, while the launch was achieved by an American rocket. In 1990, almost 30 years since the first search for lunar X-rays, the Moon was indeed imaged at X-ray wavelengths by Rosat. The X-rays were emitted by the Sun, and were scattered by atoms in the surface layers on the Moon. Rosat detected and catalogued about 100 000 separate X-ray sources, i.e. many more than the number of stars one can see with the naked eye.

In July 1999, the Chandra X-ray Observatory was launched from the Shuttle. (Its original name was AXAF, The Advanced X-ray Astrophysics Facility, but it was then renamed in honour of Subrahmanyan Chandrasekhar, the great Indian astrophysicist.) Chandra was put into an orbit which takes it from 9980 to 140 000 km above the Earth, so that for much of the time it is above the terrestrial radiation belts. It is a hundred times more sensitive than any other X-ray telescope, and quickly showed its

Table 31.4 *Selected list of ultra violet telescopes (all are Earth orbiters)*

Name	Agency	Launch	End	Distance, km
OAO 2	NASA	21Aug 1972	1980	713×724
Copernicus (OAO 3)	NASA	7 Dec 1968	Feb 1973	749×758
IUE	ESA, NASA	26 Jan 1978	30 Sept 1996	$32\,050 \times 52\,254$
ANS	SRON	30 Aug 1974	June 1976	266×1176
EUVE	NASA	7 June 1992	30 Jan 2002	515×527
FUSE	NASA, CNES	24 June 1999	12 July 2007	752×767
GALEX	NASA	28 Apr 2003	–	691×697
Kasistsat 4	Korean	27 Sept 2003	–	675×695

(ANS = Astronomical Netherlands Satellite)

potential; its first success was the detection of a brilliant ring inside the Crab Nebula.

The X-ray Multi-Mirror Mission (XMM-Newton) of ESA was launched on 10 December 1999 from an Ariane rocket at Kourou in South America. It is designed to perform detailed spectroscopy of cosmic X-ray sources over a broad band of energies ranging from 0.1 keV to 10 keV. There are three highly-nested grazing incidence mirror modules of type Wolter 1, coupled to reflection grating spectrometers and X-ray CCD cameras. It is now known as the Newton Satellite in honour of the scientist who first showed that light was composed of many different colours (wavelengths).

Newton and Chandra are both powerful in their own ways. Chandra is optimised to provide the sharpest X-ray images, whilst Newton is the most sensitive for X-ray spectra.

ULTRAVIOLET ASTRONOMY

The existence of ultraviolet radiation from the Sun was demonstrated in 1801 by J. Ritter, in producing a spectrum with a prism and noting the darkening of paper soaked in sodium chloride held in the region beyond the violet. This was possible because ultraviolet radiation of between 300 and 400 nm can penetrate the Earth's atmosphere. Most research in ultraviolet astronomy has to be carried out from altitude – initially by rockets, and nowadays by satellites.

The first ultraviolet spectrograph was launched from White Sands, on 28 June 1946, by a V2 rocket. It crashed, and the film was lost, but on 10 October 1946 a V2-soared to 88 km, and R. Tousey recorded the first recoverable solar ultraviolet spectra from above the atmosphere.

Ultraviolet observations were subsequently made from the American satellites of the OAO (Orbiting Astronomical Observatory) series. OAO 1 (8 April 1966) failed; OAO 2 (7 December 1968) operated until 23 February 1973; it moved in an orbit ranging between 770 and 780 km above the ground. Its 11 ultraviolet telescopes viewed 1930 objects. Discoveries included a hydrogen cloud round a comet (Tago–Sato–Kosaka), magnetic fields of stars and ultraviolet radiation from a supernova. OAO 3 (the Copernicus satellite) was launched on 21 August 1972, and discovered many

new sources, including supernova remnants and pulsars. It operated until late 1980.

However, the most successful ultraviolet satellite to date has been the IUE (International Ultra-violet Explorer). It was launched on 26 January 1978, with an estimated lifetime of five years; in fact it continued to operate until deliberately switched off on 30 September 1996. It has provided material for more scientific papers than any other satellite.

The IUE carried a 45-cm aperture telescope, which could feed two spectrographs (one in the 190–320 nm range, the other 115–200 nm). Although the IUE was not designed to produce ultraviolet pictures, it did provide a complete survey of ultraviolet spectra for virtually every kind of astronomical object – hot stars, cool stars, variable stars, nebulæ, the interstellar medium, extra-galactic objects and members of the Solar System, including Halley's Comet. It moved in a synchronous orbit, at a distance of 42 164 km and an inclination of 34°. Its discoveries included many stars with magnetic fields and surface activity, measurements of stellar winds, mapping of low-density bubbles of gas around the Sun and nearby stars, and measurements of the composition of planetary nebulæ.

The work of the IUE is now continued using the Hubble Space Telescope, which, in addition to its visible instruments, also carries on-board cameras and spectrographs that are sensitive to the ultraviolet.

Observations at extreme ultraviolet wavelengths really began with the Rosat satellite, in 1990. Previously, it had – wrongly – been thought that no worthwhile observations in this part of the electromagnetic spectrum could be made, because of strong absorption by neutral hydrogen in the interstellar medium. The first object to be studied by Rosat was the binary system WFC1 made up of a very hot white dwarf, HZ43, which has a temperature of 200 000 °C and is one of the hottest stars known; it has a red dwarf companion. Rosat also obtained the first EUV record of the Moon.

The brightest star in the sky at EUV wavelengths is ε Canis Majoris (Adhara). It is almost 500 light-years away, and had the interstellar medium been as opaque in EUV as was originally thought, it would have meant that radiations from that range would have been blocked. However, it now seems that the Solar System lies in a relatively dense cool part of the interstellar medium, and is surrounded by more rarefied material at a higher temperature – so that the atoms of hydrogen are ionised and do not absorb EUV.

The Extreme Ultra-Violet Explorer satellite (EUVE) was launched on 7 June 1992, into a 530-km Earth orbit; it was designed to study radiation at wavelengths from 70 and 760 Å. It carried four grazing incidence telescopes and recorded over 600 sources, ranging from Solar System objects (notably Mars, Jupiter and Io) to cool stars with hot coronæ, white dwarfs with temperatures up to 100 000 °C, and the photospheres and stellar winds of very hot stars. The nuclei of some active galaxies were also recorded. By 1993, EUVE had completed a full survey of the sky in this region of the electromagnetic spectrum.

The Far-Ultraviolet Spectroscopic Explorer (FUSE) was launched from Cape Canaveral on 24 June 1999. It records spectra of astronomical objects at the far-ultraviolet end of the electromagnetic spectrum (90 to 120 nm), which is nearly as energetic as X-rays. The FUSE equipment was ten thousand times more sensitive than that of the Copernicus satellite, which operated from 1972 to 1980; FUSE continued to transmit until 18 October 2007.

NASA's Galaxy Evolution Explorer Mission, GALEX, was launched on 28 April 2003 on a three-stage, solid-fuel Pegasus XI rocket from the Kennedy Space Center in Florida. It was 16.9 m long and 1.3 m in diameter. Its orbital altitude was 690 km, inclination 29°, period 96 min. There were two very sensitive detectors, one in the far ultraviolet to reveal stars less than ten million years old and the other in the near ultraviolet to detect stars less than 100 million years old The planned lifetime of GALEX was no more than three years, but it lasted for much longer than that and within six years had imaged over 500 million objects across two-thirds of the sky. It detected star formation in unexpected regions of the universe, and also made some surprising discoveries – notably the immensely long 'tail' of Mira Ceti.

INFRARED ASTRONOMY

Infrared astronomy has become a vitally important part of modern research. The original discovery of infrared radiation from the Sun was made in 1801 by William Herschel, by placing a thermometer beyond the red end of the solar spectrum. The range extends from a micrometre (0.001 mm) to several hundreds of micrometres, beyond which comes the microwave region.

Some wavelengths of infrared can penetrate through to the Earth's surface, particularly the near infrared towards the red end of the visible spectrum. Water vapour is the main enemy of the infrared astronomer, so that the best sites are at high altitude, with low humidity – such as the top of Mauna Kea in Hawaii, and the lofty Cerro Parañal in the Atacama Desert of northern Chile; balloons have been used to carry infrared equipment, but of course satellites are of great importance.

Infrared radiations can penetrate dust clouds, and so provide us with information about star-forming regions and central regions of the Galaxy. For example, deep inside the Orion Nebula there are powerful stars which can be detected only in the infrared; at optical wavelengths the nebulosity forms an impenetrable screen.

The first major survey of the infrared sky was undertaken in the 1960s by G. Neugebauer and R. Leighton. They discovered 6000 discrete sources. Today there are large infrared telescopes, such as UKIRT (the United Kingdom Infra-red Telescope), on Mauna Kea, above most of the atmospheric water vapour. The aperture is 380 cm. The mirror is thin; theoretically a telescope designed for infrared work need not be so accurate as an optical telescope, although in fact the UKIRT is so good that it can be used at optical wavelengths as well as in infrared. Also on Mauna Kea is the NASA IRTF (Infra-Red Telescope Facility) with an aperture of 300 cm, and there are also other large telescopes designed to study this part of the electromagnetic spectrum. Of course, the Hubble Space Telescope, above the atmosphere, is very effective in the infrared.

Of infrared satellites, one of the most successful is IRAS (the Infra-Red Astronomical Satellite), which launched on 25 January 1983; the reflector had an aperture of 60 cm, enclosed in a cooling vessel containing 100 litres of liquid helium, holding the temperature not far above absolute zero (–273 °C). IRAS operated for months, and could have picked up the radiation from a speck of dust several kilometres away. There were eight staggered rows of detectors, and observations were made in four bands (12, 25, 60 and 100 μm).

IRAS continued to function until 21 November 1985 when it ran out of coolant. By then it had surveyed 97% of the sky and had made many major discoveries. For example, it found 245 389 discrete infrared sources, increasing the number of known sources by more than 100; it found the first dust ring in the Solar System, which takes the form of a torus and lies at a distance about the same as that of the main asteroid belt, although inclined at an angle of 10° to the main plane of the Solar System; it obtained observations of comets, and detected the first cometary dust-tail in the infrared (Comet Tempel 2; the tail was 30 000 000 km long), and it made studies of infrared 'cirrus' due to cool dust clouds in the Galaxy which, when displayed on computerised maps, shows up as wispy clouds resembling cirrus clouds – hence the name. They are made up of tiny dust grains, mainly graphite.

The most spectacular discovery made by IRAS was that of clouds of cool material around some stars, notably Vega, Fomalhaut and the southern β Pictoris. In fact the cloud around Vega, the first to be found, was detected fortuitously; the main investigators, H. Aumann and R. Gillet, discovered it while calibrating the on-board equipment.

IRAS also discovered many AGNs, and starburst galaxies (galaxies undergoing massive bouts of star formation) which are radiating strongly in the infrared region of the spectrum.

Other major satellites included ISO (the Infra-red Space Observatory), an ESA satellite which was launched on 17 November 1995 and continued to operate until May 1998. It observed the universe at wavelengths from 25 to 240 μm. Its detectors were about a thousand times more sensitive than those of IRAS, and achieved angular resolutions about a hundred times better (these figures relate to its observations at a wavelength of 12 μm). Among its many achievements, it discovered water in the atmospheres of planets and of Titan, and in star-forming regions such as the Orion Nebula; it detected newly forming stars at very early stages of

Table 31.5 *Selected list of infrared telescopes*

		Launch	Term	Orbit
IRAS	NASA	25 Jan 1983	21 Nov 1985	EO, 889–70 500 km
ISO	ESA	17 Nov 1995	16 May 1998	EO, 1000–70 500 km
Spitzer	NASA	25 Aug 2003	–	SO, 0.98–1.02 a.u.
Akari	JAXA	21 Feb 2006	–	EO, 587– 610 km
Herschel	ESA, NASA	14 May 2009	–	Second Lagrangian point

EO = Earth orbit. SO = solar orbit.

development. It investigated the properties of ultra-luminous infra-red galaxies and, much nearer home, examined Comet Hale–Bopp.

The Spitzer Space Telescope (once called SIRTF, the Space Infra-Red. Telescope Facility) has been immensely successful. Its instruments were kept cold by liquid helium, expected to last for five years after the launch on 25 August 2003; actually it did not run out until 15 May 2009. This ended the main programme, but parts of the IRAC (Infrared Array Camera) remained operable. Its main instruments were IRAC, IRS (Infrared Spectrograph) and MIPS (Multiband Imaging Photometer for Spitzer).

The SST provided the first direct images of extrasolar planets. In 2004, it detected a faintly glowing body that may be the youngest star ever seen, a bright spot in the centre of a core of gas and dust known as L1014 – possibly a nearby star development with the fledgling star collecting gas and dust from the cloud around it. Spitzer has also shown that the bar through the centre of the Milky Way Galaxy is more substantial than was previously thought. Spitzer made many thousands of valuable observations before its coolant ran out, and it was still providing some data in 2010.

The Herschel Space Observatory, formerly known as FIRST (Far Infrared and Submillimetre Telescope) has three detectors: PACS (Photodetecting Array Camera and Spectrometer), an imaging camera and low-resolution spectrometer covering wavelengths from 55 to 210 μm, SPIRE (Spectral and Photometric Imaging Receiver, an imaging camera and low-resolution spectrometer covering 194 to 672 micrometres wavelength) and HIFI (Heterodyne Instrument for the Far Infrared), a detector which can electronically separate radiation of different wavelengths. In July 2009, Herschel reached its destination, the second Lagrangian point, and became fully operational. It will study the coolest and 'dustiest' regions in space.

Infrared studies can provide information which could not be obtained in any other way. It is fortunate that considerable amounts of work can be carried out from ground level, though we have to depend very largely upon space satellites.

A selected list of infrared satellites is given in Table 32.5.

MICROWAVE ASTRONOMY

This part of the electromagnetic spectrum extends from 0.3 mm to 30 cm, which is longer than infrared but shorter than radio radiation: sub-millimetre from 0.3 mm, millimetre from 1 mm.

As noted earlier the microwave background at 3 °C above absolute zero was detected in 1964 by A. Penzias and R. Wilson

at the Bell Telephone Laboratories in Holmdel, New Jersey, fortuitously, when they were calibrating a 7.35-cm wavelength receiver built for satellite communications. It is assumed to be the last manifestation of the Big Bang in which the universe was created; the overall temperature has now fallen to 2.7 °C above absolute zero, agreeing with theory. The first slight irregularities in the microwave background were detected in April 1992 with COBE, the Cosmic Background Explorer satellite. These were ripples or rarefied wisps of material, said to be the largest and most ancient structures known in the universe. This discovery was hailed as being among the most important of modern times, and confirmed that it was indeed possible for the initial material to collect into galaxies of the kind we know today.

The COBE satellite was succeeded by WMAP, the Wilkinson Microwave Anisotropy Probe, named after one of its original planners, David Todd Wilkinson (1935–2002). It was launched on 30 June 2001 by NASA, from Cape Canaveral, on a Delta II rocket, and on 1 October reached the Second Lagrangian Point, 1.5 million kilometres from Earth. Its main objective, like COBE's, was to measure temperature differences in the CMB, but it was far more sensitive than COBE. The primary reflecting mirrors make up a pair of Gregorian 1.4 m × 1.6 m dishes facing opposite directions, which focus the incoming signal on to a pair of 0.9 m × 1 m secondary mirrors. Results announced in 2008 gave the Hubble Constant as 71.9 km Mpc s^{-1}, age of the universe 13.6 thousand million years, baryonic content 0.0456 (i.e. 4.56%), cold dark matter content 0.228 (i.e. 22.8%).

The largest ground-based submillimetre telescopes are the JCMT (James Clerk Maxwell Telescope) on Mauna Kea, the SEST (Swedish Submillimetre Telescope) at La Silla in Chile, and the CSO (California Institute of Technology Sub-millimeter Telescope) on Mauna Kea. The first two have 15-m segmented mirrors, and the CSO has a 10.4-m segmented mirror.

RADIO ASTRONOMY

Radio astronomy began in 1931, when the American radio engineer Karl Jansky detected radio waves from the Milky Way. The discovery was fortuitous; Jansky was investigating 'static' on behalf of the Bell Telephone Company, using a homemade aerial nicknamed the 'Merry go-Round' (part of the mounting was made from a dismantled Ford car). Jansky's first paper was published in 1932, but caused surprisingly little interest, and Jansky never followed up his discovery as he might have been expected to do. The first intentional radio telescope

Table 31.6 *Selected list of radio observatories*

Location	Name	Latitude			Longtitude			Altitude (m)
		°	′	″	°	′	″	
Arcetri (Florence), Italy	Astrophysical Observatory	43 N	45	14.4	11 E	15	18	184
Arecibo, Puerto Rico	Arecibo Observatory, Cornell University	18 N	20	36.6	66 W	45	12	496
Big Pine, California, USA	Owens Valley Radio Observatory	37 N	13	54	118 W	16	54	1236
Bochum, Germany	Radio Telescope Station	51 N	25	43.0	07 E	11	30	160
Boulder, Colorado, USA	High Altitude Observatory	40 N	04	42.0	105 W	16	30	1692
Cambridge, England	Mullard Radio Astronomical Obs	52 N	09	45.0	00 E	02	24	26
Cassel, California, USA	Hat Creek Radio Astronomical Observatory	40 N	49	06	121 W	28	24	1043
Cebreros, Spain	Deep Space Station	40 N	27	18	04 E	22	00	789
Columbia, South Carolina, USA	University of South Carolina Radio Observatory	33 N	59	48	81 W	01	54	127
Crimea, Ukraine	Crimean Astrophysical Observatory	44 N	43	42.0	34 E	01	00	550
Culgoora, New South Wales, Australia	Australian Telescope National Facility	30 S	18	54	149 E	33	42	217
Delaware, Ohio, USA	Ohio State Observatory	40 N	15	04.7	83 W	02	54	282
Dwingeloo, Holland	Dwingeloo Radio Observatory	52 N	48	48	06 E	23	48	25
Effelsberg, Germany	Max Planck Institute of Radio Astronmy	50 N	31	36	06 E	53	06	369
Eschweiler, Germany	Stockert (Bonn University)	50 N	34	14.0	06 E	43	24	435
Port Irwin, California, USA	Goldstone Complex	35 N	23	24	116 W	50	54	1036
Gaurtbdnur, India	Gauribidanur Radio Observatory	13 N	36	12	77 E	26	06	686
Green Bank, West Virginia, USA	National Radio Astronomical Observatory	38 N	26	17.0	79 W	50	12	823
Harestua, Norway	University of Oslo Observatory	60 N	12	30.0	10 E	45	30	585
Hartebeeshoek, South Africa	Hartebeeshoek Radio Astronomical Observatory	25 S	53	24	27 E	41	06	1391
Hoskinstown, New South Wales, Australia	Molongo Radio Observatory	35 S	22	18	149 E	25	24	732
Jodrell Bank, England	Nuffield Radio Astronomical Laboratory	53 N	14	11.0	02 W	18	24	70
Kitt Peak, Arizona, USA	National Radio Astronomical Observatory	31 N	57	11.0	111 W	36	48	1920
Kunming, China	Yunnan Observatory	25 N	01	30	102 E	47	18	1940
Mauna Kea, Hawaii, USA	Caltech Submillimetre Observatory	19 S	49	36	155 W	28	18	4060
Mitaka, Japan	National Astronomical Observatory	35 N	40	18	139 E	32	30	58
Nagoya, Japan	Nagoya University Radio Astronomical Laboratory	35 N	08	54	136 E	58	24	75
Nançay, France	Radio Observatory of Nançay	47 N	22	48.0	02 E	11	48	150
Nederhorst den Berg, Holland	Radio Astronomical Observatory	52 N	14	03.0	05 E	04	36	0
Parker, New South Wales, Australia	Australian National Radio Observatory	33 S	00	00.4	148 E	15	42	392
Pulkovo Russia	Astronomical Observatory of the Academy of Sciences	59 N	46	05.0	30 E	19	24	70
Richmond Hill, Ontario, Canada	David Dunlap Observatory	43 N	51	44.0	79 W	25	12	244
St Michael, France	National Centre of Scientitie Research	43 N	55	00.0	05 E	42	30	614
Socorro, New Mexico, USA	National Radio Astronomical Observatory	34 N	04	42	107 W	37	06	81
Sugar Grove, West Virginia, USA	Naval Research Laboratory Radio Station	38 N	31	12	79 E	16	24	705

Table 31.6 (cont.)

Location	Name	Latitude			Longtitude			Altitude (m)
		°	′	″	°	′	″	
Tidbinbilla, Australia	Deep Space Station	35 S	24	06	148 E	58	48	656
Tokya, Japan	Tokyo Observatory at Mitaka	35 N	40	18.2	139 E	32	24	70
Tremsdorf, Germany	Tremsdorf Radio Astronomical Observatory	52 N	17	06	13 E	08	12	35
Uchinoura, Japan	Kagoshima Space Centre	31 N	13	42	131 E	04	00	228
Washington, District of Columbia, USA	Radio Astronomical Observatory National Laboratory	38 N	49	16.6	77 W	01	36	30
Westerbork, Holland	Westerbork Radio Astronomical Observatory	52 N	55	00	06 E	036	18	16
Westford, Massachusetts, USA	Haystack Observatory	42 N	37	24	71 W	029	18	146

was made by the American amateur Grote Reber, whose first paper appeared in 1940; Reber's 'telescope' was a 9.5-m dish. At that time Reber was the only radio astronomer in the world.

Radio waves from the Sun were first detected in 1942 by a British team led by J. S. Hey; originally the effects were thought to be due to German jamming of British radar. However, in February 1946, it was established that the giant sunspot of that month was a strong radio source.

At Jodrell Bank, in Cheshire, work began in 1945, when radar was used to measure meteor trails; in the same year, radar echoes from the Moon were detected by an American team led by J. H. de Witt and, independently, by a Hungarian team led by Z. Bay. The first Jodrell Bank radio telescope was a fixed 66-m paraboloid, due to B. Lovell (now Sir Bernard Lovell).

In 1946, the first discrete radio source beyond the Solar System was detected, Cygnus A, and in 1948 M. Ryle and F. G. Smith (now Sir Francis Graham-Smith) detected the supernova remnant Cassiopeia A. In 1949, the first optical identifications of sources beyond the Solar System were made: Taurus A (the Crab Nebula), Virgo A (the galaxy M 87) and Centaurus A (the galaxy NGC 5128, C 77). Radio waves from Jupiter were detected in 1955 by B. F. Burke and K. Franklin; the first quasar was identified in 1963 by its radio emissions, and the first pulsar in 1967, by Jocelyn Bell Burnell.

In the years after the war, large radio telescopes were built; that at Dwingeloo in Holland dates back to 1956 – its dish is 25 m across. The Lovell Telescope at Jodrell Bank, with a diameter of 76 m (250 feet), came into operation in 1957; originally it was known as the Mark I and from 1970, after major modifications, as the Mark IA. A second large telescope, the Mark II, was completed at Jodrell Bank in 1964. It is designed to operate at the shorter centimetre wavelengths beyond the range of the Lovell Telescope itself. The bowl of the Mark II is elliptical, measuring 38.2 × 25.4 m: it can be used at wavelengths down to 3 cm.

The largest steerable 'dish' is at the Max Planck Institute at Effelsberg in Germany; it was completed in 1971, and has a diameter of 100 m. The largest non-steerable dish is at Arecibo, Puerto Rico; it is 304.8 m across, built in a natural hollow in the ground, and was completed in 1963. The largest array is the VLA (Very Large Array), 80 km west of Socorro in New Mexico; it is Y-shaped, each arm being 20.9 km long, with 27 movable antenna, each 25 m across. It was completed in 1981.

Radio telescopes can be used in conjunction. In England there is the Multi-Element Radio-Linked micrometer Network (MERLIN), with six observing stations which together form a powerful telescope with a positive aperture of over 216 km; MERLIN has a maximum resolution, at 6 cm wavelength, of 40 milliarcsec, which is about 20 times better that can usually be achieved by ground-based telescopes and is comparable with the Hubble Space Telescope. Such a power is equivalent to measuring the diameter of a £1 coin from a distance of 100 km.

The base telescope is at Jodrell Bank, either the Lovell Telescope or the Mark II. Some way away are 25 m dishes at Tabley and Darnhall. A third 25-m dish is at Knockin in Shropshire. More distant is the 25 m telescope at the Royal Signals and Radar Establishment at Defford, and finally there is the 32-m dish at Cambridge. Apart from Defford, all the MERLIN telescopes can work at wavelengths as short as 13 mm, achieving a resolution of 0.04″.

The LOFAR (Low Frequency Array), centred in Holland, was opened on 12 June 2010 by Queen Béatrix of the Netherlands. It was designed and built by the Netherlands Institute for Radio Astronomy, directed by Rene Vermeulen. The 25 000 antennæ are spread over 36 fields in Holland, Germany, Sweden, France and England. Glass fibres connect the antennæ with a supercomputer at the University of Gröningen, simulating a radio telescope of unequalled size. It will be used for studies of the ways in which distant galaxies take shape, how the early Universe was lit up, to map magnetised structures across the sky, and much more.

In Australia, the Commonwealth Scientific and Industrial Research Organisation CSIRO operates the Australia Telescope, which comprises eight radio-receiving antennæ. Six of these are at the Paul Wild Observatory near Narrabri, New South Wales. The other antennæ are the 64-m Parkes radio telescope and the 22 m antenna at Mopra, near Coonabarabran, again in New South Wales. Some of these can be networked with other radio telescopes in various parts of the world in order to make very detailed pictures of small areas of the sky. It has even been extended into space, with the Japanese orbiting radio telescope Halca, launched in 1997. This is known as Very Long Baseline Interferometry.

A selected list of radio astronomy observatories is given in Table 31.6.

32 · The history of astronomy

To give every date of importance in the history of astronomy would be a mammoth undertaking. What I have therefore tried to do is to make a judicious selection, separating out purely space-research advances and discoveries.

It is impossible to say just when astronomy began, but even the earliest men capable of coherent thought must have paid attention to the various objects to be seen in the sky, so that it may be fair to say that astronomy is as old as Homo sapiens. Among the earliest peoples to make systematic studies of the stars were the Mesopotamians, the Egyptians and the Chinese, all of whom drew up constellation patterns. (There have also been suggestions that the constellations we use as a basis today were first worked out in Crete, but this is speculation only.) It seems that some constellation-systems date back to 3000 BC, probably earlier, but of course all dates in these very ancient times are uncertain.

The first essential among ancient civilisations was the compilation of a good calendar. Probably the first reasonably accurate value of the length of the year (365 days) was given by the Egyptians. (The first recorded monarch of all Egypt was Menes. who seems to have reigned around 3100 BC; he was eventually killed by a hippopotamus – possibly the only sovereign ever to have met with such a fate!) They paid great attention to the star Sirius (Sothis), because of its 'heliacal rising', or date when it could first be seen in the dawn sky, gave a reliable clue to the time of the annual flooding of the Nile, upon which the Egyptian economy depended. The Pyramids are, of course, astronomically aligned, and arguments about the methods by which they were constructed still rage as fiercely as ever.

Obviously the Egyptians had no idea of the scale of the universe, and they believed the flat Earth to be all important. So too did the Chinese, who also made observations. It has been maintained that a conjunction of the five naked-eye planets recorded during the reign of the Emperor Chuan Hsü refers to either 2449 or 2446 BC. There is also the legend of the Court Astronomers, Hsi and Ho, who were executed in 2136 BC (or, according to some authorities, 2159 BC) for their failure to predict a total solar eclipse; since the Chinese believed eclipses to be due to attacks on the Sun by a hungry dragon, this was clearly a matter of extreme importance! However, this legend is discounted by modern scholars.

The earliest data collectors were the Assyrians; all students of ancient history know of the Library of Ashurbanipal (668–626 BC). This included the 'Venus Tablet', discovered by Sir Henry Layard and deciphered in 1911 by F. X. Kugler. It claims that when Venus appears, 'rains will be in the heavens'; when it returns after an absence of three months 'hostility will be in the land; the crops will prosper'. Early attempts at drawing up tables of the movements of the Moon and planets may well date from pre-Greek times, largely for astrological reasons; until relatively modern times astrology was

regarded as a true science, and all the ancient astronomers (even Ptolemy) were also astrologers.

Babylonian astronomy continued well into Greek times, and some of the astronomers such as Naburiannu (about 500 BC) and Kidinnu (about 380 BC) may have made great advances; but we know relatively little about them, and reliable dating begins only with the rise of Greek science.

The Greeks transformed astronomy from 'stargazing' to true science. It did not happen quickly; Thales, the first of the great philosophers, was born about 624 BC, while Ptolemy, the last, died about AD 180, so that in his time Ptolemy was as remote from Thales as we are from the Crusades. Yet during that period the Greeks made remarkable progress. There were two vital problems to be solved: the shape of the Earth, and its status in the universe. The Earth was originally believed to be flat, and to be motionless in the central position, with all other bodies moving round it in circular orbits. The first of these problems was solved at a comparatively early stage, but the second was not - at least, not by most of the philosophers. A few, notably Aristarchus (c. 280 BC) were bold enough to maintain that the Sun lay in the centre of the planetary system, but he could give no firm proof, and he did not have many supporters. Even Ptolemy believed in a spherical, motionless Earth.

We know a great deal about ancient astronomy, and we have Ptolemy to thank. He wrote a book which summarises most of the scientific knowledge of the time, and this book has come down to us by way of its Arab translation; we call it the *Almagest*. It was probably written around AD 1050 and translated *c.* AD 1150.

Little progress was made in the following centuries, though there were some interesting Indian writings (Aryabhāta, fifth century AD), and in AD 570 Isidorus, Bishop of Seville, was the first to draw a definite distinction between astronomy and astrology. The revival of astronomy was due to the Arabs. In 813 Al-Ma' mūn founded the Baghdad school of astronomy, and various star catalogues were drawn up, the most notable being that of Al-Sûfî (born about 903). During this period two supernovæ were observed by Chinese astronomers; the star of 1006 (in Lupus) and 1054 (in Taurus, the remnant of which is today seen as the Crab Nebula).

The improved 'Alphonsine Tables' of planetary motions were published in 1270 by order of Alphonso X of Castile. In 1433, Ulūgh Beigh set up an elaborate observatory at Samarkand, but unfortunately he was a firm believer in astrology, and was told that his eldest son Abdallatif was destined to kill him. He therefore banished his son, who duly returned at the head of an army and had Ulūgh Beigh murdered. This marked the end of the Arab school of astronomy, and subsequent developments were mainly European. Some of the important dates in the history of astronomy are as follows:

1460s	Cardinal Nicholas Bessarion stimulated the renewal of Greek astronomy in Europe.
1543	Publication of Copernicus' book *De Revolutionibus Orbium Cælestium*. This sparked off the 'Copernican revolution', which was not really complete until the publication of Newton's *Principia* in 1687.
1545–59	First telescope built by Leonard Digges?
1572	Tycho Brahe observed the supernova in Cassiopeia.
1576–96	Tycho worked at Hven, drawing up the best star catalogue of pre-telescopic times.
1600	Giordano Bruno burned at the stake in Rome, partly because of his defence of the theory that the Earth revolves round the Sun.
1603	Publication of Johann Bayer's star catalogue, *Uranometria*.
1604	Appearance of the last supernova to be observed in our Galaxy (Kepler's Star, in Ophiuchus).
1608	Telescope built by H. Lippershey, in Holland.
1609	First telescopic lunar map, drawn by Thomas Harriot. Serious telescopic work begun by Galileo, who made a series of spectacular discoveries in 1609–10 (phases of Venus, satellites of Jupiter, stellar nature of the Milky Way). Publication of Kepler's first two laws of planetary motion.
1618	Publication of Kepler's third law of planetary motion.
1627	Publication by Kepler of improved planetary tables (the Rudolphine Tables).
1631	First transit of Mercury observed by Gassendi (following Kepler's prediction of it).
1632	Publication of Galileo's *Dialogue*, which amounted to a defence of the Copernican system. He presented it as a fact rather than as a theory. In 1633 he was condemned by the Inquisition in Rome, and was forced into a completely hollow recantation. Founding of the first official observatory (the tower observatory at Leiden, Holland).
1637	Founding of the first national observatory (Copenhagen, Denmark).
1638	Identification of the first variable star (Mira Ceti, by Phocylides Holwarda in Holland).
1639	First transit of Venus observed (by Horrocks and Crabtree, in England, following Horrocks' prediction of it).
1647	Publication of Hevelius' map of the Moon.
1651	Publication of Riccioli's map of the Moon, introducing the modern-type lunar nomenclature.
1655	Discovery of Saturn's main satellite, Titan, by C. Huygens, who announced the correct explanation of Saturn's ring system in the same year.
1656	Founding of the second Copenhagen Observatory.
1659	Markings on Mars seen for the first time (by Huygens).
1663	First description of the principle of the reflecting telescope, by the Scottish mathematician James Gregory.
1665	Newton's pioneering experiments on light and gravitation, carried out at Woolsthorpe in Lincolnshire while Cambridge University was temporarily closed because of the Plague.
1666	First observation of the Martian polar caps, by G. D. Cassini.
1667	Founding of the Paris Observatory, with Cassini as Director. (It was virtually in action by 1671.)
1668	First reflector made, by Newton. (This is the probable date. It was presented to the Royal Society in 1671 and still exists.)
1675	Founding of the Royal Greenwich Observatory. Velocity of light measured, by O. Rømer (Denmark).
1676	First serious attempt at cataloguing the southern stars, by Edmond Halley from St Helena.
1685	First astronomical observations made from South Africa (Father Guy Tachard, at the Cape).
1689	Publication of Newton's *Principia*, finally proving the truth of the theory that the Sun is the centre of the Solar System
1704	Publication of Newton's other major work, *Opticks*.
1705	Prediction of the return of a comet, by Halley (for 1758).
1723	Construction of the first really good reflecting telescope (a 6 in, by Hadley).
1725	Publication of the final version of the star catalogue by Flamsteed, drawn up at Greenwich. (Publication was posthumous.)
1728	Discovery of the aberration of light, by James Bradley.
1750	First extensive catalogue of the southern stars by Lacaille at the Cape. (His observations extended from 1750 to 1752.) Wright's theory of the origin of the Solar System.
1758	First observation of a comet at a predicted return (Halley's Comet, discovered on 25 December by Palitzsch. Perihelion occurred in 1759). Principle of the achromatic refractor discovered by Dollond. (Previously described, by Chester Moor Hall in 1729, but his basic theory was erroneous, and his discovery had been forgotten.)
1761	Discovery of the atmosphere of Venus, during the transit of that year, by M. V. Lomonosov in Russia.
1762	Completion of a new star catalogue by James Bradley; it contained the measured positions of 60 000 stars.
1767	Founding of the *Nautical Almanac*, by Nevil Maskelyne.

1769	Observations of the transit of Venus made from many stations all over the world, including Tahiti (the expedition commanded by James Cook).
1774	First recorded astronomical observation by William Herschel.
1779	Founding of Johann Schröter's private observatory at Lilienthal, near Bremen.
1781	Publication of Charles Messier's catalogue of clusters and nebulæ. Discovery of the planet Uranus, by William Herschel.
1783	First explanation of the variations of Algol, by Goodricke. (Algol's variability had been discovered by Montanari in 1669.)
1784	First Cepheid variable discovered; δ Cephei itself by Goodricke.
1786	First reasonably correct description of the shape of the Galaxy given, by William Herschel.
1789	Completion of Herschel's great reflector with a mirror 49 inch (124.5 cm) in diameter and a focal length of 40 ft (12.2 m).
1796	Publication of Laplace's 'Nebular Hypothesis' of the origin of the Solar System.
1799	Great Leonid meteor shower observed by W. Humboldt.
1800	Infrared radiation from the Sun detected by W. Herschel.
1801	First asteroid discovered (Ceres, by G. Piazzi at Palermo).
1802	Second asteroid discovered (Pallas, by H. Olbers). Existence of binary star systems established by W. Herschel. Dark lines in the solar spectrum observed by W. H. Wollaston.
1804	Third asteroid discovered (Juno, by K. Harding).
1807	Fourth asteroid discovered (Vesta, by Olbers).
1814–18	Founding of the Calton Hill Observatory, Edinburgh.
1815	Fraunhofer's first detailed map of the solar spectrum (324 lines).
1820	Foundation of the Royal Astronomical Society.
1821	Arrival of F. Fallows at the Cape as Director of the first observatory in South Africa. Founding of the Paramatta Observatory by Sir Thomas Brisbane, Governor of New South Wales. (This was the first Australian observatory. It was dismantled in 1847.)
1822	First calculated return of a short-period comet (Encke's recovered by Rümker at Paramatta).
1824	First telescope to be mounted equatorially, with clock drive (the Dorpat refractor, made by Fraunhofer).
1827	First calculation of the orbit of a binary star (ξ Ursæ Majoris, by F. Savary).
1829	Completion of the Royal Observatory at the Cape.
1834–8	First really exhaustive survey of the southern stars, carried out by John Herschel at Feldhausen (Cape).
1835	Second predicted return of Halley's Comet.
1837	Publication of the famous lunar map by W. Beer and Mädler. Publication of the first good catalogue of double stars (W. Struve's *Mensuræ Micrometricæ*).
1838	First announcement of the distance of a star (61 Cygni, by F. W. Bessel).
1839	Pulkovo Observatory completed.
1840	First attempt to photograph the Moon (by J. W. Draper).
1842	Important total solar eclipse, from which it was inferred that the corona and prominences are solar rather than lunar. First attempt to photograph totality (by G. A. Majocci), though he recorded only the partial phase.
1843	First daguerreotype of the solar spectrum obtained (by Draper).
1844	Founding of the Harvard College Observatory (first official observatory in the United States). The 15-inch refractor was installed in 1847.
1845	Completion of Lord Rosse's 72-inch reflector at Birr Castle, and the discovery with it of the spiral forms of galaxies ('spiral nebulæ'). Daguerreotype of the Sun taken by A. Fizeau and L. Foucault, in France. Discovery of the fifth asteroid (Astræa, by K. Hencke).
1846	Discovery of Neptune, by J. Galle and H. D'Arrest at Berlin, from the prediction by U. J. J. Le Verrier. The large satellite of Neptune (Triton) was discovered by W. Lassell in the same year.
1850	First photograph of a star (Vega, from Harvard College Observatory). Castor was also photographed and the image was extended, though the two components were not shown separately. Discovery of Saturn's Crêpe Ring (W. Bond, at Harvard).
1851	First photograph of a total solar eclipse (by M. Berkowski). H. Schwabe's discovery of the solar cycle established by W. Humboldt.
1857	James Clerk Maxwell proved that Saturn's rings must be composed of discrete particles. Founding of the Sydney Observatory. First good photograph of a double star (Mizar, with Alcor, by Bond, Whipple and Black).
1858	First photograph of a comet (Donati's, photographed by Usherwood).
1859	Explanation of the absorption lines in the solar spectrum given by G. Kirchhoff and R. Bunsen. Discovery of the Sun's differential rotation (by R. Carrington).

1859 Total solar eclipse. Final demonstration that the corona and prominences are solar rather than lunar.

1861–2 Publication of Kirchhoff's map of the solar spectrum.

1862 Construction of the first great refractors, including the Newall 25 inch made by Cooke. (It was for many years at Cambridge, and is now in Athens.) Discovery of the companion of Sirius (by A. Clark, at Washington). Completion of the *Bonner Durchmusterung*.

1863 A. Secchi's classification of stellar spectra published.

1864 W. Huggins' first results in his studies of stellar spectra. First spectroscopic examination of a comet (Tempel's by Donati). First spectroscopic proof that 'nebulæ' are gaseous (by Huggins). Founding of the Melbourne Observatory. (The 'Great Melbourne Reflector' completed 1869.)

1866 Association between comets and meteors established (by G. V Schiaparelli). Great Leonid meteor shower. Announcement by J. Schmidt of an alteration in the lunar crater Linné. (Though the reality of change is now discounted, regular lunar observation dates from this time.)

1867 Studies of 'Wolf–Rayet' stars by M. Wolf and G. Rayet, at Paris.

1868 First description of the method of observing the solar prominences at times of non-eclipse (independently by P. Janssen and J. N. Lockyer). Publication of a detailed map of the solar spectrum, by A. Ångström.

1870 First photograph of a solar prominence (by C. Young).

1872 First photograph of the spectrum of a star (Vega, by H. Draper, son of J. W. Draper).

1874 Transit of Venus; solar parallax redetermined. (Another transit occurred in 1882, but the overall results were disappointing.) Founding of observatories at Meudon (France) and Adelaide (Australia).

1876 First use of dry gelatine plates in stellar photography; spectrum of Vega photographed by Huggins.

1877 Discovery of the two satellites of Mars (by A. Hall, at Washington). Observations of the 'canals' of Mars (by Schiaparelli, at Milan).

1878 Publication of the elaborate lunar map by J. Schmidt (from Athens). Completion of the Potsdam Astrophysical Observatory.

1879 Founding of the Brisbane Observatory.

1880 First good photograph of a gaseous nebula (M 42, by H. Draper).

1882 D. Gill's classic photograph of the Great Comet of 1882, showing so many stars that the idea of stellar cataloguing by photography was born.

1885 Founding of the Tokyo Observatory. (An earlier naval observatory in Tokyo had been established in 1874.) Supernova in M 31, the Andromeda Galaxy (S Andromedæ). This was the only recorded extragalactic supernova to reach the fringe of naked-eye visibility until 1987.

1886 Photograph of M 31 (the Andromeda Galaxy) by I. Roberts showing spiral structure. (A better photograph was obtained by him in 1888.)

1887 Completion of the Lick 36 in refractor.

1888 Publication of J. L. E. Dreyer's *New General Catalogue* of clusters and nebulæ. H. Vogel's first spectrographic measurements of the radial velocities of stars.

1889 Spectrum of M 31 photographed by J. Scheiner, from Potsdam. Discovery at Harvard of the first spectroscopic binaries (ζ Ursæ Majoris and β Aurigæ). First photographs of the Milky Way taken (by E. E. Barnard).

1890 Foundation of the British Astronomical Association. Unsuccessful attempts to detect radio waves from the Sun, by T. Edison. (Sir Oliver Lodge was equally unsuccessful in 1896.) Publication of the Draper Catalogue of stellar spectra.

1891 Completion of the Arequipa southern station of Harvard College Observatory. Spectrohcliograph invented by G. E. Hale. First photographic discovery of an asteroid (by Wolf, from Heidelberg).

1892 First photographic discovery of a comet (by E. E. Barnard).

1893 Completion of the 28-inch Greenwich refractor.

1894 Founding of the Lowell Observatory at Flagstaff, in Arizona.

1896 Publication of the first lunar photographic atlas (Lick), Founding of the Perth Observatory. Completion of the Meudon 33-inch (83-cm) refractor. Completion of the new Royal Observatory at Blackford Hill, Edinburgh. Discovery of the predicted Companion to Procyon (by Schaeberle).

1897 Completion of the Yerkes Observatory.

1898 Discovery of the first asteroid to come well within the orbit of Mars (433 Eros, discovered by C. G. Witt at Berlin).

1899 Spectrum of the Andromeda Galaxy (M 31) photographed by Schelner.

1900 Publication of Bumham's catalogue of 1290 double stars. Horizontal refractor, of 49-inch aperture, focal length 197 ft (60 m), shown at the Paris Exhibition. (It was never used for astronomical research.)

1905 Founding of the Mount Wilson Observatory (California).

1908 Giant and dwarf stellar divisions described by E. Hertzsprung (Denmark). Completion of the Mount Wilson 60-inch reflector. Fall of the Siberian meteorite.

1912 Studies of short-period variables in the Small Magellanic Cloud, by Miss H. Leavitt, leading on to the period–luminosity law of Cepheids.

1913 Founding of the Dominion Astrophysical Observatory, Victoria (British Columbia). H. N. Russell's theory of stellar evolution announced.

1915 W. S. Adams' studies of Sirius B, leading to the identification of white dwarf stars.

1917 Completion of the 100-inch Hooker reflector at Moon Wilson (the largest until 1948).

1918 Studies by H. Shapley leading him to the first accurate estimate of the size of the Galaxy.

1919 Publication of Barnard's catalogue of dark nebulæ.

1920 The Red Shifts in the spectra of galaxies announced by V. M. Slipher.

1923 Proof given (by E. Hubble) that the galaxies are the independent systems rather than parts of our Milky Way system. Invention of the spectrohelioscope, by Hale.

1925 Establishment of the Yale Observatory at Johannesburg. (It was finally dismantled in 1952, its work done.)

1927 Completion of the Boyden Observatory at Bloemfontein, South Africa.

1930 Discovery of Pluto, by Clyde Tombaugh at Flagstaff. Invention of the Schmidt camera, by Bernhard Schmidt (Estonia).

1931 First experiments by K. Jansky at Holmdel, New Jersey, with an improvised aerial, leading on to the founding of radio astronomy. Jansky published his first results in 1932, and in 1933 found that the radio emission definitely came from the Milky Way.

1932 Discovery of carbon dioxide in the atmosphere of Venus (by T. Dunham).

1933–5 Completion of the David Dunlap Observatory near Toronto (Canada).

1937 First intentional radio telescope built (by Grote Reber); it was a 'dish', 31 ft (9.4 m) in diameter.

1938 New (and correct) theory of stellar energy proposed by H. Bethe and, independently, by C. von Weizsäcker.

1942 Solar radio emission detected by M. H. Hey and his colleagues (27–28 February). The emission had previously been attributed to intentional jamming by the Germans!

1944 Suggestion, by H. C. van de Hulst, that interstellar hydrogen must emit radio waves at a wavelength of 21.2 cm.

1945 Thermal radiation from the Moon detected at radio wavelengths (by R. H. Dicke).

1945–6 First radar contact with the Moon, by Z. Bay (Hungary) and independently by the US Army Signal Corps Laboratory.

1946 Work at Jodrell Bank begun (radar reflections from the Giacobinid meteor trails, 10 October). Beginning of radio astronomy in Australia (solar work by a team led by E. G. Bowen). Identification of the radio source Cygnus A by Hey, Parsons and Phillips.

1947–8 Photoelectric observations of variable stars in the infrared carried out by Lenouvel, using a Lallemand electronic telescope.

1948 Completion of the 200-inch Hale reflector at Palomar (USA). Identification of the radio source Cassiopeia A, by M. Ryle and F. G. Smith.

1949 Identification of further radio sources: Taurus A (the Crab Nebula), Virgo A (M 87), and Centaurus A (NGC 5128). These were the first radio sources beyond the Solar System to be identified with optical objects.

1950 M 31 (the Andromeda Galaxy) detected at radio wavelengths by M. Ryle. F. G. Smith and B. Elsmore. Funds for the building of the great Jodrell Bank radio telescope obtained by Sir Bernard Lovell.

1951 Discovery by H. Ewen and E. Purcell of the 21-cm emission from interstellar hydrogen, thus confirming van de Hulst's prediction. Optical identification of Cygnus A and Cassiopeia A (by W. Baade and R. Minkowski, using the Palomar reflector, from the positions given by Smith).

1952 W. Baade's announcement of an error in the Cepheid luminosity scale, showing that the galaxies are about twice as remote as had been previously thought. Electronic images of Saturn and θ Orionis obtained by A. Lallemand and M. Duchesne (Paris). Tycho's supernova of 1572 identified at radio wavelengths by R. Hanbury Brown and C. Hazard.

1953 I. Shklovskii explains the radio emission from the Crab Nebula as being due to synchroton radiation.

1955 Completion of the 250-ft radio 'dish' at Jodrell Bank. First detection of radio emissions from Jupiter (by B. F. Burke and K. Franklin). Construction of a radio interferometer by M. Ryle, and also the completion of the second Cambridge catalogue of radio sources. (The third Cambridge catalogue was completed in 1959.)

1958 Observations of a red event in the lunar crater Alphonsus, by N. A. Kozyrev (Crimean Astrophysical Observatory, USSR). Venus detected at radio wavelengths (by Mayer).

1959 Radar contact with the Sun (V. Eshleman, at the Stanford Research Institute, USA).

1960	Aperture synthesis method developed by M. Ryle and A. Hewish.
1961	Completion of the Parkes radio telescope, 330 km west of Sydney.
1962	Thermal radio emission detected from Mercury, by W. E. Howard and colleagues, using the 85-ft radio telescope at Michigan. First radar contact with Mercury (Kotelnikov, USSR). First X-ray source detected (in Scorpius). Sugar Grove fiasco; the US attempt to build a 600-ft fully steerable radio 'dish'. Work had begun in 1959, and when discontinued had cost $96 000 000.
1963	Announcement by P. van de Kamp, of a planet attending Barnard's Star (later found to be spurious). Identification of quasars (M. Schmidt, Palomar).
1965–6	Identification of the 3 K microwave radiation, as a result of theoretical work by Dicke and experiments by A. Penzias and R. Wilson.
1967	Completion of the 98-inch Isaac Newton reflector at Herstmonceux. Identification of the first pulsar, CP 1919, by Jocelyn Bell at Cambridge.
1968	Identification of the Vela pulsar (Large, Vaughan and Mills).
1969	First optical identification of a pulsar; the pulsar in the Crab Nebula by Cocke, Taylor and Disney at the Steward Observatory, USA.
1970	Completion of the 100-m radio 'dish' at Bonn (Germany). Completion of the large reflectors for Kitt Peak (Arizona) and Cerro Tololo (Chile); each 158-inch (401-cm) aperture. First large reflector to be erected on Mauna Kea, Hawaii; an 88-inch (224-cm).
1973	Opening of the Sutherland station of the South African Astronomical Observatories.
1974	Completion of the 153-inch (389-cm) reflector at the Siding Spring Observatory, Australia.
1976	Completion of the 236-inch (600-cm) reflector at Mount Semirodriki (USSR).
1977	Optical identification of the Vela pulsar (at Siding Spring). Discovery of Chiron (by C. Kowal, USA). Discovery of the rings of Uranus.
1978	Completion of the new Russian underground neutrino telescope. Discovery of Charon, the satellite of Pluto (J. Christy, USA). Discovery of the first satellite of an asteroid (Herculina). Rings of Uranus recorded from Earth (Matthews, Neugebauer and Nicholson). Discovery of X-rays from SS Cygni (HEAO 1).
1979	Official opening of the observatory at La Palma. Pluto and Charon recorded separately (D. Bonneau and F. Foy, Mauna Kea, thereby confirming Charon's independent existence). First comet observed to hit the Sun.
1980	Discovery of the first scintar (SS 433).
1981	Five asteroids contacted by radar from Arecibo, including two Apollos (Apollo itself, and Quetzalcoatl).
1982	Discovery of the remote quasar PKS 2000–330 (Wright and Launcey, Parkes). Recovery of Halley's Comet.
1983	Discovery of the fastest-vibrating pulsar, PKS 1937+215 in Vulpecula: period 1.557 806 449 022 milliseconds – twenty times shorter than the Crab pulsar. It spins 642 times per second.
1984	Isaac Newton Telescope installed on La Palma.
1986	Return of Halley's Comet.
1987	Completion of the William Herschel telescope at La Palma. Completion of the James Clerk Maxwell telescope on Mauna Kea. Supernova seen in the Large Cloud of Magellan.
1988	Completion of the Australia Telescope (radio astronomy network). Collapse of the Green Bank radio telescope.
1989	New Technology Telescope (NTT) brought into action at the European Southern Observatory, La Silla. Identification of the 'Great Wall' of galaxies.
1990	First brown dwarf identified by M. Hawkins. First surface details on a star (Betelgeux) detected from La Palma Observatory. First light on the Keck Telescope (Mauna Kea). White spot discovered on Saturn (24 September). End of the Pluto–Charon mutual phenomena (24 September). Sir Francis Graham-Smith retires as Astronomer Royal.
1991	Professor Arnold Wolfendale appointed Astronomer Royal. First really reliable measurement made of the distance of the Large Magellanic Cloud. Outburst of Halley's Comet (12 February). Fall of the Glatton Meteorite (5 May). First space image obtained of an asteroid, Gaspra (13 November).
1992	Completion of the Keck I telescope on Mauna Kea, Hawaii. Discovery of the first Kuiper Belt object (D. Jewitt and J. Luu 31 August).
1993	Discovery of the first asteroidal satellite, Dactyl (25 August). Start of SETI, the Search for Extra-terrestrial Intelligence (12 October). Galileo exonerated of heresy by the Pope (30 October). (Galileo had been condemned for heresy on 22 June 1633!)
1994	Hooker telescope on Mt Wilson reopened. SETI cancelled by US Congress (14 March). Impact of Comet Shoemaker–Levy 9 on Jupiter (17–22 July). Discovery of the nearest galaxy, 80 000 light-years away (Ibata, 4 August).

1995   Professor Sir Martin Rees succeeds Sir Arnold Wolfendale as Astronomer Royal (January). First light on the VATT (Mount Graham) (July). Mayor and Queloz announce the discovery of a planet round 51 Pegasi (October). First maps of Vesta (HST) (November). Galileo probe impacts Jupiter (7 December).

1996   Hipparcos catalogue completed (February). HST sends back images of star-forming regions in M16 (Eagle Nebula) (February). First surface details recorded on Pluto (HST) (March). Dedication of Keck II Telescope (8 May).

1997   First images of Mars from Pathfinder (4 July). Dedication of Hobby–Eberly Telescope (8 October).

1998   First light on Antu (first mirror of VLT) (May). Closure of the Royal Greenwich Observatory (31 October).

1999   First images released from Suburu. telescope (January). First light on Kueyen (second mirror of VLT) (March). New mirror installed at the Rosse telescope, Birr (22 June). Inauguration of the Gemini North telescope on Mauna Kea (25–27 June). Total solar eclipse seen from Cornwall and Devon (11 August). Brilliant Leonid meteor shower (18 November). December 1999 reopening of the Rosse Telescope at Birr.

2000   First light on Melipal (third mirror of VLT) (January 26). Asteroid Eros mapped from close range by the Shoemaker probe (February). First detection of an isolated black hole.

2001   After-glow of gamma-ray burster observed by Antu and 1.4-m Danish telescope (31 January). First images from the Gemini North telescope (March). Publication of results from the Boomerang experiment (April). Recovery of the lost asteroid Albert (1 May). First light on the new MMT (Mono-Mirror Telescope, 19 May). IAU General Assembly (August). First results from the new Green Bank Telescope (August). Diameters of six stars measured with Antu and Melipal.

2002   Outburst of V838 Monocerotis discovered (6 January). Dedication of the Gemini South telescope at Cerro Pachon (18 January). Possible identification of the first quark stars. Violent eruption on Io, near the volcano Surt, detected by the Keck telescope (20 February). Discovery of the Silverpit impact crater, in the sea, 130 km from Hull. Discovery of Quaoar (4 June). Fall of Peruvian meteorite (2 September). Death of Grote Reber (20 December). Clouds near Titan's south pole observed by Keck II and Gemini North.

2003   Devastating bush fire at Mount Stromlo Observatory (18 January). Asteroid Hermes recovered by B. Skiff (15 October). Discovery of the first starless galaxy, HVC 127–41–330.3. Discovery of Sedna (14 November). Most powerful solar flare recorded (4 November).

2004   Discovery of McNeil's Nebula (January). N. Chafoor and J. Zarnecki suggest that there may be waves in Titan's lakes (April). First extra-solar planet found by the gravitational lensing method (April). SuperWASP inaugurated at La Palma (April). Arecibo telescope upgraded by Arecibo L–Band Feed Display (ALFA) (21 April). Completion of the restoration of Mount Stromlo Observatory (Oct). Violent storms on Uranus detected by the Keck II telescope.

2005   Proper motion of M 33 detected by use of the Versorgungsanstalt des Bundes und der Länder (VBL). New maps made of the Silverpit crater. Lyrid meteor storm (24 April). Occultation of a star by Pluto observed from Las Campanas (10 July). Supermassive black hole in M 31 suspected from observations with the HST. First light with the Large Binocular Telescope (12 October).

2006   Outburst of RS Ophiuchi (12 February). Discovery of a second red spot on Jupiter (24 February). Two planemos identified in a star-forming region (4 August). Discovery of McNaught's Comet (7 August).

2007   Saturn's Hexagon discovered (27 March). First light on the Telescopio Gran Canarias (18 July). Tail of Encke's Comet ripped off by a coronal mass ejection (20 April). First 42 dishes of the Allen Telescope Array activated (11 October).

2008   Asteroid 2008 TC3 impacted in the Sudan (7 October).

2009   Impact scar on Jupiter discovered by A. Wesley (20 July). Discovery of Saturn's Phœbe ring (6 October).

2010   Using the VLT, astronomers identify the most distant galaxy known: UD Fy 38135539, distance 13.3 thousand million light-years.

HISTORY OF SPACE RESEARCH

It is no longer possible to separate what may be called 'pure astronomy' from space research. The Space Age began on 4 October 1957 with the launch of Russia's first artificial satellite, Sputnik 1. The following list has been restricted to the more 'astronomical' events. Full lists of all lunar and planetary probes are not given here, as they will be found elsewhere in this book.

Pre-1957

c. 150   Lucian of Samosata's *True History* about a journey to the Moon – possibly the first of all science-fiction stories.

1232   Military rockets used in a battle between the Chinese and the Mongols.

1610   John Wilkins writes *Discovery of a New World*, in which he proposes a mechanical 'flying chariot' in which to fly to the Moon.

c. 1805   Colonel William Congreve, of Woolwich Arsenal, develops the first efficient – though inaccurate – large solid fuel rockets, to be used against Napoleon's invasion forces.

1865   Publication of Jules Verne's novel *From the Earth to the Moon*.

1881   Early rocket design by N. I. Kibaltchitch. (Unwisely, he made the bomb used to kill the tzar of Russia, and was predictably executed.)

1891   Public lecture about space-flight by the eccentric German inventor Hermann Ganswindt.

1895   First scientific papers about space-flight by K. E. Tsiolkovskii. He published important papers in 1903 and subsequent years. The Russians refer to him as 'the father of space-flight'.

1919   Monograph, *A Method of Reaching Extreme Altitudes*, published by R. H. Goddard in America. This included a suggestion of sending a small vehicle to the Moon, and adverse press comments made Goddard disinclined to expose himself to further ridicule.

1924   Publication of *The Rocket into Interplanetary Space*, by H. Oberth. This was the first truly scientific account of space-research techniques.

1926   First liquid-propelled rocket launched by Goddard.

1927   Formation of the German rocket group, *Verein für Raumschiffahrt*.

1931   First European firing of a liquid-propelled rocket (Winkler, in Germany).

1937   First rocket tests at the Baltic research station at Peenemünde. One of the leaders of the team was Wernher von Braun.

1942   First firing of the A4 rocket (better known as the V2) from Peenemünde. In 1944–5, many V2s fell upon southern England.

1945   White Sands proving ground established in New Mexico. Idea of synchronous artificial satellites for communications purposes proposed by Arthur C. Clarke.

1949   First step-rocket fired from White Sands; it reached an altitude of almost 400 km. Rocket testing ground established at Cape Canaveral, Florida.

1955   Announcement of the US 'Vanguard' project for launching artificial satellites.

The Space Age

1957   4 October: launching of the first artificial satellite Sputnik I (USSR).

1958   First successful US artificial satellite (Explorer 1) led by Wehrner von Braun. Instruments carried in it were responsible for the detection of the Van Allen radiation zones surrounding the Earth.

1959   First lunar probes; Lunas 1, 2 and 3 (all USSR). Luna 1 by-passed the Moon, Luna 2 crash-landed there, and Luna 3 went on a round trip sending back pictures of the Moon's far side.

1960   First television weather satellite (Tiros 1, USA).

1961   First attempted Venus probe (USSR); contact with it lost. First manned space-flight (Yuri Gagarin, USSR). First manned US space-flight (Shepard; sub-orbital).

1962   First American to orbit the Earth (Glenn). First British built satellite (Ariel 1) launched from Cape Canaveral. First transatlantic television pictures relayed by satellite (Telstar). First attempted Mars probe (USSR; contact lost). First successful planetary probe: Mariner 2 to Venus.

1963   First occasion when two manned space-craft were in orbit simultaneously (Nikolayev and Popovich, USSR). First space-woman: Valentina Tereshkova-Nikolayeva (USSR).

1964   First good close-range photographs of the Moon (Ranger 7, USA).

1965   First 'space-walk' (Leonov, USSR). First successful Mars probe (Mariner 4, USA).

1966   First soft landing on the Moon by an automatic probe (Luna 9, USSR). First landing of a probe on Venus (Venera 3, USSR), though contact with it was lost. First soft landing of an American probe on the Moon (Surveyor 1). First circum-lunar probe (Luna 10, USSR). First really good close-range lunar pictures (Orbiter 1, USA).

1967   First soft landing of an unmanned probe on Venus (Venera 4, USSR).

1968   First recovery of a circum-lunar probe (Zond 5, USSR). First manned Apollo orbital flight (Apollo 7: Schirra, Cunningham, Eisele, USA). First manned flight round the Moon: Apollo 8 (Borman, Lovell, Anders, USA).

1969   First testing of the lunar module in orbit round the Moon (Apollo 10: Stafford, Cernan, Young, USA). 21 July, first lunar landing (Apollo 11; Armstrong, Aldrin, USA).

1970   First Chinese and Japanese artificial satellites.

1971   First capsule landed on Mars, from the USSR probe Mars 2.

1971–2   First detailed pictures of Martian volcanoes, obtained from the probe Mariner 9, which entered orbit round the planet (USA).

1972   End of the Apollo programme, with Apollo 17 (Cernan, Schmitt, Evans, USA).

1973–4   Operational 'life' of the US space-station Skylab, manned by three successive three-man crews, and from which much pioneer astronomical work was carried out.

1973   First close-range information from Jupiter (including pictures) obtained from the fly-by probe Pioneer 10. (Pioneer 11 repeated the experiments in 1974.)

The Space Age

Pioneer 10 was also the first probe to escape from the Solar System, while Pioneer 11 was defined to be the first probe to by-pass Saturn (in 1979) (USA).

1974 First pictures of the cloud-tops of Venus from close range, from the two-planet probe Mariner 10; the probe then encountered Mercury, and sent back the first pictures of the cratered surface.

1975 First pictures received from the surface of Venus, from the Russian probes Venera 9 and Venera 10.

1976 First successful soft landings on Mars (Vikings 1 and 2) sending back direct pictures and information from the surface of the planet (USA).

1977 Launching of Voyagers 1 and 2 to the outer planets (USA). Death of Wernher von Braun.

1978 Launch of two Pioneer probes to Venus – the first American attempts to put a vehicle into orbit round the planet and to land capsules there. Launching of the X-ray 'Einstein Observatory'; this operated successfully for over two years.

1979 Fly-by of Jupiter by Voyagers 1 and 2 (USA). Decay of Skylab in the Earth's atmosphere (11 July).

1980 Voyager 1 fly-by of Saturn, obtaining data from Titan and other satellites during its pass.

1981 Successful Voyager 2 pass of Saturn (USA).

1982 Landings of Veneras 13 and 14 on Venus, obtaining improved pictures and data (USSR). Longest space-mission undertaken by the Russians (2 months in orbit; Berezevoy and Lebedev in Salyut 7).

1983 Launch of IRAS (Infra-Red Astronomical Satellite).

1986 Voyager 2 fly-by of Uranus. Shuttle disaster, with the destruction of the Challenger. Five probes to Halley's Comet; two Japanese, two Russian, and one European (Giotto). 20 February, launch of Mir space station.

1987 Longest space mission (326 days) completed by Yuri Romanenko, on the space-station Mir.

1988 Longest space mission (366 days on the Mir station) completed by Titov and Manorov. Failure of the Phobos 1 probe (contact lost, 29 August).

1989 Failure of the Phobos 2 probe (contact lost, 29 March). Magellan probe to Venus launched from the Shuttle *Atlantis* (5 May). Hipparcos astrometric satellite launched (8 August). Voyager 2 passed Neptune (24 August). Galileo probe launched to Jupiter (18 October).

1990 First Japanese launch to the Moon (Muses-A, 24 January). Hubble Space Telescope launched from the Shuttle Discovery (25 April). Launch of Ulysses (6 October).

1991 Gamma-Ray Observatory (Arthur Holly Compton Observatory) launched, 5 April. First survey of the sky in extreme ultraviolet (Rosat satellite). Spectacular images of Venus sent back by the Magellan probe. Launch of Yohkoh (30 August).

1992 Ulysses (solar polar probe) flew past Jupiter (8 February). Giotto encountered Comet Grigg–Skjellerup (10 July). Discovery in April of slight variations in the microwave background radiation (COBE). First space-walk by three astronauts simultaneously (13 May). Repair of Intelsat-6 satellite. Mars Observer launched (25 September). Pioneer Venus ceased transmitting (9 October).

1993 First gravitational wave test using satellites (21 March–12 April). Japanese probe Hiten impacted on the Moon (10 April), Hipparcos completed an all-sky survey at extreme ultraviolet wavelengths (18 November). Repair to the Hubble Space Telescope by astronauts (2–13 December). COBE switched off (23 December).

1994 Launch of lunar probe Clementine (24 January); it entered lunar orbit in February, but failed to complete its programme by going on to a rendezvous with the asteroid Geographos. Mars Observer lost (2 August). Decay of Magellan probe in Venus' atmosphere (10 October).

1995 Launch of Infra-red Space Observatory (ISO) (1 November). Launch of Solar and Heliospheric Observatory (SOHO) (2 December).

1996 Loss of Cluster mission (June). The IUE (International Ultra-violet Explorer) shut down after 18 years (30 September). Launch of Mars Global Surveyor (7 November). Launch of Pathfinder to Mars (4 December).

1997 The Hubble Space Telescope reactivated after second servicing mission (19 February). Pioneers 6, 7 and 8 cease to be funded and tracked (31 March). Collision between Mir and Progress rocket (25 June). Near Earth Asteroid Rendezvous (NEAR) pass of asteroid Mathilde (June 27). Pathfinder lands on Mars (July 4). Mars Global Surveyor arrives in Martian orbit (September 11). Last full transmission from Pathfinder (September 27). Launch of Cassini mission to Saturn (October 15).

1998 Launch of Lunar Prospector (6 January). NEAR passes Earth at 148 000 km (23 January). End of ISO mission (8 April). Cassini passes Venus at 284 km (26 April). Launch of Nozomi, Japanese Mars probe (3 July). Launch of Mars Climate Orbiter (11 December). NEAR (Shoemaker probe) passes Eros, and sends back images (23 December).

1999 Launch of Mars Polar Lander (3 January). Launch of Stardust probe to Comet P/Wild 2 (15 January). Cassini passes Venus at 600 km (24 June). Launch of Chandra X-ray probe (23 July). Crash of Prospector on the Moon (30 August). Loss of Mars Climate Orbiter (3 September). Leonid meteor storm (18 November). Launch of Shenzhou, Chinese spacecraft in preparation for a manned vehicle (21 November). Mars Polar Lander lands on Mars

The Space Age

(3 December) but no contact re-established. Launch of Newton X-ray satellite (10 December 10). Third servicing mission to the Hubble Space Telescope (19–27 December).

2000   Launch of Stardust probe (7 February). Galileo flies past Io at 200 km (22 February). Compton satellite brought down in the Pacific (4 June). Cluster launches (4 June and 9 August). Cassini passes Jupiter at 10 000 000 km (30 December).

2001   Last signals from NEAR on Eros (28 February). Re-entry of Mir (23 March). Launch of Mars Odyssey (7 April). First space tourist, Dennis Tito (6 May). Launch of WMAP (June 30). Launch of Genesis probe (8 August). Deep Space probe flew past Borrelly's Comet (22 September).

2003   Launch of Spirit to Mars (10 June), followed by launch of Opportunity (7 July). Foundation of JAXA, Japan (1 October). Launch of Hayabusa to 25143 Itokawa (Oct). Launch of first Chinese taikonaut, Yang Li-wei, Shenzhou-5 (15 October). Loss of Beagle 2 (25 December).

2005   Launch of Mars Reconnaissance Orbiter (12 April). Flight of Shenzhou 6, carrying taikonauts Fei Junlong and Nie Haisheng (12–17 October). Hayabusa lands briefly on Itokawa (November).

2006   Stardust probe lands in Utah (3 January). Launch of Akari, JAXA infrared satellite (21 February). European Space Agency probe SMART-1 impacts on Moon (4 September). South African space agency founded.

2007   Old Chinese satellite Feng Yun 1c destroyed by JAXA, scattering débris into orbit (11 January). Launch of GLAST (12 June). Phoenix space-craft lands on Mars (4 August); transmitted until 2 November. Launch of Japanese lunar probe Kaguya (14 September); entered lunar orbit 4 October. Launch of Chinese lunar probe, Chang-ē 1 (24 October).

2008   Rogue US satellite destroyed by NASA (10 February). Launch of Indian lunar probe Chandrayaan-1 (22 October).

2009   Herschel and Planck Observatories launched from Ariane-5; both reached L2 point in July. First images from Lunar Reconnaissance Orbiter (23 June). First images of Apollo craft on the Moon imaged by the Lunar Reconnaissance Orbiter (LRO) (July). MESSENGER probe passes Mercury at 228 km (29 September); cross impacts the Moon (9 October). Third MESSENGER pass of Mercury, setting course for Mercury orbit (6 October).

2009   1 March, China's Chang-ē 1 lunar orbiter deliberately crashed; impact 1.5° S, 52.36° E.

2010   The SOFIA (Stratospheric Observatory for Infrared Astronomy) is a 7475 P aircraft modified to carry a 2.5-m telescope. The first images were obtained during the flight of 25–26 May 2010. It can fly up to 13 700 m. As the Pioneer spacecraft leave the Solar System it has been discovered that they are not moving as they ought to. They should slow down under the influence of the Sun's gravity, but they are slowing down more than expected. This has been called the Pioneer Anomaly.

Add 4 astronomers who died recently in the astronomers section at the back. This section is not indexed so not problem just to add these in the right place alphabetically.

Alpher,   Ralph. 1921–2007. American cosmologist of Russian descent. He made vital contributions to nucleosynthesis – the way in which complex elements are formed from lighter elements – and this relates to conditions following the Big Bang. He worked closely with George Gamow, and for one paper persuaded Hans Bethe to add his name, so that it became the Alpher-Bethe-Gamow (alpha-beta-gamma) paper. He never received full credit for his outstanding achievements. For example, he was the first to predict the temperature of the cosmic background radiation.

Gordon,   William. 1918–2010. American astronomer. He graduated from Cornell University in 1958 and planned and designed the 305-m (1000 ft) Aricebo telescope in Puerto Rico, still the largest radio "dish" in the world. He was Director there until 1965, and was professor at Rice University until 1986.

Marsden,   Brian Geoffrey. 1937–2010. English astronomer. One of the most influential comet investigators of modern times. He was born and educated in Cambridge after which he studied at New College, Oxford, took his Ph.D. at Yale University in America and then went to the Smithsonian Institution where he remained until he retired in 2006. He correctly predicted the returns of several 'lost' comets, including Swift-Tuttle in 1992. Asteroid 1877 is named after him.

Sandage,   Allan Rex. 1926–2010. Leading American cosmologist, best known for his work on the size and age of the Universe. He worked closely with Hubble, and continued to work after Hubble's death. After spells at CalTech and Palomar he joined the Carnegie Observatories in Pasadena in 1952 and spent the rest of his career there. Asteroid 9963 is named after him.

2010   The world's first spaceport. Space tourism came a step nearer on 22 October 2010, when tycoon Richard Branson and his team completed the main runway at Spaceport America, some way from the town of Las Cruces. The spacecraft VSS Enterprise, 18 metres long, is designed to take paying passengers on suborbital flights. A ticket will cost a mere $200,000!

33 · Astronomers

Selecting a limited number of astronomers for short biographical notes may be somewhat invidious. However, the list given here includes most of the great pioneers and researchers. No astronomers still living at the time of writing are included. All dates are AD unless otherwise stated.

Abul Wafa, Mohammed. 959–88. Last of the famous Baghdad school of astronomers. He wrote a book called *Almagest*, a summary of Ptolemy's great work, also called the *Almagest*, in Arabic.

Adams, John Couch. 1819–92. English astronomer, born in Lidcot, Cornwall. He graduated brilliantly from Cambridge in 1843, but had already formulated a plan to search for a new planet by studying the perturbations of Uranus. By 1845, his results were ready, but no quick search was made, and the actual discovery was due to calculations by U. Le Verrier. Later he became Director of the Cambridge Observatory, and worked upon lunar acceleration, the orbit of the Leonid meteor shower, and upon various other investigations.

Airy, George Biddell. 1801–92. English astronomer. Born in Northumberland, he graduated from Cambridge 1823 and was Professor of Astronomy there (1826–35). On becoming Astronomer Royal (1835–81) he totally reorganised the Greenwich Observatory and raised it to its present eminence. He re-equipped the Observatory and ensured that the best use was made of its instruments; it is ironical that he is probably best remembered for his failure to instigate a prompt search for Neptune when receiving Adams' calculations.

Aitken, Robert Grant. 1864–1951. American astronomer, born in Jackson, California. In 1895 he joined the staff of Lick Observatory, and specialized in double star work; he discovered 31 000 new pairs, and wrote a standard book on the subject. From 1930 until his retirement in 1935 he was Director of the Lick Observatory.

Albategnius. *c.* 850–929. This is the Latinised form of the name of the Arab prince Al Battani. Born in Batan, Mesopotamia; he drew up improved tables of the Sun and Moon, and found a more accurate value for the precession of the equinoxes. His *Movements of the Stars* enabled Hevelius, in the seventeenth century, to discover the secular variation in the Moon's motion. Albategnius was also a pioneer mathematician; in trigonometry, he introduced the use of sines.

Alcock, George. 1919–2000. Outstanding English amateur astronomer and school teacher. He lived in Peterborough. Using binoculars only, he discovered five comets and three novæ.

Alphonso X. 1223–84. King of Castile. At Toledo he assembled many of the leading astronomers of the world, and drew up the famous Alphonsine Tables, which remained the standard for three centuries.

Alfvén, Hannes. 1908–95. Swedish physicist (Nobel Laureate, 1970) and the founder of the science of magnetohydrodynamics. One of his achievements was to prove the existence of an overall galactic magnetic field.

Alhazen (Abu Ali al Hassan). 987–1038. Arab mathematician, born in Basra. He went to Cairo, where he made his observations and also wrote the first important book on optics since the time of Ptolemy.

Allen, Clabon. 1905–88. Australian astronomer who specialised in studies of the Sun. For 20 years he lived in London, and was Director of the University of London Observatory. His book *Astrophysical Quantities* (1955) is a classic.

Allen, David. 1946–94. Cambridge-born, he graduated from the university there, and concentrated upon infrared astronomy, in which he made many major contributions. He went to Australia, and worked with the Anglo–Australian Telescope at Coonabarabran. He was an outstanding writer of popular books as well as technical works. Sadly, he died of cancer when still in his forties.

Al-Ma'mūn, Abdalla. ?–833. Often referred to as Almanon. He was Caliph of Baghdad, son of Harun al Raschid; he collected and translated many Greek and Persian works, and built a major observatory in 829.

Alpher, Ralph. 1921–2007. American cosmologist of Russian descent. He made vital contributions to nucleosynthesis – the way in which complex elements are formed from lighter elements – and this of course relates to conditions following the Big Bang. He worked closely with George Gamow, and for one paper Gamow persuaded Hans Bethe to add his name, so that it became the Alpha–Beta–Gamma ($\alpha\beta\gamma$) paper. Somehow or other Alpher has never really received full credit or his outstanding achievements – for example, he was the first to predict the temperature of the cosmic background radiation.

Al-Sūfī. 903–86. A Persian nobleman, who compiled an invaluable catalogue of 1018 stars, giving their approximate positions, magnitudes and colours.

Ambartsunian, Viktor Amazaspovich. 1909–95. Outstanding Russian astronomer; he set up the astrophysics programme at Leningrad University before Stalin's purges, and managed to survive, though many of his colleagues were executed. After the war he became Director of the Byurakan Observatory in Armenia. He made many original contributions; he introduced the concept of stellar associations, and was among the first to realise that T Tauri stars are very young, and are in the pre-Main Sequence stage.

Anaxagoras. 500–428 BC. Born in Clazomenæ, Ionia. In Athens he became a friend of Pericles, and it was because of this friendship that he was merely banished, rather than being condemned to death, for teaching that the Moon contains plains, valleys and mountains, while the Sun is a blazing stone larger than the Peloponnesus (the peninsula upon which Athens stands).

Anaximander. _c._ 611–547 BC. Greek philosopher, born in Miletus. He believed the Earth to be a cylinder, suspended freely in the centre of a spherical universe. He attempted to draw up a map of the world, and introduced the gnomon into Greece.

Anaximenes. _c._ 585–525 BC. Greek philosopher, born in Miletus. He believed the Sun to be hot because of its quick motion round the Earth, that the stars were too remote to send us detectable heat, and the stars were fastened on to a crystal sphere.

Ångström, Anders. 1814–74. Swedish physicist, who graduated from Uppsala. He mapped the solar spectrum, and was the first to examine the spectra of auroræ. The Ångström unit (100-millionth part of a centimetre) is named in his honour.

Antoniadi, Eugenios. 1870–1944. Greek astronomer, who spent most of his life in France and became a naturalised Frenchman. He worked mainly at the Juvisy Observatory (with Camille Flammarion) and at Meudon, near Paris, where he used the 83-cm, refractor to make classic observations of the planets. Before the Space Age, his maps of Mars and Mercury were regarded as the standard works. He died in Occupied France during World War II.

Apian, Peter Bienewitz. 1495–1552, born in Leisnig, Saxony. He became professor of mathematics at Ingolstädt. He observed five comets, and was the first to note that their tails always point away from the Sun. The 1531 comet is known to be Halley's, and his observations of it enabled Edmond Halley to identify it with the comets of 1607 and 1682.

Apollonius. _c._ 250–200 BC. Born in Perga, Asia Minor, but lived in Alexandria. An expert mathematician, he was one of the first to develop the theory of epicycles to represent the movement of the Sun, Moon and planets.

Arago, François Jean Dominique. 1786–1853. Director of the Paris Observatory from 1830. He made many important contributions, including a recognition of the importance of photography in astronomy. He made an exhaustive study of the great total solar eclipse of 1842, and maintained (correctly!) that the Sun is wholly gaseous.

Argelander, Friedrich Wilhelm August. 1799–1875. German astronomer, who became Director of the Bonn Observatory 1836. Here he drew up his atlas of the northern heavens (the Bonn Durchmusterung), containing the positions of 324 198 stars down to the ninth magnitude. This standard work was published in 1863.

Aristarchus. _c._ 310–30BC. Greek astronomer, born in Samos. One of the first (quite possibly the very first) to maintain that the Earth moves round the Sun, and he also tried to measure the relative distances of the Sun and Moon by a method which was sound in theory, though inaccurate in practice.

Aristotle. 384–322 BC. He believed in a finite, spherical universe. He further developed the theory of concentric spheres, and gave the first practical proofs that the Earth cannot be flat.

Axford, Ian. 1933–2010. Sir Ian Axford was one of New Zealand's greatest astronomers; he specialised in all aspects of solar research, planetary stronomy and plasma physics. He spent some time in Germany (Max Planck Institut) and the USA (University of California), though most of his career was spent in New Zealand.

Baade, Wilhelm Heinrich Walter. 1893–1959. German astronomer. In 1920, while assistant at Hamburg Observatory, he discovered the unique asteroid 944 Hidalgo. In 1931, he went to America, and joined the staff of Mount Wilson. In 1952, his work upon the two classes of 'Cepheid' short-period variables enabled him to show that the galaxies are approximately twice as remote as had previously been thought.

Bailey, Solon Irving. 1854–1931. American astronomer, born in New Hampshire. He joined the Harvard staff in 1879, and was for many years in charge of the Harvard southern station at Arequipa, Peru. His studies of globular clusters led to the discovery of 'cluster variables', now known as RR Lyræ stars.

Banneker, Benjamin. 1731–1806. African-American astronomer born in Maryland, the son of a slave. He was self educated, but took up astronomy and proved to be a good mathematician. In 1791, he became a technical assistant at the Federal District of Washington, as a calculator and surveyor.

Bappu, Manali Kallat Vainu. 1927–82. One of the most distinguished of Indian astronomers; he was Director of the Kodaikanal Observatory, and modernised it. He was a specialist in stellar spectroscopy.

Barnard, Edward Emerson. 1857–1923. American astronomer, born in Nashville, Tennessee. He was self taught, but joined the staff at Lick Observatory in 1888, moving to Yerkes in 1897. He was a renowned comet-hunter; he discovered the fifth satellite of Jupiter (Amalthea) and the swift-moving star in Ophiuchus, now called Barnard's Star. He also specialised in studies of dark nebulæ.

Barrow, Isaac. 1630–77. English mathematician. He made various important contributions to science, but is perhaps best known because, in 1669, he resigned his post as Lucasian Professor at Cambridge so that his pupil Isaac Newton could succeed him.

Bayer, Johann. 1572–1625. German astronomer; a lawyer by profession and an amateur in science. He is remembered for his 1603 star catalogue, in which he introduced the system of allotting Greek letters to the stars in each constellation – the system is still in use today.

Beer, Wilhelm. 1797–1850. A Berlin banker, who set up a private observatory and collaborated with von Mädler in the great map of the Moon published in 1837–8. This map remained the standard for many years. Beer was the brother of Meyerbeer, the famous composer.

Belopolsky, Aristarch. 1854–1934. Russian astronomer, who went from Moscow to the Pulkova Observatory in 1888. He became Director of the Observatory in 1916, but resigned in 1918. He specialised in spectroscopic astronomy and in studies of variable stars.

Bessel, Friedrich Wilhelm. 1784–1846. German astronomer. He went to Lilienthal as assistant to Schröter, but in 1810 became Director of the Königsberg Observatory, retaining the post until his death. He determined the position of 75 000 stars by reducing Bradley's observations; was the first to obtain a parallax value for a star (61 Cygni, in 1838), and predicted the positions of the then unknown companions of Sirius and Procyon.

Bethe, Hans Albrecht.1906–2005. German astrophysicist who played a fundamental rôle in solving the problems of solar and stellar energy. In 1967, he was award the Nobel Prize for Physics. Most of his life was spent in the United States.

Biela, Wilhelm von. 1782–1856. Austrian army officer, and amateur astronomer, who is remembered for his discovery (in 1826) of the now-defunct periodical comet which bears his name.

Bode, Johann Elert. 1747–1826. German astronomer, Born in Hamburg. Appointed director of Berlin Observatory in 1772. In the same year he drew attention to the 'law' of planetary distances, which had been discovered by Titius of Wittenberg; rather unfairly, perhaps, this is known as Bode's law. He published a star catalogue, did much to popularise astronomy, and for 50 years edited the *Berlin Astronomisches Jahrbuch*.

Bolton, John. 1922–93. Bolton was a Yorkshireman, born in Sheffield, but spent most of his career in Australia. He was a pioneer of radio astronomy, and was the first to identify the Crab Nebula as a radio source. He spent some years as Director of the National Radio Astronomy Observatory in Australia, retiring in 1971.

Bond, George Phillips. 1825–65. Son of W. C. Bond, Born in Massachusetts. In 1859, succeeded his father as Director of the Harvard Observatory. He was a pioneer of planetary and cometary photography, and was the first to assert, upon truly scientific principles, that Saturn's rings could not be solid.

Boyd, Sir Robert. 1922–2004. A leader in space-science research, and established a new discipline of space science at University College in London. He was a specialist in studies of the upper atmosphere.

Bond, William Cranch. 1789–1859. American astronomer, born in Maine. He began his career as a watchmaker, but his fame as an amateur astronomer led to his appointment as Director of the newly-founded Harvard Observatory. In 1848 he discovered Saturn's satellite Hyperion, and in 1850 he discovered Saturn's Crepe Ring. He was also a pioneer of astronomical photography.

Bouvard, Alexis. 1767–1843. A shepherd boy, born in a hut at Chamonix. He went to Paris, taught himself mathematics, and was appointed assistant to Laplace. He made contributions to lunar theory and drew up tables of the motions of the outer planets, as well as discovering several comets.

Bradley, James. 1692–1762. English astronomer (Astronomer Royal, 1742–62). He was educated in Gloucestershire, and entered the Ministry, becoming Vicar of Bridstow in 1719; in 1721, he went to Oxford as Professor of Astronomy, and remained there until his appointment to Greenwich, mainly on the recommendation of his close friend Halley. He discovered the aberration of light and the nutation of the Earth's axis, but his greatest work was his catalogue of the positions of 60 000 stars.

Brorsen, Theodor. 1819–95. Danish astronomer, who discovered several comets, and in 1854 made the first scientific observations of the Gegenschein.

Brown, Ernest William. 1866–1938. English astronomer, who graduated from Cambridge and then went to USA. His chief work was on lunar theory, and his tables of the Moon's motion are still recognised as the standard.

Bruck, Hermann. 1905–2000. German astronomer, who spent most of his life in Britain. He was a pioneer in the automation of telescopes. Between 1957 and 1975 he was Director of the Royal Observatory, Edinburgh, which he modernised. He was also Astronomer Royal for Scotland.

Burbidge, Geoffrey. (1925–2010). English astrophysicist. Working with his wife Margaret, English astronomer Fred Hoyle, and American physicist William Fowler, he worked out how 'heavy' elements are synthesised inside stars. Their paper published in 1957 is regarded as one of the most important of the twentieth century. He was however a strong supporter of the now-rejected steady-state theory of the universe. The latter part of his life was spent at San Diego in the United States.

Burnham, Sherburne Wesley. 1838–1921. American astronomer, who began as an amateur and then went successively to Lick (1888) and Yerkes (1897). He specialised in double-star work, and discovered over 1300 new pairs. His *General Catalogue of Double Stars* remains a standard reference work.

Campbell, William Wallace. 1862–1938. American astronomer, born in Ohio. He joined the staff at Lick Observatory in 1891, and was Director from 1900 until his retirement in 1930. His main work was in spectroscopy; he discovered 339 spectroscopic binaries (among them Capella), and determined the radial velocities of stars and of 125 nebulæ, as well as carrying out spectroscopic observations of the planets.

Cannon, Annie Jump. 1863–1941. Outstanding American woman astronomer, born in Delaware. In 1896, she joined the staff at Harvard College Observatory, where she worked unceasingly on the classification of stellar spectra; the present system is due largely to her. She also discovered five novæ and over 300 variable stars. From 1938, she was William Cranch Bond Astronomer.

Carrington, Richard Christopher. 1826–75. English amateur astronomer, who had his observatory at Redhill, Surrey. He concentrated upon the Sun, and made many contributions, including the first observation of a solar flare and the independent discovery of Spörer's law concerning the distribution of sunspots throughout a cycle.

Cassini, Giovanni Domenico. 1625–1712. Italian astronomer; Professor of Astronomy, Bologna 1650–69, when he went to Paris as the first Director of the observatory there. He discovered four of Saturn's satellites as well as the main division in the rings; he drew up new tables of Jupiter's satellites, made pioneer observations of Mars, and made the first reasonably good measurement of the distance of the Sun.

Cassini, Jacques J. 1677–1756. Son of G. D. Cassini; born in Paris. He succeeded his father as Director of the Paris Observatory. He confirmed Halley's discovery of the proper motions of certain stars, and played an important part in measuring an arc of meridian from Dunkirk to the Pyrenees in order to determine the figure of the Earth.

Challis, James. 1803–62. English astronomer; Professor of Astronomy at Cambridge from 1836. He accomplished much useful work, but, unfortunately, is remembered as the man who failed to discover Neptune before the success by Galle and D'Arrest at Berlin.

Chandrasekhar, Subrahmanyan. 1910–95. Indian astrophysicist, born at Lahore; he graduated from Cambridge in 1933. In 1937 he emigrated to Chicago, and remained in the United States, becoming a US citizen. He made major contributions to astrophysics, and showed that there is a maximum possible mass for a white dwarf star – the Chandrasekhar limit. In 1983 he was awarded the Nobel Prize for Physics.

Charlier, Carl Vilhelm Ludwig. 1862–1934. Swedish cosmologist; Professor of Astronomy at Lund from 1897. He accomplished outstanding work with regard to the distribution of stars in our Galaxy.

Chelomei, Vladimir. 1914–84. Russian rocket designer. With Kororolev he designed the Proton rocket widely used in Soviet launches.

Christie, William Henry Mahoney. 1845–1922. English astronomer (Astronomer Royal 1881–1910). He modernised the Greenwich Observatory, and fully maintained its great reputation, achieved under Airy.

Clairaut, Alexis Claude. 1713–65. French mathematical genius who published his first important paper at the age of 12. He studied the motion of the Moon, and worked out the perihelion passage of Halley's Comet in 1759 to within a month of the actual date.

Clarke, Arthur Charles. 1917–2008. English scientist, inventor, a science-fiction author. He went to Sri Lanka in 1956, largely to pursue his interest in scuba diving, and remained there. He was a prolific author, and had an immense influence in popularising astronomy and astronautics. In 1945, he was the first to suggest the use of communications satellites.

Clavius, Christopher Klau. 1537–1612. German Jesuit mathematical teacher, who laid down the calendar reform of 1582 at the request of Pope Gregory.

Collinder, Per. 1890–1975. Swedish astronomer of Uppsala, best known for his catalogue of open star clusters, published in 1931.

Cooper, Gordon. 1927–2004. American astronaut, one of the 'original seven' chosen by NASA.

Copernicus, Nicolaus. 1473–1543. The Latinised name of Nikołaj Kopernik, born in Toruń, Poland. He entered the Church, and became Canon of Frombork. He had a varied career, including medicine and also the defence of his country against the Teutonic Knights, but is remembered for his great book *De Revolutionbus Orbium Cælestium*, finally published during the last days of his life. It was this book which revived the heliocentric theory according to which the Earth moves round the Sun, and sparked off the 'Copernican revolution', which came to its end with the work of Newton more than a century later.

Curtis, Heber Doust. 1872–1942. American astronomer who worked at Lick, Allegheny and Michigan observatories. He was an outstanding spectroscopist, and in 1920 took part in the 'Great Debate' with Shapley about the size of the Galaxy and the status of the resolvable nebulæ; Curtis was wrong about the size of the Galaxy, but correct in maintaining that the spiral nebulæ were independent galaxies. He also played a major rôle in the establishment of the McMath–Hulbert Observatory, renowned for its solar research.

D'Arrest, Heinrich Ludwig. 1822–75. German astronomer, born in Berlin. While Assistant at the Berlin Observatory, he joined Galle in the successful search for Neptune. From 1857, he worked at Copenhagen Observatory. He specialised in comet and asteroid work, and also published improved positions for about 2000 nebulæ.

Darwin, George Howard. 1845–1912. Son of Charles Darwin. From 1883, he was Professor of Astronomy at Cambridge, and drew up his famous, though now rejected, tidal theory of the origin of the planets. He was knighted in 1906.

Dawes, William Rutter. 1799–1868. English clergyman, and a keen-eyed amateur observer who specialised in observations of the Sun, planets and double stars. He discovered Saturn's Crepe Ring independently of Bond.

Delambre, Jean-Baptiste Joseph. 1749–1822. French astronomer, best remembered for his work on the history of the science but also a skilled computer of planetary tables.

De la Rue, Warren. 1815–89. English astronomer, born in Guernsey. He was a pioneer of astronomical photography; in 1852 he obtained the first good photographs of the Moon, and in 1857 of the Sun. His photographs of the total solar eclipse of 1860 finally proved that the prominences are solar rather than lunar.

De Vaucouleurs, Gerard. 1918–95. Outstanding French planetary observer, who specialised in research concerning Mars; he was also active in cosmological research – although the value which he derived for the Hubble constant is now known to be much too high. He was the author of many popular books as well as technical works.

Delaunay, Charles. 1816–72. French astronomer, who specialised in studies of the Moon's motion. He became Director of the Paris Observatory in 1870, but was drowned in a boating accident two years later.

Democritus. *c.* 460–360 BC. Greek philosopher, born in Abdera, Thrace. He adopted Leucippus' atomic theory, and was the first to claim that the Milky Way is made up of stars.

Descartes, René. 1596–1650. French astronomer, author of the theory that matter originates as vortices in an all-pervading ether. He also made great improvements in optics. His books were published in Holland, but he died in Sweden.

De Sitter, Willem. 1872–1934. Dutch astronomer and cosmologist, born in Friesland; he went to the Cape, and from 1908 was Professor of Astronomy at Leiden. He studied the motions of Jupiter's satellites and also the rotation of the Sun, but is best remembered for his pioneer work in relativity theory. The 'De Sitter universe', finite but unbounded, was calculated to be 2000 million light-years in radius and to contain 80 000 million galaxies.

Deslandres, Henri Alexander. 1853–1948. French astronomer (originally an Army officer); from 1907, Director of the Meudon Observatory, and from 1927, Director of the Paris Observatory also. He was a pioneer spectroscopist, and developed the spectroheliograph independently of Hale.

Dollond, John. 1706–61. English optician who reinvented the achromatic lens in 1758 and thus improved refractors beyond all recognition.

Donati, Giovanni Battista. 1826–73. Italian astronomer who discovered the great comet of 1858, and was the first to obtain the spectrum of a comet (Tempel's of 1864). From 1859, Director of the observatory at Florence, in 1872 he was largely responsible for the creation of the now-celebrated observatory at Arcetri.

Doxsey, Roger. 1947–2009. American astronomer who led the team in charge of maintaining the orbiting Hubble Space Telescope (HST). He was nicknamed 'the heart and soul of Hubble'.

Dreyer, John Louis Emil. 1852–1926. Danish astronomer, born in Copenhagen, who went to Ireland as astronomer to Lord Rosse at Birr Castle and became Director of the Armagh Observatory in 1882. He was a great astronomical historian, but is best remembered for his *New General Catalogue of Clusters and Nebulæ* (the NGC), still regarded as a standard work. In 1916 he retired from Armagh and went to Oxford, where he lived for the rest of his life.

Dyson, Frank Watson. 1868–1939. English astronomer (Astronomer Royal, 1910–33). A great administrator as well as an energetic observer of eclipses; he also carried out important work in the field of astrophysics and stellar motions.

Eddington, Sir Arthur Stanley. 1882–1945. English astronomer, born in Kendal. After working at Manchester, Cambridge and Greenwich, he was appointed Professor of Astronomy at Cambridge in 1913. He was a pioneer of the theory of the evolution and constitution of the stars, and an outstanding promoter of relativity; in 1919, he confirmed Einstein's prediction of the displacement of star positions near the eclipsed Sun. He was knighted in 1930. In addition to his outstanding work, Eddington was one of the best of all writers of popular scientific books, and was a splendid broadcaster.

Einstein, Albert. 1879–1955. German Jew, whose name will be remembered as long as Newton's; in 1905, he laid down the Special Theory of Relativity, and from 1915 to 17 he developed the General Theory. In 1933 he left Germany, fearing persecution of the Jews, and settled in the USA.

Elger, Thomas Gwyn. 1838–97. English amateur astronomer; he was first Director of the Lunar Section of the British Astronomical Association, and in 1895 published an excellent outline map of the Moon.

Empedocles of Agrigentum. *c.* 490–50 BC. He believed the Sun to be a reflection of fire, but is credited with being the first to maintain that light has a finite velocity.

Encke, Johann Franz. 1791–1865. German astronomer; from 1825, Director of the Berlin Observatory. He was responsible for compiling the star maps which enabled Galle and D'Arrest to locate Neptune. In 1818, he computed the orbit of a faint comet, and successfully predicted its return; this was Encke's Comet, which has the shortest period of any known comet (3.3 years).

Eratosthenes, *c.* 276–196 BC. Greek philosopher, born in Cyrene; he became Librarian at Alexandria, and made a remarkably accurate measurement of the circumference of the Earth.

Eudoxus, *c.* 408–355 BC. Greek astronomer, born in Cnidus. He went to Athens, and attended lectures by Plato. Finally he settled in Sicily. He developed the theory of concentric spheres – the first truly scientific attempt to explain the movements of the celestial bodies.

Euler, Leonhard. 1707–83. Brilliant Swiss mathematician, born in Basle. He pioneered studies of the lunar theory, the movements of planets, comets, and the tides. He lost his sight in 1766, but this did not stop him from working; he undertook the complicated calculations mentally.

Fabricius, David. 1564–1617. Dutch minister and amateur astronomer, who observed Mira Ceti in 1596 (though without recognising it as a variable) and made pioneer telescopic observations, notably of the Sun. In 1617, he announced from the pulpit that he knew the identity of a member of his congregation who had stolen one of his geese – and he was presumably correct, since he was assassinated before he could divulge the name of the culprit!

Fabricius, Johann. 1587–1616. Son of David Fabricius, and also a pioneer observer of the Sun by telescopic means; he discovered sunspots independently of Galileo and Scheiner.

Fallows, Fearon. 1789–1831. English astronomer, born in Cumberland. He went to South Africa in 1821 as the first Director of the Cape Observatory. He established the observatory, working under almost incredible difficulties, but the primitive living conditions undermined his health. The reduction of his Cape observations was undertaken by Airy.

Fauth, Philipp Johann Heinrich. 1867–1943. German astronomer, who compiled a large map of the Moon. Unfortunately, he believed in the absurd theory that the Moon is ice-covered, and this influenced all his work.

Feoktistov, Konstantin. 1926–2009. Russian cosmonaut, who flew in Voskhod (1962) with Komarov and Yegorov; they were the first cosmonauts not to wear space-suits. His poor health excluded him from further missions, but he continued technical work, and played a leading role in designing Mir and Soyuz. He worked closely with Korolev.

Ferguson, James. 1710–76. Scottish populariser of astronomy, who began life as a shepherd-boy but whose books gained great influence. He was also one of the first to suggest an evolutionary origin of the Solar System.

Flammarion, Camille. 1842–1925. French astronomer renowned both for his observations of Mars and for his popular books. He set up his own observatory at Juvisy, and founded the Société Astronomique de France.

Flamsteed, John. 1646–1720. English astronomer (Astronomer Royal, 1675–1720, though at first the title was 'unofficial'). His main work was the compilation of a new star catalogue, the final version of which was published posthumously. Flamsteed was also Rector of Burstow, Surrey.

Fleming, Wilhelmina. 1857–1911. Scottish woman astronomer, who emigrated to America and worked at Harvard College Observatory, where she was in charge of the famous Draper star catalogue. She discovered 10 novæ and 222 variable stars.

Fontana, Francisco. 1585–1656. Italian amateur (a lawyer by profession). He left sketches of Mars and Venus, though the 'markings' which he recorded were certainly illusory.

Fowler, William Alfred. 1911–95. American astrophysicist. With Geoffrey and Margaret Burbidge and Fred Hoyle, he wrote the classic 1957 paper describing how all but the lightest elements could be synthesised inside stars. In 1983 he was awarded the Nobel Prize for Physics.

Fowler, Alfred. 1868–1940. English astronomer, whose spectroscopic work in connection with the Sun, stars and comets was of great importance.

Franklin, Kenneth. 1923–2007. American astronomer, Chief Scientist at the Hayden Planetarium from 1956 to 1984. He is best remembered as being one of the co-discoverers of radio waves from Jupiter.

Franklin-Adams, John. 1843–1912. English businessman who took up astronomy as a hobby at the age of 47, and compiled a photographic chart of the stars which is still regarded as a standard work.

Fraunhofer, Joseph von. 1787–1826. Outstanding German optical worker, orphaned in early childhood and rescued from poverty by the Elector of Bavaria. He joined the Physical and Optical Institute of Munich, and was Director from 1823. He invented the diffraction grating, constructed the best lenses in the world, and studied the dark lines in the solar spectrum (the 'Fraunhofer Lines'). He made the Dorpat refractor for Struve (the first telescope to be clock-driven) and also the Königsberg heliometer. His comparatively early death was a tragedy for science.

Gagarin, Yuri Alexeivich. 1934–68. The first man in space. He was born in Klushino (USSR). He became a pilot, and was selected to fly in Vostok 1 (12 April 1961), completing one orbit of the Earth. Subsequently, he became training director at Star City. On 27 March 1968 he was killed while on what should have been an ordinary routine flight.

Galilei, Galileo. 1564–1642. The first great telescopic observer – and also the true founder of experimental mechanics. He worked successively at Pisa, Padua and Florence. The story of his remarkable telescopic discoveries (including the satellites of Jupiter, the phases of Venus and the gibbous aspect of Mars, the starry nature of the Milky Way and many more), and of how his defence of the Copernican theory brought him into conflict with the Church, is one of the most famous in scientific history. He was condemned by the Inquisition in 1633, and was kept a virtual prisoner in his villa at Arcetri; in his last years he also lost his sight.

Galle, Johann Gottfried. 1812–1910. German astronomer, best remembered as being the first (with D'Arrest) to locate Neptune in 1846. He discovered three comets and, in 1872, while director of the Breslau Observatory, was the first to use an asteroid for measuring solar parallax.

Gan De. 4th century BC. Chinese astronomer, from the State of Qi. With Shi Shen he is believed to be the first in history known to compile a star catalogue. He made the first known observations of the movements of Jupiter, and in 364 BC may have seen one of the satellites with the naked eye.

Ganswindt, Hermann. 1856–1934. Eccentric German who was writing about space-travel as early as 1881. His 'rocket design' involved steel cartridges. filled with dynamite, but he did more or less grasp the principle of reaction.

Gascoigne, Sidney Charles Bartholomew. 1915–2010. Astronomer deeply involved with the AAT (Anglo-Australian Telescope, now the Australian Astronomical Telescope). He made many important contributions to astrophysics and instrumentation, notably the corrector which gives Ritchey–Chrétien telescopes a much wider field of good definition. He is also remembered for another reason. During observing one night with the AAT, he went out on to the catwalk to check the weather, lost his sense of direction, stepped into space and fell 6 metres on to the wooden floor below. He injured his left arm, and was lucky not to suffer really serious damage. The site is now marked by a plaque, made by Paul Lindner, 'Gascoigne's Leap', which he unveiled in 1980!

Gassendi, Pierre. 1592–1655. French mathematician and astronomer. In 1631 he made the first of all observations of a transit of Mercury.

Gauss, Karl Friedrich. 1777–1855. German mathematical genius. In 1801, he calculated the orbit of the first asteroid, Ceres, from a few observations, and enabled Olbers to recover it in the following year. He invented the 'method of least squares', known to every mathematician.

Gill, David. 1843–1914. Scottish astronomer. In 1877, he used observations of Mars to redetermine the solar parallax, and in 1879 went to South Africa as HM Astronomer at the Cape. It was his photograph of the comet of 1882 which showed him the importance of mapping the sky photographically – since his plate showed many stars as well as the comet. He was also deeply involved in cataloguing the southern stars. He was knighted in 1900.

Glushko, Valentin Petrovitch. 1908–89. Russian rocket designer, associated with Korolev designing Sputnik 1, and later oversaw the development of Mir. He was awarded the Lenin Prize.

Goddard, Robert Hutchings. 1882–1945. American pioneer of liquid-fuel rocketry. He graduated from Clark University; in 1919, he wrote his famous paper suggesting that it might be possible to send small objects to the Moon, and on 16 March 1926 launched the first liquid-fuel rocket. It lasted for 2.5 seconds and rose to 41 ft.

Goldschmidt, Hermann. 1802–66. German astronomer, who settled in Paris. Using small telescopes poked through his attic window, he discovered 14 asteroids between 1852 and 1861.

Gold, Thomas. 1920–2004. Austrian astrophysicist who spent most of his life in America and made fundamental contributions to astrophysical research. He was a strong supporter of the steady-state theory of the universe.

Goodacre, Walter. 1856–1938. English amateur astronomer who published an excellent map of the Moon in 1910.

Goodricke, John. 1764–86. Born of English parents in Holland. He was a deaf-mute, but with a brilliant brain. It was he who found that Algol is an eclipsing binary rather than true variable, and he also discovered the fluctuations of the intrinsic variable δ Cephei.

Gould, Benjamin Apthorp. 1824–96. American astronomer, who founded the *Astrophysical Journal*. From Cordoba Observatory, Argentina, he compiled the *Uranimetria Argentina*, the first major catalogue of the southern stars.

Green, Charles. 1735–71. English astronomer who went with Captain Cook to study the 1769 transit of Venus. He died on the return voyage.

Greenstein, Jesse 1909–2002. American astrophysicist who spent most of his life working at the California Institute of Technology. He was probably the first conventionally trained astronomer to take radio astronomy seriously.

Gregory, James. 1638–75. Scottish mathematician. In 1663 he described the principle of the reflecting telescope, but never actually made one.

Grimaldi, Francesco Maria. 1618–63. Italian Jesuit, who made observations of the Moon used in the lunar map compiled by his friend Riccioli.

Gruithuisen, Franz von Paula. 1771–1852. German astronomer; from 1826, Professor of Astronomy at Munich. He was an assiduous observer of the Moon and planets, but his vivid imagination tended to discredit his work; at one stage he even reported the discovery of artificial structures on the Moon. He also proposed the impact theory of lunar crater formation.

Gum, Colin. 1924–60. Australian astronomer who carried out work of vital importance in the surveying of southern radio sources. The famous 'Gum Nebula' in Vela/Puppis is named after him. He was killed in a skiing accident at Zermatt in Switzerland.

Hadley, John. 1682–1743. English astronomer; friend of Bradley. He made the first really good reflecting telescope (6-inch aperture) in 1723, and in 1731 constructed his 'reflecting quadrant', which replaced the astrolabe and the cross-staff in navigation.

Hale, George Ellery. 1868–1938. American astronomer. A pioneer solar observer, who invented the spectroheliograph and discovered the magnetic fields of sunspots. In 1897, he became Director of Yerkes Observatory, and transferred to Mount Wilson in 1905; he master-minded the building of the 60-inch and 100-inch reflectors, as well as the Yerkes refractor. He was mainly responsible for the building of the Palomar 200-inch reflector, unfortunately not completed in his lifetime.

Hall, Asaph. 1829–1907. American astronomer, noted for his planetary work. At Washington, in 1877, he discovered the two satellites of Mars. From 1896, he was Professor of Astronomy at Harvard.

Halley, Edmond. 1656–1742. English astronomer (Astronomer Royal, 1720–42). Though best known for his prediction of the return of the great comet which now bears his name, Halley accomplished much other valuable work; he catalogued the southern stars from St Helena, studied star clusters and nebulæ, and discovered the proper motions of some of the bright stars. More importantly, he was responsible for the writing of Newton's *Principia* and personally financed its publication.

Harding, Karl Ludwig. 1765–1834. German astronomer who was at first assistant to Schröter and then was appointed Professor of Astronomy at Göttingen. In 1804, he discovered the third asteroid, Juno.

Haro, Guillermo. 1900–90. Possibly Mexico's most famous astronomer, celebrated for his studies of flare stars. He noted bright nebulæ with spectra showing emission lines, independently of G. Herbig; these are now known as Herbig–Haro Objects. He was Director of the Mexican Institute of Astronomy.

Harriot, Thomas. 1560–1621. English scholar, and friend to Sir Walter Raleigh. He made the first telescopic drawing of the Moon, and completed it some months before Galileo began his work.

Harrison, John. 1693–1776. English clockmaker who invented the marine chronometer, which revolutionised navigation. Several of his original chronometers are now on display in London.

Hartmann, Johannes Franz. 1865–1936. German astronomer. Director Göttingen Observatory (1909–21), when he went to Argentina to superintend the National Observatory there. His important work was connected with stellar and nebular radial velocities, in the course of which he discovered interstellar absorption lines in the spectrum of δ Orionis.

Hatfield, Henry. 1921–2010. Educated at Dulwich College, he joined the Navy in 1938 and became a leading hydrographer; he rewrote the Admiralty's official manual. As an amateur astronomer, using home-made equipment (notably a fine 12-inch reflector) he wrote the immensely valuable *Photographic Lunar Atlas*. As a solar observer, he literally built his house in Sevenoaks round his spectrohelioscope! While a naval midshipman in 1943, he unintentionally bombed Genoa Cathedral, fortunately without doing any substantial damage (one unexploded shell can still be seen there).

Hay, William Thompson. 1888–1949. 'Will Hay' was probably the only skilled amateur astronomer who was by profession a stage and screen comedian! In 1933 he discovered the famous white spot on Saturn – the most prominent ever seen on that planet.

Heis, Eduard. 1806–77. German astronomer; Professor at Münster from 1852. He was a leading authority on the Zodiacal Light, meteors and variable stars, and published a valuable star catalogue. He was renowned for his keen eyesight, and is said to have counted 19 naked-eye stars in the Pleiades.

Hencke, Karl Ludwig. 1793–1866. German amateur astronomer; postmaster at Driessen. In 1845, after 15 years' search, he discovered the fifth asteroid, Astræa.

Henderson, Thomas. 1798–1844. Scottish astronomer. 1832-3 HM Astronomer at the Cape. While there, he made the measurements which enabled him to measure the parallax of α Centauri. In 1834, he became the first Astronomer Royal for Scotland.

Heraclides of Pontus. *c.* 388–315 BC. He declared that the apparent daily rotation of the sky is due to the real rotation of the Earth. He also discovered that Mercury and Venus revolve round the Sun, not round the Earth.

Heraclitus of Ephesus. Born *c.* 544 BC. He took fire to be the principal element, and maintained that the diameter of the Sun was about one foot.

Herschel, Friedrich Wilhelm (always known as William Herschel). 1738–1822. Probably the greatest observer of all time. He was born in Hanover, but spent most of his life in England. He was the best telescope maker of his day, and in 1781 became famous by his discovery of the planet Uranus. He made innumerable discoveries of double stars, clusters and nebulæ; he found that many doubles are physically associated or binary systems, and he was the first to give a reasonable idea of the shape of the Galaxy. He was knighted in 1816, and received every honour that the scientific world could bestow. George III appointed him King's Astronomer (not Astronomer Royal).

Herschel, Caroline Lucretia. 1750–1848. William Herschel's sister, and constant assistant in his astronomical work. She discovered eight comets.

Herschel, John Frederick William. 1792–1871. William Herschel's son. He graduated from Cambridge 1813 and from 1832 to 1838 took a large telescope to the Cape to make the first really systematic observation of the southern heavens. He discovered 3347 double stars and 525 nebulæ, and may be said to have completed his father's pioneering work.

Hertzsprung, Ejnar. 1873–1967. Danish astronomer, who worked successively at Frederiksberg, Copenhagen, Göttingen, Mount Wilson and Leiden (Director Leiden Observatory from 1935). In 1905, he discovered the giant and dwarf subdivisions of late-type stars, and this led on to the compilation of H–R or Hertzsprung–Russell diagrams, which are of fundamental importance in astronomy.

Hevelius. 1611 87. The Latinised name of Johannes Hewelcke of Danzig (now Gdańsk). From his private observatory he drew up a catalogue of 1500 stars, and observed planets, the Moon and comets, using the unwieldy long-focus, small-aperture refractors of his day. His observatory was burned down in 1679, but he promptly constructed another. His original map of the Moon has been lost; tradition says that the copper engraving was melted down and made into a teapot after his death.

Hind, John Russell. 1823–95. English astronomer, who discovered 11 asteroids, the 1848 nova in Ophiuchus, and his 'variable nebula' round T Tauri. He also computed many cometary orbits, and from 1853 was superintendent of the Nautical Almanac.

Hipparchus. 140 BC. Great Greek astronomer, who lived in Rhodes. He drew up a star catalogue, later augmented by Ptolemy. Among his many discoveries was that of precession; he also constructed trigonometric tables. Unfortunately, all his original works have been lost.

Hoffmeister, Cuno. 1892–1967. He was born in Sonneberg, Thuringia, and after working at two German observatories he founded the Sonneberg Observatory, in 1925. He was a specialist in variable-star work, and discovered almost 10 000 new variables. He was also an authority on meteoritic astronomy.

Hooke, Robert. 1653–1703. English scientific genius, contemporary with (though no friend of!) Newton. He built various astronomical instruments, and made some useful observations, including sketches of lunar craters.

Horrocks, Jeremiah. 1619–41. English astronomer who, with his friend Crabtree, was the first to observe a transit of Venus (1639). He also worked on lunar theory. His early death was a great tragedy for science.

Howse, Derek. 1919–98. Essentially a Naval officer (Lieutenant-Commander, who won the DSC during the war), Howse became Keeper of Astronomy and Navigation at the National Maritime Museum at Greenwich, retiring in 1982. He was the author of many books and papers on all aspects of navigational astronomy.

Hoyle, Fred. 1915–2001. Outstanding English astronomer long associated with Cambridge University. He was the first to realise that chemical elements are created inside stars by nucleosynthesis, and he made fundamental contributions to astrophysics. In many ways he was unconventional, and he believed that life on Earth was brought here by way of a comet; he supported the steady-state theory of the universe, and scornfully coined the term 'Big Bang'. He was a noted speaker and broadcaster, and also the author of several science-fiction novels

Hubble, Edwin Powell. 1889–1953. American astronomer, who served in the Army during World War I and was also a qualified lawyer. In 1923, using the Mount Wilson 100-inch reflector, he discovered short-period variables in the Andromeda Spiral, and proved the Spiral to be an independent galaxy. He also established the velocity–distance relationship known as Hubble's law.

Huggins, William. 1824–1910. Pioneer English spectroscopist, who had his private observatory at Tulse Hill, near London. Pioneer of stellar spectroscopy; he established that the irresolvable nebulæ are gaseous; he was the first to determine stellar radial motions by means of the Doppler shifts in their spectral lines, and carried out important solar and planetary work. He was knighted in 1897.

Humason, Milton La Salle. 1891–1972. Born in Minnesota, he was mainly self-taught, but joined the staff of Mount Wilson Observatory in 1920, and from then on worked closely with Hubble, studying the forms, spectra, radial motions and nature of the galaxies; he also photographed the spectra of supernovæ in external systems. In 1919, he carried out a photographic search for a trans-Neptunian planet at the request of W. H. Pickering, who had made independent calculations similar to Lowell's. Humason took several plates, but failed to locate the planet. When the plates were re-examined years later, after Pluto had been discovered at Flagstaff, it was found that Humason had recorded the planet twice – but once the image was masked by a star, and on the other occasion it fell on a flaw in the plate!

Hutchison, Robert. 1928–2007. English world expert on meteorites. He led meteor hunting expeditions in China and Australia. For much of his career he worked at the Natural History Museum in London.

Huygens, Christiaan. 1629–95. Dutch astronomer; probably the best telescopic observer of his time. He discovered Saturn's brightest satellite (Titan) in 1655, and was the first to realise that the curious appearance of the planet was due to a system of rings. He was also the first to see markings on Mars. His activities extended into many fields of science; in particular, he invented the first practical pendulum clock.

Hypatia. Late 4th century AD. The first true woman astronomer daughter of Theon, Professor of Mathematics at Alexandria. She joined the Alexandrian Library and became immensely influential as a teacher; she wrote many books, compiled astronomical tables, and also invented the astrolabe. She was a supporter of Orestes, the Roman governor of Egypt and was murdered by a mob sent by Cyril, head of the Christian Church in Alexandria.

Innes, Robert Thorburn Ayton. 1861–1933. Scottish astronomer who emigrated first to Australia (becoming a wine merchant) and then went to South Africa, as Director of the Observatory at Johannesburg. He specialised in double-star work, discovering more than 1500 new pairs; he also discovered Proxima Centauri, the nearest star beyond the Sun.

Janssen, Pierre Jules César, 1824–1907. French astronomer, who specialised in solar work (in 1870 he escaped from the besieged city of Paris by balloon to study a total eclipse). Independently of Lockyer, he discovered the means of observing the Sun's chromosphere and prominences without waiting for an eclipse. From 1876 he was Director of the Meudon Observatory, and in 1904 published an elaborate solar atlas, containing more than 8000 photographs. The square at the entrance to the Meudon Observatory is still called the Place Janssen, and his statue is to be seen there.

Jansky, Karl Guthe. 1905–49. American radio engineer, of Czech descent. He joined the Bell Telephone Laboratories, and was using an improvised aerial to investigate problems of static when he detected radio waves, which he subsequently showed to come from the Milky Way. This was, in fact, the beginning of radio astronomy; but for various reasons Jansky paid little attention to it after 1937, and virtually abandoned the problem.

Jeans, Sir James Hopwood. 1877–1946. English astronomer. He elaborated the plausible but now rejected theory of the tidal origin of the planets, but his major work was in connection with stellar constitution, in which he made notable advances. He was also an expert writer of popular scientific books, and was famous as a lecturer and broadcaster.

Jeffreys, Sir Harold. 1892–1989. Though primarily a geophysicist, Jeffreys also made many important advances in astronomy, and it was he who first showed that the giant planets are not miniature suns. His great book, *The Earth; its Origin, History and Physical Constitution* (1924) was immensely influential.

Jones, Sir Harold Spencer. 1890–1960. English astronomer (Astronomer Royal 1933–55). A Cambridge graduate, who was HM Astronomer at the Cape from 1923 until his appointment to Greenwich. From the Cape he carried out much important work, mainly in connection with star catalogues and stellar radial velocities. While Astronomer Royal he redetermined the solar parallax by means of the world-wide observations of Eros, and published several excellent popular books as well as technical papers. He played a major rôle in the removal of the main equipment from Greenwich to the new site at Herstmonceux, in Sussex, and himself transferred to Herstmonceux in 1948, though it was not until 1958 that the move was completed. He was knighted in 1943.

Kant, Immanuel. 1724–1804. German philosopher, remembered astronomically for proposing a theory of the origin of the Solar System which had some points of resemblance to Laplace's later Nebular Hypothesis.

Kapteyn, Jacobus Corneleus. 1851–1922. Dutch astronomer and cosmologist. His most celebrated discovery was that of 'star-streaming'.

Kepler, Johannes. 1571–1630. German astronomer, born in Württemberg. He was the last assistant to Tycho Brahe, and after Tycho's death used the mass of observations to establish his three laws of planetary motion. He observed the 1604 supernova, and also several comets, as well as making improvements to the refracting telescope, but his main achievements were theoretical. He ranks with Copernicus and Galileo as one of the main figures in the story of the 'Copernican revolution'.

Ketakar, Venkatesh. ?–1930. Indian astronomer. In 1911, using mathematical methods rather different from those of Lowell, he gave a prediction of the orbit and position of 'Planet X' (Pluto) which was not very different from Lowell's.

Kirch, Gottfried. 1639–1710. German astronomer; Director of the Berlin Observatory from 1705. He was one of the earliest of systematic observers of comets, star-clusters and variable stars. In 1686, he discovered the variability of χ Cygni.

Kirchhoff, Gustav Robert. 1824–87. Professor of Physics at Heidelberg. One of the greatest of German physicists, he explained the dark lines in the Sun's spectrum. His great map of the solar spectrum was published from Berlin in 1860.

Kirkwood, Daniel. 1814–95. American astronomer; an authority on asteroids and meteors. He drew attention to gaps in the asteroid belt, known today as the Kirkwood Gaps; they are due to the gravitational influence of Jupiter.

Kondratyuk, Yuri. 1897–1942. Ukrainian space writer; with Tsiolkovskii and Tsander he founded the Society for Studies in Interplanetary Travel.

Korolev, Sergei Pavlovitch.1906–66. Russian rocket designer who spent six years in one of Stalin's concentration camps but subsequently, as 'Chief Designer', played the vital rôle in the Sputnik and Vostok programmes. He died during an operation, and in Soviet rocketry was irreplaceable.

Kuiper, Gerard P. 1905–73. Dutch–American astronomer, who made notable advances in planetary and lunar work and was deeply involved with the programmes of sending probes beyond the Earth. The first crater to be identified on Mercury from Mariner 10 was named in his honour.

Kulik, Leonid. 1883–1942. Russian scientist, trained as a forester, who achieved fame because of his work in meteorite research. In particular, he led several expeditions to study the Tunguska object of 1908. He died in a German prison camp in 1942.

Lacaille, Nicolas Louis de. 1713–62. French astronomer who went to the Cape to draw up the first good southern-star catalogue.

Lagrange, Joseph Louis de. 1736–1813. French mathematical genius, and author of the classic *Mécanique Analytique*. He wrote numerous astronomical papers dealing, among other topics, with the Moon's libration and the stability of the Solar System.

Laplace, Pierre Simon. 1749–1827. French mathematician who made great advances in dynamical astronomy. In 1796, he wrote *Systeme du Monde*, in which he outlined his nebular hypothesis of the origin of the planets. This was discarded in its original form, but modern theories have many points of resemblance to it.

Lassell, William. 1799–1880. English astronomer who discovered Triton, Neptune's larger satellite, and (independently of Bond) Hyperion, the seventh satellite of Saturn as well as two satellites of Uranus (Ariel and Umbriel). He set up 24- and 48-inch reflectors in Malta and discovered 600 nebulæ.

Leavitt, Henrietta Swan. 1868–1921. American woman astronomer, best remembered for her observations of Cepheids in the Small Magellanic Cloud (1912), based on photographs taken in South America; these led on to the discovery of the vital period–luminosity law for Cepheids. She also discovered four novæ, several asteroids, and over 2400 variable stars.

Lemaître, Georges. 1894–1966. Belgian priest, who was a leading mathematician; from 1927, Professor at Louvain University. His most important paper, leading to what is now called the 'Big Bang' theory of the universe, appeared in 1927, but did not become well known until publicised by Eddington three years later. During World War I Lemaître served in the Belgian Army, and won the Croix du Guerre.

Le Monnier, Pierre Charles. 1715–99. French astronomer who was concerned in star cataloguing. He observed the planet Uranus several times, but did not check his observations, and missed the chance of a classic discovery. It was said that he never failed to quarrel with anyone whom he met!

Le Verrier, Urbain Jean Joseph. 1811–77. French astronomer, whose calculations led in 1846 to the discovery of Neptune. He was an authority on meteors, and in 1867 computed the orbit of the Leonids. He also developed solar and planetary theory, and believed in the existence of a planet (Vulcan) closer to the Sun than Mercury – now known to be a myth. He was forced to resign the Directorship of the Paris Observatory in 1870 because of his irritability, but was reinstated on the death by drowning of his successor, Delaunay.

Levin, Boris Yuljevich. 1912–89. Born in Moscow. At first he was concerned with meteors, but then worked with O. Schmidt in developing his theories of the origin of the Solar System. He also made valuable contributions to cometary astronomy.

Lexell, Anders John. 1740–84. Finnish astronomer, born in Abö. He became Professor of Mathematics at St Petersburg. He discovered the periodical comet of 1770 (now lost), and was one of the first to prove that the object discovered by Herschel in 1781 was a planet rather than a comet.

Lindsay, Eric Mervyn. 1907–74. Irish astronomer, Director of the Armagh Observatory from 1936 until his death. His main work was in connection with the Magellanic Clouds and with quasars. He had close connections with the Boyden Observatory in South Africa (where he had previously been assistant astronomer) and forged close links between it, Harvard, Dunsink (Dublin) and Armagh. He was an excellent lecturer on popular astronomy, and founded the Armagh Planetarium in 1966.

Lockyer, Sir Joseph Norman. 1836–1920. English astronomer, and an independent discoverer of the method of studying the solar chromosphere and prominences at times of non-eclipse. He was knighted in 1897. He founded the Norman Lockyer Observatory at Sidmouth in Devon, which still exists and is open to the public by arrangement, and was also founder of the periodical *Nature*.

Lohrmann, Wilhelm Gotthelf. 1796–1840. German land surveyor, who began an elaborate lunar map but was unable to complete it owing to ill health. The map was completed 40 years later by Julius Schmidt.

Lomonosov, Mikhail. 1711–65. Russian astronomer; he was also termed 'the founder of Russian literature'. His father was a fisherman. In 1735, he went to the University of St Petersburg, and then to Marburg in Germany to study chemistry. On his return to Russia in 1741, he insulted some of his colleagues at the St Petersburg Academy and was imprisoned for several months, during which time he wrote two of his most famous poems. However, he later became Professor of Chemistry at St Petersburg, and in 1746 became a Secretary of State. He drew up the first accurate map of the Russian Empire, described a 'solar furnace', and investigated electrical phenomena. He also studied auroræ. In 1761, he observed the transit of Venus, and rightly concluded that Venus has a considerable atmosphere. His most important contribution was his championship of the Copernican theory and of Newton's theories, neither of which had really taken root in Russia before Lomonosov's work.

Lowell, Percival. 1855–1916. American astronomer who founded the Lowell Observatory at Flagstaff, Arizona, in 1894.

He paid great attention to Mars, and believed the 'canals' to be artificial waterways. His calculations led to the discovery of the planet Pluto, though the planet was not actually found until 1930 – by Clyde Tombaugh, at the Lowell Observatory. Lowell himself was a great astronomer who did much for science, and it is regrettable that he is today remembered mainly because of his erroneous theories about the Martian canals.

Lucian of Samosata. *c*.120–180. Greek satirist who wrote a story about a Moon voyage. His travellers were hurled Moonward by a waterspout!

Lyot, Bernard. 1897–1953. Great French astronomer; Director of the Meudon Observatory. He made many advances in instrumental techniques, and invented the coronagraph, which enables the inner corona to be studied at times of non-eclipse. He died suddenly while taking part in an eclipse expedition to Africa.

Lyttleton, Raymond. 1911–95. Cambridge astronomer; a strong advocate of the steady-state theory of the universe. He believed comets to be 'flying gravel banks'. He made very important contributions to astrophysics.

Maclear, Sir Thomas. 1794–1879. Irish astronomer, who in 1833 succeeded Henderson as HM Astronomer at the Cape. He made an accurate measurement of an arc of meridian as well as verifying Henderson's parallax of α Centauri; he also studied comets and nebulæ. He was knighted in 1860.

Mädler, Johann Heinrich von. 1794–1874. German astronomer who was the main observer in the great lunar map by himself and Beer, published in 1837–8 – a map which remained the standard for several decades. In 1840, he left his Berlin home to become Director of the Dorpat Observatory in Estonia. He erroneously believed that η Tauri (Alcyone) was the star lying at the centre of the Galaxy. He retired in 1865, and spent his last years in Hanover.

Maraldi, Giacomo Filippo. 1665–1729. Italian astronomer; nephew of G. D. Cassini. He was renowned for his observations of the planets, particularly Mars, and assisted his uncle at the Paris Observatory.

Maskelyne, Nevil. 1732–1811. English astronomer (Astronomer Royal 1765–1811). Educated at Cambridge; he then went to St Helena, at the suggestion of Bradley, to observe the transit of Venus, and decided to make a serious study of navigation. During his régime as Astronomer Royal he founded the *Nautical Almanac*.

Masursky, Harold. 1923–90. A leading planetary geologist who was closely involved with the NASA missions, playing a key role in their planning. He spent his entire career with the US Geological Survey.

McCrea, Sir William. 1904–99. Much of McCrea's career was spent in mathematical departments, and his first astronomical appointment – at the University of Sussex – came in 1966. However, his interest in astronomy dated back much further. In 1929, he was able to confirm that hydrogen is dominant in the atmosphere of the Sun, and he made many contributions to astrophysics following his graduation from Cambridge in 1923. He was also active in the field of theoretical cosmology.

McVittie, George. 1904–88. British astronomer, born in Smyrna (his mother was Greek). He graduated from Edinburgh, and worked in England and America. He was a pioneer in studies of relativity, and made many important contributions to cosmology.

Méchain, Pierre Francois Andre. 1744–1805. French astronomer who discovered eight comets (1781–99).

Menzel, Donald H. 1901–76. American astronomer, celebrated for his research into problems of the Sun and planets as well as in stellar studies. He was also an excellent lecturer, and a skilled writer of popular books.

Messier, Charles. 1730–1817. French astronomer, interested mainly in comets. Though he discovered 13 comets, he is remembered chiefly because of his catalogue of star-clusters and nebulæ, published in 1781.

Michell, John. 1725–93. English clergyman, and an amateur astronomer who made the first suggestion that many double stars may be physically associated or binary systems.

Milne, Edward Arthur. 1896–1950. English astronomer who graduated from Cambridge and then went successively to Manchester and Oxford. He made important contributions to astrophysics, and developed his theory of 'kinematic relativity', which was for a time regarded as an alternative to general Einsteinian relativity.

Minkowski, Rudolf. 1895–1976. German astronomer, who went to Mount Wilson in 1935 and remained there. He was one of the leading authorities on novæ and planetary nebulæ, and after the war became a pioneer in the new science of radio astronomy. His studies of rapidly moving gases in radio galaxies led to the rejection of the 'colliding galaxies' theory.

Mitchell, Maria. 1812–88. America's first woman astronomer and director of the Vassar College Observatory and Professor of Astronomy. She was the first woman to be elected to the American Academy of Arts and Sciences; she also discovered a comet.

Montanari, Geminiario. 1633–87. Italian astronomer who worked at Bologna and then at Padua. In 1669, he discovered the variability of Algol.

Morrison, Philip. 1915–2005. American astronomer best remembered for his contributions to the search for extraterrestrial intelligence (SETI). In 1959, with Cocconi, he wrote the pioneer paper proposing the potential of microwaves in interstellar communications.

Mutch, Thomas. 1921–1980. American astronomer and geologist, author of an important book about the geology of the Moon. He was also leader of the Viking Imaging Lander Team when landings were made on Mars. In October 1980 he was killed during a mountaineering holiday.

Nevill, Edmund Neison. 1851–1940. English astronomer, who published an important book and map concerning the Moon in 1876; he wrote under the name of Neison. He was Director of the Natal Observatory at Durban in South Africa (1882–1910), returning to England when the Observatory was closed.

Newcomb, Simon. 1835–1909. American astronomer, for some years head of the American Nautical Almanac office. His chief work was in mathematical astronomy, to which he made valuable contributions. He is also remembered as the man who proved to his own satisfaction that no heavier-than-air machine could ever fly!

Newton, Sir Isaac. 1643–1727. Probably the greatest of all astronomers. To list all his contributions here would be pointless; suffice to say that his *Principia*, published in 1687, has been described as the greatest mental effort ever made by one man. In addition to his scientific work, he sat briefly in Parliament, and served as Master of the Mint. He was knighted in 1705, and on his death was buried in Westminster Abbey.

Nikolayev, Andriyan. 1929–2004. Russian Cosmonaut who flew on two space-flights, Vostok 3 and Soyuz 9. He was the first person to make a television broadcast from space (1962). Vostok 3 was part of the first dual space-flight, with Pavel Popovich on Vostok 4.

Oberth, Hermann. 1894–1989. Romanian physicist and mathematician who wrote the first serious technical book about space research. He worked with von Braun at Peenemunde and the US.

Olbers, Heinrich Wilhelm Matthias. 1758–1840. German doctor, also a skilled amateur astronomer; established his private observatory in Bremen. He discovered two of the first four asteroids (Pallas and Vesta) and rediscovered the first (Ceres); he carried out important work in connection with cometary orbits, and discovered a periodical comet which has a period of 69.5 years, and last returned in 1956. Olbers also wrote about his celebrated paradox: 'Why is it dark at night?'

Oort, Jan Hendrick. 1900–92. Outstanding Dutch astronomer who interpreted Kapteyn's two streams of stars as evidence that the Galaxy is rotating. With C. A. Muller, he confirmed the detection by Ewen and Purcell of the 21-cm background radiation. He proposed that comets come from a huge spherical shell surrounding the Solar System, now known as the Oort Cloud.

Paczynski, Bohdan. 1940–2007. Polish Astronomer, a leading authority on theories of stellar evolutions and gamma-ray bursts.

Parmenides of Elea. Second half of the sixth century BC. He believed the stars to be of compressed fire, but agreed that the Earth was spherical, and in equilibrium because it was equidistant from all points on the sphere representing the universe.

Penston, Michael Victor. 1943–90. He was born in London, but spent much of his career at Cambridge. He made many contributions to astrophysics, and in 1983 was able to 'weigh' a black hole in the centre of the galaxy NGC 4151. He described himself as one of the LAGS – Lovers of Active Galaxies! Sadly, he died in 1990 after a long and brave fight against cancer.

Perrine, Charles Dillon. 1867–1951. American astronomer who discovered two of Jupiter's satellites as well as nine comets. He worked at the Lick Observatory until 1909, when he became Director of the Cordoba Observatory in Argentina, where he constructed a 30-inch reflector and made many observations of southern galaxies. He also planned a major star catalogue, but was politically unpopular, and after a narrow escape from assassination he retired (1936).

Peters, Christian Heinrich Friedrich. 1813–90. Danish astronomer who emigrated to America in 1848. He discovered 48 asteroids.

Piazzi, Giuseppe. 1746–1826. Italian astronomer, who became Director of the Palermo Observatory in Sicily. During the compilation of a star catalogue he discovered the first asteroid, Ceres (on 1 January 1801, the first day of the new century).

Pickering, Edward Charles. 1846–1919. American astronomer; for 43 years, from 1876, Director of the Harvard College Observatory. He concentrated upon photometry, variable stars and, above all, stellar spectra; in the famous Draper Catalogue, the stars were classified according to their spectra. During his régime the Harvard Observatory was modernised and a southern outstation was set up at Arequipa in Peru.

Pickering, William Henry. 1858–1938. Brother of E. C. Pickering, who worked with him at Harvard and also served for a while as astronomer in charge of the Arequipa out-station. In 1898, he discovered Saturn's ninth satellite, Phœbe. He made extensive studies of the Moon and Mars, mainly from the Harvard Station in Jamaica, which was set up in 1900. Independently of Lowell, he calculated the position of Pluto.

Plutarch. *c*. 46–120. Greek biographer, mentioned here because of his authorship of *De Facie in Orbe Lunæ* – On the Face in the Orb of the Moon – in which he claimed that the Moon is a world of mountains and valleys.

Pond, John. 1767–1836. English astronomer (Astronomer Royal, 1811–35). Though an excellent and painstaking astronomer, Pond was handicapped by ill health during the latter part of his régime at Greenwich, and was eventually asked to resign. He tried unsuccessfully to obtain star distances by the parallax method.

Pons, Jean Louis. 1761–1831. French astronomer, whose first post at an observatory (Marseilles) was that of caretaker! He was self-taught and concentrated on hunting for comets; he found 36 in all, and ended his career as Director of the Museum Observatory in Florence.

Popovich, Pavel. 1930–2009. Russian cosmonaut who took part in the first dual flight – Nikolayvich in Vostok 3 and Popovich in Vostok 4. After his flight he continued to be active in the field of space research.

Proctor, Richard Anthony. 1837–88. English astronomer who was an excellent cosmologist but is best known for his many popular books. In 1881, he emigrated to America, and remained there for the rest of his life. Proctor paid considerable attention to the planets, and constructed a map of Mars.

Ptolemy (Claudius Ptolemæus). *c*. 120–180. The 'Prince of Astronomers', who lived and worked in Alexandria. Nothing is known about his life, but his great work has come down to us through its Arab translation (the *Almagest*). Ptolemy's star catalogue was based on that of Hipparchus but with many contributions of his own; he also brought the geocentric system to its highest state of perfection, so that it is always known as the Ptolemaic theory. He constructed a reasonable map of the Mediterranean world, and even showed Britain, though it is true that he joined Scotland on to England in a back-to-front position.

Purbach, Georg von. 1423–61. Austrian astronomer who became a professor at Vienna in 1450. He founded a new school of astronomy, compiled tables, and began to write an *Epitome of Astronomy* based on Ptolemy's *Almagest*. After Purbach's premature death, the book was completed by his friend and pupil Regiomontanus.

Pythagoras. *c*. 572–500 BC. The great Greek geometer, mentioned here because he was one of the very first to maintain that the Earth is spherical rather than flat. He seems also to have studied the movements of the planets.

Rahe, Jurgen. 1940–97. German astronomer who emigrated to the United States and joined NASA, becoming Director of the Solar System Exploration Division; he also acted as a staff member of the California Institute of Technology. He was responsible for the overall general management, budget and strategic planning for many missions, including the Galileo probe to Jupiter. He was killed by a freak accident, when a tree fell on his car as he was driving near his home at Potomac in Maryland.

Ramsden, Jesse. 1735–1800. English maker of astronomical instruments. His meridian circles were the first to have their cross-hairs lit through the hollow axis.

Rayet, Georges Antoine. 1839–1906. French astronomer; in 1867, with Wolf, drew attention to the Wolf–Rayet stars, which have bright lines in their spectra. He went from Paris to Bordeaux, and became Director of the observatory there.

Redman, Richard Oliver. 1905–75. English astronomer who graduated from Cambridge. He made extensive studies of the Sun, stellar velocities, galactic rotation and the photometry of galaxies. In 1937, he went to the Radcliffe Observatory, Pretoria, and designed the spectrograph for the 74-inch reflector. In 1947, he returned to Cambridge as Professor of Astrophysics and Director of the Observatories. Many programmes were carried through, and Redman also devoted much time and energy to the planning and construction of the 153-inch Anglo–Australian telescope at Siding Spring.

Regiomontanus. 1436–76. The Latinised name of Johann Müller, Purbach's pupil. He completed the *Epitome of Astronomy*, and at Nürnberg set up a printing press, publishing the first printed astronomical ephemerides. He died in Rome, where he had been invited to help in reforming the calendar.

Reinmuth, Karl. 1892–1979. He spent his entire career at the Königstuhl Observatory in Heidelberg, concentrating upon asteroids and comets. He discovered over 250 asteroids, among them the 1932 Earth-grazer, Apollo.

Rhæticus, Georg Joachim. 1514–76. German astronomer, who became Professor of Astronomy at Wittenberg in 1536. An early convert to the Copernican system, he visited Copernicus at Frombork, and persuaded him to send his great book for publication.

Riccioli, Giovanni Battista. 1598–1671. Italian Jesuit astronomer who taught at Padua and Bologna. He was a pioneer telescopic observer, and drew up a lunar map, inaugurating the system of nomenclature which is still in use. Oddly enough, he never accepted the Copernican system.

Ridley, Harold Bytham. 1919–95. Born in Outer London, he was one of Britain's outstanding amateur astronomers; he specialised in meteor work, and obtained many meteor spectra. He was for many years Director of the Meteor Section of the British Astronomical Association (and was its President from 1976 to 1978).

Robinson, Thomas Romney. 1792–1882. Irish astronomer who was Director of the Armagh Observatory from 1823 to his death. He published the Armagh catalogue of over 5000 stars, and made many other contributions; he also invented the cup anemometer. It is on record that when the railway company planned to build a line close to Armagh, Robinson managed to have it diverted, since he maintained that the trains would shake his telescopes!

Rømer, Ole. 1644–1710. Danish astronomer. In 1675 he used the eclipse times of Jupiter's satellites to make an accurate measurement of the velocity of light. In 1681 he became Director of the Copenhagen Observatory. Among his numerous inventions are the transit instrument and the meridian circle.

Ronan, Colin Alastair. 1920–95. Colin Ronan was essentially an astronomical historian. During the war he served with the Army, and rose to the rank of Major; he made a very important contribution to the war effort, inventing a new method of blooming lenses to increase light transmission. He wrote many books, and was an outstanding lecturer; he was for many years Director of the Historical Section of the British Astronomical Association (President 1989–91). He was an active researcher, and produced plausible evidence that the telescope was invented in England over half a century before Galileo's time.

Rosse, 3rd Earl of. 1800–67. Irish amateur astronomer, who in 1845 completed the building of a 72-inch reflector and erected it at his home at Birr Castle. The 72-inch with its metal mirror, was much the largest telescope ever built up to that time, and has now (2010) been brought back into use. His greatest discovery was that some 'nebulæ' (galaxies) are spiral in form.

Rosse, 4th Earl of. 1840–1908. Continued his father's work, and was also the first man to measure the tiny quantity of heat coming from the Moon.

Rowland, Henry Augustus. 1848–1901. American scientist; Professor of Physics in Baltimore from 1876. His great map of the solar spectrum was published in 1895–7; it showed 20 000 absorption lines.

Rudolph, Arthur. 1906–1966. German expert on space medicine. He made immensely valuable contributions and went to America, but it was then found that during the war he had carried out inhumane experiments upon prisoners. He had to leave America, and retired to Germany

Runcorn, S. Keith. 1922–98. Born in Southport, he graduated in engineering from Cambridge in 1942. After a period working at radar research, he joined Manchester University, and carried out major researches into palæomagnetism and all other aspects of planetary magnetic phenomena.

Russell, Henry Norris. 1877–1957. American astronomer; Director, Princeton Observatory from 1908. He devoted much of his energy to studies of stellar constitution and evolution, and independently of Hertzsprung he discovered the giant and dwarf subdivisions of stars of late-spectral type. This led on to the H–R or Hertzsprung–Russell diagram, in which luminosity (or the equivalent) is plotted against spectral type.

Rutherford, Lewis Morris. 1816–92. American barrister, who gave up his profession to devote himself to astronomy. A pioneer in astronomical photography, his pictures of the Moon were outstanding; his ruled solar gratings for solar spectra were the best of their time.

Ryle, Sir Martin. 1918–84. British pioneer of radar and radio astronomy. He graduated from Oxford, and then went to Cambridge as Professor of Radio Astronomy. In 1972, he succeeded Woolley as Astronomer Royal. He developed the technique of aperture synthesis, and made many very important contributions.

Sagan, Carl. 1934–96. Born in Brooklyn, he spent all his career in the United States. He made major contributions to astrophysics, and was also a leading member of the Planetary Society, but is perhaps best remembered for his popular works, and in particular the best-selling television series 'Cosmos'. Sadly, he died of cancer when still in his early fifties. The Pathfinder station on the surface of Mars was named in his honour.

Scheiner, Christoph. 1575–1650. German Jesuit who was for some time Professor of Mathematics in Rome. He discovered sunspots independently of his contemporaries, and wrote a book, *Rosa Ursina*, which contains solar drawings and observations for the years 1611–25. He was unfriendly towards Galileo, and played a rather discreditable part in the events leading up to Galileo's trial and condemnation.

Schiaparelli, Giovanni Virginio. 1835–1910. Italian astronomer, who graduated from Turin and became Director of the Brera Observatory in Milan. He discovered the connection between meteors and comets, but his most famous work was in connection with the planets. It was he who first drew attention to the 'canal network' on Mars, in 1877.

Schirra, Walter. 1923–2007. One of the 'original seven' American astronauts in Project Mercury. He made three trips into space.

Schlesinger, Frank. 1871–1943. American astronomer, Born in. New York, his main work was in connection with stellar parallaxes. He was Director successively of the Yale and Allegheny observatories, and was responsible for the Yale 'southern station' in Johannesburg. His major works, *General Catalogue of Parallaxes* and its supplement, dealt with more than 2000 stars. He also pioneered the determination of star positions by using wide-angle cameras.

Schmidt, Julius (actually Johann Friedrich Julius). 1825–84. German astronomer, who became Director of the Athens Observatory in 1858 and spent most of his life in Greece. He concentrated upon lunar work, producing an elaborate map (based on Lohrmann's early work) and making great improvements in selenography. He drew attention to the alleged change in the lunar crater Linné, in 1866. He also discovered the outburst of the recurrent nova T Coronæ, in 1866.

Schönfeld, Eduard. 1828–91. German astronomer who collaborated with Argelander in preparing the *Bonn Durchmusterung* and later extended it to the southern hemisphere.

Schramm, David. 1945–1997. American astronomer, working mainly at Chicago. He was a specialist in theories of the early stages of the universe. He was killed when his aircraft crashed during a solo flight.

Schröter, Johann Hieronymus. 1745–1816. Chief magistrate of Lilienthal, near Bremen. He set up a private observatory, and made outstanding observations of the Moon and planets. Unfortunately, many of his notebooks were lost in 1813, with the destruction of his observatory by the invading French troops.

Schwabe, Heinrich. 1789–1875. German apothecary, who became a noted amateur astronomer concentrating on the Sun. His great discovery was that of the 11-year sunspot cycle.

Schwarzschild, Karl. 1873–1916. German astronomer, who worked successively at Vienna, Göttingen and (as Observatory Director) Potsdam. His early work dealt with photometry, but he was also a pioneer of theoretical astrophysics. Military service during World War I broke his health and led to his premature death.

Schwarzschild, Martin. 1912–97. German astronomer, who spent much of his career at Princeton University in the United States. He carried out much original research with regard to stellar structure and evolution.

Secchi, Angelo. 1818–78. Italian Jesuit astronomer; one of the great pioneers of stellar spectroscopy, classifying the stars into four types (a system superseded later by that of Harvard). He was also an authority in solar work, and his planetary observations were equally outstanding.

Seyfert, Carl. 1911–60. American astronomer who concentrated upon studies of galaxies. In 1942, he drew attention to those galaxies with very condensed nuclei, now always termed Seyfert galaxies.

Shajn, Grigorij Abramovich. 1892–1956. After serving in the Russian Army during World War I he then joined the staff of the Pulkovo Observatory, working upon meteoric astronomy. In 1924 he became Director of the Simeis Observatory, and began his work on stellar spectroscopy. During most of his latter years he was concerned mainly with the varied distribution of the faint galactic nebulæ.

Shapley, Harlow. 1885–1972. Great American astronomer who began his main work at Princeton under H. N. Russell. In 1914, he advanced the pulsation theory of Cepheid variables, and was soon able to use the variables in globular clusters to give the first accurate picture of the shape and size of the Galaxy. In 1921, he became Director of the Harvard College Observatory. In later years he concentrated upon studies of galaxies and upon the international aspect of astronomy. He was also an excellent lecturer, and a skilled writer of popular books.

Shi Shen. 4th century BC. Chinese astronomer who worked with Gan De. He made the first known observations of sunspots and realised that they were solar phenomena.

Shepard, Alan Bartlett. 1928–1988. The first American in space – a sub-orbital 'hop'. In 1971, he went to the Moon as commander of Apollo 14. He retired from NASA in 1974 and became a business executive.

Shklovskii, Iosif. 1916–85. He graduated from Moscow University, and became a professor there, later heading the radio astronomy department at the Sternberg Institute. He specialised in extraterrestrial radio sources, and showed that the emission from the Crab Nebula is synchrotron radiation. With Carl Sagan, he wrote about 'intelligent life in the universe'.

Shoemaker, Eugene. 1928–97. American astronomer geologist, who became the leading expert on meteoritics and impact craters; he was deeply involved in all the earlier planetary missions. He was also a devoted hunter of comets and near-Earth asteroids. He was killed in a car accident in Australia; subsequently, his ashes were scattered on the surface of the Moon.

Smyth, William Henry. 1788–1865. English naval officer, rising to the rank of Admiral, who in 1830 established a private observatory at Bedford and made numerous astronomical observations. He is best remembered for his famous book, *Cycle of Celestial Objects*.

Smyth, Piazzi (actually Charles Piazzi). 1819–1900. The son of Admiral Smyth. Astronomer Royal for Scotland from 1844 until his death. He was a skilled astronomer who carried out much valuable work, including spectroscopic examinations of the Zodiacal Light, but he was also an eccentric who wrote a large and totally valueless volume about the significance of the Great Pyramid.

Sosigenes. Greek astronomer, who flourished about 46 BC. He was entrusted by Julius Cæsar with the reform of the calendar. Nothing is known about his life.

South, James. 1785–1867. English amateur astronomer, who founded a private observatory, first in Southwark then at Campden Hill, and collaborated with John Herschel in studies of double stars. In 1822, he observed an occultation of a star by Mars, and the virtually instantaneous disappearance convinced him that the Martian atmosphere must be extremely tenuous.

Singer, Siegfried Friedrich. 1924–2005. Austrian physicist who emigrated to America and helped to design some of the instruments in the early satellites.

Spitzer, Lyman. 1914–97. American astrophysicist who influenced several generations of researchers through his writing and lecturing as well as his own personal contributions. He was founder and first Director of the Princeton Plasma Physics Laboratory, the leader of the group developing the Copernicus satellite.

Spörer, Friedrich Wilhelm Gustav. 1822–95. German astronomer who joined the staff at the Potsdam Observatory. He concentrated mainly upon the Sun, and discovered the variation in latitude of spot zones over the course of a solar cycle (Spörer's Law).

Steavenson, William Herbert. 1894–1975. Steavenson was never a professional astronomer; he was a medical doctor whose practice was in Outer London. He was an expert observer, and one of the few amateurs to serve as President of the Royal Astronomical Society (1957–9). His knowledge of telescopes and optics was encyclopædic, and his advice was often sought by professional astronomers.

Stibbs, Walter. 1919–2010. (Douglas) Walter Stibbs was an Australian astronomer, born in Sidney. He graduated from the university there and then went to Mount Stromlo, where he concentrated on variable stars and wrote an important book with the director, Richard Wooley. In 1959, he went to Scotland as Director of the Observatory at St Andrews University, and remained there until he retired in 1989. In addition to teaching, he continued with his own researches and oversaw the completion of the 38-inch Gregory telescope. On retirement, he returned to Australia and died on 12 April 2010. He was incidentally a good musician and a fine athlete. He was undoubtedly eccentric; either you liked him immensely (as I did) or you would remain on an entirely different wavelength.

Stock, Jürgen 1923–2004. German astronomer who went to America and chose the site for the Cerro Tololo observatory, becoming its first director. He was one of the astronomers who was largely responsible for placing the main European Southern Observatory telescopes in Chile rather than in South Africa.

Strømgren, Bengt. 1908–87. Born in Sweden, but raised in Denmark. He became Director of the Royal Copenhagen Observatory in succession to his father Elias. He went to America to direct the Yerkes and McDonald Observatories during the 1950s, returning to Denmark in 1967. He was particularly known for his studies of H.II (ionised hydrogen) regions round hot stars often called Strømgren spheres.

Struve, Friedrich Georg Wilhelm. 1793–1864. German astronomer, born in Altona. He went to Dorpat in Estonia and in 1818 became Director of the observatory. Using the 9-inch Fraunhofer refractor – the first telescope to be clock-driven – he began his classic work on double stars. In 1839, he went to Pulkova, to become director of the new observatory set up by Tsar Nicholas. Here he continued his double-star work, and his *Mensuræ Micrometricæ* gives details of over 3000 pairs. Strove also measured the parallax of Vega; his value was announced in 1840.

Struve, Otto (Wilhelm). 1819–1905. Son of F. G. W. Struve; born in Dorpat, he became assistant to his father, accompanying him to Pulkova. He continued his father's work, and became a leading authority on double stars. He succeeded to the directorship of Pulkova Observatory in 1861, retiring in 1889 and returning to Germany

Struve, Karl Hermann. 1854–1920. Son of Otto Struve; born in Pulkova, later becoming assistant to his father. His main work was in connection with planetary satellites. In 1895 he went to Königsberg, and in 1904 became Director of the Berlin Observatory, which was reorganised by him and transferred to Babelsberg during his term of office.

Struve, Gustav Wilhelm Ludwig. 1858–1920. Son of Otto Struve, and brother of Karl. He too was born at Pulkova and acted as assistant to his father. He went to Dorpat in 1886, and from 1894 was Director of the Kharkov Observatory. He was concerned mainly with statistical astronomy and with the motion of the Sun.

Struve, Otto. 1897–1963. Son of Gustav; often known as Otto Struve II. He was born in Kharkov, and fought during World War I; joined the White Army under Wrangel and Derrikin, and after their defeat reached Constantinople, where he worked as a labourer. Finally, he was offered a post at the Yerkes Observatory, where he arrived in 1921. He spent the rest of his life in America; in 1932 he became Director at Yerkes, after which he founded the McDonald Observatory in Texas and was its director from 1939 to 1947, when he became Chairman of the Department of Astronomy at Chicago. In 1959, he began a new career as the first Director of the National Radio Astronomy Observatory, but ill health forced his resignation in 1962. He was a brilliant astrophysicist, dealing mainly with spectroscopic binaries, stellar rotation and interstellar matter; he was also an author of popular books. He is (so far!) the last of the famous Struve astronomers. It is notable that four were in succession awarded the Gold Medal of the Royal Astronomical Society – a sequence unique in astronomical history.

Stuhlinger, Ernst. 1913–2008. One of Wernher von Braun's team, building the V2 missiles at Peenemünde. He subsequently went to America and played an important part in the NASA space programme. He retired in 1975 to take up a post at the University of Alabama.

Swift, Lewis. 1820–1913. American astronomer who specialised in hunting for comets and nebulæ. He found 13 comets (including the Great Comet of 1862) and 900 nebulæ.

Tempel, Ernest Wilhelm. 1821–89. German astronomer, who became Director of the Arcetri Observatory. In 1859, he discovered the nebula in the Pleiades; he also discovered six asteroids and several comets, including the comet of 1865–6, which is associated with the Leonid meteors.

Thales. *c*. 624–547 BC. The first of the great Greek philosophers. He believed the Earth to be flat, and floating in an ocean, but he was a pioneer mathematician and observer, and successfully predicted the eclipse of 585 BC which stopped the war between the Lydians and the Medes.

Timocharis. *c*. 280 BC. Greek astronomer who made some accurate measurements of star positions; one of these (of Spica) enabled Hipparchus, 150 years later, to demonstrate the precession of the equinoxes.

Tombaugh, Clyde. 1906–97. While still a young amateur, Tombaugh was called to the Lowell Observatory to search for 'Planet X', and in 1930 he discovered Pluto. He remained at the Lowell Observatory for many years, and searched for further planets and minor Earth satellites, although without success. During the war he worked at White Sands, developing telescopic methods of tracking ballistic missiles. He then went to Las Cruces University, where he remained for the rest of his career, latterly as Professor Emeritus; he was a tireless and inspiring teacher.

Tsander, Friedrikh Arturovich. 1887–1931. Latvian space scientist. In 1931 he designed a liquid-fuel rocket, but died before its successful maiden flight in 1933.

Tsiolkovskii, Konstantin Eduardovich. 1857–1935. Russian 'father of space-travel'; became a mathematics teacher, and during the 1890s was publishing material that was decades ahead of its time, though its significance was not appreciated until much later.

Turner, Herbert Hall. 1861–1930. English astronomer who played an important role in the organisation and preparation of the International Astrographic Chart. In 1903 he discovered Nova Geminorum.

Tycho Brahe. 1546–1601. The great Danish observer – probably the best of pre-telescopic times. He studied the supernova of 1572 and from 1576 to 1596 worked at his observatory at Hven, an island in the Baltic, making amazingly accurate measurements of star positions and the movements of the planets, particularly Mars. His observatory – Uraniborg – became a scientific centre, but Tycho was haughty and tactless (during his student days he had part of his nose sliced off in a duel, and made himself a replacement out of gold, silver and wax!) and after quarrels with the Danish Court he left Hven and went to Prague as Imperial Mathematician to the Holy Roman Emperor, Rudolph II. Here he was joined by Kepler, who acted as his assistant. When Tycho died, Kepler came into possession of part of the Hven observations, and used them to argue that the Earth moves round the Sun – something which Tycho himself could never accept.

Ulugh Beigh (more properly Ulugbek). 1394–1449. Grandson of the Oriental conqueror Tamerlane. About 1420, he established an observatory at Samarkand, which became an astronomical centre; he compiled a star catalogue, and compiled tables of the Moon and planets. He was assassinated in 1449.

Valier, Max. 1895–1930. Austrian rocket pioneer; colleague of Oberth and von Braun. He did much to promote the concept of space-flight. He was killed when an alcohol-fuelled rocket exploded on his test-bench in Berlin.

Van Allen, James. 1914–2006. American Space scientist who made fundamental contributions to research; it was his work that led to the discovery of radiation belts round the Earth, confirmed by the first US space probe. These belts are now named the Van Allen Belts.

Van de Kamp, Peter. 1901–1995. Dutch astronomer, who emigrated to America and became Director of the Sproule Observatory; with the 24-inch telescope there he studied Barnard's Star and believed that he had detected the presence of a planet, but his results were found to be spurious, due to instrument errors.

Van Flandern, Tom. 1940–2009. American astronomer specialising in celestia mechanics. In 1969 he received his PhD from Yale, and worked at the US Naval Observatory until 1982. His theories were varied, and frequently unconventional.

Van Maanen, Adriaan. 1884–1947. Dutch astronomer, who emigrated to America and joined the Mount Wilson staff in 1912. He specialised in stellar parallaxes and proper motions, and accomplished much valuable work, though his alleged detection of movements in the spiral arms of galaxies later proved to be erroneous. He also discovered the white dwarf still known as Van Maanen's Star.

Vehrenberg, Hans. 1910–91. German amateur astronomer, whose superb stellar photographs, contained in his *Atlas Stellarum*, are widely used.

Verbiest, Ferdinand. 1623–88. Belgian Jesuit astronomer. He became a missionary to China, and was appointed chief astronomer at the Imperial Observatory at the time of the Qing emperor K'ang His. He corrected the Chinese calendar, compiled tables of eclipses and designed and built six splendid new instruments for the Observatory, including two large armillary spheres – made of brass, with bronze dragons forming the supports!

Vogel, Hermann Carl. 1842–1907. German astronomer, born and educated in Leipzig. He went to Potsdam in 1874, and concentrated upon stellar spectroscopy, pioneering research into spectroscopic binaries. In 1883, he published the first catalogue of stellar spectra.

Von Braun, Wernher. 1912–1977. The greatest figure in the early years of space research. He developed the V2 rockets at Peenemünde, but after the war went to America and masterminded the US satellite programme as well playing a vital role in the Apollo programme.

Von Weizsacker, Carl Friedrich. 1912–2007. German physicist and philosopher. He was a pioneer in studies of solar and stellar energy and, with Bethe, was the first to realise that the basic 'fuel' is hydrogen.

Walther, Bernard. 1430–1504. (sometimes mis-spelled 'Walter'.) German amateur astronomer who lived in Nuremburg; he financed Regiomontanus' equipment, and carried on the work when Regiomontanus died. He was a very accurate observer, whose measurements of star and planetary positions were of great value to later astronomers.

Wan-Hu. *c.* 1500. Chinese official of the Ming Dynasty. According to legend, he tied 47 rockets to a chair, sat in the chair, and ordered his servants to light all 47 at once. They did. The results were predictable. If the story is true, Wan-Hu must be regarded as the first casualty of astronautics!

Wargentin, Pehr Vilhelm. 1717–83. Swedish astronomer, and Director of the Stockholm Observatory. His best work was in the preparation of accurate tables of Jupiter's satellites.

Webb, Thomas William. 1806–85. Vicar of Hardwicke in Herefordshire. He was an excellent observer, but is best remembered for his book *Celestial Objects for Common Telescopes*.

Whipple, Fred. 1906–2004. Whipple made major contributions to astronomy, but his best known theory was the 'dirty snowball' theory of comets now known to be correct. This was only one of his major achievements. During the war, he advocated the use of 'window', fragments of silver paper dropped by British bomber aircraft to confuse the German radar. It worked – I know, because I took part in the programme myself!

Whitrow, Gerald. 1912–2000. English astronomer; much of his career was spent at Imperial College, London. He was a specialist in all problems concerning time, and wrote many books and papers.

Wilkins, Hugh Percy. 1896–1960. Welsh amateur astronomer (by profession a civil servant) who concentrated upon lunar observation, and produced a 300-inch map of the Moon. He was for many years Director of the Lunar Section of the British Astronomical Association.

Wilkins, John. 1614–1672. As warden of Wadham College, Oxford, he inspired the young (Sir) Christopher Wren and Robert Hooke to undertake major astronomical researchers.

Wolf, Maximilian Franz Joseph Cornelius. 1863–1932. (Better known as Max Wolf.) German astronomer, who was born and lived in Heidelberg. He studied comets, and discovered his periodical comet in 1884; he was the first to hunt for asteroids photographically. He also carried out research into dark nebulæ, and discovered well over 1000.

Woolley, Sir Richard van der Riet. 1906–1986. Pioneer astrophysicist, who graduated from Cambridge; he succeeded Spencer Jones as Astronomer Royal, and upon retirement returned to his native South Africa to become Director of the South African National Observatories.

Wren, Sir Christopher. 1632–1723. Though his enduring fame lay in architecture, Wren's reputation until he was about 40 lay in astronomy. Mathematical astronomy and the telescopic study of Saturn were his main areas of interest, and he held the Chairs of Astronomy at Gresham College, London, and then at Oxford University before he built St Paul's Cathedral.

Wright, Thomas. 1711–85. Born near Durham, He trained as a clockmaker, though he afterwards taught mathematics. He is remembered for his book published in 1750, in which he suggested that the Galaxy is disc-shaped. He also believed Saturn's rings to be composed of small particles.

Wynne, Charles. 1911–1999. English optical worker who improved techniques in lens making, and was involved in practically all the major telescope projects of his time and produced an accurate account of space research methods.

Xenophanes. *c.* 570–478 BC. Greek philosopher, born in Colophon. His astronomical theories sound strange today; an infinitely thick flat Earth, a new Sun each day, and celestial bodies which – apart from the Moon – were made of fire!

Zach, Franz Xaver von. 1754–1832. Hungarian baron who became renowned as an amateur astronomer. He published tables of the Sun and Moon, and was one of the chief organisers of the 'Celestial Police' who banded together to hunt for the supposed planet between Mars and Jupiter. He became Director of the Seeberg Observatory at Gotha, and did much for international co-operation among astronomers.

Zhang Heng. *c.* 78–139. Chinese astronomer, mathematician, statesman and inventor. He was born in Nanyang, and became a civil servant; eventually being appointed Chief Astronomer to the Emperor. His activities were widespread; for example he compiled a star catalogue, made the first seismometer and built the first water-powered armillary sphere. He also invented the odometer.

Zwicky, Fritz. 1898–1974. Swiss astronomer; born in Bulgaria, but remained a Swiss citizen throughout his life. He graduated from Zurich, and in 1925 went to the California Institute of Technology, where he remained permanently, becoming Professor of Astrophysics from 1942 until his retirement in 1968. He became famous for his studies of galaxies and intergalactic matter; he predicted the existence of neutron stars (1934) and even black holes. He discovered many supernovæ in external galaxies, and masterminded a catalogue of compact galaxies. He was also active in the development of astronomical instrumentation, and was a pioneer worker with Schmidt telescopes. He received the Gold Medal of the Royal Astronomical Society in 1973.

34 · Glossary

Aberration of starlight. As light does not move infinitely fast, but at a rate of practically $300\,000$ km s^{-1}, and as the Earth is moving round the Sun at an average velocity of 25 km s^{-1}, the stars appear to be shifted slightly from their true positions. The best analogy is to picture a man walking along in a rainstorm, holding an umbrella. If he wants to keep himself dry, he will have to slant the umbrella forward; similarly, starlight seems to reach us 'from an angle'. Aberration may affect a star's position by up to 20.5 seconds of arc.

Ablation. The erosion of a surface by friction or vaporisation.

Absolute magnitude. The apparent magnitude that a star would have if it could be observed from a standard distance of 10 parsecs (32.6 light-years).

Absolute zero. The coldest theoretically possible temperature: − 273.16 °C.

Accretion disc. A disc structure which forms round a spinning object when material falls on to it from beyond.

Achromatic object-glass. An object-glass which has been corrected so as to eliminate chromatic aberration or false colour as much as possible.

Aerolite. A meteorite whose main composition is stony.

Airglow. The light produced and emitted by the Earth's atmosphere (excluding meteor trains, thermal radiation, lightning and auroræ).

Albedo. The reflecting power of a planet or other non-luminous body. The Moon is a poor reflector; its albedo is a mere 7% on average.

Alfvén wave. A low-frequency travelling oscillation of the ions and the magnetic field of a plasma. It has been likened to a wave travelling along a stretched string; the magnetic field line tension is analogous to string tension, and any disturbance in the magnetic field propagates along the field line. There seems to be a link between these waves and the high temperature of the solar corona.

Alpha particle. The nucleus of a helium atom, made up of two protons and two neutrons.

Altazimuth mounting for a telescope. A mounting on which the telescope may swing freely in any direction.

Altitude. The angular distance given in degrees of a celestial body above the horizon.

Analemma. The figure-of-eight shape resulting if the Sun's position in the sky is recorded at the same time of day throughout the year.

Ångström unit. One hundred-millionth part of a centimetre.

Anorthosite. An igneous rock composed largely of anorthite, a calcium-rich plagioclase feldspar. It is rich in aluminium and calcium, and is the main constituent of the ancient highland crust of the Moon.

Apastron. The orbital positions of the two members of a binary star system when at their greatest separation.

Aphelion. The furthest distance of a planet or other body from the Sun in its orbit.

Apogee. The furthest point in the orbit of a natural or artificial satellite from its parent planet.

Apparent magnitude. The apparent brightness of a celestial body. The lower the magnitude, the brighter the object: thus the Sun is approximately −27, the Pole Star +2, and the faintest stars detectable by modern techniques around +30.

Appulse. The apparent close approach of one celestial object to another. If one object covers the other, the appulse becomes an occultation.

Apsides. The two points of an elliptical orbit lying closest to and furthest from the centre of mass of the system.

Areography. The official name for 'the geography of Mars'.

Armillary sphere. A celestial globe in which the celestial sphere is represented by a skeletal framework of intersecting circles, with the Earth in the central position.

Array. An arrangement of a number of linked radio antennæ.

Ascending node. The point where an orbiting body crosses the plane of reference for its orbit, moving from south to north (the opposite point is known as the descending node).

Asterism. A pattern of stars which does not rank as a separate constellation.

Asteroids. One of the names for the minor planet swarm.

Astrobleme. A very old, very eroded crater, frequently of impact origin.

Astrograph. An astronomical telescope designed specially for astronomical photography.

Astrolabe. An ancient instrument used to measure the altitudes of celestial bodies.

Astronomical unit. The mean distance between the Earth and the Sun. It is equal to $149\,598\,500$ km.

Aurora. Auroræ are 'polar lights'; Aurora Borealis (northern) and Aurora Australis (southern). They occur in the Earth's upper atmosphere, and are caused by charged particles emitted by the Sun.

Azimuth. The bearing of an object in the sky, measured from north (0°) through east, south and west.

Bailly's beads. Brilliant points seen along the edge of the Moon just before and just after a total solar eclipse. They are caused by the sunlight shining through valleys at the Moon's limb.

Barycentre. The centre of gravity of the Earth–Moon system. Because the Earth is 81 times as massive as the Moon, the barycentre lies well inside the Earth's globe.

Basalt. A dark grey fine-grained volcanic rock, with a silica content of from 44 and 50%; basalt is the most widespread volcanic rock found on the surfaces of the terrestrial planets.

Billion. (American) One thousand million. (British) One million million. The American version is now generally used.

Binary star. A stellar system made up of two stars, genuinely associated, and moving round their common centre of gravity. The revolution periods range from millions of years for very widely separated visual pairs down to less than half an hour for pairs in which the components are almost in contact with each other. With very close pairs, the components cannot be seen separately, but may be detected by spectroscopic methods.

Black body. An idealised body which absorbs all the radiation falling on it.

Black hole. A region round a very small, very massive, collapsed star from which not even light can escape.

Blazars. Objects such as quasars, which show violent variations in light output.

BL Lacertæ objects. Variable objects which are powerful emitters of infrared radiation, and are very luminous and remote. They are of the same nature as quasars.

Bode's law. A mathematical relationship linking the distances of the planets from the Sun. It may or may not be genuinely significant. Strictly speaking, it should be called Titius' law, since it was discovered by J. D. Titius some years before J. E. Bode popularised it in 1772.

Bolide. A brilliant exploding meteor.

Bolometer. An instrument for measuring the total amount of energy received from a source of electromagnetic radiation.

Bow shock. The edge of the magnetosphere of a planetary body, where the solar wind is deflected.

Caldera (pl. calderæ). A large depression, usually found at the summit of a shield volcano, due to the withdrawal of magma from below.

Carbon stars. Red stars of spectral types R and N with unusually carbon-rich atmospheres.

Carbonaceous chondrites. Primitive stony meteorites (ærclites), containing carbonaceous compounds and hydrated silicates.

Cassegrain reflector. A reflecting telescope in which the secondary mirror is convex; the light is passed back through a hole in the main mirror. Its main advantage is that it is more compact than the Newtonian reflector.

Celestial sphere. An imaginary sphere surrounding the Earth, whose centre is the same as that of the Earth's globe.

Cepheid. A short-period variable star, very regular in behaviour; the name comes from the prototype star, δ Cephei. Cepheids are astronomically important because there is a definite law linking their variation periods with their real luminosities, so that their distances may be obtained by sheer observation.

Čerenkov radiation. Electromagnetic radiation produced when electrically charged particles move through a medium at a velocity greater than the velocity of light in that medium.

Chandrasekhar limit. The maximum possible mass limit for a white dwarf star: 1.4 times the mass of the Sun.

Chondrite. Stony asteroite containing chondrules. Chondrites make up over 90% of stony meteorites (aerolites).

Chondrules. Spherical incursions found in chondrites. They are composed mainly of pyroxene and olivine, with some glass.

Chromatic aberration. A defect in all lenses, due to the fact that light is a mixture of all wavelengths – and these wavelengths are refracted unequally, so that different colours are brought to different focal points and false colour is produced round a bright object such as a star. The fault may be reduced by making the lens a compound arrangement, using different kinds of glasses.

Chromosphere. That part of the Sun's atmosphere which lies above the bright surface or photosphere.

Circumpolar star. A star which never sets. For instance, Ursa Major (the Great Bear) is circumpolar as seen from England; Crux Australis (the Southern Cross) is circumpolar as seen from New Zealand.

Cluster variables. An obsolete name for the stars now known as RR Lyræ variables.

Cœlostat. An optical instrument making use of two mirrors, one of which is fixed, while the other is movable and is mounted parallel to the Earth's axis; as the Earth rotates, the light from the star (or other object being observed) is caught by the rotatable mirror and is reflected in a fixed direction on to the second mirror. The result is that the eyepiece of the instrument need not move at all.

Collapsar. The end product of a very massive star, which has collapsed and has surrounded itself with a black hole.

Colour index. The difference between a star's visual magnitude and its photographic magnitude. The redder the star, the greater the positive value of the colour index; bluish stars have negative colour indices. For stars of type A0, the colour index is zero.

Colures. Great circles on the celestial sphere.

Commensurability. A property of two orbits in which the period of one orbit is equal to, or a simple fraction of, the period of the other.

Conduction. A method of heat transfer in which the heat is transferred through solids by molecular impact.

Conjunction. (1) A planet is said to be in conjunction with a star, or with another planet, when the two bodies are apparently close together in the sky. (2) For the inferior planets, Mercury and Venus, inferior conjunction occurs when the planet is approximately between the Earth and the Sun; superior conjunction, when the planet is on the far side of the Sun and the three bodies are again lined up. Planets beyond the Earth's orbit can never come to inferior conjunction, for obvious reasons.

Convection. Heat transfer within a flowing material, in which hot material from lower levels rises because it is less dense. Cooler material at higher levels then sinks. The overall motion thus generated is known as a convection cell.

Corona. The outermost part of the Sun's atmosphere, made up of very tenuous gas. It is visible with the naked eye only during a total solar eclipse.

Coronagraph. A device used for studying the inner corona at times of non-eclipse.

Cosmic rays. High-velocity particles reaching the Earth from outer space. The heavier cosmic-ray particles are broken up when they enter the upper atmosphere.

Cosmic year. The time taken for the Sun to complete one revolution round the centre of the Galaxy: about 225 000 000 years.

Cosmogony. The study of the origin and evolution of the universe.

Cosmology. The study of the universe considered as a whole.

Counterglow. The English name for the sky-glow more generally called by its German name of the Gegenschein.

Culmination. The maximum altitude of a celestial body above the horizon.

Cusp. The pointed extremity of a crescent shape.

Cytherean. Relating to the planet Venus (an alternative to *Venusian*).

Dall–Kirkham telescope. A form of Cassegrain telescope using an ellipsoidal primary mirror and a spherical secondary mirror.

Dawes limit. The practical limit for the resolving power of a telescope; it is $4.56/d$ arcsec where d is the aperture of the telescope in inches, and $11.6/d$ where d is the aperture of the telescope in centimetres.

Day, sidereal. The interval between successive meridian passages, or culminations, of the same star: 23h 56m 4s .091.

Day, solar. The mean interval between successive meridian passages of the Sun: 24h 3 m 56s .555 of mean sidereal time. It is longer than the sidereal day because the Sun seems to move eastward against the stars at an average rate of approximately one degree per day.

Decametric radiation. Low-frequency radio waves, with wavelengths of tens or hundreds of metres.

Declination. The angular distance of a celestial body north or south of the celestial equator. It corresponds to latitude on the Earth.

Dewcap. An open tube fitted to the upper end of a refracting telescope. Its rôle is to prevent condensation upon the object-glass.

Dichotomy. The exact half-phase of the Moon or an inferior planet.

Diffraction grating. A device used for splitting up light – it consists of a polished metallic surface upon which thousands of parallel lines are ruled. It may be regarded as an alternative to the prism.

Direct motion. Movement of revolution or rotation in the same sense as that of the Earth.

Doppler effect. The apparent change in wavelength of the light from a luminous body which is in motion relative to the observer. With an approaching object, the wavelength is apparently shortened, and the spectral lines are shifted to the blue end of the spectral band; with a receding body there is a red shift, since the wavelength is apparently lengthened.

Double star. A star made up of two components – either genuinely associated (binary systems) or merely lined up by chance (optical pairs).

Driving clock. A mechanism for driving a telescope round at a rate which compensates for the axial rotation of the Earth, so that the object under observation remains fixed in the field of view.

Dune. An elongated mound of sand produced by wind activity. Dunes are found, for example, on Mars as well as on the Earth.

Dwarf novæ. A term sometimes applied to the U Geminorum (or SS Cygni) variable stars.

Earthshine. The faint luminosity on the night side of the Moon, frequently seen when the Moon is in its crescent phase. It is due to light reflected on to the Moon from the Earth.

Eclipse, lunar. The passage of the Moon through the shadow cast by the Earth. Lunar eclipses may be either total or partial. At some eclipses, totality may last for approximately $1\frac{3}{4}$ hours, though most are shorter.

Eclipse, solar. The blotting-out of the Sun by the Moon, so that the Moon is then directly between the Earth and the Sun. Total eclipses can last for over 7 minutes under exceptionally favourable circumstances. In a partial eclipse, the Sun is incompletely covered. In an annular eclipse, exact alignment occurs when the Moon is in the far part of its orbit, and so appears smaller than the Sun; a ring of sunlight is left showing round the dark body of the Moon. Strictly speaking, a solar 'eclipse' is the occultation of the Sun by the Moon.

Eclipsing variable (or eclipsing binary). A binary star in which one component is regularly occulted by the other, so that the total light which we receive from the system is reduced. The prototype eclipsing variable is Algol (β Persei).

Ecliptic. The apparent yearly path of the Sun among the stars. It is more accurately defined as the projection of the Earth's orbit on to the celestial sphere.

Electron. Part of an atom; a fundamental particle carrying a negative electric charge.

Electron density. The number of free electrons per unit volume of space. In interstellar space the value is around 30 000 per cubic metre.

Elongation. The angular distance of a planet from the Sun, or of a satellite from its primary planet.

Ephemeris. A table showing the predicted positions of a celestial body such as a comet, asteroid or planet.

Epoch. A date chosen for reference purposes in quoting astronomical data.

Equator, celestial. The projection of the Earth's equator on to the celestial sphere.

Equatorial mounting for a telescope. A mounting in which the telescope is set up on an axis which is parallel with the axis of the Earth. This means that one movement only (east to west) will suffice to keep an object in the field of view.

Equinox. The equinoxes are the two points at which the ecliptic cuts the celestial equator. The vernal equinox or First Point of Aries now lies in the constellation of Pisces; the Sun crosses it about 21 March each year. The autumnal equinox is known as the First Point of Libra; the Sun reaches it about 22 September yearly.

Escape velocity. The minimum velocity which an object must have in order to escape from the surface of a planet, or other celestial body, without being given any extra impetus.

Evection. An inequality in the Moon's motion, due to slight changes in the shape of the lunar orbit.

Event horizon. The 'boundary' of a black hole. No light can escape from inside the event horizon.

Exosphere. The outermost part of the atmosphere of a planetary body.

Extinction. The apparent reduction in brightness of a star or planet when low down in the sky, so that more of its light is absorbed by the Earth's atmosphere. With a star 1° above the horizon, extinction amounts to 3 magnitudes.

Eyepiece (or Ocular). The lens, or combination of lenses, at the eye-end of a telescope. It is responsible for magnifying the image of the object under study. With a positive eyepiece (for instance, a Ramsden, Orthoscopic or Monocentric) the image plane lies between the eyepiece and the object-glass (or main mirror); with a negative eyepiece (such as a Huyghenian or Tolles) the image plane lies inside the eyepiece. A Barlow lens is concave, and is mounted in a short tube which may be placed between the eyepiece and the object-glass (or mirror). It increases the effective focal length of the telescope, thereby providing increased magnification.

Faculæ. Bright, temporary patches on the surface of the Sun.

Field star. A star which is seen close to a stellar cluster, but is not a cluster member. It may be much closer or much more remote.

Filar micrometer. A device used for measuring very small angular distances as seen in the eyepiece of a telescope.

Finder. A small, wide-field telescope attached to a larger one, used for sighting purposes.

Fireball. A brilliant meteor. There is no set definition, but a meteor with a magnitude of brighter than −5 will be classed as a fireball.

Flares, solar. Brilliant eruptions in the outer part of the Sun's atmosphere. Normally they can be detected only by spectroscopic means (or the equivalent), though a few have been seen in integrated light. They are made up of hydrogen, and emit charged particles which may later reach the Earth, producing magnetic storms and displays of auroræ. Flares are generally, though not always, associated with sunspot groups.

Flare stars. Faint red dwarf stars which show sudden, short-lived increases in brilliancy, due to intense flares above their surfaces,

Flash spectrum. The sudden change-over from dark to bright lines in the Sun's spectrum, just before the onset of totality in a solar eclipse. The phenomenon is due to the fact that at this time the Moon has covered up the bright surface of the Sun, so that the chromosphere is shining 'on its own'.

Flocculi. Patches of the Sun's surface, observable with spectroscopic equipment. They are of two main kinds: bright (calcium) and dark (hydrogen).

Forbidden lines. Lines in the spectrum of a celestial body which do not appear under normal conditions, but may be seen in bodies where conditions are exceptional.

Fraunhofer lines. The dark absorption lines in the spectrum of the Sun or any other star.

Galaxies. Systems made up of stars, nebulæ and interstellar matter. Many, though by no means all, are spiral in form.

Galaxy, the. The system of which our Sun is a member. It contains approximately 100 000 million stars, and is a rather loose spiral.

Gamma-rays. Radiation of extremely short wavelength.

Gauss. Unit of measurement of a magnetic field. (The Earth's field, at the surface, is on average about 0.3 to 0.6 gauss.)

Gegenschein (Counterglow). A very faint glow seen exactly opposite to the Sun against an extremely dark sky. It is due to sunlight backscattered off millimetre-sized particles (asteroid débris) orbiting in the main plane of the Solar System. Its high angle of reflection distinguishes it from Zodiacal Light near the Sun.

Geocorona. The outermost part of the Earth's atmosphere, made up of a halo of hydrogen gas extending to around 15 Earth radii.

Geodesy. The study of the shape, size, mass and other characteristics of the Earth.

Gibbous phase. The phase of the Moon or planet when between half and full.

Geosynchronous orbit. An orbit round the Earth at an altitude of 35 900 km, where the period will be the same as the Earth's sidereal rotation period – 23h 56m 4s .1, assuming that the orbit is circular and lies in the plane of the Earth's equator.

Glitch. A sudden change in the rotation period of a pulsar, due probably to starquakes.

Globules. Small dark patches inside gaseous nebulæ. They are probably embryo stars.

Gnomon. In a sundial, it is a pointer whose function is to cast the Sun's shadow on to the dial. The gnomon always points to the celestial pole.

Graben. A downfaulted block of crust on a planetary surface bounded by a pair of normal faults.

Great circle. A circle on the surface of a sphere whose plane passes through the centre of that sphere.

Green Flash. Sudden, brief green light seen as the last segment of the Sun disappears below the horizon. It is purely an effect of the Earth's atmosphere. Venus has also been known to show a Green Flash.

Gregorian reflector. A telescope in which the secondary mirror is concave, and placed beyond the focus of the main mirror. The image is erect. Few Gregorian telescopes are in use nowadays.

H.I and H.II regions. Clouds of hydrogen in the Galaxy. In H.I regions the hydrogen is neutral; in H.II regions the hydrogen is ionised, and the presence of hot stars will make the cloud shine as a nebula.

Halo, galactic. The spherical-shaped cloud of stars round the main part of the Galaxy.

Hayashi track. The evolutionary track of a proto-star on the HR diagram – before it joins the Main Sequence.

Heliacal rising. The rising of a star or planet at the same time as the Sun, though the term is generally used to denote the time when the object is first detectable in the dawn sky.

Heliosphere. The area round the Sun extending to between 50 and 100 a.u. where the Sun's influence is dominant. The boundary, where the solar wind merges with the interstellar medium, is called the heliopause.

Herbig–Haro object. A nebulous object associated with a newly forming star.

Herschelian reflector. An obsolete type of telescope in which the main mirror is tilted, thus removing the need for a secondary mirror.

Hertzsprung gap. A region in the HR diagram, between the giant branch and the Main Sequence, containing comparatively few stars, because this stage in the evolution of a star is short.

Hertzsprung–Russell diagram (usually known as the HR diagram). A diagram in which stars are plotted according to the spectral types and their absolute magnitudes.

High-velocity star. A star travelling at more than around 65 km s^{-1} in relation to the Sun. These are old stars, which do not share the Sun's motion round the galactic centre, but travel in more elliptical orbits.

Horizon. The great circle on the celestial sphere which is everywhere 90 degrees from the observer's zenith.

Hour angle (of a celestial object). The time which has elapsed since the object crossed the meridian. If RA = right ascension of the object and LST = the local sidereal time, the hour angle = LST – RA.

Hour circle. A great circle on the celestial sphere, passing through both celestial poles. The zero hour circle coincides with the observer's meridian.

Hubble Constant. A measure of the rate at which galaxies recede from one another over vast distances. The current value is about 70 km s^{-1} megaparsec^{-1}.

Igneous rock. A rock formed by the crystallisation of a magma.

Inferior planets. Mercury and Venus, whose distances from the Sun are less than that of the Earth.

Infrared radiation. Radiation with wavelength longer than that of visible light (approximately 7500 Ångströms).

Interferometer, stellar. An instrument for measuring star diameters. The principle is based upon light interference.

Ion. An atom which has lost or gained one or more of its planetary electrons, and so has respectively a positive or negative charge.

Ionosphere. The region of the Earth's atmosphere lying above the stratosphere.

Irradiation. The effect which makes very brilliant bodies appear larger than they really are.

Julian day. A count of the days, starting from 12 noon on 1 January 4713 BC. Thus 1 January 1977 was Julian Day 2 443 145. (The name 'Julian' has nothing to do with Julius Cæsar. The system was invented in 1582 by the mathematician Scaliger, who named it in honour of his father, Julius Scaliger.)

Kelvin scale. A scale of temperature. 1 K is equal to 1 °C, but the Kelvin scale starts at absolute zero (–273.16 °C).

Kepler's laws of planetary motion. These were laid down by Johannes Kepler, from 1609 to 1618. They are: (1) The planets move in elliptical orbits, with the Sun occupying one focus. (2) The radius vector, or imaginary line joining the centre of the planet to the centre of the Sun, sweeps out equal areas in equal times. (3) With a planet, the square of the sidereal period is proportional to the cube of the mean distance from the Sun.

Kiloparsec. One thousand parsecs (3260 light-years).

Kirkwood gaps. Gaps in the main asteroid belt, where the periods would be commensurate with that of Jupiter – so that Jupiter keeps these areas 'swept clear'.

KREEP. Basaltic rocks found on the Moon, rich in potassium (K), rare earth elements (REE) and phosphorus (P).

Light echo. Really a type of reflection. When an object such as a supernova brightens rapidly, its light may be reflected off interstellar nebulosity; light from the initial flash reaches the observer first, while the light from the nebulosity (the *echo*) has travelled slightly further and arrives later. Because the light from the flash has only travelled forward as well away from the source of the flash, it gives the illusion of an echo expanding at a speed greater than that of light (*superluminal* velocity) which is impossible.

Neutron star. The remnant of a massive star which has exploded as a supernova. Neutron stars send out rapidly varying radio emissions, and are therefore called 'pulsars'. Only two (the Crab and Vela pulsars) have as yet been identified with optical objects.

Newtonian reflector. A reflecting telescope in which the light is collected by a main mirror, reflected on to a smaller flat mirror set at an angle of 45°, and thence to the side of the tube.

Nocturnal. An instrument for telling the time at night, by using the positions of the Pointers in Ursa Major (Dubhe and Merak) relative to the Pole Star.

Nodes. The points at which the orbit of the Moon, a planet or a comet cuts the plane of the ecliptic; south to north (ascending node) or north to south (descending node).

Nova. A star which suddenly flares up to many times its normal brilliancy, remaining bright for a relatively short time before fading back to obscurity.

Nutation. A slow, slight 'nodding' of the Earth's axis, due to the gravitational pull of the Moon on the Earth's equatorial bulge.

Object-glass (or objective). The main lens of a refracting telescope.

Objective prism. A small prism placed in front of the object-glass of a telescope. It produces small-scale spectra of the stars in the field of view.

Obliquity of the ecliptic. The angle between the ecliptic and the plane of the Earth's orbit. Its present value is 23° 26' 45", but it can range between 21° 55' and 24° 18'.

Occultation. The covering-up of one celestial body by another.

Ocular. Alternative name for a telescope eyepiece.

Olivine. A silicate of magnesium and iron: $(Mg,Fe)SiO_4$.

Oort Cloud. An assumed spherical shell of comets surrounding the Solar System, at a range of around one light-year.

Opposition. The position of a planet when exactly opposite to the Sun in the sky; the Sun, the Earth and the planet are then approximately lined up.

Orbit. The path of a celestial object.

Orrery. A model showing the Sun and the planets, capable of being moved mechanically so that the planets move round the Sun at their correct relative speeds.

Outgassing. The process by which volatiles within a planet gradually escape through the surface.

Palæomagnetism. The study of the fossil or remnant magnetisation of rocks of all ages.

Palimpsests. High-albedo circular features on the surfaces of some planetary bodies, notably Callisto and Ganymede. Low-walled, bright-floored circular structures on the Moon (mainly on the hemisphere turned away from Earth) have also been referred to as palimpsests.

Parallax, trigonometrical. The apparent shift of an object when observed from two different directions.

Parsec. The distance at which a star would have a parallax of one second of arc: 3.26 light-years, 206 265 astronomical units, or 30.857 million million kilometres.

Penumbra. (1) The area of partial shadow to either side of the main cone of shadow cast by the Earth. (2) The lighter part of a sunspot.

Periastron. In a binary star system, the point of closest approach of the two components.

Perigee. The position of the Moon in its orbit when closest to the Earth.

Perihelion. The position in the orbit of a planet or other body when closest to the Sun.

Perturbations. The disturbances in the orbit of a celestial body produced by the gravitational effects of other bodies.

Phases. The apparent changes in shape of the Moon and the inferior planets from new to full. Mars may show a gibbous phase, but with the other planets there are no appreciable phases as seen from Earth.

Photoelectric cell. An electronic device; light falling on the cell produces an electric current, the strength of which depends upon the intensity of the light.

Photoelectric photometer. A photoelectric cell used together with a telescope for measuring the magnitudes of celestial bodies.

Photometer. An instrument used to measure the intensity of light from any particular source.

Photometry. The measurement of the intensity of light.

Photon. The smallest 'unit' of light.

Photosphere. The bright surface of the Sun.

Pioneer Anomaly. One peculiar effect has yet to be explained.. Unmanned space-craft in the outer Solar System are not moving quite as they ought to do. Both Pioneers are escaping from the Solar System, and are slowing down because of the Sun's gravitational pull, but they are slowing down slightly more than expected. The effect is known as the Pioneer Anomaly. Other space-craft such as Galileo and Ulysses have also been affected. All sorts of explanations have been proposed, but we have to admit that at present (2010) the Pioneer Anomaly remains a complete mystery.

Planetary nebula. A small, dense, hot star surrounded by a shell of gas. The name is ill-chosen, since planetary nebulæ are neither planets nor nebulæ!

Planetoid. Obsolete name for an asteroid (minor planet).

Planemo. A celestial object with mass greater than an asteroid, but smaller than that of a star. Many are free-floating in space. (Planemo = PLANEtary Mass Object.)

Plasma. An ionised gas: a mixture of electrons and atomic nuclei.

Plerion. A supernova remnant with no shell structure. One such object is the Crab Nebula.

Polar binary. A binary; the white dwarf component is a magnetar, pushing out the inner accretion disc, only allowing matter to fall down its magnetic poles. (Example: DQ Herculis, nova 19.34.)

Poles, celestial. The north and south points of the celestial sphere.

Populations, stellar. Two main types of star regions: I (in which the brightest stars are hot and bluish) and II (in which the brightest stars are old red giants).

Position angle. The apparent direction of one object with reference to another, measured from the north point of the main object through east, south and west.

Poynting–Robertson effect. (Named for J. H. Poynting and H. P. Robertson.) A process making a very small particle in the Solar System spiral slowly inward. The 'drag' is a component of radiation pressure tangential to the motion of the particle. It was first described by Poynting in 1903, and correctly interpreted by Robertson in 1937.

Precession. The apparent slow movement of the celestial poles. This also means a shift of the celestial equator, and hence of the equinoxes; the vernal equinox moves by 50 seconds of arc yearly, and has moved out of Aries into Pisces. Precession is due to the pull of the Moon and Sun on the Earth's equatorial bulge.

Prime Meridian. The meridian on the Earth's surface which passes through the Airy Transit Circle at Greenwich Observatory. It is taken as longitude 0°.

Prominences. Masses of glowing gas rising from the surface of the Sun. They are made up chiefly of hydrogen.

Proper motion, stellar. The individual movement of a star on the celestial sphere.

Proton. A fundamental particle with a positive electric charge. The nucleus of the hydrogen atom is made up of a single proton.

Proto-planet. A body forming by the accretion of material, which will ultimately develop into a planet.

Proto-star. The earliest stage in the formation of a star.

Pulsar. A rotating neutron star, often a strong radio source. Not all pulsars can be detected by radio, since the radiation is emitted in beams, and it depends upon whether these beams sweep over the Earth.

Purkinje effect. This is due to the change of the colour sensitivity of the human eye. If two sources are equal, and are then dimmed, the redder of the two will appear the fainter.

Quadrant. An ancient astronomical instrument used for measuring the apparent positions of celestial bodies.

Quadrature. The position of the Moon or a planet when at right angles to the Sun as seen from the Earth.

Quantum. The amount of energy possessed by one photon of light.

Quasar. The core of a very powerful, remote active galaxy. The term QSO (quasi-stellar object) is also used.

Radial velocity. The movement of a celestial body toward or away from the observer; positive if receding, negative if approaching.

Radiant. The point in the sky from which the meteors of any particular shower seem to radiate.

Regolith. A layer of loose rock and mineral grains on the surface of a planetary body. With the addition of organic material it becomes a soil.

Regression of the nodes. The nodes of the Moon's orbit move slowly westward, making one complete revolution in 18.6 years. This regression is caused by the gravitational pull of the Sun.

Retardation. The difference in the time of moonrise between one night and the next.

Retrograde motion. Orbital or rotational movement in the sense opposite to that of the Earth's motion.

Reversing layer. The gaseous layer above the Sun's photosphere.

Right ascension. The angular distance of a celestial body from the vernal equinox, measured eastward. It is usually given in hours, minutes and seconds of time, so that the right ascension is the time-difference between the culmination of the vernal equinox and the culmination of the body.

Roche limit. The distance from the centre of a planet within which a second body would be broken up by the planet's gravitational pull. Note, however, that this would be the case only for a body that had no appreciable gravitational cohesion.

Saros. The period after which the Earth, Moon and Sun return to almost the same relative positions: 18 years 11.3 days. The Saros may be used in eclipse prediction, since it is usual for an eclipse to be followed by a similar eclipse exactly one Saros later.

Schmidt camera (or Schmidt telescope). An instrument that collects its light by means of a spherical mirror; a correcting plate is placed at the top of the tube. It is a purely photographic instrument.

Schwarzschild radius. The radius that a body must have if its escape velocity is to be equal to the velocity of light.

Scintillation. Twinkling of a star; it is due to the Earth's atmosphere. Planets may also show scintillation when low in the sky.

Secular acceleration of the Moon. The apparent speeding-up of the Moon in its orbit as measured over a long period of time, caused by the gradual slowing of the Earth's rotation (by 0.000 000 02 second per day).

Selenography. The study of the Moon's surface.

Selenology. The lunar equivalent of geology.

Sextant. An instrument used for measuring the altitude of a celestial object.

Seyfert galaxies. Galaxies with relatively small, bright nuclei and weak spiral arms. Some of them are strong radio emitters.

Sidereal period. The revolution period of a planet round the Sun, or of a satellite round its primary planet.

Sidereal time. The local time reckoned according to the apparent rotation of the celestial sphere. When the vernal equinox crosses the observer's meridian, the sidereal time is 0 hours.

Solar nebula. The cloud of interstellar gas and dust from which the Solar System was formed – around 5000 million years ago.

Solar wind. A flow of atomic particles streaming out constantly from the Sun in all directions.

Solstices. The times when the Sun is at its maximum declination of approximately 23½ degrees; around 22 June (summer solstice, with the Sun in the northern hemisphere of the sky) and 22 December (winter solstice, Sun in the southern hemisphere).

Specific gravity. The density of any substance, taking that of water as 1. For instance, the Earth's specific gravity is 5.5, so that the Earth 'weighs' 5.5 times as much as an equal volume of water would do.

Speckle interferometry. A technique designed to reduce the blurring of star images due to turbulence in the Earth's atmosphere.

Spectroheliograph. An instrument used for photographing the Sun in the light of one particular wavelength only. The visual equivalent of the spectroheliograph is the spectrohelioscope.

Spectroscopic binary. A binary system whose components are too close together to be seen individually, but which can be studied by means of spectroscopic analysis.

Speculum. The main mirror of a reflecting telescope.

Spherical aberration. Blurring of a telescope image; it is due to the fact that the lens (or mirror) does not bring the light-rays falling on its edge and on its centre to exactly the same focal point.

Starburst galaxy. A galaxy in which there is an exceptionally high rate of star formation.

Stellar wind. A continuous outflow of particles from a star, resulting in loss of mass.

Superior planets. All the planets lying beyond the orbit of the Earth in the Solar System (that is to say, all the principal planets apart from Mercury and Venus).

Superluminal motion. The apparent movement of material at a velocity greater than that of light. It is purely a geometrical effect.

Supernova. A colossal stellar outburst, involving (1) the total destruction of the white dwarf member of a binary system, or (2) the collapse of a very massive star.

Synchronous rotation. If the rotation period of a planetary body is equal to its orbital period, the rotation is said to be synchronous (or captured). The Moon and most planetary satellites have synchronous rotation.

Synchrotron radiation. Radiation emitted by a charged particle travelling at almost the velocity of light, moving in a strong magnetic field.

Synodic period. The interval between successive oppositions of a superior planet.

Syzygy. The position of the Moon in its orbit when new or full.

Tektites. Small, glassy objects found in a few localised areas of the Earth. They are not now believed to be meteoritic.

Terminator. The boundary between the day- and night-hemispheres of the Moon or a planet.

Thermocouple. An instrument used for measuring very small amounts of heat.

Torus. A three-dimensional ring shape, such as that of a quoit (or a doughnut).

Transit. (1) The passage of a celestial body across the observer's meridian. (2) The projection of Mercury or Venus against the face of the Sun.

Transit instrument. A telescope mounted so that it can move only in declination; it is kept pointing to the meridian, and is used for timing the passages of stars across the meridian. Transit instruments were once the basis of all practical timekeeping. The Airy transit instrument at Greenwich is accepted as the zero for all longitudes on the Earth.

Troposphere. The lowest part of the Earth's atmosphere; its top lies at an average height of about 11 km. Above it lies the stratosphere; and above the stratosphere come the ionosphere and the exosphere.

Twilight. The state of illumination when the Sun is below the horizon by less than 18°.

ULIRG. Ultra-luminous infrared galaxy

Umbra. (1) The main cone of shadow cast by the Earth. (2) The darkest part of a sunspot.

Van Allen zones. Zones of charged particles around the Earth. There are two main zones; the outer (made up chiefly of electrons) and the inner (made up chiefly of protons).

Variable stars. Stars which change in brilliancy over short periods. They are of various types.

Variation. An inequality in the Moon's motion, due to the fact that the pull of the Sun on the Moon is not constant for all positions in the lunar orbit.

White dwarf. A very small, very dense star which has used up its nuclear energy, and is in a very late stage of its evolution.

Widmanstätten patterns. If an iron meteorite is cut, polished and then etched with acid, characteristic figures of the iron crystals appear; these are the Widmanstätten patterns.

Wilson effect. Perspective effect of a sunspot near the solar limb; the limbward penumbra is broadened relative to the penumbra on the opposite side, indicating that the 'spot' is a depression. Not all sunspots show the effect, however.

WIMP. An electrically neutral, weakly interacting massive particle. As yet WIMPs are theoretical concepts only.

Wolf–Rayet stars. Very hot, greenish-white stars that are surrounded by expanding gaseous envelopes. Their spectra show bright (emission) lines.

Wormhole. A hypothetical tunnel-like structure in the fabric of spacetime.

Yarkovsky effect. Named after the Russian engineer I. O. Yarkovsky, who proposed it in 1900. It produces a slight change in the orbit of an asteroidal body. The surface is heated by the Sun by day, and cools at night. The heated side emits more radiation, producing a slight acceleration. The asteroid 6489 Golevka has been studied for 12 years, and the effect has caused a shift of 15 km.; Golevka was discovered in 1991 by E. Helin; and is irregular in shape, with a mean diameter of 0.53 km. It is a Mars-crosser, and in 2046 will pass the Earth at a range of 7.6 million miles. The orbital period is 3.95 years; escape velocity 0.0003 km s^{-1}.

Year. (1) Sidereal: the period taken for the Earth to complete one journey round the Sun (365.26 days). (2) Tropical: the interval between successive passages of the Sun across the vernal equinox (365.24 days). (3) Anomalistic: the interval between successive perihelion passages of the Earth (365.26 days; slightly less than 5 minutes longer than the sidereal year, because the position of the perihelion point moves along the Earth's orbit by about 11 seconds of arc every year). (4) Calendar: the mean length of the year according to the Gregorian calendar (365.24 days, or 365d 5h) 49m 12s).

Zenith. The observer's overhead point (altitude 90°).

Zenith distance. The angular distance of a celestial object from the Zenith.

Zenithal hourly rate. The number of naked-eye shower meteors which would be seen by an observer under ideal conditions, with the meteor radiant at the zenith. In practice, these conditions are never attained.

Zodiac. A belt stretching round the sky, 8° to either side of the ecliptic, in which the Sun, Moon and principal planets are to be found at any time. (As such, they will always be observed against one of 13 constellation)

Zodiacal Light. A cone of light rising from the horizon and stretching along the ecliptic; visible only when the Sun is a little way below the horizon and thus will often be obscured. It is due to thinly spread interplanetary material near the main plane of the Solar System reflecting sunlight.

Zone of avoidance. The sky region near the plane of the Milky Way in which few galaxies can be seen by twilight because of obscuration by interstellar dust in our Galaxy.

Zürich number (or Wolf number). A measure of solar activity. It is expressed as $R = k(10g + f)$, where R is the Zürich number, g is the number of spot-groups and f is the number of individual spots; k is a constant, usually with a value of around 1, depending upon the observer and the instrument used.

Index